Universitext

Juan Ramón Ruíz-Tolosa
Enrique Castillo

From Vectors
to Tensors

 Springer

Authors

Juan Ramón Ruíz-Tolosa
Enrique Castillo
Universidad Cantabria
Depto. Matemática Aplicada
Avenida de los Castros
39005 Santander, Spain

castie@unican.es

Library of Congress Control Number: 20041114044

Mathematics Subject Classification (2000): 15A60, 15A72

ISBN 3-540-22887-X Springer Berlin Heidelberg New York

Springer-Verlag is a part of Springer Science+Business Media
springeronline.com

© Springer-Verlag Berlin Heidelberg 2005
Printed in Germany

Cover Design: *Erich Kirchner, Heidelberg*
Printed on acid-free paper   41/3142XT 5 4 3 2 1 0

To the memory of Bernhard Riemann and Albert Einstein

Preface

It is true that there exist many books dedicated to linear algebra and somewhat fewer to multilinear algebra, written in several languages, and perhaps one can think that no more books are needed. However, it is also true that in algebra many new results are continuously appearing, different points of view can be used to see the mathematical objects and their associated structures, and different orientations can be selected to present the material, and all of them deserve publication.

Under the leadership of Juan Ramón Ruíz-Tolosa, Professor of multilinear algebra, and the collaboration of Enrique Castillo, Professor of applied mathematics, both teaching at an engineering school in Santander, a tensor textbook has been born, written from a practical point of view and free from the esoteric language typical of treatises written by algebraists, who are not interested in descending to numerical details. The balance between following this line and keeping the rigor of classical theoretical treatises has been maintained throughout this book.

The book assumes a certain knowledge of linear algebra, and is intended as a textbook for graduate and postgraduate students and also as a consultation book. It is addressed to mathematicians, physicists, engineers, and applied scientists with a practical orientation who are looking for powerful tensor tools to solve their problems.

The book covers an existing chasm between the classic theory of tensors and the possibility of solving tensor problems with a computer. In fact, the computational algebra is formulated in matrix form to facilitate its implementation on computers.

The book includes 197 examples and end-of-chapter exercises, which makes it specially suitable as a textbook for tensor courses. This material combines classic matrix techniques together with novel methods and in many cases the questions and problems are solved using different methods. They confirm the applied orientation of the book.

A computer package, written in Mathematica, accompanies this book, available on: http://personales.unican.es/castie/tensors. In it, most of the novel methods developed in the book have been implemented. We note that existing general computer software packages (Mathematica, Mathlab, etc.) for tensors are very poor, up to the point that some problems cannot be dealt with using computers because of the lack of computer programs to perform these operations.

The main contributions of the book are:

1. The book employs a new technique that permits one to extend (stretch) the tensors, as one-column matrices, solve on these matrices the desired problems, and recover the initial format of the tensor (condensation). This technique, applied in all chapters, is described and used to solve matrix equations in Chapter 1.

2. An important criterion is established in Chapter 2 for all the components of a tensor to have a given ordering, by the definition of a unique canonical tensor basis. This permits the mentioned technique to be applied.

3. In Chapter 3, factors are illustrated that have led to an important confusion in tensor books due to inadequate notation of tensors or tensor operations.

4. In addition to dealing with the classical topics of tensor books, new tensor concepts are introduced, such as the rotation of tensors, the transposer tensor, the eigentensors, and the permutation tensor structure, in Chapter 5.

5. A very detailed study of generalized Kronecker deltas is presented in Chapter 8.

6. Chapter 10 is devoted to mixed exterior algebras, analyzing the problem of change-of-basis and the exterior product of this kind of tensors.

7. In Chapter 11 the rules for the "Euclidean contraction" are given in detail. This chapter ends by introducing the geometric concepts to tensors.

8. The orientation and polar tensors in Euclidean spaces are dealt with in Chapter 12.

9. In Chapter 13 the Gram matrices $G(r)$ are established to connect exterior tensors.

10. Chapter 14 is devoted to Euclidean tensors in $E^n(\mathbb{R})$, affine geometric tensors (homographies), and some important tensors in physics and mechanics, such as the stress and strain tensors, the elastic tensor and the inertial moment tensor. It is shown how tensors allow one to solve very interesting practical problems.

In summary, the book is not a standard book on tensors because of its orientation, the many novel contributions included in it, the careful notation and the stretching–condensing techniques used for most of the transformations used in the book. We hope that our readers enjoy reading this book, discover a new world, and acquire stimulating ideas for their applications and new contributions and research.

The authors want to thank an anonimous French referee for the careful reading of the initial manuscript, and to Jeffrey Boys for the copyediting of the final manuscript.

Santander,   *Juan Ramón Ruíz-Tolosa*
September 30, 2004   *Enrique Castillo*

Contents

Part I Basic Tensor Algebra

1 Tensor Spaces 3
  1.1 Introduction 3
  1.2 Dual or reciprocal coordinate frames in affine Euclidean spaces 3
  1.3 Different types of matrix products 8
    1.3.1 Definitions....................................... 8
    1.3.2 Properties concerning general matrices 10
    1.3.3 Properties concerning square matrices 11
    1.3.4 Properties concerning eigenvalues and eigenvectors... 12
    1.3.5 Properties concerning the Schur product........... 13
    1.3.6 Extension and condensation of matrices 13
    1.3.7 Some important matrix equations................. 17
  1.4 Special tensors 26
  1.5 Exercises 30

2 Introduction to Tensors................................... 33
  2.1 Introduction 33
  2.2 The triple tensor product linear space 33
  2.3 Einstein's summation convention 36
  2.4 Tensor analytical representation 37
  2.5 Tensor product axiomatic properties 38
  2.6 Generalization... 40
  2.7 Illustrative examples 41
  2.8 Exercises 46

3 Homogeneous Tensors 47
  3.1 Introduction 47
  3.2 The concept of homogeneous tensors 47
  3.3 General rules about tensor notation 48
  3.4 The tensor product of tensors 50

3.5   Einstein's contraction of the tensor product 54
3.6   Matrix representation of tensors 56
      3.6.1   First-order tensors 56
      3.6.2   Second-order tensors 57
3.7   Exercises 61

4   Change-of-basis in Tensor Spaces 65
4.1   Introduction 65
4.2   Change of basis in a third-order tensor product space 65
4.3   Matrix representation of a change-of-basis in tensor spaces ... 67
4.4   General criteria for tensor character 69
4.5   Extension to homogeneous tensors 72
4.6   Matrix operation rules for tensor expressions 74
      4.6.1   Second-order tensors (matrices) 74
      4.6.2   Third-order tensors 77
      4.6.3   Fourth-order tensors 78
4.7   Change-of-basis invariant tensors: Isotropic tensors 80
4.8   Main isotropic tensors 80
      4.8.1   The null tensor 80
      4.8.2   Zero-order tensor (scalar invariant) 80
      4.8.3   Kronecker's delta 80
4.9   Exercises 106

5   Homogeneous Tensor Algebra: Tensor Homomorphisms 111
5.1   Introduction 111
5.2   Main theorem on tensor contraction 111
5.3   The contracted tensor product and tensor homomorphisms ... 113
5.4   Tensor product applications 119
      5.4.1   Common simply contracted tensor products 119
      5.4.2   Multiply contracted tensor products 120
      5.4.3   Scalar and inner tensor products 120
5.5   Criteria for tensor character based on contraction 122
5.6   The contracted tensor product in the reverse sense: The
      quotient law 124
5.7   Matrix representation of permutation homomorphisms 127
      5.7.1   Permutation matrix tensor product types in K^n 127
      5.7.2   Linear span of precedent types 129
      5.7.3   The isomers of a tensor 137
5.8   Matrices associated with simply contraction homomorphisms . 141
      5.8.1   Mixed tensors of second order ($r = 2$): Matrices...... 141
      5.8.2   Mixed tensors of third order ($r = 3$) 141
      5.8.3   Mixed tensors of fourth order ($r = 4$) 142
      5.8.4   Mixed tensors of fifth order ($r = 5$) 143
5.9   Matrices associated with doubly contracted homomorphisms . 144
      5.9.1   Mixed tensors of fourth order ($r = 4$) 144

Contents   XI

5.9.2   Mixed tensors of fifth order $(r = 5)$ 145
5.10   Eigentensors 159
5.11   Generalized multilinear mappings 165
5.11.1   Theorems of similitude with tensor mappings 167
5.11.2   Tensor mapping types 168
5.11.3   Direct n-dimensional tensor endomorphisms 169
5.12   Exercises 183

Part II Special Tensors

6   Symmetric Homogeneous Tensors: Tensor Algebras 189
6.1   Introduction 189
6.2   Symmetric systems of scalar components 189
6.2.1   Symmetric systems with respect to an index subset .. 190
6.2.2   Symmetric systems. Total symmetry 190
6.3   Strict components of a symmetric system 191
6.3.1   Number of strict components of a symmetric system
          with respect to an index subset 191
6.3.2   Number of strict components of a symmetric system . 192
6.4   Tensors with symmetries: Tensors with branched symmetry,
        symmetric tensors 193
6.4.1   Generation of symmetric tensors 194
6.4.2   Intrinsic character of tensor symmetry: Fundamental
          theorem of tensors with symmetry 197
6.4.3   Symmetric tensor spaces and subspaces. Strict
          components associated with subspaces 204
6.5   Symmetric tensors under the tensor algebra perspective 206
6.5.1   Symmetrized tensor associated with an arbitrary
          pure tensor 210
6.5.2   Extension of the symmetrized tensor associated with
          a mixed tensor 210
6.6   Symmetric tensor algebras: The $\otimes_S$ product 212
6.7   Illustrative examples 214
6.8   Exercises 220

7   Anti-symmetric Homogeneous Tensors, Tensor and Inner
    Product Algebras 225
7.1   Introduction 225
7.2   Anti-symmetric systems of scalar components 225
7.2.1   Anti-symmetric systems with respect to an index
          subset 226
7.2.2   Anti-symmetric systems. Total anti-symmetry 228
7.3   Strict components of an anti-symmetric system and with
        respect to an index subset 228

7.3.1   Number of strict components of an anti-symmetric
        system with respect to an index subset 229
7.3.2   Number of strict components of an anti-symmetric
        system 229
7.4   Tensors with anti-symmetries: Tensors with branched
      anti-symmetry; anti-symmetric tensors 230
7.4.1   Generation of anti-symmetric tensors 232
7.4.2   Intrinsic character of tensor anti-symmetry:
        Fundamental theorem of tensors with anti-symmetry . 236
7.4.3   Anti-symmetric tensor spaces and subspaces. Vector
        subspaces associated with strict components 243
7.5   Anti-symmetric tensors from the tensor algebra perspective . . 246
7.5.1   Anti-symmetrized tensor associated with an
        arbitrary pure tensor 249
7.5.2   Extension of the anti-symmetrized tensor concept
        associated with a mixed tensor 249
7.6   Anti-symmetric tensor algebras: The $\otimes_H$ product 252
7.7   Illustrative examples 253
7.8   Exercises 265

8 Pseudotensors; Modular, Relative or Weighted Tensors 269
8.1   Introduction 269
8.2   Previous concepts of modular tensor establishment 269
8.2.1   Relative modulus of a change-of-basis 269
8.2.2   Oriented vector space 270
8.2.3   Weight tensor 270
8.3   Axiomatic properties for the modular tensor concept 270
8.4   Modular tensor characteristics 271
8.4.1   Equality of modular tensors 272
8.4.2   Classification and special denominations 272
8.5   Remarks on modular tensor operations: Consequences 272
8.5.1   Tensor addition 272
8.5.2   Multiplication by a scalar 274
8.5.3   Tensor product 275
8.5.4   Tensor contraction 276
8.5.5   Contracted tensor products 276
8.5.6   The quotient law. New criteria for modular tensor
        character 277
8.6   Modular symmetry and anti-symmetry 280
8.7   Main modular tensors 291
8.7.1 ϵ systems, permutation systems or Levi-Civita
        tensor systems 291
8.7.2   Generalized Kronecker deltas: Definition 293
8.7.3   Dual or polar tensors: Definition 301
8.8   Exercises 310

Part III Exterior Algebras

9 **Exterior Algebras:**
   Totally Anti-symmetric Homogeneous Tensor Algebras 315
   9.1   Introduction and Definitions 315
         9.1.1   Exterior product of two vectors.................... 315
         9.1.2   Exterior product of three vectors 317
         9.1.3   Strict components of exterior vectors. Multivectors... 318
   9.2   Exterior product of r vectors: Decomposable multivectors 319
         9.2.1   Properties of exterior products of order r:
                 Decomposable multivectors or exterior vectors....... 321
         9.2.2   Exterior algebras over $V^n(K)$ spaces: Terminology ... 323
         9.2.3   Exterior algebras of order r=0 and r=1 324
   9.3   Axiomatic properties of tensor operations in exterior algebras 324
         9.3.1   Addition and multiplication by an scalar 324
         9.3.2   Generalized exterior tensor product: Exterior
                 product of exterior vectors 325
         9.3.3   Anti-commutativity of the exterior product $\wedge$ 331
   9.4   Dual exterior algebras over $V_*^n(K)$ spaces................. 331
         9.4.1   Exterior product of r linear forms over $V_*^n(K)$....... 332
         9.4.2   Axiomatic tensor operations in dual exterior
                 Algebras $\bigwedge_{n*}^{(r)}(K)$. Dual exterior tensor product 333
         9.4.3   Observation about bases of primary and dual
                 exterior spaces 334
   9.5   The change-of-basis in exterior algebras.................... 337
         9.5.1   Strict tensor relationships for $\bigwedge_n^{(r)}(K)$ algebras 338
         9.5.2   Strict tensor relationships for $\bigwedge_{n*}^{(r)}(\mathbb{R})$ algebras 339
   9.6   Complements of contramodular and comodular scalars 341
   9.7   Comparative tables of algebra correspondences 342
   9.8   Scalar mappings: Exterior contractions 342
   9.9   Exterior vector mappings: Exterior homomorphisms......... 345
         9.9.1   Direct exterior endomorphism 350
   9.10 Exercises 383

10 **Mixed Exterior Algebras** 387
   10.1 Introduction 387
         10.1.1 Mixed anti-symmetric tensor spaces and their strict
                 tensor components 387
         10.1.2 Mixed exterior product of four vectors.............. 390
   10.2 Decomposable mixed exterior vectors...................... 394
   10.3 Mixed exterior algebras: Terminology...................... 397
         10.3.1 Exterior basis of a mixed exterior algebra 397
         10.3.2 Axiomatic tensor operations in the $\bigwedge_n^{(p,q)}(K)$ algebra 398

10.4   Exterior product of mixed exterior vectors 399
10.5   Anti-commutativity of the $\bigwedge$ mixed exterior product 403
10.6   Change of basis in mixed exterior algebras 404
10.7   Exercises 409

Part IV Tensors over Linear Spaces with Inner Product

---

11 Euclidean Homogeneous Tensors 413
11.1   Introduction 413
11.2   Initial concepts 413
11.3   Tensor character of the inner vector's connection in a
        $PSE^n(\mathbb{R})$ space 416
11.4   Different types of the fundamental connection tensor 418
11.5   Tensor product of vectors in $E^n(\mathbb{R})$ (or in $PSE^n(\mathbb{R})$)....... 421
11.6   Equivalent associated tensors: Vertical displacements of
        indices. Generalization 422
        11.6.1 The quotient space of isomers 426
11.7   Changing bases in $E^n(\mathbb{R})$: Euclidean tensor character criteria 427
11.8   Symmetry and anti-symmetry in Euclidean tensors 430
11.9   Cartesian tensors 433
        11.9.1 Main properties of Euclidean $E^n(\mathbb{R})$ spaces in
                orthonormal bases 433
        11.9.2 Tensor total Euclidean character in orthonormal
                bases   434
        11.9.3 Tensor partial Euclidean character in orthonormal
                bases... 436
        11.9.4 Rectangular Cartesian tensors 436
11.10   Euclidean and pseudo-Euclidean tensor algebra 451
        11.10.1 Euclidean tensor equality 451
        11.10.2 Addition and external product of Euclidean
                (pseudo-Euclidean) tensors....................... 451
        11.10.3 Tensor product of Euclidean (pseudo-Euclidean)
                tensors 452
        11.10.4 Euclidean (pseudo-Euclidean) tensor contraction..... 452
        11.10.5 Contracted tensor product of Euclidean or
                pseudo-Euclidean tensors 455
        11.10.6 Euclidean contraction of tensors of order $r = 2$ 457
        11.10.7 Euclidean contraction of tensors of order $r = 3$ 457
        11.10.8 Euclidean contraction of tensors of order $r = 4$ 457
        11.10.9 Euclidean contraction of indices by the Hadamard
                product 458
11.11   Euclidean tensor metrics............................... 482
        11.11.1 Inner connection.............................. 483
        11.11.2 The induced fundamental metric tensor 484

　　　　11.11.3　Reciprocal and orthonormal basis 486
　　11.12　Exercises 504

12 Modular Tensors over $E^n(\mathbb{R})$ Euclidean Spaces 511
　　12.1　Introduction 511
　　12.2　Diverse cases of linear space connections 511
　　12.3　Tensor character of $\sqrt{|G|}$ 512
　　12.4　The orientation tensor: Definition 514
　　12.5　Tensor character of the orientation tensor 514
　　12.6　Orientation tensors as associated Euclidean tensors 515
　　12.7　Dual or polar tensors over $E^n(\mathbb{R})$ Euclidean spaces 516
　　12.8　Exercises 525

13 Euclidean Exterior Algebra 529
　　13.1　Introduction 529
　　13.2　Euclidean exterior algebra of order $r = 2$ 529
　　13.3　Euclidean exterior algebra of order r $(2 < r < n)$ 532
　　13.4　Euclidean exterior algebra of order r=n 535
　　13.5　The orientation tensor in exterior bases 535
　　13.6　Dual or polar tensors in exterior bases 536
　　13.7　The cross product as a polar tensor in generalized Cartesian
　　　　　coordinate frames 538
　　13.8　$\sqrt{|G|}$ geometric interpretation in generalized Cartesian
　　　　　coordinate frames 539
　　13.9　Illustrative examples 540
　　13.10　Exercises 576

Part V Classic Tensors in Geometry and Mechanics

14 Affine Tensors 581
　　14.1　Introduction and Motivation 581
　　14.2　Euclidean tensors in $E^n(\mathbb{R})$ 582
　　　　　14.2.1　Projection tensor 582
　　　　　14.2.2　The momentum tensor 586
　　　　　14.2.3　The rotation tensor 587
　　　　　14.2.4　The reflection tensor 590
　　14.3　Affine geometric tensors. Homographies 597
　　　　　14.3.1　Preamble 597
　　　　　14.3.2　Definition and representation 599
　　　　　14.3.3　Affinities 600
　　　　　14.3.4　Homothecies 604
　　　　　14.3.5　Isometries 606
　　　　　14.3.6　Product of isometries 623
　　14.4　Tensors in Physics and Mechanics 626

14.4.1   The stress tensor S 628

14.4.2   The strain tensor Γ 630

14.4.3   Tensor relationships between S and Γ. Elastic tensor . 635

14.4.4   The inertial moment tensor 647

14.5   Exercises 655

Bibliography... 659

Index... 663

Part I

Basic Tensor Algebra

1

Tensor Spaces

1.1 Introduction

In this chapter we give some concepts that are required in the remaining chapters of the book. This includes the concepts of reciprocal coordinate frames, contravariant and covariant coordinates of a vector, some formulas for changes of basis, etc.

We also introduce different types of matrix products, such as the ordinary, the tensor or the Schur products, together with their main properties, that will be used extensively to operate and simplify long expressions throughout this book.

Since we extend and condense tensors very frequently, i.e., we represent tensors as vectors to take full advantage of vector theory and tools, and then we recover their initial tensor representation, we present the corresponding extension and condensation operators that permit moving from one of these representations to the other, and vice versa.

These operators are used initially to solve some important matrix equations that are introduced, together with some interesting applications.

Finally, the chapter ends with a section devoted to special tensors that are used to solve important physical problems.

1.2 Dual or reciprocal coordinate frames in affine Euclidean spaces

Let $E^n(\mathbb{R})$ be an n-dimensional affine linear space over the field $\mathbb{R}$ equipped with an inner connection (inner or dot product), $< \cdot, \cdot >$, and let $\{\vec{e}_\alpha\}$ be a basis of $E^n(\mathbb{R})$. The vector $\vec{V}$ with components $\{x^\alpha\}$ in the initial basis $\{\vec{e}_\alpha\}$, i.e., the vector $\vec{V} = \sum_{\alpha=1}^{n} x^\alpha \vec{e}_\alpha$ will be represented in the following by the symbolic matrix expression

$$\vec{V} = ||\vec{e}_\alpha||X \equiv [\vec{e}_1 \vec{e}_2 \cdots \vec{e}_n] \begin{bmatrix} x^1 \\ x^2 \\ \vdots \\ x^n \end{bmatrix}. \tag{1.1}$$

In this book vector matrices will always be represented as row matrices, denoted by $|| \cdot ||$, and component matrices always as column matrices, denoted by $[\cdot]$. So, when referring to columns of vectors or rows of components, we must use the corresponding transpose matrices.

To every pair of vectors, $\{\vec{V}, \vec{W}\}$, the connection assigns a real number (a scalar), given by the matrix relation

$$< \vec{V}, \vec{W} > = X^t G Y, \tag{1.2}$$

where X and Y are the column matrices with the coordinates of vectors $\vec{V}$ and $\vec{W}$, respectively, and G is the *Gram matrix* of the connection, which is given by

$$G_n = [g_{\alpha\beta}] = [< \vec{e}_\alpha, \vec{e}_\beta >]; \quad g_{\alpha\beta} \in \mathbb{R}; \quad G = G^t, \; |G| \neq 0. \tag{1.3}$$

As is well known, if a new basis is selected, all the mathematical objects associated with the linear space change representation.

So, if

$$||\vec{\hat{e}}_i||_{1,n} = ||\vec{e}_\alpha||_{1,n} C_{n,n} \tag{1.4}$$

is the matrix representation of the change-of-basis, and the subindices refer to the matrix dimensions (row and columns, respectively), a vector $\vec{V}$ can be written as

$$\vec{V} = ||\vec{e}_\alpha||X_{n,1}$$

and also as

$$\vec{V} = ||\vec{\hat{e}}_i||\hat{X}_{n,1},$$

where the initial $X_{n,1}$ and new $\hat{X}_{n,1}$ components are related by

$$X_{n,1} = C\hat{X}_{n,1}. \tag{1.5}$$

It is obvious that any change-of-basis can be performed with the only constraint of having an associated C non-singular matrix ($|C| \neq 0$).

However, there exists a *very special* change-of-basis that is associated with the matrix G

$$C \equiv G^{-1}, \tag{1.6}$$

for which the resulting new basis will not be denoted by $\{\vec{\hat{e}}_i\}$, but with the special notation $\{\vec{e}^{*\alpha}\}$, and it will be called the reciprocal or dual basis of $\{\vec{e}_\alpha\}$.

The vector $\vec{V} = ||\vec{e}_\alpha||X$ with components $\{x^\alpha\}$ in the initial basis now has the components $\{x_\alpha\}$, that is

$$\vec{V} = ||\vec{e}^{*\alpha}||X^* = [\vec{e}^{*1}\vec{e}^{*2}\cdots\vec{e}^{*n}]\begin{bmatrix} x_1 \\ x_2 \\ \vdots \\ x_n \end{bmatrix}.$$

Hence, taking into account (1.6), expression (1.5) leads to

$$X = G^{-1}X^* \Leftrightarrow X^* = GX \qquad (1.7)$$

and from (1.4) we get

$$||\vec{e}^{*\alpha}|| = ||\vec{e}_\alpha||G^{-1} \Leftrightarrow [\vec{e}^{*1}\vec{e}^{*2}\cdots\vec{e}^{*n}] = [\vec{e}_1\vec{e}_2\cdots\vec{e}_n]G^{-1}. \qquad (1.8)$$

Equation (1.7) gives the relation between the *contravariant coordinates*, X, of vector $\vec{V}$ in the initial frame and the *covariant coordinates*, X^*, of the same vector $\vec{V}$, when it is referred to a new frame that is the reciprocal or dual of the initial frame. In short, in a punctual affine space we make use of two frames *simultaneously*:

1. The $(O-\{\vec{e}_\alpha\})$ initial or primary (contravariant coordinates).
2. The $(O-\{\vec{e}^{*\alpha}\})$ reciprocal (covariant coordinates) (in spheric three-dimensional geometry it is the *polar* trihedron of the given one).

Following the exposition, assume that the coordinates of two vectors $\vec{V}$ and $\vec{W}$ are given and that their dot product is sought after.

1. If the two vectors are given in contravariant coordinates, we use the expression (1.2):
$$< \vec{V}, \vec{W} > = X^t G Y.$$

2. If $\vec{V}$ is given in contravariant coordinates (column matrix X) and $\vec{W}$ is given in covariant coordinates (column matrix Y^*), and at this time the *heterogeneous connection* is not known, expression (1.7) can be obtained by writing $\vec{W}$ in contravariant coordinates, $Y = G^{-1}Y^*$ and using expression (1.2):

$$< \vec{V}, \vec{W} > = X^t G(G^{-1}Y^*) = X^t Y^* = X^t I Y^*. \qquad (1.9)$$

The *surprising result* is that with data vectors in *contra-covariant coordinates* the heterogeneous connection matrix is the identity matrix I, and the result can be obtained by a *direct* product of the data coordinates. From this result, one can begin to understand that the simultaneous use of data in contra and cova forms can greatly facilitate tensor operations.

3. If $\vec{V}$ is given in covariant coordinates (X^*) and $\vec{W}$ in contravariant coordinates (matrix Y), proceeding in a similar way with vector $\vec{V}$, and using (1.7), one gets

$$< \vec{V}, \vec{W} > = (G^{-1}X^*)^t GY = (X^*)^t (G^{-1})^t GY = (X^*)^t G^{-1}GY = (X^*)^t IY, \tag{1.10}$$

where once more we observe that *cova-contravariant* data imply a unit connection matrix I.

4. Finally, if one has *cova-covariant* data, that is, $\vec{V}(X^*)$ and $\vec{W}(Y^*)$, the result will be

$$\begin{aligned} < \vec{V}, \vec{W} > &= X^t GY = (G^{-1}X^*)^t G(G^{-1}Y^*) \\ &= (X^*)^t G^{-1}GG^{-1}Y^* = (X^*)^t G^{-1}Y^*, \end{aligned} \tag{1.11}$$

which discovers that for a *reciprocal* frame, the *Gram matrix* is $G^* \equiv G^{-1}$, that is,

$$< \vec{V}, \vec{W} > = (X^*)^t G^{-1} Y^*. \tag{1.12}$$

Example 1.1 (Change of basis). The Gram matrices associated with linear spaces equipped with inner products (Euclidean, pre-Euclidean, etc.) when changing bases, transform in a "congruent" way, i.e.: $\hat{G} = C^t GC$. The proof is as follows.

Proof: By definition we have

$$G = ||\vec{e}_i||^t \bullet ||\vec{e}_j|| = \begin{bmatrix} \vec{e}_1 \\ \vec{e}_2 \\ \vdots \\ \vec{e}_n \end{bmatrix} \bullet [\, \vec{e}_1 \ \vec{e}_2 \ \cdots \ \vec{e}_n \,] = \begin{bmatrix} \vec{e}_1 \bullet \vec{e}_1 & \vec{e}_1 \bullet \vec{e}_2 & \cdots & \vec{e}_1 \bullet \vec{e}_n \\ \vec{e}_2 \bullet \vec{e}_1 & \vec{e}_2 \bullet \vec{e}_2 & \cdots & \vec{e}_2 \bullet \vec{e}_n \\ \cdots & \cdots & \cdots & \cdots \\ \vec{e}_n \bullet \vec{e}_1 & \vec{e}_n \bullet \vec{e}_2 & \cdots & \vec{e}_n \bullet \vec{e}_n \end{bmatrix}. \tag{1.13}$$

If the scalar $\vec{e}_i \bullet \vec{e}_j$ is denoted by g_{ij}, we have $G \equiv [g_{ij}]$, and since $g_{ij} = \vec{e}_i \bullet \vec{e}_j = \vec{e}_j \bullet \vec{e}_i = g_{ji}$ we get $G = G^t$.

If in the linear space we consider the change-of-basis $||\hat{\vec{e}}_i|| = ||\vec{e}_i||C$, then we have

$$\hat{G} = ||\hat{\vec{e}}_i||^t \bullet ||\hat{\vec{e}}_j|| = (||\vec{e}_i||C)^t \bullet (||\vec{e}_j||C) = C^t (||\vec{e}_i||^t \bullet ||\vec{e}_i||)C$$

and using (1.13), we finally get $\hat{G} = C^t GC$, which is the desired result.

□

Next, an example is given to clarify the above material.

Example 1.2 (Linear operator and scalar product). Assume that in the affine linear space $E^n(\mathbb{R})$ referred to the basis $\{\vec{e}_\alpha\}$, a given linear operator (of associated matrix T given in the cited basis) transforms the vectors in the affine linear space into vectors of the same space. In this situation, one performs a change-of-basis in $E^n(\mathbb{R})$ (with given associated matrix C). We are interested in finding the matrix M associated with the linear operator, such that taking vectors in contravariant coordinates of the *initial basis* returns the transformed vectors in "covariant coordinates" of the *new basis*.

We have the following well-known relations:

1. In the initial frame of reference, when changing bases, the Gram matrix (see the change-of-basis for the bilinear forms) satisfies

$$\hat{G} = C^t G C. \tag{1.14}$$

2. It is known that the linear operator operates in $E^n(\mathbb{R})$ as

$$Y = TX. \tag{1.15}$$

3. According to (1.5) the change-of-basis for vectors leads to

$$\begin{aligned} X &= C\hat{X} \\ Y &= C\hat{Y}, \end{aligned} \tag{1.16}$$

and entering with (1.7) for the vector $\vec{W}$ in the new basis gives

$$(\hat{Y})^* = \hat{G}\hat{Y} \overset{\text{because of (1.14)}}{=} (C^t G C)\hat{Y} \overset{\text{because of (1.16)}}{=} (C^t G C)(C^{-1}Y) = C^t G Y \overset{\text{because of (1.15)}}{=} C^t G(TX). \tag{1.17}$$

Thus, we get $(\hat{Y})^* = MX$ with $M = C^t G T$, which is the sought after result. $\square$

Finally, we examine in some detail how the axes scales of the reference frame $\{\vec{e}^{*i}\}$ are the "dual" or "reciprocal" of the given reference frame $\{\vec{e}_i\}$.

Equation (1.8):

$$[\vec{e}^{*1} \vec{e}^{*2} \cdots \vec{e}^{*n}] = [\vec{e}_1 \vec{e}_2 \cdots \vec{e}_n] G^{-1}$$

declares that the director vector associated with the main direction $\overline{OX}_i^*$ (in the dual reference frame $(O - X_1^*, X_2^*, \ldots, X_n^*)$) is

$$\vec{e}^{*i} = g^{1i}\vec{e}_1 + g^{2i}\vec{e}_2 + \cdots + g^{ii}\vec{e}_i + \cdots + g^{ni}\vec{e}_n, \tag{1.18}$$

where $[g^{ij}] \equiv G^{-1}$ is the inverse of G and symmetric, and then

$$g^{ij} = \frac{G^{ij}}{|G|}, \tag{1.19}$$

where G^{ij} is the adjoint of g_{ij} in G.

The modulus of the vector $\vec{e}^{*i}$ is

$$|\vec{e}^{*i}| = \sqrt{< \vec{e}^{*i}, \vec{e}^{*i} >} = \sqrt{g^{ii}} = \sqrt{\frac{G^{ii}}{|G|}}, \tag{1.20}$$

which gives the *scales* of each main direction $\overline{OX}_i^*$, in the reciprocal system, which are the reciprocal of the scales of the *fundamental system* (contravariant) when G is diagonal.

Since $< \vec{e}^{*i}, \vec{e}_j > = 0; \forall i \neq j$, all $\vec{e}^{*i}$ has a direction that is *orthogonal* to all remaining vectors $\vec{e}_j$ ($j \neq i$). All this recalls the properties of the "polar trihedron" of a given trihedron in spheric geometry.

Remark 1.1. If the reference frame is orthogonalized but not orthonormalized, that is, if

$$G \equiv \begin{vmatrix} g_{11} & 0 & \cdots & 0 \\ 0 & g_{22} & \cdots & 0 \\ \cdots & \cdots & \ddots & \vdots \\ 0 & 0 & \cdots & g_{nn} \end{vmatrix}$$

expression (1.20) becomes

$$|\vec{e}^{*i}| = \sqrt{\frac{G^{ii}}{|G|}} = \sqrt{\frac{\frac{|G|}{g^{ii}}}{|G|}} = \frac{1}{\sqrt{g_{ii}}}. \tag{1.21}$$

□

1.3 Different types of matrix products

1.3.1 Definitions

In this section the most important matrix products are defined.

Definition 1.1 (Inner, ordinary or scalar matrix product). *Consider the following matrices:*

$$A_{m,h} \equiv [a_{i\alpha}]; \quad B_{h,n} \equiv [b_{\beta j}]; \quad P_{m,n} \equiv [p_{ij}],$$

where the matrix subindices and the values within brackets refer to their dimensions and the corresponding elements, respectively.

We say that the matrix P is the inner, ordinary or scalar product of matrices A and B, and it is denoted by A • B, iff (if and only if)

$$P = A \bullet B \Rightarrow p_{ij} = \sum_{\alpha=1}^{\alpha=h} a_{i\alpha} b_{\alpha j}; \quad i = 1, 2, \ldots, m; \quad j = 1, 2, \ldots, n.$$

Definition 1.2 (External product of matrices). *Consider the following matrices:*

$$A_{m,h} \equiv [a_{ij}]; \quad P_{m,h} \equiv [p_{ij}],$$

and the scalar $\lambda \in K$. We say that the matrix P is the external product of the scalar λ and the matrix A, and is denoted by $\lambda \circ A$, iff

$$P = \lambda \circ A \Rightarrow p_{ij} = \lambda a_{ij}.$$

Definition 1.3 (Kronecker, direct or tensor product of matrices). *Consider the following matrices:*

$$A_{m,n} \equiv [a_{\alpha\beta}]; \quad B_{p,q} \equiv [b_{\gamma\delta}]; \quad P_{mp,nq} \equiv [p_{ij}].$$

We say that the matrix P is the Kronecker, direct or tensor product of matrices A and B and it is denoted by $A \otimes B$, iff

$$P = A \otimes B \Rightarrow p_{ij} = a_{\alpha\beta}b_{\gamma\delta} = a_{\lfloor \frac{i-1}{p}\rfloor+1,\lfloor \frac{j-1}{q}\rfloor+1}b_{i-\lfloor \frac{i-1}{p}\rfloor p,j-\lfloor \frac{j-1}{q}\rfloor q}, \quad (1.22)$$

where $i = 1,2,\ldots,mp$, ; $j = 1,2,\ldots,nq$, $\lfloor x \rfloor$ is the integer part of x, with an order fixed by convention and represented by means of "submatrices":

$$P_{mp,nq} = A_{m,n} \otimes B_{p,q}.$$

$$P_{mp,nq} = \begin{bmatrix} a_{11} \circ B & a_{12} \circ B & \cdots & a_{1n} \circ B \\ \hline a_{21} \circ B & a_{22} \circ B & \cdots & a_{2n} \circ B \\ \hline \cdots & \cdots & \cdots & \cdots \\ \hline a_{m1} \circ B & a_{m2} \circ B & \cdots & a_{mn} \circ B \end{bmatrix},$$

where each partition has p rows and q columns.

It is interesting to know what the row i and column j are, where the factor $a_{\alpha\beta}b_{\gamma\delta}$ appears in the tensor product $A_{m,n} \otimes B_{p,q}$, i.e., what is the corresponding element p_{ij}. Its row and column are given by

$$i = (\alpha-1)p + \gamma; \quad j = (\beta-1)q + \delta. \quad (1.23)$$

Similarly, the reverse transformation, according to (1.22) is

$$\alpha = \left\lfloor \frac{i-1}{p} \right\rfloor + 1; \quad \beta = \left\lfloor \frac{j-1}{q} \right\rfloor + 1; \quad \gamma = i - \left\lfloor \frac{i-1}{p} \right\rfloor p; \quad \delta = j - \left\lfloor \frac{j-1}{q} \right\rfloor q. \quad (1.24)$$

Some authors call this product the total product of matrices, which causes confusion with the total product of linear spaces.

Definition 1.4 (Hadamard or Schur product of matrices). *Consider the following matrices:*

$$A_{m,n} \equiv [a_{ij}]; \quad B_{m,n} \equiv [b_{ij}]; \quad P_{m,n} \equiv [p_{ij}].$$

We say that the matrix P is the Hadamard or Schur product of matrices A and B, and it is denoted by $A\square B$, iff

$$P_{m,n} = A_{m,n}\square B_{m,n} \Rightarrow p_{ij} = a_{ij}b_{ij}; \quad i = 1,2,\ldots,m; \quad j = 1,2,\ldots,n.$$

1.3.2 Properties concerning general matrices

The properties of the sum $+$ and the ordinary product $\bullet$ of matrices, which are perfectly developed in the linear algebra of matrices, are not developed here. Conversely, the most important properties of other products are given.

The most important properties of these products for general matrices are:

1. $A \otimes (B \otimes C) = (A \otimes B) \otimes C$ (associativity of $\otimes$).
2. $\begin{cases} A \otimes (B+C) = A \otimes B + A \otimes C \text{ (right distributivity of } \otimes). \\ (A+B) \otimes C = A \otimes C + B \otimes C \text{ (left distributivity of } \otimes). \end{cases}$
3. $(A \otimes B)^t = A^t \otimes B^t$ (be aware of the order invariance).
4. $(A \otimes B)^* = A^* \otimes B^*$, where $X^* = (\tilde{X}^t)$ (complex fields).
5. Relation between scalar and tensor products. Let $A_{m,n}, B_{p,q}, C_{n,r}$ and $F_{q,s}$ be four data matrices. Then, we have

$$(A \otimes B) \bullet (C \otimes F) = (A \bullet C) \otimes (B \bullet F).$$

In fact, we have

$$A_{m,n} \otimes B_{p,q} = P_{mp,nq}$$
$$C_{n,r} \otimes F_{q,s} = Q_{nq,rs},$$

so that the scalar product $P \bullet Q$ is possible:

$$(A_{m,n} \otimes B_{p,q}) \bullet (C_{n,r} \otimes F_{q,s}) = P_{mp,nq} \bullet Q_{nq,rs} = R_{mp,rs}.$$

In addition, we have

$$A \bullet C = A_{m,n} \bullet C_{n,r} = P'_{m,r}$$

and

$$B \bullet F = B_{p,q} \bullet F_{q,s} = Q'_{p,s}$$

and then

$$(A \bullet C) \otimes (B \bullet F) = P'_{m,r} \otimes Q'_{p,s} = R_{mp,rs},$$

where these formulas aim only to justify the dimensions of the data matrices.
6. Generalization of the relations between scalar and tensor products:

$$(A_1 \otimes B_1) \bullet (A_2 \otimes B_2) \bullet \cdots \bullet (A_k \otimes B_k) = (A_1 \bullet A_2 \bullet \cdots \bullet A_k) \otimes (B_1 \bullet B_2 \bullet \cdots \bullet B_k).$$

This is how one moves from several tensor products to a single one. This is possible only when the dimensions of the corresponding matrices allow the inner product.
7. There is another way of generalizing Property 5, which follows. Consider now the product

$$P = (A_1 \otimes B_1 \otimes C_1) \bullet (A_2 \otimes B_2 \otimes C_2).$$

Assuming that the matrix dimensions allow the products, and using Properties 1 and 5, one gets

$$P = [(A_1 \otimes B_1) \otimes C_1] \bullet [(A_2 \otimes B_2) \otimes C_2] = [(A_1 \otimes B_1) \bullet (A_2 \otimes B_2)] \otimes (C_1 \bullet C_2)$$

and using again Property 5 to the bracket on the second member, we have

$$P = [(A_1 \bullet A_2) \otimes (B_1 \bullet B_2)] \otimes (C_1 \bullet C_2)$$

and using Property 1, the result is

$$P = (A_1 \bullet A_2) \otimes (B_1 \bullet B_2) \otimes (C_1 \bullet C_2).$$

In summary, the following relation holds:

$$(A_1 \otimes B_1 \otimes C_1) \bullet (A_2 \otimes B_2 \otimes C_2) = (A_1 \bullet A_2) \otimes (B_1 \bullet B_2) \otimes (C_1 \bullet C_2),$$

which after generalization leads to

$$(A_1 \otimes A_2 \otimes \cdots \otimes A_k) \bullet (B_1 \otimes B_2 \otimes \cdots \otimes B_k) = (A_1 \bullet B_1) \otimes (A_2 \bullet B_2) \otimes \cdots \otimes (A_k \bullet B_k).$$

8. If we denote by A^k the product $A \bullet A \bullet \cdots \bullet A$ and by $A^{[k]}$ the product $A \otimes A \otimes \cdots \otimes A$, with $k \in \mathbb{N}$, we have

$$A_{m,n}, B_{n,q} \Leftrightarrow (A \bullet B)^{[k]} = A^{[k]} \bullet B^{[k]}.$$

We remind the reader that $(A \bullet B)^k \neq A^k \bullet B^k$, unless A and B commute.

9.

$$\begin{aligned}
\text{rank } (A \otimes B) &= (\text{rank } A)(\text{rank } B) \\
&= (\text{rank } B)(\text{rank } A) \\
&= \text{rank } (B \otimes A). \quad\quad (1.25)
\end{aligned}$$

1.3.3 Properties concerning square matrices

Next we consider only square matrices, that is, of the form $A_{m,m}$ and $B_{p,p}$. The most important properties of these matrices are:

1. $(A \otimes I_p) \bullet (I_m \otimes B) = (I_m \otimes B) \bullet (A \otimes I_p) = A \otimes B$.
2. $\det(A \otimes B) = (\det A)^p (\det B)^m = (\det B)^m (\det A)^p = \det(B \otimes A)$.
3. trace $(A \otimes B) = (\text{trace } A)(\text{trace } B) = (\text{trace } B)(\text{trace } A) = \text{trace } (B \otimes A)$.
4. $(A \otimes B)^{-1} = A^{-1} \otimes B^{-1}$, where one must be aware of the order, and A and B must be regular matrices.
5. Remembering the meaning of the notation A^k and $A^{[k]}$ introduced in Property 8 above, Property 6 of that section for square matrices becomes

$$(A \otimes B)^k = A^k \otimes B^k.$$

1.3.4 Properties concerning eigenvalues and eigenvectors

Let $\{\lambda_i | i = 1, 2, \ldots, m\}$ and $\{\mu_i | i = 1, 2, \ldots, p\}$ be the sets of eigenvalues of $A_{m,m}$, and $B_{p,p}$, respectively. If v_i (column matrix) is an eigenvector of A_m, of eigenvalue λ_i and w_j (column matrix) is an eigenvector of B_p, of eigenvalue μ_j, that is, if $A_m \bullet v_i = \lambda_i \circ v_i$ and $B_p \bullet w_j = \mu_j \circ w_j$, then we have:

1. The set of eigenvalues of the matrix $A \otimes B$ is the set

$$\{\lambda_i \mu_j | i = 1, 2, \ldots, m; j = 1, 2, \ldots, p\}. \tag{1.26}$$

2. The set of eigenvalues of the matrix $Z = (A \otimes I_p) + (I_m \otimes B)$ is the set

$$\{\lambda_i + \mu_j | i = 1, 2, \ldots, m; j = 1, 2, \ldots, p\}. \tag{1.27}$$

Remark 1.2. The matrix A can be replaced by the matrix A^t and the matrix B by the matrix B^t.   □

3. The set of eigenvectors of the matrix $A \otimes B$ is the set

$$\{\vec{v}_i \otimes \vec{w}_j | i = 1, 2, \ldots, m; j = 1, 2, \ldots, p\}.$$

Proof.

$$(A \otimes B) \bullet (v_i \otimes w_j) = (A \bullet v_i) \otimes (B \bullet w_j) = (\lambda_i \circ v_i) \otimes (\mu_j \circ w_j) = (\lambda_i \mu_j) \circ (v_i \otimes w_j),$$

which shows that $\vec{v}_i \otimes \vec{w}_j$ are the eigenvectors of $A \otimes B$.

Example 1.3 (Eigenvalues). Consider the tridiagonal symmetric matrix

$$A_{n,n} = \begin{bmatrix} 2 & -1 & 0 & 0 & \cdots & 0 & 0 & 0 \\ -1 & 2 & -1 & 0 & \cdots & 0 & 0 & 0 \\ 0 & -1 & 2 & -1 & \cdots & 0 & 0 & 0 \\ \cdots & \cdots & \cdots & \cdots & \cdots & \cdots & \cdots & \cdots \\ 0 & 0 & 0 & 0 & \cdots & -1 & 2 & -1 \\ 0 & 0 & 0 & 0 & \cdots & 0 & -1 & 2 \end{bmatrix}$$

of order n, which is also called *finite difference matrix of order n*, and let I_n be the unit matrix. The matrix

$$L_{n^2,n^2} = (A_{n,n} \otimes I_n) + (I_n \otimes A_{n,n})$$

is called the *Laplace discrete bidimensional matrix*.

Since the eigenvalues of matrix $A_{n,n}$ are

$$4\sin^2\left(\frac{\pi i}{2(n+1)}\right); \quad i = 1, 2, \ldots, n$$

and in this case $A = B = A_{n,n}$, according to the Property 2 above, the set of eigenvalues of L_{n^2,n^2} is

$$\{\lambda_{ij}\} \equiv \left\{4\left[\sin^2\left(\frac{\pi i}{2(n+1)}\right) + \sin^2\left(\frac{\pi j}{2(n+1)}\right)\right]; \; i, j = 1, 2, \ldots, n\right\}.$$

□

1.3.5 Properties concerning the Schur product

Some important properties of the Schur product are:

1. Associativity,
$$(A\square B)\square C = A\square(B\square C).$$

2. Commutativity,
$$A\square B = B\square A.$$

3. For matrices $A_{m,n} = [a_{ij}]$, the matrix
$$N_{m,n} = \begin{bmatrix} 1 & 1 & \cdots & 1 \\ \cdots & \cdots & \cdots & \cdots \\ 1 & 1 & \cdots & 1 \end{bmatrix}; \quad \text{with } n_{ij} = 1; \ i = 1, 2, \ldots, m; \ j = 1, 2, \ldots, n$$

   is the "unit" element of the Schur product "$\square$".
4. For matrices $A_{mn} = [a_{ij}](\forall a_{ij} \neq 0)$, there exists an "inverse matrix" for the product $\square$ (Abelian group), $\forall a_{ij} \in K$.
5. Distributivity of $\square$ with respect to $+$,
$$A\square(B+C) = A\square B + A\square C$$
$$(A+B)\square C = A\square C + B\square C.$$

6. Schur product transpose,
$$(A\square B)^t = A^t\square B^t.$$

7. Other properties.
$$(A\square B)^* = A^*\square B^*; \quad A^* = (\tilde{A}^t).$$

1.3.6 Extension and condensation of matrices

Next, we consider $\{A_{m,n}\}$ the linear space $K^{m\times n}(+, \circ)$, from the point of view of a manifold.

We shall refer vectors in this space to its "canonical basis" $\mathcal{B} = \{E_{ij}\}$, which consists of the simplest matrices of $K^{m\times n}$, that is,

$$\mathcal{B} = \{(E_{ij})_{mn}\} \equiv \{E_{11}, E_{12}, \cdots, E_{1n}, E_{21}, E_{22}, \cdots, E_{mn}\},$$

where $(E_{ij})_{mn} = [k_{\alpha\beta}]$ and

$$k_{\alpha\beta} = \begin{cases} 1 \text{ if } \alpha = i \text{ and } \beta = j \\ 0 \text{ otherwise.} \end{cases}$$

When choosing the basis $\mathcal{B}$, matrix $A_{m,n}$ is expressed as a linear manifold spanned by $\mathcal{B}$. If a "matrix form" is adopted to notate the linear manifold, we get

$$A_{m,n} = [E_{11}|E_{12}|\cdots|E_{ij}|\cdots|E_{mn}] \begin{bmatrix} a_{11} \\ \hline a_{12} \\ \hline \vdots \\ \hline a_{1n} \\ \hline a_{21} \\ \hline a_{22} \\ \hline \vdots \\ \hline a_{2n} \\ \hline \vdots \\ \hline a_{m1} \\ \hline \vdots \\ \hline a_{mn} \end{bmatrix} \equiv ||E_{ij}||X_{mn,1},$$

where all the elements of matrix A_{mn} appear "stacked" in a column matrix X according to the ordering criteria imposed by the given basis $\mathcal{B}$, and the matrix product must be understood in symbolic form and as products of blocks. When one desires a given matrix $A_{m,n}$ in this form, the English language texts write:

"obtained by stacking the elements of the rows of $A_{m,n}$ in sequence."

However, we want to note that it is not necessary to express this result in words; one can use the universal language of linear algebra.

Extension of a matrix

Given a matrix $T_{m,n}$, and calling $\sigma = m \cdot n$ (not a prime number) the dimension of the linear space $T_{m,n}(K^\sigma)$ of matrices, we define by "extension" the mapping

$$E : K^{m \times n} \to K^\sigma,$$

such that $\forall T_{m,n} \in K^{m \times n} : E(T_{m,n}) = T_{\sigma,1}$ with $T_{\sigma,1} \in K^\sigma$, that is, the "stacked" view is replaced by "stack and extend the given matrix and write it in column form".

The "stacked" column matrix $T_{\sigma,1}$, associated with matrix $T_{m,n}$ can be obtained by

$$(T_{\sigma,1})^t = [1 \quad 1 \quad \cdots \quad 1]_{1,m^2} \bullet [D_{m^2} \bullet (I_m \otimes T_{m,n})], \tag{1.28}$$

where the diagonal matrix D_{m^2} is such that $d_{ii} \in \{0,1\}$ with $d_{ii} = 1$ if $i = 1 + (m+1)(\alpha - 1); \alpha = 1, 2, \ldots, m$.

If $\mathcal{B}_{m^2} = \{E_{11}, E_{12}, \ldots, E_{ij}, \ldots, E_{mm}\}$ is the canonical basis of matrices $\mathbb{R}^{m \times m}$, we have that the matrix D_{m^2} in block form is

$$D_{m^2,m^2} = \left[\begin{array}{c|c|c|c} E_{11} & \Omega & \cdots & \Omega \\ \hline \Omega & E_{22} & \cdots & \Omega \\ \hline \Omega & \Omega & \cdots & E_{mm} \end{array}\right], \qquad (1.29)$$

that is, we have an alternative way of obtaining the matrix D_{m^2,m^2} to be used in the formula for the stacked X matrix.

However, we shall use even simpler expressions.

If $\mathcal{B}'_m = \{E_i\}_{1\le i\le m}$ is the canonical basis of matrices in $\mathbb{R}^{m\times 1}$, we have the following:

Extension formula:

$$(T_{\sigma,1})^t = [\, E_1^t \;\; |E_2^t \;\; | \;\; \cdots \;\; | \;\; E_m^t \,]_{1,m^2} \bullet (I_m \otimes T_{m,n}) = J_{1,m^2} \bullet (I_m \otimes T_{m,n}),$$
$$(1.30)$$

because

$$[\,1 \;\; 1 \;\; \cdots \;\; 1\,]_{1,m^2} \bullet D_{m^2} = [\, E_1^t \;\; |E_2^t \;\; | \;\; \cdots \;\; | \;\; E_m^t \,]_{1,m^2}\,,$$

where we have denoted by J_{1,m^2} the matrix $[\, E_1^t \;\; |E_2^t \;\; | \;\; \cdots \;\; | \;\; E_m^t \,]_{1,m^2}$, and then

$$T_{\sigma,1} = (I_m \otimes T_{m,n}^t) \bullet J_{1,m^2}^t. \qquad (1.31)$$

Condensation of a matrix

Similarly, given a "stacked" matrix, $T_{\sigma,1}$ we can be interested in its "condensation", that is, recover its original format $T_{m,n}$ as a matrix.

Since we know that $\sigma = m \cdot n$, we define as "condensation" the mapping

$$\mathcal{C} : K^\sigma \to K^{m\times n}$$

such that $\forall T_{\sigma,1} \in K^\sigma : \mathcal{C}(T_{\sigma,1}) = T_{m,n}$ with $T_{m,n} \in K^{m\times n}$.

Condensation formula:

$$T_{m,n} = \left(I_m \otimes T_{\sigma,1}^t\right) \bullet \left(\left[\begin{array}{c} E_1 \\ \hline E_2 \\ \hline \vdots \\ \hline E_m \end{array}\right] \otimes I_n\right) = \left(I_m \otimes T_{\sigma,1}^t\right) \bullet \left(J_{1,m^2}^t \otimes I_n\right).$$

$$sizes \;:\quad (m, m^2 n) \qquad (m^2 n, n)$$

$$(1.32)$$

Example 1.4 (Extension of a matrix). Consider the matrix

$$A_{3,4} = \begin{bmatrix} a_{11} & a_{12} & a_{13} & a_{14} \\ a_{21} & a_{22} & a_{23} & a_{24} \\ a_{31} & a_{32} & a_{33} & a_{34} \end{bmatrix};$$

where $m = 3$ and $n = 4$, then

$$I_3 \otimes A_{3,4} = \begin{bmatrix} A_{3,4} & | & \Omega_{3,4} & | & \Omega_{3,4} \\ -- & + & -- & + & -- \\ \Omega_{3,4} & | & A_{3,4} & | & \Omega_{3,4} \\ -- & + & -- & + & -- \\ \Omega_{3,4} & | & \Omega_{3,4} & | & A_{3,4} \end{bmatrix}_{9,12}.$$

Next, we obtain the diagonal matrix $D_{m^2,m^2} \equiv D_{9,9}$. In this case we have

α	1	2	3
i	1	5	9

and then

$$D_{9,9} \bullet (I_3 \otimes A_{3,4}) = \begin{bmatrix} \begin{smallmatrix} 1 & 0 & 0 \\ 0 & 0 & 0 \\ 0 & 0 & 0 \end{smallmatrix} & | & \Omega & | & \Omega \\ -- & + & -- & + & -- \\ \Omega & | & \begin{smallmatrix} 0 & 0 & 0 \\ 0 & 1 & 0 \\ 0 & 0 & 0 \end{smallmatrix} & | & \Omega \\ -- & + & -- & + & -- \\ \Omega & | & \Omega & | & \begin{smallmatrix} 0 & 0 & 0 \\ 0 & 0 & 0 \\ 0 & 0 & 1 \end{smallmatrix} \end{bmatrix} \bullet (I_3 \otimes A_{3,4})$$

$$= \begin{bmatrix} \begin{smallmatrix} a_{11} & a_{12} & a_{13} & a_{14} \\ 0 & 0 & 0 & 0 \\ 0 & 0 & 0 & 0 \end{smallmatrix} & | & \Omega & | & \Omega \\ -- & + & -- & + & -- \\ \Omega & | & \begin{smallmatrix} 0 & 0 & 0 & 0 \\ a_{21} & a_{22} & a_{23} & a_{24} \\ 0 & 0 & 0 & 0 \end{smallmatrix} & | & \Omega \\ -- & + & -- & + & -- \\ \Omega & | & \Omega & | & \begin{smallmatrix} 0 & 0 & 0 & 0 \\ 0 & 0 & 0 & 0 \\ a_{31} & a_{32} & a_{33} & a_{34} \end{smallmatrix} \end{bmatrix}_{9,12}.$$

Thus, in summary:

$$X^t = \begin{bmatrix} 1 & 1 & \cdots & 1 \end{bmatrix}_{1,9} \bullet (D_{9,9} \bullet (I_3 \otimes A_{3,4})) \equiv [a_{11}a_{12} \ldots a_{14}a_{21} \ldots a_{33}a_{34}].$$

□

1.3.7 Some important matrix equations

In this section, after introducing some concepts, we state and solve some important matrix equations, i.e., some equations where the unknowns are matrices.

There are a long list of references on Matrix equations (see some of them in the references)[1]

We call a *transposition matrix* of order n, every matrix resulting from exchanging any two rows of the unit matrix I_n, and leaving the remaining rows unchanged. The transposition matrices are *always* regular ($|P| \neq 0$, symmetric ($P = P^t$), involutive ($P = P^{-1}$) and orthogonal ($P^{-1} = P^t$).

We call a *permutation matrix* the scalar or tensor product of several "transposition matrices" (in the second case they can be of different order $(P = P_{m,m} \otimes P_{n,n})$.

Next, we solve the following equations.

Matrix tensor product commuters equation

Consider the equation

$$B \otimes A = P_1 \bullet (A \otimes B) \bullet P_2, \tag{1.33}$$

where $P_1 \in \{\text{permutations of } I_{mp}\}$ and $P_2 \in \{\text{permutations of } I_{nq}\}$ are the unknown matrices.

Note that in general $A \otimes B \neq B \otimes A$, where $[A \otimes B]_{mp,nq}$, i.e., the tensor product is not commutative. Thus, since direct reversal of the tensor product is not permitted, Equation (1.33) allows us to find two correction matrices P_1 and P_2 for reversing the tensor product; these will be called "transposer matrices" due to reasons to be explained in Chapter 5, on tensor morphisms.

We shall give two different expressions for the solution matrices P_1 and P_2.

The first solution is as follows. The permutation matrices P_1 and P_2 (orthogonal matrices $P_1^{-1} = P_1^t; P_2^{-1} = P_2^t$) that solve Equation (1.33), for the products $A_{m,n} \otimes B_{p,q}$ and $B_{p,q} \otimes A_{m,n}$ are as follows:

$$P_1(m, p) \equiv P_{mp,mp} = [p_{1ij}],$$

where

$$p_{1ij} = \begin{cases} 1 \text{ if } i = \left(1 + \left\lfloor \frac{i-1}{p} \right\rfloor\right) + \left[(j-1) - p \left\lfloor \frac{i-1}{p} \right\rfloor\right] m \\ 0 \text{ otherwise} \end{cases} ; \; i, j = 1, \ldots, mp,$$

$$\tag{1.34}$$

where $\lfloor x \rfloor$ is the integer part of x, and

---

[1] Ruíz-Tolosa and Castillo [48] have generalized these equations to tensor equations.

$$P_2(n,q) \equiv P_{nq,nq} = [p_{2ij}],$$

where

$$p_{2ij} = \begin{cases} 1 \text{ if } j = \left(1 + \left\lfloor \frac{i-1}{q} \right\rfloor\right) + \left[(i-1) - q\left\lfloor \frac{i-1}{q} \right\rfloor\right] n \\ 0 \text{ otherwise} \end{cases} ; \; i,j = 1, \ldots, nq,$$

(1.35)

which shows that

$$P_2(n,q) = P_1^t(n,q). \tag{1.36}$$

Remark 1.3. It is interesting to check that the matrices P_1 and P_2 do not depend on the elements of A and B in (1.33), but only on their dimensions.□
□

Example 1.5 (Commuting the tensor product). Consider the particular case $A_{3,3}$ and $B_{3,3}$ ($m = n = p = q = 3$), with $A_{3,3} = [a_{ij}]; B = [b_{ij}]$.
   Applying the indicated formulas, one obtains $P_1 \equiv P_{mp,mp} \equiv P_{9,9}$ with $p_{ij} = 0$, with the exception of

row	1	4	7	2	5	8	3	6	9
column	1	2	3	4	5	6	7	8	9

that is, in the positions

$$(i,j) \equiv (1,1),(4,2),(7,3),(2,4),(5,5),(8,6),(3,7),(6,8),(9,9),$$

which take a value of 1, and $P_2 \equiv P_{nq,nq} \equiv P_{9,9}$ with $p_{ij} = 0$, with the exception of

row	1	2	3	4	5	6	7	8	9
column	1	4	7	2	5	8	3	6	9

that leads to a value of 1 in positions

$$(i,j) \equiv (1,1),(2,4),(3,7),(4,2),(5,5),(6,8),(7,3),(8,6),(9,9).$$

As one can see, the results are identical, and then $P = P_1 = P_2$, where P is symmetric, involutive and orthogonal; thus, we get

$$\begin{bmatrix} b_{11} & b_{12} & b_{13} \\ b_{21} & b_{22} & b_{23} \\ b_{31} & b_{32} & b_{33} \end{bmatrix} \otimes \begin{bmatrix} a_{11} & a_{12} & a_{13} \\ a_{21} & a_{22} & a_{23} \\ a_{31} & a_{32} & a_{33} \end{bmatrix}$$

$$= P \bullet \left(\begin{bmatrix} a_{11} & a_{12} & a_{13} \\ a_{21} & a_{22} & a_{23} \\ a_{31} & a_{32} & a_{33} \end{bmatrix} \otimes \begin{bmatrix} b_{11} & b_{12} & b_{13} \\ b_{21} & b_{22} & b_{23} \\ b_{31} & b_{32} & b_{33} \end{bmatrix} \right) \bullet P \tag{1.37}$$

with

$$P = \left[\begin{array}{ccc|ccc|ccc}
1 & 0 & 0 & 0 & 0 & 0 & 0 & 0 & 0 \\
0 & 0 & 0 & 1 & 0 & 0 & 0 & 0 & 0 \\
0 & 0 & 0 & 0 & 0 & 0 & 1 & 0 & 0 \\
\hline
0 & 1 & 0 & 0 & 0 & 0 & 0 & 0 & 0 \\
0 & 0 & 0 & 0 & 1 & 0 & 0 & 0 & 0 \\
0 & 0 & 0 & 0 & 0 & 0 & 0 & 1 & 0 \\
\hline
0 & 0 & 1 & 0 & 0 & 0 & 0 & 0 & 0 \\
0 & 0 & 0 & 0 & 0 & 1 & 0 & 0 & 0 \\
0 & 0 & 0 & 0 & 0 & 0 & 0 & 0 & 1 \\
\end{array}\right].$$

$\square$

For the case with $p = m, q = n$, that is, matrices A_{mn} and B_{mn} that have the same number of rows and columns, we have $P_1 = P_2^t$ and since they are orthogonal, the matrices $B \otimes A$ and $A \otimes B$ are "similar inside the orthogonal group"

$$P^{-1} \bullet (A \otimes B) \bullet P \equiv P^t \bullet (A \otimes B) \bullet P = B \otimes A.$$

As a final result of the analysis of the matrices P_1 and P_2, that appear in Formula (1.33), we shall propose a second and simple general expression of such matrices.

Let $\mathcal{B}_1 = \{E_{11}, E_{12}, \ldots, E_{ij}, \ldots, E_{mp}\}$ be the canonical basis, with $m \times p$ matrices, of the $\mathbb{R}^{m \times p}$ matrix linear space.

Let $\mathcal{B}_2 = \{E'_{11}, E'_{12}, \ldots, E'_{k\ell}, \ldots, E'_{nq}\}$ be the canonical basis with $n \times q$ matrices, in the $\mathbb{R}^{n \times q}$ matrix linear space.

Matrices P_1 and P_2 will be represented by blocks:

$$P_1 \equiv P_{mp,mp} = \left[\begin{array}{c|c|c|c}
E_{11} & E_{21} & \cdots & E_{m1} \\
\hline
E_{12} & E_{22} & \cdots & E_{m2} \\
\cdots & \cdots & \cdots & \cdots \\
\hline
E_{1p} & E_{2p} & \cdots & E_{mp}
\end{array}\right]. \qquad (1.38)$$

$$P_2 \equiv P_{nq,nq} = \left[\begin{array}{c|c|c|c}
E'_{11} & E'_{21} & \cdots & E'_{n1} \\
\hline
E'_{12} & E'_{22} & \cdots & E'_{n2} \\
\cdots & \cdots & \cdots & \cdots \\
\hline
E'_{1q} & E'_{2q} & \cdots & E'_{nq}
\end{array}\right]^t. \qquad (1.39)$$

Special attention must be given to the "block ordering" inside matrices P_1 and P_2; *it is not* the canonical order but the transpose. As an example, all these results will be applied to the previous application related to Formula (1.33).

Example 1.6 (Commuting the tensor product). Consider again the matrices $A_{3,3}$ and $B_{3,3}$ in Example 1.5. The matrices P_1 and P_2 that solve our application ($m = n = p = q = 3$) now have a direct construction:

$$P_1 = P_2 = P = \begin{bmatrix} E_{11} & E_{21} & E_{31} \\ \hline E_{12} & E_{22} & E_{32} \\ \hline E_{13} & E_{23} & E_{33} \end{bmatrix} = \left[\begin{array}{ccc|ccc|ccc} 1 & 0 & 0 & 0 & 0 & 0 & 0 & 0 & 0 \\ 0 & 0 & 0 & 1 & 0 & 0 & 0 & 0 & 0 \\ 0 & 0 & 0 & 0 & 0 & 0 & 1 & 0 & 0 \\ \hline 0 & 1 & 0 & 0 & 0 & 0 & 0 & 0 & 0 \\ 0 & 0 & 0 & 0 & 1 & 0 & 0 & 0 & 0 \\ 0 & 0 & 0 & 0 & 0 & 0 & 0 & 1 & 0 \\ \hline 0 & 0 & 1 & 0 & 0 & 0 & 0 & 0 & 0 \\ 0 & 0 & 0 & 0 & 0 & 1 & 0 & 0 & 0 \\ 0 & 0 & 0 & 0 & 0 & 0 & 0 & 0 & 1 \end{array}\right], \quad (1.40)$$

which evidently coincides with the P in Example 1.5, which was obtained after using complicated subindex relationships.   □

Linear equation in a single matrix variable (case 1)

Consider the matrix equation

$$A_{n,n} \bullet X_{n,m} + X_{n,m} \bullet B_{m,m} = C_{n,m}, \quad (1.41)$$

in which the unknown is the matrix $X_{n,m}$.

To solve this equation, we proceed to write it in an equivalent form, using the "tensor product"

$$[A_{n,n} \otimes I_{m,m} + I_{n,n} \otimes B_{m,m}^t] \bullet x_{nm,1} = c_{nm,1} <> M \bullet x = c \quad (1.42)$$

with

$$x = [x_{11}, x_{12}, \ldots, x_{1m}, x_{21}, x_{22}, \ldots, x_{2m}, x_{n1}, \ldots, x_{nm}]^t$$

and

$$c = [c_{11}, c_{12}, \ldots, c_{1m}, c_{21}, c_{22}, \ldots, c_{2m}, c_{n1}, \ldots, c_{nm}]^t = (I_n \otimes C_{n,m}^t) \bullet J_{1,n^2}^t,$$

where we have used (1.31).

Now we present equation (1.42) as a matrix equation, in the usual form.

The solution x (and then X) *is unique* if $|M_{mn,mn}| \neq 0$, that is $x = M^{-1} \bullet c$. Then, a unique solution exists if

$$M_{mn,mn} = A_{n,n} \otimes I_m + I_n \otimes B_{m,m}^t$$

is non-singular.

Once x is obtained, we must use (1.32) to obtain the condensed matrix sought after, $X_{n,m}$.

Example 1.7 (Equivalent matrices). Given the two matrices

$$A = \begin{pmatrix} -2 & 4 & -4 \\ 3 & 0 & 4 \\ 2 & -1 & 3 \end{pmatrix} \quad \text{and} \quad B = \begin{pmatrix} 2 & 1 & 1 \\ 1 & 0 & 2 \\ 1 & 1 & -1 \end{pmatrix}, \quad (1.43)$$

obtain the most general matrix C such that $AC = CB$.

If a matrix C exists such that $AC = CB$, then we have

$$AC - CB = \Omega,$$

which is of the form (1.41); thus, after stretching the matrix C, we get (see Equation (1.42)):

$$M \bullet c = (A \otimes I_3 - I_3 \otimes B^t) \bullet c = \left(\begin{array}{ccc|ccc|ccc} -2 & 0 & 0 & 4 & 0 & 0 & -4 & 0 & 0 \\ 0 & -2 & 0 & 0 & 4 & 0 & 0 & -4 & 0 \\ 0 & 0 & -2 & 0 & 0 & 4 & 0 & 0 & -4 \\ \hline 3 & 0 & 0 & 0 & 0 & 0 & 4 & 0 & 0 \\ 0 & 3 & 0 & 0 & 0 & 0 & 0 & 4 & 0 \\ 0 & 0 & 3 & 0 & 0 & 0 & 0 & 0 & 4 \\ \hline 2 & 0 & 0 & -1 & 0 & 0 & 3 & 0 & 0 \\ 0 & 2 & 0 & 0 & -1 & 0 & 0 & 3 & 0 \\ 0 & 0 & 2 & 0 & 0 & -1 & 0 & 0 & 3 \end{array}\right) \left(\begin{array}{c} c_{11} \\ c_{12} \\ c_{13} \\ \hline c_{21} \\ c_{22} \\ c_{23} \\ \hline c_{31} \\ c_{32} \\ c_{33} \end{array}\right)$$

$$- \left(\begin{array}{ccc|ccc|ccc} 2 & 1 & 1 & 0 & 0 & 0 & 0 & 0 & 0 \\ 1 & 0 & 1 & 0 & 0 & 0 & 0 & 0 & 0 \\ 1 & 2 & -1 & 0 & 0 & 0 & 0 & 0 & 0 \\ \hline 0 & 0 & 0 & 2 & 1 & 1 & 0 & 0 & 0 \\ 0 & 0 & 0 & 1 & 0 & 1 & 0 & 0 & 0 \\ 0 & 0 & 0 & 1 & 2 & -1 & 0 & 0 & 0 \\ \hline 0 & 0 & 0 & 0 & 0 & 0 & 2 & 1 & 1 \\ 0 & 0 & 0 & 0 & 0 & 0 & 1 & 0 & 1 \\ 0 & 0 & 0 & 0 & 0 & 0 & 1 & 2 & -1 \end{array}\right) \left(\begin{array}{c} c_{11} \\ c_{12} \\ c_{13} \\ \hline c_{21} \\ c_{22} \\ c_{23} \\ \hline c_{31} \\ c_{32} \\ c_{33} \end{array}\right) = \Omega,$$

$$\tag{1.44}$$

where c_{ij} are the elements of matrix C and the block representation has been used for illustrating the relation of the new matrix to matrices A and B.

Whence

$$\left(\begin{array}{ccc|ccc|ccc} -4 & -1 & -1 & 4 & 0 & 0 & -4 & 0 & 0 \\ -1 & -2 & -1 & 0 & 4 & 0 & 0 & -4 & 0 \\ -1 & -2 & -1 & 0 & 0 & 4 & 0 & 0 & -4 \\ \hline 3 & 0 & 0 & -2 & -1 & -1 & 4 & 0 & 0 \\ 0 & 3 & 0 & -1 & 0 & -1 & 0 & 4 & 0 \\ 0 & 0 & 3 & -1 & -2 & 1 & 0 & 0 & 4 \\ \hline 2 & 0 & 0 & -1 & 0 & 0 & 1 & -1 & -1 \\ 0 & 2 & 0 & 0 & -1 & 0 & -1 & 3 & -1 \\ 0 & 0 & 2 & 0 & 0 & -1 & -1 & -2 & 4 \end{array}\right) \left(\begin{array}{c} c_{11} \\ c_{12} \\ c_{13} \\ \hline c_{21} \\ c_{22} \\ c_{23} \\ \hline c_{31} \\ c_{32} \\ c_{33} \end{array}\right) = \Omega. \tag{1.45}$$

The orthogonal set to the linear subspace spanned by the rows of the square matrix (1.45), that is, the solution of (1.45), is the linear subspace spanned by the set of vectors

$$\{(4,8,8,17,7,7,9,0,0)^t, (1,-1,1,1,1,0,0,1,0)^t, (7,5,-13,5,1,10,0,0,9)^t\}.$$

This implies that the most general matrix C that satisfies equation $AC = CB$ is

$$C = \begin{pmatrix} 4\rho_1+\rho_2+7\rho_3 & 8\rho_1-\rho_2+5\rho_3 & 8\rho_1+\rho_2-13\rho_3 \\ 17\rho_1+\rho_2+5\rho_3 & 7\rho_1+\rho_2+\rho_3 & 7\rho_1+10\rho_3 \\ 9\rho_1 & \rho_2 & 9\rho_3, \end{pmatrix}, \qquad (1.46)$$

where ρ_1, ρ_2 and ρ_3 are arbitrary real constants. Its determinant is

$$|C| = (9\rho_3 - \rho_2)(90\rho_1^2 - 9\rho_1\rho_2 - \rho_2^2 - 9\rho_1\rho_3 - 9\rho_2\rho_3 - 18\rho_3^2),$$

and thus, the most general change-of-basis matrix that transforms matrix A into matrix B, by the similarity transformation, $C^{-1}AC = B$, is that given by (1.46) subject to

$$(9\rho_3 - \rho_2)(90\rho_1^2 - 9\rho_1\rho_2 - \rho_2^2 - 9\rho_1\rho_3 - 9\rho_2\rho_3 - 18\rho_3^2) \neq 0.$$

□

Linear equation in a single matrix variable (case 2)

Similarly, if the equation is

$$A_{m,n} \bullet X_{n,p} \bullet B_{p,q} = C_{m,q}, \qquad (1.47)$$

the corresponding usual equation is

$$[A_{m,n} \otimes B_{p,q}^t] \bullet x_{np,1} = c_{mq,1}. \qquad (1.48)$$

Again, once x is obtained, we must use (1.32) to obtain the condensed matrix sought after, $X_{n,p}$.

Linear equation in two matrix variables

Consider the matrix equation

$$A_{m,p} \bullet X_{p,q} + Y_{m,r} \bullet B_{r,q} = C_{m,q}, \qquad (1.49)$$

where the unknown matrices are $X_{p,q}$ and $Y_{m,r}$.

To solve this equation we proceed to write it stretched in an equivalent form, using the "tensor product":

$$(A_{m,p} \otimes I_q) \bullet x_{pq,1} + (I_m \otimes B_{q,r}^t) \bullet y_{mr,1} = c_{mq,1} \qquad (1.50)$$

with

$$x = [x_{11}, x_{12}, \ldots, x_{1q}, x_{21}, x_{22}, \ldots, x_{2q}, \ldots, x_{pq}]^t$$

$$y = [y_{11}, y_{12}, \ldots, x_{1r}, y_{21}, y_{22}, \ldots, y_{2r}, \ldots, y_{mr}]^t$$

and

$$c = [c_{11}, c_{12}, \ldots, c_{1q}, c_{21}, c_{22}, \ldots, c_{2q}, \ldots, c_{mq}]^t.$$

Finally, Equation (1.50) can be written as

$$\left(A_{m,p} \otimes I_q | (I_m \otimes (B^t)_{q,r}\right) \begin{pmatrix} x_{pq,1} \\ - \\ y_{mr,1} \end{pmatrix} = c_{mq,1}, \tag{1.51}$$

which is the same equation (1.49) but written in the usual form. Then, the solution (x, y) can be obtained by solving a linear system of equations.

Example 1.8 (Equivalent matrices). Given the two matrices

$$A = \begin{pmatrix} -2 & 4 & -4 \\ 3 & 0 & 4 \\ 2 & -1 & 3 \end{pmatrix} \quad \text{and} \quad B = \begin{pmatrix} 2 & 1 & 1 \\ 1 & 0 & 2 \end{pmatrix}, \tag{1.52}$$

obtain the most general matrices X and Y such that $AX + YB = \Omega$.

Since this expression is of the form (1.49), after stretching matrices X and Y, one gets (see Equation (1.51)):

$$\left(\begin{array}{ccc|ccc|ccc|cc|cc|cc} -2 & 0 & 0 & 4 & 0 & 0 & -4 & 0 & 0 & 2 & 1 & 0 & 0 & 0 & 0 \\ 0 & -2 & 0 & 0 & 4 & 0 & 0 & -4 & 0 & 1 & 0 & 0 & 0 & 0 & 0 \\ 0 & 0 & -2 & 0 & 0 & 4 & 0 & 0 & -4 & 1 & 2 & 0 & 0 & 0 & 0 \\ \hline 3 & 0 & 0 & 0 & 0 & 0 & 4 & 0 & 0 & 0 & 0 & 2 & 1 & 0 & 0 \\ 0 & 3 & 0 & 0 & 0 & 0 & 0 & 4 & 0 & 0 & 0 & 1 & 0 & 0 & 0 \\ 0 & 0 & 3 & 0 & 0 & 0 & 0 & 0 & 4 & 0 & 0 & 1 & 2 & 0 & 0 \\ \hline 2 & 0 & 0 & -1 & 0 & 0 & 3 & 0 & 0 & 0 & 0 & 0 & 0 & 2 & 1 \\ 0 & 2 & 0 & 0 & -1 & 0 & 0 & 3 & 0 & 0 & 0 & 0 & 0 & 1 & 0 \\ 0 & 0 & 2 & 0 & 0 & -1 & 0 & 0 & 3 & 0 & 0 & 0 & 0 & 1 & 2 \end{array}\right) \begin{pmatrix} x_{11} \\ x_{12} \\ x_{13} \\ x_{21} \\ x_{22} \\ x_{23} \\ x_{31} \\ x_{32} \\ x_{33} \\ - \\ y_{11} \\ y_{12} \\ y_{21} \\ y_{22} \\ y_{31} \\ y_{32} \end{pmatrix} = \Omega,$$

and solving the resulting homogeneous system and condensing X and Y one finally gets

$$X = \begin{pmatrix} -\rho_3 - 2\rho_4 - 4\rho_7 & -\rho_4 - 4\rho_6 & -2\rho_3 - \rho_4 - 4\rho_5 \\ \rho_1 + 2\rho_2 - 2\rho_3 - 4\rho_4 + \rho_7 & \rho_2 - 2\rho_4 + \rho_6 & 2\rho_1 + \rho_2 - 4\rho_3 - 2\rho_4 + \rho_5 \\ 3\rho_7 & 3\rho_6 & 3\rho_5 \end{pmatrix}$$

$$Y = \begin{pmatrix} -4\rho_2 + 6\rho_4 & -4\rho_1 + 6\rho_3 \\ 3\rho_4 & 3\rho_3 \\ \rho_2 & \rho_1 \end{pmatrix},$$

□

where $\rho_1, \rho_2, \ldots, \rho_7$ are arbitrary real constants.

Example 1.9 (One application to probability). Assume that $\Sigma_{n,n}$ is the variance–covariance matrix of the n-dimensional random variable X, then, the variance–covariance matrix of the n-dimensional random variable $Y_{n,1} = C_{n,n}X_{n,1}$ is $\Sigma^*_{n,n} = C_{n,n}\Sigma_{n,n}(C^t)_{n,n}$. If we look for $\Sigma_{n,n} = I_n$, it must be

$$\Sigma^* = CC^t \Leftrightarrow I_n C^t - C^{-1}\Sigma^* = \Omega.$$

In order to obtain all change-of-basis matrices C leading to this result, we initially solve the equation

$$I_n X + Y\Sigma^* = \Omega,$$

which is of the form (1.49) and then it can be written as

$$\left(I_n \otimes I_n \middle| (I_n \otimes (\Sigma^*_{n,n})^t)\right) \begin{pmatrix} x_{nn,1} \\ -- \\ y_{nn,1} \end{pmatrix} = c_{nn,1}, \qquad (1.53)$$

from which matrices X and Y can be obtained. Next, it suffices to impose the condition

$$Y = \left((-X)^t\right)^{-1}. \qquad (1.54)$$

As an example, consider the matrix

$$\Sigma^* = \begin{pmatrix} 3 & 1 & -1 \\ 1 & 4 & 0 \\ -1 & 0 & 2 \end{pmatrix}$$

then, we get

$$X = \begin{pmatrix} -\rho_7 + \rho_8 + 3\rho_9 & 4\rho_8 + \rho_9 & 2\rho_7 - \rho_9 \\ -\rho_4 + \rho_5 + 3\rho_6 & 4\rho_5 + \rho_6 & 2\rho_4 - \rho_6 \\ -\rho_1 + \rho_2 + 3\rho_3 & 4\rho_2 + \rho_3 & 2\rho_1 - \rho_3 \end{pmatrix}; \quad Y = \begin{pmatrix} \rho_9 & \rho_8 & \rho_7 \\ \rho_6 & \rho_5 & \rho_4 \\ \rho_3 & \rho_2 & \rho_1 \end{pmatrix}.$$

where the ρ_s are arbitrary real numbers that must satisfy (1.54). $\qquad \square$

Linear equation in several matrix variables

Finally we mention that Equations (1.41) and (1.49) can be immediately generalized to

$$\sum_{i=1}^{I}(A_i)_{m,p_i} \bullet (X_i)_{p_i,q} + \sum_{j=1}^{J}(Y_j)_{m,r_j} \bullet (B_j)_{r_j,q} = C_{m,q} \qquad (1.55)$$

leading to the following system of linear equations, which generalizes (1.51):

$$\left((A_1)_{m,p_1} \otimes I_{q,q}| \cdots |(A_I)_{m,p_I} \otimes I_{q,q}|I_{m,m} \otimes (B_1)^t_{r_1,q}| \cdots |I_{m,m} \otimes (B_J)^t_{r_J,q}\right)$$

$$\bullet \begin{pmatrix} (x_1)_{p_1q,1} \\ \vdots \\ (x_I)_{p_Iq,1} \\ --- \\ (y_1)_{mr_1,1} \\ \vdots \\ (y_J)_{mr_J,1} \end{pmatrix} = c_{mq,1}. \qquad (1.56)$$

The Schur-tensor product equation

Consider the following matrix equation, which allows us to replace a tensor product by a Schur product:

$$A_{m,n} \square B_{m,n} = Q_{m,m^2} \bullet (A_{m,n} \otimes B_{m,n}) \bullet P_{n^2,n}, \qquad (1.57)$$

where P and Q, the unknowns, are never square matrices. Note that this is a direct relationship among the "three matrix products".

Solution matrices Q_{m,m^2} and $P_{n^2,n}$ are, respectively, given by

$$Q_{m,m^2} = [q_{ij}]; \quad q_{ij} \in \{0,1\} \text{ with } q_{ij} = \begin{cases} 1 \text{ if } j = i(m+1) - m \\ 0 \text{ otherwise} \end{cases}$$

and

$$P_{n^2,n} = [p_{ij}]; \quad p_{ij} \in \{0,1\} \text{ with } p_{ij} = \begin{cases} 1 \text{ if } i = j(n+1) - n \\ 0 \text{ otherwise.} \end{cases}$$

Nevertheless, and following the previous criterion of having a faster formulation for matrices Q_{m,m^2} and $P_{n^2,n}$ in Formula (1.57) we propose the following block alternative:

$$Q_{m,m^2} = [\, E_{11} \mid E_{22} \mid \cdots \mid E_{m,m} \,], \qquad (1.58)$$

where $\mathcal{B}_{m^2} \equiv \{E_{ij}\}$ is the canonical basis of the matrix linear space $\mathbb{R}^{m,m}$, and

$$P_{n^2,n} = \begin{bmatrix} E'_{11} \\ -- \\ E'_{22} \\ -- \\ \vdots \\ -- \\ E'_{nn} \end{bmatrix}, \qquad (1.59)$$

where $\mathcal{B}'_{n^2} = \{E'_{ij}\}$ is the canonical basis of the matrix linear space $\mathbb{R}^{n,n}$.

Remark 1.4. It is interesting to check that the matrices Q and P do not depend on the elements of A and B in (1.57), but only on their dimensions.   □

We end this section by mentioning another interesting relationship. The matrix D_{m^2}, which appeared in the matrix "stacking" process, Formula (1.29), is also $D_{m^2} = Q_{m,m^2}^t \otimes Q_{m,m^2}$ (be aware of the matrix composition law $\otimes$, tensor product of the matrix blocks).

Example 1.10 (Replacing a tensor product by Schur product). Returning to the case $A_{3,3}$ and $B_{3,3}$ of a previous example, we have

$$A_{3,3} \square B_{3,3} = Q_{3,9} \bullet (A_{3,3} \otimes B_{3,3}) \bullet P_{9,3},$$

where $Q = P^t$ or $P = Q^t$ and

$$Q = \begin{bmatrix} 1 & 0 & 0 & 0 & 0 & 0 & 0 & 0 & 0 \\ 0 & 0 & 0 & 0 & 1 & 0 & 0 & 0 & 0 \\ 0 & 0 & 0 & 0 & 0 & 0 & 0 & 0 & 1 \end{bmatrix},$$

which can be checked easily, using the previous formulas.   □

As a consequence of (1.57), a relation between dot and tensor products can be obtained for the particular case $p = m, q = n$. In fact, we know that

$$A_{m,n} \square B_{m,n} = Q_{m,m^2} \bullet (A_{m,n} \otimes B_{m,n}) \bullet P_{n^2,n}$$
$$B_{m,n} \square A_{m,n} = Q_{m,m^2} \bullet (B_{m,n} \otimes A_{m,n}) \bullet P_{n^2,n}$$

and applying the commutative Property 2 to the left-hand member and equaling the right-hand members, we get

$$Q_{m,m^2} \bullet (A_{m,n} \otimes B_{m,n}) \bullet P_{n^2,n} = Q_{m,m^2} \bullet (B_{m,n} \otimes A_{m,n}) \bullet P_{n^2,n}.$$

1.4 Special tensors

In this section we study the case of special tensors defined in the usual Euclidean space with an orientation to the treatment of physical problems and its main branches, mechanics, hydraulics, etc.

In the following we assume that our Euclidean space $E^3(\mathbb{R})$, whether or not an affine space, has been orthonormalized, that is, the basic vectors $\{\vec{e}_i\}$ satisfy the constraint

$$\vec{e}_i \bullet \vec{e}_j = \delta_{ij} \text{ (Kronecker delta)},$$

and then, the corresponding Gram matrix associated with the dot product is $G_3 \equiv I_3$.

Only at the very end will these tensors be established for "oblique" reference frames, non-orthonormalized and with arbitrary G_3, that will satisfy only the following conditions:

1. $G_3 = G_3^t$
2. $\exists C_0 \ |C_0| \neq 0$ such that $C_0^t G_3 C_0 = I_3$.

Next, the following matrix representation for vectors is used:

$$\vec{u} = ||\vec{e}_i||U, \quad \vec{v} = ||\vec{e}_j||V \quad \text{and} \quad \vec{w} = ||e_i||W$$

and the following product will be particularized to this case:

1. Dot product of vectors:

$$\vec{u} \bullet \vec{v} = <\vec{u}, \vec{v}> = U^t V = [u^1 u^2 u^3] \begin{bmatrix} v^1 \\ v^2 \\ v^3 \end{bmatrix} = u^1 v^1 + u^2 v^2 + u^3 v^3,$$

meaning the scalar value $\vec{u} \bullet \vec{v} = |\vec{u}||\vec{v}| \cos \theta$, which proves that $\vec{u} \bullet \vec{v} = \vec{v} \bullet \vec{u}$.

2. Cross product of vectors:

$$\vec{u} \wedge \vec{v} = \begin{vmatrix} \vec{e}_1 & \vec{e}_2 & \vec{e}_3 \\ u^1 & u^2 & u^3 \\ v^1 & v^2 & v^3 \end{vmatrix},$$

meaning

$$\begin{cases} \vec{u} \wedge \vec{v} \perp \vec{u} \text{ and } \vec{u} \wedge \vec{v} \perp \vec{v} \\ |\vec{u} \wedge \vec{v}| = |\vec{u}||\vec{v}| \sin \theta \\ \text{Dextrorsum sense.} \end{cases}$$

In addition we have $\vec{v} \wedge \vec{u} = -\vec{u} \wedge \vec{v}$.

3. Scalar triple product:

$$\vec{u} \bullet (\vec{v} \wedge \vec{w}) = \begin{vmatrix} u^1 & u^2 & u^3 \\ v^1 & v^2 & v^3 \\ w^1 & w^2 & w^3 \end{vmatrix},$$

which is denoted by $[\vec{u}, \vec{v}, \vec{w}]$, and which mean the volume of the parallelepiped with concurrent edges $\vec{u}, \vec{v}, \vec{w}$.

In addition we have

$$\vec{u} \bullet (\vec{v} \wedge \vec{w}) = \vec{v} \bullet (\vec{w} \wedge \vec{u}) = \vec{w} \bullet (\vec{u} \wedge \vec{v}).$$

4. Vector triple product:

$$\vec{u} \wedge (\vec{v} \wedge \vec{w}) = \begin{vmatrix} \vec{v} & \vec{w} \\ \vec{u} \bullet \vec{v} & \vec{u} \bullet \vec{w} \end{vmatrix} = (\vec{u} \bullet \vec{w})\vec{v} - (\vec{u} \bullet \vec{v})\vec{w},$$

which is called the "back cab rule".

5. The "cosines law" (for plane triangles): Let $\vec{w} = \vec{u} + \vec{v}$, then we have (see Figure 1.1)

$$|\vec{w}|^2 = \vec{w} \bullet \vec{w} = (\vec{u} + \vec{v}) \bullet (\vec{u} + \vec{v}) = \vec{u} \bullet \vec{u} + \vec{v} \bullet \vec{v} + 2\vec{u} \bullet \vec{v},$$

$$|\vec{w}|^2 = |\vec{u}|^2 + |\vec{v}|^2 + 2|\vec{u}||\vec{v}| \cos(\pi - \alpha) = |\vec{u}|^2 + |\vec{v}|^2 - 2|\vec{u}||\vec{v}| \cos \alpha.$$

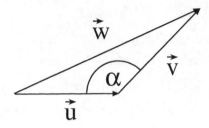

Fig. 1.1. Illustration of the cosines law for plane triangles.

6. Vector product of four vectors:

$$(\vec{u}\wedge\vec{v})\wedge(\vec{w}\wedge\vec{z}) = \begin{vmatrix} \vec{w} & \vec{z} \\ (\vec{u}\wedge\vec{v})\bullet\vec{w} & (\vec{u}\wedge\vec{v})\bullet\vec{z} \end{vmatrix} = ((\vec{u}\wedge\vec{v})\bullet\vec{z})\,\vec{w} - ((\vec{u}\wedge\vec{v})\bullet\vec{w})\,\vec{z}.$$

7. Scalar product of four vectors:

$$(\vec{u}\wedge\vec{v})\bullet(\vec{w}\wedge\vec{z}) = \begin{vmatrix} \vec{u}\bullet\vec{w} & \vec{u}\bullet\vec{z} \\ \vec{v}\bullet\vec{w} & \vec{v}\bullet\vec{z} \end{vmatrix}.$$

8. Cosines law (for spherical triangles):
   Applying the previous property to the case

$$\begin{cases} |\vec{u}| = |\vec{v}| = |\vec{z}| = 1 \\ \vec{w} = \vec{u}, \end{cases}$$

the following vector relation is obtained:

$$(\vec{u}\wedge\vec{v})\bullet(\vec{u}\wedge\vec{z}) = (\vec{v}\bullet\vec{z}) - (\vec{u}\bullet\vec{v})(\vec{u}\bullet\vec{z}),$$

which interpreted on the spherical triangle in Figure 1.2, allows us to write

$$\vec{u}\bullet\vec{v} = \cos\beta; \quad \vec{u}\bullet\vec{z} = \cos\gamma; \quad \vec{v}\bullet\vec{z} = \cos\alpha$$

$$(\vec{u}\wedge\vec{v})\bullet(\vec{w}\wedge\vec{z}) = (\sin\beta)(\sin\gamma)\cos\hat{A},$$

where $\hat{A}$ is the dihedron associated with the faces $(\vec{u},\vec{v})$ and $(\vec{u},\vec{z})$ of the trihedron with vertex O, which taken to the vector relation, leads to

$$\sin\beta\sin\gamma\cos\hat{A} = \cos\alpha - \cos\beta\cos\gamma,$$

which relates the three face angles of the trihedron to one of the dihedron (the one with edge $\vec{u}$).

Finally, we dedicate some lines to the case of a non-orthonormalized reference frame $\{\vec{e}_i\}$, i.e., $(G_3 \neq I_3)$. In this case:

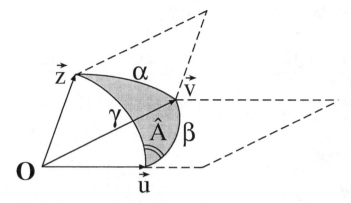

Fig. 1.2. Illustration of the cosines law for spherical triangles.

1. Dot product of vectors:

$$\vec{u} \bullet \vec{v} = <\vec{u}, \vec{v}> = U^t G V = [u^1 u^2 u^3] \begin{bmatrix} g_{11} & g_{12} & g_{13} \\ g_{21} & g_{22} & g_{23} \\ g_{31} & g_{32} & g_{33} \end{bmatrix} \begin{bmatrix} v^1 \\ v^2 \\ v^3 \end{bmatrix},$$

where the data are in contravariant coordinates.

2. Cross product of vectors:

$$\vec{z} = \vec{u} \wedge \vec{v} = \frac{1}{\sqrt{|G|}} \begin{vmatrix} \vec{e}_1 & \vec{e}_2 & \vec{e}_3 \\ u_1^* & u_2^* & u_3^* \\ v_1^* & v_2^* & v_3^* \end{vmatrix},$$

where $\sqrt{|G|}$ is the volume of the oblique solid with edges $[\vec{e}_1, \vec{e}_2, \vec{e}_3]$, the data tensor is given in *covariant* coordinates and returns vector $\vec{z}$ in contravariant coordinates. The covariant coordinates that appear in the $\vec{z}$ expression, as is very well known, can be obtained from

$$U^* = GU; \quad V^* = GV.$$

The expression

$$\sin^2 \tau = \frac{|G|}{g_{11} g_{22} g_{33}}$$

is called the "trihedron sine".

An alternative, when the data are in contravariant coordinates, but the output is in covariant coordinates, is to use the formula

$$\vec{z} = \vec{u} \wedge \vec{v} = \sqrt{|G|} \begin{vmatrix} \vec{e}^{*1} & \vec{e}^{*2} & \vec{e}^{*3} \\ u^1 & u^2 & u^3 \\ v^1 & v^2 & v^3 \end{vmatrix}.$$

3. Scalar triple product:

$$[\vec{u}, \vec{v}, \vec{w}] = \vec{u} \bullet (\vec{v} \wedge \vec{w}) = \sqrt{|G|} \begin{vmatrix} u^1 & u^2 & u^3 \\ v^1 & v^2 & v^3 \\ w^1 & w^2 & w^3 \end{vmatrix},$$

which is a scalar, with data in contravariant coordinates. The following expression is also valid:

$$[\vec{u}, \vec{v}, \vec{w}] = \vec{u} \bullet (\vec{v} \wedge \vec{w}) = \frac{1}{\sqrt{|G|}} \begin{vmatrix} u_1^* & u_2^* & u_3^* \\ v_1^* & v_2^* & v_3^* \\ w_1^* & w_2^* & w_3^* \end{vmatrix},$$

which is the same scalar but with data in covariant coordinates.

1.5 Exercises

1.1. In the affine Euclidean vector space $E^n(\mathbb{R})$ with reference to a basis $\{\vec{e}_\alpha\}$, the Gram connection matrix is given and is denoted by G.

In that space, we consider a quadratic form ϕ that is represented in matrix form as

$$\phi(\vec{V}) = X^t F X,$$

where F is regular and symmetric.

Assuming that in this space a change-of-basis is performed with associated matrix C, we look for the new matrix $\hat{F}$ associated with ϕ, but referred to the dual basis $\{\vec{e}^{*i}\}$ of the new basis

$$\phi(\vec{V}) = (\hat{X}^*)^t \hat{F} \hat{X}^*.$$

1.2. Check whether or not the Properties 6 and 8 of Section 1.3.2, which refer to matrices $A_{m,n}, B_{n,q}$

$$(A \otimes B)^k = A^k \otimes B^k \quad \text{and} \quad (A \bullet B)^{[k]} = A^{[k]} \bullet B^{[k]},$$

are extensible to negative exponents.

1.3. Given the following matrices:

$$A_{m,n} \equiv A_{2,3} = \begin{bmatrix} 1 & 0 & -1 \\ 2 & 1 & 3 \end{bmatrix} \text{ and } B_{p,q} \equiv B_{4,2} = \begin{bmatrix} 2 & 1 \\ 3 & -1 \\ 0 & 4 \\ 5 & 2 \end{bmatrix}$$

1. Determine by direct methods the matrices Q_1 and Q_2 defined by

$$Q_1 = A \otimes B; \quad Q_2 = B \otimes A.$$

2. Determine matrices P_1 and P_2 such that $Q_2 = P_1 \bullet Q_1 \bullet P_2$, applying Formulas (1.38) and (1.39).
3. Solve the previous questions using the computer.

1.4. As a particular application of the extension-condensation procedure:

1. Extend the matrix
$$A_{3,4} = \begin{bmatrix} 2 & 3 & 5 & 1 \\ 4 & 3 & 2 & -1 \\ -1 & 2 & 1 & 3 \end{bmatrix}$$

using the "extension rule".
2. Condense the matrix
$$T^t_{\sigma,1} = \begin{bmatrix} 2 & 4 & -1 & 3 & 3 & 2 & 5 & 2 & 1 & 1 & -1 & 3 \end{bmatrix}$$

using the "condensation rule" and format $B_{4,3}$.
3. Solve the previous questions using the computer.

1.5. Given the matrices
$$A_{m,n} \equiv A_{2,3} = \begin{bmatrix} 1 & 0 & -1 \\ 2 & 1 & 3 \end{bmatrix} \text{ and } B_{m,n} \equiv B_{2,3} = \begin{bmatrix} -1 & 2 & 5 \\ 3 & 0 & 4 \end{bmatrix}.$$

1. Determine the matrix $C_{2,3} = A_{2,3} \square B_{2,3}$, which is the Schur product of the given matrices.
2. Build, using Formulas (1.58) and (1.59), the matrices $Q_{m,m^2} \equiv Q_{2,4}$ and $p_{n^2,n} \equiv P_{9,3}$ such that

$$A_{2,3} \square B_{2,3} = Q_{2,4} \bullet (A_{2,3} \otimes B_{2,3}) \bullet P_{9,3}.$$

3. Solve the previous questions using the computer.

Introduction to Tensors

2.1 Introduction

This chapter introduces the tensor concept and the corresponding axiomatic properties and places special emphasis on the ordering criteria for the canonical basis, which will play an important role throughout the book to avoid a lot of the confusion that exists in many published books.

To facilitate the comprehension of the new concept we deal first with the triple tensor product linear space referred only to primal bases, and later we generalize the concept for k vectors and the case of simultaneous primal and dual bases.

Next, the important Einstein convention for repeated indices is introduced and the axiomatic properties of tensors discussed.

The chapter ends with an illustrative example from physics.

2.2 The triple tensor product linear space

Consider three given linear spaces, which will be denoted by $U^m(K), V^n(K)$ and $W^p(K)$, where m, n and p are their respective dimensions and $K(+, \cdot)$ is the field of associated scalars, by definition *the same* for all given spaces.

Assume that each of these spaces is referred to their respective bases, which will be denoted by $\{\vec{e_i}\}, \{\vec{\epsilon_j}\}$ and $\{\vec{\eta_k}\}$, respectively, in order to emphasize their different nature.

We know that

$$\vec{u} = ||\vec{e_i}|| U \equiv [\, \vec{e_1} \quad \vec{e_2} \quad \cdots \quad \vec{e_m} \,] \begin{bmatrix} u^1 \\ u^2 \\ \vdots \\ u^m \end{bmatrix} ; \forall \vec{u} \in U^m(K) \qquad (2.1)$$

$$\vec{v} = ||\vec{\epsilon}_j||V \equiv [\vec{\epsilon}_1 \quad \vec{\epsilon}_2 \quad \cdots \quad \vec{\epsilon}_n] \begin{bmatrix} v^1 \\ v^2 \\ \vdots \\ v^n \end{bmatrix}; \quad \forall \vec{v} \in V^n(K) \qquad (2.2)$$

$$\vec{w} = ||\vec{\eta}_k||W \equiv [\vec{\eta}_1 \quad \vec{\eta}_2 \quad \cdots \quad \vec{\eta}_p] \begin{bmatrix} w^1 \\ w^2 \\ \vdots \\ w^p \end{bmatrix}; \quad \forall \vec{w} \in W^p(K), \qquad (2.3)$$

where

$$U \equiv \begin{bmatrix} u^1 \\ u^2 \\ \vdots \\ u^m \end{bmatrix}; \quad V \equiv \begin{bmatrix} v^1 \\ v^2 \\ \vdots \\ v^n \end{bmatrix}; \quad W \equiv \begin{bmatrix} w^1 \\ w^2 \\ \vdots \\ w^p \end{bmatrix},$$

and $u^i, v^j, w^k \in K$.

Note the upper position of the indices, which refer to contravariant coordinates.

Definition 2.1 (tensor product). *The "tensor product" of the three given linear spaces, is a new linear space that will be denoted by $U^m \otimes V^n \otimes W^p(K)$, and its vectors will be called "tensors", built with the following criteria:*

1. *The tensor product of all possible terms of the form*

$$(u^i\vec{\epsilon}_i) \otimes (v^j\vec{\epsilon}_j) \otimes (w^k\vec{\eta}_k); i = 1, 2, \ldots, m; \; j = 1, 2, \ldots, n; \; k = 1, 2, \ldots, p$$

are constructed and, by multiplying the scalars u^i, v^j and w^k as elements of K, one writes the tensor product as a function of the basic vectors in the form

$$(u^i\vec{\epsilon}_i) \otimes (v^j\vec{\epsilon}_j) \otimes (w^k\vec{\eta}_k) = u^i v^j w^k \vec{\epsilon}_i \otimes \vec{\epsilon}_j \otimes \vec{\eta}_k. \qquad (2.4)$$

2. *Next, one proceeds to "sum" them for all possible combinations of indices i, j and k using the symbol "+" for the sum of the new linear space. In summary, a new set of vectors $\vec{t} \in U^m \otimes V^n \otimes U^p(K)$ appears with the expression*

$$\vec{t} = \sum_{\substack{1 \leq i \leq m \\ 1 \leq j \leq n \\ 1 \leq k \leq p}} u^i v^j w^k \vec{\epsilon}_i \otimes \vec{\epsilon}_j \otimes \vec{\eta}_k \equiv \sum_{\substack{1 \leq i \leq m \\ 1 \leq j \leq n \\ 1 \leq k \leq p}} t^{ijk} \vec{\pi}_{ijk}, \qquad (2.5)$$

where $t^{ijk} \equiv u^i v^j w^k$ and $\vec{\pi}_{ijk} \equiv \vec{\epsilon}_i \otimes \vec{\epsilon}_j \otimes \vec{\eta}_k$, assigning to each different product $\vec{\pi}_{ijk}$ the role of a basic vector, creating the basis $\{\vec{\pi}_{ijk}\}$, where $\vec{\pi}_{ijk} \in U^m \otimes V^n \otimes W^p(K)$.

We note that the sum + must be interpreted as

$$\vec{u} + \vec{v} = \sum_{\substack{1 \le i \le m \\ 1 \le j \le n \\ 1 \le k \le p}} (u^{ijk} + v^{ijk})\vec{e}_i \otimes \vec{\epsilon}_j \otimes \vec{\eta}_k$$

and the product for a scalar as

$$\lambda \circ \vec{u} = \sum_{\substack{1 \le i \le m \\ 1 \le j \le n \\ 1 \le k \le p}} (\lambda \cdot u^{ijk})\vec{e}_i \otimes \vec{\epsilon}_j \otimes \vec{\eta}_k.$$

As a consequence of assuming that the vectors $\{\vec{\pi}_{ijk}\}$ are linearly independent, i.e., a basis of the new linear space, we have

$$\sigma = dim[U^m \otimes V^n \otimes W^p(K)] = m \cdot n \cdot p. \tag{2.6}$$

3. Finally, it is very important that the different products

$$\{\vec{e}_i \otimes \vec{\epsilon}_j \otimes \vec{\eta}_k; i = 1, 2, \ldots, m; \ j = 1, 2, \ldots, n; \ k = 1, 2, \ldots, p\},$$

the basic vectors of $U^m \otimes V^n \otimes W^p$, be "ordered" by means of an agreed upon criteria, for the mentioned basis to be unique.
We use the following criterion:
The vector $\vec{e}_i \otimes \vec{\epsilon}_j \otimes \vec{\eta}_k$ in this ordering will be ahead of vector $\vec{e}_{i_1} \otimes \vec{\epsilon}_{j_1} \otimes \vec{\eta}_{k_1}$ iff at least one of the following conditions holds:
  a) $i < i_1$.
  b) For $i = i_1$, $j < j_1$.
  c) For $i = i_1$ and $j = j_1$, $k < k_1$.
This rule implies the order (from left to right and from top to bottom):

$$
\begin{array}{llll}
\vec{e}_1 \otimes \vec{\epsilon}_1 \otimes \vec{\eta}_1 & \vec{e}_1 \otimes \vec{\epsilon}_1 \otimes \vec{\eta}_2 & \ldots & \vec{e}_1 \otimes \vec{\epsilon}_1 \otimes \vec{\eta}_p \\
\vec{e}_1 \otimes \vec{\epsilon}_2 \otimes \vec{\eta}_1 & \vec{e}_1 \otimes \vec{\epsilon}_2 \otimes \vec{\eta}_2 & \ldots & \vec{e}_1 \otimes \vec{\epsilon}_2 \otimes \vec{\eta}_p \\
\ldots & \ldots & \ldots & \ldots \\
\vec{e}_1 \otimes \vec{\epsilon}_n \otimes \vec{\eta}_1 & \vec{e}_1 \otimes \vec{\epsilon}_n \otimes \vec{\eta}_2 & \ldots & \vec{e}_1 \otimes \vec{\epsilon}_n \otimes \vec{\eta}_p \\
\vec{e}_2 \otimes \vec{\epsilon}_1 \otimes \vec{\eta}_1 & \vec{e}_2 \otimes \vec{\epsilon}_1 \otimes \vec{\eta}_2 & \ldots & \vec{e}_2 \otimes \vec{\epsilon}_1 \otimes \vec{\eta}_p \\
\vec{e}_2 \otimes \vec{\epsilon}_2 \otimes \vec{\eta}_1 & \vec{e}_2 \otimes \vec{\epsilon}_2 \otimes \vec{\eta}_2 & \ldots & \vec{e}_2 \otimes \vec{\epsilon}_2 \otimes \vec{\eta}_p \\
\ldots & \ldots & \ldots & \ldots \\
\vec{e}_2 \otimes \vec{\epsilon}_n \otimes \vec{\eta}_1 & \vec{e}_2 \otimes \vec{\epsilon}_n \otimes \vec{\eta}_2 & \ldots & \vec{e}_2 \otimes \vec{\epsilon}_n \otimes \vec{\eta}_p \\
\ldots & \ldots & \ldots & \ldots \\
\vec{e}_m \otimes \vec{\epsilon}_1 \otimes \vec{\eta}_1 & \vec{e}_m \otimes \vec{\epsilon}_1 \otimes \vec{\eta}_2 & \ldots & \vec{e}_m \otimes \vec{\epsilon}_1 \otimes \vec{\eta}_p \\
\vec{e}_m \otimes \vec{\epsilon}_2 \otimes \vec{\eta}_1 & \vec{e}_m \otimes \vec{\epsilon}_2 \otimes \vec{\eta}_2 & \ldots & \vec{e}_m \otimes \vec{\epsilon}_2 \otimes \vec{\eta}_p \\
\ldots & \ldots & \ldots & \ldots \\
\vec{e}_m \otimes \vec{\epsilon}_n \otimes \vec{\eta}_1 & \vec{e}_m \otimes \vec{\epsilon}_n \otimes \vec{\eta}_2 & \ldots & \vec{e}_m \otimes \vec{\epsilon}_n \otimes \vec{\eta}_p.
\end{array}
$$

Remark 2.1 (Very important). The reader could think that *other simpler vector symbols*, could be adopted for the basic vectors of the linear space $U^m \otimes V^n \otimes W^p$, for example, $\{\vec{\beta}_h;\ h = 1, 2, \ldots, \sigma\}$, and numbering with t^h, the corresponding component; so that $t^{111}\ \vec{e}_1 \otimes \vec{\epsilon}_1 \otimes \vec{\eta}_1$ would be $t^1 \vec{\beta}_1$, $t^{112}\ \vec{e}_1 \otimes \vec{\epsilon}_1 \otimes \vec{\eta}_2$ would be $t^2 \vec{\beta}_2$, and $t^{mnp}\ \vec{e}_m \otimes \vec{\epsilon}_n \otimes \vec{\eta}_p$ would be $t^\sigma \vec{\beta}_\sigma$, that is, by means of $\vec{t} = t^i \vec{\beta}_i$. If this is done, the linear space $U^m \otimes V^n \otimes W^p(K)$ would appear with the classic notation of any other linear space, in which changes of basis could be performed, and working with homomorphisms or assigning an inner connection (dot product), etc. could be possible, as is done in algebras. However, if this is adopted, a reader of this book, with the exception of this chapter, *would not comprehend* why the vectors $\vec{t} \in U^m \otimes V^n \otimes W^p(K)$ are called *tensors*. In other words, the adopted notation, which is not casual, will later be of *decisive* importance. $\qquad\square$

It is interesting to know the existing relation between the position n that occupies a component in the ordered set of all of them, and the indices $i_1, i_2, \ldots, i_k$ of the given component.

The direct problem is: assuming that the indices $i_1, i_2, \ldots, i_k$ of a given component and the ranges $r_1, r_2, \ldots, r_k$ of each index are known, determine n, the position that occupies such component in the ordered set.

The solution to this problem is

$$n = i_k + r_k((i_{k-1} - 1) + r_{k-1}((i_{k-2} - 1) + \cdots (\cdots + r_2((i_1 - 1)))))$$.

The tensor ranges are precisely the dimensions $n_1, n_2, \ldots, n_k$ of each linear space factor.

The reverse problem consists of assuming that the position n (position of a given component) and the ranges $r_1, r_2, \ldots, r_k$ of all indices are known, determine the indices $i_1, i_2, \ldots, i_k$ of the given component.

The indices $i_1, i_2, \ldots, i_k$ can be obtained by means of the following algorithm:

1. $q = n - 1$.
2. Repeat with j from k to 2, by -1: $i_j = \text{mod}[q, r_j] + 1$; $q = \text{quotient}[q, r_j]$, where quotient$[q, r]$ and mod$[q, r]$ are the quotient and the rest, respectively, when dividing q by r.
3. $i_1 = q + 1$.

2.3 Einstein's summation convention

From now on, in expressions such as $\vec{u} = \sum\limits_{i=1}^{i=m} u^i \vec{e}_i$, in which common indices appear in different positions with a *summation* symbol to be developed by giving the index i its full range from 1 to m, the summation symbol and the corresponding range $\sum\limits_{i=1}^{i=m}$ will be suppressed, since the sum can be "assumed" by the superindices and subindices with the same letter and their positions, and its range by the dimension of the linear space $(\vec{e}_i \in U^m(K))$.

Similarly, the expression

$$, v^i = t^{ij} z_j \tag{2.7}$$

if the ranges are $i = 1, 2, \ldots, g$ and $j = 1, 2, \ldots, h$, is in fact

$$v^i = t^{i1} \cdot z_1 + t^{i2} \cdot z_2 + \cdots + t^{ih} \cdot z_h; \ (operations + and \cdot of \ K). \tag{2.8}$$

The repeated indices in superindex and subindex positions, must be interpreted with Einstein's criterion, and are called "dummy indices" because if we change its name

$$v^i = t^{i\alpha} z_\alpha, \tag{2.9}$$

the expression (2.8) remains unchanged.

The non-repeated indices (*in the same term* of the equality) are called "free indices". For example, in the given expression (2.7), i is the free index and j is the dummy index.

Based on this convention, in the following the relation (2.5) will be written as

$$\vec{t} = u^i v^j w^k \vec{e}_i \otimes \vec{\epsilon}_j \otimes \vec{\eta}_k \equiv t^{ijk} \vec{\pi}_{ijk}, \tag{2.10}$$

where $i = 1, 2, \ldots, m;\ j = 1, 2, \ldots, n;\ k = 1, 2, \ldots, p$ are dummy indices.

2.4 Tensor analytical representation

We follow the exposition with another confusing situation. Since the set $U^m \otimes V^n \otimes W^p(K)$ has been given the structure of a linear space over K, it has an associated sum $(+)$ and an external product $(\circ)$ for its vectors, and then we can consider a tensor as

$$\begin{aligned}
\vec{a} &= \lambda_0 \vec{t} + \mu_0 \vec{z} + \nu_0 \vec{r} \\
&\equiv \left[\lambda u^i_t v^j_t w^k_t + \mu u^i_z v^j_z w^k_z + \nu u^i_r v^j_r w^k_r \right] \vec{e}_i \otimes \vec{\epsilon}_j \otimes \vec{\eta}_k \\
&= a^{ijk} \vec{e}_i \otimes \vec{\epsilon}_j \otimes \vec{\eta}_k,
\end{aligned} \tag{2.11}$$

which is different from the vector $u^i_a v^j_a w^k_a \vec{e}_i \otimes \vec{\epsilon}_j \otimes \vec{\eta}_k$, i.e., *there are no* vectors $u^i_a \vec{e}_i \in U^m(K), v^j_a \in V^n(K)$ and $w^k_a \in W^p(K)$ from which tensor product $\vec{a}$ arises.

According to the Einstein convention, $\vec{a}$ will be denoted by

$$\vec{a} = a^{ijk} \vec{e}_i \otimes \vec{\epsilon}_j \otimes \vec{\eta}_k.$$

The conclusion is that in the linear space $U^m \otimes V^n \otimes W^p(K)$, which has vectors that are *all* tensors, there exist some of them that are tensor products of the vectors of the "factor" linear spaces (some authors call these decomposable tensors) and there exist other vectors that do not come from tensor products (non-decomposable tensors), but from linear combinations of them.

For all of them the following analytical expression holds:

$$t = t^{ijk}\vec{e}_i \otimes \vec{\epsilon}_j \otimes \vec{\eta}_k; \quad \forall \vec{t} \in U^m \otimes V^n \otimes W^p(K). \tag{2.12}$$

2.5 Tensor product axiomatic properties

Next, some of the axiomatic properties of tensors that have been used in the above paragraphs are declared.

If the composition laws for the linear spaces are denoted by

$$\{U^m(K); (+', \circ')\}; \quad \{V^n(K); (+'', \circ'')\}; \quad \{W^p(K); (+''', \circ''')\},$$

we have:

1. *Associativity of $\otimes$*: Si $\vec{u} \in U^m(K), \vec{v} \in V^n(K)$ and $\vec{w} \in W^p(K)$

$$\vec{z} = \vec{u} \otimes \vec{v} \otimes \vec{w} = (\vec{u} \otimes \vec{v}) \otimes \vec{w} = \vec{u} \otimes (\vec{v} \otimes \vec{w}), \tag{2.13}$$

   that is, the triple product can be directly calculated or in steps, first $\vec{u} \otimes \vec{v}$ and then this by $\vec{w}$, or calculate first $\vec{v} \otimes \vec{w}$ and next the product $\vec{u} \otimes (\vec{v} \otimes \vec{w})$. The result is always the unique vector $\vec{z}$.

2. *Distributivity of $\otimes$ with respect to any of the sums $+', +'', +'''$*:

$$(\vec{u}_1 +' \vec{u}_2) \otimes \vec{v} \otimes \vec{w} = \vec{u}_1 \otimes \vec{v} \otimes \vec{w} + \vec{u}_2 \otimes \vec{v} \otimes \vec{w}$$
$$\vec{u} \otimes (\vec{v}_1 +'' \vec{v}_2) \otimes \vec{w} = \vec{u} \otimes \vec{v}_1 \otimes \vec{w} + \vec{u} \otimes \vec{v}_2 \otimes \vec{w} \tag{2.14}$$
$$\vec{u} \otimes \vec{v} \otimes (\vec{w}_1 +''' \vec{w}_2) = \vec{u} \otimes \vec{v} \otimes \vec{w}_1 + \vec{u} \otimes \vec{v} \otimes \vec{w}_2.$$

3. *Associativity of $\otimes$ with respect to any of the external products*: $\circ', \circ'', \circ'''$:

$$(\lambda \circ' \vec{u}) \otimes \vec{v} \otimes \vec{w} = \lambda \circ (\vec{u} \otimes \vec{v} \otimes \vec{w})$$
$$\vec{u} \otimes (\mu \circ'' \vec{v}) \otimes \vec{w} = \mu \circ (\vec{u} \otimes \vec{v} \otimes \vec{w}) \tag{2.15}$$
$$\vec{u} \otimes \vec{v} \otimes (\nu \circ''' \vec{w}) = \nu \circ (\vec{u} \otimes \vec{v} \otimes \vec{w}).$$

   For example

$$(\lambda \circ' \vec{u}) \otimes (\mu \circ'' \vec{v}) \otimes \vec{w} = \vec{u} \otimes ((\lambda\mu) \circ'' \vec{v}) \otimes \vec{w} = (\lambda\mu) \circ (\vec{u} \otimes \vec{v} \otimes \vec{w}).$$

4. *Axiom of the linear space "tensor product" basis*: We assume that the set $\beta \equiv \{\vec{e}_i \otimes \vec{\epsilon}_j \otimes \vec{\eta}_k\}$ includes each of the vectors in the canonical basis of the linear space $U^m \otimes V^n \otimes W^p(K)$, which *by convention* are given the indicated notation, in this axiom (see Definition 2.1 point 2). The dimension of β is σ:

$$\sigma = dim[U^m \otimes V^n \otimes W^p(K)] = m \cdot n \cdot p.$$

Similarly, point 3 of Definition 2.1 indicates the way in which the tensor bases vector subindices must be ordered.

If this criterion is accepted and the subindices of each vector in the tensor basis are consecutively read, as if they were *quantities*, the set of those quantities must appear strictly ordered with respect to the natural order, if the basis is correctly established. This criterion was already used for matrices when we read them by rows.

Example 2.1 (Tensor product). Given the "tensor product" linear space by its notation

$$U^3 \otimes V^2 \otimes W^4(\mathbb{R}),$$

1. Obtain the dimension σ of the linear space, that is, the number of linearly independent vectors that must belong to a given basis.
2. Notate a generic vector (tensor) of the canonical basis $\mathcal{B}$.
3. Write in a sequential order the subindices associated with the canonical basis vectors.
4. Sort these subindices in the usual matrix form for the components of tensors of third order $(r = 3)$.

Solution:

1. The required dimension is $\sigma = 3 \times 2 \times 4 = 24$.
2. A generic vector of the canonical basis is

$$\beta = \{\vec{e}_i \otimes \vec{e}_j \otimes \vec{\eta}_k\}; \quad 1 \le i \le 3; \ 1 \le j \le 2; \ 1 \le k \le 4.$$

3. The required subindices are

1	111	2	112	3	113	4	114	5	121	6	122	7	123	8	124
9	211	10	212	11	213	12	214	13	221	14	222	15	223	16	224
17	311	18	312	19	313	20	314	21	321	22	322	23	323	24	324

Note that, when the indices are read "as quantities" the set is ordered.
4. The required representation in matrix form is a block column matrix:

$$\beta \equiv \begin{bmatrix} 111 & 112 & 113 & 114 \\ 121 & 122 & 123 & 124 \\ -- & -- & -- & -- \\ 211 & 212 & 213 & 214 \\ 221 & 222 & 223 & 224 \\ -- & -- & -- & -- \\ 311 & 312 & 313 & 314 \\ 321 & 322 & 323 & 324 \end{bmatrix},$$

where i is the number of the row that occupies the block, j is the row in each block, and k is the column in each of the blocks.

□

2.6 Generalization

At this point, it is convenient to point out that if at the beginning of this chapter some of the selected factor linear spaces were given with respect to their dual basis instead of the fundamental or initial basis, for example, $V_*^n(K)$, i.e.

$$\vec{v} = ||\vec{\epsilon}^{*j}||V^* = [\,\vec{\epsilon}^{*1} \quad \vec{\epsilon}^{*2} \quad \cdots \quad \vec{\epsilon}^{*n}\,] \begin{bmatrix} v_1^* \\ v_2^* \\ \vdots \\ v_n^* \end{bmatrix}; \;\; \vec{v} \in V_*^n(K),$$

this would cause changes in the notation and the relations (2.2), (2.4),(2.5), (2.10), etc. For example, the last would appear as

$$\vec{t} = u^i v_j^* w^k \vec{e}_i \otimes \vec{\epsilon}^{*j} \otimes \vec{\eta}_k = t_{\;\;o\,j\,o}^{i\,o\,k}\, \vec{\pi}_{\;\;i\,o\,k}^{o\,j\,o} \tag{2.16}$$

and the "analytical expression" of the tensor, relation (2.12), would be[1]

$$t = t_{\;\;o\,j\,o}^{i\,o\,k}\, \vec{e}_i \otimes \vec{\epsilon}^{*j} \otimes \vec{\eta}_k; \;\; \forall \vec{t} \in U^m \otimes V_*^n \otimes W^p(K). \tag{2.17}$$

The development of the rest of the theory would be exactly the same with the exception of the notation.

Finally, consider r linear spaces:

$$\{V_1^{n_1}(K); (+', \circ')\}, \;\; \{V_2^{n_2}(K); (+'', \circ'')\}, \;\; \cdots \;\;, \;\; \{V_r^{n_r}(K); (+', \circ^r)\},$$

the vectors of which are arbitrary, but have associated the same field of scalars $K(+, \circ)$.

We shall denote by $V_1^{n_1} \otimes V_2^{n_2} \otimes \cdots \otimes V_r^{n_r}(K)$ a new linear space, with sum "$+$" and external product "$\circ$", called the "tensor product" of the given spaces, the vectors $\vec{t}$ of which are called *tensors*, and are referred to a basis β of the given space, which is denoted by

$$\beta = \{\vec{e}_{1i_1} \otimes \vec{e}_2^{*i_2} \otimes \cdots \otimes \vec{e}_{r-1}^{*i_{r-1}} \otimes \vec{e}_{ri_r}\}, \tag{2.18}$$

with the ranges: $1 \le i_1 \le n_1; 1 \le i_2 \le n_2; \ldots, 1 \le i_r \le n_r$ and the previously established order.

The dimension of this linear space is

$$\sigma = dim[V_1^{n_1} \otimes V_2^{n_2} \otimes \cdots \otimes V_r^{n_r}(K)] = \prod_{h=1}^{h=r} n_h. \tag{2.19}$$

Any tensor in this space will be named a "tensor of order r" and will usually be presented by its "analytical expression" or "tensor expression", according to Einstein's convention:

---

[1] Note that we use here for the first time the semaphoric notation for tensor components, that will be explained in detail in Chapter 3, Section 3.3.

$$\vec{t} = t^{i_1 \circ i_3}_{\circ i_2 \circ} \cdots {}^{\circ \ i_r}_{i_{r-1} \circ} \ \vec{e}_{1i_1} \otimes \vec{e}_2^{*i_2} \otimes \cdots \otimes \vec{e}_{r-1}^{*i_{r-1}} \otimes \vec{e}_{ri_r}, \qquad (2.20)$$

where all scalars $t^{i_1 \circ i_3}_{\circ i_2 \circ} \cdots {}^{\circ \ i_r}_{i_{r-1} \circ} \in K$ are data.

It is convenient to get used to a correct reading of Formula (2.20): "heterogeneous tensor of order r, contra-cova-contra-···-cova-contravariant".

We end this section by giving a general matrix expression for the tensor product of vectors.

Consider r vectors $\vec{v}_i \in V_i^{n_i}(K)$; $1 \le i \le r$, each belonging to one of the linear space factors of the tensor space $V_1^{n_1} \otimes V_2^{n_2} \otimes \cdots \otimes V_r^{n_r}(K)$, with dimension σ given by Formula (2.19), assuming each linear space $V_i^{n_i}$ to be referred to its basis $\{\vec{e}_{i\alpha}\}$; we know that

$$\vec{v}_i = [\vec{e}_{i1} \quad \vec{e}_{i2} \quad \cdots \quad \vec{e}_{in_i}] \begin{bmatrix} x^1 \\ x^2 \\ \vdots \\ x^{n_i} \end{bmatrix} \equiv ||\vec{e}_\alpha|| X_{n_i},$$

where the set of column matrices $\{ X_{n_1} \quad X_{n_2} \quad \cdots \quad X_{n_r} \}$ are the data vectors for which the tensor product is sought after. We propose as a matrix formula for calculating the product, the expression

$$X_\sigma = X_{n_1} \otimes X_{n_2} \otimes \cdots \otimes X_{n_r}, \qquad (2.21)$$

where X_σ is the column matrix associated with the "stretching" of the tensor product components of the data vectors, and $\sigma = n_1 \cdot n_2 \cdot \ldots \cdot n_r$.

2.7 Illustrative examples

This section includes some examples to clarify the concepts introduced so far.

Example 2.2 (Physical relationships). In physics, mainly in classical mechanics, the punctual gravitational mass is studied. To this end, we denote by $L(\mathbb{R})$, $T(\mathbb{R})$ and $M(\mathbb{R})$ the unidimensional linear spaces, all of them heterogeneous, corresponding to the physical magnitudes space, time and mass (we accept negative virtual masses), respectively. These are the fundamental linear spaces, by convention. The corresponding bases are *the vectors cm* (centimeter), *sec* (second) and *g* (mass gram).

The dual linear spaces, reciprocal to the fundamental spaces, are $L_*(\mathbb{R})$, $T_*(\mathbb{R})$ and $M_*(\mathbb{R})$ and their respective bases are cm^{-1}, sec^{-1} and g^{-1}.

The following tensor spaces will be called velocity V, acceleration A, and force F:

$$V = L \otimes T_*(\mathbb{R}); \quad A = L \otimes T_* \otimes T_*(\mathbb{R}); \quad F = M \otimes L \otimes T_* \otimes T_*(\mathbb{R}).$$

In this example we solve the following problems:

1. Relate the bases of $L(\mathbb{R})$ and $L_*(\mathbb{R})$.
2. Find the bases of the tensor linear spaces V, A and F.
3. If in the fundamental linear spaces $L(\mathbb{R}), T(\mathbb{R})$ and $M(\mathbb{R})$ we take as new basic vectors m (meter), h (hour) and kg (mass kilogram), respectively, give the change-of-basis equations for the vectors $\vec{v} \in V(\mathbb{R})$, $\vec{t} \in T(\mathbb{R})$ and $\vec{f} \in F(\mathbb{R})$.
4. Solve questions 2 and 3 for the following tensor spaces (we assume that the reader knows elemental mechanics):
   a) Pressure.
   b) Moment of inertia.
   c) Power.
5. Another more complex physical example of a sum of force tensor spaces is the called "Lorentz force". Consider a punctual particle with an electrical charge of $q = 6$ coulombs, that moves in the tridimensional space by a conducting filament in the form of a helix Γ with constant velocity $\vec{v} = 2$ m/sec (ascendent), and assume that in the medium an electrical field acts with intensity $\vec{E} = 8\vec{i}$ newton/coulomb, and also a magnetic field of intensity $\vec{B} = 5\vec{k}$ tesla. We wish to determine the tensor $\vec{f}$ and its modulus $|\vec{f}|$ when the particle is at the point $H \equiv (3, 4, z_0)$.
   We have the following data: the vector equation of the "Lorentz force",

$$\vec{f} = q(\vec{E} + \vec{V} \wedge \vec{B})$$

and the Γ helix equations

$$x^2 + y^2 = 25 \tag{2.22}$$

$$z = \arcsin\left(\frac{y}{5}\right). \tag{2.23}$$

Solution: Table 2.1 gives the fundamental linear spaces and the corresponding bases.

Table 2.1. The fundamental linear spaces and the corresponding bases for Example 2.2.

Space	Basis $\mathcal{B}$	Vector	Dimension	Dual	Reciprocal basis	Name
$L(\mathbb{R})$	$\{\vec{e}_1 \equiv \overline{cm}\}$	$\vec{\ell} = \ell\vec{e}_1$	$n_1 = 1$	$L_*(\mathbb{R})$	$\{\vec{e}_1^* \equiv \overline{cm}^{-1}\}$	space
$T(\mathbb{R})$	$\{\vec{\tau}_1 \equiv \overline{sec}\}$	$\vec{t} = t\vec{\tau}_1$	$n_2 = 1$	$T_*(\mathbb{R})$	$\{\vec{e}^{*1} \equiv \overline{sec}^{-1}\}$	time
$M(\mathbb{R})$	$\{\vec{\epsilon}_1 \equiv \overline{g}\}$	$\vec{m} = m\vec{\epsilon}_1$	$n_3 = 1$	$M_*(\mathbb{R})$	$\{\vec{\epsilon}^{*1} \equiv \overline{g}^{-1}\}$	gravitational mass
$Q(\mathbb{R})$	$\{\vec{\eta}_1 \equiv \overline{U.E.E.}\}$	$\vec{q} = q\vec{\eta}_1$	$n_4 = 1$	$Q_*(\mathbb{R})$	$\{\vec{\eta}^{*1} \equiv \overline{U.E.E.}^{-1}\}$	electric mass

1. From $\vec{e} \cdot \vec{e}^{*1} = 1 \quad \Rightarrow \quad \overline{cm} \cdot \vec{e}^{*1} = 1 \quad \Rightarrow \quad \vec{e}^{*1} = \frac{1}{cm} = \overline{cm}^{-1}$ where the dot refers to the dot product.

2. The basis of the velocity tensor space $V(\mathbb{R})$ is

$$\mathcal{B}_V = \{\vec{e}_1 \otimes \vec{\tau}^{*1}\} \equiv \{cm \otimes sec^{-1}\}.$$

The basis of the acceleration tensor space $A(\mathbb{R})$ is

$$\mathcal{B}_A = \{\vec{e}_1 \otimes \vec{\tau}^{*1} \otimes \vec{\tau}^{*1}\} \equiv \{cm \otimes sec^{-1} \otimes sec^{-1}\} \equiv \{cm \otimes sec^{-2}\}.$$

Note that $\{cm \otimes sec^{-2}\}$ is just a new name for $\{cm \otimes sec^{-1} \otimes sec^{-1}\}$ and cannot be interpreted literally, i.e., as a basic vector of a tensor product linear space of two linear spaces.
The basis of the force tensor space $F(\mathbb{R})$ is

$$\mathcal{B}_F = \{\vec{\epsilon}_1 \otimes \vec{e}_1 \otimes \vec{\tau}^{*1} \otimes \vec{\tau}^{*1}\} \equiv \{g \otimes cm \otimes sec^{-1} \otimes sec^{-1}\}$$
$$\equiv \{g \otimes cm \otimes sec^{-2}\} \equiv \{dyn\}.$$

The same comment applies here for the literal interpretation of $\{g \otimes cm \otimes sec^{-2}\}$ and $\{dyn\}$.

3. From the international system (SI) of physical units, we know that the basis changes to be considered are:
   a) In $L(\mathbb{R})$: $1m = 10^2 cm$.
   b) In $T(\mathbb{R})$: $1h = 60^2 sec$.
   c) In $M(\mathbb{R})$: $1kg = 10^3 g$.
   The expression in $T(\mathbb{R})$ induces in the reciprocal space $T_*(\mathbb{R})$ the change $1h^{-1} = \frac{1}{60^2} sec^{-1} = 60^{-2} sec^{-1}$.
   Then, the new bases associated with the tensor spaces are
   a) In $V(\mathbb{R})$: $\hat{\mathcal{B}}_V = \{\widehat{\vec{e}_1} \otimes \widehat{\vec{\tau}}^{*1}\} \equiv \{m \otimes h^{-1}\}$; the relationship is $m \otimes h^{-1} = 10^2 cm \otimes 60^{-2} sec^{-1} = \frac{1}{36} cm \otimes sec^{-1}$.
   b) In $A(\mathbb{R})$: $\hat{\mathcal{B}}_A = \{\widehat{\vec{e}_1} \otimes \widehat{\vec{\tau}}^{*1} \otimes \widehat{\vec{\tau}}^{*1}\} \equiv \{m \otimes h^{-1} \otimes h^{-1}\}$; the relationship is $m \otimes h^{-2} = m \otimes h^{-1} \otimes h^{-1} = 10^2 cm \otimes 60^{-2} sec^{-1} \otimes 60^{-2} sec^{-1} = \frac{1}{129600} cm \otimes sec^{-2}$.
   c) In $F(\mathbb{R})$: $\hat{\mathcal{B}}_F = \{\widehat{\vec{\epsilon}_1} \otimes \widehat{\vec{e}_1} \otimes \widehat{\vec{\tau}}^{*1} \otimes \widehat{\vec{\tau}}^{*1}\} \equiv \{kg \otimes m \otimes h^{-2}\}$; the relationship is $kg \otimes m \otimes h^{-2} = 10^3 grs \otimes 10^2 cm \otimes \left(\frac{1}{60^2} sec^{-1}\right)^2 = \frac{10}{1296} g \otimes cm \otimes sec^{-2} \equiv \frac{10}{1296} dyn$.

4. From the corresponding physical concepts, the tensor spaces are:

$$\text{Pressure } P(\mathbb{R}) = \left(\frac{force}{surface}\right) = M \otimes L \otimes T_* \otimes T_* \otimes L_* \otimes L_*(\mathbb{R}).$$

$$\text{Moment of inertia } I(\mathbb{R}) = M \otimes L \otimes L(\mathbb{R}).$$

$$\text{Power } W(\mathbb{R}) = (force \otimes velocity) = M \otimes L \otimes T_* \otimes T_* \otimes L_* \otimes T_*(\mathbb{R}).$$

The bases of these tensor spaces in the CGS system become:
Initial basis of the pressure tensor space:

$$\beta_P = \{\vec{e}_1 \otimes \vec{e}_1 \otimes \vec{\tau}^{*1} \otimes \vec{\tau}^{*1} \otimes \vec{e}^{*1} \otimes \vec{e}^{*1}\}$$
$$\equiv \{g \otimes cm \otimes sec^{-1} \otimes sec^{-1} \otimes cm^{-1} \otimes cm^{-1}\}$$
$$\equiv \{dyn \otimes cm^{-2}\} \equiv \{bar\}.$$

Initial basis of the moment of inertia tensor space:

$$\beta_I = \{\vec{e}_1 \otimes \vec{e}_1 \otimes \vec{e}_1\} \equiv \{g \otimes cm^2\}.$$

Initial basis of the power tensor space:

$$\beta_W = \{\vec{e}_1 \otimes \vec{e}_1 \otimes \vec{\tau}^{*1} \otimes \vec{\tau}^{*1} \otimes \vec{e}_1 \otimes \vec{\tau}^{*1}\}$$
$$\equiv \{g \otimes cm \otimes sec^{-2} \otimes cm \otimes sec^{-1}\} \equiv \{(dyn \otimes cm) \otimes sec^{-1}\}$$
$$\equiv \{erg \otimes sec^{-1}\}.$$

The change-of-basis relations in the fundamental linear spaces have been already mentioned in question 3.
The new basis of the pressure tensor space is

$$\hat{\beta}_P = \{\widehat{\vec{e}_1} \otimes \widehat{\vec{e}_1} \otimes \widehat{\vec{\tau}}^{*1} \otimes \widehat{\vec{\tau}}^{*1} \otimes \widehat{\vec{e}}^{*1} \otimes \widehat{\vec{e}}^{*1}\} \equiv \{kg \otimes m \otimes h^{-2} \otimes m^{-2}\}$$

with the following equivalence:

$$kg \otimes m \otimes h^{-2} \otimes m^{-2} = 1000g \otimes 10^2 cm \otimes \left(\frac{1}{60^2}sec^{-1}\right)^2 \otimes \left(\frac{1}{100}cm^{-1}\right)^2$$
$$= \frac{1}{1296000}g \otimes cm \otimes sec^{-2} \otimes cm^{-2}$$
$$= \frac{1}{1296000}dyn \otimes cm^{-2} \equiv \frac{1}{1296000}bar.$$

The new basis of the moment of inertia tensor space is

$$\hat{\beta}_I = \{\widehat{\vec{e}_1} \otimes \widehat{\vec{e}_1} \otimes \widehat{\vec{e}_1}\} \equiv \{kg \otimes m^2\}$$

and the equivalence is

$$kg \otimes m^2 = 10^3 g \otimes (10^2 cm) = 10^7 g \otimes cm^2.$$

The new basis of the power tensor space is

$$\hat{\beta}_W = \{\widehat{\vec{e}_1} \otimes \widehat{\vec{e}_1} \otimes \widehat{\vec{\tau}}^{*1} \otimes \widehat{\vec{\tau}}^{*1} \otimes \widehat{\vec{e}_1} \otimes \widehat{\vec{\tau}}^{*1}\} \equiv \{kg \otimes m \otimes h^{-2} \otimes m \otimes h^{-1}\}$$

with the following equivalence

$$kg \otimes m^2 \otimes h^{-3} = 1000g \otimes (10^2 cm)^2 \otimes \left(\frac{1}{60^2}sec^{-1}\right)^3$$
$$= \frac{10}{6^6}dyn \otimes cm \otimes sec^{-1} = \frac{10}{6^6}erg \otimes sec^{-1}.$$

5. In this question the ordinary punctual geometric affine space $(n = 3)$ is the support space of the tensors, in this case "vectors".

First, the coordinate of point H will be determined:

Since $z_0 = \arcsin\left(\frac{4}{5}\right)$, from the curve Γ we obtain $H\left(3, 4, \arcsin\frac{4}{5}\right)$.

Next, we establish the velocity vector $\vec{v}$ at H:

Differentiating Γ we get

$$\pi_1 = x\,dx + y\,dy = 0 \tag{2.24}$$

$$\pi_2 = dz = \frac{dy}{\sqrt{5^2 + y^2}}, \tag{2.25}$$

which implies

$$\frac{dx}{-y\sqrt{25 + y^2}} = \frac{dy}{x\sqrt{25 + y^2}} = \frac{dz}{x},$$

that is, the direction vectors of the tangent at H are

$$-4\sqrt{25 + 4^2}, 3\sqrt{25 + 4^2}, 3$$

and then, those of the unit tangent vector at H become

$$\frac{-4\sqrt{41}}{\sqrt{1034}}, \frac{3\sqrt{41}}{\sqrt{1034}}, \frac{3}{\sqrt{1034}}$$

and since the velocity vector has the previous direction and modulus 2m/sec, we have

$$\vec{V}_H = \frac{1}{\sqrt{1034}}\left(-8\sqrt{41}\vec{i} + 6\sqrt{41}\vec{j} + 6\vec{k}\right).$$

Thus, the vector $\vec{V} \wedge \vec{B}$ at H is

$$(\vec{V} \wedge \vec{B})_H = \frac{1}{\sqrt{1034}}\begin{vmatrix} \vec{i} & \vec{j} & \vec{k} \\ -8\sqrt{41} & 6\sqrt{41} & 6 \\ 0 & 0 & 5 \end{vmatrix} = \frac{10\sqrt{41}}{\sqrt{1034}}(3\vec{i} + 4\vec{j}).$$

The requested tensor, applying the "Lorentz force", is the vector

$$\vec{f} = 6\left[8\vec{i} + \frac{10\sqrt{41}}{\sqrt{1034}}(3\vec{i} + 4\vec{j})\right] = \left(48 + \frac{180\sqrt{41}}{\sqrt{1034}}\right)\vec{i} + \frac{240\sqrt{41}}{\sqrt{1034}}\vec{j}$$

with modulus:

$$|\vec{f}| = \sqrt{\left(48 + \frac{180\sqrt{41}}{\sqrt{1034}}\right)^2 + \left(\frac{240\sqrt{41}}{\sqrt{1034}}\right)^2} \text{ newtons.}$$

□

2.8 Exercises

2.1. Given the vectors $\vec{u}, \vec{v}$ and $\vec{w}$ in the following linear spaces:

$$\vec{u} = 2\vec{e}_1 + 3\vec{e}_2 - 4\vec{e}_3 \in U^3(\mathbb{R})$$
$$\vec{v} = 5\vec{\epsilon}_1 - 2\vec{\epsilon}_2 + 4\vec{\epsilon}_3 - 3\vec{\epsilon}_4 \in V^4(\mathbb{R})$$
$$\vec{w} = \vec{\eta}_1 - 2\vec{\eta}_2 + 7\vec{\eta}_3 \in W^3(\mathbb{R}).$$

1. Let $\vec{p}$ be the vector "tensor product" of the given vectors

$$\vec{p} \in U^3 \otimes V^4 \otimes W^3(\mathbb{R}).$$

   Give the vector $\vec{p}$, completely developed, expressing it in the basis associated with the linear space "tensor product" by means of the adequate ordering criterion.
2. Consider the component $p_{\circ\circ\circ}^{233}\vec{e}_2 \otimes \vec{\epsilon}_3 \otimes \vec{\eta}_3$ the indices of which are given. Obtain by a direct method the value of the component $p_{\circ\circ\circ}^{233}$.
3. Obtain the position n that this component occupies in the natural ordering of the vector $\vec{p}$, checking that it is in agreement with the answer to question 1.
4. Reciprocally, obtain by a direct method the indices and the value of the component that occupies position $n' = 27$, checking that it is in agreement with the answer to question 1.
5. Solve the previous questions using the computer.

2.2. A particle of mass m, electric charge q and velocity v is thrown orthogonally to a uniform magnetic field of intensity B. Then, the particle describes a circular trajectory of radius r in the plane π of the throw, orthogonal to the field.

Assuming equilibrium between the centripetal and the Lorentz force (due to B):

1. Determine B and establish the notation of the tensor space B, as a product of the fundamental linear spaces $L(\mathbb{R}), T(\mathbb{R}), M(\mathbb{R}), Q(\mathbb{R})$ and their duals $L_*(\mathbb{R}), T_*(\mathbb{R}), M_*(\mathbb{R}), Q_*(\mathbb{R})$.
2. Perform the expression $L(\text{centimeter}), T(\text{second}), M(\text{mass gram}), Q(\text{mass U.E.E. electrostatic unit of charge}), B(\text{gauss})$ to $L(\text{meter}), T(\text{second}), M(\text{mass kilogram}), Q(\text{coulomb}), B(\text{tesla})$.

2.3. Consider a certain electric line going from East to West on the earth. Assuming that the current intensity is $I = 20$ amperes, obtain the force per unit length (newtons/meter) to which the filament of the line is subjected by the action of the earth's magnetic field ($B \sim 1$ gauss).

We assume that the "Lorentz force" is $\vec{f} = q\vec{V} \wedge \vec{B}$ and that 1 tesla $=1$ weber/m$^2 = 10^4$ gauss.

3

Homogeneous Tensors

3.1 Introduction

This chapter introduces the homogeneous tensors and the semaphoric notation used in the book, which is justified to avoid important errors when operating with tensors. We suggest that the reader makes an effort to understand the reasons leading to this special notation.

It also defines the tensor product of tensors as a particular case of that for vectors through the stretched representation of tensors.

Next, the tensor product known as Einstein's contraction of tensor products together with the matrix representation of tensors is introduced, and the latter is discussed in more detail for tensors of first and second order.

3.2 The concept of homogeneous tensors

Once we know what tensors are, by now simple vectors belonging to the tensor product linear space $V_1^{n_1} \otimes V_2^{n_2} \otimes \cdots \otimes V_r^{n_r}(K)$, with a general expression given in (2.20), we consider the case where the "factor spaces" are a *unique* linear space $V^n(K)$, sometimes referred to its primal basis $\{\vec{e}_i\}$ and other times to the dual basis $\{\vec{e}^{*i}\}$, which are simultaneously being considered. Consequently, we have the following expressions, which are characteristic of homogeneous tensors:

$$\overset{\longleftarrow \qquad r \text{ times} \qquad \longrightarrow}{V^n \otimes V_*^n \otimes V^n \otimes \cdots \otimes V_*^n \otimes V^n(K),} \qquad (3.1)$$
$$\underset{\longleftarrow \qquad r \text{ times} \qquad \longrightarrow}{}$$

which is the notation used for a homogeneous tensor product space. Its dimension is

$$\sigma = dim \left[\overset{r}{\underset{1}{\otimes}} V^n(K) \right] = n^r \qquad (3.2)$$

and any given tensor can be written as

$$\vec{t} = t^{iok}_{ojo} \cdots {}^{or}_{qo} \; \vec{e}_i \otimes \vec{e}^{*j} \otimes \vec{e}_k \otimes \cdots \otimes \vec{e}^{*q} \otimes \vec{e}_r. \tag{3.3}$$

Of course, they can be combined linearly, by means of the sum $(+)$ and the external product $(\circ)$, as in any linear space.

Example 3.1 (Homogeneous tensor). A homogeneous tensor, of third order $(r = 3)$,

$$\vec{t} = t^{iok}_{ojo} \; \vec{e}_i \otimes \vec{e}^{*j} \otimes \vec{e}_k$$

of the linear space $V^3 \otimes V^3_* \otimes V^3(K)$ is usually given as

$$T = \left\{t^{iok}_{ojo}\right\} = \left\{ \begin{bmatrix} t^{1o1}_{o1o} & t^{1o2}_{o1o} & t^{1o3}_{o1o} & | & t^{2o1}_{o1o} & t^{2o2}_{o1o} & t^{2o3}_{o1o} & | & t^{3o1}_{o1o} & t^{3o2}_{o1o} & t^{3o3}_{o1o} \\ t^{1o1}_{o2o} & t^{1o2}_{o2o} & t^{1o3}_{o2o} & | & t^{2o1}_{o2o} & t^{2o2}_{o2o} & t^{2o3}_{o2o} & | & t^{3o1}_{o2o} & t^{3o2}_{o2o} & t^{3o3}_{o2o} \\ t^{1o1}_{o3o} & t^{1o2}_{o3o} & t^{1o3}_{o3o} & | & t^{2o1}_{o3o} & t^{2o2}_{o3o} & t^{2o3}_{o3o} & | & t^{3o1}_{o3o} & t^{3o2}_{o3o} & t^{3o3}_{o3o} \end{bmatrix} \right\}$$

$$= \left\{ \begin{bmatrix} -1 & 4 & 0 & | & -1 & 4 & 0 & | & 8 & 7 & 6 \\ 2 & 1 & 3 & | & 2 & 1 & 2 & | & -1 & 2 & 3 \\ 5 & 6 & 7 & | & 6 & 9 & 5 & | & 5 & 5 & 9 \end{bmatrix} \right\}, \tag{3.4}$$

where i, j and k are the matrix, the row and the column, respectively, with $\sigma = n^r = 3^3 = 27$ components.

In order to check if the criterion for the indices i, j, k employed by each author when given the tensor data "satisfies", or not, the "basic order convention" (axiom 4, Section 2.5), it is interesting to present the tensor completely developed as a vector. □

3.3 General rules about tensor notation

It is probable that the reader is already aware that in the case of homogeneous tensors, the simple knowledge of the notation of the component of a tensor (t^{oj}_{io}) permits the knowledge of the notation of the corresponding basic vector companion $(\vec{e}^{*i} \otimes \vec{e}_j)$ and reciprocally.

In the following some rules that must be satisfied by a correct notation are indicated. The set of scalars that constitute a tensor is usually denoted by an uppercase letter $T, A, B, \ldots$, etc. This symbol is equated, by means of the "set of" symbol $\{\}$, to the corresponding lower case letter, which carries to its right semaphoric columns"

$$T = \{t^{\bullet}_{o} {}^{\circ}_{\bullet} \cdots {}^{\circ}_{\bullet} {}^{\bullet}_{o}\}$$

with two places per column, which can be occupied by a free or dummy index. The number of columns coincides with the order r of the tensor, and in each column *only* one of the two indices can appear, and there cannot be columns without an index.

Depending on the order r simple tensors receive different names:

Number of columns (order r)	Name of tensor
0	scalar
1	vector
2	matrix

In the "components" each upper index is called a "contravariant index", and each lower index is called a "covariant index".

Given the tensor T, using the notation (in the example $t^{i\,o\,k}_{o\,j\,o}$) one counts the number of indices and finds its order r in Example 3.1 $r = 3$.

Given all scalars of a tensor T (see Example 3.1), the power of T can be found by counting the number of scalars it has, and since it must be $\sigma = Pot\ T = n^r$ (in our example $n^3 = 27$), we discover the dimension n of the linear space factor $V^3(\mathbb{R})$ and then we can reconstruct:

1. The notation of the tensor product linear space $\left(\overset{3}{\underset{1}{\otimes}} V^3(\mathbb{R}) \right)$ to which the tensor belongs.

2. The basic vector associated with the given component, placing the basic vectors in an r-tensor product (for $r = 3 : \vec{e} \otimes \vec{e} \otimes \vec{e}$) and associating them with the same indices as the component, in the reverse position (i.e., lower for upper and upper for lower). If the index in the basic vector is an upper index, it will carry an *asterisk* to remind us that the basis of the corresponding space $V^3(\mathbb{R})$ is the dual one; in our example we get

$$\text{basic vector of } t^{i\,o\,k}_{o\,j\,o} \equiv \vec{e}_i \otimes \vec{e}^{*j} \otimes \vec{e}_k.$$

Remark 3.1. The fact that the simple knowledge of the components of a tensor, with their indices, allows us to know its notation as a vector, has given rise to the omission of the tensor product linear space to which it belongs, and the companion basic vector that follows each component. This motivates the use in tensor books of "scalar packages" without any vectors at all, when in reality tensors are vectors.

There is a reason why the component indices are located in a position reversed to that of basic vectors; it is simply to *satisfy Einstein's convention*, of the sum associated with the linear manifold generated by the basis.   □

Finally, we call to the reader's attention the fact that some authors "stack" the indices for typographical reasons or other notation criteria. For example, $t^{\alpha\beta}_{\gamma\delta}$. This is a *very dangerous* notation, especially if one uses matrix methods to operate tensors.

In this form, one does not know if the tensor $t^{\alpha\beta}_{\gamma\delta}$ is

$$t^{\alpha\beta o o}_{o o \gamma\delta} \quad \text{or} \quad t^{\alpha o \beta o}_{o \gamma o \delta} \quad \text{or} \quad t^{o\alpha\beta o}_{\gamma o o \delta}, \text{ etc.}$$

Since the basic vectors corresponding to each notation *are different*, problems will arise when performing "change-of-basis".

To prevent possible stacks of indices, we use the symbol "∘", when the upper or lower index of the semaphoric column is not occupied, as for example $t^{\circ \alpha \circ \circ}_{\gamma \circ \beta \delta}$.

Other authors present Formula (3.3), the tensor or analytic expression of a homogeneous tensor as

$$\vec{t} = t^{ijk}_{\circ\circ\circ} \cdots {}^{\ell\,\circ\,\circ}_{omn} \cdots {}^{\circ\circ}_{qr}\, \vec{e}_i \otimes \vec{e}_j \otimes \vec{e}_k \otimes \cdots \vec{e}_\ell \otimes \vec{e}^{*m} \otimes \vec{e}^{*n} \otimes \cdots \otimes \vec{e}^{*q} \otimes \vec{e}^{*r}$$

in which all contravariant indices appear stacked ahead and then all covariant indices also stacked.

It is true that expressions

$$u^i v^*_j w^k \vec{e}_i \otimes \vec{e}^{*j} \otimes \vec{e}_k \equiv u^i w^k v^*_j \vec{e}_i \otimes \vec{e}^{*j} \otimes \vec{e}_k$$

are identical, with respect to the tensor product of vectors, because the field is commutative, but the expression

$$u^i v^*_j w^k \vec{e}_i \otimes \vec{e}^{*j} \otimes \vec{e}_k \equiv u^i w^k v^*_j \vec{e}_i \otimes \vec{e}_k \otimes \vec{e}^{*j}$$

alters the basis of the space $\overset{3}{\underset{1}{\otimes}} V^n(K)$ and the ordering convention that is axiomatic. Thus, these "simplifications" will not be used in this book.

3.4 The tensor product of tensors

When the concept of "tensor product linear space" of three given linear spaces was introduced at the beginning of Chapter 2, and later in the generalization of this concept, it was indicated that the nature of vectors in the "factor" linear spaces was not relevant, i.e., it is indifferent.

Now it is the adequate place for assuming that such "factor" spaces are linear spaces of *tensor products*, guaranteeing that the resulting tensor product linear space is another tensor linear space. We shall see this in detail. First, we must note that when operating this product, the factor tensors must be notated with *different* indices to impose the axiomatic condition of coming from different tensor spaces.

Consider the heterogeneous tensors of orders $r_1 = 3$ and $r_2 = 2$

$$t^{i \circ k}_{\circ j \circ} \equiv \vec{e}_i \otimes \vec{e}^{*j} \otimes \vec{\delta}_k \text{ and } \vec{z} = z^{\circ \ell}_{h \circ}\, \vec{\alpha}^{*h} \otimes \vec{\beta}_\ell,$$

that belong to the different tensor product spaces

$$\vec{t} \in U^m \otimes V^n_* \otimes W^p(K) \text{ and } \vec{z} \in R^s_* \otimes S^x(K).$$

The space $(U^m \otimes V^n_* \otimes W^p) \otimes (R^s_* \otimes S^x)(K)$ will be called the "tensor product of tensors" linear space, a heterogeneous tensor linear space, which by axiom arises with the following form:

$$\vec{p} = p^{iok o \ell}_{ojoho}\ \vec{e}_i \otimes \vec{\epsilon}^{*j} \otimes \vec{\delta}_k \otimes \vec{\alpha}^{*h} \otimes \vec{\beta}_\ell \qquad (3.5)$$

and is of order $r = 5$, where each $p^{iok o \ell}_{ojoho} = t^{iok}_{ojo} \cdot z^{o\ell}_{ho}$ can be obtained as a product of scalars, both belonging to K.

The indices ranges are evidently $1 \le i \le m$, $1 \le j \le n$, $1 \le k \le p$, $1 \le h \le s$, and $1 \le \ell \le x$, and the dimension of the new space already with the correct notation, is

$$\sigma = dim[U^m \otimes V^n_* \otimes W^p \otimes R^s_* \otimes S^x(K)] = m \cdot n \cdot p \cdot s \cdot x \text{ (product in } \mathbb{N}).$$

Due to the same reasons previously indicated, in the space $U^m \otimes V^n_* \otimes W^p \otimes R^s_* \otimes S^x(K)(+, \circ)$ there will exist tensors that are linear combinations of the previous ones:

$$a^{iok o \ell}_{ojoho}\ \vec{e}_i \otimes \vec{\epsilon}^{*j} \otimes \vec{\delta}_k \otimes \vec{\alpha}^{*h} \otimes \vec{\beta}_\ell.$$

Although they belong to the "tensor product of tensors" linear space, they *are not* products of tensors, but tensors.

The order of a tensor in the product space is the sum of the orders of the factor tensors: $r = r_1 + r_2$; in our case $r = 3 + 2 = 5$.

Though other textbooks take time to "show" the distributive character of the tensor product of tensors, with respect to the sum $(+)$, and the associative character with respect to the product by a scalar $(\circ)$, we will not insist on this, after the view with which the beginning of this section was treated, of seeing the tensor product of tensors as a simple tensor product of vectors, and knowing that it will share the same axioms and properties.

Example 3.2 (Linear and quadratic forms).

1. Obtain the tensor products $d\phi \otimes d\psi$ and $d\psi \otimes d\phi$ of the linear differential forms

$$d\phi = dx^1 + 5dx^2 + 7dx^3$$
$$d\psi = dx^1 - dx^2$$

   supplying their matrix expressions and their developed analytical expressions.

2. Find the scalar associated with the vector $(7\vec{e}_1 + 3\vec{e}_2 + \vec{e}_3) \otimes (2\vec{e}_1 + 5\vec{e}_2 - 21\vec{e}_3)$ by means of the quadratic form

$$\Phi = (4\vec{\epsilon}^1 - 6\vec{\epsilon}^2 - 9\vec{\epsilon}^3) \otimes (\vec{\epsilon}^1 + 17\vec{\epsilon}^2 - 6\vec{\epsilon}^3).$$

Solution:

1. *First product:*

$$d\phi \otimes d\psi = \begin{bmatrix} dx^1 & dx^2 & dx^3 \end{bmatrix} \begin{bmatrix} 1 \\ 5 \\ 7 \end{bmatrix} \otimes \begin{bmatrix} 1 & -1 & 0 \end{bmatrix} \begin{bmatrix} dx^1 \\ dx^2 \\ dx^3 \end{bmatrix}$$

$$= \begin{bmatrix} dx^1 & dx^2 & dx^3 \end{bmatrix} \begin{bmatrix} 1 & -1 & 0 \\ 5 & -5 & 0 \\ 7 & -7 & 0 \end{bmatrix}_\otimes \begin{bmatrix} dx^1 \\ dx^2 \\ dx^3 \end{bmatrix}$$

$$= dx^1 \otimes dx^1 - dx^1 \otimes dx^2 + 5dx^2 \otimes dx^1$$
$$- 5dx^2 \otimes dx^2 + 7dx^3 \otimes dx^1 - 7dx^3 \otimes dx^2. \qquad (3.6)$$

where the $\otimes$ operator appears as a subindex to refer to a quadratic form of tensor products.

Second product:

$$d\psi \otimes d\phi = \begin{bmatrix} dx^1 & dx^2 & dx^3 \end{bmatrix} \begin{bmatrix} 1 \\ -1 \\ 0 \end{bmatrix} \otimes \begin{bmatrix} 1 & 5 & 7 \end{bmatrix} \begin{bmatrix} dx^1 \\ dx^2 \\ dx^3 \end{bmatrix}$$

$$= \begin{bmatrix} dx^1 & dx^2 & dx^3 \end{bmatrix} \begin{bmatrix} 1 & 5 & 7 \\ -1 & -5 & -7 \\ 0 & 0 & 0 \end{bmatrix}_\otimes \begin{bmatrix} dx^1 \\ dx^2 \\ dx^3 \end{bmatrix}$$

$$= dx^1 \otimes dx^1 + 5dx^1 \otimes dx^2 + 7dx^1 \otimes dx^3 - dx^2 \otimes dx^1$$
$$- 5dx^2 \otimes dx^2 - 7dx^2 \otimes dx^3. \qquad (3.7)$$

We note that they are different tensors, but they belong to the same tensor space.

2. We shall present two different methods for solving this problem:

(a) Applying to each vector its linear form, and then the product of forms (product of scalars), yields

$$\rho_1 = \begin{bmatrix} 7 & 3 & 1 \end{bmatrix} \begin{bmatrix} 4 \\ -6 \\ -9 \end{bmatrix} = 1; \quad \rho_2 = \begin{bmatrix} 2 & 5 & -21 \end{bmatrix} \begin{bmatrix} 1 \\ 17 \\ -6 \end{bmatrix} = 2 + 85 + 126 = 213;$$

$$\rho_1 \otimes \rho_2 = \rho_1 \rho_2 = 1 \times 213 = 213.$$

(b) Finding the vector $\vec{V}_1 \otimes \vec{V}_2$ in its linear space $V \otimes V(\mathbb{R})$, and determining the quadratic form in the dual tensor space (in dual bases) $V_* \otimes V_*(\mathbb{R})$, and finally, applying the quadratic form to the vector, yields

$$\vec{V} = \vec{V}_1 \otimes \vec{V}_2 = 14\vec{e}_1 \otimes \vec{e}_1 + 35\vec{e}_1 \otimes \vec{e}_2 - 147\vec{e}_1 \otimes \vec{e}_3 + 6\vec{e}_2 \otimes \vec{e}_1$$
$$+ 15\vec{e}_2 \otimes \vec{e}_2 - 63\vec{e}_2 \otimes \vec{e}_3 + 2\vec{e}_3 \otimes \vec{e}_1 + 5\vec{e}_3 \otimes \vec{e}_2 - 21\vec{e}_3 \otimes \vec{e}_3$$

$$\vec{F} = \vec{F}' \otimes \vec{F}'^2 = 4\vec{\epsilon}^1 \otimes \vec{\epsilon}^1 + 68\vec{\epsilon}^1 \otimes \vec{\epsilon}^2 - 24\vec{\epsilon}^1 \otimes \vec{\epsilon}^3 - 6\vec{\epsilon}^2 \otimes \vec{\epsilon}^1$$
$$- 102\vec{\epsilon}^2 \otimes \vec{\epsilon}^2 + 36\vec{\epsilon}^2 \otimes \vec{\epsilon}^3 - 9\vec{\epsilon}^3 \otimes \vec{\epsilon}^1 - 153\vec{\epsilon}^3 \otimes \vec{\epsilon}^2 + 54\vec{\epsilon}^3 \otimes \vec{\epsilon}^3.$$

Since the two tensor spaces are expressed in dual bases, the connection matrix (the Gram matrix) is $G \equiv I_{3\times3} = I_9$, and then

$$\rho = \vec{F}(\vec{V}) = \vec{F} \bullet \vec{V} = [\,14\ 35\ -147\ 6\ 15\ -63\ 2\ 5\ -21\,]I_9 \begin{bmatrix} 4 \\ 68 \\ -24 \\ -6 \\ -102 \\ 36 \\ -9 \\ -153 \\ 54 \end{bmatrix} = 213.$$

Note: There are other possible methods for solving this problem, which will be used later in this book.

□

Example 3.3 (Tensor product). Obtain the tensor product of $\vec{t}$ and $\vec{s}$, where

1. $\vec{t}$ and $\vec{s}$ are:

$$\vec{t} = 2\vec{e}_1 \otimes \vec{e}_2 \otimes \vec{e}^{*1} - 4\vec{e}_1 \otimes \vec{e}_1 \otimes \vec{e}^{*1} + 3\vec{e}_2 \otimes \vec{e}_1 \otimes \vec{e}^{*2} \qquad (3.8)$$
$$\vec{s} = -4\vec{e}_1 \otimes \vec{e}^{*2} + 3\vec{e}_2 \otimes \vec{e}^{*1}. \qquad (3.9)$$

2. Idem, by assuming a commutative tensor algebra.

Solution: Obviously we perform the product directly and we give the ordered tensor analytical expression, as follows.

1. In the first case one gets:

$$\vec{t} \otimes \vec{s} = 16\vec{e}_1 \otimes \vec{e}_1 \otimes \vec{e}^{*1} \otimes \vec{e}_1 \otimes \vec{e}^{*2} - 12\vec{e}_1 \otimes \vec{e}_1 \otimes \vec{e}^{*1} \otimes \vec{e}_2 \otimes \vec{e}^{*1}$$
$$- 8\vec{e}_1 \otimes \vec{e}_2 \otimes \vec{e}^{*1} \otimes \vec{e}_1 \otimes \vec{e}^{*2} + 6\vec{e}_1 \otimes \vec{e}_2 \otimes \vec{e}^{*1} \otimes \vec{e}_2 \otimes \vec{e}^{*1}$$
$$-12\vec{e}_2 \otimes \vec{e}_1 \otimes \vec{e}^{*2} \otimes \vec{e}_1 \otimes \vec{e}^{*2} + 9\vec{e}_2 \otimes \vec{e}_1 \otimes \vec{e}^{*2} \otimes \vec{e}_2 \otimes \vec{e}^{*1}. \; (3.10)$$

2. In the second case one gets:

$$\vec{t} \otimes \vec{s} = 16\vec{e}_1 \otimes \vec{e}_1 \otimes \vec{e}_1 \otimes \vec{e}^{*1} \otimes \vec{e}^{*2} - 12\vec{e}_1 \otimes \vec{e}_1 \otimes \vec{e}_2 \otimes \vec{e}^{*1} \otimes \vec{e}^{*1}$$
$$-8\vec{e}_1 \otimes \vec{e}_1 \otimes \vec{e}_2 \otimes \vec{e}^{*1} \otimes \vec{e}^{*2} - 12\vec{e}_1 \otimes \vec{e}_1 \otimes \vec{e}_2 \otimes \vec{e}^{*2} \otimes \vec{e}^{*2}$$
$$+6\vec{e}_1 \otimes \vec{e}_2 \otimes \vec{e}_2 \otimes \vec{e}^{*1} \otimes \vec{e}^{*1} + 9\vec{e}_1 \otimes \vec{e}_2 \otimes \vec{e}_2 \otimes \vec{e}^{*1} \otimes \vec{e}^{*2}. \; (3.11)$$

□

3.5 Einstein's contraction of the tensor product

When considering the tensor product of two tensors, both vectors of the same "tensor product of vectors" linear space, for example $U^m \otimes V_*^n(K)$, we know that the previously proposed axiom forces the change of subindices both in the scalar factors and in their bases, i.e., if $a_{oj}^{i\,o}\, \vec{e}_i \otimes \vec{e}^{*j}, b_{oj}^{i\,o}\, \vec{e}_i \otimes \vec{e}^{*j} \in U^m \otimes V_*^n(K)$, they must be considered as $a_{oj}^{i\,o}\, \vec{e}_i \otimes \vec{e}^{*j}$, $b_{o\ell}^{k\,o}\, \vec{e}_k \otimes \vec{e}^{*\ell}$ in order for them to be multiplied as tensors. Of course the result will be "external", i.e.

$$p_{oj\,o\,\ell}^{i\,o\,k\,o}\, \vec{e}_i \otimes \vec{e}^{*j} \otimes \vec{e}_k \otimes \vec{e}^{*\ell} \notin U^m \otimes V_*^n(K) \tag{3.12}$$

where

$$p_{oj\,o\,\ell}^{i\,o\,k\,o} = a_{oj}^{i\,o} \cdot b_{o\,\ell}^{k\,o}; \quad i,k \in \{1,2,\dots,m\}; \quad j,\ell \in \{1,2,\dots,n\}.$$

In other words, the "tensor product of tensors" is never *internal*, even in the case of both factors coming from the same linear space.

We shall apply *Einstein's contraction* to certain indices of a *mixed homogeneous tensor* (with contravariant and covariant indices), with the purpose of obtaining a new scalar system with the tensor character but lower order. This operation is known as "tensor contraction".

Successive or simultaneous contractions over pairs of indices of *different valency* in a mixed homogeneous tensor, lead to another tensor. The proof will be given in Chapter 5.

The use of the contraction operation, not only with indices of different valency, but also chosen in different factors, is called a "contracted tensor product". In this way we can reduce the tensor order until reaching the initial order of the *factors*, resulting in an internal product that leads to an internal tensor or linear algebra, with the aim of distinguishing it from general multilineal algebras.

Example 3.4 (Tensor contraction of a tensor). Consider the tensor

$$\vec{t} \in \overset{5}{\underset{1}{\otimes}} V^n(K); \quad \vec{t} = t_{ookh\ell}^{ij\,ooo}\vec{e}_i \otimes \vec{e}_j \otimes \vec{e}^{*k} \otimes \vec{e}^{*h} \otimes \vec{e}^{*\ell}.$$

We will contract the indices 2 and 4 (j contravariant, with h covariant), to get

$$\vec{u} \in \overset{3}{\underset{1}{\otimes}} V^n(K); \quad \vec{u} = C\binom{2}{4}(\vec{t}) = t_{ookal}^{i\,\alpha\,ooo}\, \vec{e}_i \otimes \vec{e}^{*k} \otimes \vec{e}^{*\ell}, \tag{3.13}$$

where

$$u_{ok\ell}^{i\,oo} \equiv t_{ook\alpha\ell}^{i\,\alpha\,ooo} \equiv t_{ook1\ell}^{i\,1\,ooo} + t_{ook2\ell}^{i\,2\,ooo} + \cdots + t_{ookn\ell}^{i\,n\,ooo} \tag{3.14}$$

and the free indices vary in $i,k,\ell \in \{1,2,\cdots,n\}$ (n^3 components). $\square$

Example 3.5 (Contracted tensor product). Consider the homogeneous tensors:

$$\vec{s}, \vec{t} \in \overset{2}{\underset{1}{\otimes}} V^n(K); \quad \text{or } S, T \in \mathbb{R}^{n \times n} \text{ (two matrices)},$$

where

$$\vec{s} = s^{i \circ}_{\circ j} \, \vec{e}_i \otimes \vec{e}^{*j}; \quad \vec{t} = t^{k \circ}_{\circ \ell} \, \vec{e}_k \otimes \vec{e}^{*\ell}.$$

The tensor product resulting from contracting indices (the j index of the first covariant factor, with the k index of the second contravariant factor) is

$$\vec{p} = C\binom{2}{3}(\vec{s} \otimes \vec{t}); \quad p^{i \circ \alpha \circ}_{\circ \alpha \circ \ell} \equiv s^{i \circ}_{\circ \alpha} \cdot t^{\alpha \circ}_{\circ \ell} \equiv p^{i \circ}_{\circ \ell},$$

that is, $P \in \mathbb{R}^{n \times n}$, i.e., it is a matrix, where

$$p^{i \circ}_{\circ \ell} = s^{i \circ}_{\circ 1} \cdot t^{1 \circ}_{\circ \ell} + s^{i \circ}_{\circ 2} \cdot t^{2 \circ}_{\circ \ell} + \cdots + s^{i \circ}_{\circ n} \cdot t^{n \circ}_{\circ \ell}; \quad i, \ell \in \{1, 2, \cdots, n\}. \quad (3.15)$$

Note that this contraction is no more than the classic matrix product (the inner product), which converts the linear space $p^{i \circ}_{\circ \ell}{}_{\ell} \, \vec{e}_i \otimes \vec{e}^{*\ell}$ to which the data belong, into a *linear algebra*, the linear algebra of matrices.

We invite the reader to study the result of contracting the indices i of the first factor with the ℓ of the second factor. □

Finally, we simply mention the contraction of "pure" tensors (totally contravariant or totally covariant) and also the contraction of indices of the *same valency* in mixed tensors.

In this case *it is not assured* that the resulting system of scalars will be a tensor; there are cases in which one obtains a tensor, and others in which it is not a tensor. Due to this reason, when one needs to know if the resulting system is a tensor, other techniques to be explained later in this book must be used.

Theorem 3.1 (Contractions as homomorphisms). *Tensor contractions are none other than homomorphisms (linear applications) of a linear space in another linear space of smaller dimension. We use the notation:*

$$C: \overset{p+q}{\underset{1}{\otimes}} V^n(K) \longrightarrow \overset{[(p-1)+(q-1)]}{\underset{1}{\otimes}} V^n(K). \quad (3.16)$$

□

Proof. Consider the data $\lambda, \mu \in K$ and $\vec{s}, \vec{t} \in V^n \otimes V^n_* \otimes V^n_* \otimes V^n(K)$. We shall contract indices 1 and 2 of a linear combination of these tensors (which is another tensor of the same space):

$$C\binom{1}{2}(\lambda\vec{s}+\mu\vec{t}) = C\binom{1}{2}[\lambda s^{i\,o\,o\,\ell}_{o\,j\,k\,o}\,\vec{e}_i\otimes\vec{e}^{*j}\otimes\vec{e}^{*k}\otimes\vec{e}_\ell$$

$$+\mu t^{i\,o\,o\,\ell}_{o\,j\,k\,o}\,\vec{e}_i\otimes\vec{e}^{*j}\otimes\vec{e}^{*k}\otimes\vec{e}_\ell]$$

$$= C\binom{1}{2}[\lambda s^{i\,o\,o\,\ell}_{o\,j\,k\,o}+\mu t^{i\,o\,o\,\ell}_{o\,j\,k\,o}]\,\vec{e}_i\otimes\vec{e}^{*j}\otimes\vec{e}^{*k}\otimes\vec{e}_\ell$$

$$= [\lambda s^{\alpha\,o\,o\,\ell}_{o\,\alpha\,k\,o}+\mu t^{\alpha\,o\,o\,\ell}_{o\,\alpha\,k\,o}]\,\vec{e}^{*k}\otimes\vec{e}_\ell$$

$$= \lambda s^{\alpha\,o\,o\,\ell}_{o\,\alpha\,k\,o}\,\vec{e}^{*k}\otimes\vec{e}_\ell+\mu t^{\alpha\,o\,o\,\ell}_{o\,\alpha\,k\,o}\,\vec{e}^{*k}\otimes\vec{e}_\ell$$

$$= \lambda C\binom{1}{2}(\vec{s})+\mu C\binom{1}{2}(\vec{t}). \tag{3.17}$$

The contracted tensor of the linear combination is the linear combination of the contracted tensors. Thus, it is an homomorphism of the linear space $V^n\otimes V^n_*\otimes V^n_*\otimes V^n(K)$ into the linear space $V^n_*\otimes V^n(K)$:

$$C\binom{1}{2}: V^n\otimes V^n_*\otimes V^n_*\otimes V^n(K)\longrightarrow V^n_*\otimes V^n(K).$$

3.6 Matrix representation of tensors

Some authors call scalars tensors of order zero. In this section we deal first with vectors, which are tensors of order one, and later with matrices, which are tensors of order two.

3.6.1 First-order tensors

It is convenient to introduce the reader to the conventions of the matrix notation of vectors, spaces with connections, bilinear forms, etc. in such a way that their relations to tensor notation, Einstein's convention, etc. be so logical and simple that we may move quickly from one formulation to the other, at will.

The basic vector matrices will be presented as row matrices, with "hat" or not, depending on whether or not they are a new basis or the initial basis, respectively.

Scalars always will appear as a column matrix, with the same meaning; "$\|\ \|$" is the symbol used for row matrices, and upper case letters X, Y, Z are symbols for scalar column matrices. The basis of the linear space $V^n(K)$ is denoted by

$$\|\vec{e}_j\| \equiv [\vec{e}_1\ \ \vec{e}_2\ \ \cdots\ \ \vec{e}_j\ \ \cdots\ \ \vec{e}_n],$$

the new basis

$$\|\vec{\hat{e}}_j\| \equiv [\vec{\hat{e}}_1\ \ \vec{\hat{e}}_2\ \ \cdots\ \ \vec{\hat{e}}_j\ \ \cdots\ \ \vec{\hat{e}}_n]$$

and the new basis of the dual space $V^n_*(K)$ by

$$||\vec{\hat{e}}^{*i}|| \equiv [\vec{\hat{e}}^{*1} \quad \vec{\hat{e}}^{*2} \quad \cdots \quad \vec{\hat{e}}^{*i} \quad \cdots \quad \vec{\hat{e}}^{*n} \,].$$

Examples of vectors are

$$\vec{V} = ||\vec{e}_j||X \equiv [\,\vec{e}_1 \ \vec{e}_2 \ \cdots \ \vec{e}_n\,] \begin{bmatrix} x^1 \\ x^2 \\ \vdots \\ x^n \end{bmatrix} \equiv x^1\vec{e}_1 + x^2\vec{e}_2 + \cdots + x^n\vec{e}_n$$

$$\equiv x^j\vec{e}_j \text{ (tensor notation).} \quad (3.18)$$

$$\vec{W} = y^i\vec{e}_i \equiv ||\vec{e}_i||Y \equiv [\,\vec{e}_1 \ \vec{e}_2 \ \cdots \ \vec{e}_n\,] \begin{bmatrix} y^1 \\ y^2 \\ \vdots \\ y^n \end{bmatrix}. \quad (3.19)$$

$$\vec{z} = ||\vec{e}^{*i}||Z^* \equiv [\,\vec{e}^{*1} \ \vec{e}^{*2} \ \cdots \ \vec{e}^{*n}\,] \begin{bmatrix} z_1 \\ z_2 \\ \vdots \\ z_n \end{bmatrix} \equiv z_1\vec{e}^{*1} + z_2\vec{e}^{*2} + \cdots + z_n\vec{e}^{*n}$$

$$\equiv z_i\vec{e}^{*i} \text{ (tensor notation)} \quad (3.20)$$

$$\vec{W} = y^i\vec{e}_i = \hat{y}^i\vec{\hat{e}}_i \equiv ||\vec{\hat{e}}_i||\hat{Y} \equiv [\,\vec{\hat{e}}_1 \ \vec{\hat{e}}_2 \ \cdots \ \vec{\hat{e}}_n\,] \begin{bmatrix} \hat{y}^1 \\ \hat{y}^2 \\ \vdots \\ \hat{y}^n \end{bmatrix}. \quad (3.21)$$

3.6.2 Second-order tensors

Though there is a wide range of tensor relations where matrices are involved, we will pay particular attention to the change-of-basis in linear spaces and to its double matrix and tensor notation.

The classic change-of-basis in a linear space $V^n(K)$ can be notated as

$$||\vec{\hat{e}}_p|| = ||\vec{e}_h||C, \ |C| \neq 0 \quad (3.22)$$

$$[\,\vec{\hat{e}}_1 \ \vec{\hat{e}}_2 \ \cdots \ \vec{\hat{e}}_n\,] = [\,\vec{e}_1 \ \vec{e}_2 \ \cdots \ \vec{e}_n\,] \begin{bmatrix} c_{\circ 1}^{1\circ} & c_{\circ 2}^{1\circ} & \cdots & c_{\circ n}^{1\circ} \\ c_{\circ 1}^{2\circ} & c_{\circ 2}^{2\circ} & \cdots & c_{\circ n}^{2\circ} \\ \vdots & \vdots & \cdots & \vdots \\ c_{\circ 1}^{n\circ} & c_{\circ 2}^{n\circ} & \cdots & c_{\circ n}^{n\circ} \end{bmatrix}. \quad (3.23)$$

where the semaphoric columns of matrix C have been selected to satisfy the matrix notation convention (first index for the row and second index for the column) and the symbolic Einstein convention in the expression that follows.

Identifying the column p in both members of (3.22), one gets

$$\vec{\acute{e}}_p = c_{op}^{1o}\vec{e}_1 + c_{op}^{2o}\vec{e}_2 + \cdots + c_{op}^{no}\vec{e}_n = c_{op}^{ho}\vec{e}_h \quad \text{(tensor notation)}.$$

Thus, the tensor notation of (3.22) is

$$\vec{\acute{e}}_p = c_{op}^{ho}\vec{e}_h, \tag{3.24}$$

where it must be $|C| \neq 0$.

The conclusion is that the change-of-basis notations in $V^n(K)$ correspond as

$$C <> c_{op}^{ho}, \tag{3.25}$$

i.e., a homogeneous second-order $(r = 2)$ tensor contra-covariant, with n^2 components.

With respect to the change-of-basis of the "representation" of the vector $\vec{V}$, Formula (3.18), we have

$$\vec{V} = ||\vec{e}_h||X \quad \text{in the initial basis} \tag{3.26}$$

$$\vec{V} = ||\vec{\acute{e}}_p||\hat{X} \quad \text{in the new basis} \tag{3.27}$$

and substituting (3.22) in the last expression one gets

$$\vec{V} = ||\vec{e}_h||C\hat{X},$$

which is again the expression of $\vec{V}$ in the initial basis, which requires

$$||\vec{e}_h||X \equiv ||\vec{e}_h||C\hat{X} \longrightarrow X = C\hat{X}, \tag{3.28}$$

a matrix relation between the initial and new components of the vector, the tensor representation of which, on account of (3.25), becomes

$$x^h = c_{o\alpha}^{ho}\,\hat{x}^\alpha \tag{3.29}$$

in which one detects with clarity "a contracted tensor product", where α is the dummy index, and h is the free index.

In order to have a tensor representation for transposed matrices, we exchange the semaphore columns of the tensor (see expression (3.25)) and get[1]

$$C^t <> c_{po}^{oh}. \tag{3.30}$$

Next we study the inversion of matrices. Assume that based on the matrix relation (3.22) we obtain the initial basis. For the following proof and for the sake of simplicity we change the name of the free index p to

---

[1] Note that as in the matrix notation we exchange rows and column indices to obtain the transpose matrix, in tensor notation, we exchange semaphoric columns.

$$||\vec{\bar{e}}_p|| = ||\vec{\bar{e}}_h||C \longrightarrow ||\vec{\bar{e}}_h|| = ||\vec{\bar{e}}_q||C^{-1} \tag{3.31}$$

$$||\vec{\bar{e}}_h|| = ||\vec{\bar{e}}_q|| \begin{bmatrix} \gamma_{\circ 1}^{1\circ} & \gamma_{\circ 2}^{1\circ} & \cdots & \gamma_{\circ n}^{1\circ} \\ \gamma_{\circ 1}^{2\circ} & \gamma_{\circ 2}^{2\circ} & \cdots & \gamma_{\circ n}^{2\circ} \\ \vdots & \vdots & \cdots & \vdots \\ \gamma_{\circ 1}^{n\circ} & \gamma_{\circ 2}^{n\circ} & \cdots & \gamma_{\circ n}^{n\circ} \end{bmatrix} \tag{3.32}$$

and identifying the h column of the two members of (3.32), we get

$$\vec{\bar{e}}_h = \gamma_{\circ h}^{1\circ}\vec{\bar{e}}_1 + \gamma_{\circ h}^{2\circ}\vec{\bar{e}}_2 + \cdots + \gamma_{\circ h}^{n\circ}\vec{\bar{e}}_n = \gamma_{\circ h}^{q\circ}\vec{\bar{e}}_q \text{ (tensor notation)},$$

so that the tensor notation of (3.31) becomes

$$\vec{\bar{e}}_h = \gamma_{\circ h}^{q\circ}\vec{\bar{e}}_q. \tag{3.33}$$

Consequently, the correspondence of matrix and tensor notations is

$$C^{-1} <> \gamma_{\circ h}^{q\circ} \text{ (provisional)}. \tag{3.34}$$

The notation used for the elements of C^{-1} must satisfy the definition of inverse matrix, so that taking into account the equivalences (3.25) and (3.34), we have

$$C^{-1}C = I_n \Leftrightarrow \gamma_{\circ h}^{q\circ} \cdot c_{\circ p}^{h\circ},$$

a contracted tensor product (h dummy), from which we get $\delta_{\circ p}^{q\circ}$ (Kronecker delta), that is the tensor notation of matrix I_n.

So, we must have

$$\gamma_{\circ h}^{q\circ} \cdot c_{\circ p}^{h\circ} = \delta_{\circ p}^{q\circ}. \tag{3.35}$$

From now on, we use Greek letters, those corresponding to the Roman alphabet for the components of the inverse matrix; so that if $M(m)$ is the notation for a matrix M, and its components m, its inverse will be denoted $M^{-1}(\mu)$, which implies (by convention) that (3.35) is satisfied.

Due to reasons of convenient tensor notation, we return the index q of (3.34) to its initial notation p (since it was altered just with the purpose of developing the already presented proof).

Concluding, the correspondence between matrix and tensor notations is given by[2]

$$C^{-1} <> \gamma_{\circ h}^{p\circ}, \tag{3.36}$$

so that for inverting a matrix equation, such as (3.24) it suffices to move to the other member the letter c, replace it with γ, and exchange indices:

---

[2] Note that the Greek letter γ has been used to denote the inverse of c, the corresponding Roman letter, as indicated above.

$$\vec{e}_h = \gamma^{p\,\circ}_{\circ\,h}\vec{e}_p, \qquad (3.37)$$

which is Equation (3.33) with the desired notation.

In summary, if $A <> a^{i\,\circ}_{\circ\,j}$, we have

$$A^t <> a^{\circ\,i}_{j\,\circ}; \quad A^{-1} <> \alpha^{j\,\circ}_{\circ\,i}; \quad (A^{-1})^t <> \alpha^{\circ\,j}_{i\,\circ}. \qquad (3.38)$$

An important case remains to be analyzed. Consider two linear spaces $V^n(K)$ and $V^n_*(K)$ which are "mutually connected" (they are dual), and have both the same dimension. We know that to any pair of vectors $\vec{V} \in V^n(K)$ and $\vec{W} \in V^n_*(K)$, chosen from each of the n-dimensional spaces, corresponds a scalar $\rho \in K$, denoted by $\rho = <\vec{V}, \vec{W}>$. We know that if the bases of the primal and dual spaces are $\{\vec{e}_i\}$ and $\{\vec{\epsilon}_j\}$, respectively, the matrix representation of the connection is

$$\rho = <\vec{V}, \vec{W}> = X^t G Y; \quad G = [g_{ij}]; \quad |G| \neq 0, \qquad (3.39)$$

where $g_{ij} = <\vec{e}_i, \vec{\epsilon}_j>$, and G is the Gram connection matrix with respect to the given bases (usually data).

If in a special situation, the linear spaces $V^n(K)$ and $V^n_*(K)$ are referred to *dual bases* $(\{\vec{e}_i\}, \{\vec{e}^{*j}\})$, which implies a matrix $G \equiv I_n$, we perform a change-of-basis in the primal space $V^n(K)$, that is

$$||\vec{\hat{e}}_k|| = ||\vec{e}_i||C, \qquad (3.40)$$

then the following question would arise: What change-of-basis must be used in the secondary or dual space $V^n_*(K)$:

$$||\vec{\hat{e}}^{*\ell}|| = ||\vec{e}^{*j}||\Gamma \qquad (3.41)$$

for the new bases $\{\vec{\hat{e}}_k\}$ and $\{\vec{\hat{e}}^{*\ell}\}$ of the corresponding spaces to remain *dual bases*?

We know that

$$\vec{V} = ||\vec{e}_i||X = ||\vec{\hat{e}}_k||\hat{X}; \quad \vec{W} = ||\vec{e}^{*j}||Y^* = ||\vec{\hat{e}}^{*\ell}||\hat{Y}^*$$

and that matrices $X, \hat{X}$ and $Y^*, \hat{Y}^*$ are related by

$$X = C\hat{X} \qquad (3.42)$$

and

$$Y^* = \Gamma\hat{Y}^*, \qquad (3.43)$$

respectively.

Since the bases of these linear spaces are *dual bases* before and after the bases changes, we must have

$$\rho = \langle \vec{V}, \vec{W} \rangle = X^t I_n Y^* = \hat{X}^t \hat{I}_n \hat{Y}^*; \quad \forall \vec{V}, \vec{W} \qquad (3.44)$$

and substituting into (3.44) Equations (3.42) and (3.43) one gets

$$(C\hat{X})^t I_n (\Gamma \hat{Y}^*) = \hat{X}^t I_n \hat{Y}^*; \quad \forall \vec{V}, \vec{W}$$

$$\hat{X}^t (C^t I_n \Gamma) \hat{Y}^* \equiv \hat{X}^t I_n \hat{Y}^*; \quad \forall \hat{X}, \hat{Y}^*,$$

which implies $C^t I_n \Gamma = I_n$ or

$$\Gamma = (C^t)^{-1} \equiv (C^{-1})^t. \qquad (3.45)$$

Then, if in the primary linear space $V^n(K)$ a change-of-basis (3.24) is performed, in the dual space $V_*^n(K)$, according to (3.38) and (3.45), the following change-of-basis must be done:

$$\vec{\hat{e}}^{*p} = \gamma^{\circ p}_{h\circ} \, \vec{e}^{*h} \longrightarrow \gamma^{\circ p}_{h\circ} <> \left(C^{-1}\right)^t. \qquad (3.46)$$

Obtaining the "initial" dual basis from (3.46) we finally have

$$\vec{e}^{*h} = c^{\circ h}_{p\circ} \, \vec{\hat{e}}^{*p}. \qquad (3.47)$$

Formulas (3.24), (3.25) and (3.46), together with

$$\vec{e}_h = \gamma^{p\circ}_{\circ h} \, \vec{\hat{e}}_p \longrightarrow \gamma^{p\circ}_{\circ h} <> C^{-1} \qquad (3.48)$$

$$\vec{e}^{*h} = c^{\circ h}_{p\circ} \, \vec{\hat{e}}^{*p} \longrightarrow c^{\circ h}_{p\circ} <> C^t \qquad (3.49)$$

will be of decisive importance in the following chapters.

3.7 Exercises

3.1. Solve the following items:

1. Answer question 1 of Example 3.3 using Formula (2.21). Determine first matrices $X_{8,1}$ of $\vec{t}$ and $X_{4,1}$ of $\vec{s}$.
   After obtaining the "extended" components of $\vec{t} \otimes \vec{s}$, the corresponding tensor basic vectors must be added to them, using the "ordered basis" criterion.
2. Solve the previous question using the computer.

3.2. Perform the following contractions:

1. $a_\circ^i b^{\circ j}_{i\circ} c^\circ_j$, where $i \in \{1, 2, 3\}$ and $j \in \{1, 2\}$.
2. $a^\circ_i b^{i\circ}_{\circ j} c^{j\circ}_{\circ k} d^k_\circ$, where $i, j, k \in \{1, 2, 3\}$.

3. $a_{i\,i}^{\circ\circ}b_{\circ\circ}^{ij}c_{j}^{\circ}$, where $i \in \{1,2,3\}$ and $j \in \{1,2\}$ (be aware of the indices of the first factor).

3.3. Notate the following expressions with the minimum possible number of dummy indices:

1. $\left(g_{ij}^{\circ\circ}x_{\circ}^{i}x_{\circ}^{j} + h_{pq}^{\circ\circ}x_{\circ}^{p}x_{\circ}^{q}\right)$, where $i,j,p,q \in \{1,2,3,\ldots,n\}$.

2. $\left(a_{ij}^{\circ\circ}b_{\circ\circ}^{jk}c_{k\ell}^{\circ\circ} + a_{ip}^{\circ\circ}b_{\circ\circ}^{pq}c_{q\ell}^{\circ\circ}\right)$, where $i,j,k,\ell,p,q \in \{1,2,3,\ldots,n\}$.

3. Is it licit to propose $\delta_{ij}^{\circ\circ}a_{\circ}^{i}a_{\circ}^{j}a_{\circ}^{k} = \delta_{ik}^{\circ\circ}a_{\circ}^{i}a_{\circ}^{k}a_{\circ}^{j}$; $i,j,k \in \{1,2,3\ldots,n\}$, where $[\delta_{ij}^{\circ\circ}]$ is the Kronecker square matrix I_n?

3.4. Consider the homogeneous tensors A and B, $A \in U^3 \otimes U_*^3(\mathbb{R})$ with basis $\{\vec{e}_\alpha \otimes \vec{e}^{*\beta}\}$ and $B \in U_*^3 \otimes U^3 \otimes U^3(\mathbb{R})$ with basis $\{\vec{e}^{*\gamma} \otimes \vec{e}_\delta \otimes \vec{e}_\eta\}$.
Their components are represented by the following matrices:

$$A = [a_{\circ\beta}^{\alpha\circ}] \equiv \begin{bmatrix} 2 & 5 & -1 \\ 7 & -4 & 0 \\ 3 & 6 & 1 \end{bmatrix} ; \quad B = [b_{\gamma\circ\circ}^{\circ\delta\eta}] \equiv \begin{bmatrix} 1 & 2 & 5 \\ 4 & 2 & 1 \\ 0 & 3 & 3 \\ \hline 2 & 5 & -1 \\ 6 & 2 & 7 \\ 1 & 1 & -2 \\ \hline 3 & 0 & 2 \\ 4 & -1 & 3 \\ 6 & 5 & 4 \end{bmatrix} ,$$

where γ is the block row, δ is the row of each block and η is the column of each block.

1. Solve the following examples of the extension operation:
    a) Obtain the "extended" components of tensors A and B (matrices $A_{9,1}$ and $B_{27,1}$).
    b) Obtain the "extended" components (column matrix $P_{243,1}$) of tensor $P = A \otimes B$, as a tensor product of the extended vectors $A_{9,1}$ and $B_{27,1}$, using Formula (2.21).
    The components of $P_{243,1}$ will be notated in tensor format $(p_{\circ\beta\gamma\circ\circ}^{\alpha\circ\circ\delta\eta})$ using the axiomatic order.
    c) Let $F \in U^3 \otimes U_*^3 \otimes U^3$ be the order $r = 3$ tensor with components $f_{\circ\gamma\circ}^{\alpha\circ\eta}$, resulting from the contraction of indices 2 and 4, that is, the contracted tensor product $A \otimes B$:

$$F = \mathcal{C}\binom{2}{4}(P) \quad \text{with} \quad f_{\circ\gamma\circ}^{\alpha\circ\eta} = p_{\circ\theta\gamma\circ\circ}^{\alpha\circ\circ\theta\eta}.$$

Obtain the "extended" matrix $F_{27,1}$ of tensor F with its components in strict order.

d) Obtain the matrix representation $F = [f_{\circ\gamma\circ}^{\alpha\circ\eta}]$, using the "condensation" of $F_{27,1}$ until the same format than the one in tensor B is obtained.

e) Let $K \in U^3(\mathbb{R})$ be the order $r = 1$ tensor (a vector), resulting from the new contraction of indices 1 and 2 of F, that is, the second contraction of indices 1 and 3 of $A \otimes B$:

$$K = C\binom{1}{2}(F) = C\left(\begin{array}{c|c} 2 & 1 \\ 4 & 3 \end{array}\right)(A \otimes B), \text{ with } k_{\circ}^{\eta} = f_{\circ\phi\circ}^{\phi\circ\eta} = p_{\circ\theta\phi\circ\circ}^{\phi\circ\circ\theta\eta}.$$

f) Obtain the "extended" matrix K_3, of tensor K.

g) Explain if the following statement is correct: "K is a doubly *contracted product* tensor of $A \otimes B$".

2. Solve the previous questions using the computer.

3.5. In the Euclidean space $E^4(\mathbb{R})$ referred to dual bases $(G \equiv I_4)$, a change-of-basis with associated matrix

$$C = \begin{bmatrix} 1 & 1 & 1 & 1 \\ 1 & 2 & 3 & 4 \\ 1 & 3 & 6 & 10 \\ 1 & 4 & 10 & 20 \end{bmatrix}$$

is performed.

Determine:

1. a) The new dual basis, as a function of the initial basis.

   b) The new contravariant and covariant coordinates of a vector $\vec{V}$ with

   initial contravariant coordinates $X = \begin{bmatrix} -1 \\ 1 \\ 0 \\ 3 \end{bmatrix}$.

   c) The new Gram connection matrix $\hat{G}$.

2. Solve the previous questions using the computer.

4

Change-of-basis in Tensor Spaces

4.1 Introduction

This chapter first discusses the tensor criteria that allows us to determine when a set of scalars is a tensor. For reasons of simplicity, we deal first with third-order tensors and obtain a matrix representation for the change-of-basis for tensors in stretched form (column matrix form) that reveals the sought after tensor criteria.

Later the method is generalized to the case of k vectors and extended to mixed homogeneous tensors, i.e., expressed in mixed contravariant and covariant components.

Some useful rules to operate tensor expressions in matrix form are given for tensors of orders 2, 3 and 4.

Finally, the chapter ends with the tensors that are invariant to changes of basis, that is, isotropic tensors, which include the null tensor and Kronecker's delta tensor.

4.2 Change of basis in a third-order tensor product space

We have reached the crucial point where the tensor concept must be clearly established. Up to now (Chapter 2) the linear space tensor product $U^m \otimes V^n \otimes W^p(K)$ (ignoring how it was generated) has been a simple space, of dimension $\sigma = m \cdot n \cdot p$, the basis of which has a peculiar notation and order $\{\vec{e}_i \otimes \vec{\epsilon}_j \otimes \vec{\eta}_k\}$, and the algebraic operations of which (sum, scalar–vector product and contracted tensor product of its vectors) are known.

A change-of-basis in this linear space would be given by means of an arbitrary square matrix of entities over K, $Z_{\sigma,\sigma}$ of "order σ", regular $|Z_{\sigma,\sigma}| \neq 0$, that relate the new basis to the initial; in matrix form:

$$||\vec{\hat{e}}_i \otimes \vec{\hat{\epsilon}}_j \otimes \vec{\hat{\eta}}_k|| = ||\vec{e}_i \otimes \vec{\epsilon}_j \otimes \vec{\eta}_k||Z_{\sigma,\sigma} \qquad (4.1)$$

where $1 \leq i \leq m$; $1 \leq j \leq n$; $1 \leq k \leq p$.

However, from now on and by *convention* (*Axiom 5* of the absolute tensor character) only the changes of basis coming from the "linear space history" *are licit*, as is to be proved below.

We consider only the changes of basis $Z_{\sigma,\sigma}$ associated with changes of basis in the factor linear spaces with which the tensor product space was built.

Thus, in the linear space $U^m(K)$ we perform the change

$$
\begin{aligned}
&\text{Matrix notation: } ||\vec{\hat{e}}_p|| = ||\vec{e}_i||C_m; \; ||\vec{e}_i|| = ||\vec{\hat{e}}_p||C_m^{-1}\\
&\text{Tensor notation: } \quad \vec{\hat{e}}_p = c_{op}^{i\,o}\,\vec{e}_i; \qquad \vec{e}_i = \gamma_{oi}^{po}\,\vec{\hat{e}}_p
\end{aligned}
\tag{4.2}
$$

in the linear space $V^n(K)$, the change

$$
\begin{aligned}
&\text{Matrix notation: } ||\vec{\hat{\epsilon}}_q|| = ||\vec{\epsilon}_j||R_n; \; ||\vec{\epsilon}_j|| = ||\vec{\hat{\epsilon}}_q||R_n^{-1}\\
&\text{Tensor notation: } \quad \vec{\hat{\epsilon}}_q = r_{oq}^{j\,o}\,\vec{\epsilon}_j; \qquad \vec{\epsilon}_j = \rho_{oj}^{qo}\,\vec{\hat{\epsilon}}_q
\end{aligned}
\tag{4.3}
$$

and in the linear space $W^p(K)$, the change

$$
\begin{aligned}
&\text{Matrix notation: } ||\vec{\hat{\eta}}_r|| = ||\vec{\eta}_k||S_p; \; ||\vec{\eta}_k|| = ||\vec{\hat{\eta}}_r||S_p^{-1}\\
&\text{Tensor notation: } \quad \vec{\hat{\eta}}_r = s_{or}^{k\,o}\,\vec{\eta}_k; \qquad \vec{\eta}_k = \sigma_{ok}^{ro}\,\vec{\hat{\eta}}_r
\end{aligned}
\tag{4.4}
$$

Table 4.1 summarizes the previous changes.

If $\vec{t} \in U^m \otimes V^n \otimes W^p(K)$ is an arbitrary tensor, according to (2.12), in the initial and the new basis it can be written as

$$
\vec{t} = t_{ooo}^{ijk}\vec{e}_i \otimes \vec{\epsilon}_j \otimes \vec{\eta}_k = \hat{t}_{ooo}^{pqr}\vec{\hat{e}}_p \otimes \vec{\hat{\epsilon}}_q \otimes \vec{\hat{\eta}}_r
\tag{4.5}
$$

Table 4.1. Changes of basis in the factor linear spaces.

Space $U^m(K)$		
	Direct	Inverse
Matrix notation	$||\vec{\hat{e}}_p|| = ||\vec{e}_i||C_m$	$||\vec{e}_i|| = ||\vec{\hat{e}}_p||C_m^{-1}$
Tensor notation	$\vec{\hat{e}}_p = c_{op}^{i\,o}\vec{e}_i$	$\vec{e}_i = \gamma_{oi}^{po}\vec{\hat{e}}_p$
Space $V^n(K)$		
	Direct	Inverse
Matrix notation	$||\vec{\hat{\epsilon}}_q|| = ||\vec{\epsilon}_j||R_n$	$||\vec{\epsilon}_j|| = ||\vec{\hat{\epsilon}}_q||R_n^{-1}$
Tensor notation	$\vec{\hat{\epsilon}}_q = r_{oq}^{j\,o}\vec{\epsilon}_j$	$\vec{\epsilon}_j = \rho_{oj}^{qo}\vec{\hat{\epsilon}}_q$
Space $W^p(K)$		
	Direct	Inverse
Matrix notation	$||\vec{\hat{\eta}}_r|| = ||\vec{\eta}_k||S_p$	$||\vec{\eta}_k|| = ||\vec{\hat{\eta}}_r||S_p^{-1}$
Tensor notation	$\vec{\hat{\eta}}_r = s_{or}^{k\,o}\vec{\eta}_k$	$\vec{\eta}_k = \sigma_{ok}^{ro}\vec{\hat{\eta}}_r$

and substituting into the first equality of (4.5) Equations (4.2)–(4.4) one gets

$$\vec{t} = t^{ijk}_{\circ\circ\circ}(\gamma^{po}_{\circ i}\,\vec{\hat{e}}_p) \otimes (\rho^{qo}_{\circ j}\,\vec{\hat{\epsilon}}_q) \otimes (\sigma^{ro}_{\circ k}\,\vec{\hat{\eta}}_r) \equiv \hat{t}^{pqr}_{\circ\circ\circ}\vec{\hat{e}}_p \otimes \vec{\hat{\epsilon}}_q \otimes \vec{\hat{\eta}}_r,$$

from which

$$\vec{t} = t^{ijk}_{\circ\circ\circ}\gamma^{po}_{\circ i}\,\rho^{qo}_{\circ j}\,\sigma^{ro}_{\circ k}\,\vec{\hat{e}}_p \otimes \vec{\hat{\epsilon}}_q \otimes \vec{\hat{\eta}}_r \equiv \hat{t}^{pqr}_{\circ\circ\circ}\vec{\hat{e}}_p \otimes \vec{\hat{\epsilon}}_q \otimes \vec{\hat{\eta}}_r,$$

which leads to

$$\hat{t}^{pqr}_{\circ\circ\circ} = t^{ijk}_{\circ\circ\circ}\gamma^{po}_{\circ i}\,\rho^{qo}_{\circ j}\,\sigma^{ro}_{\circ k}; \quad 1 \le i,p \le m; \quad 1 \le j,q \le n; \quad 1 \le k,r \le p, \quad (4.6)$$

where i, j, k are the dummy indices, and p, q, r the free indices. This is the formula that allows us to find the new tensor component $\hat{t}^{pqr}_{\circ\circ\circ}$, once the initial components and the change-of-basis matrix $C_m^{-1}, R_n^{-1}, S_p^{-1}$ in the "factor" linear spaces are known.

From a tensor point of view, the licit (of tensor character) change-of-basis in the linear space $U^m \otimes V^n \otimes W^p(K)$, is given by the expression (4.6); however, since we are looking for the matrix $Z_{\sigma,\sigma}$ that represents such a change-of-basis, we continue the study.

4.3 Matrix representation of a change-of-basis in tensor spaces

According to (3.28) and (3.29), the relationship between the vector components and the change-of-basis matrix in a linear space is

$$x^h = c^{ho}_{\circ\alpha}\,\hat{x}^\alpha$$

or in matrix form

$$X = C\hat{X}. \tag{4.7}$$

If we write the initial components as a function of the new ones from (4.6), using the rules in Formulas (3.36) and (3.37) we obtain

$$t^{ijk}_{\circ\circ\circ} = \hat{t}^{pqr}_{\circ\circ\circ}c^{io}_{\circ p}\,r^{jo}_{\circ q}\,s^{ko}_{\circ r} = [c^{io}_{\circ p}\,r^{jo}_{\circ q}\,s^{ko}_{\circ r}]\,\hat{t}^{pqr}_{\circ\circ\circ}. \tag{4.8}$$

If now we assume that $T_{\sigma,1} = [t^{ijk}_{\circ\circ\circ}]$ and $\hat{T}_{\sigma,1} = [\hat{t}^{pqr}_{\circ\circ\circ}]$ are the respective column matrices of the extended tensor components, in the initial and the new bases, respectively, ordered according to the *required* criterion (axiom 4, Chapter 2), imposed by the bases $\{\vec{e}_i \otimes \vec{\epsilon}_j \otimes \vec{\eta}_k\}$ and $\{\vec{\hat{e}}_p \otimes \vec{\hat{\epsilon}}_q \otimes \vec{\hat{\eta}}_r\}$, Formula (4.8) appears in matrix form as

$$T_{\sigma,1} = \overset{\overset{\textstyle Z_\sigma}{\longleftarrow\qquad\longrightarrow}}{[c^{io}_{\circ p}\,r^{jo}_{\circ q}\,s^{ko}_{\circ r}]}\;\hat{T}_{\sigma,1} \;\equiv\; T_{\sigma,1} = Z_\sigma\hat{T}_{\sigma,1}, \tag{4.9}$$

the matrix "construction" of which is that of Formula (4.7).

After the corresponding study, one arrives at the conclusion that Z_σ is the square matrix

$$Z_\sigma = C_m \otimes R_n \otimes S_p, \qquad (4.10)$$

where the sign "$\otimes$" is the Kronecker product, direct product or *tensor product of matrices*; that is the reason why we keep its "double" meaning.

In case some readers cannot clearly "see" matrices $Z_\sigma, T_{\sigma,1}, \hat{T}_{\sigma,1}$ in Formulas (4.9) and (4.10), Formula (4.9), $T = Z\hat{T}$, will be analyzed in detail for the particular case $m = 2; n = 3; p = 4$, that is, with the change-of-basis matrices C_2, R_3 and S_4, and $\sigma = 2 \times 3 \times 4 = 24$. We will give some details of how (4.9), $T_{\sigma,1} = Z_\sigma \hat{T}_{\sigma,1}$, is developed. Once the matrices are operated one gets (4.8):

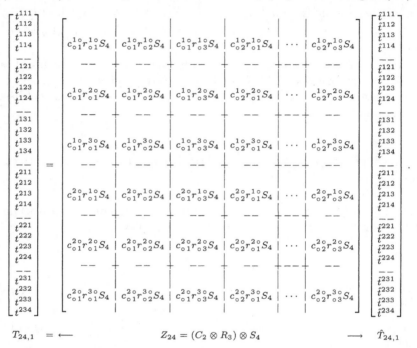

$$T_{24,1} \quad = \longleftarrow \qquad Z_{24} = (C_2 \otimes R_3) \otimes S_4 \qquad \longrightarrow \qquad \hat{T}_{24,1}$$

With Formulas (4.9) and (4.10) we finish the explanation about the matrix representation of the tensor relation (4.8).

If we look for the matrix expression of the tensor relation (4.6), it suffices to obtain $\hat{T}_{\sigma,1}$ from (4.9) to get $\hat{T}_{\sigma,1} = Z_\sigma^{-1} T_{\sigma,1}$, and replacing (4.10) and using the Kronecker product properties, we have

$$\hat{T}_{\sigma,1} = (C_m \otimes R_n \otimes S_p)^{-1} T_{\sigma,1} \Rightarrow \hat{T}_{\sigma,1} = C_m^{-1} \otimes R_n^{-1} \otimes S_p^{-1} T_{\sigma,1}. \qquad (4.11)$$

Expression (4.11) is the matrix representation of (4.6). Our desire of arriving at matrix notation is none other than to build computer programs able to perform tensor operations.

At this point we can clarify the central question, stated at the beginning of this chapter, of what are the matrices Z_σ associated with changes of basis of an arbitrary linear space $V^\sigma(K)$, of dimension $\sigma \in \mathbb{N}$, that allow us to classify its vectors as tensors. In reality its quality is in no way *arbitrary*. The answer is that if σ is an arbitrary number, vectors in $\vec{t} \in V^\sigma(K)$ can be considered as tensors of order $r = 1$ because *any* change-of-basis $Z_\sigma \equiv C_\sigma$, implies the relation

$$\hat{t}^p = t^i \gamma^{p\,o}_{\;o\,i}$$

or in matrix form

$$\hat{T}_{\sigma,1} = C_\sigma^{-1} \cdot T_{\sigma,1},$$

which reveals the satisfaction of relations (4.6) and (4.11) for $r = 1$.

If $\sigma = n_1 \cdot n_2 \neq$ prime (a decomposable natural number) it is additionally possible to notate the basis of $V^\sigma(K)$ as $\{\vec{e}_i \otimes \vec{\epsilon}_j\}$ with $1 \leq i \leq n_1$; $1 \leq j \leq n_2$ and perform change-of-basis, the matrices of which satisfy $Z_\sigma = C_{n_1} \otimes C_{n_2}$, with which we can see $V^\sigma(K)$ as a linear space of second order tensors, $r = 2$, because we have

$$\hat{t}^{pq} = t^{ij} \; \gamma_1{}^{p\,o}_{\;o\,i} \; \gamma_2{}^{q\,o}_{\;o\,j}$$

or in matrix form

$$\hat{T}_{\sigma,1} = (C_{n_1}^{-1} \otimes C_{n_2}^{-1}) \bullet T_{\sigma,1},$$

which proves that the relations (4.6) and (4.11) are satisfied for $r = 2$.

If σ permits a factor decomposition as $\sigma = n_1 \cdot n_2 \cdot n_3$, there is another possibility of notating the basis of $V^\sigma(K)$ as $\{\vec{e}_i \otimes \vec{\epsilon}_j \otimes \vec{\eta}_k\}$ with $1 \leq i \leq n_1$; $1 \leq j \leq n_2$; $1 \leq k \leq n_3$ and performing changes of basis the square matrices of which satisfy $Z_\sigma = C_{n_1} \otimes C_{n_2} \otimes C_{n_3}$ so that we can give $V^\sigma(K)$ a "third $r = 3$ order tensor" linear space character, since the following holds:

$$\hat{t}^{pqr}_{\;ooo} = t^{ijk}_{\;ooo} \; \gamma_1{}^{p\,o}_{\;o\,i} \; \gamma_2{}^{q\,o}_{\;o\,j} \; \gamma_3{}^{r\,o}_{\;o\,k}$$

or

$$\hat{T}_{\sigma,1} = (C_{n_1}^{-1} \otimes C_{n_2}^{-1} \otimes C_{n_3}^{-1}) \bullet T_{\sigma,1},$$

which are relations (4.6) and (4.11) for the present case. The conclusion is that the same linear space $V^\sigma(K)$ can be considered as a *source* of different tensors.

Now we are in a good position to establish the tensor character criterion in all its possible extensions.

4.4 General criteria for tensor character

From this point on, we shall establish a new tensor notation with the aim of adapting it to what is common in calculus and tensor analysis textbooks. Some examples of it will be given in the second part of the book.

For the "factor" linear spaces we shall use the notation $V_i^{n_i}(K)$, where $i, n_i \in \mathbb{N}$, $I_r \equiv \{1, 2, \ldots, r\}$ is the range of i, and n_i is the dimension of

$V_i^{n_i}(K)$, r the total number of linear space "factors", of arbitrary individual nature, but with the same associated scalar field K.

For the basic vectors (and their components) of the "factor" linear spaces, Greek letters $\{\alpha, \beta, \gamma, \delta, \mu, \ldots\}$ will be used for the "initial", and Roman letters $\{i, j, k, m, \ldots\}$ for the "new", that is

$$t^{iok}_{ojo}, \quad t^{\alpha o\gamma}_{o\beta o}, \quad x^\alpha \vec{e}_\alpha, \quad y^j \vec{e}_j, \text{ etc.}$$

With this convention we *avoid the hats in the analytical expressions of tensors*.

The tensor product linear space will be notated as

$$\overset{i=r}{\underset{i=1}{\otimes}} V_i^{n_i}(K); \quad \sigma \equiv dim \left(\overset{r}{\underset{1}{\otimes}} V_i \right) = n_1 \cdot n_2 \cdot n_3 \cdot \cdots \cdot n_r. \tag{4.12}$$

The subindices $\beta_i, \delta_i \in I_{n_i} \equiv \{1, 2, 3, \ldots, n_i\}$, will have the respective ranges, according to the i value.

The initial and new bases of $\overset{r}{\underset{1}{\otimes}} V_i^{n_i}(K)$ will be

$$B(\text{initial}) \equiv \{\vec{e}_{1\beta_1} \otimes \vec{e}_{2\beta_2} \otimes \cdots \otimes \vec{e}_{r\beta_r}\} \tag{4.13}$$

$$B'(\text{new}) \equiv \{\vec{e}_{1j_1} \otimes \vec{e}_{2j_2} \otimes \cdots \otimes \vec{e}_{rj_r}\}. \tag{4.14}$$

Axiom 4, referring to the ordering of indices, also holds.

The changes of basis performed in each of the factor linear spaces will be, for the basic vectors and components

$$\vec{e}_{1j_1} = c_1{}^{\beta_1 \circ}_{\circ j_1} \vec{e}_{1\beta_1}; \quad \vec{e}_{1\beta_1} = \gamma_1{}^{j_1 \circ}_{\circ \beta_1} \vec{e}_{1j_1}; \quad x_1^{\beta_1} = c_1{}^{\beta_1 \circ}_{\circ j_1} \cdot x^{j_1}; \tag{4.15}$$

$$\vec{e}_{ij_i} = c_i{}^{\beta_i \circ}_{\circ j_i} \vec{e}_{i\beta_i}; \quad \vec{e}_{i\beta_i} = \gamma_i{}^{j_i \circ}_{\circ \beta_i} \vec{e}_{ij_i}; \quad x_i^{\beta_i} = c_i{}^{\beta_i \circ}_{\circ j_i} \cdot x^{j_i}; \tag{4.16}$$

$$\vec{e}_{rj_r} = c_r{}^{\beta_r \circ}_{\circ j_r} \vec{e}_{r\beta_r}; \quad \vec{e}_{r\beta_r} = \gamma_r{}^{j_r \circ}_{\circ \beta_r} \vec{e}_{rj_r}; \quad x_r^{\beta_r} = c_r{}^{\beta_r \circ}_{\circ j_r} \cdot x^{j_r}. \tag{4.17}$$

Finally, the tensor relations between the two bases B and B' of $\overset{r}{\underset{1}{\otimes}} V_i^{n_i}(K)$, once replaced (4.15)–(4.17) and grouped are:

Direct relation:

$$\vec{e}_{1j_1} \otimes \vec{e}_{2j_2} \otimes \cdots \cdots \otimes \vec{e}_{rj_r} = c_1{}^{\beta_1 \circ}_{\circ j_1} c_2{}^{\beta_2 \circ}_{\circ j_2} \cdots c_r{}^{\beta_r \circ}_{\circ j_r} \vec{e}_{1\beta_1} \otimes \vec{e}_{2\beta_2} \otimes \cdots \otimes \vec{e}_{r\beta_r} \tag{4.18}$$

of matrix expression (tensor change-of-basis)

$$\|\vec{e}_{1j_1} \otimes \vec{e}_{2j_2} \otimes \cdots \cdots \otimes \vec{e}_{rj_r}\| = \|\vec{e}_{1\beta_1} \otimes \vec{e}_{2\beta_2} \otimes \cdots \otimes \vec{e}_{r\beta_r}\|(C_1 \otimes C_2 \otimes \cdots \otimes C_r). \tag{4.19}$$

Inverse relation:

Table 4.2. Changes of basis performed in each of the different factor linear spaces for the basic vectors and components.

Direct	Inverse	Components
$\vec{e}_{1j_1} = c_1{}^{\beta_1}{}_{\circ}{}^{\circ}{}_{j_1}\,\vec{e}_{1\beta_1}$	$\vec{e}_{1\beta_1} = \gamma_1{}^{j_1}{}_{\circ}{}^{\circ}{}_{\beta_1}\,\vec{e}_{1j_1}$	$x_1^{\beta_1} = c_1{}^{\beta_1}{}_{\circ}{}^{\circ}{}_{j_1} \cdot x^{j_1}$
$\cdots$	$\cdots$	$\cdots$
$\vec{e}_{ij_i} = c_i{}^{\beta_i}{}_{\circ}{}^{\circ}{}_{j_i}\,\vec{e}_{i\beta_i}$	$\vec{e}_{i\beta_i} = \gamma_i{}^{j_i}{}_{\circ}{}^{\circ}{}_{\beta_i}\,\vec{e}_{ij_i}$	$x_i^{\beta_i} = c_i{}^{\beta_i}{}_{\circ}{}^{\circ}{}_{j_i} \cdot x^{j_i}$
$\cdots$	$\cdots$	$\cdots$
$\vec{e}_{rj_r} = c_r{}^{\beta_r}{}_{\circ}{}^{\circ}{}_{j_r}\,\vec{e}_{r\beta_r}$	$\vec{e}_{r\beta_r} = \gamma_r{}^{j_r}{}_{\circ}{}^{\circ}{}_{\beta_r}\,\vec{e}_{rj_r}$	$x_r^{\beta_r} = c_r{}^{\beta_r}{}_{\circ}{}^{\circ}{}_{j_r} \cdot x^{j_r}$

$$\vec{e}_{1\beta_1} \otimes \vec{e}_{2\beta_2} \otimes \cdots\cdots \otimes \vec{e}_{r\beta_r} = \gamma_1{}^{j_1}{}_{\circ}{}^{\circ}{}_{\beta_1}\,\gamma_2{}^{j_2}{}_{\circ}{}^{\circ}{}_{\beta_2} \cdots \gamma_r{}^{j_r}{}_{\circ}{}^{\circ}{}_{\beta_r}\,\vec{e}_{1j_1} \otimes \vec{e}_{2j_2} \otimes \cdots \otimes \vec{e}_{rj_r} \quad (4.20)$$

of matrix expression

$$||\vec{e}_{1\beta_1} \otimes \vec{e}_{2\beta_2} \otimes \cdots\cdots \otimes \vec{e}_{r\beta_r}|| = ||\vec{e}_{1j_1} \otimes \vec{e}_{2j_2} \otimes \cdots \otimes \vec{e}_{rj_r}||(C_1^{-1} \otimes C_2^{-1} \otimes \cdots \otimes C_r^{-1}). \tag{4.21}$$

These changes of basis produce on the "components", two relations (direct and inverse), known as *tensor character criteria*.

Let $\vec{t} \in \overset{r}{\underset{1}{\otimes}} V_i^{n_i}(K)$ be a vector of the tensor product linear space that, expressed with respect to bases $\mathcal{B}$ and $\mathcal{B}'$, becomes

$$t^{\beta_1\beta_2\cdots\beta_r}{}_{\circ\ \circ\ \circ\ \circ}\,\vec{e}_{1\beta_1} \otimes \vec{e}_{2\beta_2} \otimes \cdots\cdots \otimes \vec{e}_{r\beta_r} \text{ (with } \sigma \text{ components)}, \tag{4.22}$$

$$t^{j_1j_2\cdots j_r}{}_{\circ\ \circ\ \circ\ \circ}\,\vec{e}_{1j_1} \otimes \vec{e}_{2j_2} \otimes \cdots\cdots \otimes \vec{e}_{rj_r} \text{ (with } \sigma \text{ components)}. \tag{4.23}$$

The relation between the components to be obtained permits qualifying these vectors $\vec{t}$ as "absolute tensors of order r".

Preferential relation in tensor algebra:

$$t^{j_1j_2\cdots j_r}{}_{\circ\ \circ\ \circ\ \circ} = t^{\beta_1\beta_2\cdots\beta_r}{}_{\circ\ \circ\ \circ\ \circ}\,\gamma_1{}^{j_1}{}_{\circ}{}^{\circ}{}_{\beta_1}\,\gamma_2{}^{j_2}{}_{\circ}{}^{\circ}{}_{\beta_2} \cdots \gamma_r{}^{j_r}{}_{\circ}{}^{\circ}{}_{\beta_r} \tag{4.24}$$

or *preferential relation in tensor algebra that uses matrix representations*:

$$t^{\beta_1\beta_2\cdots\beta_r}{}_{\circ\ \circ\ \circ\ \circ} = t^{j_1j_2\cdots j_r}{}_{\circ\ \circ\ \circ\ \circ}\,c_1{}^{\beta_1}{}_{\circ}{}^{\circ}{}_{j_1}\,c_2{}^{\beta_2}{}_{\circ}{}^{\circ}{}_{j_2} \cdots c_r{}^{\beta_r}{}_{\circ}{}^{\circ}{}_{j_r}. \tag{4.25}$$

A *necessary and sufficient* condition for the vectors $\vec{t} \in \overset{r}{\underset{1}{\otimes}} V_i^{n_i}(K)$ to be tensors is that they satisfy conditions (4.24) or (4.25).

We remind the reader that the indices $\beta_1, \beta_2, \cdots, \beta_r$ in (4.24) and $j_1, \cdots, j_r$ in (4.25) are "dummy" indices (Einstein's convention) and that their ranges are

$$\beta_i, j_i \in I_{n_i} \equiv \{1, 2, \cdots, n_i\} \text{ for } i \in I_r \equiv \{1, 2, \cdots, r\}. \tag{4.26}$$

Conversely, indices $j_1, j_2, \cdots, j_r$ in (4.24) and $\beta_1, \beta_2, \cdots, \beta_r$ in (4.25) are "free" indices, and their ranges, those in (4.26).

We also remind the reader that the matrix expression of (4.25) is

$$T_{\sigma,1} = [C_1 \otimes C_2 \otimes \cdots \otimes C_r] \bullet \hat{T}_{\sigma,1}, \tag{4.27}$$

where the symbol $\otimes$ is used to represent the Kronecker or direct product of matrices.

Formulas (4.24) or (4.25) constitute the fifth formal axiom, which together with the other four axioms established in Section 2.5, Formulas (2.13), (2.14), (2.15) and in Definition 2.1, point 3, order axiom of basic vectors, constitute the fundamental frame for the formal definition of absolute tensors, also called "multilineal applications", "multilineal morphisms", "multilineal operators", etc., by other authors who consider these entities as transformations.

We add that the criterion defended by the authors of this book consists of maintaining the concept of a tensor as a vector, because this concept is "more primitive" than the concept of transformation in abstract algebra.

4.5 Extension to homogeneous tensors

In Section 3.2 we clarified the analytical expression of homogeneous tensors.

As in this case there exists only one generator linear space $V^n(K)$, in primary basis $\{\vec{e}_\alpha\}$, $\{\vec{e}_i\}$ (initial or new), or the dual space $V_*^n(K)$ in the corresponding dual bases $\{\vec{e}^{*\alpha}\}$, $\{\vec{e}^{*i}\}$, here we have only the changes of basis given by Formulas (3.37) and (3.47), that is

$$\vec{e}_\alpha = \gamma_{\circ\alpha}^{i\circ} \vec{e}_i \text{ or } \vec{e}^{*\beta} = c_{j\circ}^{\circ\beta} \vec{e}^{*j}. \tag{4.28}$$

When substituting these expressions into the tensors' analytical expression, as was done in Section 4.2, one arrives at the corresponding tensor character criteria for homogeneous tensors, which are simple applications of the type (4.18) to (4.27).

Let

$$\overset{\longleftarrow \qquad r \text{ times} \qquad \longrightarrow}{V^n \otimes V_*^n \otimes V^n \otimes \cdots \otimes V_*^n \otimes V^n(K)} \tag{4.29}$$

be the "homogeneous tensor product" linear space associated with a *mixed* r-order tensor (with contravariant and covariant indices) with dimension $\sigma = n^r$, and $\beta_i, j_i \in I_n = \{1, 2, \cdots, n\}$. The relation between the bases is:

Direct relation: New basis as a function of the initial basis

$$\vec{e}_{j_1} \otimes \vec{e}^{*j_2} \otimes \vec{e}_{j_3} \otimes \cdots \otimes \vec{e}^{*j_{r-1}} \otimes \vec{e}_{j_r}$$
$$= c_{\circ j_1}^{\beta_1 \circ} \gamma_{\beta_2 \circ}^{\circ j_2} c_{\circ j_3}^{\beta_3 \circ} \cdots \gamma_{\beta_{r-1} \circ}^{\circ j_{r-1}} c_{\circ j_r}^{\beta_r \circ} \vec{e}_{\beta_1} \otimes \vec{e}^{*\beta_2} \otimes \vec{e}_{\beta_3} \otimes \cdots \otimes \vec{e}^{*\beta_{r-1}} \otimes \vec{e}_{\beta_r}$$
$$\tag{4.30}$$

and the matrix expression of the change-of-basis (4.30) is

$$||\vec{e}_{j_1} \otimes \vec{e}^{*j_2} \otimes \vec{e}_{j_3} \otimes \cdots \otimes \vec{e}^{*j_{r-1}} \otimes \vec{e}_{j_r}||$$
$$= ||\vec{e}_{\beta_1} \otimes \vec{e}^{*\beta_2} \otimes \vec{e}_{\beta_3} \otimes \cdots \otimes \vec{e}^{*\beta_{r-1}} \otimes \vec{e}_{\beta_r}||(C \otimes (C^{-1})^t \otimes C \otimes \cdots \otimes (C^{-1})^t \otimes C).$$

(4.31)

Inverse relation: Initial basis as a function of the new basis

$$\vec{e}_{\beta_1} \otimes \vec{e}^{*\beta_2} \otimes \vec{e}_{\beta_3} \otimes \cdots \otimes \vec{e}^{*\beta_{r-1}} \otimes \vec{e}_{\beta_r}$$
$$= \gamma_{\circ\,\beta_1}^{j_1\,\circ} c_{j_2\,\circ}^{\circ\,\beta_2} \gamma_{\circ\,\beta_3}^{j_3\,\circ} \cdots c_{j_{r-1}\,\circ}^{\circ\,\beta_{r-1}} \gamma_{\circ\,\beta_r}^{j_r\,\circ} \vec{e}_{j_1} \otimes \vec{e}^{*j_2} \otimes \vec{e}_{j_3} \otimes \cdots \otimes \vec{e}^{*j_{r-1}} \otimes \vec{e}_{j_r}$$

(4.32)

and the matrix expression of the change-of-basis (4.30) is

$$||\vec{e}_{\beta_1} \otimes \vec{e}^{*\beta_2} \otimes \vec{e}_{\beta_3} \otimes \cdots \otimes \vec{e}^{*\beta_{r-1}} \otimes \vec{e}_{\beta_r}||$$
$$= ||\vec{e}_{j_1} \otimes \vec{e}^{*j_2} \otimes \vec{e}_{j_3} \otimes \cdots \otimes \vec{e}^{*j_{r-1}} \otimes \vec{e}_{j_r}||(C^{-1} \otimes C^t \otimes C^{-1} \otimes \cdots \otimes C^t \otimes C^{-1}),$$

(4.33)

so that the "tensor character criteria" for a homogeneous tensor becomes

$$t^{j_1\,\circ\,j_3}_{\circ\,j_2\,\circ}\cdots{}^{\circ\,j_r}_{j_{r-1}\,\circ} = t^{\beta_1\,\circ\,\beta_3}_{\circ\,\beta_2\,\circ}\cdots{}^{\circ\,\beta_r}_{\beta_{r-1}\,\circ}\gamma_{\circ\,\beta_1}^{j_1\,\circ}c_{j_2\,\circ}^{\circ\,\beta_2}\gamma_{\circ\,\beta_3}^{j_3\,\circ}\cdots c_{j_{r-1}\,\circ}^{\circ\,\beta_{r-1}}\gamma_{\circ\,\beta_r}^{j_r\,\circ}$$ (4.34)

$$t^{\beta_1\,\circ\,\beta_3}_{\circ\,\beta_2\,\circ}\cdots{}^{\circ\,\beta_r}_{\beta_{r-1}\,\circ} = t^{j_1\,\circ\,j_3}_{\circ\,j_2\,\circ}\cdots{}^{\circ\,j_r}_{j_{r-1}\,\circ}c_{\circ\,j_1}^{\beta_1\,\circ}\gamma_{\beta_2\,\circ}^{\circ\,j_2}c_{\circ\,j_3}^{\beta_3\,\circ}\cdots\gamma_{\beta_{r-1}\,\circ}^{\circ\,j_{r-1}}c_{\circ\,j_r}^{\beta_r\,\circ}.$$ (4.35)

We remind the reader that $\beta_i, j_i \in I_n \equiv \{1, 2, \cdots, n\}$. When the number of indices is reduced we shall use Roman and Greek letters *without* subindices.

The matrix expressions of (4.34) and (4.35) are, respectively

$$\hat{T}_{\sigma,1} = (C^{-1} \otimes C^t \otimes C^{-1} \otimes \cdots \otimes C^t \otimes C^{-1}) \bullet T_{\sigma,1} \qquad (4.36)$$

$$T_{\sigma,1} = (C \otimes (C^{-1})^t \otimes C \otimes \cdots \otimes (C^{-1})^t \otimes C) \bullet \hat{T}_{\sigma,1}. \qquad (4.37)$$

Before finishing this topic, we call to the reader's attention on the "formal rules" that all tensor change-of-basis equations must satisfy.

Any homogeneous tensor change-of-basis equation of type (4.34) must satisfy:

Rule 1. The general tensor component with the indices in the new basis (free indices) will appear on the left hand side, and the general tensor component with the indices in the initial basis (dummy indices) will appear on the right-hand side, and always as the first factor.

Rule 2. The change-of-basis terms will appear on the right-hand, as second, third, etc., in the same order as the corresponding free indices on the left hand side.

Rule 3. The free indices columns, in each change-of-basis term, will appear ahead of the dummy indices columns.

Rule 4. The tensor component dummy indices on the right-hand will appear in the reverse position (position of Einstein's convention) to the one corresponding in each change-of-basis factor.

These rules are given for the reader to satisfy them.

4.6 Matrix operation rules for tensor expressions

We study in some detail the matrix processes and their tensor correspondences for certain tensors, in the following order:

1. Second-order tensors (matrices).
2. Third-order tensors.
3. Fourth-order tensors.

We understand that they require some attention because they are the most frequently used tensors.

4.6.1 Second-order tensors (matrices)

Among all the linear spaces treated in algebra books, the chapter dedicated to homomorphisms, of an initial or primary linear space $(V^m(K))$ into another final or secondary linear space $(W^n(K))$ and the problems generated in them by the changes of basis $(C_m$ in the primary and Γ_n in the secondary), deserves special attention.

We conclude that if T_{nm} is the matrix associated with the homomorphism before the change, which belongs to the linear space $\tau^{n \times m}(K)$ of matrices, T_{nm} also suffers a change (Z_1) and the new matrix associated with the homomorphism $\hat{T}_{nm}$, is related to the initial one by the expression

$$\hat{T}_{nm} = \Gamma_n^{-1} T_{nm} C_m. \tag{4.38}$$

In the particular case of endomorphisms $(V^n(K) \equiv W^n(K))$ the square matrices associated with the endomorphism (linear operator) are related by

$$\hat{T}_n = C^{-1} T C, \tag{4.39}$$

an expression that raises all the theory about eigenvalues and eigenvectors.

Another unavoidable chapter of algebra textbooks is that of dual linear spaces with a connection matrix G that, for different linear spaces changes of basis, transforms by the relation

$$\hat{G}_n = \Gamma^t G_n C. \tag{4.40}$$

If the linear spaces are coincident (autodual), we enter into the Euclidean spaces with

$$\hat{G}_n = C^t G C; \quad \text{with } (G = G^t) \tag{4.41}$$

or into the Hermitian spaces with

$$\hat{H}_n = C^* H_n C; \quad \text{with } (H = H^*; \ C^* = (\tilde{C}^t)), \tag{4.42}$$

with applications to bilinear and quadratic forms, etc.

In order to generalize formulas (4.38) to (4.42), etc. we will say that in algebra, there exist square matrices $T_n, G_n, H_n, \ldots$ of order n, that represent abstract entities, such that when changing basis in its proper linear space, it will be done with an "algorithm" of the type

$$\hat{T} = PTQ; \quad |P| \neq 0; \ |Q| \neq 0. \tag{4.43}$$

We will prove that when a matrix T changes basis, according to model (4.43), it is behaving as a "tensor", that is, all possible changes of the model "have tensor character".

Next, for the sake of exposition simplicity, we show the tensor character of Formula (4.43) for $n = 2$.

Consider matrices $P = \begin{bmatrix} a & b \\ c & d \end{bmatrix}$ and $Q = \begin{bmatrix} e & f \\ g & h \end{bmatrix}$ where $|P| \neq 0, |Q| \neq 0$.

We use (4.43) for finding the image matrices of the canonical basis matrices of $\tau^2(K)$:

$$E_{11} \equiv \begin{bmatrix} 1 & 0 \\ 0 & 0 \end{bmatrix}; \ E_{12} \equiv \begin{bmatrix} 0 & 1 \\ 0 & 0 \end{bmatrix}; \ E_{21} \equiv \begin{bmatrix} 0 & 0 \\ 1 & 0 \end{bmatrix}; \ E_{22} \equiv \begin{bmatrix} 0 & 0 \\ 0 & 1 \end{bmatrix};$$

$$\hat{E}_{11} = P \begin{bmatrix} 1 & 0 \\ 0 & 0 \end{bmatrix} Q \equiv \begin{bmatrix} a & b \\ c & d \end{bmatrix} \begin{bmatrix} 1 & 0 \\ 0 & 0 \end{bmatrix} \begin{bmatrix} e & f \\ g & h \end{bmatrix} = \begin{bmatrix} ae & af \\ ce & cf \end{bmatrix} <> \begin{bmatrix} ae \\ af \\ ce \\ cf \end{bmatrix};$$

similarly

$$\hat{E}_{12} = P \begin{bmatrix} 0 & 1 \\ 0 & 0 \end{bmatrix} Q \equiv \begin{bmatrix} a & b \\ c & d \end{bmatrix} \begin{bmatrix} 0 & 1 \\ 0 & 0 \end{bmatrix} \begin{bmatrix} e & f \\ g & h \end{bmatrix} = \begin{bmatrix} ag & ah \\ cg & ch \end{bmatrix} <> \begin{bmatrix} ag \\ ah \\ cg \\ ch \end{bmatrix};$$

$$\hat{E}_{21} = P \begin{bmatrix} 0 & 0 \\ 1 & 0 \end{bmatrix} Q \equiv \begin{bmatrix} a & b \\ c & d \end{bmatrix} \begin{bmatrix} 0 & 0 \\ 1 & 0 \end{bmatrix} \begin{bmatrix} e & f \\ g & h \end{bmatrix} = \begin{bmatrix} be & bf \\ de & df \end{bmatrix} <> \begin{bmatrix} be \\ bf \\ de \\ df \end{bmatrix};$$

$$\hat{E}_{22} = P \begin{bmatrix} 0 & 0 \\ 0 & 1 \end{bmatrix} Q \equiv \begin{bmatrix} a & b \\ c & d \end{bmatrix} \begin{bmatrix} 0 & 0 \\ 0 & 1 \end{bmatrix} \begin{bmatrix} e & f \\ g & h \end{bmatrix} = \begin{bmatrix} bg & bh \\ dg & dh \end{bmatrix} <> \begin{bmatrix} bg \\ bh \\ dg \\ dh \end{bmatrix}.$$

Because of the linear space theory we know they are the columns of the change-of-basis matrix Z_4^{-1} in $\tau^2(K)$. Thus, we have

$$
Z_4^{-1} = \begin{bmatrix} ae & ag & be & bg \\ af & ah & bf & bh \\ ce & cg & de & dg \\ cf & ch & df & dh \end{bmatrix} = \begin{bmatrix} a \begin{bmatrix} e & g \\ f & h \end{bmatrix} & b \begin{bmatrix} e & g \\ f & h \end{bmatrix} \\ c \begin{bmatrix} e & g \\ f & h \end{bmatrix} & d \begin{bmatrix} e & g \\ f & h \end{bmatrix} \end{bmatrix}
$$

$$
= \begin{bmatrix} a & b \\ c & d \end{bmatrix} \otimes \begin{bmatrix} e & g \\ f & h \end{bmatrix} = P \otimes Q^t, \tag{4.44}
$$

which leads to

$$
Z_4 = (P \otimes Q^t)^{-1} = P^{-1} \otimes (Q^{-1})^t, \tag{4.45}
$$

and then Equation (4.43), the model for (4.38) to (4.41), etc, can be translated by stacking matrices T_n and $\hat{T}_n$ (generalized to order n) "in columns", $T_{\sigma,1}$ and $\hat{T}_{\sigma,1}$, respectively and using the equation

$$
\hat{T}_{\sigma,1} = Z_\sigma^{-1} T_{\sigma,1}; \quad Z_\sigma^{-1} = P \otimes Q^t; \quad Z_\sigma = P^{-1} \otimes (Q^{-1})^t. \tag{4.46}
$$

However, according to Formula (4.10), expression (4.46) is a change-of-basis matrix of tensor construction, which is exactly what we wanted to show.

In summary, in linear algebras, the matrices associated with homomorphisms, linear operators, bilinear forms, quadratic forms, dot products, Hermitian products, etc. when changing bases behave as second-order tensors. Thus, as a general conclusion, in algebra are all tensors.

If we extend (4.43) to four consecutive changes of basis, we have

$$
\hat{T} = (P_2 \bullet P_1) T (Q_1 \bullet Q_2) \tag{4.47}
$$

and using Formula (4.46) we get

$$
\hat{T}_{\sigma,1} = Z_\sigma^{-1} \bullet T_{\sigma,1} = \left[(P_2 \bullet P_1) \otimes (Q_1 \bullet Q_2)^t \right] \bullet T_{\sigma,1},
$$

and due to Property 5 of the Kronecker product $\otimes$ (Chapter 1) we obtain

$$
Z_\sigma^{-1} = (P_2 \bullet P_1) \otimes (Q_1 \bullet Q_2)^t = (P_2 \bullet P_1) \otimes (Q_2^t \bullet Q_1^t) \Rightarrow Z_\sigma^{-1} = (P_2 \otimes Q_2^t) \bullet (P_1 \otimes Q_1^t). \tag{4.48}
$$

Generalizing (4.47) and (4.48), we get

$$
\hat{T} = (P_h \bullet \cdots \bullet P_2 \bullet P_1) T (Q_1 \bullet Q_2 \bullet \cdots \bullet Q_h), \tag{4.49}
$$

which is equivalent to

$$
\hat{T}_{\sigma,1} = Z_\sigma^{-1} T_{\sigma,1}
$$

with

$$
Z_\sigma^{-1} = (P_h \otimes Q_h^t) \bullet \cdots \bullet (P_2 \otimes Q_2^t) \bullet (P_1 \otimes Q_1^t). \tag{4.50}
$$

Formulas (4.49) and (4.50) end our comments on Equation (4.43).

Finally, a normal mode of operating change-of-basis homogeneous tensor expressions by means of matrices is the following.

Let $C = [c_{o\,i}^{\alpha\,o}]$ and $D = [d_{o\,j}^{\beta\,o}]$ be the change-of-basis matrices, in the "factor" spaces, and consider the tensor expression

$$t_{o\,o}^{\alpha\beta} = t_{o\,o}^{i\,j}\, c_{o\,i}^{\alpha\,o}\, d_{o\,j}^{\beta\,o}. \tag{4.51}$$

We prepare the expression in such way that the indices to be contracted have their columns in consecutive order, without modifying the tensor indices:

$$t_{o\,o}^{\alpha\beta} = c_{o\,i}^{\alpha\,o}\, t_{o\,o}^{i\,j}\, d_{j\,o}^{o\,\beta}, \tag{4.52}$$

where it can be noted that it has been necessary to alter the index columns of the matrix $[d_{o\,j}^{\beta\,o}]$. Thus, by means of (4.52), Equation (4.51) transforms to matrix form as

$$T = C\hat{T}D^t \tag{4.53}$$

and, if we write it with the ordering in Formula (4.43), we get

$$\hat{T} = C^{-1}T(D^{-1})^t.$$

Formulas (4.46) can be applied, that is, if we work in the linear space $T^{n^2}(K) \equiv T^{\sigma}(K)$, the matrix operation becomes

$$\hat{T}_{\sigma,1} = Z_{\sigma}^{-1}T_{\sigma,1}; \quad Z_{\sigma}^{-1} = C^{-1} \otimes \left[(D^{-1})^t\right]^t$$

and operating one gets

$$\hat{T}_{\sigma,1} = Z_{\sigma}^{-1}T_{\sigma,1}; \quad Z_{\sigma} = C \otimes D, \tag{4.54}$$

which is consistent with the general theory (Formula (4.11)), confirming the correctness of the matrix interpretation of the tensor expression (4.51) with the help of the matrix expression (4.53).

4.6.2 Third-order tensors

Consider the homogeneous tensor, given by the change-of-basis tensor expression

$$t_{o\,j\,o}^{i\,o\,k} = t_{o\,\beta\,o}^{\alpha\,o\,\gamma}\, \gamma_{o\,\alpha}^{i\,o}\, c_{j\,o}^{o\,\beta}\, \gamma_{o\,\gamma}^{k\,o} \tag{4.55}$$

with

$$C = [c_{o\,i}^{\alpha\,o}] \equiv [c_{o\,j}^{\beta\,o}]; \quad C^{-1} = [\gamma_{o\,\alpha}^{i\,o}] \equiv [\gamma_{o\,\gamma}^{k\,o}].$$

Let

1. (i, α) be the row indices of a matrix of submatrices,
2. (j, β) be the row indices of a submatrix,

3. (k, γ) be the column indices of a submatrix.

In this way, the submatrices can be operated either in the data tensor $(t^{\alpha\circ\gamma}_{\circ\beta\circ})$ or in the new output tensor $(t^{i\circ k}_{\circ j\circ})$.

We contract first the row and column indices of the submatrices of (4.55):

$$t^{i\circ k}_{\circ j\circ} = \gamma^{i\circ}_{\circ\alpha} \left(c^{\circ\beta}_{j\circ} t^{\alpha\circ\gamma}_{\circ\beta\circ} \gamma^{\circ k}_{\gamma\circ} \right) \equiv \gamma^{i\circ}_{\circ\alpha} (t'^{\alpha\circ k}_{\circ j\circ}), \qquad (4.56)$$

where in matrix form, observing the change of columns of matrices $[c^{\beta\circ}_{\circ j}]$ and $[\gamma^{k\circ}_{\circ\gamma}]$, the result is

$$[t'^{\alpha\circ k}_{\circ j\circ}] = C^t [T^\alpha](C^{-1})^t = \begin{bmatrix} T^1 \\ T^2 \\ \vdots \\ T^n \end{bmatrix}' \quad \text{(submatrices)}.$$

Next, we operate the rest in matrix form:

$$t^{i\circ k}_{\circ j\circ} = \gamma^{i\circ}_{\circ\alpha} t'^{\alpha\circ k}_{\circ j\circ} \qquad (4.57)$$

taking into account that in (4.57), the dummy index is a *row index of a matrix of submatrices*

$$\hat{T} = [t^{i\circ k}_{\circ j\circ}] = C^{-1} \odot \begin{bmatrix} T^1 \\ T^2 \\ \vdots \\ T^n \end{bmatrix}', \qquad (4.58)$$

where the product "$\odot$" indicates that we proceed to multiply the row elements of C^{-1}, by *blocks* of the block column matrix, and add them, as a sort of *extension* of the classic product of matrices. We proceed exactly the same as if the first matrix were made of blocks and the second of elements. In fact we extend its use when both factors are made of blocks, using the classic product of matrices for the block among them.

4.6.3 Fourth-order tensors

Consider the tensor with expression

$$t^{ij\ell\circ}_{\circ\circ\circ m} = t^{\alpha\beta\lambda\circ}_{\circ\circ\circ\mu} \gamma^{i\circ}_{\circ\alpha} \gamma^{j\circ}_{\circ\beta} \gamma^{\ell\circ}_{\circ\lambda} c^{\circ\mu}_{m\circ}. \qquad (4.59)$$

Let

1. (i, α) be the row indices of a matrix of submatrices,
2. (j, β) be the column indices of a matrix of submatrices,
3. (ℓ, λ) be the row indices of a submatrix,

4. (m, μ) be the column indices of a submatrix.

Following the previous process, we contract first the submatrix indices (λ and μ):

$$t'^{\alpha\beta\ell\,\circ}_{\;\circ\circ\circ m} = [\gamma^{\ell\,\circ}_{\circ\lambda}][t^{\alpha\beta\lambda\circ}_{\circ\circ\circ\mu}][c^{\mu\,\circ}_{\circ m}]$$

and in matrix form

$$t'^{\alpha\beta\ell\,\circ}_{\;\circ\circ\circ m} = C^{-1}[T^{\alpha\beta}]C = [T^{\alpha\beta}]' \tag{4.60}$$

next, we contract the row and column matrix of block indices (α, β):

$$\hat{T} = [t^{ij\ell\,\circ}_{\circ\circ\circ m}] = [\gamma^{i\,\circ}_{\circ\alpha}] \odot [T^{\alpha\beta}]' \odot [\gamma^{\circ\,j}_{\beta\circ}] \tag{4.61}$$

that is

$$\hat{T} = C^{-1} \odot [T^{\alpha\beta}]' \odot (C^{-1})^t, \tag{4.62}$$

where the special product $\odot$, previously defined, appears playing again the same role.

The reader can consult the exercises proposed in this book to improve the understanding of these concepts.

To end this section, we point out with emphasis, a property of the Kronecker product for square matrices $(A_n \otimes B_m)$, the Property 6, relations (1.38) and (1.39) which will be used with some frequency in the treatment of tensors, up to a point that seems to arise from tensor relations, for example (4.43) and (4.46) and (4.53) and (4.54) of the present chapter, and which deserves to be cited with a proper name; we propose the name property of "condensation of products $\bullet$ and $\otimes$".

We cite this property from the matrix point of view:

"The following matrix equations are equivalent":

$$A_n \bullet X_{n,m} \bullet B_m^t = C_{n,m} <> [A_n \otimes B_m] \bullet x = c,$$

where x and c are column–matrices, "stretched" from $X_{n,m}$ and $C_{n,m}$, respectively.

Now, we see the problem from the tensor point of view, with another construction. Consider the linear spaces $V^n(K), W^m(K)$ and $V^\sigma(K)$, where $\sigma = n \cdot m$, and let $K^{n \times m}$ be the linear space of matrices $n \times m$ over K. Let P_n, Q_m^t and $Z_\sigma = (P \otimes Q^t)^{-1}$ be change-of-basis square matrices, in their respective linear spaces.

According to the property of "condensation of matrix products" the following relations correspond univocally:

$$\hat{T}_{n,m} = P \bullet T_{n,m} \bullet Q \Leftrightarrow \hat{T}_{\sigma,1} = (P \otimes Q^t) \bullet T_{\sigma,1}, \tag{4.63}$$

where $T_{\sigma,1}, \hat{T}_{\sigma,1} \in V^\sigma(K)$ and $T_{n,m}, \hat{T}_{n,m} \in K^{n \times m}$ are matrices, with the same scalars as the vectors $T_{\sigma,1}, \hat{T}_{\sigma,1}$ but condensed and ordered, according to the usual matrix disposition.

4.7 Change-of-basis invariant tensors: Isotropic tensors

It is convenient to make a comment on certain homogeneous tensors with a special behavior. When performing a change-of-basis in the linear space $V^n(\mathbb{R})$, this produces in the associated linear spaces of tensor powers a change in their tensor components. However, there exist some tensors the components of which *do not* change, even though their basis is changed. Their components take always the same values; thus, when given the tensor data it is not necessary to mention with respect to which basis $\{\vec{e}_\alpha\}$ in $V^n(\mathbb{R})$ they are defined, information that is needed for the standard tensors. These tensors are known as isotropic tensors.

4.8 Main isotropic tensors

In this section we mention the most important isotropic tensors.

4.8.1 The null tensor

As is well known, any tensor space, also called "tensor power", is above all a linear space, with sum of tensors and exterior product (scalar-tensor) established over the same field K as the factor space $V^n(K)$; as such there exists a neutral tensor for the sum, the null tensor, which will be denoted by $\bar{0}$, or Ω; if the order and valency of the tensor space are those of the tensor with components: $t^{i_1 i_2 \cdots i_p \, \circ \, \circ \, \cdots \, \circ}_{\circ \, \circ \, \cdots \, \circ \, j_1 j_2 \cdots j_q}$, the null tensor is Ω, with components $\omega^{i_1 i_2 \cdots i_p \, \circ \, \circ \, \cdots \, \circ}_{\circ \, \circ \, \cdots \, \circ \, j_1 j_2 \cdots j_q} = 0$; Ω has n^r zeroes and has the property that its tensor product for any other tensor in the tensor space is the null tensor of the tensor product space, that is, $A \otimes \Omega = \Omega'$.

4.8.2 Zero-order tensor (scalar invariant)

A zero-order tensor is a tensor such that its unique component is any of the scalars in the field K over which the factor linear spaces $V^n(K)$ are defined.

Obviously, when performing changes of basis in $V^n(K)$, the components of this tensor are invariant.

4.8.3 Kronecker's delta

To start with and due to reasons that *a posteriori* will become clear, we define this tensor as a system of scalars of second order over $V^n(K)$ as

$$\delta(\alpha\beta) = \begin{cases} 1 \text{ if } \alpha = \beta \\ 0 \text{ if } \alpha \neq \beta \end{cases} \alpha, \beta \in I_n = \{1, 2, \cdots, n\}; \{0, 1\} \in K \text{ the neutral elements.}$$

and also as a matrix $[\delta(\alpha\beta)] = I_n$ (unit matrix).

When examining the tensor character of the system, we arrive at the following conclusions:

1. $\delta(ij) \neq \delta(\alpha\beta)\gamma^{i\,\circ}_{\circ\alpha}\gamma^{j\,\circ}_{\circ\beta}$ (it is not a contravariant tensor).

2. $\delta(ij) \neq \delta(\alpha\beta)c^{\circ\alpha}_{i\,\circ}c^{\circ\beta}_{j\,\circ}$ (it is not a covariant tensor).

3. $\delta(ij) = \delta(\alpha\beta)\gamma^{i\,\circ}_{\circ\alpha}c^{\circ\beta}_{j\,\circ} \rightarrow \delta^{i\,\circ}_{\circ j} = \delta^{\alpha\,\circ}_{\circ\beta}\gamma^{i\,\circ}_{\circ\alpha}c^{\circ\beta}_{j\,\circ}$ (it is a contra-covariant tensor).

4. $\delta(ij) = \delta(\alpha\beta)c^{\circ\alpha}_{i\,\circ}\gamma^{j\,\circ}_{\circ\beta} \rightarrow \delta^{\circ j}_{i\,\circ} = \delta^{\circ\beta}_{\alpha\,\circ}c^{\circ\alpha}_{i\,\circ}\gamma^{j\,\circ}_{\circ\beta}$ (it is a cova-contravariant tensor).

In conclusion, the Kronecker delta is a tensor that presents the *anomaly* of being *simultaneously* a contra-covariant and a cova-contravariant tensor, so that as a vector it permits the following formulations:

$$D \equiv \vec{d} = \delta^{\alpha\,\circ}_{\circ\beta}\vec{e}_\alpha \otimes \vec{e}^{*\beta} = \vec{e}_1 \otimes \vec{e}^{*1} + \vec{e}_2 \otimes \vec{e}^{*2} + \cdots + \vec{e}_n \otimes \vec{e}^{*n}$$

$$D \equiv \vec{d} = \delta^{\circ\beta}_{\alpha\,\circ}\vec{e}^{*\alpha} \otimes \vec{e}_\beta = \vec{e}^{*1} \otimes \vec{e}_1 + \vec{e}^{*2} \otimes \vec{e}_2 + \cdots \vec{e}^{*n} \otimes \vec{e}_n.$$

A matrix justification of the statements in 1, 2, 3 and 4 are the following:

1. $[\delta(\alpha\beta)\gamma^{i\,\circ}_{\circ\alpha}\gamma^{j\,\circ}_{\circ\beta}] = [\gamma^{i\,\circ}_{\circ\alpha}][\delta(\alpha\beta)][\gamma^{\circ j}_{\beta\,\circ}] = C^{-1}I_n(C^{-1})^t \neq I_n = [\delta(ij)]$.

2. $[\delta(\alpha\beta)c^{\circ\alpha}_{i\,\circ}c^{\circ\beta}_{j\,\circ}] = [c^{\circ\alpha}_{i\,\circ}][\delta(\alpha\beta)][c^{\beta\,\circ}_{\circ j}] = C^tI_n(C^t)^t \neq I_n = [\delta(ij)]$.

3. $[\delta(\alpha\beta)\gamma^{i\,\circ}_{\circ\alpha}c^{\circ\beta}_{j\,\circ}] = [\gamma^{i\,\circ}_{\circ\alpha}][\delta(\alpha\beta)][c^{\beta\,\circ}_{\circ j}] = C^{-1}I_n(C^t)^t = I_n = [\delta(ij)]$.

4. $[\delta(\alpha\beta)c^{\circ\alpha}_{i\,\circ}\gamma^{j\,\circ}_{\circ\beta}] = [c^{\circ\alpha}_{i\,\circ}][\delta(\alpha\beta)][\gamma^{\circ j}_{\beta\,\circ}] = C^tI_n(C^{-1})^t = I_n = [\delta(ij)]$.

Example 4.1 (Some questions about the tensor product). We propose the following questions:

1. What condition must two vectors $\vec{V}$ and $\vec{W} \in \mathbb{R}^n$ satisfy for their tensor product to satisfy the condition $\vec{V} \otimes \vec{W} = \vec{W} \otimes \vec{V}$?

2. Let $\vec{t} = t^{ij}_{\circ\circ}\vec{e}_i \otimes \vec{e}_j$ be an arbitrary tensor of $V^n \otimes V^n(\mathbb{R})$. What is the necessary and sufficient condition for the existence of two vectors $\vec{V}$ and $\vec{W} \in V^n(\mathbb{R})$ such that $\vec{V} \otimes \vec{W} = \vec{t}$, for $n = 2$?

3. Idem but for $n = 3$.

4. Let $\vec{t} = t^{i\,\circ}_{\circ j}\vec{e}_i \otimes \vec{e}^*_j$ be a mixed tensor of $\mathbb{R}^n \times \mathbb{R}^n_*(\mathbb{R})$. Can we say that the determinant $D(t^{i\,\circ}_{\circ j})$ has tensor character? How is it transformed?

Solution:

1. We show that $\vec{V} \otimes \vec{W} = \vec{W} \otimes \vec{V}$ is equivalent to $\vec{W} = \lambda\vec{V}$.

Necessary:

We know that $\vec{V} \otimes \vec{W} = x^i y^j \vec{e}_i \otimes \vec{e}_j$ and $\vec{W} \otimes \vec{V} = y^i x^j \vec{e}_i \otimes \vec{e}_j$, so that $\vec{V} \otimes \vec{W} = \vec{W} \otimes \vec{V}$ implies $x^i y^j = y^i x^j$, which is equivalent to

$$\frac{y^i}{x^i} = \frac{y^j}{x^j} = \lambda \Rightarrow \vec{W} = \lambda \vec{V},$$

then it is true that $\vec{V} \otimes \vec{W} = \vec{W} \otimes \vec{V}$ if they are parallel.

Sufficient:

If $\vec{W} = \lambda \vec{V}$ we have

$$\vec{V} \otimes \vec{W} = \vec{V} \otimes \lambda \vec{V} = \lambda(\vec{V} \otimes \vec{V}) \tag{4.64}$$
$$\vec{W} \otimes \vec{V} = \lambda \vec{V} \otimes \vec{V} = \lambda(\vec{V} \otimes \vec{V}), \tag{4.65}$$

that is, $\vec{V} \otimes \vec{W} = \vec{W} \otimes \vec{V}$.

2. *Necessary:*

Let $\vec{V} = x^1 \vec{e}_1 + x^2 \vec{e}_2$ and $\vec{W} = y^1 \vec{e}_1 + y^2 \vec{e}_2$ from which

$$\vec{V} \otimes \vec{W} = x^1 y^1 \vec{e}_1 \otimes \vec{e}_1 + x^1 y^2 \vec{e}_1 \otimes \vec{e}_2 + x^2 y^1 \vec{e}_2 \otimes \vec{e}_1 + x^2 y^2 \vec{e}_2 \otimes \vec{e}_2$$

Since in addition we have

$$\vec{t} = t_{oo}^{11} \vec{e}_1 \otimes \vec{e}_1 + t_{oo}^{12} \vec{e}_1 \otimes \vec{e}_2 + t_{oo}^{21} \vec{e}_2 \otimes \vec{e}_1 + t_{oo}^{22} \vec{e}_2 \otimes \vec{e}_2,$$

we must also have:

$$t_{oo}^{11} = x^1 y^1; \quad t_{oo}^{12} = x^1 y^2; \quad t_{oo}^{21} = x^2 y^1; \quad t_{oo}^{22} = x^2 y^2;$$

and eliminating x^1, y^1, x^2 and y^2, we obtain

$$\begin{cases} t_{oo}^{11} \cdot t_{oo}^{22} = x^1 y^1 x^2 y^2 \\ t_{oo}^{12} \cdot t_{oo}^{21} = x^1 y^2 x^2 y^1 \end{cases} \Rightarrow t_{oo}^{11} t_{oo}^{22} - t_{oo}^{12} t_{oo}^{21} = 0 \Rightarrow \begin{vmatrix} t_{oo}^{11} & t_{oo}^{12} \\ t_{oo}^{21} & t_{oo}^{22} \end{vmatrix} = 0.$$

Then, the necessary condition for tensor $\vec{t}$ to be a tensor product of vectors is that "the associated matrix of scalars be of rank $= 1$".

Sufficient:

Assume that the rank of the matrix of $\vec{t}$ is one, i.e. $\vec{t} \neq \vec{\Omega}$:

$$\text{rank} \begin{bmatrix} t^{11}_{\text{oo}} & t^{12}_{\text{oo}} \\ t^{21}_{\text{oo}} & t^{22}_{\text{oo}} \end{bmatrix} = 1 \;\Rightarrow\; \begin{vmatrix} t^{11}_{\text{oo}} & t^{12}_{\text{oo}} \\ t^{21}_{\text{oo}} & t^{22}_{\text{oo}} \end{vmatrix} = 0; \Rightarrow \; \frac{t^{11}_{\text{oo}}}{t^{21}_{\text{oo}}} = \frac{t^{12}_{\text{oo}}}{t^{22}_{\text{oo}}} = \frac{1}{\lambda}$$

$$\Rightarrow \begin{cases} t^{21}_{\text{oo}} = \lambda t^{11}_{\text{oo}} \\ t^{22}_{\text{oo}} = \lambda t^{12}_{\text{oo}} \end{cases}$$

which implies

$$\begin{aligned}
\vec{t} &= t^{11}_{\text{oo}}\vec{e}_1 \otimes \vec{e}_1 + t^{12}_{\text{oo}}\vec{e}_1 \otimes \vec{e}_2 + \lambda t^{11}_{\text{oo}}\vec{e}_2 \otimes \vec{e}_1 + \lambda t^{12}_{\text{oo}}\vec{e}_2 \otimes \vec{e}_2 \\
&= (t^{11}_{\text{oo}}\vec{e}_1 + \lambda t^{11}_{\text{oo}}\vec{e}_2) \otimes \vec{e}_1 + (t^{12}_{\text{oo}}\vec{e}_1 + \lambda t^{12}_{\text{oo}}\vec{e}_2) \otimes \vec{e}_2 \\
&= (\vec{e}_1 + \lambda\vec{e}_2) \otimes (t^{11}_{\text{oo}}\vec{e}_1) + (\vec{e}_1 + \lambda\vec{e}_2) \otimes (t^{12}_{\text{oo}}\vec{e}_2) \\
&= (\vec{e}_1 + \lambda\vec{e}_2) \otimes (t^{11}_{\text{oo}}\vec{e}_1 + \lambda t^{12}_{\text{oo}}\vec{e}_2).
\end{aligned} \tag{4.66}$$

Then we have $\vec{t} = \vec{X} \otimes \vec{Y}$.

3. *Necessary*:

Let

$$\vec{V} = x^i\vec{e}_1; \; \vec{W} = x^j\vec{e}_j, 1 \le i,j \le 3 \;\Rightarrow\; \begin{cases} \vec{V} \otimes \vec{W} = x^i y^j \vec{e}_i \otimes \vec{e}_j \\ \vec{t} \quad = t^{ij}_{\text{oo}}\vec{e}_i \otimes \vec{e}_j \end{cases}$$

The identification $\vec{V} \otimes \vec{W} = \vec{t}$, leads to

$$\begin{cases} x^1 y^1 = t^{11}_{\text{oo}} \\ x^1 y^2 = t^{12}_{\text{oo}} \\ x^1 y^3 = t^{13}_{\text{oo}} \end{cases} \begin{cases} x^2 y^1 = t^{21}_{\text{oo}} \\ x^2 y^2 = t^{22}_{\text{oo}} \\ x^2 y^3 = t^{23}_{\text{oo}} \end{cases} \begin{cases} x^3 y^1 = t^{31}_{\text{oo}} \\ x^3 y^2 = t^{32}_{\text{oo}} \\ x^3 y^3 = t^{33}_{\text{oo}} \end{cases}$$

that tensorialy speaking is a system of n^2 equations $x^i y^j = t^{ij}_{\text{oo}}$, in our case of $n^2 = 9$ equations, with $x^1, x^2, x^3, y^1, y^2, y^3$, six unknowns, so that there must be some compatibility conditions for it to hold.

We proceed to develop a gradual elimination technique of the unknowns, as follows:

$$\begin{cases} t^{11}_{\text{oo}}t^{22}_{\text{oo}} = x^1 y^1 x^2 y^2 \\ t^{12}_{\text{oo}}t^{21}_{\text{oo}} = x^1 y^2 x^2 y^1 \end{cases} \Rightarrow t^{11}_{\text{oo}}t^{22}_{\text{oo}} - t^{12}_{\text{oo}}t^{21}_{\text{oo}} = 0 \;\Rightarrow\; \frac{t^{11}_{\text{oo}}}{t^{21}_{\text{oo}}} = \frac{t^{12}_{\text{oo}}}{t^{22}_{\text{oo}}}$$

and also

$$\begin{cases} t^{22}_{\text{oo}}t^{13}_{\text{oo}} = x^2 y^2 x^1 y^3 \\ t^{23}_{\text{oo}}t^{12}_{\text{oo}} = x^2 y^3 x^1 y^2 \end{cases} \Rightarrow t^{22}_{\text{oo}}t^{13}_{\text{oo}} - t^{23}_{\text{oo}}t^{12}_{\text{oo}} = 0 \;\Rightarrow\; \frac{t^{12}_{\text{oo}}}{t^{22}_{\text{oo}}} = \frac{t^{13}_{\text{oo}}}{t^{23}_{\text{oo}}},$$

which leads to

$$\frac{t^{11}_{oo}}{t^{21}_{oo}} = \frac{t^{12}_{oo}}{t^{22}_{oo}} = \frac{t^{13}_{oo}}{t^{23}_{oo}}.$$

In a similar way we establish

$$\frac{t^{11}_{oo}}{t^{31}_{oo}} = \frac{t^{12}_{oo}}{t^{32}_{oo}} = \frac{t^{13}_{oo}}{t^{33}_{oo}},$$

which leads to

$$\text{rank} \begin{bmatrix} t^{11}_{oo} & t^{12}_{oo} & t^{13}_{oo} \\ t^{21}_{oo} & t^{22}_{oo} & t^{23}_{oo} \\ t^{31}_{oo} & t^{32}_{oo} & t^{33}_{oo} \end{bmatrix} = 1.$$

Sufficient:

Let rank $\begin{bmatrix} t^{11}_{oo} t^{12}_{oo} t^{13}_{oo} \\ t^{21}_{oo} t^{22}_{oo} t^{23}_{oo} \\ t^{31}_{oo} t^{32}_{oo} t^{33}_{oo} \end{bmatrix} = 1$, with $\vec{t} = t^{ij}_{oo} \vec{e}_i \otimes \vec{e}_j \neq \vec{\Omega}, 1 \leq i, j \leq 3$.

There exists $\beta^i, \beta^j, \beta^k \ldots, \beta^n \in K$; $\forall i, j, k$, we have

$$\frac{t^{ij}_{oo}}{\beta^j} = \frac{t^{ik}_{oo}}{\beta^k} = \cdots = \frac{t^{in}_{oo}}{\beta^n} = \alpha^i; \text{ for a given } i$$

$$\frac{t^{ji}_{oo}}{\beta^i} = \frac{t^{jk}_{oo}}{\beta^k} = \cdots = \frac{t^{jn}_{oo}}{\beta^n} = \alpha^j; \text{ for a given } j$$

etc.

Then, the following relation holds: $t^{ij}_{oo} = \alpha^i \beta^j$. This allows us to write the matrix expression:

$$\vec{t} = \begin{bmatrix} \vec{e}_1 & \vec{e}_2 & \vec{e}_3 \end{bmatrix} \begin{bmatrix} t^{11}_{oo} & t^{12}_{oo} & t^{13}_{oo} \\ t^{21}_{oo} & t^{22}_{oo} & t^{23}_{oo} \\ t^{31}_{oo} & t^{32}_{oo} & t^{33}_{oo} \end{bmatrix} \begin{bmatrix} \vec{e}_1 \\ \vec{e}_2 \\ \vec{e}_3 \end{bmatrix}$$

$$= \begin{bmatrix} \vec{e}_1 & \vec{e}_2 & \vec{e}_3 \end{bmatrix} \begin{bmatrix} \alpha^1\beta^1 & \alpha^1\beta^2 & \alpha^1\beta^3 \\ \alpha^2\beta^1 & \alpha^2\beta^2 & \alpha^2\beta^3 \\ \alpha^3\beta^1 & \alpha^3\beta^2 & \alpha^3\beta^3 \end{bmatrix} \begin{bmatrix} \vec{e}_1 \\ \vec{e}_2 \\ \vec{e}_3 \end{bmatrix}$$

$$= \begin{bmatrix} \vec{e}_1 & \vec{e}_2 & \vec{e}_3 \end{bmatrix} \left(\begin{bmatrix} \alpha^1 \\ \alpha^2 \\ \alpha^3 \end{bmatrix} \otimes \begin{bmatrix} \beta_1 & \beta_2 & \beta_3 \end{bmatrix} \right) \begin{bmatrix} \vec{e}_1 \\ \vec{e}_2 \\ \vec{e}_3 \end{bmatrix}$$

$$= \left([\vec{e}_1 \ \ \vec{e}_2 \ \ \vec{e}_3] \begin{bmatrix} \alpha^1 \\ \alpha^2 \\ \alpha^3 \end{bmatrix} \right) \otimes \left([\beta_1 \ \ \beta_2 \ \ \beta_3] \begin{bmatrix} \vec{e}_1 \\ \vec{e}_2 \\ \vec{e}_3 \end{bmatrix} \right)$$

$$= [\alpha^1 \vec{e}_1 + \alpha^2 \vec{e}_2 + \alpha^3 \vec{e}_3] \otimes [\beta^1 \vec{e}_1 + \beta^2 \vec{e}_2 + \beta^3 \vec{e}_3] = \vec{V} \otimes \vec{W},$$

which proves the sufficiency.

4. We know that t satisfies the relation

$$t_{oj}^{io} = t_{o\beta}^{\alpha o} \gamma_{o\alpha}^{io} c_{jo}^{o\beta};$$

and taking determinants and considering Formula (4.36), we get

$$D(t_{oj}^{io}) = D(t_{o\beta}^{\alpha o}) D(\gamma_{o\alpha}^{io}) D(c_{jo}^{o\beta}) = D(t_{o\beta}^{\alpha o}) |C^{-1}| \cdot |C^t|$$

$$= D(t_{o\beta}^{\alpha o}) |C^{-1}| \cdot |C| = D(t_{o\beta}^{\alpha o}) \text{ (real scalar)}.$$

So that

$$D(t_{oj}^{io}) = D(t_{o\beta}^{\alpha o})$$

is an absolute tensor of order zero (invariant), which does not get transformed by a change-of-basis since it is invariant.

□

Example 4.2 (Change of basis). In a linear space $V_1^2(\mathbb{R})$ consider the initial and new bases $(\mathcal{B}_1, \mathcal{B}_1')$ related as

$$\vec{\hat{e}}_1 = 2\vec{e}_1 - \vec{e}_2$$
$$\vec{\hat{e}}_2 = 3\vec{e}_1 + \vec{e}_2. \tag{4.67}$$

In another linear space $V_2^3(\mathbb{R})$, their respective bases $(\mathcal{B}_2, \mathcal{B}_2')$ are given by

$$\vec{\hat{c}}_1 = \vec{c}_1 - 2\vec{c}_2$$
$$\vec{\hat{c}}_2 = \vec{c}_1 + 4\vec{c}_3$$
$$\vec{\hat{c}}_3 = \vec{c}_2 - \vec{c}_3. \tag{4.68}$$

Let $\vec{t} \in V_1^2 \otimes V_2^3(\mathbb{R})$ be the tensor

$$\vec{T} = 2\vec{e}_1 \otimes \vec{c}_1 - \vec{e}_1 \otimes \vec{c}_3 - 3\vec{e}_2 \otimes \vec{c}_2.$$

1. Obtain the tensor components, in the new basis associated with the "tensor product" linear space $V_1^2 \otimes V_2^3(\mathbb{R})$.

2. Express the tensor $\vec{t}$ in its new canonical basis.

Solution:

1. *First matrix process (direct process):*

The dimension of the tensor space is $\sigma = 2 \times 3 = 6$, and the tensor components are

$$T_{6,1} \equiv \begin{bmatrix} t^{11} \\ t^{12} \\ t^{13} \\ t^{21} \\ t^{22} \\ t^{23} \end{bmatrix} = \begin{bmatrix} 2 \\ 0 \\ -1 \\ 0 \\ -3 \\ 0 \end{bmatrix}.$$

The direct changes of basis are

$$\vec{e}_i = c_{\circ i}^{\alpha \circ}\, \vec{e}_\alpha \text{ in } V_1^2(\mathbb{R})$$

and

$$\vec{c}_j = d_{\circ j}^{\beta \circ}\, \vec{c}_\beta \text{ in } V_2^3(\mathbb{R})$$

and the tensor relation of the change-of-basis is

$$t^{ij} = t^{\alpha\beta}\, \gamma_{\circ\alpha}^{i\,\circ}\, \delta_{\circ\beta}^{j\,\circ} \text{ (a \textit{totally} contravariant tensor)} \tag{4.69}$$

and the matrix information for the changes of basis is

$$\| \vec{\tilde{e}}_1 \ \ \vec{\tilde{e}}_2 \| = \| \vec{e}_1 \ \ \vec{e}_2 \| C \equiv \| \vec{e}_1 \ \ \vec{e}_2 \| \begin{bmatrix} 2 & 3 \\ -1 & 1 \end{bmatrix} \text{ for } V_1^2(\mathbb{R})$$

$$\| \vec{c}_1 \ \vec{c}_2 \ \vec{c}_3 \| = \| \vec{\tilde{c}}_1 \ \vec{\tilde{c}}_2 \ \vec{\tilde{c}}_3 \| D^{-1} \equiv \| \vec{\tilde{c}}_1 \ \vec{\tilde{c}}_2 \ \vec{\tilde{c}}_3 \| \begin{bmatrix} 1 & 1 & 0 \\ -2 & 0 & 1 \\ 0 & 4 & -1 \end{bmatrix} \text{ for } V_2^3(\mathbb{R}).$$

Whence $C^{-1} = \frac{1}{5}\begin{bmatrix} 1 & -3 \\ 1 & 2 \end{bmatrix}$, and the matrix expression of (4.69) is (see the theory)

$$\hat{T}_{6,1} = (C^{-1} \otimes D^{-1}) \bullet T_{6,1} = \left(\frac{1}{5}\begin{bmatrix} 1 & -3 \\ 1 & 2 \end{bmatrix} \otimes \begin{bmatrix} 1 & 1 & 0 \\ -2 & 0 & 1 \\ 0 & 4 & -1 \end{bmatrix} \right) \bullet \begin{bmatrix} 2 \\ 0 \\ -1 \\ 0 \\ -3 \\ 0 \end{bmatrix}$$

$$= \frac{1}{5}\begin{bmatrix} 1 \cdot \begin{bmatrix} 1 & 1 & 0 \\ -2 & 0 & 1 \\ 0 & 4 & -1 \end{bmatrix} & (-3) \cdot \begin{bmatrix} 1 & 1 & 0 \\ -2 & 0 & 1 \\ 0 & 4 & -1 \end{bmatrix} \\ 1 \cdot \begin{bmatrix} 1 & 1 & 0 \\ -2 & 0 & 1 \\ 0 & 4 & -1 \end{bmatrix} & 2 \cdot \begin{bmatrix} 1 & 1 & 0 \\ -2 & 0 & 1 \\ 0 & 4 & -1 \end{bmatrix} \end{bmatrix} \bullet \begin{bmatrix} 2 \\ 0 \\ -1 \\ 0 \\ -3 \\ 0 \end{bmatrix}$$

$$= \frac{1}{5}\begin{bmatrix} 1 & 1 & 0 & -3 & -3 & 0 \\ -2 & 0 & 1 & 6 & 0 & -3 \\ 0 & 4 & -1 & 0 & -12 & 3 \\ 1 & 1 & 0 & 2 & 2 & 0 \\ -2 & 0 & 1 & -4 & 0 & 2 \\ 0 & 4 & -1 & 0 & 8 & -2 \end{bmatrix} \bullet \begin{bmatrix} 2 \\ 0 \\ -1 \\ 0 \\ -3 \\ 0 \end{bmatrix}, \tag{4.70}$$

which implies

$$
\hat{T}_{6,1} = \frac{1}{5}
\begin{bmatrix}
11 \\ -5 \\ 37 \\ -4 \\ -5 \\ -23
\end{bmatrix}
\equiv
\begin{bmatrix}
11/5 \\ -1 \\ 37/5 \\ -4/5 \\ -1 \\ -23/5
\end{bmatrix},
$$

that is

$$
\vec{t} = \frac{11}{5}\vec{e}_1 \otimes \vec{c}_1 - \vec{e}_1 \otimes \vec{c}_2 + \frac{37}{5}\vec{e}_1 \otimes \vec{c}_3 - \frac{4}{5}\vec{e}_2 \otimes \vec{c}_1 - \vec{e}_2 \otimes \vec{c}_2 - \frac{23}{5}\vec{e}_2 \otimes \vec{c}_3. \quad (4.71)
$$

2. *Second matrix process (classic matrix method):*
We start again from the relation (4.69), and as its entities are scalars of the field K, they can be sorted as desired. The indices of the tensor $(t^{\alpha\beta})$ must not be modified.
First we pass the matrix $\gamma^{i\,o}_{o\,\alpha}$ ahead, for the dummy indices α to appear in the *contraction* position (of the matrix product, because both are matrices). Then, we have

$$
t^{ij}_{oo} = (\gamma^{i\,o}_{o\,\alpha}\, t^{\alpha\beta}_{o\,o})\, \delta^{j\,o}_{o\,\beta}, \quad (4.72)
$$

and since the indicators β are not "consecutive" (they are separated by the "semaphoric column" j), the matrix $[\delta^{j\,o}_{o\,\beta}]$ is transposed, resulting in

$$
t^{ij}_{oo} = \gamma^{i\,o}_{o\,\alpha}\, t^{\alpha\beta}_{o\,o}\, \delta^{o\,j}_{\beta o}, \quad (4.73)
$$

where $\alpha \equiv \{1,2\}$, $\beta \equiv \{1,2,3\}$ and the matrix $[t^{\alpha\beta}_{o\,o}]$ is precisely matrix $T_{6,1}$ after condensation:

$$
T = [t^{\alpha\beta}_{o\,o}] \equiv
\begin{bmatrix}
2 & 0 & -1 \\
0 & -3 & 0
\end{bmatrix},
$$

so that (4.73) permits the operation of (4.69) in matrix form by means of its double contraction, as

$$
\hat{T} = C^{-1}T(D^{-1})^t = \frac{1}{5}
\begin{bmatrix} 1 & -3 \\ 1 & 2 \end{bmatrix}
\begin{bmatrix} 2 & 0 & -1 \\ 0 & -3 & 0 \end{bmatrix}
\begin{bmatrix} 1 & -2 & 0 \\ 1 & 0 & 4 \\ 0 & 1 & -1 \end{bmatrix}
$$

$$
= \frac{1}{5}
\begin{bmatrix} 11 & -5 & 37 \\ -4 & -5 & -23 \end{bmatrix}
\equiv
\begin{bmatrix} 11/5 & -1 & 37/5 \\ -4/5 & -1 & -23/5 \end{bmatrix},
$$

which is $\hat{T}_{6,1}$ after condensation.
The answer is identical to that in (4.71), since the $t^{ij} \in \hat{T}$ coincide.

□

Example 4.3 (Index contraction). In the linear space $\mathbb{R}^3$ we consider the change-of-basis given by the new vectors $\vec{e}_1(-1,0,-1), \vec{e}_2(1,1,0)$ and $\vec{e}_3(0,0,3)$.

In the homogeneous tensor space $\mathbb{R}^3 \otimes \mathbb{R}_*^3 \otimes \mathbb{R}^3$ consider the tensor $\vec{t}$ with components

$$
T = [t^{\alpha\circ\gamma}_{\circ\beta\circ}] =
\begin{bmatrix}
t^{1\circ1}_{\circ1\circ} & t^{1\circ2}_{\circ1\circ} & t^{1\circ3}_{\circ1\circ} \\
t^{1\circ1}_{\circ2\circ} & t^{1\circ2}_{\circ2\circ} & t^{1\circ3}_{\circ2\circ} \\
t^{1\circ1}_{\circ3\circ} & t^{1\circ2}_{\circ3\circ} & t^{1\circ3}_{\circ3\circ} \\
\hline
t^{2\circ1}_{\circ1\circ} & t^{2\circ2}_{\circ1\circ} & t^{2\circ3}_{\circ1\circ} \\
t^{2\circ1}_{\circ2\circ} & t^{2\circ2}_{\circ2\circ} & t^{2\circ3}_{\circ2\circ} \\
t^{2\circ1}_{\circ3\circ} & t^{2\circ2}_{\circ3\circ} & t^{2\circ3}_{\circ3\circ} \\
\hline
t^{3\circ1}_{\circ1\circ} & t^{3\circ2}_{\circ1\circ} & t^{3\circ3}_{\circ1\circ} \\
t^{3\circ1}_{\circ2\circ} & t^{3\circ2}_{\circ2\circ} & t^{3\circ3}_{\circ2\circ} \\
t^{3\circ1}_{\circ3\circ} & t^{3\circ2}_{\circ3\circ} & t^{3\circ3}_{\circ3\circ}
\end{bmatrix}
=
\begin{bmatrix}
1 & 0 & -1 \\
2 & 3 & 0 \\
-1 & 2 & 0 \\
\hline
2 & -1 & 1 \\
0 & 0 & 0 \\
2 & 0 & 1 \\
\hline
0 & 0 & 1 \\
5 & 1 & 2 \\
0 & 0 & 0
\end{bmatrix},
$$

where α refers to block row, β to row, and γ to column of each block.

1. Find the new components $t^{i\circ k}_{\circ j \circ}$ of the tensor in its new canonical basis.
2. Contract the indices (β, γ) before the change.
3. Contract the indices (j, k) after the change.
4. Is the contracted tensor the same? (Call it S, if it is unique.)

Solution:

1. The change-of-basis matrix is $C = \begin{bmatrix} -1 & 1 & 0 \\ 0 & 1 & 0 \\ -1 & 0 & 3 \end{bmatrix}$, and from it we obtain the matrices

$$
C^t = \begin{bmatrix} -1 & 0 & -1 \\ 1 & 1 & 0 \\ 0 & 0 & 3 \end{bmatrix} \text{ and } C^{-1} = \begin{bmatrix} -1 & 1 & 0 \\ 0 & 1 & 0 \\ -1/3 & 1/3 & 1/3 \end{bmatrix},
$$

which will be used later. From Formula (4.34), we obtain the change-of-basis tensor relation for this homogeneous tensor (of $\sigma = 3^3 = 27$):

$$
t^{i\circ k}_{\circ j \circ} = t^{\alpha\circ\gamma}_{\circ\beta\circ} \gamma^{i\circ}_{\circ\alpha} c^{\beta}_{j\circ} \gamma^{k\circ}_{\circ\gamma}. \tag{4.74}
$$

First matrix process (direct process):
The matrix interpretation of (4.74), according to Formula (4.36), is

$$
\hat{T}_{27,1} = (C^{-1} \otimes C^t \otimes C^{-1}) \bullet T_{27,1} \equiv Z \bullet T_{27,1}. \tag{4.75}
$$

We start by fixing $Z = (C^{-1} \otimes C^t) \otimes C^{-1}$, that is

$$
Z = \begin{bmatrix}
(-1)\begin{bmatrix} -1 & 0 & -1 \\ 1 & 1 & 0 \\ 0 & 0 & 3 \end{bmatrix} & (1)\begin{bmatrix} -1 & 0 & -1 \\ 1 & 1 & 0 \\ 0 & 0 & 3 \end{bmatrix} & (0)\begin{bmatrix} -1 & 0 & -1 \\ 1 & 1 & 0 \\ 0 & 0 & 3 \end{bmatrix} \\[2em]
(0)\begin{bmatrix} -1 & 0 & -1 \\ 1 & 1 & 0 \\ 0 & 0 & 3 \end{bmatrix} & (1)\begin{bmatrix} -1 & 0 & -1 \\ 1 & 1 & 0 \\ 0 & 0 & 3 \end{bmatrix} & (0)\begin{bmatrix} -1 & 0 & -1 \\ 1 & 1 & 0 \\ 0 & 0 & 3 \end{bmatrix} \\[2em]
(-1/3)\begin{bmatrix} -1 & 0 & -1 \\ 1 & 1 & 0 \\ 0 & 0 & 3 \end{bmatrix} & (1/3)\begin{bmatrix} -1 & 0 & -1 \\ 1 & 1 & 0 \\ 0 & 0 & 3 \end{bmatrix} & (1/3)\begin{bmatrix} -1 & 0 & -1 \\ 1 & 1 & 0 \\ 0 & 0 & 3 \end{bmatrix}
\end{bmatrix} \otimes C^{-1}
$$

or equivalently

$$
Z = \begin{bmatrix}
1 & 0 & 1 & -1 & 0 & -1 & 0 & 0 & 0 \\
-1 & -1 & 0 & 1 & 1 & 0 & 0 & 0 & 0 \\
0 & 0 & -3 & 0 & 0 & 3 & 0 & 0 & 0 \\
0 & 0 & 0 & -1 & 0 & -1 & 0 & 0 & 0 \\
0 & 0 & 0 & 1 & 1 & 0 & 0 & 0 & 0 \\
0 & 0 & 0 & 0 & 0 & 3 & 0 & 0 & 0 \\
\frac{1}{3} & 0 & \frac{1}{3} & -\frac{1}{3} & 0 & -\frac{1}{3} & -\frac{1}{3} & 0 & -\frac{1}{3} \\
-\frac{1}{3} & -\frac{1}{3} & 0 & \frac{1}{3} & \frac{1}{3} & 0 & \frac{1}{3} & \frac{1}{3} & 0 \\
0 & 0 & -1 & 0 & 0 & 1 & 0 & 0 & 1
\end{bmatrix} \otimes \begin{bmatrix} -1 & 1 & 0 \\ 0 & 1 & 0 \\ -\frac{1}{3} & \frac{1}{3} & \frac{1}{3} \end{bmatrix},
$$

which leads to the matrix

$$
\left[\begin{array}{ccccccccc|ccccccccc|ccccccccc}
-1 & 1 & 0 & 0 & 0 & 0 & -1 & 1 & 0 & 1 & -1 & 0 & 0 & 0 & 0 & 1 & -1 & 0 & 0 & 0 & 0 & 0 & 0 & 0 & 0 & 0 & 0 \\
0 & 1 & 0 & 0 & 0 & 0 & 0 & 1 & 0 & 0 & -1 & 0 & 0 & 0 & 0 & 0 & -1 & 0 & 0 & 0 & 0 & 0 & 0 & 0 & 0 & 0 & 0 \\
-\frac{1}{3} & \frac{1}{3} & \frac{1}{3} & 0 & 0 & 0 & -\frac{1}{3} & \frac{1}{3} & \frac{1}{3} & \frac{1}{3} & -\frac{1}{3} & -\frac{1}{3} & 0 & 0 & 0 & \frac{1}{3} & -\frac{1}{3} & -\frac{1}{3} & 0 & 0 & 0 & 0 & 0 & 0 & 0 & 0 & 0 \\ \hline
1 & -1 & 0 & 1 & -1 & 0 & 0 & 0 & 0 & -1 & 1 & 0 & -1 & 1 & 0 & 0 & 0 & 0 & 0 & 0 & 0 & 0 & 0 & 0 & 0 & 0 & 0 \\
0 & -1 & 0 & 0 & -1 & 0 & 0 & 0 & 0 & 0 & 1 & 0 & 0 & 1 & 0 & 0 & 0 & 0 & 0 & 0 & 0 & 0 & 0 & 0 & 0 & 0 & 0 \\
\frac{1}{3} & -\frac{1}{3} & -\frac{1}{3} & \frac{1}{3} & -\frac{1}{3} & -\frac{1}{3} & 0 & 0 & 0 & -\frac{1}{3} & \frac{1}{3} & \frac{1}{3} & -\frac{1}{3} & \frac{1}{3} & \frac{1}{3} & 0 & 0 & 0 & 0 & 0 & 0 & 0 & 0 & 0 & 0 & 0 & 0 \\ \hline
0 & 0 & 0 & 0 & 0 & 0 & 3 & -3 & 0 & 0 & 0 & 0 & 0 & 0 & 0 & -3 & 3 & 0 & 0 & 0 & 0 & 0 & 0 & 0 & 0 & 0 & 0 \\
0 & 0 & 0 & 0 & 0 & 0 & 0 & -3 & 0 & 0 & 0 & 0 & 0 & 0 & 0 & 0 & 3 & 0 & 0 & 0 & 0 & 0 & 0 & 0 & 0 & 0 & 0 \\
0 & 0 & 0 & 0 & 0 & 0 & 1 & -1 & -1 & 0 & 0 & 0 & 0 & 0 & 0 & -1 & 1 & 1 & 0 & 0 & 0 & 0 & 0 & 0 & 0 & 0 & 0 \\ \hline
0 & 0 & 0 & 0 & 0 & 0 & 0 & 0 & 0 & 1 & -1 & 0 & 0 & 0 & 0 & 1 & -1 & 0 & 0 & 0 & 0 & 0 & 0 & 0 & 0 & 0 & 0 \\
0 & 0 & 0 & 0 & 0 & 0 & 0 & 0 & 0 & 0 & -1 & 0 & 0 & 0 & 0 & 0 & -1 & 0 & 0 & 0 & 0 & 0 & 0 & 0 & 0 & 0 & 0 \\
0 & 0 & 0 & 0 & 0 & 0 & 0 & 0 & 0 & \frac{1}{3} & -\frac{1}{3} & -\frac{1}{3} & 0 & 0 & 0 & \frac{1}{3} & -\frac{1}{3} & -\frac{1}{3} & 0 & 0 & 0 & 0 & 0 & 0 & 0 & 0 & 0 \\ \hline
0 & 0 & 0 & 0 & 0 & 0 & 0 & 0 & 0 & -1 & 1 & 0 & -1 & 1 & 0 & 0 & 0 & 0 & 0 & 0 & 0 & 0 & 0 & 0 & 0 & 0 & 0 \\
0 & 0 & 0 & 0 & 0 & 0 & 0 & 0 & 0 & 0 & 1 & 0 & 0 & 1 & 0 & 0 & 0 & 0 & 0 & 0 & 0 & 0 & 0 & 0 & 0 & 0 & 0 \\
0 & 0 & 0 & 0 & 0 & 0 & 0 & 0 & 0 & -\frac{1}{3} & \frac{1}{3} & \frac{1}{3} & -\frac{1}{3} & \frac{1}{3} & \frac{1}{3} & 0 & 0 & 0 & 0 & 0 & 0 & 0 & 0 & 0 & 0 & 0 & 0 \\ \hline
0 & 0 & 0 & 0 & 0 & 0 & 0 & 0 & 0 & 0 & 0 & 0 & 0 & 0 & 0 & -3 & 3 & 0 & 0 & 0 & 0 & 0 & 0 & 0 & 0 & 0 & 0 \\
0 & 0 & 0 & 0 & 0 & 0 & 0 & 0 & 0 & 0 & 0 & 0 & 0 & 0 & 0 & 0 & 3 & 0 & 0 & 0 & 0 & 0 & 0 & 0 & 0 & 0 & 0 \\
0 & 0 & 0 & 0 & 0 & 0 & 0 & 0 & 0 & 0 & 0 & 0 & 0 & 0 & 0 & -1 & 1 & 1 & 0 & 0 & 0 & 0 & 0 & 0 & 0 & 0 & 0 \\ \hline
-\frac{1}{3} & \frac{1}{3} & 0 & 0 & 0 & 0 & -\frac{1}{3} & \frac{1}{3} & 0 & \frac{1}{3} & -\frac{1}{3} & 0 & 0 & 0 & 0 & \frac{1}{3} & -\frac{1}{3} & 0 & \frac{1}{3} & -\frac{1}{3} & 0 & 0 & 0 & 0 & \frac{1}{3} & -\frac{1}{3} & 0 \\
0 & -\frac{1}{3} & 0 & 0 & 0 & 0 & 0 & \frac{1}{3} & 0 & 0 & -\frac{1}{3} & 0 & 0 & 0 & 0 & 0 & -\frac{1}{3} & 0 & 0 & \frac{1}{3} & 0 & 0 & 0 & 0 & 0 & -\frac{1}{3} & 0 \\
-\frac{1}{9} & \frac{1}{9} & \frac{1}{9} & 0 & 0 & 0 & -\frac{1}{9} & \frac{1}{9} & \frac{1}{9} & \frac{1}{9} & -\frac{1}{9} & -\frac{1}{9} & 0 & 0 & 0 & \frac{1}{9} & -\frac{1}{9} & -\frac{1}{9} & \frac{1}{9} & -\frac{1}{9} & -\frac{1}{9} & 0 & 0 & 0 & \frac{1}{9} & -\frac{1}{9} & -\frac{1}{9} \\ \hline
\frac{1}{3} & -\frac{1}{3} & 0 & \frac{1}{3} & -\frac{1}{3} & 0 & 0 & 0 & 0 & -\frac{1}{3} & \frac{1}{3} & 0 & -\frac{1}{3} & \frac{1}{3} & 0 & 0 & 0 & 0 & -\frac{1}{3} & \frac{1}{3} & 0 & -\frac{1}{3} & \frac{1}{3} & 0 & 0 & 0 & 0 \\
0 & -\frac{1}{3} & 0 & 0 & -\frac{1}{3} & 0 & 0 & 0 & 0 & 0 & \frac{1}{3} & 0 & 0 & \frac{1}{3} & 0 & 0 & 0 & 0 & 0 & \frac{1}{3} & 0 & 0 & \frac{1}{3} & 0 & 0 & 0 & 0 \\
\frac{1}{9} & -\frac{1}{9} & -\frac{1}{9} & \frac{1}{9} & -\frac{1}{9} & -\frac{1}{9} & 0 & 0 & 0 & -\frac{1}{9} & \frac{1}{9} & \frac{1}{9} & -\frac{1}{9} & \frac{1}{9} & \frac{1}{9} & 0 & 0 & 0 & -\frac{1}{9} & \frac{1}{9} & \frac{1}{9} & -\frac{1}{9} & \frac{1}{9} & \frac{1}{9} & 0 & 0 & 0 \\ \hline
0 & 0 & 0 & 0 & 0 & 0 & 1 & -1 & 0 & 0 & 0 & 0 & 0 & 0 & 0 & -1 & 1 & 0 & 0 & 0 & 0 & 0 & 0 & 0 & -1 & 1 & 0 \\
0 & 0 & 0 & 0 & 0 & 0 & 0 & -1 & 0 & 0 & 0 & 0 & 0 & 0 & 0 & 0 & 1 & 0 & 0 & 0 & 0 & 0 & 0 & 0 & 0 & 1 & 0 \\
0 & 0 & 0 & 0 & 0 & 0 & \frac{1}{3} & -\frac{1}{3} & -\frac{1}{3} & 0 & 0 & 0 & 0 & 0 & 0 & -\frac{1}{3} & \frac{1}{3} & \frac{1}{3} & 0 & 0 & 0 & 0 & 0 & 0 & -\frac{1}{3} & \frac{1}{3} & \frac{1}{3}
\end{array} \right]
$$

$$(4.76)$$

Returning to (4.75), and taking into account (4.76), we get

$$\hat{T}_{27,1} = Z \bullet \begin{bmatrix} t^{1o1}_{o1o} \\ t^{1o2}_{o1o} \\ \vdots \\ t^{1o3}_{o3o} \\ \hline t^{2o1}_{o1o} \\ t^{2o2}_{o1o} \\ \vdots \\ t^{2o3}_{o3o} \\ \hline t^{3o1}_{o1o} \\ t^{3o2}_{o1o} \\ \vdots \\ t^{3o3}_{o3o} \end{bmatrix} = Z \bullet \begin{bmatrix} 1 \\ 0 \\ -1 \\ 2 \\ 3 \\ 0 \\ -1 \\ 2 \\ 0 \\ \hline 2 \\ -1 \\ 1 \\ 0 \\ 0 \\ 0 \\ 2 \\ 0 \\ 1 \\ \hline 0 \\ 0 \\ 1 \\ 5 \\ 1 \\ 2 \\ 0 \\ 0 \\ 0 \end{bmatrix} = \begin{bmatrix} 7 \\ 3 \\ 4/3 \\ -3 \\ -4 \\ -1/3 \\ -15 \\ -6 \\ -4 \\ \hline 5 \\ 1 \\ 1 \\ -3 \\ -1 \\ -2/3 \\ -6 \\ 0 \\ -1 \\ \hline 7/3 \\ 1 \\ 1/3 \\ -7/3 \\ -1 \\ -2/9 \\ -5 \\ -2 \\ -4/3 \end{bmatrix},$$

which after condensing $\hat{T}_{27,1}$ into three matrices of 3^2 elements, leads to the block matrix

$$\hat{T} = [t^{iok}_{ojo}] = \begin{bmatrix} 7 & 3 & 4/3 \\ -3 & -4 & -1/3 \\ -15 & -6 & -4 \\ \hline 5 & 1 & 1 \\ -3 & -1 & -2/3 \\ -6 & 0 & -1 \\ \hline 7/3 & 1 & 1/3 \\ -7/3 & -1 & -2/9 \\ -5 & -2 & -4/3 \end{bmatrix}, \tag{4.77}$$

which contains the new tensor components.

Second matrix process (classic matrix method):

We start by conveniently ordering and associating the factors appearing in the tensor Equation (4.74) to get the matrix ordering required by the method. From Equation (4.74) we get

$$t^{iok}_{ojo} = \gamma^{io}_{oa}(t^{ao\gamma}_{o\beta o}c^{o\beta}_{jo}\gamma^{ko}_{o\gamma})$$

and then we operate first the associated factors, preparing them first, for the dummy indices to be contracted to appear consecutively (we have already used this technique in the previous problem):

$$t^{\alpha o k}_{ojo} = t^{\alpha o \gamma}_{o\beta o} c^{o\beta}_{jo} \gamma^{ko}_{o\gamma} = c^{o\beta}_{jo} t^{\alpha o \gamma}_{o\beta o} \gamma^{ok}_{\gamma o}$$

and operating them as matrices we get

$$[t^{1ok}_{ojo}] = C^t[t^{1o\gamma}_{o\beta o}](C^{-1})^t = \begin{bmatrix} -1 & 0 & -1 \\ 1 & 1 & 0 \\ 0 & 0 & 3 \end{bmatrix}\begin{bmatrix} 1 & 0 & -1 \\ 2 & 3 & 0 \\ -1 & 2 & 0 \end{bmatrix}\begin{bmatrix} -1 & 0 & -\frac{1}{3} \\ 1 & 1 & \frac{1}{3} \\ 0 & 0 & \frac{1}{3} \end{bmatrix} = \begin{bmatrix} -2 & -2 & -\frac{1}{3} \\ 0 & 3 & -\frac{1}{3} \\ 9 & 6 & \frac{1}{3} \end{bmatrix}$$

$$[t^{2ok}_{ojo}] = C^t[t^{2o\gamma}_{o\beta o}](C^{-1})^t = \begin{bmatrix} -1 & 0 & -1 \\ 1 & 1 & 0 \\ 0 & 0 & 3 \end{bmatrix}\begin{bmatrix} 2 & -1 & 1 \\ 0 & 0 & 0 \\ 2 & 0 & 1 \end{bmatrix}\begin{bmatrix} -1 & 0 & -\frac{1}{3} \\ 1 & 1 & \frac{1}{3} \\ 0 & 0 & \frac{1}{3} \end{bmatrix} = \begin{bmatrix} 5 & 1 & 1 \\ -3 & -1 & -2/3 \\ -6 & 0 & -1 \end{bmatrix}$$

$$[t^{3ok}_{ojo}] = C^t[t^{3o\gamma}_{o\beta o}](C^{-1})^t = \begin{bmatrix} -1 & 0 & -1 \\ 1 & 1 & 0 \\ 0 & 0 & 3 \end{bmatrix}\begin{bmatrix} 0 & 0 & 1 \\ 5 & 1 & 2 \\ 0 & 0 & 0 \end{bmatrix}\begin{bmatrix} -1 & 0 & -\frac{1}{3} \\ 1 & 1 & \frac{1}{3} \\ 0 & 0 & \frac{1}{3} \end{bmatrix} = \begin{bmatrix} 0 & 0 & -\frac{1}{3} \\ -4 & -1 & -\frac{1}{3} \\ 0 & 0 & 0 \end{bmatrix}.$$

It remains to operate in matrix form the expression

$$t^{iok}_{ojo} = \gamma^{io}_{o\alpha} t^{\alpha ok}_{ojo}.$$

To this end, we multiply matrix $[\gamma^{io}_{o\alpha}] \equiv C^{-1}$ by the block matrix $[t^{\alpha ok}_{ojo}]$ by means of the product "$\odot$" defined in Formula (4.58):

$$\hat{T} = [t^{iok}_{ojo}] = \begin{bmatrix} -1 & 1 & 0 \\ 0 & 1 & 0 \\ -\frac{1}{3} & \frac{1}{3} & \frac{1}{3} \end{bmatrix} \odot \left[\begin{array}{ccc} -2 & -2 & -\frac{1}{3} \\ 0 & 3 & -\frac{1}{3} \\ 9 & 6 & \frac{1}{3} \\ \hline 5 & 1 & 1 \\ -3 & -1 & -2/3 \\ -6 & 0 & -1 \\ \hline 0 & 0 & -\frac{1}{3} \\ -4 & -1 & -\frac{1}{3} \\ 0 & 0 & 0 \end{array} \right]$$

$$= \begin{bmatrix} (-1)\begin{bmatrix} -2 & -2 & -\frac{1}{3} \\ 0 & 3 & -\frac{1}{3} \\ 9 & 6 & \frac{1}{3} \end{bmatrix} + (1)\begin{bmatrix} 5 & 1 & 1 \\ -3 & -1 & -2/3 \\ -6 & 0 & -1 \end{bmatrix} + (0)\begin{bmatrix} 0 & 0 & -\frac{1}{3} \\ -4 & -1 & -\frac{1}{3} \\ 0 & 0 & 0 \end{bmatrix} \\ (0)\begin{bmatrix} -2 & -2 & -\frac{1}{3} \\ 0 & 3 & -\frac{1}{3} \\ 9 & 6 & \frac{1}{3} \end{bmatrix} + (1)\begin{bmatrix} 5 & 1 & 1 \\ -3 & -1 & -2/3 \\ -6 & 0 & -1 \end{bmatrix} + (0)\begin{bmatrix} 0 & 0 & -\frac{1}{3} \\ -4 & -1 & -\frac{1}{3} \\ 0 & 0 & 0 \end{bmatrix} \\ (-\frac{1}{3})\begin{bmatrix} -2 & -2 & -\frac{1}{3} \\ 0 & 3 & -\frac{1}{3} \\ 9 & 6 & \frac{1}{3} \end{bmatrix} + (\frac{1}{3})\begin{bmatrix} 5 & 1 & 1 \\ -3 & -1 & -2/3 \\ -6 & 0 & -1 \end{bmatrix} + (\frac{1}{3})\begin{bmatrix} 0 & 0 & -\frac{1}{3} \\ -4 & -1 & -\frac{1}{3} \\ 0 & 0 & 0 \end{bmatrix} \end{bmatrix}.$$

Once the previously obtained matrices $[t^{\alpha ok}_{ojo}]$ have been replaced, after operating one gets the block matrix already obtained in (4.77).

2. Next we contract indices (β, γ) before the change.

We write the contraction of (β, γ) in tensor $t^{\alpha \circ \gamma}_{\circ \beta \circ} \equiv C\left(\begin{array}{c}\gamma\\\beta\end{array}\right)\vec{t} \equiv s^{\alpha}$, so that in matrix form

$$S = [s^{\alpha}] = [C\left(\begin{array}{c}\gamma\\\beta\end{array}\right)\vec{t}] \equiv [t^{\alpha\circ\theta}_{\circ\theta\circ}] = [t^{\alpha\circ 1}_{\circ 1 \circ} + t^{\alpha\circ 2}_{\circ 2 \circ} + t^{\alpha\circ 3}_{\circ 3 \circ}]$$

$$= \left[\begin{array}{c} \text{trace } [t^{1\circ\gamma}_{\circ\beta\circ}] \\ \text{trace } [t^{2\circ\gamma}_{\circ\beta\circ}] \\ \text{trace } [t^{3\circ\gamma}_{\circ\beta\circ}] \end{array}\right] \equiv \left[\begin{array}{c} 1+3+0 \\ 2+0+1 \\ 0+1+0 \end{array}\right] = \left[\begin{array}{c} 4 \\ 3 \\ 1 \end{array}\right].$$

3. Next, we contract indices (j, k) after the change

$$\hat{S} = [s^i] = [C\left(\begin{array}{c}k\\j\end{array}\right)\vec{t}] \equiv [t^{i\circ\theta}_{\circ\theta\circ}] = [t^{i\circ 1}_{\circ 1 \circ} + t^{i\circ 2}_{\circ 2 \circ} + t^{i\circ 3}_{\circ 3 \circ}]$$

$$= \left[\begin{array}{c} \text{trace } [t^{1\circ k}_{\circ j \circ}] \\ \text{trace } [t^{2\circ k}_{\circ j \circ}] \\ \text{trace } [t^{3\circ k}_{\circ j \circ}] \end{array}\right] \equiv \left[\begin{array}{c} 7-4-4 \\ 5-1-1 \\ 7/3-1-4/3 \end{array}\right] = \left[\begin{array}{c} -1 \\ 3 \\ 0 \end{array}\right].$$

4. Since tensor $\bar{s}$ is a vector (first-order tensor), and it is expressed in different bases we cannot say anything, based on the fact that its components are different.

Next, we check if they are related by the change-of-basis in $\mathbb{R}^3$. We know that the change-of-basis in $\mathbb{R}^3$, acts on the vector components as (see (3.28)):

$$X = C\hat{X}.$$

Since in our case we have

$$C\hat{S} = \left[\begin{array}{ccc} -1 & 1 & 0 \\ 0 & 1 & 0 \\ -1 & 0 & 3 \end{array}\right]\left[\begin{array}{c} -1 \\ 3 \\ 0 \end{array}\right] = \left[\begin{array}{c} 4 \\ 3 \\ 1 \end{array}\right] \equiv S,$$

we are looking at a *single tensor* S, expressed in two different bases (as it corresponds to every tensor contraction).

$\square$

Example 4.4 (Index contractions). Consider the homogeneous product tensor space, of fourth order ($r = 4$), of the factor linear space $V^2(\mathbb{R})$, $V^2 \otimes V^2 \otimes V^2 \otimes V^2_*(\mathbb{R})$ and in it a tensor $\vec{t}$, the components of which with respect to the canonical basis $\{\vec{e}_\alpha \otimes \vec{e}_\beta \otimes \vec{e}_\lambda \otimes \vec{e}^{*\mu}\}$ are:

$$
[t^{\alpha\beta\lambda\circ}_{\circ\circ\circ\mu}] = \left[\begin{array}{cc|cc} 2 & -3 & 0 & 1 \\ 0 & 1 & 7 & -4 \\ \hline 6 & 0 & 4 & 1 \\ -5 & 3 & 0 & -8 \end{array}\right],
$$

where α is the row of the submatrices matrix, β is the column of the submatrices matrix, λ is the row of each submatrix, and μ is the column of each submatrix.

1. Write the tensor $\vec{t}$ as a sum of 2^{r-2} tensor products of basic vectors, over "matrices" as scalars.
2. Find the new components of tensor $\vec{t}$, assuming that the linear space $V^2(\mathbb{R})$ on which it is defined, experiments a change-of-basis given by

$$
\begin{aligned}
\hat{e}_1 &= 2\vec{e}_1 + 3\vec{e}_2 \\
\hat{e}_2 &= \vec{e}_1 + 2\vec{e}_2.
\end{aligned}
$$

3. Contract the last two indices of the tensor before and after the change, showing that the resulting contracted tensor is the same.
4. Solve question 2 but for the following tensor (it does not satisfy the ordering axiom 4 given in Chapter 2):

$$
T' = [t^{\alpha\beta\lambda\circ}_{\circ\circ\circ\mu}] = \left[\begin{array}{cc|cc} 2 & -3 & 0 & 1 \\ 0 & 1 & 7 & -4 \\ \hline 6 & 0 & 4 & 1 \\ -5 & 3 & 0 & -8 \end{array}\right],
$$

where α is the row of each submatrix, β is the column of each submatrix, λ is the submatrix row, and μ is the submatrix column.

Solution:

1. When developing the tensor expression of tensor $\vec{t}$, in the canonical basis of $V^2 \otimes V^2 \otimes V^2 \otimes V_*^2(\mathbb{R})$, we get

$$
r = 4; \quad n = 2; \quad \sigma = 2^4 = 16; \quad \sigma' = 2^{4-2} = 2^2 = 4 \text{ summands.}
$$

$$
\vec{t} = t^{111\circ}_{\circ\circ\circ 1}\, \vec{e}_1 \otimes \vec{e}_1 \otimes \vec{e}_1 \otimes \vec{e}^{*1} + t^{111\circ}_{\circ\circ\circ 2}\, \vec{e}_1 \otimes \vec{e}_1 \otimes \vec{e}_1 \otimes \vec{e}^{*2}
$$

$$
+ t^{112\circ}_{\circ\circ\circ 1}\, \vec{e}_1 \otimes \vec{e}_1 \otimes \vec{e}_2 \otimes \vec{e}^{*1} + \cdots + t^{222\circ}_{\circ\circ\circ 1}\, \vec{e}_2 \otimes \vec{e}_2 \otimes \vec{e}_2 \otimes \vec{e}^{*1}
$$

$$
+ t^{222\circ}_{\circ\circ\circ 2}\, \vec{e}_2 \otimes \vec{e}_2 \otimes \vec{e}_2 \otimes \vec{e}^{*2},
$$

and associating the summands (the last two common basic vectors), we have

$$\vec{t} = [t^{111o}_{ooo1}\vec{e}_1 \otimes \vec{e}_1 + t^{121o}_{ooo1}\vec{e}_1 \otimes \vec{e}_2 + t^{211o}_{ooo1}\vec{e}_2 \otimes \vec{e}_1 + t^{221o}_{ooo1}\vec{e}_2 \otimes \vec{e}_2] \otimes \vec{e}_1 \otimes \vec{e}^{*1}$$

$$+ [t^{111o}_{ooo2}\vec{e}_1 \otimes \vec{e}_1 + t^{121o}_{ooo2}\vec{e}_1 \otimes \vec{e}_2 + t^{211o}_{ooo2}\vec{e}_2 \otimes \vec{e}_1 + t^{221o}_{ooo2}\vec{e}_2 \otimes \vec{e}_2] \otimes \vec{e}_1 \otimes \vec{e}^{*2}$$

$$+ [t^{112o}_{ooo1}\vec{e}_1 \otimes \vec{e}_1 + t^{122o}_{ooo1}\vec{e}_1 \otimes \vec{e}_2 + t^{212o}_{ooo1}\vec{e}_2 \otimes \vec{e}_1 + t^{222o}_{ooo1}\vec{e}_2 \otimes \vec{e}_2] \otimes \vec{e}_2 \otimes \vec{e}^{*1}$$

$$+ [t^{112o}_{ooo2}\vec{e}_1 \otimes \vec{e}_1 + t^{122o}_{ooo2}\vec{e}_1 \otimes \vec{e}_2 + t^{212o}_{ooo2}\vec{e}_2 \otimes \vec{e}_1 + t^{222o}_{ooo2}\vec{e}_2 \otimes \vec{e}_2] \otimes \vec{e}_2 \otimes \vec{e}^{*2}$$

$$= \left([\vec{e}_1 \quad \vec{e}_2]\begin{bmatrix} 2 & 0 \\ 6 & 4 \end{bmatrix}\begin{bmatrix} \vec{e}_1 \\ \vec{e}_2 \end{bmatrix}\right)\vec{e}_1 \otimes \vec{e}^{*1} + \left([\vec{e}_1 \quad \vec{e}_2]\begin{bmatrix} -3 & 1 \\ 0 & 1 \end{bmatrix}\begin{bmatrix} \vec{e}_1 \\ \vec{e}_2 \end{bmatrix}\right)\vec{e}_1 \otimes \vec{e}^{*2}$$

$$+ \left([\vec{e}_1 \quad \vec{e}_2]\begin{bmatrix} 0 & 7 \\ -5 & 0 \end{bmatrix}\begin{bmatrix} \vec{e}_1 \\ \vec{e}_2 \end{bmatrix}\right)\vec{e}_2 \otimes \vec{e}^{*1} + \left([\vec{e}_1 \quad \vec{e}_2]\begin{bmatrix} 1 & -4 \\ 3 & -8 \end{bmatrix}\begin{bmatrix} \vec{e}_1 \\ \vec{e}_2 \end{bmatrix}\right)\vec{e}_2 \otimes \vec{e}^{*2},$$

that *briefly and symbolically* can be written as

$$\vec{t} = X^1\vec{e}_1 \otimes \vec{e}^{*1} + X^2\vec{e}_1 \otimes \vec{e}^{*2} + X^3\vec{e}_2 \otimes \vec{e}^{*1} + X^4\vec{e}_2 \otimes \vec{e}^{*2},$$

where

$$X^1 = \begin{bmatrix} 2 & 0 \\ 6 & 4 \end{bmatrix}; \quad X^2 = \begin{bmatrix} -3 & 1 \\ 0 & 1 \end{bmatrix}; \quad X^3 = \begin{bmatrix} 0 & 7 \\ -5 & 0 \end{bmatrix}; \quad X^4 = \begin{bmatrix} 1 & -4 \\ 3 & -8 \end{bmatrix}$$

are the components with respect to the basis $\vec{e}_\alpha \otimes \vec{e}^{*\beta}$. This practice is very dubious because it omits information on the tensor.

2. The data tensor components, in the canonical basis are the data scalars directly read with the criterion imposed by the axiomatic ordering of the basis

$$T_{16,1} = \begin{bmatrix} t^{111o}_{ooo1} \\ t^{111o}_{ooo2} \\ t^{112o}_{ooo1} \\ t^{112o}_{ooo2} \\ t^{121o}_{ooo1} \\ t^{121o}_{ooo2} \\ t^{122o}_{ooo1} \\ t^{122o}_{ooo2} \\ t^{211o}_{ooo1} \\ t^{211o}_{ooo2} \\ t^{212o}_{ooo1} \\ t^{212o}_{ooo2} \\ t^{221o}_{ooo1} \\ t^{221o}_{ooo2} \\ t^{222o}_{ooo1} \\ t^{222o}_{ooo2} \end{bmatrix} = \begin{bmatrix} 2 \\ -3 \\ 0 \\ 1 \\ 0 \\ 1 \\ 7 \\ -4 \\ 6 \\ 0 \\ -5 \\ 3 \\ 4 \\ 1 \\ 0 \\ -8 \end{bmatrix}.$$

From (4.34), the tensor expression to be considered is

$$t^{ij\ell o}_{ooom} = t^{\alpha\beta\lambda o}_{ooo\mu}\,\gamma^{io}_{o\alpha}\,\gamma^{jo}_{o\beta}\,\gamma^{\ell o}_{o\lambda}\,c^{o\mu}_{mo} \qquad (4.78)$$

and the data matrices are

$$C \equiv [c^{\alpha o}_{oi}] = \begin{bmatrix} 2 & 1 \\ 3 & 2 \end{bmatrix}; \quad C^{-1} \equiv [\gamma^{io}_{o\alpha}] = \begin{bmatrix} 2 & -1 \\ -3 & 2 \end{bmatrix}; \quad C^{t} \equiv [c^{o\mu}_{mo}] = \begin{bmatrix} 2 & 3 \\ 1 & 2 \end{bmatrix}.$$

First matrix process (direct process):

The matrix interpretation of (4.78), according to Formula (4.36), is

$$\hat{T}_{16,1} = [t^{ij\ell o}_{ooom}] = \left(C^{-1} \otimes C^{-1} \otimes C^{-1} \otimes C^{t}\right) \bullet T_{16,1}, \qquad (4.79)$$

where

$$C^{-1} \otimes C^{-1} = \begin{bmatrix} 2 & -1 \\ -3 & 2 \end{bmatrix} \otimes \begin{bmatrix} 2 & -1 \\ -3 & 2 \end{bmatrix} = \begin{bmatrix} 4 & -2 & -2 & 1 \\ -6 & 4 & 3 & -2 \\ -6 & 3 & 4 & -2 \\ 9 & -6 & -6 & 4 \end{bmatrix}$$

$$C^{-1} \otimes C^{t} = \begin{bmatrix} 2 & -1 \\ -3 & 2 \end{bmatrix} \otimes \begin{bmatrix} 2 & 3 \\ 1 & 2 \end{bmatrix} = \begin{bmatrix} 4 & 6 & -2 & -3 \\ 2 & 4 & -1 & -2 \\ -6 & -9 & 4 & 6 \\ -3 & -6 & 2 & 4 \end{bmatrix}.$$

The change-of-basis matrix Z^{-1} in the tensor space is

$$Z^{-1} = (C^{-1} \otimes C^{-1}) \otimes (C^{-1} \otimes C^{t}) = \begin{bmatrix} 4 & -2 & -2 & 1 \\ -6 & 4 & 3 & -2 \\ -6 & 3 & 4 & -2 \\ 9 & -6 & -6 & 4 \end{bmatrix} \otimes \begin{bmatrix} 4 & 6 & -2 & -3 \\ 2 & 4 & -1 & -2 \\ -6 & -9 & 4 & 6 \\ -3 & -6 & 2 & 4 \end{bmatrix}$$

that is

$$Z^{-1} = \left[\begin{array}{rrrr|rrrr|rrrr|rrrr}
16 & 24 & -8 & -12 & -8 & -12 & 4 & 6 & -8 & -12 & 4 & 6 & 4 & 6 & -2 & -3 \\
8 & 16 & -4 & -8 & -4 & -8 & 2 & 4 & -4 & -8 & 2 & 4 & 2 & 4 & -1 & -2 \\
-24 & -36 & 16 & 24 & 12 & 18 & -8 & -12 & 12 & 18 & -8 & -12 & -6 & -9 & 4 & 6 \\
-12 & -24 & 8 & 16 & 6 & 12 & -4 & -8 & 6 & 12 & -4 & -8 & -3 & -6 & 2 & 4 \\
\hline
-24 & -36 & 12 & 18 & 16 & 24 & -8 & -12 & 12 & 18 & -6 & -9 & -8 & -12 & 4 & 6 \\
-12 & -24 & 6 & 12 & 8 & 16 & -4 & -8 & 6 & 12 & -3 & -6 & -4 & -8 & 2 & 4 \\
36 & 54 & -24 & -36 & -24 & -36 & 16 & 24 & -18 & -27 & 12 & 18 & 12 & 18 & -8 & -12 \\
18 & 36 & -12 & -24 & -12 & -24 & 8 & 16 & -9 & -18 & 6 & 12 & 6 & 12 & -4 & -8 \\
\hline
-24 & -36 & 12 & 18 & 12 & 18 & -6 & -9 & 16 & 24 & -8 & -12 & -8 & -12 & 4 & 6 \\
-12 & -24 & 6 & 12 & 6 & 12 & -3 & -6 & 8 & 16 & -4 & -8 & -4 & -8 & 2 & 4 \\
36 & 54 & -24 & -36 & -18 & -27 & 12 & 18 & -24 & -36 & 16 & 24 & 12 & 18 & -8 & -12 \\
18 & 36 & -12 & -24 & -9 & -18 & 6 & 12 & -12 & -24 & 8 & 16 & 6 & 12 & -4 & -8 \\
\hline
36 & 54 & -18 & -27 & -24 & -36 & 12 & 18 & -24 & -36 & 12 & 18 & 16 & 24 & -8 & -12 \\
18 & 36 & -9 & -18 & -12 & -24 & 6 & 12 & -12 & -24 & 6 & 12 & 8 & 16 & -4 & -8 \\
-54 & -81 & 36 & 54 & 36 & 54 & -24 & -36 & 36 & 54 & -24 & -36 & -24 & -36 & 16 & 24 \\
-27 & -54 & 18 & 36 & 18 & 36 & -12 & -24 & 18 & 36 & -12 & -24 & -12 & -24 & 8 & 16
\end{array}\right].$$

Then, relation (4.79) leads to

$$
\hat{T}_{16,1} = Z^{-1} \bullet
\begin{bmatrix}
2 \\
-3 \\
0 \\
1 \\
-- \\
0 \\
1 \\
7 \\
-4 \\
-- \\
6 \\
0 \\
-5 \\
3 \\
-- \\
4 \\
1 \\
0 \\
-8
\end{bmatrix}
=
\begin{bmatrix}
-64 \\
-44 \\
89 \\
62 \\
-- \\
77 \\
57 \\
-98 \\
-76 \\
-- \\
98 \\
63 \\
-131 \\
-84 \\
-- \\
-107 \\
-74 \\
123 \\
88
\end{bmatrix},
$$

which, transformed to the data tensor format, becomes

$$
\hat{T} = [t_{\circ\circ\circ m}^{ij\ell\circ}] =
\left[
\begin{array}{cc|cc}
-64 & -44 & 77 & 57 \\
89 & 62 & -98 & -76 \\
\hline
98 & 63 & -107 & -74 \\
-131 & -84 & 123 & 88
\end{array}
\right]. \tag{4.80}
$$

Second matrix process (classic matrix method):

Starting from the Formula (4.78), we contract the indices λ and μ. First we operate the parenthesis

$$
t_{\circ\circ\circ m}^{ij\ell\circ} = \gamma_{\circ\alpha}^{i\circ}\, \gamma_{\circ\beta}^{j\circ}\, (t_{\circ\circ\circ\mu}^{\alpha\beta\lambda\circ}\, \gamma_{\circ\lambda}^{\ell\circ}\, c_{m\circ}^{\circ\mu}) = t'^{\alpha\beta}\gamma_{\circ\alpha}^{i\circ}\, \gamma_{\circ\beta}^{j\circ}.
$$

Next, we prepare it to be operated by means of matrix products

$$
t_{\circ\circ\circ\mu}^{\alpha\beta\lambda\circ}\, \gamma_{\circ\lambda}^{\ell\circ}\, c_{m\circ}^{\circ\mu} = \gamma_{\circ\lambda}^{\ell\circ}\, t_{\circ\circ\circ\mu}^{\alpha\beta\lambda\circ}\, c_{\circ m}^{\mu\circ} \Rightarrow [t_{\circ\circ\circ m}^{\alpha\beta\ell\circ}] = C^{-1}[t_{\circ\circ\circ\mu}^{\alpha\beta\lambda\circ}]C
$$

$$
[t_{\circ\circ\circ m}^{11\ell\circ}] = C^{-1}[t_{\circ\circ\circ\mu}^{11\lambda\circ}]C =
\begin{bmatrix} 2 & -1 \\ -3 & 2 \end{bmatrix}
\begin{bmatrix} 2 & -3 \\ 0 & 1 \end{bmatrix}
\begin{bmatrix} 2 & 1 \\ 3 & 2 \end{bmatrix}
=
\begin{bmatrix} -13 & -10 \\ 21 & 16 \end{bmatrix}
$$

$$
[t_{\circ\circ\circ m}^{12\ell\circ}] = C^{-1}[t_{\circ\circ\circ\mu}^{12\lambda\circ}]C =
\begin{bmatrix} 2 & -1 \\ -3 & 2 \end{bmatrix}
\begin{bmatrix} 0 & 1 \\ 7 & -4 \end{bmatrix}
\begin{bmatrix} 2 & 1 \\ 3 & 2 \end{bmatrix}
=
\begin{bmatrix} 4 & 5 \\ -5 & -8 \end{bmatrix}
$$

$$
[t_{\circ\circ\circ m}^{21\ell\circ}] = C^{-1}[t_{\circ\circ\circ\mu}^{21\lambda\circ}]C =
\begin{bmatrix} 2 & -1 \\ -3 & 2 \end{bmatrix}
\begin{bmatrix} 6 & 0 \\ -5 & 3 \end{bmatrix}
\begin{bmatrix} 2 & 1 \\ 3 & 2 \end{bmatrix}
=
\begin{bmatrix} 25 & 11 \\ -38 & -16 \end{bmatrix}
$$

$$[t^{22\ell\,o}_{\,o\,o\,o\,m}] = C^{-1}[t^{22\lambda\,o}_{\,o\,o\,o\,\mu}]C = \begin{bmatrix} 2 & -1 \\ -3 & 2 \end{bmatrix}\begin{bmatrix} 4 & 1 \\ 0 & -8 \end{bmatrix}\begin{bmatrix} 2 & 1 \\ 3 & 2 \end{bmatrix} = \begin{bmatrix} 46 & 28 \\ -81 & -50 \end{bmatrix}$$

and we prepare the contraction of the *submatrix indices* (α, β)

$$T' = [t'^{\,ij}_{\,o\,o}]; \quad t'^{\,ij}_{\,o\,o} = t'^{\,\alpha\beta}_{\,o\,o}\,\gamma^{\,i\,o}_{\,o\,\alpha}\,\gamma^{\,j\,o}_{\,o\,\beta} = \gamma^{\,i\,o}_{\,o\,\alpha}\,t'^{\,\alpha\beta}_{\,o\,o}\,\gamma^{\,o\,j}_{\,\beta\,o}.$$

The last expression, interpreted in matrix terms, becomes

$$T' = [t'^{\,ij}_{\,o\,o}] = C^{-1} \odot \left[\begin{array}{cc|cc} -13 & -10 & 4 & 5 \\ 21 & 16 & -5 & -8 \\ \hline 25 & 11 & 46 & 28 \\ -38 & -16 & -81 & -51 \end{array}\right] \odot (C^{-1})^t$$

$$= \left(\begin{bmatrix} 2 & -1 \\ -3 & 2 \end{bmatrix} \odot \left[\begin{array}{cc|cc} -13 & -10 & 4 & 5 \\ 21 & 16 & -5 & -8 \\ \hline 25 & 11 & 46 & 28 \\ -38 & -16 & -81 & -50 \end{array}\right] \right) \odot \begin{bmatrix} 2 & -3 \\ -1 & 2 \end{bmatrix}$$

$$= \begin{bmatrix} 2\begin{bmatrix} -13 & -10 \\ 21 & 16 \end{bmatrix} - \begin{bmatrix} 25 & 11 \\ -38 & -16 \end{bmatrix} & 2\begin{bmatrix} 4 & 5 \\ -5 & -8 \end{bmatrix} - \begin{bmatrix} 46 & 28 \\ -81 & -50 \end{bmatrix} \\ (-3)\begin{bmatrix} -13 & -10 \\ 21 & 16 \end{bmatrix} + 2\begin{bmatrix} 25 & 11 \\ -38 & -16 \end{bmatrix} & (-3)\begin{bmatrix} 4 & 5 \\ -5 & -8 \end{bmatrix} + 2\begin{bmatrix} 46 & 28 \\ -81 & -50 \end{bmatrix} \end{bmatrix}$$

$$\odot \begin{bmatrix} 2 & -3 \\ -1 & 2 \end{bmatrix},$$

that is,

$$T' = \left[\begin{array}{cc|cc} -51 & -31 & -38 & -18 \\ 80 & 48 & 71 & 34 \\ \hline 89 & 52 & 80 & 41 \\ -139 & -80 & -147 & -76 \end{array}\right] \odot \begin{bmatrix} 2 & -3 \\ -1 & 2 \end{bmatrix}$$

$$= \begin{bmatrix} 2\begin{bmatrix} -51 & -31 \\ 80 & 48 \end{bmatrix} - \begin{bmatrix} -38 & -18 \\ 71 & 34 \end{bmatrix} & (-3)\begin{bmatrix} -51 & -31 \\ 80 & 48 \end{bmatrix} + 2\begin{bmatrix} -38 & -18 \\ 71 & 34 \end{bmatrix} \\ 2\begin{bmatrix} 89 & 52 \\ -139 & -80 \end{bmatrix} - \begin{bmatrix} 80 & 41 \\ -147 & -76 \end{bmatrix} & (-3)\begin{bmatrix} 89 & 52 \\ -139 & -80 \end{bmatrix} + 2\begin{bmatrix} 80 & 41 \\ -147 & -76 \end{bmatrix} \end{bmatrix}$$

$$= \left[\begin{array}{cc|cc} -64 & -44 & 77 & 57 \\ 89 & 62 & -98 & -76 \\ \hline 98 & 63 & -107 & -74 \\ -131 & -84 & 123 & 88 \end{array}\right],$$

which coincides with the $\hat{T}$ obtained by the direct method.

3. We perform the contraction $C\begin{pmatrix} \lambda \\ \mu \end{pmatrix}\vec{t} = u^{\alpha\beta\theta\,o}_{\,o\,\theta}$ of indices (λ, μ), to obtain

$$[u^{\alpha\beta}] = \begin{bmatrix} 2+1 & 0-4 \\ 6+3 & 4-8 \end{bmatrix} \equiv \begin{bmatrix} 3 & -4 \\ 9 & -4 \end{bmatrix},$$

where the last matrix is the matrix of all traces, and in the new basis we have

$$[u^{ij}] = \begin{bmatrix} -64+62 & 77-76 \\ 98-84 & -107+88 \end{bmatrix} \equiv \begin{bmatrix} -2 & 1 \\ 14 & -19 \end{bmatrix},$$

where the last matrix is also the matrix of all traces, but the new traces. Since $u^{ij} = u^{\alpha\beta} \gamma^{i\,\circ}_{\circ\alpha}\gamma^{j\,\circ}_{\circ\beta} = \gamma^{i\,\circ}_{\circ\alpha}\, u^{\alpha\beta}\, \gamma^{\circ j}_{\beta\circ}$, the matrix interpretation must be

$$[u^{ij}] = C^{-1}[u^{\alpha\beta}](C^{-1})^t = \begin{bmatrix} 2 & -1 \\ -3 & 2 \end{bmatrix}\begin{bmatrix} 3 & -4 \\ 9 & -4 \end{bmatrix}\begin{bmatrix} 2 & -3 \\ -1 & 2 \end{bmatrix}$$

$$= \begin{bmatrix} -3 & -4 \\ 9 & 4 \end{bmatrix}\begin{bmatrix} 2 & -3 \\ -1 & 2 \end{bmatrix} = \begin{bmatrix} -2 & 1 \\ 14 & -19 \end{bmatrix},$$

which is the one obtained by direct contraction.

4. Since this tensor is not given in the canonical basis of the tensor space, we proceed to position the tensor components in the order forced by the cited canonical basis and we assign the values corresponding to the results in point 4.

Once we know the tensor by its components in the canonical basis of the tensor space, we proceed exactly as we did in the first question.

After finishing the process, we recover the format imposed by point 4.

Components (in canonical order)							
$t^{111\circ}_{\circ\circ\circ1}$	$t^{111\circ}_{\circ\circ\circ2}$	$t^{112\circ}_{\circ\circ\circ1}$	$t^{112\circ}_{\circ\circ\circ2}$	$t^{121\circ}_{\circ\circ\circ1}$	$t^{121\circ}_{\circ\circ\circ2}$	$t^{122\circ}_{\circ\circ\circ1}$	$t^{122\circ}_{\circ\circ\circ2}$
2	0	6	4	−3	1	0	1
$t^{211\circ}_{\circ\circ\circ1}$	$t^{211\circ}_{\circ\circ\circ2}$	$t^{212\circ}_{\circ\circ\circ1}$	$t^{212\circ}_{\circ\circ\circ2}$	$t^{221\circ}_{\circ\circ\circ1}$	$t^{221\circ}_{\circ\circ\circ2}$	$t^{222\circ}_{\circ\circ\circ1}$	$t^{222\circ}_{\circ\circ\circ2}$
0	7	−5	0	1	−4	3	−8

Thus, we have the following tensor, defined with respect to the canonical basis:

$$[t'^{\alpha\beta\lambda\circ}_{\ \circ\circ\circ\mu}] = \left[\begin{array}{cc|cc} 2 & 0 & -3 & 1 \\ 6 & 4 & 0 & 1 \\ \hline 0 & 7 & 1 & -4 \\ -5 & 0 & 3 & -8 \end{array}\right], \tag{4.81}$$

where α is the row of the block matrix, β is the column of the block matrix, λ is the submatrix row, and μ is the submatrix column. We observe that it coincides with the submatrices X_i of question 1.

First matrix process (direct process):
Since we already have Z^{-1}, we also have

$$\hat{T}'_{16,1} = Z^{-1} \bullet T'_{16,1} = Z^{-1} \bullet \begin{bmatrix} 2 \\ 0 \\ 6 \\ 4 \\ -- \\ -3 \\ 1 \\ 0 \\ 1 \\ -- \\ 0 \\ 7 \\ -5 \\ 0 \\ -- \\ 1 \\ -4 \\ 3 \\ -8 \end{bmatrix} \begin{bmatrix} -152 \\ -99 \\ 274 \\ 173 \\ -- \\ 220 \\ 145 \\ -393 \\ -250 \\ -- \\ 281 \\ 182 \\ -491 \\ -309 \\ -- \\ -410 \\ -268 \\ 708 \\ 448 \end{bmatrix} = \quad,$$

and in the new canonical basis

$$\hat{T}' = [t'^{ij\ell o}_{ooom}] = \begin{bmatrix} -152 & -99 & 220 & 145 \\ 274 & 173 & -393 & -250 \\ -- & -- & -- & -- \\ 281 & 182 & -410 & -268 \\ -491 & -309 & 708 & 448 \end{bmatrix}.$$

It only remains to recover the ordering imposed by the statement in point 4. We take the t'^{1o}_{o1} in each submatrix, and build the matrix $\begin{bmatrix} -152 & 220 \\ 281 & -410 \end{bmatrix}$.

Next, all the t'^{1o}_{o2} in each submatrix, and we obtain $\begin{bmatrix} -99 & 145 \\ 182 & -268 \end{bmatrix}$, and

so on with t'^{2o}_{o1} and t'^{2o}_{o2}, to obtain $\begin{bmatrix} 274 & -393 \\ -491 & 708 \end{bmatrix}$ and $\begin{bmatrix} 173 & -250 \\ -309 & 448 \end{bmatrix}$.

Hence, we get

$$\hat{T} = [t^{ij\ell o}_{ooom}] = \begin{bmatrix} -152 & 220 & -99 & 145 \\ 281 & -410 & 182 & -268 \\ -- & -- & -- & -- \\ 274 & -393 & 173 & -250 \\ -491 & 708 & -309 & 448 \end{bmatrix},$$

where i is the submatrix row, j is the submatrix column, ℓ is the row of the block matrix, and m is the column of the block matrix.

Second matrix procedure (classic matrix method)

Proceeding as in the first question, we have:

Third and fourth contraction, over the data in 4)

$$[t^{11\ell\,\circ}_{\circ\circ\circ m}] = C^{-1}[t^{11\lambda\circ}_{\circ\circ\circ\mu}]C = \begin{bmatrix} 2 & -1 \\ -3 & 2 \end{bmatrix}\begin{bmatrix} 2 & 0 \\ 6 & 4 \end{bmatrix}\begin{bmatrix} 2 & 1 \\ 3 & 2 \end{bmatrix} = \begin{bmatrix} -16 & -10 \\ 36 & 22 \end{bmatrix}$$

$$[t^{12\ell\,\circ}_{\circ\circ\circ m}] = C^{-1}[t^{12\lambda\circ}_{\circ\circ\circ\mu}]C = \begin{bmatrix} 2 & -1 \\ -3 & 2 \end{bmatrix}\begin{bmatrix} -3 & 1 \\ 0 & 1 \end{bmatrix}\begin{bmatrix} 2 & 1 \\ 3 & 2 \end{bmatrix} = \begin{bmatrix} -9 & -4 \\ 15 & 7 \end{bmatrix}$$

$$[t^{21\ell\,\circ}_{\circ\circ\circ m}] = C^{-1}[t^{21\lambda\circ}_{\circ\circ\circ\mu}]C = \begin{bmatrix} 2 & -1 \\ -3 & 2 \end{bmatrix}\begin{bmatrix} 0 & 7 \\ -5 & 0 \end{bmatrix}\begin{bmatrix} 2 & 1 \\ 3 & 2 \end{bmatrix} = \begin{bmatrix} 52 & 33 \\ -83 & -52 \end{bmatrix}$$

$$[t^{22\ell\,\circ}_{\circ\circ\circ m}] = C^{-1}[t^{22\lambda\circ}_{\circ\circ\circ\mu}]C = \begin{bmatrix} 2 & -1 \\ -3 & 2 \end{bmatrix}\begin{bmatrix} 1 & -4 \\ 3 & -8 \end{bmatrix}\begin{bmatrix} 2 & 1 \\ 3 & 2 \end{bmatrix} = \begin{bmatrix} -2 & -1 \\ -6 & -5 \end{bmatrix}.$$

The first contraction is

$$\begin{bmatrix} 2 & -1 \\ -3 & 2 \end{bmatrix} \odot \left[\begin{array}{cc|cc} -16 & -10 & -9 & -4 \\ 36 & 22 & 15 & 7 \\ \hline 52 & 33 & -2 & -1 \\ -83 & -52 & -6 & -5 \end{array}\right],$$

that is

$$\left[\begin{array}{cc} \begin{bmatrix} -32 & -20 \\ 72 & 44 \end{bmatrix} + \begin{bmatrix} -52 & -33 \\ 83 & 52 \end{bmatrix} & \begin{bmatrix} -18 & -8 \\ 30 & 14 \end{bmatrix} + \begin{bmatrix} 2 & 1 \\ 6 & 5 \end{bmatrix} \\ \begin{bmatrix} 48 & 30 \\ -108 & -66 \end{bmatrix} + \begin{bmatrix} 104 & 66 \\ -166 & -104 \end{bmatrix} & \begin{bmatrix} 27 & 12 \\ -45 & -21 \end{bmatrix} + \begin{bmatrix} -4 & -2 \\ -12 & -10 \end{bmatrix} \end{array}\right],$$

and we get

$$\left[\begin{array}{cc|cc} -84 & -53 & -16 & -7 \\ 155 & 96 & 36 & 19 \\ \hline 152 & 96 & 23 & 10 \\ -274 & -170 & -57 & -31 \end{array}\right].$$

The second contraction is

$$[t'^{ij\ell\,\circ}_{\circ\circ\circ m}] = \left[\begin{array}{cc|cc} -84 & -53 & -16 & -7 \\ 155 & 96 & 36 & 19 \\ \hline 152 & 96 & 23 & 10 \\ -274 & -170 & -57 & -31 \end{array}\right] \odot \begin{bmatrix} 2 & -3 \\ -1 & 2 \end{bmatrix},$$

that is

$$\left[\begin{array}{cc} \begin{bmatrix} -168 & -106 \\ 310 & 192 \end{bmatrix} + \begin{bmatrix} 16 & 7 \\ -36 & -19 \end{bmatrix} & \begin{bmatrix} 252 & 159 \\ -465 & -288 \end{bmatrix} + \begin{bmatrix} -32 & -14 \\ 72 & 38 \end{bmatrix} \\ \begin{bmatrix} 304 & 192 \\ -548 & -340 \end{bmatrix} + \begin{bmatrix} -23 & -10 \\ 57 & 31 \end{bmatrix} & \begin{bmatrix} -456 & -288 \\ 822 & 510 \end{bmatrix} + \begin{bmatrix} 46 & 20 \\ -114 & -62 \end{bmatrix} \end{array}\right],$$

which leads to

$$\left[\begin{array}{cc|cc} -152 & -99 & 220 & 145 \\ 274 & 173 & -393 & -250 \\ \hline 281 & 182 & -410 & -268 \\ -491 & -309 & 708 & 448 \end{array}\right],$$

which coincides with the $\hat{T}'$ obtained by the direct method.

Finally, we reorder it to the desired format, as was done in (a).

□

Example 4.5 (Permutation of indices). In the tensor space $\mathbb{R}^3 \otimes \mathbb{R}^3_* \otimes \mathbb{R}^3$ we consider a tensor $\vec{t}$ with components $t^{\alpha \circ \gamma}_{\circ \beta \circ}$ given by

$$T \equiv [t^{\alpha \circ \gamma}_{\circ \beta \circ}] = \begin{bmatrix} 1 & 0 & -1 \\ 2 & 3 & 0 \\ -1 & 2 & 0 \\ \hline 2 & -1 & 1 \\ 0 & 1 & 0 \\ 2 & 0 & 1 \\ \hline 0 & 0 & 1 \\ 5 & 1 & 2 \\ 1 & 0 & 0 \end{bmatrix}.$$

If a tensor is of third order, for it to be directly read, it is necessary to refer the tensor space to its canonical basis, which implies the following assignment of indices: the first index will always be the row indicator of a column matrix of submatrices, the second index will be the row indicator of each submatrix, the third index will be the column indicator of each submatrix. This quality is absolute, that is, inalterable for the tensor as data.

1. Given the numerical components that correspond to the cited indices, indicate the matrix notation associated with all semaphoric column permutations of the given tensor (use letters $\bar{u}, \bar{v}, \bar{w}, \bar{r}, \bar{s}$, for each of the resulting tensors).
2. Certain authors, in performing their tensor expression operations, use the trick previously mentioned, with the advantage of being able to solve them with matrix products. Assuming that the considered permutations are $t^{\alpha \circ \gamma}_{\circ \beta \circ}$ and $u^{\gamma \alpha \circ}_{\circ \circ \beta}$, give the notation of the respective linear spaces to which they belong, together with all vectors in the respective canonical basis in order.
3. Are $\vec{t}$ and $\vec{u}$ the same tensor? Justify why the cited authors think that these permutations do not alter their data tensor, when they operate them.

Solution:

1. The required notation is

$$t^{\alpha \circ \gamma}_{\circ \beta \circ}; \ u^{\gamma \alpha \circ}_{\circ \circ \beta}; \ v^{\circ \gamma \alpha}_{\beta \circ \circ}; w^{\alpha \gamma \circ}_{\circ \circ \beta}; \ r^{\gamma \circ \alpha}_{\circ \beta \circ}; \ s^{\circ \alpha \gamma}_{\beta \circ \circ},$$

where

$$U \equiv [u^{\gamma\alpha\circ}_{\;\circ\circ\beta}] = \begin{bmatrix} 1 & 2 & -1 \\ 2 & 0 & 2 \\ 0 & 5 & 1 \\ \hline 0 & 3 & 2 \\ -1 & 1 & 0 \\ 0 & 1 & 0 \\ \hline -1 & 0 & 0 \\ 1 & 0 & 1 \\ 1 & 2 & 0 \end{bmatrix} \;;\; V \equiv [v^{\circ\gamma\alpha}_{\beta\circ\circ}] = \begin{bmatrix} 1 & 2 & 0 \\ 0 & -1 & 0 \\ -1 & 1 & 1 \\ \hline 2 & 0 & 5 \\ 3 & 1 & 1 \\ 0 & 0 & 2 \\ \hline -1 & 2 & 1 \\ 2 & 0 & 0 \\ 0 & 1 & 0 \end{bmatrix} \;;$$

$$W \equiv [w^{\alpha\gamma\circ}_{\;\circ\circ\beta}] = \begin{bmatrix} 1 & 2 & -1 \\ 0 & 3 & 2 \\ -1 & 0 & 0 \\ \hline 2 & 0 & 2 \\ -1 & 1 & 0 \\ 1 & 0 & 1 \\ \hline 0 & 5 & 1 \\ 0 & 1 & 0 \\ 1 & 2 & 0 \end{bmatrix} \;;\; R \equiv [r^{\gamma\circ\alpha}_{\;\circ\beta\circ}] = \begin{bmatrix} 1 & 2 & 0 \\ 2 & 0 & 5 \\ -1 & 2 & 1 \\ \hline 0 & -1 & 0 \\ 3 & 1 & 1 \\ 2 & 0 & 0 \\ \hline -1 & 1 & 1 \\ 0 & 0 & 2 \\ 0 & 1 & 0 \end{bmatrix} \;;$$

$$S \equiv [s^{\circ\alpha\gamma}_{\beta\circ\circ}] = \begin{bmatrix} 1 & 0 & -1 \\ 2 & -1 & 1 \\ 0 & 0 & 1 \\ \hline 2 & 3 & 0 \\ 0 & 1 & 0 \\ 5 & 1 & 2 \\ \hline -1 & 2 & 0 \\ 2 & 0 & 1 \\ 1 & 0 & 0 \end{bmatrix}.$$

Some of the tensors T, U, V, W, R and S are called isomers, and will we analyzed in Chapter 5.

2. Consider the tensor $\vec{t} \in \mathbb{R}^3 \otimes \mathbb{R}^3_* \otimes \mathbb{R}^3$, such that its basis of 27 vectors is

$$\beta_1 \equiv \{\vec{e}_1 \otimes \vec{e}^{*1} \otimes \vec{e}_1, \vec{e}_1 \otimes \vec{e}^{*1} \otimes \vec{e}_2, \vec{e}_1 \otimes \vec{e}^{*1} \otimes \vec{e}_3, \vec{e}_1 \otimes \vec{e}^{*2} \otimes \vec{e}_1, \vec{e}_1 \otimes \vec{e}^{*2} \otimes \vec{e}_2,$$
$$\vec{e}_1 \otimes \vec{e}^{*2} \otimes \vec{e}_3, \vec{e}_1 \otimes \vec{e}^{*3} \otimes \vec{e}_1, \vec{e}_1 \otimes \vec{e}^{*3} \otimes \vec{e}_2, \vec{e}_1 \otimes \vec{e}^{*3} \otimes \vec{e}_3, \vec{e}_2 \otimes \vec{e}^{*1} \otimes \vec{e}_1,$$
$$\cdots, \vec{e}_2 \otimes \vec{e}^{*3} \otimes \vec{e}_3, \vec{e}_3 \otimes \vec{e}^{*1} \otimes \vec{e}_1, \cdots, \vec{e}_3 \otimes \vec{e}^{*3} \otimes \vec{e}_3, \}.$$

For the case of tensor $\vec{u} \in \mathbb{R}^3 \otimes \mathbb{R}^3 \otimes \mathbb{R}^3_*$, its basis of 27 vectors is

$$\beta_2 \equiv \{\vec{e}_1 \otimes \vec{e}_1 \otimes \vec{e}^{*1}, \vec{e}_1 \otimes \vec{e}_1 \otimes \vec{e}^{*2}, \vec{e}_1 \otimes \vec{e}_1 \otimes \vec{e}^{*3}, \vec{e}_1 \otimes \vec{e}_2 \otimes \vec{e}^{*1}, \vec{e}_1 \otimes \vec{e}_2 \otimes \vec{e}^{*2},$$
$$\vec{e}_1 \otimes \vec{e}_2 \otimes \vec{e}^{*3}, \vec{e}_1 \otimes \vec{e}_3 \otimes \vec{e}^{*1}, \vec{e}_1 \otimes \vec{e}_3 \otimes \vec{e}^{*2}, \vec{e}_1 \otimes \vec{e}_3 \otimes \vec{e}^{*3}, \vec{e}_2 \otimes \vec{e}_1 \otimes \vec{e}^{*1},$$
$$\cdots, \vec{e}_2 \otimes \vec{e}_3 \otimes \vec{e}^{*3}, \vec{e}_3 \otimes \vec{e}_1 \otimes \vec{e}^{*1}, \cdots, \vec{e}_3 \otimes \vec{e}_3 \otimes \vec{e}^{*3}\}.$$

3. Evidently $\vec{t} \neq \vec{u}$, due to the previously established reasons. However, it is true that a problem stated in a given linear space can be solved in another linear space by means of a certain isomorphism; however, the resulting solutions in the second space must be placed in the first linear space, by the inverse isomorphism. That is, the found tensor after performing index

permutations and contractions, *is not* the sought after tensor, but the corresponding one. If the order of a tensor is greater than 4, one needs to state the morphism in matrix form, since otherwise it can be impossible to retrace the steps.

□

Example 4.6 (Changes of basis of tensor and non tensor nature). In the linear space of matrices, two independent changes of basis are performed with associated matrices

$$C_1 = \frac{1}{2} \begin{bmatrix} 8 & 24 & -2 & -6 \\ 16 & 64 & -4 & -16 \\ -3 & -9 & 1 & 3 \\ -6 & -24 & 2 & 8 \end{bmatrix} \quad \text{and} \quad C_2 = \begin{bmatrix} 1 & -1 & -1 & -1 \\ 1 & 2 & -1 & -1 \\ 1 & 2 & 3 & -1 \\ 1 & 2 & 3 & 4 \end{bmatrix}.$$

Determine whether or not some of them has second-order tensor character.

Solution:

We start with the first case (matrix C_1):
We know that if $\sigma = 2 \times 2 = 4$, the relation (4.45) establishes

$$Z_4 = (P \otimes Q^t)^{-1} \equiv P^{-1} \otimes (Q^{-1})^t.$$

Assuming that

$$P^{-1} = \frac{1}{2} \begin{bmatrix} \alpha & \beta \\ \gamma & \delta \end{bmatrix} \quad \text{and} \quad (Q^{-1})^t = \begin{bmatrix} \epsilon & \eta \\ \rho & \lambda \end{bmatrix},$$

we must have $C_1 \equiv Z_4$:

$$C_1 = \frac{1}{2} \begin{bmatrix} 8 & 24 & -2 & -6 \\ 16 & 64 & -4 & -16 \\ -3 & -9 & 1 & 3 \\ -6 & -24 & 2 & 8 \end{bmatrix} = \frac{1}{2} \begin{bmatrix} \alpha & \beta \\ \gamma & \delta \end{bmatrix} \otimes \begin{bmatrix} \epsilon & \eta \\ \rho & \lambda \end{bmatrix}$$

$$= \frac{1}{2} \begin{bmatrix} \alpha \begin{bmatrix} \epsilon & \eta \\ \rho & \lambda \end{bmatrix} & \beta \begin{bmatrix} \epsilon & \eta \\ \rho & \lambda \end{bmatrix} \\ \gamma \begin{bmatrix} \epsilon & \eta \\ \rho & \lambda \end{bmatrix} & \delta \begin{bmatrix} \epsilon & \eta \\ \rho & \lambda \end{bmatrix} \end{bmatrix},$$

and taking the last block B_{22} (of elements prime "among them"), we must have

$$\delta \begin{bmatrix} \epsilon & \eta \\ \rho & \lambda \end{bmatrix} \equiv \begin{bmatrix} 1 & 3 \\ 2 & 8 \end{bmatrix},$$

which requires $\delta = 1$ and $\begin{bmatrix} \epsilon & \eta \\ \rho & \lambda \end{bmatrix} = \begin{bmatrix} 1 & 3 \\ 2 & 8 \end{bmatrix}$. Taking block B_{11}, and taking into account the previous result, we have

$$\alpha \begin{bmatrix} \epsilon & \eta \\ \rho & \lambda \end{bmatrix} \equiv \begin{bmatrix} 8 & 24 \\ 16 & 64 \end{bmatrix} \Rightarrow \alpha = 8 \text{ and } \begin{bmatrix} \epsilon & \eta \\ \rho & \lambda \end{bmatrix} = \begin{bmatrix} 1 & 3 \\ 2 & 8 \end{bmatrix}.$$

With block B_{12} we get $\beta \begin{bmatrix} \epsilon & \eta \\ \rho & \lambda \end{bmatrix} \equiv \begin{bmatrix} -2 & -6 \\ -4 & -16 \end{bmatrix} \Rightarrow \beta = -2$, and finally, block B_{21} gives the parameter $\gamma = -3$.

Hence

$$P^{-1} = \frac{1}{2} \begin{bmatrix} \alpha & \beta \\ \gamma & \delta \end{bmatrix} \equiv \frac{1}{2} \begin{bmatrix} 8 & -2 \\ -3 & 1 \end{bmatrix} \text{ and } (Q^{-1})^t \equiv \begin{bmatrix} 1 & 3 \\ 2 & 8 \end{bmatrix},$$

from which we obtain

$$P = \begin{bmatrix} 1 & 2 \\ 3 & 8 \end{bmatrix} \text{ and } Q = \frac{1}{2} \begin{bmatrix} 8 & -2 \\ -3 & 1 \end{bmatrix}.$$

Then, $Z_4^{-1} = P \otimes Q^t$. But, we check that $P \bullet Q = \begin{bmatrix} 1 & 0 \\ 0 & 1 \end{bmatrix}$, whence $P = Q^{-1}$, and then $Z_4^{-1} = Q^{-1} \otimes Q^t$, and $C_1^{-1} = Q^{-1} \otimes Q^t$, which proves that C_1 is a change-of-basis of tensor nature.

Because of the condensation formula we know that

$$\hat{T}_{4,1} = (Q^{-1} \otimes Q^t) \bullet T_{4,1} <> \hat{T} = Q^{-1}TQ,$$

that finally discovers that the change-of-basis C_1 corresponds to a similar matrix in $\mathbb{R}^{2\times 2}$.

Next we deal with the second case (matrix C_2):

Since the "matrix equivalence" ($\hat{T} = PTQ$) keeps the rank of matrices T and $\hat{T}$ (because P and Q are regular), and the equivalent equation is, by condensation, $\hat{T}_{\sigma,1} = (P \otimes Q^t) \bullet T_{\sigma,1}$, we will give a new solution to this problem.

We choose a matrix $H \in \mathbb{R}^{2\times 2}$; $H = \begin{bmatrix} 4 & -2 \\ 2 & -8 \end{bmatrix}$ with rank 2, and we change the basis by means of the matrix C_2

$$H_{\sigma,1} = C_2 \hat{H}_{\sigma,1} \Rightarrow \begin{bmatrix} 4 \\ -2 \\ 2 \\ -8 \end{bmatrix} = \begin{bmatrix} 1 & -1 & -1 & -1 \\ 1 & 2 & -1 & -1 \\ 1 & 2 & 3 & -1 \\ 1 & 2 & 3 & 4 \end{bmatrix} \begin{bmatrix} \hat{x} \\ \hat{y} \\ \hat{z} \\ \hat{t} \end{bmatrix},$$

and solving the system of equations we obtain $\hat{H}_{\sigma,1} \equiv \begin{bmatrix} \hat{x} \\ \hat{y} \\ \hat{z} \\ \hat{t} \end{bmatrix} = \begin{bmatrix} 1 \\ -2 \\ 1 \\ -2 \end{bmatrix}$ and

after condensation: $\hat{H} = \begin{bmatrix} 1 & -2 \\ 1 & -2 \end{bmatrix}$.

Since the rank of $\hat{H}$ is 1, different from the rank of H, which is 2, the matrix C_2 does not maintain the rank of *all* matrices after a change-of-basis. Then C_2 is not a change-of-basis for second-order tensors, and if we proceed as in the first case, some incompatibilities would appear during the identification process.

□

Example 4.7 (Changes of basis of non-tensor nature). In the linear space of square matrices of order 3 over the real numbers, $\mathbb{R}^{3\times3}$, a change-of-basis is performed of the associated matrix

$$
C = \begin{bmatrix}
2 & 0 & 0 & 0 & 0 & 0 & 0 & 0 & 0 \\
0 & 3 & 0 & 0 & 0 & 0 & 0 & 0 & 0 \\
0 & 0 & 1 & 0 & 0 & 0 & 0 & 0 & 0 \\
0 & 0 & 0 & 2 & 0 & 0 & 0 & 0 & 0 \\
0 & 0 & 0 & 0 & -1 & 0 & 0 & 0 & 0 \\
0 & 0 & 0 & 0 & 0 & 1 & 0 & 0 & 0 \\
0 & 0 & 0 & 0 & 0 & 0 & 2 & 0 & 0 \\
0 & 0 & 0 & 0 & 0 & 0 & 0 & -1 & 0 \\
0 & 0 & 0 & 0 & 0 & 0 & 0 & 0 & 1
\end{bmatrix}.
$$

We want to know if such change is of tensor nature for matrices of $\mathbb{R}^{3\times3}$.

Solution: we solve the problem by means of two different techniques.

First resolution method:

We choose in $\mathbb{R}^{3\times3}$, the singular matrix ($|X| = 0$):

$$
X = \begin{bmatrix}
3 & 2 & 3 \\
0 & 1 & 3 \\
2 & 1 & 1
\end{bmatrix}
$$

and then we perform a change-of-basis.
Inverting C and knowing that $\hat{X}_{\sigma,1} = C^{-1}X_{\sigma,1}$ we have

$$
\hat{X}_{9,1} = \begin{bmatrix}
1/2 & 0 & 0 & 0 & 0 & 0 & 0 & 0 & 0 \\
0 & 1/3 & 0 & 0 & 0 & 0 & 0 & 0 & 0 \\
0 & 0 & 1 & 0 & 0 & 0 & 0 & 0 & 0 \\
0 & 0 & 0 & 1/2 & 0 & 0 & 0 & 0 & 0 \\
0 & 0 & 0 & 0 & -1 & 0 & 0 & 0 & 0 \\
0 & 0 & 0 & 0 & 0 & 1 & 0 & 0 & 0 \\
0 & 0 & 0 & 0 & 0 & 0 & 1/2 & 0 & 0 \\
0 & 0 & 0 & 0 & 0 & 0 & 0 & -1 & 0 \\
0 & 0 & 0 & 0 & 0 & 0 & 0 & 0 & 1
\end{bmatrix}
\begin{bmatrix}
3 \\ 2 \\ 3 \\ 0 \\ 1 \\ 3 \\ 2 \\ 1 \\ 1
\end{bmatrix}
=
\begin{bmatrix}
3/2 \\ 2/3 \\ 3 \\ 0 \\ -1 \\ 3 \\ 1 \\ -1 \\ 1
\end{bmatrix}
$$

and condensing $\hat{X}_{9,1}$, we have

$$\hat{X} = \begin{bmatrix} 3/2 & 2/3 & 3 \\ 0 & -1 & 3 \\ 1 & -1 & 1 \end{bmatrix}.$$

Since we want to know if two regular matrices P and Q exist, such that $C = P^{-1} \otimes (Q^{-1})^t$, that is, $\hat{X}_{\sigma,1} = P^{-1} \otimes (Q^{-1})^t X$, which implies $\hat{X}_{\sigma,1} = (P \otimes Q^t)^{-1} X_{\sigma,1}$, the problem consists (due to the condensation properties) of knowing if P and Q satisfy $\hat{X} = PXQ$, that is, if $\hat{X}$ and X are "equivalent matrices". They are not, because the rank of X is 2, and the rank of $\hat{X}$ is 3, with determinants 0 and 8, respectively. Consequently, C is not a change-of-basis of a tensor nature.

Second resolution method:

We want to know if

$$C = P^{-1} \otimes (Q^{-1})^t, \tag{4.82}$$

where P and Q are of order 3 and regular.

If

$$P^{-1} \equiv \begin{bmatrix} p_{11} & p_{12} & p_{13} \\ p_{21} & p_{22} & p_{23} \\ p_{31} & p_{32} & p_{33} \end{bmatrix}$$

then

$$P^{-1} \otimes (Q^{-1})^t \equiv \begin{bmatrix} p_{11}(Q^{-1})^t & p_{12}(Q^{-1})^t & p_{13}(Q^{-1})^t \\ p_{21}(Q^{-1})^t & p_{22}(Q^{-1})^t & p_{23}(Q^{-1})^t \\ p_{31}(Q^{-1})^t & p_{32}(Q^{-1})^t & p_{33}(Q^{-1})^t \end{bmatrix},$$

and identifying with matrix C we must have

$$\begin{bmatrix} 2 & 0 & 0 \\ 0 & 3 & 0 \\ 0 & 0 & 1 \end{bmatrix} \equiv p_{11}(Q^{-1})^t; \quad \begin{bmatrix} 2 & 0 & 0 \\ 0 & -1 & 0 \\ 0 & 0 & 1 \end{bmatrix} \equiv p_{22}(Q^{-1})^t; \quad \begin{bmatrix} 2 & 0 & 0 \\ 0 & -1 & 0 \\ 0 & 0 & 1 \end{bmatrix} \equiv p_{33}(Q^{-1})^t \tag{4.83}$$

and obtaining $(Q^{-1})^t$ from the first equation in (4.83) and substituting it into the second yields

$$(Q^{-1})^t = \frac{1}{p_{11}} \begin{bmatrix} 2 & 0 & 0 \\ 0 & 3 & 1 \\ 0 & 0 & 1 \end{bmatrix} \Rightarrow \begin{bmatrix} 2 & 0 & 0 \\ 0 & -1 & 0 \\ 0 & 0 & 1 \end{bmatrix} \equiv \frac{p_{22}}{p_{11}} \begin{bmatrix} 2 & 0 & 0 \\ 0 & 3 & 0 \\ 0 & 0 & 1 \end{bmatrix}.$$

Since no rational number satisfies $2 = \frac{p_{22}}{p_{11}} \cdot 2$ and $-1 = \frac{p_{22}}{p_{11}} \cdot 3$, identification (4.82) is impossible. Thus, *C is not of tensor nature.* □

4.9 Exercises

4.1. Consider the absolute tensor space $U^3 \otimes V^2 \otimes W^2(\mathbb{R})$.

In the first linear space $U^3(\mathbb{R})$ we perform the change-of-basis

$$\vec{\hat{e}} = 2\vec{e}_1 - \vec{e}_2 + \vec{e}_3; \quad \vec{\hat{e}}_2 = \vec{e}_1 + 5\vec{e}_3; \quad \vec{\hat{e}}_3 = \vec{e}_1 + 2\vec{e}_2 + 3\vec{e}_3.$$

In the second space $V^2(\mathbb{R})$ a change-of-basis is performed such that it satisfies the relations

$$\vec{\epsilon}_1 = 2\vec{\hat{\epsilon}}_1 - \vec{\hat{\epsilon}}_2; \quad \vec{\epsilon}_2 = -3\vec{\hat{\epsilon}}_1 + 2\vec{\hat{\epsilon}}_2.$$

In the third space $W^2(\mathbb{R})$ we perform the change-of-basis given by the matrix expression

$$[\vec{\hat{\eta}}_1 \quad \vec{\hat{\eta}}_2] = [\vec{\eta}_1 \quad \eta_2] \begin{bmatrix} 2 & 3 \\ 1 & 2 \end{bmatrix}.$$

A tensor T in the initial basis of the tensor space is given by

$$T = 2\vec{e}_1 \otimes \vec{\epsilon}_1 \otimes \eta_1 - 5\vec{e}_2 \otimes \vec{\epsilon}_2 \otimes \eta_1 + 7\vec{e}_3 \otimes \vec{\epsilon}_1 \otimes \eta_1 + 6\vec{e}_3 \otimes \vec{\epsilon}_2 \otimes \eta_2.$$

1. Find the new tensor components in tensor notation, using the general expression for a change-of-basis (Formula (4.6)).
2. Find the extended matrix T_σ of the tensor and determine the new components $\hat{T}_\sigma$, using the corresponding matrix Z_σ according to Formula (4.11) (direct method).
3. Condense $\hat{T}_\sigma$, determining $\hat{T}$ by means of its matrix representation, as a tensor of order $r = 3$.
4. Solve the previous questions using the computer.

4.2. In the linear space $V^3 \otimes V^3_* \otimes V^3(\mathbb{R})$ a tensor T, is given by its matrix representation

$$[t^{\alpha \circ \gamma}_{\circ \beta \circ}] = \begin{bmatrix} -1 & 0 & 0 \\ 0 & 3 & 2 \\ 1 & 2 & -1 \\ \hline 3 & 5 & 1 \\ 0 & 1 & 0 \\ 2 & 2 & -2 \\ \hline 1 & 1 & 4 \\ 0 & 0 & 1 \\ 2 & 0 & 1 \end{bmatrix}.$$

A change-of-basis is performed in the linear space $V^3(\mathbb{R})$ given by

$$[\vec{\hat{e}}_i] = [\vec{e}_\alpha] \begin{bmatrix} 1 & 1 & 0 \\ 0 & 0 & 3 \\ -1 & 0 & -1 \end{bmatrix}.$$

1. Calculate $\hat{T}$, using the classical methods in Formula (4.56).
2. Given the components of the tensor $Z = \mathcal{C}\binom{\alpha}{\beta}(T)$ obtained by contracting the indices 1 and 2 of T.

3. Perform the change-of-basis on the tensor T, giving $\hat{T}_\sigma$, extended in the new basis.
4. Give $\hat{T}$, by condensation of the previous tensor.
5. Give the new components of tensor $\hat{Z}$, contracting before the change-of-basis.
6. Give the new components of tensor $\hat{Z}$, contracting after the change-of-basis.
7. Solve all the previous questions but the first, using the computer.

4.3. Consider the homogeneous tensor $T \in V_*^3 \otimes V^3 \otimes V^3 \otimes V^3(\mathbb{R})$ of fourth order, the components of which in the basis $\{\vec{e}_\alpha\}$ associated with $V^3(\mathbb{R})$ are

$$[t^{\circ\beta\gamma\delta}_{\alpha\circ\circ\circ}] = \begin{bmatrix} 1 & 0 & 1 & 1 & 2 & -1 & 3 & 0 & 2 \\ 1 & 2 & 3 & -1 & 0 & 0 & 0 & -1 & 1 \\ -2 & 1 & 0 & 1 & 0 & 1 & 1 & 0 & 1 \\ 5 & 0 & 1 & -2 & -1 & 5 & 4 & 0 & 2 \\ 1 & 2 & 0 & 1 & 0 & 1 & 1 & 1 & 1 \\ 1 & -3 & 1 & 1 & 3 & 1 & 1 & 1 & 1 \\ 2 & 0 & -1 & 3 & -1 & 0 & 1 & -1 & 2 \\ 1 & 0 & 2 & 1 & 5 & 2 & 4 & 2 & 2 \\ 0 & 1 & 0 & 0 & 6 & 1 & 1 & 0 & 1 \end{bmatrix},$$

where α is the block row, β is the block column, γ is the row of each block and δ is the column of each block.

Consider the second-order tensor $U \in V^3 \otimes V^3(\mathbb{R})$, the components of which in the basis $\{\vec{e}_i\}$ of $V^3(\mathbb{R})$ are

$$[u^{mn}_{\circ\circ}] \equiv \begin{bmatrix} 3 & -2 & 0 \\ 1 & 1 & 1 \\ -2 & 1 & 0 \end{bmatrix} \text{ where } [\vec{\hat{e}}_1 \quad \vec{\hat{e}}_3 \quad \vec{\hat{e}}_3] = [\vec{e}_1 \quad \vec{e}_3 \quad \vec{e}_3] \begin{bmatrix} 1 & 1 & 0 \\ 1 & 0 & 1 \\ 0 & 1 & 1 \end{bmatrix}.$$

1. a) Calculate the components of the tensor $\hat{T}$ in the new basis using the classical methods given in Formula (4.62). Give the matrix representation of $[t^{\circ jk\ell}_{i\circ\circ\circ}]$.

   b) Solve the previous question by the direct method, determining $\hat{T}_\sigma$ and proceeding to its condensation.

   c) Determine the tensor U in the initial basis, by the direct method, giving its matrix representation $[u^{\mu\nu}_{\circ\circ}]$.

   d) Give tensors A, B and D, resulting from the contractions $A = \mathcal{C}\binom{\beta}{\alpha}(T)$, $B = \mathcal{C}\binom{\gamma}{\alpha}(T)$ and $D = \mathcal{C}\binom{\delta}{\alpha}(T)$ in the initial basis.

   e) Give tensors $\hat{A}, \hat{B}$ and $\hat{D}$ resulting from the previous contractions but on $\hat{T}$.

2. Solve the previous questions using the computer.

4.4. Consider a second-order homogeneous tensor $T \in V^3 \otimes V^3_*(\mathbb{R})$, the matrix representation of which, in the basis $\{\vec{e}_\alpha\}$ de $V^3(\mathbb{R})$ is

$$[t^{\alpha\circ}_{\circ\beta}] = \begin{bmatrix} 2 & 2 & 0 \\ 2 & 2 & 0 \\ 0 & 0 & 1 \end{bmatrix}.$$

1. Give the matrix representation of $\hat{T}$, in each of the new bases and permutations of $\{\vec{e}_\alpha\}$.
2. Determine a new basis $\{\vec{e}_i\}$ of $V^3(\mathbb{R})$, such that in it, the matrix representation of $\hat{T}$ be a diagonal matrix of increasing scalars: $\hat{t}^{1\circ}_{\circ 1} \leq \hat{t}^{2\circ}_{\circ 2} \leq \hat{t}^{3\circ}_{\circ 3}$.

5

Homogeneous Tensor Algebra: Tensor Homomorphisms

5.1 Introduction

The chapter starts by presenting the main theorem on tensor contraction, which ensures that a contraction of a tensor product when applied to indices of different valency leads to a tensor.

It continues by presenting the contracted tensor products as homomorphisms and applies them to different tensor products as particular cases. Some tensor criteria motivated by the contraction are also discussed, including the well-known quotient law criterion.

Next, a detailed study of the matrix representation of the permutation tensors and some simple and double contracted homomorphisms is performed.

The chapter ends with a novel theory of eigentensors and generalized multilinear mappings.

5.2 Main theorem on tensor contraction

Though in Section 3.5 the contraction of tensor products has already been mentioned, and in Theorem 3.1 any contraction of mixed tensors has been examined from the homomorphism point of view, that is, of linear mappings of a primary linear space (tensor space) into another secondary linear space, one can have doubts about whether or not the resulting "range" space would be a simple linear space, or would also be a tensor space.

Fortunately, this doubt is positively resolved, because the "homomorphic" image of a tensor space is another tensor space.

Remark: the word "homomorphic" *always* has the sense of a mixed tensor "contracted from two indices of different valency".

Next, we prove this property with the required emphasis, and later it will be enunciated as a theorem.

Consider a mixed homogeneous tensor $\vec{t} \in \left(\overset{3}{\underset{1}{\otimes}} V^n \right) \otimes \left(\overset{2}{\underset{1}{\otimes}} V^n_* \right) (K)$ of order $r = 5$,

$$\vec{t} = t^{\alpha\beta\gamma\circ\circ}_{\circ\circ\circ\lambda\mu} \vec{e}_\alpha \otimes \vec{e}_\beta \otimes \vec{e}_\gamma \otimes \vec{e}^{*\lambda} \otimes \vec{e}^{*\mu} \tag{5.1}$$

and denote by $S(\alpha, \gamma, \mu)$ the "system of scalars" resulting from the contraction of indices 2 and 4 of different valency (β and λ):

$$s(\alpha, \gamma, \mu) = t^{\alpha\theta\gamma\circ\circ}_{\circ\circ\circ\theta\mu} . \tag{5.2}$$

In detail, we have

$$s(\alpha, \gamma, \mu) = t^{\alpha 1 \gamma\circ\circ}_{\circ\circ\circ 1 \mu} + t^{\alpha 2 \gamma\circ\circ}_{\circ\circ\circ 2 \mu} + \cdots + t^{\alpha n \gamma\circ\circ}_{\circ\circ\circ n \mu} . \tag{5.3}$$

The system $S(\alpha, \gamma, \mu)$ is called a system of scalars because one cannot anticipate if it is a tensor. The power of the set $S(\alpha, \gamma, \mu)$ is n^3, because we have three free indices.

Next, we perform a change-of-basis in the $\left(\overset{3}{\underset{1}{\otimes}} V^n \right) \otimes \left(\overset{2}{\underset{1}{\otimes}} V^n_* \right) (K)$ tensor space. Since its vectors are homogeneous tensors, we have

$$t^{ijk\circ\circ}_{\circ\circ\circ\ell m} = t^{\alpha\beta\gamma\circ\circ}_{\circ\circ\circ\lambda\mu} \gamma^{i\circ}_{\circ\alpha} \gamma^{j\circ}_{\circ\beta} \gamma^{k\circ}_{\circ\gamma} c^{\circ\lambda}_{\ell\circ} c^{\circ\mu}_{m\circ}. \tag{5.4}$$

The indices (j, ℓ) are contracted. Preparing Expression (5.4) and calling the set of scalars in the left-hand side $s(i, k, m)$, we get

$$t^{ijk\circ\circ}_{\circ\circ\circ\ell m} = t^{\alpha\beta\gamma\circ\circ}_{\circ\circ\circ\lambda\mu} \gamma^{i\circ}_{\circ\alpha} \gamma^{k\circ}_{\circ\gamma} (\gamma^{j\circ}_{\circ\beta} c^{\circ\lambda}_{\ell\circ}) c^{\circ\mu}_{m\circ} \tag{5.5}$$

$$s(i, k, m) = t^{\alpha\beta\gamma\circ\circ}_{\circ\circ\circ\lambda\mu} \gamma^{i\circ}_{\circ\alpha} \gamma^{k\circ}_{\circ\gamma} (\gamma^{x\circ}_{\circ\beta} c^{\circ\lambda}_{x\circ}) c^{\circ\mu}_{m\circ}. \tag{5.6}$$

The expression $(\gamma^{x\circ}_{\circ\beta} c^{\circ\lambda}_{x\circ})$ is the "product" of matrices $C^{-1} \odot C^t$ but executed by "multiplying row by row" (not by column, due to the position of x); but this is the same as $C^{-1} \cdot C = I_n$. So, that

$$\gamma^{x\circ}_{\circ\beta} c^{\circ\lambda}_{x\circ} = \delta^{\circ\lambda}_{\beta\circ}, \tag{5.7}$$

and replacing (5.7) into (5.6) we obtain

$$s(i, k, m) = (t^{\alpha\beta\gamma\circ\circ}_{\circ\circ\circ\lambda\mu} \delta^{\circ\lambda}_{\beta\circ}) \gamma^{i\circ}_{\circ\alpha} \gamma^{k\circ}_{\circ\gamma} c^{\circ\mu}_{m\circ},$$

where the product in parentheses is the contraction of (β, λ)

$$s(i, k, m) = t^{\alpha\theta\gamma\circ\circ}_{\circ\circ\circ\theta\mu} \gamma^{i\circ}_{\circ\alpha} \gamma^{k\circ}_{\circ\gamma} c^{\circ\mu}_{m\circ} \tag{5.8}$$

and, on account of (5.2), the previous expression can be written as

$$s(i, k, m) = s(\alpha, \gamma, \mu)\gamma_{o\alpha}^{i\,o}\gamma_{o\gamma}^{ko}c_{m\,o}^{o\,\mu}, \tag{5.9}$$

which declares that the system of scalars $S(\alpha, \gamma, \mu)$ satisfies the tensor criteria, that is, it is a tensor. Whence

$$s(\alpha, \gamma, \mu) \equiv s_{o\,o\,\mu}^{\alpha\gamma\,o}.$$

This proof can be repeated over other two indices with *different valency*. We leave this for the reader to do.

If, by error, we were to choose two indices with the same valency, when reaching Expression (5.6) products of the type $(\gamma_{o\beta}^{x\,o}\gamma_{o\lambda}^{x\,o})$ or $(c_{x\,o}^{o\beta}c_{x\,o}^{o\lambda})$ would appear that *are not* the Kronecker delta, making the proof invalid. We understand that this expression can be generalized to tensors of order superior to $r = 5$, and proceed to state the "tensor contraction" general theorem.

Theorem 5.1 (Fundamental theorem of tensor contraction). *The contraction with respect to indices of different valency in mixed tensors of order* r, *is a sufficient condition for obtaining another homogeneous tensor of order* $(r - 2)$. □

5.3 The contracted tensor product and tensor homomorphisms

In Section 3.4 we have dealt with tensor product of tensors, and in Section 3.5 the contracted tensor product concept was defined. Since in that definition the conditions of "tensor contraction" are satisfied, Theorem 3.1 guarantees that the *contracted tensor products* can be considered as simple homomorphisms (Formula (3.16)), that transform tensors from a tensor space into tensors of another space by the action of a *contracted tensor homomorphism.*

This point of view will be exploited at the end of this chapter, more precisely, on tensors of simple order, and it will be executed using the matrix expression

$$T_{\sigma'} = H_{n^{r-2},n^r} \bullet T_\sigma; \quad \text{with } \sigma = n^r; \quad \sigma' = n^{r-2}. \tag{5.10}$$

Nevertheless, before ending, we want to point out the analytical representation of the contracted tensor product, in the classic mode.

Given the tensors $\vec{t} = t_{o\,o\,\gamma}^{\alpha\beta\,o}\vec{e}_\alpha \otimes \vec{e}_\beta \otimes \vec{e}^{*\gamma}$ and $\vec{v} = v_{\lambda\mu}^{o\,o}\vec{e}^{*\lambda} \otimes \vec{e}^{*\mu}$, we look for the *contracted tensor product* tensor $\vec{p} = C\left(_{\lambda}^{\alpha}\right)(\vec{t} \otimes \vec{v})$, with $\vec{p} = p_{o\,\gamma\mu}^{\beta\,o\,o}\vec{e}_\beta \otimes \vec{e}^{*\gamma} \otimes \vec{e}^{*\mu}$.

This can be done in two different forms:

1. We obtain the tensor product tensor

$$\vec{w} = \vec{t} \otimes \vec{v} = w^{\alpha\beta\circ\circ\circ}_{\circ\circ\gamma\lambda\mu}\vec{e}_\alpha \otimes \vec{e}_\beta \otimes \vec{e}^{*\gamma} \otimes \vec{e}^{*\lambda} \otimes \vec{e}^{*\mu} \tag{5.11}$$

with the condition

$$w^{\alpha\beta\circ\circ\circ}_{\circ\circ\gamma\lambda\mu} = t^{\alpha\beta\circ}_{\circ\circ\gamma} \cdot v^{\circ\circ}_{\lambda\mu} \tag{5.12}$$

and then we contract

$$\vec{p} = C\begin{pmatrix}\alpha\\\lambda\end{pmatrix}\vec{w} = w^{\theta\beta\circ\circ\circ}_{\circ\circ\gamma\theta\mu}\vec{e}_\beta \otimes \vec{e}^{*\gamma} \otimes \vec{e}^{*\mu},$$

where we also have

$$w^{\theta\beta\circ\circ\circ}_{\circ\circ\gamma\theta\mu} = w^{\alpha\beta\circ\circ\circ}_{\circ\circ\gamma\lambda\mu} \cdot \delta^{\circ\lambda}_{\alpha\circ}, \tag{5.13}$$

where $\delta^{\circ\lambda}_{\alpha\circ}$ is the Kronecker tensor.

2. The second form is used by certain authors, who prefer a direct execution of the product and the contraction *simultaneously*, based on matrix representations:

$$p^{\beta\circ\circ}_{\circ\gamma\mu} = (t^{\alpha\beta\circ}_{\circ\circ\gamma}) \cdot \delta^{\circ\lambda}_{\alpha\circ} \cdot (v^{\circ\circ}_{\lambda\mu}). \tag{5.14}$$

Evidently, (5.14) is the result of replacing (5.12) into (5.13), because $\delta^{\circ\lambda}_{\alpha\circ} \equiv \delta^{\lambda\circ}_{\circ\alpha}$ is symmetric, and then, both methods lead to the same result.

Example 5.1 (Matrix associated with an operator). Consider two linear spaces $V^n(K)$ and $W^p(K)$. In the first space we consider a linear operator T_1 with associated matrix A_n in the basis $\{\vec{e}_i\}$ of the given space. Similarly, another linear operator T_2, with associated matrix B_p, transforms the vectors of $W^p(K)$ in the basis $\{\vec{\epsilon}_j\}$. We look for the matrix associated with the operator T defined to transform vectors in the tensor space $V \otimes W(K)$, in such way that

$$T(\vec{V} \otimes \vec{W}) = T_1(\vec{V}) \otimes T_2(\vec{W}).$$

Will $A_n \otimes B_p$ be the T operator matrix? That is, will "the tensor product homomorphism" be the homomorphisms' tensor product?

Solution: For a homomorphism to be correctly defined we need to know the image vectors of all basic vectors that will constitute the columns of the operator associated matrix.

The basis of our tensor space is $\beta = \{\vec{e}_i \otimes \vec{e}_j\}$, with $i = 1, 2, \ldots, n$ and $j = 1, 2, \ldots, p$.

The sought after matrix T is a square matrix of $n \times p$ rows and columns, because in the basis there exist $n \times p$ vectors the images of which are to be studied.

Applying the formula proposed in the statement to an arbitrary basic vector, we have

$$T(\vec{e}_i \otimes \vec{e}_j) = T_1(\vec{e}_1) \otimes T_2(\vec{e}_j)$$

$$= \left(\begin{bmatrix} \vec{e}_1 & \vec{e}_2 & \cdots & \vec{e}_n \end{bmatrix} \begin{bmatrix} a_{1i} \\ a_{2i} \\ \vdots \\ a_{ni} \end{bmatrix} \right) \otimes \left(\begin{bmatrix} \vec{e}_1 & \vec{e}_2 & \cdots & \vec{e}_p \end{bmatrix} \begin{bmatrix} b_{1j} \\ b_{2j} \\ \vdots \\ b_{pj} \end{bmatrix} \right)^t$$

$$= \begin{bmatrix} \vec{e}_1 & \vec{e}_2 & \cdots & \vec{e}_n \end{bmatrix} \begin{bmatrix} a_{1i} \\ a_{2i} \\ \vdots \\ a_{ni} \end{bmatrix} \otimes \begin{bmatrix} b_{1j} & b_{2j} & \cdots & b_{pj} \end{bmatrix} \begin{bmatrix} \vec{e}_1 \\ \vec{e}_2 \\ \vdots \\ \vec{e}_p \end{bmatrix}$$

$$= \left(\begin{bmatrix} \vec{e}_1 & \vec{e}_2 & \cdots & \vec{e}_n \end{bmatrix} \begin{bmatrix} a_{1i}b_{1j} & a_{1i}b_{2j} & \cdots & a_{1i}b_{pj} \\ a_{2i}b_{1j} & a_{2i}b_{2j} & \cdots & a_{2i}b_{pj} \\ \cdots & \cdots & \cdots & \cdots \\ a_{ni}b_{1j} & a_{ni}b_{2j} & \cdots & a_{ni}b_{pj} \end{bmatrix} \right) \otimes \begin{bmatrix} \vec{e}_1 \\ \vec{e}_2 \\ \vdots \\ \vec{e}_p \end{bmatrix}$$

$$= a_{1i}b_{1j}\vec{e}_1 \otimes \vec{e}_1 + a_{1i}b_{2j}\vec{e}_1 \otimes \vec{e}_2 + \cdots + a_{hi}b_{kj}\vec{e}_h \otimes \vec{e}_k$$
$$+ \cdots + a_{ni}b_{pj}\vec{e}_n \otimes \vec{e}_p$$

with $h = 1, 2, \ldots, n$; $k = 1, 2, \ldots, p$.

Assigning now values to the indices (i, j), according to the axiomatic ordering criterion for the basis $\mathcal{B} = \{\vec{e}_i \otimes \vec{e}_j\}$ and placing the image vectors in consecutive columns, the matrix $T_{n \times p}$ is obtained, which is the solution to the problem, and the columns of which correspond to

$$T(\vec{e}_1 \otimes \vec{e}_1) \quad T(\vec{e}_1 \otimes \vec{e}_2) \quad \cdots \quad T(\vec{e}_h \otimes \vec{e}_k) \quad \cdots \quad T(\vec{e}_n \otimes \vec{e}_p)$$

$$T = \begin{bmatrix} a_{11}b_{11} & a_{11}b_{12} & \cdots & a_{1h}b_{1k} & \cdots & a_{1n}b_{1p} \\ \vdots & \vdots & \vdots & \vdots & \vdots & \vdots \\ a_{21}b_{11} & a_{21}b_{12} & \cdots & a_{2h}b_{1k} & \cdots & a_{2n}b_{1p} \\ \vdots & \vdots & \vdots & \vdots & \vdots & \vdots \\ a_{h1}b_{11} & a_{h1}b_{12} & \cdots & a_{hh}b_{1k} & \cdots & a_{hn}b_{1p} \\ \vdots & \vdots & \vdots & \vdots & \vdots & \vdots \\ a_{n1}b_{11} & a_{n1}b_{12} & \cdots & a_{nh}b_{1k} & \cdots & a_{nn}b_{1p} \\ \vdots & \vdots & \vdots & \vdots & \vdots & \vdots \end{bmatrix},$$

a square matrix of order $n \times p$.

Assigning particular values to n and p (for example $n = 2; p = 3$) we immediately detect the following block construction:

$$T = \begin{bmatrix} a_{11}B & \cdots & a_{1h}B & \cdots & a_{1n}B \\ \cdots & \cdots & \cdots & \cdots & \cdots \\ a_{h1}B & \cdots & a_{hh}B & \cdots & a_{hn}B \\ \cdots & \cdots & \cdots & \cdots & \cdots \\ a_{n1}B & \cdots & a_{nh}B & \cdots & a_{nn}B \end{bmatrix} = A \otimes B.$$

The conclusion is that the proposed theorem in our statement: "the tensor product homomorphism (T) is the tensor product of the given homomorphisms $(T_1 \otimes T_2)$" is correct.   □

Example 5.2 (Change of basis). In the geometric affine ordinary space $E^2(\mathbb{R})$ we consider two bases: the initial basis of the unit classic vectors of a rectangular system XOY (on the OX axis and on the OY axis), and the new basis of the unit vectors on the OX axis and on the bisectrix of the XOY quadrant.

The new unit basic vectors $||\hat{\vec{e}}_i||$ referred to the initial basic vectors $||\vec{e}_\alpha||$, are

$$\hat{\vec{e}}_1 = \vec{e}_1; \quad \hat{\vec{e}}_2 = \frac{\sqrt{2}}{2}\vec{e}_1 + \frac{\sqrt{2}}{2}\vec{e}_2.$$

Determine the new components as a function of the initial ones in the following cases:

1. For a tensor of first order, i.e. the vector v^α.
2. For a mixed tensor of second order, i.e. the matrix $t^{\alpha\,\circ}_{\circ\,\beta}$.
3. For a mixed tensor of third order, $t^{\alpha\,\circ\,\gamma}_{\circ\,\beta\,\circ}$.
4. Solve the second question using the homomorphism (contracted product) $y^\alpha_\circ = t^{\alpha\,\circ}_{\circ\,\theta}x^\theta_\circ$.

Solution:

The change-of-basis can be written in matrix form as

$$||\hat{\vec{e}}_i|| = ||\vec{e}_\alpha||C \rightarrow [\hat{\vec{e}}_1 \quad \hat{\vec{e}}_2] = [\vec{e}_1 \quad \vec{e}_2]\begin{bmatrix} 1 & \frac{\sqrt{2}}{2} \\ 0 & \frac{\sqrt{2}}{2} \end{bmatrix},$$

and then

$$C = \begin{bmatrix} 1 & \frac{\sqrt{2}}{2} \\ 0 & \frac{\sqrt{2}}{2} \end{bmatrix}; \quad C^{-1} = \begin{bmatrix} 1 & -1 \\ 0 & \sqrt{2} \end{bmatrix}; \quad C^t = \begin{bmatrix} 1 & 0 \\ \frac{\sqrt{2}}{2} & \frac{\sqrt{2}}{2} \end{bmatrix}.$$

1. The tensor analytical equation of the vector is $v^i_\circ = v^\alpha_\circ\gamma^{i\,\circ}_{\circ\,\alpha}$, and in matrix form

$$[v^i_\circ] = [\gamma^{i\,\circ}_{\circ\,\alpha}][v^\alpha_\circ] \rightarrow \begin{bmatrix} \hat{v}^1 \\ \hat{v}^2 \end{bmatrix} = C^{-1}\begin{bmatrix} v^1 \\ v^2 \end{bmatrix} \tag{5.15}$$

$$\begin{bmatrix} \hat{v}^1 \\ \hat{v}^2 \end{bmatrix} = \begin{bmatrix} 1 & -1 \\ 0 & \sqrt{2} \end{bmatrix}\begin{bmatrix} v^1 \\ v^2 \end{bmatrix} = \begin{bmatrix} v^1 - v^2 \\ \sqrt{2}v^2 \end{bmatrix}.$$

2. The tensor analytical equation of the vector is $t^{i\,\circ}_{\circ\,j} = t^{\alpha\,\circ}_{\circ\,\beta}\gamma^{i\,\circ}_{\circ\,\alpha}c^{\circ\,\beta}_{j\,\circ}$ (classic matrix method), and in matrix form

$$[t^{i\,\circ}_{\circ\,j}] = [\gamma^{i\,\circ}_{\circ\,\alpha}][t^{\alpha\,\circ}_{\circ\,\beta}][c^{\beta\,\circ}_{\circ\,j}],$$

that is,

$$\begin{bmatrix} \hat{t}^{1o}_{o1} & \hat{t}^{1o}_{o2} \\ \hat{t}^{2o}_{o1} & \hat{t}^{2o}_{o2} \end{bmatrix} = C^{-1} \begin{bmatrix} t^{1o}_{o1} & t^{1o}_{o2} \\ t^{2o}_{o1} & t^{2o}_{o2} \end{bmatrix} (C^t)^t = \begin{bmatrix} 1 & -1 \\ 0 & \sqrt{2} \end{bmatrix} \begin{bmatrix} t^{1o}_{o1} & t^{1o}_{o2} \\ t^{2o}_{o1} & t^{2o}_{o2} \end{bmatrix} \begin{bmatrix} 1 & \frac{\sqrt{2}}{2} \\ 0 & \frac{\sqrt{2}}{2} \end{bmatrix}$$

$$= \begin{bmatrix} (t^{1o}_{o1} - t^{2o}_{o1}) & \frac{\sqrt{2}}{2}[(t^{1o}_{o1} - t^{2o}_{o1}) + (t^{1o}_{o2} - t^{2o}_{o2})] \\ \sqrt{2}t^{2o}_{o1} & (t^{2o}_{o1} + t^{2o}_{o2}) \end{bmatrix}.$$

3. The tensor analytical equation of the vector (direct method) is

$$t^{iok}_{ojo} = t^{\alpha o \gamma}_{o\beta o} \gamma^{io}_{o\alpha} c^{o\beta}_{jo} \gamma^{ko}_{o\gamma}; \quad \sigma = n^\tau = 2^3 = 8.$$

and its "extended" matrix expression

$$\hat{T}_{\sigma,1} = Z^{-1}_{\sigma,\sigma} T_{\sigma,1} \rightarrow \hat{T}_{8,1} = (C^{-1} \otimes C^t \otimes C^{-1}) \bullet T_{8,1}$$

$$Z^{-1}_{\sigma} = C^{-1} \otimes C^t \otimes C^{-1} = \begin{bmatrix} 1 & -1 \\ 0 & \sqrt{2} \end{bmatrix} \otimes \begin{bmatrix} 1 & 0 \\ \frac{\sqrt{2}}{2} & \frac{\sqrt{2}}{2} \end{bmatrix} \otimes \begin{bmatrix} 1 & -1 \\ 0 & \sqrt{2} \end{bmatrix};$$

$$\hat{T}_{8,1} = \begin{bmatrix} \hat{t}^{1o1}_{o1o} \\ \hat{t}^{1o2}_{o1o} \\ \hat{t}^{1o1}_{o2o} \\ \hat{t}^{1o2}_{o2o} \\ \hat{t}^{2o1}_{o1o} \\ \hat{t}^{2o2}_{o1o} \\ \hat{t}^{2o1}_{o2o} \\ \hat{t}^{2o2}_{o2o} \end{bmatrix} = Z^{-1}_{\sigma,\sigma} \bullet T_{8,1}$$

$$= \begin{bmatrix} 1 & -1 & 0 & 0 & -1 & 1 & 0 & 0 \\ 0 & \sqrt{2} & 0 & 0 & 0 & -\sqrt{2} & 0 & 0 \\ \frac{\sqrt{2}}{2} & -\frac{\sqrt{2}}{2} & \frac{\sqrt{2}}{2} & -\frac{\sqrt{2}}{2} & -\frac{\sqrt{2}}{2} & \frac{\sqrt{2}}{2} & -\frac{\sqrt{2}}{2} & \frac{\sqrt{2}}{2} \\ 0 & 1 & 0 & 1 & 0 & -1 & 0 & -1 \\ 0 & 0 & 0 & 0 & \sqrt{2} & -\sqrt{2} & 0 & 0 \\ 0 & 0 & 0 & 0 & 0 & 2 & 0 & 0 \\ 0 & 0 & 0 & 0 & 1 & -1 & 1 & -1 \\ 0 & 0 & 0 & 0 & 0 & \sqrt{2} & 0 & \sqrt{2} \end{bmatrix} \bullet \begin{bmatrix} t^{1o1}_{o1o} \\ t^{1o2}_{o1o} \\ t^{1o1}_{o2o} \\ t^{1o2}_{o2o} \\ t^{2o1}_{o1o} \\ t^{2o2}_{o1o} \\ t^{2o1}_{o2o} \\ t^{2o2}_{o2o} \end{bmatrix},$$

that is,

$$\hat{t}^{1o1}_{o1o} = t^{1o1}_{o1o} - t^{1o2}_{o1o} - t^{2o1}_{o1o} + t^{2o2}_{o1o}$$

$$\hat{t}^{1o2}_{o1o} = \sqrt{2}t^{1o2}_{o1o} - \sqrt{2}t^{2o2}_{o1o}$$

$$\hat{t}^{1o1}_{o2o} = \frac{\sqrt{2}}{2}(t^{1o1}_{o1o} - t^{1o2}_{o1o} + t^{1o1}_{o2o} - t^{1o2}_{o2o} - t^{2o1}_{o1o} + t^{2o2}_{o1o} - t^{2o1}_{o2o} + t^{2o2}_{o2o})$$

$$\hat{t}^{1o2}_{o2o} = t^{1o2}_{o1o} + t^{1o2}_{o2o} - t^{2o2}_{o1o} - t^{2o2}_{o2o}$$

$$\hat{t}^{2o1}_{o1o} = \sqrt{2}t^{2o1}_{o1o} - \sqrt{2}t^{2o2}_{o1o}$$

$$\hat{t}^{2o2}_{o1o} = 2t^{2o2}_{o1o}$$

$$\hat{t}^{2o1}_{o2o} = t^{2o1}_{o1o} - t^{2o2}_{o1o} + t^{2o1}_{o2o} - t^{2o2}_{o2o}$$

$$\hat{t}^{2o2}_{o1o} = \sqrt{2}t^{2o2}_{o1o} + \sqrt{2}t^{2o2}_{o2o}.$$

4. The given tensor homomorphism can be interpreted in matrix form as
In the initial basis $\{\vec{e}_\alpha\}$:

$$\begin{bmatrix} y^1 \\ y^2 \end{bmatrix} = \begin{bmatrix} t^{1o}_{o1} & t^{1o}_{o2} \\ t^{2o}_{o1} & t^{2o}_{o2} \end{bmatrix} \begin{bmatrix} x^1 \\ x^2 \end{bmatrix}, \tag{5.16}$$

and in the new basis $\{\hat{\vec{e}}_i\}$:

$$\begin{bmatrix} \hat{y}^1 \\ \hat{y}^2 \end{bmatrix} = \begin{bmatrix} \hat{t}^{1o}_{o1} & \hat{t}^{1o}_{o2} \\ \hat{t}^{2o}_{o1} & \hat{t}^{2o}_{o2} \end{bmatrix} \begin{bmatrix} \hat{x}^1 \\ \hat{x}^2 \end{bmatrix}. \tag{5.17}$$

Applying the relation (5.15) to matrices X and Y, we have

$$\begin{bmatrix} \hat{x}^1 \\ \hat{x}^2 \end{bmatrix} = C^{-1} \begin{bmatrix} x^1 \\ x^2 \end{bmatrix}, \tag{5.18}$$

and

$$\begin{bmatrix} \hat{y}^1 \\ \hat{y}^2 \end{bmatrix} = C^{-1} \begin{bmatrix} y^1 \\ y^2 \end{bmatrix}, \tag{5.19}$$

and substituting (5.18) and (5.19) into (5.17), we get

$$C^{-1} \begin{bmatrix} y^1 \\ y^2 \end{bmatrix} = \begin{bmatrix} \hat{t}^{1o}_{o1} & \hat{t}^{1o}_{o2} \\ \hat{t}^{2o}_{o1} & \hat{t}^{2o}_{o2} \end{bmatrix} C^{-1} \begin{bmatrix} x^1 \\ x^2 \end{bmatrix}$$

and substituting into the left-hand side of (5.16) the result is

$$C^{-1}\begin{bmatrix} t^{1\circ}_{\circ 1} & t^{1\circ}_{\circ 2} \\ t^{2\circ}_{\circ 1} & t^{2\circ}_{\circ 2} \end{bmatrix}\begin{bmatrix} x^1 \\ x^2 \end{bmatrix} = \begin{bmatrix} \hat{t}^{1\circ}_{\circ 1} & \hat{t}^{1\circ}_{\circ 2} \\ \hat{t}^{2\circ}_{\circ 1} & \hat{t}^{2\circ}_{\circ 2} \end{bmatrix}C^{-1}\begin{bmatrix} x^1 \\ x^2 \end{bmatrix},$$

and for this to be valid for any matrix X, we must have

$$C^{-1}\begin{bmatrix} t^{1\circ}_{\circ 1} & t^{1\circ}_{\circ 2} \\ t^{2\circ}_{\circ 1} & t^{2\circ}_{\circ 2} \end{bmatrix} = \begin{bmatrix} \hat{t}^{1\circ}_{\circ 1} & \hat{t}^{1\circ}_{\circ 2} \\ \hat{t}^{2\circ}_{\circ 1} & \hat{t}^{2\circ}_{\circ 2} \end{bmatrix}C^{-1}$$

or

$$\begin{bmatrix} \hat{t}^{1\circ}_{\circ 1} & \hat{t}^{1\circ}_{\circ 2} \\ \hat{t}^{2\circ}_{\circ 1} & \hat{t}^{2\circ}_{\circ 2} \end{bmatrix} = C^{-1}\begin{bmatrix} t^{1\circ}_{\circ 1} & t^{1\circ}_{\circ 2} \\ t^{2\circ}_{\circ 1} & t^{2\circ}_{\circ 2} \end{bmatrix}C$$

and operating we finally get

$$\begin{bmatrix} \hat{t}^{1\circ}_{\circ 1} & \hat{t}^{1\circ}_{\circ 2} \\ \hat{t}^{2\circ}_{\circ 1} & \hat{t}^{2\circ}_{\circ 2} \end{bmatrix} = \begin{bmatrix} 1 & -1 \\ 0 & \sqrt{2} \end{bmatrix}\begin{bmatrix} t^{1\circ}_{\circ 1} & t^{1\circ}_{\circ 2} \\ t^{2\circ}_{\circ 1} & t^{2\circ}_{\circ 2} \end{bmatrix}\begin{bmatrix} 1 & \frac{\sqrt{2}}{2} \\ 0 & \frac{\sqrt{2}}{2} \end{bmatrix}$$

$$\begin{bmatrix} \hat{t}^{1\circ}_{\circ 1} & \hat{t}^{1\circ}_{\circ 2} \\ \hat{t}^{2\circ}_{\circ 1} & \hat{t}^{2\circ}_{\circ 2} \end{bmatrix} = \begin{bmatrix} (t^{1\circ}_{\circ 1} - t^{2\circ}_{\circ 1}) & \frac{\sqrt{2}}{2}[(t^{1\circ}_{\circ 1} - t^{2\circ}_{\circ 1}) + (t^{1\circ}_{\circ 2} - t^{2\circ}_{\circ 2})] \\ \sqrt{2}t^{2\circ}_{\circ 1} & (t^{2\circ}_{\circ 1} + t^{2\circ}_{\circ 2}) \end{bmatrix}.$$

$\square$

5.4 Tensor product applications

In this section, some important tensor products applications are discussed.

5.4.1 Common simply contracted tensor products

First, we mention the contracted tensor product of first-order tensors.

Consider the tensors $\vec{x} = x^{\alpha}_{\circ}\vec{e}_{\alpha} \in V^n(K)$, $\vec{y} = y^{\beta}_{\circ}\vec{e}^{*\beta} \in V^n_*(K)$; their contracted tensor product is

$$p = C\begin{pmatrix} \alpha \\ \beta \end{pmatrix}(\vec{x} \otimes \vec{y}) = C\begin{pmatrix} \alpha \\ \beta \end{pmatrix}(x^{\alpha}_{\circ}y^{\circ}_{\beta}\vec{e}_{\alpha} \otimes \vec{e}^{*\beta}) = x^{\theta}y_{\theta} = x^1y_1 + x^2y_2 + \cdots + x^ny_n,$$

$$(5.20)$$

which is the classic dot product for geometric vectors or the classic inner product for first-order matrices.

Second, we mention the contracted tensor product of second-order tensors, known as the "interior product" or "classic product" of matrices.

Consider the tensors $\vec{a} = a^{\alpha\circ}_{\circ\beta}\vec{e}_{\alpha} \otimes \vec{e}^{*\beta}$ and $\vec{b} = b^{\gamma\circ}_{\circ\delta}\vec{e}_{\gamma} \otimes \vec{e}^{*\delta}$. Let $\vec{c}$ be their contracted tensor product of indices 2 and 3:

$$\vec{c} = \mathcal{C}\begin{pmatrix}\gamma\\\beta\end{pmatrix}(\vec{a} \otimes \vec{b}) = (a^{\alpha\circ}_{\circ\theta} \cdot b^{\theta\circ}_{\circ\delta})\vec{e}_\alpha \otimes \vec{e}^{*\delta},$$

where

$$c^{\alpha\circ}_{\circ\delta} = a^{\alpha\circ}_{\circ\theta} \cdot b^{\theta\circ}_{\circ\delta}, \tag{5.21}$$

which is the analytical tensor expression of the classic matrix product, of both matrices, as tensors.

Remark 5.1. In reality, the discovery of this idea occurred in the reverse order; first, Kronecker established the interior and tensor products of matrices, and then, under the name of Einstein's contraction, this concept was extended to tensors. □

5.4.2 Multiply contracted tensor products

It is obvious that when contracting a tensor product of tensors of certain orders, the resulting tensor can be a mixed tensor, with indices *not only of different valency* but coming from *different factors*; we can then continue contracting more indices, following the same criteria as the first time.

If we do this, we will obtain another tensor and we could practice contractions successively when the following two conditions are satisfied: (a) the indices must be of different valency, and (b) of different factor-tensor.

Evidently, this concept can be extended to products of *three or more tensors*, satisfying the associative law by operating the tensors two by two, and satisfying the index conditions.

On the other hand, the result of the contractions can be a zero-order tensor, that is, a scalar, which obviously is invariant under changes of basis. This is the reason why zero-order tensors are called "invariants".

5.4.3 Scalar and inner tensor products

Certain authors use the term "scalar product of tensors" for the totally contracted product of two tensors A and B, which allow it, and denote it by $A \bullet B = k$. The result is a zero-order tensor (a scalar). In this way, but based on a third fundamental tensor, we will later establish the tensor spaces with a interior connection.

It is also convenient to mention that, as a consequence of the concept of contracted tensor product, when selecting a tensor space of mixed tensors which contravariant and covariant indices coincide ($p = q$), the tensor product of *two* arbitrary tensors of this space can be contracted p times, leading to a contracted tensor product, that is, another tensor of the same space.

In such cases, some authors talk about a "tensor space with an interior product". For example, for $p = q = 2$, we would have

$$(t)^{\alpha\beta\circ\circ}_{\circ\circ\gamma\delta} \bullet (t')^{\alpha\beta\circ\circ}_{\circ\circ\gamma\delta} = (t'')^{\alpha\beta\circ\circ}_{\circ\circ\gamma\delta}$$

leading to the *interior product of tensors* concept, and the concept of associated linear algebras.

Next, some illustrative examples of contractions will be given.

Example 5.3 (Multiple contractions). Consider the tensors

$$\vec{t} = t_{\alpha\beta\circ}^{\circ\circ\gamma}\vec{e}^{*\alpha} \otimes \vec{e}^{*\beta} \otimes \vec{e}_{\gamma}; \quad \vec{x} = x^{\lambda}\vec{e}_{\lambda} \text{ and } \vec{u} = u_{\circ\nu}^{\mu\circ}\vec{e}_{\mu} \otimes \vec{e}^{*\nu}$$

and the tensor $P = \vec{t} \otimes \vec{x} \otimes \vec{u}$ with components

$$P_{\alpha\beta\circ\circ\circ\nu}^{\circ\circ\gamma\lambda\mu\circ} = t_{\alpha\beta\circ}^{\circ\circ\gamma} \cdot x_{\circ}^{\lambda} \cdot u_{\circ\nu}^{\mu\circ}.$$

We want to perform the following multiple contractions:

1. *Double:*

$$\vec{p}_1 = C\begin{pmatrix} \lambda & \mu \\ \alpha & \beta \end{pmatrix}(\vec{t} \otimes \vec{x} \otimes \vec{u}) \tag{5.22}$$

$$\vec{p}_2 = C\begin{pmatrix} \lambda & \gamma \\ \alpha & \nu \end{pmatrix}(\vec{t} \otimes \vec{x} \otimes \vec{u}) \tag{5.23}$$

$$\vec{p}_3 = C\begin{pmatrix} \mu & \gamma \\ \alpha & \nu \end{pmatrix}(\vec{t} \otimes \vec{x} \otimes \vec{u}), \tag{5.24}$$

which lead to

$$\vec{p}_1 \rightarrow p_{\circ\nu}^{\gamma\circ} = t_{\theta\phi\circ}^{\circ\circ\gamma}x_{\circ}^{\theta}a_{\circ\nu}^{\phi\circ} \tag{5.25}$$

$$\vec{p}_2 \rightarrow p_{\beta\circ}^{\circ\mu} = t_{\theta\beta\circ}^{\circ\circ\phi}x_{\circ}^{\theta}a_{\circ\phi}^{\mu\circ} \tag{5.26}$$

$$\vec{p}_3 \rightarrow p_{\beta\circ}^{\circ\lambda} = t_{\theta\beta\circ}^{\circ\circ\phi}x_{\circ}^{\lambda}a_{\circ\phi}^{\theta\circ}. \tag{5.27}$$

2. *Triple:*

$$\vec{p}_4 = C\begin{pmatrix} \lambda & \mu & \gamma \\ \alpha & \beta & \nu \end{pmatrix}(\vec{t} \otimes \vec{x} \otimes \vec{u}) \tag{5.28}$$

$$\vec{p}_5 = C\begin{pmatrix} \mu & \lambda & \gamma \\ \alpha & \beta & \nu \end{pmatrix}(\vec{t} \otimes \vec{x} \otimes \vec{u}), \tag{5.29}$$

which lead to the scalars

$$p_4 = t_{\theta\phi\circ}^{\circ\circ w}x_{\circ}^{\theta}a_{\circ w}^{\phi\circ} \tag{5.30}$$

$$p_5 = t_{\theta\phi\circ}^{\circ\circ w}x_{\circ}^{\phi}a_{\circ w}^{\theta\circ}. \tag{5.31}$$

□

5.5 Criteria for tensor character based on contraction

In Section 4.5 the tensor criteria for homogeneous tensors were established with respect to changes of basis in tensor spaces. However, next we will establish other tensor criteria based on tensor contraction.

We present them as theorems, and in the proof of the third we will examine in detail its necessarity and sufficiency.

Theorem 5.2 (First elemental criterion for tensor character). *The necessary and sufficient condition for a system of scalars $s(\alpha_1, \alpha_2, \cdots, \alpha_r)$ of order r (the α_j are indices) to be a pure homogeneous tensor, of order r, totally contravariant, is that the expression "totally r-contracted product":*

$$s(\alpha_1, \alpha_2, \ldots, \alpha_r) x_{\alpha_1} \cdot x_{\alpha_2} \ldots x_{\alpha_r}; \quad \forall \vec{x} = x_{\alpha_j} \vec{e}^{*\alpha_j} \in V_*^n(K),$$
$$j \in I_r = \{1, 2, \ldots, r\}; \quad \alpha_j \in I_n = \{1, 2, \ldots, n\} \qquad (5.32)$$

be a escalar, that is, be invariant with respect to changes of basis in $V_*^n(K)$. □

Theorem 5.3 (Second elemental criterion for tensor character). *The necessary and sufficient condition for a system of scalars $s(\alpha_1, \alpha_2, \ldots, \alpha_r)$ of order r to be a pure homogeneous tensor, of order r, totally covariant, is that the expression "totally r-contracted product":*

$$s(\alpha_1, \alpha_2, \ldots, \alpha_r) x^{\alpha_1} \cdot x^{\alpha_2} \ldots x^{\alpha_r}; \quad \forall \vec{x} = x^{\alpha_j} \vec{e}_{\alpha_j} \in V^n(K),$$
$$j \in I_r = \{1, 2, \ldots, r\}; \quad \alpha_j \in I_n = \{1, 2, \ldots, n\} \qquad (5.33)$$

be a escalar, that is, be invariant with respect to changes of basis in $V^n(K)$. □

Theorem 5.4 (General criterion for homogeneous tensor character). *The necessary and sufficient condition for a system of scalars $s(\alpha_1, \alpha_2, \cdots, \alpha_r)$ of order r to be a mixed homogeneous tensor, of order r, p-contravariant and q-covariant $(p + q = r)$, is that the expression "totally r-contracted product":*

$$s(\alpha_1, \alpha_2, \cdots, \alpha_p, \alpha_{p+1}, \ldots, \alpha_{p+q}) x_{\alpha_1} \cdot x_{\alpha_2} \ldots x_{\alpha_p} \cdot x^{\alpha_{p+1}} \cdot x^{\alpha_{p+2}} \ldots x^{\alpha_{p+q}};$$
$$\forall \vec{x} = x_{\alpha_j} \vec{e}^{*\alpha_j} \in V_*^n(K); \quad j \in I_p = \{1, 2, \ldots, p\}; \; \alpha_j \in I_n$$
$$\forall \vec{x} = x^{\alpha_k} \vec{e}_{\alpha_k} \in V^n(K) \quad k \in I_q = \{p+1, p+2, \ldots, p+q\}; \; \alpha_k \in I_n$$
$$(5.34)$$

be a escalar, that is, be invariant with respect to changes of basis in $V^n(K)$ and the corresponding changes of basis "in dual bases" in $V_*^n(K)$. □

Proof.

Necessarity:

Let $\vec{t} = t^{\alpha\beta\,\circ}_{\circ\,\circ\,\gamma}\vec{e}_\alpha \otimes \vec{e}_\beta \otimes \vec{e}^{*\gamma}$ be a mixed tensor of third order $(r = 3)$, $p = 2$ times contravariant and $q = 1$ covariant, with $p + q = 2 + 1 = 3 = r$.

Consider the vectors $\vec{x} = x_\lambda \vec{e}^{*\lambda}, \vec{y} = y_\mu \vec{e}^{*\mu}$ and $\vec{z} = z^\nu \vec{e}_\nu$, where $\vec{x}, \vec{y} \in V_*^n(K)$ and $\vec{z} \in V^n(K)$.

If we execute the r-contracted tensor product:

$$p = C \begin{pmatrix} \alpha & \beta & \lambda \\ \lambda & \mu & \nu \end{pmatrix} (\vec{t} \otimes \vec{x} \otimes \vec{y} \otimes \vec{z}) \tag{5.35}$$

we get

$$p = t^{\theta\phi\,\circ}_{\circ\,\circ\,w} \cdot x_\theta \cdot y_\phi \cdot z^w = \text{scalar (zero-order tensor)}, \tag{5.36}$$

which proves that if $\vec{t}$ is a tensor, the theorem holds.

Sufficiency:

Consider now a system of scalars such that

$$p = s(\alpha, \beta, \gamma) \cdot x_\alpha \cdot y_\beta \cdot z^\gamma \tag{5.37}$$

for any pair of vectors $\vec{x} = x_\alpha \vec{e}^{*\alpha}, \vec{y} = y_\beta \vec{e}^{*\beta} \in V_*^n(K)$ and for all $\vec{z} = z^\gamma \vec{e}_\gamma \in V^n(K)$ where p is a given scalar.

We perform a change-of-basis in the linear space $V^n(K)$, and in the dual space $V_*^n(K)$ in which we choose the dual reciprocal basis of the one selected in $V^n(K)$.

Since p is a fixed scalar, the relation (5.37) is also satisfied in the new basis, that is, the p remains invariant for any new vector:

$$p = s(i, j, k) \cdot x_i \cdot y_j \cdot z^k; \quad \forall \vec{x} = x_i \vec{e}^{*i}, \vec{y} = y_j \vec{e}^{*j} \in V_*^n(K) \text{ and } \forall \vec{z} = z^k \vec{e}_k \in V^n(K). \tag{5.38}$$

Using the change-of-basis relations (3.46) and (3.24):

$$\vec{e}^{*i} = \gamma^{\circ\,i}_{\alpha\,\circ} \vec{e}^{*\alpha} \text{ in } V_*^n(K),$$

$$\vec{e}_k = c^{\gamma\,\circ}_{\circ\,k} \vec{e}_\gamma \text{ in } V^n(K)$$

we get the expressions that directly relate the vector components, in the initial and new bases:

$$x_\alpha = \gamma^{\circ\,i}_{\alpha\,\circ} x_i; \quad y_\beta = \gamma^{\circ\,j}_{\beta\,\circ} y_j,$$

for vectors of $V_*^n(K)$, and

$$z^\gamma = c^{\gamma\,\circ}_{\circ\,k} z^k,$$

for the vector of $V^n(K)$.

Transposing these equalities one gets

$$x_\alpha = x_i \gamma_{\circ\alpha}^{i\circ}; \quad y_\beta = y_j \gamma_{\circ\beta}^{j\circ}; \quad z^\gamma = z^k c_{k\circ}^{\circ\gamma} \tag{5.39}$$

and replacing (5.39) and (5.37) we obtain

$$p = s(\alpha, \beta, \gamma)(x_i \gamma_{\circ\alpha}^{i\circ})(y_j \gamma_{\circ\beta}^{j\circ})(z^k c_{k\circ}^{\circ\gamma}),$$

which is operated as

$$p = (x_i \cdot y_j \cdot z^k)(s(\alpha, \beta, \gamma)\gamma_{\circ\alpha}^{i\circ}\gamma_{\circ\beta}^{j\circ}c_{k\circ}^{\circ\gamma}). \tag{5.40}$$

Equating the constant p in (5.38) and (5.40), we get

$$p = (x_i \cdot y_j \cdot z^k)s(i,j,k) = (x_i \cdot y_j \cdot z^k)\left(s(\alpha, \beta, \gamma)\gamma_{\circ\alpha}^{i\circ}\gamma_{\circ\beta}^{j\circ}c_{k\circ}^{\circ\gamma}\right)$$

and since the previous relation must hold for all x_i, y_j, z^k, it must be

$$s(i,j,k) = s(\alpha, \beta, \gamma)\gamma_{\circ\alpha}^{i\circ}\gamma_{\circ\beta}^{j\circ}c_{k\circ}^{\circ\gamma}, \tag{5.41}$$

which shows that the system of scalars $s(\alpha, \beta, \gamma)$ satisfies the general tensor character criterion, Formula (4.34), so that the system of scalars must be notated as

$$s(\alpha, \beta, \gamma) = t_{\circ\circ\gamma}^{\alpha\beta\circ} \text{ or } s(i,j,k) = t_{\circ\circ k}^{ij\circ},$$

which proves its tensor character. Obviously, the necessity and the sufficiency have been proved only for $r = 3$, but we have preferred this simple case, which clearly reveals the process followed, to the general case with the generic r, which hides the demonstration process under the confused complexity of subindices.

We close this part, dedicated to tensor product contraction, simple or multiple, of homogeneous tensors, by pointing out that its treatment can be considered in the wider frame of absolute tensors, that is, of heterogeneous tensors established on diverse factor linear spaces, studied in Chapters 2 and 3, and in Chapter 4, where the absolute tensor character criteria for them were established, Formulas (4.24), (4.25), (4.34) and (4.35).

However, the most frequent use of contraction occurs in the homogeneous tensor algebra, which justifies the decision made in this chapter.

5.6 The contracted tensor product in the reverse sense: The quotient law

Theorem 5.5 (Quotient law). *Consider the system of scalars $S(\alpha_1, \ldots, \alpha_r)$ of order r. A sufficient condition for such a system to be considered a homogeneous tensor is that its p-contracted tensor product by a generic (arbitrary) homogeneous tensor $\vec{b}$ of order r', called a "test tensor", lead to another tensor of order $(r + r' - 2p)$.* □

Proof. We state the proof for a concrete case.

Let $s(\alpha, \beta, \gamma, \delta)$ be the data system of scalars, of order $r = 4$, and let $\vec{b} = b^{\lambda\,o\,o}_{o\,\mu\nu}\vec{e}_\lambda \otimes \vec{e}^{*\mu} \otimes \vec{e}^{*\nu}$ be the "test" tensor, of order $r' = 3$. As a consequence of their doubly contracted ($p = 2$) product we arrive at the set of scalars $h^{\alpha\,o\,o}_{o\,\delta\mu}$, which is a known tensor, of order $r + r' - 2p = 4 + 3 - 2 \times 2 = 3$.

Since $\vec{h}$ is a tensor, due to the tensor criteria we have

$$h^{i\,o\,o}_{o\,dm} = h^{\alpha\,o\,o}_{o\,\delta\mu}\gamma^{i\,o}_{o\,\alpha}c^{o\,\delta}_{d\,o}c^{o\,\mu}_{m\,o}. \tag{5.42}$$

In addition we have

$$h^{\alpha\,o\,o}_{o\,\delta\mu} = C\left(\begin{array}{c|c}\lambda & \gamma \\ \beta & \nu\end{array}\right)\left(s(\alpha, \beta, \gamma, \delta) \otimes b^{\lambda\,o\,o}_{o\,\mu\nu}\right), \tag{5.43}$$

a relation stated in the initial basis of $V^n(K)$, and also

$$h^{i\,o\,o}_{o\,dm} = C\left(\begin{array}{c|c}\ell & k \\ j & n\end{array}\right)\left(s(i, j, k, d) \otimes b^{\ell\,o\,o}_{o\,mn}\right), \tag{5.44}$$

stated in the final basis of $V^n(K)$.

Executing the contraction indicated in (5.43) and (5.44) and using the Kronecker deltas, we get the relations

$$h^{\alpha\,o\,o}_{o\,\delta\mu} = s(\alpha, \beta, \gamma, \delta)\delta^{\beta\,o}_{o\,\lambda}\delta^{o\,\nu}_{\gamma\,o}b^{\lambda\,o\,o}_{o\,\mu\nu} \tag{5.45}$$

$$h^{i\,o\,o}_{o\,dm} = s(i, j, k, d)\delta^{j\,o}_{o\,\ell}\delta^{o\,n}_{k\,o}b^{\ell\,o\,o}_{o\,mn} \tag{5.46}$$

and since $\vec{b}$ is a tensor (the "test" tensor), we state its tensor character criterion in the form (4.35), leading to

$$b^{\lambda\,o\,o}_{o\,\mu\nu} = b^{\ell\,o\,o}_{o\,mn}c^{\lambda\,o}_{o\,\ell}\gamma^{o\,m}_{\mu\,o}\gamma^{o\,n}_{\nu\,o} \tag{5.47}$$

and replacing (5.47) into (5.45), we get

$$h^{\alpha\,o\,o}_{o\,\delta\mu} = s(\alpha, \beta, \gamma, \delta)\delta^{\beta\,o}_{o\,\lambda}\delta^{o\,\nu}_{\gamma\,o}b^{\ell\,o\,o}_{o\,mn}c^{\lambda\,o}_{o\,\ell}\gamma^{o\,m}_{\mu\,o}\gamma^{o\,n}_{\nu\,o}. \tag{5.48}$$

Finally, substituting (5.46) and (5.48) into the left- and right-hand sides of (5.42), respectively, we get

$$s(i, j, k, d)\delta^{j\,o}_{o\,\ell}\delta^{o\,n}_{k\,o}b^{\ell\,o\,o}_{o\,mn} = \left[s(\alpha, \beta, \gamma, \delta)\delta^{\beta\,o}_{o\,\lambda}\delta^{o\,\nu}_{\gamma\,o}b^{\ell\,o\,o}_{o\,mn}c^{\lambda\,o}_{o\,\ell}\gamma^{o\,m}_{\mu\,o}\gamma^{o\,n}_{\nu\,o}\right]\gamma^{i\,o}_{o\,\alpha}c^{o\,\delta}_{d\,o}c^{o\,\mu}_{m\,o},$$

and conveniently grouping the factors we obtain

$$\left[s(i, j, k, d)\delta^{j\,o}_{o\,\ell}\delta^{o\,n}_{k\,o}\right]b^{\ell\,o\,o}_{o\,mn} = \left[s(\alpha, \beta, \gamma, \delta)\gamma^{i\,o}_{o\,\alpha}(\delta^{\beta\,o}_{o\,\lambda}c^{\lambda\,o}_{o\,\ell})(\gamma^{o\,m}_{\mu\,o}c^{o\,\mu}_{m\,o})(\delta^{o\,\nu}_{\gamma\,o}\gamma^{o\,n}_{\nu\,o}), c^{o\,\delta}_{d\,o}\right]b^{\ell\,o\,o}_{o\,mn}$$

and executing the indicated contractions:

$$\left[s(i,j,k,d)\delta^{j\circ}_{\circ\ell}\delta^{\circ n}_{k\circ}\right]b^{\ell\;\circ\;\circ}_{\circ mn} = \left[s(\alpha,\beta,\gamma,\delta)\gamma^{i\;\circ}_{\circ\alpha}(c^{\beta\circ}_{\circ\ell})\cdot 1\cdot(\gamma^{\circ n}_{\gamma\circ})c^{\circ\delta}_{do}\right]b^{\ell\;\circ\;\circ}_{\circ mn}.$$

Finally, passing everything to the left-hand side and taking common factors, the result is

$$\left[s(i,j,k,d)\delta^{j\circ}_{\circ\ell}\delta^{\circ n}_{k\circ} - s(\alpha,\beta,\gamma,\delta)\gamma^{i\;\circ}_{\circ\alpha}c^{\beta\circ}_{\circ\ell}\gamma^{\circ n}_{\gamma\circ}c^{\circ\delta}_{do}\right]b^{\ell\;\circ\;\circ}_{\circ mn} = 0. \tag{5.49}$$

Since the "test" tensor $\vec{b} \neq \vec{\Omega}$ (it is not the null tensor), their components $b^{\ell\;\circ\;\circ}_{\circ mn} \neq 0$, which forces the null factor to be the bracketed term in (5.49)

$$s(i,j,k,d)\delta^{j\circ}_{\circ\ell}\delta^{\circ n}_{k\circ} = s(\alpha,\beta,\gamma,\delta)\gamma^{i\;\circ}_{\circ\alpha}c^{\beta\circ}_{\circ\ell}\gamma^{\circ n}_{\gamma\circ}c^{\circ\delta}_{do}. \tag{5.50}$$

Next, we isolate the factor $s(i,j,k,d)$ on the left-hand side of (5.50). To this end, we multiply both members by the Kronecker delta $\delta^{\ell\circ}_{\circ j}$, inverse of $\delta^{j\circ}_{\circ\ell}$:

$$s(i,j,k,d)(\delta^{\ell\circ}_{\circ j}\delta^{j\circ}_{\circ\ell})\delta^{\circ n}_{k\circ} = s(\alpha,\beta,\gamma,\delta)\gamma^{i\;\circ}_{\circ\alpha}(\delta^{\ell\circ}_{\circ j}c^{\beta\circ}_{\circ\ell})\gamma^{\circ n}_{\gamma\circ}c^{\circ\delta}_{do}$$

or

$$s(i,j,k,d)(1)\delta^{\circ n}_{k\circ} = s(\alpha,\beta,\gamma,\delta)\gamma^{i\;\circ}_{\circ\alpha}(c^{\circ\beta}_{\ell\circ}\delta^{\ell\circ}_{\circ j})\gamma^{\circ n}_{\gamma\circ}c^{\circ\delta}_{do}$$

contracting the grouped product, and multiplying both members by $\delta^{\circ k}_{n\circ}$, the inverse of $\delta^{\circ n}_{k\circ}$, we get

$$s(i,j,k,d)(\delta^{\circ k}_{n\circ}\delta^{\circ n}_{k\circ}) = s(\alpha,\beta,\gamma,\delta)\gamma^{i\;\circ}_{\circ\alpha}c^{\circ\beta}_{\ell\circ}(\delta^{\circ k}_{n\circ}\gamma^{\circ n}_{\gamma\circ})c^{\circ\delta}_{do}$$

or

$$s(i,j,k,d)(1) = s(\alpha,\beta,\gamma,\delta)\gamma^{i\;\circ}_{\circ\alpha}c^{\circ\beta}_{\ell\circ}(\gamma^{\circ k}_{\circ\gamma}\delta^{\circ k}_{n\circ})c^{\circ\delta}_{do},$$

and contracting the grouped product, we finally get

$$s(i,j,k,d) = s(\alpha,\beta,\gamma,\delta)\gamma^{i\;\circ}_{\circ\alpha}c^{\circ\beta}_{\ell\circ}\gamma^{k\circ}_{\circ\gamma}c^{\circ\delta}_{do}. \tag{5.51}$$

This last expression indicates that the set of scalars $s(\alpha,\beta,\gamma,\delta)$ is a tensor, since it satisfies a concrete tensor criterion. In addition, it shows us *its whole nature*. In reality it is

$$s(\alpha,\beta,\gamma,\delta) = s^{\alpha\circ\gamma\circ}_{\circ\beta\circ\delta}. \tag{5.52}$$

The theorem that has been proved is called the "quotient law", a disputed title, that some impute to a simple conception of this relation among tensors, such as

If $X \cdot T_1 = T_2 \rightarrow X = \frac{T_2}{T_1} \rightarrow X = T_2 \cdot T_1^{-1}$, which is certainly simple, and at least justifies its name.

This theorem, which is frequently used in solving tensor analysis theoretical problems and also in practical exercises, to detect whether or not a system of scalars is a tensor, has severe limitations that it is convenient to point out.

On one hand, one must be lucky when choosing the test tensor because, if after an unfortunate selection, the contraction does not lead to a tensor, no conclusion can be drawn, because of the sufficient character of the theorem. So, another test tensor must be selected and so on.

On the other hand, frequently, after the contraction is performed with the selected "test" tensor, we have great difficulties in proving that the result is *another* tensor, arriving at a new problem that can be even more complex than the initial one.

Consequently, the most frequent applications of the "quotient law" are those in which the contracted product is an *invariant*, which it is well known to be a zero-order tensor.

5.7 Matrix representation of permutation homomorphisms

We say that a tensor is the "permutation tensor of a given tensor" if it has the same associated scalars as the given tensor, but in different *positions*; one possibility of building a permutation tensor of a given tensor is to create with a different name a tensor with at least a changed index but with the same scalars:

$$\forall t^{\alpha \circ \gamma \circ}_{\circ \beta \circ \delta} \equiv u^{\alpha' \circ \gamma' \circ}_{\circ \beta' \circ \delta'},$$

where $(\alpha', \beta', \gamma', \delta')$ is one of the possible permutations of $(\alpha, \beta, \gamma, \delta)$.

Consider the linear space $K^\sigma, \sigma = n^r$, i.e., the linear space of matrices $T_{\sigma,1} \in K^\sigma$, "extensions" of the homogeneous tensors of a generic type $t^{\alpha_1 \circ \alpha_3}_{\circ \alpha_2 \circ} \cdots {}^{\circ}_{\alpha_r}$, defined over the "factor" linear space $V^n(K)$. We will study the permutation homomorphisms $P : K^\sigma \to K^\sigma$, the associated square matrix of which, P_{n^r}, is a permutation of the unit matrix I_{n^r} and which transforms by means of the following matrix equation:

$$P_{n^r} \bullet T_{\sigma,1} = T'_{\sigma,1}. \tag{5.53}$$

These transformations maintain the tensor dimension σ, together with its scalars, though obviously they change them in position. We will study two different types of homomorphisms P.

5.7.1 Permutation matrix tensor product types in K^n

Consider the tensor

$$T = \begin{bmatrix} t^{\alpha \beta}_{\circ \circ} \end{bmatrix} = \begin{bmatrix} a & b & c & d \\ e & f & g & h \\ m & n & p & q \\ r & s & t & u \end{bmatrix}$$

and the tensor

$$T' = \left[u^{\gamma\delta}_{\circ\circ}\right] = \begin{bmatrix} d & c & b & a \\ q & p & n & m \\ h & g & f & e \\ u & t & s & r \end{bmatrix},$$

which obviously is a permutation of T, where $\sigma = 4^2 = 16$.

We build the corresponding matrix extensions of T and T' ($T_{\sigma,1} = T_{16,1}$ and $T'_{\sigma,1} = T'_{16,1}$), and we observe that the permutation matrix that relates both is

$$T'_{16,1} = P \cdot T_{16,1},$$

where

$$T'_{16,1} = \begin{bmatrix} d \\ c \\ b \\ a \\ q \\ p \\ n \\ m \\ h \\ g \\ f \\ e \\ u \\ t \\ s \\ r \end{bmatrix} ; \quad T_{16,1} = \begin{bmatrix} a \\ b \\ c \\ d \\ e \\ f \\ g \\ h \\ m \\ n \\ p \\ q \\ r \\ s \\ t \\ u \end{bmatrix}$$

$$P \equiv P_{16,16} = \left[\begin{array}{c|c|c|c} \begin{smallmatrix} 0&0&0&1 \\ 0&0&1&0 \\ 0&1&0&0 \\ 1&0&0&0 \end{smallmatrix} & \Omega & \Omega & \Omega \\ \hline \Omega & \Omega & \begin{smallmatrix} 0&0&0&1 \\ 0&0&1&0 \\ 0&1&0&0 \\ 1&0&0&0 \end{smallmatrix} & \Omega \\ \hline \Omega & \begin{smallmatrix} 0&0&0&1 \\ 0&0&1&0 \\ 0&1&0&0 \\ 1&0&0&0 \end{smallmatrix} & \Omega & \Omega \\ \hline \Omega & \Omega & \Omega & \begin{smallmatrix} 0&0&0&1 \\ 0&0&1&0 \\ 0&1&0&0 \\ 1&0&0&0 \end{smallmatrix} \end{array} \right].$$

An analysis of P discovers that in this case

$$P = \begin{bmatrix} 1&0&0&0 \\ 0&0&1&0 \\ 0&1&0&0 \\ 0&0&0&1 \end{bmatrix} \otimes \begin{bmatrix} 0&0&0&1 \\ 0&0&1&0 \\ 0&1&0&0 \\ 1&0&0&0 \end{bmatrix},$$

i.e., the permutation matrix is the tensor product of two permutation matrices that operate in the linear space K^4, which reveals that some P have this type of construction.

5.7.2 Linear span of precedent types

In Example 4.5 of Chapter 4, we considered the five permutation tensors U, V, W, R and S of a given tensor $T = [t^{\alpha\,\circ\,\gamma}_{\circ\,\beta\,\circ}]$ of third order ($r = 3, n = 3$ and $\sigma = n^r = 27$). We will examine what type of construction has the permutation homomorphism matrix that applies $P_{(1)} : [t^{\alpha\,\circ\,\gamma}_{\circ\,\beta\,\circ}] \to [u^{\gamma\,\alpha\,\circ}_{\circ\,\circ\,\beta}]$, that is, $U_{27,1} = P_{(1)} \bullet T_{27,1}$. The solution matrices in this case are (see Example 4.5)

$$
T_{27,1} = \begin{bmatrix} 1 \\ 0 \\ -1 \\ 2 \\ 3 \\ 0 \\ -1 \\ 2 \\ 0 \\ \hline 2 \\ -1 \\ 1 \\ 0 \\ 1 \\ 0 \\ 2 \\ 0 \\ 1 \\ \hline 0 \\ 0 \\ 1 \\ 5 \\ 1 \\ 2 \\ 1 \\ 0 \\ 0 \end{bmatrix} ; \quad U_{27,1} = \begin{bmatrix} 1 \\ 2 \\ -1 \\ 2 \\ 0 \\ 2 \\ 0 \\ 5 \\ 1 \\ \hline 0 \\ 3 \\ 2 \\ -1 \\ 1 \\ 0 \\ 0 \\ 1 \\ 0 \\ \hline -1 \\ 0 \\ 0 \\ 1 \\ 0 \\ 1 \\ 1 \\ 2 \\ 0 \end{bmatrix} ; \quad \text{where } \beta = \{E_i\} \equiv \left\{ \begin{bmatrix} 1 \\ 0 \\ 0 \end{bmatrix} \begin{bmatrix} 0 \\ 1 \\ 0 \end{bmatrix} \begin{bmatrix} 0 \\ 0 \\ 1 \end{bmatrix} \right\}.
$$

We have that

$$
P_{(1)} \equiv P_{27} = E_1 \otimes I_3 \otimes I_3 \otimes E_1^t + E_2 \otimes I_3 \otimes I_3 \otimes E_2^t + E_3 \otimes I_3 \otimes I_3 \otimes E_3^t, \quad (5.54)
$$

that is, a matrix written as a linear combination of tensor products.

With respect to the permutation tensor V:

$$
P_{(2)} : [t^{\alpha\,\circ\,\gamma}_{\circ\,\beta\,\circ}] \to [v^{\circ\,\gamma\,\alpha}_{\beta\,\circ\,\circ}],
$$

that is,

$$
V_{27,1} = P_{(2)} \bullet T_{27,1}.
$$

The solution matrices are in this case, the $T_{27,1}$ matrix previously cited, and matrices

$$V_{27,1} = \begin{bmatrix} 1 \\ 2 \\ 0 \\ 0 \\ -1 \\ 0 \\ -1 \\ 1 \\ 1 \\ \hline 2 \\ 0 \\ 5 \\ 3 \\ 1 \\ 1 \\ 0 \\ 0 \\ 2 \\ \hline -1 \\ 2 \\ 1 \\ 2 \\ 0 \\ 0 \\ 0 \\ 1 \\ 0 \end{bmatrix}$$

$$P_{(2)} \equiv P'_{27} = E_1^t \otimes I_3 \otimes I_3 \otimes E_1 + E_2^t \otimes I_3 \otimes I_3 \otimes E_2 + E_3^t \otimes I_3 \otimes I_3 \otimes E_3. \quad (5.55)$$

The permutation matrix $P_{(2)}$ is of the type $P_{(1)}$, that is, a linear combination of tensor products, and $P_{(2)}$ is $P_{(1)}^t$.

For the permutation tensor W:

$$P_{(3)} : [t^{\alpha \circ \gamma}_{\circ \beta \circ}] \rightarrow [w^{\alpha \gamma \circ}_{\circ \circ \beta}],$$

that is

$$W_{27,1} = P_{(3)} \bullet T_{27,1}$$

the solution is given by the matrices

$$W_{27,1} = \begin{bmatrix} 1 \\ 2 \\ -1 \\ 0 \\ 3 \\ 2 \\ -1 \\ 0 \\ 0 \\ \hline 2 \\ 0 \\ 2 \\ -1 \\ 1 \\ 0 \\ 1 \\ 0 \\ 1 \\ \hline 0 \\ 5 \\ 1 \\ 0 \\ 1 \\ 0 \\ 1 \\ 2 \\ 0 \end{bmatrix}$$

$$P_{(3)} = I_3 \otimes \left[E_1 \otimes I_3 \otimes E_1^t + E_2 \otimes I_3 \otimes E_2^t + E_3 \otimes I_3 \otimes E_3^t \right], \qquad (5.56)$$

also a linear combination of tensor products. $P_{(3)}$ is symmetric.

Next, we analyze the permutation tensor R:

$$P_{(4)} : [t^{\alpha \circ \gamma}_{\circ \beta \circ}] \to [r^{\gamma \circ \alpha}_{\circ \beta \circ}],$$

that is,

$$R_{27,1} = P_{(4)} \bullet T_{27,1},$$

the solution of which is

$$R_{27} = \begin{bmatrix} 1 \\ 2 \\ 0 \\ 2 \\ 0 \\ 5 \\ -1 \\ 2 \\ 1 \\ \hline 0 \\ -1 \\ 0 \\ 3 \\ 1 \\ 1 \\ 2 \\ 0 \\ 0 \\ \hline -1 \\ 1 \\ 1 \\ 0 \\ 0 \\ 2 \\ 0 \\ 1 \\ 0 \end{bmatrix} ; \quad \beta \equiv \text{ basis of } \mathbb{R}^3 \equiv \{E_1, E_2, E_3\} \equiv \left\{ \begin{bmatrix} 1 \\ 0 \\ 0 \end{bmatrix} \begin{bmatrix} 0 \\ 1 \\ 0 \end{bmatrix} \begin{bmatrix} 0 \\ 0 \\ 1 \end{bmatrix} \right\}$$

$$P_{(4)} = \sum_{1 \le i,j \le 3} (E_i \otimes E_j^t) \otimes I_3 \otimes (E_i^t \otimes E_j), \qquad (5.57)$$

also a sum of tensor products. $P_{(4)}$ is symmetric.

We arrive at the last permutation S of Example 4.5, i.e.,

$$P_{(5)} : [t^{\alpha \circ \gamma}_{\circ \beta \circ}] \to [s^{\circ \alpha \gamma}_{\beta \circ \circ}],$$

that is,

$$S_{27,1} = P_{(5)} \bullet T_{27,1},$$

with solution matrices

$$S_{27,1} = \begin{bmatrix} 1 \\ 0 \\ -1 \\ 2 \\ -1 \\ 1 \\ 0 \\ 0 \\ 1 \\ -- \\ 2 \\ 3 \\ 0 \\ 0 \\ 1 \\ 0 \\ 5 \\ 1 \\ 2 \\ -- \\ -1 \\ 2 \\ 0 \\ 2 \\ 0 \\ 1 \\ 1 \\ 0 \\ 0 \end{bmatrix}$$

with

$$P_{(5)} = \left[E_1 \otimes I_3 \otimes E_1^t + E_2 \otimes I_3 \otimes E_2^t + E_3 \otimes I_3 \otimes E_3^t \right] \otimes I_3 = P_{(5)}^t. \quad (5.58)$$

We end the section dedicated to permutation homomorphisms by citing the model of this type of matrices which will be called "transposer" since it operates over second-order homogeneous tensors, the square matrices $a^{\alpha\beta}_{\circ\circ}, b^{\alpha\circ}_{\circ\beta}, c^{\circ\beta}_{\alpha\circ}$ or $d^{\circ\circ}_{\alpha\beta}$, with $r = 2; n = n; \sigma = n^2$, *transposing them.*

Whence

$$H(a^{\alpha\beta}_{\circ\circ}) = a^{\beta\alpha}_{\circ\circ}; \quad H(b^{\alpha\circ}_{\circ\beta}) = b^{\circ\alpha}_{\beta\circ}, \ldots, \text{ etc.}$$

is the matrix called a "transposition matrix" in Section 1.3.7, Formula (1.38). Here we present a generalization, in its usual mode of permutation homomorphism:

$$T'_{\sigma,1} = P_{n^2} \bullet T_{\sigma,1},$$

where $T_{\sigma,1}$ is the extension matrix that is to be transposed.

The permutation "transposer" is the block matrix:

$$P_{n^2} = P = \begin{bmatrix} E_{11} & | & E_{21} & | & \cdots & | & E_{n1} \\ -- & + & -- & + & -- & + & -- \\ E_{12} & | & E_{22} & | & \cdots & | & E_{n2} \\ -- & + & -- & + & -- & + & -- \\ \cdots & | & \cdots & | & \cdots & | & \cdots \\ -- & + & -- & + & -- & + & -- \\ E_{1n} & | & E_{2n} & | & \cdots & | & E_{nn} \end{bmatrix}, \quad (5.59)$$

where $\mathcal{B} = \{E_{ij}\}$ is the canonical basis of the tensor space $K^{n\times n}$ of square matrices of order n (noting the block ordering inside P_{n^2})

The reader can test its effect using it in the exercises.

With respect to the permutation type "transposer", responds to the expression

$$P_{n^2} = \sum E_{ii} \otimes E_{ii} + \sum_{i<j;i,j\in\{1,2,\cdots,n\}} (E_{ij} \otimes E_{ji} + E_{ji} \otimes E_{ij}). \qquad (5.60)$$

Example 5.4 (Permutation homomorphisms). Consider the linear space $\tau^{27}(\mathbb{R})$ as a tensor product of $\mathbb{R}^3 \otimes \mathbb{R}^{*3} \otimes \mathbb{R}^3$. Let $T \in \tau$ be a tensor of components

$$t^{\alpha \circ \gamma}_{\circ \beta \circ} = \begin{bmatrix} 1 & 2 & 0 & 2 & 0 & 5 & -1 & 2 & 1 \\ 0 & -1 & 0 & 3 & 1 & 1 & 2 & 0 & 0 \\ -1 & 1 & 1 & 0 & 0 & 2 & 0 & 1 & 0 \end{bmatrix},$$

where α is the row, β is the column, and γ is the matrix.

Let $\widehat{e}_1(-1,0,-1), \widehat{e}_2(1,1,0), \widehat{e}_3(0,0,3)$ be a change of the canonical basis of $\mathbb{R}^3$ that produces the corresponding change-of-basis of tensor nature in τ.

Determine the new components of tensor T, using the permutation homomorphisms, to execute the change-of-basis on the tensor ordered according to the axiom.

Solution: It is evident that the assigning of subindices in the statement does not correspond to the *axiomatic* order for the canonical basis of $\mathbb{R}^3 \otimes \mathbb{R}^{*3} \otimes \mathbb{R}^3$, which requires (see the theory and Example 2.1, question 4) that the matrix index (γ) must be the first and the column index (β) must be the last. So, before executing the change-of-basis, we must find the *fundamental* tensor $(t')^{\gamma \alpha \circ}_{\circ \circ \beta}$, which, subject to adequate permutation, provides the given data.

Tensor $(t')^{\gamma \alpha \circ}_{\circ \circ \beta}$ is the one that must be subject to the change-of-basis, given by the theory, and obviously the permutation must be undone in order to find the sought after tensor $t^{i \circ k}_{\circ j \circ}$.

Let $T'_{27,1}$, be the stretched version of $(t')^{\gamma \alpha \circ}_{\circ \circ \beta}$, and $T_{27,1}$, the stretched version of $t^{\alpha \circ \gamma}_{\circ \beta \circ}$ (data).

The permutation relation between them is

$$P_{(2)} \cdot T'_{27,1} = T_{27,1},$$

where $P_{(2)}$ (Formula (5.55)) is

$$P_{(2)} = E_1^t \otimes I_9 \otimes E_1 + E_2^t \otimes I_9 \otimes E_2 + E_3^t \otimes I_9 \otimes E_3$$

$$= [1\ 0\ 0] \otimes I_9 \otimes \begin{bmatrix} 1 \\ 0 \\ 0 \end{bmatrix} + [0\ 1\ 0] \otimes I_9 \otimes \begin{bmatrix} 0 \\ 1 \\ 0 \end{bmatrix} + [0\ 0\ 1] \otimes I_9 \otimes \begin{bmatrix} 1 \\ 0 \\ 0 \end{bmatrix}.$$

Then, $P_{(2)}$ becomes

$$P_{(2)} =
\begin{bmatrix}
\begin{smallmatrix}1&0&0\\0&0&0\\0&0&0\\0&1&0\\0&0&0\\0&0&0\\0&0&1\\0&0&0\\0&0&0\end{smallmatrix} & \Omega & \Omega &
\begin{smallmatrix}0&0&0\\1&0&0\\0&0&0\\0&0&0\\0&1&0\\0&0&0\\0&0&0\\0&0&1\\0&0&0\end{smallmatrix} & \Omega & \Omega &
\begin{smallmatrix}0&0&0\\0&0&0\\1&0&0\\0&0&0\\0&0&0\\0&1&0\\0&0&0\\0&0&0\\0&0&1\end{smallmatrix} & \Omega & \Omega \\[1ex]
\Omega & \begin{smallmatrix}1&0&0\\0&0&0\\0&0&0\\0&1&0\\0&0&0\\0&0&0\\0&0&1\\0&0&0\\0&0&0\end{smallmatrix} & \Omega & \Omega &
\begin{smallmatrix}0&0&0\\1&0&0\\0&0&0\\0&0&0\\0&1&0\\0&0&0\\0&0&0\\0&0&1\\0&0&0\end{smallmatrix} & \Omega & \Omega &
\begin{smallmatrix}0&0&0\\0&0&0\\1&0&0\\0&0&0\\0&0&0\\0&1&0\\0&0&0\\0&0&0\\0&0&1\end{smallmatrix} & \Omega \\[1ex]
\Omega & \Omega & \begin{smallmatrix}1&0&0\\0&0&0\\0&0&0\\0&1&0\\0&0&0\\0&0&0\\0&0&1\\0&0&0\\0&0&0\end{smallmatrix} & \Omega & \Omega &
\begin{smallmatrix}0&0&0\\1&0&0\\0&0&0\\0&0&0\\0&1&0\\0&0&0\\0&0&0\\0&0&1\\0&0&0\end{smallmatrix} & \Omega & \Omega &
\begin{smallmatrix}0&0&0\\0&0&0\\1&0&0\\0&0&0\\0&0&0\\0&1&0\\0&0&0\\0&0&0\\0&0&1\end{smallmatrix}
\end{bmatrix}.$$

Since $P_{(2)}$ is orthogonal, $P_{(2)}^{-1} \equiv P_{(2)}^t$,

$$P_{(2)}^{-1} =
\begin{bmatrix}
\begin{smallmatrix}100000000\\000100000\\000000100\end{smallmatrix} & \Omega & \Omega \\[1ex]
\Omega & \begin{smallmatrix}100000000\\000100000\\000000100\end{smallmatrix} & \Omega \\[1ex]
\Omega & \Omega & \begin{smallmatrix}100000000\\000100000\\000000100\end{smallmatrix} \\[1ex]
\begin{smallmatrix}010000000\\000010000\\000000010\end{smallmatrix} & \Omega & \Omega \\[1ex]
\Omega & \begin{smallmatrix}010000000\\000010000\\000000010\end{smallmatrix} & \Omega \\[1ex]
\Omega & \Omega & \begin{smallmatrix}010000000\\000010000\\000000010\end{smallmatrix} \\[1ex]
\begin{smallmatrix}001000000\\000001000\\000000001\end{smallmatrix} & \Omega & \Omega \\[1ex]
\Omega & \begin{smallmatrix}001000000\\000001000\\000000001\end{smallmatrix} & \Omega \\[1ex]
\Omega & \Omega & \begin{smallmatrix}001000000\\000001000\\000000001\end{smallmatrix}
\end{bmatrix}.$$

Returning to the initial permutation relations, we get

$$
T'_{27,1} = P^{-1}_{(2)} \cdot T_{27,1} = P^{-1}_{(2)}
\begin{bmatrix}
1 \\ 2 \\ 0 \\ 0 \\ -1 \\ 0 \\ -1 \\ 1 \\ 1 \\ \hline 2 \\ 0 \\ 5 \\ 3 \\ 1 \\ 1 \\ 0 \\ 0 \\ 2 \\ \hline -1 \\ 2 \\ 1 \\ 2 \\ 0 \\ 0 \\ 0 \\ 1 \\ 0
\end{bmatrix}
=
\begin{bmatrix}
1 \\ 0 \\ -1 \\ 2 \\ 3 \\ 0 \\ -1 \\ 2 \\ 0 \\ \hline 2 \\ -1 \\ 1 \\ 0 \\ 1 \\ 0 \\ 2 \\ 0 \\ 1 \\ \hline 0 \\ 0 \\ 1 \\ 5 \\ 1 \\ 2 \\ 1 \\ 0 \\ 0
\end{bmatrix}.
$$

Since $T'_{27,1}$ is the stretched version of the fundamental tensor $(t')^{\gamma\alpha\,\circ}_{\quad\circ\,\beta}$, according to the Formula (4.36) the corresponding is $\widehat{T}'_{27,1}$, the stretched version of $(t')^{k\,i\,\circ}_{\circ\,\circ\,j}$, with expression

$$
\widehat{T}'_{27,1} = Z^{-1} \cdot T'_{27,1}; \text{ with } Z^{-1} = C^{-1} \otimes C^{-1} \otimes C^t.
$$

In our case is

$$
C = \begin{bmatrix} -1 & 1 & 0 \\ 0 & 1 & 0 \\ -1 & 0 & 3 \end{bmatrix}; \quad
C^{-1} = \begin{bmatrix} -1 & 1 & 0 \\ 0 & 1 & 0 \\ -1/3 & 1/3 & 1/3 \end{bmatrix}; \quad
C^t = \begin{bmatrix} -1 & 0 & -1 \\ 1 & 1 & 0 \\ 0 & 0 & 3 \end{bmatrix}.
$$

Then, the matrix associated with the indicated change-of-basis is

$$
Z^{-1} = (C^{-1} \otimes C^{-1}) \otimes C^t =
\left[
\begin{array}{ccc|ccc|ccc}
1 & -1 & 0 & -1 & 1 & 0 & 0 & 0 & 0 \\
0 & -1 & 0 & 0 & 1 & 0 & 0 & 0 & 0 \\
1/3 & -1/3 & -1/3 & -1/3 & 1/3 & 1/3 & 0 & 0 & 0 \\
\hline
0 & 0 & 0 & -1 & 1 & 0 & 0 & 0 & 0 \\
0 & 0 & 0 & 0 & 1 & 0 & 0 & 0 & 0 \\
0 & 0 & 0 & -1/3 & 1/3 & 1/3 & 0 & 0 & 0 \\
\hline
1/3 & -1/3 & 0 & -1/3 & 1/3 & 0 & -1/3 & 1/3 & 0 \\
0 & -1/3 & 0 & 0 & 1/3 & 0 & 0 & 1/3 & 0 \\
1/9 & -1/9 & -1/9 & -1/9 & 1/9 & 1/9 & -1/9 & 1/9 & 1/9
\end{array}
\right] \otimes C^t
$$

$$Z^{-1} = \left[
\begin{array}{ccccccccc|ccccccccc|ccccccccc}
-1 & 0 & -1 & 1 & 0 & 1 & 0 & 0 & 0 & 1 & 0 & 1 & -1 & 0 & -1 & 0 & 0 & 0 & & & & & & & & & \\
1 & 1 & 0 & -1 & -1 & 0 & 0 & 0 & 0 & -1 & -1 & 0 & 1 & 1 & 0 & 0 & 0 & 0 & & & & & & & & & \\
0 & 0 & 3 & 0 & 0 & -3 & 0 & 0 & 0 & 0 & 0 & -3 & 0 & 0 & 3 & 0 & 0 & 0 & & & & & & & & & \\
0 & 0 & 0 & 1 & 0 & 1 & 0 & 0 & 0 & 0 & 0 & 0 & -1 & 0 & -1 & 0 & 0 & 0 & & & & \Omega & & & & & \\
0 & 0 & 0 & -1 & -1 & 0 & 0 & 0 & 0 & 0 & 0 & 0 & 1 & 1 & 0 & 0 & 0 & 0 & & & & & & & & & \\
0 & 0 & 0 & 0 & 0 & -3 & 0 & 0 & 0 & 0 & 0 & 0 & 0 & 0 & 3 & 0 & 0 & 0 & & & & & & & & & \\
-\frac{1}{3} & 0 & -\frac{1}{3} & \frac{1}{3} & 0 & \frac{1}{3} & \frac{1}{3} & 0 & \frac{1}{3} & \frac{1}{3} & 0 & \frac{1}{3} & -\frac{1}{3} & 0 & -\frac{1}{3} & -\frac{1}{3} & 0 & -\frac{1}{3} & & & & & & & & & \\
\frac{1}{3} & \frac{1}{3} & 0 & -\frac{1}{3} & -\frac{1}{3} & 0 & -\frac{1}{3} & -\frac{1}{3} & 0 & -\frac{1}{3} & -\frac{1}{3} & 0 & \frac{1}{3} & \frac{1}{3} & 0 & \frac{1}{3} & \frac{1}{3} & 0 & & & & & & & & & \\
0 & 0 & 1 & 0 & 0 & -1 & 0 & 0 & -1 & 0 & 0 & -1 & 0 & 0 & 1 & 0 & 0 & 1 & & & & & & & & & \\
\hline
 & & & & & & & & & 1 & 0 & 1 & -1 & 0 & -1 & 0 & 0 & 0 & & & & & & & & & \\
 & & & & & & & & & -1 & -1 & 0 & 1 & 1 & 0 & 0 & 0 & 0 & & & & & & & & & \\
 & & & & & & & & & 0 & 0 & -3 & 0 & 0 & 3 & 0 & 0 & 0 & & & & & & & & & \\
 & & & & & & & & & 0 & 0 & 0 & -1 & 0 & -1 & 0 & 0 & 0 & & & & & & & & & \\
 & & & \Omega & & & & & & 0 & 0 & 0 & 1 & 1 & 0 & 0 & 0 & 0 & & & & \Omega & & & & & \\
 & & & & & & & & & 0 & 0 & 0 & 0 & 0 & 3 & 0 & 0 & 0 & & & & & & & & & \\
 & & & & & & & & & \frac{1}{3} & 0 & \frac{1}{3} & -\frac{1}{3} & 0 & -\frac{1}{3} & -\frac{1}{3} & 0 & -\frac{1}{3} & & & & & & & & & \\
 & & & & & & & & & -\frac{1}{3} & -\frac{1}{3} & 0 & \frac{1}{3} & \frac{1}{3} & 0 & \frac{1}{3} & \frac{1}{3} & 0 & & & & & & & & & \\
 & & & & & & & & & 0 & 0 & -1 & 0 & 0 & 1 & 0 & 0 & 1 & & & & & & & & & \\
\hline
-\frac{1}{3} & 0 & -\frac{1}{3} & \frac{1}{3} & 0 & \frac{1}{3} & 0 & 0 & 0 & \frac{1}{3} & 0 & \frac{1}{3} & -\frac{1}{3} & 0 & -\frac{1}{3} & 0 & 0 & 0 & \frac{1}{3} & 0 & \frac{1}{3} & -\frac{1}{3} & 0 & -\frac{1}{3} & 0 & 0 & 0 \\
\frac{1}{3} & \frac{1}{3} & 0 & -\frac{1}{3} & -\frac{1}{3} & 0 & 0 & 0 & 0 & -\frac{1}{3} & -\frac{1}{3} & 0 & \frac{1}{3} & \frac{1}{3} & 0 & 0 & 0 & 0 & -\frac{1}{3} & -\frac{1}{3} & 0 & \frac{1}{3} & \frac{1}{3} & 0 & 0 & 0 & 0 \\
0 & 0 & 1 & 0 & 0 & -1 & 0 & 0 & 0 & 0 & 0 & -1 & 0 & 0 & 1 & 0 & 0 & 0 & 0 & 0 & -1 & 0 & 0 & 1 & 0 & 0 & 0 \\
0 & 0 & 0 & \frac{1}{3} & 0 & \frac{1}{3} & 0 & 0 & 0 & 0 & 0 & 0 & -\frac{1}{3} & 0 & -\frac{1}{3} & 0 & 0 & 0 & 0 & 0 & 0 & -\frac{1}{3} & 0 & -\frac{1}{3} & 0 & 0 & 0 \\
0 & 0 & 0 & -\frac{1}{3} & -\frac{1}{3} & 0 & 0 & 0 & 0 & 0 & 0 & 0 & \frac{1}{3} & \frac{1}{3} & 0 & 0 & 0 & 0 & 0 & 0 & 0 & \frac{1}{3} & \frac{1}{3} & 0 & 0 & 0 & 0 \\
0 & 0 & 0 & 0 & 0 & -1 & 0 & 0 & 0 & 0 & 0 & 0 & 0 & 0 & 1 & 0 & 0 & 0 & 0 & 0 & 0 & 0 & 0 & 1 & 0 & 0 & 0 \\
-\frac{1}{9} & 0 & -\frac{1}{9} & \frac{1}{9} & 0 & \frac{1}{9} & 0 & 0 & \frac{1}{9} & \frac{1}{9} & 0 & \frac{1}{9} & -\frac{1}{9} & 0 & -\frac{1}{9} & -\frac{1}{9} & 0 & -\frac{1}{9} & \frac{1}{9} & 0 & \frac{1}{9} & -\frac{1}{9} & 0 & -\frac{1}{9} & -\frac{1}{9} & 0 & -\frac{1}{9} \\
\frac{1}{9} & \frac{1}{9} & 0 & -\frac{1}{9} & -\frac{1}{9} & 0 & -\frac{1}{9} & -\frac{1}{9} & 0 & -\frac{1}{9} & -\frac{1}{9} & 0 & \frac{1}{9} & \frac{1}{9} & 0 & \frac{1}{9} & \frac{1}{9} & 0 & -\frac{1}{9} & -\frac{1}{9} & 0 & \frac{1}{9} & \frac{1}{9} & 0 & \frac{1}{9} & \frac{1}{9} & 0 \\
0 & 0 & \frac{1}{3} & 0 & 0 & -\frac{1}{3} & 0 & 0 & -\frac{1}{3} & 0 & 0 & -\frac{1}{3} & 0 & 0 & \frac{1}{3} & 0 & 0 & \frac{1}{3} & 0 & 0 & -\frac{1}{3} & 0 & 0 & \frac{1}{3} & 0 & 0 & \frac{1}{3}
\end{array}
\right]$$

$$\hat{T}'_{27,1} = Z^{-1}
\begin{bmatrix}
1 \\ 0 \\ -1 \\ 2 \\ 3 \\ 0 \\ -1 \\ 2 \\ 0 \\ -- \\ 2 \\ -1 \\ 1 \\ 0 \\ 1 \\ 0 \\ 2 \\ 0 \\ 1 \\ -- \\ 0 \\ 0 \\ 1 \\ 5 \\ 1 \\ 2 \\ 1 \\ 0 \\ 0
\end{bmatrix}
=
\begin{bmatrix}
5 \\ -4 \\ -6 \\ 2 \\ -4 \\ 0 \\ 1/3 \\ -1 \\ -1 \\ -- \\ 3 \\ 0 \\ -3 \\ 0 \\ 1 \\ 0 \\ 0 \\ 2/3 \\ 0 \\ -- \\ -1/3 \\ 2/3 \\ -1 \\ -5/3 \\ 2/3 \\ 2 \\ -2/3 \\ 4/9 \\ 0
\end{bmatrix}.$$

Once the change-of-basis has been performed, we must return to the data permutation:

$$P_{(2)} \cdot \hat{T}'_{27,1} = \hat{T}_{27,1}$$

yielding

$$
\hat{T}_{27} =
\begin{bmatrix}
5 \\
3 \\
-1/3 \\
-4 \\
0 \\
2/3 \\
-6 \\
-3 \\
-1 \\
\hline
2 \\
0 \\
-5/3 \\
-4 \\
1 \\
2/3 \\
0 \\
0 \\
2 \\
\hline
1/3 \\
0 \\
-2/3 \\
-1 \\
2/3 \\
4/9 \\
-1 \\
0 \\
0
\end{bmatrix}
$$

which after its condensation leads to

$$
t^{iok}_{\ ojo} =
\left[
\begin{array}{rrr|rrr|rrr}
5 & 3 & -1/3 & 2 & 0 & -5/3 & 1/3 & 0 & -2/3 \\
-4 & 0 & 2/3 & -4 & 1 & 2/3 & -1 & 2/3 & 4/9 \\
-6 & -3 & -1 & 0 & 0 & 2 & -1 & 0 & 0
\end{array}
\right],
$$

where i is the row index, j is the column index, and k is the matrix index. □

5.7.3 The isomers of a tensor

We give the name "isomers" to certain tensors that come from permutations of a given tensor; they are the isomeric tensors of such a tensor.

For pure tensors (totally contravariant or covariant) the permutation of a partial number (or all) of its indices, strictly between them, leads to an isomer.

In other words, not all permutation tensors coming from a *pure* tensor are isomers of such a tensor, since some of them do not come from altering the indices. If the tensor is a mixed tensor, they are the tensors coming from permuting partially or totally: (a) only the contravariant indices among them, without altering the covariant indices, and (b) only the covariant indices among them, without altering the contravariant indices.

In Example 4.5, which was examined in Section 5.7.2, the tensor $R = [r^{\gamma\,o\,\alpha}_{\ o\,\beta\,o}]$ is an isomer of tensor $T = [t^{\alpha\,o\,\gamma}_{\ o\,\beta\,o}]$. Similarly, the tensor $U = [u^{\gamma\alpha\,o}_{\ \ o\,o\,\beta}]$ is an isomer of tensor $W = [w^{\alpha\gamma\,o}_{\ \ o\,o\,\beta}]$.

Example 5.5 (Rotation tensor). Consider a given pure tensor (totally con-travariant or covariant) of order r: $t_{\circ\circ\circ\circ\cdots\circ}^{\alpha\beta\gamma\delta\cdots\rho}$. We define as the "rotation ten-sor" of the given tensor any of its isomers that *do not* maintain indices in the same positions as the initial one.

We will denote such a tensor $(t_{\circ\circ\circ\circ\cdots\circ}^{\alpha\beta\gamma\delta\cdots\rho})^{R(k)}$ where $k \in Z; k \neq 0; |k| < r$ is the "rotation index".

By extension, we define as the rotation tensor of a given mixed tensor all those isomers that do not maintain dummy indices of the *same valency*, in the same positions.

These rotation tensors carry the notation $(t_{\circ\beta\circ\circ\epsilon\cdots\circ}^{\alpha\circ\gamma\delta\circ\cdots\rho})^{R(k,k')}$ with two index-parameters k, k', where $k, k' \in Z; k, k' \neq 0; |k| < p; |k'| < q$ with $p + q = r$, where p and q are the contravariant and covariant orders of the given tensor, respectively.

1. Determine the rotation tensor associated with a tensor of order $(r = 2)$ over the linear space $V^2(\mathbb{R})$, and do the same over the linear space $V^3(\mathbb{R})$.

2. Determine the rotation tensors associated with a tensor of order $(r = 3)$ over the linear space $V^2(\mathbb{R})$, and do the same over the linear space $V^3(\mathbb{R})$.

3. Determine the rotation tensors associated with a tensor of order $(r = 4)$ over the linear space $V^2(\mathbb{R})$.

Solution:

1. Case $r = 2, n = 2$. Let $\vec{t} = t_{\alpha\beta}^{\circ\circ}\vec{e}^{*\alpha} \otimes \vec{e}^{*\beta} \Rightarrow (t_{\alpha\beta}^{\circ\circ})^{R(1)} = t_{\beta\alpha}^{\circ\circ}\vec{e}^{*\alpha} \otimes \vec{e}^{*\beta}$.

   Then, $[t_{\alpha\beta}^{\circ\circ}] = \begin{bmatrix} a_1 & b_1 \\ c_1 & d_1 \end{bmatrix} \Rightarrow [t_{\alpha\beta}^{\circ\circ}]^{R(1)} = \begin{bmatrix} a_1 & c_1 \\ b_1 & d_1 \end{bmatrix}$, which is known as the "transposed" matrix of the given matrix.

   Case $r = 2, n = 3$. In this case we have

   $$[t_{\alpha\beta}^{\circ\circ}] = \begin{bmatrix} a_1 & b_1 & c_1 \\ d_1 & e_1 & f_1 \\ g_1 & h_1 & i_1 \end{bmatrix} \Rightarrow [t_{\alpha\beta}^{\circ\circ}]^{R(1)} = \begin{bmatrix} a_1 & d_1 & g_1 \\ b_1 & e_1 & h_1 \\ c_1 & f_1 & i_1 \end{bmatrix}$$

   which also is the transposed matrix of the given matrix.

2. Case $r = 3, n = 2$. Let $\vec{t} = t_{\alpha\beta\gamma}^{\circ\circ\circ}\vec{e}^{*\alpha} \otimes \vec{e}^{*\beta} \otimes \vec{e}^{*\gamma}$. Since there are *two* rotations $\beta\gamma\alpha$ and $\gamma\alpha\beta$ we have $[t_{\alpha\beta\gamma}^{\circ\circ\circ}]^{R(1)}$ and $[t_{\alpha\beta\gamma}^{\circ\circ\circ}]^{R(2)}$, or, if one prefers the notation $[t_{\alpha\beta\gamma}^{\circ\circ\circ}]^{R(1)}$ and $[t_{\alpha\beta\gamma}^{\circ\circ\circ}]^{R(-1)}$, the "first rotation" and its opposite.

Let $[t^{\circ\circ\circ}_{\alpha\beta\gamma}] = \begin{bmatrix} a_1\, b_1 \\ c_1\, d_1 \\ -\ - \\ a_2\, b_2 \\ c_2\, d_2 \end{bmatrix}$ be the data tensor, with α the row submatrix in-

dex, β the row index of each submatrix, and γ the column index of each
submatrix (axiomatic order).

The correspondences in both rotations are

	First rotation	Second rotation
Initial	Transformed	Transformed
$t^{\circ\circ\circ}_{\alpha\beta\gamma}$	$t^{\circ\circ\circ}_{\beta\gamma\alpha}$	$t^{\circ\circ\circ}_{\gamma\alpha\beta}$
$t^{\circ\circ\circ}_{112}$	$t^{\circ\circ\circ}_{121}$	$t^{\circ\circ\circ}_{211}$
$t^{\circ\circ\circ}_{121}$	$t^{\circ\circ\circ}_{211}$	$t^{\circ\circ\circ}_{112}$
$t^{\circ\circ\circ}_{122}$	$t^{\circ\circ\circ}_{221}$	$t^{\circ\circ\circ}_{212}$
$t^{\circ\circ\circ}_{211}$	$t^{\circ\circ\circ}_{112}$	$t^{\circ\circ\circ}_{121}$
$t^{\circ\circ\circ}_{212}$	$t^{\circ\circ\circ}_{122}$	$t^{\circ\circ\circ}_{221}$
$t^{\circ\circ\circ}_{221}$	$t^{\circ\circ\circ}_{212}$	$t^{\circ\circ\circ}_{122}$

and then

$$[t^{\circ\circ\circ}_{\alpha\beta\gamma}]^{R(1)} = \begin{bmatrix} a_1 & a_2 \\ b_1 & b_2 \\ - & - \\ c_1 & c_2 \\ d_1 & d_2 \end{bmatrix}$$

and

$$[t^{\circ\circ\circ}_{\alpha\beta\gamma}]^{R(2)} = \begin{bmatrix} a_1 & c_1 \\ a_2 & c_2 \\ - & - \\ b_1 & d_1 \\ b_2 & d_2 \end{bmatrix},$$

which are the "transposed" (beware of the word) tensors of tensors ($r = 3, n = 2$).

Case $r = 3, n = 3$. Let

$$[t^{\circ\circ\circ}_{\alpha\beta\gamma}] = \begin{bmatrix} a_1\, b_1\, c_1 \\ d_1\, e_1\, f_1 \\ g_1\, h_1\, i_1 \\ -\ -\ - \\ a_2\, b_2\, c_2 \\ d_2\, e_2\, f_2 \\ g_2\, h_2\, i_2 \\ -\ -\ - \\ a_3\, b_3\, c_3 \\ d_3\, e_3\, f_3 \\ g_3\, h_3\, i_3 \end{bmatrix}$$

be the data tensor. In this case we have

$$[t^{\ \circ\ \circ\ \circ}_{\alpha\beta\gamma}]^{R(1)} = \begin{bmatrix} a_1\, a_2\, a_3 \\ b_1\, b_2\, b_3 \\ c_1\, c_2\, c_3 \\ \hline d_1\, d_2\, d_3 \\ e_1\, e_2\, e_3 \\ f_1\, f_2\, f_3 \\ \hline g_1\, g_2\, g_3 \\ h_1\, h_2\, h_3 \\ i_1\, i_2\, i_3 \end{bmatrix}$$

and

$$[t^{\ \circ\ \circ\ \circ}_{\alpha\beta\gamma}]^{R(2)} = \begin{bmatrix} a_1\, d_1\, g_1 \\ a_2\, d_2\, g_2 \\ a_3\, d_3\, g_3 \\ \hline b_1\, e_1\, h_1 \\ b_2\, e_2\, h_2 \\ b_3\, e_3\, h_3 \\ \hline c_1\, f_1\, i_1 \\ c_2\, f_2\, i_2 \\ c_3\, f_3\, i_3 \end{bmatrix}.$$

3. Case $r = 4, n = 2$. The data tensor is $\vec{t} = t^{\ \circ\ \circ\ \circ\ \circ}_{\alpha\beta\gamma\delta}\vec{e}^{*\alpha} \otimes \vec{e}^{*\beta} \otimes \vec{e}^{*\gamma} \otimes \vec{e}^{*\delta}$, where

$$[t^{\ \circ\ \circ\ \circ\ \circ}_{\alpha\beta\gamma\delta}] = \left[\begin{array}{cc|cc} a_1\, b_1 & a_2\, b_2 \\ c_1\, d_1 & c_2\, d_2 \\ \hline a_3\, b_3 & a_4\, b_4 \\ c_3\, d_3 & c_4\, d_4 \end{array} \right].$$

In this case $\alpha\beta\gamma\delta$ has the three rotations: $\beta\gamma\delta\alpha$, $\gamma\delta\alpha\beta$ and $\delta\alpha\beta\gamma$. Then, for the rotation $R(1)$, since $t^{\ \circ\ \circ\ \circ\ \circ}_{\alpha\beta\gamma\delta} \to t^{\ \circ\ \circ\ \circ\ \circ}_{\beta\gamma\delta\alpha}$, we have

$$[t^{\ \circ\ \circ\ \circ\ \circ}_{\alpha\beta\gamma\delta}]^{R(1)} = \left[\begin{array}{cc|cc} a_1\, a_3 & c_1\, c_3 \\ b_1\, b_3 & d_1\, d_3 \\ \hline a_2\, a_4 & c_2\, c_4 \\ b_2\, b_4 & d_2\, d_4 \end{array} \right].$$

For the rotation $R(2)$, since $t^{\ \circ\ \circ\ \circ\ \circ}_{\alpha\beta\gamma\delta} \to t^{\ \circ\ \circ\ \circ\ \circ}_{\gamma\delta\alpha\beta}$, we have

$$[t^{\ \circ\ \circ\ \circ\ \circ}_{\alpha\beta\gamma\delta}]^{R(2)} = \left[\begin{array}{cc|cc} a_1\, a_2 & b_1\, b_2 \\ a_3\, a_4 & b_3\, b_4 \\ \hline c_1\, c_2 & d_1\, d_2 \\ c_3\, c_4 & d_3\, d_4 \end{array} \right].$$

and, finally, for the rotation $R(3)$, since $t^{\ \circ\ \circ\ \circ\ \circ}_{\alpha\beta\gamma\delta} \to t^{\ \circ\ \circ\ \circ\ \circ}_{\delta\alpha\beta\gamma}$, we obtain

$$[t^{\ \circ\ \circ\ \circ\ \circ}_{\alpha\beta\gamma\delta}]^{R(3)} = \left[\begin{array}{cc|cc} a_1\, c_1 & a_3\, c_3 \\ a_2\, c_2 & a_4\, c_4 \\ \hline b_1\, d_1 & b_3\, d_3 \\ b_2\, d_2 & b_4\, d_4 \end{array} \right].$$

$\square$

5.8 Matrices associated with simply contraction homomorphisms

We give the name "simply contraction homomorphism" to a homomorphism that operates according to the equation

$$T'_{\sigma'} = H \bullet T_{\sigma,1} \tag{5.61}$$

and that apply the mixed tensor (because the contraction is assumed to be of indices of different valency), into another tensor of *smaller* dimension $\sigma' < \sigma$, by a tensor contraction of *two* indices.

We will construct the matrices H for the usual cases, that is, for tensors of orders 2, 3, 4 and 5.

5.8.1 Mixed tensors of second order ($r = 2$): Matrices.

This is the case of $T = [t^{\circ\,\beta}_{\alpha\circ}]$ or $T = [t^{\alpha\,\circ}_{\circ\,\beta}]$, with $r = 2; \sigma = n^2; n = \dim V^n(K); \sigma' = n^0 = 0$.

Thus, the result of the contraction is a scalar, which is called the "matrix trace".

Assuming that $\{E_{ij}\}$ is the canonical basis of the matrices of order n, the fundamental equation (5.61) is in this case

$$\rho = H_{1,n^2}(\alpha, \beta) \bullet T_{\sigma,1} = ([1\ 1\ \cdots\ 1]_{1,n} \bullet [\,E_{11}\,|\,E_{22}\,|\,\cdots\,|\,E_{nn}\,]) \bullet T_{\sigma,1}$$
$$= [\,E^t_1\,|\,E^t_2\,|\,\cdots\,|\,E^t_n\,] \bullet T_{\sigma,1}; \rho \in K; T'_{\sigma'} = \rho. \tag{5.62}$$

The notation of H declares its number of rows and columns, together with the indices to be contracted.

$\mathcal{B} = \{E_i\}$ is the canonical basis of the linear space $V^n(K)$:

$$\mathcal{B} = \left\{ \begin{bmatrix} 1 \\ 0 \\ \vdots \\ 0 \end{bmatrix} \begin{bmatrix} 0 \\ 1 \\ \vdots \\ 0 \end{bmatrix} \cdots \begin{bmatrix} 0 \\ 0 \\ \vdots \\ 1 \end{bmatrix} \right\}.$$

5.8.2 Mixed tensors of third order ($r = 3$)

These are tensors of the type $T = t^{\alpha\,\circ\,\circ}_{\circ\,\beta\gamma}; T = t^{\alpha\,\circ\,\gamma}_{\circ\,\beta\circ}\cdots$, etc. with $\dim V^n(K) = n; r = 3; \sigma = n^3; \sigma' = n$.

There are three possible models:

Model 1. $T = t^{\alpha\beta\circ}_{\circ\,\circ\,\gamma}$ or $T = t^{\alpha\,\circ\,\gamma}_{\circ\,\beta\,\circ}$

$$T'_{\sigma'} = u^\alpha = H_{n,n^3}(\beta, \gamma) \bullet T_{\sigma,1} = (I_n \otimes ([1\,1\cdots 1]_{1,n} \bullet [\,E_{11}\,|\,E_{22}\,|\,\cdots\,|\,E_{nn}\,])) \bullet T_{\sigma,1}. \tag{5.63}$$

Model 2. $T = t_{\circ\beta\circ}^{\alpha\circ\gamma}$ or $T = t_{\alpha\circ\circ}^{\circ\beta\gamma}$

$$T'_{\sigma'} = v^\gamma = H_{n,n^3}(\alpha,\beta) \bullet T_{\sigma,1} = (([1\,1\cdots 1]_{1,n} \bullet [E_{11}\,|\,E_{22}\,|\cdots|\,E_{nn}]) \otimes I_n) \bullet T_{\sigma,1}. \tag{5.64}$$

Model 3. $T = t_{\circ\circ\gamma}^{\alpha\beta\circ}$ or $T = t_{\alpha\circ\circ}^{\circ\beta\gamma}$

$$\begin{aligned}
T'_{\sigma'} = z^\beta &= H_{n,n^3}(\alpha,\gamma) \bullet T_{\sigma,1} \\
&= [I_n \otimes ([1\,1\cdots 1]_{1,n} \bullet E_{11})|I_n \otimes ([1\,1\cdots 1]_{1,n} \\
&\quad \bullet E_{22})|\ldots|I_n \otimes ([1\,1\cdots 1]_{1,n} \bullet E_{nn})] \bullet T_{\sigma,1}. \tag{5.65}
\end{aligned}$$

Formulas (5.63) to (5.65) can be written in a simpler form:

$$u^\alpha = H_{n,n^3}(\beta,\gamma) \bullet T_{\sigma,1} = \left(I_n \otimes [E_1^t \mid E_2^t \mid \cdots \mid E_n^t]\right) \bullet T_{\sigma,1} \tag{5.66}$$

$$v^\gamma = H_{n,n^3}(\alpha,\beta) \bullet T_{\sigma,1} = \left([E_1^t \mid E_2^t \mid \cdots \mid E_n^t] \otimes I_n\right) \bullet T_{\sigma,1} \tag{5.67}$$

$$z^\beta = H_{n,n^3}(\alpha,\gamma) \bullet T_{\sigma,1} = \left[I_n \otimes E_1^t|\, I_n \otimes E_2^t|\cdots|I_n \otimes E_n^t\right] \bullet T_{\sigma,1}. \tag{5.68}$$

5.8.3 Mixed tensors of fourth order ($r = 4$)

These are tensors of the type

$$T = t_{\circ\beta\circ\circ}^{\alpha\circ\gamma\delta}; \quad T = t_{\circ\circ\circ\delta}^{\alpha\beta\gamma\circ}, \cdots \text{ etc. };\, \sigma \equiv n^r = n^4;\, \sigma' \equiv n^{r-2} = n^{4-2} = n^2.$$

Possibilities for the contraction: $\binom{4}{2} = 6$ models.

Model 1. $T = t_{\circ\beta\circ\circ}^{\alpha\circ\gamma\delta}$; The fundamental equation (5.61) in this case is

$$T'_{\sigma'} = H_{n^2,n^4}(\alpha,\beta) \bullet T_{\sigma,1} = \left([E_1^t \mid E_2^t \mid \cdots \mid E_n^t] \otimes I_n \otimes I_n\right) \bullet T_{\sigma,1} \tag{5.69}$$

Model 2. $T = t_{\circ\circ\gamma\circ}^{\alpha\beta\circ\delta}$

$$T'_{\sigma'} = H_{n^2,n^4}(\alpha,\gamma) \bullet T_{\sigma,1} = \left([I_n \otimes E_1^t \mid I_n \otimes E_2^t \mid \cdots \mid I_n \otimes E_n^t] \otimes I_n\right) \bullet T_{\sigma,1}. \tag{5.70}$$

Model 3. $T = t_{\circ\circ\circ\delta}^{\alpha\beta\gamma\circ}$

$$T'_{\sigma'} = H_{n^2,n^4}(\alpha,\delta) \bullet T_{\sigma,1} = [I_n \otimes I_n \otimes E_1^t|I_n \otimes I_n \otimes E_2^t|\cdots|I_n \otimes I_n \otimes E_n^t] \bullet T_{\sigma,1}. \tag{5.71}$$

Model 4. $T = t_{\circ\circ\gamma\circ}^{\alpha\beta\circ\delta}$

$$T'_{\sigma'} = H_{n^2,n^4}(\beta,\gamma) \bullet T_{\sigma,1} = \left(I_n \otimes [E_1^t \mid E_2^t \mid \cdots \mid E_n^t] \otimes I_n\right) \bullet T_{\sigma,1}. \tag{5.72}$$

Model 5. $T = t_{\circ\circ\circ\delta}^{\alpha\beta\gamma\circ}$

$$T'_{\sigma'} = H_{n^2,n^4}(\beta,\delta) \bullet T_{\sigma,1} = \left(I_n \otimes [I_n \otimes E_1^t \mid I_n \otimes E_2^t \mid \cdots \mid I_n \otimes E_n^t]\right) \bullet T_{\sigma,1}. \tag{5.73}$$

Model 6. $T = t_{\circ\circ\circ\delta}^{\alpha\beta\gamma\circ}$

$$T'_{\sigma'} = H_{n^2,n^4}(\gamma,\delta) \bullet T_{\sigma,1} = \left(I_n \otimes I_n \otimes [E_1^t \mid E_2^t \mid \cdots \mid E_n^t]\right) \bullet T_{\sigma,1}. \tag{5.74}$$

$T'_{\sigma',1}$ must be given in "condensed" form (as a square matrix).

5.8.4 Mixed tensors of fifth order ($r = 5$)

We present rules for the sequence of formation of matrices representing contractions in tensors of order 5 ($r = 5$) associated with linear spaces $\mathbb{R}^n$ of basis

$$\{E_i\} \equiv \left\{ \begin{bmatrix} 1 \\ 0 \\ 0 \\ \vdots \\ 0 \end{bmatrix} \begin{bmatrix} 0 \\ 1 \\ 0 \\ \vdots \\ 0 \end{bmatrix} \cdots \begin{bmatrix} 0 \\ 0 \\ 0 \\ \vdots \\ 1 \end{bmatrix} \right\}.$$

We notate the morphism matrix using power indices and parentheses that declare the indices to be contracted.

Contractions of two indices, resulting tensors of order $r = 3$.

Tensor dimensions of the "stretched" tensors $T_{\sigma,1}$ and $T_{\sigma'}$: $\sigma = n^5$; $\sigma' = n^3$.

There exist $\binom{5}{2} = \frac{5 \times 4}{2} = 10$ models. Operation: $T'_{\sigma',1} = H_{\sigma',\sigma} \bullet T_{\sigma,1}$

Model 1. $t^{\alpha \circ \gamma \delta \epsilon}_{\circ \beta \circ \circ \circ}$

$$H_{\sigma',\sigma} \equiv H_{n^3,n^5}(\alpha,\beta) \equiv \left[E_1^t \mid E_2^t \mid \cdots \mid E_n^t \right] \otimes I_n \otimes I_n \otimes I_n.$$

Model 2. $t^{\alpha \beta \circ \delta \epsilon}_{\circ \circ \gamma \circ \circ}$

$$H_{\sigma',\sigma} \equiv H_{n^3,n^5}(\alpha,\gamma) \equiv [I_n \otimes E_1^t \mid I_n \otimes E_2^t \mid \cdots \mid I_n \otimes E_n^t] \otimes I_n \otimes I_n.$$

Model 3. $t^{\alpha \beta \gamma \circ \epsilon}_{\circ \circ \circ \delta \circ}$

$$H_{\sigma',\sigma} \equiv H_{n^3,n^5}(\alpha,\delta)$$
$$\equiv [I_n \otimes I_n \otimes E_1^t \mid I_n \otimes I_n \otimes E_2^t \mid \cdots \mid I_n \otimes I_n \otimes E_n^t] \otimes I_n.$$

Model 4. $t^{\alpha \beta \gamma \delta \circ}_{\circ \circ \circ \circ \epsilon}$

$$H_{\sigma',\sigma} \equiv H_{n^3,n^5}(\alpha,\epsilon)$$
$$\equiv [I_n \otimes I_n \otimes I_n \otimes E_1^t \mid I_n \otimes I_n \otimes I_n \otimes E_2^t \mid \cdots \mid I_n \otimes I_n \otimes I_n \otimes E_n^t].$$

Model 5. $t^{\alpha \beta \circ \delta \epsilon}_{\circ \circ \gamma \circ \circ}$

$$H_{\sigma',\sigma} \equiv H_{n^3,n^5}(\beta,\gamma) \equiv I_n \otimes [E_1^t \mid E_2^t \mid \cdots \mid E_n^t] \otimes I_n \otimes I_n.$$

Model 6. $t^{\alpha \beta \gamma \circ \epsilon}_{\circ \circ \circ \delta \circ}$

$$H_{\sigma',\sigma} \equiv H_{n^3,n^5}(\beta,\delta) \equiv I_n \otimes [I_n \otimes E_1^t \mid I_n \otimes E_2^t \mid \cdots \mid I_n \otimes E_n^t] \otimes I_n.$$

Model 7. $t^{\alpha\beta\gamma\delta\circ}_{\circ\circ\circ\circ\epsilon}$

$$H_{\sigma',\sigma} \equiv H_{n^3,n^5}(\beta,\epsilon)$$
$$\equiv I_n \otimes [I_n \otimes I_n \otimes E_1^t \,|\, I_n \otimes I_n \otimes E_2^t \,|\cdots|\, I_n \otimes I_n \otimes E_n^t].$$

Model 8. $t^{\alpha\beta\gamma\circ\epsilon}_{\circ\circ\circ\delta\circ}$

$$H_{\sigma',\sigma} \equiv H_{n^3,n^5}(\gamma,\delta) \equiv I_n \otimes I_n \otimes \left[E_1^t \,|\, E_2^t \,|\cdots|\, E_n^t\right] \otimes I_n.$$

Model 9. $t^{\alpha\beta\gamma\delta\circ}_{\circ\circ\circ\circ\epsilon}$

$$H_{\sigma',\sigma} \equiv H_{n^3,n^5}(\gamma,\epsilon) \equiv I_n \otimes I_n \otimes \left[I_n \otimes E_1^t \,|\, I_n \otimes E_2^t \,|\cdots|\, I_n \otimes E_n^t\right].$$

Model 10. $t^{\alpha\beta\gamma\delta\circ}_{\circ\circ\circ\circ\epsilon}$

$$H_{\sigma',\sigma} \equiv H_{n^3,n^5}(\delta,\epsilon) \equiv I_n \otimes I_n \otimes I_n \otimes \left[E_1^t \,|\, E_2^t \,|\cdots|\, E_n^t\right].$$

$T'_{\sigma',1}$ must be given in "condensed" form (as a column-matrix of submatrices).

5.9 Matrices associated with doubly contracted homomorphisms

5.9.1 Mixed tensors of fourth order ($r = 4$)

We look for the tensor resulting from a homogeneous mixed tensor that accepts a double contraction, that is, has at least two contravariant indices and other two covariant indices; $\sigma = n^4; \sigma' = n^0 = 1$. The resulting tensor after the double contraction always is a scalar.

The possibilities for the contraction are: $\binom{4}{2} = 6$ models. Let $\rho \in K$.

Model 1. $H_{1,n^4}(\alpha, \beta | \gamma, \delta)$ means that we contract first indices (α, β) and then, indices (γ, δ):

$$\rho = H_{1,n^4}(\alpha, \beta | \gamma, \delta) \bullet T_{\sigma,1} = ([E_1^t \,|\, E_2^t \,|\cdots|\, E_n^t] \otimes [E_1^t \,|\, E_2^t \,|\cdots|\, E_n^t]) \bullet T_{\sigma,1}.$$
$$(5.75)$$

Model 2. $H_{1,n^4}(\alpha, \gamma | \beta, \delta)$.

$$\rho = H_{1,n^4}(\alpha, \gamma | \beta, \delta) \bullet T_{\sigma,1}$$
$$= [E_1^t \otimes E_1^t | E_1^t \otimes E_2^t |\cdots| E_1^t \otimes E_n^t |\cdots| E_n^t \otimes E_1^t | E_n^t \otimes E_2^t |\cdots| E_n^t \otimes E_n^t] \bullet T_{\sigma,1}.$$
$$(5.76)$$

Model 3. $H_{1,n^4}(\alpha,\delta|\beta,\gamma)$.

$$\rho = H_{1,n^4}(\alpha,\delta|\beta,\gamma) \bullet T_{\sigma,1}$$
$$= [[E_1^t|E_2^t|\cdots|E_n^t] \otimes E_1^t|[E_1^t|E_2^t|\cdots|E_n^t] \otimes E_2^t|\cdots|[E_1^t|E_2^t|\cdots|E_n^t] \otimes E_n^t] \bullet T_{\sigma,1}.$$
$$(5.77)$$

Model 4. $H_{1,n^4}(\beta,\gamma|\alpha,\delta)$

$$\rho = H_{1,n^4}(\beta,\gamma|\alpha,\delta) \bullet T_{\sigma,1} \equiv H_{1,n^4}(\alpha,\delta|\beta,\gamma) \bullet T_{\sigma,1}. \qquad (5.78)$$

Model 5. $H_{1,n^4}(\beta,\delta|\alpha,\gamma)$

$$\rho = H_{1,n^4}(\beta,\delta|\alpha,\gamma) \bullet T_{\sigma,1} \equiv H_{1,n^4}(\alpha,\gamma|\beta,\delta) \bullet T_{\sigma,1}. \qquad (5.79)$$

Model 6. $H_{1,n^4}(\gamma,\delta|\alpha,\beta)$

$$\rho = H_{1,n^4}(\gamma,\delta|\alpha,\beta) \bullet T_{\sigma,1} \equiv H_{1,n^4}(\alpha,\beta|\gamma,\delta) \bullet T_{\sigma,1}. \qquad (5.80)$$

As a mapping of the simple contraction formulas (5.72) and those of extension and condensation (1.30) and (1.32), respectively, we propose that reader establish the direct relation between the classic product of matrices $(A \bullet B)$ where $A = [a_{\circ\beta}^{\alpha\circ}]$ and $B = [b_{\circ\delta}^{\gamma\circ}]$) and its tensor product $(A \otimes B)$, simplifying the resulting expression.

We remind the reader that the classic product of matrices $(A \bullet B)$ is a contracted tensor product.

5.9.2 Mixed tensors of fifth order $(r = 5)$

The contraction of four indices leads to tensors of order $(r = 1)$, that is, vectors. The dimensions of the "extended" tensors $T_{\sigma,1}$ and $T_{\sigma'}$ are $\sigma = n^5$ and $\sigma' = n$.

There exist $\binom{5}{2} \times \binom{3}{2} = \frac{5\times4}{2} \times 3 = 30$ models of double contraction.

Model 1. $t_{\circ\beta\circ\delta\circ}^{\alpha\circ\gamma\circ\epsilon}$

$$H_{\sigma',\sigma} \equiv H_{n,n^5}(\alpha,\beta|\gamma,\delta) = H_{n,n^3}(\gamma,\delta) \bullet H_{n^3,n^5}(\alpha,\beta)$$
$$= \left(\left[E_1^t\,|\,E_2^t\,|\cdots|\,E_n^t\right] \otimes I_n\right) \bullet \left(\left[E_1^t\,|\,E_2^t\,|\cdots|\,E_n^t\right] \otimes I_n \otimes I_n \otimes I_n\right).$$

Model 2. $t_{\circ\beta\circ\circ\epsilon}^{\alpha\circ\gamma\delta\circ}$

$$H_{\sigma',\sigma} \equiv H_{n,n^5}(\alpha,\beta|\gamma,\epsilon) = H_{n,n^3}(\gamma,\epsilon) \bullet H_{n^3,n^5}(\alpha,\beta)$$
$$= \left(\left[I_n \otimes E_1^t\,|\,I_n \otimes E_2^t\,|\cdots|\,I_n \otimes E_n^t\right]\right) \bullet \left(\left[E_1^t\,|\,E_2^t\,|\cdots|\,E_n^t\right] \otimes I_n \otimes I_n \otimes I_n\right).$$

Model 3. $t_{\circ\beta\circ\circ\epsilon}^{\alpha\circ\gamma\delta\circ}$

$$H_{\sigma',\sigma} \equiv H_{n,n^5}(\alpha,\beta|\delta,\epsilon) = H_{n,n^3}(\delta,\epsilon) \bullet H_{n^3,n^5}(\alpha,\beta)$$
$$= \left(I_n \otimes \left[E_1^t\,|\,E_2^t\,|\cdots|\,E_n^t\right]\right) \bullet \left(\left[E_1^t\,|\,E_2^t\,|\cdots|\,E_n^t\right] \otimes I_n \otimes I_n \otimes I_n\right).$$

In a similar form the remaining models can be obtained.

Example 5.6 (Tensor contraction). Contract all indices of the tensor

$$\vec{t} = (2\vec{e}_1 - 3\vec{e}_2) \otimes (5\vec{e}_1 + \vec{e}_2) \otimes (4\vec{e}^{*1} + \vec{e}^{*2}) \otimes (\vec{e}^{*1} - 2\vec{e}^{*2}).$$

Solution: We solve the problem using four different methods.

First method:

We decide to execute the contractions of each pair of contravariant factors with the corresponding covariant factors. There exist two possibilities:

1. We contract factor 1 with factor 3 and factor 2 with factor 4. The connection Gram matrix is I_2, because they are in dual bases:

$$\rho = [2 \;\; -3]I_2 \begin{bmatrix} 4 \\ 1 \end{bmatrix} \otimes [5 \;\; 1]I_2 \begin{bmatrix} 1 \\ -2 \end{bmatrix} = (8-3) \times (5-2) = 5 \times 3 = 15.$$

2. We contract factor 1 with factor 4 and factor 2 with factor 3.

$$\rho = [2 \;\; -3]I_2 \begin{bmatrix} 1 \\ -2 \end{bmatrix} \otimes [5 \;\; 1]I_2 \begin{bmatrix} 4 \\ 1 \end{bmatrix} = (2+6) \times (20+1) = 8 \times 21 = 168.$$

Second method:

We decide to associate the contravariant indices between them, and also the covariant indices between them; then, we execute the contraction, to obtain the unique result:

$$\vec{t} = \left([\vec{e}_1 \;\; \vec{e}_2] \begin{bmatrix} 2 \\ -3 \end{bmatrix} \otimes \left([\vec{e}_1 \;\; \vec{e}_2] \begin{bmatrix} 5 \\ 1 \end{bmatrix} \right)^t \right)$$

$$\otimes \left([\vec{e}^{*1} \;\; \vec{e}^{*2}] \begin{bmatrix} 4 \\ 1 \end{bmatrix} \otimes \left([\vec{e}^{*1} \;\; \vec{e}^{*2}] \begin{bmatrix} 1 \\ -2 \end{bmatrix} \right)^t \right)$$

$$= \left([\vec{e}_1 \;\; \vec{e}_2] \begin{bmatrix} 2 \\ -3 \end{bmatrix} \otimes [5 \;\; 1] \begin{bmatrix} \vec{e}_1 \\ \vec{e}_2 \end{bmatrix} \right) \otimes \left([\vec{e}^{*1} \;\; \vec{e}^{*2}] \begin{bmatrix} 4 \\ 1 \end{bmatrix} \otimes [1 \;\; -2] \begin{bmatrix} \vec{e}^{*1} \\ \vec{e}^{*2} \end{bmatrix} \right)$$

$$= \left([\vec{e}_1 \;\; \vec{e}_2] \begin{bmatrix} 10 & 2 \\ -15 & -3 \end{bmatrix} \begin{bmatrix} \vec{e}_1 \\ \vec{e}_2 \end{bmatrix} \right) \otimes \left([\vec{e}^{*1} \;\; \vec{e}^{*2}] \begin{bmatrix} 4 & -8 \\ 1 & -2 \end{bmatrix} \begin{bmatrix} \vec{e}^{*1} \\ \vec{e}^{*2} \end{bmatrix} \right)$$

$$= \left([\vec{e}_1 \otimes \vec{e}_1 \;\; \vec{e}_1 \otimes \vec{e}_2 \;\; \vec{e}_2 \otimes \vec{e}_1 \;\; \vec{e}_2 \otimes \vec{e}_2] \begin{bmatrix} 10 \\ 2 \\ -15 \\ -3 \end{bmatrix} \right)$$

$$\otimes \left([\vec{e}^{*1} \otimes \vec{e}^{*1} \;\; \vec{e}^{*1} \otimes \vec{e}^{*2} \;\; \vec{e}^{*2} \otimes \vec{e}^{*1} \;\; \vec{e}^{*2} \otimes \vec{e}^{*2}] \begin{bmatrix} 4 \\ -8 \\ 1 \\ -2 \end{bmatrix} \right).$$

The Gram matrix G between both dual tensor spaces is $G \equiv I_4$, because they are in dual bases:

$$\rho = [\,10 \quad 2 \quad -15 \quad -3\,]I_4 \begin{bmatrix} 4 \\ -8 \\ 1 \\ -2 \end{bmatrix} = 40 - 16 - 15 + 6 = 15.$$

Third method:

We decide to execute the contraction using its definition. To this end, we need to know the tensor, with the axiomatic ordering of its components.

Executing the last tensor product indicated in the previous method, we obtain

$$\vec{t} = [\,\vec{e}_1 \otimes \vec{e}_1 \;\; \vec{e}_1 \otimes \vec{e}_2 \;\; \vec{e}_2 \otimes \vec{e}_1 \;\; \vec{e}_2 \otimes \vec{e}_2\,] \begin{bmatrix} 10 \\ 2 \\ -15 \\ -3 \end{bmatrix} \otimes [\,4 \;\; -8 \;\; 1 \;\; 2\,] \begin{bmatrix} \vec{e}^{*1} \otimes \vec{e}^{*1} \\ \vec{e}^{*1} \otimes \vec{e}^{*2} \\ \vec{e}^{*2} \otimes \vec{e}^{*1} \\ \vec{e}^{*2} \otimes \vec{e}^{*2} \end{bmatrix}$$

$$= [\,\vec{e}_1 \otimes \vec{e}_1 \;\; \vec{e}_1 \otimes \vec{e}_2 \;\; \vec{e}_2 \otimes \vec{e}_1 \;\; \vec{e}_2 \otimes \vec{e}_2\,] \begin{bmatrix} 40 & -80 & 10 & -20 \\ 8 & -16 & 2 & -4 \\ -60 & 120 & -15 & 30 \\ -12 & 24 & -3 & 6 \end{bmatrix} \begin{bmatrix} \vec{e}^{*1} \otimes \vec{e}^{*1} \\ \vec{e}^{*1} \otimes \vec{e}^{*2} \\ \vec{e}^{*2} \otimes \vec{e}^{*1} \\ \vec{e}^{*2} \otimes \vec{e}^{*2} \end{bmatrix} .$$

This matrix expression leads to the desired fourth-order tensor $T = [t^{\alpha\beta\circ\circ}_{\circ\circ\gamma\delta}]$. We develop it by rows, in order to get the axiomatic ordering:

$$\vec{t} = (\vec{e}_1 \otimes \vec{e}_1) \otimes (40\vec{e}^{*1} \otimes \vec{e}^{*1} - 80\vec{e}^{*1} \otimes \vec{e}^{*2} + 10\vec{e}^{*2} \otimes \vec{e}^{*1} - 20\vec{e}^{*2} \otimes \vec{e}^{*2})$$
$$+ (\vec{e}_1 \otimes \vec{e}_2) \otimes (8\vec{e}^{*1} \otimes \vec{e}^{*1} - 16\vec{e}^{*1} \otimes \vec{e}^{*2} + 2\vec{e}^{*2} \otimes \vec{e}^{*1} - 4\vec{e}^{*2} \otimes \vec{e}^{*2})$$
$$+ (\vec{e}_2 \otimes \vec{e}_1) \otimes (-60\vec{e}^{*1} \otimes \vec{e}^{*1} + 120\vec{e}^{*1} \otimes \vec{e}^{*2} - 15\vec{e}^{*2} \otimes \vec{e}^{*1} + 30\vec{e}^{*2} \otimes \vec{e}^{*2})$$
$$+ (\vec{e}_2 \otimes \vec{e}_2) \otimes (-12\vec{e}^{*1} \otimes \vec{e}^{*1} + 24\vec{e}^{*1} \otimes \vec{e}^{*2} - 3\vec{e}^{*2} \otimes \vec{e}^{*1} + 6\vec{e}^{*2} \otimes \vec{e}^{*2}).$$

Since the first two factors refer to row and column of each submatrix (first and second tensor indices), we finally get the tensor matrix expression, with the correct ordering

$$[t^{\alpha\beta\circ\circ}_{\circ\circ\gamma\delta}] = \left[\begin{array}{cc|cc} 40 & -80 & 8 & -16 \\ 10 & -20 & 2 & -4 \\ \hline -60 & 120 & -12 & 24 \\ -15 & 30 & -3 & 6 \end{array} \right],$$

where α is the row of submatrices, β the column of submatrices, γ the row of each submatrix, and δ the column of each submatrix.

Next, we start with the contractions. There exist two possibilities:

1. We first contract α with γ, and then β with δ:

$$[u^{\beta\circ}_{\circ\delta}] = C\binom{\alpha}{\gamma} [t^{\alpha\beta\circ\circ}_{\circ\circ\gamma\delta}] = [t^{\theta\beta\circ\circ}_{\circ\circ\theta\delta}] = t^{1\beta\circ\circ}_{\circ\circ1\delta} + t^{2\beta\circ\circ}_{\circ\circ2\delta}$$

$$\rho = C\begin{pmatrix}\beta\\\delta\end{pmatrix}[u^{\beta\circ}_{\circ\delta}] = [t^{1\theta\circ\circ}_{\circ\circ1\theta}] + [t^{2\theta\circ\circ}_{\circ\circ2\theta}] = (t^{11\circ\circ}_{\circ\circ11} + t^{12\circ\circ}_{\circ\circ12}) + (t^{21\circ\circ}_{\circ\circ21} + t^{22\circ\circ}_{\circ\circ22})$$

$$\rho = 40 - 16 - 15 + 6 = 46 - 31 = 15.$$

2. We first contract α with δ, and then β with γ:

$$[v^{\beta\circ}_{\circ\gamma}] = C\begin{pmatrix}\alpha\\\delta\end{pmatrix}[t^{\alpha\beta\circ\circ}_{\circ\circ\gamma\delta}] = [t^{\theta\beta\circ\circ}_{\circ\circ\gamma\theta}] = t^{1\beta\circ\circ}_{\circ\circ\gamma1} + t^{2\beta\circ\circ}_{\circ\circ\gamma2}$$

$$\rho' = C\begin{pmatrix}\beta\\\gamma\end{pmatrix}[v^{\beta\circ}_{\circ\gamma}] = [v^{\theta\circ}_{\circ\theta}] = [t^{1\theta\circ\circ}_{\circ\circ\theta1}] + [t^{2\theta\circ\circ}_{\circ\circ\theta2}] = (t^{11\circ\circ}_{\circ\circ11} + t^{12\circ\circ}_{\circ\circ21}) + (t^{21\circ\circ}_{\circ\circ12} + t^{22\circ\circ}_{\circ\circ22})$$

$$\rho' = 40 + 2 + 120 + 6 = 168.$$

Fourth method:

We use the direct homomorphism on the components $T_\sigma (\sigma = 2\times2\times2\times2 = 16)$, that is, the tensor components in a "column matrix".

There are two models to be considered:

1. The homomorphism model (2) of double contraction, Formula (5.76):

$$\rho = H_{1,16}(\alpha, \gamma|\beta, \delta) \bullet T_{16,1} = [E^t_1 \otimes E^t_1 | E^t_1 \otimes E^t_2 | E^t_2 \otimes E^t_1 | E^t_2 \otimes E^t_2 |] \bullet T_{16,1}$$

$$= [[1\ 0] \otimes [1\ 0] | [1\ 0] \otimes [0\ 1] | [0\ 1] \otimes [1\ 0] | [0\ 1] \otimes [0\ 1]] \bullet T_{16,1}$$

$$= [1\ 0\ 0\ 0\ 0\ 1\ 0\ 0\ |\ 0\ 0\ 1\ 0\ 0\ 0\ 0\ 1\,] \bullet \begin{bmatrix} 40 \\ -80 \\ 10 \\ -20 \\ 8 \\ -16 \\ 2 \\ -4 \\ -60 \\ 120 \\ -15 \\ 30 \\ -12 \\ 24 \\ -3 \\ 6 \end{bmatrix}$$

$$= 40 - 16 - 15 + 6 = 15.$$

2. The homomorphism model (3), Formula (5.77):

$$\rho' = H_{1,16}(\alpha, \delta|\beta, \gamma) \bullet T_{16,1} = [[E^t_1|E^t_2] \otimes E^t_1 | [E^t_1|E^t_2] \otimes E^t_2] \bullet T_{16,1}$$

$$= [[1\ \ 0\ \ 0\ \ 1] \otimes [1\ \ 0] | [1\ \ 0\ \ 0\ \ 1] \otimes [0\ \ 1]] \bullet T_{16,1}$$

$$= [1\ 0\ 0\ 0\ 0\ 0\ 1\ 0\ |\ 0\ 1\ 0\ 0\ 0\ 0\ 0\ 1] \bullet \begin{bmatrix} 40 \\ -80 \\ 10 \\ -20 \\ 8 \\ -16 \\ 2 \\ -4 \\ -60 \\ 120 \\ -15 \\ 30 \\ -12 \\ 24 \\ -3 \\ 6 \end{bmatrix}$$

$$= 40 + 2 + 120 + 6 = 168.$$

□

Example 5.7 (Contractions). Consider the tensor $\vec{t} = \vec{v}_1 \otimes \vec{v}_2 \otimes \vec{f}^3 \otimes \vec{f}^4$ defined over $\mathbb{R}^2$, where the factor vectors are

$$\vec{v}_1 = \vec{e}_1 + \vec{e}_2; \quad \vec{v}_2 = 2\vec{e}_1 - \vec{e}_2; \quad \vec{f}^3 = 2\vec{e}^{*1} + \vec{e}^{*2}; \quad \vec{f}^4 = 3\vec{e}^{*1};$$

1. Obtain the totally developed analytical expression of the tensor expressed in its corresponding tensor basis.
2. Execute all possible simple contractions, indicating which of the obtained systems of scalars have tensor character.
3. Express the resulting tensors in the previous question, as a function of the vectors $\vec{v}_1, \vec{v}_2, \vec{f}^3, \vec{f}^4$.

Solution:

1. We develop the tensor product

$$\begin{aligned}
\vec{t} &= (\vec{v}_1 \otimes \vec{v}_2) \otimes (\vec{f}^3 \otimes \vec{f}^4) \\
&= (2\vec{e}_1 \otimes \vec{e}_1 - \vec{e}_1 \otimes \vec{e}_2 + 2\vec{e}_2 \otimes \vec{e}_1 - \vec{e}_2 \otimes \vec{e}_2) \otimes (6\vec{e}^{*1} \otimes \vec{e}^{*1} + 3\vec{e}^{*2} \otimes \vec{e}^{*1}) \\
&= 12\vec{e}_1 \otimes \vec{e}_1 \otimes \vec{e}^{*1} \otimes \vec{e}^{*1} + 6\vec{e}_1 \otimes \vec{e}_1 \otimes \vec{e}^{*2} \otimes \vec{e}^{*1} - 6\vec{e}_1 \otimes \vec{e}_2 \otimes \vec{e}^{*1} \otimes \vec{e}^{*1} \\
&\quad - 3\vec{e}_1 \otimes \vec{e}_2 \otimes \vec{e}^{*2} \otimes \vec{e}^{*1} + 12\vec{e}_2 \otimes \vec{e}_1 \otimes \vec{e}^{*1} \otimes \vec{e}^{*1} + 6\vec{e}_2 \otimes \vec{e}_1 \otimes \vec{e}^{*2} \otimes \vec{e}^{*1} \\
&\quad - 6\vec{e}_2 \otimes \vec{e}_2 \otimes \vec{e}^{*1} \otimes \vec{e}^{*1} - 3\vec{e}_2 \otimes \vec{e}_2 \otimes \vec{e}^{*2} \otimes \vec{e}^{*1}
\end{aligned}$$

and in matrix form

$$[t^{\alpha\beta\circ\circ}_{\circ\circ\gamma\delta}] = \left[\begin{array}{cc|cc} 12 & 0 & -6 & 0 \\ 6 & 0 & -3 & 0 \\ \hline 12 & 0 & -6 & 0 \\ 6 & 0 & -3 & 0 \end{array}\right],$$

where α is the matrix row indicator, β is the matrix column indicator, γ the row indicator of each submatrix, and δ the column indicator of each submatrix, that is, according to the basis axiomatic ordering.

2. Contraction is an operation that can be applied to any *system of scalars* of r indices and n^r components, with $n, r \geq 2$. The result is another system of scalars of order $(r-2)$ and n^{r-2} components, so that the cited operation is defined for such sets independently of whether they are or not tensors. According to this, we separate those contractions over tensor indices of the same valency that do not guarantee a resulting tensor from contractions executed over indices of different valency, in which case it is guaranteed that the contracted system is a tensor.

The contractions of tensor $\vec{t}$, *with no tensor character*, are (be aware of the special notation used for these type of non-tensor contractions)

$$\mathcal{C}(\alpha, \beta) - \mathcal{C}(\gamma, \delta)-, \text{ that is}$$

$$[a^{\circ\circ}_{\gamma\delta}] = \mathcal{C}(\alpha, \beta)[t^{\alpha\beta\circ\circ}_{\circ\circ\gamma\delta}] = [t^{11\circ\circ}_{\circ\circ\gamma\delta} + t^{22\circ\circ}_{\circ\circ\gamma\delta}] :$$

$$a^{\circ\circ}_{11} = t^{11\circ\circ}_{\circ\circ11} + t^{22\circ\circ}_{\circ\circ11} = 12 - 6 = 6$$

$$a^{\circ\circ}_{12} = t^{11\circ\circ}_{\circ\circ12} + t^{22\circ\circ}_{\circ\circ12} = 0 + 0 = 0$$

$$a^{\circ\circ}_{21} = t^{11\circ\circ}_{\circ\circ21} + t^{22\circ\circ}_{\circ\circ21} = 6 - 3 = 3$$

$$a^{\circ\circ}_{22} = t^{11\circ\circ}_{\circ\circ22} + t^{22\circ\circ}_{\circ\circ22} = 0 + 0 = 0$$

$$\Rightarrow [a^{\circ\circ}_{\gamma\delta}] = \begin{bmatrix} 6 & 0 \\ 3 & 0 \end{bmatrix}$$

$$[b^{\circ\circ}_{\alpha\beta}] = \mathcal{C}(\gamma, \delta)[t^{\alpha\beta\circ\circ}_{\circ\circ\gamma\delta}] = [t^{\alpha\beta\circ\circ}_{\circ\circ11} + t^{\alpha\beta\circ\circ}_{\circ\circ22}] :$$

$$b^{\circ\circ}_{11} = t^{11\circ\circ}_{\circ\circ11} + t^{11\circ\circ}_{\circ\circ22} = 12 + 0 = 12$$

$$b^{\circ\circ}_{12} = t^{12\circ\circ}_{\circ\circ21} + t^{12\circ\circ}_{\circ\circ22} = -6 + 0 = -6$$

$$b^{\circ\circ}_{21} = t^{21\circ\circ}_{\circ\circ11} + t^{21\circ\circ}_{\circ\circ22} = 12 + 0 = 12$$

$$b^{\circ\circ}_{22} = t^{22\circ\circ}_{\circ\circ11} + t^{22\circ\circ}_{\circ\circ22} = -6 + 0 = -6$$

$$\Rightarrow [b^{\circ\circ}_{\alpha\beta}] = \begin{bmatrix} 12 & -6 \\ 12 & -6 \end{bmatrix}.$$

The contractions with a *tensor nature* of tensor $\vec{t}$ are

$$-\mathcal{C}\begin{pmatrix}\alpha\\\gamma\end{pmatrix} - \mathcal{C}\begin{pmatrix}\alpha\\\delta\end{pmatrix} - \mathcal{C}\begin{pmatrix}\beta\\\gamma\end{pmatrix} - \mathcal{C}\begin{pmatrix}\beta\\\delta\end{pmatrix} -.$$

We execute them using two different procedures:

1. Direct procedure, according to the contraction definition:

$$[c_{\circ\,\delta}^{\beta\circ}] = C\binom{\alpha}{\gamma}[t_{\circ\,\circ\,\gamma\delta}^{\alpha\beta\circ\circ}] = [t_{\circ\,\circ\,1\delta}^{1\beta\circ\circ}] + [t_{\circ\,\circ\,2\delta}^{2\beta\circ\circ}]$$

$$\left.\begin{aligned}
c_{\circ1}^{1\circ} &= t_{\circ\circ11}^{11\circ\circ} + t_{\circ\circ21}^{21\circ\circ} = 12 + 6 = 18 \\[2mm]
c_{\circ2}^{1\circ} &= t_{\circ\circ12}^{11\circ\circ} + t_{\circ\circ22}^{21\circ\circ} = 0 + 0 = 0 \\[2mm]
c_{\circ1}^{2\circ} &= t_{\circ\circ11}^{12\circ\circ} + t_{\circ\circ21}^{22\circ\circ} = -6 - 3 = -9 \\[2mm]
c_{\circ2}^{2\circ} &= t_{\circ\circ12}^{12\circ\circ} + t_{\circ\circ22}^{22\circ\circ} = 0 + 0 = 0
\end{aligned}\right\} \Rightarrow [c_{\circ\,\delta}^{\beta\circ}] = \begin{bmatrix} 18 & 0 \\ -9 & 0 \end{bmatrix};$$

$$[d_{\circ\,\gamma}^{\beta\circ}] = C\binom{\alpha}{\delta}[t_{\circ\,\circ\,\gamma\delta}^{\alpha\beta\circ\circ}] = [t_{\circ\,\circ\,\gamma1}^{1\beta\circ\circ}] + [t_{\circ\,\circ\,\gamma2}^{2\beta\circ\circ}]$$

$$\left.\begin{aligned}
d_{\circ1}^{1\circ} &= t_{\circ\circ11}^{11\circ\circ} + t_{\circ\circ12}^{21\circ\circ} = 12 + 0 = 12 \\[2mm]
d_{\circ2}^{1\circ} &= t_{\circ\circ21}^{11\circ\circ} + t_{\circ\circ22}^{21\circ\circ} = 6 + 0 = 6 \\[2mm]
d_{\circ1}^{2\circ} &= t_{\circ\circ11}^{12\circ\circ} + t_{\circ\circ12}^{22\circ\circ} = -6 + 0 = -6 \\[2mm]
d_{\circ2}^{2\circ} &= t_{\circ\circ21}^{12\circ\circ} + t_{\circ\circ22}^{22\circ\circ} = -3 + 0 = -3
\end{aligned}\right\} \Rightarrow [d_{\circ\,\gamma}^{\beta\circ}] = \begin{bmatrix} 12 & 6 \\ -6 & -3 \end{bmatrix};$$

$$[f_{\circ\,\delta}^{\alpha\circ}] = C\binom{\beta}{\gamma}[t_{\circ\,\circ\,\gamma\delta}^{\alpha\beta\circ\circ}] = [t_{\circ\,\circ\,1\delta}^{\alpha1\circ\circ}] + [t_{\circ\,\circ\,2\delta}^{\alpha2\circ\circ}]$$

$$\left.\begin{aligned}
f_{\circ1}^{1\circ} &= t_{\circ\circ11}^{11\circ\circ} + t_{\circ\circ21}^{12\circ\circ} = 12 - 3 = 9 \\[2mm]
f_{\circ2}^{1\circ} &= t_{\circ\circ12}^{11\circ\circ} + t_{\circ\circ22}^{12\circ\circ} = 0 + 0 = 0 \\[2mm]
f_{\circ1}^{2\circ} &= t_{\circ\circ11}^{21\circ\circ} + t_{\circ\circ21}^{22\circ\circ} = 12 - 3 = 9 \\[2mm]
f_{\circ2}^{2\circ} &= t_{\circ\circ12}^{21\circ\circ} + t_{\circ\circ22}^{22\circ\circ} = 0 + 0 = 0
\end{aligned}\right\} \Rightarrow [f_{\circ\,\delta}^{\alpha\circ}] = \begin{bmatrix} 9 & 0 \\ 9 & 0 \end{bmatrix};$$

$$[g_{\circ\,\gamma}^{\alpha\circ}] = C\binom{\beta}{\delta}[t_{\circ\,\circ\,\gamma\delta}^{\alpha\beta\circ\circ}] = [t_{\circ\,\circ\,\gamma1}^{\alpha1\circ\circ}] + [t_{\circ\,\circ\,\gamma2}^{\alpha2\circ\circ}]$$

$$\left.\begin{array}{l} g^{1o}_{o1} = t^{11oo}_{oo11} + t^{12oo}_{oo12} = 12 + 0 = 12 \\[2mm] g^{1o}_{o2} = t^{11oo}_{oo211} + t^{12oo}_{oo22} = 6 + 0 = 6 \\[2mm] g^{2o}_{o1} = t^{21oo}_{oo11} + t^{22oo}_{oo12} = 12 + 0 = 12 \\[2mm] g^{2o}_{o2} = t^{21oo}_{oo21} + t^{22oo}_{oo22} = 6 + 0 = 6 \end{array}\right\} \Rightarrow [g^{\alpha o}_{o\gamma}] = \begin{bmatrix} 12 & 6 \\ 12 & 6 \end{bmatrix}.$$

2. Procedure based on the use of the simple contraction homomorphisms and of order $r = 4$. $\mathcal{C}\left(^{\alpha}_{\gamma}\right) \rightarrow$ Model (2), Formula (5.70):

$$T'_4 = H_{4,16}(\alpha,\gamma) \bullet T_{16,1} = \left([I_2 \otimes E^t_1 | I_2 \otimes E^t_2] \otimes I_2\right) \bullet T_{16,1}$$

$$= \left[\left[\begin{bmatrix} 1 & 0 \\ 0 & 1 \end{bmatrix} \otimes [1 \ 0]\right]\Bigg|\left[\begin{bmatrix} 1 & 0 \\ 0 & 1 \end{bmatrix} \otimes [0 \ 1]\right] \otimes \begin{bmatrix} 1 & 0 \\ 0 & 1 \end{bmatrix}\right] \bullet T_{16,1}$$

$$= \left(\begin{bmatrix} 1 & 0 & 0 & 0 & 0 & 1 & 0 & 0 \\ 0 & 0 & 1 & 0 & 0 & 0 & 0 & 1 \end{bmatrix} \otimes \begin{bmatrix} 1 & 0 \\ 0 & 1 \end{bmatrix}\right) \bullet T_{16,1}$$

$$= \begin{bmatrix} 1 & 0 & 0 & 0 & 0 & 0 & 0 & 0 & 0 & 0 & 1 & 0 & 0 & 0 & 0 & 0 \\ 0 & 1 & 0 & 0 & 0 & 0 & 0 & 0 & 0 & 0 & 0 & 1 & 0 & 0 & 0 & 0 \\ 0 & 0 & 0 & 0 & 1 & 0 & 0 & 0 & 0 & 0 & 0 & 0 & 0 & 0 & 1 & 0 \\ 0 & 0 & 0 & 0 & 0 & 1 & 0 & 0 & 0 & 0 & 0 & 0 & 0 & 0 & 0 & 1 \end{bmatrix} \bullet \begin{bmatrix} 12 \\ 0 \\ 6 \\ 0 \\ -6 \\ 0 \\ -3 \\ 0 \\ 12 \\ 0 \\ 6 \\ 0 \\ -6 \\ 0 \\ -3 \\ 0 \end{bmatrix}$$

$$= \begin{bmatrix} 12 + 6 \\ 0 + 0 \\ -6 - 3 \\ 0 + 0 \end{bmatrix} = \begin{bmatrix} 18 \\ 0 \\ -9 \\ 0 \end{bmatrix},$$

and after condensation the result is

$$[c^{\beta o}_{o \delta}] = \begin{bmatrix} 18 & 0 \\ -9 & 0 \end{bmatrix}.$$

$\mathcal{C}\left(^{\alpha}_{\delta}\right) \rightarrow$ Model (3), Formula (5.71):

$$T''_4 = H_{4,16}(\alpha,\delta) \bullet T_{16,1} = [I_2 \otimes I_2 \otimes E^t_1 | I_2 \otimes I_2 \otimes E^t_2] \bullet T_{16,1}$$

$$= \left[\left[\begin{bmatrix} 1 & 0 & 0 & 0 \\ 0 & 1 & 0 & 0 \\ 0 & 0 & 1 & 0 \\ 0 & 0 & 0 & 1 \end{bmatrix} \otimes [1 \ 0]\right]\Bigg|\left[\begin{bmatrix} 1 & 0 & 0 & 0 \\ 0 & 1 & 0 & 0 \\ 0 & 0 & 1 & 0 \\ 0 & 0 & 0 & 1 \end{bmatrix} \otimes [0 \ 1]\right]\right] \bullet T_{16,1}$$

$$
= \begin{bmatrix} 1 & 0 & 0 & 0 & 0 & 0 & 0 & 0 & 0 & 1 & 0 & 0 & 0 & 0 & 0 & 0 \\ 0 & 0 & 1 & 0 & 0 & 0 & 0 & 0 & 0 & 0 & 0 & 1 & 0 & 0 & 0 & 0 \\ 0 & 0 & 0 & 0 & 1 & 0 & 0 & 0 & 0 & 0 & 0 & 0 & 0 & 1 & 0 & 0 \\ 0 & 0 & 0 & 0 & 0 & 0 & 1 & 0 & 0 & 0 & 0 & 0 & 0 & 0 & 0 & 1 \end{bmatrix} \bullet \begin{bmatrix} 12 \\ 0 \\ 6 \\ 0 \\ -6 \\ 0 \\ -3 \\ 0 \\ 12 \\ 0 \\ 6 \\ 0 \\ -6 \\ 0 \\ -3 \\ 0 \end{bmatrix}
$$

$$
= \begin{bmatrix} 12+0 \\ 8+0 \\ -6+0 \\ -3+0 \end{bmatrix} = \begin{bmatrix} 12 \\ 6 \\ -6 \\ -3 \end{bmatrix},
$$

and after condensation we get

$$
[d^{\beta\circ}_{\circ\gamma}] = \begin{bmatrix} 12 & 6 \\ -6 & -3 \end{bmatrix}.
$$

$\mathcal{C}\binom{\beta}{\gamma} \to$ Model (4), Formula (5.72):

$$
T_4''' = H_{4,16}(\beta,\gamma) \bullet T_{16,1} = [I_2 \otimes [E_1^t | E_2^t] \otimes I_2] \bullet T_{16,1}
$$

$$
= \left[\begin{bmatrix} 1 & 0 \\ 0 & 1 \end{bmatrix} \otimes [1 \ \ 0 \ \ 0 \ \ 1] \otimes \begin{bmatrix} 1 & 0 \\ 0 & 1 \end{bmatrix} \right] \bullet T_{16,1}
$$

$$
= \left(\begin{bmatrix} 1 & 0 & 0 & 1 & 0 & 0 & 0 & 0 \\ 0 & 0 & 0 & 0 & 1 & 0 & 0 & 1 \end{bmatrix} \otimes \begin{bmatrix} 1 & 0 \\ 0 & 1 \end{bmatrix} \right) \bullet T_{16,1}
$$

$$
= \begin{bmatrix} 1 & 0 & 0 & 0 & 0 & 0 & 1 & 0 & 0 & 0 & 0 & 0 & 0 & 0 & 0 & 0 \\ 0 & 1 & 0 & 0 & 0 & 0 & 0 & 1 & 0 & 0 & 0 & 0 & 0 & 0 & 0 & 0 \\ 0 & 0 & 0 & 0 & 0 & 0 & 0 & 0 & 1 & 0 & 0 & 0 & 0 & 0 & 1 & 0 \\ 0 & 0 & 0 & 0 & 0 & 0 & 0 & 0 & 0 & 1 & 0 & 0 & 0 & 0 & 0 & 1 \end{bmatrix} \bullet \begin{bmatrix} 12 \\ 0 \\ 6 \\ 0 \\ -6 \\ 0 \\ -3 \\ 0 \\ 12 \\ 0 \\ 6 \\ 0 \\ -6 \\ 0 \\ -3 \\ 0 \end{bmatrix}
$$

$$
= \begin{bmatrix} 12-3 \\ 0+0 \\ 12-3 \\ 0+0 \end{bmatrix} = \begin{bmatrix} 9 \\ 0 \\ 9 \\ 0 \end{bmatrix},
$$

and after condensation we get

$$
[f^{\alpha\circ}_{\circ\delta}] = \begin{bmatrix} 9 & 0 \\ 9 & 0 \end{bmatrix}.
$$

$\mathcal{C}\binom{\beta}{\gamma} \to$ Model (5), Formula (5.73):

$$T_4^{IV} = H_{4,16}(\beta, \delta) \bullet T_{16,1} = \left(I_2 \otimes \left[I_2 \otimes E_1^t \middle| I_2 \otimes E_2^t\right]\right) \bullet T_{16,1}$$

$$= \left(\begin{bmatrix} 1 & 0 \\ 0 & 1 \end{bmatrix} \otimes \left[\begin{bmatrix} 1 & 0 \\ 0 & 1 \end{bmatrix} \otimes [1 \quad 0] \middle| \begin{bmatrix} 1 & 0 \\ 0 & 1 \end{bmatrix} \otimes [0 \quad 1]\right]\right) \bullet T_{16,1}$$

$$= \left[\begin{bmatrix} 1 & 0 \\ 0 & 1 \end{bmatrix} \otimes \begin{bmatrix} 1 & 0 & 0 & 0 & 0 & 1 & 0 & 0 \\ 0 & 0 & 1 & 0 & 0 & 0 & 0 & 1 \end{bmatrix}\right] \bullet T_{16,1}$$

$$= \begin{bmatrix} 1 & 0 & 0 & 0 & 0 & 1 & 0 & 0 & 0 & 0 & 0 & 0 & 0 & 0 & 0 & 0 \\ 0 & 0 & 1 & 0 & 0 & 0 & 0 & 1 & 0 & 0 & 0 & 0 & 0 & 0 & 0 & 0 \\ 0 & 0 & 0 & 0 & 0 & 0 & 0 & 0 & 1 & 0 & 0 & 0 & 0 & 1 & 0 & 0 \\ 0 & 0 & 0 & 0 & 0 & 0 & 0 & 0 & 0 & 1 & 0 & 0 & 0 & 0 & 0 & 1 \end{bmatrix} \bullet \begin{bmatrix} 12 \\ 0 \\ 6 \\ 0 \\ -6 \\ 0 \\ -3 \\ 0 \\ 12 \\ 0 \\ 6 \\ 0 \\ -6 \\ 0 \\ -3 \\ 0 \end{bmatrix}$$

$$= \begin{bmatrix} 12 + 0 \\ 6 + 0 \\ 12 + 0 \\ 6 + 0 \end{bmatrix} = \begin{bmatrix} 12 \\ 6 \\ 12 \\ 6 \end{bmatrix},$$

and condensing yields

$$\left[g_{\circ\,\gamma}^{\alpha\,\circ}\right] = \begin{bmatrix} 12 & 6 \\ 12 & 6 \end{bmatrix}.$$

3. We will express each of the tensors previously obtained in a developed analytical form, and later we will try to factorize each of them, as a function of the factors $\vec{v}_1, \vec{v}_2, \vec{f}^3, \vec{f}^4$. Then

$$\vec{c} = c_{\circ\,\delta}^{\beta\,\circ}\vec{e}_\beta \otimes \vec{e}^{*\delta} = 18\vec{e}_1 \otimes \vec{e}^{*1} - 9\vec{e}_2 \otimes \vec{e}^{*1} = (2\vec{e}_1 - \vec{e}_2) \otimes (9\vec{e}^{*1})$$

and according to the statement data:

$$\vec{c} = \vec{v}_2 \otimes 3\vec{f}^4 = 3(\vec{v}_2 \otimes \vec{f}^4)$$

$$\vec{d} = d_{\circ\,\gamma}^{\beta\,\circ}\vec{e}_\beta \otimes \vec{e}^{*\gamma} = 12\vec{e}_1 \otimes \vec{e}^{*1} + 6\vec{e}_1 \otimes \vec{e}^{*2} - 6\vec{e}_2 \otimes \vec{e}^{*1} - 3\vec{e}_2 \otimes \vec{e}^{*2}$$

$$= (12\vec{e}_1 - 6\vec{e}_2) \otimes \vec{e}^{*1} + (6\vec{e}_1 - 3\vec{e}_2) \otimes \vec{e}^{*2}$$

$$= 2(2\vec{e}_1 - \vec{e}_2) \otimes 3\vec{e}^{*1} + (2\vec{e}_1 - \vec{e}_2) \otimes 3\vec{e}^{*2}$$

$$= (2\vec{e}_1 - \vec{e}_2) \otimes 3(2\vec{e}^{*1} + \vec{e}^{*2})$$

$$= 3(2\vec{e}_1 - \vec{e}_2) \otimes (2\vec{e}^{*1}1 + \vec{e}^{*2})$$

$$= 3(\vec{v}_2 \otimes \vec{f}^3)$$

$$\vec{f} = f_{\circ\,\delta}^{\alpha\,\circ}\vec{e}_\alpha \otimes \vec{e}^{*\delta}$$

$$= 9\vec{e}_1 \otimes \vec{e}^{*1} + 9\vec{e}_2 \otimes \vec{e}^{*1}$$
$$= 3(\vec{e}_1 + \vec{e}_2) \otimes 3\vec{e}^{*1} = 3(\vec{v}_1 \otimes \vec{f}^4)$$
$$\vec{g} = g^{\alpha\circ}_{\circ\gamma}\vec{e}_\alpha \otimes \vec{e}^{*\gamma} = 12\vec{e}_1 \otimes \vec{e}^{*1} + 6\vec{e}_1 \otimes \vec{e}^{*2} + 12\vec{e}_2 \otimes \vec{e}^{*1} + 6\vec{e}_2 \otimes \vec{e}^{*2}$$
$$= 12(\vec{e}_1 + \vec{e}_2) \otimes \vec{e}^{*1} + 6(\vec{e}_1 + \vec{e}_2) \otimes \vec{e}^{*2}$$
$$= 6(\vec{e}_1 + \vec{e}_2) \otimes (2\vec{e}^{*1} + \vec{e}^{*2}) = 6(\vec{v}_1 \otimes \vec{f}^3).$$

$\square$

Example 5.8 (Contracted tensor product). Consider the two tensors $\vec{a}$ and $\vec{b}$ given by their components with respect to the canonical basis of the linear space $\mathbb{R}^3$:

$$[a^{\alpha\beta}_{\circ\circ}] = \begin{bmatrix} 1 & 3 & 0 \\ 0 & 0 & -1 \\ 2 & -2 & 1 \end{bmatrix} ; \quad [b^{\gamma\circ\epsilon}_{\circ\delta\circ}] = \left[\begin{array}{ccc} 1 & -1 & 1 \\ 2 & 3 & 0 \\ 0 & 4 & 5 \\ \hline 1 & 0 & 1 \\ 0 & 2 & 1 \\ 3 & 1 & -1 \\ \hline 0 & 3 & -2 \\ 5 & 1 & 0 \\ 3 & -1 & 2 \end{array}\right] .$$

1. Obtain *all* possible contracted tensor products with both tensors.
2. Determine the type of homomorphism that directly relates two of the contracted products with the other two.

Solution:

1. A tensor product tensor is

$$t^{\alpha\beta\gamma\circ\epsilon}_{\circ\circ\circ\delta\circ} = a^{\alpha\beta}_{\circ\circ} \otimes b^{\gamma\circ\epsilon}_{\circ\delta\circ}.$$

There are two possible tensor contractions: $C\binom{\alpha}{\delta} - C\binom{\beta}{\delta}$, because contractions $C\binom{\gamma}{\delta}$ and $C\binom{\epsilon}{\delta}$ correspond to indices of the same factor.

$$[u^{\beta\gamma\epsilon}_{\circ\circ\circ}] = C\binom{\alpha}{\delta}[t^{\alpha\beta\gamma\circ\epsilon}_{\circ\circ\circ\delta\circ}] = [t^{\theta\beta\gamma\circ\epsilon}_{\circ\circ\circ\theta\circ}] = [a^{\theta\beta}_{\circ\circ} \cdot b^{\gamma\circ\epsilon}_{\circ\theta\circ}].$$

If $[a^{\theta\beta}_{\circ\circ}]^t = [a^{\beta\theta}_{\circ\circ}]$, for $\gamma = 1$, we get

$$[u^{\beta 1\epsilon}_{\circ\circ\circ}] = [a^{\beta\theta}_{\circ\circ}] \cdot [b^{1\circ\epsilon}_{\circ\theta\circ}] = \begin{bmatrix} 1 & 0 & 2 \\ 3 & 0 & -2 \\ 0 & -1 & 1 \end{bmatrix} \begin{bmatrix} 1 & -1 & 1 \\ 2 & 3 & 0 \\ 0 & 4 & 5 \end{bmatrix} = \begin{bmatrix} 1 & 7 & 11 \\ 3 & -11 & -7 \\ -2 & 1 & 5 \end{bmatrix},$$

for $\gamma = 2$:

$$[u^{\beta 2\epsilon}_{\circ\circ\circ}] = [a^{\beta\theta}_{\circ\circ}] \cdot [b^{2\circ\epsilon}_{\circ\theta\circ}] = \begin{bmatrix} 1 & 0 & 2 \\ 3 & 0 & -2 \\ 0 & -1 & 1 \end{bmatrix} \begin{bmatrix} 1 & 0 & 1 \\ 0 & 2 & 1 \\ 3 & 1 & -1 \end{bmatrix} = \begin{bmatrix} 7 & 2 & -1 \\ -3 & -2 & 5 \\ 3 & -1 & -2 \end{bmatrix},$$

and for $\gamma = 3$:

$$[u_{\circ\circ\circ}^{\beta 3 \epsilon}] = [a_{\circ\circ}^{\beta\theta}] \cdot [b_{\circ\theta\circ}^{3\circ\epsilon}] = \begin{bmatrix} 1 & 0 & 2 \\ 3 & 0 & -2 \\ 0 & -1 & 1 \end{bmatrix} \begin{bmatrix} 0 & 3 & -2 \\ 5 & 1 & 0 \\ 3 & -1 & 2 \end{bmatrix} = \begin{bmatrix} 6 & 1 & 2 \\ -6 & 11 & -10 \\ -2 & -2 & 2 \end{bmatrix}.$$

So that letting $\beta = 1$, and assigning to ϵ the values $1, 2, 3$ in the above three matrices we arrive at

$$[u_{\circ\circ\circ}^{1\gamma\epsilon}] = \begin{bmatrix} 1 & 7 & 11 \\ 7 & 2 & -1 \\ 6 & 1 & 2 \end{bmatrix}.$$

Similarly, for $\beta = 2$ and ϵ taking values $1, 2, 3$, we get

$$[u_{\circ\circ\circ}^{2\gamma\epsilon}] = \begin{bmatrix} 3 & -11 & -7 \\ -3 & -2 & 5 \\ -6 & 11 & -10 \end{bmatrix}.$$

Finally, for $\beta = 3$ and values $1, 2, 3$ we obtain

$$[u_{\circ\circ\circ}^{3\gamma\epsilon}] = \begin{bmatrix} -2 & 1 & 5 \\ 3 & -1 & -2 \\ -2 & -2 & 2 \end{bmatrix}.$$

Then, the first contracted product is

$$[u_{\circ\circ\circ}^{\beta\gamma\epsilon}] = \begin{bmatrix} 1 & 7 & 11 \\ 7 & 2 & -1 \\ 6 & 1 & 2 \\ \hline 3 & -11 & -7 \\ -3 & -2 & 5 \\ -6 & 11 & -10 \\ \hline -2 & 1 & 5 \\ 3 & -1 & -2 \\ -2 & -2 & 2 \end{bmatrix},$$

and the second is

$$[v_{\circ\circ\circ}^{\alpha\gamma\epsilon}] = \mathcal{C}\binom{\beta}{\delta}[t_{\circ\circ\circ\circ\delta\circ}^{\alpha\beta\gamma\circ\epsilon}] = [t_{\circ\circ\circ\theta\circ}^{\alpha\theta\gamma\circ\epsilon}] = [a_{\circ\circ}^{\alpha\theta} \cdot b_{\circ\theta\circ}^{\gamma\circ\epsilon}].$$

For $\gamma = 1$, we get

$$[v_{\circ\circ\circ}^{\alpha 1 \epsilon}] = [a_{\circ\circ}^{\alpha\theta} \cdot b_{\circ\theta\circ}^{1\circ\epsilon}] = \begin{bmatrix} 1 & 3 & 0 \\ 0 & 0 & -1 \\ 2 & -2 & 1 \end{bmatrix} \bullet \begin{bmatrix} 1 & -1 & 1 \\ 2 & 3 & 0 \\ 0 & 4 & 5 \end{bmatrix} = \begin{bmatrix} 7 & 8 & 1 \\ 0 & -4 & -5 \\ -2 & -4 & 7 \end{bmatrix},$$

for $\gamma = 2$:

$$[v_{\circ\circ\circ}^{\alpha 2 \epsilon}] = [a_{\circ\circ}^{\alpha\theta} \cdot b_{\circ\theta\circ}^{2\circ\epsilon}] = \begin{bmatrix} 1 & 3 & 0 \\ 0 & 0 & -1 \\ 2 & -2 & 1 \end{bmatrix} \bullet \begin{bmatrix} 1 & 0 & 1 \\ 0 & 2 & 1 \\ 3 & 1 & -1 \end{bmatrix} = \begin{bmatrix} 1 & 6 & 4 \\ -3 & -1 & 1 \\ 5 & -3 & -1 \end{bmatrix},$$

and for $\gamma = 3$:

$$[v^{\alpha 3 \epsilon}_{\circ\circ\circ}] = [a^{\alpha\theta}_{\circ\circ} \cdot b^{3\circ\epsilon}_{\circ\theta\circ}] = \begin{bmatrix} 1 & 3 & 0 \\ 0 & 0 & -1 \\ 2 & -2 & 1 \end{bmatrix} \bullet \begin{bmatrix} 0 & 3 & -2 \\ 5 & 1 & 0 \\ 3 & -1 & 2 \end{bmatrix} = \begin{bmatrix} 15 & 6 & -2 \\ -3 & 1 & -2 \\ -7 & 3 & -2 \end{bmatrix}.$$

Doing exactly the same as in the previous case, $\beta = 1, 2, 3$, and ϵ successively equal to $1, 2, 3$ in each jump of β, we obtain the second contracted product

$$[v^{\alpha\gamma\epsilon}_{\circ\circ\circ}] = \begin{bmatrix} 7 & 8 & 1 \\ 1 & 6 & 4 \\ 15 & 6 & -2 \\ \hline 0 & -4 & -5 \\ -3 & -1 & 1 \\ -3 & 1 & -2 \\ \hline -2 & -4 & 7 \\ 5 & -3 & -1 \\ -7 & 3 & -2 \end{bmatrix}.$$

Another tensor product is $p^{\gamma\circ\epsilon\alpha\beta}_{\circ\delta\circ\circ\circ} = b^{\gamma\circ\epsilon}_{\circ\delta\circ} \otimes a^{\alpha\beta}_{\circ\circ}$.

There are two possible tensor contractions (of contracted product): $C\binom{\alpha}{\delta} - C\binom{\beta}{\delta}$. Thus, we have

$$[w^{\gamma\epsilon\beta}_{\circ\circ\circ}] = C\binom{\alpha}{\delta}[p^{\gamma\circ\epsilon\alpha\beta}_{\circ\delta\circ\circ\circ}] = [p^{\gamma\circ\epsilon\theta\beta}_{\circ\theta\circ\circ\circ}] = [b^{\gamma\circ\epsilon}_{\circ\theta\circ} \cdot a^{\theta\beta}_{\circ\circ}].$$

For $\gamma = 1$, taking into account that $[b^{1\epsilon\circ}_{\circ\circ\theta}] = [b^{1\circ\epsilon}_{\circ\theta\circ}]^t$, the result is

$$[w^{1\epsilon\beta}_{\circ\circ\circ}] = [b^{1\circ\epsilon}_{\circ\theta\circ} \cdot a^{\theta\beta}_{\circ\circ}] = [b^{1\epsilon\circ}_{\circ\circ\theta}] \cdot [a^{\theta\beta}_{\circ\circ}] = \begin{bmatrix} 1 & 2 & 0 \\ -1 & 3 & 4 \\ 1 & 0 & 5 \end{bmatrix} \bullet \begin{bmatrix} 1 & 3 & 0 \\ 0 & 0 & -1 \\ 2 & -2 & 1 \end{bmatrix}$$

$$= \begin{bmatrix} 1 & 3 & -2 \\ 7 & -11 & 1 \\ 11 & -7 & 5 \end{bmatrix}.$$

For $\gamma = 2$, we get

$$[w^{2\epsilon\beta}_{\circ\circ\circ}] = [b^{2\circ\epsilon}_{\circ\theta\circ} \cdot a^{\theta\beta}_{\circ\circ}] = [b^{2\epsilon\circ}_{\circ\circ\theta}] \cdot [a^{\theta\beta}_{\circ\circ}] = \begin{bmatrix} 1 & 0 & 3 \\ 0 & 2 & 1 \\ 1 & 1 & -1 \end{bmatrix} \bullet \begin{bmatrix} 1 & 3 & 0 \\ 0 & 0 & -1 \\ 2 & -2 & 1 \end{bmatrix}$$

$$= \begin{bmatrix} 7 & -3 & 3 \\ 2 & -2 & -1 \\ -1 & 5 & -2 \end{bmatrix},$$

and for $\gamma = 3$, is

$$[w^{3\epsilon\beta}_{\circ\circ\circ}] = [b^{3\circ\epsilon}_{\circ\theta\circ} \cdot a^{\theta\beta}_{\circ\circ}] = [b^{3\epsilon\circ}_{\circ\circ\theta}] \cdot [a^{\theta\beta}_{\circ\circ}] = \begin{bmatrix} 0 & 5 & 3 \\ 3 & 1 & -1 \\ -2 & 0 & 2 \end{bmatrix} \bullet \begin{bmatrix} 1 & 3 & 0 \\ 0 & 0 & -1 \\ 2 & -2 & 1 \end{bmatrix}$$

$$= \begin{bmatrix} 6 & -6 & -2 \\ 1 & 11 & -2 \\ 2 & -10 & 2 \end{bmatrix}.$$

So that the present contraction becomes

$$[w^{\gamma\epsilon\beta}_{\circ\circ\circ}] = \begin{bmatrix} 1 & 3 & -2 \\ 7 & -11 & 1 \\ 11 & -7 & 5 \\ \hline 7 & -3 & 3 \\ 2 & -2 & -1 \\ -1 & 5 & -2 \\ \hline 6 & -6 & -2 \\ 1 & 11 & -2 \\ 2 & -10 & 2 \end{bmatrix}.$$

Finally, the following contraction remains to be calculated:

$$[s^{\gamma\epsilon\alpha}_{\circ\circ\circ}] = C\binom{\beta}{\delta}[p^{\gamma\circ\epsilon\alpha\beta}_{\circ\delta\circ\circ\circ}] = [p^{\gamma\circ\epsilon\alpha\theta}_{\circ\theta\circ\circ\circ}] = [b^{\gamma\circ\epsilon}_{\circ\theta\circ} \cdot a^{\alpha\theta}_{\circ\circ}].$$

This time we will transpose the matrices associated with both factors, in order to be able to execute them in matrix form.
For $\gamma = 1$, we obtain

$$[s^{1\epsilon\alpha}_{\circ\circ\circ}] = [b^{1\circ\epsilon}_{\circ\theta\circ} \cdot a^{\alpha\theta}_{\circ\circ}] = [b^{1\epsilon\circ}_{\circ\circ\theta}] \cdot [a^{\theta\alpha}_{\circ\circ}] = \begin{bmatrix} 1 & 2 & 0 \\ -1 & 3 & 4 \\ 1 & 0 & 5 \end{bmatrix} \bullet \begin{bmatrix} 1 & 0 & 2 \\ 3 & 0 & -2 \\ 0 & -1 & 1 \end{bmatrix}$$

$$= \begin{bmatrix} 7 & 0 & -2 \\ 8 & -4 & -4 \\ 1 & -5 & 7 \end{bmatrix}.$$

For $\gamma = 2$:

$$[s^{2\epsilon\alpha}_{\circ\circ\circ}] = [b^{2\circ\epsilon}_{\circ\theta\circ} \cdot a^{\alpha\theta}_{\circ\circ}] = [b^{2\epsilon\circ}_{\circ\circ\theta}] \cdot [a^{\theta\alpha}_{\circ\circ}] = \begin{bmatrix} 1 & 0 & 3 \\ 0 & 2 & 1 \\ 1 & 1 & -1 \end{bmatrix} \bullet \begin{bmatrix} 1 & 0 & 2 \\ 3 & 0 & -2 \\ 0 & -1 & 1 \end{bmatrix}$$

$$= \begin{bmatrix} 1 & -3 & 5 \\ 6 & -1 & -3 \\ 4 & 1 & -1 \end{bmatrix},$$

and for $\gamma = 3$:

$$[s^{3\epsilon\alpha}_{\circ\circ\circ}] = [b^{3\circ\epsilon}_{\circ\theta\circ} \cdot a^{\alpha\theta}_{\circ\circ}] = [b^{3\epsilon\circ}_{\circ\circ\theta}] \cdot [a^{\theta\alpha}_{\circ\circ}] = \begin{bmatrix} 0 & 5 & 3 \\ 3 & 1 & -1 \\ -2 & 0 & 2 \end{bmatrix} \bullet \begin{bmatrix} 1 & 0 & 2 \\ 3 & 0 & -2 \\ 0 & -1 & 1 \end{bmatrix}$$

$$= \begin{bmatrix} 15 & -3 & -7 \\ 6 & 1 & 3 \\ -2 & -2 & -2 \end{bmatrix},$$

which yields the contracted tensor

$$[s^{\gamma\epsilon\alpha}_{\circ\circ\circ}] = \begin{bmatrix} 7 & 0 & -2 \\ 8 & -4 & -4 \\ 1 & -5 & 7 \\ \hline 1 & -3 & 5 \\ 6 & -1 & -3 \\ 4 & 1 & -1 \\ \hline 15 & -3 & -7 \\ 6 & 1 & 3 \\ -2 & -2 & -2 \end{bmatrix}.$$

2. A careful examination of the tensor $[w^{\gamma\epsilon\beta}_{\circ\circ\circ}]$ reveals that it is a certain permutation of $[u^{\beta\gamma\epsilon}_{\circ\circ\circ}]$ and, since all dummy indices change position, it is a rotation. Compared with the Example 5.5 of rotation tensors, we finally establish that $[w^{\gamma\epsilon\beta}_{\circ\circ\circ}] = [u^{\beta\gamma\epsilon}_{\circ\circ\circ}]^{R(1)}$.

Similarly, we establish that $[s^{\gamma\epsilon\alpha}_{\circ\circ\circ}] = [v^{\alpha\gamma\epsilon}_{\circ\circ\circ}]^{R(1)}$, an interesting relation, which enables us to avoid half of the operations in the previous question.

□

5.10 Eigentensors

Given an arbitrary tensor, T, we examine what possible tensors exist of a given order, r, that in a *contracted tensor product* with the given tensor, become a tensor that is λ times ($\lambda \in K$) the initial tensor, that is, the following tensor equation is satisfied, with T and $r = 3$:

$$C\left(\begin{array}{c|c} \alpha & \phi \\ \theta & \beta \end{array}\right)(T^{\alpha\circ\circ\delta}_{\circ\beta\gamma\circ} \otimes X^{\circ\phi\circ}_{\theta\circ w}) = \lambda X^{\circ\phi\circ}_{\theta\circ w}. \tag{5.81}$$

First case:

Data tensor: $A = [a^{\alpha\circ}_{\circ\beta}]$, of second order, over $n = dim V^2(K) = 2$.

Test tensor $r = 1$: vector $X = [x^\theta] \equiv \begin{bmatrix} x^1 \\ x^2 \end{bmatrix}$

According to (5.81), we must have

$$C\left(\begin{array}{c} \theta \\ \beta \end{array}\right)[A \otimes X] = [a^{\alpha\circ}_{\circ\beta} \cdot \delta^{\beta\circ}_{\circ\theta} \cdot x^\theta_\circ] = \left[a^{\alpha\circ}_{\circ\theta} \cdot x^\theta_\circ\right] = A \bullet X = \lambda X \tag{5.82}$$

and the relation (5.82) leads to the classic relation

$$[A - \lambda I] \bullet X = \Omega, \tag{5.83}$$

which is solved in algebras with the eigenvalues and eigenvectors associated with matrix A, for the eigenvalues λ_1 and λ_2 of the characteristic polynomial.

We do not insist on this, since we assume that it is well known by the reader. Let A_1 and A_2 be the matrices of eigenvectors associated with the eigenvalues λ_1 and λ_2 (assuming they coexist in K); we assume from now on that they are *known*.

The solutions in this first case are

$$X_1 = A_1 \text{ arbitrary eigenvector of the matrix } A, \text{ associated with } \lambda_1.$$

$$(5.84)$$

$$X_2 = A_2 \text{ arbitrary eigenvector of the matrix } A, \text{ associated with } \lambda_2.$$

Second case:

Data tensor: $A = [a^{\alpha\circ}_{\circ\beta}]$, of second order, over $n = dim\, V^2(K) = 2$.

Test tensor $r = 2$: matrix $X = [x^{\gamma\circ}_{\circ\delta}] \equiv \begin{bmatrix} x & y \\ z & t \end{bmatrix}$.

According to (5.81), the first term must be

Let $P = [p^{\alpha\circ\gamma\circ}_{\circ\beta\circ\delta}] = A \otimes X$; There are several possible contractions:

First possible contraction

$$[q^{\alpha\circ}_{\circ\delta}] = C\begin{pmatrix} \gamma \\ \beta \end{pmatrix} P \equiv C\begin{pmatrix} \gamma \\ \beta \end{pmatrix} \left[p^{\alpha\circ\gamma\circ}_{\circ\beta\circ\delta}\right] \qquad (5.85)$$

Equation (5.85) is stated by "extension":

$$q_{\sigma'} = H_{n^2,n^4}(\beta,\gamma) P_{\sigma,1} \qquad (5.86)$$

with the help of the homomorphism (5.72).

The details are

$$A = \begin{bmatrix} a^{1\circ}_{\circ 1} & a^{1\circ}_{\circ 2} \\ a^{2\circ}_{\circ 1} & a^{2\circ}_{\circ 2} \end{bmatrix}; \quad X = \begin{bmatrix} x & y \\ z & t \end{bmatrix}; \quad P = A \otimes X$$

$$P = \left[\begin{array}{cc|cc} a^{1\circ}_{\circ 1}x & a^{1\circ}_{\circ 1}y & a^{1\circ}_{\circ 2}x & a^{1\circ}_{\circ 2}y \\ a^{1\circ}_{\circ 1}z & a^{1\circ}_{\circ 1}t & a^{1\circ}_{\circ 2}z & a^{1\circ}_{\circ 2}t \\ \hline a^{2\circ}_{\circ 1}x & a^{2\circ}_{\circ 1}y & a^{2\circ}_{\circ 2}x & a^{2\circ}_{\circ 2}y \\ a^{2\circ}_{\circ 1}z & a^{2\circ}_{\circ 1}t & a^{2\circ}_{\circ 2}z & a^{2\circ}_{\circ 2}t \end{array} \right],$$

which in our case is $n = 2; \sigma = n^4 = 2^4 = 16; \sigma' = n^2 = 2^2 = 4$, and then, (5.72) leads to

$$H_{4,16}(\beta,\gamma) = I_2 \otimes [E^t_1 | E^t_2] \otimes I_2 \equiv \begin{bmatrix} 1 & 0 \\ 0 & 1 \end{bmatrix} \otimes [1 \ \ 0 \ \ 0 \ \ 1] \otimes \begin{bmatrix} 1 & 0 \\ 0 & 1 \end{bmatrix}$$

$$= \begin{bmatrix} 1 & 0 \\ 0 & 1 \end{bmatrix} \otimes \begin{bmatrix} 1 & 0 & 0 & 0 & 0 & 0 & 1 & 0 \\ 0 & 1 & 0 & 0 & 0 & 0 & 0 & 1 \end{bmatrix}$$

$$= \begin{bmatrix} 1 & 0 & 0 & 0 & 0 & 0 & 1 & 0 & 0 & 0 & 0 & 0 & 0 & 0 & 0 & 0 \\ 0 & 1 & 0 & 0 & 0 & 0 & 0 & 1 & 0 & 0 & 0 & 0 & 0 & 0 & 0 & 0 \\ 0 & 0 & 0 & 0 & 0 & 0 & 0 & 0 & 1 & 0 & 0 & 0 & 0 & 0 & 1 & 0 \\ 0 & 0 & 0 & 0 & 0 & 0 & 0 & 0 & 0 & 1 & 0 & 0 & 0 & 0 & 0 & 1 \end{bmatrix}$$

and (5.86) gives

$$q_4 = H_{4,16}(\beta,\gamma) \bullet P_{16} = \begin{bmatrix} a^{1\circ}_{\circ 1}x + a^{1\circ}_{\circ 2}z \\ a^{1\circ}_{\circ 1}y + a^{1\circ}_{\circ 2}t \\ a^{2\circ}_{\circ 1}x + a^{2\circ}_{\circ 2}z \\ a^{2\circ}_{\circ 1}y + a^{2\circ}_{\circ 2}t \end{bmatrix}$$

and once condensed, we identify with the right-hand of (5.81):

$$\begin{bmatrix} a^{1\circ}_{\circ 1}x + a^{1\circ}_{\circ 2}z & a^{1\circ}_{\circ 1}y + a^{1\circ}_{\circ 2}t \\ a^{2\circ}_{\circ 1}x + a^{2\circ}_{\circ 2}z & a^{2\circ}_{\circ 1}y + a^{2\circ}_{\circ 2}t \end{bmatrix} = \lambda \begin{bmatrix} x & y \\ z & t \end{bmatrix}$$

and passing all terms to the left-hand side leads to the matrix system:

$$\begin{cases} [A - \lambda I] \bullet \begin{bmatrix} x \\ z \end{bmatrix} = \begin{bmatrix} 0 \\ 0 \end{bmatrix} \\ [A - \lambda I] \bullet \begin{bmatrix} y \\ t \end{bmatrix} = \begin{bmatrix} 0 \\ 0 \end{bmatrix} \end{cases},$$

the solutions of which are the eigenvalues and eigenvectors of the classic, which has been solved in the first case.

Thus, the solution *matrices*, built by blocks are the following:

$X_1 = [A_1|\mu A_1]$ automatrix associated with λ_1

$X_2 = [A_2|\nu A_2]$ automatrix associated with λ_2
$\quad ; \quad \forall \mu, \nu \in K.$ (5.87)

Second possible contraction

$$[q^{\circ\gamma}_{\beta\circ}] = C\begin{pmatrix} \alpha \\ \delta \end{pmatrix} [p^{\alpha\circ\gamma\circ}_{\circ\beta\circ\delta}],$$ (5.88)

which once stretched leads to the new $q_{\sigma'}$:

$$q_{\sigma'} = H_{4,16}(\alpha,\delta) \cdot P_\sigma.$$ (5.89)

With the help of the homomorphism (5.71) we obtain

$$H_{4,16}(\alpha,\delta) = [I_4 \otimes E^t_1 | I_4 \otimes E^t_2]$$
$$= \left[\begin{bmatrix} 1 & 0 & 0 & 0 \\ 0 & 1 & 0 & 0 \\ 0 & 0 & 1 & 0 \\ 0 & 0 & 0 & 1 \end{bmatrix} \otimes [1 \ \ 0] \ \middle| \ \begin{bmatrix} 1 & 0 & 0 & 0 \\ 0 & 1 & 0 & 0 \\ 0 & 0 & 1 & 0 \\ 0 & 0 & 0 & 1 \end{bmatrix} \otimes [0 \ \ 1] \right]$$

$$
= \begin{bmatrix} 1 & 0 & 0 & 0 & 0 & 0 & 0 & 0 & 0 & 1 & 0 & 0 & 0 & 0 & 0 & 0 \\ 0 & 0 & 1 & 0 & 0 & 0 & 0 & 0 & 0 & 0 & 0 & 1 & 0 & 0 & 0 & 0 \\ 0 & 0 & 0 & 0 & 1 & 0 & 0 & 0 & 0 & 0 & 0 & 0 & 0 & 1 & 0 & 0 \\ 0 & 0 & 0 & 0 & 0 & 0 & 1 & 0 & 0 & 0 & 0 & 0 & 0 & 0 & 0 & 1 \end{bmatrix}
$$

and (5.89) gives

$$
q_4 = H_{4,16}(\alpha, \delta) P_{16} = \begin{bmatrix} a_{\circ 1}^{1\circ} x + a_{\circ 1}^{2\circ} y \\ a_{\circ 1}^{1\circ} z + a_{\circ 1}^{2\circ} t \\ a_{\circ 2}^{1\circ} x + a_{\circ 2}^{2\circ} y \\ a_{\circ 2}^{1\circ} z + a_{\circ 2}^{2\circ} t \end{bmatrix},
$$

which once condensed and according to (5.88) leads to $[q_{\beta\circ}^{\circ\gamma}]$:

$$
[q_{\beta\circ}^{\circ\gamma}] = \begin{bmatrix} a_{\circ 1}^{1\circ} x + a_{\circ 1}^{2\circ} y & a_{\circ 1}^{1\circ} z + a_{\circ 1}^{2\circ} t \\ a_{\circ 2}^{1\circ} x + a_{\circ 2}^{2\circ} y & a_{\circ 2}^{1\circ} z + a_{\circ 2}^{2\circ} t \end{bmatrix}.
$$

According to (5.81) matrix $[q_{\beta\circ}^{\circ\gamma}]$ must be equal to $\lambda X \equiv \lambda[x_{\circ\delta}^{\gamma\circ}]$, which requires transposing one of them, then

$$
[q_{\beta\circ}^{\circ\gamma}]^t = \lambda X; \quad \begin{bmatrix} a_{\circ 1}^{1\circ} x + a_{\circ 1}^{2\circ} y & a_{\circ 2}^{1\circ} x + a_{\circ 2}^{2\circ} y \\ a_{\circ 1}^{1\circ} z + a_{\circ 1}^{2\circ} t & a_{\circ 2}^{1\circ} z + a_{\circ 2}^{2\circ} t \end{bmatrix} = \lambda \begin{bmatrix} x & y \\ z & t \end{bmatrix}
$$

and passing all terms to the left-hand side, and adequately sorting the equations, yields the matrix system

$$
\begin{cases} [A^t - \lambda I] \bullet \begin{bmatrix} x \\ y \end{bmatrix} = \begin{bmatrix} 0 \\ 0 \end{bmatrix} \\ [A^t - \lambda I] \bullet \begin{bmatrix} z \\ t \end{bmatrix} = \begin{bmatrix} 0 \\ 0 \end{bmatrix} \end{cases},
$$

the solutions of which are the same eigenvalues λ_1 and λ_2 as in possibility (a), but the eigenvectors A_1' and A_2' are those corresponding to matrix A^t. So,

$$
\begin{bmatrix} x \\ y \end{bmatrix} = A_1' \rightarrow [x \quad y] = A_1'^t \text{ eigenvector of } \lambda_1
$$

$$
\begin{bmatrix} z \\ t \end{bmatrix} = \mu A_1' \rightarrow [z \quad t] = \mu A_1'^t \text{ eigenvector of } \lambda_1
$$

and similarly A_2' and $\nu A_2'$ for $\lambda = \lambda_2$.

Finally, we give the following matrices, built by blocks as left solutions:

$$X_1 = \left[\begin{array}{c} A_1'^t \\ ---- \\ \mu A_1'^t \end{array} \right] \quad \text{automatrix associated with } \lambda_1$$

$$\left. \vphantom{\begin{array}{c} A \\ A \\ A \\ A \\ A \\ A \end{array}} \right\} \quad \forall \mu, \nu \in K, \qquad (5.90)$$

$$X_2 = \left[\begin{array}{c} A_2'^t \\ ---- \\ \nu A_2'^t \end{array} \right] \quad \text{automatrix associated with } \lambda_2$$

which satisfy

$$X_1 \bullet A = \lambda_1 X_1 \text{ and } X_2 \bullet A = \lambda_2 X_2.$$

Third case:

Finally, we will study the autotensor of order $r = 3$.

Data tensor: $A = [a^{\alpha\,\circ}_{\circ\,\beta}]$, of second order, over n.

Test tensor $r = 3$: (tensor of order 3). Among several possible choices, we select the tensor $X = [x^{\gamma\circ\epsilon}_{\circ\delta\circ}]$.

Let $P = [p^{\alpha\circ\gamma\circ\epsilon}_{\circ\beta\circ\delta\circ}] = A \otimes X$.

X is a contra–cova–contravariant tensor. The possible contraction tensor products are:

$$M = C \left(\begin{array}{c} \alpha \\ \delta \end{array} \right) P = C \left(\begin{array}{c} \alpha \\ \delta \end{array} \right) [p^{\alpha\circ\gamma\circ\epsilon}_{\circ\beta\circ\delta\circ}] = [m^{\circ\gamma\epsilon}_{\beta\circ\circ}], \quad \text{cova-contra-contravariant}$$

$$N = C \left(\begin{array}{c} \epsilon \\ \beta \end{array} \right) P = C \left(\begin{array}{c} \epsilon \\ \beta \end{array} \right) [p^{\alpha\circ\gamma\circ\epsilon}_{\circ\beta\circ\delta\circ}] = [n^{\alpha\gamma\circ}_{\circ\circ\delta}], \quad \text{contra-contra-covavariant}$$

$$Q = C \left(\begin{array}{c} \gamma \\ \beta \end{array} \right) P = C \left(\begin{array}{c} \gamma \\ \beta \end{array} \right) [p^{\alpha\circ\gamma\circ\epsilon}_{\circ\beta\circ\delta\circ}] = [q^{\alpha\circ\epsilon}_{\circ\delta\circ}], \quad \text{contra-cova-contravariant}$$

So, the only valid option is the third one. Since the dimensions of the tensors to be contracted and contracted are, respectively, for $n = 2 : \sigma = 2^3 \times 2^2 = 32$ and $\sigma' = \sigma/2^2 = 32/4 = 8$, the following tensor equations must be satisfied

$$Q = C \left(\begin{array}{c} \gamma \\ \beta \end{array} \right) P = C \left(\begin{array}{c} \gamma \\ \beta \end{array} \right) [A \otimes X] = \lambda X. \qquad (5.91)$$

We start from

$$X = [x^{\gamma\circ\epsilon}_{\circ\delta\circ}] = \left[\begin{array}{cc} a & b \\ c & d \\ - & - \\ e & f \\ g & h \end{array} \right] ; \quad A = [a^{\alpha\circ}_{\circ\beta}] = \left[\begin{array}{cc} a^{1\circ}_{\circ1} & a^{1\circ}_{\circ2} \\ a^{2\circ}_{\circ1} & a^{2\circ}_{\circ2} \end{array} \right]$$

Having performed the contraction, the fundamental relation (5.91) can be stated as

$$Q = [q^{\alpha \circ \epsilon}_{\circ \delta \circ}] = \begin{bmatrix} q^{1 \circ 1}_{\circ 1 \circ} & q^{1 \circ 2}_{\circ 1 \circ} \\ q^{1 \circ 1}_{\circ 2 \circ} & q^{1 \circ 2}_{\circ 2 \circ} \\ -- & -- \\ q^{2 \circ 1}_{\circ 1 \circ} & q^{2 \circ 2}_{\circ 1 \circ} \\ q^{2 \circ 1}_{\circ 2 \circ} & q^{2 \circ 2}_{\circ 2 \circ} \end{bmatrix} = \begin{bmatrix} a^{1 \circ}_{\circ 1}a + a^{1 \circ}_{\circ 2}e & a^{1 \circ}_{\circ 1}b + a^{1 \circ}_{\circ 2}f \\ a^{1 \circ}_{\circ 1}c + a^{1 \circ}_{\circ 2}g & a^{1 \circ}_{\circ 1}d + a^{1 \circ}_{\circ 2}h \\ ---- & ---- \\ a^{2 \circ}_{\circ 1}a + a^{2 \circ}_{\circ 2}e & a^{2 \circ}_{\circ 1}b + a^{2 \circ}_{\circ 2}f \\ a^{2 \circ}_{\circ 1}c + a^{2 \circ}_{\circ 2}g & a^{2 \circ}_{\circ 1}d + a^{2 \circ}_{\circ 2}h \end{bmatrix} = \lambda \begin{bmatrix} a & b \\ c & d \\ -- \\ e & f \\ g & h \end{bmatrix}$$

$$(5.92)$$

passing all terms to the left-hand side, and grouping adequately the equations, we obtain the systems

$$\begin{cases} [A - \lambda I] \bullet \begin{bmatrix} a \\ e \end{bmatrix} = \begin{bmatrix} 0 \\ 0 \end{bmatrix}; & [A - \lambda I] \bullet \begin{bmatrix} b \\ f \end{bmatrix} = \begin{bmatrix} 0 \\ 0 \end{bmatrix} \\ [A - \lambda I] \bullet \begin{bmatrix} c \\ g \end{bmatrix} = \begin{bmatrix} 0 \\ 0 \end{bmatrix}; & [A - \lambda I] \bullet \begin{bmatrix} d \\ h \end{bmatrix} = \begin{bmatrix} 0 \\ 0 \end{bmatrix} \end{cases}.$$

that can be summarized as

$$[A - \lambda I] \bullet \begin{bmatrix} a & b & c & d \\ e & f & g & h \end{bmatrix} = \Omega_{2,4}.$$

Their interpretation is evident: the matrix solution appears as a permutation of X, and the columns of such a matrix, must be eigenvectors of the eigenvalue λ_1 for X_1, or, for the solution X_2, eigenvectors of the eigenvalue λ_2.

Built by blocks they are

$$X_1 = \begin{bmatrix} \begin{bmatrix} 1 & \mu \\ \nu & \rho \end{bmatrix} \otimes [1 \ \ 0]A_1 \\ ------------ \\ \begin{bmatrix} 1 & \mu \\ \nu & \rho \end{bmatrix} \otimes [0 \ \ 1]A_1 \end{bmatrix}_{4 \times 2}$$

$$X_2 = \begin{bmatrix} \begin{bmatrix} 1 & \mu' \\ \nu' & \rho' \end{bmatrix} \otimes [1 \ \ 0]A_2 \\ ------------ \\ \begin{bmatrix} 1 & \mu \\ \nu & \rho \end{bmatrix} \otimes [0 \ \ 1]A_2 \end{bmatrix}_{4 \times 2} \quad ; \forall \mu, \nu, \dots \rho, \mu', \nu', \dots \rho' \in K.$$

$$(5.93)$$

The reader has now enough tools and experience to solve again the problem using the direct homomorphism model 5 in Section 5.8.4, on P_σ. that is, the tensor components of $A \otimes X$ in a column matrix. Then, it can be checked that the resulting matrix $Q_{\sigma'} = H_{\sigma',\sigma} \bullet P_\sigma$ is the stretched expression of the matrix Q in (5.92). Then, the solution, that must be (5.93), can be obtained.

5.11 Generalized multilinear mappings

We analyze here the mapping of a linear space absolute direct product
$\begin{pmatrix} r \\ \times V_i^{n_i} \\ 1 \end{pmatrix} (K)$ into an arbitrary linear space $W^m(K)$.

As is well known, we call this an "absolute total" linear space or "total product" linear space, which is denoted by

$$V_1^{n_1} \times V_2^{n_2} \times V_r^{n_r}(K) \text{ or } \begin{pmatrix} r \\ \times V_i^{n_i} \\ 1 \end{pmatrix} (K) \tag{5.94}$$

to a linear space, the vectors of which are r-tuples of vectors chosen one per each factor linear space and in order:

$$(\vec{v}_1, \vec{v}_2, \cdots, \vec{v}_r) \in \begin{pmatrix} r \\ \times V_i^{n_i} \\ 1 \end{pmatrix} (K); \quad \vec{v}_i \in V_i^{n_i}(K) \tag{5.95}$$

and its dimension $n = n_1 + n_2 + \cdots + n_r$.

Next, we establish two formal axioms that must be satisfied by the generalized multilineal mappings:

1. F is a mapping that associates with each r-tuple of vectors in $\begin{pmatrix} r \\ \times V_i^{n_i} \\ 1 \end{pmatrix} (K)$,

   a vector $\vec{w} \in W^m(K)$:

$$F : \begin{pmatrix} r \\ \times V_i^{n_i} \\ 1 \end{pmatrix} (K) \to W^m(K) \tag{5.96}$$

   for all r-tuple it is

$$F(\vec{v}_1, \vec{v}_2, \ldots, \vec{v}_r) = \vec{w} \in W^m(K). \tag{5.97}$$

2. This mapping is multilinear:

$$F(\vec{v}_1, \vec{v}_2, \ldots, \vec{v}_h' + \vec{v}_h'', \ldots, \vec{v}_r) = F(\vec{v}_1, \vec{v}_2, \ldots, \vec{v}_h', \ldots, \vec{v}_r)$$
$$+ F(\vec{v}_1, \vec{v}_2, \ldots, \vec{v}_h'', \ldots, \vec{v}_r) \tag{5.98}$$

$$F(\vec{v}_1, \vec{v}_2, \ldots, \lambda \vec{v}_h, \ldots, \vec{v}_r) = \lambda F(\vec{v}_1, \vec{v}_2, \cdots, \vec{v}_h, \ldots, \vec{v}_r); \quad 1 \le h \le r. \tag{5.99}$$

Based on these axioms, we will establish how data are presented and what the operative formulas are for practical use. First, we select bases for the intervening linear spaces, and thus, to the vectors of components:

$$\vec{v}_1 = \vec{e}_{\beta_1} x^{\beta_1 \circ}_{\circ\ 1} \quad \text{with} \quad \vec{v}_1 \in V_1^{n_1}(K) \quad \text{and} \quad 1 \le \beta_1 \le n_1$$
$$\vec{v}_2 = \vec{e}_{\beta_2} x^{\beta_2 \circ}_{\circ\ 2} \quad \text{with} \quad \vec{v}_2 \in V_2^{n_2}(K) \quad \text{and} \quad 1 \le \beta_2 \le n_2$$
$$\cdots \qquad \cdots \qquad \cdots \qquad \cdots \qquad \cdots$$
$$\vec{v}_i = \vec{e}_{\beta_i} x^{\beta_i \circ}_{\circ\ i} \quad \text{with} \quad \vec{v}_i \in V_i^{n_i}(K) \quad \text{and} \quad 1 \le \beta_i \le n_i \qquad (5.100)$$
$$\cdots \qquad \cdots \qquad \cdots \qquad \cdots \qquad \cdots$$
$$\vec{v}_r = \vec{e}_{\beta_r} x^{\beta_r \circ}_{\circ\ r} \quad \text{with} \quad \vec{v}_r \in V_r^{n_r}(K) \quad \text{and} \quad 1 \le \beta_r \le n_r,$$

where $\forall x^{\beta_i \circ}_{\circ\ i}$ is *data*.

When introducing these data in (5.97), on account of (5.98) and (5.99), we obtain

$$\vec{w} = F(\vec{v}_1, \vec{v}_2, \ldots, \vec{v}_r) = x^{\beta_1 \circ}_{\circ\ 1} \cdot x^{\beta_2 \circ}_{\circ\ 2} \cdot \cdots \cdot x^{\beta_r \circ}_{\circ\ r} F(\vec{e}_{\beta_1}, \vec{e}_{\beta_2}, \ldots, \vec{e}_{\beta_r}). \quad (5.101)$$

This expression with contracted dummy indices has a total of $\sigma = n_1 \cdot n_2 \cdot \cdots \cdot n_r$ summands, which correspond with the possibilities of the r-tuples $(\vec{e}_{\beta_1}, \vec{e}_{\beta_2}, \ldots, \vec{e}_{\beta_r})$.

Assume now that the σ basic mappings:

$$F(\vec{e}_{\beta_1}, \vec{e}_{\beta_2}, \ldots, \vec{e}_{\beta_r}) = \vec{w}(\beta_1, \beta_2, \ldots, \beta_r); \quad \vec{w}(\beta_1, \beta_2, \ldots, \beta_r) \in W^m(K) \quad (5.102)$$

are given (again data).

We also assume that vectors $\vec{w}(\beta_1, \beta_2, \cdots, \beta_r)$ are data of the following form.

If the basis of the linear space $W^m(K)$ is $\{\vec{\epsilon}_k\}_1^m$, expressing the vector $\vec{w}(\beta_1, \beta_2, \ldots, \beta_r)$ as a *vector covariant tensor*:

$$\vec{w}(\beta_1, \beta_2, \ldots, \beta_r) = w^{1 \circ \circ \cdots \circ}_{\circ \beta_1 \beta_2 \cdots \beta_r} \vec{\epsilon}_1 + w^{2 \circ \circ \cdots \circ}_{\circ \beta_1 \beta_2 \cdots \beta_r} \vec{\epsilon}_2 + \cdots$$
$$+ w^{k \circ \circ \cdots \circ}_{\circ \beta_1 \beta_2 \cdots \beta_r} \vec{\epsilon}_k + \cdots + w^{m \circ \circ \cdots \circ}_{\circ \beta_1 \beta_2 \cdots \beta_r} \vec{\epsilon}_m, \quad (5.103)$$

where the vector coefficients are mounted with the corresponding covariant tensors, the m covariant tensors are *the data that characterize* the mapping $F(\vec{e}_{\beta_1}, \vec{e}_{\beta_2}, \ldots, \vec{e}_{\beta_r}) = \vec{w}(\beta_1, \beta_2, \ldots, \beta_r)$. (In reality $\vec{w}(\beta_1, \beta_2, \ldots, \beta_r)$ is a vector covariant tensor built with vectors of $W^m(K)$, instead of scalars of K; the reader can see this by executing the sum indicated in (5.103) by separate summands, and then grouping them *into a single entity*.

Assuming that F is delivered as indicated, in (5.103), and entering it in (5.101) we obtain the image of the stated multilinear mapping, by means of the final calculation formula:

$$\vec{w} = x^{\beta_1 \circ}_{\circ\ 1} \cdot x^{\beta_2 \circ}_{\circ\ 2} \cdot \cdots \cdot x^{\beta_r \circ}_{\circ\ r} \left(w^{1 \circ \circ \cdots \circ}_{\circ \beta_1 \beta_2 \cdots \beta_r} \vec{\epsilon}_1 + w^{2 \circ \circ \cdots \circ}_{\circ \beta_1 \beta_2 \cdots \beta_r} \vec{\epsilon}_2 + \cdots + w^{m \circ \circ \cdots \circ}_{\circ \beta_1 \beta_2 \cdots \beta_r} \vec{\epsilon}_m \right),$$
$$(5.104)$$

which is built with m contracted products of the contravariant components of the data vectors by the covariant components of the multilinear mapping F.

One perfectly detects that in Formula (5.104) the notation used has a free index in the *interior* of the coefficients (the index h of $w^{h\ \circ\ \circ\ \cdots\ \circ}_{\circ\ \beta_1\beta_2\cdots\beta_r}$) but it is useful for the calculation; it is the "vector" index of the basis $\{\vec{e}_h\}$ of $W^m(K)$.

If in Formula (5.104) we take as fixed, for example, the vectors $(\vec{v}_2)_0, (\vec{v}_3)_0,$ $\ldots, (\vec{v}_r)_0$, leaving as dummy the $\vec{v}_1$, since they are constant during all the multilinear mappings F all $(x^{\beta_h\ \circ}_{\ \circ\ h})_0; 2 \leq h \leq r$ the multilinear mapping degenerates into a homomorphism H_1 that applies $H_1 : V_1^{n_1}(K) \to W^m(K)$; similarly, if we fix as constant other vectors $\vec{v}_h$ with the exception of a given vector. This is the way most authors *define* multilinear mappings, which in the authors present opinion is correct, but not useful from a practical point of view, because none of them arrives at a concrete expression, like the one in (5.104).

5.11.1 Theorems of similitude with tensor mappings

Theorem 5.6 (Similitude). *There exists a univocal correspondence between the σ r-tuples $(\vec{e}_{\beta_1}, \vec{e}_{\beta_2}, \ldots, \vec{e}_{\beta_r}); 1 \leq \beta_i \leq n_i; i \in I_r$ that appear in Formula (5.101) and the σ basic tensor products, of the basis $\mathcal{B}' = \{\vec{e}_{\beta_1} \otimes \vec{e}_{\beta_2} \otimes \ldots \otimes \vec{e}_{\beta_r}\}$ of the tensor space $V_1^{n_1} \otimes V_2^{n_2} \otimes \cdots \otimes V_r^{n_r}(K) \equiv \left(\overset{r}{\underset{1}{\otimes}} V_i^{n_i}\right)(K)$*

$$(\vec{e}_{\beta_1}, \vec{e}_{\beta_2}, \cdots, \vec{e}_{\beta_r}) \overset{\longrightarrow}{\underset{\longleftarrow}{}} \vec{e}_{\beta_1} \otimes \vec{e}_{\beta_2} \otimes \cdots \otimes \vec{e}_{\beta_r}. \tag{5.105}$$

$\square$

It should be surprising for any reader the evidence of the above theorem's final expression. Next, we give a second theorem that is based on the one above.

Theorem 5.7 (Similitude). *There exists a unique multilinear mapping:*

$$F' : \left(\overset{r}{\underset{1}{\otimes}} V_i^{n_i}\right)(K) \to W^m(K),$$

such that

$$F'(\vec{v}_1 \otimes \vec{v}_2 \otimes \cdots \otimes \vec{v}_r) = F(\vec{v}_1, \vec{v}_2, \ldots, \vec{v}_r) = \vec{w}; \quad \vec{w} \in W^m(K); \quad \forall \vec{v}_i \in V_i^{n_i}(K). \tag{5.106}$$

$\square$

So that the problem of solving images by means of the multilinear mapping $F : \left(\overset{r}{\underset{1}{\times}} V_i^{n_i}\right)(K) \to W^m(K)$ can be solved indistinctly, with the tensor multilinear morphism $F' : \left(\overset{r}{\underset{1}{\otimes}} V_i^{n_i} \right)(K) \to W^m(K)$, by simply changing the notation with the help of Formula (5.105).

Finally, we consider two tensor spaces: the tensor space

$$A \equiv [V_1^{n_1} \otimes V_2^{n_2} \otimes \cdots \otimes V_r^{n_r}(K)] \otimes [V_1^{n_1} \otimes V_2^{n_2} \otimes \cdots \otimes V_r^{n_r}(K)]^*$$

and the tensor space B, the set of all (tensor) multilinear *endomorphisms* that operate inside the tensor space $V_1^{n_1} \otimes V_2^{n_2} \otimes \cdots \otimes V_r^{n_r}(K)$:

$$B = \mathcal{ML}\left[V_1^{n_1} \otimes V_2^{n_2} \otimes \cdots \otimes V_r^{n_r}(K), V_1^{n_1} \otimes V_2^{n_2} \otimes \cdots \otimes V_r^{n_r}(K)\right].$$

Theorem 5.8 (Similitude). *There exists a unique isomorphism Φ*

$$\Phi : \left(\overset{r}{\underset{1}{\otimes}} V_i^{n_i}(K) \right) \otimes \left(\overset{r}{\underset{1}{\otimes}} V_i^{n_i}(K) \right)^* \underset{\leftarrow}{\overset{\to}{}} \mathcal{ML}\left[\left(\overset{r}{\underset{1}{\otimes}} V_i^{n_i} \right)(K), \left(\overset{r}{\underset{1}{\otimes}} V_i^{n_i} \right)(K) \right]$$

(5.107)

such that with each tensor $(\vec{v}_1 \otimes \vec{v}_2 \otimes \cdots \otimes \vec{v}_r) \otimes (\vec{u}_1 \otimes \vec{u}_2 \otimes \cdots \otimes \vec{u}_r)^* \in A$ it associates a tensor multilinear endomorphism

$$T_{(\vec{v}_1 \otimes \vec{v}_2 \otimes \cdots \otimes \vec{v}_r) \otimes (\vec{u}_1 \otimes \vec{u}_2 \otimes \cdots \otimes \vec{u}_r)^*} \in B,$$

that transforms the multivectors $\vec{w} = \vec{w}_1 \otimes \vec{w}_2 \otimes \cdots \otimes \vec{w}_r \in \left(\overset{r}{\underset{1}{\otimes}} V_i^{n_i}(K) \right)$ into the following form:

$$T(\vec{w}) = [(\vec{w}_1 \otimes \vec{w}_2 \otimes \cdots \otimes \vec{w}_r) \bullet (\vec{u}_1 \otimes \vec{u}_2 \otimes \cdots \otimes \vec{u}_r)^*](\vec{v}_1 \otimes \vec{v}_2 \otimes \cdots \otimes \vec{v}_r).$$

(5.108)

□

Theorems 5.7 and 5.8 will be proved by means of concrete models in the proposed examples, so that the interested reader will be able to obtain the general proofs.

5.11.2 Tensor mapping types

If we reconstruct Formula (5.101) adapted for generalized tensor mapping or as a mapping of the correspondence (5.105):

$$\vec{w} = F(\vec{t}) = F(t^{\beta_1 \beta_2 \cdots \beta_r}_{\circ \; \circ \; \ldots \; \circ} \vec{e}_{\beta_1} \otimes \vec{e}_{\beta_2} \otimes \cdots \otimes \vec{e}_{\beta_r}) = t^{\beta_1 \beta_2 \cdots \beta_r}_{\circ \; \circ \; \ldots \; \circ} F(\vec{e}_{\beta_1} \otimes \vec{e}_{\beta_2} \otimes \cdots \otimes \vec{e}_{\beta_r})$$

(5.109)

and we do the same with (5.102) and (5.103):

$$F(\vec{e}_{\beta_1} \otimes \vec{e}_{\beta_2} \otimes \cdots \otimes \vec{e}_{\beta_r}) = \vec{w}(\beta_1, \beta_2, \cdots, \beta_r),$$ (5.110)

the development of the tensor mapping is performed using the same expression (5.104) but with these changes.

It is obvious that in (5.104) the tensor coefficients $w_{\circ\,\beta_1\beta_2\cdots\beta_r}^{h\,\circ\,\circ\,\cdots\,\circ}$ with $1 \leq h \leq m$, can be in some cases *symmetric*, or *anti-symmetric* for the covariant subindices leading to the existence of *tensor mappings F -symmetric and F-anti-symmetric*.

The tensor F-anti-symmetric mappings will be studied in later chapters.

It must be clarified, however, that the tensor mapping type F is completely *independent* of the tensor type over which it is applied, in other words, for example *it is not necessary* to transform symmetric tensors with symmetric mappings.

5.11.3 Direct n-dimensional tensor endomorphisms

We study here the particular case of tensor mappings. Consider the tensor space $\overset{r}{\underset{1}{\otimes}} V_i^n(K) \equiv V_1^n \otimes V_2^n \otimes \cdots \otimes V_r^n(K)$ tensor product of r n-dimensional linear spaces of dimension $\sigma = n^r$, over the same field K. We assume that in each of the linear spaces $V_i^n(K)$ acts an endomorphism of associated square matrix H_i of order n, which transforms the vectors $\vec{v}_i \in V_i^n(K)$ in $H_i(\vec{v}_i) = \vec{w}_i \in V_i^n(K)$.

We look for the heterogeneous tensor endomorphism H_σ, which applies the prototype multivector $\vec{v}_1 \otimes \vec{v}_2 \otimes \cdots \otimes \vec{v}_r \in \overset{r}{\underset{1}{\otimes}} V_i^n(K)$ on the image multivector $\vec{w}_1 \otimes \vec{w}_2 \otimes \cdots \otimes \vec{w}_r \in \overset{r}{\underset{1}{\otimes}} V_i^n(K)$, that is,

$$H_\sigma(\vec{v}) = \vec{w} \Leftrightarrow H_\sigma(\vec{v}_1 \otimes \vec{v}_2 \otimes \cdots \otimes \vec{v}_r) = \vec{w}_1 \otimes \vec{w}_2 \otimes \cdots \otimes \vec{w}_r. \tag{5.111}$$

We solve the problem in a direct form until we find H_σ. Later, the result will be related with the formulas in Section 5.11.

If we notate in tensor form the individual endomorphisms, if $\vec{v}_i = x_{(i)}^{\;\alpha_i}{}_\circ \vec{e}_{\alpha_i}$ and $\vec{w}_i = y_{(i)}^{\;\beta_j}{}_\circ \vec{e}_{\beta_j}$ with $\alpha_i, \beta_j \in I_n; \; i, j \in I_r$, the result is

$$y_{(i)}^{\;\beta_j}{}_\circ = h_{(i)}^{\;\beta_j\;\circ}{}_{\circ\;\alpha_j} x_{(i)}^{\;\alpha_j}{}_\circ. \tag{5.112}$$

Replacing in $\vec{w} = \vec{w}_1 \otimes \vec{w}_2 \otimes \cdots \otimes \vec{w}_r$ the expression of each vector, we arrive at

$$\vec{w} = (y_{(1)}^{\;\beta_1}{}_\circ \vec{e}_{\beta_1}) \otimes (y_{(2)}^{\;\beta_2}{}_\circ \vec{e}_{\beta_2}) \otimes \cdots \otimes (y_{(r)}^{\;\beta_r}{}_\circ \vec{e}_{\beta_r})$$

$$= \left(y_{(1)}^{\;\beta_1}{}_\circ\, y_{(2)}^{\;\beta_2}{}_\circ \cdots y_{(r)}^{\;\beta_r}{}_\circ \right) \vec{e}_{\beta_1} \otimes \vec{e}_{\beta_2} \otimes \cdots \otimes \vec{e}_{\beta_r} \tag{5.113}$$

and replacing (5.112) we get

$$\vec{w} = \left[\left(h_{(1)}^{\;\beta_1\;\circ}{}_{\circ\;\alpha_1} x_{(1)}^{\;\alpha_1}{}_\circ \right) \left(h_{(2)}^{\;\beta_2\;\circ}{}_{\circ\;\alpha_2} x_{(2)}^{\;\alpha_2}{}_\circ \right) \cdots \left(h_{(r)}^{\;\beta_r\;\circ}{}_{\circ\;\alpha_r} x_{(r)}^{\;\alpha_r}{}_\circ \right) \right]$$

$$\vec{e}_{\beta_1} \otimes \vec{e}_{\beta_2} \otimes \cdots \otimes \vec{e}_{\beta_r}, \tag{5.114}$$

which after operating and grouping yields

$$\vec{w} = H_\sigma(\vec{v}) = H_\sigma(\vec{v}_1 \otimes \vec{v}_2 \otimes \cdots \otimes \vec{v}_r)$$
$$= \left(x_{(1)\ \circ}^{\ \ \alpha_1} x_{(2)\ \circ}^{\ \ \alpha_2} \cdots x_{(r)\ \circ}^{\ \ \alpha_r} \right) \left(h_{(1)\ \circ\ \alpha_1}^{\ \beta_1\ \circ} h_{(2)\ \circ\ \alpha_2}^{\ \beta_2\ \circ} \cdots h_{(r)\ \circ\ \alpha_r}^{\ \beta_r\ \circ} \right)$$
$$\vec{e}_{\beta_1} \otimes \vec{e}_{\beta_2} \otimes \cdots \otimes \vec{e}_{\beta_r}. \tag{5.115}$$

If we write

$$h_{\ \circ\ \alpha_1\ \circ\ \alpha_2\cdots\ \circ\ \alpha_r}^{\ \beta_1\ \circ\ \beta_2\ \circ\ \cdots\beta_r\ \circ} = h_{(1)\ \circ\ \alpha_1}^{\ \beta_1\ \circ} h_{(2)\ \circ\ \alpha_2}^{\ \beta_2\ \circ} \cdots h_{(r)\ \circ\ \alpha_r}^{\ \beta_r\ \circ}, \tag{5.116}$$

Expression (5.115) becomes

$$\vec{w} = H_\sigma(\vec{v}_1 \otimes \vec{v}_2 \otimes \cdots \otimes \vec{v}_r)$$
$$= \left(x_{(1)\ \circ}^{\ \ \alpha_1} x_{(2)\ \circ}^{\ \ \alpha_2} \cdots x_{(r)\ \circ}^{\ \ \alpha_r} \right) \left(h_{\ \circ\ \alpha_1\ \circ\ \alpha_2\cdots\ \circ\ \alpha_r}^{\ \beta_1\ \circ\ \beta_2\ \circ\ \cdots\beta_r\ \circ} \right)$$
$$\vec{e}_{\beta_1} \otimes \vec{e}_{\beta_2} \otimes \cdots \otimes \vec{e}_{\beta_r}. \tag{5.117}$$

Since $x_{(j)\ \circ}^{\ \ \alpha_j}$ are the vector data $\vec{v}_j$ and $h_{(j)\ \circ\ \alpha_j}^{\ \beta_j\ \circ}$ are the endomorphism data inside each $V_j^n(K)$, Formula (5.117) solves the problem stated in this section.

In matrix form, expression (5.116) is solved in the matrix

$$H_\sigma = H_1 \otimes H_2 \otimes \cdots \otimes H_r. \tag{5.118}$$

If the column matrix $V_{\sigma,1}$ is an extension of the components of $\vec{v}_1 \otimes \vec{v}_2 \otimes \cdots \otimes \vec{v}_r$, and the column matrix $W_{\sigma,1}$ is an extension of the components of $\vec{w}_1 \otimes \vec{w}_2 \otimes \cdots \otimes \vec{w}_r$, then, expression (5.117) leads to the endomorphism (in matrix form)

$$W_{\sigma,1} = H_\sigma \bullet V_{\sigma,1}. \tag{5.119}$$

If we consider

$$m \equiv n^r; \vec{\epsilon}_k \equiv \vec{e}_{\beta_1} \otimes \vec{e}_{\beta_2} \otimes \cdots \otimes \vec{e}_{\beta_r},$$

with $1 \le k \le m$ and finally $h_{\ \circ\ \alpha_1\ \circ\ \alpha_2\cdots\ \circ\ \alpha_r}^{\ \beta_1\ \circ\ \beta_2\ \circ\ \cdots\beta_r\ \circ} \equiv w_{\ \circ\ \beta_1\ \beta_2\cdots\beta_r}^{k\ \circ\ \circ\ \cdots\ \circ}$, the tensor equation (5.117) represents a variant of Formula (5.104).

One can easily conclude that Formulas (5.104) and (5.119) can be applied to tensors in $\overset{r}{\underset{1}{\otimes}} V_i^n(K)$ not coming from tensor products, as it was indicated in Formula (5.109) and will be in the following formulas.

Example 5.9 (Proof of Theorem 5.7). In this example we prove the tensor similitude Theorem 5.7 for the homogeneous case with the help of tensor and matrix tools.

Consider the homogeneous linear space "total product" (initial space):

$$\left(\overset{r}{\underset{1}{\times}} (V^n)(K) \right) \equiv V^n \times V^n \times V^n \cdots \times V^n(K)$$

of dimension $(r \cdot n)$. Let the r-tuple $(\vec{v}_1, \vec{v}_2, \cdots, \vec{v}_r)$ be one of its vectors, where $\forall \vec{v}_i \in V^n(K)$, and consider the final linear space $W^m(K)$. The vectors $\vec{v}_i = x^{\alpha_i \, \circ}_{\circ \, i} \vec{e}_{\alpha_i}$ are given by its components $(x^{\alpha_i \, \circ}_{\circ \, i})$.

Consider a multilinear mapping F that applies the initial space on the final space by means of the following report, the coefficients f of which are data tensors:

$$F : \left(\begin{array}{c} r \\ \times (V^n)(K) \\ 1 \end{array} \right) \rightarrow W^m(K)$$

$$F(\vec{v}_1, \vec{v}_2, \ldots, \vec{v}_r) = f^{\beta \ \circ \ \circ \ \cdots \ \circ}_{\circ \, \alpha_1 \alpha_2 \cdots \alpha_r} x^{\alpha_1 \circ}_{\circ \, 1} x^{\alpha_2 \circ}_{\circ \, 2} \cdots x^{\alpha_r \circ}_{\circ \, r} \vec{e}_\beta, \tag{5.120}$$

where $\{\vec{e}_\beta\}$ is the basis of the linear space $W^m(K)$, with $\alpha_i \in I_n; 1 \leq \beta \leq m$.

Developing the sum associated with index β in (5.120) to obtain its matrix expression, we get

$$F(\vec{v}_1, \vec{v}_2, \ldots, \vec{v}_r) = f^{1 \ \circ \ \circ \ \cdots \ \circ}_{\circ \, \alpha_1 \alpha_2 \cdots \alpha_r} x^{\alpha_1 \circ}_{\circ \, 1} x^{\alpha_2 \circ}_{\circ \, 2} \cdots x^{\alpha_r \circ}_{\circ \, r} \vec{e}_1$$

$$+ f^{2 \ \circ \ \circ \ \cdots \ \circ}_{\circ \, \alpha_1 \alpha_2 \cdots \alpha_r} x^{\alpha_1 \circ}_{\circ \, 1} x^{\alpha_2 \circ}_{\circ \, 2} \cdots x^{\alpha_r \circ}_{\circ \, r} \vec{e}_2 + \cdots + f^{m \ \circ \ \circ \ \cdots \ \circ}_{\circ \, \alpha_1 \alpha_2 \cdots \alpha_r} x^{\alpha_1 \circ}_{\circ \, 1} x^{\alpha_2 \circ}_{\circ \, 2} \cdots x^{\alpha_r \circ}_{\circ \, r} \vec{e}_m$$

$$= [\vec{e}_1 \vec{e}_2 \cdots \vec{e}_m] \bullet \begin{bmatrix} f^{1 \circ \circ \cdots \circ}_{\circ 1 1 \cdots 1} & f^{1 \circ \circ \cdots \circ}_{\circ 1 1 \cdots 2} & \cdots & f^{1 \ \circ \ \circ \ \cdots \ \circ}_{\circ \, \alpha_1 \alpha_2 \cdots \alpha_r} & \cdots & f^{1 \circ \circ \cdots \circ}_{\circ n n \cdots n} \\ f^{2 \circ \circ \cdots \circ}_{\circ 1 1 \cdots 1} & f^{2 \circ \circ \cdots \circ}_{\circ 1 1 \cdots 2} & \cdots & f^{2 \ \circ \ \circ \ \cdots \ \circ}_{\circ \, \alpha_1 \alpha_2 \cdots \alpha_r} & \cdots & f^{2 \circ \circ \cdots \circ}_{\circ n n \cdots n} \\ \cdots & \cdots & & \cdots & & \cdots \\ f^{m \circ \circ \cdots \circ}_{\circ 1 1 \cdots 1} & f^{m \circ \circ \cdots \circ}_{\circ 1 1 \cdots 2} & \cdots & f^{m \ \circ \ \circ \ \cdots \ \circ}_{\circ \, \alpha_1 \alpha_2 \cdots \alpha_r} & \cdots & f^{m \circ \circ \cdots \circ}_{\circ n n \cdots n} \end{bmatrix}$$

$$\bullet \begin{bmatrix} x^{1 \circ}_{\circ 1} x^{1 \circ}_{\circ 2} \cdots x^{1 \circ}_{\circ r} \\ x^{1 \circ}_{\circ 1} x^{1 \circ}_{\circ 2} \cdots x^{2 \circ}_{\circ r} \\ \cdots \cdots \cdots \cdots \cdots \cdots \\ x^{n \circ}_{\circ 1} x^{n \circ}_{\circ 2} \cdots x^{n \circ}_{\circ r} \end{bmatrix}, \tag{5.121}$$

the symbolic matrix expression of which, with declaration of the sizes of the matrices appearing (with $\sigma = n^r$) is

$$F(\vec{v}_1, \vec{v}_2, \cdots, \vec{v}_r) = [\vec{e}_1 \vec{e}_2 \cdots \vec{e}_m] H_{m,\sigma} \left[x^{\alpha_1 \circ}_{\circ \, 1} x^{\alpha_2 \circ}_{\circ \, 2} \cdots x^{\alpha_r \circ}_{\circ \, r} \right]_{\sigma, 1}, \tag{5.122}$$

which is the matrix expression of the multilinear mapping F.

Next, we will discover a multilinear tensor morphism F'. Remembering that

$$\vec{v}_1 \otimes \vec{v}_2 \otimes \cdots \otimes \vec{v}_r = x^{\alpha_1 \circ}_{\circ \, 1} x^{\alpha_2 \circ}_{\circ \, 2} \cdots x^{\alpha_r \circ}_{\circ \, r} \vec{e}_{\alpha_1} \otimes \vec{e}_{\alpha_2} \otimes \cdots \otimes \vec{e}_{\alpha_r},$$

and applying Theorems 5.6 and 5.7 we choose the following equality:

$$F'(\vec{e}_{\alpha_1} \otimes \vec{e}_{\alpha_2} \otimes \cdots \otimes \vec{e}_{\alpha_r}) \equiv F(\vec{e}_{\alpha_1}, \vec{e}_{\alpha_2}, \cdots, \vec{e}_{\alpha_r})$$

and then

$$F'(\vec{v}_1 \otimes \vec{v}_2 \otimes \cdots \otimes \vec{v}_r) = F'(x_{\circ\ 1}^{\alpha_1\circ} x_{\circ\ 2}^{\alpha_2\circ} \cdots x_{\circ\ r}^{\alpha_r\circ} \vec{e}_{\alpha_1} \otimes \vec{e}_{\alpha_2} \otimes \cdots \otimes \vec{e}_{\alpha_r})$$

$$= F'(\vec{e}_{\alpha_1} \otimes \vec{e}_{\alpha_2} \otimes \cdots \otimes \vec{e}_{\alpha_r})(x_{\circ\ 1}^{\alpha_1\circ} x_{\circ\ 2}^{\alpha_2\circ} \cdots x_{\circ}^{\alpha_r})$$

$$= F(\vec{e}_{\alpha_1}, \vec{e}_{\alpha_2}, \cdots, \vec{e}_{\alpha_r})(x_{\circ\ 1}^{\alpha_1\circ} x_{\circ\ 2}^{\alpha_2\circ} \cdots x_{\circ}^{\alpha_r}). \quad (5.123)$$

If now we apply (5.122) to the vectors $(\vec{e}_{\alpha_1}, \vec{e}_{\alpha_2}, \ldots, \vec{e}_{\alpha_r})$ in matrix form we get

$$F(\vec{e}_{\alpha_1}, \vec{e}_{\alpha_2}, \ldots, \vec{e}_{\alpha_r}) = [\vec{\epsilon}_1, \vec{\epsilon}_2, \ldots, \vec{\epsilon}_m] H_{m,\sigma} \bullet [E_{\alpha_1} \otimes E_{\alpha_2} \otimes \cdots \otimes E_{\alpha_r}]_{\sigma,1}, \quad (5.124)$$

where $\{E_{\alpha_i}\}$ is the matrix canonical basis of $V^n(K)$.

The matrix

$$H'_{m,\sigma} = H_{m,\sigma}$$
$$\bullet [E_1 \otimes E_1 \otimes \cdots \otimes E_1| \cdots |E_{\alpha_1} \otimes E_{\alpha_2} \otimes \cdots \otimes E_{\alpha_r}| \ldots |E_n \otimes E_n \otimes \cdots \otimes E_n]_{\sigma,\sigma} \quad (5.125)$$

represents the operator F', and then the final expression for Formula (5.123) is

$$F'(\vec{v}_1 \otimes \vec{v}_2 \otimes \cdots \otimes \vec{v}_r) = [\vec{\epsilon}_1, \vec{\epsilon}_2, \ldots, \vec{\epsilon}_n] H'_{m,\sigma} \left[x_{\circ\ 1}^{\alpha_1\circ} x_{\circ\ 2}^{\alpha_2\circ} \cdots x_{\circ}^{\alpha_r} \right]_{\sigma,1}. \quad (5.126)$$

Developing Equation (5.125) one gets

$$H'_{m,\sigma} = H_{m,\sigma} \bullet I_n \equiv H_{m,\sigma},$$

which proves our theorem.

□

Example 5.10 (Confirmation of Theorem 5.7). We wish to prove the similitude Theorem 5.7 by means of the following model. Consider two linear spaces $U^m(K)$ and $V^n(K)$ referred to their bases $\{\vec{e}_{\alpha_1}\}_1^m$ and $\{\vec{e}_{\alpha_2}\}_1^n$, respectively, and the two vectors

$$\vec{u}(x^1, x^2, \ldots, x^m) \in U^m(K) \text{ and } \vec{v}(y^1, y^2, \ldots, y^n) \in V^n(K).$$

Consider also another linear space $W^{m \times n}(K)$ referred to a basis $\{\vec{\epsilon}_k\}_1^{m \times n}$, and a bilinear mapping:

$$F: U^m \times V^n(K) \to W^{m \times n}(K),$$

which transforms the vector duples of the "direct product" space $U^m \times V^n(K)$, into vectors of $W^{m \times n}(K)$ by means of

$$\vec{w} \equiv F(\vec{u}, \vec{v}) = [x^1 x^2 \cdots x^m] \begin{bmatrix} \vec{w}_{11} & \vec{w}_{12} & \cdots & \vec{w}_{1n} \\ \vec{w}_{21} & \vec{w}_{22} & \cdots & \vec{w}_{2n} \\ \cdots & \cdots & \cdots & \cdots \\ \vec{w}_{m1} & \vec{w}_{m2} & \cdots & \vec{w}_{mn} \end{bmatrix} \begin{bmatrix} y^1 \\ y^2 \\ \vdots \\ y^n \end{bmatrix}, \quad (5.127)$$

where each of the vectors $\vec{w}_{\alpha\beta}$ in this matrix comes from $W^{m \times n}(K)$ and can be written as

$$\vec{w}_{\alpha\beta} = w^{1\,o\,o}_{\,o\,\alpha\beta}\vec{\epsilon}_1 + w^{2\,o\,o}_{\,o\,\alpha\beta}\vec{\epsilon}_2 + \cdots + w^{m \times n\,o\,o}_{\quad\,\,o\,\,\alpha\beta}\vec{\epsilon}_{m \times n}; \quad 1 \le \alpha \le m; \ 1 \le \beta \le n. \tag{5.128}$$

1. Give a matrix expression of the image vector $\vec{w}$.
2. Prove the existence of the mapping $F'(\vec{u} \otimes \vec{v})$ in Theorem 5.7.
3. Answer questions 1 and 2 for the particular case

$$m = 2; \quad n = 3; \quad \vec{u}(2, -1); \quad \vec{v}(3, 2, 1);$$

$$\vec{w}_{11} = 2\vec{\epsilon}_2 - 3\vec{\epsilon}_3; \quad \vec{w}_{12} = \vec{0}; \quad \vec{w}_{13} = 5\vec{\epsilon}_1 + 2\vec{\epsilon}_2 - \vec{\epsilon}_4 + \vec{\epsilon}_6;$$

$$\vec{w}_{21} = \vec{\epsilon}_1 + \vec{\epsilon}_6; \quad \vec{w}_{22} = \vec{\epsilon}_2 - \vec{\epsilon}_5; \quad \vec{w}_{23} = \vec{\epsilon}_1 + \vec{\epsilon}_2 - \vec{\epsilon}_3 - \vec{\epsilon}_4 + \vec{\epsilon}_6.$$

Solution:

1. Expression (5.127) can be written in tensor form as

$$\vec{w} = F(\vec{u}, \vec{v}) = x^{\alpha}_{o} y^{\beta}_{o} \vec{w}_{\alpha\beta}, \tag{5.129}$$

and developing the sums associated with the dummy indices α and β, and writing them as a matrix product and, as required, representing the *vector* matrices as row matrices, we finally get

$$\vec{w} = F(\vec{u}, \vec{v}) = [\vec{w}_{11}\vec{w}_{12} \cdots \vec{w}_{1n}\vec{w}_{21}\vec{w}_{22} \cdots \vec{w}_{2n} \cdots \vec{w}_{m1}\vec{w}_{m2} \cdots \vec{w}_{mn}] \begin{bmatrix} x^1_o y^1_o \\ x^1_o y^2_o \\ \vdots \\ x^1_o y^n_o \\ x^2_o y^1_o \\ x^2_o y^2_o \\ \vdots \\ x^2_o y^n_o \\ \vdots \\ x^m_o y^1_o \\ x^m_o y^2_o \\ \vdots \\ x^m_o y^n_o \end{bmatrix},$$

which is the answer to the first question.

2. Substituting vectors $\vec{w}_{\alpha\beta}$ in Formula (5.128) into the last expression and grouping in matrix form yields

$$\vec{w} = F(\vec{u}, \vec{v})$$

$$= [\vec{\epsilon}_1 \vec{\epsilon}_2 \cdots \vec{\epsilon}_{m \times n}] \begin{bmatrix} w^{1 o o}_{o 11} & w^{1 o o}_{o 12} & \cdots & w^{1 o o}_{o \alpha \beta} & \cdots & w^{1 o o}_{o mn} \\ w^{2 o o}_{o 11} & w^{2 o o}_{o 12} & \cdots & w^{2 o o}_{o \alpha \beta} & \cdots & w^{2 o o}_{o mn} \\ \cdots & \cdots & \cdots & \cdots & \cdots & \cdots \\ w^{m \times n o o}_{o \ \ 11} & w^{m \times n o o}_{o \ \ 12} & \cdots & w^{m \times n o o}_{o \ \ \alpha \beta} & \cdots & w^{m \times n o o}_{o \ \ mn} \end{bmatrix} \begin{bmatrix} x^1_o y^1_o \\ x^1_o y^2_o \\ \vdots \\ x^1_o y^n_o \\ x^2_o y^1_o \\ x^2_o y^2_o \\ \vdots \\ x^2_o y^n_o \\ \vdots \\ x^m_o y^1_o \\ x^m_o y^2_o \\ \vdots \\ x^m_o y^n_o \end{bmatrix}.$$

$$(5.130)$$

Note that Expression (5.130) is the matrix expression of a multilinear mapping $F'(\vec{u} \otimes \vec{v})$ by means of the *central data matrix*, which "stacks" tensor F'.

Consequently $F(\vec{u}, \vec{v}) = F'(\vec{u} \otimes \vec{v}) = \vec{w}$, which is Theorem 5.7, answering the second question.

3. Next, we illustrate this numerically.

$$m \cdot n = 2 \times 3 = 6; \quad \vec{u} = [\vec{e}_1 \ \ \vec{e}_2] \begin{bmatrix} 2 \\ -1 \end{bmatrix}; \quad \vec{v} = [\vec{e}'_1 \ \ \vec{e}'_2 \ \ \vec{e}'_3] \begin{bmatrix} 3 \\ 2 \\ 1 \end{bmatrix},$$

where $\vec{u} \in U^2(\mathbb{R})$ and $\vec{v} \in V^3(\mathbb{R})$.

Let $\vec{z} = \vec{u} \otimes \vec{v} = \left([\vec{e}_1 \ \ \vec{e}_2] \begin{bmatrix} 2 \\ -1 \end{bmatrix} \right) \otimes \left([\vec{e}'_1 \ \ \vec{e}'_2 \ \ \vec{e}'_3] \begin{bmatrix} 3 \\ 2 \\ 1 \end{bmatrix} \right);$ by extension we get

$$\vec{z} = ([\vec{e}_1 \ \ \vec{e}_2] \otimes [\vec{e}'_1 \ \ \vec{e}'_2 \ \ \vec{e}'_3]) \bullet \left(\begin{bmatrix} 2 \\ -1 \end{bmatrix}^t \otimes \begin{bmatrix} 3 \\ 2 \\ 1 \end{bmatrix}^t \right)^t$$

$$= [\, \vec{e}_1 \otimes \vec{e}_1' \quad \vec{e}_1 \otimes \vec{e}_2' \quad \vec{e}_1 \otimes \vec{e}_3' \quad \vec{e}_2 \otimes \vec{e}_1' \quad \vec{e}_2 \otimes \vec{e}_2' \quad \vec{e}_2 \otimes \vec{e}_3' \,] \bullet \begin{bmatrix} 6 \\ 4 \\ 2 \\ -3 \\ -2 \\ -1 \end{bmatrix}.$$

The vector $\vec{z} = \vec{u} \otimes \vec{v} \in U^2 \otimes V^3(\mathbb{R})$ will be useful later.
Using Formula (5.127), we obtain

$$\vec{w} = F(\vec{u}, \vec{v}) = [\, 2 \quad -1 \,] \begin{bmatrix} 2\vec{\epsilon}_2 - 3\vec{\epsilon}_3 & \vec{0} & 5\vec{\epsilon}_1 + 2\vec{\epsilon}_2 - \vec{\epsilon}_4 + \vec{\epsilon}_6 \\ \vec{\epsilon}_1 + \vec{\epsilon}_6 & \vec{\epsilon}_2 - \vec{\epsilon}_5 & \vec{\epsilon}_1 + \vec{\epsilon}_2 - \vec{\epsilon}_3 - \vec{\epsilon}_4 + \vec{\epsilon}_6 \end{bmatrix} \begin{bmatrix} 3 \\ 2 \\ 1 \end{bmatrix}$$

$$= [\, 2 \quad -1 \,] \begin{bmatrix} 0 & 0 & 5 \\ 1 & 0 & 1 \end{bmatrix} \begin{bmatrix} 3 \\ 2 \\ 1 \end{bmatrix} \vec{\epsilon}_1 + [\, 2 \quad -1 \,] \begin{bmatrix} 2 & 0 & 2 \\ 0 & 1 & 1 \end{bmatrix} \begin{bmatrix} 3 \\ 2 \\ 1 \end{bmatrix} \vec{\epsilon}_2$$

$$+ [\, 2 \quad -1 \,] \begin{bmatrix} -3 & 0 & 0 \\ 0 & 0 & -1 \end{bmatrix} \begin{bmatrix} 3 \\ 2 \\ 1 \end{bmatrix} \vec{\epsilon}_3 + [\, 2 \quad -1 \,] \begin{bmatrix} 0 & 0 & -1 \\ 0 & 0 & -1 \end{bmatrix} \begin{bmatrix} 3 \\ 2 \\ 1 \end{bmatrix} \vec{\epsilon}_4$$

$$+ [\, 2 \quad -1 \,] \begin{bmatrix} 0 & 0 & 0 \\ 0 & -1 & 0 \end{bmatrix} \begin{bmatrix} 3 \\ 2 \\ 1 \end{bmatrix} \vec{\epsilon}_5 + [\, 2 \quad -1 \,] \begin{bmatrix} 0 & 0 & 1 \\ 1 & 0 & 1 \end{bmatrix} \begin{bmatrix} 3 \\ 2 \\ 1 \end{bmatrix} \vec{\epsilon}_6$$

$$= 6\vec{\epsilon}_1 + 13\vec{\epsilon}_2 - 17\vec{\epsilon}_3 - \vec{\epsilon}_4 + 2\vec{\epsilon}_5 - 2\vec{\epsilon}_6;$$

$$\vec{w} = [\, \vec{\epsilon}_1 \quad \vec{\epsilon}_2 \quad \cdots \quad \vec{\epsilon}_6 \,] \begin{bmatrix} 6 \\ 13 \\ -17 \\ -1 \\ 2 \\ -2 \end{bmatrix},$$

which answers the first question. Next, we build the central matrix of the multilinear mapping F', the structure of which has been given in Formula (5.130), Thus, we arrange the data vector components $\vec{w}_{\alpha\beta}$ as columns, and then, we apply the mentioned formula

$$\vec{w} = F'(\vec{z}) = F'(\vec{u} \otimes \vec{v})$$

$$= [\, \vec{\epsilon}_1 \quad \vec{\epsilon}_2 \quad \cdots \quad \vec{\epsilon}_6 \,] \begin{bmatrix} 0 & 0 & 5 & 1 & 0 & 1 \\ 2 & 0 & 2 & 0 & 1 & 1 \\ -3 & 0 & 0 & 0 & 0 & -1 \\ 0 & 0 & -1 & 0 & 0 & -1 \\ 0 & 0 & 0 & 0 & -1 & 0 \\ 0 & 0 & 1 & 1 & 0 & 1 \end{bmatrix} \bullet \begin{bmatrix} 6 \\ 4 \\ 2 \\ -3 \\ -2 \\ -1 \end{bmatrix}$$

$$= [\, \vec{e}_1 \otimes \vec{e}_1'\ \ \vec{e}_1 \otimes \vec{e}_2'\ \ \vec{e}_1 \otimes \vec{e}_3'\ \ \vec{e}_2 \otimes \vec{e}_1'\ \ \vec{e}_2 \otimes \vec{e}_2'\ \ \vec{e}_2 \otimes \vec{e}_3' \,] \begin{bmatrix} 6 \\ 13 \\ -17 \\ -1 \\ 2 \\ -2 \end{bmatrix} , \quad (5.131)$$

which gives the answer to the second question.

As one can see the result is the same as the one obtained in the first question, which is in agreement with the tensor similitude Theorem 5.7.

$\square$

Example 5.11 (Proof of Theorem 5.8). Consider the linear spaces $V^n(K)$ and its dual $V_*^n(K)$ referred to the reciprocal bases $\{\vec{e}_\beta\}$ and $\{\vec{e}^{*\alpha}\}$.

Consider also the linear space of all linear operators T that transform vectors inside $V^n(K)$, that is,

$$T : V^n(K) \to V^n(K); \ \ T \in \mathcal{L}[V^n(K), V^n(K)],$$

where $\mathcal{L}$ refers to *linear* operators and $V^n(K), V^n(K)$ to the endomorphism *initial* and *final* linear spaces, respectively.

Let $\{\vec{e}_{\alpha\beta}\}$ be the canonical basis of $\mathcal{L}[V^n(K), V^n(K)]$,

$$\vec{e}_{\alpha\beta} = \begin{bmatrix} 0 & 0 & \cdots & 0 & \cdots & 0 \\ \cdots & \cdots & \cdots & \cdots & \cdots & \cdots \\ 0 & 0 & \cdots & 1 & \cdots & 0 \\ \cdots & \cdots & \cdots & \cdots & \cdots & \cdots \\ 0 & 0 & \cdots & 0 & \cdots & 0 \end{bmatrix}_{n \times n} ,$$

with a *one* in the position associated with row α and column β and zero otherwise; there exist n^2 basic vectors.

Consider two data vectors

$$\vec{u} = ||\vec{e}^{*\alpha}|| \begin{bmatrix} u_1^{\circ} \\ u_2^{\circ} \\ \vdots \\ u_n^{\circ} \end{bmatrix}, \vec{u}^* \in V_*^n(K) \text{ and } \vec{v} = ||\vec{e}_\beta|| \begin{bmatrix} v_{\circ}^1 \\ v_{\circ}^2 \\ \vdots \\ v_{\circ}^n \end{bmatrix}, \vec{v} \in V^n(K).$$

If we build in matrix form the vector $\vec{v} \otimes \vec{u}^* \in V^n \otimes V_*^n(K)$ and make the matrix of order $n \times n$ of the product equal to $\Phi(\vec{v} \otimes \vec{u}^*)$, we apply the tensor space $V^n \otimes V_*^n(K)$ in the space $\mathcal{L}[V^n(K), V^n(K)]$. Show that the endomorphism transforms the vectors as stated in the tensor similitude Theorem 5.8.

Solution: We calculate the vector $\vec{v} \otimes \vec{u}^*$:

$$\vec{v} \otimes \vec{u}^* = \left([\vec{e}_1 \vec{e}_2 \cdots \vec{e}_n] \begin{bmatrix} v_\circ^1 \\ v_\circ^2 \\ \vdots \\ v_\circ^n \end{bmatrix} \right) \otimes \left([\vec{e}^{*1}\vec{e}^{*2}\cdots \vec{e}^{*n}] \begin{bmatrix} u_1^\circ \\ u_2^\circ \\ \vdots \\ u_n^\circ \end{bmatrix} \right)^t$$

$$= [\vec{e}_1\vec{e}_2\cdots\vec{e}_n] \left(\begin{bmatrix} v_\circ^1 \\ v_\circ^2 \\ \vdots \\ v_\circ^n \end{bmatrix} \bullet \begin{bmatrix} u_1^\circ & u_2^\circ & \cdots & u_n^\circ \end{bmatrix} \right) \begin{bmatrix} \vec{e}^{*1} \\ \vec{e}^{*2} \\ \vdots \\ \vec{e}^{*n} \end{bmatrix}_\otimes$$

$$= [\vec{e}_1\vec{e}_2\cdots\vec{e}_n] \left(\begin{bmatrix} v_\circ^1 u_1^\circ & v_\circ^1 u_2^\circ & \cdots & v_\circ^1 u_n^\circ \\ v_\circ^2 u_1^\circ & v_\circ^2 u_2^\circ & \cdots & v_\circ^2 u_n^\circ \\ \cdots & \cdots & \cdots & \cdots \\ v_\circ^n u_1^\circ & v_\circ^n u_2^\circ & \cdots & v_\circ^n u_n^\circ \end{bmatrix} \right) \begin{bmatrix} \vec{e}^{*1} \\ \vec{e}^{*2} \\ \cdots \\ \vec{e}^{*n} \end{bmatrix}_\otimes,$$

where the $\otimes$ operator appears as a subindex to refer to a quadratic form of tensor products.

Following the stated conditions, we have

$$\Phi(\vec{v} \otimes \vec{u}^*) = \begin{bmatrix} v_\circ^1 u_1^\circ & v_\circ^1 u_2^\circ & \cdots & v_\circ^1 u_n^\circ \\ v_\circ^2 u_1^\circ & v_\circ^2 u_2^\circ & \cdots & v_\circ^2 u_n^\circ \\ \cdots & \cdots & \cdots & \cdots \\ v_\circ^n u_1^\circ & v_\circ^n u_2^\circ & \cdots & v_\circ^n u_n^\circ \end{bmatrix}, \tag{5.132}$$

which gives the endomorphism matrix. Next, following Theorem 5.8 we examine how the vectors $\vec{w} \in V^n(K)$ (since in this example there exists only one space as primary and dual factors) are transformed.

We call the matrix in (5.132) $T_{\vec{v}\otimes\vec{u}^*}$, and transforming a vector $\vec{w} \in V^n(K)$ with the operator T we get

$$T_{\vec{v}\otimes\vec{u}^*}(\vec{w}) = \begin{bmatrix} v_\circ^1 u_1^\circ & v_\circ^1 u_2^\circ & \cdots & v_\circ^1 u_n^\circ \\ v_\circ^2 u_1^\circ & v_\circ^2 u_2^\circ & \cdots & v_\circ^2 u_n^\circ \\ \cdots & \cdots & \cdots & \cdots \\ v_\circ^n u_1^\circ & v_\circ^n u_2^\circ & \cdots & v_\circ^n u_n^\circ \end{bmatrix} \begin{bmatrix} w_\circ^1 \\ w_\circ^2 \\ \vdots \\ w_\circ^n \end{bmatrix}$$

$$= \begin{bmatrix} (u_1^\circ w_\circ^1 + u_2^\circ w_\circ^2 + \cdots + u_n^\circ w_\circ^n)v_\circ^1 \\ (u_1^\circ w_\circ^1 + u_2^\circ w_\circ^2 + \cdots + u_n^\circ w_\circ^n)v_\circ^2 \\ \cdots \\ (u_1^\circ w_\circ^1 + u_2^\circ w_\circ^2 + \cdots + u_n^\circ w_\circ^n)v_\circ^n \end{bmatrix}$$

$$= (u_1^{\circ} w_{\circ}^1 + u_2^{\circ} w_{\circ}^2 + \cdots + u_n^{\circ} w_{\circ}^n) \begin{bmatrix} v_{\circ}^1 \\ v_{\circ}^2 \\ \vdots \\ v_{\circ}^n \end{bmatrix}$$

$$= (\vec{w} \bullet \vec{u}^{*}) \begin{bmatrix} v_{\circ}^1 \\ v_{\circ}^2 \\ \vdots \\ v_{\circ}^n \end{bmatrix}. \tag{5.133}$$

The tensor conclusion of (5.133) is that

$$T_{\vec{v} \otimes \vec{u}^{*}}(\vec{w}) = (\vec{w} \bullet \vec{u}^{*})\vec{v}. \tag{5.134}$$

From (5.132) to (5.133) we conclude that the equality

$$\Phi(\vec{v} \otimes \vec{u}^{*}) = T_{\vec{v} \otimes \vec{u}^{*}}$$

has the property (5.134) and then, Theorem 5.8 has been proved with the present model.

The isomorphism character is detected if we apply (5.132) to the vectors $\vec{e}_i$ and $\vec{e}^{*j}$ of matrices $\begin{bmatrix} 0 \\ 0 \\ \vdots \\ 1 \\ \vdots \\ 0 \end{bmatrix}$, with the 1 in row i, and $\begin{bmatrix} 0 \\ 0 \\ \vdots \\ 1 \\ \vdots \\ 0 \end{bmatrix}$, with the 1 in the row j, respectively:

$$\Phi(\vec{e}_i \otimes \vec{e}^{*j}) = \begin{bmatrix} 0 & 0 & \cdots & 0 & \cdots & 0 \\ 0 & 0 & \cdots & 0 & \cdots & 0 \\ \cdots & \cdots & \cdots & \cdots & \cdots & \cdots \\ 0 & 0 & \cdots & 1 & \cdots & 0 \\ \cdots & \cdots & \cdots & \cdots & \cdots & \cdots \\ 0 & 0 & \cdots & 0 & \cdots & 0 \end{bmatrix} \equiv \vec{\epsilon}_{ij}.$$

Thus, it is shown that this multilinear endomorphism associates the basis of $V^n \otimes V_*^n(K)$ with the basis of $\mathcal{L}[V^n(K), V^n(K)]$, and then, in this particular case it is an isomorphism.

□

Example 5.12 (Total and tensor products). Consider the *total product* homogeneous linear space

$$\left(\begin{array}{c} 3 \\ \times V_k^{n_0} \\ 1 \end{array}\right)(\mathbb{R}) \equiv V_1^{n_0} \times V_2^{n_0} \times V_3^{n_0}(\mathbb{R})$$

of dimension $n = 3n_0 = 9$, and let the tuple $(\vec{v}_1, \vec{v}_2, \vec{v}_3)$ be one of its vectors, where $\vec{v}_k \in V_k^3(\mathbb{R})$. The bases for each factor linear space will be denoted by $\{\vec{e}_i(k)\}; 1 \le i, k \le 3$, and thus, we have

$$\vec{v}_k = ||\vec{e}(k)|| X_k; \quad X_k = \begin{bmatrix} x^1(k) \\ x^2(k) \\ x^3(k) \end{bmatrix}, \quad \forall x^i(k) \in \mathbb{R}.$$

The basis of the total product linear space $\left(\begin{array}{c} 3 \\ \times V_k^3 \\ 1 \end{array}\right)(\mathbb{R})$ will be notated

$$B = \{\vec{e}_1(1) \ \vec{e}_2(1) \ \vec{e}_3(1), \vec{e}_1(2) \ \vec{e}_2(2) \ \vec{e}_3(2), \vec{e}_1(3) \ \vec{e}_2(3) \ \vec{e}_3(3)\}$$

and therefore, the matrix representation of the 3-tuple $(\vec{v}_1, \vec{v}_2, \vec{v}_3)$ results

$$(\vec{v}_1, \vec{v}_2, \vec{v}_3) = ||B|| X,$$

where X is the block column matrix

$$X = \begin{bmatrix} X_1 \\ X_2 \\ X_3 \end{bmatrix}.$$

Three morphisms "$f(k)$" apply each space factor $V_k^3(\mathbb{R})$ into a linear space $W^m(\mathbb{R})$ of dimension $m = 4$ and basis $\{\varepsilon_\ell\}_1^4$.

The matrix representation of a vector $\vec{w} \in W^4(\mathbb{R})$ is $\vec{w} = ||\vec{\varepsilon}_\ell|| Y$. Assuming that the associated matrix representation, relative to such bases, of morphisms $f(k)$ are the data matrix $H_{4,3}(k)$, the morphism matrix representations become

$$Y_k = H_{4,3}(k) \bullet X_k; \quad 1 \le k \le 3.$$

Finally, let us build an homomorphism f that applies the initial total product linear space into the final space $W^m(\mathbb{R})$:

$$f : \left(\begin{array}{c} 3 \\ \times V_k^3 \\ 1 \end{array}\right)(\mathbb{R}) \to W^m(\mathbb{R}); \quad f(\vec{v}_1, \vec{v}_2, \vec{v}_3) = \vec{w}$$

which matrix representation is:

$$Y = H_{m,n} \bullet X;$$

where $H_{m,n} = [H(1) \ H(2) \ H(3)]$ is built with $H_{4,3}(k)$ matrices as blocks.

Assuming now that the data matrices are:

$$H(1) = \begin{bmatrix} 5 & 3 & 4 \\ 1 & 5 & 3 \\ 2 & 4 & 3 \\ 4 & 2 & 3 \end{bmatrix}; \quad H(2) = \begin{bmatrix} 8 & -7 & 4 \\ 4 & 5 & -2 \\ -4 & 2 & 0 \\ 2 & 3 & 6 \end{bmatrix}; \quad H(3) = \begin{bmatrix} 9 & 0 & 6 \\ -6 & -3 & 9 \\ 7 & 2 & -4 \\ 10 & -1 & 11 \end{bmatrix};$$

1. Give the representation of morphism f specifying the components of each matrix.
2. If
$$(\vec{v}_1, \vec{v}_2, \vec{v}_3) = (2\vec{e}_1 + 3\vec{e}_2 - \vec{e}_3, 5\vec{e}_2 - 2\vec{e}_3, \vec{e}_1 - \vec{e}_2 + 3\vec{e}_3) \qquad (5.135)$$
find the image vector $\vec{w} = f(\vec{v}_1, \vec{v}_2, \vec{v}_3)$.
3. In $\begin{pmatrix} 3 \\ \times V_k^3 \\ 1 \end{pmatrix}$ ($\mathbb{R}$), consider the multivector relation

$$(2\vec{e}_1, 2\vec{e}_2 - 5\vec{e}_3, 2\vec{e}_3) = (\vec{e}_1, 2\vec{e}_2, \vec{e}_3) + (\vec{e}_1, -5\vec{e}_3, \vec{e}_3).$$

Using this relation, examine if f is a *multilinear transformation* for the addition of the total product linear space.
4. Find a basis of the null space relative to morphism f, verifying that the dimension of the resulting basis is coherent with the dimension of the range space.
5. Based on the knowledge we already have on f, build a multilinear mapping

$$F: \begin{pmatrix} 3 \\ \times V_k^3 \\ 1 \end{pmatrix} (\mathbb{R}) \to W^4(\mathbb{R}).$$

To get it, one must answer the following questions:
(a) Determine matrix $M_{n,\sigma}$ where $\sigma = n_0^r = 3^3 = 27$; the matrix columns X of $M_{n,\sigma}$ are the matrix representations of the σ 3-tuples $(\vec{e}_{\beta_1}(1), \vec{e}_{\beta_2}(2), \vec{e}_{\beta_3}(3)), \forall \beta_i, 1 \le \beta_i \le 3$, in the B basis.
(b) Set condition

$$F(\vec{e}_{\beta_1}, \vec{e}_{\beta_2}, \vec{e}_{\beta_3}) = f(\vec{e}_{\beta_1}(1), \vec{e}_{\beta_2}(2), \vec{e}_{\beta_3}(3)), \forall \beta_i, 1 \le \beta_i \le 3,$$

through the matrix relation:

$$H(F)_{m,\sigma} = H_{m,n} \bullet M_{n,\sigma}.$$

Give the matrix $H_{m,\sigma}$ associated with the multilinear application F.
(c) Determine matrix $X_{\sigma,1}$ as the representation of multivector $(\vec{v}_1, \vec{v}_2, \vec{v}_3)$ given in (5.135), but now with the appropriate components as shown in formula (5.122).
6. Determine the image vector $\vec{w}' = F(\vec{v}_1, \vec{v}_2, \vec{v}_3)$, in accordance with matrix equation $Y'_{m,1} = H_{m,\sigma} \bullet X_{\sigma,1}$.
7. Determine if $\vec{w}' = 3\vec{w}$.

Solution:

1. The matrix representation of morphism f is:

$$\begin{bmatrix} y^1 \\ y^2 \\ y^3 \\ y^4 \end{bmatrix} = \begin{bmatrix} 5 & 3 & 4 & 8 & -7 & 4 & 9 & 0 & 6 \\ 1 & 5 & 3 & 4 & 5 & -2 & -6 & -3 & 9 \\ 2 & 4 & 3 & -4 & 2 & 0 & 7 & 2 & -4 \\ 4 & 2 & 3 & 2 & 3 & 6 & 10 & -1 & 11 \end{bmatrix} \begin{bmatrix} x^1(1) \\ x^2(1) \\ x^3(1) \\ x^1(2) \\ x^2(2) \\ x^3(2) \\ x^1(3) \\ x^2(3) \\ x^3(3) \end{bmatrix}$$

2. Since

$$\begin{bmatrix} y^1 \\ y^2 \\ y^3 \\ y^4 \end{bmatrix} = \begin{bmatrix} 5 & 3 & 4 & 8 & -7 & 4 & 9 & 0 & 6 \\ 1 & 5 & 3 & 4 & 5 & -2 & -6 & -3 & 9 \\ 2 & 4 & 3 & -4 & 2 & 0 & 7 & 2 & -4 \\ 4 & 2 & 3 & 2 & 3 & 6 & 10 & -1 & 11 \end{bmatrix} \begin{bmatrix} 2 \\ 3 \\ -1 \\ 0 \\ 5 \\ -2 \\ 1 \\ -1 \\ 3 \end{bmatrix} = \begin{bmatrix} -1 \\ 67 \\ 16 \\ 58 \end{bmatrix}$$

we have

$$\vec{w} = f(\vec{v}_1, \vec{v}_2, \vec{v}_3) = -\vec{\varepsilon}_1 + 67\vec{\varepsilon}_2 + 16\vec{\varepsilon}_3 + 58\vec{\varepsilon}_4.$$

3. Since f is a morphism, we have

$$f(2\vec{e}_1, 2\vec{e}_2 - 5\vec{e}_3, 2\vec{e}_3) = f\big((\vec{e}_1, 2\vec{e}_2, \vec{e}_3) + (\vec{e}_1, -5\vec{e}_3, \vec{e}_3)\big)$$
$$= f(\vec{e}_1, 2\vec{e}_2, \vec{e}_3) + f(\vec{e}_1, -5\vec{e}_3, \vec{e}_3)$$

If f were a multilinear mapping, it should be

$$f(2\vec{e}_1, 2\vec{e}_2 - 5\vec{e}_3, 2\vec{e}_3) = f(2\vec{e}_1, 2\vec{e}_2, 2\vec{e}_3) + f(2\vec{e}_1, -5\vec{e}_3, 2\vec{e}_3)$$

so, f is not a multilinear mapping.

4. A basis of the null space is

$$B_N = \begin{bmatrix} -1 & 0 & -176 & -766 & -768 \\ -1 & 0 & -82 & 196 & 442 \\ 2 & 0 & 0 & 0 & 0 \\ 0 & 0 & 405 & 207 & -1 \\ 0 & 0 & 302 & 46 & -118 \\ 0 & 0 & 0 & 0 & 424 \\ 0 & -2 & 0 & 212 & 0 \\ 0 & 13 & 848 & 0 & 0 \\ 0 & 3 & 0 & 0 & 0 \end{bmatrix}$$

As matrix $H(2)$ has rank 4, this is the rank of matrix $H_{4,9}$. Thus, we have

dim(null space)+dim(range space) $= 5+4 = 9 = $ dim(total product space).

5. (a) The matrix representations of $(\vec{e}_1(1), \vec{e}_1(2), \vec{e}_1(3)), (\vec{e}_1(1), \vec{e}_1(2), \vec{e}_2(3))$ in the B basis are:

$$
\begin{bmatrix} 1 \\ 0 \\ 0 \\ 1 \\ 0 \\ 0 \\ 1 \\ 0 \\ 0 \end{bmatrix},
\begin{bmatrix} 1 \\ 0 \\ 0 \\ 1 \\ 0 \\ 0 \\ 0 \\ 1 \\ 0 \end{bmatrix}
$$

so that following this we get matrix $M_{n,\sigma}$:

$$
M_{9,27} = \begin{bmatrix}
1 & 1 & 1 & 1 & 1 & 1 & 1 & 1 & 1 & 0 & 0 & 0 & 0 & 0 & 0 & 0 & 0 & 0 & 0 & 0 & 0 & 0 & 0 & 0 & 0 & 0 & 0 \\
0 & 0 & 0 & 0 & 0 & 0 & 0 & 0 & 0 & 1 & 1 & 1 & 1 & 1 & 1 & 1 & 1 & 1 & 0 & 0 & 0 & 0 & 0 & 0 & 0 & 0 & 0 \\
0 & 0 & 0 & 0 & 0 & 0 & 0 & 0 & 0 & 0 & 0 & 0 & 0 & 0 & 0 & 0 & 0 & 0 & 1 & 1 & 1 & 1 & 1 & 1 & 1 & 1 & 1 \\
1 & 1 & 1 & 0 & 0 & 0 & 0 & 0 & 0 & 1 & 1 & 1 & 0 & 0 & 0 & 0 & 0 & 0 & 1 & 1 & 1 & 0 & 0 & 0 & 0 & 0 & 0 \\
0 & 0 & 0 & 1 & 1 & 1 & 0 & 0 & 0 & 0 & 0 & 0 & 1 & 1 & 1 & 0 & 0 & 0 & 0 & 0 & 0 & 1 & 1 & 1 & 0 & 0 & 0 \\
0 & 0 & 0 & 0 & 0 & 0 & 1 & 1 & 1 & 0 & 0 & 0 & 0 & 0 & 0 & 1 & 1 & 1 & 0 & 0 & 0 & 0 & 0 & 0 & 1 & 1 & 1 \\
1 & 0 & 0 & 1 & 0 & 0 & 1 & 0 & 0 & 1 & 0 & 0 & 1 & 0 & 0 & 1 & 0 & 0 & 1 & 0 & 0 & 1 & 0 & 0 & 1 & 0 & 0 \\
0 & 1 & 0 & 0 & 1 & 0 & 0 & 1 & 0 & 0 & 1 & 0 & 0 & 1 & 0 & 0 & 1 & 0 & 0 & 1 & 0 & 0 & 1 & 0 & 0 & 1 & 0 \\
0 & 0 & 1 & 0 & 0 & 1 & 0 & 0 & 1 & 0 & 0 & 1 & 0 & 0 & 1 & 0 & 0 & 1 & 0 & 0 & 1 & 0 & 0 & 1 & 0 & 0 & 1
\end{bmatrix}
$$

(b) The relation $H(F)_{m,\sigma} = H_{m,n} \bullet M_{n,\sigma}$, that in this case is

$$
H(F)_{4,27} = H_{4,9} \bullet M_{9,27}
$$

becomes

$$
\begin{bmatrix}
22 & 13 & 19 & 7 & -2 & 4 & 18 & 9 & 15 & 20 & 11 & 17 & 5 & -4 & 2 & 16 & 7 & 13 & 21 & 12 & 18 & 6 & -3 & 3 & 17 & 8 & 14 \\
-1 & 2 & 14 & 0 & 3 & 15 & -7 & -4 & 8 & 3 & 6 & 18 & 4 & 7 & 19 & -30 & 12 & 1 & 4 & 16 & 2 & 5 & 17 & -5 & -2 & 10 \\
5 & 0 & -6 & 11 & 6 & 0 & 9 & 4 & -2 & 7 & 2 & -4 & 13 & 8 & 2 & 11 & 6 & 0 & 6 & 1 & -5 & 12 & 7 & 1 & 10 & 5 & -1 \\
16 & 5 & 17 & 17 & 6 & 18 & 20 & 9 & 21 & 14 & 3 & 15 & 15 & 4 & 16 & 18 & 7 & 19 & 15 & 4 & 16 & 16 & 5 & 17 & 19 & 8 & 20
\end{bmatrix}
$$

(c) Applying formula (2.21) one gets

$$X_{27,1} = X_1 \otimes X_2 \otimes X_3 = \begin{bmatrix} 2 \\ 3 \\ -1 \end{bmatrix} \otimes \begin{bmatrix} 0 \\ 5 \\ -2 \end{bmatrix} \otimes \begin{bmatrix} 1 \\ -1 \\ 3 \end{bmatrix} = \begin{pmatrix} 0 \\ 0 \\ 0 \\ 10 \\ -10 \\ 30 \\ -4 \\ 4 \\ -12 \\ 0 \\ 0 \\ 0 \\ 15 \\ -15 \\ 45 \\ -6 \\ 6 \\ -18 \\ 0 \\ 0 \\ 0 \\ -5 \\ 5 \\ -15 \\ 2 \\ -2 \\ 6 \end{pmatrix}$$

6. The matrix representation of F is:

$$Y'_{4,1} = H_{4,27} \bullet X_{27,1} = \begin{bmatrix} -57 \\ 762 \\ 153 \\ 663 \end{bmatrix}$$

and then
$$\vec{w}\,' = -57\vec{\varepsilon}_1 + 762\vec{\varepsilon}_2 + 153\vec{\varepsilon}_3 + 663\vec{\varepsilon}_4.$$

7. It is clear that

$$\begin{aligned} \vec{w}' &= 3(-19\vec{\varepsilon}_1 + 254\vec{\varepsilon}_2 + 51\vec{\varepsilon}_3 + 221\vec{\varepsilon}_4) \\ &\neq 3(-\vec{\varepsilon}_1 + 67\vec{\varepsilon}_2 + 16\vec{\varepsilon}_3 + 58\vec{\varepsilon}_4.) \\ &= 3\vec{w} \end{aligned}$$

□

5.12 Exercises

5.1. In the tensor space $\overset{4}{\underset{1}{\otimes}} R^2_*$, we consider the totally covariant homogeneous tensor T, given by its matrix representation:

$$[t^{\circ\,\circ\,\circ\,\circ}_{\alpha\beta\gamma\delta}] = \begin{bmatrix} a_1 & b_1 & | & a_2 & b_2 \\ c_1 & d_1 & | & c_2 & d_2 \\ - & - & + & - & - \\ a_3 & b_3 & | & a_4 & b_4 \\ c_3 & d_3 & | & c_4 & d_4 \end{bmatrix}.$$

Obtain the "permutation" matrices P_1, P_2 and P_3 associated with the three rotation isomers (1),(2),(3) mentioned in Example 5.5, point 3, that transform the tensor $T_{\sigma,1}$ into its "extended" isomers.

5.2. Consider the homogeneous tensors P, Q and D (the last is the Kronecker delta), all of them associated with the linear space $V^n(\mathbb{R})$. Determine if the tensors A, B, C, contracted products of the data tensors, are their isomers:

$$A: \ \delta^{\circ\,\beta}_{\alpha\,\circ}p^{\circ\,\circ\,\lambda}_{\beta\gamma\circ}; \quad B: \ \delta^{\circ\,\beta}_{\alpha\,\circ}q^{\circ\,\circ\,\circ}_{\gamma\beta\lambda}\delta^{\lambda\circ}_{\circ\,\mu}; \quad C: \ \delta^{\circ\,\beta}_{\alpha\,\circ}\delta^{\circ\,\beta}_{\alpha\,\circ}\delta^{\circ\,\lambda}_{\beta\circ}\delta^{\circ\,\lambda}_{\gamma\circ}.$$

5.3. Two tensors T of order $r_1 = 2$ and U of order $r_2 = 3$ are defined over a certain linear space $V^3(\mathbb{R})$ referred to a certain basis $\{\vec{e}_\alpha\}$. Their matrix representations are

$$[t^{\alpha\beta}_{\circ\circ}] = \begin{bmatrix} 2 & 3 & 0 \\ -2 & -1 & 1 \\ 1 & 0 & -1 \end{bmatrix} \quad (\alpha \text{ row}, \beta \text{ column})$$

$$[u^{\gamma\circ\mu}_{\circ\lambda\circ}] = \begin{bmatrix} 0 & 0 & 1 & | & 2 & 1 & 0 & | & 0 & 1 & 3 \\ 1 & 3 & 5 & | & 1 & 0 & -1 & | & 1 & -1 & 4 \\ 0 & 2 & 0 & | & -1 & 1 & 4 & | & 6 & 1 & 0 \end{bmatrix},$$

where γ is the row block, λ is the column of each block and μ is the block column. (beware of the matrix block disposition of this tensor).

1. Determine, as contractions of the tensors $Q_1 = T \otimes U$ and $Q_2 = U \otimes T$ (of order $r = 5$), the contracted products that follow:

$$A: \ t^{\theta\beta}_{\circ\circ}u^{\gamma\circ\mu}_{\circ\theta\circ}; \quad B: \ t^{\alpha\theta}_{\circ\circ}u^{\gamma\circ\mu}_{\circ\theta\circ}; \quad F: \ u^{\gamma\circ\mu}_{\circ\theta\circ}t^{\theta\beta}_{\circ\circ}; \quad G: \ u^{\gamma\circ\mu}_{\circ\theta\circ}t^{\alpha\theta}_{\circ\circ}.$$

Note: the matrix representations of tensors A, B, F, G must have the same ordering criterion as the one given in the statement for tensors of order $r = 3$.

2. Since the tensor U does not satisfy the correct axiomatic ordering in its matrix representation, give the matrix P of the permutation that transforms $U_{\sigma,1}$ in the isomer $U'_{\sigma,1}$ the condensation of which leads to tensor U' with the correct ordering.

3. Examine if P is an orthogonal matrix.

4. Give A', B', F', G', the correct contracted products, with the usual matrix representation.

5. Give the matrices $H_{A'}, H_{B'}, H_{F'}, H_{G'}$ corresponding to the contraction homomorphisms executed in the previous question, over the "extended" tensors.

6. If we recover the isomers from tensors A', B', F', G' by means of matrix P^{-1} (inverse permutation), do we get the results of question 1?. Check this result.

7. We perform a change-of-basis, in the linear space $V^3(\mathbb{R})$ of matrix

$$C = \begin{bmatrix} 1 & 0 & 0 \\ 1 & 2 & 0 \\ 1 & 2 & 3 \end{bmatrix}.$$

Give the new tensors $\hat{T}$ and $\hat{U}$ that would present this statement.

8. Solve for $\hat{T}$ and $\hat{U}$, questions 1 to 6.

5.4. Consider the tensor $T \in V^3 \otimes V_*^3(\mathbb{R})$, with matrix representation

$$[t^{\alpha\,\circ}_{\circ\,\beta}] = \begin{bmatrix} 7 & 4 & -1 \\ 4 & 7 & -1 \\ -4 & -4 & 4 \end{bmatrix}.$$

Give two right-autotensors A and B, contracted $(T \otimes A) = \lambda A$ and contracted $(T \otimes B) = \mu B$, where A is of order $(r = 2)$ symmetric and B of order $(r = 3)$.

5.5. Consider the linear space $V^3(\mathbb{R})$ referred to the basis $\{\vec{e}_\alpha\}$. We take a particular vector $(\vec{V}_1, \vec{V}_2, \vec{V}_3) \in \overset{3}{\underset{1}{\times}} V^3(\mathbb{R})$ belonging to the *total* product linear space, the matrix of which associated with the basis $\{\vec{e}_\alpha\}$ is $[X_1 \quad X_2 \quad X_3] =$
$$\begin{bmatrix} 1 & 4 & 2 \\ -1 & 1 & 5 \\ 2 & 3 & 1 \end{bmatrix}.$$

A multilinear transformation $F : \overset{3}{\underset{1}{\times}} V^3(\mathbb{R}) \to W^4(\mathbb{R})$ that applies the total product linear space in $W^4(\mathbb{R})$, is given by (5.104):

$$F[(\vec{V}_1, \vec{V}_2, \vec{V}_3)] = \vec{W} \in W^4(K),$$

which results from the total contraction of the four covariant tensors of order $(r = 3)$ that appear as vector components of

$$F^{h\,\circ\,\circ\,\circ}_{\circ\,\alpha\beta\gamma}\vec{e}_h = (\alpha-1)\vec{\epsilon}_1 + (\alpha-\beta+2)\vec{\epsilon}_2 + (\beta-\gamma-3)\vec{\epsilon}_3 + (\gamma+4)\vec{\epsilon}_4; \quad 1 \le \alpha, \beta, \gamma \le 3,$$

with the vector $\vec{V}_1 \otimes \vec{V}_2 \otimes \vec{V}_3 \in \overset{3}{\underset{1}{\otimes}} V^3(\mathbb{R})$.

Give the image vector $\vec{W}$ of the multilinear mapping.

5.6. In the linear space $V^3(\mathbb{R})$ referred to a certain basis $\{\vec{e}_\alpha\}$, we consider three linear operators with associated matrices:

$$H_1 = \begin{bmatrix} 1 & 1 & 0 \\ 1 & -2 & 0 \\ 0 & 1 & -1 \end{bmatrix} ; \quad H_2 = \begin{bmatrix} 2 & -2 & 3 \\ 1 & 1 & 1 \\ 1 & 3 & -1 \end{bmatrix} ; \quad H_3 = \begin{bmatrix} 1 & 1 & 1 \\ 1 & 2 & 2 \\ 1 & 2 & -2 \end{bmatrix} .$$

1. Obtain the eigenvalues of H_1, H_2, H_3 in increasing order. They will be notated as $(\lambda_1, \lambda_2, \lambda_3)$, (μ_1, μ_2, μ_3) and (ν_1, ν_2, ν_3), for H_1, H_2 and H_2, respectively.
2. Obtain the eigenvectors $(X_1, X_2, X_3)_{H_i}$ associated with each operator, giving their components in columns.
3. If $\vec{W}_i = H_i(\vec{V}_i)$, $1 \le i \le 3$ are the images of the vectors:

$$\vec{V}_1 = ||\vec{e}_\alpha|| \begin{bmatrix} 4 \\ 5 \\ 2 \end{bmatrix} ; \quad \vec{V}_2 = ||\vec{e}_\alpha|| \begin{bmatrix} 3 \\ -1 \\ 2 \end{bmatrix} ; \quad \vec{V}_3 = ||\vec{e}_\alpha|| \begin{bmatrix} 1 \\ -1 \\ 1 \end{bmatrix} ,$$

determine the multivector $\vec{V}_1 \otimes \vec{V}_2 \otimes \vec{V}_3 \in \overset{3}{\underset{1}{\otimes}} V^3(\mathbb{R})$ and the multivector

$\vec{W}_1 \otimes \vec{W}_2 \otimes \vec{W}_3 \in \overset{3}{\underset{1}{\otimes}} V^3(\mathbb{R})$.

4. Obtain the matrix H_σ associated with the direct endomorphism that transforms $H_\sigma(\vec{V}_1 \otimes \vec{V}_2 \otimes \vec{V}_3) = \vec{W}_1 \otimes \vec{W}_2 \otimes \vec{W}_3$.
5. Determine the eigenvalues of H_σ.
6. Determine the eigenvectors of H_σ (remember Section 1.3.4).
7. Solve questions 4, 5 and 6 using the computer and assuming that the solutions of 1, 2 and 3 are known.

Part II

Special Tensors

6

Symmetric Homogeneous Tensors: Tensor Algebras

6.1 Introduction

This chapter is devoted to symmetric homogeneous tensors that are initially defined.

Because of their symmetry, the number of data components required for their definition can be substantially reduced, from the concept of strict components (a minimum set of data), which is explained and a formula given for determining this number.

The problem of generating symmetric tensors from a given tensor, and the tensor nature of symmetry are also discussed.

Next, the symmetrization operator is extended to the case of mixed tensors, and a new interior symmetric product for exclusive use of interior symmetric tensor algebras is introduced.

The chapter ends with some illustrative examples that clarify the established concepts.

6.2 Symmetric systems of scalar components

Though the concept of systems of scalars has already been used in Formulas (5.2), (5.32), (5.33), etc., we fix here their fundamental conditions. Let $S(\alpha_1, \alpha_2, \ldots, \alpha_r)$ be a system of scalar components $s(\alpha_1, \alpha_2, \ldots, \alpha_r)$, of order r, defined over a certain linear space $V^n(K)$; we say that such a system is symmetric with respect to two indices (i, j), if it satisfies

$$\rho = s(\alpha_1, \alpha_2, \ldots, \alpha_i, \ldots, \alpha_j, \ldots, \alpha_r)$$
$$= s(\alpha_1, \alpha_2, \ldots, \alpha_j, \ldots, \alpha_i, \ldots, \alpha_r); \ \rho \in K; \ \alpha_h \in I_n = \{1, 2, \ldots, n\} \subset N,$$

that is, all given components share the same value ρ.

6.2.1 Symmetric systems with respect to an index subset

The previous concept can be extended to k indices $(k \le r)$, of the total set I_r of indices. Let

$$I_k = \{i, j, \dots, k\} \subset I_r = \{1, 2, \dots, i, \dots, j, \dots, r\} \tag{6.1}$$

if the following equalities hold:

$$
\begin{aligned}
\rho &= s(\alpha_1, \alpha_2, \dots, \alpha_i, \dots, \alpha_j, \dots, \alpha_k, \dots, \alpha_r) \\
&= s(\alpha_1, \alpha_2, \dots, \alpha_j, \dots, \alpha_i, \dots, \alpha_k, \dots, \alpha_r) \\
&= \dots\dots\dots\dots\dots\dots\dots\dots\dots\dots\dots\dots\dots\dots \\
&= s(\alpha_1, \alpha_2, \dots, \alpha_k, \dots, \alpha_j, \dots, \alpha_i, \dots, \alpha_r),
\end{aligned}
\tag{6.2}
$$

where all $k!$ permutations of the subset I_k of selected indices are considered, we say that the system of components is symmetric with respect to the index subset I_k.

6.2.2 Symmetric systems. Total symmetry

If $I_k \equiv I_r$, we say that the system is totally symmetric, and better, simply that it is *symmetric*.

Next, we propose a particular example for $(n = 4, r = 3)$, grouping all components of the system $S(\alpha, \beta, \gamma)$ with the same scalar value $\rho \in K$, and ordering them according to the axiomatic ordering criteria. Since the number of components is $\sigma = n^r = 4^3 = 64$, we obtain

$$
\begin{aligned}
&s(1\ \ 1\ \ 1) \\
&s(1\ \ 1\ \ 2) = s(1\ \ 2\ \ 1) = s(2\ \ 1\ \ 1) \\
&s(1\ \ 1\ \ 3) = s(1\ \ 3\ \ 1) = s(3\ \ 1\ \ 1) \\
&s(1\ \ 1\ \ 4) = s(1\ \ 4\ \ 1) = s(4\ \ 1\ \ 1) \\
&s(1\ \ 2\ \ 2) = s(2\ \ 1\ \ 2) = s(2\ \ 2\ \ 1) \\
&s(1\ \ 2\ \ 3) = s(1\ \ 3\ \ 2) = s(2\ \ 1\ \ 3) \\
&\qquad\quad = s(2\ \ 3\ \ 1) = s(3\ \ 1\ \ 2) = s(3\ \ 2\ \ 1) \\
&s(1\ \ 2\ \ 4) = s(1\ \ 4\ \ 2) = s(2\ \ 1\ \ 4) \\
&\qquad\quad = s(2\ \ 4\ \ 1) = s(4\ \ 1\ \ 2) = s(4\ \ 2\ \ 1) \\
&s(1\ \ 3\ \ 3) = s(3\ \ 1\ \ 3) = s(3\ \ 3\ \ 1) \\
&s(1\ \ 3\ \ 4) = s(1\ \ 4\ \ 3) = s(3\ \ 1\ \ 4) \\
&\qquad\quad = s(3\ \ 4\ \ 1) = s(4\ \ 1\ \ 3) = s(4\ \ 3\ \ 1) \\
&s(1\ \ 4\ \ 4) = s(4\ \ 1\ \ 4) = s(4\ \ 4\ \ 1) \\
&s(2\ \ 2\ \ 2) \\
&s(2\ \ 2\ \ 3) = s(2\ \ 3\ \ 2) = s(3\ \ 2\ \ 2)
\end{aligned}
$$

$$s(2 \quad 2 \quad 4) = s(2 \quad 4 \quad 2) = s(4 \quad 2 \quad 2)$$
$$s(2 \quad 3 \quad 4) = s(2 \quad 4 \quad 3) = s(3 \quad 2 \quad 4)$$
$$= s(3 \quad 4 \quad 2) = s(4 \quad 2 \quad 3) = s(4 \quad 3 \quad 2)$$
$$s(2 \quad 4 \quad 4) = s(4 \quad 2 \quad 4) = s(4 \quad 4 \quad 2)$$
$$s(3 \quad 3 \quad 2) = s(3 \quad 2 \quad 3) = s(3 \quad 3 \quad 2)$$
$$s(3 \quad 3 \quad 3)$$
$$s(3 \quad 3 \quad 4) = s(3 \quad 4 \quad 3) = s(4 \quad 3 \quad 3)$$
$$s(3 \quad 4 \quad 4) = s(4 \quad 3 \quad 4) = s(4 \quad 4 \quad 3)$$
$$s(4 \quad 4 \quad 4).$$

Its total number is:

- With three equal indices $s(\alpha, \alpha, \alpha)$: $N_1 = \binom{4}{1} = 4 \times 1 = 4$.
- With two equal indices $s(\alpha, \beta, \alpha)$: $N_2 = \left[\binom{4+3-1}{3} - \binom{4}{1} - \binom{4}{3}\right]\frac{3!}{2!} = 12 \times 3 = 36$.
- With different indices $s(\alpha, \beta, \gamma)$: $N_3 = \binom{4}{3}\frac{3!}{1!} = 4 \times 6 = 24$.

   Total: $4 + 36 + 24 = 64 \equiv \sigma$.

6.3 Strict components of a symmetric system

Definition 6.1 (Strict components of a symmetric system). *We define this as the maximal set of* different value *components, ordered according to the axiomatic ordering, that can have an arbitrary symmetric system of scalars (with partial or total symmetry).*

   In the example in Section 6.2.2 the strict components are the first column of the given table (one can notice that the rest of columns do not have a total ordering character).

6.3.1 Number of strict components of a symmetric system with respect to an index subset

Consider the system $S(\alpha_1, \alpha_2, \ldots, \alpha_i, \ldots, \alpha_j, \ldots, \alpha_k, \ldots, \alpha_r)$ defined with respect to a linear space $V^n(K)$ and let k be the number of indices of the subset $I_k = \{i, j, \ldots, k\}$ the components of which are symmetric.

   The number of valid realizations for the non-symmetric $(r - k)$ indices is given by the following number of variations with repetitions:

$$VR_{n,r-k} = n^{r-k}. \tag{6.3}$$

The number of strict realizations for the symmetric k indices is given by the combinations with repetition:

$$CR_{n,k} = \binom{n+k-1}{k}.$$ (6.4)

Thus, the total number of realizations for the strict components is

$$CR_{n,k} \cdot VR_{n,r-k} = \binom{n+k-1}{k} n^{r-k}.$$ (6.5)

Obviously, the total number of components is, as usual, the dimension of the tensor space, i.e., $\sigma = n^r$.

6.3.2 Number of strict components of a symmetric system

To find this number and since now the r indices of the system present symmetry, we let $k = r$ in Formula (6.5).

Thus, the number of strict components of the system of scalars with total symmetry of indices is

$$CR_{n,r} = \binom{n+r-1}{r}.$$ (6.6)

Number of symmetric components associated with each strict component of a symmetric system

If the strict component has α repeated indices, β repeated indices, γ repeated indices, ..., δ repeated indices, that is, its pattern is

$$s(\alpha, \alpha, \ldots, \alpha, \beta, \beta, \ldots, \beta, \gamma, \gamma, \ldots, \gamma, \ldots, \delta, \delta, \ldots, \delta),$$

the resulting number is the permutations with repetition

$$P_r^{\alpha,\beta,\gamma,\ldots,\delta} = \frac{r!}{\alpha!\beta!\gamma!\cdots\delta!},$$ (6.7)

where $\alpha + \beta + \gamma + \cdots + \delta = r$.

Finally, we want to point out that in a symmetric system of scalars $S(\alpha_1, \alpha_2, \ldots, \alpha_r)$ of order r, over a linear space $V^n(K)$, the number c of strict components with θ different indices ($1 \leq \theta \leq r$), is

$$c = \binom{r-1}{\theta-1} \cdot \binom{n}{\theta}.$$ (6.8)

Thus, in the example in Section 6.2.2 the number "c" of strict components of $\theta = 2$ different indices is the cardinality of the set

$$\{s(1\ 1\ 2), s(1\ 1\ 3), s(1\ 1\ 4), s(1\ 2\ 2), \ldots, s(3\ \ 3\ \ 4), s(3\ 4\ 4)\}.$$

Applying Formula (6.8), we have

$$c = \binom{3-1}{2-1} \cdot \binom{4}{2} = \binom{2}{1} \cdot \binom{4}{2} = 2 \times 6 = 12.$$ (6.9)

Remark 6.1. Expression (6.8) comes from the expansion formula of the combinations with repetition, $CR_{m,n}$. □

6.4 Tensors with symmetries: Tensors with branched symmetry, symmetric tensors

The symmetry criteria previously established for systems of scalar components are directly applicable to homogeneous tensors, with the following limitations, to be justified later:

1. Tensors with symmetry with respect to the subset I_k of indices ($2 \leq k \leq r$), must present the subset I_k among the p contravariant indices or among the q covariant indices ($p+q = r$); that is, all indices belonging to I_k must be of the *same valency*. Some authors define these tensors as "tensors with branched symmetry".

   This is a sufficient condition for the symmetry with respect to the indices of I_k to remain invariant under changes of basis of tensor nature, but it is *not necessary*, as we shall see when the Kronecker delta tensor system is presented; this can present stable symmetries between indices of different valency (a very particular case).

2. By definition, symmetric tensors (with respect to all their indices), are pure tensors, that is, totally contravariant or totally covariant. Some books even give them a special notation.

Consider a homogeneous tensor of order r totally contravariant, notated $t^{\alpha_1 \alpha_2 \cdots \alpha_r}_{\circ \ \circ \ \cdots \ \circ}$; if it is symmetric, it can be expressed as

$$t^{(\alpha_1 \alpha_2 \cdots \alpha_r)}_{\circ \ \circ \ \cdots \ \circ} \text{ with } \alpha_1 \leq \alpha_2 \leq \cdots \leq \alpha_r, \tag{6.10}$$

which is precisely the adopted notation for its strict components.

Similarly, the notation $t^{\circ \ \circ \ \cdots \ \circ}_{(\alpha_1 \alpha_2 \cdots \alpha_r)}$ is used for the covariant symmetric tensor.

The number of strict components, number of components associated with θ different indices, etc., can be obtained with the general formulas established for the systems of scalars.

Finally, we point out that some authors extend the total symmetry to mixed tensors, in which case it is assumed that the symmetry of the p contravariant indices and the symmetry of the q covariant indices are *independent*. Then, the expression for the strict components becomes

$$t^{(\alpha_1 \alpha_2 \cdots \alpha_p) \ \circ \ \ \circ \ \cdots \ \circ}_{\ \circ \ \circ \ \cdots \ \circ \ (\alpha_{p+1} \alpha_{p+2} \cdots \alpha_{p+q})} \tag{6.11}$$

with

$$\alpha_1 \leq \alpha_2 \leq \cdots \leq \alpha_p; \quad \alpha_{p+1} \leq \alpha_{p+2} \leq \cdots \leq \alpha_{p+q}.$$

So, for the case of a mixed tensor over the linear space $\mathbb{R}^2$ of order $r = 4$, contra-contra-cova-covariant, ($p = q = 2$) symmetric the strict components would be

$$t^{(11)\,\circ\,\circ}_{\ \ \ \circ\,\circ\,(11)}$$

$$t^{(11)\,\circ\,\circ}_{\ \ \ \circ\,\circ\,(12)} = t^{11\circ\circ}_{\circ\circ 21}$$

$$t^{(11)\,\circ\,\circ}_{\ \ \ \circ\,\circ\,(22)}$$

$$t^{(12)\,\circ\,\circ}_{\ \ \ \circ\,\circ\,(11)} = t^{21\circ\circ}_{\circ\circ 11}$$

$$t^{(12)\,\circ\,\circ}_{\ \ \ \circ\,\circ\,(12)} = t^{12\circ\circ}_{\circ\circ 21} = t^{21\circ\circ}_{\circ\circ 12} = t^{21\circ\circ}_{\circ\circ 21}$$

$$t^{(12)\,\circ\,\circ}_{\ \ \ \circ\,\circ\,(22)} = t^{21\circ\circ}_{\circ\circ 22}$$

$$t^{(22)\,\circ\,\circ}_{\ \ \ \circ\,\circ\,(11)}$$

$$t^{(22)\,\circ\,\circ}_{\ \ \ \circ\,\circ\,(12)} = t^{22\circ\circ}_{\circ\circ 21}$$

$$t^{(22)\,\circ\,\circ}_{\ \ \ \circ\,\circ\,(22)}$$

with a total number of $\sigma = n^r = 2^4 = 16$ components, of which 9 would be strict.

It is convenient in this section, in which we consider *mixed tensors*, to clarify the number of strict components when the branching is mixed, that is, it involves not only one part of the p contravariant indices (k_1, of the p indices, are symmetric), but also at the same time one part of the covariant indices (k_2, of the q indices, are symmetric).

In such a case we have

$$2 \leq k_1 \leq p; \quad 2 \leq k_2 \leq q; \quad p + q = r \text{ (order of the mixed tensor)}.$$

The number of strict components is

$$\sigma' = \binom{n + k_1 - 1}{k_1} n^{p-k_1} \cdot \binom{n + k_2 - 1}{k_2} n^{q-k_2};$$

or

$$\sigma' = \binom{n + k_1 - 1}{k_1} \cdot \binom{n + k_2 - 1}{k_2} n^{r-(k_1+k_2)}.$$

If the mixed tensor is symmetric, then $k_1 = p$ and $k_2 = q$. Thus, the number of strict components is

$$\sigma' = \binom{n + p - 1}{p} \cdot \binom{n + q - 1}{q},$$

which is the formula used in the mixed example that has been given ($n = p = q = 2$).

6.4.1 Generation of symmetric tensors

We discuss two different ways of generating symmetric tensors.

By means of decomposable symmetric tensors

We remind the reader of the beginning of Section 2.2 where we have given the bases for the tensor product of vectors concept. To center our objective, we assume that we are in the homogeneous tensor space $V^n \otimes V^n \otimes V^n(K)$ of order $r = 3$, and that we choose three arbitrary vectors $\vec{u}, \vec{v}, \vec{w}, \in V^n(K)$ the tensor product of which satisfies Formulas (2.4) and (2.5), which for the homogeneous case are

$$\vec{u} \otimes \vec{v} \otimes \vec{w} = t_{\circ\circ\circ}^{ijk} \vec{e}_i \otimes \vec{e}_j \otimes \vec{e}_k, \text{ with } t_{\circ\circ\circ}^{ijk} = x^i y^j z^k; \quad \forall \vec{u}, \vec{v}, \vec{w}, \in V^n(K). \quad (6.12)$$

As is well known, the linear combinations of these "decomposable" tensors generate all other tensors (not decomposable) of the tensor space.

If we restrict ourselves now to vectors $\vec{u}, \vec{v}, \vec{w}, \in V^n(K)$ such that

$$\vec{u} \equiv \vec{v} \equiv \vec{w} \Rightarrow X \equiv Y \equiv Z \Rightarrow x^i \equiv y^i \equiv z^i; \quad x^j \equiv y^j \equiv z^j; \quad x^k \equiv y^k \equiv z^k; \quad (6.13)$$

the result is that the set of decomposable tensors, generated by $\vec{v} \in V^n(K)$, that is, the new condition (6.12) with this constraint leads to

$$\vec{v} \otimes \vec{v} \otimes \vec{v} = t_{\circ\circ\circ}^{ijk} \vec{e}_i \otimes \vec{e}_j \otimes \vec{e}_k \text{ con } t_{\circ\circ\circ}^{ijk} = x^i x^j x^k; \quad \forall \vec{v} \in V^n(K), \quad (6.14)$$

generating a tensor subspace $S_m(\vec{v}) \subset V^n \otimes V^n \otimes V^n(K)$ of *symmetric tensors*, as can be observed in a particular case ($n = 3, r = 3$), when building the matrix representation of their components, that leads to the following decomposable tensor:

$$[t_{\circ\circ\circ}^{ijk}] = \begin{bmatrix} t_{\circ\circ\circ}^{111} & t_{\circ\circ\circ}^{112} & t_{\circ\circ\circ}^{113} \\ t_{\circ\circ\circ}^{121} & t_{\circ\circ\circ}^{122} & t_{\circ\circ\circ}^{123} \\ t_{\circ\circ\circ}^{131} & t_{\circ\circ\circ}^{132} & t_{\circ\circ\circ}^{133} \\ \hline t_{\circ\circ\circ}^{211} & t_{\circ\circ\circ}^{212} & t_{\circ\circ\circ}^{213} \\ t_{\circ\circ\circ}^{221} & t_{\circ\circ\circ}^{222} & t_{\circ\circ\circ}^{223} \\ t_{\circ\circ\circ}^{231} & t_{\circ\circ\circ}^{232} & t_{\circ\circ\circ}^{233} \\ \hline t_{\circ\circ\circ}^{311} & t_{\circ\circ\circ}^{312} & t_{\circ\circ\circ}^{313} \\ t_{\circ\circ\circ}^{321} & t_{\circ\circ\circ}^{322} & t_{\circ\circ\circ}^{323} \\ t_{\circ\circ\circ}^{331} & t_{\circ\circ\circ}^{332} & t_{\circ\circ\circ}^{333} \end{bmatrix} \equiv \begin{bmatrix} (x^1)^3 & (x^1)^2 x^2 & (x^1)^2 x^3 \\ (x^1)^2 x^2 & x^1(x^2)^2 & x^1 x^2 x^3 \\ (x^1)^2 x^3 & x^1 x^2 x^3 & x^1(x^3)^2 \\ \hline (x^1)^2 x^2 & x^1(x^2)^2 & x^1 x^2 x^3 \\ x^1(x^2)^2 & (x^2)^3 & (x^2)^2 x^3 \\ x^1 x^2 x^3 & (x^2)^2 x^3 & x^2(x^3)^2 \\ \hline (x^1)^2 x^3 & x^1 x^2 x^3 & x^1(x^3)^2 \\ x^1 x^2 x^3 & (x^2)^2 x^3 & x^2(x^3)^2 \\ x^1(x^3)^2 & x^2(x^3)^2 & (x^3)^3 \end{bmatrix}$$

with i the index of the block row, j the index of each submatrix row, and k the index of each submatrix column.

Before continuing, we point out that the rank of each submatrix is *one*, which declares that, in fact, it is a decomposable tensor, generated by the tensor product of a *single* vector.

When observing the value of the different components, they can be classified as

$$t_{\circ\circ\circ}^{(111)} = (x^1)^3$$

$$t_{\circ\circ\circ}^{(112)} = t_{\circ\circ\circ}^{121} = t_{\circ\circ\circ}^{211} = (x^1)^2 x^2$$

$$t_{\circ\circ\circ}^{(113)} = t_{\circ\circ\circ}^{131} = t_{\circ\circ\circ}^{311} = (x^1)^2 x^3$$

$$t_{\circ\circ\circ}^{(122)} = t_{\circ\circ\circ}^{212} = t_{\circ\circ\circ}^{221} = x^1 (x^2)^2$$

$$t_{\circ\circ\circ}^{(123)} = t_{\circ\circ\circ}^{132} = t_{\circ\circ\circ}^{213} = t_{\circ\circ\circ}^{231} = t_{\circ\circ\circ}^{312} = t_{\circ\circ\circ}^{321} = x^1 x^2 x^3$$

$$t_{\circ\circ\circ}^{(133)} = t_{\circ\circ\circ}^{313} = t_{\circ\circ\circ}^{331} = x^1 (x^3)^2$$

$$t_{\circ\circ\circ}^{(222)} = (x^2)^3$$

$$t_{\circ\circ\circ}^{(223)} = t_{\circ\circ\circ}^{232} = t_{\circ\circ\circ}^{322} = (x^2)^2 x^3$$

$$t_{\circ\circ\circ}^{(233)} = t_{\circ\circ\circ}^{323} = t_{\circ\circ\circ}^{332} = x^2 (x^3)^2$$

$$t_{\circ\circ\circ}^{(333)} = (x^3)^3.$$

This table is a declaration of symmetric tensor of $\sigma = n^r = 3^3 = 27$ components, with $\binom{n+r-1}{r} = \binom{3+3-1}{3} = \binom{5}{3} = 10$ strict components.

Then, we have perfectly established a procedure for generating *symmetric tensor subspaces*

$$S_m(\vec{v}) \subset (\otimes V^n)_s^r(K) \subset (\otimes V^n)^r(K),$$

that belong to a tensor space, by each one of the chosen vectors $\vec{v} \in V^n(K)$. The sum subspace of the subspace generated by several vectors $\vec{v}, \vec{u}, \vec{w}$, etc. or the linear combination of them, leads to other *symmetric tensor* subspaces that *cannot be generated by decomposable tensors*.

By means of isomer tensors

Assume that an arbitrary pure tensor $\vec{t}(n, r)$ is given, $t_{\circ\ \circ\ \dots\ \circ}^{\alpha_1 \alpha_2 \cdots \alpha_r}$ and that we decide to associate it with another tensor $\vec{u}(n, r)$ using the rules:

1. We choose a determined set of indices I_k (of the same valency) of the given tensor.
2. We notate the new tensor $\vec{u}(n, r)$ by placing the selected indices in parentheses:

$$u_{\circ\ \circ\ \ \circ\ \ \dots\ \circ}^{\alpha_1 \alpha_2 (\leftarrow I_k \rightarrow) \cdots \alpha_r}.$$

This notation will be the one used for the strict components of the new tensor $\vec{u}(n, r)$.

3. Considering all permutations $(k!)$ of the indices in I_k, we determine all the corresponding isomer tensors of $\vec{t}$.

4. Once they have been found, we build the strict components of $\vec{u}(n,r)$ by means of the formula

$$u^{\alpha_1 \alpha_2 (\leftarrow I_k \rightarrow) \cdots \alpha_r}_{\circ\ \circ\ \ \ \circ\ \ \cdots\ \circ}$$

$$= \frac{1}{k!} \left[t^{\alpha_1 \alpha_2 \alpha_3 \cdots \alpha_{r-1} \alpha_r}_{\circ\ \circ\ \circ\ \cdots\ \circ\ \circ} + t^{\alpha_1 \alpha_2 \alpha_4 \alpha_3 \cdots \alpha_r}_{\circ\ \circ\ \circ\ \circ\ \cdots\ \circ} + \cdots + t^{\alpha_1 \alpha_2 \alpha_{r-1} \cdots \alpha_3 \alpha_r}_{\circ\ \circ\ \circ\ \ \cdots\ \circ\ \circ} \right], \quad (6.15)$$

where the right-hand factor includes all the $k!$ isomers as summands.

The tensor $\vec{u}(n,r)$ with these strict components is a *symmetric tensor* with respect to the set of indices I_k. Some authors present this from the point of view of an "isomerization" endomorphism H, that directly transforms the tensor $\vec{t}(n,r)$ into the tensor $\vec{u}(n,r)$.

We give some illustrative examples:

1. Consider the data tensor $g^{\circ\ \circ}_{\alpha\beta}$; $I_k \equiv I_r = [\alpha, \beta]$ (all).
   The symmetric tensor associated by this technique is the tensor with components

   $$g'^{\circ\circ}_{(\alpha\beta)} = \frac{1}{2!} [g^{\circ\ \circ}_{\alpha\beta} + g^{\circ\ \circ}_{\beta\alpha}].$$

2. Consider the data tensor $W^{\alpha\beta\gamma\delta\epsilon\lambda}_{\circ\circ\circ\circ\circ\circ}$. An associated symmetric tensor with respect to the indices $I_k = [\beta\gamma\delta]$ is the tensor with strict components

   $$u^{\alpha(\beta\gamma\delta)\epsilon\lambda}_{\circ\ \circ\circ\circ\ \circ\circ}$$

   $$= \frac{1}{3!} \left[w^{\alpha\beta\gamma\delta\epsilon\lambda}_{\circ\circ\circ\circ\circ\circ} + w^{\alpha\beta\delta\gamma\epsilon\lambda}_{\circ\circ\circ\circ\circ\circ} + w^{\alpha\gamma\beta\delta\epsilon\lambda}_{\circ\circ\circ\circ\circ\circ} + w^{\alpha\gamma\delta\beta\epsilon\lambda}_{\circ\circ\circ\circ\circ\circ} + w^{\alpha\delta\beta\gamma\epsilon\lambda}_{\circ\circ\circ\circ\circ\circ} + w^{\alpha\delta\gamma\beta\epsilon\lambda}_{\circ\circ\circ\circ\circ\circ} \right].$$

Evidently, if the "isomerization" technique were to be applied (by error) to a symmetric data tensor, one would obtain as the associate the initial data tensor.

In the next chapter we will see how an analogous technique, but with anti-symmetric tensors, leads to the exterior tensor algebras.

6.4.2 Intrinsic character of tensor symmetry: Fundamental theorem of tensors with symmetry

The obvious question a reader can ask himself at this point is if all tensors with symmetry with respect to certain components, maintain it when performing licit changes of basis, that is, of a tensor nature.

The answer to this question is formulated in the following theorem:

Theorem 6.1 (Intrinsic character of symmetry). *The symmetry of k indices $(2 \le k \le p$ or $2 \le k \le q, p + q = r)$ of the same valency in the components of a homogeneous tensor has intrinsic character. The symmetry in the components of two or more indices of different valency in a homogeneous tensor has no intrinsic character in the general case.* $\square$

Proof.

Sufficiency: Consider the case of indices of the same valency. Consider the homogeneous tensor $t^{\alpha\beta\gamma\circ\circ}_{\circ\circ\circ\lambda\mu}$ of fifth order $(r = 5)$, defined over the linear space $V^n(K)$, which satisfies the property

$$t^{\alpha\beta\gamma\circ\circ}_{\circ\circ\circ\lambda\mu} = t^{\beta\alpha\gamma\circ\circ}_{\circ\circ\circ\lambda\mu}; \forall \alpha_j \in I_k = [\alpha, \beta] \text{ and } j \in I_5, \tag{6.16}$$

that is, it is symmetric with respect to the two first contravariant indices.

Consider the change-of-basis in $V^n(K), \|\vec{e}_i\| = \|\vec{e}_\alpha\|[c^{\alpha\circ}_{\circ\,i}]$, which produces the change of tensor basis in the space $(\otimes V^n)^3(\otimes V^n_*)^2(K)$; we indicate its action over the two components, in the new and the initial bases

$$t^{ijko\,\circ}_{\circ\circ\circ\ell m} = t^{\alpha\beta\gamma\circ\circ}_{\circ\circ\circ\lambda\mu}\gamma^{i\,\circ}_{\circ\alpha}\gamma^{j\,\circ}_{\circ\beta}\gamma^{k\circ}_{\circ\gamma}c^{\circ\lambda}_{\ell\circ}c^{\circ\,\mu}_{m\circ} \tag{6.17}$$

$$t^{jiko\,\circ}_{\circ\circ\circ\ell m} = t^{\beta\alpha\gamma\circ\circ}_{\circ\circ\circ\lambda\mu}\gamma^{j\,\circ}_{\circ\beta}\gamma^{i\,\circ}_{\circ\alpha}\gamma^{k\circ}_{\circ\gamma}c^{\circ\lambda}_{\ell\circ}c^{\circ\,\mu}_{m\circ}. \tag{6.18}$$

Taking into account (6.16), and the commutativity of the scalars $\gamma^{i\,\circ}_{\circ\alpha}$ with $\gamma^{j\,\circ}_{\circ\beta}$ in the field K, it is evident that the two right-hand members of (6.17) and (6.18) are equal, and then

$$t^{ijko\,\circ}_{\circ\circ\circ\ell m} = t^{jiko\,\circ}_{\circ\circ\circ\ell m}; \; \forall i,j \in I_5 \text{ and } \forall i_h \in I_2 = [i,j]. \tag{6.19}$$

Similar conclusions can be established by the reader for any other pair of contravariant or covariant indices of the analyzed tensor $t^{\alpha\beta\gamma\circ\circ}_{\circ\circ\circ\lambda\mu}$, by following a similar process.

The analyzed model can be used for any generalization of the fundamental hypothesis.

Necessarity: Consider the case of indices of different valency. Consider the homogeneous tensor $t^{\alpha\beta\gamma\circ\circ}_{\circ\circ\circ\lambda\mu}$ of fifth order $(r = 5)$, defined over the linear space $V^n(K)$, that satisfies the property

$$t^{\alpha\beta\gamma\circ\circ}_{\circ\circ\circ\lambda\mu} = t^{\alpha\lambda\gamma\circ\circ}_{\circ\circ\circ\beta\mu}; \; \forall \alpha_j \in I_k = [\beta, \lambda] \text{ and } j \in I_5. \tag{6.20}$$

When performing the previous change-of-basis in $V^n(K)$, the linear space $(\otimes V^n)^3(\otimes V^n_*)^2(K)$ suffers the corresponding tensor change, where the new components of those in (6.20), are related to the initial ones by

$$t^{ijko\,\circ}_{\circ\circ\circ\ell m} = t^{\alpha\beta\gamma\circ\circ}_{\circ\circ\circ\lambda\mu}\gamma^{i\,\circ}_{\circ\alpha}\gamma^{j\,\circ}_{\circ\beta}\gamma^{k\circ}_{\circ\gamma}c^{\circ\lambda}_{\ell\circ}c^{\circ\,\mu}_{m\circ} \tag{6.21}$$

$$t^{i\ell ko\,\circ}_{\circ\circ\circ jm} = t^{\alpha\lambda\gamma\circ\circ}_{\circ\circ\circ\beta\mu}\gamma^{i\,\circ}_{\circ\alpha}\gamma^{\ell\circ}_{\circ\lambda}\gamma^{k\circ}_{\circ\gamma}c^{\circ\beta}_{j\circ}c^{\circ\,\mu}_{m\circ}, \tag{6.22}$$

which proves that $\gamma_{o\beta}^{j\,o}c_{\ell o}^{o\lambda} \neq \gamma_{o\lambda}^{\ell o}c_{jo}^{o\beta}$, as can be appreciated in a simple numerical counterexample.

Let $n = r = 3$ and

$$[c_{o\,i}^{\alpha o}] = C = \begin{bmatrix} -1 & 1 & 0 \\ 0 & 1 & 0 \\ -1 & 0 & 3 \end{bmatrix} ; [\gamma_{o\alpha}^{i\,o}] = C^{-1} = \begin{bmatrix} -1 & 1 & 0 \\ 0 & 1 & 0 \\ -1/3 & 1/3 & 1/3 \end{bmatrix} ;$$

$$[c_{i\,o}^{o\alpha}] = C^t = \begin{bmatrix} -1 & 0 & -1 \\ 1 & 1 & 0 \\ 0 & 0 & 3 \end{bmatrix} .$$

If we choose $j = 2, \ell = 3, \beta = 1, \lambda = 2$, then we have

$$\gamma_{o\beta}^{j\,o}c_{\ell o}^{o\lambda} = \gamma_{o1}^{2o}(\text{ of } C^{-1})c_{3o}^{o2}(\text{ of } C^t) = 0 \cdot 0 = 0$$

$$\gamma_{o\lambda}^{\ell o}c_{jo}^{o\beta} = \gamma_{o2}^{3o}c_{2o}^{o1} = 1/3 \cdot 1 = 1/3,$$

which implies $\gamma_{o1}^{2o}c_{3o}^{o2} \neq \gamma_{o2}^{3o}c_{2o}^{o1}$.

If we choose $j = 1, \ell = 2, \beta = 3, \lambda = 2$, then we have

$$\gamma_{o3}^{1o}c_{2o}^{o2} = 0 \cdot 1 = 0; \quad \gamma_{o2}^{2o}c_{1o}^{o3} = 1 \cdot (-1) = -1,$$

and then $\gamma_{o3}^{1o}c_{2o}^{o2} \neq \gamma_{o2}^{2o}c_{1o}^{o3}$.

Then, we conclude that in general $t_{ooo\ell m}^{ijkoo} \neq t_{ooojm}^{i\ell koo}$, that is, the tensor in the new basis does not keep the symmetry.

However, if the data tensor were to have a great number of components $t_{ooo\lambda\mu}^{\alpha\beta\gamma oo} = 0$, these could in some case absorb the indicated inequalities, and maintain the symmetries of indices of different covariance, because of its nullity, that is, due to qualities of the concrete tensor that are special or proper (Kronecker delta, for example).

Example 6.1 (Strict components).

1. Consider the symmetric covariant tensor $T_1 = t_{\alpha\beta\gamma\delta\epsilon}^{\circ\circ\circ\circ\circ}$ over the linear space $V^n(\mathbb{R})$.
   a) Determine the number of components (N_1).
   b) Determine the maximum number of strict components (N_2).

2. Consider the mixed tensor $T_2 = t_{\alpha\beta\gamma\delta\epsilon oooo}^{\circ\circ\circ\circ\circ\pi\rho\sigma\tau}$ with $q = 5$ covariant indices and $p = 4$ contravariant indices, over the linear space $V^n(\mathbb{R})$, which is partially symmetric with respect to the covariant indices $(\beta\delta)$ and with respect to the contravariant indices $(\pi\sigma\tau)$.
   a) Obtain the number of components (N_3).
   b) Obtain the maximum number of strict components (N_4).

Solution:

1. Consider the tensor $T_1 = t^{\circ\circ\circ\circ\circ}_{\;\alpha\beta\gamma\delta\epsilon}$.

   a) Since the order is $r = 5$, the number of components is $N_1 = \sigma = n^r = n^5$.

   b) The number of strict components is $N_2 = \sigma' = CR_{n,r} = \binom{n+r-1}{r}$.

2. Consider the mixed tensor $T_2 = t^{\circ\circ\circ\circ\circ\pi\rho\sigma\tau}_{\;\alpha\beta\gamma\delta\epsilon\circ\circ\circ\circ}$.

   a) The order of the tensor is $r = p + q = 4 + 5 = 9$, and then $N_3 = \sigma'' = VR_{n,r} = n^r = n^9$.

   b) We consider the following possibilities for the different indices:

   i. The possibilities for index α are $CR_{n,1} \equiv C_{n,1} = \binom{n}{1} = n$.

   ii. The possibilities for indices $(\beta\delta)$ are $CR_{n,2} = \binom{n+2-1}{2} = \binom{n+1}{2} = \frac{n(n+1)}{2}$.

   iii. The possibilities for index γ are $CR_{n,1} \equiv C_{n,1} = \binom{n}{1} = n$.

   iv. The possibilities for index ϵ are $CR_{n,1} \equiv C_{n,1} = \binom{n}{1} = n$.

   v. The possibilities for indices $(\pi\sigma\tau)$ are $CR_{n,3} = \binom{n+3-1}{3} = \binom{n+2}{3} = \frac{n(n+1)(n+2)}{6}$.

   vi. The possibilities for index ρ are $CR_{n,1} \equiv C_{n,1} = \binom{n}{1} = n$.

   Thus, the number of strict components for this case is

   $$N_4 = n \cdot \frac{n(n+1)}{2} \cdot n \cdot n \cdot \frac{n(n+1)(n+2)}{6} \cdot n = \frac{1}{12}n^6(n+1)^2(n+2).$$

   $\square$

Example 6.2 (Total and strict components).

1. Given the symmetric contravariant tensor $s^{\alpha\beta\gamma}_{\circ\circ\circ}$ of order $r = 3$, over the linear space $V^3(\mathbb{R})$, $n = 3$, do the following:

   a) Give the number of components, total and strict, using different Roman letters for the last ones.

   b) Represent by a matrix the general aspect of these tensors.

2. Given a symmetric contravariant tensor $s^{\alpha\beta\gamma\delta}_{\circ\circ\circ\circ}$ of order $r = 4$, over the linear space $V^2(\mathbb{R})$, $n = 2$:

   a) Give the number of components, total and strict, using different Roman letters for the last ones.

   b) Represent by a matrix the general aspect of these tensors.

Solution:

1. Consider the tensor $s^{\alpha\beta\gamma}_{\circ\circ\circ}$.

   a) The total number of components is $\sigma = VR_{n,r} = n^r = 3^3 = 27$.
   Similarly, the number of strict components is $\sigma' = CR_{n,r} = \binom{n+r-1}{r} = \binom{3+3-1}{3} = \binom{5}{3} = 10$, and their values are

$$s^{(111)}_{\circ\circ\circ} = a$$

$$s^{(112)}_{\circ\circ\circ} = s^{121}_{\circ\circ\circ} = s^{211}_{\circ\circ\circ} = b$$

$$s^{(113)}_{\circ\circ\circ} = s^{131}_{\circ\circ\circ} = s^{311}_{\circ\circ\circ} = c$$

$$s^{(122)}_{\circ\circ\circ} = s^{212}_{\circ\circ\circ} = s^{221}_{\circ\circ\circ} = d$$

$$s^{(123)}_{\circ\circ\circ} = s^{132}_{\circ\circ\circ} = s^{213}_{\circ\circ\circ} = s^{231}_{\circ\circ\circ} = s^{312}_{\circ\circ\circ} = s^{321}_{\circ\circ\circ} = e$$

$$s^{(133)}_{\circ\circ\circ} = s^{313}_{\circ\circ\circ} = s^{331}_{\circ\circ\circ} = f$$

$$s^{(222)}_{\circ\circ\circ} = g$$

$$s^{(223)}_{\circ\circ\circ} = s^{232}_{\circ\circ\circ} = s^{322}_{\circ\circ\circ} = h$$

$$s^{(233)}_{\circ\circ\circ} = s^{323}_{\circ\circ\circ} = s^{332}_{\circ\circ\circ} = i$$

$$s^{(333)}_{\circ\circ\circ} = j.$$

   b) Thus, its most general aspect is

$$[s^{\alpha\beta\gamma}_{\circ\circ\circ}] = \left[\begin{array}{ccc|ccc|ccc} a & b & c \\ b & d & e \\ c & e & f \\ \hline b & d & e \\ d & g & h \\ e & h & i \\ \hline c & e & f \\ e & h & i \\ f & i & j \end{array}\right].$$

2. Consider the tensor $s^{\alpha\beta\gamma\delta}_{\circ\circ\circ\circ}$.

   a) The total number of components is $\sigma = VR_{n,r} = n^r = 2^4 = 16$,
   and the number of strict components is $\sigma' = CR_{n,r} = \binom{n+r-1}{r} = \binom{2+4-1}{4} = \binom{5}{4} = 5$, and their values are

$$s^{(1111)}_{\circ\circ\circ\circ} = a$$

$$s^{(1112)}_{\circ\circ\circ\circ} = s^{1121}_{\circ\circ\circ\circ} = s^{1211}_{\circ\circ\circ\circ} = s^{2111}_{\circ\circ\circ\circ} = b$$

$$s^{(1122)}_{\circ\,\circ\,\circ\,\circ} = s^{1212}_{\circ\circ\circ\circ} = s^{2112}_{\circ\circ\circ\circ} = s^{2121}_{\circ\circ\circ\circ} = s^{1221}_{\circ\circ\circ\circ} = s^{2211}_{\circ\circ\circ\circ} = c$$

$$s^{(1222)}_{\circ\,\circ\,\circ\,\circ} = s^{2122}_{\circ\circ\circ\circ} = s^{2212}_{\circ\circ\circ\circ} = s^{2221}_{\circ\circ\circ\circ} = d$$

$$s^{(2222)}_{\circ\,\circ\,\circ\,\circ} = e.$$

b) Thus, its most general aspect is

$$[s^{\alpha\beta\gamma\delta}_{\circ\,\circ\,\circ\,\circ}] = \left[\begin{array}{cc|cc} a & b & b & c \\ b & c & c & d \\ \hline b & c & c & d \\ c & d & d & e \end{array}\right].$$

□

Example 6.3 (Symmetric and anti-symmetric tensors). Let $t^{\circ\circ\circ}_{ijk}$ be the components of a tensor that belongs to the tensor space $(\otimes V^n_*)^3(\mathbb{R})$.

1. Show that if the equality $t^{\circ\circ\circ}_{jki} = \lambda t^{\circ\circ\circ}_{ijk}$ holds for some values of the scalar $\lambda \in \mathbb{R}$, such values are fixed. Obtain these values.
2. Show that the equality proposed in the previous question is intrinsic.
3. What can we say for the particular case $t^{\circ\circ}_{ji} = \lambda t^{\circ\circ}_{ij}$?

Solution:

1. If we apply the property successively, by permuting the indices in a circle, from $t^{\circ\circ\circ}_{jki} = \lambda t^{\circ\circ\circ}_{ijk}$ we obtain

$$t^{\circ\circ\circ}_{kij} = \lambda t^{\circ\circ\circ}_{jki} = \lambda(\lambda t^{\circ\circ\circ}_{ijk}) = \lambda^2 t^{\circ\circ\circ}_{ijk},$$

and permuting again, the result is

$$t^{\circ\circ\circ}_{ijk} = \lambda^2 t^{\circ\circ\circ}_{jki} = \lambda^2(\lambda t^{\circ\circ\circ}_{ijk}) = \lambda^3 t^{\circ\circ\circ}_{ijk} \Rightarrow \lambda^3 = 1.$$

Since $K \equiv \mathbb{R}$, the only solution is $\lambda = 1$, because the other two $\lambda = 1_{2\pi/3} \equiv j$ and $\lambda = 1_{4\pi/3} = j^2$, belong to the field of the complex numbers.

2. The intrinsic character can be immediately shown. If we start from

$$t^{\circ\circ\circ}_{\beta\gamma\alpha} = \lambda t^{\circ\circ\circ}_{\alpha\beta\gamma} \tag{6.23}$$

and take into account Formulas (4.35), we get

$$t^{\circ\circ\circ}_{\alpha\beta\gamma} = t^{\circ\circ\circ}_{ijk}\gamma^{\circ i}_{\alpha\circ}\gamma^{\circ j}_{\beta\circ}\gamma^{\circ k}_{\gamma\circ}; \quad t^{\circ\circ\circ}_{\beta\gamma\alpha} = t^{\circ\circ\circ}_{jki}\gamma^{\circ j}_{\beta\circ}\gamma^{\circ k}_{\gamma\circ}\gamma^{\circ i}_{\alpha\circ}, \tag{6.24}$$

and replacing (6.24) into (6.23), yields

$$t^{\circ\circ\circ}_{jki}\gamma^{\circ j}_{\beta\circ}\gamma^{\circ k}_{\gamma\circ}\gamma^{\circ i}_{\alpha\circ} = \lambda t^{\circ\circ\circ}_{ijk}\gamma^{\circ i}_{\alpha\circ}\gamma^{\circ j}_{\beta\circ}\gamma^{\circ k}_{\gamma\circ},$$

and since $\mathbb{R}$ is commutative, the result is $t^{\circ\circ\circ}_{jki} = \lambda t^{\circ\circ\circ}_{ijk}$, which is Equation (6.23) in the new basis, which proves the tensor character of the property.

3. In the given particular case, we similarly arrive at $\lambda^2 = 1$, with roots $\lambda_1 = 1$ and $\lambda_2 = -1$, which implies $t^{\circ\circ}_{ji} = t^{\circ\circ}_{ij}$ or $t^{\circ\circ}_{ji} = -t^{\circ\circ}_{ij}$, that is, only two classes of tensors satisfying it exist: (a) symmetric tensors, and (b) anti-symmetric tensors.

□

Example 6.4 (Different classes of systems). Let T be a tensor belonging to the tensor space $V^n_*(\otimes V^n)^4 \otimes V^n_*(K)$, with components $t^{\circ j k \ell p \circ}_{i \circ \circ \circ \circ q}$, that is symmetric with respect to the set of indices $(k\ell p)$.

Classify the following systems of scalars as tensors:

1. Contracted tensor of T with respect to the indices (i, j).
2. Contracted tensor of T with respect to the indices (p, q).
3. Contracted tensor of T with respect to the indices (ℓ, p).
4. Doubly contracted tensor of T with respect to the indices (i, p) and (j, q).
5. Doubly contracted tensor of T with respect to the indices (i, q) and (j, k).

Solution:

1. The contracted tensor of T with respect to the indices (i, j) is

$$C\binom{j}{i}(T) = [t^{\circ\theta k \ell p \circ}_{\theta \circ \circ \circ \circ q}] = [u^{k\ell p\circ}_{\circ\circ\circ q}],$$

which is a symmetric tensor with respect to the contravariant indices.

2. The contracted tensor of T with respect to the indices (p, q) is

$$C\binom{p}{q}(T) = [t^{\circ j k \ell \theta \circ}_{i \circ \circ \circ \circ \theta}] = [v^{\circ j k \ell}_{i \circ \circ \circ}],$$

which is a symmetric tensor with respect to the set $(k\ell)$.

3. The contracted tensor of T with respect to the indices (ℓ, p) is

$$C(\ell, p)(T) = [t^{\circ j k \theta \theta \circ}_{i \circ \circ \circ \circ q}] = s_1[i, j, k, q],$$

which is a system of scalar components.

4. The doubly contracted tensor of T with respect to the indices (i, p) and (j, q) is

$$C\left(\begin{array}{c|c} p & j \\ i & q \end{array}\right)(T) = [t^{\circ \phi k \ell \theta \circ}_{\theta \circ \circ \circ \circ \phi}] = [w^{k\ell}_{\circ\circ}],$$

which is a symmetric contravariant tensor.

5. The doubly contracted tensor of T with respect to the indices (i, q) and (j, k) is

$$C(i, q | j, k)(T) = [t^{\circ \phi \phi \ell p \circ}_{\theta \circ \circ \circ \circ \theta}] = s_2[\ell, p],$$

which is a symmetric system of scalar components.

□

6.4.3 Symmetric tensor spaces and subspaces. Strict components associated with subspaces

Consider as a tensor space of reference the tensor space of homogeneous tensors of order r totally contravariant over the linear space $V^n(K)$, notated: $(\otimes V^n)^r(K)$. Obviously inside it there exist tensor subspaces, and some of them of k-symmetric (partial or totally symmetric) tensors.

In particular, we denote by $(\otimes V^n)^r_s(K)$ the subspace sum of all symmetric tensor subspaces. We point out that such subspace is not only a vector subspace but a *tensor* subspace.

For the sake of simplicity, we examine this only for $r = 2$.

Let $\vec{u}, \vec{v} \in (\otimes V^n)^2_s(K)$ be two symmetric tensors:

$$\vec{u} = u^{\alpha\beta}_{\circ\circ}\vec{e}_\alpha \otimes \vec{e}_\beta; \ \ \text{with } u^{\alpha\beta}_{\circ\circ} = u^{\beta\alpha}_{\circ\circ} \text{ and } u^{ij}_{\circ\circ} = u^{\alpha\beta}_{\circ\circ}\gamma^{i\circ}_{\circ\alpha}\gamma^{j\circ}_{\circ\beta} \tag{6.25}$$

$$\vec{v} = v^{\alpha\beta}_{\circ\circ}\vec{e}_\alpha \otimes \vec{e}_\beta; \ \ \text{with } v^{\alpha\beta}_{\circ\circ} = v^{\beta\alpha}_{\circ\circ} \text{ and } v^{ij}_{\circ\circ} = v^{\alpha\beta}_{\circ\circ}\gamma^{i\circ}_{\circ\alpha}\gamma^{j\circ}_{\circ\beta} \tag{6.26}$$

1. The sum subspace contains the vectors of the form

$$\vec{w} = \vec{u} + \vec{v} = (u^{\alpha\beta}_{\circ\circ} + v^{\alpha\beta}_{\circ\circ})\vec{e}_\alpha \otimes \vec{e}_\beta = w^{\alpha\beta}_{\circ\circ}\vec{e}_\alpha \otimes \vec{e}_\beta$$

and since

$$w^{\alpha\beta}_{\circ\circ} = u^{\alpha\beta}_{\circ\circ} + v^{\alpha\beta}_{\circ\circ} = u^{\beta\alpha}_{\circ\circ} + v^{\beta\alpha}_{\circ\circ} = w^{\beta\alpha}_{\circ\circ} \tag{6.27}$$

and

$$w^{ij}_{\circ\circ} = u^{ij}_{\circ\circ} + v^{ij}_{\circ\circ} = (u^{\alpha\beta}_{\circ\circ} + v^{\alpha\beta}_{\circ\circ})\gamma^{i\circ}_{\circ\alpha}\gamma^{j\circ}_{\circ\beta} = w^{\alpha\beta}_{\circ\circ}\gamma^{i\circ}_{\circ\alpha}\gamma^{j\circ}_{\circ\beta} \tag{6.28}$$

the result is that the subspace $S_m(\vec{w}) = S_m(\vec{u} + \vec{v}) \subset (\otimes V^n)^2_s(K)$.

2. If $\vec{u} \in (\otimes V^n)^2_s$ and $\lambda \in K$:

$$\vec{z} = \lambda\vec{u} = \lambda u^{\alpha\beta}_{\circ\circ}\vec{e}_\alpha \otimes \vec{e}_\beta = z^{\alpha\beta}_{\circ\circ}\vec{e}_\alpha \otimes \vec{e}_\beta \ \text{con} \ z^{\alpha\beta}_{\circ\circ} = \lambda u^{\alpha\beta}_{\circ\circ} = \lambda u^{\beta\alpha}_{\circ\circ} = z^{\beta\alpha}_{\circ\circ} \tag{6.29}$$

and

$$z^{ij}_{\circ\circ} = \lambda u^{ij}_{\circ\circ} = (\lambda u^{\alpha\beta}_{\circ\circ})\gamma^{i\circ}_{\circ\alpha}\gamma^{j\circ}_{\circ\beta} = z^{\alpha\beta}_{\circ\circ}\gamma^{i\circ}_{\circ\alpha}\gamma^{j\circ}_{\circ\beta} \tag{6.30}$$

so that the set $(\otimes V^n)^2_s(K)$ has the structure not only of *linear* space, but of *tensor* subspace contained in $(\otimes V^n)^2(K)$.

The generalization to $(\otimes V^n)^r_s(K) \subset (\otimes V^n)^r(K)$ is immediate.

It is convenient to study here certain limitations that come from executing algebraic operations with symmetric tensors, together with others that come from using them with other tensor.

1. The first limitation refers to the fact that some authors consider the set of strict components associated with a symmetric tensor as components of

a vector associated with the symmetric tensor, endowing to such components with the corresponding "basic vectors", with the added constraint of having *repeated and only increasing* indices.

Then, if in particular we refer to the symmetric tensor

$$\vec{u} = u^{\alpha\beta\gamma}_{\circ\circ\circ}\vec{e}_{\alpha} \otimes \vec{e}_{\beta} \otimes \vec{e}_{\gamma} \in (\otimes V^4)^3_s(K),$$

the basis of which $\mathcal{B} = \{\vec{e}_{\alpha} \otimes \vec{e}_{\beta} \otimes \vec{e}_{\gamma}\}$ has 64 basic vectors, in agreement with its dimension $\sigma = n^r = 4^3 = 64$, the "associated" corresponding vector is

$$\vec{u}_0 = u^{(\alpha\beta\gamma)}_{\circ\circ\circ}\vec{e}_{\alpha} \otimes \vec{e}_{\beta} \otimes \vec{e}_{\gamma}$$

formed with the strict components of the previous tensor, with dimension $\sigma_0 = CR_{4,3} = \binom{4+3-1}{3} = \frac{6!}{3!3!} = 20$ and its basis

$$\mathcal{B}_0 = \{\vec{e}_1 \otimes \vec{e}_1 \otimes \vec{e}_1, \vec{e}_1 \otimes \vec{e}_1 \otimes \vec{e}_2, \ldots, \vec{e}_4 \otimes \vec{e}_4 \otimes \vec{e}_4\}$$

contains only 20 basic vectors, of the 64 basic vectors that form the tensor basis.

The apparent conclusion is that such authors consider as important vectors of the type $\vec{u}_0$ arising from each tensor of the symmetric tensor subspaces in $(\otimes V^n)^r_s$, which lead to the set $S(\vec{u}_0, \vec{v}_0, \vec{w}_0, \ldots)$ which is a distinguished subspace of the subspace $(\otimes V^n)^r_s$, to which they even endow a "product" to build commutative algebras that are isomorphic to those of polynomials. Certain authors have covered this subject at length. However, we want to point out that though it is true that the set of "associated" vectors $S(\vec{u}_0, \vec{v}_0, \vec{w}_0, \ldots)$ is a *linear space*, in the tensor subspace of symmetric tensors $(S(\vec{u}_0) \subset (V^n)^r_s)$, of dimension $\sigma_0 = CR_{n,r} = \binom{n+r-1}{r}$, *it is not* a tensor subspace, precisely because the ordering *without repetition* of their basic tensor products goes against Axiom 4 of Section 2.5, which forces the number of components of a (pure homogeneous) tensor-vector to be n^r, a quantity that is never equal to $\binom{n+r-1}{r}$.

Since the vectors of the subspace $S(\vec{u}_0, \vec{v}_0, \vec{w}_0, \ldots)$ *are not tensors*, there is no interest in studying them in the present book, though we understand their detailed development in other books under other different points of view.

2. The second limitation refers to the use of symmetric tensors as test tensors in the quotient law, to find if a data system of scalar components is a tensor. This is a common practice in many books of tensor analysis.

Assume that one desires to analyze the possible tensor nature of a given system $S(\alpha, \beta)$ of order $r = 2$, over a linear space $V^n(K)$, applying the quotient law.

Since we suspect that $S(\alpha, \beta)$ could be a pure covariant tensor, we must choose an *arbitrary* contravariant test tensor $\vec{t} = t^{\alpha\beta}_{\circ\circ}\vec{e}_{\alpha} \otimes \vec{e}_{\beta}$ for the contracted tensor of both to be an invariant, that is, if the initial and the new bases of $V^n(K)$ are $\{\vec{e}_{\alpha}\}$ and $\{\vec{e}_i\}$ we must have

$$t^{ij}_{\circ\circ} = t^{\alpha\beta}_{\circ\circ}\gamma^{i\circ}_{\circ\alpha}\gamma^{j\circ}_{\circ\beta} \rightarrow t^{\alpha\beta}_{\circ\circ} = t^{ij}_{\circ\circ}c^{\alpha\circ}_{\circ i}c^{\beta\circ}_{\circ j}, \qquad (6.31)$$

with $s(i,j)t^{ij} = I_0 \in K$ and $s(\alpha\beta)t^{\alpha\beta} = I_0 \in K$, and taking into account (6.31), transposing the order the result is

$$s(i,j)t^{ij} = s(\alpha,\beta)c^{\circ\alpha}_{i\circ}c^{\circ\beta}_{j\circ}t^{ij}_{\circ\circ} \rightarrow (s(i,j) - s(\alpha,\beta)c^{\circ\alpha}_{i\circ}c^{\circ\beta}_{j\circ})t^{ij} = 0 \quad (6.32)$$

and since the t^{ij} are in general non-null, it must be $s(i,j) = s(\alpha,\beta)c^{\circ\alpha}_{i\circ}c^{\circ\beta}_{j\circ}$, and because of the tensor relation (4.34), we finally get

$$s^{\circ\circ}_{ij} = s^{\circ\circ}_{\alpha\beta}c^{\circ\alpha}_{i\circ}c^{\circ\beta}_{j\circ}. \qquad (6.33)$$

However, if we use as test tensor a symmetric tensor, which adds the symmetry conditions (it is not fully arbitrary), we have

$$t^{ij}_{\circ\circ} = t^{ji}_{\circ\circ} \text{ and } t^{\alpha\beta}_{\circ\circ} = t^{\beta\alpha}_{\circ\circ}. \qquad (6.34)$$

Grouping Equations (6.32), by strict components, on account of (6.34), yields

$$\left([s(i,j) + s(j,i)] - [s(\alpha,\beta) + s(\beta,\alpha)]c^{\circ\alpha}_{i\circ}c^{\circ\beta}_{j\circ}\right)t^{ij}_{\circ\circ} = 0. \qquad (6.35)$$

Since the $t^{ij}_{\circ\circ}$ are not null, they would require again

$$s(i,j) + s(j,i) = [s(\alpha,\beta) + s(\beta,\alpha)]c^{\circ\alpha}_{i\circ}c^{\circ\beta}_{j\circ}, \qquad (6.36)$$

which would prove the tensor nature of the system $[s(i,j) + s(j,i)]$ but *not* the tensor nature of the first summand $s(i,j)$ isolatedly.

6.5 Symmetric tensors under the tensor algebra perspective

As will be shown, a simple counterexample forces us to accept that the symmetric tensor algebra of symmetric tensors, *does not* exist, that is, the tensor product $\vec{t}_1 \otimes \vec{t}_2$ of two symmetric tensors, is not in general another symmetric tensor. We give the following counterexample.

Consider two tensors of order $r = 2$ and $n = 3$, both contravariant and symmetric:

$$[a^{ij}_{\circ\circ}] = \begin{bmatrix} a_1 & x_1 & y_1 \\ x_1 & b_1 & z_1 \\ y_1 & z_1 & c_1 \end{bmatrix} \text{ and } [b^{kl}_{\circ\circ}] = \begin{bmatrix} a_2 & x_2 & y_2 \\ x_2 & b_2 & z_2 \\ y_2 & z_2 & c_2 \end{bmatrix}.$$

Since $\forall a_{\circ\circ}^{ij} = a_{\circ\circ}^{ji}$ and $\forall b_{\circ\circ}^{k\ell} = b_{\circ\circ}^{\ell k}$ we will represent the components of the tensor of order $r = 4$ and $n = 3$ in matrix form as

$$P \equiv [p_{\circ\circ\circ\circ}^{ijk\ell}] = [a_{\circ\circ}^{ij}] \otimes [b_{\circ\circ}^{k\ell}], \tag{6.37}$$

the tensor product of the given tensors.

The matrix P is represented by nine blocks:

$$P = [p_{\circ\circ\circ\circ}^{ijk\ell}] = \left[\begin{array}{ccc|ccc|ccc} a_1a_2 & a_1x_2 & a_1y_2 & x_1a_2 & x_1x_2 & x_1y_2 & y_1a_2 & y_1x_2 & y_1y_2 \\ a_1x_2 & a_1b_2 & a_1z_2 & x_1x_2 & x_1b_2 & x_1z_2 & y_1x_2 & y_1b_2 & y_1z_2 \\ a_1y_2 & a_1z_2 & a_1c_2 & x_1y_2 & x_1z_2 & x_1c_2 & y_1y_2 & y_1z_2 & y_1c_2 \\ \hline x_1a_2 & x_1x_2 & x_1y_2 & b_1a_2 & b_1x_2 & b_1y_2 & z_1a_2 & z_1x_2 & z_1y_2 \\ x_1x_2 & x_1b_2 & x_1z_2 & b_1x_2 & b_1b_2 & b_1z_2 & z_1x_2 & z_1b_2 & z_1z_2 \\ x_1y_2 & x_1z_2 & x_1c_2 & b_1y_2 & b_1z_2 & b_1c_2 & z_1y_2 & z_1z_2 & z_1c_2 \\ \hline y_1a_2 & y_1x_2 & y_1y_2 & z_1a_2 & z_1x_2 & z_1y_2 & c_1a_2 & c_1x_2 & c_1y_2 \\ y_1x_2 & y_1b_2 & y_1z_2 & z_1x_2 & z_1b_2 & z_1z_2 & c_1x_2 & c_1b_2 & c_1z_2 \\ y_1y_2 & y_1z_2 & y_1c_2 & z_1y_2 & z_1z_2 & z_1c_2 & c_1y_2 & c_1z_2 & c_1c_2 \end{array}\right],$$

$$\tag{6.38}$$

where (axiomatic ordering) i is the index of the row block, j is the index of the column block, k is the row of each submatrix, and ℓ is the column of each submatrix.

Next, we see some of the tensor strict components and their permutations.

If the tensor were symmetric, we should have $p_{\circ\circ\circ\circ}^{(ijk\ell)} = p_{\circ\circ\circ\circ}^{kji\ell}$, but for $i = 1, j = 2, k = 3, \ell = 3$ the result is $p_{\circ\circ\circ\circ}^{1233} = x_1c_2 \neq p_{\circ\circ\circ\circ}^{3213} = z_1y_2$, that is, the equality does not hold. Similarly, for $i = 3, j = 2, k = 1, \ell = 2$ we have $p_{\circ\circ\circ\circ}^{3212} = z_1x_2 \neq p_{\circ\circ\circ\circ}^{1232} = x_1z_2$, that is, the equality does not hold, and $p_{\circ\circ\circ\circ}^{(iijj)} = p_{\circ\circ\circ\circ}^{jiji}$, which for $i = 2; j = 3$ leads to $p_{\circ\circ\circ\circ}^{2233} = b_1c_2 \neq p_{\circ\circ\circ\circ}^{3232} = z_1z_2$, which also fails to hold, etc.

Consequently, P is a symmetric matrix, but *is not* a symmetric tensor of order 4.

Example 6.5 (Kronecker tensor product). Consider the symmetric matrices $A_2 \equiv [a_{\circ\circ}^{ij}] = \begin{bmatrix} 1 & 3 \\ 3 & 4 \end{bmatrix}$ and $B_2 \equiv [b_{\circ\circ}^{k\ell}] = \begin{bmatrix} 2 & -1 \\ -1 & 5 \end{bmatrix}$ as two contravariant tensors, of order $r = 2$, over the linear space $\mathbb{R}^2 (n = 2)$.

We wish to determine:

1. The matrix direct tensor product or the Kronecker tensor product, $C_4 = A \otimes B$, of both matrices.
2. Considering the matrices A and B as matrix representations of the corresponding tensors

$$\vec{a} = \vec{e}_1 \otimes \vec{e}_1 + 3\vec{e}_1 \otimes \vec{e}_2 + 3\vec{e}_2 \otimes \vec{e}_1 + 4\vec{e}_2 \otimes \vec{e}_2 \tag{6.39}$$

$$\vec{b} = 2\vec{e}_1 \otimes \vec{e}_1 - \vec{e}_1 \otimes \vec{e}_2 - \vec{e}_2 \otimes \vec{e}_1 + 5\vec{e}_2 \otimes \vec{e}_2, \tag{6.40}$$

determine the representation matrix of the tensor product of both tensors.

3. Prepare a warning comment with respect to the operators "$\otimes$".

Solution:

1. The matrix direct tensor product or the Kronecker tensor product of both matrices is

$$C_4 = A_2 \otimes B_2 = \begin{bmatrix} 2 & -1 & 6 & -3 \\ -1 & 5 & -3 & 15 \\ 6 & -3 & 8 & -4 \\ -3 & 15 & -4 & 20 \end{bmatrix},$$

which obviously, is a symmetric matrix (it is a symmetric tensor of order $r = 2$ over the linear space $\mathbb{R}^4$).

2. Since the tensor product of the tensors $\vec{a}$ and $\vec{b}$ is

$$\vec{p} = \vec{a} \otimes \vec{b} = p^{ijk\ell}_{\circ\circ\circ\circ}\vec{e}_i \otimes \vec{e}_j \otimes \vec{e}_k \otimes \vec{e}_\ell; \text{ con } p^{ijk\ell}_{\circ\circ\circ\circ} = a^{ij}_{\circ\circ} \cdot b^{k\ell}_{\circ\circ},$$

the matrix representation of the tensor $\vec{p}$ can be written as

$$P = [p^{ijk\ell}_{\circ\circ\circ\circ}] \equiv \left[\begin{array}{cc|cc} p^{1111}_{\circ\circ\circ\circ} & p^{1112}_{\circ\circ\circ\circ} & p^{1211}_{\circ\circ\circ\circ} & p^{1212}_{\circ\circ\circ\circ} \\ p^{1121}_{\circ\circ\circ\circ} & p^{1122}_{\circ\circ\circ\circ} & p^{1221}_{\circ\circ\circ\circ} & p^{1222}_{\circ\circ\circ\circ} \\ \hline p^{2111}_{\circ\circ\circ\circ} & p^{2112}_{\circ\circ\circ\circ} & p^{2211}_{\circ\circ\circ\circ} & p^{2212}_{\circ\circ\circ\circ} \\ p^{2121}_{\circ\circ\circ\circ} & p^{2122}_{\circ\circ\circ\circ} & p^{2221}_{\circ\circ\circ\circ} & p^{2222}_{\circ\circ\circ\circ} \end{array} \right]$$

$$= \left[\begin{array}{cc|cc} 2 & -1 & 6 & -3 \\ -1 & 5 & -3 & 15 \\ \hline 6 & -3 & 8 & -4 \\ -3 & 15 & -4 & 20 \end{array} \right],$$

which is a tensor of order $r = 4$, $n = 2$, i.e., over $\mathbb{R}^2$.

However, according to what was established in Example 6.2, question 2(a), the matrix P has not the properties of a symmetric tensor of order $r = 4$, because for example

$$p^{1211}_{\circ\circ\circ\circ} = 6 \neq p^{1112}_{\circ\circ\circ\circ} = -1 \Rightarrow p^{1211}_{\circ\circ\circ\circ} \neq p^{1112}_{\circ\circ\circ\circ}$$

$$p^{1122}_{\circ\circ\circ\circ} = 5 \neq p^{1212}_{\circ\circ\circ\circ} = -3 \Rightarrow p^{1122}_{\circ\circ\circ\circ} \neq p^{1212}_{\circ\circ\circ\circ}$$

$$p^{1212}_{\circ\circ\circ\circ} = -3 \neq p^{2211}_{\circ\circ\circ\circ} = 8 \Rightarrow p^{1212}_{\circ\circ\circ\circ} \neq p^{2211}_{\circ\circ\circ\circ}.$$

Thus, $\vec{p}$ is *not* a symmetric tensor (it is a tensor of order $r = 4$, over $n = 2$, *non-symmetric*).

We immediately conclude that the tensor product of symmetric tensors is not in general symmetric.

3. The major warning of this exercise is that the tensor product or the Kronecker product of symmetric matrices is a symmetric matrix, while the tensor product of symmetric matrices that represent symmetric tensors *is not* a symmetric tensor, though we use the same symbol "$\otimes$" and operating procedure (the matrices in the previous paragraphs 1 and 2 are identical) for both products. In other words, the operator "$\otimes$" of matrices has *not* the *same* properties as the operator "$\otimes$" for tensors.

□

Example 6.6 (Tensor products).

1. Show that the Kronecker tensor product of two square matrices, both symmetric is a symmetric matrix.
2. Show that the Kronecker tensor product of two anti-symmetric matrices is a symmetric matrix.
3. Show that the contracted tensor product, of two symmetric matrices (classic product of matrices) is not in general a symmetric matrix. In which cases is it symmetric?

Solution:

1. Consider the equalities

$$A_n = A_n^t; \; B_m = B_m^t. \tag{6.41}$$

Multiplying them in order and using the Kronecker product we obtain

$$A_n \otimes B_m = A_n^t \otimes B_m^t = (A_n \otimes B_m)^t \Rightarrow A_n \otimes B_m \text{ is symmetric.} \tag{6.42}$$

2. Consider the equalities

$$A_n = -A_n^t; \; B_m = -B_m^t. \tag{6.43}$$

Multiplying them in order and using the Kronecker product we obtain

$$A_n \otimes B_m = (-A_n^t) \otimes (-B_m^t) = (A_n \otimes B_m)^t \Rightarrow A_n \otimes B_m \text{ is symmetric.} \tag{6.44}$$

3. First we have

$$C\begin{pmatrix} j \\ k \end{pmatrix}[P_{\circ\circ k\ell}^{ij\circ\circ}] = C\begin{pmatrix} j \\ k \end{pmatrix}[A \otimes B] = C\begin{pmatrix} j \\ k \end{pmatrix}[a_{\circ\circ}^{ij} \otimes b_{k\ell}^{\circ\circ}] = [a_{\circ\circ}^{i\theta} \otimes b_{\theta\ell}^{\circ\circ}] = A_n \bullet B_n$$

and since $A_n = A_n^t$ and $B_n = B_n^t$, the properties of the classic product of matrices lead to

$$A \bullet B = A^t \bullet B^t = (B \bullet A)^t \neq (A \bullet B)^t \Rightarrow A \bullet B \neq (A \bullet B)^t, \tag{6.45}$$

unless the product $A \bullet B$ commutes. If $B \bullet A = A \bullet B$ then, returning to (6.45) we obtain

$$A \bullet B = A^t \bullet B^t = (B \bullet A)^t = (A \bullet B)^t,$$

which proves the symmetry of $A \bullet B$.

□

6.5.1 Symmetrized tensor associated with an arbitrary pure tensor

In Section 6.4.1 we have already developed a technique that allows us to associate with an arbitrary pure tensor, another tensor which is symmetric with respect to a set I_k of its indices. Here we consider only one associated tensor that corresponds to $I_k \equiv I_r$, that is, to selecting all $r!$ isomers corresponding to the set $I_r = \{\alpha_1, \alpha_2, \ldots, \alpha_r\}$ of all tensor indices.

The resulting symmetric tensor will be called a "symmetrized" form of the given tensor. To illustrate, we give a detailed example.

Let $T = t^{\alpha_1 \alpha_2 \cdots \alpha_r}_{\circ \ \circ \ \circ \cdots \ \circ}$ be the given tensor over the linear space $V^n(K)$, and consider the $r!$ permutations of its indices, which generate the corresponding isomers of the given tensor.

We denote by $S(T)^{\alpha_1, \alpha_2, \cdots, \alpha_r}$ the "symmetrized" tensor of the given tensor T the components of which are established using the formula

$$S(T)^{\alpha_1 \alpha_2 \cdots \alpha_r}_{\circ \ \circ \ \circ \cdots \ \circ} = \frac{1}{r!} \left[t^{\alpha_1 \alpha_2 \cdots \alpha_r}_{\circ \ \circ \ \circ \cdots \ \circ} + t^{\alpha_2 \alpha_1 \cdots \alpha_r}_{\circ \ \circ \ \circ \cdots \ \circ} + t^{\alpha_3 \alpha_1 \cdots \alpha_r}_{\circ \ \circ \ \circ \cdots \ \circ} + \cdots + t^{\alpha_r \alpha_{r-1} \cdots \alpha_1}_{\circ \ \circ \ \circ \cdots \ \circ} \right],$$
(6.46)

the summands of which are all the $r!$ isomer tensors.

This is a symmetric tensor, defined over the same tensor space $(\otimes V^n)^r(K)$.

6.5.2 Extension of the symmetrized tensor associated with a mixed tensor

If the data tensor is mixed, it is possible to associate with it a symmetrized tensor by establishing all isomers that result from considering all permutations of its p contravariant indices, independently of the corresponding permutations of its q covariant indices. If the data tensor is $T = t^{\alpha_1 \alpha_2 \cdots \alpha_p \ \circ \ \circ \cdots \ \circ}_{\circ \ \circ \ \cdots \ \circ \ \beta_1 \beta_2 \cdots \beta_q}$, of order $r = p + q$, the number of isomers will be $p!q!$, and the symmetrized tensor

$$S(T)^{\alpha_1 \alpha_2 \cdots \alpha_p \ \circ \ \circ \cdots \ \circ}_{\circ \ \circ \ \cdots \ \circ \ \beta_1 \beta_2 \cdots \beta_q} = \frac{1}{p!q!} \left[t^{\alpha_1 \cdots \alpha_p \ \circ \ \circ \cdots \ \circ}_{\circ \ \cdots \ \circ \ \beta_1 \cdots \beta_q} + t^{\alpha_2 \cdots \alpha_p \ \circ \ \circ \cdots \ \circ}_{\circ \ \cdots \ \circ \ \beta_1 \cdots \beta_q} + \cdots + t^{\alpha_p \cdots \alpha_1 \ \circ \ \circ \cdots \ \circ}_{\circ \ \cdots \ \circ \ \beta_q \cdots \beta_1} \right],$$
(6.47)

where the right-hand sum extends to all $p!q!$ isomer tensors.

In order to determine the number of strict components of the symmetrized tensor, we proceed with an ordered criterion, in this mixed case, separating the permutations corresponding to different valencies.

For example consider the tensor $t^{\alpha \beta \circ \circ}_{\circ \circ \gamma \delta}$ of order $r = 4$ with $p = 2$ and $q = 2$, over a $V^2(\mathbb{R})$. Then, the strict components of the symmetrized tensor will be sorted as

$$[11] [11]$$
$$[11] [12] = 1121$$
$$[11] [22]$$
$$[12] [11] = 2111$$

$$[12]\,[12] = 1221 = 2112 = 2121$$
$$[12]\,[22] = 2122$$
$$[22]\,[11]$$
$$[22]\,[12] = 2221$$
$$[22]\,[22]\,,$$

showing that, we have nine strict components of the $\sigma = n^r = 2^4 = 16$ total components.

Example 6.7 (Symmetrized tensors).

1. Obtain the symmetrized tensor of the tensor T given in Example 4.5, assuming it is totally contravariant, using the general formula.
2. Obtain the symmetrized tensor of the mixed tensor A:

$$[A^{\alpha\beta\circ\circ}_{\circ\circ\gamma\delta}] = \begin{bmatrix} 1 & 2 & -2 & 1 \\ -3 & 4 & 0 & 6 \\ \hline 5 & -3 & -2 & -5 \\ -1 & 4 & -1 & 3 \end{bmatrix}.$$

Solution:

1. From Exercise 4.5 we extract the following numerical data of tensor T and its isomers:

$$T = [t^{\alpha\beta\gamma}_{\circ\circ\circ}] = \begin{bmatrix} 1 & 0 & -1 \\ 2 & 3 & 0 \\ -1 & 2 & 0 \\ 2 & -1 & 1 \\ 0 & 1 & 0 \\ 2 & 0 & 1 \\ 0 & 0 & 1 \\ 5 & 1 & 2 \\ 1 & 0 & 0 \end{bmatrix} ; \quad U = [t^{\gamma\alpha\beta}_{\circ\circ\circ}] = \begin{bmatrix} 1 & 2 & -1 \\ 2 & 0 & 2 \\ 0 & 5 & 1 \\ 0 & 3 & 2 \\ -1 & 1 & 0 \\ 0 & 1 & 0 \\ -1 & 0 & 0 \\ 1 & 0 & 1 \\ 1 & 2 & 0 \end{bmatrix} ;$$

$$V = [t^{\beta\gamma\alpha}_{\circ\circ\circ}] = \begin{bmatrix} 1 & 2 & 0 \\ 0 & -1 & 0 \\ -1 & 1 & 1 \\ 2 & 0 & 5 \\ 3 & 1 & 1 \\ 0 & 0 & 2 \\ -1 & 2 & 1 \\ 2 & 0 & 0 \\ 0 & 1 & 0 \end{bmatrix} ; W = [t^{\alpha\gamma\beta}_{\circ\circ\circ}] = \begin{bmatrix} 1 & 2 & -1 \\ 0 & 3 & 2 \\ -1 & 0 & 0 \\ 2 & 0 & 2 \\ -1 & 1 & 0 \\ 1 & 0 & 1 \\ 0 & 5 & 1 \\ 0 & 1 & 0 \\ 1 & 2 & 0 \end{bmatrix} ;$$

$$R = [t^{\gamma\beta\alpha}_{\circ\circ\circ}] = \begin{bmatrix} 1 & 2 & 0 \\ 2 & 0 & 5 \\ -1 & 2 & 1 \\ 0 & -1 & 0 \\ 3 & 1 & 1 \\ 2 & 0 & 0 \\ -1 & 1 & 1 \\ 0 & 0 & 2 \\ 0 & 1 & 0 \end{bmatrix} ; S = [t^{\beta\alpha\gamma}_{\circ\circ\circ}] = \begin{bmatrix} 1 & 0 & -1 \\ 2 & -1 & 1 \\ 0 & 0 & 1 \\ 2 & 3 & 0 \\ 0 & 1 & 0 \\ 5 & 1 & 2 \\ -1 & 2 & 0 \\ 2 & 0 & 1 \\ 1 & 0 & 0 \end{bmatrix}.$$

The symmetrized tensor is

$$S(T) = \frac{1}{6}[T + U + V + W + R + S] = \frac{1}{6}\begin{bmatrix} 6 & 8 & -4 \\ 8 & 4 & 10 \\ -4 & 10 & 4 \\ 8 & 4 & 10 \\ 4 & 6 & 2 \\ 10 & 2 & 6 \\ -4 & 10 & 4 \\ 10 & 2 & 6 \\ 4 & 6 & 0 \end{bmatrix} = \frac{1}{3}\begin{bmatrix} 3 & 4 & -2 \\ 4 & 2 & 5 \\ -2 & 5 & 2 \\ 4 & 2 & 5 \\ 2 & 3 & 1 \\ 5 & 1 & 3 \\ -2 & 5 & 2 \\ 5 & 1 & 3 \\ 2 & 3 & 0 \end{bmatrix}.$$

2. We have four isomers:

$$[a^{\alpha\beta o o}_{o o \gamma\delta}] = \begin{bmatrix} 1 & 2 & -2 & 1 \\ -3 & 4 & 0 & 6 \\ 5 & -3 & -2 & -5 \\ -1 & 4 & -1 & 3 \end{bmatrix} ; \quad [a^{\beta\alpha o o}_{o o \gamma\delta}] = \begin{bmatrix} 1 & 2 & 5 & -3 \\ -3 & 4 & -1 & 4 \\ -2 & 1 & -2 & -5 \\ 0 & 6 & -1 & 3 \end{bmatrix} ;$$

$$[a^{\alpha\beta o o}_{o o \delta\gamma}] = \begin{bmatrix} 1 & -3 & -2 & 0 \\ 2 & 4 & 1 & 6 \\ 5 & -1 & -2 & -1 \\ -3 & 4 & -5 & 3 \end{bmatrix} ; \quad [a^{\beta\alpha o o}_{o o \delta\gamma}] = \begin{bmatrix} 1 & -3 & 5 & -1 \\ 2 & 4 & -3 & 4 \\ -2 & 0 & -2 & -1 \\ 1 & 6 & -5 & 3 \end{bmatrix}.$$

Then, $S(A)$ is

$$[s^{\alpha\beta o o}_{o o \gamma\delta}] = S(A) = \frac{1}{4}\begin{bmatrix} 4 & -2 & 6 & -3 \\ -2 & 16 & -3 & 20 \\ 6 & -3 & -8 & -12 \\ -3 & 20 & -12 & 12 \end{bmatrix}.$$

□

Remark 6.2. The resulting symmetric tensor $[s^{\alpha\beta o o}_{o o \gamma\delta}]$ *does not* satisfy the symmetric conditions required in Example 6.2, question 2(a). However, we must not panic, because such conditions are for a totally contravariant tensor $[s^{\alpha\beta\gamma\delta}_{o o o o}]$, and our tensor is a mixed tensor.   □

6.6 Symmetric tensor algebras: The $\otimes_S$ product

Once the concept of "symmetrization" of an arbitrary homogeneous tensor has been established, it is possible to define a new tensor product for exclusive use with symmetric tensors, the product of which is another symmetric tensor, and since the sum of two symmetric tensors is another symmetric tensor, this will supply a good foundation for an interior symmetric tensor algebra, that is, an algebra structure for symmetric tensors.

This new product will be called a "symmetric tensor product" and will be denoted by "$\otimes_S$".

Let A and B be two symmetric homogeneous tensors of orders r_a and r_b, respectively, both over the linear space $V^n(K)$. If $P = A \otimes B$, of order $r = r_a + r_b$, the new product is defined by the following formula

$$A \otimes_S B = S[P] = S(A \otimes B) \text{ (symmetrized tensor of } A \otimes B). \qquad (6.48)$$

For the sake of illustration, we give one example of the new product.

Let $\vec{a}, \vec{b} \in V^2 \otimes V^2(K)$ be two contravariant symmetric tensors, and let $\sigma = n^r = 2^2 = 4$ be the dimension of the tensor space to which they belong. Their matrix representations are:

$$A = [a_{\circ\circ}^{\alpha\beta}] \equiv \begin{bmatrix} a_1 & b_1 \\ b_1 & c_1 \end{bmatrix}; \quad B = [b_{\circ\circ}^{\gamma\delta}] \equiv \begin{bmatrix} a_2 & b_2 \\ b_2 & c_2 \end{bmatrix}; \quad r = r_a + r_b = 2 + 2 = 4.$$

$$A \otimes_S B = \begin{bmatrix} a_1 & b_1 \\ b_1 & c_1 \end{bmatrix} \otimes_S \begin{bmatrix} a_2 & b_2 \\ b_2 & c_2 \end{bmatrix} = S[a_{\circ\circ}^{\alpha\beta}] \otimes [b_{\circ\circ}^{\gamma\delta}]] = S[p_{\circ\circ\circ\circ}^{\alpha\beta\gamma\delta}]$$

$$= S \left(\begin{bmatrix} a_1a_2 & a_1b_2 & | & b_1a_2 & b_1b_2 \\ a_1b_2 & a_1c_2 & | & b_1b_2 & b_1c_2 \\ -- & -- & + & -- & -- \\ b_1a_2 & b_1b_2 & | & c_1a_2 & c_1b_2 \\ b_1b_2 & b_1c_2 & | & c_1b_2 & c_1c_2 \end{bmatrix} \right)$$

$$= \frac{1}{4!} \left[\left(p_{\circ\circ\circ\circ}^{\alpha\beta\gamma\delta} + p_{\circ\circ\circ\circ}^{\alpha\beta\delta\gamma} + p_{\circ\circ\circ\circ}^{\alpha\gamma\beta\delta} + p_{\circ\circ\circ\circ}^{\alpha\gamma\delta\beta} + p_{\circ\circ\circ\circ}^{\alpha\delta\gamma\beta} + p_{\circ\circ\circ\circ}^{\alpha\delta\beta\gamma} \right) \right.$$
$$+ \left(p_{\circ\circ\circ\circ}^{\beta\alpha\gamma\delta} + p_{\circ\circ\circ\circ}^{\beta\alpha\delta\gamma} + p_{\circ\circ\circ\circ}^{\beta\gamma\alpha\delta} + p_{\circ\circ\circ\circ}^{\beta\gamma\delta\alpha} + p_{\circ\circ\circ\circ}^{\beta\delta\gamma\alpha} + p_{\circ\circ\circ\circ}^{\beta\delta\alpha\gamma} \right)$$
$$+ \left(p_{\circ\circ\circ\circ}^{\gamma\alpha\beta\delta} + p_{\circ\circ\circ\circ}^{\gamma\alpha\delta\beta} + p_{\circ\circ\circ\circ}^{\gamma\beta\alpha\delta} + p_{\circ\circ\circ\circ}^{\gamma\beta\delta\alpha} + p_{\circ\circ\circ\circ}^{\gamma\delta\beta\alpha} + p_{\circ\circ\circ\circ}^{\gamma\delta\alpha\beta} \right)$$
$$+ \left. \left(p_{\circ\circ\circ\circ}^{\delta\alpha\beta\gamma} + p_{\circ\circ\circ\circ}^{\delta\alpha\gamma\beta} + p_{\circ\circ\circ\circ}^{\delta\beta\alpha\gamma} + p_{\circ\circ\circ\circ}^{\delta\beta\gamma\alpha} + p_{\circ\circ\circ\circ}^{\delta\gamma\beta\alpha} + p_{\circ\circ\circ\circ}^{\delta\gamma\alpha\beta} \right) \right]$$

$$= \begin{bmatrix} a_1a_2 & \frac{a_1b_2+b_1a_2}{2} & | & \frac{a_1b_2+b_1a_2}{2} & \frac{4b_1b_2+a_1c_2+c_1a_2}{6} \\ \frac{a_1b_2+b_1a_2}{2} & \frac{4b_1b_2+a_1c_2+c_1a_2}{6} & | & \frac{4b_1b_2+a_1c_2+c_1a_2}{6} & \frac{b_1c_2+c_1b_2}{2} \\ ------ & ------ & + & ------ & ------ \\ \frac{a_1b_2+b_1a_2}{2} & \frac{4b_1b_2+a_1c_2+c_1a_2}{6} & | & \frac{4b_1b_2+a_1c_2+c_1a_2}{6} & \frac{b_1c_2+c_1b_2}{2} \\ \frac{4b_1b_2+a_1c_2+c_1a_2}{6} & \frac{b_1c_2+c_1b_2}{2} & | & \frac{b_1c_2+c_1b_2}{2} & c_1c_2 \end{bmatrix}.$$

It can be easily proved that the symmetric tensor product has the following properties:

1. It is commutative:
$$A \otimes_S B = B \otimes_S A. \qquad (6.49)$$

2. It is associative:
$$(A \otimes_S B) \otimes_S C = A \otimes_S (B \otimes_S C). \qquad (6.50)$$

3. It is distributive with respect to the tensor sum "+":
$$(A + B) \otimes_S C = A \otimes_S C + B \otimes_S C. \qquad (6.51)$$
$$A \otimes_S (B + C) = A \otimes_S B + A \otimes_S C. \qquad (6.52)$$

4. Due to the previously given properties, it is obvious that the symmetric tensor product, together with the sum, endow the set of symmetric homogeneous tensors with the character of a *symmetric tensor algebra*.

6.7 Illustrative examples

Example 6.8 (The rotation tensor and the symmetry). A known homogeneous contravariant tensor T, of third order is related to another tensor W, by the expression

$$t^{ijk}_{\circ\circ\circ} = w^{jik}_{\circ\circ\circ} + w^{ijk}_{\circ\circ\circ}.$$

It is also known that W is a contravariant tensor of third order and symmetric with respect to the last two indices.

1. Prove that the data tensor T is symmetric with respect to the first two indices.
2. Prove that the rotation tensor of the given tensor T shares the same symmetry.
3. Determine the generic component $w^{ijk}_{\circ\circ\circ}$ of the tensor W as a function of the components of T and of its isomers.

□

Solution:

1. Exchanging i and j in the given expression, we have

$$t^{ijk}_{\circ\circ\circ} = w^{jik}_{\circ\circ\circ} + w^{ijk}_{\circ\circ\circ} \tag{6.53}$$

$$t^{jik}_{\circ\circ\circ} = w^{ijk}_{\circ\circ\circ} + w^{jik}_{\circ\circ\circ} \tag{6.54}$$

and subtracting (6.53) from (6.54) gives

$$t^{jik}_{\circ\circ\circ} - t^{ijk}_{\circ\circ\circ} = 0 \quad \Rightarrow \quad t^{jik}_{\circ\circ\circ} = t^{ijk}_{\circ\circ\circ}.$$

2. Consider the isomer tensors of the given tensor, coming from the circular permutations of its indices (rotation tensors):

Rotation(1). (the index i passes to j || the index j passes to k || the index k passes to i)

Rotation(2). (the index i passes to k || the index j passes to i || the index k passes to j)

Thus, from $t^{ijk}_{\circ\circ\circ} = w^{jik}_{\circ\circ\circ} + w^{ijk}_{\circ\circ\circ}$ we obtain the rotation isomer(1):

$$t^{jki}_{\circ\circ\circ} = w^{kji}_{\circ\circ\circ} + w^{jki}_{\circ\circ\circ}. \tag{6.55}$$

Exchanging j and k in Expression (6.55) gives

$$t^{kji}_{\circ\circ\circ} = w^{jki}_{\circ\circ\circ} + w^{kji}_{\circ\circ\circ}, \tag{6.56}$$

and subtracting (6.55) from (6.56) we get

$$t^{kji}_{\circ\circ\circ} - t^{jki}_{\circ\circ\circ} = 0 \quad \Rightarrow \quad t^{kji}_{\circ\circ\circ} = t^{jki}_{\circ\circ\circ}.$$

Similarly, rotation(2) is

$$t^{kij}_{\circ\circ\circ} = w^{ikj}_{\circ\circ\circ} + w^{kij}_{\circ\circ\circ}. \tag{6.57}$$

Exchanging k and i in Expression (6.57)gives

$$t^{ikj}_{\circ\circ\circ} = w^{kij}_{\circ\circ\circ} + w^{ikj}_{\circ\circ\circ}, \tag{6.58}$$

and subtracting (6.57) from (6.58) we obtain

$$t^{ikj}_{\circ\circ\circ} - t^{kij}_{\circ\circ\circ} = 0 \quad \Rightarrow \quad t^{ikj}_{\circ\circ\circ} = t^{kij}_{\circ\circ\circ}.$$

3. Considering the rotations:

$$t^{ijk}_{\circ\circ\circ} = w^{jik}_{\circ\circ\circ} + w^{ijk}_{\circ\circ\circ} \tag{6.59}$$

$$t^{jki}_{\circ\circ\circ} = w^{kji}_{\circ\circ\circ} + w^{jki}_{\circ\circ\circ} \tag{6.60}$$

$$t^{kij}_{\circ\circ\circ} = w^{ikj}_{\circ\circ\circ} + w^{kij}_{\circ\circ\circ} \tag{6.61}$$

executing the operation (6.59) + (6.61) − (6.60) and on account of the symmetry of the last two indices of tensor W, we have

$$t^{ijk}_{\circ\circ\circ} - t^{jki}_{\circ\circ\circ} + t^{kij}_{\circ\circ\circ} = (w^{jki}_{\circ\circ\circ} + w^{ijk}_{\circ\circ\circ}) - (w^{kij}_{\circ\circ\circ} + w^{jki}_{\circ\circ\circ}) + (w^{ijk}_{\circ\circ\circ} + w^{kij}_{\circ\circ\circ}) = 2w^{ijk}_{\circ\circ\circ},$$

from which

$$w^{ijk}_{\circ\circ\circ} = \frac{1}{2}[t^{ijk}_{\circ\circ\circ} - t^{jki}_{\circ\circ\circ} + t^{kij}_{\circ\circ\circ}].$$

Example 6.9 (Products of symmetric contraction). Let A and B be two given homogeneous tensors, both over the linear space $V^n(K)$. Show that the constant $d \in K$, resulting from the contracted tensor product

$$d = a^{\circ\circ}_{ij} b^i_{\circ} b^j_{\circ}$$

can also be obtained by using a certain symmetric tensor $\vec{c} = c^{\circ\circ}_{ij} \vec{e}^{*i} \otimes \vec{e}^{*j}$; as

$$d = c^{\circ\circ}_{ij} b^i_{\circ} b^j_{\circ}.$$

□

Solution: We know that any covariant homogeneous tensor of second order can be written as the sum of a symmetric tensor and an anti-symmetric tensor (matrices theory). Thus, in our case we can write

$$d = a^{\circ\circ}_{ij}b^i_{\circ}b^j_{\circ} = \left(\frac{a^{\circ\circ}_{ij} + a^{\circ\circ}_{ji}}{2} + \frac{a^{\circ\circ}_{ij} - a^{\circ\circ}_{ji}}{2}\right)b^i_{\circ}b^j_{\circ}$$

$$= \left(\frac{a^{\circ\circ}_{ij} + a^{\circ\circ}_{ji}}{2}\right)b^i_{\circ}b^j_{\circ} + \frac{1}{2}\left(a^{\circ\circ}_{ij}b^i_{\circ}b^j_{\circ} - a^{\circ\circ}_{ji}b^i_{\circ}b^j_{\circ}\right)$$

but, because of the commutativity of the product in the field K, we have

$$d = \left(\frac{a^{\circ\circ}_{ij} + a^{\circ\circ}_{ji}}{2}\right)b^i_{\circ}b^j_{\circ} + \frac{1}{2}\left(a^{\circ\circ}_{ij}b^i_{\circ}b^j_{\circ} - a^{\circ\circ}_{ji}b^j_{\circ}b^i_{\circ}\right) = c^{\circ\circ}_{ij}b^i_{\circ}b^j_{\circ} + \frac{1}{2}(d - d),$$

and then, $d = c^{\circ\circ}_{ij}b^i_{\circ}b^j_{\circ}$, where $c^{\circ\circ}_{ij} = \frac{1}{2}(a^{\circ\circ}_{ij} + a^{\circ\circ}_{ji})$ is a symmetric tensor (the symmetrized tensor of A).

Example 6.10 (Strict components of a symmetric mixed tensor). Consider the symmetric mixed tensor, with components $t^{\circ\circ k}_{ijo}$ over a linear space $V^n(K)$. Calculate the number of strict components.

Solution: For the symmetry to have tensor character it must be that of the covariant indices. From the formulation in Section 6.4, we have

$$p = 1; \quad k_1 = 0; \quad q = 2; \quad k_2 = 2; \quad r = p + q = 3;$$

$$\sigma' = \binom{n + k_1 - 1}{k_1}\binom{n + k_2 - 1}{k_2}n^{r-(k_1+k_2)}$$

$$= \binom{n + 0 - 1}{0}\binom{n + 2 - 1}{2}n^{3-(0+2)} = \binom{n+1}{2}\cdot n = \frac{n^2(n+1)}{2}.$$

□

Example 6.11 (Main tensors associated with a given tensor). Consider the homogeneous contravariant tensor $\vec{t}$, of second order, over a linear space $V^3(K)$, the components of which are

$$[t^{ij}_{\circ\circ}] = \begin{bmatrix} 1 & 3 & -2 \\ 0 & 1 & 3 \\ -1 & 0 & 2 \end{bmatrix}.$$

1. Give the components $t_r{}^{ij}_{\circ\circ}$ of the tensor rotation of the given tensor.

2. Give the tensor components $t_s{}_{\circ\circ}^{\;ij}$ of the symmetrized tensor of the given tensor.

3. Give a null trace tensor $t_\circ{}_{\circ\circ}^{\;ij}$ associated with the given tensor, by subtracting a "scalar" tensor.

4. Give a null trace tensor $t_{so}{}_{\circ\circ}^{\;ij}$ associated with the symmetrized tensor, $t_s{}_{\circ\circ}^{\;ij}$.

Solution:

1. The components of the rotation tensor are:

$$t_r{}_{\circ\circ}^{\;ij} = [t_{\circ\circ}^{ji}] = \begin{bmatrix} 1 & 0 & -1 \\ 3 & 1 & 0 \\ -2 & 3 & 2 \end{bmatrix}$$

2. The tensor components of the symmetrized tensor of the given tensor are:

$$[t_s{}_{\circ\circ}^{\;ij}] = \frac{1}{2}[t_{\circ\circ}^{ij} + t_{\circ\circ}^{ji}] = \begin{bmatrix} 1 & 3/2 & -3/2 \\ 3/2 & 1 & 3/2 \\ -3/2 & 3/2 & 2 \end{bmatrix}.$$

3. The components of the null trace tensor associated with the given tensor by subtracting a "scalar" tensor are:

$$[t_\circ{}_{\circ\circ}^{\;ij}] = [t_{\circ\circ}^{ij}] - \left(\frac{t_{\circ\circ}^{\alpha\alpha}}{3}\right)[\delta_{\circ\circ}^{ij}]$$

$$= \begin{bmatrix} 1 & 3 & -2 \\ 0 & 1 & 3 \\ -1 & 0 & 2 \end{bmatrix} - \frac{1+1+2}{3}\begin{bmatrix} 1 & 0 & 0 \\ 0 & 1 & 0 \\ 0 & 0 & 1 \end{bmatrix} = \begin{bmatrix} -1/3 & 3 & -2 \\ 0 & -1/3 & 3 \\ -1 & 0 & 2/3 \end{bmatrix}.$$

4. The components of the null trace tensor associated with the symmetrized tensor are:

$$[t_{so}{}_{\circ\circ}^{\;ij}] = [t_s{}_{\circ\circ}^{\;ij}] - \left(\frac{t_s{}_{\circ\circ}^{\;\alpha\alpha}}{3}\right)[\delta_{\circ\circ}^{ij}]$$

$$= \begin{bmatrix} 1 & 3/2 & -3/2 \\ 3/2 & 1 & 3/2 \\ -3/2 & 3/2 & 2 \end{bmatrix} - \frac{4}{3}\begin{bmatrix} 1 & 0 & 0 \\ 0 & 1 & 0 \\ 0 & 0 & 1 \end{bmatrix} = \begin{bmatrix} -1/3 & 3/2 & -3/2 \\ 3/2 & -1/3 & 3/2 \\ -3/2 & 3/2 & 2/3 \end{bmatrix}.$$

□

Example 6.12 (Symmetric total contractions). Let A and B be two homogeneous tensor homomorphisms, of second order, covariant and *symmetric*, that transform homogeneous contravariant tensors of first order (vectors), all of them defined over a linear space $V^n(K)$, and such that they satisfy the tensor equations

$$(a_{ij}^{\circ\circ} - kb_{ij}^{\circ\circ})u_\circ^i = 0; \quad \forall i,j \in I_n \tag{6.62}$$

$$(a_{ij}^{\circ\circ} - k'b_{ij}^{\circ\circ})v_\circ^i = 0; \quad \forall i,j \in I_n, \tag{6.63}$$

where $k, k' \in K$; $0 \neq k \neq k' \neq 0$; $\vec{u}, \vec{v} \in V^n(K)$; $\vec{u} \neq \vec{v}$.

218 6 Symmetric Homogeneous Tensors: Tensor Algebras

1. Show that all the following tensor total contractions are null:

$$a^{\circ\circ}_{ij}v^i_\circ v^j_\circ = b^{\circ\circ}_{ij}v^i_\circ v^j_\circ = 0.$$

2. Give the value of the constant k, as a function of the components of tensors A, B and $\vec{u}$.

3. Idem, of the constant k', as a function of the components of the tensors A, B and $\vec{v}$.

4. Prove that in reality k and k' are different eigenvalues and non-null, functions only of A and B.

Solution:

1. Executing the contracted tensor products of Equation (6.62) by the tensor $\vec{v}$, and of Equation (6.63) by the tensor $\vec{u}$, we have

$$(a^{\circ\circ}_{ij} - kb^{\circ\circ}_{ij})u^i_\circ v^j_\circ = 0; \quad \forall i, j \in I_n \tag{6.64}$$

$$(a^{\circ\circ}_{ij} - k'b^{\circ\circ}_{ij})v^i_\circ u^j_\circ = 0; \quad \forall i, j \in I_n \tag{6.65}$$

and exchanging indices i, j in (6.65) and subtracting (6.65) from (6.64) we have

$$(a^{\circ\circ}_{ij} - kb^{\circ\circ}_{ij} - a^{\circ\circ}_{ji} + k'b^{\circ\circ}_{ji})u^i_\circ v^j_\circ = 0.$$

Remembering that because of the symmetry $a^{\circ\circ}_{ji} = a^{\circ\circ}_{ij}$ and $b^{\circ\circ}_{ji} = b^{\circ\circ}_{ij}$ we can write

$$(k' - k)b^{\circ\circ}_{ij}u^i_\circ v^j_\circ = 0 \quad \Rightarrow \quad b^{\circ\circ}_{ij}u^i_\circ v^j_\circ = 0, \tag{6.66}$$

and finally, substituting (6.66) into (6.64), we have $a^{\circ\circ}_{ij}u^i_\circ v^j_\circ = 0$. Whence

$$a^{\circ\circ}_{ij}u^i_\circ v^j_\circ = b^{\circ\circ}_{ij}u^i_\circ v^j_\circ = 0.$$

2. From Equation (6.62), giving values to j we get the system of n equations

$$j = 1 \rightarrow k = \frac{a^{\circ\circ}_{i1}u^i_\circ}{b^{\circ\circ}_{i1}u^i_\circ} = \frac{a^{\circ\circ}_{11}u^1_\circ + a^{\circ\circ}_{21}u^2_\circ + \cdots + a^{\circ\circ}_{n1}u^n_\circ}{b^{\circ\circ}_{11}u^1_\circ + b^{\circ\circ}_{21}u^2_\circ + \cdots + b^{\circ\circ}_{n1}u^n_\circ}$$

$$j = 2 \rightarrow k = \frac{a^{\circ\circ}_{i2}u^i_\circ}{b^{\circ\circ}_{i2}u^i_\circ} = \frac{a^{\circ\circ}_{12}u^1_\circ + a^{\circ\circ}_{22}u^2_\circ + \cdots + a^{\circ\circ}_{n2}u^n_\circ}{b^{\circ\circ}_{12}u^1_\circ + b^{\circ\circ}_{22}u^2_\circ + \cdots + b^{\circ\circ}_{n2}u^n_\circ} \tag{6.67}$$

$$\cdots\cdots\cdots\cdots\cdots\cdots\cdots\cdots\cdots\cdots\cdots\cdots\cdots$$

$$j = n \rightarrow k = \frac{a^{\circ\circ}_{in}u^i_\circ}{b^{\circ\circ}_{in}u^i_\circ} = \frac{a^{\circ\circ}_{1n}u^1_\circ + a^{\circ\circ}_{2n}u^2_\circ + \cdots + a^{\circ\circ}_{nn}u^n_\circ}{b^{\circ\circ}_{1n}u^1_\circ + b^{\circ\circ}_{2n}u^2_\circ + \cdots + b^{\circ\circ}_{nn}u^n_\circ},$$

the solution of which is

$$k = \frac{a_{i1}^{\circ\circ}u_{\circ}^{i}}{b_{i1}^{\circ\circ}u_{\circ}^{i}} = \frac{a_{i2}^{\circ\circ}u_{\circ}^{i}}{b_{i2}^{\circ\circ}u_{\circ}^{i}} = \cdots = \frac{a_{in}^{\circ\circ}u_{\circ}^{i}}{b_{in}^{\circ\circ}u_{\circ}^{i}}. \tag{6.68}$$

3. Operating Equation (6.63) in a similar way, we obtain the solution

$$k' = \frac{a_{i1}^{\circ\circ}v_{\circ}^{i}}{b_{i1}^{\circ\circ}v_{\circ}^{i}} = \frac{a_{i2}^{\circ\circ}v_{\circ}^{i}}{b_{i2}^{\circ\circ}v_{\circ}^{i}} = \cdots = \frac{a_{in}^{\circ\circ}v_{\circ}^{i}}{b_{in}^{\circ\circ}v_{\circ}^{i}}. \tag{6.69}$$

4. The system (6.67) can be notated as

$$(a_{11}^{\circ\circ} - kb_{11}^{\circ\circ})u_{\circ}^{1} + (a_{21}^{\circ\circ} - kb_{21}^{\circ\circ})u_{\circ}^{2} + \cdots + (a_{n1}^{\circ\circ} - kb_{n1}^{\circ\circ})u_{\circ}^{n} = 0$$

$$(a_{12}^{\circ\circ} - kb_{12}^{\circ\circ})u_{\circ}^{1} + (a_{22}^{\circ\circ} - kb_{22}^{\circ\circ})u_{\circ}^{2} + \cdots + (a_{n2}^{\circ\circ} - kb_{n2}^{\circ\circ})u_{\circ}^{n} = 0$$

$$\tag{6.70}$$

$$\cdots\cdots\cdots\cdots\cdots\cdots\cdots\cdots\cdots\cdots\cdots\cdots\cdots\cdots\cdots$$

$$(a_{1n}^{\circ\circ} - kb_{1n}^{\circ\circ})u_{\circ}^{1} + (a_{2n}^{\circ\circ} - kb_{2n}^{\circ\circ})u_{\circ}^{2} + \cdots + (a_{nn}^{\circ\circ} - kb_{nn}^{\circ\circ})u_{\circ}^{n} = 0,$$

a homogeneous system of n equations, the compatibility of which requires a null determinant of the coefficient matrix:

$$|A^{t} - kB^{t}| = 0 \quad \rightarrow \quad |A - kB| = 0,$$

which, if matrices A or B are regular, leads to

$$|A^{-1}||A - kB| = 0 \quad \rightarrow \quad |I_{n} - kA^{-1}B| = 0 \quad \rightarrow \quad \left|\frac{1}{k}I_{n} - A^{-1}B\right| = 0$$

or

$$|A - kB||B^{-1}| = 0 \quad \rightarrow \quad |AB^{-1} - kI_{n}| = 0,$$

where both are characteristic polynomials, showing that k is an eigenvalue. If A and B were singular tensors, we should analyze the system (6.70) by removing the redundant equations.

Obviously with k', starting from equation (6.63), one arrives at exactly the same conclusions.

Also the k' are eigenvalues (with $k \neq k'$) of the same characteristic polynomials.

$\square$

Example 6.13 (Matrix algorithms do not keep tensor symmetry). Let A and B be the contravariant and symmetric homogeneous tensors, of order $r = 2$

$$A \equiv [a^{ij}_{oo}] = \begin{bmatrix} a_1 & x_1 & y_1 \\ x_1 & b_1 & z_1 \\ y_1 & z_1 & c_1 \end{bmatrix} \quad \text{and} \quad B \equiv [b^{ij}_{oo}] = \begin{bmatrix} a_2 & x_2 & y_2 \\ x_2 & b_2 & z_2 \\ y_2 & z_2 & c_2 \end{bmatrix}.$$

Obtain the matrix $S = A \otimes B - B \otimes A$ and make some comments about the resulting tensor.

Solution: Based on the following determinants:

$$(1) = \begin{vmatrix} a_1 & a_2 \\ x_1 & x_2 \end{vmatrix}; \quad (2) = \begin{vmatrix} a_1 & a_2 \\ y_1 & y_2 \end{vmatrix}; \quad (3) = \begin{vmatrix} a_1 & a_2 \\ z_1 & z_2 \end{vmatrix}; \quad (4) = \begin{vmatrix} a_1 & a_2 \\ b_1 & b_2 \end{vmatrix};$$

$$(5) = \begin{vmatrix} a_1 & a_2 \\ c_1 & c_2 \end{vmatrix}; \quad (6) = \begin{vmatrix} b_1 & b_2 \\ c_1 & c_2 \end{vmatrix}; \quad (7) = \begin{vmatrix} x_1 & x_2 \\ y_1 & y_2 \end{vmatrix}; \quad (8) = \begin{vmatrix} x_1 & x_2 \\ b_1 & b_2 \end{vmatrix};$$

$$(9) = \begin{vmatrix} x_1 & x_2 \\ z_1 & z_2 \end{vmatrix}; \quad (10) = \begin{vmatrix} x_1 & x_2 \\ c_1 & c_2 \end{vmatrix}; \quad (11) = \begin{vmatrix} y_1 & y_2 \\ b_1 & b_2 \end{vmatrix}; \quad (12) = \begin{vmatrix} b_1 & b_2 \\ z_1 & z_2 \end{vmatrix};$$

$$(13) = \begin{vmatrix} y_1 & y_2 \\ z_1 & z_2 \end{vmatrix}; \quad (14) = \begin{vmatrix} y_1 & y_2 \\ c_1 & c_2 \end{vmatrix}; \quad (15) = \begin{vmatrix} z_1 & z_2 \\ c_1 & c_2 \end{vmatrix};$$

we have

$$S = A \otimes B - B \otimes A = \left[\begin{array}{ccc|ccc|ccc} 0 & (1) & (2) & -(1) & 0 & (7) & -(2) & -(7) & 0 \\ (1) & (4) & (3) & 0 & (8) & (9) & -(7) & (11) & (13) \\ (2) & (3) & (5) & (7) & (9) & (10) & 0 & (13) & (14) \\ \hline -(1) & 0 & (7) & -(4) & -(8) & -(11) & -(3) & -(9) & -(13) \\ 0 & (8) & (9) & -(8) & 0 & (12) & -(9) & -(12) & 0 \\ (7) & (9) & (10) & -(11) & (12) & (6) & -(13) & 0 & (15) \\ \hline -(2) & -(7) & 0 & -(3) & -(9) & -(13) & -(5) & -(10) & -(14) \\ -(7) & (11) & (13) & -(9) & -(12) & 0 & -(10) & -(6) & -(15) \\ 0 & (13) & (14) & -(13) & 0 & (15) & -(14) & -(15) & 0 \end{array} \right].$$

The required comment could be: S is a symmetric matrix and a second-order symmetric tensor. However, *it is not* a contravariant *symmetric* fourth-order tensor, which is what it should be, as the difference of two contravariant fourth-order tensors.

□

6.8 Exercises

6.1. Consider a tensor of $p = 3$ contravariant indices, $q = 2$ covariant indices and covariance $(1-2-1-1-2)$, built over a linear space of dimension $n = 4$, that is symmetric with respect to all the contravariant indices.

1. Obtain the total number of components (N_1).
2. Obtain the maximum number of strict components (N_2).
3. Obtain the number of strict components with repeated contravariant index (N_3).
4. Obtain the number of strict components for which the contravariant indices have two repeated indices (N_4).
5. Obtain the number of strict components with all contravariant coordinates indices being different (N_5).
6. Obtain the number of components with repeated covariant indices (N_6).

6.2. Consider a symmetric contravariant tensor $t_{\circ\circ\circ}^{\alpha\beta\gamma}$ of order $(r = 3)$, over the linear space $\mathbb{R}^4$ (dimension $n = 4$).

1. Obtain the number of total components (N_1) and the number of strict components (N_2).
2. Give values to the strict components in axiomatic order by means of the Roman letter in order, so that they correspond to:

$$a\,b\,c\,d\,e\,f\,g\,h\,i\,j\,k\,\ell\,m\,n\,o\,p\,q\,r\,s\,t\,u.$$

3. Execute a matrix representation of the general pattern of this model of tensors, with the given values.

6.3. Determine if the following tensor list over the linear space $V^n(\mathbb{R})$, presents total or partial symmetries, specifying in the second case with respect to which indices:

1. $t_{\circ\circ\gamma\delta}^{\alpha\beta\circ\circ} = \begin{vmatrix} \alpha & \gamma \\ \delta & \beta \end{vmatrix}$.

2. $t_{\alpha\circ\gamma}^{\circ\beta\circ} = \alpha + \beta + \gamma - \alpha\beta\gamma$.

3. $t_{\circ\circ\circ}^{\alpha\beta\gamma} = \alpha\beta - \gamma$.

4. $t_{\circ\beta\circ\delta\circ}^{\alpha\circ\gamma\circ\lambda} = \alpha + \beta + \gamma + \delta + \lambda - \beta\gamma$.

5. $t_{\alpha\beta\gamma\circ\circ}^{\circ\circ\circ\delta\lambda} = \alpha\beta\gamma(1 + \delta\lambda)$.

6. Obtain the number of strict components of each of the above tensors.

6.4. In the linear space $V^3 \otimes V^3 \otimes V^3(\mathbb{R})$ a tensor T is given by its usual matrix representation

$$[t_{\circ\circ\circ}^{\alpha\beta\gamma}] = \begin{bmatrix} -1 & 0 & 0 \\ 0 & 3 & 2 \\ 1 & 2 & -1 \\ \hline 3 & 5 & 1 \\ 0 & 1 & 0 \\ 2 & 2 & -2 \\ \hline 1 & 1 & 4 \\ 0 & 0 & 1 \\ 2 & 0 & 1 \end{bmatrix}.$$

1. Using the permutation tensors, $P_{(1)}, P_{(2)}, P_{(3)}, P_{(4)}$ and $P_{(5)}$ established in Section 5.7.2, extract the isomer tensors of T in the clasic matrix representation using the notation U, V, W, R and S.
2. Obtain the *Symmetrized* tensor associated with tensor T, which will be denoted by $S(T)$.
3. Execute the triple contraction of all indices in $S(T)$, obtaining the invariant k.
4. A change-of-basis is performed in the linear space $V^3(\mathbb{R})$ of associated matrix $C = \begin{bmatrix} 1 & 0 & 1 \\ 0 & 1 & 1 \\ 1 & 1 & 0 \end{bmatrix}$. Obtain the symmetrized tensor $S(T)$ giving its matrix representation.
5. We repeat the triple contraction over $\hat{S}(T)$, to obtain $\hat{k}$.
6. Is $k = \hat{k}$? If it remains invariant, would it be a "particular tensor" contraction of $S(T)$?
7. Answer questions 2 and 4 using the computer for the symmetrization.

6.5. 1. We wish to know the number of different strict components of a tensor T of order $(r = 3)$ totally contravariant and symmetric, built over $V^3(\mathbb{R})$.
2. It is known that when executing over this tensor all possible contractions, $C(\alpha\beta)$, $C(\alpha\gamma)$ and $C(\beta\gamma)$, one always obtains the null system of scalars Ω. According to this, how many of the strict components of T are really independent?
3. Extend the previous question to the general case: $T \in \overset{r}{\underset{1}{\otimes}} V^n(\mathbb{R})$ symmetric, and such that all simple contractions lead to Ω.

6.6. Let $A, B \in V^3 \otimes V^3(\mathbb{R})$ be two symmetric contravariant tensors of second order $(r = 2)$, the matrix representation of which is

$$[a^{\alpha\beta}_{\circ\circ}] = \begin{bmatrix} 0 & 4 & 2 \\ 4 & 9 & 7 \\ 2 & 7 & 12 \end{bmatrix} \quad \text{and} \quad [b^{\alpha\beta}_{\circ\circ}] = \begin{bmatrix} -1 & 2 & 0 \\ 2 & 3 & 1 \\ 0 & 1 & 4 \end{bmatrix}.$$

1. Obtain the matrix representation of its "interior symmetric" tensor product $P = A \otimes_S B$.
2. Find the tensors $T = A \otimes_S (A \bullet B)$ and $U = B \otimes_S (A \bullet B)$.
3. Find the tensor $W = (A + B) \otimes_S (A \bullet B)$.
4. Check the distributivity of the operator "$\otimes_S$" with respect to "+", knowing that $W = T + U$.

6.7. Consider a tensor $T \in V^3 \otimes V^3_* \otimes V^3_*(\mathbb{R})$ defined over a linear space $V^3(\mathbb{R})$ referred to the basis $\{\vec{e}_\alpha\}$, the matrix representation of which is

$$[t^{\alpha\,\circ\,\circ}_{\circ\,\beta\gamma}] = \begin{bmatrix} 1 & -1 & 4 \\ -1 & -4 & 0 \\ 4 & 0 & 3 \\ \hline -1 & -4 & 0 \\ -4 & 2 & -2 \\ 0 & -2 & -3 \\ \hline 4 & 0 & 3 \\ 0 & -2 & -3 \\ 2 & -3 & 5 \end{bmatrix},$$

where α is the block row, β is the row of each block, and γ is the column of each block.

Let $\{\vec{e}_i\}$ be a basis of $V^3(\mathbb{R})$ related to the initial one by the change-of-basis matrix:

$$C = \begin{bmatrix} 0 & 1 & 1 \\ 1 & 1 & 1 \\ 1 & 1 & 0 \end{bmatrix}.$$

1. Study the possible "partial" symmetries of the components of T in the initial basis.
2. Are there tensor symmetries, i.e., intrinsic, among those analyzed in the previous question?
3. Is T a totally symmetric tensor?
4. Perform the change-of-basis and give a matrix representation of T in the new basis: $[\hat{t}^{i\,\circ\,\circ}_{\circ\,jk}]$.
5. Answer again questions 1, 2 and 3.

7

Anti-symmetric Homogeneous Tensors, Tensor and Inner Product Algebras

7.1 Introduction

This chapter is devoted to anti-symmetric homogeneous tensors that are initially defined.

Because of their anti-symmetry, the number of data components required for their definition can be substantially reduced, arising from the concept of strict components (a minimum set of data), which is explained and a formula given for determining this number.

The problem of generating anti-symmetric tensors from a given tensor, and the tensor nature of anti-symmetry are also discussed.

Next, the anti-symmetrization operator is extended to the case of mixed tensors, and a new interior anti-symmetric product for exclusive use of interior anti-symmetric tensor algebras is introduced.

The chapter ends with some illustrative examples that clarify the established concepts.

7.2 Anti-symmetric systems of scalar components

Definition 7.1 (Anti-symmetric systems of scalar components). *Consider a system of scalar components $S(\alpha_1, \alpha_2, \ldots, \alpha_r)$ of order r, defined with respect to a certain linear space $V^n(K)$ in a certain basis. We say that such a system is anti-symmetric with respect to the indices (i, j), if the following holds:*

$$\left\{ \begin{array}{l} s(\alpha_1, \alpha_2, \ldots, \alpha_i, \ldots, \alpha_j, \ldots, \alpha_r) = \rho \\ s(\alpha_1, \alpha_2, \ldots, \alpha_j, \ldots, \alpha_i, \ldots, \alpha_r) = -\rho \end{array} \right\} ; \rho \in K; \alpha_h \in I_n = \{1, 2, \ldots, n\} \subset \mathbb{N} \tag{7.1}$$

that is, the values ρ of the cited components are of opposing sign.

7.2.1 Anti-symmetric systems with respect to an index subset

The above concept can be extended to k indices $(2 \le k \le r)$, of the total set I_r of indices. Let

$$I_k = \{i, j, \ldots, k\} \subset I_r = \{1, 2, \ldots, i, \ldots, j, \ldots, r\} \in I_n \qquad (7.2)$$

be the set of indices that satisfy Definition 7.1. We emphasize now two concepts that will be very important throughout this chapter.

The first refers to the consequences of Definition 7.1 if the set I_k, that is, the index set in the definition, were to have *some repeated index*. If the index i were to appear repeated in I_k, this would mean that Definition 7.1 should satisfy

$$\left\{ \begin{array}{l} s(\alpha_1, \alpha_2, \ldots, \alpha_i, \ldots, \alpha_i, \ldots, \alpha_r) = \rho \\ s(\alpha_1, \alpha_2, \ldots, \alpha_i, \ldots, \alpha_i, \ldots, \alpha_r) = -\rho \end{array} \right\}$$

when exchanging them, and because of the identity of the left-hand members, we will have $\rho = -\rho$, and since $\rho \in K$ (field of characteristic $\ne 2$), this implies $\rho = 0$, and

$$s(\alpha_1, \alpha_2, \ldots, \alpha_i, \ldots, \alpha_i, \ldots, \alpha_r) = 0. \qquad (7.3)$$

Consequently, in this type of tensor we only study the components set with anti-symmetric indices that are different, because one knows that otherwise, the component is *null*.

The second important concept motivated by Definition 7.1 is the concept of "evenness of a permutation", of a subset of *different* natural numbers.

Definition 7.2 (Permutation). *Consider a collection of certain different natural numbers, assumed sorted in the natural ordering. An alteration of the given ordering, results in a new grouping that is called a "permutation" of the initial one.*

Definition 7.3 (Transposition). *If we exchange two elements of the initial set, keeping all the remaining elements in their initial positions, the permutation receives the special name of "transposition".*

For example, in Formula (7.1), the second collection is a transposition of the first one.

If several elements of the initial collection are altered, the resulting permutation has τ transpositions. The total number τ of transpositions that are associated with each permutation, can be counted as indicated in the following example.

Example 7.1 (Transpositions). Consider the initial collection $P_0 = \{2, 3, 5, 6, 7, 8\}$ and the permutation $P = \{8, 3, 7, 5, 2, 6\}$. We wish to find the corresponding value of τ.

Comparing P with P_0, we modify P as follows:

1. We proceed with a transposition of element 8 with element 2, to position element 2 in the first position, since this is its position in P_0, and we start counting the number of transpositions to be executed ($\tau = 1$). We are in permutation $P_1 = \{2, 3, 7, 5, 8, 6\}$.
2. Now we proceed with the following place; since in P_1 element 3 is in its place (remember P_0), we move to the third place, and proceed to exchange element 7 with element 5, and we already have ($\tau = 2$) transpositions. We are in the permutation $P_2 = \{2, 3, 5, 7, 8, 6\}$.
3. Element 6 is not yet in the fourth place, so that we exchange element 7 with element 6; and we have ($\tau = 3$) transpositions. We are in permutation $P_3 = \{2, 3, 5, 6, 8, 7\}$.
4. Finally, we exchange elements 8 and 7, and have ($\tau = 4$) transpositions. We are in permutation $P_4 = \{2, 3, 5, 6, 7, 8\} \equiv P_0$.

Since we have arrived at P_0 in an organized way (that was our aim), we say that P is the "product of $\tau = 4$ transpositions" of the collection P_0.

Next, we assign the permutation P a sign, according to the formula:

$$P \rightarrow \tau(\text{number of transpositions with respect to } P_0) \rightarrow \text{sign} = (-1)^\tau \quad (7.4)$$

Obviously, the collection P_0, as the main permutation, also has a sign: $(-1)^0 = +1$.

If the sign of $(-1)^\tau$ is $+$, we say that the permutation P is of class *even*. If the sign of $(-1)^\tau$ is $-$, we say that the permutation P is of class *odd*.

We point out again that, obviously, have classified only the permutations of *different* anti-symmetric indices.

Returning to Relation (7.2), and summarizing all the previous work, we propose the following:

- If $h \in I_k$ and the component *has* that index repeated we have

$$s(\alpha_1, \alpha_2, \ldots, \alpha_h, \ldots, \alpha_h, \ldots, \alpha_r) = 0. \quad (7.5)$$

- If the analyzed components *have no* repeated indices $h \in I_k$, then their $k!$ components take the value

$$s(\alpha_1, \alpha_2, \ldots, \alpha_k, \ldots, \alpha_j, \ldots, \alpha_i, \ldots, \alpha_r) = (-1)^\tau \rho, \quad (7.6)$$

that is:
$+\rho$ if the permutation of I_k is even ($\tau = $ even)
$-\rho$ if the permutation of I_k is odd ($\tau = $ odd)
The result is that $k!/2$ components take the value ρ and the other $k!/2$ take the value $-\rho$, which completes the $k!$ *analyzed* components (the reader must be aware that the indices α_h can vary with $h \notin I_k$).

□

7.2.2 Anti-symmetric systems. Total anti-symmetry

If $I_k \equiv I_r$, we say that the system is totally anti-symmetric, and better, simply an anti-symmetric system of scalars.

We give now a particular example of the last system $(n = 4, r = 3)$, grouping the system (α, β, γ) by components with the same scalar $\rho \in K$ in nine classes, according to the axiomatic ordering. The number of components is $\sigma = n^r = 4^3 = 64$, that are grouped as:

$$\text{Class } 1: s(1\ 1\ 1) = s(1\ 1\ 2) = s(1\ 2\ 1) = \cdots = s(1\ 2\ 2)$$
$$= \cdots = \cdots s(2\ 2\ 2) = s(4\ 4\ 4) = 0.$$
$$\text{Class } 2: s(1\ 2\ 3) = s(2\ 3\ 1) = s(3\ 1\ 2) = \rho_1.$$
$$\text{Class } 3: s(1\ 3\ 2) = s(2\ 1\ 3) = s(3\ 2\ 1) = -\rho_1.$$
$$\text{Class } 4: s(1\ 2\ 4) = s(2\ 4\ 1) = s(4\ 1\ 2) = \rho_2.$$
$$\text{Class } 5: s(1\ 4\ 2) = s(2\ 1\ 4) = s(4\ 2\ 1) = -\rho_2.$$
$$\text{Class } 6: s(1\ 3\ 4) = s(3\ 4\ 1) = s(4\ 1\ 3) = \rho_3.$$
$$\text{Class } 7: s(1\ 4\ 3) = s(3\ 1\ 4) = s(4\ 3\ 1) = -\rho_3.$$
$$\text{Class } 8: s(2\ 3\ 4) = s(3\ 4\ 2) = s(4\ 2\ 3) = \rho_4.$$
$$\text{Class } 9: s(2\ 4\ 3) = s(3\ 2\ 4) = s(4\ 3\ 2) = -\rho_4.$$

The total number of components is:

- Of repeated indices (null): $VR_{n,r} - V_{n,r} = n^r - \frac{n!}{(n-r)!} = 4^3 - \frac{4!}{(4-3)!} = 64 - \frac{4!}{1!} = 40$.
- Of different indices: $V_{n,r} = \frac{n!}{(n-r)!} = \frac{n!}{(4-3)!} = \frac{4!}{1!} = 24$.

Total: $40 + 24 = 64 \equiv \sigma$.

An evident conclusion is that there cannot be anti-symmetric systems of scalars with $r > n$, because this would force *all* components to have repeated indices, and then *all* components would be null.

7.3 Strict components of an anti-symmetric system and with respect to an index subset

Definition 7.4 (Strict components of an anti-symmetric system). *In the systems with anti-symmetry (partial or total), we will define strict components to be the maximum number of non-null components of different absolute value (we ignore the sign) that can be present in the system, sorted according to the axiomatic criteria.*

In the example examined in Section 7.2.2 the strict components are:

$$\{s(1\ 2\ 3), s(1\ 2\ 4), s(1\ 3\ 4), s(2\ 3\ 4)\}.$$

This concept is different from the number of different components (as some books propose), which in our case would be nine, and not four, that is, the set

$$\{0, \rho_1, -\rho_1, \rho_2, -\rho_2, \rho_3, -\rho_3, \rho_4, -\rho_4\}.$$

7.3.1 Number of strict components of an anti-symmetric system with respect to an index subset

Next, we discuss the number of strict components of an anti-symmetric system with respect to an index subset.

Consider the system $S(\alpha_1, \alpha_2, \ldots, \alpha_i, \ldots, \alpha_j, \ldots, \alpha_k, \ldots, \alpha_r)$ defined with respect to the linear space $V^n(K)$ and let k be the number of indices of the subset $I_k = \{i, j, \ldots, k\}$ the components of which are anti-symmetric.

The number of valid realizations for the non anti-symmetric $(r-k)$ indices is the following number of variations with repetition:

$$VR_{n,r-k} = n^{r-k}. \tag{7.7}$$

The number of strict realizations for the k anti-symmetric indices is the number of combinations

$$C_{n,k} = \binom{n}{k}, \tag{7.8}$$

thus, the total number of possibilities for the strict components is

$$Cn, k \cdot VR_{n,r-k} = \binom{n}{k} n^{r-k}. \tag{7.9}$$

Obviously, the total number of components is, as always, the dimension of the tensor space: $\sigma = n^r$.

Finally, the maximum number of different values is

$$d = 2 \binom{n}{k} n^{r-k} + 1, \tag{7.10}$$

which is double (7.9) due to the reverse sign, plus one (the zero value). In the example of Section 7.2.2, it is the number of classes $(n = 4, r = k = 3)$.

7.3.2 Number of strict components of an anti-symmetric system

Obviously, it is enough to let $k = r$ in Formula (7.9), because now the r indices of the system present anti-symmetry.

The number of strict components of the system of scalars with total anti-symmetry of indices is

$$C_{n,r} = \binom{n}{r}. \tag{7.11}$$

Number of anti-symmetric components associated with each strict component in an anti-symmetric system

Evidently, this number is the number of permutations of the r different indices of the strict component. Thus, we have

$$P_r = r! \tag{7.12}$$

components associated with each strict component; half with sign $+$ (the even permutations) and the other half with sign $-$ (the odd permutations).

The maximum number of non-null components in an anti-symmetric system is

$$C_{n,r} \cdot P_r = \binom{n}{r} \cdot r! = \frac{n!}{(n-r)!r!} \cdot r! = \frac{n!}{(n-r)!} = V_{n.r}, \tag{7.13}$$

i.e., the variations of n elements taken "r in r". Then, the number of null components is

$$VR_{n,r} - V_{n,r} = n^r - \frac{n!}{(n-r)!}. \tag{7.14}$$

7.4 Tensors with anti-symmetries: Tensors with branched anti-symmetry; anti-symmetric tensors

The previously established anti-symmetry criteria for systems of scalar components are directly applicable to homogeneous tensors, with the following limitations, to be justified later:

1. Tensors with anti-symmetry with respect to the subset of indices I_k ($2 \le k \le r$) must present the subset I_k inside the set of the p contravariant indices or inside the set of the q covariant indices ($p + q = r$), that is, all indices in I_k must be of the *same valency*. Some authors define these tensors as "tensors with branched anti-symmetry".

   This is a sufficient condition for the anti-symmetry with respect to the indices of I_k to remain invariant under changes of basis of tensor nature, but it is *not necessary*, as we shall see for certain particular tensor systems, that can present stable anti-symmetries between indices of different valency.

2. By definition, anti-symmetric tensors (with respect to all their indices), are pure tensors, that is, totally contravariant or totally covariant. Some books even give them a special notation. Certain authors call these tensors "multivectors", "r-vectors", "polivectors", etc.

Consider a homogeneous tensor of order r that is totally contravariant, notated $t^{\alpha_1 \alpha_2 \cdots \alpha_r}_{\circ \ \circ \ \cdots \ \circ}$. If it is anti-symmetric, it can be expressed as

$$t^{(\alpha_1 \alpha_2 \cdots \alpha_r)}_{\circ \ \circ \ \cdots \ \circ} \text{ with } \alpha_1 \le \alpha_2 \le \cdots \le \alpha_r, \tag{7.15}$$

which is precisely the adopted notation for its strict components.

Similarly, we use the notation $t_{(\alpha_1 \alpha_2 \cdots \alpha_r)}^{\circ \ \circ \ \cdots \ \circ}$ for the covariant anti-symmetric tensor.

The number of strict components can be obtained with the general formulas established for the systems of scalars.

We recall here the formulas applicable to pure tensors (totally contravariant or covariant) of order r, with k anti-symmetric indices ($2 \leq k \leq r$), over $V^n(K)$. The number of strict components is

$$\sigma' = \binom{n}{k} n^{r-k} \tag{7.16}$$

and the maximum number of different values is

$$d = \binom{n}{k} n^{r-k} + 1. \tag{7.17}$$

If the pure tensor is anti-symmetric, then $k = r$, and the number of strict components is

$$\sigma' = \binom{n}{r} \tag{7.18}$$

and the maximum number of different values is

$$d = 2\binom{n}{r} + 1. \tag{7.19}$$

Finally, we point out that some authors extend the total anti-symmetry to mixed tensors; in such cases it is assumed that the anti-symmetry of the p contravariant indices and the anti-symmetry of the q covariant indices are *independent*. Then, the expression for the strict components becomes

$$t^{(\alpha_1 \alpha_2 \cdots \alpha_p) \ \circ \quad \circ \ \cdots \ \circ}_{\quad \circ \ \circ \ \cdots \ \circ \ (\alpha_{p+1} \alpha_{p+2} \cdots \alpha_{p+q})} \tag{7.20}$$

with $\alpha_1 < \alpha_2 < \cdots < \alpha_p$; $\alpha_{p+1} < \alpha_{p+2} < \cdots < \alpha_{p+q}$; $p + q = r$.

So, for the case of a mixed tensor over the linear space $\mathbb{R}^3$ of order $r = 4$, contra-contra-cova-covariant ($p = q = 2$) anti-symmetric the strict components would be

$$t^{(12) \circ \circ}_{\circ \circ (12)} = t^{21 \circ \circ}_{\circ \circ 21} = a; \quad t^{12 \circ \circ}_{\circ \circ 21} = t^{21 \circ \circ}_{\circ \circ 12} = -a;$$

$$t^{(12) \circ \circ}_{\circ \circ (13)} = t^{21 \circ \circ}_{\circ \circ 31} = b; \quad t^{12 \circ \circ}_{\circ \circ 31} = t^{21 \circ \circ}_{\circ \circ 13} = -b;$$

$$t^{(12) \circ \circ}_{\circ \circ (23)} = t^{21 \circ \circ}_{\circ \circ 32} = c; \quad t^{12 \circ \circ}_{\circ \circ 32} = t^{21 \circ \circ}_{\circ \circ 23} = -c;$$

$$t^{(13) \circ \circ}_{\circ \circ (12)} = t^{31 \circ \circ}_{\circ \circ 21} = d; \quad t^{13 \circ \circ}_{\circ \circ 21} = t^{31 \circ \circ}_{\circ \circ 12} = -d;$$

$$t^{(13) \circ \circ}_{\circ \circ (13)} = t^{31 \circ \circ}_{\circ \circ 31} = e; \quad t^{13 \circ \circ}_{\circ \circ 31} = t^{31 \circ \circ}_{\circ \circ 13} = -e;$$

$$t^{(13) \circ \circ}_{\circ \circ (23)} = t^{31 \circ \circ}_{\circ \circ 32} = f; \quad t^{13 \circ \circ}_{\circ \circ 32} = t^{31 \circ \circ}_{\circ \circ 23} = -f;$$

$$t^{(23) \circ \circ}_{\circ \circ (12)} = t^{32 \circ \circ}_{\circ \circ 21} = g; \quad t^{23 \circ \circ}_{\circ \circ 21} = t^{32 \circ \circ}_{\circ \circ 12} = -g;$$

$$t^{(23) \circ \circ}_{\circ \circ (13)} = t^{32 \circ \circ}_{\circ \circ 31} = h; \quad t^{23 \circ \circ}_{\circ \circ 31} = t^{32 \circ \circ}_{\circ \circ 13} = -h;$$

$$t^{(23) \circ \circ}_{\circ \circ (23)} = t^{32 \circ \circ}_{\circ \circ 32} = i; \quad t^{23 \circ \circ}_{\circ \circ 32} = t^{32 \circ \circ}_{\circ \circ 23} = -i;$$

with a total of $\sigma = n^r = 3^4 = 81$ components, of which 9 are strict, those in the first column of the table. The maximum number of components that are different from zero is $4 \times 9 = 36$.

It is convenient in this section, in which we consider *mixed tensors*, to clarify the number of strict components when the branching is mixed, that is, it involves not only one part of the p contravariant indices (k_1, of the p indices, are anti-symmetric), but also at the same time to one part of the covariant indices (k_2, of the q indices, are anti-symmetric).

In such a case we have

$$2 \leq k_1 \leq p; \quad 2 \leq k_2 \leq q; \quad p + q = r \text{ (order of the mixed tensor)}$$

and the number of strict components is

$$\sigma' = \binom{n}{k_1} n^{p-k_1} \cdot \binom{n}{k_2} n^{q-k_2}; \tag{7.21}$$

or

$$\sigma' = \binom{n}{k_1} \cdot \binom{n}{k_2} n^{r-(k_1+k_2)}.$$

If the mixed tensor is anti-symmetric, then $k_1 = p$ and $k_2 = q$; thus, the number of strict components becomes

$$\sigma' = \binom{n}{p} \cdot \binom{n}{q}, \tag{7.22}$$

which is the formula used in the mixed example ($n = 3, p = q = 2$), in the first answer. The maximum number of components different from zero is

$$c = \sigma' \cdot p! \cdot q! = \binom{n}{p} \cdot \binom{n}{q} p! q!, \tag{7.23}$$

which is the formula used in the mixed example, in its second answer.

7.4.1 Generation of anti-symmetric tensors

By combining decomposable anti-symmetric tensors

We remind the reader again of the beginning of Section 2.2 where we have given the bases for the tensor product of vectors concept. To center our objective, we assume that we are in the homogeneous tensor space $V^n \otimes V^n \otimes V^n(K)$ of order $r = 3$, and that we choose three arbitrary vectors and *in this order* $\vec{u}, \vec{v}, \vec{w}, \in V^n(K)$ and that we consider the algebraic sum of *all* tensor products that can be executed with them as factors, endowing to each product the sign ($+$ or $-$) depending on the factor $(-1)^\tau$, where τ is the number of transpositions of the permutation of factors $\vec{u} \otimes \vec{v} \otimes \vec{w}$ that are being considered.

From this criterion, the following tensor, a linear combination of the composed tensors, arises:

$$t^{ijk}_{\circ\circ\circ}\vec{e}_i \otimes \vec{e}_j \otimes \vec{e}_k = \sum_{1}^{6}(-1)^{\tau}\vec{u} \otimes \vec{v} \otimes \vec{w} = (-1)^0 \vec{u} \otimes \vec{v} \otimes \vec{w} + (-1)^2 \vec{v} \otimes \vec{w} \otimes \vec{u}$$
$$+ (-1)^2\vec{w} \otimes \vec{u} \otimes \vec{v} + (-1)^1 \vec{v} \otimes \vec{u} \otimes \vec{w} + (-1)^1\vec{w} \otimes \vec{v} \otimes \vec{u}$$
$$+(-1)^1\vec{u} \otimes \vec{w} \otimes \vec{v},$$

which operating, leads to

$$t^{ijk}_{\circ\circ\circ}\vec{e}_i \otimes \vec{e}_j \otimes \vec{e}_k = (\vec{u} \otimes \vec{v} \otimes \vec{w} + \vec{v} \otimes \vec{w} \otimes \vec{u} + \vec{w} \otimes \vec{u} \otimes \vec{v}) - (\vec{v} \otimes \vec{u} \otimes \vec{w}$$
$$+ \vec{w} \otimes \vec{v} \otimes \vec{u} + \vec{u} \otimes \vec{w} \otimes \vec{v}). \tag{7.24}$$

Taking into account that

$$\vec{u} = x^i\vec{e}_i; \quad \vec{v} = y^j\vec{e}_j; \quad \vec{w} = z^k\vec{e}_k; \quad \text{with } (\vec{u} \neq \vec{v} \neq \vec{w})$$

and based on Formula (6.12), we have

$$\vec{u} \otimes \vec{v} \otimes \vec{w} = a^{ijk}_{\circ\circ\circ}\vec{e}_i \otimes \vec{e}_j \otimes \vec{e}_k, \text{ with } a^{ijk}_{\circ\circ\circ} = x^i y^j z^k; \quad \forall \vec{u}, \vec{v}, \vec{w} \in V^n(K). \tag{7.25}$$

Substituting (7.25) for each summand into (7.24), we get

$$t^{ijk}_{\circ\circ\circ}\vec{e}_i \otimes \vec{e}_j \otimes \vec{e}_k = [(x^i y^j z^k + x^j y^k z^i + x^k y^i z^j) - (x^j y^i z^k + x^k y^j z^i + x^i y^k z^j)]\vec{e}_i \otimes \vec{e}_j \otimes \vec{e}_k,$$

which for $n = r = 3$ requires (after identification)

$$t^{ijk}_{\circ\circ\circ} = \begin{vmatrix} x^i & y^i & z^i \\ x^j & y^j & z^j \\ x^k & y^k & z^k \end{vmatrix}; \quad \forall i, j, k \in \{1, 2, 3\}. \tag{7.26}$$

Formula (7.26) totally declares how the components, created by the established definition of the new tensor, are. The first thing to be observed is that if two or more indices of the set $\{i, j, k\}$ are repeated, we must have $t^{ijk}_{\circ\circ\circ} = 0$, because the matrix has identical rows. Thus, the non-null components are those generated by indices $i \neq j \neq k$.

The first triplet of distinct indices is $\{1, 2, 3\}$, leading to the component

$$t^{(123)}_{\circ\circ\circ} = \begin{vmatrix} x^1 & y^1 & z^1 \\ x^2 & y^2 & z^2 \\ x^3 & y^3 & z^3 \end{vmatrix} \equiv |U \quad V \quad W| = \Delta. \tag{7.27}$$

The permutations $\{2, 3, 1\}, \{3, 1, 2\}, \{3, 2, 1\}$, etc. lead to the alteration of the rows of Δ, and as a consequence, the sign will change or not, but the absolute value will be the same. The conclusion is that

$$t^{ijk}_{\circ\circ\circ} = 0 \text{ if two or more indices are repeated.}$$

$$t^{(123)}_{\circ\ \circ\ \circ} = t^{231}_{\circ\circ\circ} = t^{312}_{\circ\circ\circ} = \Delta. \tag{7.28}$$

$$t^{213}_{\circ\circ\circ} = t^{321}_{\circ\circ\circ} = t^{132}_{\circ\circ\circ} = -\Delta.$$

The previous classification clearly declares an anti-symmetric homogeneous contravariant tensor of $\sigma = n^r = 3^3 = 27$ components, with $\sigma' = \binom{n}{r} = \binom{3}{3} = 1$ strict component (of value Δ).

The maximum number of *non-null* components, Formula (7.13), is

$$V_{n,r} = \frac{n!}{(n-r)!} = \frac{3!}{0!} = 6,$$

such that half of them are of value $+\Delta$, and the other half $-\Delta$. Then, $n^r - V_{n,r} = n^r - \frac{n!}{(n-r)!} = 27 - 6 = 21$ components are null.

We generate a set of *anti-symmetric tensors* per each triplet of selected vectors $(\vec{u}, \vec{v}, \vec{w})$, such as can be observed in the particular case being analyzed. The matrix representation of the studied tensor is

$$[t^{ijk}_{\circ\circ\circ}] =
\begin{bmatrix}
t^{111}_{\circ\circ\circ} & t^{112}_{\circ\circ\circ} & t^{113}_{\circ\circ\circ} \\
t^{121}_{\circ\circ\circ} & t^{122}_{\circ\circ\circ} & t^{123}_{\circ\circ\circ} \\
t^{131}_{\circ\circ\circ} & t^{132}_{\circ\circ\circ} & t^{133}_{\circ\circ\circ} \\
t^{211}_{\circ\circ\circ} & t^{212}_{\circ\circ\circ} & t^{213}_{\circ\circ\circ} \\
t^{221}_{\circ\circ\circ} & t^{222}_{\circ\circ\circ} & t^{223}_{\circ\circ\circ} \\
t^{231}_{\circ\circ\circ} & t^{232}_{\circ\circ\circ} & t^{233}_{\circ\circ\circ} \\
t^{311}_{\circ\circ\circ} & t^{312}_{\circ\circ\circ} & t^{313}_{\circ\circ\circ} \\
t^{321}_{\circ\circ\circ} & t^{322}_{\circ\circ\circ} & t^{323}_{\circ\circ\circ} \\
t^{331}_{\circ\circ\circ} & t^{332}_{\circ\circ\circ} & t^{333}_{\circ\circ\circ}
\end{bmatrix}
=
\begin{bmatrix}
0 & 0 & 0 \\
0 & 0 & \Delta \\
0 & -\Delta & 0 \\
0 & 0 & -\Delta \\
0 & 0 & 0 \\
\Delta & 0 & 0 \\
0 & \Delta & 0 \\
-\Delta & 0 & 0 \\
0 & 0 & 0
\end{bmatrix}
; \text{ with } \Delta = |U \quad V \quad W|,$$

$$\tag{7.29}$$

which is the pattern associated with an anti-symmetric contravariant tensor $(n = r = 3)$.

We have perfectly established a procedure for generating *anti-symmetric tensors* in a tensor space $(\otimes V^n)^r_h(K)$. For each r-tuple $\{\vec{u}, \vec{v}, \vec{w}, \cdots, \vec{z}\}_r \subset V^n(K); \vec{u} \neq \vec{v} \neq \vec{w} \neq \cdots \neq \vec{z}$ of selected vectors, they generate an anti-symmetric decomposable tensor.

The sum of several tensors generated by some different vector r-tuples, or the linear combination of them, leads to an *anti-symmetric tensor* subspace, S_h, that *is not generated only by decomposable tensors*:

$$S_h(\vec{u}, \vec{v}, \vec{w}, \ldots, \vec{z}) \subseteq (\otimes V^n)^r_h(K) \subset (\otimes V^n)^r(K).$$

This subspace is contained in or coincides with the tensor space $(\otimes V^n)^r_h(K)$ of all anti-symmetric tensors of order r, which are contained in the tensor space $(\otimes V^n)^r(K)$ of order r.

The generalization to the case of selecting from $\{\vec{u}, \vec{v}, \vec{w}, \ldots, \vec{z}\}_r \subset V^n(K)$, r-vectors for generating an anti-symmetric tensor in $(\otimes V^n)_h^r(K)$, must take into account that if $r < n$, Formula (7.26) will be extended, because it is applied to as many determinants of order r, as possible choices of different r-tuples of indices $\{\alpha_1, \alpha_2, \ldots, \alpha_r\}$ can be obtained from $I_n = \{1, 2, \cdots, n\}$, with the condition $\alpha_1 \neq \alpha_2 \neq \cdots \neq \alpha_r$. Note that this formula is only for non-null components.

Thus, we arrive at a total of $V_{n,r}$ variations, and therefore at the anti-symmetric components:

$$
t^{\alpha_1 \alpha_2 \cdots \alpha_r}_{\circ \ \circ \ \cdots \ \circ} = \Delta_r \equiv
\begin{vmatrix}
x^{\alpha_1 \circ}_{\circ \ 1} & x^{\alpha_1 \circ}_{\circ \ 2} & \cdots & x^{\alpha_1 \circ}_{\circ \ r} \\
x^{\alpha_2 \circ}_{\circ \ 1} & x^{\alpha_2 \circ}_{\circ \ 2} & \cdots & x^{\alpha_2 \circ}_{\circ \ r} \\
\cdots & \cdots & \cdots & \cdots \\
x^{\alpha_r \circ}_{\circ \ 1} & x^{\alpha_r \circ}_{\circ \ 2} & \cdots & x^{\alpha_r \circ}_{\circ \ r}
\end{vmatrix},
\tag{7.30}
$$

where Δ_r is the minor of the data matrix $[\, X_1 \ \ X_2 \ \ \cdots \ \ X_r \,]$ containing the rows $\{\alpha_1, \alpha_2, \ldots, \alpha_r\}$.

Since this requires a great amount of computational time, usually we get only the strict components $t^{(\alpha_1 \alpha_2 \cdots \alpha_r)}_{\circ \ \circ \ \cdots \ \circ}$, i.e., selecting only the $\binom{n}{r}$ combinations $\{\alpha_1, \alpha_2, \ldots, \alpha_r\}$ with $\alpha_1 < \alpha_2 < \cdots < \alpha_r$ for the minor Δ_r and next, we obtain $t^{\alpha_1 \alpha_2 \cdots \alpha_r}_{\circ \ \circ \ \cdots \ \circ} = \pm t^{(\alpha_1 \alpha_2 \cdots \alpha_r)}_{\circ \ \circ \ \cdots \ \circ}$ with the sign corresponding to the associated permutation.

By means of signed isomer tensors

Assume that an arbitrary tensor $\vec{t}(n, r)$, for example the pure contravariant tensor $t^{\alpha_1 \alpha_2 \cdots \alpha_r}_{\circ \ \circ \ \cdots \ \circ}$, is given and that we decide to associate it with another tensor $\vec{u}(n, r)$ with the following criterion:

1. A given set of indices I_k (of the same valency) from the set of the r tensor *dummy indices* is chosen.
2. We notate the new tensor $\vec{u}(n, r)$ (that is, the associated tensor) by placing the selected dummy indices in parentheses:

$$
u^{\alpha_1 \alpha_2 (\leftarrow I_k \rightarrow) \cdots \alpha_r}_{\circ \ \circ \ \ \ \ \circ \ \ \cdots \circ}.
$$

This notation is that of the strict components of the new tensor $\vec{u}(n, r)$.
3. Considering all permutations $(k!)$ of the dummy indices of I_k, we determine all the corresponding isomer tensors of tensor $\vec{t}$.
4. Once they are found, we build the strict components of $\vec{u}(n, r)$ by means of the formula

$$
u^{\alpha_1 \alpha_2 (\leftarrow I_k \rightarrow) \cdots \alpha_r}_{\circ \ \circ \ \ \ \ \circ \ \ \cdots \circ} = \frac{1}{k!} \sum_1^{k!} (-1)^{\tau} t^{\alpha_1 \alpha_2 (\leftarrow \text{permutation of } I_k \rightarrow) \cdots \alpha_r}_{\circ \ \circ \ \ \ \ \ \ \ \ \ \ \ \ \ \ \ \ \ \ \cdots \circ}, \tag{7.31}
$$

where the right-hand factor includes all the $k!$ isomers with their corresponding sign as summands. The sign of each summand is the one corresponding to the number τ of transpositions of the permutation of I_k (that each summand carries).

The tensor $\vec{u}(n, r)$ with the indicated strict components is an *anti-symmetric tensor* with respect to the set of indices I_k. Some authors present this from the point of view of an "anti-symmetrization" endomorphism H, which directly transforms the given tensor $\vec{t}(n, r)$ into the associated anti-symmetric tensor $\vec{u}(n, r)$.

Next, some illustrative examples are given.

1. Consider the data tensor $g^{\circ\circ}_{\alpha\beta}$; $I_k \equiv I_r = [\alpha, \beta]$ (all indices).

   The anti-symmetric tensor associated by means of this technique is the tensor with components given by Formula (7.31):

$$g'^{\circ\circ}_{(\alpha\beta)} = \frac{1}{2!}[(-1)^0 g^{\circ\circ}_{\alpha\beta} + (-1)^1 g^{\circ\circ}_{\beta\alpha}] = \frac{1}{2!}[g^{\circ\circ}_{\alpha\beta} - g^{\circ\circ}_{\beta\alpha}]. \tag{7.32}$$

2. Given the data tensor $W^{\alpha\beta\gamma\delta\epsilon\lambda}_{\circ\circ\circ\circ\circ\circ}$, an anti-symmetric associated tensor with respect to the indices $I_k = [\beta, \gamma, \delta]$, is the following tensor with strict components $u^{\alpha(\beta\gamma\delta)\epsilon\lambda}_{\circ\,\circ\circ\circ\,\circ\circ}$ given by Formula (7.31):

$$u^{\alpha(\beta\gamma\delta)\epsilon\lambda}_{\circ\,\circ\circ\circ\,\circ\circ} = \frac{1}{3!}\left[(-1)^0 w^{\alpha\beta\gamma\delta\epsilon\lambda}_{\circ\circ\circ\circ\circ\circ} + (-1)^2 w^{\alpha\gamma\delta\beta\epsilon\lambda}_{\circ\circ\circ\circ\circ\circ} + (-1)^2 w^{\alpha\delta\beta\gamma\epsilon\lambda}_{\circ\circ\circ\circ\circ\circ}\right.$$
$$\left. + (-1)^1 w^{\alpha\gamma\beta\delta\epsilon\lambda}_{\circ\circ\circ\circ\circ\circ} + (-1)^1 w^{\alpha\delta\gamma\beta\epsilon\lambda}_{\circ\circ\circ\circ\circ\circ} + (-1)^1 w^{\alpha\beta\delta\gamma\epsilon\lambda}_{\circ\circ\circ\circ\circ\circ}\right],$$

which, when operated gives the following anti-symmetric tensor with respect to I_k

$$u^{\alpha(\beta\gamma\delta)\epsilon\lambda}_{\circ\,\circ\circ\circ\,\circ\circ} = \frac{1}{6}\left[w^{\alpha\beta\gamma\delta\epsilon\lambda}_{\circ\circ\circ\circ\circ\circ} + w^{\alpha\gamma\delta\beta\epsilon\lambda}_{\circ\circ\circ\circ\circ\circ} + w^{\alpha\delta\beta\gamma\epsilon\lambda}_{\circ\circ\circ\circ\circ\circ})\right.$$
$$\left. - (w^{\alpha\gamma\beta\delta\epsilon\lambda}_{\circ\circ\circ\circ\circ\circ} + w^{\alpha\delta\gamma\beta\epsilon\lambda}_{\circ\circ\circ\circ\circ\circ} + w^{\alpha\beta\delta\gamma\epsilon\lambda}_{\circ\circ\circ\circ\circ\circ})\right]. \tag{7.33}$$

Obviously, if the anti-symmetrization technique were applied (by error) to an anti-symmetric data tensor, the result would be that the associated tensor is again the same data tensor.

Though not indicated, we have already entered in the exterior algebras in detail.

7.4.2 Intrinsic character of tensor anti-symmetry: Fundamental theorem of tensors with anti-symmetry

The obvious question a reader can ask himself is if any tensor with anti-symmetry with respect to certain components, maintains it when performing licit changes of basis, that is, of a tensor nature.

The answer to this question is formulated in the following theorem:

Theorem 7.1 (Intrinsic character of anti-symmetry). *The anti-symmetry of k indices ($2 \leq k \leq p$ or $2 \leq k \leq q, p + q = r$) of the* same *valency in the components of a homogeneous tensor has intrinsic character. The anti-symmetry in the components of two or more indices of* different *valency in a homogeneous tensor has no intrinsic character in the general case.* $\square$

Proof.

Sufficiency: Consider the case of indices of the same valency. Consider the homogeneous tensor $t_{\circ\circ\circ\lambda\mu}^{\alpha\beta\gamma\circ\circ}$ of fifth order ($r = 5$), defined over the linear space $V^n(K)$, which satisfies the property:

$$t_{\circ\circ\circ\lambda\mu}^{\beta\alpha\gamma\circ\circ} = -t_{\circ\circ\circ\lambda\mu}^{\alpha\beta\gamma\circ\circ}; \ \forall \alpha_j \in I_k = [\alpha, \beta] \text{ and } j \in I_5, \quad (7.34)$$

where $\alpha_1 = \alpha$, $\alpha_2 = \beta$, $\alpha_3 = \gamma$, $\alpha_4 = \lambda$, $\alpha_5 = \mu$, and $\alpha_1 \neq \alpha_2$.

Consider the change-of-basis in $V^n(K), \|\vec{e}_i\| = \|\vec{e}_\alpha\|[c_{\circ i}^{\alpha\circ}]$, which produces the change of tensor basis in the space $(\otimes V^n)^3 (\otimes V_*^n)^2(K)$; we indicate its action over the two components that appear in (7.34):

$$t_{\circ\circ\circ\ell m}^{jik\circ\circ} = t_{\circ\circ\circ\lambda\mu}^{\beta\alpha\gamma\circ\circ}\gamma_{\circ\beta}^{j\circ}\gamma_{\circ\alpha}^{i\circ}\gamma_{\circ\gamma}^{k\circ}c_{\ell\circ}^{\circ\lambda}c_{m\circ}^{\circ\mu} \quad (7.35)$$

$$-t_{\circ\circ\circ\ell m}^{ijk\circ\circ} = -t_{\circ\circ\circ\lambda\mu}^{\alpha\beta\gamma\circ\circ}\gamma_{\circ\alpha}^{i\circ}\gamma_{\circ\beta}^{j\circ}\gamma_{\circ\gamma}^{k\circ}c_{\ell\circ}^{\circ\lambda}c_{m\circ}^{\circ\mu}. \quad (7.36)$$

Taking into account (7.34), and the commutativity of the scalars $\gamma_{\circ\alpha}^{i\circ}$ with $\gamma_{\circ\beta}^{j\circ}$ in the field K, it is evident that the two right-hand members of (7.35) and (7.36) are equal, and then

$$\text{If } \forall i \neq j, \ t_{\circ\circ\circ\ell m}^{jik\circ\circ} = -t_{\circ\circ\circ\ell m}^{ijk\circ\circ}; \quad \forall \hat{\alpha}_h \in \hat{I}_k = [i, j]; \ \forall h \in I_5. \quad (7.37)$$

Following a similar process, analogous conclusions can be established by the reader for any other pair of contravariant or covariant indices of the analyzed tensor $t_{\circ\circ\circ\lambda\mu}^{\alpha\beta\gamma\circ\circ}$.

The analyzed model can be used for any generalization of the fundamental hypothesis to a larger set of indices of I_k.

Necessarity: Consider the case of indices of different valency. Consider the homogeneous tensor $t_{\circ\circ\circ\lambda\mu}^{\alpha\beta\gamma\circ\circ}$ of fifth order ($r = 5$), defined over the linear space $V^n(K)$, which satisfies the property

$$t_{\circ\circ\circ\beta\mu}^{\alpha\lambda\gamma\circ\circ} = -t_{\circ\circ\circ\lambda\mu}^{\alpha\beta\gamma\circ\circ}; \ ; \forall \alpha_j \in I_k = [\beta, \lambda], \beta \neq \lambda, \text{ and } j \in I_5. \quad (7.38)$$

When performing the change-of-basis in $V^n(K)$ mentioned in the sufficiency proof, the linear space $(\otimes V^n)^3(\otimes V_*^n)^2(K)$ suffers the corresponding tensor change, where the new components replacing those in (7.38) are related to the initial ones by

$$t^{i\ell k o o}_{o o o j m} = t^{\alpha\lambda\gamma o o}_{o o o \beta \mu}\gamma^{i o}_{o\alpha}\gamma^{\ell o}_{o\lambda}\gamma^{k o}_{o\gamma}c^{o\beta}_{j o}c^{o\mu}_{m o} \tag{7.39}$$

$$-t^{i j k o o}_{o o o \ell m} = -t^{\alpha\beta\gamma o o}_{o o o \lambda \mu}\gamma^{i o}_{o\alpha}\gamma^{j o}_{o\beta}\gamma^{k o}_{o\gamma}c^{o\lambda}_{\ell o}c^{o\mu}_{m o}, \tag{7.40}$$

proving that $\gamma^{\ell o}_{o\lambda}c^{o\beta}_{j o} \neq \gamma^{j o}_{o\beta}c^{o\lambda}_{\ell o}$, as can be appreciated in a simple numerical counterexample.

Let $n = r = 3$ and

$$[c^{\alpha o}_{o i}] = C = \begin{bmatrix} -1 & 1 & 0 \\ 0 & 1 & 0 \\ -1 & 0 & 3 \end{bmatrix}; [\gamma^{i o}_{o\alpha}] = C^{-1} = \begin{bmatrix} -1 & 1 & 0 \\ 0 & 1 & 0 \\ -1/3 & 1/3 & 1/3 \end{bmatrix};$$

$$[c^{o\alpha}_{i o}] = C^t = \begin{bmatrix} -1 & 0 & -1 \\ 1 & 1 & 0 \\ 0 & 0 & 3 \end{bmatrix};$$

if we choose $j = 2, \ell = 3, \beta = 1$, and $\lambda = 2$, then we have

$$\gamma^{\ell o}_{o\lambda}c^{o\beta}_{j o} = \gamma^{3 o}_{o 2}(\text{of } C^{-1})c^{o 1}_{2 o}(\text{of } C^t) = 1/3 \cdot 1 = 1/3$$

$$\gamma^{j o}_{o\beta}c^{o\lambda}_{\ell o} = \gamma^{2 o}_{o 1}c^{o 2}_{3 o} = 0 \cdot 0 = 0,$$

which implies $\gamma^{\ell o}_{o\lambda}c^{o\beta}_{j o} \neq \gamma^{j o}_{o\beta}c^{o\lambda}_{\ell o}$.

If we choose $j = 1, \ell = 2, \beta = 3$ and $\lambda = 2$, the result is

$$\gamma^{\ell o}_{o\lambda}c^{o\beta}_{j o} = \gamma^{2 o}_{o 2}c^{o 3}_{1 o} = 1 \cdot (-1) = -1,$$

$$\gamma^{j o}_{o\beta}c^{o\lambda}_{\ell o} = \gamma^{1 o}_{o 3}c^{o 2}_{2 o} = 0 \cdot 1 = 0,$$

so that again we have $\gamma^{\ell o}_{o\lambda}c^{o\beta}_{j o} \neq \gamma^{j o}_{o\beta}c^{o\lambda}_{\ell o}$.

We conclude that in general $t^{i\ell k o o}_{o o o j m} \neq -t^{i j k o o}_{o o o \ell m}$, that is, the tensor in the new basis does not keep the anti-symmetry of indices of different valency, that existed in the initial basis.

However, if the data tensor were to have a great number of null components, $t^{\alpha\beta\gamma o o}_{o o o \lambda \mu} = 0$, these could in some cases absorb the indicated inequalities, and maintain the anti-symmetries of indices of different valency, because of its nullity, that is, due to qualities that are special or proper to the concrete tensor.

Example 7.2 (Strict components).

1. Consider the anti-symmetric covariant tensor $T_1 = t^{o o o o o}_{\alpha\beta\gamma\delta\epsilon}$, over the linear space $V^n(\mathbb{R})$.

a) Determine the number of components (N_1).
b) Determine the number of strict components (N_2).
c) Determine the maximum number of strict components that are different (N_3).
d) Determine the maximum number of non-null components (N_4).
e) Determine the minimum number of null components (N_5).

2. Consider the mixed tensor $T_2 = t^{\circ\circ\circ\circ\circ\pi\rho\sigma\tau}_{\alpha\beta\gamma\delta\epsilon\circ\circ\circ\circ}$, with $q = 5$ covariant indices and $p = 4$ contravariant indices, over the linear space $V^n(\mathbb{R})$, which is partially anti-symmetric, with respect to the covariant indices $k_2 = [\beta, \delta]$ and contravariant indices $k_1 = [\pi, \sigma, \tau]$.

a) Determine the number of components (N_1).
b) Determine the number of strict components (N_2).
c) Determine the maximum number of strict components that are different (N_3).
d) Determine the maximum number of non-null components (N_4).
e) Determine the minimum number of null components (N_5).

Solution:

1. a) Since the order is $r = 5$, the number of components is $N_1 = \sigma = n^r = n^5$ (we assume $n \geq 5$).
   b) Using Formula (7.18), we have
   $$N_2 = \binom{n}{r} = \binom{n}{5} = \frac{n(n-1)(n-2)(n-3)(n-4)}{5!}.$$
   c) Using Formula (7.19), we have
   $$N_3 = 2\binom{n}{r} + 1 = 2\binom{n}{5} + 1.$$
   d) Based on Formula (7.13), we can write
   $$N_4 = V_{n,r} = \frac{n!}{(n-r)!} = \frac{n(n-1)(n-2)(n-3)(n-4)(n-5)!}{(n-5)!}$$
   $$= n(n-1)(n-2)(n-3)(n-4).$$
   e) Because of (7.14) we have
   $$N_5(\text{null}) = N_1 - N_4 = n^r - \frac{n!}{(n-r)!} = n^5 - n(n-1)(n-2)(n-3)(n-4).$$

2. Consider the mixed tensor $T_2 = t^{\circ\circ\circ\circ\circ\pi\rho\sigma\tau}_{\alpha\beta\gamma\delta\epsilon\circ\circ\circ\circ}$.
   a) Since the order of this tensor is $r = p + q = 4 + 5 = 9$, the result is $N_1 = n^r = n^9$.

b) We use Formula (7.21):

$$N_2 = \binom{n}{k_1}\binom{n}{k_2} n^{r-(k_1+k_2)} = \binom{n}{3}\binom{n}{2} n^{9-(3+2)} = \frac{1}{12} n^6 (n-1)^2 (n-2).$$

c) We use an extension of the previous formula

$$N_3 = 2\binom{n}{3} \cdot 2\binom{n}{2} n^{9-5} + 1 = \frac{1}{3} n^6 (n-1)^2 (n-2) + 1.$$

d) We use a formula that is an extension of (7.13) to mixed indices

$$N_4 = V_{n,k_1} \cdot V_{n.k_2} n^{r-(k_1+k_2)} = \frac{n!}{(n-k_1)!} \frac{n!}{(n-k_2)!} n^{r-(k_1+k_2)}$$

$$= \frac{n!}{(n-3)!} \frac{n!}{(n-2)!} n^{9-5} = n^6 (n-1)^2 (n-2).$$

e) Obviously, we have

$$N_5 = N_1 - N_4 = n^9 - n^6 (n-1)^2 (n-2)$$
$$= n^6 [n^3 - (n^3 - 4n^2 + 5n - 2)] = n^6 [4n^2 - 5n + 2].$$

□

Example 7.3 (Total components and strict components).

1. Consider the anti-symmetric contravariant tensor $H^{\alpha\beta\gamma}_{\circ\circ\circ}$ of order $r = 3$, over the linear space $V^3(\mathbb{R})$, $n = 3$.
   a) Obtain the number of total and strict components, using different Roman letters for the non-null components.
   b) Give the matrix representation of these tensors.
2. Given a contravariant anti-symmetric tensor $H^{\alpha\beta\gamma\delta}_{\circ\circ\circ\circ}$ of order $r = 4$, over the linear space $V^5(\mathbb{R})$, $n = 5$, give the number of total and strict components, using different Roman letters for the non-null components.

Solution:

1. Consider the tensor $H^{\alpha\beta\gamma}_{\circ\circ\circ}$.
   a) The total number of components is $\sigma = V R_{n,r} = n^r = 3^3 = 27$. Similarly, the number of strict components is $\sigma' = \binom{n}{r} = \binom{3}{3} = 1$, and the values of the non-null components are

$$h^{(123)}_{\circ\circ\circ} = h^{231}_{\circ\circ\circ} = h^{312}_{\circ\circ\circ} = a; \quad h^{213}_{\circ\circ\circ} = h^{321}_{\circ\circ\circ} = h^{132}_{\circ\circ\circ} = -a.$$

b) Then, its most general pattern is

$$[H^{\alpha\beta\gamma}_{\circ\circ\circ}] = \begin{bmatrix} 0 & 0 & 0 \\ 0 & 0 & a \\ 0 & -a & 0 \\ 0 & 0 & -a \\ 0 & 0 & 0 \\ a & 0 & 0 \\ 0 & a & 0 \\ -a & 0 & 0 \\ 0 & 0 & 0 \end{bmatrix}.$$

2. Consider the tensor $H^{\alpha\beta\gamma\delta}_{\circ\circ\circ\circ}$.

a) The total number of components is $\sigma = VR_{n,r} = n^r = 5^4 = 625$, the number of strict components is $\sigma' = \binom{n}{r} = \binom{5}{4} = \binom{5}{4} = 5$, and the values of the non-null components (a total of 120) are

$$h^{(1234)}_{\circ\circ\circ\circ} = h^{1342}_{\circ\circ\circ\circ} = h^{1423}_{\circ\circ\circ\circ} = h^{2143}_{\circ\circ\circ\circ} = h^{2314}_{\circ\circ\circ\circ} = h^{2431}_{\circ\circ\circ\circ} = h^{3124}_{\circ\circ\circ\circ}$$
$$= h^{3241}_{\circ\circ\circ\circ} = h^{3412}_{\circ\circ\circ\circ} = h^{4132}_{\circ\circ\circ\circ} = h^{4213}_{\circ\circ\circ\circ} = h^{4321}_{\circ\circ\circ\circ} = a$$

$$h^{1243}_{\circ\circ\circ\circ} = h^{1324}_{\circ\circ\circ\circ} = h^{1432}_{\circ\circ\circ\circ} = h^{2134}_{\circ\circ\circ\circ} = h^{2341}_{\circ\circ\circ\circ} = h^{2413}_{\circ\circ\circ\circ} = h^{3142}_{\circ\circ\circ\circ}$$
$$= h^{3214}_{\circ\circ\circ\circ} = h^{3421}_{\circ\circ\circ\circ} = h^{4123}_{\circ\circ\circ\circ} = h^{4231}_{\circ\circ\circ\circ} = h^{4312}_{\circ\circ\circ\circ} = -a$$

$$h^{(1235)}_{\circ\circ\circ\circ} = h^{1352}_{\circ\circ\circ\circ} = h^{1523}_{\circ\circ\circ\circ} = h^{2153}_{\circ\circ\circ\circ} = h^{2315}_{\circ\circ\circ\circ} = h^{2531}_{\circ\circ\circ\circ} = h^{3125}_{\circ\circ\circ\circ}$$
$$= h^{3251}_{\circ\circ\circ\circ} = h^{3512}_{\circ\circ\circ\circ} = h^{5132}_{\circ\circ\circ\circ} = h^{5213}_{\circ\circ\circ\circ} = h^{5321}_{\circ\circ\circ\circ} = b$$

$$h^{1253}_{\circ\circ\circ\circ} = h^{1325}_{\circ\circ\circ\circ} = h^{1532}_{\circ\circ\circ\circ} = h^{2135}_{\circ\circ\circ\circ} = h^{2351}_{\circ\circ\circ\circ} = h^{2513}_{\circ\circ\circ\circ} = h^{3152}_{\circ\circ\circ\circ}$$
$$= h^{3215}_{\circ\circ\circ\circ} = h^{3521}_{\circ\circ\circ\circ} = h^{5123}_{\circ\circ\circ\circ} = h^{5231}_{\circ\circ\circ\circ} = h^{5312}_{\circ\circ\circ\circ} = -b$$

$$h^{(1245)}_{\circ\circ\circ\circ} = h^{1452}_{\circ\circ\circ\circ} = h^{1524}_{\circ\circ\circ\circ} = h^{2154}_{\circ\circ\circ\circ} = h^{2415}_{\circ\circ\circ\circ} = h^{2541}_{\circ\circ\circ\circ} = h^{4125}_{\circ\circ\circ\circ}$$
$$= h^{4251}_{\circ\circ\circ\circ} = h^{4512}_{\circ\circ\circ\circ} = h^{5142}_{\circ\circ\circ\circ} = h^{5214}_{\circ\circ\circ\circ} = h^{5421}_{\circ\circ\circ\circ} = c$$

$$h^{1254}_{\circ\circ\circ\circ} = h^{1425}_{\circ\circ\circ\circ} = h^{1542}_{\circ\circ\circ\circ} = h^{2145}_{\circ\circ\circ\circ} = h^{2451}_{\circ\circ\circ\circ} = h^{2514}_{\circ\circ\circ\circ} = h^{4152}_{\circ\circ\circ\circ}$$
$$= h^{4215}_{\circ\circ\circ\circ} = h^{4521}_{\circ\circ\circ\circ} = h^{5124}_{\circ\circ\circ\circ} = h^{5241}_{\circ\circ\circ\circ} = h^{5412}_{\circ\circ\circ\circ} = -c$$

$$h^{(1345)}_{\circ\circ\circ\circ} = h^{1453}_{\circ\circ\circ\circ} = h^{1534}_{\circ\circ\circ\circ} = h^{3154}_{\circ\circ\circ\circ} = h^{3415}_{\circ\circ\circ\circ} = h^{3541}_{\circ\circ\circ\circ} = h^{4135}_{\circ\circ\circ\circ}$$
$$= h^{4351}_{\circ\circ\circ\circ} = h^{4513}_{\circ\circ\circ\circ} = h^{5143}_{\circ\circ\circ\circ} = h^{5314}_{\circ\circ\circ\circ} = h^{5431}_{\circ\circ\circ\circ} = d$$

$$h^{1354}_{\circ\circ\circ\circ} = h^{1435}_{\circ\circ\circ\circ} = h^{1543}_{\circ\circ\circ\circ} = h^{3145}_{\circ\circ\circ\circ} = h^{3451}_{\circ\circ\circ\circ} = h^{3514}_{\circ\circ\circ\circ} = h^{4153}_{\circ\circ\circ\circ}$$
$$= h^{4315}_{\circ\circ\circ\circ} = h^{4531}_{\circ\circ\circ\circ} = h^{5134}_{\circ\circ\circ\circ} = h^{5341}_{\circ\circ\circ\circ} = h^{5413}_{\circ\circ\circ\circ} = -d$$

$$h^{(2345)}_{\circ\circ\circ\circ} = h^{2453}_{\circ\circ\circ\circ} = h^{2534}_{\circ\circ\circ\circ} = h^{3254}_{\circ\circ\circ\circ} = h^{3425}_{\circ\circ\circ\circ} = h^{3542}_{\circ\circ\circ\circ} = h^{4235}_{\circ\circ\circ\circ}$$

242 7 Anti-symmetric Homogeneous Tensors, Tensor and Inner Product Algebras

$$= h^{4352}_{\circ\circ\circ\circ} = h^{4523}_{\circ\circ\circ\circ} = h^{5243}_{\circ\circ\circ\circ} = h^{5324}_{\circ\circ\circ\circ} = h^{5432}_{\circ\circ\circ\circ} = e$$

$$h^{2354}_{\circ\circ\circ\circ} = h^{2435}_{\circ\circ\circ\circ} = h^{2543}_{\circ\circ\circ\circ} = h^{3245}_{\circ\circ\circ\circ} = h^{3452}_{\circ\circ\circ\circ} = h^{3524}_{\circ\circ\circ\circ} = h^{4253}_{\circ\circ\circ\circ}$$

$$= h^{4325}_{\circ\circ\circ\circ} = h^{4532}_{\circ\circ\circ\circ} = h^{5234}_{\circ\circ\circ\circ} = h^{5342}_{\circ\circ\circ\circ} = h^{5423}_{\circ\circ\circ\circ} = -e.$$

□

Example 7.4 (Anti-symmetric system). A system of scalars of second order, defined over the linear space $V^n(K)$ is given, such that its components satisfy the condition

$$bs(\alpha, \beta) + cs(\beta, \alpha) = 0, \quad \forall \alpha, \beta,$$

where b and c are different and constant scalars.

Ignoring the tensor nature of the proposed system $S(\alpha, \beta)$ what intrinsic property has this system?

Solution: Four different cases for the scalars b and c are to be considered:

1. $b = 0$ and $c \neq 0$. Then, $cs(\beta, \alpha) = 0 \to s(\beta, \alpha) = 0 \to S(\alpha, \beta) \equiv \Omega$
2. $b \neq 0$ and $c = 0$. Then, $bs(\alpha, \beta) = 0 \to s(\alpha, \beta) = 0 \to S(\alpha, \beta) \equiv \Omega$.
3. $b = -c \neq 0$ (since they are different). Using the assumption and the fundamental condition we get $-cs(\alpha, \beta) + cs(\beta, \alpha) = 0 \to c(s(\beta, \alpha) - s(\alpha, \beta)) = 0$, which requires a *symmetric system*, $s(\beta, \alpha) = s(\alpha, \beta) \to$ symmetric system.
4. $b \neq -c$ ($b \neq 0$ and $c \neq 0$). We start from the relation

$$bs(\alpha, \beta) + cs(\beta, \alpha) = 0; \tag{7.41}$$

and changing the index notation we obtain

$$bs(\beta, \alpha) + cs(\alpha, \beta) = 0, \tag{7.42}$$

and adding (7.41) and (7.42) we get

$$(b + c)s(\alpha, \beta) + (b + c)s(\beta, \alpha) = 0 \to (b + c)(s(\alpha, \beta) + s(\beta, \alpha)) = 0,$$

which requires (since $b + c \neq 0$): $s(\beta, \alpha) = -s(\alpha, \beta)$, that is, $s(\alpha, \beta)$ is an *anti-symmetric system*.

□

Example 7.5 (Different classes of systems). Let T be a tensor belonging to the tensor space $V^n_*(\otimes V^n)^4 \otimes V^n_*(K)$, with components $t^{\circ jk\ell po}_{i\circ\circ\circ\circ q}$, anti-symmetric with respect to a set of indices $[k, \ell, p]$.

Classify the following systems of scalars as tensors:

1. Contracted tensor of T with respect to the indices (i, j).
2. Contracted tensor of T with respect to the indices (p, q).
3. Contracted tensor of T with respect to the indices (ℓ, p).
4. Doubly contracted tensor of T with respect to the indices (i, p) and (j, q).
5. Doubly contracted tensor of T with respect to the indices (i, q) and (j, k).

Solution:

1. The contracted tensor of T with respect to the indices (i,j) is

$$C\binom{j}{i}(T) = [t^{\circ\theta k\ell p\circ}_{\theta\circ\circ\circ\circ q}] = [u^{k\ell p\circ}_{\circ\circ\circ q}],$$

which is an anti-symmetric tensor with respect to its contravariant indices.

2. The contracted tensor of T with respect to the indices (p,q) is

$$C\binom{p}{q}(T) = [t^{\circ j k\ell\theta\circ}_{i\circ\circ\circ\circ\theta}] = [v^{\circ j k\ell}_{i\circ\circ\circ}],$$

which is an anti-symmetric tensor with respect to the set $(k\ell)$.

3. The contracted tensor of T with respect to the indices (ℓ,p) is

$$\mathcal{C}(\ell,p)(T) = [t^{\circ j k\theta\theta\circ}_{i\circ\circ\circ\circ q}] = s_1[i,j,k,q],$$

which is a system of scalar components.

4. The doubly contracted tensor of T with respect to the indices (i,p) and (j,q) is

$$C\left(\begin{array}{c|c}p & j\\i & q\end{array}\right)(T) = [t^{\circ\phi k\ell\theta\circ}_{\theta\circ\circ\circ\circ\phi}] = [w^{k\ell}_{\circ\circ}],$$

which is an anti-symmetric contravariant tensor of order 2.

5. The doubly contracted tensor of T with respect to the indices (i,q) and (j,k) is

$$\mathcal{C}(i,q|j,k)(T) = [t^{\circ\phi\phi\ell p\circ}_{\theta\circ\circ\circ\circ\theta}] = s_2[\ell,p],$$

which is an anti-symmetric system of scalar components.

$\square$

7.4.3 Anti-symmetric tensor spaces and subspaces. Vector subspaces associated with strict components

Consider the tensor space of homogeneous tensors of order r totally contravariant, over the linear space $V^n(K)$, notated: $(\otimes V^n)^r(K)$, as the tensor space of reference. Obviously inside it tensor subspaces exist, and some of them are k-anti-symmetric (partial or totally anti-symmetric) tensors.

In particular, we denote by $(\otimes V^n)^r_h(K)$ the subspace sum of all anti-symmetric tensor subspaces. We point out that such a subspace is not only a linear subspace but a *tensor* subspace.

For the sake of simplicity, we examine this only for $r = 2$.

Let $\vec{u}, \vec{v} \in (\otimes V^n)^2_h(K)$ be two anti-symmetric tensors:

$$\vec{u} = u^{\alpha\beta}_{\circ\circ}\vec{e}_\alpha \otimes \vec{e}_\beta; \quad \text{with } u^{\beta\alpha}_{\circ\circ} = -u^{\alpha\beta}_{\circ\circ} \text{ and } u^{ij}_{\circ\circ} = u^{\alpha\beta}_{\circ\circ}\gamma^{i\circ}_{\circ\alpha}\gamma^{j\circ}_{\circ\beta} \tag{7.43}$$

$$\vec{v} = v^{\alpha\beta}_{\circ\circ}\vec{e}_\alpha \otimes \vec{e}_\beta; \quad \text{with } v^{\beta\alpha}_{\circ\circ} = -v^{\alpha\beta}_{\circ\circ} \text{ and } v^{ij}_{\circ\circ} = v^{\alpha\beta}_{\circ\circ}\gamma^{i\circ}_{\circ\alpha}\gamma^{j\circ}_{\circ\beta} \tag{7.44}$$

1. The sum subspace contains the vectors of the form:

$$\vec{w} = \vec{u} + \vec{v} = (u_{\circ\circ}^{\alpha\beta} + v_{\circ\circ}^{\alpha\beta})\vec{e}_\alpha \otimes \vec{e}_\beta = w_{\circ\circ}^{\alpha\beta}\vec{e}_\alpha \otimes \vec{e}_\beta$$

and since

$$w_{\circ\circ}^{\beta\alpha} = u_{\circ\circ}^{\beta\alpha} + v_{\circ\circ}^{\beta\alpha} = -u_{\circ\circ}^{\alpha\beta} + (-v_{\circ\circ}^{\alpha\beta}) = -(u_{\circ\circ}^{\alpha\beta} + v_{\circ\circ}^{\alpha\beta}) = -w_{\circ\circ}^{\beta\alpha} \quad (7.45)$$

and

$$w_{\circ\circ}^{ij} = u_{\circ\circ}^{ij} + v_{\circ\circ}^{ij} = (u_{\circ\circ}^{\alpha\beta} + v_{\circ\circ}^{\alpha\beta})\gamma_{\circ\alpha}^{i\circ}\gamma_{\circ\beta}^{j\circ} = w_{\circ\circ}^{\alpha\beta}\gamma_{\circ\alpha}^{i\circ}\gamma_{\circ\beta}^{j\circ}, \quad (7.46)$$

the result is that the subspace $S_h(\vec{w}) = S_h(\vec{u} + \vec{v}) \subset (\otimes V^n)_h^2(K)$.

2. If $\vec{u} \in (\otimes V^n)_h^2$ and $\lambda \in K$:

$$\vec{z} = \lambda\vec{u} = \lambda u_{\circ\circ}^{\alpha\beta}\vec{e}_\alpha \otimes \vec{e}_\beta = z_{\circ\circ}^{\alpha\beta}\vec{e}_\alpha \otimes \vec{e}_\beta \text{ with } z_{\circ\circ}^{\beta\alpha} = \lambda u_{\circ\circ}^{\beta\alpha} = -\lambda u_{\circ\circ}^{\alpha\beta} = -z_{\circ\circ}^{\alpha\beta}$$

$$(7.47)$$

and

$$z_{\circ\circ}^{ij} = \lambda u_{\circ\circ}^{ij} = (\lambda u_{\circ\circ}^{\alpha\beta})\gamma_{\circ\alpha}^{i\circ}\gamma_{\circ\beta}^{j\circ} = z_{\circ\circ}^{\alpha\beta}\gamma_{\circ\alpha}^{i\circ}\gamma_{\circ\beta}^{j\circ}, \quad (7.48)$$

so that the set $(\otimes V^n)_h^2(K)$ has the structure not only of a *linear* space, but of a *tensor* subspace contained in $(\otimes V^n)^2(K)$.

The generalization to $(\otimes V^n)_h^r(K) \subset (\otimes V^n)^r(K)$ is immediate.

It is convenient to study here a certain convention that comes from performing algebraic operations inside the set of anti-symmetric tensors. Such a convention refers to the fact that diverse authors consider the set of strict components associated with an anti-symmetric tensor as the components of a *vector* associated with the anti-symmetric tensor, endowing to such components of the corresponding basic vectors created *ad hoc* special notation with the added constraint of having exclusively *non-repeated and only increasing* indices.

With the aim of clarifying the convention, in particular we refer to the anti-symmetric tensor

$$\vec{u} = u_{\circ\circ\circ}^{\alpha\beta\gamma}\vec{e}_\alpha \otimes \vec{e}_\beta \otimes \vec{e}_\gamma \in (\otimes V^4)_h^3(K). \quad (7.49)$$

The basis $\mathcal{B} = \{\vec{e}_\alpha \otimes \vec{e}_\beta \otimes \vec{e}_\gamma\}$ of the tensor space $(\otimes V^4)_h^3(K)$, has 64 basic vectors, in agreement with its dimension $\sigma = n^r = 4^3 = 64$, and the associated corresponding vector is (*in principle*)

$$\vec{u}_0' = u_{\circ\circ\circ}^{(\alpha\beta\gamma)}\vec{e}_\alpha \otimes \vec{e}_\beta \otimes \vec{e}_\gamma$$

formed with the strict components of the previous tensor, with dimension $\sigma_0 = \binom{n}{r} = \binom{n}{3} = 4$ and its basis would be

$$\mathcal{B}_0' = \{\vec{e}_1 \otimes \vec{e}_2 \otimes \vec{e}_3, \vec{e}_1 \otimes \vec{e}_2 \otimes \vec{e}_4, \vec{e}_1 \otimes \vec{e}_3 \otimes \vec{e}_4, \vec{e}_2 \otimes \vec{e}_3 \otimes \vec{e}_4\}.$$

The development of the anti-symmetric tensor given in (7.49), *sorted by strict components* is

$$
\begin{aligned}
\vec{u} = &+u^{(123)}_{\circ\circ\circ}[(\vec{e}_1 \otimes \vec{e}_2 \otimes \vec{e}_3 + \vec{e}_2 \otimes \vec{e}_3 \otimes \vec{e}_1 + \vec{e}_3 \otimes \vec{e}_1 \otimes \vec{e}_2) \\
&-(\vec{e}_1 \otimes \vec{e}_3 \otimes \vec{e}_2 + \vec{e}_2 \otimes \vec{e}_1 \otimes \vec{e}_3 + \vec{e}_3 \otimes \vec{e}_2 \otimes \vec{e}_1)] \\
&+u^{(124)}_{\circ\circ\circ}[(\vec{e}_1 \otimes \vec{e}_2 \otimes \vec{e}_4 + \vec{e}_2 \otimes \vec{e}_4 \otimes \vec{e}_1 + \vec{e}_4 \otimes \vec{e}_1 \otimes \vec{e}_2) \\
&-(\vec{e}_1 \otimes \vec{e}_4 \otimes \vec{e}_2 + \vec{e}_2 \otimes \vec{e}_1 \otimes \vec{e}_4 + \vec{e}_4 \otimes \vec{e}_2 \otimes \vec{e}_1)] \\
&+u^{(134)}_{\circ\circ\circ}[(\vec{e}_1 \otimes \vec{e}_3 \otimes \vec{e}_4 + \vec{e}_3 \otimes \vec{e}_4 \otimes \vec{e}_1 + \vec{e}_4 \otimes \vec{e}_1 \otimes \vec{e}_3) \\
&-(\vec{e}_1 \otimes \vec{e}_4 \otimes \vec{e}_3 + \vec{e}_3 \otimes \vec{e}_1 \otimes \vec{e}_4 + \vec{e}_4 \otimes \vec{e}_3 \otimes \vec{e}_1)] \\
&+u^{(234)}_{\circ\circ\circ}[(\vec{e}_2 \otimes \vec{e}_3 \otimes \vec{e}_4 + \vec{e}_3 \otimes \vec{e}_4 \otimes \vec{e}_2 + \vec{e}_4 \otimes \vec{e}_2 \otimes \vec{e}_3) \\
&-(\vec{e}_2 \otimes \vec{e}_4 \otimes \vec{e}_3 + \vec{e}_3 \otimes \vec{e}_2 \otimes \vec{e}_4 + \vec{e}_4 \otimes \vec{e}_3 \otimes \vec{e}_2)].
\end{aligned} \tag{7.50}
$$

To each of the "vector" brackets accompanying the strict components corresponds a vector, which is notated as

$$
\begin{aligned}
\vec{e}_1 \wedge \vec{e}_2 \wedge \vec{e}_3 &\equiv [\vec{e}_1 \otimes \vec{e}_2 \otimes \vec{e}_3 + \cdots - \vec{e}_3 \otimes \vec{e}_2 \otimes \vec{e}_1] \\
\vec{e}_1 \wedge \vec{e}_2 \wedge \vec{e}_4 &\equiv [\vec{e}_1 \otimes \vec{e}_2 \otimes \vec{e}_4 + \cdots - \vec{e}_4 \otimes \vec{e}_2 \otimes \vec{e}_1] \\
\vec{e}_1 \wedge \vec{e}_3 \wedge \vec{e}_4 &\equiv [\vec{e}_1 \otimes \vec{e}_3 \otimes \vec{e}_4 + \cdots - \vec{e}_4 \otimes \vec{e}_3 \otimes \vec{e}_1] \\
\vec{e}_2 \wedge \vec{e}_3 \wedge \vec{e}_4 &\equiv [\vec{e}_2 \otimes \vec{e}_3 \otimes \vec{e}_4 + \cdots - \vec{e}_4 \otimes \vec{e}_3 \otimes \vec{e}_2].
\end{aligned} \tag{7.51}
$$

The sign criterion from the interior of each bracket comes evidently from the sign criterion $(-1)^\tau$, used for the indices of the anti-symmetric tensors.

Consequently, the vector $\vec{u}_0$ associated with tensor $\vec{u}$ is

$$
\vec{u}_0 = u^{(\alpha\beta\gamma)}_{\circ\circ\circ}\vec{e}_\alpha \wedge \vec{e}_\beta \wedge \vec{e}_\gamma \text{ with } \alpha < \beta < \gamma. \tag{7.52}
$$

The basis of the linear space just created, which holds vector $\vec{u}_0$, is

$$
\mathcal{B}_0 = \{\vec{e}_1 \wedge \vec{e}_2 \wedge \vec{e}_3, \ \vec{e}_1 \wedge \vec{e}_2 \wedge \vec{e}_4, \ \vec{e}_1 \wedge \vec{e}_3 \wedge \vec{e}_4, \ \vec{e}_2 \wedge \vec{e}_3 \wedge \vec{e}_4\} \tag{7.53}
$$

of dimension $\sigma_0 = \binom{n}{r} = \binom{4}{3} = 4$.

The vector $\vec{u}_0$ once developed can be written as

$$
\vec{u}_0 = u^{123}_{\circ\circ\circ}\vec{e}_1 \wedge \vec{e}_2 \wedge \vec{e}_3 + u^{124}_{\circ\circ\circ}\vec{e}_1 \wedge \vec{e}_2 \wedge \vec{e}_4 + u^{134}_{\circ\circ\circ}\vec{e}_1 \wedge \vec{e}_3 \wedge \vec{e}_4 + u^{234}_{\circ\circ\circ}\vec{e}_2 \wedge \vec{e}_3 \wedge \vec{e}_4. \tag{7.54}
$$

As can be seen, the indices of the strict components are the subindices of its basic vector.

It is sufficient to know the strict components of an anti-symmetric tensor $\vec{u}$, to "build" the associated vector $\vec{u}_0$ and reciprocally. In fact $\vec{u}_0 = \vec{u}$ if we substitute Formulas (7.51) in $\vec{u}_0$.

This justifies the notation

$$
V^4 \wedge V^4 \wedge V^4(K) \equiv (\wedge V^4)^3(K) \tag{7.55}
$$

for the linear space of basis $\mathcal{B}_0$.

As will be seen in Chapter 9, we have entered the subject of exterior algebras. Before ending this section, it is convenient to insist with emphasis, that $\vec{u}_0$ *is not* a tensor, in spite of the "provocative" notation of Formula (7.54), and that the linear space $(\wedge V^4)^3(K)$ is not a tensor space of the already established tensor spaces.

The dimension of the exterior linear space $\sigma_0 = \binom{n}{r}$ can never be equal to $\sigma = VR_{n,r} = n^r$, which is the typical dimension of a tensor space, because the condition $\alpha_1 < \alpha_2 < \alpha_3 < \cdots < \alpha_r$ of (7.52) forbids the typical constraint of anti-symmetric tensors $\alpha_1 \neq \alpha_2 \neq \alpha_3 \neq \cdots \neq \alpha_r$.

However, though they are not tensors, following most authors, they will be dealt with in depth in Chapter 9, because of their very important consequences.

We end here our explanation, initiated with Formula (7.49), the convention that univocally relates each anti-symmetric tensor in $(\otimes V^n)_h^r(K)$ to a vector of the "exterior linear space" $(\wedge V^n)^r(K)$, which is the notation received by the proper anti-symmetric tensor space $(\otimes V^n)_h^r(K)$, when in it we adopt a special basis *of type* $\mathcal{B}_0$, and then its dimension *is modified* (initially $\sigma = n^r$ and finally $\sigma' = \binom{n}{r}$).

7.5 Anti-symmetric tensors from the tensor algebra perspective

As will be shown, a simple counterexample forces us to accept that the anti-symmetric tensor algebra of anti-symmetric tensors *does not* exist, that is, the tensor product $\vec{t}_1 \otimes \vec{t}_2$ of two anti-symmetric tensors, is not in general another anti-symmetric tensor.

We give the following counterexample.

Consider two tensors of order $r = 2$ and $n = 3$, both contravariant:

$$[a_{\circ\circ}^{ij}] = \begin{bmatrix} 0 & x_1 & -y_1 \\ -x_1 & 0 & z_1 \\ y_1 & -z_1 & 0 \end{bmatrix} \text{ with } x_1 \cdot y_1 \cdot z_1 \neq 0;$$

$$[b_{\circ\circ}^{k\ell}] = \begin{bmatrix} 0 & x_2 & -y_2 \\ -x_2 & 0 & z_2 \\ y_2 & -z_2 & 0 \end{bmatrix} \text{ with } x_2 \cdot y_2 \cdot z_2 \neq 0,$$

that are anti-symmetric, because $\forall a_{\circ\circ}^{ij} = -a_{\circ\circ}^{ji}$ and $\forall b_{\circ\circ}^{k\ell} = -b_{\circ\circ}^{\ell k}$.

Next, we represent in matrix form the components of the tensor of order $r = 4$ and $n = 3$:

$$P \equiv [p_{\circ\circ\circ\circ}^{ijk\ell}] = [a_{\circ\circ}^{ij}] \otimes [b_{\circ\circ}^{k\ell}], \tag{7.56}$$

which is the tensor product of the given tensors.

The matrix P is represented by nine blocks:

$$P = [p_{\circ\circ\circ\circ}^{ijk\ell}] = \left[\begin{array}{ccc|ccc|ccc} 0 & 0 & 0 & 0 & x_1x_2 & -x_1y_2 & 0 & -x_2y_1 & y_1y_2 \\ 0 & 0 & 0 & -x_1x_2 & 0 & x_1z_2 & x_2y_1 & 0 & -y_1z_2 \\ 0 & 0 & 0 & x_1y_2 & -x_1z_2 & 0 & -y_1y_2 & y_1z_2 & 0 \\ \hline 0 & -x_1x_2 & x_1y_2 & 0 & 0 & 0 & 0 & x_2z_1 & -y_2z_1 \\ x_1x_2 & 0 & -x_1z_2 & 0 & 0 & 0 & -x_2z_1 & 0 & z_1z_2 \\ -x_1y_2 & x_1z_2 & 0 & 0 & 0 & 0 & y_2z_1 & -z_1z_2 & 0 \\ \hline 0 & x_2y_1 & -y_1y_2 & 0 & -x_2z_1 & y_2z_1 & 0 & 0 & 0 \\ -x_2y_1 & 0 & y_1z_2 & x_2z_1 & 0 & -z_1z_2 & 0 & 0 & 0 \\ y_1y_2 & -y_1z_2 & 0 & -y_2z_1 & z_1z_2 & 0 & 0 & 0 & 0 \end{array}\right]$$

$$(7.57)$$

Since $n = 3$ and $r = 4$, that is, $n < r$, in the component $p_{\circ\circ\circ\circ}^{ijk\ell}$ there will *always be* a repeated index, so that, if P were an anti-symmetric tensor then $[p_{\circ\circ\circ\circ}^{ijk\ell}] \equiv \Omega$, that is, all its components would be null, i.e., $p_{\circ\circ\circ\circ}^{1111} = p_{\circ\circ\circ\circ}^{1233} = p_{\circ\circ\circ\circ}^{3122} = p_{\circ\circ\circ\circ}^{3212} = \cdots = p_{\circ\circ\circ\circ}^{3333} = 0$.

However, in the obtained matrix P this is not true, because $p_{\circ\circ\circ\circ}^{1212} = x_1x_2 \neq 0$; $p_{\circ\circ\circ\circ}^{1313} = y_1y_2 \neq 0$; $p_{\circ\circ\circ\circ}^{2113} = x_1y_2 \neq 0$; etc.

Thus, P is a symmetric matrix, but *it is not* the matrix representation of a contravariant anti-symmetric tensor of order $r = 4$, over a $V^3(K)$.

Example 7.6 (The Kronecker tensor product). Consider the anti-symmetric matrices $A_2 \equiv [a_{\circ\circ}^{ij}] = \begin{bmatrix} 0 & 2 \\ -2 & 0 \end{bmatrix}$ and $B_2 \equiv [b_{\circ\circ}^{k\ell}] = \begin{bmatrix} 0 & 3 \\ -3 & 0 \end{bmatrix}$, as two contravariant tensors, of order $r = 2$, over the linear space $\mathbb{R}^2 (n = 2)$.

1. Determine the matrix direct tensor product or the Kronecker tensor product, $C_4 = A \otimes B$, of both matrices.
2. Considering the matrices A and B as matrix representations of the corresponding tensors:

$$\vec{a} = 2\vec{e}_1 \otimes \vec{e}_2 - 2\vec{e}_2 \otimes \vec{e}_1 \tag{7.58}$$

$$\vec{b} = 3\vec{e}_1 \otimes \vec{e}_2 - 3\vec{e}_2 \otimes \vec{e}_1, \tag{7.59}$$

determine the representation matrix of the tensor product of both tensors.
3. Prepare a warning comment with respect to the operators "$\otimes$".

Solution:

1. The matrix direct tensor product or the Kronecker tensor product of both matrices is

$$C_4 = A \otimes B = \begin{bmatrix} 0 & 0 & 0 & 6 \\ 0 & 0 & -6 & 0 \\ 0 & -6 & 0 & 0 \\ 6 & 0 & 0 & 0 \end{bmatrix},$$

which obviously, is a symmetric matrix (it is a symmetric tensor of order $r = 2$ over the linear space $\mathbb{R}^4$).

2. The tensor product of the tensors $\vec{a}$ and $\vec{b}$ is

$$\vec{p} = \vec{a} \otimes \vec{b} = p_{\circ\circ\circ\circ}^{ijk\ell}\vec{e}_i \otimes \vec{e}_j \otimes \vec{e}_k \otimes \vec{e}_\ell; \text{ con } p_{\circ\circ\circ\circ}^{ijk\ell} = a_{\circ\circ}^{ij} \cdot b_{\circ\circ}^{k\ell}.$$

Thus, the matrix representation of the tensor $\vec{p}$ becomes

$$P = [p_{\circ\circ\circ\circ}^{ijk\ell}] \equiv \begin{bmatrix} p_{\circ\circ\circ\circ}^{1111} & p_{\circ\circ\circ\circ}^{1112} & | & p_{\circ\circ\circ\circ}^{1211} & p_{\circ\circ\circ\circ}^{1212} \\ p_{\circ\circ\circ\circ}^{1121} & p_{\circ\circ\circ\circ}^{1122} & | & p_{\circ\circ\circ\circ}^{1221} & p_{\circ\circ\circ\circ}^{1222} \\ - - - & - - - & + & - - - & - - - \\ p_{\circ\circ\circ\circ}^{2111} & p_{\circ\circ\circ\circ}^{2112} & | & p_{\circ\circ\circ\circ}^{2211} & p_{\circ\circ\circ\circ}^{2212} \\ p_{\circ\circ\circ\circ}^{2121} & p_{\circ\circ\circ\circ}^{2122} & | & p_{\circ\circ\circ\circ}^{2221} & p_{\circ\circ\circ\circ}^{2222} \end{bmatrix}$$

$$= \begin{bmatrix} 0 & 0 & | & 0 & 6 \\ 0 & 0 & | & -6 & 0 \\ - & - & + & - & - \\ 0 & -6 & | & 0 & 0 \\ 6 & 0 & | & 0 & 0 \end{bmatrix},$$

which is a tensor of order $r = 4$, $n = 2$, i.e., over $\mathbb{R}^2$.

However, the matrix P does not satisfy the properties of an anti-symmetric tensor of order $r = 4$ and $n = 2$, because since $r > n$ all its components have repeated indices (as one detects in the matrix of the P components) and then, all its components would be null, a condition that is not satisfied by P.

Then, $\vec{p}$ is not an anti-symmetric tensor of order $r = 4$ and $n = 2$ (it is a tensor of order $r = 4$ and $n = 2$ but not anti-symmetric).

We immediately conclude that the tensor product of anti-symmetric tensors is not in general anti-symmetric.

3. The major warning of this exercise leads to an analogous conclusion to that of Example 6.5: while the tensor product or Kronecker product of anti-symmetric matrices is a symmetric matrix, the tensor product of matrices that represent anti-symmetric tensors is not a tensor (neither symmetric nor anti-symmetric), though the same symbol "$\otimes$" and operating procedure (matrices in paragraphs 1 and 2 of the solution are identical for the mentioned products) be utilized. In simple words, the operation "$\otimes$" for matrices has not the same properties as the operation "$\otimes$" for tensors.

$\square$

Example 7.7 (Contracted product). Show that the contracted tensor product of two anti-symmetric matrices (classic product of matrices) is not in general a symmetric matrix. In which case is it?

Let $A_n = -A_n^t$ and $B_n = -B_n^t$ be two anti-symmetric matrices. Since

$$A_n \bullet B_n = (-A_n^t) \bullet (-B_n^t) = A_n^t \bullet B_n^t = (B_n \bullet A_n)^t \neq (A_n \bullet B_n)^t \quad (7.60)$$

the matrix $A_n \bullet B_n$ is not symmetric.

However, if $A_n \bullet B_n = B_n \bullet A_n$ the inequality of (7.60) becomes an equality:

$$A_n \bullet B_n = (B_n \bullet A_n)^t = (A_n \bullet B_n)^t \Rightarrow A_n \bullet B_n \text{ is symmetric,}$$

so that, if they commute, the product of two anti-symmetric matrices is symmetric.   □

7.5.1 Anti-symmetrized tensor associated with an arbitrary pure tensor

In Section 7.4.1 we have already developed a technique that allows us to associate with an arbitrary pure tensor, another tensor that is anti-symmetric with respect to a subset I_k of its indices. More precisely, Formula (7.31) was stipulated to build them.

Here we consider only the associated tensor that corresponds to $I_k \equiv I_r$, that is, the one corresponding to selecting *all* $r!$ isomers that correspond to the set $I_r = \{\alpha_1, \alpha_2, \cdots, \alpha_r\}$ of all the tensor indices with the sign corresponding to the factor $(-1)^\tau$.

The resulting anti-symmetric tensor will be called an "anti-symmetrized" form of the given tensor.

To illustrate, we give a detailed example. Let $T = t^{\alpha_1 \alpha_2 \cdots \alpha_r}_{\circ \ \circ \ \cdots \ \circ}$ be the given tensor over the linear space $V^n(K)$, and consider all the $r!$ permutations of its indices, which generate the corresponding isomers of the given tensor.

We denote by $H(T)^{\alpha_1, \alpha_2, \cdots, \alpha_r}$ the "anti-symmetrized" tensor of T, which is an anti-symmetric tensor, defined over the same tensor space $(\otimes V^n)^r(K)$, and the components of which can be obtained using the formula

$$H(T)^{\alpha_1 \alpha_2 \cdots \alpha_r}_{\circ \ \circ \ \cdots \ \circ} = \frac{1}{r!} \left[(-1)^{\tau_1} t^{\alpha_1 \alpha_2 \cdots \alpha_r}_{\circ \ \circ \ \cdots \ \circ} + (-1)^{\tau_2} t^{\alpha_2 \alpha_1 \cdots \alpha_r}_{\circ \ \circ \ \cdots \ \circ} \right.$$
$$\left. + (-1)^{\tau_3} t^{\alpha_3 \alpha_1 \cdots \alpha_r}_{\circ \ \circ \ \cdots \ \circ} + \cdots + (-1)^{\tau_h} t^{\alpha_r \alpha_{r-1} \cdots \alpha_1}_{\circ \ \circ \ \ \cdots \ \circ} \right]. \quad (7.61)$$

The summands are all the $h = r!$ isomer tensors.

7.5.2 Extension of the anti-symmetrized tensor concept associated with a mixed tensor

If the data tensor is mixed, it is possible to associate it with an anti-symmetrized tensor by establishing all isomers that result from considering all permutations of its p contravariant indices, independently of the corresponding permutations of its q covariant indices. If the data tensor is $T = t^{\alpha_1 \alpha_2 \cdots \alpha_p \ \circ \ \circ \ \cdots \ \circ}_{\ \circ \ \circ \ \cdots \ \circ \ \beta_1 \beta_2 \cdots \beta_q}$, of order $r = p + q$, the number of isomers will be $p! q!$, and if we denote by $\tau_1 = p_1 + q_1$ the number of transpositions of the first isomer, $\tau_2 = p_2 + q_2$ the number of transpositions of the second isomer, etc.,

$\tau_u = p_u + q_u$ the number of transpositions of the last isomer $(u = p!q!)$, the anti-symmetrized tensor is

$$H(T)^{\alpha_1 \alpha_2 \cdots \alpha_p \, \circ \, \circ \, \cdots \, \circ}_{\circ \, \circ \, \cdots \, \circ \, \beta_1 \beta_2 \cdots \beta_q} = \frac{1}{p!q!}\left[(-1)^{\tau_1} t^{\alpha_1 \cdots \alpha_p \, \circ \, \cdots \, \circ}_{\circ \, \cdots \, \circ \, \beta_1 \cdots \beta_q} + (-1)^{\tau_2} t^{\alpha_2 \cdots \alpha_p \, \circ \, \cdots \, \circ}_{\circ \, \cdots \, \circ \, \beta_1 \cdots \beta_q} \right.$$
$$\left. + \cdots + (-1)^{\tau_u} t^{\alpha_p \cdots \alpha_1 \, \circ \, \cdots \, \circ}_{\circ \, \cdots \, \circ \, \beta_q \cdots \beta_1} \right], \qquad (7.62)$$

where the right-hand sum extends to all $p!q!$ isomer tensors.

In order to determine the number of strict components of the anti-symmetrized tensor, we use Formula (7.22) and for the non-null components, Formula (7.23).

For example, consider the tensor $t^{\alpha\beta\circ\circ}_{\circ\circ\gamma\delta}$ with $p = 2$ and $q = 2$, of order $r = p + q = 4$ over the linear space $V^2(\mathbb{R})$, that is, $n = 2$, and a total of $\sigma = n^r = 2^4 = 16$ components. The anti-symmetrized tensor of the given tensor will have $N = \binom{n}{p}\binom{n}{q} = \binom{2}{2}\binom{2}{2} = 1$ strict component and $N = \binom{n}{p}\binom{n}{q}p!q! = \binom{2}{2}\binom{2}{2}2!2! = 4$ non-null components, so that

$$H(T)^{\alpha\beta\circ\circ}_{\circ\circ\gamma\delta} = \frac{1}{2!2!}[(t^{\alpha\beta\circ\circ}_{\circ\circ\gamma\delta} + t^{\beta\alpha\circ\circ}_{\circ\circ\delta\gamma} - (t^{\beta\alpha\circ\circ}_{\circ\circ\gamma\delta} + t^{\alpha\beta\circ\circ}_{\circ\circ\delta\gamma})],$$

and its matrix representation is

$$H(T)\{[12][12] = [21][21] = a; \ [12][21] = [21][12] = -a\}$$

$$[H(T)^{\alpha\beta\circ\circ}_{\circ\circ\gamma\delta}] = \begin{bmatrix} 0 & 0 & 0 & a \\ 0 & 0 & -a & 0 \\ 0 & -a & 0 & 0 \\ a & 0 & 0 & 0 \end{bmatrix},$$

where $a = \frac{1}{4}\left(t^{12\circ\circ}_{\circ\circ12} + t^{21\circ\circ}_{\circ\circ21} - t^{12\circ\circ}_{\circ\circ21} - t^{21\circ\circ}_{\circ\circ12}\right)$.

Example 7.8 (Anti-symmetrized tensors).

1. Obtain the anti-symmetrized tensor of tensor T given in Example 4.5, assuming that it is totally contravariant, using the general formula (7.61).
2. Obtain the anti-symmetrized tensor of the mixed tensor

$$A \equiv [A^{\alpha\beta\circ\circ}_{\circ\circ\gamma\delta}] = \begin{bmatrix} 1 & 2 & -2 & 1 \\ -3 & 4 & 0 & 6 \\ \hline 5 & -3 & -2 & -5 \\ -1 & 4 & -1 & 3 \end{bmatrix}.$$

Solution:

1. From Exercise 4.5 we extract the numerical data of the tensor T and of its isomers

$$T = [t^{\alpha\beta\gamma}_{\circ\circ\circ}] = \begin{bmatrix} 1 & 0 & -1 \\ 2 & 3 & 0 \\ -1 & 2 & 0 \\ 2 & -1 & 1 \\ 0 & 1 & 0 \\ 2 & 0 & 1 \\ 0 & 0 & 1 \\ 5 & 1 & 2 \\ 1 & 0 & 0 \end{bmatrix} \; ; \; U = [t^{\gamma\alpha\beta}_{\circ\circ\circ}] = \begin{bmatrix} 1 & 2 & -1 \\ 2 & 0 & 2 \\ 0 & 5 & 1 \\ 0 & 3 & 2 \\ -1 & 1 & 0 \\ 0 & 1 & 0 \\ -1 & 0 & 0 \\ 1 & 0 & 1 \\ 1 & 2 & 0 \end{bmatrix} \; ; \quad (7.63)$$

$$V = [t^{\beta\gamma\alpha}_{\circ\circ\circ}] = \begin{bmatrix} 1 & 2 & 0 \\ 0 & -1 & 0 \\ -1 & 1 & 1 \\ 2 & 0 & 5 \\ 3 & 1 & 1 \\ 0 & 0 & 2 \\ -1 & 2 & 1 \\ 2 & 0 & 0 \\ 0 & 1 & 0 \end{bmatrix} \; ; W = [t^{\alpha\gamma\beta}_{\circ\circ\circ}] = \begin{bmatrix} 1 & 2 & -1 \\ 0 & 3 & 2 \\ -1 & 0 & 0 \\ 2 & 0 & 2 \\ -1 & 1 & 0 \\ 1 & 0 & 1 \\ 0 & 5 & 1 \\ 0 & 1 & 0 \\ 1 & 2 & 0 \end{bmatrix} \; ; \quad (7.64)$$

$$R = [t^{\gamma\beta\alpha}_{\circ\circ\circ}] = \begin{bmatrix} 1 & 2 & 0 \\ 2 & 0 & 5 \\ -1 & 2 & 1 \\ 0 & -1 & 0 \\ 3 & 1 & 1 \\ 2 & 0 & 0 \\ -1 & 1 & 1 \\ 0 & 0 & 2 \\ 0 & 1 & 0 \end{bmatrix} \; ; \; S = [t^{\beta\alpha\gamma}_{\circ\circ\circ}] = \begin{bmatrix} 1 & 0 & -1 \\ 2 & -1 & 1 \\ 0 & 0 & 1 \\ 2 & 3 & 0 \\ 0 & 1 & 0 \\ 5 & 1 & 2 \\ -1 & 2 & 0 \\ 2 & 0 & 1 \\ 1 & 0 & 0 \end{bmatrix} . \quad (7.65)$$

The anti-symmetrized tensor is

$$H(T) = \frac{1}{6}[(T+U+V)-(W+R+S)] = \frac{1}{6}\begin{bmatrix} 0 & 0 & 0 \\ 0 & 0 & -6 \\ 0 & 6 & 0 \\ 0 & 0 & 6 \\ 0 & 0 & 0 \\ -6 & 0 & 0 \\ 0 & -6 & 0 \\ 6 & 0 & 0 \\ 0 & 0 & 0 \end{bmatrix} = \begin{bmatrix} 0 & 0 & 0 \\ 0 & 0 & -1 \\ 0 & 1 & 0 \\ 0 & 0 & 1 \\ 0 & 0 & 0 \\ -1 & 0 & 0 \\ 0 & -1 & 0 \\ 1 & 0 & 0 \\ 0 & 0 & 0 \end{bmatrix} .$$

2. There are four isomers

$$[a^{\alpha\beta\circ\circ}_{\circ\circ\gamma\delta}] = \begin{bmatrix} 1 & 2 & -2 & 1 \\ -3 & 4 & 0 & 6 \\ 5 & -3 & -2 & -5 \\ -1 & 4 & -1 & 3 \end{bmatrix} \; ; \; [a^{\beta\alpha\circ\circ}_{\circ\circ\gamma\delta}] = \begin{bmatrix} 1 & 2 & 5 & -3 \\ -3 & 4 & -1 & 4 \\ -2 & 1 & -2 & -5 \\ 0 & 6 & -1 & 3 \end{bmatrix} \; ;$$

$$[a^{\alpha\beta\circ\circ}_{\circ\circ\delta\gamma}] = \begin{bmatrix} 1 & -3 & -2 & 0 \\ 2 & 4 & 1 & 6 \\ 5 & -1 & -2 & -1 \\ -3 & 4 & -5 & 3 \end{bmatrix} \; ; \; [a^{\beta\alpha\circ\circ}_{\circ\circ\delta\gamma}] = \begin{bmatrix} 1 & -3 & 5 & -1 \\ 2 & 4 & -3 & 4 \\ -2 & 0 & -2 & -1 \\ 1 & 6 & -5 & 3 \end{bmatrix} ,$$

so that

$$H(A) = \frac{1}{2!2!}\left[(a^{\alpha\beta\circ\circ}_{\circ\circ\gamma\delta} + a^{\beta\alpha\circ\circ}_{\circ\circ\delta\gamma}) - (a^{\beta\alpha\circ\circ}_{\circ\circ\gamma\delta} + a^{\alpha\beta\circ\circ}_{\circ\circ\delta\gamma})\right].$$

$$[h^{\alpha\beta\circ\circ}_{\circ\circ\gamma\delta}] = H(A) = \frac{1}{4}\begin{bmatrix} 0 & 0 & 0 & 3 \\ 0 & 0 & -3 & 0 \\ \hline 0 & -3 & 0 & 0 \\ 3 & 0 & 0 & 0 \end{bmatrix} = \begin{bmatrix} 0 & 0 & 0 & \frac{3}{4} \\ 0 & 0 & -\frac{3}{4} & 0 \\ 0 & -\frac{3}{4} & 0 & 0 \\ \frac{3}{4} & 0 & 0 & 0 \end{bmatrix};$$

□

7.6 Anti-symmetric tensor algebras: The $\otimes_H$ product

Once the concept of "anti-symetrization" of an arbitrary homogeneous tensor has been established, it is possible to define a new tensor product for exclusive use with anti-symmetric tensors, the product of which is another anti-symmetric tensor, and since the sum of two anti-symmetric tensors is another anti-symmetric tensor, establish a good foundation for an anti-symmetric tensor algebra, that is, for anti-symmetric tensors.

This new product will receive the name "anti-symmetric tensor product" and will be notated "$\otimes_H$".

Let A and B be two anti-symmetric homogeneous tensors, of orders r_a and r_b, respectively, both over the linear space $V^n(K)$; if $P = A \otimes B$, of order $r = r_a + r_b$, we define the "anti-symmetric tensor product" by means of the following formula:

$$A \otimes_H B = H[P] = H(A \otimes B) \text{ (anti-symmetrized tensor of } A \otimes B). \quad (7.66)$$

Next, we give an example of the new product.

Let $\vec{a}, \vec{b} \in V^2 \otimes V^2(K)$ be two anti-symmetric contravariant tensors, and let $\sigma = n^r = 2^2 = 4$ be the dimension of the tensor space to which they belong. The matrix representation of our data is

$$A = [a^{\alpha\beta}_{\circ\circ}] \equiv \begin{bmatrix} 0 & b_1 \\ -b_1 & 0 \end{bmatrix}; \quad B = [b^{\gamma\delta}_{\circ\circ}] \equiv \begin{bmatrix} 0 & b_2 \\ -b_2 & 0 \end{bmatrix}; r = r_a + r_b = 2 + 2 = 4.$$

$$A \otimes_H B = \begin{bmatrix} 0 & b_1 \\ -b_1 & 0 \end{bmatrix} \otimes_H \begin{bmatrix} 0 & b_2 \\ -b_2 & 0 \end{bmatrix} = H[[a^{\alpha\beta}_{\circ\circ}] \otimes [b^{\gamma\delta}_{\circ\circ}]] = H[p^{\alpha\beta\gamma\delta}_{\circ\circ\circ\circ}]$$

$$= H\left(\begin{bmatrix} 0 & 0 & 0 & b_1 b_2 \\ 0 & 0 & 1 & -b_1 b_2 & 0 \\ \hline 0 & -b_1 b_2 & 0 & 0 \\ b_1 b_2 & 0 & 0 & 0 \end{bmatrix}\right)$$

$$= \frac{1}{4!}\left[\left(p^{\alpha\beta\gamma\delta}_{\circ\circ\circ\circ} - p^{\alpha\beta\delta\gamma}_{\circ\circ\circ\circ} - p^{\alpha\gamma\beta\delta}_{\circ\circ\circ\circ} + p^{\alpha\gamma\delta\beta}_{\circ\circ\circ\circ} - p^{\alpha\delta\gamma\beta}_{\circ\circ\circ\circ} + p^{\alpha\delta\beta\gamma}_{\circ\circ\circ\circ}\right)\right.$$

$$+ \left(-p^{\beta\alpha\gamma\delta}_{\circ\circ\circ\circ} + p^{\beta\alpha\delta\gamma}_{\circ\circ\circ\circ} + p^{\beta\gamma\alpha\delta}_{\circ\circ\circ\circ} - p^{\beta\gamma\delta\alpha}_{\circ\circ\circ\circ} + p^{\beta\delta\gamma\alpha}_{\circ\circ\circ\circ} - p^{\beta\delta\alpha\gamma}_{\circ\circ\circ\circ}\right)$$

$$+ \left.\left(p^{\gamma\alpha\beta\delta}_{\circ\circ\circ\circ} - p^{\gamma\alpha\delta\beta}_{\circ\circ\circ\circ} - p^{\gamma\beta\alpha\delta}_{\circ\circ\circ\circ} + p^{\gamma\beta\delta\alpha}_{\circ\circ\circ\circ} + p^{\gamma\delta\beta\alpha}_{\circ\circ\circ\circ} - p^{\gamma\delta\alpha\beta}_{\circ\circ\circ\circ}\right)\right.$$

$$+ \left(-p_{\circ\circ\circ\circ}^{\delta\alpha\beta\gamma} + p_{\circ\circ\circ\circ}^{\delta\alpha\gamma\beta} + p_{\circ\circ\circ\circ}^{\delta\beta\alpha\gamma} - p_{\circ\circ\circ\circ}^{\delta\beta\gamma\alpha} + p_{\circ\circ\circ\circ}^{\delta\gamma\beta\alpha} - p_{\circ\circ\circ\circ}^{\delta\gamma\alpha\beta} \right) \Big]$$

$$= \left[\begin{array}{cc|cc} 0 & 0 & 0 & 0 \\ 0 & 0 & 0 & 0 \\ \hline 0 & 0 & 0 & 0 \\ 0 & 0 & 0 & 0 \end{array} \right] \equiv \Omega.$$

The obtained result was predictable; an anti-symmetric tensor $[p_{\circ\circ\circ\circ}^{ijk\ell}]$ of order $r = 4$ and $n = 2$, has $r > n$, and then all components of P have repeated indices, which implies that all $[p_{\circ\circ\circ\circ}^{ijk\ell}]$ must be null.

It can be easily proved that the anti-symmetric tensor product has the following properties:

1. It is commutative:

$$A \otimes_H B = B \otimes_H A. \tag{7.67}$$

2. It is associative:

$$(A \otimes_H B) \otimes_H C = A \otimes_H (B \otimes_H C). \tag{7.68}$$

3. It is distributive for the sum "+" of tensors:

$$(A + B) \otimes_H C = A \otimes_H C + B \otimes_H C. \tag{7.69}$$

$$A \otimes_H (B + C) = A \otimes_H B + A \otimes_H C. \tag{7.70}$$

4. Based on the above properties, it is obvious that the anti-symmetric tensor product, together with the sum, endow the set of homogeneous anti-symmetric tensors with the character of an *anti-symmetric tensor algebra*.

7.7 Illustrative examples

Example 7.9 (The rotation tensor and the anti-symmetry). Consider a homogeneous contravariant tensor T, of third order, that is related to another tensor W, by means of the expression

$$t_{\circ\circ\circ}^{ijk} = -w_{\circ\circ\circ}^{ijk} + w_{\circ\circ\circ}^{jik}.$$

It is also known that the unknown tensor W is of third order, contravariant and anti-symmetric with respect to the last two indices.

1. Show that the tensor T is anti-symmetric with respect to the first two indices.
2. Show that the rotation tensors of the given tensor T also satisfy the same anti-symmetry.
3. Determine the generic component $w_{\circ\circ\circ}^{ijk}$ of the tensor W as a function of the components of T and its isomers.

Solution:

1. Exchanging i and j in the given expression we have

$$t^{ijk}_{\circ\circ\circ} = w^{jik}_{\circ\circ\circ} - w^{ijk}_{\circ\circ\circ} \tag{7.71}$$

$$t^{jik}_{\circ\circ\circ} = w^{ijk}_{\circ\circ\circ} - w^{jik}_{\circ\circ\circ} \tag{7.72}$$

and adding them gives

$$t^{jik}_{\circ\circ\circ} + t^{ijk}_{\circ\circ\circ} = 0 \quad \rightarrow \quad t^{jik}_{\circ\circ\circ} = -t^{ijk}_{\circ\circ\circ}.$$

2. Consider the isomer tensors of the given tensor, resulting from the circular permutations of its indices (rotation tensors; see Example 5.5):
   Rotation 1 (the index i passes to j || the index j passes to k || the index k passes to i)
   Rotation 2 (the index i passes to k || the index j passes to i || the index k passes to j).

   Then

$$t^{ijk}_{\circ\circ\circ} = w^{jik}_{\circ\circ\circ} - w^{ijk}_{\circ\circ\circ} \tag{7.73}$$

   and we obtain the rotation isomer 1:

$$t^{jki}_{\circ\circ\circ} = w^{kji}_{\circ\circ\circ} - w^{jki}_{\circ\circ\circ}. \tag{7.74}$$

   Exchanging j and k in Expression (7.74), the result is

$$t^{kji}_{\circ\circ\circ} = w^{jki}_{\circ\circ\circ} - w^{kji}_{\circ\circ\circ} \tag{7.75}$$

   and adding (7.74) and (7.75) we get

$$t^{jki}_{\circ\circ\circ} + t^{kji}_{\circ\circ\circ} = 0 \quad \rightarrow \quad t^{kji}_{\circ\circ\circ} = -t^{jki}_{\circ\circ\circ}. \tag{7.76}$$

   Similarly, rotation 2 is

$$t^{kij}_{\circ\circ\circ} = w^{ikj}_{\circ\circ\circ} - w^{kij}_{\circ\circ\circ}. \tag{7.77}$$

   Exchanging k and i in Expression (7.77), we obtain

$$t^{ikj}_{\circ\circ\circ} = w^{kij}_{\circ\circ\circ} - w^{ikj}_{\circ\circ\circ} \tag{7.78}$$

   and adding (7.77) and (7.78) we get

$$t^{kij}_{\circ\circ\circ} + t^{ikj}_{\circ\circ\circ} = 0 \quad \rightarrow \quad t^{ikj}_{\circ\circ\circ} = -t^{kij}_{\circ\circ\circ}. \tag{7.79}$$

3. Considering the rotations

$$t^{ijk}_{\circ\circ\circ} = w^{jik}_{\circ\circ\circ} - w^{ijk}_{\circ\circ\circ} \tag{7.80}$$

$$t^{jki}_{\circ\circ\circ} = w^{kji}_{\circ\circ\circ} - w^{jki}_{\circ\circ\circ} \tag{7.81}$$

$$t^{kij}_{\circ\circ\circ} = w^{ikj}_{\circ\circ\circ} - w^{kij}_{\circ\circ\circ} \tag{7.82}$$

because of the anti-symmetry of the last two indices of W, they can be written

$$t^{ijk}_{\circ\circ\circ} = -w^{jki}_{\circ\circ\circ} - w^{ijk}_{\circ\circ\circ} \tag{7.83}$$

$$-t^{jki}_{\circ\circ\circ} = -w^{kji}_{\circ\circ\circ} + w^{jki}_{\circ\circ\circ} \tag{7.84}$$

$$t^{kij}_{\circ\circ\circ} = -w^{ijk}_{\circ\circ\circ} + w^{kji}_{\circ\circ\circ} \tag{7.85}$$

and adding (7.83), (7.84), and (7.85) we get

$$t^{ijk}_{\circ\circ\circ} - t^{jki}_{\circ\circ\circ} + t^{kij}_{\circ\circ\circ} = (-w^{jki}_{\circ\circ\circ} - w^{ijk}_{\circ\circ\circ}) + (-w^{kji}_{\circ\circ\circ} + w^{jki}_{\circ\circ\circ})$$

$$+(-w^{ijk}_{\circ\circ\circ} + w^{kji}_{\circ\circ\circ}) = -2w^{ijk}_{\circ\circ\circ};$$

and finally

$$w^{ijk}_{\circ\circ\circ} = \frac{1}{2}\left[t^{jki}_{\circ\circ\circ} - t^{ijk}_{\circ\circ\circ} - t^{kij}_{\circ\circ\circ} \right].$$

□

Example 7.10 (Contracted products of tensors with partial and reverse symmetries).

1. Consider a homogeneous contravariant symmetric tensor A of second order $(r = 2)$, and another homogeneous covariant anti-symmetric tensor B of second order $(r = 2)$, both over the linear space $V^n(K)$, obtain the doubly contracted product tensor $a^{\alpha\beta}_{\circ\circ}b^{\circ\circ}_{\alpha\beta}$ of both tensors.

2. Let A be a contravariant tensor of third order $(r = 3)$, symmetric with respect to the first two indices, and B a covariant tensor of third order $(r = 3)$ anti-symmetric with respect to the first two indices, both over the linear space $V^n(K)$. Obtain the doubly contracted tensor product C of both tensors

$$c^{i\circ}_{\circ j} = a^{\alpha\beta i}_{\circ\circ\circ}b^{\circ\circ\circ}_{\alpha\beta j}.$$

Solution:

1. Since it is a totally contracted product of homogeneous tensors of second order, the resulting tensor is a tensor of zero order (a scalar invariant). Whence

$$a^{\alpha\beta}_{\circ\circ}b^{\circ\circ}_{\alpha\beta} = c;\quad c \in K.$$

Since α and β are dummy indices with the same range, they are exchangeable, that is

$$a^{\alpha\beta}_{\circ\circ}b^{\circ\circ}_{\alpha\beta} = a^{\beta\alpha}_{\circ\circ}b^{\circ\circ}_{\beta\alpha} = c, \tag{7.86}$$

and since A is symmetric ($a^{ji}_{\circ\circ} = a^{ij}_{\circ\circ}$) and B is anti-symmetric ($b^{\circ\circ}_{ji} = -b^{\circ\circ}_{ij}$), from (7.86), we get

$$c = a^{\beta\alpha}_{\circ\circ}b^{\circ\circ}_{\beta\alpha} = a^{\alpha\beta}_{\circ\circ}(-b^{\circ\circ}_{\alpha\beta}) = -a^{\alpha\beta}_{\circ\circ}b^{\circ\circ}_{\alpha\beta} = -c.$$

Thus, $c = -c \;\Rightarrow\; c = 0$ (because $K \neq 2$).
2. If in the relation

$$c^{i\circ}_{\circ j} = a^{\alpha\beta i}_{\circ\circ\circ}b^{\circ\circ\circ}_{\alpha\beta j} \tag{7.87}$$

we exchange the dummy indices α and β, we get

$$c^{i\circ}_{\circ j} = a^{\beta\alpha i}_{\circ\circ\circ}b^{\circ\circ\circ}_{\beta\alpha j}. \tag{7.88}$$

In addition, from the symmetry of A and the anti-symmetry of B, we have

$$a^{\beta\alpha i}_{\circ\circ\circ} = a^{\alpha\beta i}_{\circ\circ\circ} \tag{7.89}$$

$$b^{\circ\circ\circ}_{\beta\alpha j} = -b^{\circ\circ\circ}_{\alpha\beta j}. \tag{7.90}$$

The product, term by term, of (7.89) and (7.90) gives

$$a^{\beta\alpha i}_{\circ\circ\circ}b^{\circ\circ\circ}_{\beta\alpha j} = -a^{\alpha\beta i}_{\circ\circ\circ}b^{\circ\circ\circ}_{\alpha\beta j} \tag{7.91}$$

and substituting (7.87) and (7.88) into (7.91), we finally obtain

$$c^{i\circ}_{\circ j} = -c^{i\circ}_{\circ j} = \;\Rightarrow\; 2c^{i\circ}_{\circ j} = 0 \;\rightarrow\; c^{i\circ}_{\circ j} = 0.$$

This result is different from the previous result, because now we have $n^r = n^2$ zeroes, which implies $C \equiv \Omega_n$.

□

Example 7.11 (Strict components of an anti-symmetric mixed tensor). Consider the anti-symmetric mixed tensor with components $t^{\circ\circ k}_{ij\circ}$ over the linear space $(V^n(K)$. Calculate the number of strict components.

Solution: Since the anti-symmetry has a tensor nature, it must be of the covariant indices. From Formula (7.21), we have

$$p = 1; \; k_1 = 0; \; q = 2; \; k_2 = 2; \; r = p + q = 1 + 2 = 3$$

$$\sigma' = \binom{n}{k_1}\binom{n}{k_2}n^{r-(k_1+k_2)} = \binom{n}{0}\binom{n}{2}n^{3-(0+2)} = \binom{n}{2}n = \frac{n^2(n-1)}{2}.$$

□

Example 7.12 (Main tensors associated with the given tensor). Consider the tensor $\vec{t}$ in Example 6.11 and assume that we know the solutions of such a example.

1. Give the components of the anti-symmetrized tensor $t_{h\,\substack{i\,j\\o\,o}}$ of the given tensor.
2. Give the components of the anti-symmetrized tensor $t_{hr\,\substack{i\,j\\o\,o}}$ of the rotation tensor of the given tensor.
3. Decompose $\vec{t}$ as the sum of a symmetric and an anti-symmetric tensor.

Solution:

1. The components of the anti-symmetrized tensor $t_{h\,\substack{i\,j\\o\,o}}$ are

$$[t_{h\,\substack{i\,j\\o\,o}}] = \frac{1}{2}[t_{\substack{i\,j\\o\,o}} - t_{\substack{j\,i\\o\,o}}] = \begin{bmatrix} 0 & 3/2 & -1/2 \\ -3/2 & 0 & 3/2 \\ 1/2 & -3/2 & 0 \end{bmatrix}.$$

2. The components of the anti-symmetrized tensor $t_{hr\,\substack{i\,j\\o\,o}}$ of the rotation tensor of $\vec{t}$ are:

$$[t_{hr\,\substack{i\,j\\o\,o}}] = \frac{1}{2}[t_{r\,\substack{i\,j\\o\,o}} - t_{r\,\substack{j\,i\\o\,o}}] = \frac{1}{2}[t_{\substack{j\,i\\o\,o}} - t_{\substack{i\,j\\o\,o}}] = \begin{bmatrix} 0 & -3/2 & 1/2 \\ 3/2 & 0 & -3/2 \\ -1/2 & 3/2 & 0 \end{bmatrix}.$$

3. In this case we have

$$\vec{t} = t_{s\,\substack{i\,j\\o\,o}}\vec{e}_i \otimes \vec{e}_j + t_{h\,\substack{i\,j\\o\,o}}\vec{e}_i \otimes \vec{e}_j$$

$$= [\vec{e}_1 \quad \vec{e}_2 \quad \vec{e}_3] \begin{bmatrix} 1 & 3/2 & -3/2 \\ 3/2 & 1 & 3/2 \\ -3/2 & 3/2 & 2 \end{bmatrix} \begin{bmatrix} \vec{e}_1 \\ \vec{e}_2 \\ \vec{e}_3 \end{bmatrix}$$

$$+ [\vec{e}_1 \quad \vec{e}_2 \quad \vec{e}_3] \begin{bmatrix} 0 & 3/2 & -1/2 \\ -3/2 & 0 & 3/2 \\ 1/2 & -3/2 & 0 \end{bmatrix} \begin{bmatrix} \vec{e}_1 \\ \vec{e}_2 \\ \vec{e}_3 \end{bmatrix}$$

$$= (\vec{e}_1 \otimes \vec{e}_1 + \vec{e}_2 \otimes \vec{e}_2 + 2\vec{e}_3 \otimes \vec{e}_3 + 3/2\vec{e}_1 \otimes \vec{e}_2 + 3/2\vec{e}_2 \otimes \vec{e}_1 - 3/2\vec{e}_1 \otimes \vec{e}_3$$
$$-3/2\vec{e}_3 \otimes \vec{e}_1 + 3/2\vec{e}_2 \otimes \vec{e}_3 + 3/2\vec{e}_3 \otimes \vec{e}_2) + (3/2\vec{e}_1 \otimes \vec{e}_2 - 3/2\vec{e}_2 \otimes \vec{e}_1$$
$$-1/2\vec{e}_1 \otimes \vec{e}_3 + 1/2\vec{e}_3 \otimes \vec{e}_1 + 3/2\vec{e}_2 \otimes \vec{e}_3 - 3/2\vec{e}_3 \otimes \vec{e}_2)$$
$$= \vec{e}_1 \otimes \vec{e}_1 + \vec{e}_2 \otimes \vec{e}_2 + 2\vec{e}_3 \otimes \vec{e}_3 + 3\vec{e}_1 \otimes \vec{e}_2 - 2\vec{e}_1 \otimes \vec{e}_3 - \vec{e}_3 \otimes \vec{e}_1 + 3\vec{e}_2 \otimes \vec{e}_3,$$

which is the data tensor $\vec{t}$. This shows that a second-order tensor is the sum of its symmetrized tensor and its anti-symmetrized tensor.

□

Example 7.13 (Analysis of tensor anti-symmetries). Study the possible anti-symmetric nature of each of the homogeneous tensors defined over a linear space $V^n(K)$, the components of which in a given basis appear as:

1. $t^{ijk\ell}_{\circ\circ\circ\circ} = i - j + k - \ell$.
2. $t^{\circ\circ\circ}_{ijk} = (i-j)(j-k)(k-i)$.
3. $t^{ij}_{\circ\circ} = (i-j)(j-i)$.
4. $t^{\circ\circ\circ\ell\circ}_{ijkom} = (i-j)k\ell m$.

Solution:

1. When letting $i = j = 1$, $k = 2$ and $\ell = 3$, the result is $t^{1123}_{\circ\circ\circ\circ} = 1-1+2-3 = -1 \neq 0$.

   Since the studied component has repeated indices, and is not null, the tensor *is not anti-symmetric*.

2. If two indices are identical ($i = j, j = k, k = i$) the component is null. Next, we study the permutations of different indices

$$t^{\circ\circ\circ}_{jik} = -t^{\circ\circ\circ}_{ijk}; \quad t^{\circ\circ\circ}_{ikj} = -t^{\circ\circ\circ}_{ijk}; \quad t^{\circ\circ\circ}_{kji} = -t^{\circ\circ\circ}_{ijk};$$

$$t^{\circ\circ\circ}_{jki} = (j-k)(k-i)(i-j) = (i-j)(j-k)(k-i) = t^{\circ\circ\circ}_{ijk}$$

$$t^{\circ\circ\circ}_{kij} = (k-i)(i-j)(j-k) = (i-j)(j-k)(k-i) = t^{\circ\circ\circ}_{ijk},$$

which shows it is a third-order *covariant anti-symmetric* tensor.

3. In this case we have

$$t^{ji}_{\circ\circ} = (j-i)(i-j) = (i-j)(j-i) = t^{ij}_{\circ\circ},$$

which shows it is a second-order *contravariant symmetric* tensor, though since $t^{ii}_{\circ\circ} = 0$, it is of null trace; *it is not anti-symmetric*.

4. In this case we have

$$t^{\circ\circ\circ\ell\circ}_{jikom} = (j-i)k\ell m = -(i-j)k\ell m = -t^{\circ\circ\circ\ell\circ}_{ijkom}.$$

If $i = j$, then $t^{\circ\circ\circ\ell\circ}_{iikom} = 0 \cdot k\ell m = 0$. It is thus a fifth-order mixed tensor, anti-symmetric with respect to its first two indices.

□

Example 7.14 (Intrinsic character of symmetry and anti-symmetry). In a linear space $V^3(\mathbb{R})$ two bases $\{\vec{e}_\alpha\}$ and $\{\vec{\hat{e}}_i\}$ are considered, which are related by

$$\vec{\hat{e}}_1 = \vec{e}_1 - \vec{e}_2; \quad \vec{\hat{e}}_2 = \vec{e}_2 - \vec{e}_3; \quad \vec{\hat{e}}_3 = \vec{e}_3.$$

Consider two homogeneous tensors $\vec{a} \in V^3 \otimes V^3(\mathbb{R})$ and $\vec{b} \in V^3 \otimes V^3_* \otimes V^3(\mathbb{R})$ the components of which in the tensor bases associated with $\{\vec{e}_\alpha\}$, are

$$[a^{\alpha\beta}_{\circ\circ}] = \begin{bmatrix} 0 & 0 & 1 \\ 0 & -1 & 0 \\ 1 & 0 & 1 \end{bmatrix} \quad \text{and} \quad [b^{\alpha\circ\gamma}_{\circ\beta\circ}] = \begin{bmatrix} 0 & 1 & -1 \\ 0 & 0 & 1 \\ 0 & -1 & 1 \\ -1 & 0 & 0 \\ 0 & 0 & 2 \\ 1 & 0 & 0 \\ 1 & 0 & 0 \\ -1 & -2 & 0 \\ -1 & 0 & 0 \end{bmatrix}.$$

1. Obtain the components of tensors $\vec{a}$ and $\vec{b}$ in the new bases associated with the basis $\{\vec{\tilde{e}}_i\}$.
2. Analyze the possible tensor symmetries or anti-symmetries of $\vec{a}$ and $\vec{b}$.
3. Obtain the contracted tensor product $\vec{p}$, with respect to the last index of $\vec{a}$ and the second index of $\vec{b}$, in the initial basis.
4. Give $\vec{p}$ in the new basis, by a direct calculation of the contraction.
5. Idem, but executing the change-of-basis.
6. Has $\vec{p}$ some symmetry or anti-symmetry?

Solution: The change-of-basis relations are

$$||\vec{\tilde{e}}_i|| = ||\vec{e}_\alpha|| C; \quad C = \begin{bmatrix} 1 & 0 & 0 \\ -1 & 1 & 0 \\ 0 & -1 & 1 \end{bmatrix}; \quad C^{-1} = \begin{bmatrix} 1 & 0 & 0 \\ 1 & 1 & 0 \\ 1 & 1 & 1 \end{bmatrix}; \quad C^t = \begin{bmatrix} 1 & -1 & 0 \\ 0 & 1 & -1 \\ 0 & 0 & 1 \end{bmatrix}.$$

1. Change of basis of tensor $\vec{a}$: we directly apply Formula (4.36), that is, we use the direct method. Since $\sigma = n^r = 3^2 = 9$, we have

$$\hat{A}_{\sigma,1} = \hat{A}_{9,1} = (C^{-1} \otimes C^{-1}) \bullet A_{9,1}$$

$$= \left(\begin{bmatrix} 1 & 0 & 0 \\ 1 & 1 & 0 \\ 1 & 1 & 1 \end{bmatrix} \otimes \begin{bmatrix} 1 & 0 & 0 \\ 1 & 1 & 0 \\ 1 & 1 & 1 \end{bmatrix} \right) \begin{bmatrix} 0 \\ 0 \\ 1 \\ 0 \\ -1 \\ 0 \\ 1 \\ 0 \\ 1 \end{bmatrix} = \begin{bmatrix} 0 \\ 0 \\ 1 \\ 0 \\ -1 \\ 0 \\ 1 \\ 0 \\ 2 \end{bmatrix}$$

and condensing $\hat{A}_{9,1}$ we obtain

$$\hat{A} \equiv [a^{ij}_{\circ\circ}] = \begin{bmatrix} 0 & 0 & 1 \\ 0 & -1 & 0 \\ 1 & 0 & 2 \end{bmatrix}.$$

Change of basis of tensor $\vec{b}$: we apply again Formula (4.36). Since $\sigma = n^r = 3^3 = 27$, we have

$$\hat{B}_{\sigma,1} = \hat{B}_{27,1} = (C^{-1} \otimes C^t \otimes C^{-1}) \bullet B_{27,1}$$

$$
= \left(\begin{bmatrix} 1 & 0 & 0 \\ 1 & 1 & 0 \\ 1 & 1 & 1 \end{bmatrix} \otimes \begin{bmatrix} 1 & -1 & 0 \\ 0 & 1 & -1 \\ 0 & 0 & 1 \end{bmatrix} \otimes \begin{bmatrix} 1 & 0 & 0 \\ 1 & 1 & 0 \\ 1 & 1 & 1 \end{bmatrix} \right)
\begin{bmatrix} 0 \\ 1 \\ -1 \\ 0 \\ 0 \\ 1 \\ 0 \\ -1 \\ 1 \\ \hline -1 \\ 0 \\ 0 \\ 0 \\ 0 \\ 0 \\ 2 \\ 1 \\ 0 \\ 0 \\ \hline 1 \\ 0 \\ 0 \\ -1 \\ -2 \\ 0 \\ -1 \\ 0 \\ 0 \end{bmatrix}
=
\begin{bmatrix} 0 \\ 1 \\ -1 \\ 0 \\ 1 \\ 1 \\ 0 \\ -1 \\ 0 \\ \hline -1 \\ 0 \\ -4 \\ -1 \\ 0 \\ 2 \\ 1 \\ 0 \\ 1 \\ \hline 1 \\ 4 \\ 0 \\ -1 \\ -2 \\ 0 \\ 0 \\ -1 \\ 0 \end{bmatrix}
$$

and condensing $\hat{B}_{27,1}$ we get

$$
\hat{B} \equiv [b^{\,i\,\circ\,k}_{\circ\,j\,\circ}] = \begin{bmatrix} 0 & 1 & -1 \\ 0 & 1 & 1 \\ 0 & -1 & 0 \\ -1 & 0 & -4 \\ -1 & 0 & 2 \\ 1 & 0 & 1 \\ 1 & 4 & 0 \\ -1 & -2 & 0 \\ 0 & -1 & 0 \end{bmatrix}.
$$

2. Analysis of tensor $\vec{a}$ in the initial basis:

$$
a^{12}_{\circ\circ} = a^{21}_{\circ\circ} = 0; \quad a^{13}_{\circ\circ} = a^{31}_{\circ\circ} = 1; \quad a^{23}_{\circ\circ} = a^{32}_{\circ\circ} = 0.
$$

Analysis of $\vec{a}$ in the final basis:

$$
\hat{a}^{12}_{\circ\circ} = \hat{a}^{21}_{\circ\circ} = 0; \quad \hat{a}^{13}_{\circ\circ} = \hat{a}^{31}_{\circ\circ} = 1; \quad \hat{a}^{23}_{\circ\circ} = \hat{a}^{32}_{\circ\circ} = 0 \quad \rightarrow a^{\alpha\beta}_{\circ\circ} = a^{\beta\alpha}_{\circ\circ}.
$$

Evidently $\vec{a}$ is a contravariant symmetric tensor.
Analysis of tensor $\vec{b}$, in the initial basis:

$$
b^{1\,\circ\,1}_{\circ\,1\,\circ} = b^{2\,\circ\,2}_{\circ\,1\,\circ} = b^{3\,\circ\,3}_{\circ\,1\,\circ} = 0; \quad b^{1\,\circ\,2}_{\circ\,1\,\circ} = -b^{2\,\circ\,1}_{\circ\,1\,\circ} = 1.
$$

$$
b^{1\,\circ\,3}_{\circ\,1\,\circ} = -b^{3\,\circ\,1}_{\circ\,1\,\circ} = -1; \quad b^{2\,\circ\,3}_{\circ\,1\,\circ} = -b^{3\,\circ\,2}_{\circ\,1\,\circ} = 0.
$$

Analysis of $\vec{b}$ in the final basis:

$$\hat{b}^{1\circ1}_{\circ1\circ} = \hat{b}^{2\circ2}_{\circ1\circ} = \hat{b}^{3\circ3}_{\circ1\circ} = 0; \quad \hat{b}^{1\circ2}_{\circ1\circ} = -\hat{b}^{2\circ1}_{\circ1\circ} = 1.$$

$$\hat{b}^{1\circ3}_{\circ1\circ} = -\hat{b}^{3\circ1}_{\circ1\circ} = -1; \quad \hat{b}^{2\circ3}_{\circ1\circ} = -\hat{b}^{3\circ2}_{\circ1\circ} = -4.$$

so that

$$b^{\alpha\circ\gamma}_{\circ\beta\circ} = -b^{\gamma\circ\alpha}_{\circ\beta\circ}; \quad \text{si } \gamma = \alpha \to b^{\alpha\circ\alpha}_{\circ\beta\circ} = 0,$$

that is, the $\vec{b}$ is a mixed anti-symmetric tensor with respect to its con-travariant indices tensor.

3. The contraction $\binom{\beta}{\epsilon}$ of the tensor product $(a^{\alpha\beta}_{\circ\circ} \otimes b^{\delta\circ\gamma}_{\circ\epsilon\circ})$ leads to

$$[p^{\alpha\delta\gamma}_{\circ\circ\circ}] = [a^{\alpha\theta}_{\circ\circ} \cdot b^{\delta\circ\gamma}_{\circ\theta\circ}] = [a^{\alpha1}_{\circ\circ} \cdot b^{\delta\circ\gamma}_{\circ1\circ} + a^{\alpha2}_{\circ\circ} \cdot b^{\delta\circ\gamma}_{\circ2\circ} + a^{\alpha3}_{\circ\circ} \cdot b^{\delta\circ\gamma}_{\circ3\circ}]. \quad (7.92)$$

For $\alpha = 1$ (first block of the tensor P matrix, called P_1), since $\delta = \{1,2,3\}$ (blocks 1,2,3 of tensor B) and $\gamma = 1$ (first column of each block of B and of P), Expression (7.92) leads to

$$[p^{1\delta1}_{\circ\circ\circ}] = a^{11}_{\circ\circ}b^{\delta\circ1}_{\circ1\circ} + a^{12}_{\circ\circ}b^{\delta\circ1}_{\circ2\circ} + a^{13}_{\circ\circ}b^{\delta\circ1}_{\circ3\circ}$$

$$= [a^{11}_{\circ\circ} \quad a^{12}_{\circ\circ} \quad a^{13}_{\circ\circ}] \begin{bmatrix} b^{\delta\circ1}_{\circ1\circ} \\ b^{\delta\circ1}_{\circ2\circ} \\ b^{\delta\circ1}_{\circ3\circ} \end{bmatrix}$$

$$= [0 \quad 0 \quad 1] \begin{bmatrix} 0 & -1 & 1 \\ 0 & 0 & -1 \\ 0 & 1 & -1 \end{bmatrix} = [0 \quad 1 \quad -1],$$

that is the first column of block P_1.

For $\alpha = 1$, $\delta = \{1,2,3\}$ and $\gamma = 2$, Expression (7.92) gives

$$[p^{1\delta2}_{\circ\circ\circ}] = [a^{11}_{\circ\circ} \quad a^{12}_{\circ\circ} \quad a^{13}_{\circ\circ}] \begin{bmatrix} b^{\delta\circ2}_{\circ1\circ} \\ b^{\delta\circ2}_{\circ2\circ} \\ b^{\delta\circ2}_{\circ3\circ} \end{bmatrix}$$

$$= [0 \quad 0 \quad 1] \begin{bmatrix} 1 & 0 & 0 \\ 0 & 0 & -2 \\ -1 & 0 & 0 \end{bmatrix} = [-1 \quad 0 \quad 0],$$

which is the second column of block P_1.

For $\alpha = 1$, $\delta = \{1,2,3\}$ and $\gamma = 3$, Expression (7.92) gives

$$[p^{1\delta3}_{\circ\circ\circ}] = [0 \quad 0 \quad 1] \begin{bmatrix} -1 & 0 & 0 \\ 1 & 2 & 0 \\ 1 & 0 & 0 \end{bmatrix} = [1 \quad 0 \quad 0],$$

which is the third column of block P_1.

So, in summary

$$P_1 = \begin{bmatrix} 0 & -1 & 1 \\ 1 & 0 & 0 \\ -1 & 0 & 0 \end{bmatrix}.$$

For $\alpha = 2$ (second block of the matrix of tensor P, called P_2); $\delta = \{1, 2, 3\}$ and $\gamma = 1$, Expression (7.92) gives

$$[p_{ooo}^{2\delta1}] = a_{oo}^{21}b_{o1o}^{\delta o1} + a_{oo}^{22}b_{o2o}^{\delta o1} + a_{oo}^{23}b_{o3o}^{\delta o1} = [a_{oo}^{21} \ \ a_{oo}^{22} \ \ a_{oo}^{23}] \begin{bmatrix} b_{o1o}^{\delta o1} \\ b_{o1o}^{\delta o1} \\ b_{o2o}^{\delta o1} \\ b_{o3o}^{\delta o1} \end{bmatrix}$$

$$= [0 \ \ -1 \ \ 0] \begin{bmatrix} 0 & -1 & 1 \\ 0 & 0 & -1 \\ 0 & 1 & -1 \end{bmatrix} = [0 \ \ 0 \ \ 1],$$

which is the first column of block P_2, and

$$[p_{ooo}^{2\delta2}] = [a_{oo}^{21} \ \ a_{oo}^{22} \ \ a_{oo}^{23}] \begin{bmatrix} b_{o1o}^{\delta o2} \\ b_{o1o}^{\delta o2} \\ b_{o2o}^{\delta o2} \\ b_{o3o}^{\delta o2} \end{bmatrix} = [0 \ \ -1 \ \ 0] \begin{bmatrix} 1 & 0 & 0 \\ 0 & 0 & -2 \\ -1 & 0 & 0 \end{bmatrix}$$

$$= [0 \ \ 0 \ \ 2],$$

which is the second column of block P_2, and

$$[p_{ooo}^{2\delta3}] = [0 \ \ -1 \ \ 0] \begin{bmatrix} -1 & 0 & 0 \\ 1 & 2 & 0 \\ 1 & 0 & 0 \end{bmatrix} = [-1 \ \ -2 \ \ 0],$$

which is the third column of block P_2.

Thus, in summary

$$P_2 = \begin{bmatrix} 0 & 0 & -1 \\ 0 & 0 & -2 \\ 1 & 2 & 0 \end{bmatrix}.$$

Finally, we calculate the third block (P_3):

$$[p_{ooo}^{3\delta1}] = [a_{oo}^{31} \ \ a_{oo}^{32} \ \ a_{oo}^{33}] \begin{bmatrix} b_{o1o}^{\delta o1} \\ b_{o1o}^{\delta o1} \\ b_{o2o}^{\delta o1} \\ b_{o3o}^{\delta o1} \end{bmatrix}$$

$$= [1 \ 0 \ 1] \begin{bmatrix} 0 & -1 & 1 \\ 0 & 0 & -1 \\ 0 & 1 & -1 \end{bmatrix} = [0 \ 0 \ 0]$$

$$[p^{3\delta 2}_{\circ\circ\circ}] = [1\ 0\ 1] \begin{bmatrix} 1 & 0 & 0 \\ 0 & 0 & -2 \\ -1 & 0 & 0 \end{bmatrix} = [0\ 0\ 0]$$

$$[p^{3\delta 3}_{\circ\circ\circ}] = [1\ 0\ 1] \begin{bmatrix} -1 & 0 & 0 \\ 1 & 2 & 0 \\ 1 & 0 & 0 \end{bmatrix} = [0\ 0\ 0]$$

and then

$$P_3 = \begin{bmatrix} 0 & 0 & 0 \\ 0 & 0 & 0 \\ 0 & 0 & 0 \end{bmatrix} \equiv \Omega,$$

so that finally, the matrix representation of the contracted tensor, in the initial basis is

$$P \equiv [p^{\alpha\delta\gamma}_{\circ\circ\circ}] = \begin{bmatrix} 0 & -1 & 1 \\ 1 & 0 & 0 \\ -1 & 0 & 0 \\ 0 & 0 & -1 \\ 0 & 0 & -2 \\ 1 & 2 & 0 \\ 0 & 0 & 0 \\ 0 & 0 & 0 \\ 0 & 0 & 0 \end{bmatrix}.$$

4. As the statement requires, we proceed again to execute the contraction, but using the matrices $\hat{A}$ and $\hat{B}$ associated with tensors $\vec{a}$ and $\vec{b}$, in the new basis.

$$\left[\hat{p}^{1\delta 1}_{\circ\circ\circ}\right] = [0\ 0\ 1] \begin{bmatrix} 0 & -1 & 1 \\ 0 & -1 & -1 \\ 0 & 1 & 0 \end{bmatrix} = [0\ 1\ 0]$$

$$\left[\hat{p}^{1\delta 2}_{\circ\circ\circ}\right] = [0\ 0\ 1] \begin{bmatrix} 1 & 0 & 4 \\ 1 & 0 & -2 \\ -1 & 0 & -1 \end{bmatrix} = [-1\ 0\ -1]$$

$$\left[\hat{p}^{1\delta 3}_{\circ\circ\circ}\right] = [0\ 0\ 1] \begin{bmatrix} -1 & -4 & 0 \\ 1 & 2 & 0 \\ 0 & 1 & 0 \end{bmatrix} = [0\ 1\ 0]; \Rightarrow \hat{P}_1 = \begin{bmatrix} 0 & -1 & 0 \\ 1 & 0 & 1 \\ 0 & -1 & 0 \end{bmatrix}$$

$$\left[\hat{p}^{2\delta 1}_{\circ\circ\circ}\right] = [0\ -1\ 0] \begin{bmatrix} 0 & -1 & 1 \\ 0 & -1 & -1 \\ 0 & 1 & 0 \end{bmatrix} = [0\ 1\ 1]$$

$$\left[\hat{p}^{2\delta 2}_{\circ\circ\circ}\right] = [0\ -1\ 0] \begin{bmatrix} 1 & 0 & 4 \\ 1 & 0 & -2 \\ -1 & 0 & -1 \end{bmatrix} = [-1\ 0\ 2]$$

$$\left[\hat{p}^{2\delta 3}_{\circ\circ\circ}\right] = [0\ -1\ 0] \begin{bmatrix} -1 & -4 & 0 \\ 1 & 2 & 0 \\ 0 & 1 & 0 \end{bmatrix} = [-1\ -2\ 0]; \Rightarrow \hat{P}_2 = \begin{bmatrix} 0 & -1 & -1 \\ 1 & 0 & -2 \\ 1 & 2 & 0 \end{bmatrix}$$

$$\left[\hat{p}^{3\delta 1}_{\circ\circ\circ}\right] = [1\ 0\ 2] \begin{bmatrix} 0 & -1 & 1 \\ 0 & -1 & -1 \\ 0 & 1 & 0 \end{bmatrix} = [0\ 1\ 1]$$

$$\left[\hat{p}^{3\delta 2}_{\circ\circ\circ}\right] = [1\ 0\ 2] \begin{bmatrix} 1 & 0 & 4 \\ 1 & 0 & -2 \\ -1 & 0 & -1 \end{bmatrix} = [-1\ 0\ 2]$$

$$\left[\hat{p}^{3\delta3}_{\circ\circ\circ}\right] = [\,1\;0\;2\,]\begin{bmatrix} -1 & -4 & 0 \\ 1 & 2 & 0 \\ 0 & 1 & 0 \end{bmatrix} = [\,-1\;-2\;0\,] \;\Rightarrow\; \hat{P}_3 = \begin{bmatrix} 0 & -1 & -1 \\ 1 & 0 & -2 \\ 1 & 2 & 0 \end{bmatrix}.$$

In conclusion:

$$\hat{P} \equiv [p^{ijk}_{\circ\circ\circ}] = \begin{bmatrix} 0 & -1 & 0 \\ 1 & 0 & 1 \\ 0 & -1 & 0 \\ 0 & -1 & -1 \\ 1 & 0 & -2 \\ 1 & 2 & 0 \\ 0 & -1 & -1 \\ 1 & 0 & -2 \\ 1 & 2 & 0 \end{bmatrix}.$$

5. We change the basis of the contracted tensor P, in the same form as we did with the data tensors, to solve the proposed question, which aims to show the correctness of the answers to the previous questions.

Since tensor P is of order $r = 3$ and totally contravariant, we apply Formula (4.36) with $\sigma = n^r = 3^3 = 27$, and get

$$\hat{P}_{\sigma,1} = \hat{P}_{27,1} = (C^{-1}\otimes C^{-1}\otimes C^{-1})\bullet P_{27,1} = \left(\begin{bmatrix} 1 & 0 & 0 \\ 1 & 1 & 0 \\ 1 & 1 & 1 \end{bmatrix} \otimes \begin{bmatrix} 1 & 0 & 0 \\ 1 & 1 & 0 \\ 1 & 1 & 1 \end{bmatrix} \otimes \begin{bmatrix} 1 & 0 & 0 \\ 1 & 1 & 0 \\ 1 & 1 & 1 \end{bmatrix}\right) \begin{bmatrix} 0 \\ -1 \\ 1 \\ 1 \\ 0 \\ 0 \\ -1 \\ 0 \\ 0 \\ \hline 0 \\ 0 \\ -1 \\ 0 \\ 0 \\ -2 \\ 1 \\ 2 \\ 0 \\ \hline 0 \\ 0 \\ 0 \\ 0 \\ 0 \\ 0 \\ 0 \\ 0 \\ 0 \end{bmatrix} = \begin{bmatrix} 0 \\ -1 \\ 0 \\ 1 \\ 0 \\ 1 \\ 0 \\ -1 \\ 0 \\ \hline 0 \\ -1 \\ -1 \\ 1 \\ 0 \\ -2 \\ 1 \\ 2 \\ 0 \\ \hline 0 \\ -1 \\ -1 \\ 1 \\ 0 \\ -2 \\ 1 \\ 2 \\ 0 \end{bmatrix}$$

Condensing $\hat{P}_{27,1}$ we again obtain

$$\hat{P} \equiv [p^{ijk}_{\circ\circ\circ}] = \begin{bmatrix} 0 & -1 & 0 \\ 1 & 0 & 1 \\ 0 & -1 & 0 \\ 0 & -1 & -1 \\ 1 & 0 & -2 \\ 1 & 2 & 0 \\ 0 & -1 & -1 \\ 1 & 0 & -2 \\ 1 & 2 & 0 \end{bmatrix},$$

which confirms the correction of the answers to the previous questions.

6. It becomes clear that in the initial basis and in the new basis, $\vec{p}$ is a homogeneous contravariant and anti-symmetric with respect to the *last two* indices tensor of third order:

$$p_{\circ\circ\circ}^{\alpha\delta\delta} = 0 \quad \text{and} \quad p_{\circ\circ\circ}^{\alpha\gamma\delta} = -p_{\circ\circ\circ}^{\alpha\delta\gamma} \text{ if } \delta \neq \gamma.$$

$\square$

7.8 Exercises

7.1. Assuming that in a homogeneous mixed anti-symmetric tensor, defined over a linear space $V^5(\mathbb{R})$ and its dual, the following strict component has the value

$$t_{(\circ 1 \circ 3 4 \circ 5 \circ)}^{(1 \circ 2 \circ \circ 3 \circ 4)} = a; \quad a \in \mathbb{R}.$$

1. How many components are sure to have the same absolute value?
2. How many components are null?
3. Give the value of the component $t_{\circ 5 \circ 4 1 \circ 3 \circ}^{3 \circ 1 \circ \circ 4 \circ 2}$.

7.2. 1. Give the general matrix representation of a homogeneous anti-symmetric tensor $T \in \left(\otimes V^4\right)_h^2 (\mathbb{R})$, the strict components of which in increasing order are a, b, c, d, e, f.
2. The same question for a tensor $T \in \left(\otimes V^4\right)_h^3 (\mathbb{R})$, the strict components of which in increasing order are a, b, c, d.

7.3. 1. Let $P_1 \in \left(\otimes V^n\right)_h^2 (\mathbb{R})$ and Q_1 be a homogeneous anti-symmetric tensor and $Q_1 \in \left(\otimes V_*^n\right)_s^2 (\mathbb{R})$ be a homogeneous symmetric tensor. If the totally contracted product of both tensor is executed, determine the tensor

$$Z = C \begin{pmatrix} \alpha & \beta \\ \gamma & \delta \end{pmatrix} p_{\circ\circ}^{\alpha\beta} q_{\gamma\delta}^{\circ\circ} = p_{\circ\circ}^{\theta\phi} q_{\theta\phi}^{\circ\circ}.$$

2. Let $P_2 \in \left(\otimes V^n\right)^3 (\mathbb{R})$ be a homogeneous, and symmetric with respect to its first two indices tensor, and $Q_2 \in \left(\otimes V_*^n\right)^3 (\mathbb{R})$ be a homogeneous and anti-symmetric with respect to its first two indices tensor.
Consider the tensor U product of the two tensors contracted with respect to its first two indices

$$u_{\circ\eta}^{\gamma\circ} = C \begin{pmatrix} \alpha & \beta \\ \delta & \epsilon \end{pmatrix} p_{\circ\circ\circ}^{\alpha\beta\gamma} q_{\delta\epsilon\eta}^{\circ\circ\circ} = p_{\circ\circ\circ}^{\theta\phi\gamma} q_{\theta\phi\eta}^{\circ\circ\circ}.$$

Determine tensor U.

7.4. Consider the tensor $T \in (\otimes V^2)^3 (\mathbb{R})$, the matrix representation of which is

$$[t_{\circ\circ\circ}^{\alpha\beta\gamma}] = \begin{bmatrix} 1 & -1 \\ 2 & 3 \\ \hline 2 & 1 \\ 5 & 3 \end{bmatrix}.$$

1. Give the representation of the anti-symmetrized tensor $H(T)$, associated with tensor T.
2. Execute in the linear space $V^2(\mathbb{R})$ a change-of-basis of associated matrix
   $C \equiv [c_{\circ i}^{\alpha\circ}] = \begin{bmatrix} 1 & 2 \\ 1 & 3 \end{bmatrix}$, and give the matrix representation of tensor T in the new basis $\hat{T} = [\hat{t}_{\circ\circ\circ}^{ijk}]$.
3. Find the matrix representation of the anti-symmetrized tensor $H(\hat{T})$, associated with $\hat{T}$, in the new basis.
4. Check the previous result, directly changing the basis of the tensor $H(T)$.
5. Answer questions 1 and 3 using the computer.

7.5. Consider the tensor $A \in V_*^3 \otimes V^3 \otimes V^3(\mathbb{R})$ by means of its matrix representation in the basis $\{\vec{e}_\alpha\}$ of $V^3(\mathbb{R})$

$$[a_{\alpha\circ\circ}^{\circ\beta\gamma}] = \begin{bmatrix} 0 & 1 & -2 \\ -1 & 0 & -3 \\ 2 & 3 & 0 \\ \hline 0 & 4 & 1 \\ -4 & 0 & -3 \\ -1 & 3 & 0 \\ \hline 0 & 2 & 5 \\ -2 & 0 & 0 \\ -5 & 0 & 0 \end{bmatrix},$$

where α is the block row, β is the row of each block, and γ is the column of each block.

Consider also a change-of-basis in $V^3(\mathbb{R})$ given by the expression

$$\|\vec{e}_i\| = \|\vec{e}_\alpha\| \begin{bmatrix} 0 & 1 & 1 \\ 1 & 1 & 1 \\ 1 & 1 & 0 \end{bmatrix}.$$

1. Study the possible partial anti-symmetries of the components of A in the initial basis.
2. If any of the anti-symmetries is of a tensor nature, execute the change-of-basis and obtain the matrix representation of the tensor A in the new basis $\hat{A} = [a_{i\circ\circ}^{\circ jk}]$.
3. Check the intrinsic anti-symmetric character discovered in the previous question.

4. If $B \in V^3(\mathbb{R})$ the vector $[b_o^\delta] = \begin{bmatrix} 2 \\ 1 \\ -3 \end{bmatrix}$ determines the contracted product

tensor

$$P = C \begin{pmatrix} \delta \\ \alpha \end{pmatrix} A \otimes B.$$

5. Determine if the tensor P is anti-symmetric.

7.6. The following general matrix representations are given:

1. Of a symmetric tensor $T_1 \in \left(\otimes V^3 \right)_s^3 (\mathbb{R})$.
2. Of an anti-symmetric tensor $T_2 \in \left(\otimes V^3 \right)_h^3 (\mathbb{R})$.

Obtain:

1. $H(T_1)$, the anti-symmetrized tensor of tensor T_1.
2. $S(T_2)$, the symmetrized tensor of tensor T_2.

8

Pseudotensors; Modular, Relative or Weighted Tensors

8.1 Introduction

First of all, we recognize that the need to create mathematical models that are more and more complex, and capable of giving mathematical support to the new theories in physics in its diverse branches (solid mechanics, elasticity, plasticity, mechanics of deformable media, mechanics of fluids, quantum mechanics, relativity, etc.) has forced us to extend the concept of absolute tensors with new contributions. This is the aim of this chapter.

More precisely, in this chapter we will extend the concept of *homogeneous tensors* to other systems of scalar components, of "close" formal characteristics, in such way that the cited absolute homogeneous tensors be integrated into the "new ones" called modular tensors, relative tensors, and also pseudotensors, as a specific part of special relevance, such as the ϵ-systems of Levi-Civita, the generalized Kronecker deltas and polar tensors.

The modular tensors are established in the same form as the absolute tensors, over real finite-dimensional linear spaces $V^n(\mathbb{R})$.

The term "pseudotensors" (from the Greek, false tensors), reminds us that they are not tensors, under the classic conception of a tensor that has been considered up to now. The terms "modular tensors" and "relative tensors", to be defined below, refer to the concept of "relative modulus" of their two bases in a linear space $V^n(\mathbb{R})$.

8.2 Previous concepts of modular tensor establishment

8.2.1 Relative modulus of a change-of-basis

Consider the linear space $V^n(\mathbb{R})$ referred to a basis $\{\vec{e}_\alpha\}, \alpha \in I_n$. If we take as a new basis $\{\vec{e}_i\}$, we know that the tensor relation between the two bases is

$$\vec{e}_i = c_{\circ\, i}^{\alpha\circ}\vec{e}_\alpha$$

and in matrix form

$$||\vec{e}_i|| = ||\vec{e}_\alpha||[c^{\alpha\,\circ}_{\,\circ\,i}], \tag{8.1}$$

where $C = [c^{\alpha\,\circ}_{\,\circ\,i}]$ is the change-of-basis matrix, with $|C| \neq 0 \in \mathbb{R}$.

The determinant $|C|$ is called the *relative modulus* of the basis $\{\vec{e}_i\}$ with respect to the basis $\{\vec{e}_\alpha\}$.

This concept also appears in the dual changes of basis. From Section 3.6.2, the result is

$$\vec{e}^{*i} = \gamma^{\circ\,i}_{\alpha\,\circ}\vec{e}^{*\alpha} \text{ where } [\gamma^{\circ\,i}_{\alpha\,\circ}] = \Gamma^t \text{ and } \Gamma \equiv C^{-1} = [\gamma^{i\,\circ}_{\circ\,\alpha}]; \tag{8.2}$$

where, obviously,

$$|\Gamma| = \frac{1}{|C|} = |C|^{-1},$$

which shows that $|C|$ and $|\Gamma|$ *always* have the same sign.

8.2.2 Oriented vector space

We say that two bases $\{\vec{e}_\alpha\}$ and $\{\vec{e}_i\}$ *have the same orientation* if the relative modulus is positive, that is, if $|C| > 0$. This property is an equivalence relation, thus all new bases remain classified with respect to the initial one *directly* or *inversely*, depending on whether or not they have the same orientation as the initial one.

If in a linear space, we consider *only* changes of basis with $|C| > 0$, we say that the linear space $V^n(\mathbb{R})$ is *oriented* and if tensors are constructed on it, by power of the tensor product, we say that they are "oriented tensors".

8.2.3 Weight tensor

An exponent w that appears in the general expression of a change-of-basis of a modular tensor will be called a weight tensor. It is a parameter that belongs to the set of relative integers

$$w \in Z; \quad Z = \{\ldots -3, -2, -1, 0, 1, 2, 3, \ldots\},$$

which will be mentioned after establishing the axiomatic properties for these tensors.

8.3 Axiomatic properties for the modular tensor concept

1. Consider a linear space $V^n(\mathbb{R})$ referred to a basis $\{\vec{e}_\alpha\}$ and its dual linear space $V^n_*(\mathbb{R})$ referred to the reciprocal basis of $\{\vec{e}^{*\alpha}\}$. Over them we build another linear space, called a modular tensor space.

2. The modular tensor system is notated:

$$V^n \otimes V^n \otimes V^n_* \otimes V^n_* \otimes V^n \otimes V^n \otimes \cdots \otimes V^n_* \otimes V^n(\mathbb{R})|_w.$$
$$\longleftarrow r \text{ factors } \longrightarrow$$

Its field of scalars is also $\mathbb{R}$, and the elements of these tensor system are notated

$$t^{\alpha\beta\circ\circ\epsilon\cdots\circ\rho}_{\circ\circ\gamma\delta\cdots\pi\circ}\vec{e}_\alpha \otimes \vec{e}_\beta \otimes \vec{e}^{*\gamma} \otimes \vec{e}^{*\delta} \otimes \vec{e}_\epsilon \otimes \cdots \otimes \vec{e}^{*\pi} \otimes \vec{e}_\rho$$

appearing as a linear combination of a basis, built as a tensor power of the basic vectors of $V^n(\mathbb{R})$ and $V^n_*(\mathbb{R})$, according to the axiomatic criteria proposed in Section 2.5. Among the r dummy indices of the scalar components and those of each basic vector of the tensor space, the Einstein convention holds.

3. In this tensor system only changes of basis that respond to the following formulation are licit:

$$t^{ij\circ\circ e\cdots\circ r}_{\circ\circ k d\circ\cdots p\circ} = |C|^w t^{\alpha\beta\circ\circ\epsilon\cdots\circ\ell}_{\circ\circ\gamma\delta\circ\cdots\pi\circ}\gamma^{i\circ}_{\circ\alpha}\gamma^{j\circ}_{\circ\beta}c^{\circ\gamma}_{k\circ}c^{\circ\delta}_{d\circ}\gamma^{e\circ\cdots}_{\circ\epsilon}\cdots c^{\circ\pi}_{p\circ}\gamma^{r\circ}_{\circ\rho}, \tag{8.3}$$

where C is the matrix of a change-of-basis $||\vec{e}_i|| = ||\vec{e}_\alpha||C$ executed in the linear space $V^n(\mathbb{R})$ and $|C|$ is its relative modulus.

Remark 8.1 (Important note). From now on, and following most authors, we will formulate these tensors with all contravariant indices (p) in sequence, and all covariant indices (q) in sequence, and after the contravariant indices when both indices exist. According to this criterion, the relation (8.3) can be written as

$$t^{i_1 i_2 \cdots i_p \circ \circ \cdots \circ}_{\circ \circ \cdots \circ j_1 j_2 \cdots j_q} = |C|^w t^{\alpha_1 \alpha_2 \cdots \alpha_p \circ \circ \cdots \circ}_{\circ \circ \cdots \circ \beta_1 \beta_2 \cdots \beta_q}\gamma^{i_1 \circ}_{\circ \alpha_1}\gamma^{i_2 \circ}_{\circ \alpha_2}\cdots\gamma^{i_p \circ}_{\circ \alpha_p}c^{\circ \beta_1}_{j_1 \circ}c^{\circ \beta_2}_{j_2 \circ}\cdots c^{\circ \beta_q}_{j_q \circ}$$
$$\tag{8.4}$$

but it is only for the sake of simplicity of notation, and *not* because the alteration of the ordering of indices be licit. □

8.4 Modular tensor characteristics

We use the term modular tensor characteristics (m.t.c.) to refer to the expression (order (r), species, weight (w)) of a modular tensor.

The order is the number of indices, the species is the ordered list of the valencies of each index in the component: cova-cova-contra-cova-...- contravariant, and the weight is the relative integer that is assigned to the tensor because of its particular properties.

With the same criterion as that in Section 8.3, some authors notate the modular tensor spaces in abbreviated form $(\otimes V^n)^p_w(\mathbb{R})$ or $(\otimes V^n)^q_w(\mathbb{R})$ or $(\otimes V^n)^{p+q}_w(\mathbb{R})$, but their modular tensor characteristics must be added in order to know exactly the position of their indices.

8.4.1 Equality of modular tensors

Two modular tensors are equal if:

1. They are associated with the same linear space $V^n(\mathbb{R})$.
2. They have the same modular tensor characteristics (m.t.c.).
3. They have identical components with respect to the same basis.

8.4.2 Classification and special denominations

Depending on the weight and order, modular tensors use a different terminology:

- If $w = 0$ we have the *absolute* or *proper* homogeneous tensors.
- If $w \neq 0$ we have the modular tensors or pseudotensors.
- If $w = 1$ and $r = n$ the tensor is called the "tensor density" of order $n = p + q$.
- If $w = -1$ and $r = n$ the tensor is called the "tensor capacity" of order $n = p + q$.
- If $w \neq 0$ and $r = 0$ the tensor is called the "pseudo-scalar of relative invariant".
- If $w = 1$ and $r = 0$ the tensor is called the "scalar density or comodular scalar".
- If $w = -1$ and $r = 0$ the tensor is called the "scalar capacity or contramodular scalar".

Evidently, the product of a comodular scalar by a contramodular scalar is an intrinsic scalar or invariant.

This terminology is commonly used by such authors as [13] and [23].

8.5 Remarks on modular tensor operations: Consequences

8.5.1 Tensor addition

The sum of two modular tensors, defined over a $V^n(\mathbb{R})$ and of identical modular tensor characteristics, is another modular tensor of identical m.t.c., the components of which relative to a basis, are the sum of the corresponding components of the tensor summands, in the same basis.

To prove it, we will prove only the satisfaction of the third axiom, Formula (8.4) for the tensor sum, assuming that axioms 1 and 2 hold.

1. Let $A, B \in (\otimes V^n)^2 \otimes V_*^n(\mathbb{R})|_w$ be two tensors of the given modular tensor space; let $S = A + B$ be its sum tensor. Consider the components: $s_{\circ\circ\gamma}^{\alpha\beta\circ} = a_{\circ\circ\gamma}^{\alpha\beta\circ} + b_{\circ\circ\gamma}^{\alpha\beta\circ}$.

Since A and B are modular, Formula (8.4) must hold for both:

$$a^{ij\circ}_{\circ\circ k} = |C|^w a^{\alpha\beta\circ}_{\circ\circ\gamma}\gamma^{i\circ}_{\circ\alpha}\gamma^{j\circ}_{\circ\beta}c^{\circ\gamma}_{k\circ}$$

$$b^{ij\circ}_{\circ\circ k} = |C|^w b^{\alpha\beta\circ}_{\circ\circ\gamma}\gamma^{i\circ}_{\circ\alpha}\gamma^{j\circ}_{\circ\beta}c^{\circ\gamma}_{k\circ}$$

and adding them we get

$$a^{ij\circ}_{\circ\circ k} + b^{ij\circ}_{\circ\circ k} = |C|^w(a^{\alpha\beta\circ}_{\circ\circ\gamma} + b^{\alpha\beta\circ}_{\circ\circ\gamma})\gamma^{i\circ}_{\circ\alpha}\gamma^{j\circ}_{\circ\beta}c^{\circ\gamma}_{k\circ},$$

that is

$$s^{ij\circ}_{\circ\circ k} = |C|^w s^{\alpha\beta\circ}_{\circ\circ\gamma}\gamma^{i\circ}_{\circ\alpha}\gamma^{j\circ}_{\circ\beta}c^{\circ\gamma}_{k\circ}.$$

In fact, we get

$$S = A + B; \quad S \in (\otimes V^n)^2(V^n_*)(\mathbb{R})|_w, \tag{8.5}$$

which proves that the sum is an internal law of composition.

2. Note that if $A, B, D \in (\otimes V^n)^2 \otimes V^n_*(\mathbb{R})|_w$, since $a^{\alpha\beta\circ}_{\circ\circ\gamma}, b^{\alpha\beta\circ}_{\circ\circ\gamma}, d^{\alpha\beta\circ}_{\circ\circ\gamma} \in \mathbb{R}, \forall \alpha, \beta, \gamma$, holds and $\mathbb{R}$ being a field, we have

$$(a^{\alpha\beta\circ}_{\circ\circ\gamma} + b^{\alpha\beta\circ}_{\circ\circ\gamma}) + d^{\alpha\beta\circ}_{\circ\circ\gamma} = a^{\alpha\beta\circ}_{\circ\circ\gamma} + (b^{\alpha\beta\circ}_{\circ\circ\gamma} + d^{\alpha\beta\circ}_{\circ\circ\gamma})$$

and then

$$(A + B) + D = A + (B + D), \tag{8.6}$$

which implies the associativity of the modular law $(+)$.

3. $\exists \Omega \in (\otimes V^n)^2 \otimes V^n_*(\mathbb{R})|_w$, that is, a zero modular tensor exists.
Since $\forall a^{\alpha\beta\circ}_{\circ\circ\gamma} \in \mathbb{R}$, there exists a system of zeroes in $\mathbb{R}$, $0^{\alpha\beta\circ}_{\circ\circ\gamma} \in \mathbb{R}$, such that

$$a^{\alpha\beta\circ}_{\circ\circ\gamma} + 0^{\alpha\beta\circ}_{\circ\circ\gamma} = 0^{\alpha\beta\circ}_{\circ\circ\gamma} + a^{\alpha\beta\circ}_{\circ\circ\gamma} = a^{\alpha\beta\circ}_{\circ\circ\gamma};$$

whence

$$A + \Omega = \Omega + A = A \tag{8.7}$$

proving that the modular law $(+)$ has a zero element.

4.

$$\forall A \in (\otimes V^n)^2 \otimes V^n_*(\mathbb{R})|_w, \exists(-A) \in (\otimes V^n)^2 \otimes V^n_*(\mathbb{R})|_w;$$

this is the consequence of the fact that in $\mathbb{R}$, $\forall a^{\alpha\beta\circ}_{\circ\circ\gamma}$ corresponds to a $(-a^{\alpha\beta\circ}_{\circ\circ\gamma}) \in \mathbb{R}$, because $\mathbb{R}$ is a field. Thus,

$$\forall \alpha, \beta, \gamma: \quad a^{\alpha\beta\circ}_{\circ\circ\gamma} + (-a^{\alpha\beta\circ}_{\circ\circ\gamma}) = 0^{\alpha\beta\circ}_{\circ\circ\gamma},$$

which leads to

$$A + (-A) = (-A) + A = \Omega, \tag{8.8}$$

which proves that the modular law $(+)$ is "cancelative".

5. $\forall A, B \in (\otimes V^n)^2 \otimes V_*^n(\mathbb{R})|_w$, since in $\mathbb{R}$ the result is that:

$$\forall \alpha, \beta, \gamma: \quad a^{\alpha\beta\circ}_{\circ\circ\gamma} + b^{\alpha\beta\circ}_{\circ\circ\gamma} = b^{\alpha\beta\circ}_{\circ\circ\gamma} + a^{\alpha\beta\circ}_{\circ\circ\gamma}$$

then we get

$$A + B = B + A, \tag{8.9}$$

which proves that the modular law $(+)$ is Abelian or commutative. Consequently, $(\otimes V^n)^2 \otimes V_*^n(\mathbb{R})|_w$, because of (8.5), (8.6), (8.7), (8.8) and (8.9), is an Abelian group for the modular sum.

8.5.2 Multiplication by a scalar

The product of a modular tensor defined over a $V^n(\mathbb{R})$ by a scalar of $\mathbb{R}$ is another modular tensor of identical m.t.c., the relative components of which with respect to a basis are the real products of the scalar by the corresponding component of the tensor factor.

We will show only the satisfaction of the third axiom, Formula (8.4) for the tensor product, assuming that axioms 1 and 2 are satisfied.

If $A \in (\otimes V^n)^2 \otimes V_*^n(\mathbb{R})|_w$ and $\lambda \in \mathbb{R}$, let $Q = \lambda A$ be the tensor product. Let the components be

$$q^{\alpha\beta\circ}_{\circ\circ\gamma} = \lambda a^{\alpha\beta\circ}_{\circ\circ\gamma}.$$

Since A is modular, it is $a^{ij\circ}_{\circ\circ k} = |C|^w a^{\alpha\beta\circ}_{\circ\circ\gamma} \gamma^{i\circ}_{\circ\alpha} \gamma^{j\circ}_{\circ\beta} c^{\circ\gamma}_{k\circ}$, and since all scalars belong to $\mathbb{R}$, multiplying both sides by λ, we get

$$\lambda a^{ij\circ}_{\circ\circ k} = |C|^w (\lambda a^{\alpha\beta\circ}_{\circ\circ\gamma}) \gamma^{i\circ}_{\circ\alpha} \gamma^{j\circ}_{\circ\beta} c^{\circ\gamma}_{k\circ}$$

or

$$q^{ij\circ}_{\circ\circ k} = |C|^w q^{\alpha\beta\circ}_{\circ\circ\gamma} \gamma^{i\circ}_{\circ\alpha} \gamma^{j\circ}_{\circ\beta} c^{\circ\gamma}_{k\circ}.$$

Thus, we obtain

$$\lambda A \in (\otimes V^n)^2 \otimes V_*^n(\mathbb{R})|_w \tag{8.10}$$

and a new "external product" is created.

Without further demonstrations we advance that $\forall A, B$ and $\forall \lambda, \mu$, we have:

Distributivity property of the external product with respect to the sum of tensors:

$$\lambda(A + B) = \lambda A + \lambda B. \tag{8.11}$$

Distributivity property of the sum of scalars with respect to the external product:

$$(\lambda + \mu)A = \lambda A + \mu A. \tag{8.12}$$

Associative property of the external product and the real product:

$$\lambda(\mu A) = (\lambda \cdot \mu)A. \tag{8.13}$$

Unitary character of the external product:

$$1 \cdot A = A, \tag{8.14}$$

that is, the scalar $1 \in \mathbb{R}$, unit element of the real product is also the "unit" element of the external product. Thus, the Abelian group $(\otimes V^n)^2 \otimes V_*(\mathbb{R})|_w$ after (8.10), (8.11), (8.12), (8.13) and (8.14) is a linear space of modular tensors, that is, a modular tensor space.

8.5.3 Tensor product

The pending question is if the tensor product of modular tensors is also modular. We will try with two modular tensors $A \in (\otimes V^n)^2 \otimes V_*(\mathbb{R})|_{w_1}$ and $B \in V^n \otimes V_*^n(\mathbb{R})|_{w_2}$.

That is, now each tensor can have different m.t.c., though they must be defined over the same $V^n(\mathbb{R})$.

In summary, they could be $A \in (V^n)_{w_1}^{r_1}$ with $r_1 = p_1 + q_1$ and $B \in (V^n)_{w_2}^{r_2}$ with $r_2 = p_2 + q_2$.

The weight of the product will be $w_p = w_1 + w_2$ and the order $r_p = r_1 + r_2$. The species would be the direct sum of the species $esp_p = esp_1 \oplus esp_2$.

Let A and B be the data tensors. Then we have

$$a_{\circ\circ k}^{ij\circ} = |C|^{w_1} a_{\circ\circ\gamma}^{\alpha\beta\circ} \gamma_{\circ\alpha}^{i\circ} \gamma_{\circ\beta}^{j\circ} c_{k\circ}^{\circ\gamma}$$

$$b_{\circ m}^{\ell\circ} = |C|^{w_2} b_{\circ\mu}^{\lambda\circ} \gamma_{\circ\lambda}^{\ell\circ} c_{m\circ}^{\circ\mu}.$$

Let $P = A \otimes B$, then multiplying them in $\mathbb{R}$, the components are

$$p_{\circ\circ k\circ m}^{ij\circ\ell\circ} = |C|^{w_1+w_2} (a_{\circ\circ\gamma}^{\alpha\beta\circ} \cdot b_{\circ\mu}^{\lambda\circ}) \gamma_{\circ\alpha}^{i\circ} \gamma_{\circ\beta}^{j\circ} c_{k\circ}^{\circ\gamma} \gamma_{\circ\lambda}^{\ell\circ} c_{m\circ}^{\circ\mu}.$$

If we call $p_{\circ\circ\gamma\circ\mu}^{\alpha\beta\circ\lambda\circ} = a_{\circ\circ\gamma}^{\alpha\beta\circ} \cdot b_{\circ\mu}^{\lambda\circ}$ the result is

$$p_{\circ\circ k\circ m}^{ij\circ\ell\circ} = |C|^{w_1+w_2} p_{\circ\circ\gamma\circ\mu}^{\alpha\beta\circ\lambda\circ} \gamma_{\circ\alpha}^{i\circ} \gamma_{\circ\beta}^{j\circ} c_{k\circ}^{\circ\gamma} \gamma_{\circ\lambda}^{\ell\circ} c_{m\circ}^{\circ\mu}$$

and we obtain

$$A \otimes B \in (\otimes V^n)^2 \otimes V_*^n \otimes V^n \otimes V_*^n(\mathbb{R})|_{w_1+w_2}, \tag{8.15}$$

which shows that it is another modular tensor.

Since the tensor product is distributive with respect to the sum, we have arrived at a modular tensor algebra.

8.5.4 Tensor contraction

As could be expected, the system resulting from the contraction with respect to two indices of the same valency has no modular tensor nature. Next, we analyze the case of indices of different valency, over a mixed modular tensor.

Let $A \in (V^n)_w^{r=2+2}$. Because of Formula (8.4), we have

$$a^{ij oo}_{ookl} = |C|^w a^{\alpha \beta oo}_{oo\gamma\lambda} \gamma^{io}_{o\alpha} \gamma^{jo}_{o\beta} c^{o\gamma}_{ko} c^{o\lambda}_{lo}. \tag{8.16}$$

Let $B = \mathcal{C}\binom{\beta}{\gamma} A$ and also $B = \mathcal{C}\binom{j}{k} A$, then

$$b^{\alpha o}_{o\lambda} = \mathcal{C}\binom{\beta}{\gamma} a^{\alpha \beta oo}_{oo\gamma\lambda} = a^{\alpha \theta oo}_{oo\theta\lambda} \tag{8.17}$$

$$b^{io}_{ol} = \mathcal{C}\binom{j}{k} a^{ij oo}_{ookl} = a^{iz oo}_{oozl}. \tag{8.18}$$

Letting $j = k = z$ in (8.16) and using (8.18) we get

$$\begin{aligned}
b^{io}_{ol} = a^{iz oo}_{oozl} &= |C|^w a^{\alpha \beta oo}_{oo\gamma\lambda} \gamma^{io}_{o\alpha} \gamma^{zo}_{o\beta} c^{o\gamma}_{zo} c^{o\lambda}_{lo} \\
&= |C|^w a^{\alpha \beta oo}_{oo\gamma\lambda} \gamma^{io}_{o\alpha} (c^{\gamma o}_{oz} \gamma^{zo}_{o\beta}) c^{o\lambda}_{lo} \\
&= |C|^w a^{\alpha \beta oo}_{oo\gamma\lambda} \gamma^{io}_{o\alpha} \delta^{\gamma o}_{o\beta} c^{o\lambda}_{lo} \\
&= |C|^w (\delta^{o\gamma}_{\beta o} a^{\alpha \beta oo}_{oo\gamma\lambda}) \gamma^{io}_{o\alpha} c^{o\lambda}_{lo} \\
&= |C|^w a^{\alpha \theta oo}_{oo\theta\lambda} \gamma^{io}_{o\alpha} c^{o\lambda}_{lo}
\end{aligned}$$

and because of (8.17), we finally obtain

$$b^{io}_{ol} = |C|^w b^{\alpha o}_{o\lambda} \gamma^{io}_{o\alpha} c^{o\lambda}_{lo}. \tag{8.19}$$

Thus, the contracted tensor, is another modular tensor with *the same weight* that the given tensor, order $(r-2)$ and species $(p-1) + (q-1)$, in perfect analogy with the absolute homogeneous tensors.

It is evident that the multiple modular contraction is also modular.

8.5.5 Contracted tensor products

As is well known, the contracted tensor product is defined in exactly the same form as for absolute homogeneous tensors. The operation consists of multiplying in tensor form two modular tensors (then, the weight of the product would be the sum of the weights of the factors: $w_p = w_a + w_b$) and then executing a contraction of indices *of different valency and tensor factor* that maintains the weight. Thus, the contracted tensor product will have weight $w_p = w_a + w_b$, order $r_p = r_a + r_b - 2$, and species the ordered sequence coming from the tensor product with the exception of the contracted indices, which obviously disappear.

Next, we analyze a model to check the modular tensor nature. Consider the tensors

$$\text{Tensor A}: a^{ij\,\circ}_{\circ\circ k} = |C|^{w_1} a^{\alpha\beta\,\circ}_{\circ\circ\gamma}\gamma^{i\,\circ}_{\circ\alpha}\gamma^{j\,\circ}_{\circ\beta}c^{\circ\gamma}_{k\circ} \tag{8.20}$$

$$\text{Tensor B}: b^{\ell\,\circ}_{\circ m} = |C|^{w_2} b^{\lambda\circ}_{\circ\mu}\gamma^{\ell\,\circ}_{\circ\lambda}c^{\circ\mu}_{m\circ}. \tag{8.21}$$

The contracted tensor product tensor is $P_c = \mathcal{C}\binom{\lambda}{\gamma}P = \mathcal{C}\binom{\lambda}{\gamma}A \otimes B$ in the initial basis, and also $P_c = \mathcal{C}\binom{\ell}{k}P = \mathcal{C}\binom{\ell}{k}A \otimes B$ in the new basis.

Whence

$$pc^{ij\,\circ}_{\circ\circ m} = p^{ij\circ z\circ}_{\circ\circ z\circ m} = a^{ij\,\circ}_{\circ\circ z}\cdot b^{z\,\circ}_{\circ m};$$

and using (8.20) and (8.21) and letting $k = \ell = z$, the result is

$$pc^{ij\,\circ}_{\circ\circ m} = (|C|^{w_1} a^{\alpha\beta\circ}_{\circ\circ\gamma}\gamma^{i\,\circ}_{\circ\alpha}\gamma^{j\,\circ}_{\circ\beta}c^{\circ\gamma}_{z\circ})(|C|^{w_2} b^{\lambda\circ}_{\circ\mu}\gamma^{z\circ}_{\circ\lambda}c^{\circ\mu}_{m\circ})$$

$$= |C|^{w_1+w_2}(a^{\alpha\beta\circ}_{\circ\circ\gamma}b^{\lambda\circ}_{\circ\mu})\gamma^{i\,\circ}_{\circ\alpha}\gamma^{j\,\circ}_{\circ\beta}(c^{\circ\gamma}_{z\circ}\gamma^{z\circ}_{\circ\lambda})c^{\circ\mu}_{m\circ}$$

$$= |C|^{w_1+w_2}(a^{\alpha\beta\circ}_{\circ\circ\gamma}b^{\lambda\circ}_{\circ\mu})\gamma^{i\,\circ}_{\circ\alpha}\gamma^{j\,\circ}_{\circ\beta}(\gamma^{\circ z}_{\lambda\circ}c^{\circ\gamma}_{z\circ})c^{\circ\mu}_{m\circ}$$

$$= |C|^{w_1+w_2}(a^{\alpha\beta\circ}_{\circ\circ\gamma}b^{\lambda\circ}_{\circ\mu})\gamma^{i\,\circ}_{\circ\alpha}\gamma^{j\,\circ}_{\circ\beta}(\delta^{\circ\gamma}_{\lambda\circ})c^{\circ\mu}_{m\circ}$$

$$= |C|^{w_1+w_2}(a^{\alpha\beta\circ}_{\circ\circ\gamma}\delta^{\gamma\circ}_{\circ\lambda}b^{\lambda\circ}_{\circ\mu})\gamma^{i\,\circ}_{\circ\alpha}\gamma^{j\,\circ}_{\circ\beta}c^{\circ\mu}_{m\circ}$$

$$= |C|^{w_1+w_2}(a^{\alpha\beta\circ}_{\circ\circ\theta}b^{\theta\circ}_{\circ\mu})\gamma^{i\,\circ}_{\circ\alpha}\gamma^{j\,\circ}_{\circ\beta}c^{\circ\mu}_{m\circ}$$

$$= |C|^{w_1+w_2}pc^{\alpha\beta\circ}_{\circ\circ\mu}\gamma^{i\,\circ}_{\circ\alpha}\gamma^{j\,\circ}_{\circ\beta}c^{\circ\mu}_{m\circ}. \tag{8.22}$$

The modular tensor character is kept after contracted tensor products.

8.5.6 The quotient law. New criteria for modular tensor character

As indicated in Section 5.6, the aim of the quotient law is to determine the tensor nature of a certain system of scalars and discover its species.

We will use two theoretical models to show that the quotient law can be extended to determine the *modular tensor character* of a given system.

Model 1. We have a system of scalars of order $r = 3$, $S(\alpha\beta\gamma)$ defined over a linear space $V^n(\mathbb{R})$ and we decide to apply to it a "test" modular tensor A, the components of which satisfy

$$a^i_\circ = |C|^{w_a} a^\alpha_\circ \gamma^{i\,\circ}_{\circ\alpha}. \tag{8.23}$$

The contracted product $S(\alpha\beta\gamma)\cdot a^\alpha_\circ$ leads to the tensor B, the components of which satisfy the relation

$$b_{jk}^{\circ\circ} = |C|^{w_b} b_{\beta\gamma}^{\circ\circ} c_{j\circ}^{\circ\beta} c_{k\circ}^{\circ\gamma}. \tag{8.24}$$

We wish to determine the tensor nature of the system $S(\alpha\beta\gamma)$ using the quotient law.

We start from

$$S(\alpha\beta\gamma)a_{\circ}^{\alpha} = b_{\beta\gamma}^{\circ\circ} \tag{8.25}$$

in the initial basis and also

$$S(ijk)a_{\circ}^{i} = b_{jk}^{\circ\circ} \tag{8.26}$$

in the new basis.

We can obtain $a_{\circ}^{\alpha}$ from expression (8.23). First, we pass the scalar $|C|^{w_a}$ to the left-hand side and write the matrix $\gamma_{\circ\alpha}^{i\circ}$ in front of the component $a_{\circ}^{\alpha}$ and get

$$|C|^{-w_a} a_{\circ}^{i} = \gamma_{\alpha\circ}^{\circ i} a_{\circ}^{\alpha};$$

next, we pass the matrix $\gamma_{\alpha\circ}^{\circ i}$ to the left-hand side

$$|C|^{-w_a} c_{i\circ}^{\circ\alpha} a_{\circ}^{i} = a_{\circ}^{\alpha}. \tag{8.27}$$

Now we substitute (8.27) into (8.25) and get

$$|C|^{-w_a} S(\alpha\beta\gamma) c_{i\circ}^{\circ\alpha} a_{\circ}^{i} = b_{\beta\gamma}^{\circ\circ} \tag{8.28}$$

and then we substitute (8.28) into (8.24), and the result is substituted into (8.26), to obtain

$$S(ijk)a_{\circ}^{i} = |C|^{w_b} (|C|^{-w_a} S(\alpha\beta\gamma) c_{i\circ}^{\circ\alpha} a_{\circ}^{i}) c_{j\circ}^{\circ\beta} c_{k\circ}^{\circ\gamma}$$

and operating and grouping in the left-hand side, we obtain

$$S(ijk)a_{\circ}^{i} - |C|^{w_b-w_a} S(\alpha\beta\gamma) c_{i\circ}^{\circ\alpha} c_{j\circ}^{\circ\beta} c_{k\circ}^{\circ\gamma} a_{\circ}^{i} = 0$$

$$(S(ijk) - |C|^{w_b-w_a} S(\alpha\beta\gamma) c_{i\circ}^{\circ\alpha} c_{j\circ}^{\circ\beta} c_{k\circ}^{\circ\gamma})a_{\circ}^{i} = 0,$$

which implies

$$S(ijk) = |C|^{w_b-w_a} S(\alpha\beta\gamma) c_{i\circ}^{\circ\alpha} c_{j\circ}^{\circ\beta} c_{k\circ}^{\circ\gamma}.$$

The last equation shows two facts:

First: $s(\alpha\beta\gamma) \equiv s_{\alpha\beta\gamma}^{\circ\circ\circ}$ is a covariant tensor of order 3.

Second: $S(\alpha\beta\gamma)$ is a modular tensor of weight $w_b - w_a$.

Thus, finally we get

$$s_{ijk}^{\circ\circ\circ} = |C|^{w_b - w_a} s_{\alpha\beta\gamma}^{\circ\circ\circ} c_{i\,\circ}^{\circ\alpha} c_{j\,\circ}^{\circ\beta} c_{k\,\circ}^{\circ\gamma}.$$

Model 2. Consider a system of scalars of order $r = 2$, defined over a linear space $V^n(\mathbb{R})$. The test tensor is the modular tensor A of zero weight, that is, the absolute tensor:

$$a_{\circ\circ}^{ij} = a_{\circ\circ}^{\alpha\beta} \gamma_{\circ\alpha}^{i\,\circ} \gamma_{\circ\beta}^{j\,\circ}. \tag{8.29}$$

The contracted product B is a modular tensor of zero order and weight $w \neq 0$, that is, a pseudoscalar. Thus, we have

$$S(\alpha\beta) \cdot a_{\circ\circ}^{\alpha\beta} = b \tag{8.30}$$

$$S(ij) \cdot a_{\circ\circ}^{ij} = b' \tag{8.31}$$

$$b' = |C|^w b. \tag{8.32}$$

Expression (8.29) can be written as

$$a_{\circ\circ}^{ij} = \gamma_{\alpha\circ}^{\circ i} \gamma_{\beta\circ}^{\circ j} a_{\circ\circ}^{\alpha\beta}$$

and also, passing matrices γ to the left-hand side

$$c_{i\,\circ}^{\circ\alpha} c_{j\,\circ}^{\circ\beta} a_{\circ\circ}^{ij} = a_{\circ\circ}^{\alpha\beta}. \tag{8.33}$$

So that substituting (8.33) into (8.30), the result is

$$S(\alpha\beta) c_{i\,\circ}^{\circ\alpha} c_{j\,\circ}^{\circ\beta} a_{\circ\circ}^{ij} = b. \tag{8.34}$$

Substituting (8.34) and (8.31) into (8.32), we have

$$S(ij) a_{\circ\circ}^{ij} = |C|^w S(\alpha\beta) c_{i\,\circ}^{\circ\alpha} c_{j\,\circ}^{\circ\beta} a_{\circ\circ}^{ij}$$

and taking common factors we obtain

$$(S(ij) - |C|^w S(\alpha\beta) c_{i\,\circ}^{\circ\alpha} c_{j\,\circ}^{\circ\beta}) a_{\circ\circ}^{ij} = 0,$$

which requires

$$S(ij) = |C|^w S(\alpha\beta) c_{i\,\circ}^{\circ\alpha} c_{j\,\circ}^{\circ\beta}, \tag{8.35}$$

showing that $S(\alpha\beta)$ is a modular covariant tensor, $S_{\alpha\beta}^{\circ\circ}$ of weight w. Then, expression (8.35) can be written as

$$S_{ij}^{\circ\circ} = |C|^w S_{\alpha\beta}^{\circ\circ} c_{i\,\circ}^{\circ\alpha} c_{j\,\circ}^{\circ\beta}.$$

This proves the validity of the quotient law for modular tensors and its validity for establishing modular tensor criteria for systems of scalars.

8.6 Modular symmetry and anti-symmetry

We simply establish the existence of tensors that are symmetric or anti-symmetric with respect to a subset I_k of their indices. We insist that in the case of mixed tensors, the subset $I_k = I_{k'} \cup I_{k''}$ must be formed by subsets $I_{k'} \subset \{\text{contravariant indices}\}$, $I_{k''} \subset \{\text{covariant indices}\}$, considered separately, that is, the symmetry or the anti-symmetry are always contemplated over indices of the same valency.

Example 8.1 (Calculus of the modular characteristics). Let $[t^{\alpha\,\circ}_{\,\circ\,\beta}]$ be a mixed tensor of $\mathbb{R}^n \times \mathbb{R}^n_*$.

1. Can we say that the determinant $|t^{\,\circ\,\beta}_{\alpha\,\circ}|$ has a tensor nature? Give its modular characteristics.
2. Idem for $[t^{\,\circ\,\circ}_{\alpha\,\beta}]$.
3. Idem for $[t^{\alpha\,\beta}_{\,\circ\,\circ}]$.

Solution:

1. We know that the homogeneous mixed tensors of order $r = 2$ change basis according to the tensor relation

$$t^{i\,\circ}_{\,\circ\,j} = t^{\alpha\,\circ}_{\,\circ\,\beta}\gamma^{i\,\circ}_{\,\circ\,\alpha}c^{\,\circ\,\beta}_{j\,\circ}.$$

If we take determinants on both sides, on account of the Binet–Cauchy theorem for determinants, the result is

$$|t^{i\,\circ}_{\,\circ\,j}| = |t^{\alpha\,\circ}_{\,\circ\,\beta}\gamma^{i\,\circ}_{\,\circ\,\alpha}c^{\,\circ\,\beta}_{j\,\circ}| = |t^{\alpha\,\circ}_{\,\circ\,\beta}||\gamma^{i\,\circ}_{\,\circ\,\alpha}||c^{\,\circ\,\beta}_{j\,\circ}|$$
$$= |t^{\alpha\,\circ}_{\,\circ\,\beta}||C^{-1}||C^t| = |t^{\alpha\,\circ}_{\,\circ\,\beta}||C|^{-1}|C| = |t^{\alpha\,\circ}_{\,\circ\,\beta}|,$$

which remains invariant, so that we have an absolute homogeneous tensor of order $r = 0$, with modular tensor characteristic (m.t.c.)

$$(order, species, weight) \equiv (0, -, 0).$$

2. Similarly

$$t^{\circ\circ}_{ij} = t^{\,\circ\,\circ}_{\alpha\beta}c^{\,\circ\,\alpha}_{i\,\circ}c^{\,\circ\,\beta}_{j\,\circ}$$

and taking determinants we get

$$|t^{\circ\circ}_{ij}| = |t^{\,\circ\,\circ}_{\alpha\beta}||c^{\,\circ\,\alpha}_{i\,\circ}||c^{\,\circ\,\beta}_{j\,\circ}| = |t^{\,\circ\,\circ}_{\alpha\beta}||C||C| = |C|^2|t^{\,\circ\,\circ}_{\alpha\beta}|,$$

which proves that it is a modular tensor of order $r = 0$, and weight 2, also called a "pseudoscalar" of weight 2, and with m.t.c. $(0, -, 2)$.

3. Finally, we have

$$t^{ij}_{\circ\circ} = t^{\alpha\beta}_{\circ\circ}\gamma^{i\circ}_{\circ\alpha}\gamma^{j\circ}_{\circ\beta}$$

and following the same process we obtain

$$|t^{ij}_{\circ\circ}| = |t^{\alpha\circ}_{\circ\beta}||\gamma^{i\circ}_{\circ\alpha}||\gamma^{j\circ}_{\circ\beta}| = |t^{\alpha\beta}_{\circ\circ}||C|^{-1}|C|^{-1} = |C|^{-2}|t^{\alpha\beta}_{\circ\circ}|,$$

which proves it is a modular tensor of order $r = 0$, and weight -2, also called a "pseudoscalar" of weight -2, and with m.t.c. $(0, -, -2)$.

□

Example 8.2 (Modular quotient law). Consider a covariant tensor $A \equiv [a^{\circ\circ}_{\lambda\mu}]$ of order $r = 2$, built over the linear space $V^n(\mathbb{R})$ in a certain basis. We denote by $A(\alpha\beta)$ the adjoint of the element $a_{\alpha\beta}$ in the determinant $|a^{\circ\circ}_{\lambda\mu}|$ and we know that $|a^{\circ\circ}_{\lambda\mu}| \in \mathbb{R}$ and $|a^{\circ\circ}_{\lambda\mu}| \neq 0$.

1. We wish to analyze the possible tensor nature of the system of scalars:

$$b(\alpha\beta) = \frac{A(\alpha\beta)}{|a^{\circ\circ}_{\lambda\mu}|}.$$

2. Solve the same problem, but assuming that A is a contra-contravariant tensor of order $r = 2$, and that $|a^{\lambda\mu}_{\circ\circ}| \neq 0$.

Solution: We set $|a^{\circ\circ}_{\lambda\mu}| = |A|$, so that $|a^{\circ\circ}_{\lambda\mu}|^{-1} = |A|^{-1}$. From the theory of determinants, we know that the inverse matrix can be defined as

$$[b(\alpha\beta)]^t = \left[\frac{A(\alpha\beta)}{|A|}\right]^t = [a^{\circ\circ}_{\alpha\beta}]^{-1} = A^{-1}, \tag{8.36}$$

so that pre-multiplying by the matrix $[a^{\circ\circ}_{\alpha\beta}]$, we obtain

$$[a^{\circ\circ}_{\alpha\beta}][b(\alpha\beta)]^t = [a^{\circ\circ}_{\alpha\beta}][a^{\circ\circ}_{\alpha\beta}]^{-1} = A \cdot A^{-1} = I_n.$$

This matrix expression, notated in tensor form is

$$a^{\circ\circ}_{\gamma\alpha}b(\alpha\beta) = \delta^{\circ\beta}_{\gamma\circ}. \tag{8.37}$$

If we now choose a contravariant tensor, of order $r = 1$, $T = t^{\gamma}_{\circ}\vec{e}_{\gamma} \in V^n(\mathbb{R})$, as the test tensor to apply the quotient law, the contracted product with the data tensor $a^{\circ\circ}_{\gamma\alpha}$ is the covariant tensor, of first order $S = s^{\circ}_{\alpha}\vec{e}^{*\alpha} \in V^n_*(\mathbb{R})$. In effect

$$t_{\circ}^{\gamma} a_{\gamma\alpha}^{\circ\circ} = s_{\alpha}^{\circ}; \ \forall T = t_{\circ}^{\gamma} \vec{e}_{\gamma} \in V^n(\mathbb{R}). \tag{8.38}$$

If we now execute the contracted product of tensor S by the system of scalars $b(\alpha\beta)$, using (8.38), we get:

$$t_{\circ}^{\gamma} a_{\gamma\alpha}^{\circ\circ} b(\alpha\beta) = s_{\alpha}^{\circ} b(\alpha\beta)$$

and applying (8.37) to the previous relation, we obtain

$$t_{\circ}^{\gamma} \delta_{\gamma\circ}^{\circ\beta} = s_{\alpha}^{\circ} b(\alpha\beta) \to s_{\alpha}^{\circ} b(\alpha\beta) = t_{\circ}^{\beta}. \tag{8.39}$$

Equation (8.39) shows, by the quotient law, that the system $b(\alpha\beta)$ is a totally contravariant homogeneous tensor of order $r = 2$. Then, we have

$$b(\alpha\beta) = b_{\circ\circ}^{\alpha\beta} = \frac{A(\alpha\beta)}{|A|}. \tag{8.40}$$

We complete this example with another thought. If we remove denominators in (8.40), we get

$$A(\alpha\beta) = |A| b_{\circ\circ}^{\alpha\beta}. \tag{8.41}$$

We have seen in Example 8.1 that $|A| = |a_{\lambda\mu}^{\circ\circ}|$ is a modular tensor of order $r = 0$ and weight 2, which reveals that the system of adjoints or cofactors of the determinant $|a_{\lambda\mu}^{\circ\circ}| = |A|$, according to (8.41), is the tensor product of the tensor analyzed in (8.40) and determinant $|A|$. Whence the set $A(\alpha\beta)$ of the adjoints resulting from such product is a modular tensor of order $r = 2$ and weight 2.

3. Since the process is similar to the process presented, though starting from

$$[b(\alpha\beta)]^t = \left[\frac{A(\alpha\beta)}{|a_{\circ\circ}^{\lambda\mu}|} \right]^t = \left[a_{\circ\circ}^{\alpha\beta} \right]^{-1}, \text{ we invite the reader to solve and com-}$$

plete it, with the meaning of the adjoint or cofactors of a contravariant tensor of order $r = 2$.

□

Example 8.3 (Pseudoscalars). Let A and B be two homogeneous tensors, of order $r = 2$, both covariant and referred to a certain basis in the linear space $V_*^n(\mathbb{R})$, and let $\lambda \in \mathbb{R}$. Consider the system of scalar components associated with the determinant:

$$s = |a_{\alpha\beta}^{\circ\circ} + \lambda b_{\alpha\beta}^{\circ\circ}|.$$

We wish to study its possible tensor nature.

Solution: Due to the tensor nature of the entities A and B, we have for the other basis of $V_*^n(\mathbb{R})$:

$$a_{ij}^{\circ\circ} = a_{\alpha\beta}^{\circ\circ} c_{i\circ}^{\circ\alpha} c_{j\circ}^{\circ\beta}$$

$$b_{ij}^{\circ\circ} = b_{\alpha\beta}^{\circ\circ} c_{i\circ}^{\circ\alpha} c_{j\circ}^{\circ\beta}.$$

Multiplying the second relation by λ and adding, we obtain

$$a_{ij}^{\circ\circ} + \lambda b_{ij}^{\circ\circ} = (a_{\alpha\beta}^{\circ\circ} + \lambda b_{\alpha\beta}^{\circ\circ}) c_{i\circ}^{\circ\alpha} c_{j\circ}^{\circ\beta}$$

and taking determinants, we get

$$\left| a_{ij}^{\circ\circ} + \lambda b_{ij}^{\circ\circ} \right| = \left| a_{\alpha\beta}^{\circ\circ} + \lambda b_{\alpha\beta}^{\circ\circ} \right| \left| c_{i\circ}^{\circ\alpha} \right| \left| c_{j\circ}^{\circ\beta} \right|$$

and, since $\left| c_{i\circ}^{\circ\alpha} \right| = \left| c_{j\circ}^{\circ\beta} \right| = |C^t| = |C|$, the previous relation leads to

$$\left| a_{ij}^{\circ\circ} + \lambda b_{ij}^{\circ\circ} \right| = |C|^2 \left| a_{\alpha\beta}^{\circ\circ} + \lambda b_{\alpha\beta}^{\circ\circ} \right|,$$

which reveals that the entity $s = \left| a_{\alpha\beta}^{\circ\circ} + \lambda b_{\alpha\beta}^{\circ\circ} \right|$ is a modular tensor of order 0 and weight 2, or better a relative invariant of weight 2, or a pseudoscalar of weight 2.

□

Remark 8.2. We observe that the case $\left| a_{\circ\beta}^{\alpha\circ} + \lambda b_{\circ\beta}^{\alpha\circ} \right|$ and the case $\left| a_{\circ\circ}^{\alpha\beta} + \lambda b_{\circ\circ}^{\alpha\beta} \right|$ lead to an absolute tensor of zero order or to an absolute invariant, and to a modular tensor of order 0 and weight -2, respectively.

□

Example 8.4 (Modular tensor nature of determinants and their cofactors). Let T be a modular tensor, of order $r = 2$ and weight $p = w(w \in Z)$, referred to a certain basis of the linear space $V^n(\mathbb{R})$. We wish to study the modular tensor nature of the system $|T|$, determinant of T, in the following cases:

1. The totally covariant tensor $T = [t_{\alpha\beta}^{\circ\circ}]$.

2. The contra-covariant tensor $T = [t_{\circ\beta}^{\alpha\circ}]$

3. The cova-contravariant tensor $T = [t_{\alpha\circ}^{\circ\beta}]$.

4. The totally contravariant tensor $T = [t_{\circ\circ}^{\alpha\beta}]$.

5. The system of scalars:
$$s(\alpha\beta) \equiv T(\alpha\beta),$$

the components of which are the adjoints or cofactors of the determinant $|T|$, for the four subcases indicated in 1, 2, 3 and 4. In all of them we assume that $|T| \neq 0$.

Solution:

1. Since it is a tensor with modular characteristics $(order, species, weight) \equiv (2, cova\text{-}covariant, w)$, we have

$$t_{ij}^{\circ\circ} = |C|^w t_{\alpha\beta}^{\circ\circ} c_{i\circ}^{\circ\alpha} c_{j\circ}^{\circ\beta}. \tag{8.42}$$

Taking into account that the scalar $|C|^w$ appears in each row (column) of the matrix $[t_{\alpha\beta}^{\circ\circ}]$, and that $|c_{i\circ}^{\circ\alpha}| = |c_{j\circ}^{\circ\beta}| = |C^t| = |C|$ when taking determinants on both sides of (8.42), one gets

$$|t_{ij}^{\circ\circ}| = |\,|C|^w t_{\alpha\beta}^{\circ\circ}\,|\,|c_{i\circ}^{\circ\alpha}||c_{j\circ}^{\circ\beta}| = |C|^{nw} \cdot |t_{\alpha\beta}^{\circ\circ}| \cdot |C| \cdot |C| = C|^{nw+2} \cdot |t_{\alpha\beta}^{\circ\circ}|. \tag{8.43}$$

In this case $|T|$ is a modular tensor of order 0 (pseudo–scalar) and weight $(nw + 2)$.

2. Now it is a tensor with modular characteristics $(order, species, weight) \equiv (2, contra\text{-}covariant, w)$, thus we have

$$t_{ij}^{\circ\circ} = |C|^w t_{\circ\beta}^{\alpha\circ} \gamma_{\circ\alpha}^{i\circ} c_{j\circ}^{\circ\beta} \tag{8.44}$$

and considering the comments made in 1, taking determinants of (8.44), we get

$$|t_{\circ j}^{i\circ}| = |C|^{nw} |t_{\circ\beta}^{\alpha\circ}||C|^{-1}|C| = |C|^{nw}|t_{\circ\beta}^{\alpha\circ}|, \tag{8.45}$$

which proves that in this case $|T|$ is a modular tensor of order 0 and weight (nw).

3. Now the modular characteristics are $(2, cova\text{-}contravariant, w)$, that is, we can write

$$t_{ij}^{\circ\circ} = |C|^w t_{\alpha\circ}^{\circ\beta} c_{i\circ}^{\circ\alpha} \gamma_{\circ\beta}^{j\circ} \tag{8.46}$$

and taking determinants we obtain

$$|t_{i\circ}^{\circ j}| = |C|^{nw} |t_{\alpha\circ}^{\circ\beta}||C||C|^{-1} = |C|^{nw}|t_{\alpha\circ}^{\circ\beta}|, \tag{8.47}$$

which shows that in the present case $|T|$ is a modular tensor of order 0 and weight (nw).

4. In this question the modular characteristics are $(2, contra\text{-}contravariant, w)$, so that the tensor change-of-basis is

$$t_{\circ\circ}^{ij} = |C|^w t_{\circ\circ}^{\alpha\beta} \gamma_{\circ\alpha}^{i\circ} \gamma_{\circ\beta}^{j\circ} \tag{8.48}$$

and taking determinants and operating yields

$$|t_{\circ\circ}^{ij}| = |C|^{nw} \cdot |t_{\circ\circ}^{\alpha\beta}| \cdot |C|^{-1} \cdot |C|^{-1} = |C|^{nw-2} \cdot |t_{\circ\circ}^{\alpha\beta}|, \tag{8.49}$$

which shows that this time T is a modular tensor of order 0 and weight $(nw - 2)$.

5. We solve only two cases, and leave the reader to solve the rest but guessing the results in advance.

Case 2: The data tensor is of characteristics $(2, contra\text{-}covariant, w)$. From the definition of inverse matrix, in tensor notation, we have

$$\frac{S(\beta\alpha)}{|t^{\lambda\circ}_{\circ\mu}|} \cdot t^{\alpha\circ}_{\circ\gamma} = \delta^{\beta\circ}_{\circ\gamma} \tag{8.50}$$

and contracting the data tensor with another homogeneous contravariant tensor $A = [a^{\gamma}_{\circ}] \in V^{n}(\mathbb{R})$, we get

$$t^{\alpha\circ}_{\circ\gamma}a^{\gamma}_{\circ} = b^{\alpha}_{\circ}, \tag{8.51}$$

with $b^{\alpha}_{\circ}$ of weight (w), and then, contracting the two sides of (8.50) with tensor A we have

$$\frac{S(\beta\alpha)}{|t^{\lambda\circ}_{\circ\mu}|} \cdot t^{\alpha\circ}_{\circ\gamma}a^{\gamma}_{\circ} = \delta^{\beta\circ}_{\circ\gamma}a^{\gamma}_{\circ},$$

which implies

$$\frac{S(\beta\alpha)}{|t^{\lambda\circ}_{\circ\mu}|} b^{\alpha}_{\circ} = a^{\beta}_{\circ}. \tag{8.52}$$

Interpreting (8.52) as a quotient law, the entity $\frac{S(\beta\alpha)}{|t^{\lambda\circ}_{\circ\mu}|}$ becomes a modular tensor of order $r = 2$, contra-covariant, of weight $(-w)$. Its tensor character equation is

$$\frac{s^{j\circ}_{\circ i}}{|t^{\ell\circ}_{\circ m}|} = |C|^{-w} \frac{s^{\beta\circ}_{\circ\alpha}}{|t^{\lambda\circ}_{\circ\mu}|} \gamma^{j\circ}_{\circ\beta} c^{\circ\alpha}_{i\circ} \tag{8.53}$$

and if one takes into account the relation (8.45)

$$|t^{\ell\circ}_{\circ m}| = |C|^{nw}|t^{\lambda\circ}_{\circ\mu}|,$$

substituting it into the left-hand side of (8.53) yields

$$\frac{s^{j\circ}_{\circ i}}{|C|^{nw}|t^{\lambda\circ}_{\circ\mu}|} = |C|^{-w} \frac{s^{\beta\circ}_{\circ\alpha}}{|t^{\lambda\circ}_{\circ\mu}|} \gamma^{j\circ}_{\circ\beta} c^{i\circ}_{\circ\alpha},$$

which after simplifying and renaming the indices, finally leads to

$$s^{i\circ}_{\circ j} = |C|^{(n-1)w} s^{\alpha\circ}_{\circ\beta} \gamma^{i\circ}_{\circ\alpha} c^{\circ\beta}_{j\circ}, \tag{8.54}$$

which proves that in this case the system S of the adjoints or cofactors of the determinant $|T| = |t^{\lambda \circ}_{\circ \mu}|$ is a modular tensor of order $r = 2$, contra–covariant, of weight $(n-1)w$.

Case 4. The data tensor is of characteristics $(2, contra\text{-}contravariant, w)$. Before examining this part, it is convenient to read the solution to Example 8.2, which can save much explanation.

We start from the definition of an inverse matrix, to formulate the first matrix expression

$$\frac{[S(\alpha\beta)]^t}{|T|}[t^{\alpha\beta}_{\circ\circ}] = I_n, \tag{8.55}$$

the tensor expression of which is

$$\frac{S(\beta\alpha)}{|T|}[t^{\alpha\gamma}_{\circ\circ}] = \delta^{\circ\gamma}_{\beta\circ}. \tag{8.56}$$

If we execute the contracted tensor product of tensor T by a first-order covariant tensor $A = [a^{\circ}_{\gamma}], A \in V^n_*(\mathbb{R})$, we get

$$t^{\alpha\gamma}_{\circ\circ}a^{\circ}_{\gamma} = b^{\alpha}_{\circ}, \tag{8.57}$$

where $b^{\alpha}_{\circ}$ is a modular contravariant tensor of first-order and weight (w).

Multiplying (8.56) by a°_{γ} and contracting yields

$$\frac{S(\beta\alpha)}{|T|}t^{\alpha\gamma}_{\circ\circ}a^{\circ}_{\gamma} = \delta^{\circ\gamma}_{\beta\circ}a^{\circ}_{\gamma}$$

and taking into account (8.57) the result is

$$\frac{S(\beta\alpha)}{|T|}b^{\alpha}_{\circ} = a^{\circ}_{\beta}. \tag{8.58}$$

Since in relation (8.58) $|T| \neq 0$, it is valid to interpret it as a modular quotient law, in which a modular co-covariant tensor of order $r = 2$ and of weight $(-w)$ is contracted with another modular contravariant tensor of first order and weight (w), to give a homogeneous covariant tensor of first order.

Then, representing $|T| = |t^{\lambda\mu}_{\circ\circ}|$, we have the tensor character equation just discovered, as the first factor of the left-hand side of (8.58), i.e.,

$$\frac{s^{\circ\circ}_{ji}}{|t^{\ell m}_{\circ\circ}|} = |C|^{-w}\frac{s^{\circ\circ}_{\beta\alpha}}{|t^{\lambda\mu}_{\circ\circ}|}c^{\circ\beta}_{j\circ}c^{\circ\alpha}_{i\circ}. \tag{8.59}$$

However, in the first part of this example, point 4, we have seen that Formula (8.49) suggests

$$|t_{\circ\,\circ}^{\ell m}| = |C|^{nw-2}|t_{\circ\,\circ}^{\lambda\mu}|,$$

so that we replace it into (8.59) to get

$$\frac{s_{j\,i}^{\circ\circ}}{|C|^{nw-2}|t_{\circ\,\circ}^{\lambda\mu}|} = |C|^{-w}\frac{s_{\beta\alpha}^{\circ\,\circ}}{|t_{\circ\,\circ}^{\lambda\mu}|}c_{j\,\circ}^{\circ\beta}c_{i\,\circ}^{\circ\alpha}$$

and operating and sorting we arrive at

$$s_{i\,j}^{\circ\circ} = |C|^{(n-1)w-2}s_{\alpha\beta}^{\circ\,\circ}c_{i\,\circ}^{\circ\alpha}c_{j\,\circ}^{\circ\beta}, \tag{8.60}$$

which reveals that the system S of the adjoints or cofactors of the determinant $|T| = |t_{\circ\,\circ}^{\lambda\mu}|$ is a modular tensor of order $r = 2$, totally covariant and of weight $((n-1)w-2)$.

$\square$

Example 8.5 (Tensor character of subsets of tensor components). Consider a tensor with modular characteristics $(3, contra\text{-}cova\text{-}contravariant, 3)$ referred to the linear space $\mathbb{R}^3$ in its canonical basis.

The matrix representation of its components is

$$T \equiv [t_{\circ\,\beta\circ}^{\alpha\circ\gamma}] = \left[\begin{array}{rrr} 1 & -1 & 1 \\ 0 & 2 & 0 \\ -1 & 1 & 3 \\ \hline 2 & 0 & -1 \\ 0 & 0 & 1 \\ 3 & 0 & 1 \\ \hline 1 & 0 & 0 \\ 1 & -1 & 1 \\ 1 & 0 & 2 \end{array}\right],$$

where α is the row block index (there are three blocks), β is the row index in each block, and γ is the column index in each block.

If we choose a new basis in $\mathbb{R}^3$, with vectors

$$\vec{e}_1(0,1,0); \quad \vec{e}_2(1,0,-1); \quad \vec{e}_3(1,0,-2),$$

1. Obtain the components of the modular tensor T in the new basis.
2. Determine the system of scalars S defined by the expression

$$s(\beta,\gamma) = t_{\circ\,\beta\circ}^{1\circ\gamma}.$$

3. Idem in the new basis

$$s(j,k) = t_{\circ\,j\circ}^{1\circ k}.$$

4. Examine the tensor nature of the system S.

5. Idem for the system P defined by

$$p(\alpha,\gamma) = t^{\alpha\circ\gamma}_{\circ 1 \circ}.$$

6. Idem for the system Q defined by

$$q(\alpha) = t^{\alpha\circ\beta}_{\circ\beta\circ}.$$

Solution: The matrices that will appear in the solution process are

$$||\vec{\hat{e}}_i|| = ||\vec{\hat{e}}_\alpha|| [c^{\alpha\circ}_{\circ i}]; \quad C \equiv [c^{\alpha\circ}_{\circ i}] = \begin{bmatrix} 0 & 1 & 1 \\ 1 & 0 & 0 \\ 0 & -1 & -2 \end{bmatrix}; \quad |C| = 1 \neq 0$$

$$C^{-1} \equiv [\gamma^{i\circ}_{\circ\alpha}] = \begin{bmatrix} 0 & 1 & 0 \\ 2 & 0 & 1 \\ -1 & 0 & -1 \end{bmatrix}; \quad C^t \equiv [c^{\circ\alpha}_{i\circ}] = \begin{bmatrix} 0 & 1 & 0 \\ 1 & 0 & -1 \\ 1 & 0 & -2 \end{bmatrix}; \quad |C|^3 = 1.$$

1. Note that $\sigma = n^r = 3^3 = 27$.

Using the "extension" formula (1.30) we have $(T_{27})^t$:

$$[1\ -1\ 1\ 0\ 2\ 0\ -1\ 1\ 3\ |\ 2\ 0\ -1\ 0\ 0\ 1\ 3\ 0\ 1\ |\ 1\ 0\ 0\ 1\ -1\ 1\ 1\ 0\ 2].$$

and according to Formula (4.36), we can write

$$\left(\hat{T}^t_{27}\right)_0 = (T_{27})^t \bullet \left[C^{-1} \otimes C^t \otimes C^{-1}\right]^t = (T_{27})^t \bullet \left[(C^{-1})^t \otimes C \otimes (C^{-1})^t\right]$$

$$= (T_{27})^t \bullet \left(\begin{bmatrix} 0 & 2 & -1 \\ 1 & 0 & 0 \\ 0 & 1 & -1 \end{bmatrix} \otimes \begin{bmatrix} 0 & 1 & 1 \\ 1 & 0 & 0 \\ 0 & -1 & -2 \end{bmatrix} \otimes \begin{bmatrix} 0 & 2 & -1 \\ 1 & 0 & 0 \\ 0 & 1 & -1 \end{bmatrix} \right)$$

$$= [0\ 1\ -1\ 0\ -4\ 3\ 0\ -11\ 7\ |\ 3\ 3\ -2\ -4\ 2\ 2\ -6\ -4\ -9\ |\ -1\ -3\ 2\ 2\ 0\ -2\ 3\ 5\ -7].$$

In addition, taking into account that it is a modular tensor, we have

$$\hat{T}^t_{27} = \left(\hat{T}^t_{27}\right)_0 \cdot |C|^3$$

$$= [0\ 1\ -1\ 0\ -4\ 3\ 0\ -11\ 7\ |\ 3\ 3\ -2\ -4\ 2\ 2\ -6\ -4\ -9\ |\ -1\ -3\ 2\ 2\ 0\ -2\ 3\ 5\ -7]$$

and using the "condensation" formula (1.32), we obtain the matrix representation of the tensor $\hat{T}$, in the new basis

$$\hat{T} = [t^{i\circ k}_{\circ j\circ}] = \begin{bmatrix} 0 & 1 & -1 \\ 0 & -4 & 3 \\ 0 & -11 & 7 \\ 3 & 3 & -2 \\ -4 & 2 & 2 \\ -6 & -4 & 9 \\ -1 & -3 & 2 \\ 2 & 0 & -2 \\ 3 & 5 & -7 \end{bmatrix}.$$

2. Consider now the system S. Since $[t_{\circ\beta\circ}^{1\circ\gamma}]$ is the first block of the tensor T, we have

$$S \equiv [s(\beta, \gamma)] = \begin{bmatrix} 1 & -1 & 1 \\ 0 & 2 & 0 \\ -1 & 1 & 3 \end{bmatrix}. \tag{8.61}$$

3. Similarly, since $[t_{\circ j \circ}^{1\circ k}]$ is the first block of $\hat{T}$, we get

$$S \equiv [s(j, k)] = \begin{bmatrix} 0 & 1 & -1 \\ 0 & -4 & 3 \\ 0 & -11 & 7 \end{bmatrix}. \tag{8.62}$$

4. Considering the tensor nature of T, and developing its dummy index α according to the Einstein convention yields

$$\begin{aligned}
t_{\circ j \circ}^{i \circ k} &= |C|^3 t_{\circ\beta\circ}^{\alpha\circ\gamma} \gamma_{\circ\alpha}^{i\circ} c_{j\circ}^{\circ\beta} \gamma_{\circ\gamma}^{k\circ} \\
&= |C|^3 t_{\circ\beta\circ}^{1\circ\gamma} \gamma_{\circ 1}^{i\circ} c_{j\circ}^{\circ\beta} \gamma_{\circ\gamma}^{k\circ} + |C|^3 t_{\circ\beta\circ}^{2\circ\gamma} \gamma_{\circ 2}^{i\circ} c_{j\circ}^{\circ\beta} \gamma_{\circ\gamma}^{k\circ} + |C|^3 t_{\circ\beta\circ}^{3\circ\gamma} \gamma_{\circ 3}^{i\circ} c_{j\circ}^{\circ\beta} \gamma_{\circ\gamma}^{k\circ}
\end{aligned}$$

and letting $i = 1$, we get

$$t_{\circ j \circ}^{1\circ k} = |C|^3 t_{\circ\beta\circ}^{1\circ\gamma} \gamma_{\circ 1}^{1\circ} c_{j\circ}^{\circ\beta} \gamma_{\circ\gamma}^{k\circ} + |C|^3 \gamma_{\circ\gamma}^{k\circ} c_{j\circ}^{\circ\beta} \left[t_{\circ\beta\circ}^{2\circ\gamma} \gamma_{\circ 2}^{1\circ} + t_{\circ\beta\circ}^{3\circ\gamma} \gamma_{\circ 3}^{1\circ} \right]$$

and taking into account (8.61) and (8.62), we finally obtain

$$s(j, k) = \gamma_{\circ 1}^{1\circ} \left[|C|^3 s(\beta, \gamma) c_{j\circ}^{\circ\beta} \gamma_{\circ\gamma}^{k\circ} \right] + |C|^3 c_{j\circ}^{\circ\beta} \gamma_{\circ\gamma}^{k\circ} \left[t_{\circ\beta\circ}^{2\circ\gamma} \gamma_{\circ 2}^{1\circ} + t_{\circ\beta\circ}^{3\circ\gamma} \gamma_{\circ 3}^{1\circ} \right]. \tag{8.63}$$

Expression (8.63) has two properties that prevent the system S from having a tensor character:

a) The factor $\gamma_{\circ 1}^{1\circ}$ is a *variable* real number dependent on the change-of-basis C performed in $V^n(\mathbb{R})$.

b) The second summand will not be null in general, unless one precisely selects a tensor such that $t_{\circ\beta\circ}^{2\circ\gamma} = t_{\circ\beta\circ}^{3\circ\gamma} = 0$, that is, having blocks (2) and (3) null.

In summary, S *does not have a tensor character*.

5. We examine the system P:

a)

$$P \equiv [t_{\circ 1 \circ}^{\alpha\circ\gamma}] = \begin{bmatrix} 1 & -1 & 1 \\ 2 & 0 & -1 \\ 1 & 0 & 0 \end{bmatrix}. \tag{8.64}$$

b)

$$\hat{P} \equiv [t_{\circ 1 \circ}^{i\circ k}] = \begin{bmatrix} 0 & 1 & -1 \\ 3 & 3 & -2 \\ -1 & -3 & 2 \end{bmatrix}. \tag{8.65}$$

c) To analyze its tensor nature, we follow a process similar to the one in question 4. Developing the dummy index β, the result is

$$t_{ojo}^{iok} = |C|^3 t_{o\beta o}^{\alpha o\gamma}\gamma_{o\alpha}^{io}c_{jo}^{o\beta}\gamma_{o\gamma}^{ko}$$

$$= |C|^3 t_{o1o}^{\alpha o\gamma}\gamma_{o\alpha}^{io}c_{jo}^{o1}\gamma_{o\gamma}^{ko} + |C|^3 t_{o2o}^{\alpha o\gamma}\gamma_{o\alpha}^{io}c_{jo}^{o2}\gamma_{o\gamma}^{ko} + |C|^3 t_{o3o}^{\alpha o\gamma}\gamma_{o\alpha}^{io}c_{jo}^{o3}\gamma_{o\gamma}^{ko}$$

which letting $j = 1$, yields

$$t_{o1o}^{iok} = |C|^3 t_{o1o}^{\alpha o\gamma}\gamma_{o\alpha}^{io}c_{1o}^{o1}\gamma_{o\gamma}^{ko} + |C|^3\gamma_{o\alpha}^{io}\gamma_{o\gamma}^{ko}\left[t_{o2o}^{\alpha o\gamma}c_{1o}^{o2} + t_{o3o}^{\alpha o\gamma}c_{1o}^{o3}\right]$$

and taking into account (8.64) and (8.65), we finally get

$$p(i,k) = c_{1o}^{o1}\left[|C|^3 p(\alpha,\gamma)\gamma_{o\alpha}^{io}\gamma_{o\gamma}^{ko}\right] + |C|^3\gamma_{o\alpha}^{io}\gamma_{o\gamma}^{ko}\left[t_{o2o}^{\alpha o\gamma}c_{1o}^{o2} + t_{o3o}^{\alpha o\gamma}c_{1o}^{o3}\right].$$
(8.66)

The discussion is the same as Expression (8.63), and the result is the same, i.e., the system P *does not have a tensor character*.
6. Next, we discuss the system Q.

a)

$$Q \equiv [q(\alpha)] = [t_{o\beta o}^{\alpha o\beta}] = \begin{bmatrix} t_{o\beta o}^{1o\beta} \\ t_{o\beta o}^{2o\beta} \\ t_{o\beta o}^{3o\beta} \end{bmatrix} = \begin{bmatrix} 1+2+3 \\ 2+0+1 \\ 1+(-1)+2 \end{bmatrix} = \begin{bmatrix} 6 \\ 3 \\ 2 \end{bmatrix} \equiv \begin{bmatrix} q^1 \\ q^2 \\ q^3 \end{bmatrix}.$$

b)

$$\hat{Q} \equiv [q(i)] = [t_{oxo}^{iox}] = \begin{bmatrix} t_{oxo}^{1ox} \\ t_{oxo}^{2ox} \\ t_{oxo}^{3ox} \end{bmatrix} = \begin{bmatrix} 0-4+7 \\ 3+2+9 \\ -1+0-7 \end{bmatrix} = \begin{bmatrix} 3 \\ 14 \\ -8 \end{bmatrix} \equiv \begin{bmatrix} \hat{q}^1 \\ \hat{q}^2 \\ \hat{q}^3 \end{bmatrix}.$$

c) We check if

$$q_o^i = |C|^w q_o^\alpha \gamma_{o\alpha}^{io}.$$
(8.67)

Since $|C|^w = 1$, we have

$$\begin{bmatrix} \hat{q}^1 \\ \hat{q}^2 \\ \hat{q}^3 \end{bmatrix} = [\gamma_{o\alpha}^{io}][q_o^\alpha] = C^{-1}\begin{bmatrix} q^1 \\ q^2 \\ q^3 \end{bmatrix} = \begin{bmatrix} 0 & 1 & 0 \\ 2 & 0 & 1 \\ -1 & 0 & -1 \end{bmatrix}\begin{bmatrix} 6 \\ 3 \\ 2 \end{bmatrix} = \begin{bmatrix} 3 \\ 14 \\ -8 \end{bmatrix},$$

which evidently coincides with that of (b), showing that, as could be expected from a contraction $[t_{o\beta o}^{\alpha o\beta}]$, the result has a tensor character. Consequently, Q as indicated by (8.67) is a modular contravariant tensor of order $r = 1$ and unknown weight; Section 8.5.4 solves the question for all tensor contractions and states that the weight is $w = 3$, i.e., the same of $t^{i \circ k}_{\circ j \circ}$; so (8.67) must be rewritten as

$$q^i_{\circ} = |C|^3 q^{\alpha}_{\circ} \gamma^{i \circ}_{\circ \alpha}.$$

□

8.7 Main modular tensors

8.7.1 ϵ systems, permutation systems or Levi-Civita tensor systems

By means of these three different names we refer to a system of scalar components defined over a linear space $V^n(\mathbb{R})$, that *in any basis* satisfies the following formal properties for their components:

1. $\epsilon(\alpha_1, \alpha_2, \ldots, \alpha_j, \ldots, \alpha_n) = 0$, if two or more indices are repeated.
2. $\epsilon(\alpha_1, \alpha_2, \ldots, \alpha_j, \ldots, \alpha_n) = 1$, if the index permutation is even.
3. $\epsilon(\alpha_1, \alpha_2, \ldots, \alpha_j, \ldots, \alpha_n) = -1$, if the index permutation is odd.

Each index α_j takes values in the set $I_n = \{1, 2, \cdots, n-1, n\}$ with total independence, that guarantee a total of $VR_{n,n} = n^n$ possibilities, that is, a system with n^n components.

A total of $n!$ components will *not* be null (since it must have different elements of I_n), and from them, due to the equivalence relation in the definition, it will be $\frac{n!}{2}$ with value 1 and $\frac{n!}{2}$ with value -1.

The proposed definition was established by Ricci, though Professor Levi-Civita popularized it in his publications. A set of important properties that we list has been obtained from the bibliography:

1. For each value of n and each linear space $V^n(\mathbb{R})$, that is, for each dimension and each field (it could be extended to another linear spaces $V^n(K)$), there exists a system ϵ.
2. The ϵ systems are systems of n^n scalar components, with totally antisymmetric character.
3. They are isotropic systems.
4. If we make the determinant of order n a "less primitive" (deducted) entity, which is the reverse of what has happened historically, we obtain the relation

$$|A| = |a^{\alpha \circ}_{\circ \beta}| = a^{1 \circ}_{\circ \alpha_1} \cdot a^{2 \circ}_{\circ \alpha_2} \cdot \cdots \cdot a^{n \circ}_{\circ \alpha_n} \epsilon(\alpha_1, \alpha_2, \ldots, \alpha_n)$$

$$= \epsilon(\alpha_1, \alpha_2, \cdots, \alpha_n) a^{1 \circ}_{\circ \alpha_1} \cdot a^{2 \circ}_{\circ \alpha_2} \cdot \cdots \cdot a^{n \circ}_{\circ \alpha_n} \quad (8.68)$$

and also

$$|A| = |a_{\alpha\,\circ}^{\,\circ\,\beta}| = a_{\alpha_1\,\circ}^{\,\circ\,1} \cdot a_{\alpha_2\,\circ}^{\,\circ\,2} \cdots \cdots a_{\alpha_n\,\circ}^{\,\circ\,n}\epsilon(\alpha_1,\alpha_2,\ldots,\alpha_n)$$

$$= \epsilon(\alpha_1,\alpha_2,\ldots,\alpha_n)a_{\alpha_1\,\circ}^{\,\circ\,1} \cdot a_{\alpha_2\,\circ}^{\,\circ\,2} \cdots \cdots a_{\alpha_n\,\circ}^{\,\circ\,n}. \qquad (8.69)$$

The justification comes obviously from the definition of the determinant.
5. Returning to the paragraph "from the existing bibliography", the authors of the present book propose another property that allows these entities to be used in determinants by means of a simple function. Remembering the properties of the function $y = sgn(x)$ (function "sign of x", over the real numbers), we have

$$\epsilon(\alpha_1,\alpha_2,\cdots,\alpha_n) = \prod_{j=1}^{n-1}\prod_{i=1}^{n-j} sgn[\alpha_{i+j} - \alpha_j], \qquad (8.70)$$

which satisfies the conditions of the definition with $\epsilon \neq 0$, and which it is very convenient for its computer implementation.

Modular tensor character of ϵ systems

As has been established in Formulas (8.1) and (8.2), we know that $C = [c_{\circ\,i}^{\alpha\circ}], C^t = [c_{i\,\circ}^{\circ\alpha}]$ and $C^{-1} = [\gamma_{\circ\alpha}^{i\circ}]$. Applying Formula (8.68) to the determinant $|C^{-1}| = |\gamma_{\circ\alpha}^{i\circ}|$, we have

$$|C^{-1}| = |\gamma_{\circ\alpha}^{i\circ}| = \epsilon(\alpha_1,\alpha_2,\ldots,\alpha_n)\gamma_{\circ\alpha_1}^{1\,\circ}\gamma_{\circ\alpha_2}^{2\,\circ}\cdots\gamma_{\circ\alpha_n}^{n\,\circ}.$$

Multiplying both members by $\epsilon(i_1,i_2,\ldots,i_n)$ and taking into account that $|C^{-1}| = |C|^{-1}$, yields

$$\epsilon(i_1,i_2,\ldots,i_n)|C|^{-1} = \epsilon(\alpha_1,\alpha_2,\ldots,\alpha_n)\left[\epsilon(i_1,i_2,\ldots,i_n)\gamma_{\circ\alpha_1}^{1\,\circ}\gamma_{\circ\alpha_2}^{2\,\circ}\cdots\gamma_{\circ\alpha_n}^{n\,\circ}\right].$$

If we take into account that when $i_1 = 1, i_2 = 2, \cdots, i_n = n$ it is $\epsilon(i_1,i_2,\ldots,i_n) = 1$ and passing $|C|$ to the right-hand, we can write

$$\epsilon(i_1,i_2,\ldots,i_n) = |C|\epsilon(\alpha_1,\alpha_2,\ldots,\alpha_n)\left[\gamma_{\circ\,\alpha_1}^{i_1\,\circ}\gamma_{\circ\,\alpha_2}^{i_2\,\circ}\cdots\gamma_{\circ\,\alpha_n}^{i_n\,\circ}\right].$$

Since the right-hand side is the expression of the change-of-basis of a modular tensor of weight 1, order n, totally contravariant species, isotropic and anti-symmetric, the result is

$$\epsilon_{\circ\,\circ\,\cdots\,\circ}^{i_1 i_2 \cdots i_n} = |C|\epsilon_{\circ\,\circ\,\cdots\,\circ}^{\alpha_1\alpha_2\cdots\alpha_n}\gamma_{\circ\,\alpha_1}^{i_1\,\circ}\gamma_{\circ\,\alpha_2}^{i_2\,\circ}\cdots\gamma_{\circ\,\alpha_n}^{i_n\,\circ}. \qquad (8.71)$$

Now, we develop another determinant according to the formula corresponding to fixed rows

$$|C^t| = |c_{i\,\circ}^{\,\circ\,\alpha}| = \epsilon(\alpha_1\alpha_2\cdots\alpha_n)c_{1\,\circ}^{\circ\,\alpha_1}c_{2\,\circ}^{\circ\,\alpha_2}\cdots c_{n\,\circ}^{\circ\,\alpha_n}.$$

Multiplying both members by $\epsilon(i_1, i_2, \ldots, i_n)$ and taking into account that $|C^t| = |C|^t = |C|$, we get

$$\epsilon(i_1, i_2, \ldots, i_n)|C| = \epsilon(\alpha_1\alpha_2\ldots\alpha_n)\left[\epsilon(i_1, i_2, \ldots, i_n)c_{1\,\circ}^{\circ\,\alpha_1}c_{2\,\circ}^{\circ\,\alpha_2}\cdots c_{n\,\circ}^{\circ\,\alpha_n}\right].$$

If we take into account that when $i_1 = 1, i_2 = 2, \ldots, i_n = n$ it is $\epsilon(i_1, i_2, \ldots, i_n) = 1$ and passing $|C|^{-1}$ to the right-hand side, we can write

$$\epsilon(i_1, i_2, \ldots, i_n) = |C|^{-1}\epsilon(\alpha_1, \alpha_2, \ldots, \alpha_n)\left[c_{i_1\,\circ}^{\;\circ\,\alpha_1}c_{i_2\,\circ}^{\;\circ\,\alpha_2}\cdots c_{i_n\,\circ}^{\;\circ\,\alpha_n}\right].$$

Since the right-hand side is the expression of the change-of-basis of a modular tensor of weight -1, order n, totally covariant species, isotropic and anti-symmetric, we have

$$\epsilon_{i_1 i_2 \cdots i_n}^{\;\circ\;\circ\;\cdots\;\circ} = |C|^{-1}\epsilon_{\alpha_1\alpha_2\cdots\alpha_n}^{\;\circ\;\circ\;\cdots\;\circ}c_{i_1\,\circ}^{\;\circ\,\alpha_1}c_{i_2\,\circ}^{\;\circ\,\alpha_2}\cdots c_{i_n\,\circ}^{\;\circ\,\alpha_n}. \tag{8.72}$$

As a conclusion, we can say that the systems ϵ are modular tensors, *simultaneously* contravariant of weight 1, and covariant of weight -1, depending on how the user wants to use them. They have a unique strict component $\epsilon_{\circ\circ\cdots\circ}^{(12\cdots n)} \equiv \epsilon_{(12\cdots n)}^{\circ\circ\cdots\circ} = 1.$

They are also called the Levi-Civita tensor density (the tensor $\epsilon_{\;\circ\;\circ\;\cdots\;\circ}^{\alpha_1\,\alpha_2\,\cdots\,\alpha_n}$) and the Levi-Civita tensor capacity (the tensor $\epsilon_{\alpha_1\alpha_2\cdots\alpha_n}^{\;\circ\;\circ\;\cdots\;\circ}$).

It cannot be forgotten that the permutation systems or Levi-Civita tensor systems are entities of a tensor linear space, and thus its complete notation as such modular anti-symmetric tensors is, respectively

$$\vec{\mathcal{E}} = \epsilon_{\;\circ\;\circ\;\circ\;\cdots\;\circ}^{\alpha_1\,\alpha_2\,\cdots\,\circ\,\alpha_n}\vec{e}_{\alpha_1} \otimes \vec{e}_{\alpha_2} \otimes \cdots \otimes \vec{e}_{\alpha_n} \quad \begin{cases} \alpha_i \in I_n; \\ \alpha_1 \neq \alpha_2 \neq \cdots \neq \alpha_n, \end{cases}$$
$$\vec{\mathcal{E}} = \epsilon_{\alpha_1\alpha_2\;\circ\;\cdots\alpha_n}^{\;\circ\;\circ\;\cdots\;\circ\;\circ}\vec{e}^{*\alpha_1} \otimes \vec{e}^{*\alpha_2} \otimes \cdots \otimes \vec{e}^{*\alpha_n}$$

8.7.2 Generalized Kronecker deltas: Definition

We give the name "generalized Kronecker delta of order $r(\leq n)$" or simply "delta of order r" to a system of scalar components of order $2r$, associated with or built over the linear space $V^n(\mathbb{R})$, with n^{2r} components, *for any basis of $V^n(\mathbb{R})$*, that satisfy the following formal properties:

1. $\delta_{\beta_1 \beta_2 \cdots \beta_r}^{\alpha_1 \alpha_2 \cdots \alpha_r} = 0$ if (a) the set of the r contravariant indices *is different* than the set of the r covariant indices, or (b) the set of the r contravariant indices is the same as the set of the r covariant indices, with repeated indices appearing in the mentioned sets.

2. $\delta^{\alpha_1 \alpha_2 \cdots \alpha_r}_{\beta_1 \beta_2 \cdots \beta_r} = 1$ if the set of the r contravariant indices is identical to the set of the r covariant indices, without repeated indices in the mentioned sets, and if their permutations are of the same parity.

3. $\delta^{\alpha_1 \alpha_2 \cdots \alpha_r}_{\beta_1 \beta_2 \cdots \beta_r} = -1$ if the set of the r contravariant indices is identical to the set of the r covariant indices, without repeated indices in the mentioned sets, if their permutations are of different parity, and $r \geq 2$.

The deltas satisfy other properties such as:

1. There exist deltas for each value of n, and for each value of r ($r \leq n$). There even exists the possibility of modifying the field of the linear space $(V^n(K))$.
2. As a consequence, the generalized Kronecker deltas of order r, are systems of n^{2r} components with anti-symmetric character with respect to the set of contravariant indices and with respect to the set of covariant indices.
3. The Kronecker deltas are isotropic systems.
4. The *classic* Kronecker delta $[\delta^{\alpha\circ}_{\circ\beta}]$, a tensor of order 2, is the generalized Kronecker delta of order $r = 1$.

Deltas of order r=n: relations with ϵ systems

We have the following relations:

1.
$$\delta^{\alpha_1 \alpha_2 \cdots \alpha_n}_{\beta_1 \beta_2 \cdots \beta_n} = \epsilon^{\alpha_1 \alpha_2 \cdots \alpha_n}_{\circ \circ \cdots \circ} \cdot \epsilon^{\circ \circ \cdots \circ}_{\beta_1 \beta_2 \cdots \beta_n} \tag{8.73}$$

2.
$$\delta^{\alpha_1 \alpha_2 \cdots \alpha_n}_{1 \ 2 \ \cdots \ n} = \epsilon^{\alpha_1 \alpha_2 \cdots \alpha_n}_{\circ \circ \cdots \circ} \tag{8.74}$$

3.
$$\delta^{1 \ 2 \ \cdots \ n}_{\beta_1 \beta_2 \cdots \beta_n} = \epsilon^{\circ \circ \cdots \circ}_{\beta_1 \beta_2 \cdots \beta_n}. \tag{8.75}$$

From these relations, the following consequences can be obtained.

In Section 8.7.1 we have proven the modular tensor character of the ϵ systems and in 8.5.3 we have established that the tensor product of modular tensors is another modular tensor.

Consequently, Formula (8.73) establishes that the generalized delta of order $r = n$ is a modular n-contravariant and n-covariant tensor of order $2n$ and zero weight, that is, it is a homogeneous mixed tensor and with anti-symmetry for each set of indices, which should be correctly notated as a tensor, i.e., for (8.73), we have

$$\delta^{\alpha_1 \alpha_2 \cdots \alpha_n}_{\beta_1 \beta_2 \cdots \beta_n} = \delta^{\alpha_1 \alpha_2 \cdots \alpha_n \ \circ \ \circ \ \cdots \ \circ}_{\circ \ \circ \ \cdots \ \circ \ \beta_1 \beta_2 \cdots \beta_n} = \epsilon^{\alpha_1 \alpha_2 \cdots \alpha_n}_{\circ \ \circ \ \cdots \ \circ} \otimes \epsilon^{\circ \ \circ \ \cdots \ \circ}_{\beta_1 \beta_2 \cdots \beta_n} \equiv \epsilon^{\alpha_1 \alpha_2 \cdots \alpha_n}_{\circ \ \circ \ \cdots \ \circ} \cdot \epsilon^{\circ \ \circ \ \cdots \ \circ}_{\beta_1 \beta_2 \cdots \beta_n}$$

and for (8.74) and (8.75)

$$\delta^{\alpha_1 \alpha_2 \cdots \alpha_n}_{1 \; 2 \; \cdots \; n} \equiv \delta^{\alpha_1 \alpha_2 \cdots \alpha_n \, \circ \circ \cdots \circ}_{\circ \; \circ \; \cdots \; \circ \; 12 \cdots n} = \epsilon^{\alpha_1 \alpha_2 \cdots \alpha_n}_{\circ \; \circ \; \cdots \; \circ}$$

$$\delta^{1 \; 2 \; \cdots \; n}_{\beta_1 \beta_2 \cdots \beta_n} \equiv \delta^{12 \cdots n \, \circ \; \circ \; \cdots \; \circ}_{\circ \circ \cdots \circ \, \beta_1 \beta_2 \cdots \beta_n} = \epsilon^{\circ \; \circ \; \cdots \; \circ}_{\beta_1 \beta_2 \cdots \beta_n}.$$

However, it is common to present them with *stacked* indices, so, we will use this convention from now on.

Based on relations (8.70) and (8.73) the following property can be justified:

$$\delta^{\alpha_1 \alpha_2 \ldots \alpha_n}_{\beta_1 \beta_2 \ldots \beta_n} = \epsilon(\alpha_1, \alpha_2, \ldots, \alpha_n) \epsilon(\beta_1, \beta_2, \ldots, \beta_n), \qquad (8.76)$$

which strictly satisfies the formal conditions of the definition of a Kronecker delta, for $\delta \neq 0$.

Deltas of order r=n: Tensor character

Though we have already justified that the Kronecker deltas of order $r = n$ are homogeneous mixed tensors, we present the proof in an analytical form.

We start from the fact that the change-of-basis in $V^n(\mathbb{R})$ produces, because of (8.71), a change of modular tensor basis of the system ϵ:

$$\epsilon^{i_1 i_2 \cdots i_n}_{\circ \; \circ \; \cdots \; \circ} = |C| \epsilon^{\alpha_1 \alpha_2 \cdots \alpha_n}_{\circ \; \circ \; \cdots \; \circ} \gamma^{i_1 \; \circ}_{\circ \; \alpha_1} \gamma^{i_2 \; \circ}_{\circ \; \alpha_2} \cdots \gamma^{i_n \; \circ}_{\circ \; \alpha_n},$$

and because of (8.72):

$$\epsilon^{\circ \; \circ \; \cdots \; \circ}_{j_1 j_2 \cdots j_n} = |C|^{-1} \epsilon^{\circ \; \circ \; \cdots \; \circ}_{\beta_1 \beta_2 \cdots \beta_n} c^{\circ \; \beta_1}_{j_1 \; \circ} c^{\circ \; \beta_2}_{j_2 \; \circ} \cdots c^{\circ \; \beta_n}_{j_n \; \circ}.$$

Multiplying term by term, taking into account the property (8.73) and with $|C| \cdot |C|^{-1} = 1$, we get

$$\delta^{i_1 i_2 \cdots i_n}_{j_1 j_2 \cdots j_n} = \delta^{\alpha_1 \alpha_2 \cdots \alpha_n}_{\beta_1 \beta_2 \cdots \beta_n} \gamma^{i_1 \; \circ}_{\circ \; \alpha_1} \gamma^{i_2 \; \circ}_{\circ \; \alpha_2} \cdots \gamma^{i_n \; \circ}_{\circ \; \alpha_n} c^{\circ \; \beta_1}_{j_1 \; \circ} c^{\circ \; \beta_2}_{j_2 \; \circ} \cdots c^{\circ \; \beta_n}_{j_n \; \circ}, \qquad (8.77)$$

which shows the homogeneous mixed tensor character of the Kronecker deltas of order $r = n$.

Having proved the mixed tensor character of the generalized Kronecker deltas, we insist again that they are entities of a tensor linear space, thus its complete notation as such mixed homogeneous tensors, considering the notation

$$\delta^{\alpha_1 \alpha_2 \cdots \alpha_r}_{\beta_1 \beta_2 \cdots \beta_r} \equiv \delta^{\alpha_1 \alpha_2 \cdots \alpha_r \; \circ \; \circ \; \cdots \; \circ}_{\circ \; \circ \; \cdots \; \circ \; \beta_1 \beta_2 \cdots \beta_r}$$

is

$$D_r = \delta^{\alpha_1 \alpha_2 \cdots \alpha_r}_{\beta_1 \beta_2 \cdots \beta_r} \vec{e}_{\alpha_1} \otimes \vec{e}_{\alpha_2} \otimes \cdots \otimes \vec{e}_{\alpha_r} \vec{e}^{*\beta_1} \otimes \vec{e}^{*\beta_2} \otimes \cdots \otimes \vec{e}^{*\beta_r},$$

where $\alpha_i, \beta_i \in I_n$; $\alpha_1 \neq \alpha_2 \neq \cdots \neq \alpha_r$ and $\beta_1 \neq \beta_2 \neq \cdots \neq \beta_r$ with a total of $n^r \cdot n^r = n^{2r}$ components.

In the particular case $r = n$, the generalized Kronecker delta becomes the tensor product of the Levi-Civita tensors:

$$D_n = \delta^{\alpha_1\,\alpha_2\,\cdots\,\alpha_n}_{\beta_1\,\beta_2\,\cdots\,\beta_n}\vec{e}_{\alpha_1}\otimes\vec{e}_{\alpha_2}\otimes\cdots\otimes\vec{e}_{\alpha_n}\vec{e}^{*\beta_1}\otimes\vec{e}^{*\beta_2}\otimes\cdots\otimes\vec{e}^{*\beta_n}$$

$$= (\epsilon^{\alpha_1\,\alpha_2\,\cdots\,\,\circ\,\alpha_n}_{\,\circ\,\,\circ\,\,\circ\,\cdots\,\,\circ}\vec{e}_{\alpha_1}\otimes\vec{e}_{\alpha_2}\otimes\cdots\otimes\vec{e}_{\alpha_n})\otimes(\epsilon^{\,\circ\,\,\circ\,\cdots\,\circ\,\,\circ}_{\alpha_1\,\alpha_2\,\circ\,\cdots\,\alpha_n}\vec{e}^{*\alpha_1}\otimes\vec{e}^{*\alpha_2}\otimes\cdots\otimes\vec{e}^{*\alpha_n})$$

according to the relation (8.73) and those established in the final part of Section 5.

Deltas of order r < n

Expressions as a function of the deltas of order n

1. Deltas of order inferior to n are obtained from the delta of order n, $(\delta^{\alpha_1\,\alpha_2\,\cdots\,\alpha_n}_{\beta_1\,\beta_2\,\cdots\,\beta_n})$ by contracting indices of different valency, as shown below:

We contract the last two indices of $\delta^{\alpha_1\,\alpha_2\,\cdots\,\alpha_{n-1}\,\alpha_n}_{\beta_1\,\beta_2\,\cdots\,\beta_{n-1}\,\beta_n}$, which guarantees the resulting tensor to be a mixed homogeneous tensor of order $(n-1)$:

$$C\binom{\alpha_n}{\beta_n}\delta^{\alpha_1\,\alpha_2\,\cdots\,\alpha_{n-1}\,\alpha_n}_{\beta_1\,\beta_2\,\cdots\,\beta_{n-1}\,\beta_n} = \delta^{\alpha_1\,\alpha_2\,\cdots\,\alpha_{n-1}\,\theta_n}_{\beta_1\,\beta_2\,\cdots\,\beta_{n-1}\,\theta_n}$$

$$= \delta^{\alpha_1\,\alpha_2\,\cdots\,\alpha_{n-1}\,1}_{\beta_1\,\beta_2\,\cdots\,\beta_{n-1}\,1} + \delta^{\alpha_1\,\alpha_2\,\cdots\,\alpha_{n-1}\,2}_{\beta_1\,\beta_2\,\cdots\,\beta_{n-1}\,2} + \cdots + \delta^{\alpha_1\,\alpha_2\,\cdots\,\alpha_{n-1}\,n}_{\beta_1\,\beta_2\,\cdots\,\beta_{n-1}\,n},$$

$$(8.78)$$

an expression with n summands in which each summand will be null if

$$\{\alpha_1,\alpha_2,\ldots\alpha_{n-1}\}\equiv\{\beta_1,\beta_2,\ldots\beta_{n-1}\}$$

without repeated indices.

There exist n summands and $(n-1)$ indices that must be different (for the *calculated* components of the first member $\delta^{\alpha_1\,\alpha_2\,\cdots\,\alpha_{n-1}\,\theta_n}_{\beta_1\,\beta_2\,\cdots\,\beta_{n-1}\,\theta_n}$ to be non–null) and the indices vary from 1 to n.

There can exist only *one* summand for which the index $\theta_n = a \in \{\alpha_1,\alpha_2,\ldots,\alpha_{n-1}\}$. Whence

$$\delta^{\alpha_1\,\alpha_2\,\cdots\,\alpha_{n-1}\,a}_{\beta_1\,\beta_2\,\cdots\,\beta_{n-1}\,a} = \delta^{\alpha_1\,\alpha_2\,\cdots\,\alpha_{n-1}}_{\beta_1\,\beta_2\,\cdots\,\beta_{n-1}} \qquad (8.79)$$

and from (8.78) and (8.79) we conclude

$$\delta^{\alpha_1\,\alpha_2\,\cdots\,\alpha_{n-1}\,\alpha_n}_{\beta_1\,\beta_2\,\cdots\,\beta_{n-1}\,\alpha_n} = 1\cdot\delta^{\alpha_1\,\alpha_2\,\cdots\,\alpha_{n-1}}_{\beta_1\,\beta_2\,\cdots\,\beta_{n-1}}. \qquad (8.80)$$

Contracting again the last two indices of different valency in $\delta^{\alpha_1\,\alpha_2\,\cdots\,\alpha_{n-1}}_{\beta_1\,\beta_2\,\cdots\,\beta_{n-1}}$, yields

$$\delta^{\alpha_1\,\alpha_2\,\cdots\,\alpha_{n-2}\,\theta_{n-1}}_{\beta_1\,\beta_2\,\cdots\,\beta_{n-2}\,\theta_{n-1}} = \delta^{\alpha_1\,\alpha_2\,\cdots\,\alpha_{n-2}\,1}_{\beta_1\,\beta_2\,\cdots\,\beta_{n-2}\,1} + \delta^{\alpha_1\,\alpha_2\,\cdots\,\alpha_{n-2}\,2}_{\beta_1\,\beta_2\,\cdots\,\beta_{n-2}\,2} + \cdots + \delta^{\alpha_1\,\alpha_2\,\cdots\,\alpha_{n-2}\,n}_{\beta_1\,\beta_2\,\cdots\,\beta_{n-2}\,n} \quad (8.81)$$

and since there exist n summands and $(n-2)$ indices that must be different, now there can exist only *two* summands in which the index $\theta_{n-1} = a, b \in \{\alpha_1, \alpha_2, \dots, \alpha_{n-1}\}$. Thus

$$\delta^{\alpha_1\,\alpha_2\cdots\alpha_{n-2}\,a}_{\beta_1\,\beta_2\cdots\beta_{n-2}\,a} = \delta^{\alpha_1\,\alpha_2\cdots\alpha_{n-2}\,b}_{\beta_1\,\beta_2\cdots\beta_{n-2}\,b} = \delta^{\alpha_1\,\alpha_2\cdots\alpha_{n-2}}_{\beta_1\,\beta_2\cdots\beta_{n-2}}. \tag{8.82}$$

From (8.81) and (8.82) we obtain

$$\delta^{\alpha_1\,\alpha_2\cdots\alpha_{n-2}\,\alpha_{n-1}}_{\beta_1\,\beta_2\cdots\beta_{n-2}\,\alpha_{n-1}} = 2\delta^{\alpha_1\,\alpha_2\cdots\alpha_{n-2}}_{\beta_1\,\beta_2\cdots\beta_{n-2}}$$

and substituting it into (8.80), we get

$$\delta^{\alpha_1\,\alpha_2\cdots\alpha_{n-2}\,\alpha_{n-1}\,\alpha_n}_{\beta_1\,\beta_2\cdots\beta_{n-2}\,\alpha_{n-1}\,\alpha_n} = 1 \cdot 2\delta^{\alpha_1\,\alpha_2\cdots\alpha_{n-2}}_{\beta_1\,\beta_2\cdots\beta_{n-2}} = 2!\,\delta^{\alpha_1\,\alpha_2\cdots\alpha_{n-2}}_{\beta_1\,\beta_2\cdots\beta_{n-2}}. \tag{8.83}$$

The relations (8.80) and (8.83) perfectly declare the generation law:

1. the second member coefficient is the factorial of the number k of contractions,
2. the order of the second member tensor is $(n-k)$, and
3. the number of contracted indices in the left-hand side is the number of contractions.

From all this, one concludes that after $(n-r)$ contractions one has arrived at the relation

$$\delta^{\alpha_1\,\alpha_2\cdots\alpha_r\,\alpha_{r+1}}_{\beta_1\,\beta_2\cdots\beta_r\,\alpha_{r+1}} = (n-r) \cdot \delta^{\alpha_1\,\alpha_2\cdots\alpha_r}_{\beta_1\,\beta_2\cdots\beta_r}, \tag{8.84}$$

which, substituted in the previous formulation leads to the final relation

$$\delta^{\alpha_1\,\alpha_2\cdots\alpha_r\,\alpha_{r+1}\cdots\alpha_{n-1}\,\alpha_n}_{\beta_1\,\beta_2\cdots\beta_r\,\alpha_{r+1}\cdots\alpha_{n-1}\,\alpha_n} = (n-r)! \cdot \delta^{\alpha_1\,\alpha_2\cdots\alpha_r}_{\beta_1\,\beta_2\cdots\beta_r}, \tag{8.85}$$

that is,

$$\delta^{\alpha_1\,\alpha_2\cdots\alpha_r}_{\beta_1\,\beta_2\cdots\beta_r} = \frac{1}{(n-r)!}\delta^{\alpha_1\,\alpha_2\cdots\alpha_r\,\alpha_{r+1}\cdots\alpha_{n-1}\,\alpha_n}_{\beta_1\,\beta_2\cdots\beta_r\,\alpha_{r+1}\cdots\alpha_{n-1}\,\alpha_n}, \tag{8.86}$$

a formula that answers the question.

Before finishing this part, we examine some interesting consequences of Formula (8.86).

If k is a natural number such that $1 \le r < k < n$ and we let $r = k$ in (8.86), we get

$$\delta^{\alpha_1\,\alpha_2\cdots\alpha_r\,\alpha_{r+1}\cdots\alpha_k}_{\beta_1\,\beta_2\cdots\beta_r\,\beta_{r+1}\cdots\beta_k} = \frac{1}{(n-k)!}\delta^{\alpha_1\,\alpha_2\cdots\alpha_k\,\alpha_{k+1}\cdots\alpha_n}_{\beta_1\,\beta_2\cdots\beta_k\,\alpha_{k+1}\cdots\alpha_n}. \tag{8.87}$$

If we continue with the contractions in (8.87) from index k up to index $r+1$, we obtain

$$\delta^{\alpha_1\,\alpha_2\cdots\alpha_r\,\alpha_{r+1}\cdots\alpha_k}_{\beta_1\,\beta_2\cdots\beta_r\,\alpha_{r+1}\cdots\alpha_k} = \frac{1}{(n-k)!}\delta^{\alpha_1\,\alpha_2\cdots\alpha_r\,\alpha_{r+1}\cdots\alpha_{n-1}\,\alpha_n}_{\beta_1\,\beta_2\cdots\beta_r\,\alpha_{r+1}\cdots\alpha_{n-1}\,\alpha_n}$$

and obtaining the last delta and substituting it into (8.86), we finally get

$$\delta^{\alpha_1 \alpha_2 \cdots \alpha_r}_{\beta_1 \beta_2 \cdots \beta_r} = \frac{(n-k)!}{(n-r)!} \delta^{\alpha_1 \alpha_2 \cdots \alpha_r \alpha_{r+1} \cdots \alpha_k}_{\beta_1 \beta_2 \cdots \beta_r \alpha_{r+1} \cdots \alpha_k}, \qquad (8.88)$$

an expression that relates the delta of order r with another delta of order k greater than r but smaller than n, which is more general than (8.86).

If in (8.86) we let $r = 1$, the classic Kronecker delta appears as a function of the generalized Kronecker delta of order n

$$\delta^{\alpha_1}_{\beta_1} = \frac{1}{(n-1)!} \delta^{\alpha_1 \alpha_2 \alpha_3 \cdots \alpha_n}_{\beta_1 \alpha_2 \alpha_3 \cdots \alpha_n}. \qquad (8.89)$$

2. *The contracted deltas.* If in (8.89) we contract all indices, we get

$$\delta^{\theta_1}_{\theta_1} = \frac{1}{(n-1)!} \delta^{\alpha_1 \alpha_2 \cdots \alpha_n}_{\alpha_1 \alpha_2 \cdots \alpha_n}$$

and developing the first index, we obtain

$$\delta^1_1 + \delta^2_2 + \cdots + \delta^n_n = \frac{1}{(n-1)!} \delta^{\alpha_1 \alpha_2 \cdots \alpha_n}_{\alpha_1 \alpha_2 \cdots \alpha_n}. \qquad (8.90)$$

Adding the deltas on the left-hand side, removing denominators and writing them in reverse order the result is

$$\delta^{\alpha_1 \alpha_2 \cdots \alpha_n}_{\alpha_1 \alpha_2 \cdots \alpha_n} = (1+1+1+\cdots+1)(n-1)! = n \cdot (n-1)! = n!. \qquad (8.91)$$

Finally, if we contract the indices on the left-hand member of (8.86), we get

$$\delta^{\alpha_1 \alpha_2 \cdots \alpha_r}_{\alpha_1 \alpha_2 \cdots \alpha_r} = \frac{1}{(n-r)!} \delta^{\alpha_1 \alpha_2 \cdots \alpha_n}_{\alpha_1 \alpha_2 \cdots \alpha_n} \qquad (8.92)$$

and considering (8.91), (8.92) becomes

$$\delta^{\alpha_1 \alpha_2 \cdots \alpha_r}_{\alpha_1 \alpha_2 \cdots \alpha_r} = \frac{n!}{(n-r)!}. \qquad (8.93)$$

Generalized Kronecker delta of order r as a function of the classical Kronecker deltas

We start this topic by defining the system of determinants, Δr, of order r, that are functions of the classic Kronecker deltas:

$$\Delta_r = \begin{vmatrix} \delta^{\alpha_1}_{\circ\ \beta_1} & \delta^{\alpha_1}_{\circ\ \beta_2} & \cdots & \delta^{\alpha_1}_{\circ\ \beta_r} \\ \delta^{\alpha_2}_{\circ\ \beta_1} & \delta^{\alpha_2}_{\circ\ \beta_2} & \cdots & \delta^{\alpha_2}_{\circ\ \beta_r} \\ \cdots & \cdots & \cdots & \cdots \\ \delta^{\alpha_r}_{\circ\ \beta_1} & \delta^{\alpha_r}_{\circ\ \beta_2} & \cdots & \delta^{\alpha_r}_{\circ\ \beta_r} \end{vmatrix} ; \ \alpha_h, \beta_h \in I_n = \{1, \cdots, n\}; \ h \in I_r = \{1, \cdots, r\},$$

$$(8.94)$$

which can also be written by means of another determinant, D_n, of order n, as

$$\Delta_r = D_n = \begin{vmatrix} \Delta_r & | & \Omega \\ -- & + & -- \\ \Omega & | & I_{n-r} \end{vmatrix}, \tag{8.95}$$

where I_{n-r} is the unitary matrix of order $(n-r)$.

Next, we replace each one of the elements of the blocks I_{n-r} and Ω in D_n, by selected classic Kronecker deltas, the values of which (1 and 0) coincide with the corresponding blocks

$$\Delta_r = D_n = \begin{vmatrix} \delta^{\alpha_1 \ \circ}_{\ \circ \ \beta_1} & \delta^{\alpha_1 \ \circ}_{\ \circ \ \beta_2} & \cdots & \delta^{\alpha_1 \ \circ}_{\ \circ \ \beta_r} & | & \delta^{\alpha_1 \ \circ}_{\ \circ \ \beta_{r+1}} & \delta^{\alpha_1 \ \circ}_{\ \circ \ \beta_{r+2}} & \cdots & \delta^{\alpha_1 \ \circ}_{\ \circ \ \beta_n} \\ \delta^{\alpha_2 \ \circ}_{\ \circ \ \beta_1} & \delta^{\alpha_2 \ \circ}_{\ \circ \ \beta_2} & \cdots & \delta^{\alpha_2 \ \circ}_{\ \circ \ \beta_r} & | & \delta^{\alpha_2 \ \circ}_{\ \circ \ \beta_{r+1}} & \delta^{\alpha_2 \ \circ}_{\ \circ \ \beta_{r+2}} & \cdots & \delta^{\alpha_2 \ \circ}_{\ \circ \ \beta_n} \\ \cdots & \cdots & \cdots & \cdots & | & \cdots & \cdots & \cdots & \cdots \\ \delta^{\alpha_r \ \circ}_{\ \circ \ \beta_1} & \delta^{\alpha_r \ \circ}_{\ \circ \ \beta_2} & \cdots & \delta^{\alpha_r \ \circ}_{\ \circ \ \beta_r} & | & \delta^{\alpha_r \ \circ}_{\ \circ \ \beta_{r+1}} & \delta^{\alpha_r \ \circ}_{\ \circ \ \beta_{r+2}} & \cdots & \delta^{\alpha_r \ \circ}_{\ \circ \ \beta_n} \\ \delta^{\alpha_{r+1} \ \circ}_{\ \ \circ \ \beta_1} & \delta^{\alpha_{r+1} \ \circ}_{\ \ \circ \ \beta_2} & \cdots & \delta^{\alpha_{r+1} \ \circ}_{\ \ \circ \ \beta_r} & | & & & & \\ \delta^{\alpha_{r+2} \ \circ}_{\ \ \circ \ \beta_1} & \delta^{\alpha_{r+2} \ \circ}_{\ \ \circ \ \beta_2} & \cdots & \delta^{\alpha_{r+2} \ \circ}_{\ \ \circ \ \beta_r} & | & & & & \\ \cdots & \cdots & \cdots & \cdots & | & & I_{n-r} & & \\ \delta^{\alpha_n \ \circ}_{\ \circ \ \beta_1} & \delta^{\alpha_n \ \circ}_{\ \circ \ \beta_2} & \cdots & \delta^{\alpha_n \ \circ}_{\ \circ \ \beta_r} & | & & & & \end{vmatrix}, \tag{8.96}$$

where $\alpha_{r+1}, \alpha_{r+2}, \ldots, \alpha_n, \beta_{r+1}, \beta_{r+2}, \ldots, \beta_n \in I_n = \{1, 2, \ldots, n\}$, and represent indices that satisfy

$$\alpha_{r+1}, \alpha_{r+2}, \ldots, \alpha_n \notin \{\alpha_1, \alpha_2, \ldots, \alpha_r\}; \quad \beta_{r+1}, \beta_{r+2}, \ldots, \beta_n \notin \{\beta_1, \beta_2, \ldots, \beta_r\}.$$

Finally, taking into account that the determinant

$$A_{n-r} = \begin{vmatrix} \delta^{\alpha_{r+1} \ \ \circ}_{\ \ \circ \ \beta_{r+1}} & \delta^{\alpha_{r+1} \ \ \circ}_{\ \ \circ \ \beta_{r+2}} & \cdots & \delta^{\alpha_{r+1} \ \ \circ}_{\ \ \circ \ \beta_n} \\ \delta^{\alpha_{r+2} \ \ \circ}_{\ \ \circ \ \beta_{r+1}} & \delta^{\alpha_{r+2} \ \ \circ}_{\ \ \circ \ \beta_{r+2}} & \cdots & \delta^{\alpha_{r+2} \ \ \circ}_{\ \ \circ \ \beta_n} \\ \cdots & \cdots & \cdots & \cdots \\ \delta^{\alpha_n \ \ \circ}_{\ \ \circ \ \beta_{r+1}} & \delta^{\alpha_n \ \ \circ}_{\ \ \circ \ \beta_{r+2}} & \cdots & \delta^{\alpha_n \ \ \circ}_{\ \ \circ \ \beta_n} \end{vmatrix},$$

when developed by the Laplace method by all fixed columns, takes as its value the number of permutations of its rows $(\alpha_{r+1}, \alpha_{r+2}, \ldots, \alpha_n)$ by $|I_{n-r}|$, the result is

$$A_{n-r} = (n-r)! \, |I_{n-r}| = (n-r)!. \tag{8.97}$$

Note that the determinant A_{n-r} is such that for *all* permutations of its *rows* $\{\alpha_{r+1}, \alpha_{r+2}, \ldots, \alpha_n\}$, it takes the value one ($|I_{n-r}|$), so that using the Laplace formula, we get a sum of $(n-r)!$ ones.

As a logical consequence, replacing $r = 0$ in (8.97), we obtain $A_n = n!$.

Relation (8.97) can be written as

$$|I_{n-r}| = \frac{A_{n-r}}{(n-r)!} \tag{8.98}$$

and substituting the determinant $|I_{n-r}|$ of Formula (8.98) into the corresponding block of (8.96), we get

$$
D_n = \frac{1}{(n-r)!}
\begin{vmatrix}
\delta^{\alpha_1\ \circ}_{\ \circ\ \beta_1} & \delta^{\alpha_1\ \circ}_{\ \circ\ \beta_2} & \cdots & \delta^{\alpha_1\ \circ}_{\ \circ\ \beta_r} & \delta^{\alpha_1\ \ \circ}_{\ \circ\ \beta_{r+1}} & \delta^{\alpha_1\ \ \circ}_{\ \circ\ \beta_{r+2}} & \cdots & \delta^{\alpha_1\ \circ}_{\ \circ\ \beta_n} \\
\delta^{\alpha_2\ \circ}_{\ \circ\ \beta_1} & \delta^{\alpha_2\ \circ}_{\ \circ\ \beta_2} & \cdots & \delta^{\alpha_2\ \circ}_{\ \circ\ \beta_r} & \delta^{\alpha_2\ \ \circ}_{\ \circ\ \beta_{r+1}} & \delta^{\alpha_2\ \ \circ}_{\ \circ\ \beta_{r+2}} & \cdots & \delta^{\alpha_2\ \circ}_{\ \circ\ \beta_n} \\
\cdots & \cdots & \cdots & \cdots & \cdots & \cdots & \cdots & \cdots \\
\delta^{\alpha_r\ \circ}_{\ \circ\ \beta_1} & \delta^{\alpha_r\ \circ}_{\ \circ\ \beta_2} & \cdots & \delta^{\alpha_r\ \circ}_{\ \circ\ \beta_r} & \delta^{\alpha_r\ \ \circ}_{\ \circ\ \beta_{r+1}} & \delta^{\alpha_r\ \ \circ}_{\ \circ\ \beta_{r+2}} & \cdots & \delta^{\alpha_r\ \circ}_{\ \circ\ \beta_n} \\
\delta^{\alpha_{r+1}\ \circ}_{\ \ \circ\ \beta_1} & \delta^{\alpha_{r+1}\ \circ}_{\ \ \circ\ \beta_2} & \cdots & \delta^{\alpha_{r+1}\ \circ}_{\ \ \circ\ \beta_r} & \delta^{\alpha_{r+1}\ \ \circ}_{\ \ \circ\ \beta_{r+1}} & \delta^{\alpha_{r+1}\ \ \circ}_{\ \ \circ\ \beta_{r+2}} & \cdots & \delta^{\alpha_{r+1}\ \circ}_{\ \ \circ\ \beta_n} \\
\delta^{\alpha_{r+2}\ \circ}_{\ \ \circ\ \beta_1} & \delta^{\alpha_{r+2}\ \circ}_{\ \ \circ\ \beta_2} & \cdots & \delta^{\alpha_{r+2}\ \circ}_{\ \ \circ\ \beta_r} & \delta^{\alpha_{r+2}\ \ \circ}_{\ \ \circ\ \beta_{r+1}} & \delta^{\alpha_{r+2}\ \ \circ}_{\ \ \circ\ \beta_{r+2}} & \cdots & \delta^{\alpha_{r+2}\ \circ}_{\ \ \circ\ \beta_n} \\
\cdots & \cdots & \cdots & \cdots & \cdots & \cdots & \cdots & \cdots \\
\delta^{\alpha_n\ \circ}_{\ \circ\ \beta_1} & \delta^{\alpha_n\ \circ}_{\ \circ\ \beta_2} & \cdots & \delta^{\alpha_n\ \circ}_{\ \circ\ \beta_r} & \delta^{\alpha_n\ \ \circ}_{\ \circ\ \beta_{r+1}} & \delta^{\alpha_n\ \ \circ}_{\ \circ\ \beta_{r+2}} & \cdots & \delta^{\alpha_n\ \circ}_{\ \circ\ \beta_n}
\end{vmatrix},
$$

$$(8.99)$$

in which the correlation of row and column indices is totally in accordance with its order.

Next, using the Laplace formula we develop the determinant D_n of (8.99) fixing its columns and expressing its development as a function of the components of the Levi-Civita ϵ system, Formulas (8.73), (8.74) and (8.75). This leads to

$$
D_n = \frac{1}{(n-r)!}
\begin{vmatrix}
\delta^{\alpha_1\ \circ}_{\ \circ\ 1} & \delta^{\alpha_1\ \circ}_{\ \circ\ 2} & \cdots & \delta^{\alpha_1\ \circ}_{\ \circ\ n} \\
\delta^{\alpha_2\ \circ}_{\ \circ\ 1} & \delta^{\alpha_2\ \circ}_{\ \circ\ 2} & \cdots & \delta^{\alpha_2\ \circ}_{\ \circ\ n} \\
\cdots & \cdots & \cdots & \cdots \\
\delta^{\alpha_n\ \circ}_{\ \circ\ 1} & \delta^{\alpha_n\ \circ}_{\ \circ\ 2} & \cdots & \delta^{\alpha_n\ \circ}_{\ \circ\ n}
\end{vmatrix}
\epsilon^{\circ\ \circ\ \cdots\ \circ\ \ \circ\ \ \ \ \ \circ\ \cdots\ \circ}_{\beta_1\beta_2\cdots\beta_r\,\alpha_{r+1}\alpha_{r+2}\cdots\alpha_n}
$$

$$
= \frac{1}{(n-r)!}\epsilon^{\alpha_1\alpha_2\cdots\alpha_r\,\alpha_{r+1}\alpha_{r+2}\cdots\alpha_n}_{\circ\ \circ\ \cdots\ \circ\ \ \circ\ \ \ \ \ \circ\ \cdots\ \circ}\cdot\epsilon^{\circ\ \circ\ \cdots\ \circ\ \ \circ\ \ \ \ \ \circ\ \cdots\ \circ}_{\beta_1\beta_2\cdots\beta_r\,\alpha_{r+1}\alpha_{r+2}\cdots\alpha_n}
$$

$$
= \frac{1}{(n-r)!}\delta^{\alpha_1\alpha_2\cdots\alpha_r\,\alpha_{r+1}\cdots\alpha_n}_{\beta_1\beta_2\cdots\beta_r\,\alpha_{r+1}\cdots\alpha_n}.
$$

$$(8.100)$$

If we now take into account Formulas (8.86) and (8.95), we obtain

$$
\Delta_r = D_n = \delta^{\alpha_1\alpha_2\cdots\alpha_r}_{\beta_1\beta_2\cdots\beta_r},
$$

$$(8.101)$$

and finally, remembering (8.94), we conclude that

$$
\delta^{\alpha_1\alpha_2\cdots\alpha_r}_{\beta_1\beta_2\cdots\beta_r} =
\begin{vmatrix}
\delta^{\alpha_1\ \circ}_{\ \circ\ \beta_1} & \delta^{\alpha_1\ \circ}_{\ \circ\ \beta_2} & \cdots & \delta^{\alpha_1\ \circ}_{\ \circ\ \beta_r} \\
\delta^{\alpha_2\ \circ}_{\ \circ\ \beta_1} & \delta^{\alpha_2\ \circ}_{\ \circ\ \beta_2} & \cdots & \delta^{\alpha_2\ \circ}_{\ \circ\ \beta_r} \\
\cdots & \cdots & \cdots & \cdots \\
\delta^{\alpha_r\ \circ}_{\ \circ\ \beta_1} & \delta^{\alpha_r\ \circ}_{\ \circ\ \beta_2} & \cdots & \delta^{\alpha_r\ \circ}_{\ \circ\ \beta_r}
\end{vmatrix}
\ ; \quad
\begin{cases}
\alpha_h, \beta_h \in I_n = \{1, 2, \ldots, n\}; \\
\qquad\quad h \in I_r = \{1, 2, \ldots, r\},
\end{cases}
$$

$$(8.102)$$

which answers the question.

Use of generalized Kronecker deltas in the anti-symmetric tensors

The components of an anti-symmetric tensor of order r, with $r \leq n$, established over the linear space $V^n(\mathbb{R})$, can be directly notated, as a function of the strict components, by a simple formulation: If the tensor is contravariant

$$t^{\alpha_1 \alpha_2 \cdots \alpha_r}_{\circ \ \circ \ \cdots \ \circ} = \delta^{\alpha_1 \alpha_2 \cdots \alpha_r}_{(\beta_1 \beta_2 \cdots \beta_r)} t^{(\beta_1 \beta_2 \cdots \beta_r)}_{\circ \ \circ \ \cdots \ \circ} \tag{8.103}$$

and if it is covariant

$$t^{\circ \ \circ \ \cdots \ \circ}_{\alpha_1 \alpha_2 \cdots \alpha_r} = \delta^{(\beta_1 \beta_2 \cdots \beta_r)}_{\alpha_1 \alpha_2 \cdots \alpha_r} t^{\circ \ \circ \ \cdots \ \circ}_{(\beta_1 \beta_2 \cdots \beta_r)}. \tag{8.104}$$

8.7.3 Dual or polar tensors: Definition

Dual or polar tensors are modular tensors the are associated *only* with given homogeneous anti-symmetric tensors.

Consider the given covariant tensor $B^{\circ \ \circ \ \cdots \ \circ}_{\beta_1 \beta_2 \cdots \beta_q}$ of order q where $q \leq n$, because otherwise, since it is totally anti-symmetric, would have all components null. We give the name "adjoint tensor" or "polar tensor" of the given tensor to the *contravariant* tensor:

$$D^{\beta_{q+1} \beta_{q+2} \cdots \beta_n}_{\circ \ \circ \ \cdots \ \circ} = \frac{1}{q!} \epsilon^{\beta_1 \beta_2 \cdots \beta_n}_{\circ \ \circ \ \cdots \ \circ} B^{\circ \ \circ \ \cdots \ \circ}_{\beta_1 \beta_2 \cdots \beta_q}, \tag{8.105}$$

an expression that in reality is a contracted tensor product in its first q indices, of the Levi-Civita contravariant tensor of order n and the given tensor.

Similarly, if the given tensor is $A^{\alpha_1 \alpha_2 \cdots \alpha_p}_{\circ \ \circ \ \cdots \ \circ}$, its "adjoint" or "polar" tensor is the *covariant* tensor

$$D^{\circ \ \circ \ \cdots \ \circ}_{\alpha_{p+1} \alpha_{p+2} \cdots \alpha_n} = \frac{1}{p!} \epsilon^{\circ \ \circ \ \cdots \ \circ}_{\alpha_1 \alpha_2 \cdots \alpha_n} A^{\alpha_1 \alpha_2 \cdots \alpha_p}_{\circ \ \circ \ \cdots \ \circ}. \tag{8.106}$$

Tensor character of polar systems

In the previous definitions we have advanced, a priori, that the defined systems of scalars are tensors. In fact, in the covariant case, the explanation following Formula (8.105) is a proof of the tensor nature of the defined entity.

However, we will show how these "polar" entities change basis, to clearly manifest their tensor character.

If we perform a change-of-basis in $V^n(\mathbb{R})$ of matrix C, the polar contravariant tensor changes as

$$D^{j_{q+1} j_{q+2} \cdots j_n}_{\circ \ \circ \ \cdots \ \circ} = |C| D^{\beta_{q+1} \beta_{q+2} \cdots \beta_n}_{\circ \ \circ \ \cdots \ \circ} \gamma^{j_{q+1} \ \circ}_{\circ \ \beta_{q+1}} \gamma^{j_{q+2} \ \circ}_{\circ \ \beta_{q+2}} \cdots \gamma^{j_n \ \circ}_{\circ \ \beta_n}, \tag{8.107}$$

which shows it is a "tensor density of order $(n - q)$", i.e., a modular tensor type.

With respect to the polar covariant tensor, we have

$$D_{\substack{\circ\;\circ\;\cdots\;\circ\\ i_{p+1} i_{p+2}\cdots i_n}} = |C|^{-1} D_{\substack{\circ\;\circ\;\cdots\;\circ\\ \alpha_{p+1}\alpha_{p+2}\cdots\alpha_n}} c_{\substack{\circ\;\;\alpha_{p+1}\\ i_{p+1}\;\circ}} c_{\substack{\circ\;\;\alpha_{p+2}\\ i_{p+2}\;\circ}} \cdots c_{\substack{\circ\;\;\alpha_n\\ i_n\;\circ}}. \qquad (8.108)$$

In this case it is a "tensor capacity of order $(n-p)$", another type of modular tensor.

Example 8.6 (Contraction of ϵ systems). Remembering the two representations of the ϵ systems, $\epsilon^{\substack{\alpha_1\alpha_2\cdots\alpha_n\\ \circ\;\circ\;\cdots\;\circ}}$ and $\epsilon_{\substack{\circ\;\circ\;\cdots\;\circ\\ \alpha_1\alpha_2\cdots\alpha_n}}$ respectively, we wish to obtain the totally contracted product of such modular tensors:

$$\epsilon^{\substack{\alpha_1\alpha_2\cdots\alpha_n\\ \circ\;\circ\;\cdots\;\circ}} \cdot \epsilon_{\substack{\circ\;\circ\;\cdots\;\circ\\ \alpha_1\alpha_2\cdots\alpha_n}}.$$

Solution: After interpreting Formula (8.73), we obtain

$$\delta^{\substack{\alpha_1\alpha_2\cdots\alpha_n\\ \beta_1\beta_2\cdots\beta_n}} = \epsilon^{\substack{\alpha_1\alpha_2\cdots\alpha_n\\ \circ\;\circ\;\cdots\;\circ}} \otimes \epsilon_{\substack{\circ\;\circ\;\cdots\;\circ\\ \beta_1\beta_2\cdots\beta_n}} \equiv \epsilon^{\substack{\alpha_1\alpha_2\cdots\alpha_n\\ \circ\;\circ\;\cdots\;\circ}} \cdot \epsilon_{\substack{\circ\;\circ\;\cdots\;\circ\\ \beta_1\beta_2\cdots\beta_n}},$$

where the last product is that of the real numbers.

Contracting all indices α_h with the corresponding β_h, we obtain

$$\delta^{\substack{\alpha_1\alpha_2\cdots\alpha_n\\ \alpha_1\alpha_2\cdots\alpha_n}} = \epsilon^{\substack{\alpha_1\alpha_2\cdots\alpha_n\\ \circ\;\circ\;\cdots\;\circ}} \cdot \epsilon_{\substack{\circ\;\circ\;\cdots\;\circ\\ \alpha_1\alpha_2\cdots\alpha_n}}$$

and taking into account Formula (8.91), the result is

$$\epsilon^{\substack{\alpha_1\alpha_2\cdots\alpha_n\\ \circ\;\circ\;\cdots\;\circ}} \cdot \epsilon_{\substack{\circ\;\circ\;\cdots\;\circ\\ \alpha_1\alpha_2\cdots\alpha_n}} = n!,$$

which is an absolute tensor of zero order (a scalar invariant).   □

Example 8.7 (Tensors with simultaneous different species). Consider a contravariant tensor $A \equiv [a_{\circ\circ}^{\lambda\mu}]$ of order $r = 2$, defined over the linear space $V^3(\mathbb{R})$, in the tensor basis associated with the basis $\{\vec{e}_\alpha\}$ of $V^3(\mathbb{R})$, and we assume that $|A| \equiv |a_{\circ\circ}^{\lambda\mu}| > 0$. We wish to analyze the possible tensor nature of the system S of scalar components defined by

$$s(\alpha\beta\gamma) = \frac{\epsilon(\alpha\beta\gamma)}{\sqrt{|a_{\circ\circ}^{\lambda\mu}|}}. \qquad (8.109)$$

Solution: Since A is the tensor of the given characteristics, according to the relation (4.34), we have

$$a_{\circ\circ}^{\ell m} = a_{\circ\circ}^{\lambda\mu} \gamma_{\circ\lambda}^{\ell\circ} \gamma_{\circ\mu}^{m\circ},$$

which in matrix form and taking determinants, becomes

$$|a_{\circ\circ}^{\ell m}| = |a_{\circ\circ}^{\lambda\mu}||C|^{-1}|C|^{-1} \Leftrightarrow \sqrt{|a_{\circ\circ}^{\ell m}|} = \sqrt{|a_{\circ\circ}^{\lambda\mu}|}|C|^{-1}. \qquad (8.110)$$

1. On the other hand, we can consider that $\epsilon(\alpha\beta\gamma) \equiv \epsilon_{\circ\circ\circ}^{\alpha\beta\gamma}$, so that from a tensor point of view, according to (8.39) it is

$$\epsilon_{\circ\circ\circ}^{ijk} = |C|\epsilon_{\circ\circ\circ}^{\alpha\beta\gamma}\gamma_{\circ\alpha}^{i\circ}\gamma_{\circ\beta}^{j\circ}\gamma_{\circ\gamma}^{k\circ}. \tag{8.111}$$

Dividing term by term the (8.111) and (8.110), we have

$$\frac{\epsilon_{\circ\circ\circ}^{ijk}}{\sqrt{|a_{\circ\circ}^{\ell m}|}} = \frac{|C|}{|C|^{-1}}\frac{\epsilon_{\circ\circ\circ}^{\alpha\beta\gamma}}{\sqrt{|a_{\circ\circ}^{\lambda\mu}|}}\gamma_{\circ\alpha}^{i\circ}\gamma_{\circ\beta}^{j\circ}\gamma_{\circ\gamma}^{k\circ}$$

and operating and taking into account expression (8.109), we get

$$s(ijk) = |C|^2 s(\alpha\beta\gamma)\gamma_{\circ\alpha}^{i\circ}\gamma_{\circ\beta}^{j\circ}\gamma_{\circ\gamma}^{k\circ}, \tag{8.112}$$

which declares that the given system is a modular contravariant tensor of third order $(r = 3)$, and weight $w = 2$.

2. If we wish we can also consider that $\epsilon(\alpha\beta\gamma) \equiv \epsilon_{\alpha\beta\gamma}^{\circ\circ\circ}$, which from a tensor point of view, according to (8.40), leads to

$$\epsilon_{ijk}^{\circ\circ\circ} = |C|^{-1}\epsilon_{\alpha\beta\gamma}^{\circ\circ\circ}c_{i\circ}^{\circ\alpha}c_{j\circ}^{\circ\beta}c_{k\circ}^{\circ\gamma}. \tag{8.113}$$

Dividing term by term the (8.113) and (8.110), we have

$$\frac{\epsilon_{ijk}^{\circ\circ\circ}}{\sqrt{|a_{\circ\circ}^{\ell m}|}} = \frac{|C|^{-1}}{|C|^{-1}}\frac{\epsilon_{\alpha\beta\gamma}^{\circ\circ\circ}}{\sqrt{|a_{\circ\circ}^{\lambda\mu}|}}c_{i\circ}^{\circ\alpha}c_{j\circ}^{\circ\beta}c_{k\circ}^{\circ\gamma}$$

and operating and taking into account expression (8.109), we get

$$s(ijk) = s(\alpha\beta\gamma)c_{i\circ}^{\circ\alpha}c_{j\circ}^{\circ\beta}c_{k\circ}^{\circ\gamma}, \tag{8.114}$$

which declares that the given system also (simultaneously) is a homogeneous covariant tensor of third order $(r = 3)$.

□

Example 8.8 (Partial contractions of ϵ systems). Consider the ϵ system, defined over a linear space $V^3(\mathbb{R})$. We wish to analyze the tensor nature of the system the components of which are defined by the doubly contracted expression

$$\epsilon(\alpha\gamma\lambda) \cdot \epsilon(\beta\gamma\lambda).$$

Solution: For the contracted product to have a tensor nature we consider only two possibilities.

1. The first factor is a modular contravariant tensor and the second is a modular covariant tensor. Multiplying (8.71) and (8.72) adapted to this case, we get

$$\epsilon^{ik\ell}_{\circ\circ\circ} \otimes \epsilon^{\circ\circ\circ}_{jmn} = \left(|C|\epsilon^{\alpha\gamma\lambda}_{\circ\circ\circ}\gamma^{i\circ}_{\circ\alpha}\gamma^{k\circ}_{\circ\gamma}\gamma^{\ell\circ}_{\circ\lambda}\right) \otimes \left(|C|^{-1}\epsilon^{\circ\circ\circ}_{\beta\mu\nu}c^{\circ\beta}_{j\circ}c^{\circ\mu}_{k\circ}c^{\circ\nu}_{\ell\circ}\right)$$

and contracting we obtain

$$\epsilon^{ik\ell}_{\circ\circ\circ}\delta^{\circ m}_{k\circ}\delta^{\circ n}_{\ell\circ}\,\epsilon^{\circ\circ\circ}_{jmn} = \epsilon^{\alpha\gamma\lambda}_{\circ\circ\circ}(\gamma^{k\circ}_{\circ\gamma}c^{\circ\mu}_{k\circ})(\gamma^{\ell\circ}_{\circ\lambda}c^{\circ\nu}_{\ell\circ})\,\epsilon^{\circ\circ\circ}_{\beta\mu\nu}\gamma^{i\circ}_{\circ\alpha}c^{\circ\beta}_{j\circ}$$

$$= \epsilon^{\alpha\gamma\lambda}_{\circ\circ\circ}\delta^{\circ\mu}_{\gamma\circ}\delta^{\circ\nu}_{\lambda\circ}\,\epsilon^{\circ\circ\circ}_{\beta\mu\nu}\gamma^{i\circ}_{\circ\alpha}c^{\circ\beta}_{j\circ}.$$

Operating the Kronecker deltas leads to

$$\epsilon^{ik\ell}_{\circ\circ\circ} \cdot \epsilon^{\circ\circ\circ}_{jk\ell} = \left(\epsilon^{\alpha\gamma\lambda}_{\circ\circ\circ} \cdot \epsilon^{\circ\circ\circ}_{\beta\gamma\lambda}\right)\gamma^{i\circ}_{\circ\alpha}c^{\circ\beta}_{j\circ}, \tag{8.115}$$

which declares that the proposed system is a homogeneous contra-covariant tensor of order $r = 2$.

2. Since the first factor is a modular covariant tensor and the second is a modular contravariant tensor, by a similar process we obtain

$$\epsilon^{\circ\circ\circ}_{ik\ell} \cdot \epsilon^{jk\ell}_{\circ\circ\circ} = \left(\epsilon^{\circ\circ\circ}_{\alpha\gamma\lambda} \cdot \epsilon^{\beta\gamma\lambda}_{\circ\circ\circ}\right)c^{\circ\alpha}_{i\circ}\gamma^{j\circ}_{\circ\beta}, \tag{8.116}$$

which shows that the proposed system also (simultaneously) is a homogeneous cova-contravariant tensor of order $(r = 2)$.

□

Example 8.9 (Total contractions of ϵ systems). Let $c^{\circ n}_{m\circ}$ be the elements of a square regular matrix of change-of-basis in $V^3(\mathbb{R})$. Obtain the scalar resulting from the multiple contraction

$$\rho = \frac{\epsilon^{\circ\circ\circ}_{\alpha\beta\gamma}\epsilon^{ijk}_{\circ\circ\circ}c^{\circ\alpha}_{i\circ}c^{\circ\beta}_{j\circ}c^{\circ\gamma}_{k\circ}}{|c^{\circ n}_{m\circ}|}. \tag{8.117}$$

Solution: Taking into account (8.72), for a modular tensor of third order, we get

$$\epsilon^{\circ\circ\circ}_{ijk} = \frac{1}{|c^{\circ n}_{m\circ}|}\epsilon^{\circ\circ\circ}_{\alpha\beta\gamma}c^{\circ\alpha}_{i\circ}c^{\circ\beta}_{j\circ}c^{\circ\gamma}_{k\circ}$$

and substituting it into (8.117) we obtain

$$\rho = \epsilon^{ijk}_{\circ\circ\circ}\left(\frac{1}{|c^{\circ n}_{m\circ}|}\epsilon^{\circ\circ\circ}_{\alpha\beta\gamma}c^{\circ\alpha}_{i\circ}c^{\circ\beta}_{j\circ}c^{\circ\gamma}_{k\circ}\right) = \epsilon^{ijk}_{\circ\circ\circ} \cdot \epsilon^{\circ\circ\circ}_{ijk}.$$

Taking into account the Example 8.6, and that the linear space is $V^3(\mathbb{R})$ $(n = 3)$, we get

$$\rho = 3! = 6; \quad \text{homogeneous tensor of order zero (scalar invariant).}$$

□

Example 8.10 (Tensor products of ϵ systems). Consider a system of scalars S, with components $s(ijk\ell mn) = \epsilon_{ooo}^{ijk} \cdot \epsilon_{\ell mn}^{ooo}$, where the ϵ systems are defined over the linear space $V^3(\mathbb{R})$.

1. Give all different values of the components of S.
2. Discuss the tensor nature of the system S.

Solution:

1. According to Formula (8.73), we know that

$$\epsilon_{ooo}^{ijk} \cdot \epsilon_{\ell mn}^{ooo} = \delta_{\ell mn}^{ijk}$$

and then, because of the generalized deltas of order $r = 3$ properties, the possible values are:
   a) If it has repeated indices in contravariant or covariant positions, as in $s(122123) = \delta_{123}^{122} = 0$ or $s(123111) = \delta_{111}^{123} = 0$, the value is *zero*.
   b) If the contravariant indices have a permutation with parity identical to that for the covariant indices (non-repeated indices), as in $s(213213) = \delta_{213}^{213} = 1$ or $s(123312) = \delta_{312}^{123} = 1$, the value is *one*.
   c) If the non-repeated indices in each valency, have permutations of different parity, as in $s(213123) = \delta_{123}^{213} = -1$ or $s(312213) = \delta_{213}^{312} = -1$, the value is *minus one*.
2. Since the tensor character of the generalized deltas has been established in Formula (8.77), the system $s(ijk\ell mn) \equiv s_{ooo\ell mn}^{ijkooo}$ is classified as a tricontra-tricovariant isotropic homogeneous tensor of sixth order $(r = 6)$ with $\sigma = n^r = 3^6 = 729$ components.

□

Example 8.11 (Relation between the ϵ systems and the alternated multilinear forms). Consider the linear space "total product" $(\times V^n(\mathbb{R}))^3$, defined over the linear space $V^n(\mathbb{R})$, the vectors of which are vector triplets of $V^n(\mathbb{R})$, so that it is not a "tensor product".

Consider also a *trilinear* mapping:

$$F : V^n(\mathbb{R}) \times V^n(\mathbb{R}) \times V^n(\mathbb{R}) \to \mathbb{R}$$

that applies the mentioned "total space" in the field of real numbers. They are called "trilinear forms".

Assuming that the linear space $V^n(\mathbb{R})$ is referred to the basis $\{\vec{e}_\alpha\}$ and given three vectors $\vec{V}, \vec{W}, \vec{Z} \in V^n(\mathbb{R})$ by its matrix representations in the given basis

$$\vec{V} = ||\vec{e}_\alpha||X; \quad \vec{W} = ||\vec{e}_\alpha||Y; \quad \vec{Z} = ||\vec{e}_\alpha||Z;$$

1. Obtain the general expression of this mapping, as a function of the contravariant components of the triplet vectors.
2. Show that the coefficients of the previous expressions are in reality the components of a tensor F, the nature of which is to be determined.
3. Show how a change-of-basis $||\vec{\tilde{e}}_i|| = ||\vec{e}_\alpha||C$ in $V^n(\mathbb{R})$, changes the tensor F (the trilinear form).
4. Assuming that $n = 3$ and that the form be anti-symmetric for each pair of vectors, study the final expression of the function in question 1.

Solution:

1. By definition, the form associates with each vector triplet a given scalar $\rho \in \mathbb{R}$. By convention, we shall distinguish with a special notation the scalars that are images of three basic vectors, as

$$F(\vec{e}_\alpha, \vec{e}_\beta, \vec{e}_\gamma) = f^{\circ\circ\circ}_{\alpha\beta\gamma}; \quad f^{\circ\circ\circ}_{\alpha\beta\gamma} \in \mathbb{R}; \quad \forall \alpha, \beta, \gamma \in I_n = \{1, 2, \ldots, n\}. \quad (8.118)$$

As is well known

$$X = \begin{bmatrix} x^1 \\ x^2 \\ \vdots \\ x^n \end{bmatrix}; \quad Y = \begin{bmatrix} y^1 \\ y^2 \\ \vdots \\ y^n \end{bmatrix}; \quad Z = \begin{bmatrix} z^1 \\ z^2 \\ \vdots \\ z^n \end{bmatrix}$$

are data, and in tensor notation $\vec{V} = x^\alpha_{\circ}\vec{e}_\alpha$, $\vec{W} = y^\beta_{\circ}\vec{e}_\beta$, $\vec{Z} = z^\gamma_{\circ}\vec{e}_\gamma$, and since F is trilinear, we have

$$\rho = F(\vec{V}, \vec{W}, \vec{Z}) = F(x^\alpha_{\circ}\vec{e}_\alpha, y^\beta_{\circ}\vec{e}_\beta, z^\gamma_{\circ}\vec{e}_\gamma) = F(\vec{e}_\alpha, \vec{e}_\beta, \vec{e}_\gamma)x^\alpha_{\circ}y^\beta_{\circ}z^\gamma_{\circ}$$

and taking into account (8.118), we get

$$\rho = F(\vec{V}, \vec{W}, \vec{Z}) = f^{\circ\circ\circ}_{\alpha\beta\gamma}x^\alpha_{\circ}y^\beta_{\circ}z^\gamma_{\circ} \quad \alpha, \beta, \gamma \in I_n. \quad (8.119)$$

The Einstein convention holds in expression (8.119), that is, α, β, γ are dummy contraction indices.

The form F is given by the collection of the n^3 coefficients $f^{\circ\circ\circ}_{\alpha\beta\gamma}$, which are data, assuming that the dual linear space is its reciprocal basis.

2. The simple fact of giving adequate notation to the intervening entities, allows us to discover that Equation (8.119) is no more than a *total contraction* (since the resulting tensor ρ is a scalar *invariant*) of the tensor product of a covariant homogeneous tensor of third order $\vec{F} = f^{\circ\circ\circ}_{\alpha\beta\gamma}\vec{e}^{*\alpha} \otimes \vec{e}^{*\beta} \otimes \vec{e}^{*\gamma}$, by three contravariant coordinate tensors of first order (vectors) $(x^\alpha_{\circ}\vec{e}_\alpha, y^\beta_{\circ}\vec{e}_\beta, z^\gamma_{\circ}\vec{e}_\gamma)$.

Remark 8.3. It is very important that the reader understands how simply the tensor algebra includes all types of homomorphisms of spaces into another, linear or "multilinear", together with the world of "forms" of linear spaces of all kinds in its field, with the simple *use* of the tensor product concept and contractions (total or partial).

One can still find many authors of tensor books who use the terms "mappings", "quotient-space", "forms", etc. continuously in his/her explanations with the help of "functional diagrams", endowing the chapters with a complexity that is not easy to understand for the uninitiated, when in reality all can be expressed by means of contravariant (vectors)-covariant (forms) contractions of the tensor (products or not) indices using the powerful Einstein–Kronecker convention.   □

3. It is known that the vectors change basis by means of the tensor expressions

$$x_\circ^i = x_\circ^\alpha \gamma_{\circ\alpha}^{i\,\circ}; \quad y_\circ^j = y_\circ^\beta \gamma_{\circ\beta}^{j\,\circ}; \quad z_\circ^k = z_\circ^\gamma \gamma_{\circ\gamma}^{k\circ}$$

and the trilinear form changes by means of the tensor expression

$$f_{ijk}^{\circ\circ\circ} = f_{\alpha\beta\gamma}^{\circ\circ\circ} c_{i\,\circ}^{\circ\alpha} c_{j\,\circ}^{\circ\beta} c_{k\circ}^{\circ\gamma}. \tag{8.120}$$

Next, we see that the trilinear forms are tensors. If we execute its tensor product, we get

$$f_{ijk}^{\circ\circ\circ} \otimes x_\circ^\ell \otimes y_\circ^m \otimes z_\circ^p = \left(f_{\alpha\beta\gamma}^{\circ\circ\circ} c_{i\,\circ}^{\circ\alpha} c_{j\,\circ}^{\circ\beta} c_{k\circ}^{\circ\gamma} \right) \otimes \left(x_\circ^\lambda \gamma_{\circ\lambda}^{i\,\circ} \right) \otimes \left(y_\circ^\mu \gamma_{\circ\mu}^{j\,\circ} \right) \otimes \left(z_\circ^\pi \gamma_{\circ\pi}^{k\circ} \right)$$

and if we proceed to its contraction, we obtain

$$f_{ijk}^{\circ\circ\circ} \left(\delta_{\ell\,\circ}^{\circ\,i} x_\circ^\ell \right) \left(\delta_{m\circ}^{\circ\,j} y_\circ^m \right) \left(\delta_{p\circ}^{\circ k} z_\circ^p \right) = f_{\alpha\beta\gamma}^{\circ\circ\circ} \left(c_{i\,\circ}^{\circ\alpha} \gamma_{\circ\lambda}^{i\,\circ} \right) x_\circ^\lambda \left(c_{j\,\circ}^{\circ\beta} \gamma_{\circ\mu}^{j\,\circ} \right) y_\circ^\mu \left(c_{k\circ}^{\circ\gamma} \gamma_{\circ\pi}^{k\circ} \right) z_\circ^\pi$$

and operating we finally get

$$f_{ijk}^{\circ\circ\circ} x_\circ^i y_\circ^j z_\circ^k = f_{\alpha\beta\gamma}^{\circ\circ\circ} \left(\delta_{\circ\lambda}^{\alpha\circ} x_\circ^\lambda \right) \left(\delta_{\circ\mu}^{\beta\circ} y_\circ^k \right) \left(\delta_{\circ\pi}^{\gamma\circ} z_\circ^\pi \right),$$

that is,

$$f_{ijk}^{\circ\circ\circ} x_\circ^i y_\circ^j z_\circ^k = f_{\alpha\beta\gamma}^{\circ\circ\circ} x_\circ^\alpha y_\circ^\beta z_\circ^\gamma = \rho,$$

which shows that the form is always $\bar{F}(\vec{V}, \vec{W}, \vec{Z}) = \rho$, ignoring the reference systems (bases) and the "entities representations".

In other words, the "trilinear forms" have tensor character.

4. As in this case $n = 3$, there is no more than a strict basic component, different from zero, that is $F(\vec{e}_1, \vec{e}_2, \vec{e}_3)$.

Thus, because of Formula (8.104), for $r = 3$ and $(\alpha_1, \alpha_2, \alpha_3) \equiv (123)$, the result is

$$f_{\alpha\beta\gamma}^{\circ\circ\circ} = F(\vec{e}_\alpha, \vec{e}_\beta, \vec{e}_\gamma) = \delta_{\alpha\beta\gamma}^{1\,2\,3} F(\vec{e}_1, \vec{e}_2, \vec{e}_3) = F(\vec{e}_1, \vec{e}_2, \vec{e}_3) \epsilon_{\alpha\beta\gamma}^{\circ\circ\circ},$$

with $\alpha, \beta, \gamma \in I_3 \equiv \{1, 2, 3\}$ and $F(\vec{e}_1, \vec{e}_2, \vec{e}_3) \in \mathbb{R}$.

As a consequence, we obtain

$$\rho = F(\vec{V}, \vec{W}, \vec{Z}) = f^{\circ\circ\circ}_{\alpha\beta\gamma} x^{\alpha}_{\circ} y^{\beta}_{\circ} z^{\gamma}_{\circ} = F(\vec{e}_1, \vec{e}_2, \vec{e}_3) \left(\epsilon^{\circ\circ\circ}_{\alpha\beta\gamma} x^{\alpha}_{\circ} y^{\beta}_{\circ} z^{\gamma}_{\circ} \right).$$

The last factor is the development of the determinant $|XYZ|$. Then, we have

$$\rho = F(\vec{V}, \vec{W}, \vec{Z}) = |XYZ| F(\vec{e}_1, \vec{e}_2, \vec{e}_3).$$

$\square$

Example 8.12 (Contractions of Kronecker deltas of order r). In the mathematical developments of some physical theories, contracted expressions of the generalized Kronecker deltas of order k ($k \leq n$), built over the linear space $V^n(\mathbb{R})$ are used. For the case $k = 5$, we wish to know the tensor expressions of the contractions in this Kronecker delta of:

1. Its five indices, as a function of the generalized delta of order 0.
2. Its four last indices, as a function of the generalized delta of order 1.
3. Its three last indices, as a function of the generalized delta of order 2.
4. Its two last indices, as a function of the generalized delta of order 3.
5. Its last index, as a function of the generalized delta of order 4.

Solution: Starting from Formula (8.88), for $r < k < n$, we have

$$\delta^{\alpha_1 \alpha_2 \cdots \alpha_r}_{\beta_1 \beta_2 \cdots \beta_r} = \frac{(n-k)!}{(n-r)!} \delta^{\alpha_1 \alpha_2 \cdots \alpha_r \alpha_{r+1} \cdots \alpha_k}_{\beta_1 \beta_2 \cdots \beta_r \alpha_{r+1} \cdots \alpha_k}. \tag{8.121}$$

1. Letting $r = 0$ and $k = 5$, we obtain

$$1 = \frac{(n-5)!}{(n-0)!} \delta^{\alpha_1 \alpha_2 \alpha_3 \alpha_4 \alpha_5}_{\alpha_1 \alpha_2 \alpha_3 \alpha_4 \alpha_5} \Rightarrow \delta^{\alpha_1 \alpha_2 \alpha_3 \alpha_4 \alpha_5}_{\alpha_1 \alpha_2 \alpha_3 \alpha_4 \alpha_5} = \frac{n!}{(n-5)!}.$$

2. Letting $r = 1$ and $k = 5$, we obtain

$$\delta^{\alpha_1}_{\beta_1} = \frac{(n-5)!}{(n-1)!} \delta^{\alpha_1 \alpha_2 \alpha_3 \alpha_4 \alpha_5}_{\beta_1 \alpha_2 \alpha_3 \alpha_4 \alpha_5} \Rightarrow \delta^{\alpha_1 \alpha_2 \alpha_3 \alpha_4 \alpha_5}_{\beta_1 \alpha_2 \alpha_3 \alpha_4 \alpha_5} = \frac{(n-1)!}{(n-5)!} \delta^{\alpha_1}_{\beta_1}.$$

3. Letting $r = 2$ and $k = 5$, we obtain

$$\delta^{\alpha_1 \alpha_2}_{\beta_1 \beta_2} = \frac{(n-5)!}{(n-2)!} \delta^{\alpha_1 \alpha_2 \alpha_3 \alpha_4 \alpha_5}_{\beta_1 \beta_2 \alpha_3 \alpha_4 \alpha_5} \Rightarrow \delta^{\alpha_1 \alpha_2 \alpha_3 \alpha_4 \alpha_5}_{\beta_1 \beta_2 \alpha_3 \alpha_4 \alpha_5} = \frac{(n-2)!}{(n-5)!} \delta^{\alpha_1 \alpha_2}_{\beta_1 \beta_2}.$$

4. Letting $r = 3$ and $k = 5$, we obtain

$$\delta^{\alpha_1 \alpha_2 \alpha_3}_{\beta_1 \beta_2 \beta_3} = \frac{(n-5)!}{(n-3)!} \delta^{\alpha_1 \alpha_2 \alpha_3 \alpha_4 \alpha_5}_{\beta_1 \beta_2 \beta_3 \alpha_4 \alpha_5} \Rightarrow \delta^{\alpha_1 \alpha_2 \alpha_3 \alpha_4 \alpha_5}_{\beta_1 \beta_2 \beta_3 \alpha_4 \alpha_5} = \frac{(n-3)!}{(n-5)!} \delta^{\alpha_1 \alpha_2 \alpha_3}_{\beta_1 \beta_2 \beta_3}.$$

5. Letting $r = 4$ and $k = 5$, we obtain

$$\delta^{\alpha_1 \alpha_2 \alpha_3 \alpha_4}_{\beta_1 \beta_2 \beta_3 \beta_4} = \frac{(n-5)!}{(n-4)!}\delta^{\alpha_1 \alpha_2 \alpha_3 \alpha_4 \alpha_5}_{\beta_1 \beta_2 \beta_3 \beta_4 \alpha_5} \Rightarrow \delta^{\alpha_1 \alpha_2 \alpha_3 \alpha_4 \alpha_5}_{\beta_1 \beta_2 \beta_3 \beta_4 \alpha_5} = \frac{(n-4)!}{(n-5)!}\delta^{\alpha_1 \alpha_2 \alpha_3 \alpha_4}_{\beta_1 \beta_2 \beta_3 \beta_4}.$$

$\square$

Example 8.13 (Systems of determinants with Kronecker deltas). Consider two systems of scalar components, S_1 and S_2, built over a certain $V^3(\mathbb{R})$:

$$s_1(\alpha\beta\gamma) = \begin{vmatrix} \delta^{1\,o}_{\,o\alpha} & \delta^{1\,o}_{\,o\beta} & \delta^{1\,o}_{\,o\gamma} \\ \delta^{2\,o}_{\,o\alpha} & \delta^{2\,o}_{\,o\beta} & \delta^{2\,o}_{\,o\gamma} \\ \delta^{3\,o}_{\,o\alpha} & \delta^{3\,o}_{\,o\beta} & \delta^{3\,o}_{\,o\gamma} \end{vmatrix} \quad \text{and} \quad s_2(\alpha\beta\gamma) = \begin{vmatrix} \delta^{\alpha\,o}_{\,o1} & \delta^{\alpha\,o}_{\,o2} & \delta^{\alpha\,o}_{\,o3} \\ \delta^{\beta\,o}_{\,o1} & \delta^{\beta\,o}_{\,o2} & \delta^{\beta\,o}_{\,o3} \\ \delta^{\gamma\,o}_{\,o1} & \delta^{\gamma\,o}_{\,o2} & \delta^{\gamma\,o}_{\,o3} \end{vmatrix},$$

where $\delta^{\alpha\,o}_{\,o\beta}$ are the components of the classic Kronecker delta tensor, over $V^3(\mathbb{R})$.

We wish to determine the tensor nature of the systems S_1 and S_2.

Solution: *Study of the system S_1.* Developing the determinant $s_1(\alpha\beta\gamma)$ by the "column" elements, and endowing the products with a sign by means of an ϵ system, we have

$$s_1(\alpha\beta\gamma) = \delta^{i\,o}_{\,o\alpha}\delta^{j\,o}_{\,o\beta}\delta^{k\,o}_{\,o\gamma}\epsilon^{o\,o\,o}_{ijk}$$

and contracting, we get

$$s_1(\alpha\beta\gamma) = \epsilon^{o\,o\,o}_{\alpha\beta\gamma},$$

which shows that the determinants of S are an expression of the generalized Kronecker delta of order 3, a modular covariant tensor of weight -1.

Study of the system S_2. Proceeding in a similar form, i.e., developing by the row elements, we get

$$s_2(\alpha\beta\gamma)\delta^{\alpha\,o}_{\,o\,i}\delta^{\beta\,o}_{\,o\,j}\delta^{\gamma\,o}_{\,o\,k}\epsilon^{ijk}_{ooo}$$

and contracting, we obtain

$$s_2(\alpha\beta\gamma) = \epsilon^{\alpha\beta\gamma}_{o\,o\,o},$$

which shows that the determinants of S_2, are an expression of the generalized Kronecker delta of order 3, a modular contravariant tensor of weight 1. $\square$

Example 8.14 (Contractions of generalized Kronecker deltas with ϵ systems). Assuming that the tensors are built over $V^n(\mathbb{R})$, prove the formula

$$\delta^{\alpha_1 \alpha_2 \cdots \alpha_r}_{\beta_1 \beta_2 \cdots \beta_r}\epsilon^{o\;o\;\cdots\;o\quad o\quad o\;\cdots\;o}_{\alpha_1 \alpha_2 \cdots \alpha_r \beta_{r+1}\beta_{r+2}\cdots\beta_n} = r!\epsilon^{o\;o\;\cdots\;o}_{\beta_1 \beta_2 \cdots \beta_n}. \tag{8.122}$$

Solution: Ignoring repeated β indices, for which the equality is trivial, because it is null on both sides, we have:

1. The indices $\beta_1, \beta_2, \ldots, \beta_r, \beta_{r+1}, \ldots, \beta_n$ are all different and fixed, that is, a fixed permutation of $I_n \equiv \{1, 2, \ldots, n\}$.

2. The indices $\alpha_1, \alpha_2, \ldots, \alpha_r$ satisfy the set equality $\{\alpha_1, \alpha_2, \ldots, \alpha_r\} = \{\beta_1, \beta_2, \ldots, \beta_r\}$, because otherwise, the $\delta^{\alpha_1 \alpha_2 \cdots \alpha_r}_{\beta_1 \beta_2 \cdots \beta_r}$ would be null, which would make impossible (8.122).

   This second constraint guarantees that there exist only $r!$ possible permutations for the set $\{\alpha_1, \alpha_2, \ldots, \alpha_r\}$ of delta superindices, and then, since (8.122) is a contraction of dummy indices $\{\alpha_1, \alpha_2, \ldots, \alpha_r\}$, the development of the left-hand side of (8.122) will have $r!$ summands; in $r!/2$ of them, the generalized delta will take the value $+1$, and in the other half, the value -1, due to the parity of the index permutations. All this leads to

$$
\delta^{\alpha_1 \alpha_2 \cdots \alpha_r}_{\beta_1 \beta_2 \cdots \beta_r} \epsilon^{\circ\circ\cdots\circ\circ\circ\cdots\circ}_{\alpha_1 \alpha_2 \cdots \alpha_r \beta_{r+1} \beta_{r+2} \cdots \beta_n} = \frac{r!}{2}(+1)\epsilon^{\circ\circ\cdots\circ\circ\circ\cdots\circ}_{\beta_1 \beta_2 \cdots \beta_r \beta_{r+1} \beta_{r+2} \cdots \beta_n}
$$

$$
+ \frac{r!}{2}(-1)\left(-\epsilon^{\circ\circ\cdots\circ\circ\circ\cdots\circ}_{\beta_1 \beta_2 \cdots \beta_r \beta_{r+1} \beta_{r+2} \cdots \beta_n}\right)
$$

$$
= r!\epsilon^{\circ\circ\cdots\circ}_{\beta_1 \beta_2 \cdots \beta_n}.
$$

$\square$

8.8 Exercises

8.1. Solve the following questions:

1. Consider the two tensors $U, W \in (\otimes V_*^n)^2 (\mathbb{R})$, Discuss the tensor nature of the system of scalars represented by the determinant of its secular equation, where $\lambda \in \mathbb{R}$ is

$$
\left| u^{\circ\circ}_{\alpha\beta} - \lambda w^{\circ\circ}_{\alpha\beta} \right|.
$$

2. The same question, but with $U, W \in (\otimes V^n)^2 (\mathbb{R})$.

8.2. Solve the following questions:

1. As a consequence of the results obtained in Example 8.4 (point 5), we propose the following. Let $U(1)$ be a system of scalars the components of which are defined and notated as

$$
u(1)(\alpha\beta) = \text{Adjoint of } t^{\circ\circ}_{\alpha\beta},
$$

   where $T = t^{\circ\circ}_{\alpha\beta}\vec{e}^{*\alpha} \otimes \vec{e}^{*\beta}$ is a data tensor. Similarly, let $U(2)$ be another system of scalars with components

$$
u(2)(\alpha\beta) = \text{Adjoint of } u(1)(\alpha\beta).
$$

   and so on.

a) Discuss the nature of the system $U(k), k \in \mathbb{N}$.

b) Discuss the nature of the determinant system $|u(k)(\alpha\beta)|$.

2. Exactly the same questions, but starting from the data tensor $T = t^{\alpha\beta}_{\circ\circ}\vec{e}_{\alpha} \otimes \vec{e}_{\beta}$.

8.3. Consider two tensors built over the linear space $V^3(\mathbb{R})$ in the basis $\{\vec{e}_{\alpha}\}$:

(a) The homogeneous tensor $U \in (\otimes V^3)^2 (\mathbb{R})$, of matrix representation

$$[u^{\alpha\beta}_{\circ\circ}] = \begin{bmatrix} 1 & 0 & -1 \\ 1 & 2 & 1 \\ 2 & 2 & 3 \end{bmatrix}$$

and

(b) the modular tensor $W \in V^3_* \otimes V^3 \otimes V^3(\mathbb{R})$ of weight $w = 2$ and matrix representation

$$[w^{\circ\lambda\mu}_{\gamma\circ\circ}] = \begin{bmatrix} 1 & -1 & 2 \\ 2 & 6 & 0 \\ 1 & 1 & 3 \\ \hline 1 & 1 & -1 \\ 0 & 5 & 2 \\ 2 & 3 & 1 \\ \hline 1 & -1 & 4 \\ 2 & 3 & 5 \\ 6 & 1 & 4 \end{bmatrix}.$$

1. Obtain the matrix representation of the tensor P, the contracted (with respect to their first indices), product of U and the tensor W in the basis $\{\vec{e}_{\alpha}\}$, i.e.,

$$P = [p^{\beta\lambda\mu}_{\circ\circ\circ}] = \mathcal{C}\begin{pmatrix} \alpha \\ \gamma \end{pmatrix} (U \otimes W).$$

2. Study the tensor nature of the system P.

3. We perform a change-of-basis $\{\vec{e}_i\}$ in $V^3(\mathbb{R})$ that diagonalizes the tensor U. Give the matrix $\hat{U} = [\hat{u}^{ij}_{\circ\circ}]$ with the characteristic values in increasing order, and give also the matrix C of change-of-basis that leads to this objective.

4. Give the new representation of the tensor P in the basis $\{\vec{e}_i\}$.

8.4. We represent with the notation $\varepsilon(\alpha\beta\gamma)$ and $\varepsilon(\lambda\mu\nu)$ the components of the Levi-Civita system ε, to avoid indicating its contravariant or covariant species.

1. Study the system $S_1 = \mathcal{C}(\gamma, \lambda)(\varepsilon(\alpha\beta\gamma) \otimes \varepsilon(\lambda\mu\nu))$ a contracted product of the indices γ and λ, in all possible cases.

2. Express, as a function of the Kronecker deltas, all cases in which the system S_1 is of a tensor nature.

3. Study the system $S_2 = C\left(\beta, \mu | \gamma, \nu\right)\left(\varepsilon(\alpha\beta\gamma) \otimes \varepsilon(\lambda\mu\nu)\right)$ a doubly contracted product, expressing the tensor cases also by means of the Kronecker deltas.

8.5. Consider the tensor $T \in V_*^3 \otimes V^3 \otimes V^3(\mathbb{R})$ built over the linear space $V^3(\mathbb{R})$ referred to the basis $\{\vec{e}_\alpha\}$, represented by the matrix

$$[t^{\circ\,\beta\gamma}_{\alpha\circ\circ}] = \begin{bmatrix} 0 & -2 & 4 \\ 2 & 0 & -6 \\ -4 & 6 & 0 \\ \hline 0 & -6 & 4 \\ 6 & 0 & -2 \\ -4 & 2 & 0 \\ \hline 0 & -4 & 2 \\ 4 & 0 & -6 \\ -2 & 6 & 0 \end{bmatrix},$$

where α is the block row, β is the row of each block, and γ is the column of each block.

1. Obtain all contractions of T:

$$A = C\begin{pmatrix} \beta \\ \alpha \end{pmatrix} T; \quad B = C\begin{pmatrix} \gamma \\ \alpha \end{pmatrix} T; \quad L = C\left(\beta, \gamma\right) T,$$

discussing its possible tensor nature.
2. We define the system of scalars P and Q by contraction of the tensors A and B with the Levi-Civita system ε

$$p(\lambda\mu) = a^\theta_{\circ}\varepsilon^{\circ\circ\circ}_{\theta\lambda\mu}; \quad q(\lambda\mu) = b^\theta_{\circ}\varepsilon^{\circ\circ\circ}_{\theta\lambda\mu}.$$

Study the tensor nature of the system Q.
3. a) Obtain the components of P.
   b) Obtain the components of Q.
4. Study the symmetry or anti-symmetry of P and Q.

Part III

Exterior Algebras

9

Exterior Algebras:
Totally Anti-symmetric Homogeneous Tensor Algebras

9.1 Introduction and Definitions

In this chapter the exterior product of vectors is introduced; first, the cases of two and three vectors, and then the general case with all its properties.

Due to the intrinsic anti-symmetry of this product, the problem of strict components is also analyzed.

Next, the axiomatic system of exterior algebras and the problem of the non-associativity of the exterior product of vectors and the need for a new associative exterior product is discussed.

Finally, dual exterior algebras over $V_*^n(K)$ spaces and other problems are studied.

To perceive as familiar and justified the formulas and the axioms to be introduced in creating these algebras, it is convenient for the reader to reread Sections 7.4.1 and 7.4.3, before commencing this chapter.

First we insist that the algebraic frame in which the algebras are to be inserted is a homogeneous tensor space of order r contravariant and anti-symmetric, $(\otimes V^n)_h^r(K)$, built over linear space $V^n(K)$. We shall create a new composition law to be called the "exterior product of vectors", that will be defined as a function of a linear combination of decomposable tensors in the given tensor space. Having established the intrinsic definition we shall pass to its representation in terms of some subsets of basic vectors of the tensor space grouped by strict components. Finally, we shall define a special basis for the generated subspace.

9.1.1 Exterior product of two vectors

Definition 9.1 (Exterior product). *Given two vectors $\vec{u}, \vec{v} \in V^n(K)$, we give the name "exterior product" of the vectors, notated $\vec{u} \wedge \vec{v}$, to the tensor in $(V^n \otimes V^n)_h(K)$:*

$$\vec{u} \wedge \vec{v} = \vec{u} \otimes \vec{v} - \vec{v} \otimes \vec{u}. \tag{9.1}$$

We immediately see that

$$\vec{u} \wedge \vec{u} = \vec{u} \otimes \vec{u} - \vec{u} \otimes \vec{u} = \vec{0}$$

and

$$\vec{v} \wedge \vec{u} = \vec{v} \otimes \vec{u} - \vec{u} \otimes \vec{v} = -(\vec{u} \otimes \vec{v} - \vec{v} \otimes \vec{u}) = -\vec{u} \wedge \vec{v}, \qquad (9.2)$$

which proves its intrinsic anti-symmetric nature, because of its definition.

If the linear space $V^n(K)$ is referred to a certain basis $\{\vec{e}_\alpha\}$, then, the tensor space $(\otimes V^n)^2(K)$ is referred to its basis $\{\vec{e}_\alpha \otimes \vec{e}_\beta\}$ where $\alpha, \beta \in I_n = \{1, 2, \ldots, n\}$, then, for the tensor $\vec{u} \wedge \vec{v}$, we have

$$
\begin{aligned}
\vec{u} \wedge \vec{v} &= (u_o^\alpha \vec{e}_\alpha) \wedge (v_o^\beta \vec{e}_\beta) \\
&= (u_o^\alpha \vec{e}_\alpha) \otimes (v_o^\beta \vec{e}_\beta) - (v_o^\beta \vec{e}_\beta) \otimes (u_o^\alpha \vec{e}_\alpha) \\
&= u_o^\alpha v_o^\beta \vec{e}_\alpha \otimes \vec{e}_\beta - v_o^\beta u_o^\alpha \vec{e}_\beta \otimes \vec{e}_\alpha \\
&= u_o^\alpha v_o^\beta (\vec{e}_\alpha \otimes \vec{e}_\beta - \vec{e}_\beta \otimes \vec{e}_\alpha) \\
&= u_o^\alpha v_o^\beta \vec{e}_\alpha \wedge \vec{e}_\beta; \quad \alpha \neq \beta,
\end{aligned}
\qquad (9.3)
$$

because $\vec{e}_\alpha \wedge \vec{e}_\beta = \vec{e}_\alpha \otimes \vec{e}_\beta - \vec{e}_\beta \otimes \vec{e}_\alpha$.

Expression (9.3) reveals the relation

$$(u_o^\alpha \vec{e}_\alpha) \wedge (v_o^\beta \vec{e}_\beta) = u_o^\alpha v_o^\beta \vec{e}_\alpha \wedge \vec{e}_\beta, \ \alpha \neq \beta, \qquad (9.4)$$

which is the *bilinear character* of the operator "$\wedge$", i.e., linear for the sum and the scalar–vector product of the space $V^n(K)$, to the left and the right of the operator "$\wedge$".

Expression (9.4) can be grouped into two summands and be written as

$$\vec{u} \wedge \vec{v} = u_o^\alpha v_o^\beta \vec{e}_\alpha \wedge \vec{e}_\beta + u_o^\beta v_o^\alpha \vec{e}_\beta \wedge \vec{e}_\alpha; \ \alpha, \beta \in I_n; \ \alpha < \beta$$

and if we take into account that in addition $\vec{e}_\alpha \wedge \vec{e}_\alpha = \vec{0}$, from the property (9.2), the result is

$$\vec{u} \wedge \vec{v} = (u_o^\alpha v_o^\beta - u_o^\beta v_o^\alpha) \vec{e}_\alpha \wedge \vec{e}_\beta; \ \alpha, \beta \in I_n; \ \alpha < \beta$$

or

$$\vec{u} \wedge \vec{v} = \begin{vmatrix} u_o^\alpha & v_o^\alpha \\ u_o^\beta & v_o^\beta \end{vmatrix} \vec{e}_\alpha \wedge \vec{e}_\beta; \ \alpha, \beta \in I_n \text{ and } \alpha < \beta. \qquad (9.5)$$

The number of determinant-components of (9.5) is that of the number of possibilities of choosing from n indices, two by two and without repetition or permutations, that is, there exist $\binom{n}{2}$ strict components of the possible n^2. Some authors call $\vec{u} \wedge \vec{v}$ a "bivector" or "2-vector".

9.1.2 Exterior product of three vectors

Definition 9.2 (Exterior product of three vectors). *Given three vectors* $\vec{u}, \vec{v}, \vec{w} \in V^n(K)$, *we give the name "exterior product" of the vectors, notated* $\vec{u} \wedge \vec{v} \wedge \vec{w}$, *to the tensor of* $(\otimes V^n)_h^3(K)$

$$\vec{u} \wedge \vec{v} \wedge \vec{w} = \vec{u} \otimes \vec{v} \otimes \vec{w} + \vec{v} \otimes \vec{w} \otimes \vec{u} + \vec{w} \otimes \vec{u} \otimes \vec{v} - \vec{v} \otimes \vec{u} \otimes \vec{w} - \vec{w} \otimes \vec{v} \otimes \vec{u} - \vec{u} \otimes \vec{w} \otimes \vec{v}. \quad (9.6)$$

It is easy to verify the intrinsic anti-symmetry of Expression (9.6), following an analogous exposition to that in Formula (9.2): $\vec{u} \wedge \vec{v} \wedge \vec{w} = \vec{v} \wedge \vec{w} \wedge \vec{u} = \vec{w} \wedge \vec{u} \wedge \vec{v} = -\vec{v} \wedge \vec{u} \wedge \vec{w}$, etc.

If the linear space $V^n(K)$ is referred to a basis $\{\vec{e}_\alpha\}$ then, as before we can establish that

$$\vec{u} \wedge \vec{v} \wedge \vec{w} = (u_\circ^\alpha \vec{e}_\alpha) \wedge (v_\circ^\beta \vec{e}_\beta) \wedge (w_\circ^\gamma \vec{e}_\gamma)$$

$$= u_\circ^\alpha v_\circ^\beta w_\circ^\gamma (\vec{e}_\alpha \otimes \vec{e}_\beta \otimes \vec{e}_\gamma + \vec{e}_\beta \otimes \vec{e}_\gamma \otimes \vec{e}_\alpha + \vec{e}_\gamma \otimes \vec{e}_\alpha \otimes \vec{e}_\beta$$

$$- \vec{e}_\beta \otimes \vec{e}_\alpha \otimes \vec{e}_\gamma - \vec{e}_\gamma \otimes \vec{e}_\beta \otimes \vec{e}_\alpha - \vec{e}_\alpha \otimes \vec{e}_\gamma \otimes \vec{e}_\beta)$$

$$= u_\circ^\alpha v_\circ^\beta w_\circ^\gamma \vec{e}_\alpha \wedge \vec{e}_\beta \wedge \vec{e}_\gamma; \quad \alpha, \beta, \gamma \in I_n; \ \alpha \neq \beta \neq \gamma. \quad (9.7)$$

The expression

$$(u_\circ^\alpha \vec{e}_\alpha) \wedge (v_\circ^\beta \vec{e}_\beta) \wedge (w_\circ^\gamma \vec{e}_\gamma) = u_\circ^\alpha v_\circ^\beta w_\circ^\gamma \vec{e}_\alpha \wedge \vec{e}_\beta \wedge \vec{e}_\gamma \quad (9.8)$$

declares the "trilinear" character of this mapping, of the "total" linear space $(\times V^n)^3(K)$ in the tensor space $(\otimes V^n)_h^3(K)$.

Grouping the summands of (9.8) into partial sums in such a way that for all of them $\alpha < \beta < \gamma$, it is possible to conclude that they can be grouped as

$$\vec{u} \wedge \vec{v} \wedge \vec{w} = u_\circ^\alpha v_\circ^\beta w_\circ^\gamma \vec{e}_\alpha \wedge \vec{e}_\beta \wedge \vec{e}_\gamma + u_\circ^\beta v_\circ^\gamma w_\circ^\alpha \vec{e}_\beta \wedge \vec{e}_\gamma \wedge \vec{e}_\alpha + u_\circ^\gamma v_\circ^\alpha w_\circ^\beta \vec{e}_\gamma \wedge \vec{e}_\alpha \wedge \vec{e}_\beta$$

$$+ u_\circ^\beta v_\circ^\alpha w_\circ^\gamma \vec{e}_\beta \wedge \vec{e}_\alpha \wedge \vec{e}_\gamma + u_\circ^\gamma v_\circ^\beta w_\circ^\alpha \vec{e}_\gamma \wedge \vec{e}_\beta \wedge \vec{e}_\alpha + u_\circ^\alpha v_\circ^\gamma w_\circ^\beta \vec{e}_\alpha \wedge \vec{e}_\gamma \wedge \vec{e}_\beta$$

with $\alpha, \beta, \gamma \in I_n$ and $\alpha < \beta < \gamma$.

Because of the anti-symmetry of the exterior product, the previous expression can be written as

$$\vec{u} \wedge \vec{v} \wedge \vec{w} = \left(u_\circ^\alpha v_\circ^\beta w_\circ^\gamma + u_\circ^\beta v_\circ^\gamma w_\circ^\alpha + u_\circ^\gamma v_\circ^\alpha w_\circ^\beta - u_\circ^\beta v_\circ^\alpha w_\circ^\gamma - u_\circ^\gamma v_\circ^\beta w_\circ^\alpha - u_\circ^\alpha v_\circ^\gamma w_\circ^\beta \right)$$

$$\vec{e}_\alpha \wedge \vec{e}_\beta \wedge \vec{e}_\gamma \quad (9.9)$$

with $\alpha, \beta, \gamma \in I_n$ and $\alpha < \beta < \gamma$, and also as

$$\vec{u} \wedge \vec{v} \wedge \vec{w} = \begin{vmatrix} u_\circ^\alpha & v_\circ^\alpha & w_\circ^\alpha \\ u_\circ^\beta & v_\circ^\beta & w_\circ^\beta \\ u_\circ^\gamma & v_\circ^\gamma & w_\circ^\gamma \end{vmatrix} \vec{e}_\alpha \wedge \vec{e}_\beta \wedge \vec{e}_\gamma, \quad \alpha < \beta < \gamma. \quad (9.10)$$

The number of determinant-components of (9.10) evidently coincides with the existing $\binom{n}{3}$ possibilities of formulating the row indices, extracted from the set I_n three by three.

We proceed to change the notation of Formula (9.6), in such way that it permits a later easy generalization of the intrinsic expression.

Setting $\vec{u} = \vec{v}_1$, $\vec{v} = \vec{v}_2$ and $\vec{w} = \vec{v}_3$, the new (9.6) can be written in terms of the Kronecker deltas, as

$$\vec{v}_1 \wedge \vec{v}_2 \wedge \vec{v}_3 = \delta^{\alpha_1 \alpha_2 \alpha_3}_{1 \ 2 \ 3} \vec{v}_{\alpha_1} \otimes \vec{v}_{\alpha_2} \otimes \vec{v}_{\alpha_3}; \quad \alpha_i \in I_n; \ i \in I_3; \ \alpha_1 \neq \alpha_2 \neq \alpha_3. \quad (9.11)$$

Similarly, we have

$$\vec{e}_{\beta_1} \wedge \vec{e}_{\beta_2} \wedge \vec{e}_{\beta_3} = \delta^{\alpha_1 \alpha_2 \alpha_3}_{(\beta_1 \beta_2 \beta_3)} \vec{e}_{\alpha_1} \otimes \vec{e}_{\alpha_2} \otimes \vec{e}_{\alpha_3}; \quad \alpha_i, \beta_j \in I_n, \ i \in I_3; \ \beta_1 < \beta_2 < \beta_3, \quad (9.12)$$

where the term in parentheses $(\beta_1 \beta_2 \beta_3)$ means "increasing order". It is a very interesting relation, that expresses the exterior bases, in terms of subsets of the tensor anti-symmetric basis of $(\otimes V^n)^3_h(K)$, as was proposed in the final part of Section 9.1.

If we substitute (9.12) into (9.10) the exterior product will appear with the new notation as

$$\vec{v}_1 \wedge \vec{v}_2 \wedge \vec{v}_3 = \begin{vmatrix} x^{\beta_1 \ \circ}_{\ \circ \ 1} & x^{\beta_1 \ \circ}_{\ \circ \ 2} & x^{\beta_1 \ \circ}_{\ \circ \ 3} \\ x^{\beta_2 \ \circ}_{\ \circ \ 1} & x^{\beta_2 \ \circ}_{\ \circ \ 2} & x^{\beta_2 \ \circ}_{\ \circ \ 3} \\ x^{\beta_3 \ \circ}_{\ \circ \ 1} & x^{\beta_3 \ \circ}_{\ \circ \ 2} & x^{\beta_3 \ \circ}_{\ \circ \ 3} \end{vmatrix} \vec{e}_{\beta_1} \wedge \vec{e}_{\beta_2} \wedge \vec{e}_{\beta_3}$$

$$= \begin{vmatrix} x^{\beta_1 \ \circ}_{\ \circ \ 1} & x^{\beta_1 \ \circ}_{\ \circ \ 2} & x^{\beta_1 \ \circ}_{\ \circ \ 3} \\ x^{\beta_2 \ \circ}_{\ \circ \ 1} & x^{\beta_2 \ \circ}_{\ \circ \ 2} & x^{\beta_2 \ \circ}_{\ \circ \ 3} \\ x^{\beta_3 \ \circ}_{\ \circ \ 1} & x^{\beta_3 \ \circ}_{\ \circ \ 2} & x^{\beta_3 \ \circ}_{\ \circ \ 3} \end{vmatrix} \delta^{\alpha_1 \alpha_2 \alpha_3}_{(\beta_1 \beta_2 \beta_3)} \vec{e}_{\alpha_1} \otimes \vec{e}_{\alpha_2} \otimes \vec{e}_{\alpha_3}; \ \beta_1 < \beta_2 < \beta_3,$$

$$(9.13)$$

that is, expressed as a contravariant anti-symmetric "decomposable" (tensor product of vectors) tensor of third order. Some authors call this exterior product a "trivector" or "3-vector".

9.1.3 Strict components of exterior vectors. Multivectors

As it can be seen in Formula (9.5), the exterior product $\vec{u} \wedge \vec{v}$ can be expressed by means of $\binom{n}{2}$ components in an exterior basis $\vec{e}_\alpha \wedge \vec{e}_\beta$, in which $\alpha < \beta$, and also by means of (9.12), that is,

$$\vec{u} \wedge \vec{v} = \begin{vmatrix} u^\alpha_{\ \circ} & v^\alpha_{\ \circ} \\ u^\beta_{\ \circ} & v^\beta_{\ \circ} \end{vmatrix} (\vec{e}_\alpha \otimes \vec{e}_\beta - \vec{e}_\beta \otimes \vec{e}_\alpha),$$

with n^2 components, of which $(n^2 - 2!\binom{n}{2}) = n)$ are null, in which we have only the constraint $\alpha \neq \beta$, as in any contravariant anti-symmetric tensor of order $r = 2$.

Analogously, in Formula (9.10), the exterior product $\vec{u} \wedge \vec{v} \wedge \vec{w}$ can be expressed by means of $\binom{n}{3}$ components with the constraint $\alpha < \beta < \gamma$, and also in (9.13) with n^3 components, of which

$$\left[n^3 - 3! \binom{n}{3} \right] = n(3n - 2)$$

are null, in which the constraint is only $\alpha_1 \neq \alpha_2 \neq \alpha_3$.

When the exterior product is expressed in the first form (9.5) and (9.10), we say that its components are "strict" and when it is expressed in the second mode, as an anti-symmetric tensor (9.3) and (9.7), we say that its components are "normal" or "ordinary".

9.2 Exterior product of r vectors: Decomposable multivectors

Generalizing (9.11) we adopt the following definition.

Definition 9.3 (Exterior product of r vectors). *We define the exterior product of r vectors, notated $\vec{v}_1 \wedge \vec{v}_2 \wedge \cdots \wedge \vec{v}_r$, with $\vec{v}_i \in V^n(K)$, by the intrinsic expression*

$$\vec{v}_1 \wedge \vec{v}_2 \wedge \cdots \wedge \vec{v}_r = \delta^{\alpha_1 \alpha_2 \cdots \alpha_r}_{1\ 2\ \dots\ r} \vec{v}_{\alpha_1} \otimes \vec{v}_{\alpha_2} \otimes \cdots \otimes \vec{v}_{\alpha_r}; \ 1 < 2 < \cdots < r;$$

$$\alpha_i \in I_n; \ i \in I_r; \ \alpha_1 \neq \alpha_2 \neq \cdots \neq \alpha_r \qquad (9.14)$$

Remark 9.1. It is possible that the reader will be surprised by the explicit appearance in (9.14) of the conditions $1 < 2 < \cdots < r$, which can appear as superfluous, because it is evident. However, we want to emphasize its importance. If, for example, in a certain case the vectors were to be $\{\vec{v}_{30}, \vec{v}_{17}, \vec{v}_{43}, \vec{v}_{12}\}$, Expression (9.14) would *force* them to be written on the left-hand side in the following ordered way:

$$\vec{v}_{12} \wedge \vec{v}_{17} \wedge \vec{v}_{30} \wedge \vec{v}_{43} = \delta^{\alpha_1 \alpha_2 \alpha_3 \alpha_4}_{12\ 17\ 30\ 43}, \ etc.$$

From it one can proceed to calculate the product in the desired order. □

Formula (9.14) says that the exterior product $\vec{v}_1 \wedge \vec{v}_2 \wedge \cdots \wedge \vec{v}_r$ is the sum of the tensor products of all permutations of the r vectors, where each product has the sign $+$ or $-$ depending on the corresponding permutation being of even or odd, with respect to the fundamental permutation.

Evidently, if the linear space $V^n(K)$ is referred to a basis $\{\vec{e}_\alpha\}$, we can generalize (9.12) giving the multivector $\vec{e}_{\beta_1} \wedge \vec{e}_{\beta_2} \wedge \cdots \wedge \vec{e}_{\beta_r}$ expressed in the basis of the anti-symmetric tensor space $(\otimes V^n)^r_h(K)$:

$$\vec{e}_{\beta_1} \wedge \vec{e}_{\beta_2} \wedge \cdots \wedge \vec{e}_{\beta_r} = \delta^{\,\alpha_1 \alpha_2 \cdots \alpha_r}_{(\beta_1\,\beta_2\cdots\beta_r)} \vec{e}_{\alpha_1} \otimes \vec{e}_{\alpha_2} \otimes \cdots \otimes \vec{e}_{\alpha_r}; \qquad (9.15)$$

$$\alpha_i, \beta_j \in I_n; \ i,j \in I_r; \ \alpha_1 \neq \alpha_2 \neq \cdots \neq \alpha_r; \ \beta_1 < \beta_2 < \cdots < \beta_r.$$

As a consequence of (9.4) and (9.8) the multilinear character of the exterior product has been established, so that if we assume that the vectors $\vec{V}_1, \vec{V}_2, \ldots, \vec{V}_r$ in terms of their components are

$$\vec{V}_1 = x^{\alpha_1}_{\circ}\vec{e}_{\alpha_1}; \ \vec{V}_2 = x^{\alpha_2}_{\circ}\vec{e}_{\alpha_2}; \ \cdots; \ \vec{V}_r = x^{\alpha_r}_{\circ}\vec{e}_{\alpha_r},$$

the multilinearity of the product leads to

$$\vec{V}_1 \wedge \vec{V}_2 \wedge \cdots \wedge \vec{V}_r = (x^{\alpha_1 \circ}_{\circ\ 1}\vec{e}_{\alpha_1}) \wedge (x^{\alpha_2 \circ}_{\circ\ 2}\vec{e}_{\alpha_2}) \wedge \cdots \wedge (x^{\alpha_r \circ}_{\circ\ r}\vec{e}_{\alpha_r})$$

$$= (x^{\alpha_1 \circ}_{\circ\ 1}x^{\alpha_2 \circ}_{\circ\ 2}\cdots x^{\alpha_r \circ}_{\circ\ r})\vec{e}_{\alpha_1} \wedge \vec{e}_{\alpha_2} \wedge \cdots \wedge \vec{e}_{\alpha_r}. \quad (9.16)$$

If in the diverse permutations of the set of indices $\{\alpha_1, \alpha_2, \ldots, \alpha_r\}$, we denote by $\{\beta_1, \beta_2, \ldots .\beta_r\}$ that having such indices totally ordered, that is

$$\{\alpha_1, \alpha_2, \ldots, \alpha_r\} \equiv \{\beta_1, \beta_2, \ldots, \beta_r\}; \quad \beta_1 < \beta_2 < \cdots < \beta_r,$$

because of the anti-symmetry of the exterior product, shown in (9.2) and (9.6), we have

$$\vec{e}_{\alpha_1} \wedge \vec{e}_{\alpha_2} \wedge \cdots \wedge \vec{e}_{\alpha_r} = \delta^{(\beta_1\,\beta_2\cdots\beta_r)}_{\alpha_1\,\alpha_2\cdots\alpha_r}\vec{e}_{\beta_1} \wedge \vec{e}_{\beta_2} \wedge \cdots \wedge \vec{e}_{\beta_r}, \qquad (9.17)$$

which, substituted into (9.16), leads to

$$\vec{V}_1 \wedge \vec{V}_2 \wedge \cdots \wedge \vec{V}_r = \left(\delta^{(\beta_1\,\beta_2\cdots\beta_r)}_{\alpha_1\,\alpha_2\cdots\alpha_r}x^{\alpha_1 \circ}_{\circ\ 1}x^{\alpha_2 \circ}_{\circ\ 2}\cdots x^{\alpha_r \circ}_{\circ\ r}\right)\vec{e}_{\beta_1} \wedge \vec{e}_{\beta_2} \wedge \cdots \wedge \vec{e}_{\beta_r}.$$
$$(9.18)$$

The first factor of the right-hand side is, by definition, the development by fixed columns of the determinant

$$A_r = |X_{r1}\, X_{r2}\, \cdots\, X_{rr}| = \begin{vmatrix} x^{\beta_1 \circ}_{\circ\ 1} & x^{\beta_1 \circ}_{\circ\ 2} & \cdots & x^{\beta_1 \circ}_{\circ\ r} \\ x^{\beta_2 \circ}_{\circ\ 1} & x^{\beta_2 \circ}_{\circ\ 2} & \cdots & x^{\beta_2 \circ}_{\circ\ r} \\ \cdots & \cdots & \cdots & \cdots \\ x^{\beta_r \circ}_{\circ\ 1} & x^{\beta_r \circ}_{\circ\ 2} & \cdots & x^{\beta_r \circ}_{\circ\ r} \end{vmatrix}$$

$$= \delta^{(\beta_1\,\beta_2\cdots\beta_r)}_{\alpha_1\,\alpha_2\cdots\alpha_r}x^{\alpha_1 \circ}_{\circ\ 1}x^{\alpha_2 \circ}_{\circ\ 2}\cdots x^{\alpha_r \circ}_{\circ\ r}. \qquad (9.19)$$

If the column matrices $X_1, X_2, \ldots, X_r \in K^{n \times 1}$ represent the data of the vectors $\vec{V}_1, \vec{V}_2, \ldots, \vec{V}_r$, the column matrices $X_{r1}, X_{r2}, \cdots, X_{rr} \in K^{r \times 1}$ represent submatrices, respectively, of the data matrices, precisely the rows $\beta_1, \beta_2, \ldots, \beta_r$ (remember their ordered character). Thus, from the data matrices X_i we can extract exactly $\binom{n}{r}$ different determinants A_r.

After these clarifications, we substitute (9.19) into (9.18) and finally get

$$\vec{V}_1 \wedge \vec{V}_2 \wedge \cdots \wedge \vec{V}_r = |\, X_{r1} \quad X_{r2} \quad \cdots \quad X_{rr}\,|\, \vec{e}_{\beta_1} \wedge \vec{e}_{\beta_2} \wedge \cdots \wedge \vec{e}_{\beta_r}. \quad (9.20)$$

Formula (9.20) can be interpreted as follows: it is a sum of $\binom{n}{r}$ summands; each of them carries a real coefficient (resulting from the value of Δ_r) and the corresponding $\binom{n}{r}$ exterior products of the basic vectors of $V^n(\mathbb{R})$, taken from r in r and totally ordered.

When selecting other vectors $\vec{W}_1, \vec{W}_2, \ldots, \vec{W}_r \in V^n(K)$, and executing their exterior product $\vec{W}_1 \wedge \vec{W}_2 \wedge \cdots \wedge \vec{W}_r$ we obtain other coefficients for the *same* $\binom{n}{r}$ exterior products of the set

$$\{\vec{e}_{\beta_1} \wedge \vec{e}_{\beta_2} \wedge \cdots \wedge \vec{e}_{\beta_r}\}.$$

This leads us to think, for the first time, that we could sum linear combinations of several exterior products that would produce the appearance of terms with the set $B_0 = \{\vec{e}_{\beta_1} \wedge \vec{e}_{\beta_2} \wedge \cdots \wedge \vec{e}_{\beta_r}\}$ of $\binom{n}{r}$ "common basic" vectors, and coefficients notated $t^{(\beta_1 \beta_2 \cdots \beta_r)}_{\circ\ \circ\ \ldots\ \circ}$ that *are not* developments of determinants Δ_r. Choosing a generic adequate notation for these exterior vectors, finally we have

$$T = t^{(\alpha_1 \alpha_2 \cdots \alpha_r)}_{\circ\ \circ\ \ldots\ \circ} \vec{e}_{\alpha_1} \wedge \vec{e}_{\alpha_2} \wedge \cdots \wedge \vec{e}_{\alpha_r}, \quad (9.21)$$

where the parentheses mean "strictly ordered", that is, $\alpha_1 < \alpha_2 < \cdots < \alpha_r$.

At this point a linear space is created, that we suspect will be of a tensor nature, of dimension $\sigma' = \binom{n}{r}$, and basis the entities of the set $B_0 = \{\vec{e}_{\alpha_1} \wedge \vec{e}_{\alpha_2} \wedge \cdots \wedge \vec{e}_{\alpha_r}\}$.

This suspicion is due to the fact that its creation has followed exactly the same steps as those followed in Chapter 1 for the tensor product $\otimes$, to give rise to the absolute tensors: first the creation of "products of vectors" (decomposable tensors) and then its linear combination.

Returning to the term in parentheses of Formula (9.21), and remembering Section 9.1.3, it is evident that $t^{(\alpha_1 \alpha_2 \cdots \alpha_r)}_{\circ\ \circ\ \ldots\ \circ}$ are the strict components of T.

Finally, we will express the exterior multivector (r-vector) $\vec{W} = \vec{V}_1 \wedge \vec{V}_2 \wedge \cdots \wedge \vec{V}_r$, as a tensor in $(\otimes V^n)^r_h(K)$ the tensor space of the contravariant anti-symmetric tensors of order r, that is, expressed by means of the basis $\mathcal{B}_1 = \{\vec{e}_{\alpha_1} \otimes \vec{e}_{\alpha_2} \otimes \cdots \otimes \vec{e}_{\alpha_r}\}$ of tensor products. To this end, we substitute Formula (9.15) into (9.20) and get

$$\vec{W} = \vec{V}_1 \wedge \vec{V}_2 \wedge \cdots \wedge \vec{V}_r$$
$$= |\, X_{r1} \quad X_{r2} \quad \cdots \quad X_{rr}\,| \left(\delta^{\alpha_1 \alpha_2 \cdots \alpha_r}_{(\beta_1 \beta_2 \cdots \beta_r)} \vec{e}_{\alpha_1} \otimes \vec{e}_{\alpha_2} \otimes \cdots \otimes \vec{e}_{\alpha_r} \right) \quad (9.22)$$

with $\alpha_i, \beta_j \in I_n$; $i, j \in I_r$; $\alpha_1 \neq \alpha_2 \neq \cdots \neq \alpha_r$.

9.2.1 Properties of exterior products of order r: Decomposable multivectors or exterior vectors

Property 1. The exterior product over $V^n(K)$ is distributive with respect to the sum of vectors of $V^n(K)$.

$$\vec{V}_1 \wedge \vec{V}_2 \wedge \cdots \wedge \vec{V}_h \wedge \cdots \wedge \vec{V}_r = \vec{V}_1 \wedge \vec{V}_2 \wedge \cdots \wedge \left(\vec{V}_h' + \vec{V}_h'' \right) \wedge \cdots \wedge \vec{V}_r$$
$$= \vec{V}_1 \wedge \vec{V}_2 \wedge \cdots \wedge \vec{V}_h' \wedge \cdots \wedge \vec{V}_r + \vec{V}_1 \wedge \vec{V}_2 \wedge \cdots \wedge \vec{V}_h'' \wedge \cdots \wedge \vec{V}_r; \ \forall h \in I_r.$$

Property 2. The exterior product over $V^n(K)$ is associative with respect to the external product of $V^n(K)$.

$$\vec{V}_1 \wedge \vec{V}_2 \wedge \cdots \wedge (\lambda \vec{V}_h) \wedge \cdots \wedge \vec{V}_r$$
$$= \vec{V}_1 \wedge (\lambda \vec{V}_2) \wedge \cdots \wedge \vec{V}_h \wedge \cdots \wedge \vec{V}_r = \lambda \left(\vec{V}_1 \wedge \vec{V}_2 \wedge \cdots \wedge \vec{V}_h \wedge \cdots \wedge \vec{V}_r \right);$$
$$\forall \lambda \in K.$$

Property 3. The exterior product over $V^n(K)$ is p-linear or multilinear on the left and right of "$\wedge$":

$$\vec{V}_1 \wedge \vec{V}_2 \wedge \cdots \wedge (\lambda \vec{V}_h' + \mu \vec{V}_h'') \wedge \cdots \wedge \vec{V}_r = \lambda \left(\vec{V}_1 \wedge \vec{V}_2 \wedge \cdots \wedge \vec{V}_h' \wedge \cdots \wedge \vec{V}_r \right)$$
$$+ \mu \left(\vec{V}_1 \wedge \vec{V}_2 \wedge \cdots \wedge \vec{V}_h'' \wedge \cdots \wedge \vec{V}_r \right); \ \forall \lambda, \mu \in K; \ \forall V_h', V_h'' \in V^n(K); \ \forall h \in I_r.$$

Property 4. The exterior product over $V^n(K)$ is anti-commutative or anti-symmetric:

$$\vec{V}_1 \wedge \vec{V}_2 \wedge \cdots \wedge \vec{V}_k \wedge \cdots \wedge \vec{V}_h \wedge \cdots \wedge \vec{V}_r$$
$$= -\vec{V}_1 \wedge \vec{V}_2 \wedge \cdots \wedge \vec{V}_h \wedge \cdots \wedge \vec{V}_k \wedge \cdots \wedge \vec{V}_r; \ \forall V_k, V_h \in V^n(K).$$

For two vectors, some authors express it as

$$\vec{V} \wedge \vec{W} + \vec{W} \wedge \vec{V} = \vec{0}.$$

Property 5. The exterior product of vectors over $V^n(K)$ if it has repeated indices is the null tensor. In effect, when applying Property 4 we get

$$\vec{W} = \vec{V}_1 \wedge \vec{V}_2 \wedge \cdots \wedge \vec{Z} \wedge \cdots \wedge \vec{Z} \wedge \cdots \wedge \vec{V}_r = -\vec{W} \Rightarrow 2\vec{W} = \vec{0} \Rightarrow \vec{W} = \vec{0}.$$

Property 6. If some factor-vector of the exterior product is a linear combination of the rest, the exterior product is the null tensor ($\vec{0}$):

$$\vec{V}_1 \wedge \vec{V}_2 \wedge \cdots \wedge \left(\lambda \vec{V}_1 + \mu \vec{V}_2 + \cdots + \rho \vec{V}_h \right) \wedge \cdots \wedge \vec{V}_h \wedge \cdots \wedge \vec{V}_r$$
$$= \lambda \left(\vec{V}_1 \wedge \vec{V}_2 \wedge \cdots \wedge \vec{V}_1 \wedge \cdots \wedge \vec{V}_h \wedge \cdots \wedge \vec{V}_r) \right)$$
$$+ \mu \left(\vec{V}_1 \wedge \vec{V}_2 \wedge \cdots \wedge \vec{V}_2 \wedge \cdots \wedge \vec{V}_h \wedge \cdots \wedge \vec{V}_r) \right)$$
$$+ \rho \left(\vec{V}_1 \wedge \vec{V}_2 \wedge \cdots \wedge \vec{V}_h \wedge \vec{V}_h \wedge \cdots \wedge \vec{V}_r) \right) = \lambda \vec{0} + \mu \vec{0} + \cdots + \rho \vec{0} = \vec{0}.$$

Property 7. If a set of vectors is linearly dependent (linked system), its exterior product is the null tensor. This sufficient condition is a simple corollary of Property 6.

Property 8. If a set of vectors is linearly independent (free system), its exterior product cannot be the null tensor.

Proof. By reduction to the absurd: If $\vec{V}_1, \vec{V}_2, \ldots, \vec{V}_r$ are linearly independent and we had $\vec{V}_1 \wedge \vec{V}_2 \wedge \cdots \wedge \vec{V}_r = \vec{0}$, according to Formula (9.22), $\vec{0} = \sum \rho \vec{e}_{\alpha_1} \otimes \vec{e}_{\alpha_2} \otimes \cdots \otimes \vec{e}_{\alpha_r}$, with some $\rho \neq 0$, which would indicate that the *basis* of the linear space $(\otimes V^n)_h^r(K)$ would have some linearly dependent vectors, which is absurd, because it would be not a basis.

This proves the necessarity and the sufficiency of Property 7.

Property 9. If the basis of $V^n(K)$ is $\{\vec{e}_\alpha\} = \{\vec{e}_1, \vec{e}_2, \ldots, \vec{e}_\alpha, \ldots, \vec{e}_n\}$, then, the basis of the linear subspace generated by the vectors T, Formula (9.21), i.e. linear combinations of r-vectors, is the set

$$\mathcal{B}_0 = \{\vec{e}_{\alpha_1} \wedge \vec{e}_{\alpha_2} \wedge \cdots \wedge \vec{e}_{\alpha_r}\}; \quad \alpha_i \in I_n; \quad \alpha_1 < \alpha_2 < \cdots < \alpha_r, \quad (9.23)$$

a total of $\binom{n}{r}$ r-vectors, formulated by the exterior products of the basic vectors of $V^n(K)$, taken "r in r", and with their subindices strictly ordered in each of the r-tuples.

9.2.2 Exterior algebras over $V^n(K)$ spaces: Terminology

The set of the exterior products of order r receive other diverse names according to different authors: multivectors of order r, r-vectors, decomposable exterior vectors, etc.

The linear combination of them, as we have seen in the Formula (9.21), leads to a set of vectors, referred to the basis $\mathcal{B}_0$, Formula (9.23), that are not exterior products of order r, because their coefficients do not respond to the formulation (9.20), and that will be notated with the typical notation of tensors, Formula (9.21). The generated linear subspace will be notated by means of the mnemonic expression

$$\bigwedge\nolimits_n^{(r)}(K) \equiv (\wedge V^n)^r(K) \equiv V^n \wedge V^n \wedge \cdots \wedge V^n(K); \quad r \leq n, \quad (9.24)$$

where the last term has r factors, $\bigwedge$ is the upper case Greek letter and it must be read as: "Exterior Linear Space of order r, over the linear space $V^n(K)$".

This subspace, of dimension $\binom{n}{r}$ is situated as indicated in the following relation:

$$\bigwedge\nolimits_n^{(r)}(K) \subset (\otimes V^n)_h^r(K) \subset (\otimes V^n)^r(K). \quad (9.25)$$

That is, the exterior space is a tensor linear space of anti-symmetric contravariant tensors of order r, that belong to the general tensor space of all homogeneous contravariant tensors of order r.

The dimension $\binom{n}{r}$ (and not n^r as it corresponds to the anti-symmetric tensors), is due to the fact that the new adopted basis, $\mathcal{B}_0$, Formula (9.23), has $\binom{n}{r}$ components, containing exterior products of the basic vectors, strictly ordered.

An arbitrary vector $\vec{t} \in \bigwedge_n^{(r)}(K)$ has the representation given by the Formula (9.21) of strict components

$$\vec{t} = t^{(\alpha_1 \alpha_2 \cdots \alpha_r)}_{\circ \ \circ \ \cdots \ \circ} \vec{e}_{\alpha_1} \wedge \vec{e}_{\alpha_2} \wedge \cdots \wedge \vec{e}_{\alpha_r}; \quad \alpha_i \in I_n; \ i \in I_r; \ \alpha_1 < \alpha_2 < \cdots < \alpha_r$$

and it is usually called an "exterior vector".

Thus, the exterior vectors of $\bigwedge_n^{(r)}(K)$ are sometimes exterior products and sometimes not.

We have not yet reached the exterior algebras, because it is necessary to treat the exterior product of exterior vectors, to establish them.

If one wants to see $\bigwedge^{(r)}(K)$, from the point of view of anti-symmetric tensors, one has only to develop each one of the exterior basic vectors, in terms of the tensor product "$\otimes$". To this end, we substitute Formula (9.15) into (9.21), with the adequate notation

$$\vec{t} = t^{(\alpha_1 \alpha_2 \cdots \alpha_r)}_{\circ \ \circ \ \cdots \ \circ} \vec{e}_{\alpha_1} \wedge \vec{e}_{\alpha_2} \wedge \cdots \wedge \vec{e}_{\alpha_r}$$

$$= t^{(\alpha_1 \alpha_2 \cdots \alpha_r)}_{\circ \ \circ \ \cdots \ \circ} \left(\delta^{\lambda_1 \lambda_2 \cdots \lambda_r}_{(\alpha_1 \alpha_2 \cdots \alpha_r)} \vec{e}_{\lambda_1} \otimes \vec{e}_{\lambda_2} \otimes \cdots \otimes \vec{e}_{\lambda_r} \right) \tag{9.26}$$

with $\alpha_i, \lambda_j \in I_n$; $i, j \in I_r$, $\lambda_1 \neq \lambda_2 \neq \cdots \neq \lambda_r$, and now $\vec{t}$ appears as an anti-symmetric tensor given by its components. In summary, after totally developing (9.26) we obtain $\binom{n}{r} r!$ *non-null* components with alternate signs $(+, -)$ due to the Kronecker delta.

9.2.3 Exterior algebras of order r=0 and r=1

In order to examine and supply coherence to all cases of exterior algebras that can arise for all possible values of r, we will comment on the cases $r = 1$ and $r = 0$. We understand that for $r = 1$, "formally"

$$\bigwedge_n^{(1)} \equiv V^n(K), \tag{9.27}$$

of dimension $\sigma = \binom{n}{1} = n^1 = n$.

For $r = 0$, we have

$$\bigwedge_n^{(0)} \equiv K, \tag{9.28}$$

of dimension $\sigma = \binom{n}{0} = n^0 = 1$. Only in these two cases do the exterior linear space and the anti-symmetric tensor space dimensions coincide.

9.3 Axiomatic properties of tensor operations in exterior algebras

9.3.1 Addition and multiplication by an scalar

We directly give the axiom and its formal properties.

Definition 9.4 (Sum of exterior vectors). *Given two exterior vectors*
$T, U \in \bigwedge_n^{(r)}(K)$

$$T = t^{(\alpha_1 \alpha_2 \cdots \alpha_r)}_{\circ \; \circ \; \cdots \; \circ} \vec{e}_{\alpha_1} \wedge \vec{e}_{\alpha_2} \wedge \cdots \wedge \vec{e}_{\alpha_r}$$

$$U = u^{(\alpha_1 \alpha_2 \cdots \alpha_r)}_{\circ \; \circ \; \cdots \; \circ} \vec{e}_{\alpha_1} \wedge \vec{e}_{\alpha_2} \wedge \cdots \wedge \vec{e}_{\alpha_r}$$

we define the vector sum of both vectors to the vector

$$W = T + U = w^{(\alpha_1 \alpha_2 \cdots \alpha_r)}_{\circ \; \circ \; \cdots \; \circ} \vec{e}_{\alpha_1} \wedge \vec{e}_{\alpha_2} \wedge \cdots \wedge \vec{e}_{\alpha_r} \in \bigwedge_n^{(r)}(K),$$

where

$$w^{(\alpha_1 \alpha_2 \cdots \alpha_r)}_{\circ \; \circ \; \cdots \; \circ} = t^{(\alpha_1 \alpha_2 \cdots \alpha_r)}_{\circ \; \circ \; \cdots \; \circ} + u^{(\alpha_1 \alpha_2 \cdots \alpha_r)}_{\circ \; \circ \; \cdots \; \circ} \in K.$$

This sum is associative, unitary, cancelative and Abelian, arising from the Abelian group of the exterior vectors.

Definition 9.5 (External product of scalar-exterior vectors). *For any exterior vector* $T \in \bigwedge_n^{(r)}(K)$ *and for all* $\lambda \in K$, *we define the exterior product of* λ *by* T, *as the vector:*

$$S = \lambda T = \lambda t^{(\alpha_1 \alpha_2 \cdots \alpha_r)}_{\circ \; \circ \; \cdots \; \circ} \vec{e}_{\alpha_1} \wedge \vec{e}_{\alpha_2} \wedge \cdots \wedge \vec{e}_{\alpha_r}$$

$$= s^{(\alpha_1 \alpha_2 \cdots \alpha_r)}_{\circ \; \circ \; \cdots \; \circ} \vec{e}_{\alpha_1} \wedge \vec{e}_{\alpha_2} \wedge \cdots \wedge \vec{e}_{\alpha_r} \in \bigwedge_n^{(r)}(K),$$

where $\lambda \cdot t^{(\alpha_1 \alpha_2 \cdots \alpha_r)}_{\circ \; \circ \; \cdots \; \circ} = s^{(\alpha_1 \alpha_2 \cdots \alpha_r)}_{\circ \; \circ \; \cdots \; \circ} \in K.$

This exterior product is distributive with respect to the sum of scalars, distributive with respect to the sum of exterior vectors, associative, i.e., $\lambda(\mu T) = (\lambda \cdot \mu)T$, and unitary, i.e., $1T = T$, and $1 \in K$ is the unit of the scalar product of the field.

In summary, we have created the exterior linear space of the exterior vectors.

9.3.2 Generalized exterior tensor product: Exterior product of exterior vectors

The topic we deal with now is certainly delicate. The main reason that forces us to be cautious is the fact that exterior vectors come from linear combinations of exterior products of vectors, thus, they are anti-symmetric tensors. As we already know (Section 7.5) since the product of anti-symmetric tensors *is not* an anti-symmetric tensor, it cannot be an exterior vector; so that we do not try to multiply *two* exterior vectors by means of the exterior product "∧".

Even worse, we will demonstrate the *non-associativity* of the exterior product of two exterior vectors. Even though fortune would have permitted us to establish an exterior product composition law of exterior vectors, attaining the anti-symmetry of the product by some "axiomatic additional arrangement", the effort would be in vain, because to build a ring and arrive at an algebra, we need the associativity and the distributivity of the established product.

We start the proof of the non-associativity.

Consider two exterior vectors $T, U \in \bigwedge_4^{(2)}(K)$ that in addition are bivectors; each one of them has order $r_1 = r_2 = 2$ and they are built over the linear space $V^4(K)$.

We consider $T = \vec{e}_1 \wedge \vec{e}_2$ and $U = \vec{e}_3 \wedge \vec{e}_4$, and assuming that "$\wedge$" is associative, we examine their exterior product

$$V = T \wedge U = (\vec{e}_1 \wedge \vec{e}_2) \wedge (\vec{e}_3 \wedge \vec{e}_4). \qquad (9.29)$$

Calculating this product by two different methods we arrive at a contradiction.

To develop the product by the first method, we use Formula (9.1), which establishes the exterior product of two vectors:

$$V = T \wedge U = T \otimes U - U \otimes T = (\vec{e}_1 \wedge \vec{e}_2) \otimes (\vec{e}_3 \wedge \vec{e}_4) - (\vec{e}_3 \wedge \vec{e}_4) \otimes (\vec{e}_1 \wedge \vec{e}_2);$$

applying again (9.1) in the parentheses, we get for V:

$$V = [\vec{e}_1 \otimes \vec{e}_2 - \vec{e}_2 \otimes \vec{e}_1] \otimes [\vec{e}_3 \otimes \vec{e}_4 - \vec{e}_4 \otimes \vec{e}_3] - [\vec{e}_3 \otimes \vec{e}_4 - \vec{e}_4 \otimes \vec{e}_3] \otimes [\vec{e}_1 \otimes \vec{e}_2 - \vec{e}_2 \otimes \vec{e}_1];$$
$$(9.30)$$

which leads to

$$\begin{aligned} V = &\vec{e}_1 \otimes \vec{e}_2 \otimes \vec{e}_3 \otimes \vec{e}_4 - \vec{e}_1 \otimes \vec{e}_2 \otimes \vec{e}_4 \otimes \vec{e}_3 - \vec{e}_2 \otimes \vec{e}_1 \otimes \vec{e}_3 \otimes \vec{e}_4 \\ &+\vec{e}_2 \otimes \vec{e}_1 \otimes \vec{e}_4 \otimes \vec{e}_3 - \vec{e}_3 \otimes \vec{e}_4 \otimes \vec{e}_1 \otimes \vec{e}_2 + \vec{e}_3 \otimes \vec{e}_4 \otimes \vec{e}_2 \otimes \vec{e}_1 \\ &+\vec{e}_4 \otimes \vec{e}_3 \otimes \vec{e}_1 \otimes \vec{e}_2 - \vec{e}_4 \otimes \vec{e}_3 \otimes \vec{e}_2 \otimes \vec{e}_1. \end{aligned} \qquad (9.31)$$

The resulting tensor V has 8 non-null components and

$$V \in \left(\otimes V^4\right)^4 (K). \qquad (9.32)$$

Now, we develop Expression (9.29) by a second method

$$\vec{V} = T \wedge U = (\vec{e}_1 \wedge \vec{e}_2) \wedge (\vec{e}_3 \wedge \vec{e}_4) = \vec{e}_1 \wedge \vec{e}_2 \wedge \vec{e}_3 \wedge \vec{e}_4. \qquad (9.33)$$

Developing the exterior product of (9.33) by means of the general formula (9.22), we have

$$V = \vec{e}_1 \wedge \vec{e}_2 \wedge \vec{e}_3 \wedge \vec{e}_4 = \begin{vmatrix} 1 & 0 & 0 & 0 \\ 0 & 1 & 0 & 0 \\ 0 & 0 & 1 & 0 \\ 0 & 0 & 0 & 1 \end{vmatrix} \delta^{\alpha_1 \alpha_2 \alpha_3 \alpha_4}_{1 \ 2 \ 3 \ 4} \vec{e}_{\alpha_1} \otimes \vec{e}_{\alpha_2} \otimes \vec{e}_{\alpha_3} \otimes \vec{e}_{\alpha_4} \quad (9.34)$$

and executing the generalized Kronecker delta we get a linear combination of the $4! = 24$ basic vectors of the anti-symmetric tensor space $(\otimes V^4)^4 K)$, which all appear with coefficients one and alternate signs (12 positive and 12 negative):

$$\vec{V} = \vec{e}_1 \otimes \vec{e}_2 \otimes \vec{e}_3 \otimes \vec{e}_4 - \vec{e}_2 \otimes \vec{e}_1 \otimes \vec{e}_3 \otimes \vec{e}_4 + \vec{e}_2 \otimes \vec{e}_3 \otimes \vec{e}_1 \otimes \vec{e}_4$$
$$+ \cdots + \vec{e}_4 \otimes \vec{e}_3 \otimes \vec{e}_2 \otimes \vec{e}_1; \quad V \in \left(\otimes V^4\right)^4 (K). \tag{9.35}$$

Tensor V in Formula (9.35) belongs to the same linear space as the tensor V obtained in Formula (9.31), but has 24 non-null components, i.e., apart from the 8 components referred to in (9.31) it has another additional 16 non-null components. Consequently, to assure that they equal has no sense.

From this contradiction we can conclude that the *assumption*, the associativity of "∧", is infeasible.

The summary of all that we have said in this section, is that we must abandon any attempt to use the operator "∧" to build the exterior product of exterior vectors.

This forces us to adopt the following scheme:

1. Create an exterior product, the notation of which will be the *new symbol*, "$\bigwedge$" (in upper case), in order not to confuse the symbols "∧" of the exterior vectors and of the exterior products of vectors, with the intention of endowing the set of exterior vectors of any order with a product.

2. This product will be useful to:
   a) multiply exterior vectors of different exterior spaces: $T \in \bigwedge_n^{(p)}(K)$ and $U \in \bigwedge_n^{(q)}(K)$ with $p, q \leq n$, that is,

   $$T \bigwedge U = V; \quad V \in \bigwedge_n^{(p+q)}(K), \quad p + q \leq n; \tag{9.36}$$

   b) multiply exterior vectors of the same exterior space: $T, U \in \bigwedge_n^{(r)}(K)$, $r \leq n$, that is,

   $$T \bigwedge U = V; \quad V \in \bigwedge_n^{(2r)}, \quad 2r \leq n. \tag{9.37}$$

3. Evidently, this exterior product of exterior vectors, when applied to case (b), will be associative, distributive with respect to the sum of exterior vectors and associative with respect to the product by a scalar. In this way, $\bigwedge_n^{(r)}(K)$ will become an exterior algebra.

The reader must remember that when in Sections 3.4 and 3.5 the tensor product of tensors was established, the factor-tensors were notated with indices of different notation, axiomatically forcing them to proceed from different tensor spaces.

To establish the product "$\bigwedge$" of exterior vectors, we proceed in an analogous mode with as many exterior factor–vectors as necessary. Normally, the exterior factor vectors are given by their strict components, but without prepared notation, as

a)

$$T = t^{(\alpha_1 \alpha_2 \cdots \alpha_p)}_{\circ \ \circ \ \cdots \ \circ} \vec{e}_{\alpha_1} \wedge \vec{e}_{\alpha_2} \wedge \cdots \wedge \vec{e}_{\alpha_p} \in \bigwedge_n^{(p)}(K);$$
$$\alpha_i, \alpha_j \in I_n; \ i,j \in I_p; \ \alpha_1 < \alpha_2 < \cdots < \alpha_p$$
$$U = u^{(\alpha_1 \alpha_2 \cdots \alpha_q)}_{\circ \ \circ \ \cdots \ \circ} \vec{e}_{\alpha_1} \wedge \vec{e}_{\alpha_2} \wedge \cdots \wedge \vec{e}_{\alpha_q} \in \bigwedge_n^{(q)}(K);$$
$$\alpha_i, \alpha_j \in I_n; \ i,j \in I_q; \ \alpha_1 < \alpha_2 < \cdots < \alpha_q; \qquad (9.38)$$

b)

$$T = t^{(\alpha_1 \alpha_2 \cdots \alpha_r)}_{\circ \ \circ \ \cdots \ \circ} \vec{e}_{\alpha_1} \wedge \vec{e}_{\alpha_2} \wedge \cdots \wedge \vec{e}_{\alpha_r} \in \bigwedge_n^{(r)}(K);$$
$$\alpha_i, \alpha_j \in I_r; \ i,j \in I_r; \ \alpha_1 < \alpha_2 < \cdots < \alpha_r$$
$$U = u^{(\alpha_1 \alpha_2 \cdots \alpha_r)}_{\circ \ \circ \ \cdots \ \circ} \vec{e}_{\alpha_1} \wedge \vec{e}_{\alpha_2} \wedge \cdots \wedge \vec{e}_{\alpha_r} \in \bigwedge_n^{(r)}(K). \qquad (9.39)$$

If necessary, the second factor, the U in Formulas (9.36) and (9.37), is changed in its index notation, to prepare it for executing the product as:
a)

$$U = u^{(\beta_1 \beta_2 \cdots \beta_q)}_{\circ \ \circ \ \cdots \ \circ} \vec{e}_{\beta_1} \wedge \vec{e}_{\beta_2} \wedge \cdots \wedge \vec{e}_{\beta_q}; \qquad (9.40)$$

b)

$$U = u^{(\beta_1 \beta_2 \cdots \beta_r)}_{\circ \ \circ \ \cdots \ \circ} \vec{e}_{\beta_1} \wedge \vec{e}_{\beta_2} \wedge \cdots \wedge \vec{e}_{\beta_r}. \qquad (9.41)$$

4. Next, we choose collections of indices $\{\gamma_1, \gamma_2, \ldots, \gamma_{p+q}\} \in I_n$ for the basic vectors of the type $(\vec{e}_{\gamma_1} \wedge \vec{e}_{\gamma_2} \wedge \cdots \wedge \vec{e}_{\gamma_{p+q}}); \gamma_1 < \gamma_2 < \cdots < \gamma_{p+q}$ with which the exterior vector product $V \in \bigwedge_n^{(p+q)}(K)$ is to be obtained. Obviously, we must choose $\binom{n}{p+q}$ collections of the given type, that is, the dimension of $V \in \bigwedge_n^{(p+q)}(K)$. Next, we define the coefficients of V in the already prepared basis

$$\mathcal{B}_0 = \{\vec{e}_{\gamma_1} \wedge \vec{e}_{\gamma_2} \wedge \cdots \wedge \vec{e}_{\gamma_{p+q}}\} \qquad (9.42)$$

or, in other words, we define the strict components of V. Remember from (9.36), that $p + q \le n$.

a) The exterior product $\bigwedge$ of exterior vectors valid for Formulas (9.36) and (9.38) is defined as

$$V = T \bigwedge U = v^{(\gamma_1 \gamma_2 \cdots \gamma_{p+q})}_{\circ \ \circ \ \cdots \ \ \circ} \vec{e}_{\gamma_1} \wedge \vec{e}_{\gamma_2} \wedge \cdots \wedge \vec{e}_{\gamma_{p+q}}$$
$$= t^{(\alpha_1 \alpha_2 \cdots \alpha_p)}_{\circ \ \circ \ \cdots \ \circ} \cdot u^{(\beta_1 \beta_2 \cdots \beta_q)}_{\circ \ \circ \ \cdots \ \circ}$$
$$\cdot \delta^{(\gamma_1 \ \gamma_2 \ \cdots \ \gamma_p \ \gamma_{p+1} \gamma_{p+2} \cdots \gamma_{p+q})}_{(\alpha_1 \alpha_2 \cdots \alpha_p) \ (\beta_1 \ \ \beta_2 \ \ \cdots \ \ \beta_q \ )} \ \vec{e}_{\gamma_1} \wedge \vec{e}_{\gamma_2} \wedge \cdots \wedge \vec{e}_{\gamma_{p+q}}, \quad (9.43)$$

where

$$\{\alpha_1, \alpha_2, \ldots, \alpha_p, \beta_1, \beta_2, \ldots, \beta_q\} \equiv \{\gamma_1, \gamma_2, \ldots, \gamma_{p+q}\};$$
$$\gamma_1 < \gamma_2 < \cdots < \gamma_{p+q}$$

and

$$(\alpha_1 < \alpha_2 < \cdots < \alpha_p) \text{ and } (\beta_1 < \beta_2 < \cdots < \beta_q) \Rightarrow \alpha_i \neq \beta_j.$$

Therefore, when using (9.43) we choose in the ordered set

$$\{\gamma_1, \gamma_2, \ldots, \gamma_{p+q}\}$$

combinations of p indices for the $\{\alpha_i\}$ and the remaining combination (that not chosen) for the $\{\beta_j\}$; we determine the sign of the generalized Kronecker delta with them, and we assign it to the scalar resulting from the product $t^{(\alpha_1 \alpha_2 \cdots \alpha_p)}_{\circ \circ \cdots \circ} \cdot u^{(\beta_1 \beta_2 \cdots \beta_q)}_{\circ \circ \cdots \circ}$.

We shall obtain a total of $\binom{n}{p+q}$ strict components, with $\binom{p+q}{p}$ algebraic summands each.

b) Now, the exterior product $\bigwedge$ for Formulas (9.37) and (9.39) is defined as

$$V = T \bigwedge U = v^{(\gamma_1 \gamma_2 \cdots \gamma_{2r})}_{\circ \circ \cdots \circ} \vec{e}_{\gamma_1} \wedge \vec{e}_{\gamma_2} \wedge \cdots \wedge \vec{e}_{\gamma_{2r}}$$

$$= t^{(\alpha_1 \alpha_2 \cdots \alpha_r)}_{\circ \circ \cdots \circ} \cdot u^{(\beta_1 \beta_2 \cdots \beta_r)}_{\circ \circ \cdots \circ}$$

$$\cdot \delta^{(\gamma_1 \gamma_2 \cdots \gamma_r \gamma_{r+1} \gamma_{r+2} \cdots \gamma_{2r})}_{(\alpha_1 \alpha_2 \cdots \alpha_r)(\beta_1 \beta_2 \cdots \beta_r)} \vec{e}_{\gamma_1} \wedge \vec{e}_{\gamma_2} \wedge \cdots \wedge \vec{e}_{\gamma_{2r}}, \qquad (9.44)$$

where $\{\alpha_1, \alpha_2, \ldots, \alpha_r, \beta_1, \beta_2, \cdots, \beta_r\} \equiv \{\gamma_1, \gamma_2, \ldots, \gamma_{2r}\}$ and $(\alpha_1 < \alpha_2 < \ldots < \alpha_r) \neq (\beta_1 < \beta_2 < \ldots < \beta_r)$.

The number of strict components is $\binom{n}{2r}$, with $\binom{2r}{r}$ summands each.

To use (9.44) we proceed in an analogous form to that for (9.43). Since we began this section with an examination of a product of exterior vectors in expression (9.29), we end by applying (9.44) to finally determine it. The data are $T, U \in \bigwedge_4^{(2)}(K)$, where

$$T = t^{(\alpha_1 \alpha_2)}_{\circ \circ} \vec{e}_{\alpha_1} \wedge \vec{e}_{\alpha_2}$$

with

$$\{t^{(12)}_{\circ \circ} = 1; \text{ if } (\alpha_1, \alpha_2) \neq (1, 2) \rightarrow t^{(13)}_{\circ \circ} = t^{(14)}_{\circ \circ} = t^{(23)}_{\circ \circ} = t^{(24)}_{\circ \circ} = t^{(34)}_{\circ \circ} = 0\}$$

and

$$U = u^{(\beta_1 \beta_2)}_{\circ \circ} \vec{e}_{\beta_1} \wedge \vec{e}_{\beta_2}$$

with

$$\{u\,_{\circ\circ}^{(34)} = 1; \text{ if } (\beta_1, \beta_2) \neq (3,4) \rightarrow u\,_{\circ\circ}^{(12)} = u\,_{\circ\circ}^{(13)} = u\,_{\circ\circ}^{(14)} = u\,_{\circ\circ}^{(23)} v = u\,_{\circ\circ}^{(24)} = 0\}$$

$$V = T \bigwedge U = (\vec{e}_1 \wedge \vec{e}_2) \bigwedge (\vec{e}_3 \wedge \vec{e}_4)$$
$$= t\,_{\circ\circ}^{(12)} \cdot u\,_{\circ\circ}^{(34)} \cdot \delta_{(1234)}^{(1234)} (\vec{e}_1 \wedge \vec{e}_2 \wedge \vec{e}_3 \wedge \vec{e}_4) + (\vec{0} + \vec{0} + \vec{0} + \vec{0} + \vec{0})$$
$$= \vec{e}_1 \wedge \vec{e}_2 \wedge \vec{e}_3 \wedge \vec{e}_4.$$

The number of strict components is $\binom{n}{2r} = \binom{4}{4} = 1$ of $\binom{2r}{r} = \binom{4}{2} = 6$ summands, all null but one.

The conclusion of this model of exterior product, the discussion of which has originated the extension, already finished, of this section is that in the space $\bigwedge_4^{(2)}(K)$, the stated product of exterior vectors leads to the following expression, which is to be discussed:

$$(\vec{e}_1 \wedge \vec{e}_2) \bigwedge (\vec{e}_3 \wedge \vec{e}_4) = \vec{e}_1 \wedge \vec{e}_2 \wedge \vec{e}_3 \wedge \vec{e}_4 \in \bigwedge_4^{(4)}, \qquad (9.45)$$

which together with the previous discussion, clearly declares the *non-associativity* of the symbol "$\wedge$" of the exterior product, because the central symbol of the left-hand side is *another* symbol.

In addition, it allows us to consider the exterior product as a "new" exterior composition law.

This criticism is addressed to many authors who employ all symbols "$\wedge$" identical in size, leading to a lack of understanding of why a new exterior product is "created", when the proper "$\wedge$" we already have is valid, *since it is associative* (the reader should imagine the criticized Formula (9.45) with all symbols of the same size).

The exterior product defined in this way has the following desired formal intrinsic properties:

it is associative:

$$(T \bigwedge U) \bigwedge S = T \bigwedge (U \bigwedge S) = T \bigwedge U \bigwedge S;$$

it is distributive:

$$P \bigwedge (T + U) \bigwedge S = P \bigwedge T \bigwedge S + P \bigwedge U \bigwedge S$$

it is associative with respect to the external product:

$$(\lambda T) \bigwedge U = T \bigwedge (\lambda U) = \lambda (T \bigwedge U)$$

leading to the exterior tensor algebra, $\bigwedge_n^{(r)}(K)$. The tensor nature of the exterior vectors manifests when they are expressed as anti-symmetric tensors, that is, when they are notated with the tensor basis of $(\otimes V^n)_h^r(K)$, which is the same as that of $(\otimes V^n)^r(K)$.

9.3.3 Anti-commutativity of the exterior product $\bigwedge$

If we consider the generalized Kronecker delta $\delta^{(\gamma_1 \gamma_2 \cdots \gamma_p \gamma_{p+1} \cdots \gamma_{p+q})}_{\;\;\alpha_1 \alpha_2 \cdots \alpha_p \;\; \beta_1 \;\; \cdots \;\; \beta_q}$ and we exchange two covariant indices (covariant indices, because the contravariant indices are "strictly ordered"), each performed transposition changes its sign. To place the index β_1 in the first position, we must use p transpositions until we reach the ordering $\beta_1 \alpha_1 \alpha_2 \cdots \alpha_p \beta_2 \cdots \beta_q$; if we do the same with the β_2 to place it in the second place, we need another p transpositions with each α_i, so that when we reach the ordering $\beta_1 \beta_2 \alpha_1 \alpha_2 \cdots \alpha_p \beta_3 \cdots \beta_q$ we have performed $p + p$ transpositions; so, we have a total of $2p$ transpositions. If we continue with this criterion, when all indices β_j are ahead of the indices α_i, i.e., $\beta_1 \beta_2, \cdots \beta_q \alpha_1 \alpha_2 \cdots \alpha_p$ we will have performed a total of $p \cdot q$ transpositions. This justifies the formula:

$$\delta^{(\gamma_1 \gamma_2 \cdots \gamma_p \gamma_{p+1} \cdots \gamma_{p+q})}_{\;\beta_1 \beta_2 \cdots \beta_q \;\; \alpha_1 \;\; \cdots \;\; \alpha_p} = (-1)^{p \cdot q} \delta^{(\gamma_1 \gamma_2 \cdots \gamma_p \gamma_{p+1} \cdots \gamma_{p+q})}_{(\alpha_1 \alpha_2 \cdots \alpha_p)\; (\beta_1 \;\; \cdots \;\; \beta_q)}. \qquad (9.46)$$

Once this has been established and remembering Formula (9.43), we have

$$U \bigwedge T = u^{(\beta_1 \beta_2 \cdots \beta_q)}_{\circ\;\circ\;\cdots\;\circ} t^{(\alpha_1 \alpha_2 \cdots \alpha_p)}_{\circ\;\circ\;\cdots\;\circ} \cdot \delta^{(\gamma_1 \gamma_2 \cdots \gamma_p \gamma_{p+1} \cdots \gamma_{p+q})}_{(\beta_1 \beta_2 \cdots \beta_q\; \alpha_1 \;\; \cdots \;\; \alpha_p\;)}\; \vec{e}_{\gamma_1} \wedge \vec{e}_{\gamma_2} \wedge \cdots \wedge \vec{e}_{\gamma_{p+q}}$$

$$= (-1)^{p \cdot q} t^{(\alpha_1 \alpha_2 \cdots \alpha_p)}_{\circ\;\circ\;\cdots\;\circ} u^{(\beta_1 \beta_2 \cdots \beta_q)}_{\circ\;\circ\;\cdots\;\circ} \delta^{(\gamma_1 \gamma_2 \cdots\; \gamma_p\; \gamma_{p+1} \cdots \gamma_{p+q})}_{(\alpha_1 \alpha_2 \cdots \alpha_p)\;(\beta_1 \;\; \cdots \;\; \beta_q\;)} \vec{e}_{\gamma_1} \wedge \vec{e}_{\gamma_2} \wedge \cdots \wedge \vec{e}_{\gamma_{p+q}}$$

$$= (-1)^{p \cdot q} T \bigwedge U, \qquad (9.47)$$

a property that is called "anti-commutativity" of the exterior product of exterior vectors.

In the particular case of two exterior vectors coming from the same $\bigwedge_n^{(r)}(K)$, Formulas (9.37) and (9.39), the anti-commutativity is stated by

$$U \bigwedge T = (-1)^{r^2} T \bigwedge U \qquad (9.48)$$

so that the exterior algebras $\bigwedge_n^{(r)}(K)$ of r "even" are Abelian.

9.4 Dual exterior algebras over $V_*^n(K)$ spaces

Some authors use the notation

$$F = f^{\circ\;\circ\;\cdots\;\circ}_{\alpha_1 \alpha_2 \cdots \alpha_r} \vec{e}^{*\alpha_1} \otimes \vec{e}^{*\alpha_2} \otimes \cdots \otimes \vec{e}^{*\alpha_r}; \quad \alpha_1 \neq \alpha_2 \neq \cdots \neq \alpha_r$$

when they refer to the homogeneous anti-symmetric covariant tensors of order r built over the linear space $V_*(\mathbb{R})$ dual of $V^n(\mathbb{R})$ and assume it is referred to a certain basis $\{\vec{e}^{*\alpha_j}\}$.

The reason for this notation is that the vectors of the dual linear spaces are called "forms", a word that starts with an "f" so that the vectors of the linear space $V_*(K)$ are the "linear forms" and obviously, the vectors of $(\otimes V_*^n)_h^r(K)$ are "the multilinear forms" of order r built over the linear space of the linear forms $V_*^n(K)$. In mathematics, "form" is a mapping of the linear space in its field of scalars, i.e., $f(\vec{v}) = \rho; \rho \in K$.

This notation is also used in strict components, typical of anti-symmetric forms:

$$F = f_{(\alpha_1 \alpha_2 \cdots \alpha_r)}^{\circ \ \circ \ \cdots \ \circ} \delta_{\beta_1 \beta_2 \cdots \beta_r}^{(\alpha_1 \alpha_2 \cdots \alpha_r)} \vec{e}^{*\beta_1} \otimes \vec{e}^{*\beta_2} \otimes \cdots \otimes \vec{e}^{*\beta_r}; \ \alpha_1 < \alpha_2 < \cdots < \alpha_r; \ \alpha_i \in I_n.$$

In this way we arrive at the exterior forms:

$$W \in \bigwedge\nolimits_{n*}^{(r)}(K), \ \ W = f_{(\alpha_1 \alpha_2 \cdots \alpha_r)}^{\circ \ \circ \ \cdots \ \circ} \vec{e}^{*\alpha_1} \wedge \vec{e}^{*\alpha_2} \wedge \cdots \wedge \vec{e}^{*\alpha_r}.$$

To end the exposition, we want to emphasize only that in this book we *do not* make *any distinction in notation* in the tensors, no matter they are $(t_{\circ \circ}^{\alpha \beta})$ contravariant or $(t_{\alpha \beta}^{\circ \circ})$ covariant. The reason is that the reader, when looking the totally covariant species of the tensor, can call it personally a "form", if he desires. What is important in books is not what each reader reads, but what is written and how.

9.4.1 Exterior product of r linear forms over $V_*^n(K)$

Definition 9.6 (Exterior product of r linear forms over $V_*^n(K)$). *We define this product as*

$$V_1^* \wedge V_2^* \wedge \cdots \wedge V_r^* = \delta_{1 \ 2 \ \cdots \ r}^{\alpha_1 \alpha_2 \cdots \alpha_r} V_{\alpha_1}^* \otimes V_{\alpha_2}^* \otimes \cdots \otimes V_{\alpha_r}^*;$$

$$1 < 2 < \cdots < r; \ \alpha_i \in I_n; \ i \in I_r; \ \alpha_i \neq \alpha_j. \quad (9.49)$$

We repeat the comment in Section 9.2, about data vectors with proper subindices different from the proposed ones, so that we remember the "sense" of the condition $1 < 2 < \cdots < r$. As is well known, the right-hand side of (9.49) is a homogeneous covariant and anti-symmetric tensor of order r.

We apply this product to subsets of r basic vectors of $V_*^n(K)$, to create the future exterior basis of $\bigwedge_{n*}^{(r)}(K)$ by means of the expression

$$\vec{e}^{*\beta_1} \wedge \vec{e}^{*\beta_2} \wedge \cdots \wedge \vec{e}^{*\beta_r} = \delta_{\alpha_1 \alpha_2 \cdots \alpha_r}^{(\beta_1 \beta_2 \cdots \beta_r)} \vec{e}^{*\alpha_1} \otimes \vec{e}^{*\alpha_2} \otimes \cdots \otimes \vec{e}^{*\alpha_r}; \ \alpha_i, \beta_j \in I_n;$$

$$i, j \in I_r; \ \alpha_i \neq \alpha_j; \ \beta_1 < \beta_2 < \cdots < \beta_r.$$

With the n basic vectors of $V_*^n(K)$ we can construct $\binom{n}{r}$ products of the proposed type in the first term of (9.50), that is, the number of basic exterior vectors for the exterior linear space $\bigwedge_{n*}^{(r)}(K)$ that is to be created. Thus, its dimension will be $\binom{n}{r}$.

If we wish to express the exterior product of the r vectors of the first member of (9.49) in terms of the basis of the exterior space to be created

$\bigwedge_n^{(r)}(K)$, and if we assume that the data vectors V_i^* are given by their respective column-matrices X_i^*, we proceed in an analogous form as that employed in Section 9.2 and arrive at

$$V_1^* \wedge V_2^* \wedge \cdots \wedge V_r^* = |\, X_{r_1}^*\ \ X_{r_2}^*\ \ \cdots\ \ X_{r_r}^*\,|_{rr}\ \bar{e}^{*\beta_1} \wedge \bar{e}^{*\beta_2} \wedge \cdots \wedge \bar{e}^{*\beta_r} \quad (9.50)$$

with the constraints in (9.20) and the following clarification.

We start from the matrix

$$[X_1^*\ \ X_2^*\ \ \cdots\ \ X_r^*]_{n,r}$$

built with the data, and extract the indicated minors in Formula (9.50) selecting the rows by numerical combinations of the n, taken of r in r, that is, taking the rows $(\beta_1\beta_2\cdots\beta_r)$ that indicate the indices $\bar{e}^{*\beta_j}$ in the proper formula (remember that $\beta_1 < \beta_2 < \cdots < \beta_r$).

If what we want is to express the mentioned product as an anti-symmetric tensor belonging to $(\otimes V_*^n)_h^r(K)$, following the process indicated in Section 9.2, we arrive at

$$W^* = V_1^* \wedge V_2^* \wedge \cdots \wedge V_r^* = |\, X_{r_1}^*\ \ X_{r_2}^*\ \ \cdots\ \ X_{r_r}^*\,|\ \delta_{\alpha_1\,\alpha_2\cdots\alpha_r}^{(\beta_1\,\beta_2\,\cdots\,\beta_r)}\, \bar{e}^{*\alpha_1} \otimes \bar{e}^{*\alpha_2} \otimes \cdots \otimes \bar{e}^{*\alpha_r}$$
$$(9.51)$$

with $\alpha_i, \beta_j \in I_n$.

9.4.2 Axiomatic tensor operations in dual exterior Algebras $\bigwedge_{n*}^{(r)}(K)$. Dual exterior tensor product

We have nothing to add to what has been already established in the exterior contravariant algebras $\bigwedge_n^{(r)}(K)$. The dual exterior products are combined by linear combinations to generate dual exterior vectors that are not products, notated

$$T = t_{(\alpha_1\,\alpha_2\cdots\alpha_r)}^{\circ\ \circ\ \ \circ\ \cdots\ \circ}\, \bar{e}^{*\alpha_1} \wedge \bar{e}^{*\alpha_2} \wedge \cdots \wedge \bar{e}^{*\alpha_r}; \quad \alpha_i \in I_n;\ i \in I_r;\ \alpha_1 < \alpha_2 < \ldots < \alpha_r.$$
$$(9.52)$$

We formulate axiomatically the sum of dual exterior vectors and the external product of scalars by exterior dual vectors, which constitutes the linear space $\bigwedge_{n*}^{(r)}(K)$. Finally, we establish the exterior product of exterior dual vectors, in which we use again the symbol "$\bigwedge$".

1.

$$T = T_p \bigwedge T_q = t_{(\gamma_1\,\gamma_2\cdots\gamma_{p+q})}^{\circ\ \circ\ \cdots\ \circ}\, \bar{e}^{*\gamma_1} \wedge \bar{e}^{*\gamma_2} \wedge \cdots \wedge \bar{e}^{*\gamma_{p+q}}$$

$$= t_{(\alpha_1\,\alpha_2\cdots\alpha_p)}^{\circ\ \circ\ \cdots\ \circ} \cdot t_{(\beta_1\,\beta_2\cdots\beta_q)}^{\circ\ \circ\ \circ\ \cdots\ \circ}$$

$$\cdot\, \delta_{(\gamma_1\ \gamma_2\cdots\ \gamma_p\ \gamma_{p+1}\gamma_{p+2}\cdots\gamma_{p+q})}^{(\alpha_1\,\alpha_2\cdots\alpha_p)\ (\beta_1\ \beta_2\ \cdots\ \beta_q\ )}\, \bar{e}^{*\gamma_1} \wedge \bar{e}^{*\gamma_2} \wedge \cdots \wedge \bar{e}^{*\gamma_{p+q}}, \quad (9.53)$$

where $\{\alpha_1, \alpha_2, \ldots, \alpha_p, \beta_1, \beta_2, \ldots, \beta_q\} \equiv \{\gamma_1, \gamma_2, \ldots, \gamma_{p+q}\}$ and $(\alpha_1 < \alpha_2 < \cdots < \alpha_p) \neq (\beta_1 < \beta_2 < \cdots < \beta_q)$.

2.

$$T = T_r \bigwedge U_r = t_{\binom{\circ \ \circ \ \cdots \ \circ}{\gamma_1 \gamma_2 \cdots \gamma_{2r}}} \bar{e}^{*\gamma_1} \wedge \bar{e}^{*\gamma_2} \wedge \cdots \wedge \bar{e}^{*\gamma_{2r}}$$

$$= t_{\binom{\circ \ \circ \ \cdots \ \circ}{\alpha_1 \alpha_2 \cdots \alpha_r}} \cdot u_{\binom{\circ \ \circ \ \cdots \ \circ}{\beta_1 \beta_2 \cdots \beta_r}}$$

$$\cdot \delta_{\binom{\alpha_1 \alpha_2 \cdots \alpha_r}{\gamma_1 \ \gamma_2 \ \cdots \ \gamma_p}\binom{\beta_1 \ \ \beta_2 \ \cdots \ \beta_r}{\gamma_{r+1}\gamma_{r+2}\cdots\gamma_{2r}}} \bar{e}^{*\gamma_1} \wedge \bar{e}^{*\gamma_2} \wedge \cdots \wedge \bar{e}^{*\gamma_{2r}}, \qquad (9.54)$$

where $\{\alpha_1, \alpha_2, \ldots, \alpha_r, \beta_1, \beta_2, \ldots, \beta_r\} \equiv \{\gamma_1, \gamma_2, \ldots, \gamma_r\}$ and $(\alpha_1 < \alpha_2 < \cdots < \alpha_r) \neq (\beta_1 < \beta_2 < \cdots < \beta_r)$.

The associative character, the distributive character with respect to the sum and the associative character with respect to the external product of the new product "$\bigwedge$" lead to the exterior dual algebra $\bigwedge_{n*}^{(r)}(K)$.

The anti-commutativity of the exterior dual vectors is analogous to the one already defined in Section 9.3.3.

9.4.3 Observation about bases of primary and dual exterior spaces

Since up to this moment we have developed two parallel theories in an implicit form, due to the notation, a comparative comment is necessary.

In principle, we have notated the linear space of the exterior covariant vectors (the forms) by means of $\bigwedge_{n*}^{(r)}(K)$. With this notation, we pretend to say that $\bigwedge_{n*}^{(r)}(K)$ is really the *dual* linear space in reciprocal or dual bases of $\bigwedge_{n}^{(r)}(K)$, the notation of which should be $[\bigwedge_{n}^{(r)}(K)]^*$.

On the other hand, if we see the exterior forms as homogeneous anti-symmetric dual tensors, then $\bigwedge_{*}^{(r)}(K)$ is the linear space $(\otimes V_*^n)_h^r(K)$, which is a part of $(\otimes V_*^n)^r(K)$. From this second point of view, the reader can ask himself: is $\bigwedge_{*}^{(r)}(K)$ a part of $(\otimes V_*^n)^r(K)$, dual and in reciprocal basis of $(\otimes V^n)^r(K)$, the correct notation of which should be $[(\otimes V^n)^r(K)]^*$?

In summary, we know that the bases of $(\otimes V^n)^r(K)$ and of the dual $(\otimes V_*^n)^r(K) \equiv [(\otimes V^n)^r(K)]^*$ are certainly reciprocal.

However, are the bases of $\bigwedge_{n}^{(r)}(K)$ and of $\bigwedge_{n*}^{(r)}(K)$ reciprocal or dual?. If this were true, then we should have

$$\text{basis of } \bigwedge_{n*}^{(r)}(K) \equiv \text{basis of } \left[\bigwedge_{n}^{(r)}(K)\right]^*.$$

If it were false, we should clarify what is the relation between these two bases.

For the sake of exposition clarity and consideration of uninitiated readers, we will define what we understand by "connection" of two dual linear spaces, together with the concept of "dual bases" or "reciprocal bases". We employ for these bases the second option, in order to avoid the risk of confusion that the word "dual" implies, because it is employed by many authors, for both dual spaces and dual bases, when these concepts are *strictly different*.

Definition 9.7 (Connection of two linear spaces). *Consider the basis of a linear space $V^n(K)$, the primal, notated $\{\vec{e}_\alpha\}$. Consider the basis of a linear space $V_*^n(K)$, the reciprocal of the previous one, notated $\{\vec{e}^{*\beta}\}$. We say that two linear spaces $V^n(K)$ and $W^n(K)$ are "connected" if we know all contracted products of the vectors of their respective bases:*

$$\vec{e}_\alpha \bullet \vec{e}_\beta = g_{\alpha\beta}; \quad g_{\alpha\beta} \in K; \quad \alpha, \beta \in I_n. \tag{9.55}$$

The components $g_{\alpha\beta}$ are usually given as data, and grouped in a matrix of order "n", called 'the "Gram matrix of the connection":

$$G \equiv G_n = [g_{\alpha\beta}]. \tag{9.56}$$

Definition 9.8 (Connection of primal and dual spaces). *We say that two linear spaces, the primal and the dual, are "connected" when we know their Gram matrix G. Other authors say that they are "dual linear spaces", though the "connection" G be not given. When nothing is said, one must understand that $G \equiv I_n$.*

Definition 9.9 (Reciprocal bases). *We say that two dual linear spaces are referred to in reciprocal bases, with respective notation $\{\vec{e}_\alpha\}$ and $\{\vec{e}^{*\beta}\}$ when the connection matrix G is I_n, that is,*

$$G = [g_{\alpha \circ}^{\circ \beta}] \equiv [\vec{e}_\alpha \bullet \vec{e}^{*\beta}] \equiv [\delta_{\alpha \circ}^{\circ \beta}] = I_n. \tag{9.57}$$

When two linear spaces, that are dual, are in "reciprocal" bases, the basis of the dual space has a reserved and specific notation $\{\vec{e}^{*\beta}\}$, for the reader to know that the matrix G is precisely I_n.

However, this convention *is not* in general *respected* and it is common to notate a basis of the dual space as $\{\vec{e}^{*\beta}\}$ though the connection G *be not* I_n, provoking the corresponding confusion. As most of the authors violate this convention, the authors of this book have also violated it in some of the previous chapters, while being aware of it.

Next, we extend and clarify the "dual" concept to the tensor spaces $(\otimes V^n)^r(K)$ and $(\otimes V_*^n)^r(K)$ that have already been introduced.

We have

$$(\vec{e}_{\gamma_1} \otimes \vec{e}_{\gamma_2} \otimes \cdots \otimes \vec{e}_{\gamma_r}) \bullet (\vec{e}^{*\eta_1} \otimes \vec{e}^{*\eta_2} \otimes \cdots \otimes \vec{e}^{*\eta_r}) = \delta_{A \circ}^{\circ B}, \tag{9.58}$$

where $A \equiv \{\gamma_1, \gamma_2, \ldots, \gamma_r\}$ is the set of the ordered subindices of the first factor and $B \equiv \{\eta_1, \eta_2, \ldots, \eta_r\}$ is the set of the ordered subindices of the second factor.

If the indices are separated with bars in the sets A and B to facilitate their comparison, we have

1. $A \equiv B \Leftrightarrow \{\gamma_1|\gamma_2|\cdots|\gamma_r\} \equiv \{\eta_1|\eta_2|\cdots|\eta_r\} \Leftrightarrow \gamma_i = \eta_i; \quad i \in I_r \Rightarrow \delta_{A \circ}^{\circ B} = 1.$

2. $A \neq B \Leftrightarrow \exists \gamma_i \neq \eta_i \Leftrightarrow \{\gamma_1|\gamma_2|\cdots|\gamma_r\} \not\equiv \{\eta_1|\eta_2|\cdots|\eta_r\} \Rightarrow \delta_{A \circ}^{\circ B} = 0.$

That is, $\delta_{A\,\circ}^{\,\circ\,B}$ behaves as a *classic* Kronecker delta, i.e., it takes only values 1 or 0.

To clarify this, we give one example. Let $n = 6; r = 4$.

If $\begin{cases} A \equiv \{\gamma_1|\gamma_2|\gamma_3|\gamma_4\} \equiv \{3|6|5|4\} \\ B \equiv \{\eta_1|\eta_2|\eta_3|\eta_4\} \equiv \{6|3|5|4\} \end{cases} \Rightarrow A \neq B \Rightarrow \delta_{A\,\circ}^{\,\circ\,B} = 0.$

If $\begin{cases} A \equiv \{\gamma_1|\gamma_2|\gamma_3|\gamma_4\} \equiv \{5|4|1|2\} \\ B \equiv \{\eta_1|\eta_2|\eta_3|\eta_4\} \equiv \{5|2|3|4\} \end{cases} \Rightarrow A \neq B \Rightarrow \delta_{A\,\circ}^{\,\circ\,B} = 0.$

If $\begin{cases} A \equiv \{\gamma_1|\gamma_2|\gamma_3|\gamma_4\} \equiv \{4|6|2|5\} \\ B \equiv \{\eta_1|\eta_2|\eta_3|\eta_4\} \equiv \{4|6|2|5\} \end{cases} \Rightarrow A \equiv B \Rightarrow \delta_{A\,\circ}^{\,\circ\,B} = 1.$

Contracting the factor tensors in (9.58) yields the product

$$\delta_{\gamma_1\,\circ}^{\,\circ\,\eta_1} \cdot \delta_{\gamma_2\,\circ}^{\,\circ\,\eta_2} \cdot \ldots \cdot \delta_{\gamma_r\,\circ}^{\,\circ\,\eta_r} \equiv \delta_{A\,\circ}^{\,\circ\,B}.$$

Thus, the given bases of the linear space $(\otimes V^n)^r(K)$ and the linear space $(\otimes V_*^n)^r(K)$ are reciprocal, and then we have dual linear spaces, in reciprocal bases. For this reason, Expression (9.58) has been formulated to be used in practice.

Next, we study the exterior spaces $\bigwedge_n^{(r)}(K)$ and $\bigwedge_{n*}^{(r)}(K)$, to determine if their respective bases are reciprocal.

Starting from Formula (9.15), which we apply with adequate index notation, we have

$$\vec{e}_{\alpha_1} \wedge \vec{e}_{\alpha_2} \wedge \cdots \wedge \vec{e}_{\alpha_r} = \delta_{(\alpha_1\alpha_2\cdots\alpha_r)}^{\gamma_1\gamma_2\cdots\gamma_r} \vec{e}_{\gamma_1} \otimes \vec{e}_{\gamma_2} \otimes \cdots \otimes \vec{e}_{\gamma_r}; \tag{9.59}$$

$$\alpha_i, \gamma_j \in I_n; \; i,j \in I_r; \; \gamma_i \neq \gamma_j; \; \alpha_1 < \alpha_2 < \cdots < \alpha_r.$$

Using now the correlative formula (9.50), the result is

$$\vec{e}^{*\beta_1} \wedge \vec{e}^{*\beta_2} \wedge \cdots \wedge \vec{e}^{*\beta_r} = \delta_{\eta_1\eta_2\cdots\eta_r}^{(\beta_1\beta_2\cdots\beta_r)} \vec{e}^{*\eta_1} \otimes \vec{e}^{*\eta_2} \otimes \cdots \otimes \vec{e}^{*\eta_r}; \; \beta_i, \eta_j \in I_n;$$

$$i,j \in I_r; \; \eta_i \neq \eta_j; \; \beta_1 < \beta_2 < \cdots < \beta_r. \tag{9.60}$$

First we determine the scalar $\rho \in K$:

$$\rho = (\vec{e}_{\alpha_1} \wedge \vec{e}_{\alpha_2} \wedge \cdots \wedge \vec{e}_{\alpha_r}) \bullet (\vec{e}^{*\beta_1} \wedge \vec{e}^{*\beta_2} \wedge \cdots \wedge \vec{e}^{*\beta_r}) \tag{9.61}$$

substituting into (9.61) expressions (9.59) and (9.60), we have

$$\rho = \left[\delta_{(\alpha_1\alpha_2\cdots\alpha_r)}^{\gamma_1\gamma_2\cdots\gamma_r} \vec{e}_{\gamma_1} \otimes \vec{e}_{\gamma_2} \otimes \cdots \otimes \vec{e}_{\gamma_r} \right] \bullet \left[\delta_{\eta_1\eta_2\cdots\eta_r}^{(\beta_1\beta_2\cdots\beta_r)} \vec{e}^{*\eta_1} \otimes \vec{e}^{*\eta_2} \otimes \cdots \otimes \vec{e}^{*\eta_r} \right]$$

$$= \left[\delta_{(\alpha_1\alpha_2\cdots\alpha_r)}^{\gamma_1\gamma_2\cdots\gamma_r} \bullet \delta_{\eta_1\eta_2\cdots\eta_r}^{(\beta_1\beta_2\cdots\beta_r)} \right] [(\vec{e}_{\gamma_1} \otimes \vec{e}_{\gamma_2} \otimes \cdots \otimes \vec{e}_{\gamma_r}) \bullet (\vec{e}^{*\eta_1} \otimes \vec{e}^{*\eta_2} \otimes \cdots \otimes \vec{e}^{*\eta_r})]$$

and considering Expression (9.58) leads to

$$\rho = \left[\delta_{(\alpha_1\alpha_2\cdots\alpha_r)}^{\gamma_1\gamma_2\cdots\gamma_r} \bullet \delta_{\eta_1\eta_2\cdots\eta_r}^{(\beta_1\beta_2\cdots\beta_r)} \right] \delta_{A\,\circ}^{\,\circ\,B}. \tag{9.62}$$

The factor that precedes $\delta_{A\,\circ}^{\circ\,B}$ in (9.62) carries the systems of indices $(\alpha_1, \alpha_2, \ldots, \alpha_r)$ and $(\beta_1, \beta_2, \ldots, \beta_r)$ both strictly ordered, and satisfying

$$\{\alpha_1, \alpha_1, \ldots, \alpha_r\} \equiv \{\gamma_1, \gamma_1, \ldots, \gamma_r\}; \quad \{\beta_1, \beta_1, \ldots, \beta_r\} \equiv \{\eta_1, \eta_1, \ldots, \eta_r\}.$$

If $\{\gamma_1 | \gamma_2 | \cdots | \gamma_r\} \neq \{\eta_1 | \eta_2 | \cdots | \eta_r\}$, then $\delta_{A\,\circ}^{\circ\,B}$ is *null*, and so they are *not* considered in Formula (9.62). Thus, the systems of useful indices are

$$\{\gamma_1 | \gamma_2 | \cdots | \gamma_r\} \equiv \{\eta_1 | \eta_2 | \cdots | \eta_r\} \Rightarrow (\alpha_1, \alpha_2, \ldots, \alpha_r) \equiv (\beta_1, \beta_2, \ldots, \beta_r).$$
$$\tag{9.63}$$

The equality of these sets implies $\delta_{A\,\circ}^{\circ\,B} = 1$, and then, expression (9.62) can be written as

$$\rho = \left(\delta_{(\alpha_1 \alpha_2 \cdots \alpha_r)}^{\gamma_1 \gamma_2 \cdots \gamma_r} \bullet \delta_{\gamma_1 \gamma_2 \cdots \gamma_r}^{(\alpha_1 \alpha_2 \cdots \alpha_r)} \right). \tag{9.64}$$

Considering that $\forall \gamma_i \neq \gamma_j$, there are only $r!$ sets $(\gamma_1, \gamma_2, \cdots, \gamma_r)$ because they are the permutations of the totally ordered set $(\alpha_1, \alpha_2, \cdots, \alpha_r)$.

The two generalized Kronecker deltas $\delta_{(\alpha_1 \alpha_2 \cdots \alpha_r)}^{\gamma_1 \gamma_2 \cdots \gamma_r}$ and $\delta_{\gamma_1 \gamma_2 \cdots \gamma_r}^{(\alpha_1 \alpha_2 \cdots \alpha_r)}$ have the same value and sign.

Then, the contraction $\delta_{(\alpha_1 \alpha_2 \cdots \alpha_r)}^{\gamma_1 \gamma_2 \cdots \gamma_r} \bullet \delta_{\gamma_1 \gamma_2 \cdots \gamma_r}^{(\alpha_1 \alpha_2 \cdots \alpha_r)}$ has $r!$ summands of value 1, which implies that (9.64) can be written as

$$\rho = \left(\delta_{(\alpha_1 \alpha_2 \cdots \alpha_r)}^{\gamma_1 \gamma_2 \cdots \gamma_r} \bullet \delta_{\gamma_1 \gamma_2 \cdots \gamma_r}^{(\alpha_1 \alpha_2 \cdots \alpha_r)} \right) = r!, \tag{9.65}$$

so that $\rho = r!$ ($r! \geq 2$ because we always have at least two indices).

The consequence is that $\rho \neq \delta_{A\,\circ}^{\circ\,B}$.

Finally, considering (9.61) and (9.65) we conclude

$$\rho = (\vec{e}_{\alpha_1} \wedge \vec{e}_{\alpha_2} \wedge \cdots \wedge \vec{e}_{\alpha_r}) \bullet (\vec{e}^{*\beta_1} \wedge \vec{e}^{*\beta_2} \wedge \cdots \wedge \vec{e}^{*\beta_r}) = r!. \tag{9.66}$$

As the contracted product of the left-hand side of (9.66) does not satisfy (9.58), that is, is different from $\delta_{A\,\circ}^{\circ\,B}$, the contracted bases in (9.66) *are not* reciprocal.

In summary: $\bigwedge_n^{(r)}(K)$ and $\bigwedge_{n*}^{(r)}(K)$ are dual spaces of exterior vectors, but their respective bases *are not reciprocal*: their Gram matrix G is not $I_{\sigma'}$, but

$$G \equiv \begin{bmatrix} r! & 0 & \cdots & 0 \\ 0 & r! & \cdots & 0 \\ \cdots & \cdots & \cdots & \cdots \\ 0 & 0 & \cdots & r! \end{bmatrix}_{\sigma'} \quad \text{and} \quad \sigma' = \binom{n}{r}. \tag{9.67}$$

9.5 The change-of-basis in exterior algebras

We shall proceed to establish how the components of an exterior vector in the linear space $\bigwedge_n^{(r)}(K)$ change bases or how an exterior form in $\bigwedge_{n*}^{(r)}(K)$ changes its basis, and two strict algebraic relations will appear, that *when satisfied* over a system of scalars $s(\alpha_1, \alpha_2, \ldots, \alpha_r)$ with strict notation $\alpha_1 < \alpha_2 < \cdots < \alpha_r$, are useful as necessary and sufficient conditions to establish its exterior tensor nature and then, that such a system must be notated either as $t^{(\alpha_1 \alpha_2 \cdots \alpha_r)}_{\circ \ \circ \ \cdots \ \circ}$,

or as $t_{(\alpha_1 \alpha_2 \cdots \alpha_r)}^{\circ \ \circ \ \cdots \ \circ}$, because it is a system of contravariant or covariant exterior vectors.

It will also be observed that the cited algebraic relations *do not have* a tensor construction, absolute or modular, as was already indicated in Section 7.4.3 close to Formula (7.55).

However, it is sufficient to express any exterior vector as a homogeneous anti-symmetric tensor and change its basis, for it to declare without any doubt its tensor character. Because of this, as *it is a priori known* that they are tensors, many authors call the vectors in $\bigwedge_n^{(r)}(K)$ and $\bigwedge_{n*}^{(r)}(K)$ "exterior tensors".

9.5.1 Strict tensor relationships for $\bigwedge_n^{(r)}(K)$ algebras

It is known that the tensor notation of change-of-basis in $V^n(K)$ is $\vec{e}_i = c^{\alpha \circ}_{\circ i} \vec{e}_\alpha$, where $C = [c^{\alpha \circ}_{\circ i}]$ is the change-of-basis matrix, and that $C^{-1} = [\gamma^{i \circ}_{\circ \alpha}]$ and $C^t = [c^{\circ \alpha}_{i \circ}]$.

According to what has been established, an exterior vector $T \in \bigwedge_n^{(r)}(K)$ is given depending on whether it is referred to the initial or the new basis, by the expressions

$$T = t^{(\alpha_1 \alpha_2 \cdots \alpha_r)}_{\circ \ \circ \ \cdots \ \circ} \vec{e}_{\alpha_1} \wedge \vec{e}_{\alpha_2} \wedge \cdots \wedge \vec{e}_{\alpha_r}; \quad \alpha_1 < \alpha_2 < \cdots < \alpha_r \tag{9.68}$$

or

$$T = t^{(i_1 i_2 \cdots i_r)}_{\circ \ \circ \ \cdots \ \circ} \vec{e}_{i_1} \wedge \vec{e}_{i_2} \wedge \cdots \wedge \vec{e}_{i_r}; \quad i_1 < i_2 < \cdots < i_r, \tag{9.69}$$

respectively. Replacing in Expression (9.68) the initial basis by the new one, using the tensor equation of the change-of-basis $\vec{e}_\alpha = \gamma^{i \circ}_{\circ \alpha} \vec{e}_i$, we have

$$T = t^{(\alpha_1 \alpha_2 \cdots \alpha_r)}_{\circ \ \circ \ \cdots \ \circ} (\gamma^{i_1 \circ}_{\circ \alpha_1} \vec{e}_{i_1}) \wedge (\gamma^{i_2 \circ}_{\circ \alpha_2} \vec{e}_{i_2}) \wedge \cdots \wedge (\gamma^{i_r \circ}_{\circ \alpha_r} \vec{e}_{i_r}).$$

When we distribute the parentheses and locate the basic vectors in the tail, then they *are not in strict order*:

$$T = t^{(\alpha_1 \alpha_2 \cdots \alpha_r)}_{\circ \ \circ \ \cdots \ \circ} (\gamma^{i'_1 \circ}_{\circ \alpha_1} \gamma^{i'_2 \circ}_{\circ \alpha_2} \cdots \gamma^{i'_r \circ}_{\circ \alpha_r}) \vec{e}_{i'_1} \wedge \vec{e}_{i'_2} \wedge \cdots \wedge \vec{e}_{i'_r}. \tag{9.70}$$

However, as in reality in Formula (9.70) the subindices (i'_j) have no values in strict order, we group the basic terms by strict components, to obtain

$$T = t^{(\alpha_1 \alpha_2 \cdots \alpha_r)}_{\circ \ \circ \ \cdots \ \circ} (\gamma^{i'_1 \circ}_{\circ \alpha_1} \gamma^{i'_2 \circ}_{\circ \alpha_2} \cdots \gamma^{i'_r \circ}_{\circ \alpha_r}) (\delta^{(i_1 i_1 \cdots i_r)}_{i'_1 i'_1 \cdots i'_r} \vec{e}_{i_1} \wedge \vec{e}_{i_2} \wedge \cdots \wedge \vec{e}_{i_r})$$

or

$$T = t^{(\alpha_1 \alpha_2 \cdots \alpha_r)}_{\circ\;\circ\;\cdots\;\circ} (\delta^{(i_1 i_1 \cdots i_r)}_{i'_1 i'_1 \cdots i'_r} \gamma^{i'_1 \;\circ}_{\;\circ\;\alpha_1} \gamma^{i'_2 \;\circ}_{\;\circ\;\alpha_2} \cdots \gamma^{i'_r \;\circ}_{\;\circ\;\alpha_r}) \vec{e}_{i_1} \wedge \vec{e}_{i_2} \wedge \cdots \wedge \vec{e}_{i_r}. \quad (9.71)$$

The parentheses declare the development of a determinant with rows $(i_1, i_2, \ldots, i_r)$ and columns $(\alpha_1, \alpha_2, \ldots, \alpha_r)$ both in strict order. Thus, expression (9.71) can be written as

$$T = t^{(\alpha_1 \alpha_2 \cdots \alpha_r)}_{\circ\;\circ\;\cdots\;\circ} \begin{vmatrix} \gamma^{i_1 \;\circ}_{\;\circ\;\alpha_1} & \gamma^{i_1 \;\circ}_{\;\circ\;\alpha_2} & \cdots & \gamma^{i_1 \;\circ}_{\;\circ\;\alpha_r} \\ \gamma^{i_2 \;\circ}_{\;\circ\;\alpha_1} & \gamma^{i_2 \;\circ}_{\;\circ\;\alpha_2} & \cdots & \gamma^{i_2 \;\circ}_{\;\circ\;\alpha_r} \\ \cdots & \cdots & \cdots & \cdots \\ \gamma^{i_r \;\circ}_{\;\circ\;\alpha_1} & \gamma^{i_r \;\circ}_{\;\circ\;\alpha_2} & \cdots & \gamma^{i_r \;\circ}_{\;\circ\;\alpha_r} \end{vmatrix} \vec{e}_{i_1} \wedge \vec{e}_{i_2} \wedge \cdots \wedge \vec{e}_{i_r}, \quad (9.72)$$

that compared with (9.69), permits us to establish

$$t^{(i_1 i_2 \cdots i_r)}_{\circ\;\circ\;\cdots\;\circ} = t^{(\alpha_1 \alpha_2 \cdots \alpha_r)}_{\circ\;\circ\;\cdots\;\circ} \begin{vmatrix} \gamma^{i_1 \;\circ}_{\;\circ\;\alpha_1} & \gamma^{i_1 \;\circ}_{\;\circ\;\alpha_2} & \cdots & \gamma^{i_1 \;\circ}_{\;\circ\;\alpha_r} \\ \gamma^{i_2 \;\circ}_{\;\circ\;\alpha_1} & \gamma^{i_2 \;\circ}_{\;\circ\;\alpha_2} & \cdots & \gamma^{i_2 \;\circ}_{\;\circ\;\alpha_r} \\ \cdots & \cdots & \cdots & \cdots \\ \gamma^{i_r \;\circ}_{\;\circ\;\alpha_1} & \gamma^{i_r \;\circ}_{\;\circ\;\alpha_2} & \cdots & \gamma^{i_r \;\circ}_{\;\circ\;\alpha_r} \end{vmatrix}, \quad (9.73)$$

which is the strict tensor relation for changes of bases in exterior algebras $\bigwedge_n^{(r)}(K)$, fulfilling the purpose of this section. The tensor T has a total of $\binom{n}{r}$ summands of the indicated type in its Formula (9.72). The determinant of (9.72) and (9.73) is the minor with rows $(i_1, i_2, \ldots, i_r)$, and columns $(\alpha_1, \alpha_2, \ldots, \alpha_r)$ of the matrix C^{-1}.

Obviously, if one wants to express the initial components of T in terms of the new components, we initiate the development by substituting the direct change-of-basis $\vec{e}_i = c^{\alpha\circ}_{\;\circ\;i} \vec{e}_\alpha$ into Expression (9.69) and it is executed in an analogous form, leading to

$$t^{(\alpha_1 \alpha_2 \cdots \alpha_r)}_{\circ\;\circ\;\cdots\;\circ} = t^{(i_1 i_2 \cdots i_r)}_{\circ\;\circ\;\cdots\;\circ} \begin{vmatrix} c^{\circ\;\alpha_1}_{i_1 \;\circ} & c^{\circ\;\alpha_2}_{i_1 \;\circ} & \cdots & c^{\circ\;\alpha_r}_{i_1 \;\circ} \\ c^{\circ\;\alpha_1}_{i_2 \;\circ} & c^{\circ\;\alpha_2}_{i_2 \;\circ} & \cdots & c^{\circ\;\alpha_r}_{i_2 \;\circ} \\ \cdots & \cdots & \cdots & \cdots \\ c^{\circ\;\alpha_1}_{i_r \;\circ} & c^{\circ\;\alpha_2}_{i_r \;\circ} & \cdots & c^{\circ\;\alpha_r}_{i_r \;\circ} \end{vmatrix}, \quad (9.74)$$

in which the determinant is the minor with columns $(\alpha_1, \alpha_2, \ldots, \alpha_r)$ and rows $(i_1, i_2, \ldots, i_r)$ of the matrix C^t, since this was the meaning of such indices in Formula (9.74).

9.5.2 Strict tensor relationships for $\bigwedge_{n*}^{(r)}(\mathbb{R})$ algebras

In an analogous form to that employed to define the exterior algebras $\bigwedge_n^{(r)}(K)$, from the totally contravariant and anti-symmetric tensors, we have established from the totally covariant and anti-symmetric tensors the "forms" exterior algebras $\bigwedge_{n*}^{(r)}(K)$:

$$T = t_{(\alpha_1 \alpha_2 \cdots \alpha_r)}^{\circ \; \circ \; \cdots \; \circ} \bar{e}^{*\alpha_1} \wedge \bar{e}^{*\alpha_2} \wedge \cdots \wedge \bar{e}^{*\alpha_r}, \tag{9.75}$$

where $T \in \bigwedge_{n*}^{(r)}(K)$ is expressed in the initial basis. If it were expressed in the new basis, we would have

$$T = t_{(i_1 i_2 \cdots i_r)}^{\circ \; \circ \; \cdots \; \circ} \bar{e}^{*i_1} \wedge \bar{e}^{*i_2} \wedge \cdots \wedge \bar{e}^{*i_r}. \tag{9.76}$$

A change-of-basis $\vec{e}_i = c_{\circ\,i}^{\alpha\circ} \vec{e}_\alpha$ in $V^n(K)$ produces in the dual linear space $V_*^n(K)$ the change $\bar{e}^{*i} = \gamma_{\alpha\circ}^{\circ\,i} \bar{e}^{*\alpha}$, and obtaining $\bar{e}^{*\alpha}$ from it, we have

$$\bar{e}^{*\alpha} = c_{i\,\circ}^{\circ\,\alpha} \bar{e}^{*i}, \tag{9.77}$$

where $C^t = [c_{i\,\circ}^{\circ\,\alpha}]$.

Substituting (9.77) into (9.75) we obtain

$$T = t_{(\alpha_1 \alpha_2 \cdots \alpha_r)}^{\circ \; \circ \; \cdots \; \circ} (c_{i_1 \,\circ}^{\circ\,\alpha_1} \bar{e}^{*i_1}) \wedge (c_{i_2 \,\circ}^{\circ\,\alpha_2} \bar{e}^{*i_2}) \wedge \cdots \wedge (c_{i_r \,\circ}^{\circ\,\alpha_r} \bar{e}^{*i_r}), \tag{9.78}$$

which operated in an analogous manner, and taking into account the change of ordering of the dummy indices leads to

$$T = t_{(\alpha_1 \alpha_2 \cdots \alpha_r)}^{\circ \; \circ \; \cdots \; \circ} (c_{i_1 \,\circ}^{\circ\,\alpha_1} c_{i_2 \,\circ}^{\circ\,\alpha_2} \cdots c_{i_r \,\circ}^{\circ\,\alpha_r}) \left(\delta_{i_1 i_2 \cdots i_r}^{i_1' i_2' \cdots i_r'} \bar{e}^{*i_1} \wedge \bar{e}^{*i_2} \wedge \cdots \wedge \bar{e}^{*i_r} \right). \tag{9.79}$$

Finally, we have a similar expression to that for the contravariant case, i.e.,

$$T = t_{(\alpha_1 \alpha_2 \cdots \alpha_r)}^{\circ \; \circ \; \cdots \; \circ} \begin{vmatrix} c_{i_1 \,\circ}^{\circ\,\alpha_1} & c_{i_1 \,\circ}^{\circ\,\alpha_1} & \cdots & c_{i_1 \,\circ}^{\circ\,\alpha_1} \\ c_{i_2 \,\circ}^{\circ\,\alpha_1} & c_{i_2 \,\circ}^{\circ\,\alpha_1} & \cdots & c_{i_2 \,\circ}^{\circ\,\alpha_1} \\ \cdots & \cdots & \cdots & \cdots \\ c_{i_r \,\circ}^{\circ\,\alpha_1} & c_{i_r \,\circ}^{\circ\,\alpha_1} & \cdots & c_{i_r \,\circ}^{\circ\,\alpha_1} \end{vmatrix} \bar{e}^{*i_1} \wedge \bar{e}^{*i_2} \wedge \cdots \wedge \bar{e}^{*i_r} \tag{9.80}$$

with $\alpha_1 < \alpha_2 < \cdots < \alpha_r$ e $i_1 < i_2 < \cdots < i_r$.

The determinant of Formula (9.80) is the minor of rows $(i_1, i_2, \cdots, i_r)$ and columns $(\alpha_1, \alpha_2, \cdots, \alpha_r)$ of the matrix C^t. Comparing (9.80) with (9.76), we obtain

$$t_{(i_1 i_2 \cdots i_r)}^{\circ \; \circ \; \cdots \; \circ} = t_{(\alpha_1 \alpha_2 \cdots \alpha_r)}^{\circ \; \circ \; \cdots \; \circ} \begin{vmatrix} c_{i_1 \,\circ}^{\circ\,\alpha_1} & c_{i_1 \,\circ}^{\circ\,\alpha_2} & \cdots & c_{i_1 \,\circ}^{\circ\,\alpha_r} \\ c_{i_2 \,\circ}^{\circ\,\alpha_1} & c_{i_2 \,\circ}^{\circ\,\alpha_2} & \cdots & c_{i_2 \,\circ}^{\circ\,\alpha_r} \\ \cdots & \cdots & \cdots & \cdots \\ c_{i_r \,\circ}^{\circ\,\alpha_1} & c_{i_r \,\circ}^{\circ\,\alpha_2} & \cdots & c_{i_r \,\circ}^{\circ\,\alpha_r} \end{vmatrix}, \tag{9.81}$$

which is the strict tensor relation for changes of bases in exterior dual algebras $\bigwedge_{n*}^{(r)}(K)$, fulfilling the purpose of this section. Tensor T has a total of $\binom{n}{r}$ summands of the indicated type in Formula (9.80).

9.6 Complements of contramodular and comodular scalars

In the case $r = n$, the exterior algebra $\bigwedge_n^{(n)}(K)$ consists of exterior vectors of a unique strict component

$$T = \tau \vec{e}_1 \wedge \vec{e}_1 \wedge \cdots \wedge \vec{e}_n \equiv t_{\circ\circ\cdots\circ}^{(12\cdots n)} \vec{e}_1 \wedge \vec{e}_2 \wedge \cdots \wedge \vec{e}_n \tag{9.82}$$

or, if it is written as a contravariant anti-symmetric tensor $T \in (\otimes V^n)^n(K)$:

$$T = t_{\circ\circ\cdots\circ}^{(12\cdots n)} \delta_{1 \ 2 \ \cdots \ n}^{(\alpha_1 \alpha_2 \cdots \alpha_n)} \vec{e}_{\alpha_1} \wedge \vec{e}_{\alpha_2} \wedge \cdots \wedge \vec{e}_{\alpha_n}; \tag{9.83}$$

$$\tau = t_{\circ\circ\cdots\circ}^{(12\cdots n)}; \quad \alpha_i \in I_n; \quad \alpha_i \neq \alpha_j; \quad i \neq j,$$

with $n!$ non-null components, half of them of value $(+\tau)$ and the other half of value $(-\tau)$.

The change-of-basis in this model of exterior algebra, $\bigwedge_n^{(n)}(K)$, the entities of which correspond to the expression

$$T = \hat{\tau} \vec{\hat{e}}_1 \wedge \vec{\hat{e}}_2 \wedge \cdots \wedge \vec{\hat{e}}_n \equiv \hat{t}_{\circ\circ\cdots\circ}^{(12\cdots n)} \vec{\hat{e}}_1 \wedge \vec{\hat{e}}_2 \wedge \cdots \wedge \vec{\hat{e}}_n; \quad \tau \equiv \hat{t}_{\circ\circ\cdots\circ}^{(12\cdots n)}$$

is stated as follows.

Because of (9.73), its relation with the initial (τ) is

$$\hat{\tau} = \hat{t}_{\circ\circ\cdots\circ}^{(12\cdots n)} = t_{\circ\circ\cdots\circ}^{(12\cdots n)} \begin{vmatrix} \gamma_{\circ\alpha_1}^{i_1\circ} & \gamma_{\circ\alpha_2}^{i_1\circ} & \cdots & \gamma_{\circ\alpha_n}^{i_1\circ} \\ \gamma_{\circ\alpha_1}^{i_2\circ} & \gamma_{\circ\alpha_2}^{i_2\circ} & \cdots & \gamma_{\circ\alpha_n}^{i_2\circ} \\ \cdots & \cdots & \cdots & \cdots \\ \gamma_{\circ\alpha_1}^{i_n\circ} & \gamma_{\circ\alpha_2}^{i_n\circ} & \cdots & \gamma_{\circ\alpha_n}^{i_n\circ} \end{vmatrix} = t_{\circ\circ\cdots\circ}^{(12\cdots n)} |C|^{-1}$$

and in abbreviated form

$$\hat{\tau} = \frac{1}{|C|}\tau. \tag{9.84}$$

We observe that the expression for the change-of-basis of the unique component of these exterior totally contravariant vectors is *the same* as that of a modular tensor of zero order and weight $w = -1$, called *contramodular scalar*" or "*scalar capacity*"; it is the case of determinants Δ_n in contravariant coordinates.

Similarly, in the exterior algebra $\bigwedge_{n*}^{(r)}(K)$ model with exterior forms of a unique component

$$T = \tau \vec{e}^{*1} \wedge \vec{e}^{*2} \wedge \cdots \wedge \vec{e}^{*n} \equiv t_{(12\cdots n)}^{\circ\circ\cdots\circ} \vec{e}^{*1} \wedge \vec{e}^{*2} \wedge \cdots \wedge \vec{e}^{*n}, \tag{9.85}$$

in the new basis they are written as

$$T = \hat{\tau} \vec{\hat{e}}^{*1} \wedge \vec{\hat{e}}^{*2} \wedge \cdots \wedge \vec{\hat{e}}^{*n} \equiv \hat{t}_{(12\cdots n)}^{\circ\circ\cdots\circ} \vec{\hat{e}}^{*1} \wedge \vec{\hat{e}}^{*2} \wedge \cdots \wedge \vec{\hat{e}}^{*n},$$

which is related to the initial basis by the relation (9.81), applied to the case $r = n$:

$$\hat{\tau} = \hat{t}\,^{\circ\circ\cdots\circ}_{(12\cdots n)} = t\,^{\circ\circ\cdots\circ}_{(12\cdots n)} \begin{vmatrix} c_{i_1\;\circ}^{\circ\;\alpha_1} & c_{i_1\;\circ}^{\circ\;\alpha_2} & \cdots & c_{i_1\;\circ}^{\circ\;\alpha_n} \\ c_{i_2\;\circ}^{\circ\;\alpha_1} & c_{i_2\;\circ}^{\circ\;\alpha_2} & \cdots & c_{i_2\;\circ}^{\circ\;\alpha_n} \\ \cdots & \cdots & \cdots & \cdots \\ c_{i_n\;\circ}^{\circ\;\alpha_1} & c_{i_n\;\circ}^{\circ\;\alpha_2} & \cdots & c_{i_n\;\circ}^{\circ\;\alpha_t n} \end{vmatrix} = t\,^{\circ\circ\cdots\circ}_{(12\cdots n)}|C^t|$$

and in abbreviated form

$$\hat{\tau} = |C|\tau, \tag{9.86}$$

which gives the change-of-basis for the totally covariant exterior forms. This expression also corresponds to that of a modular tensor of zero order and weight $w = 1$, called a *comodular scalar* or *scalar density*; this is the case of determinants Δ_n with columns in covariant coordinates.

9.7 Comparative tables of algebra correspondences

We establish in this section a comparison between the anti-symmetric tensor spaces $(\otimes V^n)^r_h(K), (\otimes V^n_*)^r_h(K)$ and the respective exterior algebras $\bigwedge_n^{(r)}(K)$, $\bigwedge_{n*}^{(r)}(K)$, which is shown in Table 9.1.

9.8 Scalar mappings: Exterior contractions

Let

$$T = t^{(\alpha_1\alpha_2\cdots\alpha_r)}_{\;\circ\;\circ\;\cdots\;\circ}\,\vec{e}_{\alpha_1} \wedge \vec{e}_{\alpha_2} \wedge \cdots \wedge \vec{e}_{\alpha_r} \tag{9.87}$$

be an arbitrary exterior vector $T \in \bigwedge_n^{(r)}(K)$, built over the linear space $V^n(K)$.

Let

$$F = f^{\;\circ\;\;\circ\;\cdots\;\;\circ}_{(\alpha_1\alpha_2\cdots\alpha_r)}\,\vec{e}^{*\alpha_1} \wedge \vec{e}^{*\alpha_2} \wedge \cdots \wedge \vec{e}^{*\alpha_r} \tag{9.88}$$

be an arbitrary exterior form, $F \in \bigwedge_{n*}^{(r)}(K)$, built over $V_*^n(K)$, the dual linear space that is assumed to be in *reciprocal basis* of that of $V^n(K)$.

In both cases it is $\alpha_1 < \alpha_2 < \cdots < \alpha_r; \alpha_i \in I_n$.

We assume that we want to know the scalar resulting from the contracted product of both entities. We perform the contraction by using two different procedures:

Procedure (a) (as if they were tensors in dual bases). The scheme to be followed is: We proceed to the "extension" of the exterior vector and the form. Remembering that the dimension of $\bigwedge_n^{(r)}(K)$ and of $[\bigwedge_n^{(r)}(K)]^*$, the dual of the previous linear space, is $\sigma = \binom{n}{r}$ we have:

Table 9.1. Comparative tables of correspondences

Concept	$(\otimes V^n)_h^r(K)$	$\bigwedge_n^{(r)}(K)$
Algebraic structure	Primal linear space of dimension $\sigma = n^r$	Linear space of dimension $\sigma' = \binom{n}{r}$
Entities in it	Its tensors are homogeneous tensors of order r contravariant and *anti-symmetric* over $V^n(K)$	Its vectors are called "exterior vectors" and are the tensors on the left but written in another basis and by "strict" components
Product of vectors	The tensor product of vectors $(V_2 \otimes V_1 \neq -V_1 \otimes V_2)$; $T = V_1 \otimes V_2 \otimes \cdots \otimes V_r$. Not all $T \in (\otimes V^n)_h^r(K)$ can be written as a product	The exterior product of vectors $(V_2 \wedge V_1 = -V_1 \wedge V_2)$; $T = V_1 \wedge V_2 \wedge \cdots \wedge V_r$. Not all $T \in \bigwedge_n^{(r)}(K)$ can be written as a product
Basis	The basis associated with $\{\vec{e}_\alpha\}$ of $V^n(K)$: $\{\vec{e}_{\alpha_1} \otimes \vec{e}_{\alpha_2} \otimes \cdots \otimes \vec{e}_{\alpha_r}\}$ $\alpha_i \in I_n$; $i \in I_r$ $\alpha_1 \neq \alpha_2 \neq \cdots \neq \alpha_r$	The basis associated with $\{\vec{e}_\alpha\}$ of $V^n(K)$: $\{\vec{e}_{\alpha_1} \wedge \vec{e}_{\alpha_2} \wedge \cdots \wedge \vec{e}_{\alpha_r}\}$ $\alpha_i \in I_n$; $i \in I_r$ $\alpha_1 < \alpha_2 < \cdots < \alpha_r$
Change of basis relation	Tensor equation for *anti-symmetric* totally contravariant tensors: $$t^{\overset{i_1 i_2 \cdots i_r}{\circ\ \circ\ \cdots\ \circ}} = t^{\overset{\alpha_1 \alpha_2 \cdots \alpha_r}{\circ\ \circ\ \cdots\ \circ}} \gamma_{\overset{\circ\ \alpha_1}{i_1\ \circ}} \gamma_{\overset{\circ\ \alpha_2}{i_2\ \circ}} \cdots \gamma_{\overset{\circ\ \alpha_r}{i_r\ \circ}}$$ $\alpha_i, i_j \in I_n; i,j \in I_r$; $i_1 \neq i_2 \neq \cdots \neq i_r$; $\alpha_1 \neq \alpha_2 \neq \cdots \neq \alpha_r$	Tensor relation for exterior vectors: $$t^{\overset{(i_1 i_2 \cdots i_r)}{\circ\ \circ\ \cdots\ \circ}} =$$ $$t^{\overset{(\alpha_1 \alpha_2 \cdots \alpha_r)}{\circ\ \circ\ \cdots\ \circ}} \begin{vmatrix} \gamma_{\overset{i_1\ \circ}{\circ\ \alpha_1}} & \gamma_{\overset{i_1\ \circ}{\circ\ \alpha_2}} & \cdots & \gamma_{\overset{i_1\ \circ}{\circ\ \alpha_r}} \\ \gamma_{\overset{i_2\ \circ}{\circ\ \alpha_1}} & \gamma_{\overset{i_2\ \circ}{\circ\ \alpha_2}} & \cdots & \gamma_{\overset{i_2\ \circ}{\circ\ \alpha_r}} \\ \cdots & \cdots & \cdots & \cdots \\ \gamma_{\overset{i_r\ \circ}{\circ\ \alpha_1}} & \gamma_{\overset{i_r\ \circ}{\circ\ \alpha_2}} & \cdots & \gamma_{\overset{i_r\ \circ}{\circ\ \alpha_r}} \end{vmatrix}$$ $\alpha_i, i_j \in I_n$; $i_1 < i_2 < \cdots < i_r$; $\alpha_1 < \alpha_2 < \cdots < \alpha_r$

Concept	$(\otimes V_*^n)_h^r(K)$	$\bigwedge_{n*}^{(r)}(K)$
Algebraic structure	Dual linear space of dimension $\sigma = n^r$	Dual linear space of dimension $\sigma' = \binom{n}{r}$
Entities in it	Its vectors are homogeneous covariant and *anti-symmetric* tensors of order r over $V_*^n(K)$	Its vectors are called "dual exterior vectors" and also "exterior forms" and are the vectors on the left but written in another basis and by "strict" components
Product of vectors	The tensor product of vectors $T = V_1^* \otimes V_2^* \otimes \cdots \otimes V_r^*$ with $V_2^* \otimes V_1^* \neq -V_1^* \otimes V_2^*$. Not all $T \in (\otimes V_*^n)_h^r(K)$ can be written as a product	The dual exterior product of vectors $T = V_1^* \wedge V_2^* \wedge \cdots \wedge V_r^*$ with $V_2^* \wedge V_1^* = -V_1^* \wedge V_2^*$. Not all $T \in \bigwedge_{n*}^{(r)}(K)$ can be written as a product
Basis	Basis associated with $\{\vec{e}_\alpha\}$ of $V_*^n(K)$: $\{\vec{e}^{*\alpha_1} \otimes \vec{e}^{*\alpha_2} \otimes \cdots \otimes \vec{e}^{*\alpha_r}\}$ $\alpha_i \in I_n$; $i \in I_r$; $\alpha_1 \neq \alpha_2 \neq \cdots \neq \alpha_r$	Basis associated with $\{\vec{e}^{*\alpha}\}$ of $V_*^n(K)$: $\{\vec{e}^{*\alpha_1} \wedge \vec{e}^{*\alpha_2} \wedge \cdots \wedge \vec{e}^{*\alpha_r}\}$; $\alpha_i \in I_n$; $i \in I_r$; $\alpha_1 < \alpha_2 < \cdots < \alpha_r$
Change of basis relation	Tensor equation for *anti-symmetric* totally covariant tensors: $$t_{\overset{\circ\ \circ\ \cdots\ \circ}{i_1 i_2 \cdots i_r}} = t_{\overset{\circ\ \circ\ \cdots\ \circ}{\alpha_1 \alpha_2 \cdots \alpha_r}} c_{\overset{\circ\ \alpha_1}{i_1\ \circ}} c_{\overset{\circ\ \alpha_2}{i_2\ \circ}} \cdots c_{\overset{\circ\ \alpha_r}{i_r\ \circ}}$$ $\alpha_i, i_j \in I_n; i,j \in I_r$; $i_1 \neq i_2 \neq \cdots \neq i_r$; $\alpha_1 \neq \alpha_2 \neq \cdots \neq \alpha_r$	Tensor relation for exterior vectors: $$t_{\overset{\circ\ \circ\ \cdots\ \circ}{(i_1 i_2 \cdots i_r)}} =$$ $$t_{\overset{\circ\ \circ\ \cdots\ \circ}{(\alpha_1 \alpha_2 \cdots \alpha_r)}} \begin{vmatrix} c_{\overset{\circ\ \alpha_1}{i_1\ \circ}} & c_{\overset{\circ\ \alpha_2}{i_1\ \circ}} & \cdots & c_{\overset{\circ\ \alpha_r}{i_1\ \circ}} \\ c_{\overset{\circ\ \alpha_1}{i_2\ \circ}} & c_{\overset{\circ\ \alpha_2}{i_2\ \circ}} & \cdots & c_{\overset{\circ\ \alpha_r}{i_2\ \circ}} \\ c_{\overset{\circ\ \alpha_1}{i_r\ \circ}} & c_{\overset{\circ\ \alpha_2}{i_r\ \circ}} & \cdots & c_{\overset{\circ\ \alpha_r}{i_r\ \circ}} \end{vmatrix}$$ $\alpha_i, i_j \in I_n; i_1 < i_2 < \cdots < i_r$; $\alpha_1 < \alpha_2 < \cdots < \alpha_r$

(a) For T we start from $T \in \bigwedge_n^{(r)}(K)$ in the basis $\{\vec{e}_{\alpha_1} \wedge \vec{e}_{\alpha_2} \wedge \cdots \wedge \vec{e}_{\alpha_r}\}$, Formula (9.87), and we arrive at $T \in W^\sigma(K)$, represented as $t^{(\alpha_1 \alpha_2 \cdots \alpha_r)} \vec{e}_{\overline{\alpha_1 \alpha_2 \cdots \alpha_r}}$ in the basis of $W^\sigma(K)$ (already stretched).

(b) For F, we start from $F \in [\bigwedge_n^{(r)}(K)]^*$ in the basis $\{\bar{e}^{*\alpha_1} \wedge \bar{e}^{*\alpha_2} \wedge \cdots \wedge \bar{e}^{*\alpha_r}\}$ and we arrive at $F \in W^{*\sigma}(K)$, represented as $f_{(\alpha_1\alpha_2\cdots\alpha_r)} \bar{e}^{* \overline{\alpha_1\alpha_2\cdots\alpha_r}}$ (already stretched) in the basis of $W^{*\sigma}(K)$.

Since we are in the case of two normal dual spaces $W^\sigma(K)$ and $W^{*\sigma}(K)$, we assume that they *are* in reciprocal bases $\{\bar{e}_{\overline{\alpha_1\alpha_2\cdots\alpha_r}}\}$ and $\{\bar{e}^{* \overline{\alpha_1\alpha_2\cdots\alpha_r}}\}$, so that the connection matrix is $G_\sigma \equiv I_\sigma$; thus, we proceed to obtain the dot product of their vectors, which will be notated with caution as $F(T)$ (the form of the vector) and will be called the "*value of the form*", that is,

$$F(T) = f_{(\alpha_1\alpha_2\cdots\alpha_r)}{}^{\circ\;\circ\;\cdots\;\circ}\, \delta^{(\alpha_1\alpha_2\cdots\alpha_r)}_{\beta_1\beta_2\cdots\beta_r}\, t^{(\beta_1\beta_2\cdots\beta_r)}{}_{\circ\;\circ\;\cdots\;\circ}$$

and then

$$F(T) = f_{(\alpha_1\alpha_2\cdots\alpha_r)}{}^{\circ\;\circ\;\cdots\;\circ}\, t^{(\alpha_1\alpha_2\cdots\alpha_r)}{}_{\circ\;\circ\;\cdots\;\circ} \in K, \tag{9.89}$$

because the Einstein convention holds.

Before continuing with procedure (b), we discuss if $F(T)$ is intrinsic, that is, if it has an invariant tensor nature.

We know that $F(T) = f_{(\alpha_1\alpha_2\cdots\alpha_r)}{}^{\circ\;\circ\;\cdots\;\circ}\, t^{(\alpha_1\alpha_2\cdots\alpha_r)}{}_{\circ\;\circ\;\cdots\;\circ}$, with $\alpha_1 < \alpha_2 < \cdots < \alpha_r$, is a sum of $\sigma = \binom{n}{r}$ main summands, over which the Einstein convention holds. Assuming that a change-of-basis in $V^n(K)$, of matrix $C = [c^{\alpha\circ}_{\circ\,i}]$ is performed, we will study the value of the form in the new basis, using the formulas (9.73) and (9.81) adequately:

$$F(T) = f_{(i_1 i_2 \cdots i_r)}{}^{\circ\;\circ\;\cdots\;\circ}\, t^{(i_1 i_2 \cdots i_r)}{}_{\circ\;\circ\;\cdots\;\circ}$$

$$= f_{(\alpha_1\alpha_2\cdots\alpha_r)}{}^{\circ\;\circ\;\cdots\;\circ} \begin{vmatrix} c^{\circ\,\alpha_1}_{i_1\,\circ} & c^{\circ\,\alpha_2}_{i_1\,\circ} & \cdots & c^{\circ\,\alpha_r}_{i_1\,\circ} \\ c^{\circ\,\alpha_1}_{i_2\,\circ} & c^{\circ\,\alpha_2}_{i_2\,\circ} & \cdots & c^{\circ\,\alpha_r}_{i_2\,\circ} \\ \cdots & \cdots & \cdots & \cdots \\ c^{\circ\,\alpha_1}_{i_r\,\circ} & c^{\circ\,\alpha_2}_{i_r\,\circ} & \cdots & c^{\circ\,\alpha_r}_{i_r\,\circ} \end{vmatrix}$$

$$\cdot\, t^{(\beta_1\beta_2\cdots\beta_r)}{}_{\circ\;\circ\;\cdots\;\circ} \begin{vmatrix} \gamma^{i_1\,\circ}_{\circ\,\beta_1} & \gamma^{i_1\,\circ}_{\circ\,\beta_2} & \cdots & \gamma^{i_1\,\circ}_{\circ\,\beta_r} \\ \gamma^{i_2\,\circ}_{\circ\,\beta_1} & \gamma^{i_2\,\circ}_{\circ\,\beta_2} & \cdots & \gamma^{i_2\,\circ}_{\circ\,\beta_r} \\ \cdots & \cdots & \cdots & \cdots \\ \gamma^{i_r\,\circ}_{\circ\,\beta_1} & \gamma^{i_r\,\circ}_{\circ\,\beta_2} & \cdots & \gamma^{i_r\,\circ}_{\circ\,\beta_r} \end{vmatrix}$$

and grouping, the result is

$$F(T) = f_{(\alpha_1\alpha_2\cdots\alpha_r)}{}^{\circ\;\circ\;\cdots\;\circ}\, t^{(\beta_1\beta_2\cdots\beta_r)}{}_{\circ\;\circ\;\cdots\;\circ} \left(\begin{vmatrix} c^{\circ\,\alpha_1}_{i_1\,\circ} & c^{\circ\,\alpha_2}_{i_1\,\circ} & \cdots & c^{\circ\,\alpha_r}_{i_1\,\circ} \\ c^{\circ\,\alpha_1}_{i_2\,\circ} & c^{\circ\,\alpha_2}_{i_2\,\circ} & \cdots & c^{\circ\,\alpha_r}_{i_2\,\circ} \\ \cdots & \cdots & \cdots & \cdots \\ c^{\circ\,\alpha_1}_{i_r\,\circ} & c^{\circ\,\alpha_2}_{i_r\,\circ} & \cdots & c^{\circ\,\alpha_r}_{i_r\,\circ} \end{vmatrix} \begin{vmatrix} \gamma^{i_1\,\circ}_{\circ\,\beta_1} & \gamma^{i_1\,\circ}_{\circ\,\beta_2} & \cdots & \gamma^{i_1\,\circ}_{\circ\,\beta_r} \\ \gamma^{i_2\,\circ}_{\circ\,\beta_1} & \gamma^{i_2\,\circ}_{\circ\,\beta_2} & \cdots & \gamma^{i_2\,\circ}_{\circ\,\beta_r} \\ \cdots & \cdots & \cdots & \cdots \\ \gamma^{i_r\,\circ}_{\circ\,\beta_1} & \gamma^{i_r\,\circ}_{\circ\,\beta_2} & \cdots & \gamma^{i_r\,\circ}_{\circ\,\beta_r} \end{vmatrix} \right)$$

$$= f_{(\alpha_1\alpha_2\cdots\alpha_r)}{}^{\circ\;\circ\;\cdots\;\circ}\, \delta^{\alpha_1\alpha_2\cdots\alpha_r}_{\beta_1\beta_2\cdots\beta_r}\, t^{(\beta_1\beta_2\cdots\beta_r)}{}_{\circ\;\circ\;\cdots\;\circ},$$

that once operated becomes

$$F(T) = f_{(i_1 i_2 \cdots i_r)}^{\circ \ \circ \ \cdots \ \circ} t^{(i_1 i_2 \cdots i_r)}_{\circ \ \circ \ \cdots \ \circ} = f_{(\alpha_1 \alpha_2 \cdots \alpha_r)}^{\circ \ \circ \ \cdots \ \circ} t^{(\alpha_1 \alpha_2 \cdots \alpha_r)}_{\circ \ \circ \ \cdots \ \circ}, \qquad (9.90)$$

which proves the scalar invariant tensor of a zero-order nature over the linear space $V^n(K)$, of the form $F(T)$ value.

Procedure (b) Now we execute the direct contraction of the exterior r-form F with the r-exterior vector T using expressions (9.87) and (9.88):

$$F \bullet T = \left[f_{(\alpha_1 \alpha_2 \cdots \alpha_r)}^{\circ \ \circ \ \cdots \ \circ} \vec{e}^{*\alpha_1} \wedge \vec{e}^{*\alpha_2} \wedge \cdots \wedge \vec{e}^{*\alpha_r} \right] \bullet \left[t^{(\alpha_1 \alpha_2 \cdots \alpha_r)}_{\circ \ \circ \ \cdots \ \circ} \vec{e}_{\alpha_1} \wedge \vec{e}_{\alpha_2} \wedge \cdots \wedge \vec{e}_{\alpha_r} \right]$$

$$= f_{(\alpha_1 \alpha_2 \cdots \alpha_r)}^{\circ \ \circ \ \cdots \ \circ} t^{(\alpha_1 \alpha_2 \cdots \alpha_r)}_{\circ \ \circ \ \cdots \ \circ} \left[(\vec{e}^{*\alpha_1} \wedge \vec{e}^{*\alpha_2} \wedge \cdots \wedge \vec{e}^{*\alpha_r}) \bullet (\vec{e}_{\alpha_1} \wedge \vec{e}_{\alpha_2} \wedge \cdots \wedge \vec{e}_{\alpha_r}) \right]$$

and remembering Formulas (9.66) and (9.67), the result is

$$F \bullet T = F(T) \cdot r! \qquad (9.91)$$

Expression (9.91) provides the title of this section, and shows again that $\bigwedge_n^{(r)}(K)$ and $\bigwedge_{n*}^{(r)}(K)$, though they are dual linear spaces, *are not* in reciprocal bases, and due to this reason, the second procedure given the result that exactly answers the question in this section, and that is not what it was expected to (we expected the same result as in procedure (a)).

9.9 Exterior vector mappings: Exterior homomorphisms

We end the present chapter by making a brief exposition of the mappings of exterior linear spaces in other linear spaces.

The exposition will follow a process analogous to that established in the final part of Chapter 5, where definitions and properties of the multilinear mappings of linear spaces "of direct product" or "total" have been developed, together with the correlative tensor ones. There is a singularity; we will restrict ourselves to homogeneous linear factor-spaces, in accordance with the concepts of the exterior algebra. We remind the reader that in Formulas (5.94) to (5.97) we introduced the multilinear mapping F, which applies the linear space "direct product" in another given linear space:

$$F : (\overset{r}{\underset{1}{\times}} V^n)(K) \to W^m(K).$$

For any r-tuple, $(\vec{v}_1, \vec{v}_2, \ldots, \vec{v}_r) \in (\overset{r}{\underset{1}{\times}} V^n)(K)$ where $\forall \vec{v}_i \in V^n(K)$, we have

$$F(\vec{v}_1, \vec{v}_2, \ldots, \vec{v}_r) = \vec{w} \in W^m(K). \qquad (9.92)$$

In addition to the formal properties given in Section 5.11 (the multilinearity of F) we establish below two further properties:

1. We say that a multilinear mapping F is "symmetric" if it satisfies

$$F(\vec{v}_1, \vec{v}_2, \ldots, \vec{v}_k, \cdots, \vec{v}_h, \ldots \vec{v}_r) = F(\vec{v}_1, \vec{v}_2, \ldots, \vec{v}_h, \ldots, \vec{v}_k, \ldots, \vec{v}_r); \quad \forall \vec{v}_h, \vec{v}_k \tag{9.93}$$

2. A multilinear mapping F is said to be "alternate" or "anti-symmetric" if it satisfies

$$F(\vec{v}_1, \vec{v}_2, \ldots, \vec{v}_k, \ldots, \vec{v}_h, \ldots \vec{v}_r) = -F(\vec{v}_1, \vec{v}_2, \ldots, \vec{v}_h, \ldots, \vec{v}_k, \ldots, \vec{v}_r); \quad \forall \vec{v}_h, \vec{v}_k. \tag{9.94}$$

In what follows, we denote by F_s and F_a the multilinear mappings that are respectively symmetric or alternate.

We remind the reader the "similitude" theorems that were treated in Section 5.11.1 for tensor morphisms. Next, we give "similitude" theorems for exterior algebras.

Theorem 9.1 (Exterior). *If F_a is an alternate multilinear mapping that applies $\left(\overset{r}{\underset{1}{\times}} V^n \right)(K)$ in $W^m(K)$, and operates in the form $F_a(\vec{v}_1, \vec{v}_2, \ldots, \vec{v}_r) = \vec{w}$, there exists a unique multilinear mapping F'_a that applies $\bigwedge_n^{(r)}(K)$ in $W^m(K)$, where*

$$F'_a(\vec{v}_1 \wedge \vec{v}_2 \wedge \cdots \wedge \vec{v}_r) = F_a(\vec{v}_1, \vec{v}_2, \ldots, \vec{v}_r) = \vec{w}.$$

$\square$

As a consequence of this theorem, when one wants to operate an alternate multilinear mapping, one can work with a mapping F'_a of exterior vectors or with a tensor anti-symmetric alternate mapping of order r in totally covariant dummy indices.

In the final part of Section 5.11.2 this circumstance was clearly shown when proposing in Formula (5.103) that the tensor coefficients $w^{h \ \circ \ \circ \ \cdots \ \circ}_{\circ \beta_1 \beta_2 \cdots \beta_r}$ with $1 \le h \le m$, of the mapping F could be symmetric tensors for the F_s and anti-symmetric for the F_a.

If we consider Formulas (5.102) and (5.103) and we notate with the indicated improvement, simultaneously grouping the basis $\vec{\epsilon}$ of the linear space $W^m(K)$, we have

$$F_a(\vec{e}_{\beta_1}, \vec{e}_{\beta_2}, \ldots, \vec{e}_{\beta_r}) = \vec{w}(\beta_1, \beta_2, \ldots, \beta_r) = w^{h \ \circ \ \circ \ \cdots \ \circ}_{\circ \beta_1 \beta_2 \cdots \beta_r} \vec{\epsilon}_h. \tag{9.95}$$

Remark 9.2. This formula proves that if $\{\vec{e}_\beta\}$ is the basis of $V^n(K)$, *the alternate forces $r \le n$.* $\square$

As the $w^{h \ \circ \ \circ \ \cdots \ \circ}_{\circ \beta_1 \beta_2 \cdots \beta_r}$ are anti-symmetric for the covariant indices, we only consider indices without repetition, and if we denote by $(\gamma_1, \gamma_2, \ldots, \gamma_r)$ such indices strictly ordered, such conditions can we written as

$$\{\gamma_1, \gamma_2, \ldots, \gamma_r\} \equiv \{\beta_1, \beta_2, \ldots, \beta_r\} \tag{9.96}$$

with $\beta_1 \neq \beta_2 \neq \cdots \neq \beta_r$ and $\gamma_1 < \gamma_2 < \cdots < \gamma_r$ from which results

$$w^{h \;\circ\;\circ\;\cdots\;\circ}_{\;\circ\;\beta_1\beta_2\cdots\beta_r} = \delta^{(\gamma_1\gamma_2\cdots\gamma_r)}_{\;\beta_1\beta_2\cdots\beta_r} w^{h \;\circ\;\circ\;\cdots\;\circ}_{\;\circ\,(\gamma_1\gamma_2\cdots\gamma_r)}. \tag{9.97}$$

Remembering expression (5.101) for linear spaces that are all homogeneous, and taking into account (9.95) and (9.97), we have

$$\begin{aligned}
\vec{w} = F_a(\vec{v}_1, \vec{v}_2, \ldots, \vec{v}_r) &= x^{\beta_1\circ}_{\;\circ\,1} x^{\beta_2\circ}_{\;\circ\,2} \cdots x^{\beta_r\circ}_{\;\circ\,r} F_a(\vec{e}_{\beta_1}, \vec{e}_{\beta_2}, \ldots, \vec{e}_{\beta_r}) \\
&= \left(x^{\beta_1\circ}_{\;\circ\,1} x^{\beta_2\circ}_{\;\circ\,2} \cdots x^{\beta_r\circ}_{\;\circ\,r} \right) w^{h \;\circ\;\circ\;\cdots\;\circ}_{\;\circ\,\beta_1\beta_2\cdots\beta_r} \vec{e}_h \\
&= \left(x^{\beta_1\circ}_{\;\circ\,1} x^{\beta_2\circ}_{\;\circ\,2} \cdots x^{\beta_r\circ}_{\;\circ\,r} \delta^{(\gamma_1\gamma_2\cdots\gamma_r)}_{\;\beta_1\beta_2\cdots\beta_r} \right) w^{h \;\circ\;\circ\;\cdots\;\circ}_{\;\circ\,(\gamma_1\gamma_2\cdots\gamma_r)} \vec{e}_h \tag{9.98} \\
&\quad \beta_1 \neq \beta_2 \neq \cdots \neq \beta_r \text{ and } \gamma_1 < \gamma_2 < \cdots < \gamma_r.
\end{aligned}$$

This expression reminds us of Formula (9.19) in which the parentheses can be interpreted as the development of a determinant, that is, the minor of the rows $\beta_1, \beta_2, \ldots, \beta_r$ of the matrix $[X_1, X_2, \ldots, X_r]_{n \times r}$ formed by the column-matrices of the *data* vectors $(\vec{v}_1, \vec{v}_2, \cdots, \vec{v}_r)$ in the basis $\{\vec{e}_\alpha\}$ of $V^n(K)$.

Thus, Expression (9.98) can be formulated as

$$\vec{w} = F_a(\vec{v}_1, \vec{v}_2, \ldots, \vec{v}_r) = \begin{vmatrix} x^{\beta_1\circ}_{\;\circ\,1} & x^{\beta_1\circ}_{\;\circ\,2} & \cdots & x^{\beta_1\circ}_{\;\circ\,r} \\ x^{\beta_2\circ}_{\;\circ\,1} & x^{\beta_2\circ}_{\;\circ\,2} & \cdots & x^{\beta_2\circ}_{\;\circ\,r} \\ \cdots & \cdots & \cdots & \cdots \\ x^{\beta_r\circ}_{\;\circ\,1} & x^{\beta_r\circ}_{\;\circ\,2} & \cdots & x^{\beta_r\circ}_{\;\circ\,r} \end{vmatrix} w^{h \;\circ\;\circ\;\cdots\;\circ}_{\;\circ\,(\gamma_1\gamma_2\cdots\gamma_r)} \vec{e}_h, \tag{9.99}$$

which satisfies the conditions (9.96) and solves the alternate mapping direct product of the given data vectors and of the data F_a, by means of the knowledge of its strict components $w^{h \;\circ\;\circ\;\cdots\;\circ}_{\;\circ\,(\gamma_1\gamma_2\cdots\gamma_r)}$.

The mapping $F_a(\vec{v}_1, \vec{v}_2, \ldots, \vec{v}_r)$ in $W^m(K)$:

$$F_a : \left(\overset{r}{\underset{1}{\times}} V^n \right)(K) \to W^m(K)$$

is usually interpreted (following the line of Theorem 9.1) by means of two options:

Option I: $H_\otimes : (\vec{v}_1 \otimes \vec{v}_2 \otimes \cdots \otimes \vec{v}_r) \to W^m(K)$.

$$\tag{9.100}$$

Option II: $H_\wedge : (\vec{v}_1 \wedge \vec{v}_2 \wedge \cdots \wedge \vec{v}_r) \to W^m(K)$.

To develop the first option ($H_\otimes$), we choose from (9.98) the equality

$$\vec{w} = F(\vec{v}_1, \vec{v}_2, \cdots, \vec{v}_r) = x^{\beta_1\circ}_{\;\circ\,1} x^{\beta_2\circ}_{\;\circ\,2} \cdots x^{\beta_r\circ}_{\;\circ\,r} \left(w^{h \;\circ\;\circ\;\cdots\;\circ}_{\;\circ\,\beta_1\beta_2\cdots\beta_r} \right) \vec{e}_h, \tag{9.101}$$

where we only require the conditions $\beta_i \in I_n$ and $r \leq n$. Grouping the $m \times n^r$ terms of the central factor, since the mapping matrix $H_\otimes$ is

$$H_\otimes = \begin{bmatrix} w_{\circ 11 \cdots 1}^{1 \circ \circ \cdots \circ} & w_{\circ 11 \cdots 2}^{1 \circ \circ \cdots \circ} & \cdots & w_{\circ nn \cdots n}^{1 \circ \circ \cdots \circ} \\ w_{\circ 11 \cdots 1}^{2 \circ \circ \cdots \circ} & w_{\circ 11 \cdots 2}^{2 \circ \circ \cdots \circ} & \cdots & w_{\circ nn \cdots n}^{2 \circ \circ \cdots \circ} \\ \cdots & \cdots & \cdots & \cdots \\ w_{\circ 11 \cdots 1}^{m \circ \circ \cdots \circ} & w_{\circ 11 \cdots 2}^{m \circ \circ \cdots \circ} & \cdots & w_{\circ nn \cdots n}^{m \circ \circ \cdots \circ} \end{bmatrix}_{m \times n^r}, \qquad (9.102)$$

in which there would be $\left[n^r - \binom{n}{r} r! \right]$ zeroes. Then, the $\vec{w}$ in (9.101) can be finally represented in matrix form as

$$H_\otimes(\vec{v}_1 \otimes \vec{v}_2 \otimes \cdots \otimes \vec{v}_r) = [\, \vec{e}_1 \quad \vec{e}_2 \quad \cdots \quad \vec{e}_m \,] \bullet H_\otimes \bullet \begin{bmatrix} x_{\circ 1}^{1 \circ} x_{\circ 2}^{1 \circ} \cdots x_{\circ r}^{1 \circ} \\ x_{\circ 1}^{1 \circ} x_{\circ 2}^{1 \circ} \cdots x_{\circ r}^{2 \circ} \\ \cdots \\ x_{\circ 1}^{\beta_1 \circ} x_{\circ 2}^{\beta_2 \circ} \cdots x_{\circ r}^{\beta_r \circ} \\ \vdots \\ x_{\circ 1}^{n \circ} x_{\circ 2}^{n \circ} \cdots x_{\circ r}^{(n-1) \circ} \\ x_{\circ 1}^{n \circ} x_{\circ 2}^{n \circ} \cdots x_{\circ r}^{n \circ} \end{bmatrix}_{n^r \times 1}.$$

$$(9.103)$$

Formula (9.103) has in the last matrix-column the components of the multivector $(\vec{v}_1 \otimes \vec{v}_2 \otimes \cdots \otimes \vec{v}_r)$ and transforms it into a vector of $W^m(K)$, so that this is the interpretation of the first option and at the same time it offers a practical answer to the proposed theorem:

$$F_a(\vec{v}_1, \vec{v}_2, \ldots, \vec{v}_r) = F_a'(\vec{v}_1 \otimes \vec{v}_2 \otimes \cdots \otimes \vec{v}_r) = H_\otimes(\vec{v}_1 \otimes \vec{v}_2 \otimes \cdots \otimes \vec{v}_r) \equiv \vec{w}. \quad (9.104)$$

The second option can be obtained by developing (9.99) in matrix form. If we take into account that the number of tensor components of the type $w_{\circ(\gamma_1 \gamma_2 \cdots \gamma_r)}^{h \circ \circ \cdots \circ}$, because they are strict in the covariant indices, is $\binom{n}{r}$, they can be grouped into the matrix

$$H_\wedge = \begin{bmatrix} w_{\circ(12 \cdots r)}^{1 \ \circ\circ\cdots\circ} & \cdots & w_{\circ(\gamma_1 \gamma_2 \cdots \gamma_r)}^{1 \ \circ \ \circ \cdots \circ} & \cdots & w_{\circ((n-r+1)\cdots(n-1)n)}^{1 \ \ \ \circ \ \ \cdots \ \circ \ \circ} \\ w_{\circ(12 \cdots r)}^{2 \ \circ\circ\cdots\circ} & \cdots & w_{\circ(\gamma_1 \gamma_2 \cdots \gamma_r)}^{2 \ \circ \ \circ \cdots \circ} & \cdots & w_{\circ((n-r+1)\cdots(n-1)n)}^{2 \ \ \ \circ \ \ \cdots \ \circ \ \circ} \\ \cdots & \cdots & \cdots & \cdots & \cdots \\ w_{\circ(12 \cdots r)}^{m \ \circ\circ\cdots\circ} & \cdots & w_{\circ(\gamma_1 \gamma_2 \cdots \gamma_r)}^{m \ \circ \ \circ \cdots \circ} & \cdots & w_{\circ((n-r+1)\cdots(n-1)n)}^{m \ \ \ \circ \ \ \cdots \ \circ \ \circ} \end{bmatrix}_{m \times \binom{n}{r}},$$

$$(9.105)$$

which is the *data* matrix of the exterior morphism $F'(\vec{v}_1 \wedge \vec{v}_2 \wedge \cdots \wedge \vec{v}_r)$ and contains its strict covariant components.

It is the right moment to return to formulate expression (9.99) in matrix form. If for the sake of simplicity we denote by Δ_r^j each one of the first factor determinants in Formula (9.99) we have

$$\vec{w} = H_{\wedge}(\vec{v}_1 \wedge \vec{v}_2 \wedge \cdots \wedge \vec{v}_r) = [\vec{\epsilon}_1 \quad \vec{\epsilon}_2 \quad \cdots \quad \vec{\epsilon}_m] \bullet H_{\wedge} \bullet \begin{bmatrix} \Delta_r^1 \\ \Delta_r^2 \\ \vdots \\ \Delta_r^{\binom{n}{r}} \end{bmatrix}_{\binom{n}{r} \times 1}, \quad (9.106)$$

which permits us to determine the image vector $F''(\vec{v}_1 \wedge \vec{v}_2 \wedge \cdots \wedge \vec{v}_r)$ directly in terms of the multivector's strict components, confirming again Theorem 9.1:

$$F_a(\vec{v}_1, \vec{v}_2, \ldots, \vec{v}_r) = F''(\vec{v}_1 \wedge \vec{v}_2 \wedge \cdots \wedge \vec{v}_r) = H_{\wedge}(\vec{v}_1 \wedge \vec{v}_2 \wedge \cdots \wedge \vec{v}_r) \equiv \vec{w}. \quad (9.107)$$

It is obvious that, since the two exterior morphisms ($H_{\otimes}$ and $H_{\wedge}$) apply the same prototype, the *data* multivector, to the same image, the $\vec{w} \in W^m(K)$, they will be related. Motivated by all the above considerations, we start from the following matrix relation:

$$[H_{\otimes}]_{m \times n^r} = [H_{\wedge}]_{m \times \binom{n}{r}} \bullet \left[\delta^{(\gamma_1 \gamma_2 \cdots \gamma_r) \cdots \; \circ \; \circ \cdots \circ}_{\circ \; \circ \cdots \circ \; \cdots \beta_1 \beta_2 \cdots \beta_r} \right]_{\binom{n}{r} \times n^r} \quad (9.108)$$

with $\gamma_1 < \gamma_2 < \cdots < \gamma_r$ and $\beta_i \in I_n$ arbitrary. The species of the generalized Kronecker delta have been separated (the strict contravariant components from the covariant components), for they are to be simultaneously used for validating $(0, 1, -1)$ and as row and column indicators of its proper matrix.

Analogously to the similitude Theorem 5.8, given in Section 5.11.1 of tensor morphisms, there can be established for the exterior morphisms another parallel theorem, that we only state, and that we invite the reader to develop, based on the trend indicated for Theorem 9.1 of exterior vectors.

We denote by A' the exterior linear space:

$$A' = \left[\bigwedge_n^{(r)}(K) \right] \bigwedge \left[\bigwedge_{n*}^{(r)}(K) \right] \quad (9.109)$$

and by B' the exterior tensor space of all exterior multilinear *endomorphisms* that operate inside $\bigwedge_n^{(r)}(K)$:

$$B' = \mathcal{ML}\left[\bigwedge_n^{(r)}(K), \bigwedge_n^{(r)}(K) \right]. \quad (9.110)$$

Theorem 9.2 (Exterior). *There exists a unique isomorphism:*

$$\Phi_{\wedge} : A' \rightleftharpoons B'$$

such that to each exterior tensor $\vec{v}_1 \wedge \vec{v}_2 \wedge \cdots \wedge \vec{v}_r) \bigwedge (\vec{u_1}^* \wedge \vec{u_2}^* \wedge \cdots \wedge \vec{u_r}^*) \in A'$ *corresponds an exterior multilinear* endomorphism:

$$T_{(\vec{v}_1 \wedge \vec{v}_2 \wedge \cdots \wedge \vec{v}_r) \bigwedge (\vec{u_1}^* \wedge \vec{u_2}^* \wedge \cdots \wedge \vec{u_r}^*)} \in B',$$

that transforms the exterior multivectors $\vec{w}' = \vec{w}_1 \wedge \vec{w}_2 \wedge \cdots \wedge \vec{w}_r \in \bigwedge_n^{(r)}(K)$ *into the form:*

$$T(\vec{w}') = [(\vec{w}_1 \wedge \vec{w}_2 \wedge \cdots \wedge \vec{w}_r) \bullet (\vec{u_1}^* \wedge \vec{u_2}^* \wedge \cdots \wedge \vec{u_r}^*)] (\vec{v}_1 \wedge \vec{v}_2 \wedge \cdots \wedge \vec{v}_r).$$
(9.111)
□

9.9.1 Direct exterior endomorphism

Having established the exterior Theorems 9.1 and 9.2, we end the explanation of the exterior morphisms by solving the particular case of a direct exterior endomorphism, associated with a known and given endomorphism:

$$H : V^n(K) \to V^n(K); \ \ \forall \vec{v}_i \in V^n(K) : \ H(\vec{v}_i) = \vec{w}_i,$$
(9.112)

where the data matrix H is square, H_n.

We look for an associated exterior operator, denoted by $P_{\sigma'}$, that operates as $P_{\sigma'}(\vec{z}) = \vec{w}$, where $\vec{z}, \vec{w} \in \bigwedge_n^{(r)}(K)$, of dimension $\sigma' = \binom{n}{r}$, and transforms $\vec{z}$ into $\vec{w}$ in the form

$$P_{\sigma'}(\vec{v}_1 \wedge \vec{v}_2 \wedge \cdots \wedge \vec{v}_r) = \vec{w}_1 \wedge \vec{w}_2 \wedge \cdots \wedge \vec{w}_r; \ \ \forall \vec{v}_i \in V^n(K).$$

Starting from $\vec{v}_i = x^{\alpha_i \circ}_{\circ \ i} \vec{e}_{\alpha_i}$, and $\vec{w}_i = y^{\beta_j \circ}_{\circ \ i} \vec{e}_{\beta_j}$, with

$$y^{\beta_j \circ}_{\circ \ i} = h^{\beta_j \ \circ}_{\circ \ \alpha_j} x^{\alpha_j \circ}_{\circ \ i},$$
(9.113)

substituting into the exterior products and remembering Formula (9.20) we get

$$\vec{v}_1 \wedge \vec{v}_2 \wedge \cdots \wedge \vec{v}_r = \left(x^{\alpha_1 \circ}_{\circ \ 1} x^{\alpha_2 \circ}_{\circ \ 2} \cdots x^{\alpha_r \circ}_{\circ \ r} \right) \vec{e}_{\alpha_1} \wedge \vec{e}_{\alpha_2} \wedge \cdots \wedge \vec{e}_{\alpha_r}$$
$$= | X_{r1} \ \ X_{r2} \ \ \cdots \ \ X_{rr} |_{rr} \, \vec{e}_{\gamma_1} \wedge \vec{e}_{\gamma_2} \wedge \cdots \wedge \vec{e}_{\gamma_r} \ (9.114)$$

with $\gamma_1 < \gamma_2 < \cdots < \gamma_r$, notated in strict order and in the same line we have

$$\vec{w}_1 \wedge \vec{w}_2 \wedge \cdots \wedge \vec{w}_r = \left(y^{\beta_1 \circ}_{\circ \ 1} \vec{e}_{\beta_1} \right) \wedge \left(y^{\beta_2 \circ}_{\circ \ 2} \vec{e}_{\beta_2} \right) \wedge \cdots \wedge \left(y^{\beta_r \circ}_{\circ \ r} \vec{e}_{\beta_r} \right).$$

Substituting the right-hand side of (9.113) into the previous expression, we get

$$\vec{w}_1 \wedge \vec{w}_2 \wedge \cdots \wedge \vec{w}_r = \left(h^{\beta_1 \ \circ}_{\circ \ \alpha_1} x^{\alpha_1 \circ}_{\circ \ 1} \right) \left(h^{\beta_2 \ \circ}_{\circ \ \alpha_2} x^{\alpha_2 \circ}_{\circ \ 2} \right) \cdots \left(h^{\beta_r \ \circ}_{\circ \ \alpha_r} x^{\alpha_r \circ}_{\circ \ r} \right) \vec{e}_{\beta_1} \wedge \vec{e}_{\beta_2} \wedge \cdots \wedge \vec{e}_{\beta_r},$$

which operated and grouped in a convenient form leads to

$$\vec{w}_1 \wedge \vec{w}_2 \wedge \cdots \wedge \vec{w}_r = \left(x^{\alpha_1 \circ}_{\circ \ 1} x^{\alpha_2 \circ}_{\circ \ 2} \cdots x^{\alpha_r \circ}_{\circ \ r} \right) \left(h^{\beta_1 \ \circ}_{\circ \ \alpha_1} h^{\beta_2 \ \circ}_{\circ \ \alpha_2} \cdots h^{\beta_r \ \circ}_{\circ \ \alpha_r} \right) \vec{e}_{\beta_1} \wedge \vec{e}_{\beta_2} \wedge \cdots \wedge \vec{e}_{\beta_r}.$$
(9.115)

As in Formula (9.115) the subindices β_j that relate the exterior products of the basic vectors $\vec{e}_{\beta_j}$ are yet totally free $(\beta_j \in I_n)$, it is convenient to write them in strict order, as it corresponds to any basis of a linear space $\bigwedge_n^{(r)}(K)$. Thus, by means of the corresponding Kronecker delta, we apply the conditions of expression (9.96) and get

$$
\begin{aligned}
&\vec{w}_1 \wedge \vec{w}_2 \wedge \cdots \wedge \vec{w}_r \\
&= \left(x_{\circ\ 1}^{\alpha_1 \circ} x_{\circ\ 2}^{\alpha_2 \circ} \cdots x_{\circ\ r}^{\alpha_r \circ} \right) \left(\delta_{\beta_1 \beta_2 \cdots \beta_r}^{(\gamma_1 \gamma_2 \cdots \gamma_r)} h_{\circ\ \alpha_1}^{\beta_1 \circ} h_{\circ\ \alpha_2}^{\beta_2 \circ} \cdots h_{\circ\ \alpha_r}^{\beta_r \circ} \right) \vec{e}_{\gamma_1} \wedge \vec{e}_{\gamma_2} \wedge \cdots \wedge \vec{e}_{\gamma_r}.
\end{aligned}
$$

$$(9.116)$$

To facilitate the development of Formula (9.116) it is convenient to replace in the scalars the notation of the strict r-tuple $(\gamma_1, \gamma_2, \ldots, \gamma_r)$ by the strict r-tuple $(\lambda_1, \lambda_2, \ldots, \lambda_r)$, which represents any of the $\binom{n}{r}$ r-tuples of type $(\gamma_1, \gamma_2, \ldots, \gamma_r)$ elegible in I_n.

We use this modification to introduce the factor $1 = \delta_{\alpha_1 \alpha_2 \cdots \alpha_r}^{(\lambda_1 \lambda_2 \cdots \lambda_r)} \cdot \delta_{(\lambda_1 \lambda_2 \cdots \lambda_r)}^{\alpha_1 \alpha_2 \cdots \alpha_r}$ adequately spread in Formula (9.116). In summary, we have

$$
\begin{aligned}
\vec{w}_1 \wedge \vec{w}_2 \wedge \cdots \wedge \vec{w}_r &= \left(\delta_{\alpha_1 \alpha_2 \cdots \alpha_r}^{(\lambda_1 \lambda_2 \cdots \lambda_r)} x_{\circ\ 1}^{\alpha_1 \circ} x_{\circ\ 2}^{\alpha_2 \circ} \cdots x_{\circ\ r}^{\alpha_r \circ} \right) \delta_{(\lambda_1 \lambda_2 \cdots \lambda_r)}^{\alpha_1 \alpha_2 \cdots \alpha_r} \\
&\cdot \left(\delta_{\beta_1 \beta_2 \cdots \beta_r}^{(\gamma_1 \gamma_2 \cdots \gamma_r)} h_{\circ\ \alpha_1}^{\beta_1 \circ} h_{\circ\ \alpha_2}^{\beta_2 \circ} \cdots h_{\circ\ \alpha_r}^{\beta_r \circ} \right) \vec{e}_{\gamma_1} \wedge \vec{e}_{\gamma_2} \wedge \cdots \wedge \vec{e}_{\gamma_r},
\end{aligned}
$$

$$(9.117)$$

where the factor-parentheses represent determinants, or more precisely, minors of the matrix

$$[X_1 \quad X_2 \quad \cdots \quad X_r]_{n \times r},$$

the columns of which are the components of the *data* vectors $[\vec{v}_1, \vec{v}_2, \ldots, \vec{v}_r]$ in the first factor, and minors of the matrix H_n of the *data* endomorphism, which operates in $V^n(K)$, in the case of the second factor. Thus, we get

$$
\begin{aligned}
&\vec{w}_1 \wedge \vec{w}_2 \wedge \cdots \wedge \vec{w}_r \\
&= \begin{vmatrix} x_{\circ\ 1}^{\lambda_1 \circ} & x_{\circ\ 2}^{\lambda_1 \circ} & \cdots & x_{\circ\ r}^{\lambda_1 \circ} \\ x_{\circ\ 1}^{\lambda_2 \circ} & x_{\circ\ 2}^{\lambda_2 \circ} & \cdots & x_{\circ\ r}^{\lambda_2 \circ} \\ \cdots & \cdots & \cdots & \cdots \\ x_{\circ\ 1}^{\lambda_r \circ} & x_{\circ\ 2}^{\lambda_r \circ} & \cdots & x_{\circ\ r}^{\lambda_r \circ} \end{vmatrix} \left(\delta_{(\lambda_1 \lambda_2 \cdots \lambda_r)}^{\alpha_1 \alpha_2 \cdots \alpha_r} h_{\circ\ \alpha_1}^{\gamma_1 \circ} h_{\circ\ \alpha_2}^{\gamma_2 \circ} \cdots h_{\circ\ \alpha_r}^{\gamma_r \circ} \right) \vec{e}_{\gamma_1} \wedge \vec{e}_{\gamma_2} \wedge \cdots \wedge \vec{e}_{\gamma_r}
\end{aligned}
$$

$$(9.118)$$

and introducing the second minor into (9.118) yields

$$
\begin{aligned}
\vec{z} &= \vec{w}_1 \wedge \vec{w}_2 \wedge \cdots \wedge \vec{w}_r \\
&= \begin{vmatrix} x_{\circ\ 1}^{\lambda_1 \circ} & x_{\circ\ 2}^{\lambda_1 \circ} & \cdots & x_{\circ\ r}^{\lambda_1 \circ} \\ x_{\circ\ 1}^{\lambda_2 \circ} & x_{\circ\ 2}^{\lambda_2 \circ} & \cdots & x_{\circ\ r}^{\lambda_2 \circ} \\ \cdots & \cdots & \cdots & \cdots \\ x_{\circ\ 1}^{\lambda_r \circ} & x_{\circ\ 2}^{\lambda_r \circ} & \cdots & x_{\circ\ r}^{\lambda_r \circ} \end{vmatrix} \begin{vmatrix} h_{\circ\ \lambda_1}^{\gamma_1 \circ} & h_{\circ\ \lambda_2}^{\gamma_1 \circ} & \cdots & h_{\circ\ \lambda_r}^{\gamma_1 \circ} \\ h_{\circ\ \lambda_1}^{\gamma_2 \circ} & h_{\circ\ \lambda_2}^{\gamma_2 \circ} & \cdots & h_{\circ\ \lambda_r}^{\gamma_2 \circ} \\ \cdots & \cdots & \cdots & \cdots \\ h_{\circ\ \lambda_1}^{\gamma_r \circ} & h_{\circ\ \lambda_2}^{\gamma_r \circ} & \cdots & h_{\circ\ \lambda_r}^{\gamma_r \circ} \end{vmatrix} \vec{e}_{\gamma_1} \wedge \vec{e}_{\gamma_2} \wedge \cdots \wedge \vec{e}_{\gamma_r},
\end{aligned}
$$

$$(9.119)$$

which gives the answer to the desired operator $P_{\sigma'}$.

In Expression (9.119) the dummy indices $(\lambda_1, \lambda_2 \cdots \lambda_r)$ appear, and take contractive values according to the Einstein convention, so that in reality *the coefficient* of $\vec{e}_{\gamma_1} \wedge \vec{e}_{\gamma_2} \wedge \cdots \wedge \vec{e}_{\gamma_r}$ has $\binom{n}{r}$ summands. Note that in the first minor, the dummy indices are indicators of row, while in the second they are indicators of column.

Example 9.1 (Associativity of the exterior product).

1. Show that the exterior product of three vectors, over the linear space $V^n(K)$: $\vec{v} \wedge \vec{v}_2 \wedge \vec{v}_3$, *is not* an associative law of composition

$$\vec{v}_1 \wedge (\vec{v}_2 \wedge \vec{v}_3) \neq (\vec{v}_1 \wedge \vec{v}_2) \wedge \vec{v}_3 \neq \vec{v}_1 \wedge \vec{v}_2 \wedge \vec{v}_3.$$

2. Show that the previous question and its answer are not coherent with the exact meaning of "exterior product".

Solution:

1. According to Formula (9.1), we have

$$\vec{v}_2 \wedge \vec{v}_3 = \vec{v}_2 \otimes \vec{v}_3 - \vec{v}_3 \otimes \vec{v}_2$$
$$\vec{v}_1 \wedge (\vec{v}_2 \wedge \vec{v}_3) = \vec{v}_1 \otimes (\vec{v}_2 \wedge \vec{v}_3) - (\vec{v}_2 \wedge \vec{v}_3) \otimes \vec{v}_1$$

and replacing the first into the second, we get

$$\vec{v}_1 \wedge (\vec{v}_2 \wedge \vec{v}_3) = \vec{v}_1 \otimes (\vec{v}_2 \otimes \vec{v}_3 - \vec{v}_3 \otimes \vec{v}_2) - (\vec{v}_2 \otimes \vec{v}_3 - \vec{v}_3 \otimes \vec{v}_2) \otimes \vec{v}_1$$
$$= \vec{v}_1 \otimes \vec{v}_2 \otimes \vec{v}_3 - \vec{v}_1 \otimes \vec{v}_3 \otimes \vec{v}_2 - \vec{v}_2 \otimes \vec{v}_3 \otimes \vec{v}_1 + \vec{v}_3 \otimes \vec{v}_2 \otimes \vec{v}_1.$$

On the other hand, if we calculate $\vec{v}_1 \wedge \vec{v}_2 \wedge \vec{v}_3$ by means of Formula (9.6) we obtain

$$\vec{v}_1 \wedge \vec{v}_2 \wedge \vec{v}_3 = \vec{v}_1 \otimes \vec{v}_2 \otimes \vec{v}_3 + \vec{v}_2 \otimes \vec{v}_3 \otimes \vec{v}_1 + \vec{v}_3 \otimes \vec{v}_1 \otimes \vec{v}_2 - \vec{v}_2 \otimes \vec{v}_1 \otimes \vec{v}_3$$
$$- \vec{v}_3 \otimes \vec{v}_2 \otimes \vec{v}_1 - \vec{v}_1 \otimes \vec{v}_3 \otimes \vec{v}_2$$
$$= \vec{v}_1 \wedge (\vec{v}_2 \wedge \vec{v}_3) + [\vec{v}_3 \otimes \vec{v}_1 \otimes \vec{v}_2 - \vec{v}_2 \otimes \vec{v}_1 \otimes \vec{v}_3$$
$$+ 2\vec{v}_2 \otimes \vec{v}_3 \otimes \vec{v}_1 - 2\vec{v}_3 \otimes \vec{v}_2 \otimes \vec{v}_1].$$

The conclusion is that $\vec{v}_1 \wedge (\vec{v}_2 \wedge \vec{v}_3) \neq \vec{v}_1 \wedge \vec{v}_2 \wedge \vec{v}_3$.

2. $\vec{v}_1 \wedge \vec{v}_2 \wedge \vec{v}_3$ must be interpreted as *a single* exterior law of composition, and not as *two* laws of composition, to which its inadequate notation leads.
   If the notation used for the exterior product of three vectors was, for example, $\wedge(\vec{v}_1, \vec{v}_2, \vec{v}_3)$, then, asking if the given expression is associative would have made no sense, because the operator symbol "$\wedge$" appears only once in this expression.

□

Example 9.2 (Change of basis in exterior algebras). Consider the contravariant anti-symmetric tensors of order $r = 2$ that constitute the tensor space $\left[V^3 \otimes V^3\right]_h^2 (\mathbb{R})$ built over $V^3(\mathbb{R})$ in the basis $\{\vec{e}_\alpha\}$.

We adopt for the tensor space the exterior basis $\{\vec{e}_{\alpha_1} \wedge \vec{e}_{\alpha_2}\}$, with the well-known conditions: $\alpha_1, \alpha_2 \in I_3; \alpha_1 < \alpha_2$, that is, we assume that we are in the exterior linear space $\bigwedge_3^{(2)}(\mathbb{R})$.

1. Obtain the expression of this basis of the exterior product or bivector $A = \vec{x} \wedge \vec{y}$, where

$$\vec{x} = 2\vec{e}_1 + 5\vec{e}_2 - 3\vec{e}_3; \quad \vec{y} = \vec{e}_1 - \vec{e}_3.$$

2. In $V^3(\mathbb{R})$ a change-of-basis $\vec{e}_i = c^{\alpha}_{\circ\ i}\vec{e}_\alpha$, of matrix

$$C = \begin{bmatrix} 1 & 2 & 1 \\ 1 & 0 & 3 \\ -1 & 1 & 2 \end{bmatrix}$$

is performed and we wish to know the expression of each of the new exterior basic vectors $\vec{e}_{i_1} \wedge \vec{e}_{i_2}$, in terms of the exterior initial basis.
3. Find the formulas that relate the components of an arbitrary exterior vector of $\bigwedge_3^{(2)}(\mathbb{R})$, in each of the mentioned bases.
4. Apply the previously found relation to obtain the new components of the bivector A.
5. Obtain the components of the tensor A, relative to the tensor space $(V^3 \otimes V^3)_h^2(\mathbb{R})$ in its initial basis.
6. The same question but in the new basis.
7. Are the formulas that relate the strict components of the tensor A in the two bases of the previous paragraph the same as those indicated in (3)?

Solution:

1. We apply the general formula (9.20).

The data vector matrix is $[X_1 X_2] \equiv \begin{bmatrix} 2 & 1 \\ 5 & 0 \\ -3 & -1 \end{bmatrix}$ and

$$A = \vec{X} \wedge \vec{Y} = \begin{vmatrix} x^{\alpha_1\,\circ}_{\circ\ 1} & x^{\alpha_1\,\circ}_{\circ\ 2} \\ x^{\alpha_2\,\circ}_{\circ\ 1} & x^{\alpha_2\,\circ}_{\circ\ 2} \end{vmatrix} \vec{e}_{\alpha_1} \wedge \vec{e}_{\alpha_2}$$

$$= \begin{vmatrix} 2 & 1 \\ 5 & 0 \end{vmatrix} \vec{e}_1 \wedge \vec{e}_2 + \begin{vmatrix} 2 & 1 \\ -3 & -1 \end{vmatrix} \vec{e}_1 \wedge \vec{e}_3 + \begin{vmatrix} 5 & 0 \\ -3 & -1 \end{vmatrix} \vec{e}_2 \wedge \vec{e}_3$$

and operating we obtain

$$A = \vec{X} \wedge \vec{Y} = -5\vec{e}_1 \wedge \vec{e}_2 + \vec{e}_1 \wedge \vec{e}_3 - 5\vec{e}_2 \wedge \vec{e}_3.$$

2. The columns of the change-of-basis matrix C are the components of the new basic vectors in $V^3(\mathbb{R})$

$$\vec{\hat{e}}_1 = \vec{e}_1 + \vec{e}_2 - \vec{e}_3, \quad \vec{\hat{e}}_2 = 2\vec{e}_1 + \vec{e}_3 \quad \text{and} \quad \vec{\hat{e}}_3 = \vec{e}_1 + 3\vec{e}_2 + 2\vec{e}_3.$$

We calculate the new basic vectors of $\bigwedge_3^{(2)}(\mathbb{R})$, again using (9.20) and obtain

$$C \equiv \begin{bmatrix} 1 & 2 & 1 \\ 1 & 0 & 3 \\ -1 & 1 & 2 \end{bmatrix}$$

$$\begin{cases} \vec{\hat{e}}_1 \wedge \vec{\hat{e}}_2 = \begin{vmatrix} 1 & 2 \\ 1 & 0 \end{vmatrix} \vec{e}_1 \wedge \vec{e}_2 + \begin{vmatrix} 1 & 2 \\ -1 & 1 \end{vmatrix} \vec{e}_1 \wedge \vec{e}_3 + \begin{vmatrix} 1 & 0 \\ -1 & 1 \end{vmatrix} \vec{e}_2 \wedge \vec{e}_3 \\ \qquad = -2\vec{e}_1 \wedge \vec{e}_2 + 3\vec{e}_1 \wedge \vec{e}_3 + \vec{e}_2 \wedge \vec{e}_3 \\[2mm] \vec{\hat{e}}_1 \wedge \vec{\hat{e}}_3 = \begin{vmatrix} 1 & 1 \\ 1 & 3 \end{vmatrix} \vec{e}_1 \wedge \vec{e}_2 + \begin{vmatrix} 1 & 1 \\ -1 & 2 \end{vmatrix} \vec{e}_1 \wedge \vec{e}_3 + \begin{vmatrix} 1 & 3 \\ -1 & 2 \end{vmatrix} \vec{e}_2 \wedge \vec{e}_3 \\ \qquad = 2\vec{e}_1 \wedge \vec{e}_2 + 3\vec{e}_1 \wedge \vec{e}_3 + 5\vec{e}_2 \wedge \vec{e}_3 \\[2mm] \vec{\hat{e}}_2 \wedge \vec{\hat{e}}_3 = \begin{vmatrix} 2 & 1 \\ 0 & 3 \end{vmatrix} \vec{e}_1 \wedge \vec{e}_2 + \begin{vmatrix} 2 & 1 \\ 1 & 2 \end{vmatrix} \vec{e}_1 \wedge \vec{e}_3 + \begin{vmatrix} 0 & 3 \\ 1 & 2 \end{vmatrix} \vec{e}_2 \wedge \vec{e}_3 \\ \qquad = 6\vec{e}_1 \wedge \vec{e}_2 + 3\vec{e}_1 \wedge \vec{e}_3 - 3\vec{e}_2 \wedge \vec{e}_3. \end{cases}$$

Summarizing, in matrix mode, the change-of-basis inside $\bigwedge_3^{(2)}(\mathbb{R})$ is

$$||\vec{\hat{e}}_i \wedge \vec{\hat{e}}_j|| = ||\vec{e}_\alpha \wedge \vec{e}_\beta|| \Gamma \rightarrow \begin{bmatrix} \vec{\hat{e}}_1 \wedge \vec{\hat{e}}_2 & \vec{\hat{e}}_1 \wedge \vec{\hat{e}}_3 & \vec{\hat{e}}_2 \wedge \vec{\hat{e}}_3 \end{bmatrix}$$

$$= \begin{bmatrix} \vec{e}_1 \wedge \vec{e}_2 & \vec{e}_1 \wedge \vec{e}_3 & \vec{e}_2 \wedge \vec{e}_3 \end{bmatrix} \begin{bmatrix} -2 & 2 & 6 \\ 3 & 3 & 3 \\ 1 & 5 & -3 \end{bmatrix}. \qquad (9.120)$$

3. We use two procedures:

Procedure (a): As we know the change-of-basis in the space $\bigwedge_3^{(2)}(\mathbb{R})$, Formula (9.120), we can deal with this question as a linear space. The matrix relation between the initial and the new components, Formula (1.5) is

$$T = \Gamma \hat{T} \rightarrow \hat{T} = \Gamma^{-1} T; \quad \begin{bmatrix} \hat{t}^{(12)}_{\circ\circ} \\ \hat{t}^{(13)}_{\circ\circ} \\ \hat{t}^{(23)}_{\circ\circ} \end{bmatrix} = \begin{bmatrix} -2 & 2 & 6 \\ 3 & 3 & 3 \\ 1 & 5 & -3 \end{bmatrix}^{-1} \begin{bmatrix} t^{(12)}_{\circ\circ} \\ t^{(13)}_{\circ\circ} \\ t^{(23)}_{\circ\circ} \end{bmatrix};$$

which, operated gives

$$\hat{t}^{(12)}_{\circ\circ} = \frac{1}{12}(-2t^{(12)}_{\circ\circ} + 3t^{(13)}_{\circ\circ} - t^{(23)}_{\circ\circ}) \qquad (9.121)$$

$$\hat{t}^{(13)}_{\circ\circ} = \frac{1}{12}(t^{(12)}_{\circ\circ} + 2t^{(23)}_{\circ\circ}) \tag{9.122}$$

$$\hat{t}^{(23)}_{\circ\circ} = \frac{1}{12}(t^{(12)}_{\circ\circ} + t^{(13)}_{\circ\circ} - t^{(23)}_{\circ\circ}). \tag{9.123}$$

Procedure (b): We will have the opportunity of checking if the calculations performed when answering the previous questions are correct, by attacking the problem again by means of the strict tensor relations formulated in (9.73) for the change-of-basis. First, we calculate the matrix

$$C^{-1} = \frac{1}{12} \begin{bmatrix} 3 & 3 & -6 \\ 5 & -3 & 2 \\ -1 & 3 & 2 \end{bmatrix}$$

and from this we take the minors in (9.73):

$$\hat{t}^{(12)}_{\circ\circ} = t^{(12)}_{\circ\circ} \frac{1}{12^2} \begin{vmatrix} 3 & 3 \\ 5 & -3 \end{vmatrix} + t^{(13)}_{\circ\circ} \frac{1}{12^2} \begin{vmatrix} 3 & -6 \\ 5 & 2 \end{vmatrix} + t^{(23)}_{\circ\circ} \frac{1}{12^2} \begin{vmatrix} 3 & -6 \\ -3 & 2 \end{vmatrix}$$

$$= \frac{1}{12}(-2t^{(12)}_{\circ\circ} + 3t^{(13)}_{\circ\circ} - t^{(23)}_{\circ\circ}) \tag{9.124}$$

$$\hat{t}^{(13)}_{\circ\circ} = t^{(12)}_{\circ\circ} \frac{1}{12^2} \begin{vmatrix} 3 & 3 \\ -1 & 3 \end{vmatrix} + t^{(13)}_{\circ\circ} \frac{1}{12^2} \begin{vmatrix} 3 & -6 \\ -1 & 2 \end{vmatrix} + t^{(23)}_{\circ\circ} \frac{1}{12^2} \begin{vmatrix} 3 & -6 \\ 3 & 2 \end{vmatrix}$$

$$= \frac{1}{12}(t^{(12)}_{\circ\circ} + 2t^{(23)}_{\circ\circ}) \tag{9.125}$$

$$\hat{t}^{(23)}_{\circ\circ} = t^{(12)}_{\circ\circ} \frac{1}{12^2} \begin{vmatrix} 5 & -3 \\ -1 & 3 \end{vmatrix} + t^{(13)}_{\circ\circ} \frac{1}{12^2} \begin{vmatrix} 5 & 2 \\ -1 & 2 \end{vmatrix} + t^{(23)}_{\circ\circ} \frac{1}{12^2} \begin{vmatrix} -3 & 2 \\ 3 & 2 \end{vmatrix}$$

$$= \frac{1}{12}(t^{(12)}_{\circ\circ} + t^{(13)}_{\circ\circ} - t^{(23)}_{\circ\circ}) \tag{9.126}$$

and since the results are coincident, we confirm the correction of the previous answers.

4. We use the Formulas (9.124) to (9.126) with the components of A obtained in question 1:

$$\hat{a}^{(12)}_{\circ\circ} = \frac{-2a^{(12)}_{\circ\circ} + 3a^{(13)}_{\circ\circ} - a^{(23)}_{\circ\circ}}{12} = \frac{-2\cdot(-5) + 3\cdot 1 - 1\cdot(-5)}{12} = \frac{3}{2}$$

$$\hat{a}^{(13)}_{\circ\circ} = \frac{a^{(12)}_{\circ\circ} + 2a^{(23)}_{\circ\circ}}{12} = \frac{(-5) + 2\cdot(-5)}{12} = -\frac{5}{4}$$

$$\hat{a}^{(23)}_{\circ\circ} = \frac{a^{(12)}_{\circ\circ} + a^{(13)}_{\circ\circ} - a^{(23)}_{\circ\circ}}{12} = \frac{(-5) + 1 - (-5)}{12} = \frac{1}{12},$$

so that

$$\hat{A} = \frac{3}{2}\vec{e}_1 \wedge \vec{e}_2 - \frac{5}{4}\vec{e}_2 \wedge \vec{e}_3 + \frac{1}{12}\vec{e}_2 \wedge \vec{e}_3. \tag{9.127}$$

5. Applying Formula (9.15) to the basic vectors of the exterior algebra $\bigwedge_3^{(2)}(\mathbb{R})$, in the bivector A, we have

$$A = -5\vec{e}_1 \wedge \vec{e}_2 + \vec{e}_1 \wedge \vec{e}_3 - 5\vec{e}_2 \wedge \vec{e}_3 = -5(\vec{e}_1 \otimes \vec{e}_2 - \vec{e}_2 \otimes \vec{e}_1)$$
$$+(\vec{e}_1 \otimes \vec{e}_3 - \vec{e}_3 \otimes \vec{e}_1) - 5(\vec{e}_2 \otimes \vec{e}_3 - \vec{e}_3 \otimes \vec{e}_2)$$

and sorting we get

$$A = -5\vec{e}_1 \otimes \vec{e}_2 + \vec{e}_1 \otimes \vec{e}_3 + 5\vec{e}_2 \otimes \vec{e}_1 - 5\vec{e}_2 \otimes \vec{e}_3 - \vec{e}_3 \otimes \vec{e}_1 + 5\vec{e}_3 \otimes \vec{e}_2,$$

so that the matrix representation of the tensor components becomes

$$A = [a^{\alpha\beta}_{\circ\circ}] \equiv \begin{bmatrix} 0 & -5 & 1 \\ 5 & 0 & -5 \\ -1 & 5 & 0 \end{bmatrix}; \text{ anti-symmetric tensor of order } r = 2.$$

6. Now we can apply two different procedures. We will use both options, because even though this complicates the example, it is more informative to the reader.

Procedure (a). Since we already have $\hat{A}$, Formula (9.127), we apply over it again Expression (9.15), which leads to

$$\hat{A} = \frac{3}{2}\vec{e}_1 \wedge \vec{e}_2 - \frac{5}{4}\vec{e}_1 \wedge \vec{e}_3 + \frac{1}{12}\vec{e}_2 \wedge \vec{e}_3$$
$$= \frac{3}{2}(\vec{e}_1 \otimes \vec{e}_2 - \vec{e}_2 \otimes \vec{e}_1) - \frac{5}{4}(\vec{e}_1 \otimes \vec{e}_3 - \vec{e}_3 \otimes \vec{e}_1) + \frac{1}{12}(\vec{e}_2 \otimes \vec{e}_3 - \vec{e}_3 \otimes \vec{e}_2)$$

and sorting, to

$$\hat{A} = \frac{3}{2}\vec{e}_1 \otimes \vec{e}_2 - \frac{5}{4}\vec{e}_1 \otimes \vec{e}_3 - \frac{3}{2}\vec{e}_2 \otimes \vec{e}_1 + \frac{1}{12}\vec{e}_2 \otimes \vec{e}_3 + \frac{5}{4}\vec{e}_3 \otimes \vec{e}_1 - \frac{1}{12}\vec{e}_3 \otimes \vec{e}_2$$

and representing the tensor components in matrix form, we finally obtain

$$\hat{A} = [\hat{a}^{ij}_{\circ\circ}] = \begin{bmatrix} 0 & 3/2 & -5/4 \\ -3/2 & 0 & 1/12 \\ 5/4 & -1/12 & 0 \end{bmatrix}.$$

Procedure (b). Now we start from the tensor components in the tensor basis as an entity $A \in (V \otimes V)^3_h(\mathbb{R})$, and it is subject to a change-of-basis of tensor nature. We know that $\sigma = n^r = 3^2 = 9$, and

$$\hat{a}^{ij}_{\circ\circ} = a^{\alpha\beta}_{\circ\circ}\gamma^{i\circ}_{\circ\alpha}\gamma^{j\circ}_{\circ\beta} \tag{9.128}$$

and writing the equation in matrix form, that is, using Formula (4.36), we have

$$\hat{A}_{\sigma,1} = \left(C^{-1} \otimes C^{-1}\right) \bullet A_{\sigma,1}$$

$$= \left(\frac{1}{12} \begin{bmatrix} 3 & 3 & -6 \\ 5 & -3 & 2 \\ -1 & 3 & 2 \end{bmatrix} \otimes \frac{1}{12} \begin{bmatrix} 3 & 3 & -6 \\ 5 & -3 & 2 \\ -1 & 3 & 2 \end{bmatrix} \right) \bullet \begin{bmatrix} 0 \\ -5 \\ 1 \\ 5 \\ 0 \\ -5 \\ -1 \\ 5 \\ 0 \end{bmatrix}$$

$$= \frac{1}{12} \begin{bmatrix} 0 \\ 18 \\ -15 \\ -18 \\ 0 \\ 1 \\ 15 \\ -1 \\ 0 \end{bmatrix} = \begin{bmatrix} 0 \\ 3/2 \\ -5/4 \\ -3/2 \\ 0 \\ 1/12 \\ 5/4 \\ -1/12 \\ 0 \end{bmatrix}$$

and condensing $\hat{A}_{\sigma,1}$, we obtain the new tensor components

$$\hat{A} = [\hat{a}^{ij}_{\circ\circ}] = \begin{bmatrix} 0 & 3/2 & -5/4 \\ -3/2 & 0 & 1/12 \\ 5/4 & -1/12 & 0 \end{bmatrix}.$$

7. The equality of the strict components of tensor A, calculated in questions 4 and 5, together with the $\hat{A}$ in the new basis, questions 6(a) and 6(b), suggests that such components (those obtained in general for a exterior vector in question 3) could be the same for a tensor change, a suggestion that is to be checked.

We start from the tensor equation (9.128) and we develop it by the classic method of matrix products (in accordance with the criterion of solving the problems by several procedures).

Preparing (9.128), for the matrix process and for arbitrary anti-symmetric tensors, we have

$$\hat{t}^{ij}_{\circ\circ} = \gamma^{i\circ}_{\circ\alpha} t^{\alpha\beta}_{\circ\circ} \gamma^{\circ j}_{\beta\circ},$$

that is,

$$[\hat{t}^{ij}_{\circ\circ}] = (C^{-1}) \begin{bmatrix} 0 & t^{(12)}_{\circ\circ} & t^{(13)}_{\circ\circ} \\ -t^{(12)}_{\circ\circ} & 0 & t^{(23)}_{\circ\circ} \\ -t^{(13)}_{\circ\circ} & -t^{(23)}_{\circ\circ} & 0 \end{bmatrix} (C^{-1})^t$$

$$= \frac{1}{12} \begin{bmatrix} 3 & 3 & -6 \\ 5 & -3 & 2 \\ -1 & 3 & 2 \end{bmatrix} \begin{bmatrix} 0 & t^{(12)}_{\circ\circ} & t^{(13)}_{\circ\circ} \\ -t^{(12)}_{\circ\circ} & 0 & t^{(23)}_{\circ\circ} \\ -t^{(13)}_{\circ\circ} & -t^{(23)}_{\circ\circ} & 0 \end{bmatrix} \frac{1}{12} \begin{bmatrix} 3 & 5 & -1 \\ 3 & -3 & 3 \\ -6 & 2 & 2 \end{bmatrix}$$

$$= \frac{1}{12} \begin{bmatrix} 0 & -2t^{(12)}_{\circ\circ} + 3t^{(13)}_{\circ\circ} - t^{(23)}_{\circ\circ} & t^{(12)}_{\circ\circ} + 2t^{(23)}_{\circ\circ} \\ 2t^{(12)}_{\circ\circ} - 3t^{(13)}_{\circ\circ} + t^{(23)}_{\circ\circ} & 0 & t^{(12)}_{\circ\circ} + t^{(13)}_{\circ\circ} - t^{(23)}_{\circ\circ} \\ -t^{(12)}_{\circ\circ} - 2t^{(23)}_{\circ\circ} & -t^{(12)}_{\circ\circ} - t^{(13)}_{\circ\circ} + t^{(23)}_{\circ\circ} & 0 \end{bmatrix},$$

and identifying with the matrix

$$\hat{t}^{ij}_{\circ\circ} = \begin{bmatrix} 0 & \hat{t}^{(12)}_{\circ\circ} & \hat{t}^{(13)}_{\circ\circ} \\ -\hat{t}^{(12)}_{\circ\circ} & 0 & \hat{t}^{(23)}_{\circ\circ} \\ -\hat{t}^{(13)}_{\circ\circ} & -\hat{t}^{(23)}_{\circ\circ} & 0 \end{bmatrix}$$

we obtain

$$\hat{t}^{(12)}_{\circ\circ} = \frac{1}{12} \left(-2t^{(12)}_{\circ\circ} + 3t^{(13)}_{\circ\circ} - t^{(23)}_{\circ\circ} \right)$$

$$\hat{t}^{(13)}_{\circ\circ} = \frac{1}{12} \left(t^{(12)}_{\circ\circ} + 2t^{(23)}_{\circ\circ} \right)$$

$$\hat{t}^{(23)}_{\circ\circ} = \frac{1}{12} \left(\hat{t}^{(12)}_{\circ\circ} + \hat{t}^{(13)}_{\circ\circ} - \hat{t}^{(23)}_{\circ\circ} \right).$$

In effect they are the same as Formulas (9.124), which proves that it is the same to execute a change-of-basis over a totally anti-symmetric tensor, as it is to execute it over the strict components of its corresponding exterior algebra. In summary, the changes of basis executed with the exterior algebra technique have a *tensor nature*, for the strict components, because the same results are obtained if the change-of-basis is performed over the anti-symmetric tensor of the respective, linear space *tensor product* with the classic homogeneous equations.

$\square$

Example 9.3 (Linked systems). Consider the non-null multivector $\vec{a}_1 \wedge \vec{a}_2 \wedge \cdots \wedge \vec{a}_p \in \bigwedge_n^{(p)}(K)$, where $\forall \vec{a}_i \in V^n(K)$.

Determine the vectors $\vec{b}_j \in V^n(K)$ that satisfy the condition

$$\vec{b}_p \wedge \vec{b}_{p-1} \wedge \cdots \wedge \vec{b}_2 \wedge \vec{b}_1 = \vec{a}_1 \wedge \vec{a}_2 \wedge \cdots \wedge \vec{a}_p; \quad i, j \in I_p.$$

Solution: It is well known that in any exterior algebra $\bigwedge_n^{(p)}(K)$ the following property is satisfied. If $\vec{a} \wedge \vec{b} \cdots \wedge \vec{c} \neq \vec{\Omega}$ then it is $\vec{x} \wedge \vec{a} \wedge \cdots \wedge \vec{c} = \vec{\Omega}$ that implies $\vec{x}$ is a linear combination of vectors $(\vec{a}, \vec{b}, \ldots, \vec{c})$.

With the help of this property, we multiply the given expression by the arbitrary vector $\vec{b}_j \in \{\vec{b}_1, \vec{b}_2, \ldots, \vec{b}_p\}$:

$$\vec{b}_j \wedge \vec{b}_p \wedge \vec{b}_{p-1} \wedge \cdots \wedge \vec{b}_j \wedge \cdots \wedge \vec{b}_2 \wedge \vec{b}_1 = \vec{b}_j \wedge \vec{a}_1 \wedge \vec{a}_2 \wedge \cdots \wedge \vec{a}_p,$$

and since the first member has two equal vectors it is null, so that $\vec{b}_j \wedge \vec{a}_1 \wedge \vec{a}_2 \wedge \cdots \wedge \vec{a}_p = \vec{\Omega}$. Thus, when applying the initially cited property, $\vec{b}_j$ is a linear combination of the remaining vectors

$$\vec{b}_j = \lambda_{\circ j}^{1\circ}\vec{a}_1 + \lambda_{\circ j}^{2\circ}\vec{a}_2 + \cdots + \lambda_{\circ j}^{p\circ}\vec{a}_p; \ \exists\lambda_j \neq 0; \ j \in I_p \qquad (9.129)$$

It remains only to determine the scalars $\lambda_{\circ j}^{i\circ}$.

Adopting the vectors $\{\vec{a}_i\}$ as a basis, which is licit because their number is p and in addition they are a free system since their multivector is non-null, we proceed to the calculation of the exterior product $\vec{b}_p \wedge \vec{b}_{p-1} \wedge \cdots \wedge \vec{b}_2 \wedge \vec{b}_1$. To this end, we first calculate $\vec{b}_1 \wedge \vec{b}_2 \wedge \cdots \wedge \vec{b}_p$ with Formula (9.20), which forces the last ordering:

$$\vec{b}_1 \wedge \vec{b}_2 \wedge \cdots \wedge \vec{b}_p = \begin{vmatrix} \lambda_{\circ 1}^{1\circ} & \lambda_{\circ 2}^{1\circ} & \cdots & \lambda_{\circ p}^{1\circ} \\ \lambda_{\circ 1}^{2\circ} & \lambda_{\circ 2}^{2\circ} & \cdots & \lambda_{\circ p}^{2\circ} \\ \cdots & \cdots & \cdots & \cdots \\ \lambda_{\circ 1}^{p\circ} & \lambda_{\circ 2}^{p\circ} & \cdots & \lambda_{\circ p}^{p\circ} \end{vmatrix} \vec{a}_1 \wedge \vec{a}_2 \wedge \cdots \wedge \vec{a}_p.$$

Next, we proceed to order the indices on the left-hand side. We execute the following transpositions: $\vec{b}_1$ is transposed with $\vec{b}_p$, $\vec{b}_2$ is transposed with $\vec{b}_{p-1}$, $\vec{b}_3$ is transposed with $\vec{b}_{p-2}$, etc. Thus, if p is even there are $p/2$ transpositions, and if it is odd, there are $(p-1)/2$. In addition, as the exterior product is alternate, we get

$$\vec{b}_p \wedge \vec{b}_{p-1} \wedge \cdots \wedge \vec{b}_2 \wedge \vec{b}_1 = \left[(-1)^{p/2} \text{ or } (-1)^{(p-1)/2}\right] \vec{b}_1 \wedge \vec{b}_2 \wedge \cdots \wedge \vec{b}_p$$

and finally, the relation becomes

$$\left[(-1)^{p/2} \text{ or } (-1)^{(p-1)/2}\right] \begin{vmatrix} \lambda_{\circ 1}^{1\circ} & \lambda_{\circ 2}^{1\circ} & \cdots & \lambda_{\circ p}^{1\circ} \\ \lambda_{\circ 1}^{2\circ} & \lambda_{\circ 2}^{2\circ} & \cdots & \lambda_{\circ p}^{2\circ} \\ \cdots & \cdots & \cdots & \cdots \\ \lambda_{\circ 1}^{p\circ} & \lambda_{\circ 2}^{p\circ} & \cdots & \lambda_{\circ p}^{p\circ} \end{vmatrix} \vec{a}_1 \wedge \vec{a}_2 \wedge \cdots \wedge \vec{a}_p = \vec{a}_1 \wedge \vec{a}_2 \wedge \cdots \wedge \vec{a}_p,$$

$$(9.130)$$

which leads to the following conclusion:

If p = even: The vectors $\vec{b}_j$ are chosen according to (9.129). Their scalars $\lambda_{\circ j}^{i\circ}$ must satisfy the condition

$$(-1)^{p/2} \begin{vmatrix} \lambda_{\circ 1}^{1\circ} & \lambda_{\circ 2}^{1\circ} & \cdots & \lambda_{\circ p}^{1\circ} \\ \lambda_{\circ 1}^{2\circ} & \lambda_{\circ 2}^{2\circ} & \cdots & \lambda_{\circ p}^{2\circ} \\ \cdots & \cdots & \cdots & \cdots \\ \lambda_{\circ 1}^{p\circ} & \lambda_{\circ 2}^{p\circ} & \cdots & \lambda_{\circ p}^{p\circ} \end{vmatrix} = 1,$$

in agreement with (9.130).

If p = odd: With respect to the $\vec{b}_j$ we choose them in the same form, with (9.129). The scalars $\lambda_{\circ j}^{i\circ}$ must satisfy the condition

$$(-1)^{(p-1)/2} \begin{vmatrix} \lambda^{1o}_{o1} & \lambda^{1o}_{o2} & \cdots & \lambda^{1o}_{op} \\ \lambda^{2o}_{o1} & \lambda^{2o}_{o2} & \cdots & \lambda^{2o}_{op} \\ \cdots & \cdots & \cdots & \cdots \\ \lambda^{po}_{o1} & \lambda^{po}_{o2} & \cdots & \lambda^{po}_{op} \end{vmatrix} = 1,$$

in agreement with (9.130).

□

Example 9.4 (Bivectors).

1. Consider the vectors $\vec{v}_1 = 2\vec{e}_1 + 3\vec{e}_2 + 4\vec{e}_4$ and $\vec{v}_2 = -\vec{e}_3$, $\vec{v}_1, \vec{v}_2 \in V^4(\mathbb{R})$, referred to a basis $\{\vec{e}_\alpha\}$.
   We wish to know the bivector $\vec{v}_1 \wedge \vec{v}_2 \in \bigwedge_4^{(2)}(\mathbb{R})$.
2. In a linear space $V^3(\mathbb{R})$ with basis $\{\vec{e}_\alpha\}$ we consider the vectors $\vec{v} = 2\vec{e}_1 - \vec{e}_2 + 2\vec{e}_3$ and $\vec{w} = \vec{e}_1 - \vec{e}_2 + \vec{e}_3$.
   a) Obtain the ordinary components of the bivector $\vec{v} \wedge \vec{w}$.
   b) Obtain the strict components of the bivector $\vec{v} \wedge \vec{w}$.

Solution:

1. The data vector matrix is $[X_1 X_2] = \begin{bmatrix} 2 & 0 \\ 3 & 0 \\ 0 & -1 \\ 4 & 0 \end{bmatrix}$, and applying Formula (9.20), we have

$$\vec{v}_1 \wedge \vec{v}_2 = \begin{vmatrix} x^{\alpha_1 o}_{o1} & x^{\alpha_1 o}_{o2} \\ x^{\alpha_2 o}_{o1} & x^{\alpha_2 o}_{o2} \end{vmatrix} \vec{e}_{\alpha_1} \wedge \vec{e}_{\alpha_2}$$

$$= \begin{vmatrix} 2 & 0 \\ 3 & 0 \end{vmatrix} \vec{e}_1 \wedge \vec{e}_2 + \begin{vmatrix} 2 & 0 \\ 0 & -1 \end{vmatrix} \vec{e}_1 \wedge \vec{e}_3 + \begin{vmatrix} 2 & 0 \\ 4 & 0 \end{vmatrix} \vec{e}_1 \wedge \vec{e}_4 + \begin{vmatrix} 3 & 0 \\ 0 & -1 \end{vmatrix} \vec{e}_2 \wedge \vec{e}_3$$
$$+ \begin{vmatrix} 3 & 0 \\ 4 & 0 \end{vmatrix} \vec{e}_2 \wedge \vec{e}_4 + \begin{vmatrix} 0 & -1 \\ 4 & 0 \end{vmatrix} \vec{e}_3 \wedge \vec{e}_4;$$

and since the dimension of $\bigwedge_4^{(2)}(\mathbb{R})$ is $\sigma' = \binom{n}{r} = \binom{4}{2} = 6$ this leads to

$$\vec{v}_1 \wedge \vec{v}_2 = -2\vec{e}_1 \wedge \vec{e}_3 - 3\vec{e}_2 \wedge \vec{e}_3 + 4\vec{e}_3 \wedge \vec{e}_4,$$

which is the exterior vector of components $(0, -2, 0, -3, 0, 4)$ in the basis

$$\{\vec{e}_1 \wedge \vec{e}_2, \vec{e}_1 \wedge \vec{e}_3, \vec{e}_1 \wedge \vec{e}_4, \vec{e}_2 \wedge \vec{e}_3, \vec{e}_2 \wedge \vec{e}_4, \vec{e}_3 \wedge \vec{e}_4\} \text{ of } \bigwedge_4^{(2)}(\mathbb{R}).$$

2. We solve this problem in two different forms.

   Solution by means of the direct procedure: We proceed to execute the exterior product of the two vectors $\vec{v}$ and $\vec{w}$ and passing it immediately to the tensor product

$$\vec{v} \wedge \vec{w} = \vec{v} \otimes \vec{w} - \vec{w} \otimes \vec{v} = (2\vec{e}_1 - \vec{e}_2 + 2\vec{e}_3) \otimes (\vec{e}_1 - \vec{e}_2 + \vec{e}_3)$$
$$- (\vec{e}_1 - \vec{e}_2 + \vec{e}_3) \otimes (2\vec{e}_1 - \vec{e}_2 + 2\vec{e}_3)$$
$$= -\vec{e}_1 \otimes \vec{e}_2 + \vec{e}_2 \otimes \vec{e}_1 + \vec{e}_2 \otimes \vec{e}_3 - \vec{e}_3 \otimes \vec{e}_2. \qquad (9.131)$$

In conclusion $(-1, 1, 0, 0, 1, -1)$ are the bivector $\vec{v} \wedge \vec{w}$ components in the basis

$$\{\vec{e}_1 \otimes \vec{e}_2, \vec{e}_2 \otimes \vec{e}_1, \vec{e}_1 \otimes \vec{e}_3, \vec{e}_3 \otimes \vec{e}_1, \vec{e}_2 \otimes \vec{e}_3, \vec{e}_3 \otimes \vec{e}_2\}$$

of the contravariant and anti-symmetric tensor subspace of order $r = 2$, $(V^3 \otimes V^3)_h^2(\mathbb{R})$, of dimension $r! \binom{n}{r} = 2\binom{3}{2} = 2 \cdot 3 = 6$.
However, the "ordinary components" are defined as a tensor of the tensor space $(V^3 \otimes V^3)^2(\mathbb{R})$ of dimension $\sigma = n^r = 3^2 = 9$, referred to its ordinary basis $\{\vec{e}_1 \otimes \vec{e}_1, \vec{e}_1 \otimes \vec{e}_2, \cdots, \vec{e}_3 \otimes \vec{e}_3\}$ and in matrix form, that leads to

$$[t_{\circ\circ}^{\alpha\beta}] = \begin{bmatrix} 0 & -1 & 0 \\ 1 & 0 & 1 \\ 0 & -1 & 0 \end{bmatrix}.$$

Solution by means of another procedure: Sorting (9.131) by *strict components*, that coincide with those of the *upper triangle* above the diagonal of the matrix $[t_{\circ\circ}^{\alpha\beta}]$, the result is

$$\vec{v} \wedge \vec{w} = (-1)(\vec{e}_1 \otimes \vec{e}_2 - \vec{e}_2 \otimes \vec{e}_1) + 0 \cdot (\vec{e}_1 \otimes \vec{e}_3 - \vec{e}_3 \otimes \vec{e}_1)$$
$$+ 1 \cdot (\vec{e}_2 \otimes \vec{e}_3 - \vec{e}_3 \otimes \vec{e}_2)$$
$$= -\vec{e}_1 \wedge \vec{e}_2 + 0\vec{e}_1 \wedge \vec{e}_3 + \vec{e}_2 \wedge \vec{e}_3$$

of components $(-1, 0, 1)$ in the basis of $\bigwedge_3^{(2)}(\mathbb{R})$.
The previous solution can be executed directly with Formula (9.20), which is to be checked.
In this case, the data vector matrix is $[X_1 X_2] = \begin{bmatrix} 2 & 1 \\ -1 & -1 \\ 2 & 1 \end{bmatrix}$ and

$$\vec{v} \wedge \vec{w} = \begin{vmatrix} x_{\circ 1}^{\alpha_1 \circ} & x_{\circ 2}^{\alpha_1 \circ} \\ x_{\circ 1}^{\alpha_2 \circ} & x_{\circ 2}^{\alpha_2 \circ} \end{vmatrix} \vec{e}_{\alpha_1} \wedge \vec{e}_{\alpha_2}$$

$$= \begin{vmatrix} 2 & 1 \\ -1 & -1 \end{vmatrix} \vec{e}_1 \wedge \vec{e}_2 + \begin{vmatrix} 2 & 1 \\ 2 & 1 \end{vmatrix} \vec{e}_1 \wedge \vec{e}_3 + \begin{vmatrix} -1 & -1 \\ 2 & 1 \end{vmatrix} \vec{e}_2 \wedge \vec{e}_3$$

$$\vec{v} \wedge \vec{w} = -\vec{e}_1 \wedge \vec{e}_2 + 0\,\vec{e}_1 \wedge \vec{e}_3 + \vec{e}_2 \wedge \vec{e}_3$$

with components $(-1, 0, 1)$, which confirms the previous processes.

$\square$

Example 9.5 (Contramodular exterior algebras $\bigwedge_n^{(n)}(\mathbb{R})$). Consider the exterior algebra $\bigwedge_n^{(n)}(\mathbb{R})$ built over the linear space $V^n(\mathbb{R})$, referred to the basis

$\{\vec{e}_\alpha\}$. Consider an exterior tensor T, $T \in \bigwedge_n^{(n)}(\mathbb{R})$ the strict component of which is $a \in \mathbb{R}; a \neq 0$.

We execute in the linear space $V^n(\mathbb{R})$ a change-of-basis of matrix C, the relative modulus of which is $a^{-1} = 1/a$.

1. Obtain the strict component of T in the new basis.
2. Give the "ordinary components" of T, in both bases.

Solution:

1. In Section 8.2.1, on modular tensors, we defined the concept of relative modulus as the determinant $|C|$, so that $|C| = a^{-1}$.

   On the other hand, the relation (9.84) gives the tensor strict relation for the change-of-basis in our exterior algebra, that is,

   $$\hat{\tau} = \hat{t}^{(12\cdots n)}_{\circ\circ\cdots\circ} = t^{(12\cdots n)}_{\circ\circ\cdots\circ}|C|^{-1} = \frac{1}{|C|}\tau,$$

   and since $\tau = t^{(12\cdots n)}_{\circ\circ\cdots\circ} = a$, the result is $\hat{\tau} = \hat{t}^{(12\cdots n)}_{\circ\circ\cdots\circ} = \frac{1}{|C|} \cdot \tau = \frac{1}{a^{-1}} \cdot a = a^2$.

2. Formula (8.103) for modular tensors, permits us to establish the components of a contravariant and anti-symmetric tensor of order r, over $V^n(K)$, in terms of its strict components

   $$t^{\alpha_1\alpha_2\cdots\alpha_r}_{\circ\;\circ\;\cdots\;\circ} = \delta^{\alpha_1\alpha_2\cdots\alpha_r}_{(\beta_1\beta_2\cdots\beta_r)}t^{(\beta_1\beta_2\cdots\beta_r)}_{\circ\;\circ\;\cdots\;\circ}.$$

   Since in our case $r = n$, we have

   $$t^{\alpha_1\alpha_2\cdots\alpha_n}_{\circ\;\circ\;\cdots\;\circ} = \delta^{\alpha_1\alpha_2\cdots\alpha_n}_{(\,1\;\;2\;\cdots\;n\,)}t^{(12\cdots n)}_{\circ\circ\cdots\circ}$$

   or

   $$t^{\alpha_1\alpha_2\cdots\alpha_n}_{\circ\;\circ\;\cdots\;\circ} = \epsilon^{\alpha_1\alpha_2\cdots\alpha_n}_{\circ\;\circ\;\cdots\;\circ}t^{(12\cdots n)}_{\circ\circ\cdots\circ} = \epsilon^{\alpha_1\alpha_2\cdots\alpha_n}_{\circ\;\circ\;\cdots\;\circ}\tau,$$

   the ordinary components of T in the initial basis, since $\tau = a$, are

   $$t^{\alpha_1\alpha_2\cdots\alpha_n}_{\circ\;\circ\;\cdots\;\circ} = \epsilon^{\alpha_1\alpha_2\cdots\alpha_n}_{\circ\;\circ\;\cdots\;\circ}a; \quad \alpha_i \in I_n; \quad \alpha_i \neq \alpha_j;$$

   with a total of $n!$ strict components (without the zero element).
   The ordinary components of T in the new basis, with $\hat{T} = a^2$, are

   $$\hat{t}^{i_1 i_2\cdots i_n}_{\circ\;\circ\cdots\;\circ} = \epsilon^{i_1 i_2\cdots i_n}_{\circ\;\circ\cdots\;\circ} \cdot a^2.$$

   As is well known, $\epsilon^{\alpha_1\alpha_2\cdots\alpha_n}_{\circ\;\circ\;\cdots\;\circ} \equiv \epsilon^{i_1 i_2\cdots i_n}_{\circ\;\circ\cdots\;\circ}$ the Levi-Civita tensor is *isotropic*.

   $\square$

Example 9.6 (Contramodular and comodular determinants). Consider the linear space $V^n(K)$ referred to a basis $\{\vec{e}_\beta\}$. We give the name "determinant" defined over $V^n(K)$ to an alternate form that applies the linear space "direct or total product" in its proper field. $\forall \vec{x}_\alpha \in V^n(K)$, where $\vec{x}_\alpha = x_{\circ\alpha}^{\beta\circ}\vec{e}_\beta$, it is

$$F_a(\vec{x}_1, \vec{x}_2, \ldots, \vec{x}_\alpha, \ldots, \vec{x}_n) = |x_{\circ\alpha}^{\beta\circ}| = \Delta; \quad \Delta \in K.$$

We wish to study the problem of change-of-basis of this tensor:

1. Facing the problem as a tensor function.
2. Based on the analysis of the strict component, contramodular or comodular of the exterior algebras $\bigwedge_n^{(r)}(K)$ and $\bigwedge_{n*}^{(r)}(K)$.

Solution:

1. Consider the initial and new bases of $V^n(\mathbb{R})$, $\{\vec{e}_\beta\}$ and $\{\hat{\vec{e}}_j\}$, respectively. A vector can be represented as $\vec{x}_\alpha = x_{\circ\alpha}^{\beta\circ}\vec{e}_\beta$ and also as $\hat{\vec{x}}_i = x_{\circ i}^{j\circ}\vec{e}_j$ depending on the considered basis.
   Each vector of the determinant (organized in columns) assuming they are contravariant vectors, changes as a contravariant tensor of order $r = 1$, according to the relation $x_{\circ i}^{j\circ} = x_{\circ\alpha}^{\beta\circ}\gamma_{\circ\beta}^{j\circ}$. Taking determinants and using the Binet–Cauchy property, we get

$$|x_{\circ i}^{j\circ}| = |x_{\circ\alpha}^{\beta\circ}||\gamma_{\circ\beta}^{j\circ}|.$$

If we set $|x_{\circ\alpha}^{\beta\circ}| = \Delta_n$ and $|x_{\circ i}^{j\circ}| = \hat{\Delta}_n$, the previous relation becomes $\hat{\Delta}_n = \Delta_n|C^{-1}|$, that is,

$$\hat{\Delta}_n = |C|^{-1}\Delta_n,$$

which proves that the determinant of contravariant vectors changes as a "tensor capacity", i.e., as a contramodular scalar (see modular tensors). Similarly, if we deal with covariant tensors of components $(x_\alpha^*, \hat{x}_i^*)$ with respect to the bases $\{\vec{e}^{*\alpha}, \hat{\vec{e}}^{*i}\}$, the tensor relation is $(x^*)_{j\circ}^{\circ i} = (x^*)_{\beta\circ}^{\circ\alpha}c_{j\circ}^{\circ\beta}$, and taking determinants and notating with one asterisk $(*)$ we have $|(x^*)_{j\circ}^{\circ i}| = |(x^*)_{\beta\circ}^{\circ\alpha}||c_{j\circ}^{\circ\beta}|$, $\hat{\Delta}^* = \Delta^*|C^t|$,

$$\hat{\Delta}^* = |C|\Delta^*,$$

which proves that the determinant of covariant vectors changes as a "tensor density", i.e., as a comodular scalar.

2. Starting from the Formula (9.20), applied to an exterior product of $\bigwedge_n^{(r)}(\mathbb{R})$, the result is

$$\vec{v}_1 \wedge \vec{v}_2 \wedge \cdots \wedge \vec{v}_n = |X_1 X_2 \cdots X_n| \vec{e}_1 \wedge \vec{e}_2 \wedge \cdots \wedge \vec{e}_n,$$

a tensor with strict component $t_{\circ\circ\cdots\circ}^{(12\cdots n)} = |X_1 X_2 \cdots X_n| = \Delta_n$.
Similarly, if we consider the Formula (9.50), applied to an exterior product of $\bigwedge_{n*}^{(n)}(K)$, the result is

$$\vec{v}_1^* \wedge \vec{v}_2^* \wedge \cdots \wedge \vec{v}_n^* = |X_1^* X_2^* \cdots X_n^*| \vec{e}^{*1} \wedge \vec{e}^{*2} \wedge \cdots \wedge \vec{e}^{*n}$$

a tensor with strict component $t_{(12\cdots n)}^{\circ\circ\cdots\circ} = |X_1^* X_2^* \cdots X_n^*| = \Delta_n^*$.
Applying to the cited components the results of the change-of-basis analyzed in Section 9.6, Formulas (9.84): $\hat{t}_{\circ\circ\cdots\circ}^{(12\cdots n)} = \frac{1}{|C|} t_{\circ\circ\cdots\circ}^{(12\cdots n)}$, and (9.86), $\hat{t}_{(12\cdots n)}^{\circ\circ\cdots\circ} = |C| t_{(12\cdots n)}^{\circ\circ\cdots\circ}$, we respectively get

$$\hat{\Delta} = \frac{1}{|C|} \Delta_n \to \hat{\Delta} = |C|^{-1} \Delta_n$$

and

$$\hat{\Delta}^* = |C| \Delta_n^*,$$

which are those previously obtained.

□

Example 9.7 (Exterior mappings). Let F be an exterior p-linear mapping

$$F \in \mathcal{L}\left[\bigwedge_n^{(p)}(K), W^m(K)\right],$$

that applies a p-exterior vector in the vector $\vec{w} \in W^m(K)$ as

$$\forall \vec{x}_i \in V^n(K), F(\vec{x}_1 \wedge \vec{x}_2 \wedge \cdots \wedge \vec{x}_i \wedge \cdots \wedge \vec{x}_p) = \vec{w}; \quad p \le n.$$

1. Show that if $\vec{x}_i = \vec{0}$, this implies $F(\vec{x}_1 \wedge \vec{x}_2 \wedge \cdots \wedge \vec{x}_p) = \vec{0}_W$.
2. Show that $F(\vec{x}_1 \wedge \vec{x}_2 \wedge \cdots \wedge (-\vec{x}_i) \wedge \cdots \wedge \vec{x}_p) = -F(\vec{x}_1 \wedge \vec{x}_2 \wedge \cdots \wedge \vec{x}_i \wedge \cdots \wedge \vec{x}_p)$.

Note: We wish intrinsic proofs, that is, without reference to bases of the linear spaces.

Solution:

1. We know that due to the tensor multilinearity of the anti-symmetric tensors (exterior vectors are anti-symmetric), the following holds:

$$F(\vec{x}_1 \wedge \vec{x}_2 \wedge \cdots \wedge \vec{x}_i \wedge \cdots \wedge \vec{x}_p) = F(\vec{x}_1 \wedge \vec{x}_2 \wedge \cdots \wedge (\vec{x}_i + \vec{0}) \wedge \cdots \wedge \vec{x}_p)$$
$$= F(\vec{x}_1 \wedge \vec{x}_2 \wedge \cdots \wedge \vec{x}_i \wedge \cdots \wedge \vec{x}_p)$$
$$+ F(\vec{x}_1 \wedge \vec{x}_2 \wedge \cdots \wedge \vec{0} \wedge \cdots \wedge \vec{x}_p).$$

Thus, it can be written that in the linear space $W^m(K)$ we have

$$\vec{w} = \vec{w} + F(\vec{x}_1 \wedge \vec{x}_2 \wedge \cdots \wedge \vec{0} \wedge \cdots \wedge \vec{x}_p),$$

which shows that the entity of $W^m(K)$, $F(\vec{x}_1 \wedge \vec{x}_2 \wedge \cdots \wedge \vec{0} \wedge \cdots \wedge \vec{x}_p)$ behaves as if it were the zero $\vec{0}_W$. But, according the properties of all linear spaces, the zero element is *unique*, so that $F(\vec{x}_1 \wedge \vec{x}_2 \wedge \cdots \wedge \vec{0} \wedge \cdots \wedge \vec{x}_p) = \vec{0}_W$.

2. Starting from the previous question, we have

$$\begin{aligned}
\vec{0}_W &= F(\vec{x}_1 \wedge \vec{x}_2 \wedge \cdots \wedge \vec{0} \wedge \cdots \wedge \vec{x}_p) \\
&= F(\vec{x}_1 \wedge \vec{x}_2 \wedge \cdots \wedge (\vec{x}_i + (-\vec{x}_i)) \wedge \cdots \wedge \vec{x}_p) \\
&= F(\vec{x}_1 \wedge \vec{x}_2 \wedge \cdots \wedge \vec{x}_i \wedge \cdots \wedge \vec{x}_p) + F(\vec{x}_1 \wedge \vec{x}_2 \wedge \cdots \wedge (-\vec{x}_i) \wedge \cdots \wedge \vec{x}_p),
\end{aligned}$$

thus, in the image linear space $W^m(K)$, the previous relation can be written as

$$\vec{0}_W = \vec{w} + F(\vec{x}_1 \wedge \vec{x}_2 \wedge \cdots \wedge (-\vec{x}_i) \wedge \cdots \wedge \vec{x}_p),$$

from which we conclude that the entity of $W^m(K)$, $F(\vec{x}_1 \wedge \vec{x}_2 \wedge \cdots \wedge (-\vec{x}_i) \wedge \cdots \wedge \vec{x}_p)$ behaves as if it were the opposite vector to the vector $\vec{w}(\vec{0}_W = \vec{w} + (-\vec{w}))$. But, since the opposite vector is *unique*, we have:

$$F(\vec{x}_1 \wedge \vec{x}_2 \wedge \cdots \wedge (-\vec{x}_i) \wedge \cdots \wedge \vec{x}_p) = -\vec{w},$$

and then

$$F(\vec{x}_1 \wedge \vec{x}_2 \wedge \cdots \wedge (-\vec{x}_i) \wedge \cdots \wedge \vec{x}_p) = -F(\vec{x}_1 \wedge \vec{x}_2 \wedge \cdots \wedge \vec{x}_i \wedge \cdots \wedge \vec{x}_p).$$

□

Example 9.8 (Products of exterior vectors). Consider the exterior tensors:

$$T = t^{(\alpha_1 \alpha_2)}_{\circ \ \circ} \vec{e}_{\alpha_1} \wedge \vec{e}_{\alpha_2}, \ \text{con} \ t^{(\alpha_1 \alpha_2)}_{\circ \ \circ} = \alpha_1 + \alpha_2,$$

and

$$S = s^{(\alpha_1 \alpha_2)}_{\circ \ \circ} \vec{e}_{\alpha_1} \wedge \vec{e}_{\alpha_2}, \ \text{con} \ s^{(\alpha_1 \alpha_2)}_{\circ \ \circ} = \alpha_1 - \alpha_2,$$

belonging to the exterior algebra $\bigwedge_4^{(2)}(\mathbb{R})$.

1. Determine the exterior tensor $W_1 = T \bigwedge S$.
2. Determine the exterior tensor $W_2 = S \bigwedge T$.
3. Determine the exterior tensor $W_3 = T \bigwedge T$.
4. Determine the exterior tensor $W_4 = S \bigwedge S$.
5. Check that it satisfies the Newton binomial formula, calculating $(T + S)^2$.

Solution: We have

$$T \bigwedge S, S \bigwedge T, T \bigwedge T, S \bigwedge S \in \bigwedge_4^{(4)}(\mathbb{R}).$$

1. Applying Formula (9.44) we have

$$W_1 = T \bigwedge S = t^{(\alpha_1 \alpha_2)}_{\;\;\circ\;\;\circ} s^{(\alpha_3 \alpha_4)}_{\;\;\circ\;\;\circ} \delta^{(\alpha_1\;\alpha_2\;\alpha_3\;\alpha_4)}_{(\alpha_1\alpha_2)(\alpha_3\alpha_4)} \vec{e}_{\alpha_1} \wedge \vec{e}_{\alpha_2} \wedge \vec{e}_{\alpha_3} \wedge \vec{e}_{\alpha_4}$$

$$= \Big[t^{(12)}_{\;\circ\;\circ} s^{(34)}_{\;\circ\;\circ} \delta^{(1\;2\;3\;4)}_{(12)(34)} + t^{(13)}_{\;\circ\;\circ} s^{(24)}_{\;\circ\;\circ} \delta^{(1\;2\;3\;4)}_{(13)(24)}$$

$$+ t^{(14)}_{\;\circ\;\circ} s^{(23)}_{\;\circ\;\circ} \delta^{(1\;2\;3\;4)}_{(14)(23)} + t^{(23)}_{\;\circ\;\circ} s^{(14)}_{\;\circ\;\circ} \delta^{(1\;2\;3\;4)}_{(23)(14)}$$

$$+ t^{(24)}_{\;\circ\;\circ} s^{(13)}_{\;\circ\;\circ} \delta^{(1\;2\;3\;4)}_{(24)(13)} + t^{(34)}_{\;\circ\;\circ} s^{(12)}_{\;\circ\;\circ} \delta^{(1\;2\;3\;4)}_{(34)(12)} \Big] \vec{e}_1 \wedge \vec{e}_2 \wedge \vec{e}_3 \wedge \vec{e}_4$$

$$= [3 \cdot (-1) \cdot (+1) + 4 \cdot (-2) \cdot (-1) + 5 \cdot (-1) \cdot (+1) + 5 \cdot (-3) \cdot (+1)$$

$$+ 6 \cdot (-2) \cdot (-1) + 7 \cdot (-1) \cdot (+1)] \vec{e}_1 \wedge \vec{e}_2 \wedge \vec{e}_3 \wedge \vec{e}_4$$

$$= (-10) \vec{e}_1 \wedge \vec{e}_2 \wedge \vec{e}_3 \wedge \vec{e}_4$$

and then
$$W_1 = (-10) \; \vec{e}_1 \wedge \vec{e}_2 \wedge \vec{e}_3 \wedge \vec{e}_4.$$

2. Applying Formula (9.48) we have

$$S \bigwedge T = (-1)^{r^2} T \bigwedge S = (-1)^{2^2} T \bigwedge S = (-1)^4 T \bigwedge S.$$

In summary

$$W_2 = S \bigwedge T = T \bigwedge S = (-10) \vec{e}_1 \wedge \vec{e}_2 \wedge \vec{e}_3 \wedge \vec{e}_4 \text{ (Abelian algebra)}.$$

3. The exterior tensor $T \bigwedge T$ is

$$W_3 = T \bigwedge T = \Big[t^{(12)}_{\;\circ\;\circ} t^{(34)}_{\;\circ\;\circ} \delta^{(1\;2\;3\;4)}_{(12)(34)} + t^{(13)}_{\;\circ\;\circ} t^{(24)}_{\;\circ\;\circ} \delta^{(1\;2\;3\;4)}_{(13)(24)}$$

$$+ t^{(14)}_{\;\circ\;\circ} t^{(23)}_{\;\circ\;\circ} \delta^{(1\;2\;3\;4)}_{(14)(23)} + t^{(23)}_{\;\circ\;\circ} t^{(14)}_{\;\circ\;\circ} \delta^{(1\;2\;3\;4)}_{(23)(14)}$$

$$+ t^{(24)}_{\;\circ\;\circ} t^{(13)}_{\;\circ\;\circ} \delta^{(1\;2\;3\;4)}_{(24)(13)} + t^{(34)}_{\;\circ\;\circ} t^{(12)}_{\;\circ\;\circ} \delta^{(1\;2\;3\;4)}_{(34)(12)} \Big] \vec{e}_1 \wedge \vec{e}_2 \wedge \vec{e}_3 \wedge \vec{e}_4$$

$$= [3 \cdot (7) \cdot (+1) + 4 \cdot (6) \cdot (-1) + 5 \cdot 5 \cdot (+1) + 5 \cdot 5 \cdot (+1)$$

$$+ 6 \cdot 4 \cdot (-1) + 7 \cdot 3 \cdot (+1)] \vec{e}_1 \wedge \vec{e}_2 \wedge \vec{e}_3 \wedge \vec{e}_4$$

$$= 44 \vec{e}_1 \wedge \vec{e}_2 \wedge \vec{e}_3 \wedge \vec{e}_4.$$

4. The exterior tensor $S \bigwedge S$ is:

$$W_4 = S \bigwedge S = \Big[s^{(12)}_{\;\circ\;\circ} s^{(34)}_{\;\circ\;\circ} \delta^{(1\;2\;3\;4)}_{(12)(34)} + s^{(13)}_{\;\circ\;\circ} s^{(24)}_{\;\circ\;\circ} \delta^{(1\;2\;3\;4)}_{(13)(24)}$$

$$+ s^{(14)}_{\;\circ\;\circ} s^{(23)}_{\;\circ\;\circ} \delta^{(1\;2\;3\;4)}_{(14)(23)} + s^{(23)}_{\;\circ\;\circ} s^{(14)}_{\;\circ\;\circ} \delta^{(1\;2\;3\;4)}_{(23)(14)}$$

$$+ s^{(24)}_{\;\circ\;\circ} s^{(13)}_{\;\circ\;\circ} \delta^{(1\;2\;3\;4)}_{(24)(13)} + s^{(34)}_{\;\circ\;\circ} s^{(12)}_{\;\circ\;\circ} \delta^{(1\;2\;3\;4)}_{(34)(12)} \Big] \vec{e}_1 \wedge \vec{e}_2 \wedge \vec{e}_3 \wedge \vec{e}_4$$

$$= [(-1) \cdot (-1) \cdot (+1) + (-2) \cdot (-2) \cdot (-1) + (-3) \cdot (-1) \cdot (+1)$$

$$+ (-1) \cdot (-3) \cdot (+1) + (-2) \cdot (-2) \cdot (-1) + (-1) \cdot (-1) \cdot (+1)]$$

$$\vec{e}_1 \wedge \vec{e}_2 \wedge \vec{e}_3 \wedge \vec{e}_4$$

$$= 0 \; \vec{e}_1 \wedge \vec{e}_2 \wedge \vec{e}_3 \wedge \vec{e}_4 = \vec{0}_4.$$

5. The Newton binomial formula for $(T+S)^2$ is

$$(T+S)^2 = (T+S)\bigwedge(T+S) = T\bigwedge T + T\bigwedge S + S\bigwedge T + S\bigwedge S$$
$$= T\bigwedge T + 2T\bigwedge S + S\bigwedge S = T^2 + 2T\bigwedge S + S^2.$$

We will prove that our tensors satisfy the Newton binomial formula. We have

$$(T+S)^2 = T\wedge T + 2T\wedge S + S\wedge S = [W_3 + 2W_1 + W_4]$$
$$= [44 + 2(-10) + 0]\vec{e}_1 \wedge \vec{e}_2 \wedge \vec{e}_3 \wedge \vec{e}_4 = 24\vec{e}_1 \wedge \vec{e}_2 \wedge \vec{e}_3 \wedge \vec{e}_4.$$

A direct calculation would have given

$$(T+S)^2 = (T+S)\bigwedge(T+S)$$
$$= [(\alpha_1+\alpha_2) + (\alpha_1-\alpha_2)]\vec{e}_{\alpha_1}\wedge\vec{e}_{\alpha_2}\bigwedge[(\alpha_3+\alpha_4) + (\alpha_3-\alpha_4)]\vec{e}_{\alpha_3}\wedge\vec{e}_{\alpha_4}$$
$$= (2\alpha_1)\vec{e}_{\alpha_1} \wedge \vec{e}_{\alpha_2}\bigwedge(2\alpha_3)\vec{e}_{\alpha_3} \wedge \vec{e}_{\alpha_4}$$
$$= 4(\alpha_1\vec{e}_{\alpha_1} \wedge \vec{e}_{\alpha_2})\bigwedge(\alpha_3\vec{e}_{\alpha_3} \wedge \vec{e}_{\alpha_4})$$
$$= 4\left[1\cdot 3\delta^{(1\,2\,3\,4)}_{(12)(34)} + 1\cdot 2\delta^{(1\,2\,3\,4)}_{(13)(24)} + 1\cdot 2\delta^{(1\,2\,3\,4)}_{(14)(23)} + 2\cdot 1\delta^{(1\,2\,3\,4)}_{(23)(14)}\right.$$
$$\left. + 2\cdot 1\delta^{(1\,2\,3\,4)}_{(24)(13)} + 3\cdot 1\delta^{(1\,2\,3\,4)}_{(34)(12)}+\right] \vec{e}_1 \wedge \vec{e}_2 \wedge \vec{e}_3 \wedge \vec{e}_4$$
$$= 24\vec{e}_1 \wedge \vec{e}_2 \wedge \vec{e}_3 \wedge \vec{e}_4,$$

which confirms the previous result

$$(T+S)^2 = 24\vec{e}_1 \wedge \vec{e}_2 \wedge \vec{e}_3 \wedge \vec{e}_4.$$

□

Example 9.9 (Condition for $T \in \bigwedge_n^{(2)}(\mathbb{R})$ to be an exterior product). Let $P = p^{(ij)}_{oo}\vec{e}_i \wedge \vec{e}_j$ be an exterior vector of the algebra $\bigwedge_n^{(2)}(\mathbb{R})$, established over the linear space $V^n(\mathbb{R})$ in the basis $\{\vec{e}_i\}$.

1. Show that for it to be decomposable, that is, for P to proceed from an exterior product $P = \vec{V} \wedge \vec{W}$, it is necessary that it satisfies the relation

$$p^{ij}_{oo}p^{k\ell}_{oo} + p^{ik}_{oo}p^{\ell j}_{oo} + p^{i\ell}_{oo}p^{jk}_{oo} = 0 \tag{9.132}$$

among the non strict components of P.

2. How many different expressions can represent the proposed condition?

Solution:

1. If $P = p_{\substack{\circ\circ}}^{(ij)} \vec{e}_i \wedge \vec{e}_j = p_{\circ\circ}^{ij}(\vec{e}_i \otimes \vec{e}_j - \vec{e}_j \otimes \vec{e}_i)$ is an exterior vector product of the form

$$P = \vec{V} \wedge \vec{W} = \begin{vmatrix} x_{\circ}^i & y_{\circ}^i \\ x_{\circ}^j & y_{\circ}^j \end{vmatrix} (\vec{e}_i \otimes \vec{e}_j - \vec{e}_j \otimes \vec{e}_i),$$

$\forall i < j$ we must have

$$p_{\circ\circ}^{ij} = \begin{vmatrix} x_{\circ}^i & y_{\circ}^i \\ x_{\circ}^j & y_{\circ}^j \end{vmatrix} \quad \text{and} \quad p_{\circ\circ}^{ji} = - \begin{vmatrix} x_{\circ}^i & y_{\circ}^i \\ x_{\circ}^j & y_{\circ}^j \end{vmatrix}. \tag{9.133}$$

We examine the conclusions to which the left-hand expression of (9.132) leads, if we substitute it in (9.133). We have

$$p_{\circ\circ}^{ij}p_{\circ\circ}^{k\ell} + p_{\circ\circ}^{ik}p_{\circ\circ}^{\ell j} + p_{\circ\circ}^{i\ell}p_{\circ\circ}^{jk} = p_{\circ\circ}^{ij}\begin{vmatrix} x_{\circ}^k & y_{\circ}^k \\ x_{\circ}^\ell & y_{\circ}^\ell \end{vmatrix} + p_{\circ\circ}^{ik}\begin{vmatrix} x_{\circ}^\ell & y_{\circ}^\ell \\ x_{\circ}^j & y_{\circ}^j \end{vmatrix} + p_{\circ\circ}^{i\ell}\begin{vmatrix} x_{\circ}^j & y_{\circ}^j \\ x_{\circ}^k & y_{\circ}^k \end{vmatrix}$$

$$= \begin{vmatrix} p_{\circ\circ}^{ij} & x_{\circ}^j & y_{\circ}^j \\ p_{\circ\circ}^{ik} & x_{\circ}^k & y_{\circ}^k \\ p_{\circ\circ}^{i\ell} & x_{\circ}^\ell & y_{\circ}^\ell \end{vmatrix} = \begin{vmatrix} \begin{vmatrix} x_{\circ}^i & y_{\circ}^i \\ x_{\circ}^j & y_{\circ}^j \end{vmatrix} & x_{\circ}^j & y_{\circ}^j \\ \begin{vmatrix} x_{\circ}^i & y_{\circ}^i \\ x_{\circ}^k & y_{\circ}^k \end{vmatrix} & x_{\circ}^k & y_{\circ}^k \\ \begin{vmatrix} x_{\circ}^i & y_{\circ}^i \\ x_{\circ}^\ell & y_{\circ}^\ell \end{vmatrix} & x_{\circ}^\ell & y_{\circ}^\ell \end{vmatrix}$$

$$= \begin{vmatrix} x_{\circ}^i y_{\circ}^j - x_{\circ}^j y_{\circ}^i & x_{\circ}^j & y_{\circ}^j \\ x_{\circ}^i y_{\circ}^k - x_{\circ}^k y_{\circ}^i & x_{\circ}^k & y_{\circ}^k \\ x_{\circ}^i y_{\circ}^\ell - x_{\circ}^\ell y_{\circ}^i & x_{\circ}^\ell & y_{\circ}^\ell \end{vmatrix} = \begin{vmatrix} x_{\circ}^i y_{\circ}^j & x_{\circ}^j & y_{\circ}^j \\ x_{\circ}^i y_{\circ}^k & x_{\circ}^k & y_{\circ}^k \\ x_{\circ}^i y_{\circ}^\ell & x_{\circ}^\ell & y_{\circ}^\ell \end{vmatrix} - \begin{vmatrix} x_{\circ}^j y_{\circ}^i & x_{\circ}^j & y_{\circ}^j \\ x_{\circ}^k y_{\circ}^i & x_{\circ}^k & y_{\circ}^k \\ x_{\circ}^\ell y_{\circ}^i & x_{\circ}^\ell & y_{\circ}^\ell \end{vmatrix}$$

$= 0 - 0 = 0$ (because their columns are proportional).

Then, if $P = \vec{V} \wedge \vec{W}$ then $\rightarrow p_{\circ\circ}^{ij}p_{\circ\circ}^{k\ell} + p_{\circ\circ}^{ik}p_{\circ\circ}^{\ell j} + p_{\circ\circ}^{i\ell}p_{\circ\circ}^{jk} = 0$.

2. If we exchange the indices i and j in (9.132), and again take into account (9.132), we obtain

$$p_{\circ\circ}^{ji}p_{\circ\circ}^{k\ell} + p_{\circ\circ}^{jk}p_{\circ\circ}^{\ell i} + p_{\circ\circ}^{j\ell}p_{\circ\circ}^{ik} = p_{\circ\circ}^{ji}p_{\circ\circ}^{k\ell} + p_{\circ\circ}^{j\ell}p_{\circ\circ}^{ik} + p_{\circ\circ}^{jk}p_{\circ\circ}^{\ell i}$$

$$= -p_{\circ\circ}^{ij}p_{\circ\circ}^{k\ell} - p_{\circ\circ}^{ik}p_{\circ\circ}^{\ell j} - p_{\circ\circ}^{i\ell}p_{\circ\circ}^{jk} = 0$$

and then

$$P = \vec{V} \wedge \vec{W} \rightarrow p^{ji}_{\circ\circ}p^{k\ell}_{\circ\circ} + p^{j\ell}_{\circ\circ}p^{ik}_{\circ\circ} + p^{jk}_{\circ\circ}p^{\ell i}_{\circ\circ} = 0.$$

If we exchange the indices j and k in (9.132), we get

$$p^{ik}_{\circ\circ}p^{j\ell}_{\circ\circ} + p^{ij}_{\circ\circ}p^{\ell k}_{\circ\circ} + p^{i\ell}_{\circ\circ}p^{kj}_{\circ\circ} = p^{ij}_{\circ\circ}p^{\ell k}_{\circ\circ} + p^{ik}_{\circ\circ}p^{j\ell}_{\circ\circ} + p^{i\ell}_{\circ\circ}p^{kj}_{\circ\circ}$$

$$= -p^{ij}_{\circ\circ}p^{k\ell}_{\circ\circ} - p^{ik}_{\circ\circ}p^{\ell j}_{\circ\circ} - p^{i\ell}_{\circ\circ}p^{jk}_{\circ\circ} = 0$$

and then

$$P = \vec{V} \wedge \vec{W} \rightarrow p^{ij}_{\circ\circ}p^{\ell k}_{\circ\circ} + p^{ik}_{\circ\circ}p^{j\ell}_{\circ\circ} + p^{i\ell}_{\circ\circ}p^{kj}_{\circ\circ} = 0.$$

If we exchange indices i and k in (9.132), it is also satisfied, which proves the intrinsic nature of the relation (9.132) for the exterior vectors of order $r = 2$, which are *decomposable*.

The concrete number of different relations is given by the different possibilities of the numerical values that can take *the indices ahead* $(ijk\ell)$ of the formula, because the order of the rest depend on them. Thus, the number of relations to be stated in each case is

$$C_{n,4} = \binom{n}{4}.$$

$\square$

Example 9.10 (Tensor and exterior algebras). Let

$$T = t^{\circ\circ\lambda\mu}_{\alpha\beta\circ\circ}\vec{e}^{*\alpha} \otimes \vec{e}^{*\beta} \otimes \vec{e}_\lambda \otimes \vec{e}_\mu, \quad U = u^{\lambda\mu}_{\circ\circ}\vec{e}_\lambda \otimes \vec{e}_\mu \text{ and } V = v^{\circ\circ}_{\lambda\mu}\vec{e}^{*\lambda} \otimes \vec{e}^{*\mu},$$

be three homogeneous tensors defined over the linear space $V^2(\mathbb{R})$, with components

$$[t^{\circ\circ\lambda\mu}_{\alpha\beta\circ\circ}] = \left[\begin{array}{cc|cc} 0 & 0 & 1 & -2 \\ 0 & 0 & -1 & 3 \\ \hline -1 & 2 & 0 & 0 \\ 1 & -3 & 0 & 0 \end{array}\right]; \quad [u^{\lambda\mu}_{\circ\circ}] = \left[\begin{array}{cc} 1 & 2 \\ -1 & 3 \end{array}\right]; \quad v^{\circ\circ}_{\lambda\mu} = \left[\begin{array}{cc} 0 & -1 \\ 1 & 2 \end{array}\right],$$

where α and β are the row and the column block, respectively, and λ and μ are the rows and the columns of each submatrix, respectively.

1. If we perform a change-of-basis in $V^2(\mathbb{R})$, defined by

$$\vec{e}_1 = \hat{\vec{e}}_1 - \hat{\vec{e}}_2; \quad \vec{e}_2 = -\hat{\vec{e}}_1 + 2\hat{\vec{e}}_2,$$

   obtain the components of the tensors T, U and V in the new basis.
2. Obtain the doubly contracted tensors of T, indicating its nature, assuming the initial basis.
3. Obtain, in the initial basis, the tensors R and S, defined as

$$r^{\lambda\mu}_{\circ\circ} = t^{\circ\circ\lambda\mu}_{\alpha\beta\circ\circ} u^{\alpha\beta}_{\circ\circ}; \quad s^{\circ\circ}_{\alpha\beta} = v^{\circ\circ}_{\lambda\mu} t^{\circ\circ\lambda\mu}_{\alpha\beta\circ\circ}.$$

4. Study the symmetries and anti-symmetries of tensor T.

5. If $\vec{a} = 5\vec{e}^{*1} + 2\vec{e}^{*2}$, obtain the conditions to be satisfied by the components of the vector $\vec{b} \in V_*^2(\mathbb{R})$ for $\vec{a} \wedge \vec{b} = s_{\alpha\beta}^{\circ\circ}$.

Solution:

1. As the given change-of-basis equations give the *inverse* information, it is necessary to invert it, to find the direct change-of-basis matrix C. We have $\sigma_1 = n^r = 2^4 = 16$, and

$$[\vec{e}_1\vec{e}_2] = [\hat{\vec{e}}_1\hat{\vec{e}}_2]\begin{bmatrix} 1 & -1 \\ -1 & 2 \end{bmatrix} \rightarrow C^{-1} = \begin{bmatrix} 1 & -1 \\ -1 & 2 \end{bmatrix};$$

$$C = \begin{bmatrix} 1 & -1 \\ -1 & 2 \end{bmatrix}^{-1} = \begin{bmatrix} 2 & 1 \\ 1 & 1 \end{bmatrix}; \quad C^t = \begin{bmatrix} 2 & 1 \\ 1 & 1 \end{bmatrix}; \quad |C| = 1.$$

The change-of-basis tensor equation is

$$t_{ijoo}^{\circ\circ\ell m} = t_{\alpha\beta\circ\circ}^{\circ\circ\lambda\mu}c_{io}^{\circ\alpha}c_{jo}^{\circ\beta}\gamma_{o\lambda}^{\ell o}\gamma_{o\mu}^{mo},$$

which in matrix form becomes

$$\hat{T}_{16,1} = (C^t \otimes C^t \otimes C^{-1} \otimes C^{-1}) \bullet T_{16,1}$$

$$= \left(\begin{bmatrix} 4 & 2 & 2 & 1 \\ 2 & 2 & 1 & 1 \\ 2 & 1 & 2 & 1 \\ 1 & 1 & 1 & 1 \end{bmatrix} \otimes \begin{bmatrix} 1 & -1 & -1 & 1 \\ -1 & 2 & 1 & -2 \\ -1 & 1 & 2 & -2 \\ 1 & -2 & -2 & 4 \end{bmatrix}\right) \bullet T_{16,1}$$

$$=
\left[\begin{array}{cccc|cccc|cccc|cccc}
4 & -4 & -4 & 4 & 2 & -2 & -2 & 2 & 2 & -2 & -2 & 2 & 1 & -1 & -1 & 1 \\
-4 & 8 & 4 & -8 & -2 & 4 & 2 & -4 & -2 & 4 & 2 & -4 & -1 & 2 & 1 & -2 \\
-4 & 4 & 8 & -8 & -2 & 2 & 4 & -4 & -2 & 2 & 4 & -4 & -1 & 1 & 2 & -2 \\
4 & -8 & -8 & 16 & 2 & -4 & -4 & 8 & 2 & -4 & -4 & 8 & 1 & -2 & -2 & 4 \\
\hline
2 & -2 & -2 & 2 & 2 & -2 & -2 & 2 & 1 & -1 & -1 & 1 & 1 & -1 & -1 & 1 \\
-2 & 4 & 2 & -4 & -2 & 4 & 2 & -4 & -1 & 2 & 1 & -2 & -1 & 2 & 1 & -2 \\
-2 & 2 & 4 & -4 & -2 & 2 & 4 & -4 & -1 & 1 & 2 & -2 & -1 & 1 & 2 & -2 \\
2 & -4 & -4 & 8 & 2 & -4 & -4 & 8 & 1 & -2 & -2 & 4 & 1 & -2 & -2 & 4 \\
\hline
2 & -2 & -2 & 2 & 1 & -1 & -1 & 1 & 2 & -2 & -2 & 2 & 1 & -1 & -1 & 1 \\
-2 & 4 & 2 & -4 & -1 & 2 & 1 & -2 & -2 & 4 & 2 & -4 & -1 & 2 & 1 & -2 \\
-2 & 2 & 4 & -4 & -1 & 1 & 2 & -2 & -2 & 2 & 4 & -4 & -1 & 1 & 2 & -2 \\
2 & -4 & -4 & 8 & 1 & -2 & -2 & 4 & 2 & -4 & -4 & 8 & 1 & -2 & -2 & 4 \\
\hline
1 & -1 & -1 & 1 & 1 & -1 & -1 & 1 & 1 & -1 & -1 & 1 & 1 & -1 & -1 & 1 \\
-1 & 2 & 1 & -2 & -1 & 2 & 1 & -2 & -1 & 2 & 1 & -2 & -1 & 2 & 1 & -2 \\
-1 & 1 & 2 & -2 & -1 & 1 & 2 & -2 & -1 & 1 & 2 & -2 & -1 & 1 & 2 & -2 \\
1 & -2 & -2 & 4 & 1 & -2 & -2 & 4 & 1 & -2 & -2 & 4 & 1 & -2 & -2 & 4
\end{array}\right]
\bullet
\begin{bmatrix} 0 \\ 0 \\ 0 \\ 0 \\ \hline 1 \\ -2 \\ -1 \\ 3 \\ \hline -1 \\ 2 \\ 1 \\ -3 \\ \hline 0 \\ 0 \\ 0 \\ 0 \end{bmatrix}
=
\begin{bmatrix} 0 \\ 0 \\ 0 \\ 0 \\ \hline 7 \\ -12 \\ -11 \\ 19 \\ \hline -7 \\ 12 \\ 11 \\ -19 \\ \hline 0 \\ 0 \\ 0 \\ 0 \end{bmatrix}
$$

which condensed gives the tensor T in the new basis

$$[t^{\circ\circ\ell m}_{ij\circ\circ}] = \left[\begin{array}{cc|cc} 0 & 0 & 7 & -12 \\ 0 & 0 & -11 & 19 \\ \hline -7 & 12 & 0 & 0 \\ 11 & -19 & 0 & 0 \end{array}\right].$$

We analyze the tensor U. The change-of-basis tensor equation is ($\sigma_2 = 2^2 = 4$):

$$u^{\ell m}_{\circ\circ} = u^{\lambda\mu}_{\circ\circ}\gamma^{\ell\circ}_{\circ\lambda}\gamma^{m\circ}_{\circ\mu}$$

and in matrix form

$$\hat{U}_{4,1} = (C^{-1} \otimes C^{-1}) \bullet U_{4,1} = \begin{bmatrix} 1 & -1 & -1 & 1 \\ -1 & 2 & 1 & -2 \\ -1 & 1 & 2 & -2 \\ 1 & -2 & -2 & 4 \end{bmatrix} \bullet \begin{bmatrix} 1 \\ 2 \\ -1 \\ 3 \end{bmatrix} = \begin{bmatrix} 3 \\ -4 \\ -7 \\ 11 \end{bmatrix},$$

which condensed gives the tensor U in the new basis $[u^{\ell m}_{\circ\circ}] = \begin{bmatrix} 3 & -4 \\ -7 & 11 \end{bmatrix}$.

Finally, for the tensor V: $v^{\circ\circ}_{\ell m} = v^{\circ\circ}_{\lambda\mu}c^{\lambda}_{\ell\circ}c^{\circ\mu}_{m\circ}$ in matrix form, with $\sigma_2 = 2^2 = 4$, we get

$$\hat{V}_{4,1} = (C^t \otimes C^t) \bullet V_{4,1} = \begin{bmatrix} 4 & 2 & 2 & 1 \\ 2 & 2 & 1 & 1 \\ 2 & 1 & 2 & 1 \\ 1 & 1 & 1 & 1 \end{bmatrix} \bullet \begin{bmatrix} 0 \\ -1 \\ 1 \\ 2 \end{bmatrix} = \begin{bmatrix} 2 \\ 1 \\ 3 \\ 2 \end{bmatrix},$$

which condensed gives the tensor V in the new basis $[v^{\circ\circ}_{\ell m}] = \begin{bmatrix} 2 & 1 \\ 3 & 2 \end{bmatrix}$.

2. The possible double contractions of the tensor T are:

  1.

$$M = C\begin{pmatrix} \lambda & \mu \\ \alpha & \beta \end{pmatrix} t^{\circ\circ\lambda\mu}_{\alpha\beta\circ\circ}$$

$$= t^{\circ\circ\theta w}_{\theta w\circ\circ} = t^{\circ\circ11}_{11\circ\circ} + t^{\circ\circ12}_{12\circ\circ} + t^{\circ\circ21}_{21\circ\circ} + t^{\circ\circ22}_{22\circ\circ} = 0 + (-2) + 1 + 0 = -1;$$

  2.

$$N = C\begin{pmatrix} \lambda & \mu \\ \beta & \alpha \end{pmatrix} t^{\circ\circ\lambda\mu}_{\alpha\beta\circ\circ}$$

$$= t^{\circ\circ\theta w}_{w\theta\circ\circ} = t^{\circ\circ11}_{11\circ\circ} + t^{\circ\circ12}_{21\circ\circ} + t^{\circ\circ21}_{12\circ\circ} + t^{\circ\circ22}_{22\circ\circ} = 0 + 2 + (-1) + 0 = 1.$$

In both cases they are invariant scalars (tensors of zero order).
We check its invariant nature by calculating the same contractions in the new basis:

  1.

$$M = C\begin{pmatrix} \ell & m \\ i & j \end{pmatrix} \hat{t}^{\circ\circ\ell m}_{ij\circ\circ}$$

$$= \hat{t}^{\circ\circ11}_{11\circ\circ} + \hat{t}^{\circ\circ21}_{12\circ\circ} + \hat{t}^{\circ\circ21}_{21\circ\circ} + \hat{t}^{\circ\circ22}_{22\circ\circ} = 0 + (-12) + 11 + 0 = -1$$

2.

$$N = C \left(\begin{array}{c|c} \ell & m \\ j & i \end{array} \right) \hat{t}_{ijoo}^{oo\ell m}$$

$$= \hat{t}_{11oo}^{oo11} + \hat{t}_{21oo}^{oo12} + \hat{t}_{12oo}^{oo21} + \hat{t}_{22oo}^{oo22} = 0 + 12 + (-11) + 0 = 1$$

with identical results.

Following the criterion of maximum information, we execute again the contractions, but with the help of computer, that is, by means of model 2 of double contraction, Formula (5.76), that applied to our case leads to the matrix of Example 5.6, fourth method (a), that is,

$$M = [1\,0\,0\,0\,0\,1\,0\,0\,|\,0\,0\,1\,0\,0\,0\,0\,1] \bullet \begin{bmatrix} 0 \\ 0 \\ 0 \\ 0 \\ - \\ 1 \\ -2 \\ -1 \\ 3 \\ - \\ -1 \\ 2 \\ 1 \\ -3 \\ - \\ 0 \\ 0 \\ 0 \\ 0 \end{bmatrix} = (-2) + 1 = -1.$$

Similarly, for the second contraction, Formula (5.77), which applied to our case leads to the matrix of Example 5.6, fourth method (b), that is,

$$N = [1\,0\,0\,0\,0\,0\,1\,0\,|\,0\,1\,0\,0\,0\,0\,0\,1] \bullet \begin{bmatrix} 0 \\ 0 \\ 0 \\ 0 \\ - \\ 1 \\ -2 \\ -1 \\ 3 \\ - \\ -1 \\ 2 \\ 1 \\ -3 \\ - \\ 0 \\ 0 \\ 0 \\ 0 \end{bmatrix} = (-1) + 2 = 1$$

with the same results.

3. We proceed to determine the first contraction tensor

$$R \equiv [r^{\lambda\mu}_{\circ\circ}] = [t^{\circ\circ\lambda\mu}_{\alpha\beta\circ\circ}]u^{\alpha\beta}_{\circ\circ} = [t^{\circ\circ\lambda\mu}_{11\circ\circ}]u^{11}_{\circ\circ} + [t^{\circ\circ\lambda\mu}_{12\circ\circ}]u^{12}_{\circ\circ} + [t^{\circ\circ\lambda\mu}_{21\circ\circ}]u^{21}_{\circ\circ} + [t^{\circ\circ\lambda\mu}_{22\circ\circ}]u^{22}_{\circ\circ},$$

which numerically gives

$$R = \begin{bmatrix} 0 & 0 \\ 0 & 0 \end{bmatrix} \cdot 1 + \begin{bmatrix} 1 & -2 \\ -1 & 3 \end{bmatrix} \cdot 2 + \begin{bmatrix} -1 & 2 \\ 1 & -3 \end{bmatrix} \cdot (-1) + \begin{bmatrix} 0 & 0 \\ 0 & 0 \end{bmatrix} \cdot 3 = \begin{bmatrix} 3 & -6 \\ -3 & 9 \end{bmatrix}.$$

Next, we analyze the second contraction-tensor

$$S = [s^{\circ\circ}_{\alpha\beta}] = v^{\circ\circ}_{\lambda\mu}[t^{\circ\circ\lambda\mu}_{\alpha\beta\circ\circ}]$$

$$= v^{\circ\circ}_{11}[t^{\circ\circ11}_{\alpha\beta\circ\circ}] + v^{\circ\circ}_{12}[t^{\circ\circ12}_{\alpha\beta\circ\circ}] + v^{\circ\circ}_{21}[t^{\circ\circ21}_{\alpha\beta\circ\circ}] + v^{\circ\circ}_{22}[t^{\circ\circ22}_{\alpha\beta\circ\circ}]$$

$$= 0\begin{bmatrix} 0 & 1 \\ -1 & 0 \end{bmatrix} + (-1)\begin{bmatrix} 0 & -2 \\ 2 & 0 \end{bmatrix} + 1\begin{bmatrix} 0 & -1 \\ 1 & 0 \end{bmatrix} + 2\begin{bmatrix} 0 & 3 \\ -3 & 0 \end{bmatrix} = \begin{bmatrix} 0 & 7 \\ -7 & 0 \end{bmatrix}.$$

4. T is a mixed tensor of order $r = 4$, with $(p = 2, q = 2)$, and anti-symmetric with respect to the covariant indices.

5. Since $S = s^{\circ\circ}_{\alpha\beta}\vec{e}^{*\alpha} \otimes \vec{e}^{*\beta}$ with $S = \begin{bmatrix} 0 & 7 \\ -7 & 0 \end{bmatrix}$, and

$$\vec{a} \wedge \vec{b} = (5\vec{e}^{*1} + 2\vec{e}^{*2}) \wedge (b_1\vec{e}^{*1} + b_2\vec{e}^{*2}) = \begin{vmatrix} 5 & b_1 \\ 2 & b_2 \end{vmatrix} \vec{e}^{*1} \wedge \vec{e}^{*2}$$

$$= (5b_2 - 2b_1)(\vec{e}^{*1} \otimes \vec{e}^{*2} - \vec{e}^{*2} \otimes \vec{e}^{*1})$$

of associated matrix $\begin{bmatrix} 0 & (5b_2 - 2b_1) \\ -(5b_2 - 2b_1) & 0 \end{bmatrix}$, identifying this with S, we get the desired condition:

$$5b_2 - 2b_1 = 7 \rightarrow \frac{b_1 + 1}{5} = \frac{b_2 - 1}{2}.$$

□

Example 9.11 (Decomposition of an exterior tensor). Given a linear space $V^n(K)$, we consider the exterior algebra $\bigwedge_n^{(p)}(K)$. Let $T \in \bigwedge_n^{(p)}(K)$ be an exterior tensor, that is, a homogeneous contravariant and anti-symmetric tensor of order $r = p$, given by its strict components and such that $T \neq \Omega_p$. We denote by $L_{(n-p)}$ the set

$$L_{(n-p)} = \{\vec{x} \in V^n(K) | T \wedge \vec{x} = \Omega_{(p+1)}\},$$

where $\Omega_{(p+1)}$ is the null tensor of the algebra $\bigwedge_n^{(p+1)}(K)$.

1. If $p = n - 1$, show that L_1 is a linear subspace of $V^n(K)$ and find their Cartesian equation.

2. Show that the dimension of $L_{(n-p)}$ satisfies dim $L_{(n-p)} \le p$.

3. If $T = \vec{V}_1 \wedge \vec{V}_2 \wedge \cdots \wedge \vec{V}_p \ne \Omega_p$, show that the necessary and sufficient condition for $\forall \vec{V}_i \in L_{(n-p)} \subset V^n(K)$; $i \in I_p$, is that dim $L_{(n-p)} = p$, and that the vectors $\vec{V}_1, \vec{V}_2, \ldots, \vec{V}_p$ satisfy the equations of $L_{(n-p)}$.
This is the fundamental decomposition theorem.

Solution:

1. We remind the reader of Properties 6, 7 and 8 of Section 9.2.1, of the r-vectors that belong to the different algebras $\bigwedge_n^{(r)}(K)$.

In particular, we point out that if in the cited exterior algebra we consider the multivector $\vec{V}_1 \wedge \vec{V}_2 \wedge \cdots \wedge \vec{V}_r \ne \Omega_r$, which requires the system $(\vec{V}_1, \vec{V}_2, \ldots, \vec{V}_r)$ in $V^n(K)$ to be free, and it is exteriorly multiplied by another vector $\vec{x} \in V^n(K)$, only two options are possible:

Option (a): The vector system $(\vec{V}_1, \vec{V}_2, \ldots, \vec{V}_r, \vec{x})$ is free in $V^n(K)$. Then, $\vec{V}_1 \wedge \vec{V}_2 \wedge \cdots \wedge \vec{V}_r \wedge \vec{x} \ne \Omega_{(r+1)}$ belongs to $\bigwedge_n^{(r+1)}(K)$ (it belongs to *another* exterior algebra).

Option (b): The vector system $(\vec{V}_1, \vec{V}_2, \cdots, \vec{V}_r, \vec{x})$ is linked in $V^n(K)$. Then $\vec{x}$ belongs to the linear subspace $\Pi \subset V^n(K)$ generated by the basis of $\Pi : \mathcal{B}_0 = \{\vec{V}_1, \vec{V}_2, \cdots, \vec{V}_r\}$.

Then, $\vec{V}_1 \wedge \vec{V}_2 \wedge \ldots \wedge \vec{V}_r \wedge \vec{x} \equiv \Omega_r$; $\Omega_r \in \bigwedge_n^{(r)}(K)$, and $\vec{x}$ is a linear combination of $\mathcal{B}_0$.

These two options are not theorems, but simple corollaries of the mentioned Properties 5, 6 and 7.

Next, we prove that the question we will solve is no more that a "sufficient" way of looking option (b).

Consider the vector $\vec{x} = x_o^1 \vec{e}_1 + x_o^2 \vec{e}_2 + \cdots + x_o^n \vec{e}_n \in V^n(K)$, let $r = n - 1$, and consider also the tensor $T \in \bigwedge_n^{(n-1)}(K)$:

$$T = t_{\circ\circ\circ\cdots}^{(234\cdots(n-1)n)} {}_\circ {}_\circ \vec{e}_2 \wedge \vec{e}_3 \wedge \vec{e}_4 \wedge \cdots \wedge \vec{e}_n$$

$$+ t_{\circ\circ\circ\cdots}^{(134\cdots(n-1)n)} {}_\circ {}_\circ \vec{e}_1 \wedge \vec{e}_3 \wedge \vec{e}_4 \wedge \cdots \wedge \vec{e}_n$$

$$+ \cdots + t_{\circ\circ\circ\cdots}^{(123\cdots(n-2)n)} {}_\circ {}_\circ \vec{e}_1 \wedge \vec{e}_2 \wedge \vec{e}_3 \wedge \cdots \wedge \vec{e}_{n-2} \wedge \vec{e}_n$$

$$+ t_{\circ\circ\circ\cdots}^{(123\cdots(n-2)(n-1))} {}_\circ {}_\circ \vec{e}_1 \wedge \vec{e}_2 \wedge \vec{e}_3 \wedge \cdots \wedge \vec{e}_{n-2} \wedge \vec{e}_{n-1}$$

also with n summands.

If we impose the condition $T \wedge \vec{x} = \Omega_n$, after *distributing* each summand of T with each summand of $\vec{x}$ to the right of T, from the n^2 resulting summands, only n are null (because the rest have repeated vectors), that is,

$$T \wedge \vec{x} = t_{\circ\circ\circ\cdots\circ}^{(234\cdots n)} x_o^1 [\vec{e}_2 \wedge \vec{e}_3 \wedge \cdots \wedge \vec{e}_n] \wedge \vec{e}_1$$

$$+ t^{(134\cdots n)}_{\circ\circ\circ\cdots\circ} x_\circ^2 [\vec{e}_1 \wedge \vec{e}_3 \wedge \cdots \wedge \vec{e}_n] \wedge \vec{e}_2 + \cdots$$

$$+ t^{(123\cdots (n-2)n)}_{\circ\circ\circ\cdots\circ\circ} x^{(n-1)}_{\circ} [\vec{e}_1 \wedge \vec{e}_2 \wedge \cdots \wedge \vec{e}_{n-2} \wedge \vec{e}_n] \wedge \vec{e}_{n-1}$$

$$+ t^{(123\cdots (n-2)(n-1))}_{\circ\circ\circ\cdots\circ\circ} x^n_{\circ} [\vec{e}_1 \wedge \vec{e}_2 \wedge \cdots \wedge \vec{e}_{n-2} \wedge \vec{e}_{n-1}] \wedge \vec{e}_n.$$

Imposing a strict order to the basic multivector products, each term takes the sign $+$ or $-$ depending on the number of transpositions, and all of them the common multivector $(\vec{e}_1 \wedge \vec{e}_2 \wedge \cdots \wedge \vec{e}_n)$, which is a common factor:

$$T \wedge \vec{x} = \Big[(-1)^{n-1} t^{(234\cdots n)}_{\circ\circ\circ\cdots\circ} x^1_{\circ} + (-1)^{n-2} t^{(134\cdots n)}_{\circ\circ\circ\cdots\circ} x^2_{\circ} + \cdots$$

$$+ (-1)^1 t^{(123\cdots (n-2)n)}_{\circ\circ\circ\cdots\circ\circ} x^{(n-1)}_{\circ} + t^{(123\cdots (n-2)(n-1))}_{\circ\circ\circ\cdots\circ\circ} x^n_{\circ} \Big] (\vec{e}_1 \wedge \vec{e}_2 \wedge \cdots \wedge \vec{e}_n).$$

After adding the condition $T \wedge \vec{x} = \Omega_n$, one of the two factors must be zero, but the multivector is the basis of the exterior algebra $\bigwedge_n^{(r)}(K)$, which cannot be zero because we have multiplied free vectors (a basis) of $V^n(K)$. Thus, the zero factor is the bracketed factor that operated leads to the expression

$$L_1 \equiv t^{(123\cdots (n-2)(n-1))}_{\circ\circ\circ\cdots\circ\circ} x^n - t^{(123\cdots (n-2)n)}_{\circ\circ\circ\cdots\circ\circ} x^{(n-1)} + \cdots$$

$$+ (-1)^{n-2} t^{(134\cdots n)}_{\circ\circ\circ\cdots\circ} x^2 + (-1)^{n-1} t^{(234\cdots n)}_{\circ\circ\circ\cdots\circ} x^1 = 0, \quad (9.134)$$

which can be interpreted in the frame of the linear space $V^n(K)$ as the Cartesian equation of the $(n-1)$-dimensional linear subspace, L_1, which is directly known as soon as an exterior tensor $T \in \bigwedge_n^{(n-1)}(K)$ is given, because with the tensor components T, we can build the first member of the Cartesian equation of L_1.

2. As in the exterior algebra $\bigwedge_n^{(n-1)}(K)$, the order is $p = n-1$ and $\dim L_1 = n-1$, and the linear subspace L_1 is the one of maximum dimension that can contain $V^n(K)$, we have that any other linear subspace $L_{(n-p)}$ being related to arbitrary tensors $T \in \bigwedge_n^{(p)}(K)$ of order p will satisfy $\dim L_{(n-p)} \le \dim L_1 = n-1 = p$, that is, $\dim L_{(n-p)} \le p$.

3. a) Necessary condition: If T is a product multivector of the free vectors $\{\vec{v}_1, \vec{v}_2, \ldots, \vec{v}_i, \ldots, \vec{v}_p\}$ and is $T \wedge \vec{x} = \Omega_{(p+1)}$, when replacing $\vec{x} = \vec{V}_i, i \in I_p$, on the left hand side, because there are repeated vectors, we get

$$T \wedge \vec{V}_i \equiv (\vec{V}_1 \wedge \vec{V}_2 \wedge \cdots \wedge \vec{V}_i \wedge \cdots \wedge \vec{V}_p) \wedge \vec{V}_i \equiv \Omega_{(p+1)},$$

which shows that as $T \wedge \vec{x} = \Omega_{(p+1)}$ leads to the Cartesian equations of $L_{(n-p)}$, then, the vectors $\vec{V}_i$ of the multivector, satisfy the equation of $L_{(n-p)}$.

In summary, to decompose an exterior tensor in products as a multi-vector, it is necessary to know a basis of the associated linear subspace, of dimension dim $L_{(n-p)} = p$ with $p \leq n - 1$.

If $p = n$, *any* exterior vector $T \in \bigwedge_n^{(n)}(K)$ can immediately be written as a multivector (n-vector).

b) Sufficient condition: As the set $L_1 \equiv \{\vec{x} \in V^n(K) | T \wedge \vec{x} = \Omega_n\}$ is a linear $(n-1)$-dimensional subspace and $T \in \bigwedge_n^{(n-1)}(K)$, the condition in (2) holds in the sufficient sense dim $L_{(n-p)} = $ dim $L_1 = p = n - 1$.

□

Example 9.12 (Building multivectors by contraction). Consider the homogeneous contravariant tensors, A and B built over the linear space $V^4(\mathbb{R})$, of components

$$[a^{\beta\gamma}_{\circ\circ}] = \begin{bmatrix} 0 & 1 & -1 & 0 \\ -1 & 0 & 0 & 1 \\ 1 & 0 & 0 & -1 \\ 0 & -1 & 1 & 0 \end{bmatrix} ;$$

$$[b^{\alpha\beta\gamma}_{\circ\circ\circ}] = \begin{bmatrix} 1 & 0 & 0 & 0 \\ 1 & 2 & 0 & 1 \\ 0 & 1 & 1 & 1 \\ 0 & 0 & 1 & 0 \\ \hline 0 & 1 & 0 & 0 \\ 1 & 0 & -1 & 0 \\ 0 & 0 & 0 & 0 \\ 0 & 0 & 0 & 0 \\ \hline 0 & 0 & 0 & 0 \\ 1 & 0 & 0 & 0 \\ 0 & 0 & 0 & 1 \\ 1 & 0 & -1 & 0 \\ \hline 0 & 0 & 0 & 0 \\ 0 & 0 & 0 & 0 \\ 0 & 0 & 0 & 0 \\ 0 & 0 & 0 & 0 \end{bmatrix} ,$$

where α is the block row index, β the row index of each block, and γ the column index of each block.

Consider also the two vectors $\vec{x}, \vec{y} \in V^n(\mathbb{R})$:

$$\vec{x} = \vec{e}_1 - \vec{e}_3 + 2\vec{e}_4 \quad \text{and} \quad \vec{y} = \vec{e}_1 + \vec{e}_2 + \vec{e}_3 + \vec{e}_4.$$

1. Write the tensor A as the sum of the minimum number of tensor products of vectors, and if possible of exterior products.
2. Obtain the ordinary and the strict components of the bivector $T = \vec{x} \wedge \vec{y}$, $T \in \bigwedge_4^{(2)}(\mathbb{R})$.
3. Let H be the tensor $H = T \otimes A$.
   a) Classify the tensor by its partial anti-symmetries.
   b) Examine if H is an exterior tensor.

c) Examine if H is a multivector. In the affirmative case give three vectors $\vec{v}_1, \vec{v}_2, \vec{v}_3 \in V^4(\mathbb{R})$ for which exterior product is H.

4. Consider a system of scalar components S, defined by

$$s(\lambda, \mu, \nu) = \delta_{\alpha\beta\gamma}^{\lambda\mu\nu} b_{\circ\circ\circ}^{\alpha\beta\gamma}.$$

Answer all the previous questions for this case.

Solution:

1. We start by grouping the tensor by opposed terms, because it is an anti-symmetric matrix:

$$\begin{aligned}
A &= [(\vec{e}_1 \otimes \vec{e}_2 - \vec{e}_2 \otimes \vec{e}_1) - (\vec{e}_1 \otimes \vec{e}_3 - \vec{e}_3 \otimes \vec{e}_1)] \\
&\quad + [(\vec{e}_2 \otimes \vec{e}_4 - \vec{e}_4 \otimes \vec{e}_2) - (\vec{e}_3 \otimes \vec{e}_4 - \vec{e}_4 \otimes \vec{e}_3)] \\
&= [\vec{e}_1 \otimes (\vec{e}_2 - \vec{e}_3) - (\vec{e}_2 - \vec{e}_3) \otimes \vec{e}_1] + [(\vec{e}_2 - \vec{e}_3) \otimes \vec{e}_4 - \vec{e}_4 \otimes (\vec{e}_2 - \vec{e}_3)] \\
&= \vec{e}_1 \wedge (\vec{e}_2 - \vec{e}_3) - \vec{e}_4 \wedge (\vec{e}_2 - \vec{e}_3) = (\vec{e}_1 - \vec{e}_4) \wedge (\vec{e}_2 - \vec{e}_3),
\end{aligned}$$

which proves A is a bivector and $A \in \bigwedge_4^{(2)}(\mathbb{R})$.

2. The data matrix is $[XY] \equiv \begin{bmatrix} 1 & 1 \\ 0 & 1 \\ -1 & 1 \\ 2 & 1 \end{bmatrix}$, and

$$\begin{aligned}
T &= \begin{bmatrix} x_\circ^\alpha & y_\circ^\alpha \\ x_\circ^\beta & y_\circ^\beta \end{bmatrix} \vec{e}_\alpha \wedge \vec{e}_\beta = \begin{vmatrix} 1 & 1 \\ 0 & 1 \end{vmatrix} \vec{e}_1 \wedge \vec{e}_2 + \begin{vmatrix} 1 & 1 \\ -1 & 1 \end{vmatrix} \vec{e}_1 \wedge \vec{e}_3 + \begin{vmatrix} 1 & 1 \\ 2 & 1 \end{vmatrix} \vec{e}_1 \wedge \vec{e}_4 \\
&\quad + \begin{vmatrix} 0 & 1 \\ -1 & 1 \end{vmatrix} \vec{e}_2 \wedge \vec{e}_3 + \begin{vmatrix} 0 & 1 \\ 2 & 1 \end{vmatrix} \vec{e}_2 \wedge \vec{e}_4 + \begin{vmatrix} -1 & 1 \\ 2 & 1 \end{vmatrix} \vec{e}_3 \wedge \vec{e}_4 \\
&= \vec{e}_1 \wedge \vec{e}_2 + 2\vec{e}_1 \wedge \vec{e}_3 - \vec{e}_1 \wedge \vec{e}_4 + \vec{e}_2 \wedge \vec{e}_3 - 2\vec{e}_2 \wedge \vec{e}_4 - 3\vec{e}_3 \wedge \vec{e}_4.
\end{aligned}$$

Thus, its strict components are $(1, 2, -1, 1, -2, -3)$ and the corresponding ordinary components can be represented as an anti-symmetric tensor of order $r = 2$, by means of the matrix

$$[t_{\circ\circ}^{\alpha\beta}] = \begin{bmatrix} 0 & 1 & 2 & -1 \\ -1 & 0 & 1 & -2 \\ -2 & -1 & 0 & -3 \\ 1 & 2 & 3 & 0 \end{bmatrix}.$$

3. $H = T \otimes A$; $\quad h_{\circ\circ\circ\circ}^{\alpha\beta\gamma\delta} = t_{\circ\circ}^{\alpha\beta} \cdot a_{\circ\circ}^{\gamma\delta}$.

a) $h_{\circ\circ\circ\circ}^{\alpha\beta\gamma\delta}$ is a homogeneous tensor of order $r = 4$, contravariant and partially anti-symmetric with respect to the indices (1,2) and (3,4).

b) In order to be an exterior tensor, all its components with repeated indices must be null, as a necessary condition. But, for example, the component $h^{1213}_{\circ\circ\circ\circ} = t^{12}_{\circ\circ} \cdot a^{13}_{\circ\circ} = 1 \cdot (-1) = -1 \neq 0$ is not null. Then, *the tensor is not an exterior tensor*, i.e., $H \notin \bigwedge^{(4)}_4(\mathbb{R})$.

c) Since $H \notin \bigwedge^{(4)}_4(\mathbb{R})$, it cannot be a multivector.

4. a) When Formula (8.103) is used with the strict components of an anti-symmetric tensor, it gives its ordinary components. If we use it over an arbitrary tensor, this tensor is "anti-symmetrized" (see Section 7.5.1). As the expression of system S responds to Formula (8.103), its notation must be

$$s^{\lambda\mu\nu}_{\circ\circ\circ} = \delta^{\lambda\mu\nu}_{\alpha\beta\gamma} b^{\alpha\beta\gamma}_{\circ\circ\circ}.$$

b) S is a homogeneous contravariant and totally anti-symmetric tensor of order $r = 3$, and then, it is an exterior vector, i.e., $S \in \bigwedge^{(3)}_4(\mathbb{R})$. Since S is an exterior vector we determine its strict components. The dimension of $\bigwedge^{(3)}_4(\mathbb{R})$ is $\sigma' = \binom{4}{3} = 4$, and

$$s^{(123)}_{\circ\circ\circ} = \delta^{(123)}_{123} b^{123}_{\circ\circ\circ} + \delta^{(123)}_{132} b^{132}_{\circ\circ\circ} + \delta^{(123)}_{231} b^{231}_{\circ\circ\circ}$$
$$+ \delta^{(123)}_{213} b^{213}_{\circ\circ\circ} + \delta^{(123)}_{312} b^{312}_{\circ\circ\circ} + \delta^{(123)}_{321} b^{321}_{\circ\circ\circ}$$
$$= 1 \cdot 0 + (-1) \cdot 1 + 1 \cdot 0 + (-1) \cdot 0 + 1 \cdot 0 + (-1) \cdot 1 = -2$$

$$s^{(124)}_{\circ\circ\circ} = \delta^{(124)}_{124} b^{124}_{\circ\circ\circ} + \delta^{(124)}_{142} b^{142}_{\circ\circ\circ} + \delta^{(124)}_{241} b^{241}_{\circ\circ\circ}$$
$$+ \delta^{(124)}_{214} b^{214}_{\circ\circ\circ} + \delta^{(124)}_{412} b^{412}_{\circ\circ\circ} + \delta^{(124)}_{421} b^{421}_{\circ\circ\circ}$$
$$= 1 \cdot 1 + (-1) \cdot 0 + 1 \cdot 0 + (-1) \cdot 0 + 1 \cdot 0 + (-1) \cdot 0 = 1$$

$$s^{(134)}_{\circ\circ\circ} = \delta^{(134)}_{134} b^{134}_{\circ\circ\circ} + \delta^{(134)}_{143} b^{143}_{\circ\circ\circ} + \delta^{(134)}_{341} b^{341}_{\circ\circ\circ}$$
$$+ \delta^{(134)}_{314} b^{314}_{\circ\circ\circ} + \delta^{(134)}_{413} b^{413}_{\circ\circ\circ} + \delta^{(134)}_{431} b^{431}_{\circ\circ\circ}$$
$$= 1 \cdot 1 + (-1) \cdot 1 + 1 \cdot 1 + (-1) \cdot 0 + 1 \cdot 0 + (-1) \cdot 0 = 1$$

$$s^{(234)}_{\circ\circ\circ} = \delta^{(234)}_{234} b^{234}_{\circ\circ\circ} + \delta^{(234)}_{243} b^{243}_{\circ\circ\circ} + \delta^{(234)}_{342} b^{342}_{\circ\circ\circ}$$
$$+ \delta^{(234)}_{324} b^{324}_{\circ\circ\circ} + \delta^{(234)}_{423} b^{423}_{\circ\circ\circ} + \delta^{(234)}_{432} b^{432}_{\circ\circ\circ}$$
$$= 1 \cdot 0 + (-1) \cdot 0 + 1 \cdot 0 + (-1) \cdot 0 + 1 \cdot 0 + (-1) \cdot 0 = 0$$

$$S = (-2)\vec{e}_1 \wedge \vec{e}_2 \wedge \vec{e}_3 + \vec{e}_1 \wedge \vec{e}_2 \wedge \vec{e}_4 + \vec{e}_1 \wedge \vec{e}_3 \wedge \vec{e}_4.$$

c) As has been established in Example 9.11, the Cartesian equation of the linear subspace of $V^4(\mathbb{R})$, in which the vectors $\vec{V}_i$ that are used to build multivectors of order $(n-1) = 4-1 = 3$, of the exterior algebra $\bigwedge_4^{(4)}(\mathbb{R})$ are contained, is given by Formula (9.134):

$$s\overset{(123)}{\underset{\circ\circ\circ}{}}x_\circ^4 - s\overset{(124)}{\underset{\circ\circ\circ}{}}x_\circ^3 + s\overset{(134)}{\underset{\circ\circ\circ}{}}x_\circ^2 - s\overset{(234)}{\underset{\circ\circ\circ}{}}x_\circ^1 = 0,$$

which numerically is

$$\Pi \equiv (-2)x^4 - x^3 + x^2 - 0x^1 = 0,$$

that is,

$$\Pi \equiv x^2 - x^3 - 2x^4 = 0,$$

a linear subspace of $V^4(\mathbb{R})$ of dimension $r \equiv (n-1) = 4-1 = 3$.
Now, we need to find three vectors $\vec{V}_1, \vec{V}_2, \vec{V}_3$ that, satisfying the Cartesian equation of the linear subspace Π, are linearly independent. In this way, the trivector will not be null.
Summarizing, we need to find a simple basis of the linear subspace Π. The most simple vector satisfying Π is $\vec{e}_1(1, 0, 0, 0)$, and then, $\vec{V}_1 = \vec{e}_1$. Another simple vector is $(\vec{e}_2 + \vec{e}_3)$ with components $(0, 1, 1, 0)$, that is, $\vec{V}_2 = \vec{e}_2 + \vec{e}_3$. Finally, the vector $(\vec{e}_2 - \vec{e}_3 + \vec{e}_4)$ of components $(0, 1, -1, 1)$ also satisfies Π. Thus $\vec{V}_3 = \vec{e}_2 - \vec{e}_3 + \vec{e}_4$.
Next, we check if this system of vectors is free, that is, a basis of the linear subspace Π; we have

$$\text{rank of } \begin{bmatrix} 1 & 0 & 0 \\ 0 & 1 & 1 \\ 0 & 1 & -1 \\ 0 & 0 & 1 \end{bmatrix} = \text{rank of } \begin{bmatrix} 1 & 0 & 0 \\ 0 & 1 & 1 \\ 0 & 1 & -1 \end{bmatrix} = 3,$$

because $\begin{vmatrix} 1 & 0 & 0 \\ 0 & 1 & 1 \\ 0 & 1 & -1 \end{vmatrix} = -2 \neq 0$, i.e., they constitute a basis.

Consequently, we propose as a simple solution (though not unique) the multivector

$$\vec{V}_1 \wedge \vec{V}_2 \wedge \vec{V}_3 \equiv \vec{e}_1 \wedge (\vec{e}_2 + \vec{e}_3) \wedge (\vec{e}_2 - \vec{e}_3 + \vec{e}_4). \tag{9.135}$$

Next, we proceed to develop the multivector, to check that it is the exterior tensor S given in 4(b).

To this end, we use Formula (9.20) over the data matrix $\begin{bmatrix} 1 & 0 & 0 \\ 0 & 1 & 1 \\ 0 & 1 & -1 \\ 0 & 0 & 1 \end{bmatrix}$.

(*Warning*: We must not use the distributive property in the apparently simple development of the given expression (9.135), because we need first to associate the sign "$\wedge$", and it *is not associative*).

$$\vec{V}_1 \wedge \vec{V}_2 \wedge \vec{V}_3 = \begin{bmatrix} 1 & 0 & 0 \\ 0 & 1 & 1 \\ 0 & 1 & -1 \end{bmatrix} \vec{e}_1 \wedge \vec{e}_2 \wedge \vec{e}_3 + \begin{bmatrix} 1 & 0 & 0 \\ 0 & 1 & 1 \\ 0 & 0 & 1 \end{bmatrix} \vec{e}_1 \wedge \vec{e}_2 \wedge \vec{e}_4$$

$$+ \begin{bmatrix} 1 & 0 & 0 \\ 0 & 1 & -1 \\ 0 & 0 & 1 \end{bmatrix} \vec{e}_1 \wedge \vec{e}_3 \wedge \vec{e}_4 + \begin{bmatrix} 0 & 1 & 1 \\ 0 & 1 & -1 \\ 0 & 0 & 1 \end{bmatrix} \vec{e}_2 \wedge \vec{e}_3 \wedge \vec{e}_4$$

$$= -2\vec{e}_1 \wedge \vec{e}_2 \wedge \vec{e}_3 + \vec{e}_1 \wedge \vec{e}_2 \wedge \vec{e}_4 + \vec{e}_1 \wedge \vec{e}_3 \wedge \vec{e}_4.$$

Then $\vec{V}_1 \wedge \vec{V}_2 \wedge \vec{V}_3 \equiv S$, that proves that the given solution is correct.

□

Example 9.13 (Number of inversion of the exterior product). Consider a free system of p vectors $\vec{V}_i; (i \in I_p; p \leq n)$ of the linear space $V^n(K)$.

1. Check that the exterior product $P \in \bigwedge_n^{(p)}(K): P = \vec{V}_1 \wedge \vec{V}_2 \wedge \cdots \wedge \vec{V}_p$ can be written as

$$P = I_0^p + I_1^p + \cdots + I_h^p + \cdots + I_N^p, \tag{9.136}$$

where $N = \binom{p}{2} = \frac{p(p-1)}{2}$ and I_h^p is the sum of all terms of the form

$$\delta_{1 \ 2 \ \cdots \ p}^{\alpha_1 \alpha_2 \cdots \alpha_p} \left(\vec{V}_{\alpha_1} \otimes \vec{V}_{\alpha_2} \otimes \cdots \otimes \vec{V}_{\alpha_p} \right),$$

with the condition that the permutation $(\alpha_1, \alpha_2, \ldots, \alpha_p)$ present strictly h and only h inversions, with respect to the "pattern" permutation $(1, 2, \ldots, p)$.
2. If we denote by (I_h^p) the number of *non-null* summands of I_h^p, find the values of (I_h^p) for $h \leq 2$.
3. Idem, but for $h = 3$.
4. Particularize for the case $p = 4$, and find the values of I_h^4 with $h = \{0, 1, 2, \cdots, N\}$. Check that the sum of all them is $\sum_{h=0}^{h=N} I_h^4 = 4!$.

Solution:

1. We check only a particular case. Consider the exterior algebra $\bigwedge_5^{(4)}(\mathbb{R})$, with $p = 4; n = 5; \sigma' = \binom{n}{p} = \binom{5}{4} = 5; N = \binom{4}{2} = 6$. Consider the exterior product

$$P = \vec{V}_1 \wedge \vec{V}_2 \wedge \vec{V}_3 \wedge \vec{V}_4.$$

The development, according to (9.136) will be

$$P = \vec{V}_1 \wedge \vec{V}_2 \wedge \vec{V}_3 \wedge \vec{V}_4$$

$$= \delta_{1 \ 2 \ 3 \ 4}^{\alpha_1 \alpha_2 \alpha_3 \alpha_4} \vec{V}_{\alpha_1} \otimes \vec{V}_{\alpha_2} \otimes \vec{V}_{\alpha_3} \otimes \vec{V}_{\alpha_4} + \delta_{1 \ 2 \ 3 \ 5}^{\alpha_1 \alpha_2 \alpha_3 \alpha_4} \vec{V}_{\alpha_1} \otimes \vec{V}_{\alpha_2} \otimes \vec{V}_{\alpha_3} \otimes \vec{V}_{\alpha_4}$$

$$+ \delta_{1 \ 2 \ 4 \ 5}^{\alpha_1 \alpha_2 \alpha_3 \alpha_4} \vec{V}_{\alpha_1} \otimes \vec{V}_{\alpha_2} \otimes \vec{V}_{\alpha_3} \otimes \vec{V}_{\alpha_4} + \delta_{1 \ 3 \ 4 \ 5}^{\alpha_1 \alpha_2 \alpha_3 \alpha_4} \vec{V}_{\alpha_1} \otimes \vec{V}_{\alpha_2} \otimes \vec{V}_{\alpha_3} \otimes \vec{V}_{\alpha_4}$$

$$+ \delta_{2 \ 3 \ 4 \ 5}^{\alpha_1 \alpha_2 \alpha_3 \alpha_4} \vec{V}_{\alpha_1} \otimes \vec{V}_{\alpha_2} \otimes \vec{V}_{\alpha_3} \otimes \vec{V}_{\alpha_4}. \tag{9.137}$$

In Table 9.2 the permutations of the strict-type sets are classified according to their number of inversions h (its total number is $p!\sigma' = 4!\cdot 5 = 120$).

Table 9.2. Classification of the permutations of the strict-type sets according to their number of inversions h.

I_h^p	1234	1235	1245	1345	2345
I_0^4	+(1234)	+(1235)	+(1245)	+(1345)	+(2345)
I_1^4	−(1243)	−(1253)	−(1254)	−(1354)	−(2354)
	−(1324)	−(1325)	−(1425)	−(1435)	−(2435)
	−(2134)	−(2135)	−(2145)	−(3145)	−(3245)
I_2^4	+(1342)	+(1352)	+(1452)	+(1453)	+(2453)
	+(1423)	+(1523)	+(1524)	+(1534)	+(2534)
	+(2143)	+(2153)	+(2154)	+(3154)	+(3254)
	+(2314)	+(2315)	+(2415)	+(3415)	+(4235)
	+(3124)	+(3125)	+(4215)	+(4135)	+(2345)
I_3^4	−(1432)	−(1532)	−(1542)	−(1543)	−(2543)
	−(2341)	−(2351)	−(2451)	−(3451)	−(3452)
	−(2413)	−(2513)	−(2514)	−(3514)	−(3524)
	−(3142)	−(3152)	−(4152)	−(4153)	−(4253)
	−(3214)	−(3215)	−(4215)	−(4315)	−(4325)
	−(4123)	−(5123)	−(5124)	−(5134)	−(5234)
I_4^4	+(2431)	+(2531)	+(2541)	+(3541)	+(3542)
	+(3241)	+(3251)	+(4251)	+(4351)	+(4352)
	+(3412)	+(3512)	+(4512)	+(4513)	+(4523)
	+(4132)	+(5132)	+(5142)	+(5143)	+(5243)
	+(4213)	+(5213)	+(5214)	+(5314)	+(5324)
I_5^4	−(4312)	−(5312)	−(5412)	−(5413)	−(5423)
	−(4231)	−(5231)	−(5241)	−(5341)	−(5342)
	−(3421)	−(3521)	−(4521)	−(4531)	−(4532)
I_6^4	+(4321)	+(5321)	+(5421)	+(5431)	+(5432)

Now, we can build the desired grouping, by simply assigning to the tensor products the signs and vector subindices that appear in *each box* of each of the *horizontal rows* in the table. We build here only I_0^4, I_5^4, I_6^4, and invite the reader to do the rest of the work.

$$I_0^4 = \left[\vec{V}_1 \otimes \vec{V}_2 \otimes \vec{V}_3 \otimes \vec{V}_4 + \vec{V}_1 \otimes \vec{V}_2 \otimes \vec{V}_3 \otimes \vec{V}_5 + \vec{V}_1 \otimes \vec{V}_2 \otimes \vec{V}_4 \otimes \vec{V}_5 \right.$$

$$+ \, \vec{V}_1 \otimes \vec{V}_3 \otimes \vec{V}_4 \otimes \vec{V}_5 + \vec{V}_2 \otimes \vec{V}_3 \otimes \vec{V}_4 \otimes \vec{V}_5 \Big]$$

$$I_5^4 = -\Big[\vec{V}_4 \otimes \vec{V}_3 \otimes \vec{V}_1 \otimes \vec{V}_2 + \vec{V}_4 \otimes \vec{V}_2 \otimes \vec{V}_3 \otimes \vec{V}_1 + \vec{V}_3 \otimes \vec{V}_4 \otimes \vec{V}_2 \otimes \vec{V}_1$$

$$+ \vec{V}_5 \otimes \vec{V}_3 \otimes \vec{V}_1 \otimes \vec{V}_2 + \vec{V}_5 \otimes \vec{V}_2 \otimes \vec{V}_3 \otimes \vec{V}_1 + \vec{V}_3 \otimes \vec{V}_5 \otimes \vec{V}_2 \otimes \vec{V}_1$$

$$+ \vec{V}_5 \otimes \vec{V}_4 \otimes \vec{V}_1 \otimes \vec{V}_2 + \vec{V}_5 \otimes \vec{V}_2 \otimes \vec{V}_4 \otimes \vec{V}_1 + \vec{V}_4 \otimes \vec{V}_5 \otimes \vec{V}_2 \otimes \vec{V}_1$$

$$+ \vec{V}_5 \otimes \vec{V}_4 \otimes \vec{V}_1 \otimes \vec{V}_3 + \vec{V}_5 \otimes \vec{V}_3 \otimes \vec{V}_4 \otimes \vec{V}_1 + \vec{V}_4 \otimes \vec{V}_5 \otimes \vec{V}_3 \otimes \vec{V}_1$$

$$+ \vec{V}_5 \otimes \vec{V}_4 \otimes \vec{V}_2 \otimes \vec{V}_3 + \vec{V}_5 \otimes \vec{V}_3 \otimes \vec{V}_4 \otimes \vec{V}_2 + \vec{V}_4 \otimes \vec{V}_5 \otimes \vec{V}_3 \otimes \vec{V}_2 \Big]$$

$$I_6^4 = \Big[\vec{V}_4 \otimes \vec{V}_3 \otimes \vec{V}_2 \otimes \vec{V}_1 + \vec{V}_5 \otimes \vec{V}_3 \otimes \vec{V}_2 \otimes \vec{V}_1 + \vec{V}_5 \otimes \vec{V}_4 \otimes \vec{V}_2 \otimes \vec{V}_1$$

$$+ \vec{V}_5 \otimes \vec{V}_4 \otimes \vec{V}_3 \otimes \vec{V}_1 + \vec{V}_5 \otimes \vec{V}_4 \otimes \vec{V}_3 \otimes \vec{V}_2 \Big] .$$

2. Since $V^n(K)$ is of dimension n, if we build the algebras $\bigwedge_n^1(K)$ with $p = 1$, it can be confused with the proper linear space $V^n(K)$ and there are no permutations of indices, because there are no exterior products. Considering the algebra $\bigwedge_n^{(2)}(K)$ with $p = 2$, this has dimension (number of basic exterior products) $\binom{n}{2}$. But, what we are asked for is how many tensor products has $\vec{V}_1 \wedge \vec{V}_2 = \vec{V}_1 \otimes \vec{V}_2 - \vec{V}_2 \otimes \vec{V}_1 \rightarrow (I_0^2) = 1$; $(I_1^2) = 1$. The answer is in total $N + 1 = \binom{2}{2} + 1 = 2$, and then $(P) = (I_0^2) + (I_1^2) = 1 + 1 = 2 = 2! \equiv p!$.

3. In $\bigwedge_n^{(3)}(K)$ with $p = 3$, we have the development of $\vec{V}_1 \wedge \vec{V}_2 \wedge \vec{V}_3$:

$$\vec{V}_1 \otimes \vec{V}_2 \otimes \vec{V}_3, \text{ with } h = 0 \text{ inversions: } (I_0^3) = 1$$

$$\vec{V}_2 \otimes \vec{V}_1 \otimes \vec{V}_3 \text{ and } \vec{V}_1 \otimes \vec{V}_3 \otimes \vec{V}_2 \text{ with } h = 1 \text{ inversions: } (I_1^3) = 2$$

$$\vec{V}_2 \otimes \vec{V}_3 \otimes \vec{V}_1 \text{ and } \vec{V}_3 \otimes \vec{V}_1 \otimes \vec{V}_2 \text{ with } h = 2 \text{ inversions: } (I_2^3) = 2$$

$$\vec{V}_3 \otimes \vec{V}_2 \otimes \vec{V}_1, \text{ with } h = 3 \text{ inversions: } (I_3^3) = 1.$$

In total $N + 1 = \binom{3}{2} + 1 = 4$ summands, thus

$$P = (I_0^3) + (I_1^3) + (I_2^3) + (I_3^3) = 1 + 2 + 2 + 1 = 6 = 3! \equiv p!$$

4. In $\bigwedge_n^{(4)}(K)$ with $p = 4$, from the development of $\vec{V}_1 \wedge \vec{V}_2 \wedge \vec{V}_3 \wedge \vec{V}_4$, counting in each box of the *first column of the table*, since it is valid that $\forall n \geq 4$, we have

$$(I_0^4) = 1; \ (I_1^4) = 3; \ (I_2^4) = 5; \ (I_3^4) = 6; \ (I_4^4) = 5; \ (I_5^4) = 3; \ (I_6^4) = 1,$$

a total of $N + 1 = \binom{4}{2} + 1 = 7$ summands, whence

$$P = (I_0^4) + (I_1^4) + (I_2^4) + (I_3^4) + (I_4^4) + (I_5^4) + (I_6^4)$$

$$= 1 + 3 + 5 + 6 + 5 + 3 + 1 = 24 = 4! \equiv p!$$

$\square$

9.10 Exercises

9.1. In the linear space $V^4(\mathbb{R})$ referred to the basis $\{\vec{e}_\alpha\}$ two vectors $\vec{V}_1, \vec{V}_2$
are represented by the matrix $[X_1, X_2] = \begin{bmatrix} -1 & 0 \\ 1 & 5 \\ -2 & 0 \\ -3 & 5 \end{bmatrix}$, and another two vectors
$\vec{W}_1, \vec{W}_2$ by $[Y_1, Y_2] = \begin{bmatrix} 1 & 2 \\ 1 & -3 \\ 2 & 4 \\ 5 & 5 \end{bmatrix}$.

1. Obtain the exterior vector $P = \vec{V}_1 \wedge \vec{V}_2$ totally developed.
2. Obtain the exterior vector $Q = \vec{W}_1 \wedge \vec{W}_2$ totally developed.
3. Let $\vec{Z} \in V^4(\mathbb{R})$ be a vector of components $[X] = \begin{bmatrix} x^1 \\ x^2 \\ x^3 \\ x^4 \end{bmatrix}$. Obtain two
   Cartesian equations of the linear subspace, which must be satisfied by
   any pair of vectors $\vec{Z}_1$ and $\vec{Z}_2$, for their exterior product to be P.
4. Add an extra condition for $\vec{Z}_1 \wedge \vec{Z}_2$ to be precisely P.
5. Solve the questions 3 and 4 for any pair of vectors $\vec{Z}_3$ and $\vec{Z}_4$ satisfying
   $\vec{Z}_3 \wedge \vec{Z}_4 = Q$.
6. Choose a set of numeric vectors $\vec{Z}_1$, $\vec{Z}_2$, $\vec{Z}_3$ and $\vec{Z}_4$ satisfying the above
   conditions, and obtain the exterior vector $\vec{Z}_1 \wedge \vec{Z}_2 \wedge \vec{Z}_3 \wedge \vec{Z}_4$.

9.2. 1. In the linear space $\mathbb{R}^4$ consider the vectors $\vec{V}_1, \vec{V}_2, \vec{V}_3$ of components
$\begin{bmatrix} 2 & 0 & 1 \\ 1 & 1 & 0 \\ 1 & 0 & 1 \\ 3 & 1 & 2 \end{bmatrix}$ and the vectors $\vec{W}_1, \vec{W}_2, \vec{W}_3$ of components $\begin{bmatrix} 1 & 1 & 1 \\ 1 & 1 & 2 \\ 1 & 0 & 1 \\ 1 & 1 & -2 \end{bmatrix}$.
Determine if the two 3-exterior vectors

$$P = \vec{V}_1 \wedge \vec{V}_2 \wedge \vec{V}_3 \text{ and } Q = \vec{W}_1 \wedge \vec{W}_2 \wedge \vec{W}_3$$

are proportional $(P = \lambda Q)$, without determining them.
2. If P and Q are not proportional, calculate the two 3-vectors $(P+Q)$ and
   $(P-Q)$.
3. Calculate the exterior products $P \wedge \vec{W}_2$ and $Q \wedge \vec{V}_2$.
4. Calculate the exterior expression $P \wedge \vec{W}_1 + Q \wedge \vec{V}_3$.
5. Determine $\left(\vec{V}_1 \wedge \vec{V}_3\right) \wedge \left(\vec{W}_1 \wedge \vec{W}_3\right)$.
6. Is there any reason that justifies the results of question 3?

9.3. Consider the exterior algebra $\bigwedge_{2n}^{(2)}(\mathbb{R})$ built over the linear space $V^{2n}(\mathbb{R})$
referred to the basis $\{\vec{e}_\alpha\}$. The power of a tensor $T \in \bigwedge_{2n}^{(2)}(\mathbb{R})$ with respect
to the exterior product of tensors, will be notated $(\bigwedge T)^k$, whence

$$\left(\bigwedge T\right)^0 = 1; \quad \left(\bigwedge T\right)^1 = T; \quad \left(\bigwedge T\right)^2 = T \bigwedge T; \quad \text{etc.}$$

1. A 2-vector $Z \in \bigwedge_{2n}^{(2)}(\mathbb{R})$ is defined by means of its strict components:

$$Z = z_{\circ\;\circ}^{(\alpha\beta)} \vec{e}_\alpha \wedge \vec{e}_\beta; \quad z_{\circ\;\circ}^{(\alpha\beta)} = 1 \text{ if } \beta = 1 + \alpha; \quad z_{\circ\;\circ}^{(\alpha\beta)} = 0 \text{ if } \beta \neq 1 + \alpha;$$

or

$$z = \vec{e}_1 \wedge \vec{e}_2 + \vec{e}_3 \wedge \vec{e}_4 + \cdots + \vec{e}_{2n-1} \wedge \vec{e}_{2n}$$

    a) Give the representation of the tensor Z, that is, the matrix $Z = [z_{\circ\;\circ}^{\alpha\beta}]_{2n}$ in the tensor space $\left(\otimes V^{2n}\right)_h^2 (\mathbb{R})$.

    b) Prove that $(\bigwedge Z)^n = f(Z)\vec{e}_1 \wedge \vec{e}_2 \wedge \cdots \wedge \vec{e}_{2n-1} \wedge \vec{e}_{2n}$ indicating what is the scalar $f(Z) \in \mathbb{R}$.

2. Consider the tensor $T = t_{\circ\;\circ}^{(\alpha\beta)} \vec{e}_\alpha \wedge \vec{e}_\beta, T \in \bigwedge_{2n}^{(2)}(\mathbb{R})$, the strict components of which are the data in this question. Give $(\bigwedge T)^n = f(T)\vec{e}_1 \wedge \vec{e}_2 \wedge \cdots \wedge \vec{e}_{2n-1} \wedge \vec{e}_{2n}$ for the cases $n = 1, n = 2$ and $n = 3$.

3. The quotients $Pf(T) = \frac{f(T)}{f(Z)}$ are polynomials of variables the strict components of T, called "Pfaffians". Give the Pfaffians $Pf(T)$ for the cases $n = 1, n = 2$ and $n = 3$.

9.4. In the linear space $V^4(\mathbb{R})$ referred to the basis $\{\vec{e}_\alpha\}$, a linear operator that transforms the vectors as $T(\vec{V}) = \vec{W}$, is defined by the matrix

$$T = \begin{bmatrix} 1 & 1 & 1 & 0 \\ 0 & 2 & 1 & 0 \\ 1 & 0 & 0 & 2 \\ 1 & -1 & 1 & 1 \end{bmatrix}.$$

1. Determine the matrix H of the direct exterior endomorphism, associated with T:

$$H : \bigwedge_4^{(3)}(\mathbb{R}) \to \bigwedge_4^{(3)}(\mathbb{R}),$$

such that $\forall \vec{V}_1, \vec{V}_2, \vec{V}_3 \in V^4(\mathbb{R})$ be $H(\vec{V}_1 \wedge \vec{V}_2 \wedge \vec{V}_3) = \vec{W}_1 \wedge \vec{W}_2 \wedge \vec{W}_3$.

2. Check the correct behavior of the operator H, using the vectors $\vec{V}_1, \vec{V}_2$ and $\vec{V}_3$ of components $[X_1 \;\; X_2 \;\; X_3] = \begin{bmatrix} 2 & 0 & 1 \\ 1 & 1 & 0 \\ 1 & 0 & 2 \\ 3 & 1 & 1 \end{bmatrix}$ and its transformed

vectors, by means of Formula (9.119) of the present chapter.

3. A change-of-basis is performed in the linear space $V^4(\mathbb{R})$ given by the matrix

$$C = \begin{bmatrix} 1 & -1 & 1 & -1 \\ 0 & 1 & -1 & 1 \\ 0 & 0 & 1 & -1 \\ 0 & 0 & 0 & 1 \end{bmatrix}.$$

Give the new matrix that represents the linear operator T.

4. Answer again questions 1 and 2.

9.5. Let $F(T) = U$ be a mapping defined in symbolic form as

$$F : \bigwedge_4^{(3)}(\mathbb{R}) \to \bigwedge_4^{(2)}(\mathbb{R}),$$

which transforms the tri-exterior vectors of $\bigwedge_4^{(3)}(\mathbb{R})$ in the bivectors of $\bigwedge_4^{(2)}(\mathbb{R})$; $\forall \vec{u}, \vec{v}, \vec{w} \in V^4(\mathbb{R})$, of basis $\{\vec{e}_\alpha\}$, such that:

$$F(\vec{u} \wedge \vec{v} \wedge \vec{w}) = \vec{u} \wedge \vec{v} + 2\vec{v} \wedge \vec{w} + 3\vec{w} \wedge \vec{u}.$$

1. Obtain the matrix associated with the given mapping F.
2. Obtain the Cartesian equation of the image linear subspace associated with the given mapping.
3. Examine the kernel linear subspace.
4. Obtain the bivector U image of the exterior vector

$$T = \vec{e}_1 \wedge \vec{e}_2 \wedge \vec{e}_3 - 2\vec{e}_1 \wedge \vec{e}_3 \wedge \vec{e}_4 + 3\vec{e}_2 \wedge \vec{e}_3 \wedge \vec{e}_4.$$

10

Mixed Exterior Algebras

10.1 Introduction

As has been indicated in Sections 7.4.1 and 7.4.3, and in the initial sections of Chapter 9, prompted by the techniques that needed this model, we developed the idea of treating the homogeneous anti-symmetric totally contravariant or totally covariant tensors from a different perspective using the strict components of those tensors, giving rise to the exterior algebras primary and dual that have already been treated. Nevertheless, though the technique still does not require it, the tensor algebra can, following a line parallel to those already treated to establish a mixed exterior algebra. Assuming the risk that any new endeavour bears, begging the pardon of those authors working in this field, who with due rights have conservative positions and remain unadventurous, and finally with the benevolence of those who can forgive any errors found, this chapter is ready to begin.

In this chapter the exterior product is extended to include the mixed exterior product of p vectors $\vec{V}_i$ and q vectors $\vec{V}_j^*$, analyzing the problem of change-of-basis and the exterior product of mixed exterior vectors.

10.1.1 Mixed anti-symmetric tensor spaces and their strict tensor components

We present in this section a brief memorandum of the diverse circumstances that particularize tensors with mixed anti-symmetry. We start with their notation, and assume that they are given by their strict components (indices between parentheses):

$$t^{(\alpha_1 \alpha_2 \cdots \alpha_p)}_{ (\alpha_{p+1} \alpha_{p+2} \cdots \alpha_{p+q})} \qquad (10.1)$$

with p contravariant and q covariant indices. Their anti-symmetry involves the totality of the p contravariant indices and *simultaneously*, the totality of the q covariant indices, but with *total independence* among them. For the sake of clarity, the already-known conditions for the exterior vectors are here repeated, assuming that the tensor is established over $V^n(K)$. They are

$$\alpha_1 < \alpha_2 < \cdots < \alpha_p; \ \alpha_{p+1} < \alpha_{p+2} < \cdots < \alpha_{p+q}; \ \alpha_i, \alpha_{p+j} \in I_n;$$
$$i \in I_p; \ (p+j) \in I_q; \ 2 \le p \le n; \ 2 \le q \le n; \ p+q = r, \qquad (10.2)$$

where r is the tensor order, which as can be seen, has a minimum of $r = 4$, and a maximum of $r = 2n$.

Though in Formula (10.1) *we give the data format*, with all indices "stacked", that is, all contravariant indices together and then, all covariant indices, we will need the *real species*, as for example

$$(t^{\alpha_1 \alpha_2 \ \circ \ \alpha_3 \ \circ \ \cdots \alpha_i \cdots \ \circ \ \cdots \alpha_p \cdots \ \circ}_{\ \circ \ \circ \ \alpha_{p+1} \ \circ \ \alpha_{p+2} \cdots \ \circ \ \cdots \alpha_{p+j} \cdots \ \circ \ \cdots \alpha_{p+q}}),$$

that is, the ordered list of the indices contra- and cova-, to be able to solve correctly all the questions that we will pose. The most commonly used formulas are:

1. The dimension of the mixed anti-symmetric tensor space

$$\left[\left(\overset{p}{\underset{1}{\otimes}} V\right) \otimes \left(\overset{q}{\underset{1}{\otimes}} V_*\right)\right]^n_h (K)$$

   is $\sigma = n^r \equiv$ (total number of components).
2. The number of strict components is $\sigma' = \binom{n}{p}\binom{n}{q}$.
3. The maximum number of components different from zero is

$$N_1 = \sigma' \cdot p! \cdot q! = \binom{n}{p}\binom{n}{q}p!q!. \qquad (10.3)$$

4. The minimum number of null components is $N_0 = n^r - \binom{n}{p}\binom{n}{q}p!q!$.
5. The "ordinary" components of the mixed anti-symmetric tensor are related to the strict components by the expression

$$t^{\alpha_1\alpha_2\cdots\alpha_p \ \circ \ \circ \ \cdots \ \circ}_{\ \circ \ \circ \ \cdots \ \circ \ \alpha_{p+1}\alpha_{p+2}\cdots\alpha_{p+q}} = \delta^{(\beta_1\beta_2\cdots\beta_p)}_{\alpha_1\alpha_2\cdots\alpha_p} \cdot \delta^{\alpha_{p+1}\alpha_{p+2}\cdots\alpha_{p+q}}_{(\beta_{p+1}\beta_{p+2}\cdots\beta_{p+q})}$$
$$\cdot t^{(\beta_1\beta_2\cdots\beta_p) \ \circ \ \circ \ \cdots \ \circ}_{\ \circ \ \circ \ \cdots \ \circ \ (\beta_{p+1}\beta_{p+2}\cdots\beta_{p+q})}, \qquad (10.4)$$

where

$$\{\beta_1\beta_2\cdots\beta_p\} = \{\alpha_1\alpha_2\cdots\alpha_p\} \subset I_n; \ \beta_1 < \beta_2 < \cdots < \beta_p; \ p+q = r$$
$$\{\beta_{p+1}\beta_{p+2}\cdots\beta_{p+q}\} = \{\alpha_{p+1}\alpha_{p+2}\cdots\alpha_{p+q}\} \subset I_n;$$
$$\beta_{p+1} < \beta_{p+2} < \cdots < \beta_{p+q}.$$

With respect to the relation between the basic vectors of a mixed exterior algebra $\bigwedge_n^{(p,q)}(K)$ and those of the corresponding mixed anti-symmetric tensor space $\left[\left(\overset{p}{\underset{1}{\otimes}}V\right)\otimes\left(\overset{q}{\underset{1}{\otimes}}V_*\right)\right]^n_h(K)$ we have for the generic case (of stretched species) the expression

$$\vec{e}_{\beta_1}\wedge\vec{e}_{\beta_2}\wedge\cdots\wedge\vec{e}_{\beta_p}\wedge\vec{e}^{*\beta_{p+1}}\wedge\vec{e}^{*\beta_{p+2}}\wedge\cdots\wedge\vec{e}^{*\beta_{p+q}}$$

$$=\delta^{\alpha_1\alpha_2\cdots\alpha_p}_{(\beta_1\beta_2\cdots\beta_p)}\cdot\delta^{(\beta_{p+1}\beta_{p+2}\cdots\beta_{p+q})}_{\alpha_{p+1}\alpha_{p+2}\cdots\alpha_{p+q}}$$

$$\vec{e}_{\alpha_1}\otimes\vec{e}_{\alpha_2}\otimes\cdots\otimes\vec{e}_{\alpha_p}\otimes\vec{e}^{*\alpha_{p+1}}\otimes\vec{e}^{*\alpha_{p+2}}\otimes\cdots\otimes\vec{e}^{*\alpha_{p+q}}, \qquad (10.5)$$

where the α_i and α_{p+j} are contraction dummy indices.

We end this section with an illustrative example of a mixed anti-symmetric tensor of order 4, $p=q=2$, $n=3$, given by its strict components, grouping its ordinary components in a block matrix and presenting it as an exterior vector of the *mixed* exterior algebra $\bigwedge^{(2,2)}_3(K)$, of *species* $[t^{\alpha\beta\circ\circ}_{\circ\circ\gamma\delta}]$. Its strict components are

$$t^{(12)\,\circ\circ}_{\circ\circ\,(12)}=a;\ \ t^{(12)\,\circ\circ}_{\circ\circ\,(13)}=b;\ \ t^{(12)\,\circ\circ}_{\circ\circ\,(23)}=c;\ \ t^{(13)\,\circ\circ}_{\circ\circ\,(12)}=d;$$

$$t^{(13)\,\circ\circ}_{\circ\circ\,(13)}=e;\ \ t^{(13)\,\circ\circ}_{\circ\circ\,(23)}=f;\ \ t^{(23)\,\circ\circ}_{\circ\circ\,(12)}=g;\ \ t^{(23)\,\circ\circ}_{\circ\circ\,(13)}=h;$$

$$t^{(23)\,\circ\circ}_{\circ\circ\,(23)}=i,$$

and some of its important numerical values are

$$r=p+q=2+2=4;\ \ \sigma=n^r=3^4=81;\ \ \sigma'=\binom{n}{p}\binom{n}{q}=\binom{3}{2}\binom{3}{2}=9.$$

$$N_1=\sigma'p!q!=9\cdot2!2!=36;\ \ N_0=81-9\cdot2!2!=45.$$

Its mixed anti-symmetric tensor notation is

$$T=t^{\alpha\beta\circ\circ}_{\circ\circ\gamma\delta}\vec{e}_\alpha\otimes\vec{e}_\beta\otimes\vec{e}^{*\gamma}\otimes\vec{e}^{*\delta}.$$

Its matrix expression by blocks is

$$[t^{\alpha\beta\circ\circ}_{\circ\circ\gamma\delta}]=\begin{bmatrix}t^{11\circ\circ}_{\circ\circ\gamma\delta}&|&\cdots&|&t^{13\circ\circ}_{\circ\circ\gamma\delta}\\ \hline t^{21\circ\circ}_{\circ\circ\gamma\delta}&|&\cdots&|&t^{23\circ\circ}_{\circ\circ\gamma\delta}\\ \hline t^{31\circ\circ}_{\circ\circ\gamma\delta}&|&\cdots&|&t^{33\circ\circ}_{\circ\circ\gamma\delta}\end{bmatrix}=\begin{bmatrix}0&0&0&0&a&b&0&d&e\\0&0&0&-a&0&c&-d&0&f\\0&0&0&-b&-c&0&-e&-f&0\\0&-a&-b&0&0&0&0&g&h\\a&0&-c&0&0&0&-g&0&i\\b&c&0&0&0&0&-h&-i&0\\0&-d&-e&0&-g&-h&0&0&0\\d&0&-f&g&0&-i&0&0&0\\e&f&0&h&i&0&0&0&0\end{bmatrix}$$

and its expression as mixed exterior vector $\in\bigwedge^{(2,2)}_3(K)$ developed only in "strict components", is

$$T=a\,\vec{e}_1\wedge\vec{e}_2\wedge\vec{e}^{*1}\wedge\vec{e}^{*2}+b\,\vec{e}_1\wedge\vec{e}_2\wedge\vec{e}^{*1}\wedge\vec{e}^{*3}+c\,\vec{e}_1\wedge\vec{e}_2\wedge\vec{e}^{*2}\wedge\vec{e}^{*3}$$

$$+d\,\vec{e}_1\wedge\vec{e}_3\wedge\vec{e}^{*1}\wedge\vec{e}^{*2}+e\,\vec{e}_1\wedge\vec{e}_3\wedge\vec{e}^{*1}\wedge\vec{e}^{*3}+f\,\vec{e}_1\wedge\vec{e}_3\wedge\vec{e}^{*2}\wedge\vec{e}^{*3}$$

$$+g\,\vec{e}_2\wedge\vec{e}_3\wedge\vec{e}^{*1}\wedge\vec{e}^{*2}+h\,\vec{e}_2\wedge\vec{e}_3\wedge\vec{e}^{*1}\wedge\vec{e}^{*3}+i\,\vec{e}_2\wedge\vec{e}_3\wedge\vec{e}^{*2}\wedge\vec{e}^{*3}$$

10.1.2 Mixed exterior product of four vectors

Following an analogous path to that employed for establishing the exterior product of vectors in the exterior algebra $\bigwedge_n^{(r)}(K)$ and in the algebra $\bigwedge_{n*}^{(r)}(K)$, we define the mixed exterior tetraproduct as follows.

Let $\vec{V_1}, \vec{V_3} \in V^n(K)$ be two arbitrary vectors in the linear space $V^n(K)$, endowed with odd notation, and $\vec{V_2^*}, \vec{V_4^*} \in V_*^n(K)$ another two arbitrary vectors, from the linear space $V_*^n(K)$, dual of the previous one, endowed with even notation.

We denote by $P_n = [2, 4, 6, \dots, 2n]$ the set of the first n *even* numbers, and similarly, $IMP_n = [1, 3, 5, \dots, (2n-1)]$ is the set of the first n *odd* numbers, where $IMP_n \cup P_n \equiv I_{2n}$.

To execute an exterior product of mixed vectors (coming from the primal and the dual) we specify first the "species" of the desired product, which fixes the notation of the exterior space in which the resulting exterior vector product is going to be, where $p = q$ and $r = p+q = 2p$. There exist χ possible species, with $\chi = \binom{r}{p}$. For example, if $p = q = 2$, there exist $\chi = \binom{4}{2} = 6$ possible species:

$$\left\{ t^{\alpha\beta\circ\circ}_{\circ\circ\gamma\delta}, t^{\alpha\circ\beta\circ}_{\circ\gamma\circ\delta}, t^{\alpha\circ\circ\beta}_{\circ\gamma\delta\circ}, t^{\circ\alpha\beta\circ}_{\gamma\circ\circ\delta}, t^{\circ\alpha\circ\beta}_{\gamma\circ\delta\circ}, t^{\circ\circ\alpha\beta}_{\gamma\delta\circ\circ} \right\}.$$

In what follows, we choose the species $\{ t^{\alpha\circ\beta\circ}_{\circ\gamma\circ\delta} \}$ that corresponds to its product exterior linear space.

For the present case, $p = q = 2, r = p+q = 2p = 4$, we call mixed exterior product of the four vectors of the given species $(V^n \wedge V_*^n \wedge V^n \wedge V_*^n)(K)$, to the vector

$$\vec{V_1} \wedge \vec{V_2^*} \wedge \vec{V_3} \wedge \vec{V_4^*} = \delta^{\alpha_1 \alpha_3}_{(\beta_1 \beta_3)} \cdot \delta^{(\beta_2 \beta_4)}_{\alpha_2 \alpha_4} \vec{V}_{\alpha_1} \otimes \vec{V}_{\alpha_2}^* \otimes \vec{V}_{\alpha_3} \otimes \vec{V}_{\alpha_4}^* \qquad (10.6)$$

with

$$\{\beta_1 \beta_3\} \subset IMP_n; \quad \{\beta_2 \beta_4\} \subset P_n; \quad \{\alpha_1 \alpha_3\} = \{\beta_1 \beta_3\};$$
$$\beta_1 < \beta_3; \quad \{\alpha_2 \alpha_4\} = \{\beta_2 \beta_4\}; \quad \beta_2 < \beta_4. \qquad (10.7)$$

To clarify, we apply expression (10.6) and conditions (10.7) to the concrete case:

$$p = q = n = 2; \quad \{\beta_1 \beta_3\} \equiv \{1, 3\}; \quad \{\beta_2 \beta_4\} \equiv \{2, 4\}$$

$$\vec{V_1} \wedge \vec{V_2^*} \wedge \vec{V_3} \wedge \vec{V_4^*} = \delta^{13}_{(13)} \cdot \delta^{(24)}_{24} \vec{V_1} \otimes \vec{V_2^*} \otimes \vec{V_3} \otimes \vec{V_4^*}$$

$$+ \delta^{13}_{(13)} \cdot \delta^{(24)}_{42} \vec{V_1} \otimes \vec{V_4^*} \otimes \vec{V_3} \otimes \vec{V_2^*}$$

$$+ \delta^{31}_{(13)} \cdot \delta^{(24)}_{24} \vec{V_3} \otimes \vec{V_2^*} \otimes \vec{V_1} \otimes \vec{V_4^*}$$

$$+ \delta^{31}_{(13)} \cdot \delta^{(24)}_{42} \vec{V_3} \otimes \vec{V_4^*} \otimes \vec{V_1} \otimes \vec{V_2^*}$$

$$= \vec{V}_1 \otimes \vec{V}_2^* \otimes \vec{V}_3 \otimes \vec{V}_4^* - \vec{V}_1 \otimes \vec{V}_4^* \otimes \vec{V}_3 \otimes \vec{V}_2^*$$
$$- \vec{V}_3 \otimes \vec{V}_2^* \otimes \vec{V}_1 \otimes \vec{V}_4^* + \vec{V}_3 \otimes \vec{V}_4^* \otimes \vec{V}_1 \otimes \vec{V}_2^*,$$

$$(10.8)$$

which clearly reveals the behavior and laws that vectors linked by the tensor product "$\otimes$" obey:

1. When we transpose indices of the same valency (of the same species) and those of the reverse valency (reverse species) do not change, the tensor product changes sign.
2. When both change, the odd indices (contravariant) and also the even indices (covariant), the final sign will depend on the *parity* of the *total* number of transpositions. If the total number of transpositions is even, the sign will be "$+$", and if it is odd, the sign will be "$-$".
3. *It is not permitted* to consider other possible alterations of the set of indices (it is axiomatically prohibited); only exchanging the odd indices (contravariant) between them or the even indices (covariant) between them are licit alterations.

Probably, the reader at this point of the exposition shares with the authors the pedagogical criterion that while we advance in the construction of this new algebra, it demands an adequate formal set of axioms.

Formula (10.6) establishes the intrinsic definition of the mixed 4-vector $\vec{V}_1 \wedge \vec{V}_2^* \wedge \vec{V}_3 \wedge \vec{V}_4^*$ as a function of the tensor product of the indicated vectors.

It is convenient to follow a process parallel to that followed in Chapter 9, establishing the calculation formula for the exterior tetraproduct, when the bases $\{\vec{e}_\alpha\}$ of the linear space $V^n(K)$ and $\{\vec{e}^{*\beta}\}$ of the dual space $V_*^n(K)$ are known, and then, vectors $\vec{V}_i$ and $\vec{V}_j^*$ are given by their components, their respective matrices X_i and X_j^*, as *data* of the given vectors in contravariant and in covariant coordinates.

Consider the following data vectors:

$$\vec{V}_1 = ||\vec{e}_{\alpha_1}|| X_1 \equiv ||\vec{e}_{\alpha_1}|| [x^{\alpha_1\,\circ}_{\,\circ\,1}]; \quad \vec{V}_3 = ||\vec{e}_{\alpha_3}|| X_3 \equiv ||\vec{e}_{\alpha_3}|| [x^{\alpha_3\,\circ}_{\,\circ\,3}] \quad \text{and}$$

$$\vec{V}_2^* = ||\vec{e}^{*\alpha_2}|| X_2^* \equiv ||\vec{e}^{*\alpha_2}|| [x^{\,\circ\,2}_{\alpha_2\circ}]; \quad \vec{V}_4^* = ||\vec{e}^{*\alpha_4}|| X_4^* \equiv ||\vec{e}^{*\alpha_4}|| [x^{\,\circ\,4}_{\alpha_4\circ}].$$

Replacing these tensor expressions into the mixed 4-vector, written according to Formula (10.6) and conditions (10.7) with adequate notation we get

$$\vec{V}_1 \wedge \vec{V}_2^* \wedge \vec{V}_3 \wedge \vec{V}_4^* = \delta^{\beta_1\beta_3}_{(\gamma_1\gamma_3)} \cdot \delta^{(\gamma_2\gamma_4)}_{\beta_2\beta_4} \vec{V}_{\beta_1} \otimes \vec{V}_{\beta_2}^* \otimes \vec{V}_{\beta_3}^* \otimes \vec{V}_{\beta_4}^*$$

$$= \delta^{\beta_1\beta_3}_{(\gamma_1\gamma_3)} \cdot \delta^{(\gamma_2\gamma_4)}_{\beta_2\beta_4} (x^{\alpha_1\,\circ}_{\,\circ\,\beta_1} \vec{e}_{\alpha_1}) \otimes (x^{\,\circ\,\beta_2}_{\alpha_2\circ} \vec{e}^{*\alpha_2}) \otimes (x^{\alpha_3\,\circ}_{\,\circ\,\beta_3} \vec{e}_{\alpha_3}) \otimes (x^{\,\circ\,\beta_4}_{\alpha_4\circ} \vec{e}^{*\alpha_4})$$

$$= \left(\delta^{\beta_1\beta_3}_{(\gamma_1\gamma_3)} x^{\alpha_1\,\circ}_{\,\circ\,\beta_1} x^{\alpha_3\,\circ}_{\,\circ\,\beta_3} \right) \left(\delta^{(\gamma_2\gamma_4)}_{\beta_2\beta_4} x^{\,\circ\,\beta_2}_{\alpha_2\circ} x^{\,\circ\,\beta_4}_{\alpha_4\circ} \right) \vec{e}_{\alpha_1} \otimes \vec{e}^{*\alpha_2} \otimes \vec{e}_{\alpha_3} \otimes \vec{e}^{*\alpha_4}$$

$$(10.9)$$

with the conditions

$$\{\gamma_1\gamma_3\} \subset IMP_n; \ \{\gamma_2\gamma_4\} \subset P_n; \ \{\beta_1\beta_3\} = \{\gamma_1\gamma_3\}; \ \gamma_1 < \gamma_3;$$
$$\{\beta_2\beta_4\} = \{\gamma_2\gamma_4\}; \ \gamma_2 < \gamma_4; \ \forall \alpha_i \in I_n, \ (10.10)$$

where $\{\gamma_1\gamma_3\}$ represents each of the *ordered combinations* that can be generated in the set IMP_n taking the natural odd numbers two by two, that is, in the set of the contravariant vectors, and $\{\gamma_2\gamma_4\}$ represents each of the *ordered combinations* that can be generated in the set P_n taking the natural even numbers two by two, that is, in the set of the covariant vectors. The number of possible elections is $\binom{n}{p} = \binom{n}{q}$ because $p = q$, in both cases.

We start the analysis of expression (10.9) for the different values of "α_i" in the particular case $p = q = n = 2$ with $I_2 = [1, 2]$.

When developing Expression (10.9) for repeated indices $\{\alpha_1\alpha_3\}$ or $\{\alpha_2\alpha_4\}$ the terms cancel, and only those that appear boldfaced in the following tables remain valid:

α_1	α_3	α_2	α_4
1	1	1	1
1	1	1	2
1	1	2	1
1	1	2	2

α_1	α_3	α_2	α_4
1	**2**	1	1
1	**2**	1	2
1	**2**	2	1
1	**2**	2	2

α_1	α_3	α_2	α_4
2	**1**	1	1
2	**1**	1	2
2	**1**	2	1
2	**1**	2	2

α_1	α_3	α_2	α_4
2	2	1	1
2	2	1	2
2	2	2	1
2	2	2	2

when they are replaced in the adopted numerical example, the result is

$$\vec{V}_1 \wedge \vec{V}_2^* \wedge \vec{V}_3 \wedge \vec{V}_4^* = \begin{vmatrix} x_{\circ 1}^{1\circ} & x_{\circ 3}^{1\circ} \\ x_{\circ 1}^{2\circ} & x_{\circ 3}^{2\circ} \end{vmatrix} \cdot \begin{vmatrix} x_{1\circ}^{\circ 2} & x_{1\circ}^{\circ 4} \\ x_{2\circ}^{\circ 2} & x_{2\circ}^{\circ 4} \end{vmatrix}$$
$$\cdot \left[\vec{e}_1 \otimes \vec{e}^{*1} \otimes \vec{e}_2 \otimes \vec{e}^{*2} - \vec{e}_1 \otimes \vec{e}^{*2} \otimes \vec{e}_2 \otimes \vec{e}^{*1} \right.$$
$$\left. - \vec{e}_2 \otimes \vec{e}^{*1} \otimes \vec{e}_1 \otimes \vec{e}^{*2} + \vec{e}_2 \otimes \vec{e}^{*2} \otimes \vec{e}_1 \otimes \vec{e}^{*1} \right], (10.11)$$

which is the expression of the mixed exterior product of our example, as a mixed tensor anti-symmetric with respect to the indices $[t_{\circ\beta\circ\delta}^{\alpha\circ\gamma\circ}]$, given with a unique strict component.

Since the matrices of the data vectors of $V^2(K)$ and $V_*^2(K)$ in the example are

$$[X_1 X_3] \equiv \begin{vmatrix} x_{\circ 1}^{1\circ} & x_{\circ 3}^{1\circ} \\ x_{\circ 1}^{2\circ} & x_{\circ 3}^{2\circ} \end{vmatrix}$$ of the primary vectors $\vec{V}_1$ and $\vec{V}_3$ and $[X_1 X_3] \equiv$

$$\begin{vmatrix} x_{1\circ}^{\circ 2} & x_{1\circ}^{\circ 4} \\ x_{2\circ}^{\circ 2} & x_{2\circ}^{\circ 4} \end{vmatrix}$$ of the dual vectors $\vec{V}_2^*$ and $\vec{V}_4^*$, we denote by

$$\Delta_2 = |X_1 X_3| \text{ and } \Delta_2^* = |X_2^* X_4^*|, \qquad (10.12)$$

the respective determinants, the product $(\Delta_2 \cdot \Delta_2^*)$ of which is the strict component.

On the other hand, considering Formula (10.5) applied to our example $(p = q = n = 2)$, we obtain the expression

$$
\begin{aligned}
\vec{e}_1 \wedge \vec{e}^{*1} \wedge \vec{e}_2 \wedge \vec{e}^{*2} &= \delta\,{}^{\alpha_1\,\alpha_3}_{(\;1\;\;2\;)}\delta\,{}^{(\;1\;\;2\;)}_{\alpha_2\,\alpha_4}\vec{e}_{\alpha_1} \otimes \vec{e}^{*\alpha_2} \otimes \vec{e}_{\alpha_3} \otimes \vec{e}^{*\alpha_4} \\
&= \vec{e}_1 \otimes \vec{e}^{*1} \otimes \vec{e}_2 \otimes \vec{e}^{*2} - \vec{e}_1 \otimes \vec{e}^{*2} \otimes \vec{e}_2 \otimes \vec{e}^{*1} \\
&\quad -\vec{e}_2 \otimes \vec{e}^{*1} \otimes \vec{e}_1 \otimes \vec{e}^{*2} + \vec{e}_2 \otimes \vec{e}^{*2} \otimes \vec{e}_1 \otimes \vec{e}^{*1}. \quad (10.13)
\end{aligned}
$$

If we substitute (10.12) and (10.13) into (10.11) we obtain the *exterior expression* of the numerical model exterior product

$$
\vec{V}_1 \wedge \vec{V}_2^* \wedge \vec{V}_3 \wedge \vec{V}_4^* = \begin{vmatrix} x^{1\circ}_{\circ 1} & x^{1\circ}_{\circ 3} \\ x^{2\circ}_{\circ 1} & x^{2\circ}_{\circ 3} \end{vmatrix} \cdot \begin{vmatrix} x^{\circ 2}_{1\circ} & x^{\circ 4}_{1\circ} \\ x^{\circ 2}_{2\circ} & x^{\circ 4}_{2\circ} \end{vmatrix} \cdot \vec{e}_1 \wedge \vec{e}^{*1} \wedge \vec{e}_2 \wedge \vec{e}^{*2}. (10.14)
$$

This model permits us to end the present section with the generalized expressions of the 4-exterior vector initially treated in (10.9).

Consider the linear space $\bigwedge_n^{2,2}(K)$ with $p = q = 2$, and two vectors $\vec{V}_1, \vec{V}_3 \in V^n(K)$ which data matrix in contravariant coordinates $(x^i_\circ)$ is

$$
A = [X_1 X_3]_{n,2} \equiv \begin{bmatrix} x^{1\circ}_{\circ 1} & x^{1\circ}_{\circ 3} \\ x^{2\circ}_{\circ 1} & x^{2\circ}_{\circ 3} \\ \cdots & \cdots \\ x^{n\circ}_{\circ 1} & x^{n\circ}_{\circ 3} \end{bmatrix}. \quad (10.15)
$$

Consider also two vectors $\vec{V}_2^*, \vec{V}_4^* \in V_*^n(K)$ which data matrix in covariant coordinates (x°_j) is

$$
B = [X_2^* X_4^*]_{n,2} \equiv \begin{bmatrix} x^{\circ 2}_{1\circ} & x^{\circ 4}_{1\circ} \\ x^{\circ 2}_{2\circ} & x^{\circ 4}_{2\circ} \\ \cdots & \cdots \\ x^{\circ 2}_{n\circ} & x^{\circ 4}_{n\circ} \end{bmatrix}. \quad (10.16)
$$

The minors will carry the notation Δ with two upper row indices and two lower column indices; they *will not* carry an asterisk if they are minors of the contravariant matrix A, and *will* carry an asterisk if they are minors of the covariant matrix B.

According to this, Expression (10.9), once contracted becomes

$$
\vec{V}_1 \wedge \vec{V}_2^* \wedge \vec{V}_3 \wedge \vec{V}_4^* = \sum_1^{\binom{n}{2}^2} \begin{vmatrix} x^{\gamma_1\circ}_{\circ 1} & x^{\gamma_1\circ}_{\circ 3} \\ x^{\gamma_3\circ}_{\circ 1} & x^{\gamma_3\circ}_{\circ 3} \end{vmatrix} \cdot \begin{vmatrix} x^{\circ 2}_{\gamma_2\circ} & x^{\circ 4}_{\gamma_2\circ} \\ x^{\circ 2}_{\gamma_4\circ} & x^{\circ 4}_{\gamma_4\circ} \end{vmatrix} \delta\,{}^{\alpha_1\,\alpha_3}_{(\gamma_1\,\gamma_3)} \cdot \delta\,{}^{(\gamma_2\,\gamma_4)}_{\alpha_2\,\alpha_4}
$$
$$
\cdot \vec{e}_{\alpha_1} \otimes \vec{e}^{*\alpha_2} \otimes \vec{e}_{\alpha_3} \otimes \vec{e}^{*\alpha_4}, \quad (10.17)
$$

where $(\gamma_1 \gamma_3)$ and $(\gamma_2 \gamma_4)$ are the combinations of I_n of order 2, which are a total of $\binom{n}{2}$ combinations. This is the 4-exterior vector product, as a mixed totally anti-symmetric tensor of the tensor space $(\vec{V} \otimes \vec{V}_* \otimes \vec{V} \otimes \vec{V}_*)_h^n(K)$.

We also accept, according to the above, the following brief notation:

$$\vec{V}_1 \wedge \vec{V}_2^* \wedge \vec{V}_3 \wedge \vec{V}_4^* = \sum_1^{\binom{n}{2}^2} \Delta_{(1\ 3)}^{(\gamma_1 \gamma_3)} \cdot \left(\Delta_{(2\ 4)}^{(\gamma_2 \gamma_4)}\right)^* \delta_{(\gamma_1 \gamma_3)}^{\alpha_1 \alpha_3} \cdot \delta_{\alpha_2 \alpha_4}^{(\gamma_2 \gamma_4)}$$
$$\vec{e}_{\alpha_1} \otimes \vec{e}^{*\alpha_2} \otimes \vec{e}_{\alpha_3} \otimes \vec{e}^{*\alpha_4} \tag{10.18}$$

with the given conditions in expression (10.10) and conditions for the minors in the lines following (10.10).

The number of non-null components is given by (10.3):

$$N_1 = \left[\binom{n}{p}p!\right]^2 = \left[\binom{n}{2}2!\right]^2 = n^2(n-1)^2.$$

Finally, it is possible to give the 4-exterior vector product, as an exterior vector of the exterior algebra $V \wedge V_* \wedge V \wedge V_*(K)$:

$$\vec{V}_1 \wedge \vec{V}_2^* \wedge \vec{V}_3 \wedge \vec{V}_4^* = \sum_1^{\binom{n}{2}^2} \Delta_{(1\ 3)}^{(\gamma_1 \gamma_3)} \cdot \left(\Delta_{(2\ 4)}^{(\gamma_2 \gamma_4)}\right)^* \vec{e}_{\gamma_1} \wedge \vec{e}^{*\gamma_2} \wedge \vec{e}_{\gamma_3} \wedge \vec{e}^{*\gamma_4}. \tag{10.19}$$

The number of strict components is given by (10.3), i.e., $\sigma' = \binom{n}{p}^2 = \binom{n}{2}^2 = \frac{1}{4}n^2(n-1)^2$.

10.2 Decomposable mixed exterior vectors: mixed exterior product of p vectors $\vec{V}_i$ and q vectors $\vec{V}_j^*$

Once we have established in detail the first mixed exterior product, the 4-vector mixed exterior product formulated with the expressions and conditions (10.6), (10.9), (10.10), (10.15), (10.16), (10.17), (10.18) and (10.19), it is the right moment for establishing the general mixed exterior product, the factors of which are p vectors $\vec{V}_i \in V^n(K)$ and q vectors $\vec{V}_j^* \in V_*^n(K)$, with $2 \leq p, q \leq n$, $p = q$ or $p \neq q$ and the "species" of which are given as data.

In the product calculation procedures *the species* is going to play an *essential* role in the establishment of the formulas. First we obtain the data matrices.

The p contravariant vectors are "stacked" in a data-matrix, previous model (10.15), the important properties of which are its power and *special numbering of its columns*.

Consider the species *"alternate" in our model*, with $p < q$:

$$\left[t^{\alpha_1 \ \circ \ \alpha_3 \ \circ \ \cdots \alpha_{(2p-1)} \ \circ \ \cdots \ \circ}_{\ \circ \ \alpha_2 \ \circ \ \alpha_4 \cdots \ \ \circ \ \ \alpha_{2p} \cdots \alpha_{2q}} \right].$$

The contravariant data-matrix A is that having in its columns the numerical components of the $\vec{V}_i, 1 \le i \le p$, contravariant vectors. The *product species* which is data, is going to decide *the numbering of the columns of the matrix* A, i.e., the columns are numbered with the column index *corresponding* to the place occupied by the contravariant indices *in the species. In our model* they are the odd numbers $IMP_p = \{1, 3, 5, \ldots, 2p-1\}$.

In the matrix B, the components of the q covariant vectors are stacked and the numbering of the columns will be that corresponding to the places occupied by the covariant indices *in the product species*. In our model they are the even numbers $P_q \equiv \{2, 4, 6, \cdots, 2p, \cdots, 2q\}$. Thus, the data matrices A and B *for the present model* are

$$A = \begin{bmatrix} x^{1\circ}_{\circ 1} & x^{1\circ}_{\circ 3} & \cdots & x^{1\ \circ}_{\circ (2p-1)} \\ x^{2\circ}_{\circ 1} & x^{2\circ}_{\circ 3} & \cdots & x^{2\ \circ}_{\circ (2p-1)} \\ \cdots & \cdots & \cdots & \cdots \\ x^{n\circ}_{\circ 1} & x^{n\circ}_{\circ 3} & \cdots & x^{n\ \circ}_{\circ (2p-1)} \end{bmatrix}_{n,p} \quad \text{and} \quad B = \begin{bmatrix} x^{\circ 2}_{1\circ} & x^{\circ 4}_{1\circ} & \cdots & x^{\circ 2q}_{1\ \circ} \\ x^{\circ 2}_{2\circ} & x^{\circ 4}_{2\circ} & \cdots & x^{\circ 2q}_{2\ \circ} \\ \cdots & \cdots & \cdots & \cdots \\ x^{\circ 2}_{n\circ} & x^{\circ 4}_{n\circ} & \cdots & x^{\circ 2q}_{n\ \circ} \end{bmatrix}_{n,q},$$

$$(10.20)$$

with $2 \le p$, $q \le n$, or with column matrix notation, as blocks:

$$A = \begin{bmatrix} X_1 X_3 \cdots X_{(2p-1)} \end{bmatrix} \quad \text{and} \quad B = \begin{bmatrix} X_2 X_4 \cdots X_{2q} \end{bmatrix}.$$

These expressions generalize those in (10.15) and (10.16).

Next, from the set I_n we extract the $\binom{n}{p}$ combinations

$$(\gamma_1 \gamma_3 \cdots \gamma_{(2p-1)}); \quad \gamma_1 < \gamma_3 < \cdots < \gamma_{(2p-1)},$$

and also from I_n we obtain the $\binom{n}{q}$ combinations

$$(\gamma_2 \gamma_4 \cdots \gamma_{2q}); \ \gamma_2 < \gamma_4 < \cdots < \gamma_{2q}, \tag{10.21}$$

which will be useful for the later formulation of the *rows* of the minor Δ and of the generalized Levi-Civita deltas that appear with the cited entities.

From such rows we extract the diverse permutations:

$$\{\alpha_1 \alpha_3 \alpha_5 \cdots \alpha_{(2p-1)}\} = \{\gamma_1 \gamma_3 \cdots \gamma_{(2p-1)}\}$$
$$\{\alpha_2 \alpha_4 \alpha_6 \cdots \alpha_{2q}\} = \{\gamma_2 \gamma_4 \cdots \gamma_{2q}\}. \tag{10.22}$$

Once these introductory conditions have been established, we proceed to the generalization, first of Expression (10.6):

$$\vec{V}_1 \wedge \vec{V}_2^* \wedge \vec{V}_3 \wedge \vec{V}_4^* \wedge \cdots \wedge \vec{V}_{(2p-1)} \wedge \vec{V}_{2p}^* \wedge \vec{V}_{(2p+2)}^* \wedge \cdots \wedge \vec{V}_{2q}^*$$

$$= \delta^{\alpha_1 \alpha_3 \cdots \alpha_{(2p-1)}}_{(\beta_1 \beta_3 \cdots \beta_{(2p-1)}))} \cdot \delta^{(\beta_2 \beta_4 \cdots \beta_{2q})}_{\alpha_2 \alpha_4 \cdots \alpha_{2q}}$$

$$\vec{V}_{\alpha_1} \otimes \vec{V}_{\alpha_2}^* \otimes \vec{V}_{\alpha_3} \otimes \vec{V}_{\alpha_4}^* \otimes \cdots \otimes \vec{V}_{\alpha_{(2p-1)}} \otimes \vec{V}_{\alpha_{2p}}^* \otimes \vec{V}_{\alpha_{(2p+2)}}^* \otimes \cdots \otimes \vec{V}_{\alpha_{2q}}^* \tag{10.23}$$

with $\binom{n}{p}p! \times \binom{n}{q}q!$ summands in its development.

Next, we generalize Expression (10.17):

$$\vec{V}_1 \wedge \vec{V}_2^* \wedge \vec{V}_3 \wedge \vec{V}_4^* \wedge \cdots \wedge \vec{V}_{(2p-1)} \wedge \vec{V}_{2p}^* \wedge \vec{V}_{(2p+2)}^* \wedge \cdots \wedge \vec{V}_{2q}^*$$

$$= \sum_1^{\binom{n}{p}\binom{n}{q}} \begin{vmatrix} x^{\gamma_1}_{\;\circ\;1}{}^{\circ} & x^{\gamma_1}_{\;\circ\;3}{}^{\circ} & \cdots & x^{\gamma_1}_{\;\circ\;(2p-1)}{}^{\circ} \\ x^{\gamma_3}_{\;\circ\;1}{}^{\circ} & x^{\gamma_3}_{\;\circ\;3}{}^{\circ} & \cdots & x^{\gamma_3}_{\;\circ\;(2p-1)}{}^{\circ} \\ \cdots & \cdots & & \cdots \\ x^{\gamma_{(2p-1)}}_{\;\circ\;1}{}^{\circ} & x^{\gamma_{(2p-1)}}_{\;\circ\;3}{}^{\circ} & \cdots & x^{\gamma_{(2p-1)}}_{\;\circ\;(2p-1)}{}^{\circ} \end{vmatrix} \cdot \begin{vmatrix} x^{\circ\;2}_{\gamma_2\circ} & x^{\circ\;4}_{\gamma_2\circ} & \cdots & x^{\circ\;2q}_{\gamma_2\circ} \\ x^{\circ\;2}_{\gamma_4\circ} & x^{\circ\;4}_{\gamma_4\circ} & \cdots & x^{\circ\;2q}_{\gamma_4\circ} \\ \cdots & \cdots & \cdots & \cdots \\ x^{\circ\;2}_{\gamma_{2q}\circ} & x^{\circ\;4}_{\gamma_{2q}\circ} & \cdots & x^{\circ\;2q}_{\gamma_{2q}\circ} \end{vmatrix}$$

$$\cdot\, \delta^{\alpha_1 \alpha_3 \cdots \alpha_{(2p-1)}}_{(\gamma_1 \gamma_3 \cdots \gamma_{(2p-1)})} \cdot \delta^{(\gamma_2 \gamma_4 \cdots \gamma_{2q})}_{\alpha_2 \alpha_4 \cdots \alpha_{2q}}$$

$$\vec{e}_{\alpha_1} \otimes \vec{e}^{*\alpha_2} \otimes \vec{e}_{\alpha_3} \otimes \vec{e}^{*\alpha_4} \otimes \cdots \otimes \vec{e}_{\alpha_{(2p-1)}} \otimes \vec{e}^{*2p} \otimes \vec{e}^{*(2p+2)} \otimes \cdots \otimes \vec{e}^{*2q},$$

$$(10.24)$$

giving rise to the $(p+q)$-exterior mixed vector as a tensor of the mixed anti-symmetric tensor algebra

$$\left[\left(\overset{p}{\underset{1}{\otimes}} V \right) \otimes \left(\overset{q}{\underset{1}{\otimes}} V_* \right) \right]^n_h (K)$$

of "alternate" species

$$\left[t^{\alpha_1 \;\circ\; \alpha_3 \;\circ\; \cdots \alpha_{(2p-1)} \;\circ\;\;\;\;\circ\;\; \cdots\;\circ}_{\;\;\circ\; \alpha_2 \;\circ\; \alpha_4 \cdots\;\;\;\;\;\;\circ\;\;\; \alpha_{2p}\alpha_{(2p+2)}\cdots\alpha_{2q}} \right].$$

We can also generalize expression (10.18), using the previous abbreviated notation:

$$\vec{V}_1 \wedge \vec{V}_2^* \wedge \vec{V}_3 \wedge \vec{V}_4^* \wedge \cdots \wedge \vec{V}_{(2p-1)} \wedge \vec{V}_{2p}^* \wedge \vec{V}_{(2p+2)}^* \wedge \cdots \wedge \vec{V}_{2q}^*$$

$$= \sum_1^{\binom{n}{p}\binom{n}{q}} \Delta^{(\gamma_1 \gamma_3 \cdots \gamma_{(2p-1)})}_{(\;1\;\;3\;\cdots\;(2p-1))} \cdot \left(\Delta^{(\gamma_2 \gamma_4 \cdots \gamma_{2q})}_{(\;2\;\;4\;\cdots\;2q)} \right)^* \delta^{\alpha_1 \alpha_3 \cdots \gamma_{(2p-1)}}_{(\gamma_1 \gamma_3 \cdots (2p-1))} \cdot \delta^{(\gamma_2 \gamma_4 \cdots \gamma_{2q})}_{\alpha_2 \alpha_4 \cdots\; 2q}$$

$$\vec{e}_{\alpha_1} \otimes \vec{e}^{*\alpha_2} \otimes \vec{e}_{\alpha_3} \otimes \vec{e}^{*\alpha_4} \otimes \cdots \otimes \vec{e}_{\alpha_{(2p-1)}} \otimes \vec{e}^{*2p} \otimes \vec{e}^{*(2p+2)} \otimes \cdots \otimes \vec{e}^{*2q}$$

$$(10.25)$$

under conditions (10.21) and (10.22).

Finally, we generalize expression (10.19):

$$\vec{V}_1 \wedge \vec{V}_2^* \wedge \vec{V}_3 \wedge \vec{V}_4^* \wedge \cdots \wedge \vec{V}_{(2p-1)} \wedge \vec{V}_{2p}^* \wedge \vec{V}_{(2p+2)}^* \wedge \cdots \wedge \vec{V}_{2q}^*$$

$$= \sum_1^{\binom{n}{p}\binom{n}{q}} \Delta^{(\gamma_1 \gamma_3 \cdots \gamma_{(2p-1)})}_{(\;1\;\;3\;\cdots\;(2p-1))} \cdot \left(\Delta^{(\gamma_2 \gamma_4 \cdots \gamma_{2q})}_{(\;2\;\;4\;\cdots\;2q)} \right)^*$$

$$\vec{e}_{\gamma_1} \wedge \vec{e}^{*\gamma_2} \wedge \vec{e}_{\gamma_3} \wedge \vec{e}^{*\gamma_4} \wedge \cdots \wedge \vec{e}_{\gamma_{(2p-1)}} \wedge \vec{e}^{*\gamma_{2p}} \wedge \vec{e}^{*\gamma_{(2p+2)}} \wedge \cdots \wedge \vec{e}^{*\gamma_{2q}}$$

$$(10.26)$$

with conditions (10.21) and (10.22).

10.3 Mixed exterior algebras: Terminology

The notation we propose is subject to the species of the mixed exterior vectors to be contained, and can be guessed after the advanced notation that has been used in the previous sections.

First, we propose a general notation, useful for any presentation of mixed exterior algebras.

The commonly used terminology will be

$$\bigwedge\nolimits_n^{(p,q)}(K) \equiv (\wedge V^n)^p \wedge (\wedge V_*^n)^q (K)$$

$$\equiv \left[\left(\overset{p}{\underset{1}{\wedge}} V \right) \wedge \left(\overset{q}{\underset{1}{\wedge}} V_* \right) \right]^n (K)$$

$$\equiv (V \wedge V_* \wedge V_* \wedge V \wedge \cdots \wedge V \wedge V_* \wedge V \wedge V)^n (K) \quad (10.27)$$

with "p" V referring to the primal linear space $V^n(K)$ and "q" V_* referring to the dual linear space $V_*^n(K)$ precisely in its concrete place. The upper index n braces all instead of bracing each one of them, when $p + q = r$ is a large natural number. The last notation *is forced*, when in any of the previous "the species" has not been mentioned. Normally we shall use any of the notations (more frequently the first one), except the last one, which it is too long, giving in addition the species by means of one of the following two procedures.

If the exterior vector is of small order r, we notate its species in tensor form as $[t^{\alpha \circ \circ \circ \lambda \mu}_{\circ \beta \gamma \delta \circ \circ}]$ with $r = 6$, the indices of which p and q are directly seen in addition to its place.

If the order r is large, we use: $1 \equiv$ as the *contravariant* index indicator, and $2 \equiv$ as *covariant* index indicator.

Thus, the previous exterior vector species has been given with the tensor component as guide, is now given as species $= [1 - 2 - 2 - 2 - 1 - 1]$, with p ones and q twos, in the foreseen order, a procedure that is simple for the computer and for large r. The univocal correspondence between both procedures of giving the species, it is evident that avoids the last terminology for the notation of mixed exterior algebras.

For the sake of simplifying the calculation formulas, it is convenient for the authors of tensor material to give *always*, when presenting the theory, a staked tensor formulation, for example, that of Expression (10.1) or the first of the terminology. When developing all their theory with this licence, the unwarned reader thinks that it is always that way, giving rise to some undesired annoyances when the examples are real, numeric and concrete, for which the order is *essential*.

10.3.1 Exterior basis of a mixed exterior algebra

As can be seen in previous expressions, such as (10.14), (10.19) and (10.26), the exterior basis of a $\bigwedge\nolimits_n^{(p,q)}(K)$ has a very clear construction. First, the number of basic vectors of the cited mixed exterior algebra is its dimension, that is, according to the group of formulas (10.3),

$$\sigma' = \binom{n}{p}\binom{n}{q}, \tag{10.28}$$

which refers to a concrete construction, from the set I_n of the $\binom{n}{p}$ groups of combinations notated $(\gamma_1 \gamma_3 \cdots \gamma_{(2p-1)})$ with $\gamma_1 < \gamma_3 < \cdots < \gamma_{(2p-1)}$ that it is possible to extract from I_n, and from the other $\binom{n}{q}$ groups of combinations notated $(\gamma_2 \gamma_4 \cdots \gamma_{2q})$ with $\gamma_2 < \gamma_4 < \cdots < \gamma_{2q}$, that it is also possible to extract from I_n. Such groups are ordered by intercalating each and every one of the odd groups γ_i, with each and every one of the even groups of indices γ_j, precisely in the order imposed by the species. With the cited groups of combinations already totally ordered, the basic set of $\bigwedge_n^{(p,q)}(K)$ with σ' vectors is built, that is,

$$B_{\wedge(p,q)} \equiv \{\vec{e}_{\gamma_1} \wedge \vec{e}^{*\gamma_2} \wedge \vec{e}_{\gamma_3} \wedge \vec{e}^{*\gamma_4} \otimes \cdots \otimes \vec{e}_{\gamma_{(2p-1)}} \otimes \vec{e}^{*\gamma_{2p}} \otimes \vec{e}^{*\gamma(2p+2)} \otimes \cdots \otimes \vec{e}^{*\gamma_{2q}}\}$$
$$(10.29)$$

assuming that the odd $\{\vec{e}_{\gamma_i}\}$ *and the even* $\{\vec{e}^{*\gamma_j}\}$ *appear in the order assigned by the species.*

This is the basis in which a mixed exterior vector must be delivered, coming or not from a mixed product of vectors, associated with their corresponding mixed strict components, which configure, in essence, any mixed exterior tensor.

10.3.2 Axiomatic tensor operations in the $\bigwedge_n^{(p,q)}(K)$ algebra

The generation of mixed exterior vectors is again stated exactly equal to the case of primal or dual exterior vectors. The sum of several different exterior products referred to the same exterior basis originates the appearance of the normal entities of a mixed exterior algebra, $T \in \bigwedge_n^{(p,q)}(K)$ and given species, that do not come from multiplying vectors; they are the proper vectors, that is, non-decomposable, of the given algebra.

Consider a concrete model as the one that is to be selected for specifying the axiomatic properties. By means of this formal support, we try to avoid the complexity that implies the general formulation.

Definition 10.1 (Sum of mixed exterior vectors). *Given the mixed exterior vectors* $T, U \in \bigwedge_n^{(3,3)}(K)$ *species* $= [1 - 2 - 2 - 2 - 1 - 1]$.

$$T = t^{(\alpha_1 \; \circ \; \circ \; \circ \; \alpha_5 \alpha_6)}_{(\; \circ \; \alpha_2 \alpha_3 \alpha_4 \; \circ \; \circ\;)} \vec{e}_{\alpha_1} \wedge \vec{e}^{*\alpha_2} \wedge \vec{e}^{*\alpha_3} \wedge \vec{e}^{*\alpha_4} \wedge \vec{e}_{\alpha_5} \wedge \vec{e}_{\alpha_6}.$$

$$U = u^{(\alpha_1 \; \circ \; \circ \; \circ \; \alpha_5 \alpha_6)}_{(\; \circ \; \alpha_2 \alpha_3 \alpha_4 \; \circ \; \circ\;)} \vec{e}_{\alpha_1} \wedge \vec{e}^{*\alpha_2} \wedge \vec{e}^{*\alpha_3} \wedge \vec{e}^{*\alpha_4} \wedge \vec{e}_{\alpha_5} \wedge \vec{e}_{\alpha_6},$$

we define as the vector sum of both, the exterior vector

$$W = T + U = w{\binom{\alpha_1 \; \circ \; \circ \; \circ \; \alpha_5 \alpha_6}{\circ \; \alpha_2 \alpha_3 \alpha_4 \; \circ \; \circ}} \vec{e}_{\alpha_1} \wedge \vec{e}^{*\alpha_2} \wedge \vec{e}^{*\alpha_3} \wedge \vec{e}^{*\alpha_4} \wedge \vec{e}_{\alpha_5} \wedge \vec{e}_{\alpha_6}$$

where

$$w{\binom{\alpha_1 \; \circ \; \circ \; \circ \; \alpha_5 \alpha_6}{\circ \; \alpha_2 \alpha_3 \alpha_4 \; \circ \; \circ}} = t{\binom{\alpha_1 \; \circ \; \circ \; \circ \; \alpha_5 \alpha_6}{\circ \; \alpha_2 \alpha_3 \alpha_4 \; \circ \; \circ}} + u{\binom{\alpha_1 \; \circ \; \circ \; \circ \; \alpha_5 \alpha_6}{\circ \; \alpha_2 \alpha_3 \alpha_4 \; \circ \; \circ}} \in K.$$

This sum is associative, unitary, cancelative and Abelian, arising the Abelian group of the mixed exterior vectors.

Definition 10.2 (External product of mixed exterior vectors). *For all mixed exterior vectors* $T \in \bigwedge_n^{(3,3)}(K)$ *and for all* $\lambda \in K$, *we define the product of* λ *and* T, *for the vector*

$$S = \lambda T = \lambda t{\binom{\alpha_1 \; \circ \; \circ \; \circ \; \alpha_5 \alpha_6}{\circ \; \alpha_2 \alpha_3 \alpha_4 \; \circ \; \circ}} \vec{e}_{\alpha_1} \wedge \vec{e}^{*\alpha_2} \wedge \vec{e}^{*\alpha_3} \wedge \vec{e}^{*\alpha_4} \wedge \vec{e}_{\alpha_5} \wedge \vec{e}_{\alpha_6}$$

$$= s{\binom{\alpha_1 \; \circ \; \circ \; \circ \; \alpha_5 \alpha_6}{\circ \; \alpha_2 \alpha_3 \alpha_4 \; \circ \; \circ}} \vec{e}_{\alpha_1} \wedge \vec{e}^{*\alpha_2} \wedge \vec{e}^{*\alpha_3} \wedge \vec{e}^{*\alpha_4} \wedge \vec{e}_{\alpha_5} \wedge \vec{e}_{\alpha_6},$$

where

$$s{\binom{\alpha_1 \; \circ \; \circ \; \circ \; \alpha_5 \alpha_6}{\circ \; \alpha_2 \alpha_3 \alpha_4 \; \circ \; \circ}} = \lambda t{\binom{\alpha_1 \; \circ \; \circ \; \circ \; \alpha_5 \alpha_6}{\circ \; \alpha_2 \alpha_3 \alpha_4 \; \circ \; \circ}} \in K.$$

The external product is distributive with respect to the sum of scalars, distributive with respect to the sum of exterior vectors, associative, i.e., $\lambda(\mu T) = (\lambda\mu)T$ and unitary, i.e., $1T = T$, and $1 \in K$ is the unit of the product of scalars in the field K.

We have created the mixed exterior linear space.

10.4 Exterior product of mixed exterior vectors

We proceed to establish the mentioned product by means of the following scheme.

First: We create a mixed exterior product, denoted by "$\bigwedge$" (in upper case), in order to avoid the possibility of confusion with the symbol "$\wedge$" that appears in the mixed exterior products of vectors and in the bases of the mixed exterior spaces.

Second: This product will be useful to:

Case (a): Multiply mixed exterior vectors of different mixed exterior spaces. Consider two mixed exterior tensors $T \in \bigwedge_n^{(p_1,q_1)}(K)$; $U \in \bigwedge_n^{(p_2,q_2)}(K)$ over $V^n(K)$ and $V_*^n(K)$, respectively, with the conditions $p_1, q_1 \leq n$; $p_2, q_2 \leq n$; $r_1 = p_1 + q_1$; $r_2 = p_2 + q_2$ and $r_1, r_2 \leq 2n$. Then, we must have

$$V = T \bigwedge U; \quad V \in \bigwedge_n^{(p,q)}(K), \tag{10.30}$$

with the conditions

$$p = p_1 + p_2; \quad q = q_1 + q_2; \quad p, q \leq n; \quad r = p + q \leq 2n, \tag{10.31}$$

which last one equality is more strict than $r \leq 4n$.

The cited expressions are necessary but not sufficient for establishing the product $V = T \bigwedge U$. Apart from knowing the mixed exterior algebra $V \in \bigwedge_n^{(p,q)}(K)$, we need, before calculating V, *the species* of the linear space $\bigwedge_n^{(p,q)}(K)$, with the aim of being able to create the basis of the mentioned mixed exterior algebra. If the species is not given, it will be taken as the union of the species of the factors.

Case (b): Multiply mixed exterior vectors, both from the same data mixed exterior linear space $\bigwedge_n^{p,q}(K)$; we must have

$$T \bigwedge U = V; \quad V \in \bigwedge_n^{(2p,2q)}(K), \quad 2p, 2q \leq n. \tag{10.32}$$

Third: The reader should remember that in the third point of Section 9.3.2 the notation was changed for tensor U, second factor of the product; thus:

Case (a): The tensors $T \in \bigwedge_n^{(p_1,q_1)}(K)$ and $U \in \bigwedge_n^{(p_2,q_2)}(K)$, respectively, are:

$$t^{\left(\begin{smallmatrix} \alpha_1 & \circ & \circ & \cdots & \circ & \alpha_{(2p_1-3)} \, \alpha_{(2p_1-1)} \\ \circ & \alpha_2 \, \alpha_4 \cdots \alpha_{2q_1} & & \circ & & \circ \end{smallmatrix}\right)} \vec{e}_{\alpha_1} \wedge \vec{e}^{*\alpha_2} \wedge \vec{e}^{*\alpha_4} \wedge \cdots \wedge \vec{e}^{*\alpha_{2q_1}} \wedge \vec{e}_{\alpha_{(2p_1-3)}} \wedge \vec{e}_{\alpha_{(2p_1-1)}},$$

$$u^{\left(\begin{smallmatrix} \circ & \alpha_1 & \circ & \cdots \alpha_{(2p_2-3)} & \circ & \alpha_{(2p_2-1)} \\ \alpha_2 & \circ & \alpha_4 \cdots & \circ & \alpha_{(2q_2)} & \circ \end{smallmatrix}\right)} \vec{e}^{*\alpha_2} \wedge \vec{e}_{\alpha_1} \wedge \vec{e}^{*\alpha_4} \wedge \cdots \wedge \vec{e}_{\alpha_{2p_2-3}} \wedge \vec{e}^{*\alpha_{(2q_2)}} \wedge \vec{e}_{\alpha_{(2p_2-1)}}, \tag{10.33}$$

with the conditions

For T: $\quad \alpha_i, \alpha_j \in I_n; \; i \in IMP_{p_1}; \; j \in P_{q_1};$

$\qquad \alpha_1 < \alpha_3 < \cdots < \alpha_{(2p_1-1)}; \; \alpha_2 < \alpha_4 < \cdots < \alpha_{(2q_1)}.$

For U: $\quad \alpha_i, \alpha_j \in I_n; \; i \in IMP_{p_2}; \; j \in P_{q_2}; \; \alpha_1 < \alpha_3 < \cdots < \alpha_{(2p_2-1)}.$

$\qquad \alpha_2 < \alpha_4 < \cdots < \alpha_{(2q_2)}.$

$$\tag{10.34}$$

We proceed to change the notation of the indices in the second factor U, as it is indicated in the third point of Section 9.3.2 for the generalized exterior product.

The mixed exterior tensor, which is the data in the second factor, will be notated U, with

$$u^{\left(\begin{smallmatrix} \circ & \beta_1 & \circ & \cdots \beta_{(2p_2-3)} & \circ & \beta_{(2p_2-1)} \\ \beta_2 & \circ & \beta_4 \cdots & \circ & \beta_{(2q_2)} & \circ \end{smallmatrix}\right)} \tag{10.35}$$

being the component associated with the basic vector

$$\vec{e}^{*\beta_2} \wedge \vec{e}_{\beta_1} \wedge \vec{e}^{*\beta_4} \wedge \cdots \wedge \vec{e}_{\beta_{2p_2-3}} \wedge \vec{e}^{*\beta_{(2q_2)}} \wedge \vec{e}_{\beta_{(2p_2-1)}}$$

with the same conditions as the α.

Case (b): The tensors $T \in \bigwedge_n^{(p,q)}(K)$ and $U \in \bigwedge_n^{(p,q)}(K)$ are:

$$t^{\left(\begin{smallmatrix} \alpha_1 & \circ & \circ & \cdots & \circ & \alpha_{(2p-3)}\,\alpha_{(2p-1)} \\ \circ & \alpha_2\,\alpha_4\cdots\alpha_{2q} & & \circ & & \circ \end{smallmatrix}\right)} \vec{e}_{\alpha_1} \wedge \vec{e}^{*\alpha_2} \wedge \vec{e}^{*\alpha_4} \wedge \cdots \wedge \vec{e}^{*\alpha_{2q}} \wedge \vec{e}_{\alpha_{(2p-3)}} \wedge \vec{e}_{\alpha_{(2p-1)}}.$$

$$u^{\left(\begin{smallmatrix} \alpha_1 & \circ & \circ & \cdots & \circ & \alpha_{(2p-3)}\,\alpha_{(2p-1)} \\ \circ & \alpha_2\,\alpha_4\cdots\alpha_{2q} & & \circ & & \circ \end{smallmatrix}\right)} \vec{e}_{\alpha_1} \wedge \vec{e}^{*\alpha_2} \wedge \vec{e}^{*\alpha_4} \wedge \cdots \wedge \vec{e}^{*\alpha_{2q}} \wedge \vec{e}_{\alpha_{(2p-3)}} \wedge \vec{e}_{\alpha_{(2p-1)}}.$$

$$(10.36)$$

and notating

$$U = u^{\left(\begin{smallmatrix} \beta_1 & \circ & \circ & \cdots & \circ & \beta_{(2p-3)}\,\beta_{(2p-1)} \\ \circ & \beta_2\,\beta_4\cdots\beta_{2q} & & \circ & & \circ \end{smallmatrix}\right)} \vec{e}_{\beta_1} \wedge \vec{e}^{*\beta_2} \wedge \vec{e}^{*\beta_4} \wedge \cdots \wedge \vec{e}^{*\beta_{2q}} \wedge \vec{e}_{\beta_{(2p-3)}} \wedge \vec{e}_{\beta_{(2p-1)}}$$

with the conditions

$$\beta_i, \beta_j \in I_n;\ i \in IMP_p;\ j \in P_q;\ \beta_1 < \beta_3 < \cdots < \beta_{(2p-1)};\ \beta_2 < \beta_4 < \cdots < \beta_{2q}.$$

$$(10.37)$$

Fourth:

Case (a): Next, we choose collections of indices $\{\gamma_1, \gamma_3, \ldots, \gamma_{(2p-1)}\} \in I_n$ and $\{\gamma_2, \gamma_4, \ldots, \gamma_{2q}\} \in I_n$ for the basic vectors of the type $(\vec{e}_{\gamma_1} \wedge \vec{e}_{\gamma_3} \wedge \cdots \wedge \vec{e}_{\gamma_{(2p-1)}})$ and for those of the type $(\vec{e}^{*\gamma_2} \wedge \vec{e}^{*\gamma_4} \wedge \cdots \wedge \vec{e}^{*\gamma_{2q}})$, respectively. Obviously, we must choose $\binom{n}{p}$ collections for the group of contravariant indices and $\binom{n}{q}$ collections for the group of covariant indices, corresponding to the dimension $\sigma' = \binom{n}{p}\binom{n}{q}$ of $V \in \bigwedge_n^{(p,q)}(K)$; the mixed exterior space V in which the product is defined.

With respect to the species of the cited exterior space, that is, the position order, it must be imposed as *additional data*. So, we must have the following data:

$$p = p_1 + p_2;\quad q = q_1 + q_2;\quad \text{species} = \{1 - 1 - 2 - 2 - \cdots - 1 - 2 - 1\}_{(p+q)},$$

which permits us to build the basis of $\bigwedge_n^{(p,q)}(K)$:

$$B_{\wedge(p,q)} \equiv \{\vec{e}_{\gamma_1} \wedge \vec{e}_{\gamma_3} \wedge \vec{e}^{*\gamma_2} \wedge \vec{e}^{*\gamma_4} \wedge \cdots \wedge \vec{e}_{\gamma_{(2p-3)}} \wedge \vec{e}^{*\gamma_{2q}} \wedge \vec{e}_{\gamma_{(2p-1)}}\} \quad (10.38)$$

Case (b): As its algebra is $\bigwedge_n^{(2p,2q)}(K)$, remembering that $2p, 2q \le n$, and that it has the species $\{1 - 1 - 2 - -2 \cdots 1 - 2 - 1\}_{(2p+2q)}$, the result is that the dimension of $\bigwedge_n^{(2p,2q)}(K)$ is $\sigma' = \binom{n}{2p}\binom{n}{2q}$, that is, the number of basic vectors and the basis with analogous disposition to that of (10.38), because we have chosen an analogous species (but with $(2p + 2q)$ indices).

Fifth: It remains only to define the coefficients of V in the already prepared basis.

Case (a): We define the mixed exterior product $\bigwedge$ for the Formula (10.30) and (10.33), as

$$V = T \bigwedge U$$

$$= v\binom{\gamma_1\,\gamma_3\,\circ\,\circ\,\cdots\gamma_{(2p-3)}\,\circ\,\gamma_{(2p-1)}}{\circ\,\circ\,\gamma_2\,\gamma_4\cdots\,\circ\,\gamma_{2q}\,\circ\,}$$

$$\vec{e}_{\gamma_1} \wedge \vec{e}_{\gamma_3} \wedge \vec{e}^{*\gamma_2} \wedge \vec{e}^{*\gamma_4} \wedge \cdots \wedge \vec{e}_{\gamma_{(2p-3)}} \wedge \vec{e}^{*\gamma_{2q}} \wedge \vec{e}_{\gamma_{(2p-1)}}$$

$$= t\binom{\alpha_1\,\circ\,\circ\,\cdots\,\circ\,\alpha_{(2p_1-3)}\,\alpha_{(2p_1-1)}}{\circ\,\alpha_2\,\alpha_4\cdots\alpha_{2q_1}\,\circ\,\circ} \cdot u\binom{\circ\,\beta_1\,\circ\,\cdots\beta_{(2p_2-3)}\,\circ\,\beta_{(2p_2-1)}}{\beta_2\,\circ\,\beta_4\cdots\,\circ\,\beta_{(2q_2)}\,\circ}$$

$$\cdot \delta\binom{\gamma_1\,\gamma_3\,\cdots\,\gamma_{(2p_1-1)}\,\gamma_{(2p_1+1)}\gamma_{(2p_1+3)}\,\cdots\,\gamma_{(2p-1)}}{(\alpha_1\,\alpha_3\cdots\alpha_{(2p_1-1)})\quad(\beta_1\quad\beta_3\quad\cdots\beta_{(2p_2-1)})} \cdot \delta\binom{\alpha_2\,\alpha_4\cdots\alpha_{2q_1})\quad(\beta_2\quad\beta_4\quad\cdots\beta_{2q_2})}{\gamma_2\,\gamma_4\cdots\,\gamma_{2q_1}\,\gamma_{(2q_1+2)}\gamma_{(2q_1+4)}\cdots\,\gamma_{2q}}$$

$$\vec{e}_{\gamma_1} \wedge \vec{e}_{\gamma_3} \wedge \vec{e}^{*\gamma_2} \wedge \vec{e}^{*\gamma_4} \wedge \cdots \wedge \vec{e}_{\gamma_{(2p-3)}} \wedge \vec{e}^{*\gamma_{2q}} \wedge \vec{e}_{\gamma_{(2p-1)}} \qquad (10.39)$$

with the conditions:

$$p = p_1 + p_2; \quad q = q_1 + q_2;$$
$$\{\alpha_1, \alpha_3, \ldots, \alpha_{(2p_1-1)}, \beta_1, \beta_3, \ldots, \beta_{(2p_2-1)}\} \equiv \{\gamma_1, \gamma_3, \ldots, \gamma_{(2p-1)}\};$$
$$\{\alpha_2, \alpha_4, \ldots, \alpha_{2q_1}, \beta_2, \beta_4, \ldots, \beta_{2q_2}\} \equiv \{\gamma_2, \gamma_4, \ldots, \gamma_{2q}\};$$
$$\gamma_1 < \gamma_3 < \ldots < \gamma_{(2p-1)}; \quad \gamma_2 < \gamma_4 < \ldots < \gamma_{2q};$$
$$(\alpha_1 < \alpha_3 < \ldots < \alpha_{(2p_1-1)}) \neq (\beta_1 < \beta_3 < \ldots < \beta_{(2p_2-1)});$$
$$(\alpha_2 < \alpha_4 < \ldots < \alpha_{2q_1}) \neq (\beta_2 < \beta_4 < \ldots < \beta_{2q_2}). \qquad (10.40)$$

When using (10.39), we choose from the ordered set $\{\gamma_1, \gamma_3, \ldots, \gamma_{(2p-1)}\}$ taken from I_n the combinations of p_1 indices for the odd $\{\alpha_i\}$ and the complementary combination (that not chosen with p_2 indices) is reserved for the even $\{\beta_j\}$ and we take these selections to the first generalized Kronecker delta, which gives it the sign. Next, we proceed in a similar form for the second delta, i.e., we choose from the ordered set $\{\gamma_2, \gamma_4, \ldots, \gamma_{2q}\}$ also taken from I_n the combinations of q_1 indices for the even $\{\alpha_i\}$ and the complementary combination (that not chosen with q_2 indices) is assigned to the odd $\{\beta_j\}$ and these selections are taken to the second generalized Kronecker delta, which gives it the sign, that is multiplied by the previous one and by the scalars t and u.

Obviously, each and every one of the selections of the first delta must be associated with each and every one of the second delta, which leads to all scalar summands of the coefficient corresponding to the basic vector of $\bigwedge_n^{(p,q)}(K)$ that is being calculated. We will obtain a total of $\sigma' = \binom{n}{p} \cdot \binom{n}{q}$ strict components, with $\binom{p}{p_1}\binom{q}{q_1}$ summands for each component.

Case (b): Finally, we define the mixed exterior product "$\bigwedge$" for Formulas (10.32) and (10.36), as

$$V = T \bigwedge U$$

$$= v_{(\circ \ \circ \ \gamma_2 \gamma_4 \cdots \ \ \ \ \circ \ \ \ \ \gamma_{4q} \ \ \circ \ \ )}^{(\gamma_1 \gamma_3 \ \circ \ \circ \ \cdots \gamma_{(4p-3)} \ \circ \ \gamma_{(4p-1)})}$$

$$\vec{e}_{\gamma_1} \wedge \vec{e}_{\gamma_3} \wedge \vec{e}^{*\gamma_2} \wedge \vec{e}^{*\gamma_4} \wedge \cdots \wedge \vec{e}_{\gamma_{(4p-3)}} \wedge \vec{e}^{*\gamma_{4q}} \wedge \vec{e}_{\gamma_{(4p-1)}} \quad (10.41)$$

$$= t_{(\circ \ \alpha_2 \alpha_4 \cdots \alpha_{2q} \ \ \circ \ \ \ \ \ \circ \ )}^{(\alpha_1 \ \circ \ \circ \ \cdots \ \circ \ \alpha_{(2p-3)} \alpha_{(2p-1)})} \cdot u_{(\circ \ \beta_2 \beta_4 \cdots \beta_{2q} \ \ \circ \ \ \ \ \ \circ \ )}^{(\beta_1 \ \circ \ \circ \ \cdots \ \circ \ \beta_{(2p-3)} \beta_{(2p-1)})}$$

$$\cdot \delta_{(\alpha_1 \alpha_3 \cdots \alpha_{(2p-1)}) \ (\beta_1 \ \ \ \beta_3 \ \ \ \cdots \beta_{(2p-1)})}^{(\gamma_1 \ \gamma_3 \ \cdots \gamma_{(2p-1)}) \ \gamma_{(2p+1)} \gamma_{(2p+3)} \cdots \gamma_{(4p-1)})}$$

$$\cdot \delta_{(\gamma_2 \ \gamma_4 \ \cdots \ \gamma_{2q} \ \gamma_{(2q+2)} \gamma_{(2q+4)} \cdots \gamma_{4q})}^{(\alpha_2 \alpha_4 \cdots \alpha_{2q}) \ (\beta_2 \ \ \ \beta_4 \ \ \ \cdots \beta_{2q})}$$

$$\vec{e}_{\gamma_1} \wedge \vec{e}_{\gamma_3} \wedge \vec{e}^{*\gamma_2} \wedge \vec{e}^{*\gamma_4} \wedge \cdots \wedge \vec{e}_{\gamma_{(4p-3)}} \wedge \vec{e}^{*\gamma_{4q}} \wedge \vec{e}_{\gamma_{(4p-1)}}.$$

with the conditions:

$$\{\alpha_1, \alpha_3, \ldots, \alpha_{(2p-1)}, \beta_1, \beta_3, \ldots, \beta_{(2p-1)}\} \equiv \{\gamma_1, \gamma_3, \ldots, \gamma_{(2p-1)}, \cdots, \gamma_{(4p-1)}\};$$

$$\{\alpha_2, \alpha_4, \ldots, \alpha_{2q}, \beta_2, \beta_4, \ldots, \beta_{2q}\} \equiv \{\gamma_2, \gamma_4, \ldots, \gamma_{2q}, \ldots, \gamma_{4q}\};$$

$$\gamma_1 < \gamma_3 < \cdots < \gamma_{(4p-1)}; \quad \gamma_2 < \gamma_4 < \cdots < \gamma_{4q};$$

$$(\alpha_1 < \alpha_3 < \cdots < \alpha_{(2p-1)}) \neq (\beta_1 < \beta_3 < \cdots < \beta_{(2p-1)});$$

$$(\alpha_2 < \alpha_4 < \cdots < \alpha_{2q}) \neq (\beta_2 < \beta_4 < \cdots < \beta_{2q}). \quad (10.42)$$

With respect to the way of using Formula (10.42) it becomes clear if we consider the sets $\{\gamma_1, \gamma_3, \ldots, \gamma_{(4p-1)}\}$ and $\{\gamma_2, \gamma_4, \ldots, \gamma_{4q}\}$ chosen in I_n, over which we choose the $\binom{n}{2p}$ combinations and $\binom{n}{2q}$, respectively, following an analogous process to that already explained for using the Formula (10.39).

10.5 Anti-commutativity of the $\bigwedge$ mixed exterior product

The alternating character of the order of the factors in the product $T \bigwedge U$ by the $U \bigwedge T$ affects the ordering of the indices of the *two* generalized Kronecker deltas in the Formula (10.39).

Applying to each delta a similar analysis to that performed in Section 9.3.3 with the mentioned delta, leads to:

Case (a):

$$\delta_1(\text{altered}) = (-1)^{p_1 p_2} \cdot \delta_1(\text{not altered})$$

$$\delta_2(\text{altered}) = (-1)^{q_1 q_2} \cdot \delta_2(\text{not altered})$$

Whence, $\delta_1 \cdot \delta_2(\text{altered}) = (-1)^{p_1 p_2 + q_1 q_2} \delta_1 \cdot \delta_2(\text{not altered})$, which leads to

$$U \bigwedge T = (-1)^{(p_1 p_2 + q_1 q_2)} T \bigwedge U. \quad (10.43)$$

Case (b):

$$U \bigwedge T = (-1)^{(p^2 + q^2)} T \bigwedge U. \quad (10.44)$$

10.6 Change of basis in mixed exterior algebras

Consider the mixed exterior algebra $\bigwedge_n^{(p,q)}(K)$ built over the linear space $V^n(K)$ and its dual $V_*^n(K)$. We execute a change-of-basis in the $V^n(K)$ of associated matrix $C = [c_{\circ\,i}^{\alpha\,\circ}]$, and in its dual the corresponding change, with associated matrix $C^t = [c_{i\,\circ}^{\circ\,\alpha}]$, where $C^{-1} = [\gamma_{\circ\,\alpha}^{i\,\circ}]$.

In summary, from Sections 9.5.1 and 9.5.2, we can establish that a mixed exterior tensor changes components according to the relation and the conditions that follow. Assuming $p > q$, and the given species, it is

$$t_{(\circ\, i_2\, i_4 \cdots i_{2q}\quad\circ\quad\ \ \circ\ )}^{(i_1\ \circ\ \circ\ \cdots\ \circ\ i_{(2p-3)}\, i_{(2p-1)})}$$

$$= t_{(\circ\ \alpha_2\, \alpha_4 \cdots \alpha_{2q}\quad\circ\quad\ \ \circ\ )}^{(\alpha_1\ \circ\ \circ\ \cdots\ \circ\ \alpha_{(2p-3)}\, \alpha_{(2p-1)})} \begin{vmatrix} \gamma_{\circ\,\alpha_1}^{i_1\ \circ} & \gamma_{\circ\,\alpha_3}^{i_1\ \circ} & \cdots & \gamma_{\circ\,\alpha_{(2p-1)}}^{i_1\quad\circ} \\ \gamma_{\circ\,\alpha_1}^{i_3\ \circ} & \gamma_{\circ\,\alpha_3}^{i_3\ \circ} & \cdots & \gamma_{\circ\,\alpha_{(2p-1)}}^{i_3\quad\circ} \\ \cdots & \cdots & \cdots & \cdots \\ \gamma_{\circ\quad\alpha_1}^{i_{(2p-1)}\ \circ} & \gamma_{\circ\quad\alpha_3}^{i_{(2p-1)}\ \circ} & \cdots & \gamma_{\circ\quad\alpha_{(2p-1)}}^{i_{(2p-1)}\quad\circ} \end{vmatrix}$$

$$\cdot \begin{vmatrix} c_{i_2\ \circ}^{\circ\,\alpha_2} & c_{i_2\ \circ}^{\circ\,\alpha_4} & \cdots & c_{i_2\ \circ}^{\circ\,\alpha_{2q}} \\ c_{i_4\ \circ}^{\circ\,\alpha_2} & c_{i_4\ \circ}^{\circ\,\alpha_4} & \cdots & c_{i_4\ \circ}^{\circ\,\alpha_{2q}} \\ \cdots & \cdots & \cdots & \cdots \\ c_{i_{2q}\ \circ}^{\circ\,\alpha_2} & c_{i_{2q}\ \circ}^{\circ\,\alpha_4} & \cdots & c_{i_{2q}\ \circ}^{\circ\,\alpha_{2q}} \end{vmatrix}, \tag{10.45}$$

where the first minor belongs to C^{-1} and the second to C^t. Since $p \neq q$, a possibility exists of the existence of components with repeated components. This means that the change only implies indices of a single valency.

Example 10.1 (Mixed exterior products). Consider the vectors $\vec{V}_1 = 2\vec{e}_1 - 2\vec{e}_3$; $\vec{V}_2 = \vec{e}_1 + \vec{e}_2$; $\vec{V}_3 = \vec{e}_2 - \vec{e}_3$, which belong to the linear space $V^3(\mathbb{R})$ and the vectors $\vec{W}_1^* = \vec{e}^{*1} - \vec{e}^{*2} + 3\vec{e}^{*3}$; $\vec{W}_2^* = 2\vec{e}^{*1} + \vec{e}^{*2}$ in the linear space $V_*^3(\mathbb{R})$.

We wish to know its mixed exterior product, knowing that the species of this product is

$$\text{species} = [-1-1-2-2-1].$$

Solution: The required product is a decomposable tensor

$$P \in (V \wedge V \wedge V_* \wedge V_* \wedge V)^3(\mathbb{R})$$

with $n = p = 3$; $q = 2$.

The data matrix in contravariant coordinates is

$$A = \begin{bmatrix} 2 & 1 & 0 \\ 0 & 1 & 1 \\ -2 & 0 & -1 \end{bmatrix}$$

and in covariant coordinates is

$$B = \begin{bmatrix} 1 & 2 \\ -1 & 1 \\ 3 & 0 \end{bmatrix}.$$

The *Blocks* to be used are:

$$\binom{n}{p} = \binom{3}{3} = 1 \longrightarrow \begin{vmatrix} 2 & 1 & 0 \\ 0 & 1 & 1 \\ -2 & 0 & -1 \end{vmatrix} = -4; \quad \binom{n}{q} = \binom{3}{2} = 3 \longrightarrow \begin{cases} \begin{vmatrix} 1 & 2 \\ -1 & 1 \end{vmatrix} = 3 \\[2mm] \begin{vmatrix} 1 & 2 \\ 3 & 0 \end{vmatrix} = -6 \\[2mm] \begin{vmatrix} -1 & 1 \\ 3 & 0 \end{vmatrix} = -3 \end{cases}$$

We use Formula (10.26) to get

$$P = \vec{V}_1 \wedge \vec{V}_2 \wedge \vec{W}_1^* \wedge \vec{W}_2^* \wedge \vec{V}_3 \tag{10.46}$$

$$= \begin{vmatrix} x^{1o}_{o1} & x^{1o}_{o3} & x^{1o}_{o5} \\ x^{2o}_{o1} & x^{2o}_{o3} & x^{2o}_{o5} \\ x^{3o}_{o1} & x^{3o}_{o3} & x^{3o}_{o5} \end{vmatrix} \begin{vmatrix} x^{o2}_{1o} & x^{o4}_{1o} \\ x^{o2}_{2o} & x^{o4}_{2o} \end{vmatrix} \vec{e}_1 \wedge \vec{e}_2 \wedge \vec{e}^{*1} \wedge \vec{e}^{*2} \wedge \vec{e}_3$$

$$+ \begin{vmatrix} x^{1o}_{o1} & x^{1o}_{o3} & x^{1o}_{o5} \\ x^{2o}_{o1} & x^{2o}_{o3} & x^{2o}_{o5} \\ x^{3o}_{o1} & x^{3o}_{o3} & x^{3o}_{o5} \end{vmatrix} \begin{vmatrix} x^{o2}_{1o} & x^{o4}_{1o} \\ x^{o2}_{3o} & x^{o4}_{3o} \end{vmatrix} \vec{e}_1 \wedge \vec{e}_2 \wedge \vec{e}^{*1} \wedge \vec{e}^{*3} \wedge \vec{e}_3$$

$$+ \begin{vmatrix} x^{1o}_{o1} & x^{1o}_{o3} & x^{1o}_{o5} \\ x^{2o}_{o1} & x^{2o}_{o3} & x^{2o}_{o5} \\ x^{3o}_{o1} & x^{3o}_{o3} & x^{3o}_{o5} \end{vmatrix} \begin{vmatrix} x^{o2}_{2o} & x^{o4}_{2o} \\ x^{o2}_{3o} & x^{o4}_{3o} \end{vmatrix} \vec{e}_1 \wedge \vec{e}_2 \wedge \vec{e}^{*2} \wedge \vec{e}^{*3} \wedge \vec{e}_3$$

and operating numerically, we obtain

$$P = \vec{V}_1 \wedge \vec{V}_2 \wedge \vec{W}_1^* \wedge \vec{W}_2^* \wedge \vec{V}_3 = (-4)(3)\vec{e}_1 \wedge \vec{e}_2 \wedge \vec{e}^{*1} \wedge \vec{e}^{*2} \wedge \vec{e}_3$$
$$+ (-4)(-6)\vec{e}_1 \wedge \vec{e}_2 \wedge \vec{e}^{*1} \wedge \vec{e}^{*3} \wedge \vec{e}_3 + (-4)(-3)\vec{e}_1 \wedge \vec{e}_2 \wedge \vec{e}^{*2} \wedge \vec{e}^{*3} \wedge \vec{e}_3$$
$$= -12\vec{e}_1 \wedge \vec{e}_2 \wedge \vec{e}^{*1} \wedge \vec{e}^{*2} \wedge \vec{e}_3 + 24\vec{e}_1 \wedge \vec{e}_2 \wedge \vec{e}^{*1} \wedge \vec{e}^{*3} \wedge \vec{e}_3$$
$$+ 12\vec{e}_1 \wedge \vec{e}_2 \wedge \vec{e}^{*2} \wedge \vec{e}^{*3} \wedge \vec{e}_3.$$

Obviously, $P \in \bigwedge_3^{(3,2)}(\mathbb{R})$ with the given species. $\quad\square$

Example 10.2 (Exterior product of mixed tensor). Consider the following two mixed exterior tensors $T \in \bigwedge_3^{(2,1)}(\mathbb{R})$, species $= [1 - 2 - 1]$:

$$T = 2\vec{e}_1 \wedge \vec{e}^{*1} \wedge \vec{e}_2 + 3\vec{e}_1 \wedge \vec{e}^{*2} \wedge \vec{e}_3 + 4\vec{e}_2 \wedge \vec{e}^{*3} \wedge \vec{e}_3;$$

and $U \in \bigwedge_3^{(1,2)}(\mathbb{R})$, species $= [2 - 1 - 2]$:

$$U = -2\vec{e}^{*1} \wedge \vec{e}_3 \wedge \vec{e}^{*2} - 3\vec{e}^{*1} \wedge \vec{e}_2 \wedge \vec{e}^{*3} - 4\vec{e}^{*2} \wedge \vec{e}_1 \wedge \vec{e}^{*3}.$$

Obtain the exterior product of both tensors, in the order $V = T \bigwedge U$ and with species $= [-1 - 2 - 1 - 2 - 1 - 2]$.

Solution: From the data we get the following information:

$$p_1 = 2; \quad q_1 = 1; \quad p_2 = 1; \quad q_2 = 2; \quad p = p_1 + p_2 = 3; \quad q = q_1 + q_2 = 3; \quad n = 3,$$

which satisfies the constraints:

$$p \leq n; \quad q \leq n; \quad r = p + q = 6 \leq 2n.$$

The notation of the first factor is

$$T = t^{(\alpha_1\, \circ\, \alpha_3)}_{(\,\circ\, \alpha_2\, \circ\,)}\vec{e}_{\alpha_1} \wedge \vec{e}^{*\alpha_2} \wedge \vec{e}_{\alpha_3} = t^{1\circ2}_{\circ1\circ}\vec{e}_1 \wedge \vec{e}^{*1} \wedge \vec{e}_2 + t^{1\circ3}_{\circ2\circ}\vec{e}_1 \wedge \vec{e}^{*2} \wedge \vec{e}_3 + t^{2\circ3}_{\circ3\circ}\vec{e}_2 \wedge \vec{e}^{*3} \wedge \vec{e}_3,$$

where $t^{1\circ2}_{\circ1\circ} = 2$; $t^{1\circ3}_{\circ2\circ} = 3$; $t^{2\circ3}_{\circ3\circ} = 4$, the rest are $t^{(\alpha_1\, \circ\, \alpha_3)}_{(\,\circ\, \alpha_2\, \circ\,)} = 0$, and the notation of the second factor is

$$U = u^{(\,\circ\, \beta_1\, \circ\,)}_{(\beta_2\, \circ\, \beta_4)}\vec{e}^{*\beta_2} \wedge \vec{e}_{\beta_1} \wedge \vec{e}^{*\beta_4}$$

$$= u^{\circ3\circ}_{1\circ2}\vec{e}^{*1} \wedge \vec{e}_3 \wedge \vec{e}^{*2} + u^{\circ2\circ}_{1\circ3}\vec{e}^{*1} \wedge \vec{e}_2 \wedge \vec{e}^{*3} + u^{\circ1\circ}_{2\circ3}\vec{e}^{*2} \wedge \vec{e}_1 \wedge \vec{e}^{*3},$$

where $u^{\circ3\circ}_{1\circ2} = -2$; $u^{\circ2\circ}_{1\circ3} = -3$; $u^{\circ1\circ}_{2\circ3} = -4$, and the rest are $u^{(\,\circ\, \beta_1\, \circ\,)}_{(\beta_2\, \circ\, \beta_4)} = 0$.

We choose the indices as follows:

$$\{\gamma_1\gamma_3\gamma_5\} \equiv \{123\} \subset I_3 \quad \text{and} \quad \{\gamma_2\gamma_4\gamma_6\} \equiv \{123\} \subset I_3.$$

Since the product $P \in \bigwedge_3^{(3,3)}(\mathbb{R})$, we extract the dimension

$$\sigma' = \binom{n}{p}\binom{n}{q} = \binom{3}{3}\binom{3}{3} = 1$$

and taking into account the data species, the basis

$$B_{\wedge(3,3)} = \{\vec{e}_1 \wedge \vec{e}^{*2} \wedge \vec{e}_3 \wedge \vec{e}^{*4} \wedge \vec{e}_5 \wedge \vec{e}^{*6}\}$$

$$P = T\bigwedge U = p^{(1\circ3\circ5\circ)}_{(\circ2\circ4\circ6)}\vec{e}_1 \wedge \vec{e}^{*2} \wedge \vec{e}_3 \wedge \vec{e}^{*4} \wedge \vec{e}_5 \wedge \vec{e}^{*6}$$

with

$$p^{(1\circ3\circ5\circ)}_{(\circ2\circ4\circ6)} = t^{\alpha_1\, \circ\, \alpha_3}_{\circ\, \alpha_2\, \circ}u^{\circ\, \beta_1\, \circ}_{\beta_2\, \circ\, \beta_4}\delta^{(\,1\quad2\quad3\,)}_{(\alpha_1\alpha_3)(\beta_1)}\delta^{(\alpha_2)(\beta_2\beta_4)}_{(\,1\quad2\quad3\,)}$$

extracted from Formula (10.39).

Table 10.1. Possible combinations of indices α, β and the corresponding products of scalar components.

$(\alpha_1\alpha_3)$	(β_1)	(α_2)	$(\beta_2\beta_4)$	$t^{\alpha_1 \,\circ\, \alpha_3}_{\circ\, \alpha_2\, \circ}\, u^{\circ\, \beta_1\, \circ}_{\beta_2\, \circ\, \beta_4}$
1 2	3	1	2 3	$t^{1\circ2}_{\circ1\circ}\, u^{\circ3\circ}_{2\circ3} = 2 \times 0 = 0$
1 3	2	1	2 3	$t^{1\circ3}_{\circ1\circ}\, u^{\circ2\circ}_{2\circ3} = 0 \times 0 = 0$
2 3	1	1	2 3	$t^{2\circ3}_{\circ1\circ}\, u^{\circ1\circ}_{2\circ3} = 0 \times (-4) = 0$
1 2	3	2	1 3	$t^{1\circ2}_{\circ2\circ}\, u^{\circ3\circ}_{1\circ3} = 0 \times 0 = 0$
1 3	2	2	1 3	$t^{1\circ3}_{\circ2\circ}\, u^{\circ2\circ}_{1\circ3} = 3 \times (-3) = -9$
2 3	1	2	1 3	$t^{2\circ3}_{\circ2\circ}\, u^{\circ1\circ}_{1\circ3} = 0 \times 0 = 0$
1 2	3	3	1 2	$t^{1\circ2}_{\circ3\circ}\, u^{\circ3\circ}_{1\circ2} = 0 \times (-2) = 0$
1 3	2	3	1 2	$t^{1\circ3}_{\circ3\circ}\, u^{\circ2\circ}_{1\circ2} = 0 \times 0 = 0$
2 3	1	3	1 2	$t^{2\circ3}_{\circ3\circ}\, u^{\circ1\circ}_{1\circ2} = 4 \times 0 = 0$

The number of summands that contain the unique component p is

$$\binom{p}{p_1}\binom{q}{q_1} = \binom{3}{2}\binom{3}{1} = 3 \times 3 = 9.$$

In Table 10.1 the possible combinations of indices α, β and the corresponding products of scalar components are shown. In it one can see that there is only one non-null component product, so that *only* the corresponding Kronecker deltas are used to finally calculate the value of p:

$$p^{(1\circ3\circ5\circ)}_{(\circ2\circ4\circ6)} = t^{1\circ3}_{\circ2\circ} u^{\circ2\circ}_{1\circ3} \delta^{(1\ 2\ 3\ )}_{(13)(2)}\delta^{(2)(13)}_{(1\ 2\ 3)} = (-9)(-1)(-1) = -9.$$

Whence

$$P = -9\vec{e}_1 \wedge \vec{e}^{*2} \wedge \vec{e}_3 \wedge \vec{e}^{*4} \wedge \vec{e}_5 \wedge \vec{e}^{*6}.$$

and according to Formula (10.43), it is

$$U\bigwedge T = (-1)^{(2\times1+1\times2)}T\bigwedge U = (-1)^4 T\bigwedge U = T\bigwedge U,$$

that proves that $\bigwedge_3^{(3,3)}(\mathbb{R})$ is an Abelian algebra.   □

Example 10.3 (Change of basis in a mixed exterior algebra). In the algebra $\bigwedge_3^{(2,1)}(\mathbb{R})$ species $= [1 - 2 - 1]$, the tensor T in the last example is given:

$$T = 2\vec{e}_1 \wedge \vec{e}^{*1} \wedge \vec{e}_2 + 3\vec{e}_1 \wedge \vec{e}^{*2} \wedge \vec{e}_3 + 4\vec{e}_2 \wedge \vec{e}^{*3} \wedge \vec{e}_3.$$

Give the developed new expression of the tensor, if in the linear space $V^3(\mathbb{R})$ over which the algebra has been defined, a change-of-basis of matrix

$$C = [c_{\circ i}^{\alpha \circ}] = \begin{bmatrix} 1 & -1 & 1 \\ 2 & 0 & 1 \\ 3 & -1 & 0 \end{bmatrix} \text{ is performed.}$$

Solution: The dimension of the exterior space $\bigwedge_3^{(2,1)}(\mathbb{R})$ is $\sigma' = \binom{3}{2}\binom{3}{1} = 3 \times 3 = 9$, which gives us the number of components to be calculated.

We shall use minors taken from the matrices

$$C^{-1} = \frac{1}{4}\begin{bmatrix} -1 & 1 & 1 \\ -3 & 3 & -1 \\ 2 & 2 & -2 \end{bmatrix} \text{ and } C^t = \begin{bmatrix} 1 & 2 & 3 \\ -1 & 0 & -1 \\ 1 & 1 & 0 \end{bmatrix}.$$

The stated question responds to Formula (10.45), which is adapted to this case as

$$t_{(\circ i_2 \circ)}^{(i_1 \circ i_3)} = t_{(\circ \alpha_2 \circ)}^{(\alpha_1 \circ \alpha_3)} \begin{vmatrix} \gamma_{\circ \alpha_1}^{i_1 \circ} & \gamma_{\circ \alpha_3}^{i_1 \circ} \\ \gamma_{\circ \alpha_1}^{i_3 \circ} & \gamma_{\circ \alpha_3}^{i_3 \circ} \end{vmatrix} \begin{vmatrix} c_{i_2 \circ}^{\circ \alpha_2} \end{vmatrix}$$

and $I_n = I_3$.

The possible indices for the new and the initial bases are ($\sigma' = 9$)

$$|(i_1 i_2 i_3)|(\alpha_1 \alpha_2 \alpha_3)|112 \ 113 \ 213|122 \ 123 \ 223|132 \ 133 \ 233|$$

and the data T tensor components

$$t_{\circ 1 \circ}^{1 \circ 2} = 2; \quad t_{\circ 2 \circ}^{1 \circ 3} = 3; \quad t_{\circ 3 \circ}^{2 \circ 3} = 4,$$

because the rest are $t_{(\circ \alpha_2 \circ)}^{(\alpha_1 \circ \alpha_3)} = 0$.

We develop in numerical form the formula only for the useful data components to get

$$\hat{t}_{(\circ 1 \circ)}^{(1 \circ 2)} = t_{(\circ 1 \circ)}^{(1 \circ 2)} \begin{vmatrix} \gamma_{\circ 1}^{1 \circ} & \gamma_{\circ 2}^{1 \circ} \\ \gamma_{\circ 1}^{2 \circ} & \gamma_{\circ 2}^{2 \circ} \end{vmatrix} \begin{vmatrix} c_{1 \circ}^{\circ 1} \end{vmatrix} + t_{(\circ 2 \circ)}^{(1 \circ 3)} \begin{vmatrix} \gamma_{\circ 1}^{1 \circ} & \gamma_{\circ 3}^{1 \circ} \\ \gamma_{\circ 1}^{2 \circ} & \gamma_{\circ 3}^{2 \circ} \end{vmatrix} \begin{vmatrix} c_{1 \circ}^{\circ 2} \end{vmatrix} + t_{(\circ 3 \circ)}^{(2 \circ 3)} \begin{vmatrix} \gamma_{\circ 2}^{1 \circ} & \gamma_{\circ 3}^{1 \circ} \\ \gamma_{\circ 2}^{2 \circ} & \gamma_{\circ 3}^{2 \circ} \end{vmatrix} \begin{vmatrix} c_{1 \circ}^{\circ 3} \end{vmatrix}$$

$$= \frac{1}{4^2} \left[2\begin{vmatrix} -1 & 1 \\ -3 & 3 \end{vmatrix} \cdot 1 + 3\begin{vmatrix} -1 & 1 \\ -3 & -1 \end{vmatrix} \cdot 2 + 4\begin{vmatrix} 1 & 1 \\ 3 & -1 \end{vmatrix} \cdot 3 \right] = -\frac{6}{4}$$

$$\hat{t}_{(\circ 1 \circ)}^{(1 \circ 3)} = t_{(\circ 1 \circ)}^{(1 \circ 2)} \begin{vmatrix} \gamma_{\circ 1}^{1 \circ} & \gamma_{\circ 2}^{1 \circ} \\ \gamma_{\circ 1}^{3 \circ} & \gamma_{\circ 2}^{3 \circ} \end{vmatrix} \begin{vmatrix} c_{1 \circ}^{\circ 1} \end{vmatrix} + t_{(\circ 2 \circ)}^{(1 \circ 3)} \begin{vmatrix} \gamma_{\circ 1}^{1 \circ} & \gamma_{\circ 3}^{1 \circ} \\ \gamma_{\circ 1}^{3 \circ} & \gamma_{\circ 3}^{3 \circ} \end{vmatrix} \begin{vmatrix} c_{1 \circ}^{\circ 2} \end{vmatrix} + t_{(\circ 3 \circ)}^{(2 \circ 3)} \begin{vmatrix} \gamma_{\circ 2}^{1 \circ} & \gamma_{\circ 3}^{1 \circ} \\ \gamma_{\circ 2}^{3 \circ} & \gamma_{\circ 3}^{3 \circ} \end{vmatrix} \begin{vmatrix} c_{1 \circ}^{\circ 3} \end{vmatrix}$$

$$= \frac{1}{4^2} \left[2\begin{vmatrix} -1 & 1 \\ 2 & 2 \end{vmatrix} \cdot 1 + 3\begin{vmatrix} -1 & 1 \\ 2 & -2 \end{vmatrix} \cdot 2 + 4\begin{vmatrix} 1 & 1 \\ 2 & -2 \end{vmatrix} \cdot 3 \right] = -\frac{14}{4}$$

$$\hat{t}_{(\circ 1 \circ)}^{(2 \circ 3)} = t_{(\circ 1 \circ)}^{(1 \circ 2)} \begin{vmatrix} \gamma_{\circ 1}^{2 \circ} & \gamma_{\circ 2}^{2 \circ} \\ \gamma_{\circ 1}^{3 \circ} & \gamma_{\circ 2}^{3 \circ} \end{vmatrix} \begin{vmatrix} c_{1 \circ}^{\circ 1} \end{vmatrix} + t_{(\circ 2 \circ)}^{(1 \circ 3)} \begin{vmatrix} \gamma_{\circ 1}^{2 \circ} & \gamma_{\circ 3}^{2 \circ} \\ \gamma_{\circ 1}^{3 \circ} & \gamma_{\circ 3}^{3 \circ} \end{vmatrix} \begin{vmatrix} c_{1 \circ}^{\circ 2} \end{vmatrix} + t_{(\circ 3 \circ)}^{(2 \circ 3)} \begin{vmatrix} \gamma_{\circ 2}^{2 \circ} & \gamma_{\circ 3}^{2 \circ} \\ \gamma_{\circ 2}^{3 \circ} & \gamma_{\circ 3}^{3 \circ} \end{vmatrix} \begin{vmatrix} c_{1 \circ}^{\circ 3} \end{vmatrix}$$

$$= \frac{1}{4^2} \left[2\begin{vmatrix} -3 & 3 \\ 2 & 2 \end{vmatrix} \cdot 1 + 3\begin{vmatrix} -3 & -1 \\ 2 & -2 \end{vmatrix} \cdot 2 + 4\begin{vmatrix} 3 & -1 \\ 2 & -2 \end{vmatrix} \cdot 3 \right] = -\frac{6}{4}$$

$$\hat{t}^{(1\circ2)}_{(\circ2\circ)} = t^{(1\circ2)}_{(\circ1\circ)}\begin{vmatrix}\gamma^{1\circ}_{\circ1}&\gamma^{1\circ}_{\circ2}\\\gamma^{2\circ}_{\circ1}&\gamma^{2\circ}_{\circ2}\end{vmatrix}\left|c^{\circ1}_{2\circ}\right| + t^{(1\circ3)}_{(\circ2\circ)}\begin{vmatrix}\gamma^{1\circ}_{\circ1}&\gamma^{1\circ}_{\circ3}\\\gamma^{2\circ}_{\circ1}&\gamma^{2\circ}_{\circ3}\end{vmatrix}\left|c^{\circ2}_{2\circ}\right| + t^{(2\circ3)}_{(\circ3\circ)}\begin{vmatrix}\gamma^{1\circ}_{\circ2}&\gamma^{1\circ}_{\circ3}\\\gamma^{2\circ}_{\circ2}&\gamma^{2\circ}_{\circ3}\end{vmatrix}\left|c^{\circ3}_{2\circ}\right|$$

$$= \frac{1}{4^2}\left[2\begin{vmatrix}-1&1\\-3&3\end{vmatrix}\cdot(-1)+3\begin{vmatrix}-1&1\\-3&-1\end{vmatrix}\cdot0+4\begin{vmatrix}1&1\\3&-1\end{vmatrix}\cdot(-1)\right] = \frac{4}{4}$$

$$\hat{t}^{(1\circ3)}_{(\circ2\circ)} = \frac{1}{4^2}\left[2\begin{vmatrix}-1&1\\2&2\end{vmatrix}\cdot(-1)+3\begin{vmatrix}-1&1\\2&-2\end{vmatrix}\cdot0+4\begin{vmatrix}1&1\\2&-2\end{vmatrix}\cdot(-1)\right] = \frac{6}{4}$$

$$\hat{t}^{(2\circ3)}_{(\circ2\circ)} = \frac{1}{4^2}\left[2\begin{vmatrix}-3&3\\2&2\end{vmatrix}\cdot(-1)+3\begin{vmatrix}-3&-1\\2&-2\end{vmatrix}\cdot0+4\begin{vmatrix}3&-1\\2&-2\end{vmatrix}\cdot(-1)\right] = \frac{10}{4}$$

$$\hat{t}^{(1\circ2)}_{(\circ3\circ)} = \frac{1}{4^2}\left[2\begin{vmatrix}-1&1\\-3&3\end{vmatrix}\cdot1+3\begin{vmatrix}-1&1\\-3&-1\end{vmatrix}\cdot1+4\begin{vmatrix}1&1\\3&-1\end{vmatrix}\cdot0\right] = \frac{3}{4}$$

$$\hat{t}^{(1\circ3)}_{(\circ3\circ)} = \frac{1}{4^2}\left[2\begin{vmatrix}-1&1\\2&2\end{vmatrix}\cdot1+3\begin{vmatrix}-1&1\\2&-2\end{vmatrix}\cdot1+4\begin{vmatrix}1&1\\2&-2\end{vmatrix}\cdot0\right] = -\frac{2}{4}$$

$$\hat{t}^{(2\circ3)}_{(\circ3\circ)} = \frac{1}{4^2}\left[2\begin{vmatrix}-3&3\\2&2\end{vmatrix}\cdot1+3\begin{vmatrix}-3&-1\\2&-2\end{vmatrix}\cdot1+4\begin{vmatrix}3&-1\\2&-2\end{vmatrix}\cdot0\right] = 0.$$

Once the components are known, we obtain the following developed expression for T:

$$T = \frac{1}{4}\left(-6\vec{e}_1\wedge\vec{e}^{*1}\wedge\vec{e}_2 - 14\vec{e}_1\wedge\vec{e}^{*1}\wedge\vec{e}_3 - 6\vec{e}_2\wedge\vec{e}^{*1}\wedge\vec{e}_3 + 4\vec{e}_1\wedge\vec{e}^{*2}\wedge\vec{e}_2\right.$$
$$\left. + 6\vec{e}_1\wedge\vec{e}^{*2}\wedge\vec{e}_3 + 10\vec{e}_2\wedge\vec{e}^{*2}\wedge\vec{e}_3 + 3\vec{e}_1\wedge\vec{e}^{*3}\wedge\vec{e}_2 - 2\vec{e}_1\wedge\vec{e}^{*3}\wedge\vec{e}_3\right).$$

□

10.7 Exercises

10.1. In the linear space $V^4(\mathbb{R})$ referred to a certain basis $\{\vec{e}_\alpha\}$ three vectors $\vec{V}_1, \vec{V}_2, \vec{V}_3$ are given by its matrix representation

$$A = \begin{bmatrix}1&1&1\\1&2&3\\2&3&1\\3&1&2\end{bmatrix}$$

and in its dual space $V^4_*(\mathbb{R})$ another three vectors $\vec{W}^*_1, \vec{W}^*_2, \vec{W}^*_3$ are given by means of the matrix

$$B = \begin{bmatrix}2&3&4\\3&4&2\\4&2&3\\1&1&1\end{bmatrix}.$$

1. Give the mixed exterior product $P_1 = \vec{V}_1 \wedge \vec{W}^*_1 \wedge \vec{V}_2$ of species= $[1-2-1]$, by its strict components.

2. Give the mixed exterior product $P_2 = \vec{W}_2^* \wedge \vec{V}_3 \wedge \vec{W}_3^*$ of species= $[2-1-2]$, by its strict components.

3. Give the exterior product of the two mixed tensors $T_1 = P_1 \wedge P_2$ of species $= [1 - 2 - 1 - 2 - 1 - 2]$.

4. Give the totally developed mixed exterior product $T_2 = \vec{V}_1 \wedge \vec{W}_1^* \wedge \vec{V}_2 \wedge \vec{W}_2^* \wedge \vec{V}_3 \wedge \vec{W}_3^*$ of species $= [1 - 2 - 1 - 2 - 1 - 2]$.

5. Is $T_1 = T_2$?

10.2. Consider two exterior tensor spaces $\bigwedge_4^{(1,2)}(\mathbb{R})$ and $\bigwedge_4^{(3,1)}(\mathbb{R})$ both built over a linear space $V^4(\mathbb{R})$ referred to a basis $\{\vec{e}_\alpha\}$.

Let P and Q be two exterior tensors, $P \in \bigwedge_4^{(1,2)}(\mathbb{R})$ and $Q \in \bigwedge_4^{(3,1)}(\mathbb{R})$. Tensor P has the non-null strict components:

$$p_{(1 \circ 2)}^{\circ 1 \circ} = 3; \quad p_{(3 \circ 4)}^{\circ 1 \circ} = 4; \quad p_{(1 \circ 3)}^{\circ 2 \circ} = 2; \quad p_{(1 \circ 4)}^{\circ 3 \circ} = 1; \quad p_{(2 \circ 3)}^{\circ 4 \circ} = -5,$$

and tensor Q has the non-null strict components:

$$q_{\circ \circ 1 \circ}^{(1 2 \circ 4)} = 1; \quad q_{\circ \circ 2 \circ}^{(1 3 \circ 2)} = 4; \quad q_{\circ \circ 2 \circ}^{(2 3 \circ 4)} = -2; \quad q_{\circ \circ 3 \circ}^{(1 2 \circ 3)} = -1; \quad q_{\circ \circ 3 \circ}^{(1 3 \circ 4)} = 2; \quad q_{\circ \circ 4 \circ}^{(2 3 \circ 4)} = 5.$$

1. Find the totally developed exterior product of the mixed tensor $T = P \wedge Q$, of components $t_{\alpha \circ \gamma \circ \circ \mu \circ}^{\circ \beta \circ \delta \lambda \circ \nu}$.

2. Over the tensor T we execute the contraction

$$U = C \begin{pmatrix} \beta & \delta \\ \mu & \gamma \end{pmatrix} T.$$

Give the tensor U.

3. In the space $V^4(\mathbb{R})$ a change-of-basis $\{\vec{e}_i\}$ of associated matrix

$$C = \begin{bmatrix} 1 & 2 & 3 & 1 \\ 1 & 3 & 3 & 2 \\ 2 & 4 & 3 & 3 \\ 1 & 1 & 1 & 1 \end{bmatrix}$$

is performed. Give the new strict components of the tensors P, Q, T and U.

10.3. Consider the set of the generalized Kronecker deltas $\delta_{\gamma\delta}^{\alpha\beta}; \alpha, \beta, \gamma, \delta \in \{1, 2, 3, 4\}$, and consider two mixed exterior vectors $D_1, D_2 \in \bigwedge_4^{(2,2)}$, of specie $d_{\circ \beta \circ \delta}^{\alpha \circ \gamma \circ}$ with respective components:

$$D_1 = (\alpha + \beta)\delta_{\gamma\delta}^{\alpha\beta} \quad \text{and} \quad D_2 = (\alpha \cdot \beta)\delta_{\gamma\delta}^{\alpha\beta}.$$

Using the computer, answer the following questions:

1. Write the tensor $D_1^2 = D_1 \wedge D_1$ totally developed.
2. Idem for $D_2^2 = D_2 \wedge D_2$.
3. Idem for tensors $A = D_1 \wedge D_2$ and $B = D_2 \wedge D_1$.

Part IV

Tensors over Linear Spaces with Inner Product

Euclidean Homogeneous Tensors

11.1 Introduction

In this chapter the problem of vertical displacements of indices in Euclidean and pseudo-Euclidean tensors is dealt with. In particular, the generated problems related to raising and lowering indices with respect to the symmetry and those associated with the "Euclidean contraction" are analyzed.

In the final part of this chapter, it is shown how geometry enters tensor spaces by means of Euclidean tensors, so that it is possible to talk, for example, about length, perimeter or angle of a tensor, and if desired, to build a tensor geometry.

11.2 Initial concepts

In Chapter 3 we have stated the study of homogeneous tensors established over a certain primal linear space $V^n(K)$ and its dual space $V_*^n(K)$. In Chapter 4 it was established how a change-of-basis $\vec{\hat{e}}_i = c_{\circ\,i}^{\alpha\circ} \vec{e}_\alpha$ of matrix $C \equiv [c_{\circ\,i}^{\alpha\circ}]$ performed in the primal, produced another change-of-basis in the dual $\vec{\hat{e}}^{*j} = \gamma_{\beta\circ}^{\circ\,j} \vec{e}^{*\beta}$ with the matrix condition $[\gamma_{\beta\circ}^{\circ\,j}] = \left([c_{\circ\,i}^{\alpha\circ}]^t \right)^{-1}$, that is, $C^* = (C^t)^{-1}$, which kept the new bases of both spaces with the reciprocal character that they initially showed.

We dedicate some paragraphs to the concept of connection in linear spaces because in the present chapter we will clearly point out the connection between both linear spaces (the so-called primal space $V^n(K)$ and the dual or secondary space $V_*^n(K)$), i.e., present in any tensor process.

Main background:

1. *Axiomatic Properties.*
   We assume here that the reader is familiar with connection in vector spaces, linear spaces with a connection and bilinear forms.

2. Field $K \equiv \mathbb{R}$.

From now on, we assume that $K \equiv \mathbb{R}$, that is, the field of real numbers, because if the associated field K were the set of the complex numbers C or the set of quaternions H, or any other arbitrary field, the axioms of the axiomatic system should be modified to adapt it to the field being considered:

(a) The bi-stability axiom of the forms with respect to the external product:

$$\begin{cases} \Phi(\lambda \circ \vec{v}, \vec{w}) = \lambda \cdot \Phi(\vec{v}, \vec{w}) \\ \Phi(\vec{v}, \mu \circ \vec{w}) = \mu \cdot \Phi(\vec{v}, \vec{w})) \end{cases} ; \quad \forall \lambda, \mu \in K; \ \ \forall \vec{v}, \vec{w} \in V^n(K).$$

(b) The real symmetric axiom of the bilinear form. We have such character if $\forall \vec{v}, \vec{w} \in V^n(K) \Rightarrow \Phi(\vec{w}, \vec{v}) = \Phi(\vec{v}, \vec{w})$.

In other words, in this book we only examine "connected tensors" over the real numbers; the spaces over which the tensors will be built will be denoted by $PSE^n(\mathbb{R})$ (primal) and $PSE^n_*(\mathbb{R})$ (dual).

3. *Existence of a unique linear space $PSE^n(\mathbb{R})$.* Since the spaces being connected are *n-dimensional and arbitrary*, we assume that from now on there exists only *one* linear space that is connected with itself, and that *simultaneously* is the primal and the dual space, so that we have $PSE^n(\mathbb{R}) \equiv PSE^n_*(\mathbb{R})$.

In other words, the universal quantifiers $\forall$, that initiate the axiomatic properties in point 1, should say for this situation: $\forall \vec{V}, \vec{W} \in PSE^n(\mathbb{R})$ is $\Phi(\vec{V}, \vec{W}) = \rho; \ \rho \in \mathbb{R}$, etc., so that we will not return to this fact again and we assume that it is already established and known.

4. *Fundamental tensor G of the connection. The Gram matrix.* Obviously, any reader who knows that our linear space $PSE^n(\mathbb{R})$ is connected by a bilinear form Φ, can ask himself what are the real numbers $\rho \in \mathbb{R}$ that are the images of any pair of basic vectors in the basis $\{\vec{e}_\alpha\}$ that has been chosen in the linear space $PSE^n(\mathbb{R})$. From now on, we notate "the inner connection" as

$$< \vec{V}, \vec{W} > = \rho. \tag{11.1}$$

In this simple way it will be understood that it is a bilinear form $\Phi(\vec{V}, \vec{W}) = \rho$ that strictly satisfies the conditions in points, 1, 2, 3 and 4, above.

Since the n^2 real numbers $< \vec{e}_\alpha, \vec{e}_\beta >, \alpha, \beta, \in [1, 2, \cdots, n]$, are the numerical data of a system of real components that will define a tensor, such scalars will carry a specific notation, so that instead of ρ, we will use

$$< \vec{e}_\alpha, \vec{e}_\beta > = g^{\circ\circ}_{\alpha\beta}; \ \forall g^{\circ\circ}_{\alpha\beta} \in \mathbb{R}. \tag{11.2}$$

They are delivered as n^2 scalars grouped as a square matrix, of order n:

$$G_n \equiv [g^{\circ\circ}_{\alpha\beta}], \tag{11.3}$$

which is called the "Gram" matrix of the connection.

The meaning of its terms was clarified in Formula (11.2).

We will see now how the general scalar $\rho = <\vec{V}, \vec{W}>$, that corresponds to two arbitrary vectors $\vec{V}, \vec{W} \in PSE^n(\mathbb{R})$ can be determined. Let $\vec{V} = x_\circ^\alpha \vec{e}_\alpha \equiv ||\vec{e}_\alpha|| X$ and $\vec{W} = y_\circ^\beta \vec{e}_\beta \equiv ||\vec{e}_\beta|| Y$. Since

$$< \vec{V}, \vec{W} > = < x_\circ^\alpha \vec{e}_\alpha, y_\circ^\beta \vec{e}_\beta >$$

is bilinear, we can take the scalars $x_\circ^\alpha y_\circ^\beta$ out of the form and since in the sums the Einstein convention holds, after using (11.2) in tensor notation we get

$$< \vec{V}, \vec{W} > = x_\circ^\alpha y_\circ^\beta < \vec{e}_\alpha, \vec{e}_\beta > = x_\circ^\alpha y_\circ^\beta g_{\alpha\beta}^{\circ\circ} \qquad (11.4)$$

and in matrix form

$$< \vec{V}, \vec{W} > = x_\circ^\alpha g_{\alpha\beta}^{\circ\circ} y_\circ^\beta = X^t G_n Y. \qquad (11.5)$$

Expression (11.5) was already obtained in Chapter 1, Formula (1.2), and in Formulas (1.3) some conditions were added for G to satisfy the axioms 3 of symmetry and 4 of regularity. These conditions are $G \equiv G^t$ and $|G| \neq 0$ (regular and symmetric Gram matrix).

In Formula (1.6) a change-of-basis $C = G^{-1}$ was proposed that led to the reciprocal basis $\{\vec{e}^{*\beta}\}$, which is the basis adopted in the proper linear space, when it wants to be seen as dual $(PSE_*^n(\mathbb{R}))$. The Gram matrix of the connection has in this case the components

$$< \vec{e}^{*\alpha}, \vec{e}^{*\beta} > = g_{\circ\circ}^{\alpha\beta}, \quad \forall g_{\circ\circ}^{\alpha\beta} \in \mathbb{R}, \qquad (11.6)$$

so that the same vector $\vec{V}$ has contravariant coordinates or covariant coordinates, depending on the adopted basis (both bases are simultaneous):

$$\vec{V} = ||\vec{e}_\alpha|| X = ||\vec{e}^{*\beta}|| X^*.$$

The relation between both components, contravariant and covariant in tensor form is

$$(x_\beta^\circ)^* = g_{\alpha\beta}^{\circ\circ} x_\circ^\alpha \qquad (11.7)$$

and also

$$x_\circ^\beta = g_{\circ\circ}^{\alpha\beta} (x_\alpha^\circ)^* \qquad (11.8)$$

or, in matrix form (and respectively):

$$X^* = G_n X \qquad (11.9)$$

$$X = G_n^{-1} X^*. \qquad (11.10)$$

Formula (11.9) was proved in Section 1.2, Formula (1.7).

The conclusion is established in Section 1.2, in which we proceeded to the calculation of the form $< \vec{V}, \vec{W} >$ in the four possible situations, depending on the data of $\vec{V}$ and $\vec{W}$, and it was shown that for:

(a) Data $\vec{V}$ and $\vec{W}$ in contravariant coordinates:

$$\rho = <\vec{V},\vec{W}> = <x_{\circ}^{\alpha}\vec{e}_{\alpha}, y_{\circ}^{\beta}\vec{e}_{\beta}>$$
$$= x_{\circ}^{\alpha} <\vec{e}_{\alpha},\vec{e}_{\beta}> y_{\circ}^{\beta} = x_{\circ}^{\alpha}g_{\alpha\beta}^{\circ\circ}y_{\circ}^{\beta} = X^{t}G_{n}Y. \qquad (11.11)$$

(b) Data $\vec{V}$ in contravariant coordinates and $\vec{W}$ in covariant coordinates:

$$\rho = <\vec{V},\vec{W}> = <x_{\circ}^{\alpha}\vec{e}_{\alpha}, y_{\beta}^{\circ}\vec{e}^{*\beta}>$$
$$= x_{\circ}^{\alpha} <\vec{e}_{\alpha},\vec{e}^{*\beta}> y_{\beta}^{\circ} = x_{\circ}^{\alpha}\delta_{\alpha\circ}^{\circ\beta}y_{\beta}^{\circ} = X^{t}I_{n}Y^{*}. \qquad (11.12)$$

(c) Data $\vec{V}$ in covariant coordinates and $\vec{W}$ in contravariant coordinates:

$$\rho = <\vec{V},\vec{W}> = <x_{\alpha}^{\circ}\vec{e}^{*\alpha}, y_{\circ}^{\beta}\vec{e}_{\beta}>$$
$$= x_{\alpha}^{\circ} <\vec{e}^{*\alpha},\vec{e}_{\beta}> y_{\circ}^{\beta} = x_{\alpha}^{\circ}\delta_{\circ\beta}^{\alpha\circ}y_{\circ}^{\beta} = (X^{*})^{t}I_{n}Y. \qquad (11.13)$$

(d) Data $\vec{V}$ and $\vec{W}$ in covariant coordinates:

$$\rho = <\vec{V},\vec{W}> = <x_{\alpha}^{\circ}\vec{e}^{*\alpha}, y_{\beta}^{\circ}\vec{e}^{*\beta}>$$
$$= x_{\alpha}^{\circ} <\vec{e}^{*\alpha},\vec{e}^{*\beta}> y_{\beta}^{\circ} = x_{\alpha}^{\circ}g_{\circ\circ}^{\alpha\beta}y_{\beta}^{\circ} = (X^{*})^{t}G_{n}^{-1}Y^{*}. \qquad (11.14)$$

It is obvious that

$$\rho = x_{\circ}^{\alpha}g_{\alpha\beta}^{\circ\circ}y_{\circ}^{\beta} = x_{\circ}^{\alpha}\delta_{\alpha\circ}^{\circ\beta}y_{\beta}^{\circ} = x_{\alpha}^{\circ}\delta_{\circ\beta}^{\alpha\circ}y_{\circ}^{\beta} = x_{\alpha}^{\circ}g_{\circ\circ}^{\alpha\beta}y_{\beta}^{\circ}. \qquad (11.15)$$

Formulas (11.12) and (11.13) imply that isotropy holds:

$$<\vec{e}_{\alpha},\vec{e}^{*\beta}> = <\vec{e}^{*\alpha},\vec{e}_{\beta}> = \delta_{\alpha\circ}^{\circ\beta} = \delta_{\circ\beta}^{\alpha\circ} \text{ (Kronecker's deltas)}, \qquad (11.16)$$

which declares that the linear space $PSE^{n}(\mathbb{R}) \equiv PSE_{*}^{n}(\mathbb{R})$ is simultaneously in reciprocal bases, and that both bases are used, one per data vector.

11.3 Tensor character of the inner vector's connection in a $PSE^{n}(\mathbb{R})$ space

As the reader can already have observed, to the n^{2} scalars $g_{\alpha\beta}^{\circ\circ}$ that constitute the Gram matrix G_{n} we have associated the title of "fundamental tensor of the connection", without a justification of its tensor nature. We prove that $G_{n} = [g(\alpha\beta)]$ is a tensor. To this end, we copy the Expressions (11.5) and (11.11), but with the notation of a system of scalars: $\rho = x_{\circ}^{\alpha}g(\alpha\beta)y_{\circ}^{\beta}$.

Since the product of vectors $(x_o^\alpha y_o^\beta)$ is a totally contravariant second-order tensor, applying the quotient law, and more precisely, by a direct application of Theorem 5.3, since ρ is an invariant scalar, the system $g(\alpha\beta)$ is a totally covariant second-order tensor $g(\alpha,\beta) \equiv g_{\alpha\beta}^{\circ\circ}$.

It is also possible that the data $\vec{V}$ and $\vec{W}$ be in covariant coordinates, in which case to the n^2 scalars $g_{\circ\circ}^{\alpha\beta}$ in the Gram matrix G_n^*, we give the same title of "fundamental tensor of the connection". The proof is very similar, we copy Formula (11.14) with the proper notation of the systems of scalar components:

$$\rho = x_\alpha^\circ g(\alpha\beta) y_\beta^\circ$$

and since $(x_\alpha^\circ y_\beta^\circ)$ is a product of covariant vectors, this product is a second-order covariant tensor. Applying the quotient law, or directly Theorem 5.2, since ρ is an invariant escalar, the system $g(\alpha\beta)$ is a totally contravariant tensor of second order:

$$g(\alpha\beta) \equiv g_{\circ\circ}^{\alpha\beta}.$$

We remind the reader that in Section 1.2 it was proven that

$$G_n^* \equiv [g_{\circ\circ}^{\alpha\beta}] = G^{-1}. \qquad (11.17)$$

Nevertheless, the direct proof can be established as follows. Since the vectors are tensors of first-order, taking the contravariant species, the result is

$$\begin{cases} x_o^i = x_o^\alpha \gamma_{o\alpha}^{i\,o} \\ x_o^j = x_o^\beta \gamma_{o\beta}^{j\,o} \end{cases} \text{ or } \begin{cases} x_o^\alpha = x_o^i c_{o\,i}^{\alpha o} \\ x_o^\beta = x_o^j c_{o\,j}^{\beta o} \end{cases}$$

On the other hand, the dot product of two vectors in the initial and new bases, formulated in tensor form, can be written as

$$\vec{V} \bullet \vec{W} = x_o^\alpha x_o^\beta g_{\alpha\beta}^{\circ\circ} = x_o^i x_o^j g_{ij}^{\circ\circ}$$

and introducing in the first member the initial equalities, we get

$$(x_o^i c_{o\,i}^{\alpha o})(x_o^j c_{o\,j}^{\beta o}) g_{\alpha\beta}^{\circ\circ} = x_o^i x_o^j g_{ij}^{\circ\circ}$$

and simplifying:

$$c_{o\,i}^{\alpha o} c_{o\,j}^{\beta o} g_{\alpha\beta}^{\circ\circ} = g_{ij}^{\circ\circ}.$$

If we sort the tensor factors, for the "format rules" established for Formulas (4.24) and (4.34) to be satisfied, we obtain

$$g_{ij}^{\circ\circ} = g_{\alpha\beta}^{\circ\circ} c_{i\,o}^{\circ\alpha} c_{j\,o}^{\circ\beta},$$

that proves that it is a totally covariant second-order tensor.

11.4 Different types of the fundamental connection tensor

In this section we extend some aspects that are convenient to clarify. If in the linear space $PSE^n(\mathbb{R})$ we execute a change-of-basis of matrix C, the fundamental tensor changes, in a congruent matrix process, to another Gram matrix by means of the expression

$$\hat{G} = C^t G C.$$

The "congruence" Theorem ensures that there exists a change-of-basis such that the new $\hat{G}$ is a diagonal matrix:

$$C^t G C = \hat{D}_\mu \equiv \begin{bmatrix} \mu_1 & 0 & \cdots & 0 \\ 0 & \mu_2 & \cdots & 0 \\ \cdots & \cdots & \cdots & \cdots \\ 0 & 0 & \cdots & \mu_n \end{bmatrix} ; \ |D_\mu| \neq 0.$$

The "Sylvester inertia law" goes further; it says that a change-of-basis can be found such that the corresponding Gram matrix D_μ, is not only diagonal, but has values $\mu = \pm 1$:

$$C_s^t G C_s = \begin{bmatrix} 1 & 0 & \cdots & 0 & 0 & 0 & \cdots & 0 \\ 0 & 1 & \cdots & 0 & 0 & 0 & \cdots & 0 \\ \cdots & \cdots & \cdots & \cdots & \cdots & \cdots & \cdots & \cdots \\ 0 & 0 & \cdots & 1 & 0 & 0 & \cdots & 0 \\ 0 & 0 & \cdots & 0 & -1 & 0 & \cdots & 0 \\ 0 & 0 & \cdots & 0 & 0 & -1 & \cdots & 0 \\ \cdots & \cdots & \cdots & \cdots & \cdots & \cdots & \cdots & \cdots \\ 0 & 0 & \cdots & 0 & 0 & 0 & \cdots & -1 \end{bmatrix} = \begin{bmatrix} I_p & \Omega \\ \Omega & -I_q \end{bmatrix}. \qquad (11.18)$$

Sylvester showed that the order n, the rank r and the *Spanish signature* σ (number of (+1) in the D_μ matrix) is an invariant for all matrices G that are "congruent". So, it is convenient to notate each matrix as $G(n, r, \sigma)$. We introduce the axiomatic convention that states that all (+1) must be ahead of all (−1) in the diagonal of D_μ, with the aim of every one to "canonize" the matrix G with the same format. The change-of-basis matrix C_s, which according to (11.18) leads to the fundamental tensor of the connection, is called the "Sylvestizer matrix of G" and the matrix

$$\hat{D}_\mu = I_\sigma \oplus (-I_{(n-\sigma)}) \equiv \begin{bmatrix} I_\sigma & \Omega \\ \Omega & -I_{n-\sigma} \end{bmatrix}_n \qquad (11.19)$$

is called the "Sylvesterian matrix" of G. Once the Sylvesterian matrix of G is known, we can denote by G the expression $G(n, r, \sigma)$ and conversely, if G is given by the notation $G(n, r, \sigma)$ we can build its Sylvesterian matrix. For example, if we write for a certain quadratic form $G(5, 3, 1)$, the reader will immediately think of the following expression, which declares its "Sylvesterian" matrix:

$$\exists C_s, |C_s| \neq 0 \text{ such that } C_s^t G C_s = \begin{bmatrix} 1 & 0 & 0 & 0 & 0 \\ 0 & -1 & 0 & 0 & 0 \\ 0 & 0 & -1 & 0 & 0 \\ 0 & 0 & 0 & 0 & 0 \\ 0 & 0 & 0 & 0 & 0 \end{bmatrix}.$$

Case 1.

The linear spaces with connection fundamental tensor G the Sylversterian matrix of which has $n = r \neq \sigma$ (that is, some (-1) in the diagonal) will be called pseudo-Euclidean spaces $\equiv PSE^n(\mathbb{R})$ (false Euclidean space). For that type of connection, we use the notations:

1. $< \vec{V}, \vec{W} > = X^t G Y$ to the bilinear form of two vectors.
2. $< \vec{V}, \vec{V} > = X^t G X$ to the quadratic from of a vector.

If $< \vec{V}, \vec{W} > = 0$, we say that the vectors $\vec{V}$ and $\vec{W}$ are *"conjugate"*.

Case 2.

On the other hand, the linear spaces with connection fundamental tensor G the Sylvesterian matrix of which has $n = r = \sigma$ (all $(+1)$ in the main diagonal), that is, $\hat{D}_\mu \equiv I_n$, will be called Euclidean spaces $\equiv E^n(\mathbb{R})$ and will have the following *exclusive* notation:

1. The bilinear form of two vectors *will not* be notated $< \vec{V}, \vec{W} >$, but $\vec{V} \bullet \vec{W} = X^t G Y$ and will be called the "dot product".
2. The quadratic form of a vector *will not* be notated $< \vec{V}, \vec{V} >$, but $\mathcal{N}(\vec{V}) = \vec{V} \bullet \vec{V}$ and will be called the "Euclidean norm of $\vec{V}$" and as it is essentially positive, its square root always exists and is called the "modulus of $\vec{V}$":

$$|\vec{V}| = \sqrt{\mathcal{N}(\vec{V})} = \sqrt{\vec{V} \bullet \vec{V}}.$$

If $\vec{V} \bullet \vec{W} = 0$ we say that the vectors $\vec{V}$ and $\vec{W}$ are *"orthogonal"*, and we notate this as $\vec{V} \perp \vec{W}$.
If $|\vec{V}| = 1$ we say that $\vec{V}$ is a "versor" or "unit vector".
Its connection fundamental tensor G will be called the "fundamental metric tensor". It is well known that this fundamental tensor defines the $\cos \alpha = \frac{\vec{V} \bullet \vec{W}}{|\vec{V}||\vec{W}|}$, etc., thus, it endows the vectors with "directions" and introduces the "metric space" character into the affine punctual spaces, reason that explains that many authors call them "metric tensors".

Case 3.

Certain authors do not require the regular character $|G| \neq 0$ to the connection tensor, it can then be $G(n, r, \sigma)$ with $r = \sigma < n$. The Sylvesterian matrix is $\hat{D}_\mu = \begin{bmatrix} I_r & \Omega \\ \Omega & \Omega \end{bmatrix}$ and the quadratic form $< \vec{V}, \vec{V} > \geq 0$; $\forall \vec{V}$.

The linear spaces with such a connection will be called pre-Euclidean spaces $\equiv PE^n(\mathbb{R})$, and the bilinear form of two vectors will be notated $< \vec{V}, \vec{W} >$, and the quadratic form $< \vec{V}, \vec{V} >$ in the same form as the pseudo-Euclidean spaces.

It is obvious that any of the mentioned linear spaces, together with their duals, can be used as support for establishing homogeneous tensors, multilinear algebras, exterior algebras, modular tensors, etc.

From all of them, we choose only Case 2 to continue in this book, so that from this point on all to be developed will be over the linear spaces that are strictly Euclidean, $E^n(\mathbb{R})$ and their simultaneous dual $E_*^n(\mathbb{R})$, endowed with the connection fundamental tensor, or metric-covariant tensor $G = [g_{\alpha\beta}^{\circ\circ}]$, where $G(n, r, \sigma)$ must satisfy $n = r = \sigma$ or the analogous condition

$$\exists C_s, |C_s| \neq 0 \text{ such that } C_s^t G C_s = I_n. \tag{11.20}$$

If it is of interest we will also work with the *same* connection fundamental tensor under its metric-contravariant expression $G^* = [g_{\circ\circ}^{\alpha\beta}]$, where $G^*(n, r, \sigma)$ must satisfy $n = r = \sigma$, or

$$\exists C_s^*, \quad (C_s^*)^t G^* C_s^* = I_n; \tag{11.21}$$

where

$$C_s^* \equiv (C_s^t)^{-1} \equiv (C_s^{-1})^t,$$

and from Formula (11.17) $G^* = G^{-1}$.

The tensor relations between the components of the metric tensors are

$$g_{\circ\circ}^{\alpha\theta} \cdot g_{\theta\beta}^{\circ\circ} = \delta_{\circ\beta}^{\alpha\circ} \equiv g_{\circ\beta}^{\alpha\circ} \tag{11.22}$$

$$g_{\theta\alpha}^{\circ\circ} \cdot g_{\circ\circ}^{\theta\beta} = \delta_{\alpha\circ}^{\circ\beta} \equiv g_{\alpha\circ}^{\circ\beta}. \tag{11.23}$$

It is impossible to try to make a summary of the diverse names used by different authors for the connection of linear spaces and even more difficult for their concrete meanings, because there are authors who with the same names as those proposed here, assume different axiomatic properties, that they do not even declare. The reader can imagine the labyrinth that results when talking about orthogonal vectors, orthogonal linear subspaces, dot product, modulus of a vector, etc., in the line of spaces that *are not* Euclidean. Another occurrence is the use of vectorial tensors (very frequent in books of physics) as the vector product, the triple vector product, the mixed vector product, etc. presenting always the same invariable formulas, because always by "definition" the connection fundamental tensor is that of matrix I_n (in such books *there are no* arbitrary changes of basis). Nevertheless, we advance some commonly used names. For spaces $PSE^n(\mathbb{R})$ of a Sylvesterian matrix with $\sigma < r \leq n$ which are Euclidian spaces (not Euclidean spaces but very similar), non-Euclidean spaces with proper names such as Minkowsky's spaces, pre-Euclidean spaces, metric spaces, etc.

For the spaces $E^n(\mathbb{R})$ of a Sylvesterian matrix with $\sigma = r = n$ Euclidean spaces, metric spaces, Cartesian spaces, "properly Euclidean spaces", "dot product" spaces, geometric spaces, OGS (ordinary geometric space), etc.

11.5 Tensor product of vectors in $E^n(\mathbb{R})$ (or in $PSE^n(\mathbb{R})$)

The existence of a unique Euclidean linear space such that the condition in Section 11.2, point 3, is $E^n(\mathbb{R}) \equiv E^n_*(\mathbb{R})$, produces the following problem. Each vector can be given in contravariant or covariant coordinates, Formulas (11.9) and (11.10).

When two factor-vectors $\vec{V}_i \in E^n(\mathbb{R}), 1 \le i \le p$; $\vec{V}^*_j \in E^n_*(\mathbb{R}), 1 \le j \le q$; $p + q = r$ are chosen, the product $\vec{P}$ belonging to the tensor space of the conditioned relation (3.1) is

$$\overset{\leftarrow \qquad r = (p+q) \text{ times} \qquad \rightarrow}{\vec{P} \in E^n \otimes E^n_* \otimes E^n \otimes \cdots \otimes E^n_* \otimes E^n(\mathbb{R}),} \qquad (11.24)$$

which, according to (3.2), has dimension

$$\sigma = dim \left[\left(\overset{p}{\underset{1}{\otimes}} E^n \right) \left(\overset{q}{\underset{1}{\otimes}} E^n_* \right) (\mathbb{R}) \right] = n^r$$

and can be notated as a mixed homogeneous tensor:

$$\begin{aligned}
P &= \vec{V}_1 \otimes \vec{V}^*_2 \otimes \vec{V}_3 \otimes \cdots \otimes \vec{V}^*_{r-1} \otimes \vec{V}_r \\
&= (x^{\alpha_1}_{\ \circ} \vec{e}_{\alpha_1}) \otimes (x^{\ \circ}_{\alpha_2} \vec{e}^{*\alpha_2}) \otimes (x^{\alpha_3}_{\ \circ} \vec{e}_{\alpha_3}) \otimes \cdots \otimes (x^{\ \circ}_{\alpha_{r-1}} \vec{e}^{*\alpha_{r-1}}) \otimes (x^{\alpha_r}_{\ \circ} \vec{e}_{\alpha_r}) \\
&= (x^{\alpha_1}_{\ \circ} x^{\ \circ}_{\alpha_2} x^{\alpha_3}_{\ \circ} \cdots x^{\ \circ}_{\alpha_{r-1}} x^{\alpha_r}_{\ \circ}) \vec{e}_{\alpha_1} \otimes \vec{e}^{*\alpha_2} \otimes \vec{e}_{\alpha_3} \otimes \cdots \vec{e}^{*\alpha_{r-1}} \otimes \vec{e}_{\alpha_r} \\
&= t^{\alpha_1 \ \circ \ \alpha_3 \cdots \ \circ \ \alpha_r}_{\ \circ \ \alpha_2 \ \circ \ \cdots \alpha_{r-1} \ \circ} \vec{e}_{\alpha_1} \otimes \vec{e}^{*\alpha_2} \otimes \vec{e}_{\alpha_3} \otimes \cdots \vec{e}^{*\alpha_{r-1}} \otimes \vec{e}_{\alpha_r}.
\end{aligned} \qquad (11.25)$$

But, *at any time* any $E^n(\mathbb{R})$ can be replaced by $E^n_*(\mathbb{R})$, because they are identical ($E^n(\mathbb{R}) \equiv E^n_*(\mathbb{R})$); this means, for example, that, since $\vec{P} \in \left(\overset{r}{\underset{1}{\otimes}} E^n \right) (\mathbb{R})$, $\vec{P}$ can *also* be given as

$$\begin{aligned}
\vec{P} &= \vec{V}_1 \otimes \vec{V}_2 \otimes \vec{V}_3 \otimes \cdots \otimes \vec{V}_{r-1} \otimes \vec{V}_r \\
&= (x^{\alpha_1}_{\ \circ} x^{\alpha_2}_{\ \circ} x^{\alpha_3}_{\ \circ} \cdots x^{\alpha_r}_{\ \circ}) \vec{e}_{\alpha_1} \vec{e}_{\alpha_2} \otimes \vec{e}_{\alpha_3} \otimes \cdots \otimes \vec{e}_{\alpha_{r-1}} \otimes \vec{e}_{\alpha_r} \qquad (11.26) \\
&= t^{\alpha_1 \alpha_2 \alpha_3 \cdots \alpha_{r-1} \alpha_r}_{\ \circ \ \circ \ \circ \ \cdots \ \circ \ \circ} \vec{e}_{\alpha_1} \vec{e}_{\alpha_2} \otimes \vec{e}_{\alpha_3} \otimes \cdots \otimes \vec{e}_{\alpha_{r-1}} \otimes \vec{e}_{\alpha_r}, \qquad (11.27)
\end{aligned}$$

422   11 Euclidean Homogeneous Tensors

where $\vec{P}$ appears as a totally contravariant tensor.

Another possibility, just to mention some more of all possible cases, is given by considering that $\vec{P} \in \left(\overset{r}{\underset{1}{\otimes}} E_*^n \right)(\mathbb{R})$:

$$P = \vec{V}_1^* \otimes \vec{V}_2^* \otimes \vec{V}_3^* \otimes \cdots \otimes \vec{V}_{r-1}^* \otimes \vec{V}_r^*$$

$$= (x_{\alpha_1}^{\circ} x_{\alpha_2}^{\circ} x_{\alpha_3}^{\circ} \cdots x_{\alpha_{r-1}}^{\circ} x_{\alpha_r}^{\circ}) \vec{e}^{*\alpha_1} \otimes \vec{e}^{*\alpha_2} \otimes \vec{e}^{*\alpha_3} \otimes \cdots \otimes \vec{e}^{*\alpha_{r-1}} \otimes \vec{e}^{*\alpha_r} \quad (11.28)$$

$$= t_{\alpha_1 \alpha_2 \alpha_3 \cdots \alpha_{r-1} \alpha_r}^{\circ \circ \circ \cdots \circ \circ} \vec{e}^{*\alpha_1} \otimes \vec{e}^{*\alpha_2} \otimes \vec{e}^{*\alpha_3} \otimes \cdots \otimes \vec{e}^{*\alpha_{r-1}} \otimes \vec{e}^{*\alpha_r}, \quad (11.29)$$

where $\vec{P}$ appears as a totally covariant tensor.

There exist a total 2^r of different possible ways of expressing $\vec{P}$, and evidently, all of them identical. We give the following equality of (11.25), (11.27) and (11.29):

$$\vec{P} = t^{\alpha_1 \ \circ \ \alpha_3 \cdots \ \circ \ \alpha_r}_{\ \circ \ \alpha_2 \ \circ \cdots \alpha_{r-1} \ \circ} \vec{e}_{\alpha_1} \otimes \vec{e}^{*\alpha_2} \otimes \vec{e}_{\alpha_3} \otimes \cdots \otimes \vec{e}^{*\alpha_{r-1}} \otimes \vec{e}_{\alpha_r}$$

$$= t^{\alpha_1 \alpha_2 \alpha_3 \cdots \alpha_{r-1} \alpha_r}_{\ \circ \ \circ \ \circ \cdots \ \circ \ \circ} \vec{e}_{\alpha_1} \otimes \vec{e}_{\alpha_2} \otimes \vec{e}_{\alpha_3} \otimes \cdots \otimes \vec{e}_{\alpha_{r-1}} \otimes \vec{e}_{\alpha_r}$$

$$= t^{\circ \ \circ \ \circ \cdots \ \circ \ \circ}_{\alpha_1 \alpha_2 \alpha_3 \cdots \alpha_{r-1} \alpha_r} \vec{e}^{*\alpha_1} \otimes \vec{e}^{*\alpha_2} \otimes \vec{e}^{*\alpha_3} \otimes \cdots \otimes \vec{e}^{*\alpha_{r-1}} \otimes \vec{e}^{*\alpha_r}$$

$$= \text{etc.}$$

This equality, which does not hold for the homogeneous tensor product, or in the modular tensor product, or in the homogeneous symmetric or anti-symmetric tensor products, declares a relation between the tensor bases that is to be studied in the following sections.

We end this section by indicating to the reader that Expressions (11.25), (11.26) and (11.28) can be executed using matrices entering into Formula (2.21) adapted to each case. For example, for (11.25) it would be

$$X_\sigma = X_1 \otimes X_2^* \otimes X_3 \otimes \cdots \otimes X_{r-1}^* \otimes X_r, \text{ with } \sigma = n^r. \quad (11.30)$$

Another warning to the reader is that the Euclidean tensor spaces

$$\left(\overset{p}{\underset{1}{\otimes}} E^n \right) \left(\overset{q}{\underset{1}{\otimes}} E_*^n \right)(\mathbb{R}), \quad \left(\overset{r}{\underset{1}{\otimes}} E^n \right)(\mathbb{R}), \quad \left(\overset{r}{\underset{1}{\otimes}} E_*^n \right)(\mathbb{R}), etc.$$

have proper tensors, that is non-decomposable tensors, that *are not* products of vectors.

We continue to remind the reader that the indicated property of $\vec{P}$, as indicated in the title of this section, is not exclusive to Euclidean spaces $E^n(\mathbb{R})$ but to all pseudo-Euclidean spaces $PSE^n(\mathbb{R})$.

11.6 Equivalent associated tensors: Vertical displacements of indices. Generalization

Before generalizing the obtained results, we will initiate and present the theoretical fundamentals for the case of *Euclidean tensors of third order*, without loss of conciseness and with simplicity of notation. We observe that all to be presented is totally extensible to pseudo-Euclidean tensors. The fact that we always move inside the same Euclidean linear space $E^n(\mathbb{R}) \equiv E_*^n(\mathbb{R})$, means that a given tensor T of order $r = 3$ belongs to a *unique* product tensor space, that is, that

$$E^n \otimes E^n \otimes E^n(\mathbb{R}) \equiv E^n \otimes E_*^n \otimes E^n(\mathbb{R}) \equiv E_*^n \otimes E_*^n \otimes E^n(\mathbb{R}) \equiv \text{ etc.} \quad (11.31)$$

giving rise to the concept of an *"associated tensor"*.

We first remind the reader about the relations between the primal and dual bases:

$$\vec{e}_\beta = g_{\theta\beta}^{\circ\circ}\vec{e}^{*\theta} = g_{\beta\theta}^{\circ\circ}\vec{e}^{*\theta} \text{ and } \vec{e}^{*\beta} = g_{\circ\circ}^{\theta\beta}\vec{e}_\theta = g_{\circ\circ}^{\beta\theta}\vec{e}_\theta \qquad (11.32)$$

corresponding to the matrix relations

$$||\vec{e}_\beta|| = ||\vec{e}^{*\theta}||G \text{ and } ||\vec{e}^{*\beta}|| = ||\vec{e}_\theta||G^{-1}. \qquad (11.33)$$

The first consequence is the appearance of the covariant and contravariant components of any vector $\vec{v} \in E^n(\mathbb{R})$, given in Formulas (11.7) and (11.8), respectively:

$$x_\beta^\circ = g_{\beta\theta}^{\circ\circ}x_\circ^\theta = g_{\theta\beta}^{\circ\circ}x_\circ^\theta \qquad (11.34)$$

$$x_\circ^\beta = g_{\circ\circ}^{\theta\beta}x_\theta^\circ = g_{\circ\circ}^{\beta\theta}x_\theta^\circ. \qquad (11.35)$$

Thus, following with the tensor T, we propose a well-known relation. Let

$$T = t_{\circ\circ\gamma}^{\alpha\beta\circ}\vec{e}_\alpha \otimes \vec{e}_\beta \otimes \vec{e}^{*\gamma} = t_{\circ\circ\circ}^{\alpha\beta\gamma}\vec{e}_\alpha \otimes \vec{e}_\beta \otimes \vec{e}_\gamma \qquad (11.36)$$

replacing $\vec{e}^{*\gamma} = g_{\circ\circ}^{\theta\gamma}\vec{e}_\theta$ from (11.32) into the left-hand side of (11.36), we get

$$T = t_{\circ\circ\gamma}^{\alpha\beta\circ}\vec{e}_\alpha \otimes \vec{e}_\beta \otimes (g_{\circ\circ}^{\theta\gamma}\vec{e}_\theta) = t_{\circ\circ\gamma}^{\alpha\beta\circ}g_{\circ\circ}^{\theta\gamma}\vec{e}_\alpha \otimes \vec{e}_\beta \otimes \vec{e}_\theta$$

and replacing the notation of the dummy index θ by γ, we have the tensor contraction

$$T = (t_{\circ\circ\theta}^{\alpha\beta\circ}g_{\circ\circ}^{\theta\gamma})\vec{e}_\alpha \otimes \vec{e}_\beta \otimes \vec{e}_\gamma,$$

which is in the same basis as that of the right-hand of (11.36). This requires that

$$t_{\circ\circ\circ}^{\alpha\beta\gamma} = t_{\circ\circ\theta}^{\alpha\beta\circ}g_{\circ\circ}^{\theta\gamma}, \qquad (11.37)$$

which represents the transformation from $t_{\circ\circ\gamma}^{\alpha\beta\circ}$ to $t_{\circ\circ\circ}^{\alpha\beta\gamma}$.

Similarly, consider another tensor relation

$$T = t_{\circ\circ\gamma}^{\alpha\beta\circ}\vec{e}_\alpha \otimes \vec{e}_\beta \otimes \vec{e}^{*\gamma} = t_{\circ\beta\gamma}^{\alpha\circ\circ}\vec{e}_\alpha \otimes \vec{e}^{*\beta} \otimes \vec{e}^{*\gamma}. \qquad (11.38)$$

Substituting in the left-hand side of (11.38) $\vec{e}_\beta = g^{\circ\circ}_{\theta\beta}\vec{e}^{*\theta}$ taken from (11.32) leads to

$$T = t^{\alpha\beta\circ}_{\circ\circ\gamma}\vec{e}_\alpha \otimes (g^{\circ\circ}_{\beta\theta}\vec{e}^{*\theta}) \otimes \vec{e}^{*\gamma} = t^{\alpha\beta\circ}_{\circ\circ\gamma}g^{\circ\circ}_{\beta\theta}\vec{e}_\alpha \otimes \vec{e}^{*\theta} \otimes \vec{e}^{*\gamma}$$

and replacing the notation of the dummy index θ by β, we have the tensor contraction

$$T = (t^{\alpha\theta\circ}_{\circ\circ\gamma}g^{\circ\circ}_{\theta\beta})\vec{e}_\alpha \otimes \vec{e}^{*\beta} \otimes \vec{e}^{*\gamma},$$

which is in the same basis as that of the right-hand of (11.38). This implies

$$t^{\alpha\circ\circ}_{\circ\beta\gamma} = t^{\alpha\theta\circ}_{\circ\circ\gamma}g^{\circ\circ}_{\theta\beta}, \tag{11.39}$$

which represents the transformation of $t^{\alpha\beta\circ}_{\circ\circ\gamma}$ to $t^{\alpha\circ\circ}_{\circ\beta\gamma}$. It is obvious that once the process is discovered, one can continue raising or lowering indices, changing the T tensor species of order $r = 3$. From all this, we conclude that similar to vectors (tensors of order 1), which present 2^1 possibilities of being represented (in contra and in covariant coordinates), the remaining tensors T of order r, present 2^r possibilities of representation.

In particular, those of order $r = 3$ present $2^3 = 8$ possibilities:

$$\{t^{\alpha\beta\gamma}_{\circ\circ\circ}, t^{\alpha\beta\circ}_{\circ\circ\gamma}, t^{\alpha\circ\gamma}_{\circ\beta\circ}, t^{\circ\beta\gamma}_{\alpha\circ\circ}, t^{\circ\circ\gamma}_{\alpha\beta\circ}, t^{\circ\beta\circ}_{\alpha\circ\gamma}, t^{\alpha\circ\circ}_{\circ\beta\gamma}, t^{\circ\circ\circ}_{\alpha\beta\gamma}, \}.$$

Generalization: all possible representations of a *Euclidean* (or *pseudo-Euclidean*) tensor T of order r are called "associated tensors".

There exists a rule for executing the vertical displacements of the indices of a mixed tensor, which is proposed in a tabular form:

For	raising	an index h	$t^{\alpha\circ\cdots\circ\cdots\lambda\circ}_{\circ\beta\cdots h\cdots\circ\mu}$	we replace the index	$t^{\alpha\circ\cdots\circ\cdots\lambda\circ}_{\circ\beta\cdots\theta\cdots\circ\mu}$
	lowering	of a data tensor	$u^{\alpha\circ\cdots h\cdots\lambda\circ}_{\circ\beta\cdots\circ\cdots\circ\mu}$	h by the dummy index θ	$u^{\alpha\circ\cdots\theta\cdots\lambda\circ}_{\circ\beta\cdots\circ\cdots\circ\mu}$

$t^{\alpha\circ\cdots\circ\cdots\lambda\circ}_{\circ\beta\cdots\theta\cdots\circ\mu}$	and it is contracted	$g^{\theta h}_{\circ\circ}$	obtaining	$t^{\alpha\circ\cdots h\cdots\lambda\circ}_{\circ\beta\cdots\circ\cdots\circ\mu} = t^{\alpha\circ\cdots\circ\cdots\lambda\circ}_{\circ\beta\cdots\theta\cdots\circ\mu}g^{\theta h}_{\circ\circ}$
$u^{\alpha\circ\cdots\theta\cdots\lambda\circ}_{\circ\beta\cdots\circ\cdots\circ\mu}$	with	$g^{\circ\circ}_{\theta h}$		$u^{\alpha\circ\cdots\circ\cdots\lambda\circ}_{\circ\beta\cdots h\cdots\circ\mu} = u^{\alpha\circ\cdots\theta\cdots\lambda\circ}_{\circ\beta\cdots\circ\cdots\circ\mu}g^{\circ\circ}_{\theta h}$

Some authors call these relations "association formulas". It is obvious that we can move several indices at the same time, if desired.

To check it, we present an illustrative example.

Let $T = t^{\alpha_1\ \circ\ \circ\ \alpha_4}_{\circ\ \alpha_2\alpha_3\ \circ}\vec{e}_{\alpha_1} \otimes \vec{e}^{*\alpha_2} \otimes \vec{e}^{*\alpha_3} \otimes \vec{e}_{\alpha_4}$ be the notation used for a Euclidean tensor, for which we wish to know the initial component associated with the basis $\{\vec{e}^{*\beta_1} \otimes \vec{e}^{*\beta_2} \otimes \vec{e}_{\beta_3} \otimes \vec{e}_{\beta_4}\}$.

Applying the rule proposed in the table, and using the data indices α_1 and α_3, as dummies, we have

$$\text{(a)} \quad T = t^{\circ\ \circ\ \beta_3\alpha_4}_{\beta_1\alpha_2\ \circ\ \circ} = g^{\circ\ \circ\ \alpha_3\beta_3}_{\alpha_1\beta_1}g^{\alpha_1\ \circ\ \circ\ \alpha_4}_{\circ\ \circ}t^{\alpha_1\ \circ\ \circ\ \alpha_4}_{\circ\ \alpha_2\alpha_3\ \circ} \tag{11.40}$$

or

$$\text{(b)} \quad T = g_{\beta_1 \alpha_1}^{\circ \ \circ} t_{\circ \ \alpha_2 \alpha_3 \ \circ}^{\alpha_1 \ \circ \ \circ \ \alpha_4} g_{\circ \ \circ}^{\alpha_3 \beta_3}. \tag{11.41}$$

In (b) we note that the metric tensors can appear in the tensor expressions as pre- or post-factors of the modified tensor component.

Nevertheless, it is observable that the results (a) and (b) come from the basis $\vec{e}^{*\beta_1} \otimes \vec{e}^{*\alpha_2} \otimes \vec{e}_{\beta_3} \otimes \vec{e}_{\alpha_4}$, which even though it has the species of the desired basis, does not coincide with it.

The sought after component is really $t_{\beta_1 \beta_2 \ \circ \ \circ}^{\circ \ \circ \ \beta_3 \beta_4}$, and in order to obtain it we propose the following solution (there are other possible solutions depending on the decision of which of them are pre- or post-factors):

$$\text{(c)} \quad t_{\beta_1 \beta_2 \ \circ \ \circ}^{\circ \ \circ \ \beta_3 \beta_4} = g_{\beta_1 \alpha_1}^{\circ \ \circ} \delta_{\beta_2 \ \circ}^{\circ \ \alpha_2} t_{\circ \ \alpha_2 \alpha_3 \ \circ}^{\alpha_1 \ \circ \ \circ \ \alpha_4} g_{\circ \ \circ}^{\alpha_3 \beta_3} \delta_{\ \alpha_4 \ \circ}^{\circ \ \beta_4} \tag{11.42}$$

and also because of Formulas (11.22) and (11.23) the result is

$$\text{(d)} \quad t_{\beta_1 \beta_2 \ \circ \ \circ}^{\circ \ \circ \ \beta_3 \beta_4} = g_{\beta_1 \alpha_1}^{\circ \ \circ} g_{\beta_2 \ \circ}^{\circ \ \alpha_2} t_{\circ \ \alpha_2 \alpha_3 \ \circ}^{\alpha_1 \ \circ \ \circ \ \alpha_4} g_{\circ \ \circ}^{\alpha_3 \beta_3} g_{\ \alpha_4 \ \circ}^{\circ \ \beta_4}, \tag{11.43}$$

where we perfectly appreciate the action of the metric-cova and metric-contra tensors, in the lowering of α_1 and raising of α_3 (as β_1 and β_3), and of the mixed-metric tensors for changing the name $(\alpha_2 \alpha_4)$ by $(\beta_2 \beta_4)$ but not the position of indices.

In fact, the studied illustrative example not only pretends to verify that it is possible to move several indices simultaneously, and by means of mixed tensors of Formulas (11.22) and (11.23) change their notation, but its real intention is to obtain a matrix process for determining the associated tensors of a given tensor.

In effect, we proceed to interpret the problem from its very beginning.

Consider the Euclidean tensor $t_{\circ \ \beta \gamma \circ}^{\alpha \circ \circ \delta}$ of order $r = 4$ over $E^n(\mathbb{R})$, which is given by its matrix representation $T = [t_{\circ \ \beta \gamma \circ}^{\alpha \circ \circ \delta}]$, the associated tensor of which $t_{\alpha \beta \circ \circ}^{\circ \circ \gamma \delta}$ is to be known numerically.

We know the fundamental metric tensor, in its four representations:

$$G_n = [g_{\alpha \beta}^{\circ \circ}]; \quad I_n = [g_{\alpha \circ}^{\circ \beta}] = [g_{\circ \beta}^{\alpha \circ}]; \quad G_n^{-1} = [g_{\circ \circ}^{\alpha \beta}].$$

Since the index ordering is decisive, we change the tensor notation, using numbered dummy indices, in such way that the resulting problem is precisely the given illustrative example.

Data: $t_{\circ \ \alpha_2 \alpha_3 \ \circ}^{\alpha_1 \ \circ \ \circ \ \alpha_4}$. Question: $t_{\beta_1 \beta_2 \ \circ \ \circ}^{\circ \ \circ \ \beta_3 \beta_4}$.

The final formula, (11.43), is going to be conveniently treated to get our objective. Let $\sigma = n^4$. Using the symmetry of all metric tensors, we propose the formula

$$t_{\beta_1 \beta_2 \ \circ \ \circ}^{\circ \ \circ \ \beta_3 \beta_4} = g_{\beta_1 \alpha_1}^{\circ \ \circ} g_{\beta_2 \ \circ}^{\circ \ \alpha_2} g_{\circ \ \circ}^{\beta_3 \alpha_3} g_{\ \alpha_4 \ \circ}^{\beta_4 \circ} t_{\circ \ \alpha_2 \alpha_3 \ \circ}^{\alpha_1 \ \circ \ \circ \ \alpha_4}. \tag{11.44}$$

Note the index ordering and the expression of the moved indices.

Denoting by $T'_{\sigma,1}$ the column matrix "extension" of the block matrix $[t_{\beta_1\beta_2\ \ \circ\ \circ}^{\ \circ\ \circ\ \beta_3\beta_4}]$ of the associated tensor and by T_σ the "extension" of the matrix $[t_{\ \circ\ \alpha_2\alpha_3\ \circ}^{\alpha_1\ \circ\ \circ\ \alpha_4}]$ of the data tensor, we write term by term the Formula (11.44) in matrix form, which can be interpreted as

$$T'_\sigma = \left(G_n \otimes I_n \otimes G_n^{-1} \otimes I_n\right) \bullet T_\sigma. \tag{11.45}$$

Once we have obtained T'_σ, we proceed to its condensation, which is indicated with the blocks in horizontal positions:

	B_{11}	B_{12}	$\cdots$	B_{1n}
	first block with the first n^2 entities	second block of the following n^2 entities	$\cdots$	n-th block following n^2
	B_{21}	B_{22}	$\cdots$	B_{2n}
$(T'_\sigma)^t =$	$(n+1)$-th block following n^2	$(n+2)$-th block following n^2	$\cdots$	$(2n)$-th block following n^2
	$\cdots$	$\cdots$	$\cdots$	$\cdots$
	B_{n1}	B_{n2}	$\cdots$	B_{nn}
	$((n-1)n+1)$-th block following n^2	$((n-1)n+2)$-th block following n^2	$\cdots$	n^2-th block last n^2

with a total of (n^2 entities)$\times$ (n^2 blocks) $= n^4 \equiv \sigma$.

Particularizing Formula (11.45) to another orders $r \neq 4$, and any selected species would be useful as a matrix process for calculating associated tensors in each case.

11.6.1 The quotient space of isomers

Let $\left(\overset{r}{\underset{1}{\otimes}} E^n\right)(\mathbb{R})$ be the unique product space of Euclidean tensors of order r. We define the following equivalent relation $\mathcal{R}$: Two Euclidean tensors are related iff they are "associate". Each class of the corresponding quotient space $\left(\overset{r}{\underset{1}{\otimes}} E^n\right)(\mathbb{R})/\mathcal{R}$ contains only all Euclidean tensors associated with $T = [t_{\circ\ \circ\ \circ\ \circ}^{\alpha_1\alpha_2\cdots\alpha_r}]$.

With respect to the expression "equivalent associated tensors" proposed in the title of this section, we show that the relation "*the tensor T is associated with tensor U*" is an equivalence relation, in the set of homogeneous tensors of order r over $E^n(\mathbb{R})$, that is, the set of vectors of the tensor space $\left(\overset{r}{\underset{1}{\otimes}} E^n\right)(\mathbb{R})$.

For the sake of simplicity we proceed with $r = 2$.

Property I. The relation is reflexive (T is associated with T):

$$t^{\alpha\beta}_{\circ\circ} = \left(t^{\alpha\theta}_{\circ\circ}g^{\circ\circ}_{\theta w}\right) g^{w\beta}_{\circ\circ}.$$

Property II. The relation is symmetric (if T is associated with U, then U is associated with T). Let T be associated with U, then

$$t^{\alpha\beta}_{\circ\circ} = g^{\alpha\theta}_{\circ\circ}g^{\beta w}_{\circ\circ}u^{\circ\circ}_{\theta w};$$

pre-multiplying by $g^{\circ\circ}_{\mu\alpha}g^{\circ\circ}_{\nu\beta}$, the result is

$$g^{\circ\circ}_{\mu\alpha}g^{\circ\circ}_{\nu\beta}t^{\alpha\beta}_{\circ\circ} = (g^{\circ\circ}_{\mu\alpha}g^{\alpha\theta}_{\circ\circ})(g^{\circ\circ}_{\nu\beta}g^{\beta w}_{\circ\circ})u^{\circ\circ}_{\theta w} = \delta^{\circ\theta}_{\mu\circ}\delta^{\circ w}_{\nu\circ}u^{\circ\circ}_{\theta w};$$

and operating and exchanging sides:

$$u^{\circ\circ}_{\mu\nu} = g^{\circ\circ}_{\mu\alpha}g^{\circ\circ}_{\nu\beta}t^{\alpha\beta}_{\circ\circ} \to U \text{ is associated with } T.$$

Property III. The relation is transitive (if T is associated with U, and U is associated with V, then T is associated with V).

If T is associated with U, then $t^{\alpha\beta}_{\circ\circ} = g^{\alpha\theta}_{\circ\circ}g^{\beta w}_{\circ\circ}u^{\circ\circ}_{\theta w}$

If U is associated with V, then $u^{\circ\circ}_{\theta w} = g^{\circ\circ}_{\theta\lambda}g^{\circ\circ}_{w\mu}v^{\lambda\mu}_{\circ\circ}$

Substituting the second into the first, we get

$$t^{\alpha\beta}_{\circ\circ} = g^{\alpha\theta}_{\circ\circ}g^{\beta w}_{\circ\circ}(g^{\circ\circ}_{\theta\lambda}g^{\circ\circ}_{w\mu}v^{\lambda\mu}_{\circ\circ}) = (g^{\alpha\theta}_{\circ\circ}g^{\circ\circ}_{\theta\lambda})(g^{\beta w}_{\circ\circ}g^{\circ\circ}_{w\mu})v^{\lambda\mu}_{\circ\circ},$$

that is, $t^{\alpha\beta}_{\circ\circ} = \delta^{\alpha\circ}_{\circ\lambda}\delta^{\beta\circ}_{\circ\mu}v^{\lambda\mu}_{\circ\circ} \to t^{\alpha\beta}_{\circ\circ} = g^{\alpha\circ}_{\circ\lambda}g^{\beta\circ}_{\circ\mu}v^{\lambda\mu}_{\circ\circ}$, and then T is associated with V.

The quotient set is the set of the diverse tensors of different order; for example of the diverse contravariant linear spaces $\left\{\left(\overset{r}{\underset{1}{\otimes}} E^n\right)(\mathbb{R})\right\}$.

11.7 Changing bases in $E^n(\mathbb{R})$: Euclidean tensor character criteria

Though the strict change-of-basis of a Euclidean tensor is the one that corresponds to it as a homogeneous tensor, already given in Chapter 4, the topic to be treated in this section is the one motivated by the additional circumstance of the existence of associated tensors.

In other words, the following problem can arise. Given a Euclidean tensor T, of order r and species χ, we perform a change-of-basis in $E^n(\mathbb{R})$ of matrix $C = [c^{\alpha o}_{o\,i}]$, and we wish to know the new components of T, knowing that the new species is $\hat{\chi}$, that is, the tensor has an initial species χ, and when changing basis, we accept that because it is Euclidean, the answer be given in the new basis, with another associated tensor of different species $\hat{\chi}$.

We analyze this situation for a tensor of order $r = 4$. Consider $T = t^{o\,o\,o\,\mu}_{\alpha\beta\lambda o}\vec{e}^{*\alpha} \otimes \vec{e}^{*\beta} \otimes \vec{e}^{*\lambda} \otimes \vec{e}_\mu$ of species $\chi = \{2 - 2 - 2 - 1\}$ over the Euclidean space $E^n(\mathbb{R})$, in which we perform the change-of-basis $\vec{e}_i = c^{\alpha o}_{o\,i}\vec{e}_\alpha$; we wish to know $T = t^{i\,o\,o\,o}_{o\,j\,\ell\,m}\vec{e}_i \otimes \vec{e}^{*j} \otimes \vec{e}^{*\ell} \otimes \vec{e}^{*m}$, of species $\hat{\chi} = \{1 - 2 - 2 - 2\}$.

Different possibilities of action:

1. We extract the associated tensor of species $\hat{\chi}$, and later, we execute the change-of-basis over this tensor.
2. We execute first the change-of-basis over that of species χ, and later we determine in the new basis, the associated tensor of species $\hat{\chi}$. The following question arises immediately: is the "order of execution" *indifferent*?

   We show that it is, no matter the process followed with respect to the result, though process 1 is more convenient, because we know directly the metric tensor G, while in process 2, as we study the associated tensors in the new basis, it is necessary to determine the new metric tensor $\hat{G}$.

1. Development of possibility 1:

   The associated tensor (already in the initial basis of $E^n(\mathbb{R})$) of the given one is

$$t^{\alpha o o o}_{o\beta\lambda\mu} = g^{\alpha\theta}_{o o}\, t^{o o o w}_{\theta\beta\lambda o}\, g^{o o}_{w\mu}, \tag{11.46}$$

   and the change-of-basis of the previous tensor is

$$t^{i o o o}_{o j \ell m} = t^{\alpha o o o}_{o\beta\lambda\mu}\gamma^{i o}_{o\alpha}c^{o\beta}_{j o}c^{o\lambda}_{\ell o}c^{o\mu}_{m o}$$

   and replacing (11.46) we obtain

$$t^{i o o o}_{o j \ell m} = g^{\alpha\theta}_{o o}\, t^{o o o w}_{\theta\beta\lambda o}\, g^{o o}_{w\mu}\gamma^{i o}_{o\alpha}c^{o\beta}_{j o}c^{o\lambda}_{\ell o}c^{o\mu}_{m o}. \tag{11.47}$$

2. Development of possibility 2:

   We determine the data tensor in the new basis of $E^n(\mathbb{R})$ as

$$t^{o o o v}_{z j \ell o} = t^{o o o \mu}_{\alpha\beta\lambda o}\, c^{o\alpha}_{z o}c^{o\beta}_{j o}c^{o\lambda}_{\ell o}\gamma^{v o}_{o\mu} \tag{11.48}$$

   and now we proceed to determine the associated tensor of the given one, with the species $\hat{\chi} = \{1 - 2 - 2 - 2\}$:

$$t^{i o o o}_{o j \ell m} = \hat{g}^{i z}_{o o}\, t^{o o o v}_{z j \ell o}\, \hat{g}^{o o}_{v m} \tag{11.49}$$

and substituting (11.48) into (11.49), we have

$$t^{i\,\circ\,\circ\,\circ}_{\circ\,j\,\ell\,m} = \hat{g}^{i\,z}_{\circ\,\circ}; t^{\circ\,\circ\,\circ\,\mu}_{\alpha\beta\lambda\circ}\, c^{\circ\,\alpha}_{z\,\circ}c^{\circ\,\beta}_{j\,\circ}c^{\circ\,\lambda}_{\ell\,\circ}\gamma^{v\,\circ}_{\circ\,\mu}\hat{g}^{\circ\,\circ}_{vm}. \tag{11.50}$$

The calculus of the new metric tensor, in contravariant and covariant cooordinates gives the tensor relations taken from (4.34):

$$\begin{aligned}\hat{g}^{i\,z}_{\circ\,\circ} &= g^{\sigma\,\rho}_{\circ\,\circ}\gamma^{i\,\circ}_{\circ\,\sigma}\gamma^{z\,\circ}_{\circ\,\rho}\\ \hat{g}^{\circ\,\circ}_{vm} &= g^{\circ\,\circ}_{\tau\,\eta}c^{\circ\,\tau}_{v\,\circ}c^{\circ\,\eta}_{m\,\circ},\end{aligned} \tag{11.51}$$

which, substituted into (11.50), leads to the expression

$$t^{i\,\circ\,\circ\,\circ}_{\circ\,j\,\ell\,m} = g^{\sigma\,\rho}_{\circ\,\circ}\gamma^{i\,\circ}_{\circ\,\alpha}\gamma^{z\,\circ}_{\circ\,\rho}t^{\circ\,\circ\,\circ\,\mu}_{\alpha\beta\lambda\circ}\,c^{\circ\,\alpha}_{z\,\circ}c^{\circ\,\beta}_{j\,\circ}c^{\circ\,\lambda}_{\ell\,\circ}\gamma^{v\,\circ}_{\circ\,\mu}\hat{g}^{\circ\,\circ}_{\tau\,\eta}c^{\circ\,\tau}_{v\,\circ}c^{\circ\,\eta}_{m\,\circ}. \tag{11.52}$$

Since in addition we have

$$\gamma^{z\,\circ}_{\circ\,\rho}\cdot c^{\circ\,\alpha}_{z\,\circ} = \delta^{\circ\,\alpha}_{\rho\,\circ}\quad\text{and}\quad \gamma^{v\,\circ}_{\circ\,\mu}\cdot c^{\circ\,\tau}_{v\,\circ} = \delta^{\circ\,\tau}_{\mu\,\circ},$$

taking them into (11.52) yields

$$\begin{aligned}t^{i\,\circ\,\circ\,\circ}_{\circ\,j\,\ell\,m} &= g^{\sigma\,\rho}_{\circ\,\circ}\gamma^{i\,\circ}_{\circ\,\alpha}(\delta^{\circ\,\alpha}_{\rho\,\circ}t^{\circ\,\circ\,\circ\,\mu}_{\alpha\beta\lambda\circ}\delta^{\circ\,\tau}_{\mu\,\circ})c^{\circ\,\beta}_{j\,\circ}c^{\circ\,\lambda}_{\ell\,\circ}g^{\circ\,\circ}_{\tau\,\eta}c^{\circ\,\eta}_{m\,\circ}\\ &= g^{\sigma\,\rho}_{\circ\,\circ}t^{\circ\,\circ\,\circ\,\tau}_{\rho\beta\lambda\circ}g^{\circ\,\circ}_{\tau\,\eta}\gamma^{i\,\circ}_{\circ\,\alpha}c^{\circ\,\beta}_{j\,\circ}c^{\circ\,\lambda}_{\ell\,\circ}c^{\circ\,\eta}_{m\,\circ},\end{aligned} \tag{11.53}$$

which is the final result for possibility 2.

If we perform a change of notation in its dummy indices, according to the correspondence:

$$\sigma\to\alpha\quad \rho\to\theta\quad \tau\to w\quad \eta\to\mu,$$

the result is

$$t^{i\,\circ\,\circ\,\circ}_{\circ\,j\,\ell\,m} = g^{\alpha\theta}_{\circ\,\circ}\,t^{\circ\,\circ\,\circ\,w}_{\theta\beta\lambda\circ}\,g^{\circ\,\circ}_{w\,\mu}\gamma^{i\,\circ}_{\circ\,\alpha}c^{\circ\,\beta}_{j\,\circ}c^{\circ\,\lambda}_{\ell\,\circ}c^{\circ\,\mu}_{m\,\circ},$$

that is coincident with (11.47), the result of process 1.

With respect to the tensor criteria of systems of scalar components over Euclidean spaces $E^n(\mathbb{R})$, and more precisely in the methods related to the quotient law for determining the tensor nature, we must point out the effect of the associated tensors.

As a conclusion, we mention *two* conditions for a set of 2^r scalar components, referred to $E^n(\mathbb{R})$, to be considered as a Euclidean tensor of order r.

1. The 2^r systems must be homogeneous tensors of order r and each one of them corresponding to the different possible species. To each one of them the quotient law would have been applied to determine its tensor nature.
2. The set of the 2^r systems must be an equivalence class with respect to the association of Euclidean tensors indicated in the previous section.

11.8 Symmetry and anti-symmetry in Euclidean tensors

Euclidean tensors can obviously present in a certain basis $\{\vec{e}_\alpha\}$ of $E^n(\mathbb{R})$, partial or total symmetries (with respect to an index set).

The situation is similar with respect to the anti-symmetry. As they are homogeneous tensors, we know that both properties have a tensor nature.

However, we raise another question, relative to an arbitrary basis.

Consider a tensor "totally covariant and symmetric". If we consider its associated tensors, raising pairs of indices, does this imply that the tensor is symmetric with respect to these contravariant indices?

Following the line of the authors that mention this, we prove that if the answer is affirmative, we will obtain a mixed symmetric tensor.

We remind the reader that only changes of indices of the *same species* between them are considered, because the symmetry of indices of different species has no tensor nature.

In effect, let $T \in \left(\overset{r}{\underset{1}{\otimes}} E^n_* \right) (\mathbb{R})$ be a symmetric covariant tensor.

Let

$$t^{\circ \ \circ \ \circ \ \circ \ \circ}_{\alpha_1 \alpha_2 \alpha_3 \dots \alpha_r} = t^{\circ \ \circ \ \circ \ \circ \ \circ}_{\alpha_1 \alpha_3 \alpha_2 \dots \alpha_r}, \ \forall \alpha_i, \alpha_j \in I_n, \ i, j \in I_r. \tag{11.54}$$

If we study

$$t^{\circ \ \alpha_2 \alpha_3 \circ \ \circ}_{\alpha_1 \ \circ \ \circ \ \dots \alpha_r} = g^{\alpha_2 \beta_2}_{\circ \ \circ} \ g^{\alpha_3 \beta_3}_{\circ \ \circ} t^{\circ \ \circ \ \circ \ \circ \ \circ}_{\alpha_1 \beta_2 \beta_3 \dots \alpha_r}$$

due to the properties of the fundamental metric tensor, it is

$$t^{\circ \ \alpha_2 \alpha_3 \circ \ \circ}_{\alpha_1 \ \circ \ \circ \ \dots \alpha_r} = g^{\alpha_3 \beta_3}_{\circ \ \circ} g^{\alpha_2 \beta_2}_{\circ \ \circ} \ t^{\circ \ \circ \ \circ \ \circ \ \circ}_{\alpha_1 \beta_2 \beta_3 \dots \alpha_r}$$

and because of the property (11.54)

$$t^{\circ \ \alpha_2 \alpha_3 \circ \ \circ}_{\alpha_1 \ \circ \ \circ \ \dots \alpha_r} = g^{\alpha_3 \beta_3}_{\circ \ \circ} g^{\alpha_2 \beta_2}_{\circ \ \circ} \ t^{\circ \ \circ \ \circ \ \circ \ \circ}_{\alpha_1 \beta_3 \beta_2 \dots \alpha_r} = t^{\circ \ \alpha_3 \alpha_2 \circ \ \circ}_{\alpha_1 \ \circ \ \circ \ \dots \alpha_r}.$$

Similarly, if the initial tensor $T \in \left(\overset{r}{\underset{1}{\otimes}} E^n \right) (\mathbb{R})$ is contravariant and symmetric, also the process of "lowering" each pair of indices, will produce the symmetry of the lowered indices.

Thus, as a new corollary we conclude that if a Euclidean tensor presents initial total covariant (contravariant) symmetry, it will present final contravariant (covariant) symmetry in the given basis, and then in any other basis.

However, if the *initial data is a mixed tensor*, Euclidean and symmetric, the symmetric character does *not* hold, in general, in its associated tensors.

The reason is clear. Assume that a mixed tensor of order $r = 4$, $p = q = 2$, is symmetric, then we have

$$t^{\alpha \beta \circ \circ}_{\circ \circ \gamma \delta} = t^{\beta \alpha \circ \circ}_{\circ \circ \gamma \delta} = t^{\alpha \beta \circ \circ}_{\circ \circ \delta \gamma} = t^{\beta \alpha \circ \circ}_{\circ \circ \delta \gamma}.$$

However, when lowering its first two indices, as is well known, the symmetry $t^{\circ\ \circ\ \circ\ \circ}_{(\alpha\beta)(\gamma\delta)}$ remains between the indices $(\alpha\beta)$ and between the $(\gamma\delta)$ but we *know* *nothing about* $t^{\circ\ \circ\ \circ\ \circ}_{\alpha(\gamma\beta)\delta}$, because though they would have the indices $(\beta\gamma)$, $t^{\alpha\beta\circ\circ}_{\circ\circ\gamma\delta}$ with vertical symmetry and keep it when lowered, in tensor form this means nothing. A counterexample will clarify this situation.

Consider the mixed symmetric tensor S of order $r = 4$ and species $(1 - 1 - 2 - 2)$, over a Euclidean space $E^2(\mathbb{R})$ of Gram matrix $G_2 = \begin{bmatrix} 1 & 1 \\ 1 & 2 \end{bmatrix}$ and $\sigma = n^r = 2^4 = 16$.

The matrix representation of the components of S is

$$S = \left[S^{\alpha\beta\circ\circ}_{\circ\circ\gamma\delta} \right] = \begin{bmatrix} a & b & | & m & p \\ b & c & | & p & q \\ - & - & + & - & - \\ m & p & | & e & d \\ p & q & | & d & f \end{bmatrix}, \tag{11.55}$$

where $a, b, c, \ldots, e, d, f \in \mathbb{N}$ with $a \neq b \neq c \neq \cdots \neq d \neq f$.

Whence

$$s^{11\circ\circ}_{\circ\circ11} = a; \quad s^{11\circ\circ}_{\circ\circ22} = c; \quad s^{22\circ\circ}_{\circ\circ11} = e; \quad s^{22\circ\circ}_{\circ\circ22} = f$$

$$s^{11\circ\circ}_{\circ\circ12} = s^{11\circ\circ}_{\circ\circ21} = b; \quad s^{12\circ\circ}_{\circ\circ11} = s^{21\circ\circ}_{\circ\circ11} = m; \quad s^{12\circ\circ}_{\circ\circ22} = s^{21\circ\circ}_{\circ\circ22} = q;$$

$$s^{22\circ\circ}_{\circ\circ12} = s^{22\circ\circ}_{\circ\circ21} = d; \quad s^{12\circ\circ}_{\circ\circ12} = s^{12\circ\circ}_{\circ\circ21} = s^{21\circ\circ}_{\circ\circ12} = s^{21\circ\circ}_{\circ\circ21} = p.$$

The expression for lowering the two indices $(\alpha\beta)$, to arrive at the totally covariant associated tensor $S^{\circ\circ\circ\circ}_{\alpha\beta\gamma\delta}$, is

$$S = \left[S^{\circ\ \circ\ \circ\ \circ}_{\beta_1\beta_2\beta_3\beta_4} \right] = g^{\circ\ \circ}_{\beta_1\ \alpha_1}\, g^{\circ\ \circ}_{\beta_2\ \alpha_2} S^{\alpha_1\alpha_2\ \circ\ \circ}_{\circ\ \circ\ \alpha_3\alpha_4}\, g^{\alpha_3\ \circ}_{\circ\ \beta_3}\, g^{\alpha_4\ \circ}_{\circ\ \beta_4} \tag{11.56}$$

obtained as the formulas (11.43) and (11.44).

We calculate (11.56) using matrices, as in Formula (11.45).

In our case, the extension S_σ of S is (in horizontal):

$$(S_{1,16}) = [\, a\ b\ b\ c\ |\ m\ p\ p\ q\ |\ m\ p\ p\ q\ |\ e\ d\ d\ f\,],$$

whence

$$M_{16,16} = [G_2 \otimes G_2 \otimes I_2 \otimes I_2]$$
$$= \left(\begin{bmatrix} 1 & 1 \\ 1 & 2 \end{bmatrix} \otimes \begin{bmatrix} 1 & 1 \\ 1 & 2 \end{bmatrix} \right) \otimes \left(\begin{bmatrix} 1 & 0 \\ 0 & 1 \end{bmatrix} \otimes \begin{bmatrix} 1 & 0 \\ 0 & 1 \end{bmatrix} \right)$$
$$= \begin{bmatrix} 1 & 1 & 1 & 1 \\ 1 & 2 & 1 & 2 \\ 1 & 1 & 2 & 2 \\ 1 & 2 & 2 & 4 \end{bmatrix} \otimes \begin{bmatrix} 1 & 0 & 0 & 0 \\ 0 & 1 & 0 & 0 \\ 0 & 0 & 1 & 0 \\ 0 & 0 & 0 & 1 \end{bmatrix},$$

which implies $\hat{S} = M_{16,16} \bullet S_{16,1}$, and operating

$$\hat{S}_{16} = \left[\begin{array}{cccc|cccc|cccc|cccc}
1\,0\,0\,0 & 1\,0\,0\,0 & 1\,0\,0\,0 & 1\,0\,0\,0 \\
0\,1\,0\,0 & 0\,1\,0\,0 & 0\,1\,0\,0 & 0\,1\,0\,0 \\
0\,0\,1\,0 & 0\,0\,1\,0 & 0\,0\,1\,0 & 0\,0\,1\,0 \\
0\,0\,0\,1 & 0\,0\,0\,1 & 0\,0\,0\,1 & 0\,0\,0\,1 \\
\hline
1\,0\,0\,0 & 2\,0\,0\,0 & 1\,0\,0\,0 & 2\,0\,0\,0 \\
0\,1\,0\,0 & 0\,2\,0\,0 & 0\,1\,0\,0 & 0\,2\,0\,0 \\
0\,0\,1\,0 & 0\,0\,2\,0 & 0\,0\,1\,0 & 0\,0\,2\,0 \\
0\,0\,0\,1 & 0\,0\,0\,2 & 0\,0\,0\,1 & 0\,0\,0\,2 \\
\hline
1\,0\,0\,0 & 1\,0\,0\,0 & 2\,0\,0\,0 & 2\,0\,0\,0 \\
0\,1\,0\,0 & 0\,1\,0\,0 & 0\,2\,0\,0 & 0\,2\,0\,0 \\
0\,0\,1\,0 & 0\,0\,1\,0 & 0\,0\,2\,0 & 0\,0\,2\,0 \\
0\,0\,0\,1 & 0\,0\,0\,1 & 0\,0\,0\,2 & 0\,0\,0\,2 \\
\hline
1\,0\,0\,0 & 2\,0\,0\,0 & 2\,0\,0\,0 & 4\,0\,0\,0 \\
0\,1\,0\,0 & 0\,2\,0\,0 & 0\,2\,0\,0 & 0\,4\,0\,0 \\
0\,0\,1\,0 & 0\,0\,2\,0 & 0\,0\,2\,0 & 0\,0\,4\,0 \\
0\,0\,0\,1 & 0\,0\,0\,2 & 0\,0\,0\,2 & 0\,0\,0\,4
\end{array}\right]
\left[\begin{array}{c}
a \\ b \\ b \\ c \\ - \\ m \\ p \\ p \\ q \\ - \\ m \\ p \\ p \\ q \\ - \\ e \\ d \\ d \\ f
\end{array}\right]
=
\left[\begin{array}{c}
a+2m+e \\ b+2p+d \\ b+2p+d \\ c+2q+f \\ \hline
a+3m+2e \\ b+3p+2d \\ b+3p+2d \\ c+3q+2f \\ \hline
a+3m+2e \\ b+3p+2d \\ b+3p+2d \\ c+3q+2f \\ \hline
a+4m+4e \\ b+4p+4d \\ b+4p+4d \\ c+4q+4f
\end{array}\right].$$

We observe that condensed, $\hat{S}_{16,1}$ is the associated tensor:

$$S' = [S^{\circ\circ\circ\circ}_{\alpha\beta\gamma\delta}] = \left[\begin{array}{cc|cc}
a+2m+e & b+2p+d & a+3m+2e & b+3p+2d \\
b+2p+d & c+2q+f & b+3p+2d & c+3q+2f \\
\hline
a+3m+2e & b+3p+2d & a+4m+4e & b+4p+4d \\
b+3p+2d & c+3q+2f & b+4p+4d & c+4q+4f
\end{array}\right].$$

$$(11.57)$$

When studying the subindices discussed in our development when the initial data tensor was a mixed tensor, we observed that, for example: $s^{\circ\circ\circ\circ}_{1211} = a + 3m + 2e = s^{\circ\circ\circ\circ}_{2111}$, that is, the "lowered" indices maintain the symmetry that they showed initially, in agreement with the theory. However, when contemplating the indices $(\beta\gamma)$, the result is $s^{\circ\ \circ\ \circ\circ}_{1(21)1} = a + 3m + 2e \neq b + 2p + d = s^{\circ\ \circ\ \circ\circ}_{1(12)1}$, and then, due to this reason and others, the associated tensor $S' = [S^{\circ\circ\circ\circ}_{\alpha\beta\gamma\delta}]$ is not symmetric, thus the symmetry of S has no tensor nature.

It is evident that with respect to the anti-symmetry the same occurs: If a tensor $T \in \left(\overset{r}{\underset{1}{\otimes}} E^n\right)(\mathbb{R})$ is anti-symmetric, $\forall \alpha_i, \alpha_j \in I_n$; $i, j \in I_r$ is

$$t^{\alpha_2\,\alpha_1\,\alpha_3\,\cdots\,\alpha_r}_{\circ\ \ \circ\ \ \circ\ \cdots\ \circ} = -t^{\alpha_1\,\alpha_2\,\alpha_3\,\cdots\,\alpha_r}_{\circ\ \ \circ\ \ \circ\ \cdots\ \circ}$$

and this implies

$$t^{\,\circ\;\;\circ\;\;\alpha_3\cdots\alpha_r}_{\;\alpha_2\,\alpha_1\,\circ\,\cdots\,\circ} = -t^{\,\circ\;\;\circ\;\;\alpha_3\cdots\alpha_r}_{\;\alpha_1\,\alpha_2\,\circ\,\cdots\,\circ}$$

however, the reverse *does not* satisfy such an implication. We do not insist further on the anti-symmetric case, that will be treated in detail later in Chapter 13 dedicated to Euclidean exterior algebras.

11.9 Cartesian tensors

Given the great importance that in particular tensor applications in solid mechanics, mechanics of continuous means, geotechnics, etc., the fact of using *exclusively* either initial or new orthonormalized bases, for the reference Euclidean space $E^n(\mathbb{R})$, has, we have decided to give the title of Cartesian tensors to the tensors that have such compelling privilege.

11.9.1 Main properties of Euclidean $E^n(\mathbb{R})$ spaces in orthonormal bases

The following properties hold:

1. Let $\{\vec{e}_\alpha\}$ be an orthonormal initial basis of $E^n(\mathbb{R})$. The fundamental metric tensor in covariant coordinates, $G \equiv [g^{\,\circ\,\circ}_{\,\alpha\,\beta}]$, satisfies the property

$$g^{\,\circ\,\circ}_{\,\alpha\,\beta} = \vec{e}_\alpha \bullet \vec{e}_\beta = \delta_{\alpha\beta}; \; \forall \alpha, \beta \in I_n,$$

so that

$$G \equiv [g^{\,\circ\,\circ}_{\,\alpha\,\beta}] = [\delta^{\,\circ\,\circ}_{\,\alpha\,\beta}] = I_n \text{ (unit matrix).} \qquad (11.58)$$

Similarly, as $E^n(\mathbb{R}) \equiv E^n_*(\mathbb{R})$, the fundamental metric tensor in contravariant coordinates is

$$G^* \equiv G^{-1} \equiv [g^{\,\alpha\beta}_{\,\circ\,\circ}] = [g^{\,\circ\,\circ}_{\,\alpha\,\beta}]^{-1} = I_n^{-1} = I_n, \qquad (11.59)$$

whence

$$g^{\,\circ\,\circ}_{\,\alpha\,\beta} = g^{\,\alpha\beta}_{\,\circ\,\circ} = \delta^{\,\alpha\,\circ}_{\,\circ\,\beta} = \delta^{\,\circ\,\beta}_{\,\alpha\,\circ}. \qquad (11.60)$$

2. The contravariant and the covariant coordinates of any vector $\vec{V} \in E^n(\mathbb{R})$ coincide.
   In effect, let $\vec{V} = x^\alpha_\circ \vec{e}_\alpha$ or $\vec{V} = x^\circ_\alpha \vec{e}^{*\alpha}$, then

$$x^\circ_\alpha = g^{\,\circ\,\circ}_{\,\alpha\,\beta} x^\beta_\circ = \delta^{\,\circ\,\circ}_{\,\alpha\,\beta} x^\beta_\circ = x^\alpha_\circ; \forall \alpha \in I_n. \qquad (11.61)$$

3. The orthonormal bases satisfy the property of being autodual or autoreciprocal: $\vec{e}_\alpha \equiv \vec{e}^{*\alpha}, \forall \alpha \in I_n,$

$$\vec{e}^{*\alpha} = g^{\,\alpha\beta}_{\,\circ\,\circ} \vec{e}_\beta = \delta^{\,\alpha\beta}_{\,\circ\,\circ} \vec{e}_\beta = \vec{e}_\alpha,$$

thus, if it is orthonormal, i.e., if

$$\vec{e}^{*\alpha} \bullet \vec{e}_\beta = \vec{e}_\alpha \bullet \vec{e}_\beta = \delta_{\alpha\beta}^{\circ\,\circ} = \delta_{\circ\,\beta}^{\alpha\,\circ}$$

then it is also autoreciprocal, i.e.,

$$\vec{e}^{*\alpha} \bullet \vec{e}_\beta = \delta_{\circ\,\beta}^{\alpha\,\circ}; \text{ (reciprocal condition)}.$$

4. The changes of basis, of initial orthonormal to final orthonormal bases are made by means of *change-of-basis orthogonal* matrices C (note the change in name, it should be "orthonormal matrices C").
Let $\vec{e}_i = c_{\circ\,i}^{\alpha\,\circ}\vec{e}_\alpha$ and $C \equiv [c_{\circ\,i}^{\alpha\,\circ}]$ be the change-of-basis in the space $E^n(\mathbb{R})$.
We know that $\vec{e}_\alpha \bullet \vec{e}_\beta = \delta_{\alpha\beta}^{\circ\,\circ}$ and that $\vec{e}_i \bullet \vec{e}_j = \delta_{ij}^{\circ\,\circ}$ because the bases must be orthonormal before and after the change. Whence

$$\vec{e}_i \bullet \vec{e}_j = \delta_{ij}^{\circ\,\circ} = (c_{\circ\,i}^{\alpha\,\circ}\vec{e}_\alpha) \bullet (c_{\circ\,j}^{\beta\,\circ}\vec{e}_\beta) = (\vec{e}_\alpha \bullet \vec{e}_\beta)c_{\circ\,i}^{\alpha\,\circ}c_{\circ\,j}^{\beta\,\circ} = \delta_{\alpha\beta}^{\circ\,\circ}c_{\circ\,i}^{\alpha\,\circ}c_{\circ\,j}^{\beta\,\circ} = c_{i\,\circ}^{\circ\,\alpha}\delta_{\alpha\beta}^{\circ\,\circ}c_{\circ\,j}^{\beta\,\circ}$$

with the correlative contraction of indices α and β.
The tensor expression $\delta_{ij}^{\circ\,\circ} = c_{i\,\circ}^{\circ\,\alpha}\delta_{\alpha\beta}^{\circ\,\circ}c_{\circ\,j}^{\beta\,\circ}$ has the matrix interpretation

$$I_n = C^t I_n C; \; C^t \bullet C = I_n \to C^t = C^{-1}, \tag{11.62}$$

which shows its ortogonal nature.
If desired, one can use a tensor procedure

$$\delta_{ij}^{\circ\,\circ} = c_{i\,\circ}^{\circ\,\alpha}\delta_{\alpha\beta}^{\circ\,\circ}c_{\circ\,j}^{\beta\,\circ} = c_{i\,\circ}^{\circ\,\alpha}(\delta_{\circ\,\beta}^{\alpha\,\circ}c_{\circ\,j}^{\beta\,\circ}) = c_{i\,\circ}^{\circ\,\alpha}c_{\circ\,j}^{\alpha\,\circ},$$

which implies

$$c_{i\,\circ}^{\circ\,\alpha} = \gamma_{\circ\,\alpha}^{i\,\circ} \to C^t = C^{-1}. \tag{11.63}$$

11.9.2 Tensor total Euclidean character in orthonormal bases

The proper Euclidean tensors, which have a tensor nature for any change-of-basis, be this or not of orthogonal matrix C, and which will be called "total" Euclidean nature, evidently can be considered in a restricted way, as Cartesian tensors, executing with them exclusively changes of basis of orthogonal matrix, in which case they exhibit its tensor nature, though in a certainly outstanding way.
Consider a tensor $T \in \left(\overset{2}{\underset{1}{\otimes}} E^n\right)(\mathbb{R})$ over the linear Euclidean space $E^n(\mathbb{R})$ of orthonormal basis $\{\vec{e}_\alpha\}$. We have

$$t_{\circ\,\circ}^{\alpha\beta} = g_{\circ\,\circ}^{\alpha\theta}g_{\circ\,\circ}^{\beta w}t_{\theta w}^{\circ\,\circ} = \delta_{\circ\,\circ}^{\alpha\theta}\delta_{\circ\,\circ}^{\beta w}t_{\theta w}^{\circ\,\circ} = t_{\alpha\beta}^{\circ\,\circ}$$

and

$$t^{\alpha\circ}_{\circ\beta} = g^{\alpha\theta}_{\circ\circ}t^{\circ\circ}_{\theta\beta} = \delta^{\alpha\theta}_{\circ\circ}t^{\circ\circ}_{\theta\beta} = t^{\circ\circ}_{\alpha\beta}$$

and also

$$t^{\circ\beta}_{\alpha\circ} = t^{\circ\circ}_{\alpha\theta}g^{\theta\beta}_{\circ\circ} = t^{\circ\circ}_{\alpha\theta}\delta^{\theta\beta}_{\circ\circ} = t^{\circ\circ}_{\alpha\beta}.$$

The conclusion is that

$$t^{\alpha\beta}_{\circ\circ} = t^{\circ\circ}_{\alpha\beta} = t^{\alpha\circ}_{\circ\beta} = t^{\circ\beta}_{\alpha\circ}, \qquad (11.64)$$

i.e., that all associated Euclidean tensors have the same values for the components of the diverse types of species, which constitutes an extension to tensors of property 2 of Section 11.9.1. They also have identical change-of-basis tensor relations:

$$\begin{aligned}
t^{ij}_{\circ\circ} &= t^{\alpha\beta}_{\circ\circ}\gamma^{i\circ}_{\circ\alpha}\gamma^{j\circ}_{\circ\beta}\\
t^{\circ\circ}_{ij} &= t^{\circ\circ}_{\alpha\beta}c^{\circ\alpha}_{i\circ}c^{\circ\beta}_{j\circ}\\
t^{i\circ}_{\circ j} &= t^{\alpha\circ}_{\circ\beta}\gamma^{i\circ}_{\circ\alpha}c^{\circ\beta}_{j\circ}\\
t^{\circ j}_{i\circ} &= t^{\circ\beta}_{\alpha\circ}c^{\circ\alpha}_{i\circ}\gamma^{j\circ}_{\circ\beta}.
\end{aligned} \qquad (11.65)$$

In effect, in agreement with Formulas (11.63) and (11.64), we have, for example with (11.65-2), that

$$t^{\circ\circ}_{ij} = t^{\circ\circ}_{\alpha\beta}c^{\circ\alpha}_{i\circ}c^{\circ\beta}_{j\circ} = t^{\circ\circ}_{\alpha\beta}\gamma^{i\circ}_{\circ\alpha}\gamma^{j\circ}_{\circ\beta} = t^{\alpha\beta}_{\circ\circ}\gamma^{i\circ}_{\circ\alpha}\gamma^{j\circ}_{\circ\beta},$$

and because of (11.65-1), we get

$$t^{\circ\circ}_{ij} = t^{\alpha\beta}_{\circ\circ}\gamma^{i\circ}_{\circ\alpha}\gamma^{j\circ}_{\circ\beta} = t^{ij}_{\circ\circ}$$

and transforming in a similar way (11.65-3) and (11.65-4) we arrive at

$$t^{\circ\circ}_{ij} = t^{ij}_{\circ\circ} = t^{i\circ}_{\circ j} = t^{\circ j}_{i\circ}, \qquad (11.66)$$

which is (11.64) but in the new basis.

Since there is only one type of components for all species it is common to give *always* the tensor in contravariant coordinates

$$T = t^{\alpha\beta}_{\circ\circ}\vec{e}_\alpha \otimes \vec{e}_\beta,$$

though the space be $(E^n_* \otimes E^n)(\mathbb{R})$, for example.

All that has been mentioned in this section for $\left(\overset{2}{\underset{1}{\otimes}} E^n\right)(\mathbb{R})$ can be generalized evidently for any Cartesian tensor in $\left(\overset{r}{\underset{1}{\otimes}} E^n\right)(\mathbb{R})$, referred to a Euclidean space $E^n(\mathbb{R})$ of orthonormal basis $\{\vec{e}_\alpha\}$.

11.9.3 Tensor partial Euclidean character in orthonormal bases

We intend to distinguish with this name the systems S of scalar components $s(\alpha_1, \alpha_2, \ldots, \alpha_r)$, such that they satisfy the change-of-basis partial tensor relation

$$s(i_1, i_2, \ldots, i_r) = s(\alpha_1, \alpha_2, \ldots, \alpha_r) c_{i_1 \; o}^{\circ \; \alpha_1} c_{i_2 \; o}^{\circ \; \alpha_2} \cdots c_{r \; o}^{\circ \alpha_r}, \tag{11.67}$$

and the *additional condition* in Formula (11.63):

$$c_{i_j \; o}^{\circ \; \alpha_j} = \gamma_{\circ \; \alpha_j}^{i_j \; \circ}, \forall \alpha_j, i_j \in I_n; \; \forall j \in I_r. \tag{11.68}$$

That is, they can be notated as partial Cartesian tensors, because they behave like that *only* for *orthogonal matrix* changes of basis in $E^n(\mathbb{R})$

$$s_{i_1 i_2 \cdots i_r}^{\circ \; \circ \; \cdots \; \circ} = s_{\alpha_1 \alpha_2 \cdots \alpha_r}^{\circ \; \circ \; \cdots \; \circ} c_{i_1 \; \circ}^{\circ \; \alpha_1} c_{i_2 \; \circ}^{\circ \; \alpha_2} \cdots c_{i_r \; \circ}^{\circ \; \alpha_r}; \tag{11.69}$$

where $C^t = [c_{i \; \circ}^{\circ \alpha}] = C^{-1}$.

These systems of scalars that manifest tensor nature, *only if* we perform changes of basis of matrix C, such that $C^t = C^{-1}$ in $E^n(\mathbb{R})$), but that do not satisfy the condition (11.69) for non-orthogonal matrices C, will be called "partial" Euclidean tensors, though as Cartesian tensors they are not different from the "total" Euclidean tensors.

The authors of this book hope the reader will have captured the great differences between both, even though there are many authors of physical applications who put all of them in the box of "Cartesian tensors" without making any more distinctions.

11.9.4 Rectangular Cartesian tensors

Rectangular Cartesian tensors are those tensors that are defined over a special Euclidean linear space, the OGS, the *ordinary geometric space*, that is, the basic vectors $\{\vec{e}_\alpha\}$ constitute n-rectangular polyhedra with the unit directional vectors, notated usually with specific letters $\vec{i}, \vec{j}, \vec{k}, \vec{\ell}$, etc., and obviously $G \equiv G^{-1} = I_n$.

Example 11.1 (Associated Euclidean tensors). Consider a linear space $V^3(\mathbb{R})$ connected with a fundamental metric tensor that, in the basis $\{\vec{e}_\alpha\}$ of such a space, is represented by the matrix

$$G \equiv [g_{\alpha\beta}^{\circ \; \circ}] = \begin{bmatrix} 1 & 1 & -1 \\ 1 & 2 & 0 \\ -1 & 0 & 3 \end{bmatrix}.$$

1. Examine if $V^3(\mathbb{R})$ is a pseudo-Euclidean space or a Euclidean space, and give it the adequate notation.

2. Give the associated tensors of the metric tensor given by its different species.
3. Consider a Euclidean tensor T defined over this space, with components
$$t^{\circ\circ\gamma}_{\alpha\beta\circ} = \alpha + \beta - 4\gamma.$$
   a) Give T by its contravariant components.
   b) Give T by its covariant components.
   c) Examine the symmetries of T, in all representations of the present example.

Solution:

1. We extract the Gram–Schmidt numbers of the matrix G:

$$N_0 = 1; \; N_1 = |g_{11}| = |1| = 1; \; N_2 = \begin{vmatrix} g_{11} & g_{12} \\ g_{21} & g_{22} \end{vmatrix} = \begin{vmatrix} 1 & 1 \\ 1 & 2 \end{vmatrix} = 1; \; N_3 = |G| = 1.$$

Then, the diagonal of the canonized matrix is $\hat{g}_{11} = \frac{N_1}{N_0} = 1; \; \hat{g}_{22} = \frac{N_2}{N_1} = 1; \; \hat{g}_{33} = \frac{N_3}{N_2} = 1.$

and the Sylvesterian matrix: $I_3 = \begin{bmatrix} 1 & 0 & 0 \\ 0 & 1 & 0 \\ 0 & 0 & 1 \end{bmatrix}$, with $n = r = \sigma = 3$.

It classifies as a positive definite quadratic form, so that it is a Euclidean space with notation $E^3(\mathbb{R})$.

2. The associated tensors of $G = [g^{\circ\circ}_{\alpha\beta}]$, are:
   a) The metric tensor in contra-covariant coordinates is

$$g^{\alpha\circ}_{\circ\beta} = g^{\alpha\theta}_{\circ\circ} \cdot g^{\circ\circ}_{\theta\beta} = \delta^{\alpha\circ}_{\circ\beta}; \quad [g^{\alpha\circ}_{\circ\beta}] = [\delta^{\alpha\circ}_{\circ\beta}] = \begin{bmatrix} 1 & 0 & 0 \\ 0 & 1 & 0 \\ 0 & 0 & 1 \end{bmatrix}.$$

   b) The metric tensor in contravariant coordinates is

$$[g^{\alpha\beta}_{\circ\circ}] = [g^{\circ\circ}_{\alpha\beta}]^{-1} = G^{-1} = \begin{bmatrix} 6 & -3 & 2 \\ -3 & 2 & -1 \\ 2 & -1 & 1 \end{bmatrix}.$$

   c) The metric tensor in cova-contravariant coordinates is

$$g^{\circ\beta}_{\alpha\circ} = g^{\circ\circ}_{\alpha\theta} \cdot g^{\theta\beta}_{\circ\circ} = \delta^{\circ\beta}_{\alpha\circ}; \quad [g^{\circ\beta}_{\alpha\circ}] = [\delta^{\circ\beta}_{\alpha\circ}] = \begin{bmatrix} 1 & 0 & 0 \\ 0 & 1 & 0 \\ 0 & 0 & 1 \end{bmatrix}.$$

3. We organize the tensor data $t^{\circ\circ\gamma}_{\alpha\beta\circ} = \alpha + \beta - 4\gamma$ in the matrix representation agreed upon in Axiom 4 of Section 2.5, where α = row block index, and β, γ = row and column indices of each block:

$$T = [t^{\circ\circ\gamma}_{\alpha\beta\circ}] = \begin{bmatrix} -2 & -6 & -10 \\ -1 & -5 & -9 \\ 0 & -4 & -8 \\ \hline -1 & -5 & -9 \\ 0 & -4 & .8 \\ 1 & -3 & -7 \\ \hline 0 & -4 & -8 \\ 1 & -3 & -7 \\ 2 & -2 & -6 \end{bmatrix}.$$

a) As is recommended in Formula (10.43) we notate the data tensor and the associated tensor with numbered dummy indices, and we relate them with the same criterion as that used in the given formula, to obtain in our case

$$t^{\beta_1\beta_2\beta_3}_{\circ\,\circ\,\circ} = g^{\beta_1\,\alpha_1}_{\circ\,\circ} \cdot g^{\beta_2\,\alpha_2}_{\circ\,\circ} \cdot g^{\beta_3\,\circ}_{\circ\,\alpha_3} \cdot t^{\circ\,\circ\,\alpha_3}_{\alpha_1\,\alpha_2\,\circ}, \qquad (11.70)$$

which can be interpreted in matrix form, as was done with (10.44):

$$T'_\sigma = (G^{-1} \otimes G^{-1} \otimes I_3) \bullet T_\sigma,$$

where $\sigma = n^r = 3^3 = 27$, that is,

$$Z'_{27,27} = \begin{bmatrix} 6 & -3 & 2 \\ -3 & 2 & -1 \\ 2 & -1 & 1 \end{bmatrix} \otimes \begin{bmatrix} 6 & -3 & 2 \\ -3 & 2 & -1 \\ 2 & -1 & 1 \end{bmatrix} \otimes \begin{bmatrix} 1 & 0 & 0 \\ 0 & 1 & 0 \\ 0 & 0 & 1 \end{bmatrix}$$

$$= \begin{pmatrix}
36 & 0 & 0 & -18 & 0 & 0 & 12 & 0 & 0 & -18 & 0 & 0 & 9 & 0 & 0 & -6 & 0 & 0 & 12 & 0 & 0 & -6 & 0 & 0 & 4 & 0 & 0 \\
0 & 36 & 0 & 0 & -18 & 0 & 0 & 12 & 0 & 0 & -18 & 0 & 0 & 9 & 0 & 0 & -6 & 0 & 0 & 12 & 0 & 0 & -6 & 0 & 0 & 4 & 0 \\
0 & 0 & 36 & 0 & 0 & -18 & 0 & 0 & 12 & 0 & 0 & -18 & 0 & 0 & 9 & 0 & 0 & -6 & 0 & 0 & 12 & 0 & 0 & -6 & 0 & 0 & 4 \\
-18 & 0 & 0 & 12 & 0 & 0 & -6 & 0 & 0 & 9 & 0 & 0 & -6 & 0 & 0 & 3 & 0 & 0 & -6 & 0 & 0 & 4 & 0 & 0 & -2 & 0 & 0 \\
0 & -18 & 0 & 0 & 12 & 0 & 0 & -6 & 0 & 0 & 9 & 0 & 0 & -6 & 0 & 0 & 3 & 0 & 0 & -6 & 0 & 0 & 4 & 0 & 0 & -2 & 0 \\
0 & 0 & -18 & 0 & 0 & 12 & 0 & 0 & -6 & 0 & 0 & 9 & 0 & 0 & -6 & 0 & 0 & 3 & 0 & 0 & -6 & 0 & 0 & 4 & 0 & 0 & -2 \\
12 & 0 & 0 & -6 & 0 & 0 & 6 & 0 & 0 & -6 & 0 & 0 & 3 & 0 & 0 & -3 & 0 & 0 & 4 & 0 & 0 & -2 & 0 & 0 & 2 & 0 & 0 \\
0 & 12 & 0 & 0 & -6 & 0 & 0 & 6 & 0 & 0 & -6 & 0 & 0 & 3 & 0 & 0 & -3 & 0 & 0 & 4 & 0 & 0 & -2 & 0 & 0 & 2 & 0 \\
0 & 0 & 12 & 0 & 0 & -6 & 0 & 0 & 6 & 0 & 0 & -6 & 0 & 0 & 3 & 0 & 0 & -3 & 0 & 0 & 4 & 0 & 0 & -2 & 0 & 0 & 2 \\
-18 & 0 & 0 & 9 & 0 & 0 & -6 & 0 & 0 & 12 & 0 & 0 & -6 & 0 & 0 & 4 & 0 & 0 & -6 & 0 & 0 & 3 & 0 & 0 & -2 & 0 & 0 \\
0 & -18 & 0 & 0 & 9 & 0 & 0 & -6 & 0 & 0 & 12 & 0 & 0 & -6 & 0 & 0 & 4 & 0 & 0 & -6 & 0 & 0 & 3 & 0 & 0 & -2 & 0 \\
0 & 0 & -18 & 0 & 0 & 9 & 0 & 0 & -6 & 0 & 0 & 12 & 0 & 0 & -6 & 0 & 0 & 4 & 0 & 0 & -6 & 0 & 0 & 3 & 0 & 0 & -2 \\
9 & 0 & 0 & -6 & 0 & 0 & 3 & 0 & 0 & -6 & 0 & 0 & 4 & 0 & 0 & -2 & 0 & 0 & 3 & 0 & 0 & -2 & 0 & 0 & 1 & 0 & 0 \\
0 & 9 & 0 & 0 & -6 & 0 & 0 & 3 & 0 & 0 & -6 & 0 & 0 & 4 & 0 & 0 & -2 & 0 & 0 & 3 & 0 & 0 & -2 & 0 & 0 & 1 & 0 \\
0 & 0 & 9 & 0 & 0 & -6 & 0 & 0 & 3 & 0 & 0 & -6 & 0 & 0 & 4 & 0 & 0 & -2 & 0 & 0 & 3 & 0 & 0 & -2 & 0 & 0 & 1 \\
-6 & 0 & 0 & 3 & 0 & 0 & -3 & 0 & 0 & 4 & 0 & 0 & -2 & 0 & 0 & 2 & 0 & 0 & -2 & 0 & 0 & 1 & 0 & 0 & -1 & 0 & 0 \\
0 & -6 & 0 & 0 & 3 & 0 & 0 & -3 & 0 & 0 & 4 & 0 & 0 & -2 & 0 & 0 & 2 & 0 & 0 & -2 & 0 & 0 & 1 & 0 & 0 & -1 & 0 \\
0 & 0 & -6 & 0 & 0 & 3 & 0 & 0 & -3 & 0 & 0 & 4 & 0 & 0 & -2 & 0 & 0 & 2 & 0 & 0 & -2 & 0 & 0 & 1 & 0 & 0 & -1 \\
12 & 0 & 0 & -6 & 0 & 0 & 4 & 0 & 0 & -6 & 0 & 0 & 3 & 0 & 0 & -2 & 0 & 0 & 6 & 0 & 0 & -3 & 0 & 0 & 2 & 0 & 0 \\
0 & 12 & 0 & 0 & -6 & 0 & 0 & 4 & 0 & 0 & -6 & 0 & 0 & 3 & 0 & 0 & -2 & 0 & 0 & 6 & 0 & 0 & -3 & 0 & 0 & 2 & 0 \\
0 & 0 & 12 & 0 & 0 & -6 & 0 & 0 & 4 & 0 & 0 & -6 & 0 & 0 & 3 & 0 & 0 & -2 & 0 & 0 & 6 & 0 & 0 & -3 & 0 & 0 & 2 \\
-6 & 0 & 0 & 4 & 0 & 0 & -2 & 0 & 0 & 3 & 0 & 0 & -2 & 0 & 0 & 1 & 0 & 0 & -3 & 0 & 0 & 2 & 0 & 0 & -1 & 0 & 0 \\
0 & -6 & 0 & 0 & 4 & 0 & 0 & -2 & 0 & 0 & 3 & 0 & 0 & -2 & 0 & 0 & 1 & 0 & 0 & -3 & 0 & 0 & 2 & 0 & 0 & -1 & 0 \\
0 & 0 & -6 & 0 & 0 & 4 & 0 & 0 & -2 & 0 & 0 & 3 & 0 & 0 & -2 & 0 & 0 & 1 & 0 & 0 & -3 & 0 & 0 & 2 & 0 & 0 & -1 \\
4 & 0 & 0 & -2 & 0 & 0 & 2 & 0 & 0 & -2 & 0 & 0 & 1 & 0 & 0 & -1 & 0 & 0 & 2 & 0 & 0 & -1 & 0 & 0 & 1 & 0 & 0 \\
0 & 4 & 0 & 0 & -2 & 0 & 0 & 2 & 0 & 0 & -2 & 0 & 0 & 1 & 0 & 0 & -1 & 0 & 0 & 2 & 0 & 0 & -1 & 0 & 0 & 1 & 0 \\
0 & 0 & 4 & 0 & 0 & -2 & 0 & 0 & 2 & 0 & 0 & -2 & 0 & 0 & 1 & 0 & 0 & -1 & 0 & 0 & 2 & 0 & 0 & -1 & 0 & 0 & 1
\end{pmatrix}$$

For the sake of convenience, we notate $T_{\sigma,1}$ horizontally:

$$T^t_{27,1} = [-2\,-6\,-10\,-1\,-5\,-9\,0\,-4\,-8\,|\,-1\,-5\,-9\,0\,-4\,-8\,1\,-3\,-7\,|\,0\,-4\,-8\,1\,-3\,-7\,2\,-2\,-6]$$
$$(11.71)$$

and we calculate $T'_{27,1}$, by means of the relation $T'_{27,1} = Z'_{27,27} \bullet T_{27,1}$. Notating $T'_{27,1}$ horizontally, we get

$$(T'_{27,1})^t = [A\,|\,B],$$

where

$$A = [\,-40\ -140\ -240\ 18\ 58\ 98\ -13\ -53\ -93\ 18\ 58\ 98\ -8\,]$$
$$B = [\,-24\ -40\ 6\ 22\ 38\ -13\ -53\ -93\ 6\ 22\ 38\ -4\ -20\ -36\,],$$

which once condensed leads to

$$[t^{\alpha\beta\gamma}_{\circ\circ\circ}] = \begin{bmatrix} -40 & -140 & -240 \\ 18 & 58 & 98 \\ -13 & -53 & -93 \\ 18 & 58 & 98 \\ -8 & -24 & -40 \\ 6 & 22 & 38 \\ -13 & -53 & -93 \\ 6 & 22 & 38 \\ -4 & -20 & -36 \end{bmatrix}.$$

b) Proceeding with a similar criterion, we have

$$t^{\circ\ \circ\ \circ}_{\beta_1\beta_2\beta_3} = g^{\circ\ \alpha_1}_{\beta_1\ \circ}\, g^{\circ\ \alpha_2}_{\beta_2\ \circ}\, g^{\circ\ \circ}_{\beta_3\alpha_3}\, t^{\circ\ \circ\ \alpha_3}_{\alpha_1\alpha_2\ \circ}, \qquad (11.72)$$

which in matrix form is

$$T''_{\sigma,1} = (I_3 \otimes I_3 \otimes G) \bullet T_{\sigma,1},$$

with $\sigma = 27$, and in our case

$$Z''_{27,27} = \begin{bmatrix} 1 & 0 & 0 \\ 0 & 1 & 0 \\ 0 & 0 & 1 \end{bmatrix} \otimes \begin{bmatrix} 1 & 0 & 0 \\ 0 & 1 & 0 \\ 0 & 0 & 1 \end{bmatrix} \otimes \begin{bmatrix} 1 & 1 & -1 \\ 1 & 2 & 0 \\ -1 & 0 & 3 \end{bmatrix}$$

$$=$$

```
 11-1 00 0 00 0 00 0 00 0 00 0 00 0 00 0 00 0
 12 0 00 0 00 0 00 0 00 0 00 0 00 0 00 0 00 0
-10 3 00 0 00 0 00 0 00 0 00 0 00 0 00 0 00 0
 00 0 11-1 00 0 00 0 00 0 00 0 00 0 00 0 00 0
 00 0 12 0 00 0 00 0 00 0 00 0 00 0 00 0 00 0
 00 0-10 3 00 0 00 0 00 0 00 0 00 0 00 0 00 0
 00 0 00 0 11-1 00 0 00 0 00 0 00 0 00 0 00 0
 00 0 00 0 12 0 00 0 00 0 00 0 00 0 00 0 00 0
 00 0 00 0-10 3 00 0 00 0 00 0 00 0 00 0 00 0
 00 0 00 0 00 0 11-1 00 0 00 0 00 0 00 0 00 0
 00 0 00 0 00 0 12 0 00 0 00 0 00 0 00 0 00 0
 00 0 00 0 00 0-10 3 00 0 00 0 00 0 00 0 00 0
 00 0 00 0 00 0 00 0 11-1 00 0 00 0 00 0 00 0
 00 0 00 0 00 0 00 0 12 0 00 0 00 0 00 0 00 0
 00 0 00 0 00 0 00 0-10 3 00 0 00 0 00 0 00 0
 00 0 00 0 00 0 00 0 00 0 11-1 00 0 00 0 00 0
 00 0 00 0 00 0 00 0 00 0 12 0 00 0 00 0 00 0
 00 0 00 0 00 0 00 0 00 0-10 3 00 0 00 0 00 0
 00 0 00 0 00 0 00 0 00 0 00 0 11-1 00 0 00 0
 00 0 00 0 00 0 00 0 00 0 00 0 12 0 00 0 00 0
 00 0 00 0 00 0 00 0 00 0 00 0-10 3 00 0 00 0
 00 0 00 0 00 0 00 0 00 0 00 0 00 0 11-1 00 0
 00 0 00 0 00 0 00 0 00 0 00 0 00 0 12 0 00 0
 00 0 00 0 00 0 00 0 00 0 00 0 00 0-10 3 00 0
 00 0 00 0 00 0 00 0 00 0 00 0 00 0 00 0 11-1
 00 0 00 0 00 0 00 0 00 0 00 0 00 0 00 0 12 0
 00 0 00 0 00 0 00 0 00 0 00 0 00 0 00 0-10 3
```

We calculate $T''_{27,1}$, using the relation $T''_{27,1} = Z''_{27,27} \bullet T_{27,1}$, and using horizontal notation for the result, we have

$$(T''_{27,1})^t = [A \mid B],$$

where

$$A = [\,2 \ {-14} \ {-28} \ 3 \ {-11} \ {-26} \ 4 \ {-8} \ {-24} \ 3 \ {-11} \ {-26} \ 4\,]$$
$$B = [\,{-8} \ {-24} \ 5 \ {-5} \ {-22} \ 4 \ {-8} \ {-24} \ 5 \ {-5} \ {-22} \ 6 \ {-2} \ {-20}\,],$$

which once condensed gives

$$[t^{\ \circ\circ\circ}_{\alpha\beta\gamma}] = \begin{bmatrix} 2 & -14 & -28 \\ 3 & -11 & -26 \\ 4 & -8 & -24 \\ \hline 3 & -11 & -26 \\ 4 & -8 & -24 \\ 5 & -5 & -22 \\ \hline 4 & -8 & -24 \\ 5 & -5 & -22 \\ 6 & -2 & -20 \end{bmatrix}.$$

c) Note that the data tensor T satisfies

$$t^{\circ\circ\gamma}_{12\circ} = t^{\circ\circ\gamma}_{21\circ}; \quad \text{in fact, if } \gamma = (1,2,3),\ t^{\circ\circ\gamma}_{12\circ} = (-1,-5,-9)$$

$$t^{\circ\circ\gamma}_{13\circ} = t^{\circ\circ\gamma}_{31\circ}; \quad \text{in fact, if } \gamma = (1,2,3),\ t^{\circ\circ\gamma}_{13\circ} = (0,-4,-8,)$$

$$t^{\circ\circ\gamma}_{23\circ} = t^{\circ\circ\gamma}_{32\circ}; \quad \text{in fact, if } \gamma = (1,2,3),\ t^{\circ\circ\gamma}_{32\circ} = (1,-3,-7).$$

Studying the tensor T in its representation $[t^{\alpha\beta\gamma}_{\ \circ\circ\circ}]$ from answer 3(a)), we get

$$t^{12\gamma}_{\circ\circ\circ} = t^{21\gamma}_{\circ\circ\circ}; \quad \text{in fact, if } \gamma = (1,2,3),\ t^{12\gamma}_{\circ\circ\circ} = (18,58,98)$$

$$t^{13\gamma}_{\circ\circ\circ} = t^{31\gamma}_{\circ\circ\circ}; \quad \text{in fact, if } \gamma = (1,2,3),\ t^{13\gamma}_{\circ\circ\circ} = (-13,-53,-93)$$

$$t^{23\gamma}_{\circ\circ\circ} = t^{32\gamma}_{\circ\circ\circ}; \quad \text{in fact, if } \gamma = (1,2,3),\ t^{23\gamma}_{\circ\circ\circ} = (6,22,38).$$

Finally, examining $T = [t^{\ \ \circ\circ\circ}_{\alpha\beta\gamma}]$ in answer 3(b)), we get

$$t^{\circ\circ\circ}_{12\gamma} = t^{\circ\circ\circ}_{21\gamma}; \quad \text{in fact, if } \gamma = (1,2,3),\ t^{\circ\circ\circ}_{12\gamma} = (3,-11,-26)$$

$$t^{\circ\circ\circ}_{13\gamma} = t^{\circ\circ\circ}_{31\gamma}; \quad \text{in fact, if } \gamma = (1,2,3),\ t^{\circ\circ\circ}_{13\gamma} = (4,-8,-24)$$

$$t^{\circ\circ\circ}_{23\gamma} = t^{\circ\circ\circ}_{32\gamma}; \quad \text{in fact, if } \gamma = (1,2,3),\ t^{\circ\circ\circ}_{23\gamma} = (5,-5,-22),$$

which finally confirms the "partially symmetric" character with respect to *the first two indices*, of the Euclidean tensor T, i.e., it presents this property not only in the analyzed representations but in all possible representations, and in any basis, because it is a tensor property.

□

Example 11.2 (Associated tensors with a change-of-basis). In a Euclidean space $E^n(\mathbb{R})$ referred to a basis $\{\vec{e}_\alpha\}$ the fundamental metric tensor is given in contravariant coordinates as

$$[g_{\circ\circ}^{\alpha\beta}] = \begin{bmatrix} 2 & 1 \\ 1 & 1 \end{bmatrix}.$$

A Euclidean tensor T of order $r = 2$

$$T = [t_{\circ\circ}^{\alpha\beta}] = \begin{bmatrix} 1 & 1 \\ 1 & 2 \end{bmatrix}$$

is considered over that space and a change-of-basis of matrix C, to the new basis of $E^2(\mathbb{R})$ $\{\vec{e}_i\}$:

$$C \equiv [c_{\circ i}^{\alpha\circ}] = \begin{bmatrix} 3 & 5 \\ 1 & 2 \end{bmatrix}$$

is performed.

1. Write of all possible forms the components of the tensor T in the initial basis.
2. Based on the data, write of all possible forms the components of the tensor T in the new basis.
3. Examine the symmetries of T.

Remark 11.1. Execute the operations using the classical method for question 1, and using the direct matrix method for question 2.

□

Solution:

1. The fundamental metric tensor in covariants is:

$$[g_{\alpha\beta}^{\circ\circ}] = [g_{\circ\circ}^{\alpha\beta}]^{-1} = \begin{bmatrix} 2 & 1 \\ 1 & 1 \end{bmatrix}^{-1} = \begin{bmatrix} 1 & -1 \\ -1 & 2 \end{bmatrix}.$$

The contra-cova components of T are:

$$t_{\circ\beta_2}^{\beta_1\circ} = g_{\circ\alpha_1}^{\beta_1\circ}g_{\beta_2\alpha_2}^{\circ\circ}t_{\circ\circ}^{\alpha_1\alpha_2} = \delta_{\circ\alpha_1}^{\beta_1\circ}t_{\circ\circ}^{\alpha_1\alpha_2}g_{\alpha_2\beta_2}^{\circ\circ};$$

$$[t_{\alpha\circ}^{\circ\beta}] = I_2 \bullet \begin{bmatrix} 1 & 1 \\ 1 & 2 \end{bmatrix}\begin{bmatrix} 1 & -1 \\ -1 & 2 \end{bmatrix}^t = \begin{bmatrix} 0 & 1 \\ -1 & 3 \end{bmatrix}.$$

The cova-contra components of T are:

$$t_{\circ\beta_2}^{\beta_1\circ} = g_{\beta_1\alpha_1}^{\circ\circ}g_{\circ\alpha_2}^{\beta_2\circ}t_{\circ\circ}^{\alpha_1\alpha_2} = g_{\beta_1\alpha_1}^{\circ\circ}t_{\circ\circ}^{\alpha_1\alpha_2}\delta_{\alpha_2\circ}^{\circ\beta_2};$$

$$[t_{\alpha\,\circ}^{\,\circ\,\beta}] = \begin{bmatrix} 1 & -1 \\ -1 & 2 \end{bmatrix}\begin{bmatrix} 1 & 1 \\ 1 & 2 \end{bmatrix}\begin{bmatrix} 1 & 0 \\ 0 & 1 \end{bmatrix} = \begin{bmatrix} 0 & -1 \\ 1 & 3 \end{bmatrix}.$$

The cova-cova components of T are:

$$t_{\beta_1\,\beta_2}^{\,\circ\,\circ} = g_{\beta_1\,\alpha_1}^{\circ\,\circ} g_{\beta_2\,\alpha_2}^{\circ\,\circ} t_{\,\circ\,\circ}^{\alpha_1\,\alpha_2} = g_{\beta_1\,\alpha_1}^{\circ\,\circ} t_{\,\circ\,\circ}^{\alpha_1\,\alpha_2} g_{\alpha_2\,\beta_2}^{\circ\,\circ};$$

$$[t_{\alpha\beta}^{\circ\circ}] = \begin{bmatrix} 1 & -1 \\ -1 & 2 \end{bmatrix}\begin{bmatrix} 1 & 1 \\ 1 & 2 \end{bmatrix}\begin{bmatrix} 1 & -1 \\ -1 & 2 \end{bmatrix}^{t} = \begin{bmatrix} 1 & -2 \\ -2 & 5 \end{bmatrix}.$$

2. The change-of-basis matrices to be used are:

$$C = [c_{\circ\,i}^{\alpha\,\circ}] \rightarrow C^{-1} = [\gamma_{\circ\alpha}^{i\,\circ}] = [\gamma_{\circ\beta}^{j\,\circ}] = \begin{bmatrix} 3 & 5 \\ 1 & 2 \end{bmatrix}^{-1} = \begin{bmatrix} 2 & -5 \\ -1 & 3 \end{bmatrix};$$

$$C^{t} = [c_{i\,\circ}^{\circ\alpha}] = [c_{j\,\circ}^{\circ\beta}] = \begin{bmatrix} 3 & 5 \\ 1 & 2 \end{bmatrix} = \begin{bmatrix} 3 & 1 \\ 5 & 2 \end{bmatrix}.$$

The new contra-contra components of T are:

$$t_{\circ\circ}^{ij} = t_{\circ\circ}^{\alpha\beta}\gamma_{\circ\alpha}^{i\,\circ}\gamma_{\circ\beta}^{j\,\circ};\ \sigma = n^{r} = 2^{2} = 4.$$

$$\hat{T} = (C^{-1} \otimes C^{-1}) \bullet T_{\sigma};$$

$$\hat{T}_{\sigma,1} = \left(\begin{bmatrix} 2 & -5 \\ -1 & 3 \end{bmatrix} \otimes \begin{bmatrix} 2 & -5 \\ -1 & 3 \end{bmatrix} \right) \bullet \begin{bmatrix} 1 \\ 1 \\ 1 \\ 2 \end{bmatrix}$$

$$= \begin{bmatrix} 4 & -10 & -10 & 25 \\ -2 & 6 & 5 & -15 \\ -2 & 5 & 6 & -15 \\ 1 & -3 & -3 & 9 \end{bmatrix}\begin{bmatrix} 1 \\ 1 \\ 1 \\ 2 \end{bmatrix} = \begin{bmatrix} 34 \\ -21 \\ -21 \\ 13 \end{bmatrix};$$

$$\hat{T} = [t_{\circ\circ}^{ij}] = \begin{bmatrix} 34 & -21 \\ -21 & 13 \end{bmatrix}.$$

The new contra-cova components of T are:

$$t_{\circ j}^{i\,\circ} = t_{\circ\beta}^{\alpha\,\circ}\gamma_{\circ\alpha}^{i\,\circ}c_{j\,\circ}^{\circ\beta};\ \hat{T}'_{\sigma} = (C^{-1} \otimes C^{t}) \bullet \begin{bmatrix} 0 \\ 1 \\ -1 \\ 3 \end{bmatrix};$$

$$\hat{T}'_{\sigma,1} = \left(\begin{bmatrix} 2 & -5 \\ -1 & 3 \end{bmatrix} \otimes \begin{bmatrix} 3 & 1 \\ 5 & 2 \end{bmatrix} \right) \bullet \begin{bmatrix} 0 \\ 1 \\ -1 \\ 3 \end{bmatrix}$$

$$= \begin{bmatrix} 6 & 2 & -15 & -5 \\ 10 & 4 & -25 & -10 \\ -3 & -1 & 9 & 3 \\ -5 & 2 & 15 & 6 \end{bmatrix}\begin{bmatrix} 0 \\ 1 \\ -1 \\ 3 \end{bmatrix} = \begin{bmatrix} 2 \\ -1 \\ -1 \\ 5 \end{bmatrix};$$

$$\hat{T} = [t_{oj}^{io}] = \begin{bmatrix} 2 & -1 \\ -1 & 5 \end{bmatrix}.$$

The new cova-contra components of T are:

$$t_{io}^{oj} = t_{\alpha o}^{o\beta} c_{io}^{o\alpha} \gamma_{o\beta}^{jo}; \quad \hat{T}_\sigma'' = (C^t \otimes C^{-1}) \bullet \begin{bmatrix} 0 \\ -1 \\ 1 \\ 3 \end{bmatrix};$$

$$\hat{T}_\sigma'' = \left(\begin{bmatrix} 3 & 1 \\ 5 & 2 \end{bmatrix} \otimes \begin{bmatrix} 2 & -5 \\ -1 & 3 \end{bmatrix} \right) \bullet \begin{bmatrix} 0 \\ -1 \\ 1 \\ 3 \end{bmatrix}$$

$$= \begin{bmatrix} 6 & -15 & 2 & -5 \\ -3 & 9 & -1 & 3 \\ 10 & -25 & 4 & -10 \\ -5 & 15 & -2 & 6 \end{bmatrix} \bullet \begin{bmatrix} 0 \\ -1 \\ 1 \\ 3 \end{bmatrix} = \begin{bmatrix} 2 \\ -1 \\ -1 \\ 1 \end{bmatrix};$$

$$\hat{T}'' = [t_{io}^{oj}] = \begin{bmatrix} 2 & -1 \\ -1 & 1 \end{bmatrix}.$$

The new cova-contra components of T are:

$$t_{ij}^{oo} = t_{\alpha\beta}^{oo} c_{io}^{o\alpha} c_{jo}^{o\beta}; \quad \hat{T}''' = (C^t \otimes C^t) \bullet \begin{bmatrix} 1 \\ -2 \\ -2 \\ 5 \end{bmatrix};$$

$$\hat{T}''' = \left(\begin{bmatrix} 3 & 1 \\ 5 & 2 \end{bmatrix} \otimes \begin{bmatrix} 3 & 1 \\ 5 & 2 \end{bmatrix} \right) \bullet \begin{bmatrix} 1 \\ -2 \\ -2 \\ 5 \end{bmatrix} = \begin{bmatrix} 9 & 3 & 3 & 1 \\ 15 & 6 & 5 & 2 \\ 15 & 5 & 6 & 2 \\ 25 & 10 & 10 & 4 \end{bmatrix} \bullet \begin{bmatrix} 1 \\ -2 \\ -2 \\ 5 \end{bmatrix} = \begin{bmatrix} 2 \\ 3 \\ 3 \\ 5 \end{bmatrix};$$

$$\hat{T}''' = [t_{ij}^{oo}] = \begin{bmatrix} 2 & 3 \\ 3 & 5 \end{bmatrix}.$$

3. The reader can easily see that

$$[t_{oo}^{\alpha\beta}]^t = [t_{oo}^{\alpha\beta}]; \quad [t_{ij}^{oo}]^t = [t_{ij}^{oo}],$$

which shows the symmetry of T.

As a practical complement, the reader can invert the order of the questions of the present exercise, by first executing the change-of-basis over the tensor $[t_{oo}^{\alpha\beta}]$ in order to determine $[t_{oo}^{ij}]$, and later proceed to the calculation of the Euclidean associated tensor of tensor $[t_{oo}^{ij}]$, that is, the $[t_{oj}^{io}], [t_{io}^{oj}]$ and $[t_{ij}^{oo}]$, and check that the results are coincident with those proposed in this exercise, as it demonstrates the theory in Section 11.7.

$\square$

Example 11.3 (Uniqueness of associated tensors). Show that if U, V and W are the associated tensors of a given Euclidean contravariant tensor $T \in \begin{pmatrix} 2 \\ \otimes E^n \\ 1 \end{pmatrix} (\mathbb{R})$ of second order there is no new associated tensor.

Solution: Let $T = t^{\alpha\beta}_{\circ\circ}\vec{e}_\alpha \otimes \vec{e}_\beta$ be the data contra-contravariant tensor, the associated tensors are

$$U = u^{\alpha\circ}_{\circ\beta} = t^{\alpha\theta}_{\circ\circ}g^{\circ\circ}_{\theta\beta} \quad \text{(contra-covariant)}$$

$$V = v^{\circ\beta}_{\alpha\circ} = g^{\circ\circ}_{\alpha\theta}t^{\theta\beta}_{\circ\circ} \quad \text{(cova-contravariant)}$$

$$W = w^{\circ\circ}_{\alpha\beta} = g^{\circ\circ}_{\alpha\theta}g^{\circ\circ}_{\beta w}t^{\theta w}_{\circ\circ} \quad \text{(cova-covariant)}.$$

If we obtain from $\vec{W}$ *another* contra–contravariant associated tensor, the result is

$$Z = z^{\alpha\beta}_{\circ\circ} = g^{\alpha\lambda}_{\circ\circ}g^{\beta\mu}_{\circ\circ}w^{\circ\circ}_{\lambda\mu} = g^{\alpha\lambda}_{\circ\circ}g^{\beta\mu}_{\circ\circ}\left(g^{\circ\circ}_{\lambda\theta}g^{\circ\circ}_{\mu w}t^{\theta w}_{\circ\circ}\right)$$
$$= \left(g^{\alpha\lambda}_{\circ\circ}g^{\circ\circ}_{\lambda\theta}\right)\left(g^{\beta\mu}_{\circ\circ}g^{\circ\circ}_{\mu w}\right)t^{\theta w}_{\circ\circ} = \delta^{\alpha\circ}_{\circ\theta}\delta^{\beta\circ}_{\circ w}t^{\theta w}_{\circ\circ} = t^{\alpha\beta}_{\circ\circ},$$

which shows that $Z \equiv T$ and that we return to the initial tensor.   □

Example 11.4 (Symmetries through pure associated tensors). In a Euclidean space $E^3(\mathbb{R})$ referred to a basis $\{\vec{e}_\alpha\}$, the fundamental metric tensor in covariant coordinates is represented by the matrix

$$G \equiv [g^{\circ\circ}_{\alpha\beta}] = \begin{bmatrix} 1 & 1 & 1 \\ 1 & 2 & 1 \\ 1 & 1 & 2 \end{bmatrix}.$$

A certain Euclidean tensor $T \in \left(E^3 \otimes E^3 \otimes E^3_*\right)(\mathbb{R})$, is given by its components

$$[t^{\alpha\beta\circ}_{\circ\circ\gamma}] = \begin{bmatrix} 6 & 8 & 7 \\ 1 & 0 & 1 \\ -1 & -1 & -3 \\ \hline 1 & 0 & 1 \\ 7 & 10 & 12 \\ 9 & 14 & 13 \\ \hline -1 & -1 & -3 \\ 9 & 14 & 13 \\ 5 & 9 & 8 \end{bmatrix},$$

where α is the block row, β is the row of each block and γ is the column of each block.

1. Verify that the space $E^3(\mathbb{R})$ is Euclidean.

2. Obtain the fundamental metric tensor in contravariant coordinates.
3. Obtain the "associated" tensor of T in contravariant coordinates.
4. Obtain the "associated" tensor of T in covariant coordinates.
5. Obtain the symmetries and anti-symmetries of T.
6. We execute a change-of-basis in $E^3(\mathbb{R})$, of matrix

$$C \equiv [c_{\circ i}^{\alpha \circ}] = \begin{bmatrix} 1 & -1 & 0 \\ 0 & 1 & -1 \\ 0 & 0 & 1 \end{bmatrix}.$$

Give the associated tensor of T in the new basis, in contravariant coordinates.

Solution:

1. The Gram–Schmidt numbers of the fundamental metric tensor matrix are

$$\Gamma_0 = 1; \quad \Gamma_1 = 1; \quad \Gamma_2 = \begin{vmatrix} 1 & 1 \\ 1 & 2 \end{vmatrix} = 1; \quad \Gamma_3 = |G| = \begin{vmatrix} 1 & 1 & 1 \\ 1 & 2 & 1 \\ 1 & 1 & 2 \end{vmatrix} = 1$$

and since $\forall \Gamma_i > 0$, the space is Euclidean.

2.
$$[g_{\circ \circ}^{\alpha \beta}] = [g_{\alpha \beta}^{\circ \circ}]^{-1} \equiv G^{-1} = \begin{bmatrix} 1 & 1 & 1 \\ 1 & 2 & 1 \\ 1 & 1 & 2 \end{bmatrix}^{-1} = \begin{bmatrix} 3 & -1 & -1 \\ -1 & 1 & 0 \\ -1 & 0 & 1 \end{bmatrix}.$$

3. To obtain the tensor in contravariant coordinates, we raise the index γ. We solve the problem by the classic method

$$t_{\circ \circ \circ}^{\alpha \beta \gamma} = t_{\circ \circ \theta}^{\alpha \beta \circ} g_{\circ \circ}^{\theta \gamma},$$

and operating, we get

$$[t_{\circ \circ \circ}^{1 \beta \gamma}] = [t_{\circ \circ \theta}^{1 \beta \circ}][g_{\circ \circ}^{\theta \gamma}] = \begin{bmatrix} 6 & 8 & 7 \\ 1 & 0 & 1 \\ -1 & -1 & -3 \end{bmatrix} \begin{bmatrix} 3 & -1 & -1 \\ -1 & 1 & 0 \\ -1 & 0 & 1 \end{bmatrix} = \begin{bmatrix} 3 & 2 & 1 \\ 2 & -1 & 0 \\ 1 & 0 & -2 \end{bmatrix}$$

$$[t_{\circ \circ \circ}^{2 \beta \gamma}] = [t_{\circ \circ \theta}^{2 \beta \circ}][g_{\circ \circ}^{\theta \gamma}] = \begin{bmatrix} 1 & 0 & 1 \\ 7 & 10 & 12 \\ 9 & 14 & 13 \end{bmatrix} \begin{bmatrix} 3 & -1 & -1 \\ -1 & 1 & 0 \\ -1 & 0 & 1 \end{bmatrix} = \begin{bmatrix} 2 & -1 & 0 \\ -1 & 3 & 5 \\ 0 & 5 & 4 \end{bmatrix}$$

$$[t_{\circ \circ \circ}^{3 \beta \gamma}] = [t_{\circ \circ \theta}^{3 \beta \circ}][g_{\circ \circ}^{\theta \gamma}] = \begin{bmatrix} -1 & -1 & 3 \\ 9 & 14 & 13 \\ 5 & 9 & 8 \end{bmatrix} \begin{bmatrix} 3 & -1 & -1 \\ -1 & 1 & 0 \\ -1 & 0 & 1 \end{bmatrix} = \begin{bmatrix} 1 & 0 & -2 \\ 0 & 5 & 4 \\ -2 & 4 & 3 \end{bmatrix}$$

whence

$$[t_{\circ \circ \circ}^{\alpha \beta \gamma}] = \begin{bmatrix} 3 & 2 & 1 \\ 2 & -1 & 0 \\ 1 & 0 & -2 \\ \hline 2 & -1 & 0 \\ -1 & 3 & 5 \\ 0 & 5 & 4 \\ \hline 1 & 0 & -2 \\ 0 & 5 & 4 \\ -2 & 4 & 3 \end{bmatrix}.$$

4. To obtain the tensor in covariant coordinates, from the data tensor, we lower the indices α and β. We solve the problem by the direct method. As the ordering is decisive in this method, we change the subindex notation to another numbered notation

$$t^{\circ\ \circ\ \circ}_{\beta_1\beta_2\beta_3} = g^{\circ\ \circ}_{\beta_1\alpha_1} g^{\circ\ \circ}_{\beta_2\alpha_2} g^{\circ\ \alpha_3}_{\beta_3\circ} t^{\alpha_1\alpha_2\ \circ}_{\circ\ \circ\ \alpha_3}$$

and we apply the Formula (11.45) adapted to our case. We have $\sigma = n^r = 3^3 = 27$ and $T'_{27,1} = (G \otimes G \otimes I_3) \bullet T_{27,1}$ with

$$(T_{27,1})^t = [A \,|\, B \,|\, C]$$

where

$$A = [\,6\ 8\ 7\ 1\ 0\ 1\ -1\ -1\ 3\,]$$
$$B = [\,1\ 0\ 1\ 7\ 10\ 12\ 9\ 14\ 13\,]$$
$$C = [\,-1\ -1\ -3\ 9\ 14\ 13\ 5\ 9\ 8\,]$$

and

$$Z_1 = G \otimes G \otimes I_3 = \begin{bmatrix} 1 & 1 & 1 \\ 1 & 2 & 1 \\ 1 & 1 & 2 \end{bmatrix} \otimes \begin{bmatrix} 1 & 1 & 1 \\ 1 & 2 & 1 \\ 1 & 1 & 2 \end{bmatrix} \otimes \begin{bmatrix} 1 & 0 & 0 \\ 0 & 1 & 0 \\ 0 & 0 & 1 \end{bmatrix}$$

$$=$$

```
1 0 0 1 0 0 1 0 0 1 0 0 1 0 0 1 0 0 1 0 0 1 0 0 1 0 0
0 1 0 0 1 0 0 1 0 0 1 0 0 1 0 0 1 0 0 1 0 0 1 0 0 1 0
0 0 1 0 0 1 0 0 1 0 0 1 0 0 1 0 0 1 0 0 1 0 0 1 0 0 1
1 0 0 2 0 0 1 0 0 1 0 0 2 0 0 1 0 0 1 0 0 2 0 0 1 0 0
0 1 0 0 2 0 0 1 0 0 1 0 0 2 0 0 1 0 0 1 0 0 2 0 0 1 0
0 0 1 0 0 2 0 0 1 0 0 1 0 0 2 0 0 1 0 0 1 0 0 2 0 0 1
1 0 0 1 0 0 2 0 0 1 0 0 1 0 0 2 0 0 1 0 0 1 0 0 2 0 0
0 1 0 0 1 0 0 2 0 0 1 0 0 1 0 0 2 0 0 1 0 0 1 0 0 2 0
0 0 1 0 0 1 0 0 2 0 0 1 0 0 1 0 0 2 0 0 1 0 0 1 0 0 2
1 0 0 1 0 0 1 0 0 2 0 0 2 0 0 2 0 0 1 0 0 1 0 0 1 0 0
0 1 0 0 1 0 0 1 0 0 2 0 0 2 0 0 2 0 0 1 0 0 1 0 0 1 0
0 0 1 0 0 1 0 0 1 0 0 2 0 0 2 0 0 2 0 0 1 0 0 1 0 0 1
1 0 0 2 0 0 1 0 0 2 0 0 4 0 0 2 0 0 1 0 0 2 0 0 1 0 0
0 1 0 0 2 0 0 1 0 0 2 0 0 4 0 0 2 0 0 1 0 0 2 0 0 1 0
0 0 1 0 0 2 0 0 1 0 0 2 0 0 4 0 0 2 0 0 1 0 0 2 0 0 1
1 0 0 1 0 0 2 0 0 2 0 0 2 0 0 4 0 0 1 0 0 1 0 0 2 0 0
0 1 0 0 1 0 0 2 0 0 2 0 0 2 0 0 4 0 0 1 0 0 1 0 0 2 0
0 0 1 0 0 1 0 0 2 0 0 2 0 0 2 0 0 4 0 0 1 0 0 1 0 0 2
1 0 0 1 0 0 1 0 0 1 0 0 1 0 0 1 0 0 2 0 0 2 0 0 2 0 0
0 1 0 0 1 0 0 1 0 0 1 0 0 1 0 0 1 0 0 2 0 0 2 0 0 2 0
0 0 1 0 0 1 0 0 1 0 0 1 0 0 1 0 0 1 0 0 2 0 0 2 0 0 2
1 0 0 2 0 0 1 0 0 1 0 0 2 0 0 1 0 0 2 0 0 4 0 0 2 0 0
0 1 0 0 2 0 0 1 0 0 1 0 0 2 0 0 1 0 0 2 0 0 4 0 0 2 0
0 0 1 0 0 2 0 0 1 0 0 1 0 0 2 0 0 1 0 0 2 0 0 4 0 0 2
1 0 0 1 0 0 2 0 0 1 0 0 1 0 0 2 0 0 2 0 0 2 0 0 4 0 0
0 1 0 0 1 0 0 2 0 0 1 0 0 1 0 0 2 0 0 2 0 0 2 0 0 4 0
0 0 1 0 0 1 0 0 2 0 0 1 0 0 1 0 0 2 0 0 2 0 0 2 0 0 4
```
$;$

$$T'_{27,1} = Z_1 \bullet T_{27,1},$$

resulting in

$$(T'_{27,1})^t = [A \,|\, B \,|\, C],$$

where

$$A = [\,36\ 53\ 49\ 53\ 77\ 75\ 49\ 75\ 67\,]$$
$$B = [\,53\ 77\ 75\ 77\ 111\ 113\ 75\ 113\ 106\,]$$
$$C = [\,49\ 75\ 67\ 75\ 113\ 106\ 67\ 106\ 93\,],$$

which, once condensed, leads to the sought after tensor

$$[t_{\alpha\beta\gamma}^{\circ\circ\circ}] = \begin{bmatrix} 36 & 53 & 49 \\ 53 & 77 & 75 \\ 49 & 75 & 67 \\ 53 & 77 & 75 \\ 77 & 111 & 113 \\ 75 & 113 & 106 \\ 49 & 75 & 67 \\ 75 & 113 & 106 \\ 67 & 106 & 93 \end{bmatrix}.$$

5. Observing the associated tensor in contravariant coordinates, we know that the indices that can present or not present symmetries are

$$t_{\circ\circ\circ}^{112} = t_{\circ\circ\circ}^{121} = t_{\circ\circ\circ}^{211} = 2;\quad t_{\circ\circ\circ}^{113} = t_{\circ\circ\circ}^{131} = t_{\circ\circ\circ}^{311} = 1;\quad t_{\circ\circ\circ}^{133} = t_{\circ\circ\circ}^{313} = t_{\circ\circ\circ}^{331} = -2;$$
$$t_{\circ\circ\circ}^{122} = t_{\circ\circ\circ}^{212} = t_{\circ\circ\circ}^{221} = -1;\quad t_{\circ\circ\circ}^{123} = t_{\circ\circ\circ}^{132} = t_{\circ\circ\circ}^{213} = t_{\circ\circ\circ}^{231} = t_{\circ\circ\circ}^{312} = t_{\circ\circ\circ}^{321} = 0;$$
$$t_{\circ\circ\circ}^{223} = t_{\circ\circ\circ}^{232} = t_{\circ\circ\circ}^{322} = 5;\quad t_{\circ\circ\circ}^{233} = t_{\circ\circ\circ}^{323} = t_{\circ\circ\circ}^{332} = 4,$$

which proves that T is a *symmetric* tensor. If desired, its symmetry can be verified over the covariant associated tensor.

6. Since we already have the tensor in contravariant coordinates, it suffices to execute the change-of-basis, which will be treated by the direct method

$$t_{\circ\circ\circ}^{ijk} = t_{\circ\circ\circ}^{\alpha\beta\gamma}\gamma_{\circ\alpha}^{i\circ}\gamma_{\circ\beta}^{j\circ}\gamma_{\circ\gamma}^{k\circ};\quad C^{-1} = \begin{bmatrix} 1 & 1 & 1 \\ 0 & 1 & 1 \\ 0 & 0 & 1 \end{bmatrix};$$

$$\hat{T}_{27,1} = \left(C^{-1}\otimes C^{-1}\otimes C^{-1}\right)\bullet T_{27,1}$$
$$T_{27,1}^t = [3212-1010-2\,|\,2-10-135054\,|\,10-2054-243]$$

$$Z_2 = C^{-1}\otimes C^{-1}\otimes C^{-1} = \begin{bmatrix} 1 & 1 & 1 \\ 0 & 1 & 1 \\ 0 & 0 & 1 \end{bmatrix}\otimes\begin{bmatrix} 1 & 1 & 1 \\ 0 & 1 & 1 \\ 0 & 0 & 1 \end{bmatrix}\otimes\begin{bmatrix} 1 & 1 & 1 \\ 0 & 1 & 1 \\ 0 & 0 & 1 \end{bmatrix}$$

$$
= \begin{pmatrix}
1 & 1 & 1 & 1 & 1 & 1 & 1 & 1 & 1 & 1 & 1 & 1 & 1 & 1 & 1 & 1 & 1 & 1 & 1 & 1 & 1 & 1 & 1 & 1 & 1 & 1 & 1 \\
0 & 1 & 1 & 0 & 1 & 1 & 0 & 1 & 1 & 0 & 1 & 1 & 0 & 1 & 1 & 0 & 1 & 1 & 0 & 1 & 1 & 0 & 1 & 1 & 0 & 1 & 1 \\
0 & 0 & 1 & 0 & 0 & 1 & 0 & 0 & 1 & 0 & 0 & 1 & 0 & 0 & 1 & 0 & 0 & 1 & 0 & 0 & 1 & 0 & 0 & 1 & 0 & 0 & 1 \\
0 & 0 & 0 & 1 & 1 & 1 & 1 & 1 & 1 & 0 & 0 & 0 & 1 & 1 & 1 & 1 & 1 & 1 & 0 & 0 & 0 & 1 & 1 & 1 & 1 & 1 & 1 \\
0 & 0 & 0 & 0 & 1 & 1 & 0 & 1 & 1 & 0 & 0 & 0 & 0 & 1 & 1 & 0 & 1 & 1 & 0 & 0 & 0 & 0 & 1 & 1 & 0 & 1 & 1 \\
0 & 0 & 0 & 0 & 0 & 1 & 0 & 0 & 1 & 0 & 0 & 0 & 0 & 0 & 1 & 0 & 0 & 1 & 0 & 0 & 0 & 0 & 0 & 1 & 0 & 0 & 1 \\
0 & 0 & 0 & 0 & 0 & 0 & 1 & 1 & 1 & 0 & 0 & 0 & 0 & 0 & 0 & 1 & 1 & 1 & 0 & 0 & 0 & 0 & 0 & 0 & 1 & 1 & 1 \\
0 & 0 & 0 & 0 & 0 & 0 & 0 & 1 & 1 & 0 & 0 & 0 & 0 & 0 & 0 & 0 & 1 & 1 & 0 & 0 & 0 & 0 & 0 & 0 & 0 & 1 & 1 \\
0 & 0 & 0 & 0 & 0 & 0 & 0 & 0 & 1 & 0 & 0 & 0 & 0 & 0 & 0 & 0 & 0 & 1 & 0 & 0 & 0 & 0 & 0 & 0 & 0 & 0 & 1 \\
0 & 0 & 0 & 0 & 0 & 0 & 0 & 0 & 0 & 1 & 1 & 1 & 1 & 1 & 1 & 1 & 1 & 1 & 1 & 1 & 1 & 1 & 1 & 1 & 1 & 1 & 1 \\
0 & 0 & 0 & 0 & 0 & 0 & 0 & 0 & 0 & 0 & 1 & 1 & 0 & 1 & 1 & 0 & 1 & 1 & 0 & 1 & 1 & 0 & 1 & 1 & 0 & 1 & 1 \\
0 & 0 & 0 & 0 & 0 & 0 & 0 & 0 & 0 & 0 & 0 & 1 & 0 & 0 & 1 & 0 & 0 & 1 & 0 & 0 & 1 & 0 & 0 & 1 & 0 & 0 & 1 \\
0 & 0 & 0 & 0 & 0 & 0 & 0 & 0 & 0 & 0 & 0 & 0 & 1 & 1 & 1 & 1 & 1 & 1 & 0 & 0 & 0 & 1 & 1 & 1 & 1 & 1 & 1 \\
0 & 0 & 0 & 0 & 0 & 0 & 0 & 0 & 0 & 0 & 0 & 0 & 0 & 1 & 1 & 0 & 1 & 1 & 0 & 0 & 0 & 0 & 1 & 1 & 0 & 1 & 1 \\
0 & 0 & 0 & 0 & 0 & 0 & 0 & 0 & 0 & 0 & 0 & 0 & 0 & 0 & 1 & 0 & 0 & 1 & 0 & 0 & 0 & 0 & 0 & 1 & 0 & 0 & 1 \\
0 & 0 & 0 & 0 & 0 & 0 & 0 & 0 & 0 & 0 & 0 & 0 & 0 & 0 & 0 & 1 & 1 & 1 & 0 & 0 & 0 & 0 & 0 & 0 & 1 & 1 & 1 \\
0 & 0 & 0 & 0 & 0 & 0 & 0 & 0 & 0 & 0 & 0 & 0 & 0 & 0 & 0 & 0 & 1 & 1 & 0 & 0 & 0 & 0 & 0 & 0 & 0 & 1 & 1 \\
0 & 0 & 0 & 0 & 0 & 0 & 0 & 0 & 0 & 0 & 0 & 0 & 0 & 0 & 0 & 0 & 0 & 1 & 0 & 0 & 0 & 0 & 0 & 0 & 0 & 0 & 1 \\
0 & 0 & 0 & 0 & 0 & 0 & 0 & 0 & 0 & 0 & 0 & 0 & 0 & 0 & 0 & 0 & 0 & 0 & 1 & 1 & 1 & 1 & 1 & 1 & 1 & 1 & 1 \\
0 & 0 & 0 & 0 & 0 & 0 & 0 & 0 & 0 & 0 & 0 & 0 & 0 & 0 & 0 & 0 & 0 & 0 & 0 & 1 & 1 & 0 & 1 & 1 & 0 & 1 & 1 \\
0 & 0 & 0 & 0 & 0 & 0 & 0 & 0 & 0 & 0 & 0 & 0 & 0 & 0 & 0 & 0 & 0 & 0 & 0 & 0 & 1 & 0 & 0 & 1 & 0 & 0 & 1 \\
0 & 0 & 0 & 0 & 0 & 0 & 0 & 0 & 0 & 0 & 0 & 0 & 0 & 0 & 0 & 0 & 0 & 0 & 0 & 0 & 0 & 1 & 1 & 1 & 1 & 1 & 1 \\
0 & 0 & 0 & 0 & 0 & 0 & 0 & 0 & 0 & 0 & 0 & 0 & 0 & 0 & 0 & 0 & 0 & 0 & 0 & 0 & 0 & 0 & 1 & 1 & 0 & 1 & 1 \\
0 & 0 & 0 & 0 & 0 & 0 & 0 & 0 & 0 & 0 & 0 & 0 & 0 & 0 & 0 & 0 & 0 & 0 & 0 & 0 & 0 & 0 & 0 & 1 & 0 & 0 & 1 \\
0 & 0 & 0 & 0 & 0 & 0 & 0 & 0 & 0 & 0 & 0 & 0 & 0 & 0 & 0 & 0 & 0 & 0 & 0 & 0 & 0 & 0 & 0 & 0 & 1 & 1 & 1 \\
0 & 0 & 0 & 0 & 0 & 0 & 0 & 0 & 0 & 0 & 0 & 0 & 0 & 0 & 0 & 0 & 0 & 0 & 0 & 0 & 0 & 0 & 0 & 0 & 0 & 1 & 1 \\
0 & 0 & 0 & 0 & 0 & 0 & 0 & 0 & 0 & 0 & 0 & 0 & 0 & 0 & 0 & 0 & 0 & 0 & 0 & 0 & 0 & 0 & 0 & 0 & 0 & 0 & 1
\end{pmatrix}
$$

$$
\hat{T}_{27,1} = Z_2 \bullet T_{27,1}
$$

$$
= [36\,30\,13\,30\,30\,14\,13\,14\,5\,|\,30\,30\,14\,30\,33\,16\,14\,16\,7\,|\,13\,14\,5\,14\,16\,7\,5\,7\,3]^t
$$

which, once condensed, leads to

$$
[t^{i\,j\,k}_{\circ\circ\circ}] =
\left[\begin{array}{ccc}
36 & 30 & 13 \\
30 & 30 & 14 \\
13 & 14 & 5 \\
\hline
30 & 30 & 14 \\
30 & 33 & 16 \\
14 & 16 & 7 \\
\hline
13 & 14 & 5 \\
14 & 16 & 7 \\
5 & 7 & 3
\end{array}\right],
$$

which shows again the symmetric nature, an invariant quality of the tensor.

□

Example 11.5 (Associated tensor in different bases). Over the Euclidean space $E^n(\mathbb{R})$, which is referred to two bases, an initial basis $\{\vec{e}_\alpha\}$ and another final basis $\{\vec{e}_i\}$, a Euclidean tensor T of order $r = 2$ is built.

1. We wish to establish all necessary tensor relations to know each and every one of the associated tensors of T in the new basis, built from each associated tensor in the initial basis. A total of 16 relations.
2. Prepare the calculus of the second, seventh and the last relations by the direct method.

Solution:

Data: Fundamental metric tensor: $G = [g^{\circ\,\circ}_{\alpha\beta}]$;

Change of basis: $C = [c^{\alpha\,\circ}_{\circ\,i}]$.

1. The desired relations are:

$$t^{ij}_{\circ\circ} = \qquad t^{\alpha\beta}_{\circ\circ}\gamma^{i\,\circ}_{\circ\alpha}\gamma^{j\,\circ}_{\circ\beta} \qquad (\text{data } t^{\alpha\beta}_{\circ\circ})$$

$$t^{ij}_{\circ\circ} = \left(g^{\beta\theta}_{\circ\circ}t^{\alpha\circ}_{\circ\theta}\right)\gamma^{i\,\circ}_{\circ\alpha}\gamma^{j\,\circ}_{\circ\beta} \quad (\text{data } t^{\alpha\circ}_{\circ\beta})$$

$$t^{ij}_{\circ\circ} = \left(g^{\alpha\theta}_{\circ\circ}t^{\circ\beta}_{\theta\circ}\right)\gamma^{i\,\circ}_{\circ\alpha}\gamma^{j\,\circ}_{\circ\beta} \quad (\text{data } t^{\circ\beta}_{\alpha\circ})$$

$$t^{ij}_{\circ\circ} = \left(g^{\alpha\theta}_{\circ\circ}g^{\beta\lambda}_{\circ\circ}t^{\circ\circ}_{\theta\lambda}\right)\gamma^{i\,\circ}_{\circ\alpha}\gamma^{j\,\circ}_{\circ\beta} \quad (\text{data } t^{\circ\circ}_{\alpha\beta})$$

$$t^{i\,\circ}_{\circ j} = \left(g^{\circ\circ}_{\beta\theta}t^{\alpha\theta}_{\circ\circ}\right)\gamma^{i\,\circ}_{\circ\alpha}c^{\circ\beta}_{j\,\circ} \quad (\text{data } t^{\alpha\beta}_{\circ\circ})$$

$$t^{i\,\circ}_{\circ j} = \qquad t^{\alpha\circ}_{\circ\beta}\gamma^{i\,\circ}_{\circ\alpha}c^{\circ\beta}_{j\,\circ} \qquad (\text{data } t^{\alpha\circ}_{\circ\beta})$$

$$t^{i\,\circ}_{\circ j} = \left(g^{\alpha\theta}_{\circ\circ}g^{\circ\circ}_{\beta\lambda}t^{\circ\lambda}_{\theta\circ}\right)\gamma^{i\,\circ}_{\circ\alpha}c^{\circ\beta}_{j\,\circ} \quad (\text{data } t^{\circ\beta}_{\alpha\circ})$$

$$t^{i\,\circ}_{\circ j} = \left(g^{\alpha\theta}_{\circ\circ}t^{\circ\circ}_{\theta\beta}\right)\gamma^{i\,\circ}_{\circ\alpha}c^{\circ\beta}_{j\,\circ} \quad (\text{data } t^{\circ\circ}_{\alpha\beta})$$

$$t^{\circ j}_{i\,\circ} = \left(g^{\circ\circ}_{\alpha\theta}t^{\theta\beta}_{\circ\circ}\right)c^{\circ\alpha}_{i\,\circ}\gamma^{j\,\circ}_{\circ\beta} \quad (\text{data } t^{\alpha\beta}_{\circ\circ})$$

$$t^{\circ j}_{i\,\circ} = \left(g^{\circ\circ}_{\alpha\theta}g^{\beta\lambda}_{\circ\circ}t^{\theta\circ}_{\circ\lambda}\right)c^{\circ\alpha}_{i\,\circ}\gamma^{j\,\circ}_{\circ\beta} \quad (\text{data } t^{\alpha\circ}_{\circ\beta})$$

$$t^{\circ j}_{i\,\circ} = \qquad t^{\circ\beta}_{\alpha\circ}c^{\circ\alpha}_{i\,\circ}\gamma^{j\,\circ}_{\circ\beta} \qquad (\text{data } t^{\circ\beta}_{\alpha\circ})$$

$$t^{\circ j}_{i\,\circ} = \left(g^{\beta\theta}_{\circ\circ}t^{\circ\circ}_{\alpha\theta}\right)c^{\circ\alpha}_{i\,\circ}\gamma^{j\,\circ}_{\circ\beta} \quad (\text{data } t^{\circ\circ}_{\alpha\beta})$$

$$t^{\circ\circ}_{ij} = \left(g^{\circ\circ}_{\alpha\theta}g^{\circ\circ}_{\beta\lambda}t^{\theta\lambda}_{\circ\circ}\right)c^{\circ\alpha}_{i\,\circ}c^{\circ\beta}_{i\,\circ} \quad (\text{data } t^{\alpha\beta}_{\circ\circ})$$

$$t^{\circ\circ}_{ij} = \left(g^{\circ\circ}_{\alpha\theta}t^{\theta\circ}_{\circ\beta}\right)c^{\circ\alpha}_{i\,\circ}c^{\circ\beta}_{j\,\circ} \quad (\text{data } t^{\alpha\circ}_{\circ\beta})$$

$$t^{\circ\circ}_{ij} = \left(g^{\circ\circ}_{\beta\theta}t^{\circ\theta}_{\alpha\circ}\right)c^{\circ\alpha}_{i\,\circ}c^{\circ\beta}_{j\,\circ} \quad (\text{data } t^{\circ\beta}_{\alpha\circ})$$

$$t^{\circ\circ}_{ij} = \qquad t^{\circ\circ}_{\alpha\beta}c^{\circ\alpha}_{i\,\circ}c^{\circ\beta}_{j\,\circ} \qquad (\text{data } t^{\circ\circ}_{\alpha\beta}).$$

2. We remind the reader of the need to use numbered indices.

Second relation: data tensor $[t^{\alpha\circ}_{\circ\beta}]$

$$t^{ij}_{\circ\circ} = \left(g^{\beta\theta}_{\circ\circ}t^{\alpha\circ}_{\circ\theta}\right)\gamma^{i\alpha}_{\circ\circ}\gamma^{j\beta}_{\circ\circ}$$

with numbered indices:

$$t^{i_1 i_2}_{\ \circ\ \circ} = \left(g^{\alpha_1\ \circ}_{\ \circ\ \beta_1}g^{\alpha_2\beta_2}_{\ \circ\ \circ}t^{\beta_1\ \circ}_{\ \circ\ \beta_2}\right)\gamma^{i_1\ \circ}_{\ \circ\ \alpha_1}\gamma^{i_2\ \circ}_{\ \circ\ \alpha_2},$$

where $T_{\sigma,1} = $ extension of $[t^{\alpha\circ}_{\circ\beta}]$ and $\hat{T}_{\sigma,1} = $ extension of $[t^{ij}_{\circ\circ}]$, we have the matrix expression:

$$\hat{T}_{\sigma,1} = \left(C^{-1}\otimes C^{-1}\right)\bullet\left[\left(I_n\otimes G^{-1}\right)\bullet T_{\sigma,1}.\right.$$

Next, we proceed to the condensation of $\hat{T}_{\sigma,1}$.

Seventh relation: data tensor: $[t^{\circ\ \beta}_{\alpha\ \circ}]$

$$t^{i\ \circ}_{\circ\ j} = \left(g^{\alpha\theta}_{\circ\circ}g^{\ \circ\circ}_{\beta\lambda}t^{\circ\lambda}_{\theta\circ}\right)\gamma^{i\ \circ}_{\circ\ \alpha}c^{\circ\beta}_{j\ \circ}$$

with numbered indices:

$$t^{i_1\ \circ}_{\ \circ\ i_2} = \left(g^{\alpha_1\beta_1}_{\ \circ\ \circ}g^{\ \circ\ \circ}_{\alpha_2\beta_2}t^{\ \circ\ \beta_2}_{\beta_1\ \circ}\right)\gamma^{i_1\ \circ}_{\ \circ\ \alpha_1}c^{\ \circ\ \alpha_2}_{i_2\ \circ},$$

where $T'_{\sigma,1} = $ extension of $[t^{\circ\ \beta}_{\alpha\ \circ}]$ and $\hat{T}'_{\sigma,1} = $ extension of $[t^{i\ \circ}_{\circ\ j}]$, we have the matrix expression

$$\hat{T}'_{\sigma,1} = \left(C^{-1}\otimes C^t\right)\bullet\left[\left(G^{-1}\otimes G\right)\bullet T'_{\sigma,1}\right].$$

Next, we proceed to the condensation of $\hat{T}'_{\sigma,1}$.

Last relation: data tensor $[t^{\circ\circ}_{\alpha\beta}]$

$$t^{\circ\circ}_{i\ j} = t^{\circ\circ}_{\alpha\beta}c^{\circ\alpha}_{i\ \circ}c^{\circ\beta}_{j\ \circ}$$

with numbered indices:

$$t^{\circ\ \circ}_{i_1 i_2} = t^{\ \circ\ \circ}_{\alpha_1\alpha_2}c^{\ \circ\ \alpha_1}_{i_1\ \circ}c^{\ \circ\ \alpha_2}_{i_2\ \circ},$$

where $T''_{\sigma,1} = $ extension of $[t^{\circ\ \circ}_{\alpha\beta}]$ and $\hat{T}''_{\sigma,1} = $ extension of $[t^{\circ\circ}_{ij}]$, we have only the change-of-basis

$$\hat{T}''_{\sigma,1} = \left(C^t\otimes C^t\right)\bullet T''_{\sigma,1}.$$

Next, we proceed to the condensation of $\hat{T}''_{\sigma,1}$.

$\square$

11.10 Euclidean tensor algebra in $\left(\overset{r}{\underset{1}{\otimes}} E^n \right)$ ($\mathbb{R}$) or pseudo-Euclidean tensor algebra in $\left(\overset{r}{\underset{1}{\otimes}} PSE^n \right)$ ($\mathbb{R}$)

It is obvious that the algebraic operations in the Euclidean tensors are based on the homogeneous tensors and in the relations motivated by the presence of the associated tensors. So, it suffices to make some brief and punctual clarifications.

In the particular case of pseudo-Euclidean tensor spaces, we must clearly indicate that we talk about "pseudo-Euclidean associated tensors".

11.10.1 Euclidean tensor equality

According to all of the previous discussion, a Euclidean tensor is perfectly defined by its order, and by the numerical components of any of its species, relative to a basis.

So, one can ask, for example, if two given Euclidean tensors

$$T \in (E^n \otimes E^n_* \otimes E^n_*)\,(\mathbb{R}) \text{ and } S \in \left(\overset{3}{\underset{1}{\otimes}} E^n_* \right)$$

are given by their components $[t^{\alpha\,\circ\,\circ}_{\circ\,\beta\gamma}]$ relative to the basis $\{\vec{e}_\alpha\}$ of $E^n(\mathbb{R})$ for the first, and respectively $[s^{\circ\,\circ\,\circ}_{i\,j\,k}]$ relative to the basis $\{\vec{e}_i\}$ of $E^n(\mathbb{R})$ for the second, the change-of-basis considered in $E^n(\mathbb{R})$ being $||\vec{e}_i|| = ||\vec{e}_\alpha||[c^{\alpha\,\circ}_{\circ\,i}]$, they represent the same Euclidean tensor $(T \equiv S)$.

We proceed to calculate

$$t^{\circ\,\circ\,\circ}_{ijk} = t^{\circ\,\circ\,\circ}_{\alpha\beta\gamma}c^{\circ\,\alpha}_{i\,\circ}c^{\circ\,\beta}_{j\,\circ}c^{\circ\,\gamma}_{k\,\circ} = g^{\circ\,\circ}_{\alpha\theta}t^{\theta\,\circ\,\circ}_{\circ\,\beta\gamma}c^{\circ\,\alpha}_{i\,\circ}c^{\circ\,\beta}_{j\,\circ}c^{\circ\,\gamma}_{k\,\circ};$$

and if

$$t^{\circ\,\circ\,\circ}_{ijk} = s^{\circ\,\circ\,\circ}_{ijk}; \;\; \forall i, j, k, \in I_n,$$

then, one can conclude that they represent the same Euclidean tensor $(T \equiv S)$.

11.10.2 Addition and external product of Euclidean (pseudo-Euclidean) tensors

Again, we indicate that the two tensors must have the same order. To sum them they must be notated by means of associated of the *same species* and in the *same basis*.

For example, the tensors mentioned in Section 11.10.1 can be added. If $U = T + S$ is the tensor sum, the covariant components of U in the new basis $\{\vec{e}_i\}$ of $E^n(\mathbb{R})$ would be (either if $T = S$ or if $T \neq S$):

$$u_{ijk}^{\circ\circ\circ} = t_{ijk}^{\circ\circ\circ} + s_{ijk}^{\circ\circ\circ}; \quad \forall i, j, k \in I_n.$$

With respect to the external product of a Euclidean tensor by a scalar, it is such that $\lambda S \in \left(\overset{3}{\underset{1}{\otimes}} E_*^N \right) (\mathbb{R})$ because the linear character of the homogeneous tensor algebra remains.

11.10.3 Tensor product of Euclidean (pseudo-Euclidean) tensors

This product can always be executed. The difference with respect to the homogeneous tensors is that due to the possibility of displacing the indices without varying the tensor, the result can be given with a great variety of species.

More precisely, if $T \in \left(\overset{r}{\underset{1}{\otimes}} E^n \right) (\mathbb{R})$ and $S \in \left(\overset{s}{\underset{1}{\otimes}} E^n \right) (\mathbb{R})$, then $T \otimes S \in \left(\overset{(r + s)}{\underset{1}{\otimes}} E^n \right) (\mathbb{R})$.

With respect to the product species, the result is that the first factor T has 2^r possibilities of associated tensors to be represented. On the other hand, the tensor S has 2^s possibilities of representation, which makes it possible to give the tensor product by means of any of the $2^{(r+s)}$ possibilities of associated tensors that have such a product.

Due to this reason, in theory it is assumed that the data tensors are both given in contravariant coordinates, as has been assumed in the present section, and that the result, obviously, is subject to the selected option, this or another one.

Next, we introduce a model to fix the concepts. Consider a tensor $T \in (E_*^n \otimes E^n \otimes E^n)(\mathbb{R})$ and another tensor $U \in (E_*^n \otimes E^n)(\mathbb{R})$ given by their components $[t_{\alpha\circ\circ}^{\circ\beta\gamma}]$ and $[u_{\alpha\circ}^{\circ\beta}]$, respectively. We wish to know the tensor product $P \otimes U$ in covariant coordinates.

As is well known, from the theory of homogeneous tensors, we must notate the second factor tensor with other indices

$$p_{\alpha\circ\circ\delta\circ}^{\circ\beta\gamma\circ\lambda} = t_{\alpha\circ\circ}^{\circ\beta\gamma} \cdot u_{\delta\circ}^{\circ\lambda} \tag{11.73}$$

and then, we proceed to determine the associated sought after tensor

$$p_{ijkd\ell}^{\circ\circ\circ\circ\circ} = g_{i\circ}^{\circ\alpha} g_{j\beta}^{\circ\circ} g_{k\gamma}^{\circ\circ} g_{d\circ}^{\circ\delta} g_{\ell\lambda}^{\circ\circ} p_{\alpha\circ\circ\delta\circ}^{\circ\beta\gamma\circ\lambda}, \tag{11.74}$$

where $g_{i\circ}^{\circ\alpha} = \delta_{i\circ}^{\circ\alpha}$ and $g_{do}^{\circ\delta} = \delta_{do}^{\circ\delta}$ are Kronecker deltas.

11.10.4 Euclidean (pseudo-Euclidean) tensor contraction

We start this section by reminding the reader of some of the properties of the Kronecker deltas:

$$\delta^{\circ j}_{i \circ} = g^{\circ\circ}_{i h} g^{h j}_{\circ\circ} = g^{\circ j}_{i \circ} \tag{11.75}$$

$$\delta^{i\circ}_{\circ j} = g^{i h}_{\circ\circ} g^{\circ\circ}_{h j} = g^{i\circ}_{\circ j} \tag{11.76}$$

and since delta is an isotropic tensor, we have

$$\delta^{\circ j}_{i \circ} = \delta^{j\circ}_{\circ i} = \delta^{\circ i}_{j \circ} = \delta^{i\circ}_{\circ j}, \tag{11.77}$$

properties that will be used when needed.

Next, we proceed to analyze the contraction of *two indices* (the first two indices) of a tensor T of order $r = 3$, built over the Euclidean space $E^n(\mathbb{R})$, in four cases of possible positions for the given indices, that is,

(a) $t^{i\circ\circ}_{\circ j k}$; (b) $t^{ij\circ}_{\circ\circ k}$; (c) $t^{\circ j\circ}_{i\circ k}$; (d) $t^{\circ\circ\circ}_{ijk}$,

that correspond to certain associated tensors of T.

We notate $u^{\circ}_k, v^{\circ}_k, w^{\circ}_k$ and r°_k to the tensors resulting after the contraction, respectively, of cases (a), (b), (c) and (d).

Case (a)

$$u^{\circ}_k = t^{\theta\circ\circ}_{\circ\theta k} \text{ Einstein's convention } = \delta^{j\circ}_{\circ i} t^{i\circ\circ}_{\circ j k}, \tag{11.78}$$

which, because of (11.75), (11.76) and (11.77), can be notated

$$u^{\circ}_k = g^{j\circ}_{\circ i} t^{i\circ\circ}_{\circ j k} \text{ Einstein's convention } = g^{\circ j}_{i \circ} t^{i\circ\circ}_{\circ j k}, \tag{11.79}$$

which is called a *"normal contraction"* by other authors, because the contracted indices are of different species.

From expression (11.79) we establish the following *mnemonic rule*: "We contract the two data tensor indices with the fundamental metric tensor, the indices of which present contrary valency."

Case (b) Since now the indices i, j are of the same species, we lower the index j, to proceed later as in case (a). For the sake of clarity we change the dummy notation of the first index

$$t^{\lambda\circ\circ}_{\circ j k} = g^{\circ\circ}_{j\theta} t^{\lambda\theta\circ}_{\circ\circ k},$$

$$v^{\circ}_k = \delta^{j\circ}_{\circ\lambda}\left(t^{\lambda\circ\circ}_{\circ j k}\right) = \delta^{j\circ}_{\circ\lambda}\left(g^{\circ\circ}_{j\theta} t^{\lambda\theta\circ}_{\circ\circ k}\right) = \left(\delta^{\circ j}_{\lambda\circ} g^{\circ\circ}_{j\theta}\right)\left(t^{\lambda\theta\circ}_{\circ\circ k}\right) = g^{\circ\circ}_{\lambda\theta} t^{\lambda\theta\circ}_{\circ\circ k}.$$

Notating the last expression with apropriate indices, gives

$$v^{\circ}_k = g^{\circ\circ}_{i j} t^{ij\circ}_{\circ\circ k}, \tag{11.80}$$

which is called a *"Euclidean contraction"* by other authors, warning the reader that the contracted indices are of the same species (this contraction is proper and exclusive to Euclidean tensors).

From expression (11.80) we establish the following *mnemonic rule*:

"We contract the two data tensor indices with the fundamental metric tensor, the indices of which present contrary valency" (*the surprise* is that it is the *same* rule).

Case (c)

$$w^{\circ}_{k} = t^{\circ\theta\circ}_{\theta\circ k} \text{ (Einstein's convention).}$$

According to properties (11.75), (11.76) and (11.77) we can also propose

$$w^{\circ}_{k} = \delta^{\circ i}_{j\circ} t^{\circ j \circ}_{i \circ k} = \delta^{i\circ}_{\circ j} t^{\circ j \circ}_{i \circ k} = g^{i\circ}_{\circ j} t^{\circ j \circ}_{i \circ k}, \tag{11.81}$$

an expression that appears in the so-called *normal contraction*.

With respect to the *mnemonic rule*, it is again the same: "We contract the two data tensor indices with the fundamental metric tensor, the indices of which present contrary valency."

Case (d) Since in this case the indices to be contracted are both covariant, we raise the index i to be again in case (a). As before, we change the notation of the j index

$$t^{i\circ\circ}_{\circ\mu k} = g^{\theta i}_{\circ\circ} t^{\circ\circ\circ}_{\theta\mu k};$$

$$r^{\circ}_{k} = \delta^{\mu\circ}_{\circ i} \left(t^{i\circ\circ}_{\circ\mu k} \right) = \delta^{\mu\circ}_{\circ i} \left(g^{\theta i}_{\circ\circ} t^{\circ\circ\circ}_{\theta\mu k} \right) = \left(\delta^{\mu\circ}_{\circ i} g^{\theta i}_{\circ\circ} \right) t^{\circ\circ\circ}_{\theta\mu k} = \left(g^{\theta i}_{\circ\circ} \delta^{\circ\mu}_{i\circ} \right) t^{\circ\circ\circ}_{\theta\mu k} = g^{\theta\mu}_{\circ\circ} t^{\circ\circ\circ}_{\theta\mu k},$$

and notating with apropriate indices

$$r^{\circ}_{k} = g^{ij}_{\circ\circ} t^{\circ\circ\circ}_{ijk}. \tag{11.82}$$

Again the Euclidean contraction appears, even though for our readers, we find the usual "practical rule".

It is obvious that if the contracted indices are not the first two indices, the same conclusions are obtained. For example, if we contract the *first and fourth* indices of the Euclidean tensor $t^{i\circ\circ\ell}_{\circ j k \circ}$, we obtain the tensor expression

$$u^{\circ\circ}_{jk} = g^{\circ\circ}_{i\ell} t^{i\circ\circ\ell}_{\circ j k \circ}. \tag{11.83}$$

When performing multiple contractions the same rules hold; if we contract indices (1, 4) and (2, 3) of the Euclidean tensor above, the tensor operation becomes

$$\rho = g^{\circ\circ}_{i\ell} g^{jk}_{\circ\circ} t^{i\circ\circ\ell}_{\circ j k \circ}; \quad \rho \in \mathbb{R}. \tag{11.84}$$

Since in the Euclidean tensors any pair of indices can be contracted, it is a good policy not to refer to the indices by name, *but by position*, as was indicated in Formula (11.84).

So we will say: contraction of the *first* with the *fourth* (1, 4), the *second* with the *third* (2, 3), etc.

Finally we give a theorem that completes this section.

Theorem 11.1 (Euclidean tensors). *When we contract certain indices in given positions, in a Euclidean tensor, the contracted tensor is* unique, *that is, it does not* depend on the associated tensor initially chosen to execute the contraction. □

Proof. As a proof of this theorem, we will study the contracted tensors, that we have obtained in the first part, in the cases (a), (b), (c) and (d), i.e., $[u_k^\circ], [v_k^\circ], [w_k^\circ]$ and $[r_k^\circ]$, because all of them come from contracting the first two indices of several associated tensors of T. We remind the reader about what was established in Section 11.10.1 with respect the equality of Euclidean tensors, and propose:

Cases (a) and (b):

$$u_k^\circ = t_{\circ\theta k}^{\theta\circ\circ} = \delta_{\circ i}^{j\circ} t_{\circ j k}^{i\circ\circ} = \delta_{\circ i}^{j\circ} \left(g_{j\theta}^{\circ\circ} t_{\circ\circ k}^{i\theta\circ} \right) = \left(\delta_{i\circ}^{\circ j} g_{j\theta}^{\circ\circ} \right) t_{\circ\circ k}^{i\theta\circ} = g_{i\theta}^{\circ\circ} t_{\circ\circ k}^{i\theta\circ} = v_k^\circ.$$

If we compare with Formula (11.80), the conclusion is

$$[u_k^\circ] = [v_k^\circ]. \tag{11.85}$$

Cases (a) and (c): We have

$$u_k^\circ = \delta_{\circ i}^{j\circ} t_{\circ j k}^{i\circ\circ} = \delta_{\circ i}^{j\circ} \left(g_{\circ\circ}^{\alpha i} g_{\lambda j}^{\circ\circ} t_{\alpha\circ k}^{\circ\lambda\circ} \right)$$

$$= \left[\left(g_{\lambda j}^{\circ\circ} \delta_{\circ i}^{j\circ} \right) g_{\circ\circ}^{i\alpha} \right] t_{\alpha\circ k}^{\circ\lambda\circ} = \left(g_{\lambda i}^{\circ\circ} g_{\circ\circ}^{i\alpha} \right) t_{\alpha\circ k}^{\circ\lambda\circ} = \delta_{\lambda\circ}^{\circ\alpha} t_{\alpha\circ k}^{\circ\lambda\circ} = t_{\theta\circ k}^{\circ\theta\circ} = w_k^\circ$$

and then

$$[u_k^\circ] = [w_k^\circ]. \tag{11.86}$$

Cases (a) and (d): If we compare with Formula (11.82) we have

$$u_k^\circ = \delta_{\circ i}^{j\circ} t_{\circ j k}^{i\circ\circ} = \delta_{\circ i}^{j\circ} \left(g_{\circ\circ}^{\theta i} t_{\theta j k}^{\circ\circ\circ} \right) = \left(\delta_{\circ i}^{j\circ} g_{\circ\circ}^{\theta i} \right) t_{\theta j k}^{\circ\circ\circ}$$

$$= \left(g_{\circ\circ}^{\theta i} \delta_{i\circ}^{\circ j} \right) t_{\theta j k}^{\circ\circ\circ} = g_{\circ\circ}^{\theta j} t_{\theta j k}^{\circ\circ\circ} = r_k^\circ,$$

and then

$$[u_k^\circ] = [r_k^\circ]. \tag{11.87}$$

The final conclusion of (11.85), (11.86) and (11.87) is that

$$[u_k^\circ] = [v_k^\circ] = [w_k^\circ] = [r_k^\circ]$$

and that all of them are the same tensor, which proves the theorem.

11.10.5 Contracted tensor product of Euclidean or pseudo-Euclidean tensors

As a consequence of all that has been established in the previous section about the Euclidean contraction, it is evident that we can execute the tensor product of two Euclidean tensors of orders r_1 and r_2, respectively, and then proceed to the contraction with the *only condition* of selecting contraction indices coming *each one from each factor*, though obviously, it is not necessary for them to have different species, because the contractions and vertical displacements of indices are permutable operations in them, due to their Euclidean nature. The order of the resulting tensor will be $(r_1 + r_2 - 2c)$, where c is the number of pairs of contracted indices.

Some authors call this contracted tensor product of Euclidean tensors the "inner product" of tensors, a name that is mentioned here but that is not used.

To end this topic, we present an illustrative model.

Consider two tensors T and U of orders $r_1 = 3$ and $r_2 = 2$, respectively, built over the same Euclidean space $E^n(\mathbb{R})$, where T and U are given by the components $[t^{\circ\,\beta\gamma}_{\alpha\circ\circ}]$ and $[u^{\circ\,\circ}_{\lambda\mu}]$, respectively.

We wish to find the contracted tensor product of the first two indices of each factor, and we do not specify the desired species for the resulting tensor.

Let $P = T \otimes U$, $P_c = \mathcal{C}(\alpha,\ \lambda)P \equiv \mathcal{C}(1,\ 4)P$ and

$$P <> p^{\circ\,\beta\gamma\circ\circ}_{\alpha\circ\circ\lambda\mu} = \left(t^{\circ\,\beta\gamma}_{\alpha\circ\circ}\right) \cdot \left(u^{\circ\,\circ}_{\lambda\mu}\right). \qquad (11.88)$$

Since P is Euclidean, to contract the indices $(1, 4)$ we apply the mnemonic rule, i.e., since α and λ are both covariant indices in the product P, the components of the contracted product P_c will be

$$P_c \equiv Q <> q^{\beta\gamma\circ}_{\circ\circ\mu} = g^{\alpha\lambda}_{\circ\circ} p^{\circ\,\beta\gamma\circ\circ}_{\alpha\circ\circ\lambda\mu}; \qquad (11.89)$$

$$q^{\beta\gamma\circ}_{\circ\circ\mu} = g^{\alpha\lambda}_{\circ\circ} \left(t^{\circ\,\beta\gamma}_{\alpha\circ\circ} \cdot u^{\circ\,\circ}_{\lambda\mu}\right). \qquad (11.90)$$

With respect to the matrix execution of the Euclidean contractions, Examples 11.4 and 11.5 illustrate diverse vertical displacements of indices over tensors, executed in matrix form, using the matrix extension T_σ of the data tensors.

If the Euclidean contraction be of indices with different valency, the matrix formula proposed in Sections 5.7 and 5.8 that operate with T_σ, could be directly applied.

If, on the contrary, they are of the same valency, once the displacement of the convenient index has been done, and since the resulting tensor T'_σ is in extended form, the matrix formulas of the mentioned sections can be applied over it.

Nevertheless, we also give alternative formulas of *direct* use.

G_0 represents the matrix of the fundamental metric tensor of $E^n(\mathbb{R})$, with the species that corresponds to the desired contraction (G, I_n, G^{-1}), T_0 represents the matrix of the data tensor with the corresponding species, "□" represents the Hadamard product, and $\{E_i\}$ is the basis of the linear space $E^n(\mathbb{R})$ *represented as column matrices*.

11.10.6 Euclidean contraction of tensors of order $r = 2$

The matrix $T_{\sigma,1}$ has been detailed in Formula (1.30).

$$\rho = ([E_1^t \;\; |E_2^t \;\; | \;\; \cdots \;\; | \;\; E_n^t] \bullet (I_n \otimes G_0))_{1,n^2} \bullet T_{\sigma,1} = G_\sigma^t \bullet T_{\sigma,1} \quad (11.91)$$

where

$$G_\sigma^t = [E_1^t \;\; |E_2^t \;\; | \;\; \cdots \;\; | \;\; E_n^t] \bullet (I_n \otimes G_0) \quad (11.92)$$

or

$$\rho = G_\sigma^t \bullet [I_n \otimes I_n] \bullet T_{\sigma,1}. \quad (11.93)$$

11.10.7 Euclidean contraction of tensors of order $r = 3$

We *notate* the input tensor in contravariant coordinates, simply as "*support notation*".

First: For the T tensor indices $(1, 2)$.
    Let

$$G_\sigma^t = [[E_1^t \;\; |E_2^t \;\; | \;\; \cdots \;\; | \;\; E_n^t] \bullet (I_n \otimes G_0)]_{1,n^2}$$

$$[u_k^\circ] = \begin{bmatrix} G_\sigma^t \bullet [(I_n \otimes I_n) \otimes E_1^t] \bullet T_\sigma \\ G_\sigma^t \bullet [(I_n \otimes I_n) \otimes E_2^t] \bullet T_\sigma \\ \cdots \\ G_\sigma^t \bullet [(I_n \otimes I_n) \otimes E_n^t] \bullet T_\sigma \end{bmatrix}_{n,1}. \quad (11.94)$$

Second: For the T tensor indices $(1, 3)$.

$$[v_k^\circ] = \begin{bmatrix} G_\sigma^t \bullet [(I_n \otimes E_1^t) \otimes I_n] \bullet T_\sigma \\ G_\sigma^t \bullet [(I_n \otimes E_2^t) \otimes I_n] \bullet T_\sigma \\ \cdots \\ G_\sigma^t \bullet [(I_n \otimes E_n^t) \otimes I_n] \bullet T_\sigma \end{bmatrix}_{n,1}. \quad (11.95)$$

Third: For the T tensor indices $(2, 3)$.

$$[w_k^\circ] = \begin{bmatrix} G_\sigma^t \bullet [E_1^t \otimes (I_n \otimes I_n)] \bullet T_\sigma \\ G_\sigma^t \bullet [E_2^t \otimes (I_n \otimes I_n)] \bullet T_\sigma \\ \cdots \\ G_\sigma^t \bullet [E_n^t \otimes (I_n \otimes I_n)] \bullet T_\sigma \end{bmatrix}_{n,1}. \quad (11.96)$$

11.10.8 Euclidean contraction of tensors of order $r = 4$

We *notate* the input tensor in contravariant coordinates, simply as "*support notation*".

First: For the T tensor indices $(1, 2)$.

$$[u_{\circ\circ}^{\alpha\beta}] = [B_1 \;\; B_2 \;\; \cdots \;\; B_n], \quad (11.97)$$

where

$$B_i = \begin{bmatrix} G_\sigma^t \bullet [(I_n \otimes I_n) \otimes (E_1^t \otimes E_i^t)] \bullet T_\sigma \\ G_\sigma^t \bullet [(I_n \otimes I_n) \otimes (E_1^t \otimes E_i^t)] \bullet T_\sigma \\ \ldots \\ G_\sigma^t \bullet [(I_n \otimes I_n) \otimes (E_1^t \otimes E_i^t)] \bullet T_\sigma \end{bmatrix}_{n,n} \quad ; \quad i = 1,2,\ldots,n.$$

Second: We propose that of indices (2, 4) as the last model

$$[v_{\circ\circ}^{\alpha\beta}] = [\, C_1 \quad C_2 \quad \cdots \quad C_n \,], \tag{11.98}$$

where

$$C_i = \begin{bmatrix} G_\sigma^t \bullet [(E_1^t \otimes I_n) \otimes (E_i^t \otimes I_n)] \bullet T_\sigma \\ G_\sigma^t \bullet [(E_2^t \otimes I_n) \otimes (E_i^t \otimes I_n)] \bullet T_\sigma \\ \ldots \\ G_\sigma^t \bullet [(E_n^t \otimes I_n) \otimes (E_i^t \otimes I_n)] \bullet T_\sigma \end{bmatrix}_{n,n} \quad ; \quad i = 1,2,\cdots,n.$$

To end this topic, we emphasize that the given contraction formulas *are not* unique.

11.10.9 Euclidean contraction of indices by the Hadamard product

As an example, we will give two models of tensors of *third* order ($r = 3$) directly contracted from the data matrix T_0, that is, without being extended, and using the Hadamard product.

We start with the contraction of indices (1, 3), which is expressed using the Hadamard products as

$$[v_k^\circ] = \begin{bmatrix} [1 \;\; 1 \;\; \cdots \;\; 1]_{1,n} \bullet \left(G_0 \square \begin{bmatrix} (E_1^t \otimes E_1^t) \bullet T_0 \\ (E_2^t \otimes E_1^t) \bullet T_0 \\ \ldots \ldots \ldots \\ (E_n^t \otimes E_1^t) \bullet T_0 \end{bmatrix} \right) \bullet \begin{bmatrix} 1 \\ 1 \\ \vdots \\ 1 \end{bmatrix} \\ [1 \;\; 1 \;\; \cdots \;\; 1]_{1,n} \bullet \left(G_0 \square \begin{bmatrix} (E_1^t \otimes E_2^t) \bullet T_0 \\ (E_2^t \otimes E_2^t) \bullet T_0 \\ \ldots \ldots \ldots \\ (E_n^t \otimes E_2^t) \bullet T_0 \end{bmatrix} \right) \bullet \begin{bmatrix} 1 \\ 1 \\ \vdots \\ 1 \end{bmatrix} \\ \cdots \cdots \cdots \cdots \cdots \cdots \cdots \cdots \cdots \cdots \cdots \\ [1 \;\; 1 \;\; \cdots \;\; 1]_{1,n} \bullet \left(G_0 \square \begin{bmatrix} (E_1^t \otimes E_n^t) \bullet T_0 \\ (E_2^t \otimes E_n^t) \bullet T_0 \\ \ldots \ldots \ldots \\ (E_n^t \otimes E_n^t) \bullet T_0 \end{bmatrix} \right) \bullet \begin{bmatrix} 1 \\ 1 \\ \vdots \\ 1 \end{bmatrix} \end{bmatrix}_{n,1}$$

$$\tag{11.99}$$

Similarly, we perform the contraction of indices (2, 3), as

$$[w_k^\circ] = \begin{bmatrix} [1 \quad 1 \quad \cdots \quad 1]_{1,n} \bullet (G_0 \square \, [(E_1^t \otimes I_n) \bullet T_0]) \bullet \begin{bmatrix} 1 \\ 1 \\ \vdots \\ 1 \end{bmatrix} \\ [1 \quad 1 \quad \cdots \quad 1]_{1,n} \bullet (G_0 \square \, [(E_2^t \otimes I_n) \bullet T_0]) \bullet \begin{bmatrix} 1 \\ 1 \\ \vdots \\ 1 \end{bmatrix} \\ \cdots \quad \cdots \quad \cdots \quad \cdots \quad \cdots \quad \cdots \quad \cdots \quad \cdots \quad \cdots \quad \cdots \quad \cdots \\ [1 \quad 1 \quad \cdots \quad 1]_{1,n} \bullet (G_0 \square \, [(E_n^t \otimes I_n) \bullet T_0]) \bullet \begin{bmatrix} 1 \\ 1 \\ \vdots \\ 1 \end{bmatrix} \end{bmatrix}_{n,1} . \tag{11.100}$$

Example 11.6 (Euclidean tensor algebra. Contractions). In a Euclidean space $E^2(\mathbb{R})$ referred to a basis $\{\vec{e}_\alpha\}$ the fundamental metric tensor is given in contravariant coordinates by the matrix

$$[g_{\circ\circ}^{\alpha\beta}] = \begin{bmatrix} 2 & 1 \\ 1 & 1 \end{bmatrix}.$$

Consider a change-of-basis $\{\vec{e}_i\}$, for which the matrix C is

$$[c_{\circ i}^{\alpha\circ}] = \begin{bmatrix} 1 & 2 \\ -1 & -1 \end{bmatrix}.$$

A tensor T built over this Euclidean space of order $r = 3$, has as a matrix expression for its components

$$[t_{\circ\beta\gamma}^{\alpha\circ\circ}] = \begin{bmatrix} 1 & 0 \\ 2 & 1 \\ 2 & 0 \\ 1 & 1 \end{bmatrix},$$

where α is the block row, β is the row of each block and γ is the column of each block.

1. Obtain the tensor U contracted with respect to the first and third indices (1, 3) of T:
   a) given by its contravariant components, in the initial basis;
   b) given by its covariant components, in the initial basis.
2. Again the previous question but in the new basis.
3. Obtain the tensor W contracted with respect to the second and third indices (2, 3) of T:
   a) given by its contravariant components, in the initial basis;
   b) given by its covariant components, in the initial basis.
4. Again the previous question but in the new basis.

5. Examine the nature and components of the system of scalar components

$$s(\alpha, \beta, \gamma, \delta, \epsilon) = g^{\alpha\beta}_{\circ\circ} t^{\gamma\circ\circ}_{\circ\delta\epsilon}.$$

6. Obtain the contravariant components of the tensor $P = G^{-1} \otimes T$ in the basis $\{\vec{e}_\alpha\}$.
7. Obtain the contracted tensor product with respect to the indices $(1, 5)$ of P in the basis $\{\vec{e}_\alpha\}$.

Solution:

1. To present the maximum information, we solve the previous questions by several methods.
   According to the definition, we know that

$$[g^{\alpha\beta}_{\circ\circ}] \equiv G^{-1} = \begin{bmatrix} 2 & 1 \\ 1 & 1 \end{bmatrix}$$

and that

$$G \equiv [g^{\circ\circ}_{\alpha\beta}] = \begin{bmatrix} -1 & -1 \\ -1 & 2 \end{bmatrix}; \quad [g^{\alpha\circ}_{\circ\beta}] = [g^{\circ\beta}_{\alpha\circ}] = I_2 = \begin{bmatrix} 1 & 0 \\ 0 & 1 \end{bmatrix}.$$

Method 1.

(a) Since the sought after contraction $(1, 3)$ is of indices of different valency, it is a "normal contraction", that is, it can be executed as a simple homogeneous tensor:

$$u^\circ_\beta = t^{\theta\circ\circ}_{\circ\beta\theta}; \quad U^* = \begin{bmatrix} u^\circ_1 \\ u^\circ_2 \end{bmatrix} = \begin{bmatrix} t^{\theta\circ\circ}_{\circ 1\theta} \\ t^{\theta\circ\circ}_{\circ 2\theta} \end{bmatrix} = \begin{bmatrix} t^{1\circ\circ}_{\circ 11} + t^{2\circ\circ}_{\circ 12} \\ t^{1\circ\circ}_{\circ 21} + t^{2\circ\circ}_{\circ 22} \end{bmatrix} = \begin{bmatrix} 1+0 \\ 2+1 \end{bmatrix} = \begin{bmatrix} 1 \\ 3 \end{bmatrix}.$$

As the resulting tensor must be in contravariant coordinates, we must raise the index β:

$$U = [u^\beta_\circ] = [g^{\theta\beta}_{\circ\circ} u^\circ_\theta] = G^{-1}U^* = \begin{bmatrix} 2 & 1 \\ 1 & 1 \end{bmatrix}\begin{bmatrix} 1 \\ 3 \end{bmatrix} = \begin{bmatrix} 5 \\ 4 \end{bmatrix}.$$

(b) It was previously established that $U^* = \begin{bmatrix} 1 \\ 3 \end{bmatrix}$.

Method 2.

(a) We perform the contraction as a homogeneous tensor, using Formula (5.65) of model 3;

$$\sigma = 2^3 = 8; \quad \sigma' = \frac{\sigma}{2^2} = \frac{8}{4} = 2.$$

$$T_2' = [u_\beta^\circ] = \left(\left[I_2 \otimes E_1^t | I_2 \otimes E_2^t \right] \right) \bullet T_8$$

$$= \left(\begin{bmatrix} 1 & 0 \\ 0 & 1 \end{bmatrix} \otimes [1 \; 0] \Big| \begin{bmatrix} 1 & 0 \\ 0 & 1 \end{bmatrix} \otimes [0 \; 1] \right) \bullet \begin{bmatrix} 1 \\ 0 \\ 2 \\ 1 \\ - \\ 2 \\ 0 \\ 1 \\ 1 \end{bmatrix}$$

$$= \begin{bmatrix} 1 & 0 & 0 & 0 & 0 & 1 & 0 & 0 \\ 0 & 0 & 1 & 0 & 0 & 0 & 0 & 1 \end{bmatrix} \bullet \begin{bmatrix} 1 \\ 0 \\ 2 \\ 1 \\ - \\ 2 \\ 0 \\ 1 \\ 1 \end{bmatrix} = \begin{bmatrix} 1 \\ 3 \end{bmatrix};$$

$$U = [u_\circ^\beta] = G^{-1} T_2' = \begin{bmatrix} 2 & 1 \\ 1 & 1 \end{bmatrix} \begin{bmatrix} 1 \\ 3 \end{bmatrix} = \begin{bmatrix} 5 \\ 4 \end{bmatrix}.$$

(b) It was previously obtained that $U^* = T_2' = [u_\beta^\circ] = \begin{bmatrix} 1 \\ 3 \end{bmatrix}$.

Method 3.

(a) The expression $\mathcal{EC} \equiv$ must be read "Euclidean contraction". We use the "mnemonic rule".

$\mathcal{EC}(1,3)[t_{\circ\beta\gamma}^{\alpha\circ\circ}] = [g_{\circ\alpha}^{\gamma\circ}][t_{\circ\beta\gamma}^{\alpha\circ\circ}]$, and obtain $G_0 = [g_{\circ\alpha}^{\gamma\circ}] = I_2$.

We also know that $T_\sigma^t = \begin{bmatrix} 1\,0\,2\,1 \,|\, 2\,0\,1\,1 \end{bmatrix}$, and because of Formula (11.92), that it is essentially (1.30) applied to G_0, which leads to

$$G_\sigma^t = [E_1^t | E_2^t] \bullet [I_2 \otimes G_0] = [1 \; 0 \; 0 \; 1] \bullet \left(\begin{bmatrix} 1 & 0 \\ 0 & 1 \end{bmatrix} \otimes \begin{bmatrix} 1 & 0 \\ 0 & 1 \end{bmatrix} \right)$$

$$= [1 \; 0 \; 0 \; 1] \bullet \begin{bmatrix} 1 & 0 & 0 & 0 \\ 0 & 1 & 0 & 0 \\ 0 & 0 & 1 & 0 \\ 0 & 0 & 0 & 1 \end{bmatrix} = [1 \; 0 \; 0 \; 1].$$

As expected, we have obtained the matrix G_σ "extended" and in horizontal form, and now we can perform the contraction using Formula (11.95):

$$[u_\beta^\circ] = \begin{bmatrix} G_\sigma^t \bullet [I_2 \otimes E_1^t \otimes I_2] \bullet T_\sigma \\ G_\sigma^t \bullet [I_2 \otimes E_2^t \otimes I_2] \bullet T_\sigma \end{bmatrix}$$

$$= \begin{bmatrix} [1 \ \ 0 \ \ 0 \ \ 1] \bullet \left(\begin{bmatrix} 1 & 0 \\ 0 & 1 \end{bmatrix} \otimes [1 \ \ 0] \otimes \begin{bmatrix} 1 & 0 \\ 0 & 1 \end{bmatrix} \right) \bullet T_\sigma \\ [1 \ \ 0 \ \ 0 \ \ 1] \bullet \left(\begin{bmatrix} 1 & 0 \\ 0 & 1 \end{bmatrix} \otimes [0 \ \ 1] \otimes \begin{bmatrix} 1 & 0 \\ 0 & 1 \end{bmatrix} \right) \bullet T_\sigma \end{bmatrix}$$

$$= \begin{bmatrix} [1 \ \ 0 \ \ 0 \ \ 1] \bullet \left(\begin{bmatrix} 1 & 0 & 0 & 0 \\ 0 & 0 & 1 & 0 \end{bmatrix} \otimes \begin{bmatrix} 1 & 0 \\ 0 & 1 \end{bmatrix} \right) \bullet T_\sigma \\ [1 \ \ 0 \ \ 0 \ \ 1] \bullet \left(\begin{bmatrix} 0 & 1 & 0 & 0 \\ 0 & 0 & 0 & 1 \end{bmatrix} \otimes \begin{bmatrix} 1 & 0 \\ 0 & 1 \end{bmatrix} \right) \bullet T_\sigma \end{bmatrix}$$

$$= \begin{bmatrix} [1 \ \ 0 \ \ 0 \ \ 1] \bullet \begin{bmatrix} 1 & 0 & 0 & 0 & 0 & 0 & 0 & 0 \\ 0 & 1 & 0 & 0 & 0 & 0 & 0 & 0 \\ 0 & 0 & 0 & 0 & 1 & 0 & 0 & 0 \\ 0 & 0 & 0 & 0 & 0 & 1 & 0 & 0 \end{bmatrix} \bullet \begin{bmatrix} 1 \\ 0 \\ 2 \\ 1 \\ - \\ 2 \\ 0 \\ 1 \\ 1 \end{bmatrix} \\ [1 \ \ 0 \ \ 0 \ \ 1] \bullet \begin{bmatrix} 0 & 0 & 1 & 0 & 0 & 0 & 0 & 0 \\ 0 & 0 & 0 & 1 & 0 & 0 & 0 & 0 \\ 0 & 0 & 0 & 0 & 0 & 0 & 1 & 0 \\ 0 & 0 & 0 & 0 & 0 & 0 & 0 & 1 \end{bmatrix} \bullet \begin{bmatrix} 1 \\ 0 \\ 2 \\ 1 \\ - \\ 2 \\ 0 \\ 1 \\ 1 \end{bmatrix} \end{bmatrix}$$

$$= \begin{bmatrix} [1 \ \ 0 \ \ 0 \ \ 1] \bullet \begin{bmatrix} 1 \\ 0 \\ 2 \\ 0 \end{bmatrix} \\ [1 \ \ 0 \ \ 0 \ \ 1] \bullet \begin{bmatrix} 2 \\ 1 \\ 1 \\ 1 \end{bmatrix} \end{bmatrix} = \begin{bmatrix} 1 \\ 3 \end{bmatrix}$$

and in contravariant coordinates:

$$U = G^{-1} U^* = \begin{bmatrix} 2 & 1 \\ 1 & 1 \end{bmatrix} \begin{bmatrix} 1 \\ 3 \end{bmatrix} = \begin{bmatrix} 5 \\ 4 \end{bmatrix}.$$

(b) It has already been found that $U^* = \begin{bmatrix} 1 \\ 3 \end{bmatrix}$.

2. The change-of-basis, for first-order tensors (vectors), is:

(a)

$$\hat{U} = [u_{\circ}^{j}] = [u_{\circ}^{\beta}][\gamma_{\circ\beta}^{j\circ}] = C^{-1} \bullet U = \begin{bmatrix} 1 & 2 \\ -1 & -1 \end{bmatrix}^{-1} \begin{bmatrix} 5 \\ 4 \end{bmatrix}$$

$$= \begin{bmatrix} -1 & -2 \\ 1 & 1 \end{bmatrix} \begin{bmatrix} 5 \\ 4 \end{bmatrix} = \begin{bmatrix} -13 \\ 9 \end{bmatrix}.$$

(b)

$$\hat{U}^* = [u_j^{\,\circ}] = [u_\beta^{\,\circ}][c_{j\,\circ}^{\,\circ\beta}] = C^t \bullet U^* = \begin{bmatrix} 1 & 2 \\ -1 & -1 \end{bmatrix}^t \begin{bmatrix} 1 \\ 3 \end{bmatrix}$$

$$= \begin{bmatrix} 1 & -1 \\ 2 & -1 \end{bmatrix}\begin{bmatrix} 1 \\ 3 \end{bmatrix} = \begin{bmatrix} -2 \\ 1 \end{bmatrix}.$$

3. *Method 1.* Since the sought after contraction (2,3) is of indices of the same valency, we have a Euclidean contraction, and it can be executed only as a Euclidean tensor

$$[w_{\,\circ}^\alpha] = \mathcal{EC}(2,3)[t_{\,\circ\,\beta\gamma}^{\alpha\,\circ\,\circ}] = [g_{\,\circ\circ}^{\beta\gamma}][t_{\,\circ\,\beta\gamma}^{\alpha\,\circ\,\circ}],$$

from which we conclude that $G_0 = [g_{\,\circ\circ}^{\beta\gamma}] = G^{-1} = \begin{bmatrix} 2 & 1 \\ 1 & 1 \end{bmatrix}$, whence

$$W \equiv [w_{\,\circ}^\alpha] = \begin{bmatrix} w_{\,\circ}^1 \\ w_{\,\circ}^2 \end{bmatrix} = \begin{bmatrix} g_{\,\circ\circ}^{11}\cdot t_{\,\circ 11}^{1\circ\circ} + g_{\,\circ\circ}^{12}\cdot t_{\,\circ 12}^{1\circ\circ} + g_{\,\circ\circ}^{21}\cdot t_{\,\circ 21}^{1\circ\circ} + g_{\,\circ\circ}^{22}\cdot t_{\,\circ 22}^{1\circ\circ} \\ g_{\,\circ\circ}^{11}\cdot t_{\,\circ 11}^{2\circ\circ} + g_{\,\circ\circ}^{12}\cdot t_{\,\circ 12}^{2\circ\circ} + g_{\,\circ\circ}^{21}\cdot t_{\,\circ 21}^{2\circ\circ} + g_{\,\circ\circ}^{22}\cdot t_{\,\circ 22}^{2\circ\circ} \end{bmatrix}$$

$$= \begin{bmatrix} 2\times 1 + 1\times 0 + 1\times 2 + 1\times 1 \\ 2\times 2 + 1\times 0 + 1\times 1 + 1\times 1 \end{bmatrix} = \begin{bmatrix} 5 \\ 6 \end{bmatrix}.$$

To find the tensor $[w_{\,\circ}^\alpha]$ in covariant coordinates, we lower the index $w_\alpha^{\,\circ} = g_{\theta\alpha}^{\circ\circ} w_{\,\circ}^\theta$:

$$W^* = G \bullet W = \begin{bmatrix} 1 & -1 \\ -1 & 2 \end{bmatrix} \bullet \begin{bmatrix} 5 \\ 6 \end{bmatrix} = \begin{bmatrix} -1 \\ 7 \end{bmatrix}.$$

Method 2. Using Formula (11.92), the matrix G_0 is stretched:

$$G_\sigma^t = [E_1^t|E_2^t] \bullet [I_2 \otimes G_0] = [\,1\ \ 0\ \ 0\ \ 1\,] \bullet \left(\begin{bmatrix} 1 & 0 \\ 0 & 1 \end{bmatrix} \otimes \begin{bmatrix} 2 & 1 \\ 1 & 1 \end{bmatrix}\right)$$

$$= [\,1\ \ 0\ \ 0\ \ 1\,] \bullet \begin{bmatrix} 2 & 1 & 0 & 0 \\ 1 & 1 & 0 & 0 \\ 0 & 0 & 2 & 1 \\ 0 & 0 & 1 & 1 \end{bmatrix} = [\,2\ \ 1\ \ 1\ \ 1\,],$$

and next, we apply the Formula (11.96):

$$W = [w_k^{\,\circ}] = \begin{bmatrix} G_\sigma^t \bullet [E_1^t \otimes (I_n \otimes I_n)] \bullet T_\sigma \\ G_\sigma^t \bullet [E_2^t \otimes (I_n \otimes I_n)] \bullet T_\sigma \end{bmatrix}$$

$$= \begin{bmatrix} [\,2\ 1\ 1\ 1\,] \bullet \left([1\ \ 0] \otimes \begin{bmatrix} 1 & 0 \\ 0 & 1 \end{bmatrix} \otimes \begin{bmatrix} 1 & 0 \\ 0 & 1 \end{bmatrix}\right) \bullet T_\sigma \\ [\,2\ \ 1\ \ 1\ \ 1\,] \bullet \left([0\ \ 1] \otimes \begin{bmatrix} 1 & 0 \\ 0 & 1 \end{bmatrix} \otimes \begin{bmatrix} 1 & 0 \\ 0 & 1 \end{bmatrix}\right) \bullet T_\sigma \end{bmatrix}$$

$$= \begin{bmatrix} \begin{bmatrix} 2 & 1 & 1 & 1 \end{bmatrix} \bullet \begin{bmatrix} 1 & 0 & 0 & 0 & 0 & 0 & 0 & 0 \\ 0 & 1 & 0 & 0 & 0 & 0 & 0 & 0 \\ 0 & 0 & 1 & 0 & 0 & 0 & 0 & 0 \\ 0 & 0 & 0 & 1 & 0 & 0 & 0 & 0 \end{bmatrix} \bullet \begin{bmatrix} 1 \\ 0 \\ 2 \\ 1 \\ \overline{2} \\ 0 \\ 1 \\ 1 \end{bmatrix} \\[2em] \begin{bmatrix} 2 & 1 & 1 & 1 \end{bmatrix} \bullet \begin{bmatrix} 0 & 0 & 0 & 0 & 1 & 0 & 0 & 0 \\ 0 & 0 & 0 & 0 & 0 & 1 & 0 & 0 \\ 0 & 0 & 0 & 0 & 0 & 0 & 1 & 0 \\ 0 & 0 & 0 & 0 & 0 & 0 & 0 & 1 \end{bmatrix} \bullet \begin{bmatrix} 1 \\ 0 \\ 2 \\ 1 \\ \overline{2} \\ 0 \\ 1 \\ 1 \end{bmatrix} \end{bmatrix}$$

$$= \begin{bmatrix} \begin{bmatrix} 2 & 1 & 1 & 1 \end{bmatrix} \bullet \begin{bmatrix} 1 \\ 0 \\ 2 \\ 1 \end{bmatrix} \\[2em] \begin{bmatrix} 2 & 1 & 1 & 1 \end{bmatrix} \bullet \begin{bmatrix} 2 \\ 0 \\ 1 \\ 1 \end{bmatrix} \end{bmatrix} = \begin{bmatrix} 5 \\ 6 \end{bmatrix}$$

$$W^* = G \bullet W = \begin{bmatrix} 1 & -1 \\ -1 & 2 \end{bmatrix} \begin{bmatrix} 5 \\ 6 \end{bmatrix} = \begin{bmatrix} -1 \\ 7 \end{bmatrix}.$$

4. We write the tensors $W = [w^\alpha_\circ]$ and $W^* = [w^\circ_\alpha]$ in the new basis, using the same matrix formulas as those used in question 2.

$$\hat{W} = C^{-1}W = \begin{bmatrix} -1 & -2 \\ 1 & 1 \end{bmatrix} \begin{bmatrix} 5 \\ 6 \end{bmatrix} = \begin{bmatrix} -17 \\ 11 \end{bmatrix};$$

$$\hat{W}^* = C^t W^* = \begin{bmatrix} 1 & -1 \\ -1 & 2 \end{bmatrix} \begin{bmatrix} -1 \\ 7 \end{bmatrix} = \begin{bmatrix} -8 \\ -9 \end{bmatrix}.$$

5. The nature of the given system of scalars $s(\alpha, \beta, \gamma, \delta, \epsilon)$ is obviously that of a Euclidean tensor of fifth order ($r = 5$):

$$p^{\alpha\beta\gamma\circ\circ}_{\circ\circ\circ\delta\epsilon} = g^{\alpha\beta}_{\circ\circ} \otimes t^{\gamma\circ\circ}_{\circ\delta\epsilon} = g^{\alpha\beta}_{\circ\circ} t^{\gamma\circ\circ}_{\circ\delta\epsilon}, \qquad (11.101)$$

because it is the tensor product of two given Euclidean tensors, $P = G^{-1} \otimes T$.

Its components will be given in matrix form as a column matrix with two blocks of four matrices: and their elements will be notated calculating the products of real numbers given by the free indices formula (11.101):

$$[p^{\alpha\beta\gamma\circ\circ}_{\circ\circ\circ\delta\epsilon}] = \begin{bmatrix} 2 & 0 & | & 4 & 0 \\ 4 & 2 & | & 2 & 2 \\ - & - & + & - & - \\ 1 & 0 & | & 2 & 0 \\ 2 & 1 & | & 1 & 1 \\ - & - & + & - & - \\ 1 & 0 & | & 2 & 0 \\ 2 & 1 & | & 1 & 1 \\ - & - & + & - & - \\ 1 & 0 & | & 2 & 0 \\ 2 & 1 & | & 1 & 1 \end{bmatrix}.$$

We obtain a total of $\sigma = n^r = 2^5 = 32$ components.

6. Replacing the notation of dummy indices by numbered indices: we raise the last two indices

$$p^{\beta_1\beta_2\beta_3\beta_4\beta_5}_{\circ\circ\circ\circ\circ} = g^{\beta_1\circ}_{\circ\alpha_1}g^{\beta_2\circ}_{\circ\alpha_2}g^{\beta_3\circ}_{\circ\alpha_3}g^{\beta_4\alpha_4}_{\circ\circ}g^{\beta_5\alpha_5}_{\circ\circ}p^{\alpha_1\alpha_2\alpha_3\circ\circ}_{\circ\circ\circ\alpha_4\alpha_5},$$

which in matrix interpretation leads to the formula

$$[p^{\alpha\beta\gamma\delta\epsilon}_{\circ\circ\circ\circ\circ}] = \left(I_2 \otimes I_2 \otimes I_2 \otimes G^{-1} \otimes G^{-1}\right) \bullet P_\sigma,$$

where

$$P^t_\sigma = \begin{bmatrix} 2\,0\,4\,2 & | & 4\,0\,2\,2 & | & 1\,0\,2\,1 & | & 2\,0\,1\,1 & | & 1\,0\,2\,1 & | & 2\,0\,1\,1 & | & 1\,0\,2\,1 & | & 2\,0\,1\,1 \end{bmatrix}$$

and

$$P'_\sigma = [p^{\alpha\beta\gamma\delta\epsilon}_{\circ\circ\circ\circ\circ}]$$

$$= \left(\left(\begin{bmatrix} 1 & 0 \\ 0 & 1 \end{bmatrix} \otimes \begin{bmatrix} 1 & 0 \\ 0 & 1 \end{bmatrix} \otimes \begin{bmatrix} 1 & 0 \\ 0 & 1 \end{bmatrix}\right) \otimes \left(\begin{bmatrix} 2 & 1 \\ 1 & 1 \end{bmatrix} \otimes \begin{bmatrix} 2 & 1 \\ 1 & 1 \end{bmatrix}\right)\right) \bullet P_\sigma$$

$$= \left(\begin{bmatrix} 1 & 0 & 0 & 0 & 0 & 0 & 0 & 0 \\ 0 & 1 & 0 & 0 & 0 & 0 & 0 & 0 \\ 0 & 0 & 1 & 0 & 0 & 0 & 0 & 0 \\ 0 & 0 & 0 & 1 & 0 & 0 & 0 & 0 \\ 0 & 0 & 0 & 0 & 1 & 0 & 0 & 0 \\ 0 & 0 & 0 & 0 & 0 & 1 & 0 & 0 \\ 0 & 0 & 0 & 0 & 0 & 0 & 1 & 0 \\ 0 & 0 & 0 & 0 & 0 & 0 & 0 & 1 \end{bmatrix} \otimes \begin{bmatrix} 4 & 2 & 2 & 1 \\ 2 & 2 & 1 & 1 \\ 2 & 1 & 2 & 1 \\ 1 & 1 & 1 & 1 \end{bmatrix}\right) \bullet P_\sigma = Z_2 \bullet P_\sigma$$

$$(P'_\sigma)^t = \begin{bmatrix} 1\,8\,10\,14\,8 & | & 22\,12\,14\,8 & | & 9\,5\,7\,4 & | & 11\,6\,7\,4 & | & 9\,5\,7\,4 & | & 11\,6\,7\,4 & | & 9\,5\,7\,4 & | & 11\,6\,7\,4 \end{bmatrix}$$

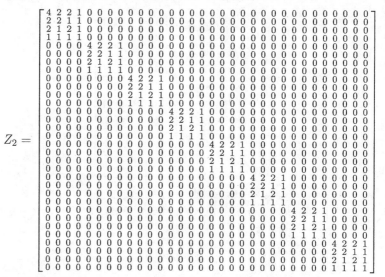

$$Z_2 =$$

and after condensing we obtain the contravariant components of $P = T \otimes U$:

$$[p^{\alpha\beta\gamma\delta\epsilon}_{\circ\circ\circ\circ\circ}] = \left[\begin{array}{cc|cc} 18 & 10 & 22 & 12 \\ 14 & 8 & 14 & 8 \\ \hline 9 & 5 & 11 & 6 \\ 7 & 4 & 7 & 4 \\ \hline 9 & 5 & 11 & 6 \\ 7 & 4 & 7 & 4 \\ \hline 9 & 5 & 11 & 6 \\ 7 & 4 & 7 & 4 \end{array}\right].$$

7. Since the indices (1,5) of the tensor product were initially the first and last indices (α, ϵ) of each factor, of the tensor product $P = G^{-1} \otimes T$, we are dealing with the determination of a contracted tensor product, that is, of what some authors call the "inner product" $(G^{-1} \cdot T)$.

In addition, as they are indices (α, ϵ) of different valency $(p^{\alpha\beta\gamma\circ\circ}_{\circ\circ\circ\delta\epsilon})$ can be contracted as a simple homogeneous tensor.

We initiate the calculus of the first components, and we express in matrix form the resulting tensor:

$$f^{\beta\gamma\circ}_{\circ\circ\delta} = p^{\theta\beta\gamma\circ\circ}_{\circ\circ\circ\delta\theta} = p^{1\beta\gamma\circ\circ}_{\circ\circ\circ\delta1} + p^{2\beta\gamma\circ\circ}_{\circ\circ\circ\delta2}$$

$$f^{11\circ}_{\circ\circ1} = p^{11\circ\circ}_{\circ\circ\circ11} + p^{211\circ\circ}_{\circ\circ\circ12} = 2 + 0 = 2$$

$$f^{11\circ}_{\circ\circ2} = p^{11\circ\circ}_{\circ\circ\circ21} + p^{211\circ\circ}_{\circ\circ\circ22} = 4 + 1 = 5$$

$$f^{12\circ}_{\circ\circ1} = p^{12\circ\circ}_{\circ\circ\circ11} + p^{212\circ\circ}_{\circ\circ\circ12} = 4 + 0 = 4$$

$$f^{12o}_{oo2} = p^{112oo}_{ooo21} + p^{212oo}_{ooo22} = 2 + 1 = 3$$

and then, we obtain

$$[f^{\beta\gamma o}_{oo\delta}] = \begin{bmatrix} 2 & 5 \\ 4 & 3 \\ - & - \\ 1 & 3 \\ 2 & 2 \end{bmatrix}.$$

It can also be obtained by contracting the blocks of G^{-1} with tensor T, and we obtain the extended tensor

$$F_\sigma = \begin{bmatrix} \begin{bmatrix} 1 & 0 \\ 2 & 1 \\ 2 & 0 \\ 1 & 1 \end{bmatrix} \begin{bmatrix} 2 \\ 1 \end{bmatrix} \\ - - - - - - - \\ \begin{bmatrix} 1 & 0 \\ 2 & 1 \\ 2 & 0 \\ 1 & 1 \end{bmatrix} \begin{bmatrix} 1 \\ 1 \end{bmatrix} \end{bmatrix} = \begin{bmatrix} 2 \\ 5 \\ 4 \\ 3 \\ - \\ 1 \\ 3 \\ 2 \\ 2 \end{bmatrix}.$$

□

Example 11.7 (Euclidean tensor algebra. Contractions). In a Euclidean space $E^2(\mathbb{R})$ referred to a basis $\{\vec{e}_\alpha\}$ the fundamental metric tensor is given by the Gram matrix

$$G \equiv [g^{oo}_{\alpha\beta}] = \begin{bmatrix} 1 & -1 \\ -1 & 2 \end{bmatrix}.$$

Consider two Euclidean tensors built over the mentioned space, with components

$$\text{Tensor } U : \quad [u^{\alphao}_{o\beta}] = \begin{bmatrix} 0 & 1 \\ -1 & 2 \end{bmatrix}.$$

$$\text{Tensor } V : \quad [v^{o\mu o}_{\lambdao\nu}] = \begin{bmatrix} 0 & 1 \\ -1 & 0 \\ - & - \\ 2 & 0 \\ 1 & 2 \end{bmatrix}.$$

1. Give the totally covariant components of $P = U \otimes V$.

2. Give the totally contravariant components of the tensor $\mathcal{EC}(1, 3)V$.
3. Obtain the tensor resulting from the contracted tensor product $\mathcal{EC}(1, 4)P$, starting from the data tensors.

4. Consider in $E^2(\mathbb{R})$ the new basis $\{\vec{e}_i\}$, and a change-of-basis given by the matrix

$$C \equiv [c^{\alpha\circ}_{\circ i}] = \begin{bmatrix} 1 & 0 \\ 1 & 2 \end{bmatrix}.$$

a) Obtain the covariant components of U in the basis $\{\vec{e}_i\}$.
b) Obtain the contravariant components of the tensor $\mathcal{E}C(1, 4)P$ in the basis $\{\vec{e}_i\}$.
c) Obtain the covariant components of P in the basis $\{\vec{e}_i\}$.

Solution: We solve the questions by the classic and direct methods.

1. We first determine the components of both factors U and V in covariant coordinates:

$$u^{\circ\circ}_{\alpha\beta} = g^{\circ\circ}_{\alpha\theta} u^{\theta\circ}_{\circ\beta}; \quad [u^{\circ\circ}_{\alpha\beta}] = G \bullet U = \begin{bmatrix} 1 & -1 \\ -1 & 2 \end{bmatrix} \begin{bmatrix} 0 & 1 \\ -1 & 2 \end{bmatrix} = \begin{bmatrix} 1 & -1 \\ -2 & 3 \end{bmatrix}$$

$$v^{\circ\circ\circ}_{\lambda\mu\nu} = g^{\circ\circ}_{\mu\theta} v^{\circ\theta\circ}_{\lambda\circ\nu},$$

which with numbered indices becomes

$$v^{\circ\ \circ\ \circ}_{\beta_1\beta_2\beta_3} = g^{\alpha_1\circ}_{\circ\ \beta_1} g^{\circ\ \circ}_{\alpha_2\beta_2} g^{\circ\ \circ}_{\circ\ \beta_3} v^{\circ\ \alpha_2\ \circ}_{\alpha_1\ \circ\ \alpha_3},$$

and in matrix form

$$V'_\sigma = (I_2 \otimes G \otimes I_2) \bullet V_\sigma = \left(\begin{bmatrix} 1 & 0 \\ 0 & 1 \end{bmatrix} \otimes \begin{bmatrix} 1 & -1 \\ -1 & 2 \end{bmatrix} \otimes \begin{bmatrix} 1 & 0 \\ 0 & 1 \end{bmatrix} \right) \bullet \begin{bmatrix} 0 \\ 1 \\ -1 \\ 0 \\ \hline 2 \\ 0 \\ 1 \\ 2 \end{bmatrix}$$

$$= \begin{bmatrix} 1 & 0 & -1 & 0 & 0 & 0 & 0 & 0 \\ 0 & 1 & 0 & -1 & 0 & 0 & 0 & 0 \\ -1 & 0 & 2 & 0 & 0 & 0 & 0 & 0 \\ 0 & -1 & 0 & 2 & 0 & 0 & 0 & 0 \\ 0 & 0 & 0 & 0 & 1 & 0 & -1 & 0 \\ 0 & 0 & 0 & 0 & 0 & 1 & 0 & -1 \\ 0 & 0 & 0 & -1 & 0 & 2 & 0 \\ 0 & 0 & 0 & 0 & 0 & -1 & 0 & 2 \end{bmatrix} \bullet \begin{bmatrix} 0 \\ 1 \\ -1 \\ 0 \\ \hline 2 \\ 0 \\ 1 \\ 2 \end{bmatrix} = \begin{bmatrix} 1 \\ 1 \\ -2 \\ -1 \\ \hline 1 \\ -2 \\ 0 \\ 4 \end{bmatrix}$$

and condensing the result, we obtain

$$[v^{\circ\circ\circ}_{\lambda\mu\nu}] = \begin{bmatrix} 1 & 1 \\ -2 & -1 \\ \hline 1 & -2 \\ 0 & 4 \end{bmatrix}.$$

So that $P = U \otimes V$ in covariant coordinates is calculated by extending the factors. Executing now its tensor product and condensing in a block matrix (of matrix blocks) the result is

$$P'_{32} = U'_4 \otimes V'_8 = \begin{bmatrix} 1 \\ -1 \\ -2 \\ 3 \end{bmatrix} \otimes \begin{bmatrix} 1 \\ 1 \\ -2 \\ -1 \\ - \\ 1 \\ -2 \\ 0 \\ 4 \end{bmatrix}$$

and once condensed we get

$$[p^{\circ\,\circ\,\circ\,\circ\,\circ}_{\alpha\beta\lambda\mu\nu}] = \begin{bmatrix} 1 & 1 & | & 1 & -2 \\ -2 & -1 & | & 0 & 4 \\ - & - & + & - & - \\ -1 & -1 & | & -1 & 2 \\ 2 & 1 & | & 0 & -4 \\ - & - & + & - & - \\ -2 & -2 & | & -2 & 4 \\ 4 & 2 & | & 0 & -8 \\ - & - & + & - & - \\ 3 & 3 & | & 3 & -6 \\ -6 & -3 & | & 0 & 12 \end{bmatrix}.$$

2. The indices of the tensor V to be contracted are both covariant indices (λ, ν), so that we need to use the Euclidean contraction. According to the mnemonic rule, we have

$$t^{\mu}_{\circ} = g^{\lambda\nu}_{\circ\circ} v^{\circ\mu\circ}_{\lambda\circ\nu},$$

and applying Formula (11.95), we obtain

$$G_0 = G^{-1} = \begin{bmatrix} 1 & -1 \\ -1 & 2 \end{bmatrix}^{-1} = \begin{bmatrix} 2 & 1 \\ 1 & 1 \end{bmatrix}$$

$$G^t_{\sigma} = (\text{extension } G_0) = \begin{bmatrix} 2 & 1 & 1 & 1 \end{bmatrix}$$

$$[t^{\mu}_{\circ}] = \begin{bmatrix} G^t_{\sigma} \bullet \left(\begin{bmatrix} 1 & 0 \\ 0 & 1 \end{bmatrix} \otimes [1 \ \ 0] \otimes \begin{bmatrix} 1 & 0 \\ 0 & 1 \end{bmatrix} \right) \bullet V_{\sigma} \\ G^t_{\sigma} \bullet \left(\begin{bmatrix} 1 & 0 \\ 0 & 1 \end{bmatrix} \otimes [0 \ \ 1] \otimes \begin{bmatrix} 1 & 0 \\ 0 & 1 \end{bmatrix} \right) \bullet V_{\sigma} \end{bmatrix}$$

$$
=
\begin{bmatrix}
[2 \;\; 1 \;\; 1 \;\; 1] \bullet
\begin{bmatrix}
1 & 0 & 0 & 0 & 0 & 0 & 0 & 0 \\
0 & 1 & 0 & 0 & 0 & 0 & 0 & 0 \\
0 & 0 & 0 & 0 & 1 & 0 & 0 & 0 \\
0 & 0 & 0 & 0 & 0 & 1 & 0 & 0
\end{bmatrix}
\bullet
\begin{bmatrix}
0 \\ 1 \\ -1 \\ 0 \\ \hline 2 \\ 0 \\ 1 \\ 2
\end{bmatrix}
\end{bmatrix}
$$

$$
=
\begin{bmatrix}
[2\,1\,1\,1] \bullet
\begin{bmatrix}
0&0&1&0&0&0&0&0 \\
0&0&0&1&0&0&0&0 \\
0&0&0&0&0&0&1&0 \\
0&0&0&0&0&0&0&1
\end{bmatrix}
\bullet
\begin{bmatrix}
0 \\ 1 \\ -1 \\ 0 \\ \hline 2 \\ 0 \\ 1 \\ 2
\end{bmatrix}
\end{bmatrix}
$$

$$
=
\begin{bmatrix}
[2\,1\,1\,1] \bullet
\begin{bmatrix} 0 \\ 1 \\ 2 \\ 0 \end{bmatrix} \\[4pt]
[2\,1\,1\,1] \bullet
\begin{bmatrix} -1 \\ 0 \\ 1 \\ 2 \end{bmatrix}
\end{bmatrix}
=
\begin{bmatrix} 3 \\ 1 \end{bmatrix}.
$$

3. We have $Q = \mathcal{E}\mathcal{C}(1,4)P$ and the indices $(1,4) \equiv (\alpha, \mu)$.

Method 1.
Classic approach. We perform the Euclidean contraction between the two factors

$$
Q <> [q^{\circ\circ\circ}_{\beta\lambda\nu}] = [u^{\alpha\circ}_{\circ\beta}g^{\circ\circ}_{\alpha\mu}v^{\circ\mu\circ}_{\lambda\circ\nu}]
$$

$$
= [u^{1\circ}_{\circ\beta}g^{\circ\circ}_{11}v^{\circ1\circ}_{\lambda\circ\nu} + u^{1\circ}_{\circ\beta}g^{\circ\circ}_{12}v^{\circ2\circ}_{\lambda\circ\nu} + u^{2\circ}_{\circ\beta}g^{\circ\circ}_{21}v^{\circ1\circ}_{\lambda\circ\nu} + u^{2\circ}_{\circ\beta}g^{\circ\circ}_{22}v^{\circ2\circ}_{\lambda\circ\nu}]
$$

Series $\beta\lambda\nu$	Summand values	$q^{\circ\circ\circ}_{\beta\lambda\nu}$
1 1 1	$0 \times 1 \times 0 + 0 \times (-1) \times (-1) + (-1) \times (-1) \times 0 + (-1) \times 2 \times (-1)$	2
1 1 2	$0 \times 1 \times 1 + 0 \times (-1) \times 0 + (-1) \times (-1) \times 1 + (-1) \times 2 \times 0$	1
1 2 1	$0 \times 1 \times 2 + 0 \times (-1) \times 1 + (-1) \times (-1) \times 2 + (-1) \times 2 \times 1$	0
1 2 2	$0 \times 1 \times 0 + 0 \times (-1) \times 2 + (-1) \times (-1) \times 0 + (-1) \times 2 \times 2$	-4
2 1 1	$1 \times 1 \times 0 + 1 \times (-1) \times (-1) + 2 \times (-1) \times 0 + 2 \times 2 \times (-1)$	-3
2 1 2	$1 \times 1 \times 1 + 1 \times (-1) \times 0 + 2 \times (-1) \times 1 + 2 \times 2 \times 0$	-1
2 2 1	$1 \times 1 \times 2 + 1 \times (-1) \times 1 + 2 \times (-1) \times 2 + 2 \times 2 \times 1$	1
2 2 2	$1 \times 1 \times 0 + 1 \times (-1) \times 2 + 2 \times (-1) \times 0 + 2 \times 2 \times 2$	6

and the sought after tensor Q is

$$
[q^{\circ\circ\circ}_{\beta\lambda\nu}] =
\left[
\begin{array}{cc}
2 & 1 \\
0 & -4 \\
\hline
-3 & -1 \\
1 & 6
\end{array}
\right].
$$

Method 2.

We determine the tensor product $P = U \otimes V$, but preparing first the indices to be contracted, with different valency.

So, we calculate $[u^{\circ\,\circ}_{\alpha\beta}]$ lowering the index α, which was already done at the beginning of question 1, where we obtained

$$[u^{\circ\,\circ}_{\alpha\beta}] = G \cdot U = \begin{bmatrix} 1 & -1 \\ -2 & 3 \end{bmatrix}.$$

Next, we execute the tensor product $P = u^{\circ\,\circ}_{\alpha\beta} \otimes v^{\circ\mu\circ}_{\lambda\circ\nu}$, operation that is performed with the extended tensor factors

$$P''_{32} = U''_4 \otimes V''_8 = \begin{bmatrix} 1 \\ -1 \\ -2 \\ 3 \end{bmatrix} \otimes \begin{bmatrix} 0 \\ 1 \\ -1 \\ 0 \\ \hline 2 \\ 0 \\ 1 \\ 2 \end{bmatrix} ;$$

$$(P''_{32})^t = [A\,|\,B\,|\,C\,|\,D],$$

where

$$A = \begin{bmatrix} 0 & 1 & -1 & 0 & 2 & 0 & 1 & 2 \end{bmatrix}$$
$$B = \begin{bmatrix} 0 & -1 & 1 & 0 & -2 & 0 & -1 & -2 \end{bmatrix}$$
$$C = \begin{bmatrix} 0 & -2 & 2 & 0 & -4 & 0 & -2 & -4 \end{bmatrix}$$
$$D = \begin{bmatrix} 0 & 3 & -3 & 0 & 6 & 0 & 3 & 6 \end{bmatrix}.$$

Finally we proceed to contract indices $(1, 4) \equiv (\alpha, \mu)$, which already are in the position of a normal homogeneous contraction, over the proper tensor product

$$Q <> [q^{\circ\,\circ\,\circ}_{\beta\lambda\nu}] = C\binom{\alpha}{\mu}[p^{\circ\,\circ\,\circ\mu\circ}_{\alpha\beta\lambda\circ\nu}] = [p^{\circ\,\circ\,\circ\theta\circ}_{\theta\beta\lambda\circ\nu}],$$

where

$$p^{\circ\,\circ\,\circ\theta\circ}_{\theta\beta\lambda\circ\nu} = p^{\circ\,\circ\,\circ1\circ}_{1\beta\lambda\circ\nu} + p^{\circ\,\circ\,\circ2\circ}_{2\beta\lambda\circ\nu}.$$

Series $\beta\lambda\nu$	Indices and summand values	$q^{\circ\,\circ\,\circ}_{\beta\lambda\nu}$
1 1 1	$(1\ 1\ 1\ 1\ 1) + (2\ 1\ 1\ 2\ 1) = 0 + 2$	2
1 1 2	$(1\ 1\ 1\ 1\ 2) + (2\ 1\ 1\ 2\ 2) = 1 + 0$	1
1 2 1	$(1\ 1\ 2\ 1\ 1) + (2\ 1\ 2\ 2\ 1) = 2 + (-2)$	0
1 2 2	$(1\ 1\ 2\ 1\ 2) + (2\ 1\ 2\ 2\ 2) = 0 + (-4)$	-4
2 1 1	$(1\ 2\ 1\ 1\ 1) + (2\ 2\ 1\ 2\ 1) = 0 - 3$	-3
2 1 2	$(1\ 2\ 1\ 1\ 2) + (2\ 2\ 1\ 2\ 2) = -1 + 0$	-1
2 2 1	$(1\ 2\ 2\ 1\ 1) + (2\ 2\ 2\ 2\ 1) = -2 + 3$	1
2 2 2	$(1\ 2\ 2\ 1\ 2) + (2\ 2\ 2\ 2\ 2) = 0 + 6$	6

As can be seen, it leads to the same result, which, due to the theorem for Euclidean tensors could not be otherwise.

4. The desired components are calculated as follows.

(a) The direct procedure in this case consists of executing the proposed change-of-basis over the tensor U in covariant coordinates.

From the beginning of question 1 we know that U in covariant coordinates is

$$[u_{\alpha\beta}^{\circ\circ}] = G \bullet U = \begin{bmatrix} 1 & -1 \\ -1 & 2 \end{bmatrix} \begin{bmatrix} 0 & 1 \\ -1 & 2 \end{bmatrix} = \begin{bmatrix} 1 & -1 \\ -2 & 3 \end{bmatrix},$$

and the tensor equation of the change-of-basis is

$$u_{ij}^{\circ\circ} = u_{\alpha\beta}^{\circ\circ} c_{i\circ}^{\circ\alpha} c_{j\circ}^{\circ\beta},$$

which interpreted in matrix form, leads to

$$\hat{U}_\sigma = (C^t \otimes C^t) \bullet U_\sigma = \left(\begin{bmatrix} 1 & 1 \\ 0 & 2 \end{bmatrix} \otimes \begin{bmatrix} 1 & 1 \\ 0 & 2 \end{bmatrix} \right) \bullet \begin{bmatrix} 1 \\ -1 \\ -2 \\ 3 \end{bmatrix}$$

$$= \begin{bmatrix} 1 & 1 & 1 & 1 \\ 0 & 2 & 0 & 2 \\ 0 & 0 & 2 & 2 \\ 0 & 0 & 0 & 4 \end{bmatrix} \bullet \begin{bmatrix} 1 \\ -1 \\ -2 \\ 3 \end{bmatrix} = \begin{bmatrix} 1 \\ 4 \\ 2 \\ 12 \end{bmatrix},$$

and condensing,

$$\hat{U} <> [u_{ij}^{\circ\circ}] = \begin{bmatrix} 1 & 4 \\ 2 & 12 \end{bmatrix}.$$

(b) The tensor $Q = \mathcal{EC}(1, 4)P \equiv \mathcal{EC}(1, 4)U \otimes V$, was already obtained in covariant coordinates in question 3:

$$Q <> [q_{\beta\lambda\nu}^{\circ\circ\circ}] = \begin{bmatrix} 2 & 1 \\ 0 & -4 \\ \hline -3 & -1 \\ 1 & 6 \end{bmatrix}.$$

We proceed to raise all indices, to have it in contravariant coordinates, and then, we proceed to change its basis, working with numbered indices

$$q_{\circ\circ\circ}^{\beta_1\beta_2\beta_3} = g_{\circ\circ}^{\beta_1\alpha_1} g_{\circ\circ}^{\beta_2\alpha_2} g_{\circ\circ}^{\beta_3\alpha_3} q_{\alpha_1\alpha_2\alpha_3}^{\circ\circ\circ},$$

which in matrix form becomes

$$Q'_\sigma = (G^{-1} \otimes G^{-1} \otimes G^{-1}) \bullet Q_\sigma$$

$$= \left(\begin{bmatrix} 2 & 1 \\ 1 & 1 \end{bmatrix} \otimes \begin{bmatrix} 2 & 1 \\ 1 & 1 \end{bmatrix} \otimes \begin{bmatrix} 2 & 1 \\ 1 & 1 \end{bmatrix} \right) \bullet \begin{bmatrix} 2 \\ 1 \\ 0 \\ -4 \\ \hline -3 \\ -1 \\ 1 \\ 6 \end{bmatrix}$$

$$
= \begin{bmatrix} 8 & 4 & 4 & 2 & 4 & 2 & 2 & 1 \\ 4 & 4 & 2 & 2 & 2 & 2 & 1 & 1 \\ 4 & 2 & 4 & 2 & 2 & 1 & 2 & 1 \\ 2 & 2 & 2 & 2 & 1 & 1 & 1 & 1 \\ 4 & 2 & 2 & 1 & 4 & 2 & 2 & 1 \\ 2 & 2 & 1 & 1 & 2 & 2 & 1 & 1 \\ 2 & 1 & 2 & 1 & 2 & 1 & 2 & 1 \\ 1 & 1 & 1 & 1 & 1 & 1 & 1 & 1 \end{bmatrix} \bullet \begin{bmatrix} 2 \\ 1 \\ 0 \\ -4 \\ -3 \\ -1 \\ 1 \\ 6 \end{bmatrix} = \begin{bmatrix} 6 \\ 3 \\ 3 \\ 1 \\ 0 \\ 1 \\ 2 \\ 2 \end{bmatrix}.
$$

Once the extended tensor product in contravariant coordinates is known, we perform the change-of-basis, where

$$
C = \begin{bmatrix} 1 & 0 \\ 1 & 2 \end{bmatrix}; \quad C^{-1} = \frac{1}{2} \begin{bmatrix} 2 & 0 \\ -1 & 1 \end{bmatrix},
$$

which leads to the tensor equation of change-of-basis

$$
[q_{ooo}^{j\ell n}] = q_{ooo}^{\beta\lambda\nu} \gamma_{o\beta}^{jo} \gamma_{o\lambda}^{\ell o} \gamma_{o\nu}^{no},
$$

which in matrix form becomes

$$
\hat{Q}'_\sigma = \left(C^{-1} \otimes C^{-1} \otimes C^{-1} \right) \bullet Q'_\sigma
$$

$$
= \left(\frac{1}{2} \begin{bmatrix} 2 & 0 \\ -1 & 1 \end{bmatrix} \otimes \frac{1}{2} \begin{bmatrix} 2 & 0 \\ -1 & 1 \end{bmatrix} \otimes \frac{1}{2} \begin{bmatrix} 2 & 0 \\ -1 & 1 \end{bmatrix} \right) \bullet \begin{bmatrix} 6 \\ 3 \\ 3 \\ 1 \\ 0 \\ 1 \\ 2 \\ 2 \end{bmatrix}
$$

$$
= \frac{1}{8} \begin{bmatrix} 8 & 0 & 0 & 0 & 0 & 0 & 0 & 0 \\ -4 & 4 & 0 & 0 & 0 & 0 & 0 & 0 \\ -4 & 0 & 4 & 0 & 0 & 0 & 0 & 0 \\ 2 & -2 & -2 & 2 & 0 & 0 & 0 & 0 \\ -4 & 0 & 0 & 0 & 4 & 0 & 0 & 0 \\ 2 & -2 & 0 & 0 & -2 & 2 & 0 & 0 \\ 2 & 0 & -2 & 0 & -2 & 0 & 2 & 0 \\ -1 & 1 & 1 & -1 & 1 & -1 & -1 & 1 \end{bmatrix} \bullet \begin{bmatrix} 6 \\ 3 \\ 3 \\ 1 \\ 0 \\ 1 \\ 2 \\ 2 \end{bmatrix} = \frac{1}{4} \begin{bmatrix} 24 \\ -6 \\ -6 \\ 1 \\ -12 \\ 4 \\ 5 \\ -1 \end{bmatrix},
$$

which, once condensed, leads to

$$
\hat{Q} <> [q_{ooo}^{j\ell n}] = \frac{1}{4} \begin{bmatrix} 24 & -6 \\ -6 & 1 \\ -12 & 4 \\ 5 & -1 \end{bmatrix}.
$$

(c) As the covariant coordinates of P in the initial basis, were calculated in question 1, it remains only to perform the change-of-basis

$$
p_{ij\ell mn}^{ooooo} = p_{\alpha\beta\lambda\mu\nu}^{ooooo} c_{io}^{o\alpha} c_{jo}^{o\beta} c_{\ell o}^{o\lambda} c_{mo}^{o\mu} c_{no}^{o\nu},
$$

where

$$(P'_{32,1})^t = [A\,|\,B\,|\,C\,|\,D],$$

where

$$A = \begin{bmatrix} 1 & 1 & -2 & -1 & 1 & -2 & 0 & 4 \end{bmatrix}$$
$$B = \begin{bmatrix} -1 & -1 & 2 & 1 & -1 & 2 & 0 & -4 \end{bmatrix}$$
$$C = \begin{bmatrix} -2 & -2 & 4 & 2 & -2 & 4 & 0 & -8 \end{bmatrix}$$
$$D = \begin{bmatrix} 3 & 3 & -6 & -3 & 3 & -6 & 0 & 12 \end{bmatrix}.$$

and in matrix form:

$$\hat{P}'_{32,1} = \left(C^t \otimes C^t \otimes C^t \otimes C^t \otimes C^t \right) \bullet P'_{32,1}$$
$$= \left(\begin{bmatrix} 1 & 0 \\ 1 & 2 \end{bmatrix}^t \otimes \begin{bmatrix} 1 & 0 \\ 1 & 2 \end{bmatrix}^t \otimes \begin{bmatrix} 1 & 0 \\ 1 & 2 \end{bmatrix}^t \otimes \begin{bmatrix} 1 & 0 \\ 1 & 2 \end{bmatrix}^t \otimes \begin{bmatrix} 1 & 0 \\ 1 & 2 \end{bmatrix}^t \right) \bullet P'_{32,1}$$
$$= [A\,|\,B\,|\,C\,|\,D]^t,$$

where

$$A = \begin{bmatrix} 2 & 4 & 2 & 12 & 6 & 8 & 16 & 32 \end{bmatrix}$$
$$B = \begin{bmatrix} 8 & 16 & 8 & 48 & 24 & 32 & 64 & 128 \end{bmatrix}$$
$$C = \begin{bmatrix} 4 & 8 & 4 & 24 & 12 & 16 & 32 & 64 \end{bmatrix}$$
$$D = \begin{bmatrix} 24 & 48 & 24 & 144 & 72 & 96 & 192 & 384 \end{bmatrix},$$

which, once condensed, gives

$$[p^{\circ\circ\circ\circ\circ\circ}_{ij\ell mn}] = \begin{bmatrix} 2 & 4 & | & 6 & 8 \\ 2 & 12 & | & 16 & 32 \\ - & - & + & - & - \\ 8 & 16 & | & 24 & 32 \\ 8 & 48 & | & 64 & 128 \\ - & - & + & - & - \\ 4 & 8 & | & 12 & 16 \\ 4 & 24 & | & 32 & 64 \\ - & - & + & - & - \\ 24 & 48 & | & 72 & 96 \\ 24 & 144 & | & 192 & 384 \end{bmatrix}.$$

□

Example 11.8 (Euclidean tensor algebra: Inner product). In a Euclidean space $E^3(\mathbb{R})$, referred to the basis $\{\vec{e}_\alpha\}$ and fundamental metric tensor, in covariant coordinates

$$G \equiv [g^{\circ\circ}_{\alpha\beta}] = \begin{bmatrix} 3 & -1 & 0 \\ -1 & 2 & 0 \\ 0 & 0 & 1 \end{bmatrix},$$

we perform a change-of-basis, of associated matrix

$$C \equiv [c_{\circ i}^{\alpha \circ}] = \begin{bmatrix} 1 & 1 & 0 \\ 1 & 2 & 1 \\ 1 & 3 & -1 \end{bmatrix}.$$

Consider the tensors T and U, built over this space, with components

$$[t_{\circ \circ}^{\alpha \beta}] = \begin{bmatrix} 1 & 0 & 1 \\ 0 & 2 & 0 \\ 0 & -1 & 0 \end{bmatrix} \quad \text{and} \quad [u_{\circ \beta}^{\alpha \circ}] = \begin{bmatrix} 0 & 1 & 0 \\ 0 & 0 & 0 \\ 1 & 0 & 3 \end{bmatrix}.$$

1. Obtain the contracted tensor of U in the basis $\{\vec{e}_\alpha\}$ and $\{\vec{e}_i\}$.
2. Obtain the tensor product $T \otimes U$ in contravariant coordinates of $\{\vec{e}_i\}$.
3. Obtain the contracted "inner product" $T \bullet U$ of $T \otimes U$ with respect to the indices second and second of each data factor, in the initial basis $\{\vec{e}_\alpha\}$.

Solution: With the aim of giving maximum information, and clarifying the theory, we proceed to solve these questions by several methods, even though solution be evident by some of them.

1. The contracted tensor of U is obtained as follows.

$$G^{-1} \equiv [g_{\circ \circ}^{\alpha \beta}] = \begin{bmatrix} 3 & -1 & 0 \\ -1 & 2 & 0 \\ 0 & 0 & 1 \end{bmatrix}^{-1} = \frac{1}{5} \begin{bmatrix} 2 & 1 & 0 \\ 1 & 3 & 0 \\ 0 & 0 & 5 \end{bmatrix}.$$

and with the mnemonic rule

$$\rho = g_{\circ \alpha}^{\beta \circ} u_{\circ \beta}^{\alpha \circ}; \quad G_0 = [g_{\circ \alpha}^{\beta \circ}] = I_3 = \begin{bmatrix} 1 & 0 & 0 \\ 0 & 1 & 0 \\ 0 & 0 & 1 \end{bmatrix}.$$

According to Formula (11.91), we have

$$G_\sigma^t = \text{extension of } G_0 = [100010001]$$

and

$$(U_\sigma)^t = \text{extension of } [u_{\circ \beta}^{\alpha \circ}]^t = [010000103]$$

so that applying the formula, we get

$$\rho = G_\sigma^t \bullet U_\sigma = \begin{bmatrix} 1 & 0 & 0 & 0 & 1 & 0 & 0 & 0 & 1 \end{bmatrix} \bullet \begin{bmatrix} 0 \\ 1 \\ 0 \\ 0 \\ 0 \\ 0 \\ 1 \\ 0 \\ 3 \end{bmatrix} = 0+0+3 = 3.$$

Since $\rho = 3$ is a scalar, it remains invariant when changing the basis, so that this also is the result of the contraction of U, under the representation $[u^{i\,\circ}_{\circ j}]$. However, following the criterion initially presented, it will be verified.

We determine the tensor U, in the new basis

$$u^{i\,\circ}_{\circ j} = u^{\alpha\,\circ}_{\circ\beta}\gamma^{i\,\circ}_{\circ\alpha}c^{\circ\beta}_{j\,\circ},$$

and in matrix form

$$\hat{U}_\sigma = [C^{-1} \otimes C^t] \bullet U_\sigma; \quad C^{-1} = \begin{bmatrix} 1 & 1 & 0 \\ 1 & 2 & 1 \\ 1 & 3 & -1 \end{bmatrix}^{-1} = \frac{1}{3}\begin{bmatrix} 5 & -1 & -1 \\ -2 & 1 & 1 \\ -1 & 2 & -1 \end{bmatrix}$$

$$\hat{U}_9 = \left(\frac{1}{3}\begin{bmatrix} 5 & -1 & -1 \\ -2 & 1 & 1 \\ -1 & 2 & -1 \end{bmatrix} \otimes \begin{bmatrix} 1 & 1 & 1 \\ 1 & 2 & 3 \\ 0 & 1 & -1 \end{bmatrix}\right) \bullet \begin{bmatrix} 0 \\ 1 \\ 0 \\ 0 \\ 0 \\ 1 \\ 0 \\ 3 \end{bmatrix} = \begin{bmatrix} 1/3 \\ 0 \\ 8/3 \\ 2/3 \\ 2 \\ -5/3 \\ -5/3 \\ -4 \\ 2/3 \end{bmatrix}.$$

Next, we contract in the new basis

$$\hat{\rho} = g^{\circ j}_{i\,\circ}u^{i\,\circ}_{\circ j}; \quad \hat{G}_0 = [g^{\circ j}_{i\,\circ}] = I_3 = \begin{bmatrix} 1 & 0 & 0 \\ 0 & 1 & 0 \\ 0 & 0 & 1 \end{bmatrix},$$

and according to Formula (11.91), we get $\hat{G}^t_\sigma$ = extension of $\hat{G}_0$ = $\begin{bmatrix} 1 & 0 & 0 & 0 & 1 & 0 & 0 & 0 & 1 \end{bmatrix}$

$$\left(\hat{U}_\sigma\right)^t \equiv \left(\hat{U}_9\right)^t = \begin{bmatrix} 1/3 & 0 & 8/3 & 2/3 & 2 & -5/3 & -5/3 & -4 & 2/3 \end{bmatrix}$$

$$\hat{\rho} = \hat{G}^t_\sigma \bullet \hat{U}_\sigma = \begin{bmatrix} 1 & 0 & 0 & 0 & 1 & 0 & 0 & 0 & 1 \end{bmatrix} \bullet \begin{bmatrix} 1/3 \\ 0 \\ 8/3 \\ 2/3 \\ 2 \\ -5/3 \\ -5/3 \\ -4 \\ 2/3 \end{bmatrix} = 3,$$

whence $\hat{\rho} = \rho = 3$.

2. The selected method for solving this question is the following. First, the tensor U in contravariant coordinates is determined and next, T and U in contravariant coordinates of the new basis and the tensor product $P = T \otimes U$ are determined.

$$u^{\alpha\beta}_{\circ\circ} = g^{\theta\beta}_{\circ\circ}u^{\alpha\circ}_{\circ\theta};$$

$$[u^{\alpha\beta}_{\circ\circ}] = [u^{\alpha\circ}_{\circ\theta}g^{\theta\beta}_{\circ\circ}] = UG^{-1} = \begin{bmatrix} 0 & 1 & 0 \\ 0 & 0 & 0 \\ 1 & 0 & 3 \end{bmatrix} \cdot \frac{1}{5}\begin{bmatrix} 2 & 1 & 0 \\ 1 & 3 & 0 \\ 0 & 0 & 5 \end{bmatrix} = \frac{1}{5}\begin{bmatrix} 1 & 3 & 0 \\ 0 & 0 & 0 \\ 2 & 1 & 15 \end{bmatrix}.$$

We change basis $[t^{\alpha\beta}_{\circ\circ}]$ and $[u^{\alpha\beta}_{\circ\circ}]$, using the classical method, and we get

$$[t^{ij}_{\circ\circ}] = [t^{\alpha\beta}_{\circ\circ}\gamma^{i\circ}_{\circ\alpha}\gamma^{j\circ}_{\circ\beta}] = [\gamma^{i\circ}_{\circ\alpha}][t^{\alpha\beta}_{\circ\circ}][\gamma^{\circ j}_{\beta\circ}] = C^{-1} \bullet T \bullet (C^{-1})^t$$

$$= \frac{1}{3}\begin{bmatrix} 5 & -1 & -1 \\ -2 & 1 & 1 \\ -1 & 2 & -1 \end{bmatrix}\begin{bmatrix} 1 & 0 & 1 \\ 0 & 2 & 0 \\ 0 & -1 & 0 \end{bmatrix}\frac{1}{3}\begin{bmatrix} 5 & -2 & -1 \\ -1 & 1 & 2 \\ -1 & 1 & -1 \end{bmatrix}$$

$$= \frac{1}{3}\begin{bmatrix} 7 & -2 & -4 \\ -3 & 1 & 2 \\ -3 & 2 & 4 \end{bmatrix}.$$

$$[u^{ij}_{\circ\circ}] = C^{-1}[u^{\alpha\beta}_{\circ\circ}](C^{-1})^t$$

$$= \frac{1}{3}\begin{bmatrix} 5 & -1 & -1 \\ -2 & 1 & 1 \\ -1 & 2 & -1 \end{bmatrix}\frac{1}{5}\begin{bmatrix} 1 & 3 & 0 \\ 0 & 0 & 0 \\ 2 & 1 & 15 \end{bmatrix}\frac{1}{3}\begin{bmatrix} 5 & -2 & -1 \\ -1 & 1 & 2 \\ -1 & 1 & -1 \end{bmatrix}$$

$$= \frac{1}{45}\begin{bmatrix} 16 & -7 & 40 \\ -10 & 10 & -25 \\ 4 & -13 & 10 \end{bmatrix}.$$

$$P = T \otimes U;$$

$$[p^{ijk\ell}_{\circ\circ\circ\circ}] = [t^{ij}_{\circ\circ}] \otimes [u^{k\ell}_{\circ\circ}] = \frac{1}{3}\begin{bmatrix} 7 & -2 & -4 \\ -3 & 1 & 2 \\ -3 & 2 & 4 \end{bmatrix} \otimes \frac{1}{45}\begin{bmatrix} 16 & -7 & 40 \\ -10 & 10 & -25 \\ 4 & -13 & 10 \end{bmatrix}$$

$$[p^{ijk\ell}_{\circ\circ\circ\circ}] = \begin{bmatrix} \frac{112}{135} & -\frac{49}{135} & \frac{56}{27} & -\frac{32}{135} & \frac{14}{135} & -\frac{16}{27} & -\frac{64}{135} & \frac{28}{135} & -\frac{32}{27} \\ -\frac{14}{27} & \frac{14}{27} & -\frac{35}{27} & \frac{4}{27} & -\frac{4}{27} & \frac{10}{27} & \frac{8}{27} & -\frac{8}{27} & \frac{20}{27} \\ \frac{28}{135} & -\frac{91}{135} & \frac{14}{27} & -\frac{8}{135} & \frac{26}{135} & -\frac{4}{27} & -\frac{16}{135} & \frac{52}{135} & -\frac{8}{27} \\ -\frac{16}{45} & \frac{7}{45} & -\frac{8}{9} & \frac{16}{135} & -\frac{7}{135} & \frac{8}{27} & \frac{32}{135} & \frac{14}{135} & \frac{16}{27} \\ \frac{2}{9} & -\frac{2}{9} & \frac{5}{9} & -\frac{2}{27} & \frac{2}{27} & -\frac{5}{27} & -\frac{4}{27} & \frac{4}{27} & -\frac{10}{27} \\ -\frac{4}{45} & \frac{13}{45} & -\frac{2}{9} & \frac{4}{135} & -\frac{13}{135} & \frac{2}{27} & \frac{8}{135} & \frac{26}{135} & \frac{4}{27} \\ -\frac{16}{45} & \frac{7}{45} & -\frac{8}{9} & \frac{32}{135} & -\frac{14}{135} & \frac{16}{27} & \frac{64}{135} & \frac{28}{135} & \frac{32}{27} \\ \frac{2}{9} & -\frac{2}{9} & \frac{5}{9} & -\frac{4}{27} & \frac{4}{27} & -\frac{10}{27} & -\frac{8}{27} & \frac{8}{27} & -\frac{20}{27} \\ -\frac{4}{45} & \frac{13}{45} & -\frac{2}{9} & \frac{8}{135} & -\frac{26}{135} & \frac{4}{27} & \frac{16}{135} & -\frac{52}{135} & \frac{8}{27} \end{bmatrix}.$$

3. We use two methods.

Method 1. We wish to know the tensor $Q = \mathcal{C}(2,4)T \otimes U \equiv T \cdot U$.

$$[T \cdot U] = [q_{\circ\circ}^{\alpha\mu}] = \left[t_{\circ\circ}^{\alpha\theta} \cdot u_{\circ\theta}^{\mu\circ}\right] = [t_{\circ\circ}^{\alpha\theta}][u_{\theta\circ}^{\circ\mu}] = T \cdot U^t$$

$$= \begin{bmatrix} 1 & 0 & 1 \\ 0 & 2 & 0 \\ 0 & -1 & 0 \end{bmatrix} \begin{bmatrix} 0 & 1 & 0 \\ 0 & 0 & 0 \\ 1 & 0 & 3 \end{bmatrix}^t = \begin{bmatrix} 1 & 0 & 1 \\ 0 & 2 & 0 \\ 0 & -1 & 0 \end{bmatrix} \begin{bmatrix} 0 & 0 & 1 \\ 1 & 0 & 0 \\ 0 & 0 & 3 \end{bmatrix}$$

$$= \begin{bmatrix} 0 & 0 & 4 \\ 2 & 0 & 0 \\ -1 & 0 & 0 \end{bmatrix}.$$

Method 2. We use the direct method (computer), and determine $P = T \otimes U$:

$$[t_{\circ\circ}^{\alpha\beta}] \otimes [u_{\circ\lambda}^{\mu\circ}] = [p_{\circ\circ\circ\lambda}^{\alpha\beta\mu\circ}] = \begin{bmatrix} 1 & 0 & 1 \\ 0 & 2 & 0 \\ 0 & -1 & 0 \end{bmatrix} \otimes \begin{bmatrix} 0 & 1 & 0 \\ 0 & 0 & 0 \\ 1 & 0 & 3 \end{bmatrix}.$$

$$[P] \equiv [p_{\circ\circ\circ\lambda}^{\alpha\beta\mu\circ}] = \left[\begin{array}{ccc|ccc|ccc} 0 & 1 & 0 & 0 & 0 & 0 & 0 & 1 & 0 \\ 0 & 0 & 0 & 0 & 0 & 0 & 0 & 0 & 0 \\ 1 & 0 & 3 & 0 & 0 & 0 & 1 & 0 & 3 \\ \hline 0 & 0 & 0 & 0 & 2 & 0 & 0 & 0 & 0 \\ 0 & 0 & 0 & 0 & 0 & 0 & 0 & 0 & 0 \\ 0 & 0 & 0 & 2 & 0 & 6 & 0 & 0 & 0 \\ \hline 0 & 0 & 0 & 0 & -1 & 0 & 0 & 0 & 0 \\ 0 & 0 & 0 & 0 & 0 & 0 & 0 & 0 & 0 \\ 0 & 0 & 0 & -1 & 0 & -3 & 0 & 0 & 0 \end{array}\right]$$

and contract the indices $(2,4) \equiv (\beta, \lambda)$, using Formula (5.73), of model 5, and $\sigma = n^r = 3^4 = 81$ and $\sigma' = \frac{n^r}{n^2} = \frac{3^4}{3^2} = 9$, $P_\sigma = P_{81} = P$ extended *as a tensor of order 4 (not as a matrix [P]):*

$$Q_9' = (I_3 \otimes [I_3 \otimes [\,1\ 0\ 0\,]|I_3 \otimes [\,0\ 1\ 0\,]|I_3 \otimes [\,0\ \ 0\ \ 1\,]]) \bullet P_{81}$$

$$= \left(\begin{bmatrix} 1 & 0 & 0 \\ 0 & 1 & 0 \\ 0 & 0 & 1 \end{bmatrix} \otimes \begin{bmatrix} 100000000 & 010000000 & 001000000 \\ 000100000 & 000010000 & 000001000 \\ 000000100 & 000000010 & 000000001 \end{bmatrix} \right) \bullet P_{81}$$

Having executed the matrix product, we obtain $Q_{9,1}'$ (extended)

$$Q_{9,1}' = [\,0\ 0\ 4\ 2\ 0\ 0\ -1\ 0\ 0\,]^t,$$

which, once condensed, leads again to

$$[Q] \equiv [q_{\circ\circ}^{\alpha\mu}] = \begin{bmatrix} 0 & 0 & 4 \\ 2 & 0 & 0 \\ -1 & 0 & 0 \end{bmatrix}.$$

$\square$

Example 11.9 (Homomorphisms of metric tensors). Consider the Euclidean space $E^4(\mathbb{R})$ referred to a basis $\{\vec{e}_\alpha\}$, such that the fundamental metric tensor expressed in covariant coordinates is

$$G \equiv [g^{\circ\circ}_{\alpha\beta}] = \begin{bmatrix} 1 & 0 & -2 & 0 \\ 0 & 1 & 0 & 0 \\ -2 & 0 & 6 & 1 \\ 0 & 0 & 1 & 1 \end{bmatrix}.$$

We wish to extend the concept of *vectorial tensor product* defined in the OGS (ordinary geometric space), in the following form. Consider three vectors $\vec{V}_1, \vec{V}_2, \vec{V}_3 \in E^4(\mathbb{R})$. We give the name "triple vector product" and notation $\bigwedge \vec{V}_1, \vec{V}_2, \vec{V}_3 \bigwedge$ to another vector with direction orthogonal to the hyperplane defined by the vectors $\vec{V}_1, \vec{V}_2$ and $\vec{V}_3$, with orientation corresponding to the corkscrew rule in the mentioned direction (rotation in the direct orientation of the trihedron $\vec{V}_1, \vec{V}_2, \vec{V}_3$) and modulus the volume of the rhombohedron of intersecting edges in the trihedron $\vec{V}_1, \vec{V}_2, \vec{V}_3$.

1. Propose by extension of the OGS, the formula of the mentioned tensor, assuming that the vectors $\vec{V}_1, \vec{V}_2$ and $\vec{V}_3$ are given by its contravariant components

$$X = \begin{bmatrix} x^1 \\ x^2 \\ x^3 \\ x^4 \end{bmatrix}, Y \text{ and } Z,$$

respectively.

2. Consider an endomorphism $T : E^4(\mathbb{R}) \longrightarrow E^4(\mathbb{R})$, such that $\forall \vec{V} \in E^4(\mathbb{R})$, $T_{\vec{V}_1, \vec{V}_2}(\vec{V}) = \vec{W}$, where $\vec{V}_1$ and $\vec{V}_2$ are two fixed vectors of coordinates

$$X_1 = \begin{bmatrix} 1 \\ 1 \\ 1 \\ 2 \end{bmatrix} \text{ and } X_2 = \begin{bmatrix} 0 \\ 1 \\ 0 \\ 1 \end{bmatrix} \text{ and } \vec{W} = \bigwedge \vec{V}_1, \vec{V}_2, \vec{V} \bigwedge,$$

where the expression means triple vector product of three vectors.
Give the matrix T associated with the given endomorphism $\xi = TX$.

3. If the previous matrix T is anti-symmetric, is the operator $T_{\vec{V}_1, \vec{V}_2}$ anti-symmetric?

4. Since any $T_{\vec{V}_1, \vec{V}_2}$ needs *two* fixed vectors of $E^4(\mathbb{R})$ to be established, a new correspondence of $E^4 \times E^4(\mathbb{R})$ in the linear space $\mathcal{L}[E^4(\mathbb{R})] \equiv$ of all operators $T_{\vec{V}_1, \vec{V}_2}$ is created.
Consider the correspondence $\Theta : E^4 \times E^4(\mathbb{R}) \longrightarrow \mathcal{L}[E^4(\mathbb{R})]$ that with any duple $(\vec{V}, X) \in E^4 \times E^4(\mathbb{R})$ associates a $T_{\vec{V}, X}$. Examine if Θ is or is not a monomorphism.

To solve this theoretical question we assume that the Euclidean space $E^4(\mathbb{R})$ is referred to an orthonormalized basis $\{\vec{e}_\beta\}$, that is, $G \equiv I_4$.

Solution.

1. The desired formula is:

$$\bigwedge \vec{V}_1, \vec{V}_2, \vec{V}_3 \bigwedge = \sqrt{|G|} \begin{vmatrix} \vec{e}^{*1} & \vec{e}^{*2} & \vec{e}^{*3} & \vec{e}^{*4} \\ x^1 & x^2 & x^3 & x^4 \\ y^1 & y^2 & y^3 & y^4 \\ z^1 & z^2 & z^3 & z^4 \end{vmatrix}$$

"copied" by extension of the dimension of the ordinary triple vector product. Note that the data input is in contravariant coordinates and *the output is in covariant coordinates* (only if G were I_4, would the output simultaneously be in contravariant coordinates):

$$|G| = \begin{vmatrix} 1 & 0 & -2 & 0 \\ 0 & 1 & 0 & 0 \\ -2 & 0 & 6 & 1 \\ 0 & 0 & 1 & 1 \end{vmatrix} = 1 \longrightarrow \sqrt{|G|} = 1.$$

2. The matrix T is obtained as follows.

$$\vec{\xi}^* \equiv \xi_\alpha^\circ \vec{e}^{*\alpha} = \begin{vmatrix} \vec{e}^{*1} & \vec{e}^{*2} & \vec{e}^{*3} & \vec{e}^{*4} \\ 1 & 1 & 1 & 2 \\ 0 & 1 & 0 & 1 \\ x^1 & x^2 & x^3 & x^4 \end{vmatrix}$$

$$= \begin{vmatrix} 1 & 1 & 2 \\ 1 & 0 & 1 \\ x^2 & x^3 & x^4 \end{vmatrix} \vec{e}^{*1} - \begin{vmatrix} 1 & 1 & 2 \\ 0 & 0 & 1 \\ x^1 & x^3 & x^4 \end{vmatrix} \vec{e}^{*2} + \begin{vmatrix} 1 & 1 & 2 \\ 0 & 1 & 1 \\ x^1 & x^2 & x^4 \end{vmatrix} \vec{e}^{*3} - \begin{vmatrix} 1 & 1 & 1 \\ 0 & 1 & 0 \\ x^1 & x^2 & x^3 \end{vmatrix} \vec{e}^{*4}$$

$$= (x^2 + x^3 - x^4)\vec{e}^{*1} + (-x^1 + x^3)\vec{e}^{*2}$$
$$+(-x^1 - x^2 + x^4)\vec{e}^{*3} + (x^1 - x^3)\vec{e}^{*4}.$$

We express this vector in contravariant coordinates, so that we "raise the index" in the Euclidean sense:

$$G^{-1} = \begin{bmatrix} 1 & 0 & -2 & 0 \\ 0 & 1 & 0 & 0 \\ -2 & 0 & 6 & 1 \\ 0 & 0 & 1 & 1 \end{bmatrix}^{-1} = \begin{bmatrix} 5 & 0 & 2 & -2 \\ 0 & 1 & 0 & 0 \\ 2 & 0 & 1 & -1 \\ -2 & 0 & -1 & 2 \end{bmatrix}$$

$$[\xi_\circ^\alpha] = [\xi_\theta^\circ g_\circ^{\theta\alpha}] = [g_\circ^{\alpha\theta}][\xi_\theta^\circ] = G^{-1}\xi^*$$

$$= \begin{bmatrix} 5 & 0 & 2 & -2 \\ 0 & 1 & 0 & 0 \\ 2 & 0 & 1 & -1 \\ -2 & 0 & -1 & 2 \end{bmatrix} \begin{bmatrix} x^2 + x^3 - x^4 \\ -x^1 + x^3 \\ -x^1 - x^2 + x^4 \\ x^1 - x^3 \end{bmatrix} = \begin{bmatrix} -4x^1 + 3x^2 + 7x^3 - 3x^4 \\ -x^1 + x^3 \\ -2x^1 + x^2 + 3x^3 - x^4 \\ 3x^1 - x^2 - 4x^3 + x^4 \end{bmatrix}$$

$$= \begin{bmatrix} -4 & 3 & 7 & -3 \\ -1 & 0 & 1 & 0 \\ -2 & 1 & 3 & -1 \\ 3 & -1 & -4 & 1 \end{bmatrix} \begin{bmatrix} x^1 \\ x^2 \\ x^3 \\ x^4 \end{bmatrix}$$

$$\xi = TX \longrightarrow \begin{bmatrix} \xi^1 \\ \xi^2 \\ \xi^3 \\ \xi^4 \end{bmatrix} = T \begin{bmatrix} x^1 \\ x^2 \\ x^3 \\ x^4 \end{bmatrix}; \quad T = \begin{bmatrix} -4 & 3 & 7 & -3 \\ -1 & 0 & 1 & 0 \\ -2 & 1 & 3 & -1 \\ 3 & -1 & -4 & 1 \end{bmatrix}.$$

3. A morphism is anti-symmetric or alternate in a Euclidean space if the dot product representation satisfies the conditions

$$\forall \vec{X}, \vec{Y} \longrightarrow \begin{cases} < T(\vec{X})|\vec{Y}> = <\vec{X}|T^*(\vec{Y})> \\ T^* = -T \end{cases}$$

that is,

$$\begin{cases} (TX)^t GY = X^t G(T^*Y) \\ T^* = -T \end{cases} \Rightarrow \begin{cases} T^t G = GT^* \\ T^* = -T \end{cases} \longrightarrow T^t G + GT = \Omega.$$

$$(11.102)$$

If $G \equiv I$, the previous condition leads to an anti-symmetric matrix $T^t = -T$.

Condition (11.102) is that of an anti-symmetric homomorphism in *any basis*. We examine T:

$$\begin{bmatrix} -4 & -1 & -2 & 3 \\ 3 & 0 & 1 & -1 \\ 7 & 1 & 3 & -4 \\ -3 & 0 & -1 & 1 \end{bmatrix} \begin{bmatrix} 1 & 0 & -2 & 0 \\ 0 & 1 & 0 & 0 \\ -2 & 0 & 6 & 1 \\ 0 & 0 & 1 & 1 \end{bmatrix} + \begin{bmatrix} 1 & 0 & -2 & 0 \\ 0 & 1 & 0 & 0 \\ -2 & 0 & 6 & 1 \\ 0 & 0 & 1 & 1 \end{bmatrix} \begin{bmatrix} -4 & 3 & 7 & -3 \\ -1 & 0 & 1 & 0 \\ -2 & 1 & 3 & -1 \\ 3 & -1 & -4 & 1 \end{bmatrix}$$

$$= \begin{bmatrix} 0 & -1 & -1 & 1 \\ 1 & 0 & -1 & 0 \\ 1 & 1 & 0 & -1 \\ -1 & 0 & 1 & 0 \end{bmatrix} + \begin{bmatrix} 0 & 1 & 1 & -1 \\ -1 & 0 & 1 & 0 \\ -1 & -1 & 0 & 1 \\ 1 & 0 & -1 & 0 \end{bmatrix} = \begin{bmatrix} 0 & 0 & 0 & 0 \\ 0 & 0 & 0 & 0 \\ 0 & 0 & 0 & 0 \\ 0 & 0 & 0 & 0 \end{bmatrix} \equiv \Omega_4.$$

Thus, though the operator does not resemble an anti-symmetric matrix, *it is a Euclidean anti-symmetric tensor*.

4. First, we examine the matrix T of the endomorphism $T_{\vec{V},\vec{X}}$, which the mapping Θ associates with the bivector $(\vec{V}, \vec{X}) \in E^4 \times E^4(\mathbb{R})$.

Consider the components of $\vec{V} <> \begin{bmatrix} v^1 \\ v^2 \\ v^3 \\ v^4 \end{bmatrix}$ and of $\vec{X} <> \begin{bmatrix} x^1 \\ x^2 \\ x^3 \\ x^4 \end{bmatrix}$.

$$\bigwedge \vec{V}, \vec{X}, \vec{Z} \bigwedge = \begin{vmatrix} \vec{e}_1 & \vec{e}_2 & \vec{e}_3 & \vec{e}_4 \\ v^1 & v^2 & v^3 & v^4 \\ x^1 & x^2 & x^3 & x^4 \\ z^1 & z^2 & z^3 & z^4 \end{vmatrix}$$

$$= \left(\begin{vmatrix} v^3 & v^4 \\ x^3 & x^4 \end{vmatrix} z^2 + \begin{vmatrix} v^4 & v^2 \\ x^4 & x^2 \end{vmatrix} z^3 + \begin{vmatrix} v^2 & v^3 \\ x^2 & x^3 \end{vmatrix} z^4 \right) \vec{e}_1 - \cdots$$

$$\cdots - \left(\begin{vmatrix} v^2 & v^3 \\ x^2 & x^3 \end{vmatrix} z^1 + \begin{vmatrix} v^3 & v^1 \\ x^3 & x^1 \end{vmatrix} z^2 + \begin{vmatrix} v^1 & v^2 \\ x^1 & x^2 \end{vmatrix} z^3 \right) \vec{e}_4,$$

then the anti-symmetric operator becomes

$$
T = \begin{bmatrix}
0 & -\begin{vmatrix} v^3 & v^4 \\ x^3 & x^4 \end{vmatrix} & \cdots & -\begin{vmatrix} v^2 & v^3 \\ x^2 & x^3 \end{vmatrix} \\
\begin{vmatrix} v^3 & v^4 \\ x^3 & x^4 \end{vmatrix} & 0 & \cdots & -\begin{vmatrix} v^3 & v^1 \\ x^3 & x^1 \end{vmatrix} \\
\begin{vmatrix} v^4 & v^2 \\ x^4 & x^2 \end{vmatrix} & \cdots & \cdots & -\begin{vmatrix} v^1 & v^2 \\ x^1 & x^2 \end{vmatrix} \\
\begin{vmatrix} v^2 & v^3 \\ x^2 & x^3 \end{vmatrix} & \cdots & \cdots & 0
\end{bmatrix}.
$$

Thus, the general matrix of T has the construction

$$
T = \begin{bmatrix}
0 & a_{12} & a_{13} & a_{14} \\
-a_{12} & 0 & a_{23} & a_{24} \\
-a_{13} & -a_{23} & 0 & a_{34} \\
-a_{14} & -a_{24} & -a_{34} & 0
\end{bmatrix}
$$

$$
= a_{12} \left[\begin{array}{cc|c} 0 & 1 & \\ -1 & 0 & \\ \hline - & - & + \; - \\ & & \mid \; \Omega \end{array}\right] + a_{13} \left[\begin{array}{ccc|c} 0 & 0 & 1 & \\ 0 & 0 & 0 & \\ -1 & 0 & 0 & \\ \hline - & - & - & + \; - \\ & & & \mid \; 0 \end{array}\right] + \cdots
$$

$$
+ a_{34} \left[\begin{array}{c|cc} \Omega & & \\ - & + & - \; - \\ \mid & 0 & 1 \\ \mid & -1 & 0 \end{array}\right],
$$

where 6 is the dimension of the basis of matrices generating T. Thus, the linear subspace of the operators T in $\mathcal{L}[E^4(\mathbb{R})]$ is of dimension 6 and since the dimension of $E^4 \times E^4$ is 8 (that of $\mathcal{L}[E^4(\mathbb{R})]$ is 16), the homomorphism Θ has a proper kernel, thus, it is *not* a *monomorphism*, because it is not injective.

(In any morphism, the kernel dimension plus the range linear subspace dimension is the dimension of the initial linear space.)

$\square$

11.11 Euclidean tensor metrics: The $\left(\overset{r}{\underset{1}{\otimes}} E^n\right)(\mathbb{R})$ space as Euclidean space

Once the "*Euclidean contraction*" has been established in detail, there arises the possibility of choosing pairs of tensors of a certain tensor space $\left(\overset{r}{\underset{1}{\otimes}} E^n\right)(\mathbb{R})$, constructed over a Euclidean space $E^n(\mathbb{R})$, and executing *totally contracted* tensor products with them, that is, creating a *dot product* associated with the tensor space, endowed with a formal rule for such a dot product to become also a *Euclidean* product, that would endow the tensor space $\left(\overset{r}{\underset{1}{\otimes}} E^n\right)(\mathbb{R})$ with a *proper fundamental metric tensor*, arising from the G of $E^n(\mathbb{R})$ as follows.

11.11.1 Inner connection in $\left(\overset{r}{\underset{1}{\otimes}} E^n\right)(\mathbb{R})$: Dot product of Euclidean tensors

Let $\{\vec{e}_\alpha\}$ be an arbitrary basis of a Euclidean linear space $E^n(\mathbb{R})$. The basis associated with the tensor space $\left(\overset{r}{\underset{1}{\otimes}} E^n\right)(\mathbb{R})$ will be

$$B = \{\vec{e}_{\alpha_1} \otimes \vec{e}_{\alpha_2} \otimes \ldots \otimes \vec{e}_{\alpha_r}\}, \tag{11.103}$$

such that $\forall \alpha_j \in I_n$; $\forall j \in I_r$, $\{\alpha_1, \alpha_2, \ldots, \alpha_r\}$ be a *totally ordered* subset of natural numbers.

The inner connection (or dot product) is defined as a "form" Φ:

$$\Phi : \left(\overset{r}{\underset{1}{\otimes}} E^n\right)(\mathbb{R}) \times \left(\overset{r}{\underset{1}{\otimes}} E^n\right)(\mathbb{R}) \longrightarrow \mathbb{R}; \forall (T, U), \Phi(T, U) = t^{\alpha_1 \alpha_2 \cdots \alpha_r}_{\circ\ \circ\ \cdots\ \circ} u^{\circ\ \circ\ \cdots\ \circ}_{\alpha_1 \alpha_2 \cdots \alpha_r} \tag{11.104}$$

that is, Φ is defined by means of a totally contracted Euclidean tensor product, with *the following criterion*: Since the dummy indices have *ordered* notation, we contract one index of the *first* factor with the *corresponding* index of the *second* factor: i.e., $(1, 1), (2, 2), (3, 3), \ldots, (r, r)$.

The tensor factor species is indifferent. We adopt, from this point on, the notation "$T \bullet U$" for the dot product $\Phi(T, U)$ of two Euclidean tensors.

We insist on the irrelevant character of the species. For example, if the two factor tensors are given, the first in contravariant coordinates and the second in covariant coordinates as in the definition formula (11.104), we show that they can be given in the reverse way.

With factors of normal contraction:

$$\begin{aligned}
T \bullet U &= t^{\alpha_1 \alpha_2 \cdots \alpha_r}_{\circ\ \circ\ \cdots\ \circ} u^{\circ\ \circ\ \cdots\ \circ}_{\alpha_1 \alpha_2 \cdots \alpha_r} \\
&= \left(t^{\circ\ \circ\ \cdots\ \circ}_{\beta_1 \beta_2 \cdots \beta_r} g^{\beta_1 \alpha_1}_{\circ\ \circ} g^{\beta_2 \alpha_2}_{\circ\ \circ} \cdots g^{\beta_r \alpha_r}_{\circ\ \circ}\right) u^{\circ\ \circ\ \cdots\ \circ}_{\alpha_1 \alpha_2 \cdots \alpha_r} \\
&= t^{\circ\ \circ\ \cdots\ \circ}_{\beta_1 \beta_2 \cdots \beta_r} \left(g^{\beta_1 \alpha_1}_{\circ\ \circ} g^{\beta_2 \alpha_2}_{\circ\ \circ} \cdots g^{\beta_r \alpha_r}_{\circ\ \circ} u^{\circ\ \circ\ \cdots\ \circ}_{\alpha_1 \alpha_2 \cdots \alpha_r}\right) \\
&= t^{\circ\ \circ\ \cdots\ \circ}_{\beta_1 \beta_2 \cdots \beta_r} u^{\beta_1 \beta_2 \cdots \beta_r}_{\circ\ \circ\ \cdots\ \circ} \tag{11.105}
\end{aligned}$$

and also of the type

$$T \bullet U = t^{\alpha_1\ \circ\ \circ\ \alpha_4 \cdots \alpha_r}_{\circ\ \alpha_2 \alpha_3\ \circ\ \cdots\ \circ} u^{\circ\ \alpha_2 \alpha_3\ \circ\ \cdots\ \circ}_{\alpha_1\ \circ\ \circ\ \alpha_4 \cdots \alpha_r}, \tag{11.106}$$

if mixed factors of complementary species are used.

With factors of Euclidean contraction, we now have the possibility of arbitrary factors

$$T \bullet U = t^{\alpha_1 \ \circ \ \alpha_3 \ \circ \ \cdots \ \circ}_{\ \circ \ \alpha_2 \ \circ \ \alpha_4 \cdots \alpha_r} g^{\ \circ \ \circ}_{\alpha_1 \beta_1} g^{\ \circ \ \circ \ \alpha_2 \beta_2}_{\ \circ \ \circ} g^{\ \circ \ \beta_3 \ \alpha_4 \ \circ}_{\alpha_3 \ \circ \ \circ \ \beta_4} \cdots g^{\ \alpha_r \ \circ}_{\ \circ \ \beta_r} u^{\beta_1 \ \circ \ \circ \ \beta_4 \cdots \beta_r}_{\ \circ \ \beta_2 \beta_3 \ \circ \ \cdots \ \circ}. \quad (11.107)$$

The formal properties are:

1. The commutativity $T \bullet U = U \bullet T$. Though the tensor product of tensors is not commutative, the commutativity is a consequence of the *total contraction*, that is, the dot product is unique, because it is a *zero-order tensor* over the field $\mathbb{R}$.

2. Positivity of the Euclidean norm. (This property *includes* the regularity that must exist in any Euclidean space): $N(\Omega) = 0 \in \mathbb{R}$.

$$\text{If } T \neq \Omega, \ N(T) = T \bullet T = t^{\alpha_1 \alpha_2 \cdots \alpha_r}_{\ \circ \ \circ \ \cdots \ \circ} g^{\ \circ \ \circ}_{\alpha_1 \beta_1} g^{\ \circ \ \circ}_{\alpha_2 \beta_2} \cdots g^{\ \circ \ \circ}_{\alpha_r \beta_r} t^{\beta_1 \beta_2 \cdots \beta_r}_{\ \circ \ \circ \ \cdots \ \circ} > 0.$$

It is a consequence of the fact that the matrix $[g^{\circ \circ}_{\alpha\beta}]$ of the fundamental metric tensor of $E^4(\mathbb{R})$ is positive definite.

3. Bilinear character of the dot product, with respect to the sum and the exterior product by scalars of the tensor space $\left(\overset{r}{\underset{1}{\otimes}} E^n\right)(\mathbb{R})$:

$$T \bullet (\alpha U + \beta V) = \alpha T \bullet U + \beta T \bullet V = \alpha U \bullet T + \beta V \bullet T$$

$$= (\alpha U + \beta V) \bullet T; \ \forall T, U, V \in \left(\overset{r}{\underset{1}{\otimes}} E^n\right)(\mathbb{R}).$$

It is an immediate consequence of the distributivity of the product of the real number with respect to the sums and of the double commutativity of the sum and the product of the real numbers.

Thus, it is established that $\left(\overset{r}{\underset{1}{\otimes}} E^n\right)(\mathbb{R})$ is also a Euclidean tensor space, so that the concepts modulus of a tensor, orthogonal tensors, $\sin(\alpha)$ and $\cos(\alpha)$ of two tensors, etc. have sense.

11.11.2 The induced fundamental metric tensor

The dimension of the tensor space $\left(\overset{r}{\underset{1}{\otimes}} E^n\right)(\mathbb{R})$, that is, the number σ of basic vectors of the basis B of the tensor space, Formula (11.103), is n^r, so that the dot products of each pair of vectors of B, are defined and notated as

$$g^{\ \circ \ \circ \ \circ \ \circ \ \cdots \ \circ \ \circ}_{\alpha_1 \beta_1 \alpha_2 \beta_2 \cdots \alpha_r \beta_r} = (\vec{e}_{\alpha_1} \otimes \vec{e}_{\alpha_2} \otimes \cdots \otimes \vec{e}_{\alpha_r}) \bullet (\vec{e}_{\beta_1} \otimes \vec{e}_{\beta_2} \otimes \cdots \otimes \vec{e}_{\beta_r})$$

$$= g^{\ \circ \ \circ}_{\alpha_1 \beta_1} \cdot g^{\ \circ \ \circ}_{\alpha_2 \beta_2} \cdot \cdots \cdot g^{\ \circ \ \circ}_{\alpha_r \beta_r}, \quad (11.108)$$

where $g^{\;\circ\;\;\circ\;\;\circ\;\;\circ\;\cdots\;\circ\;\;\circ}_{\alpha_1\beta_1\alpha_2\beta_2\cdots\alpha_r\beta_r}$ are the $(n^r)^2$ elements of a square matrix notated $G(r)$, which arise at from Formula (11.108), endowing the tensor space with a fundamental metric tensor in covariant coordinates. $G(r)$ is a matrix of order (n^r). The definition that proposes Formula (11.108) is based on Property 7 of Section 1.3.2, which refers to properties of the tensor and inner product of matrices.

Similarly, we know that *the same* Euclidean tensor space can be notated $\left(\overset{r}{\underset{1}{\otimes}} E^n_*\right)(\mathbb{R})$, as has been established in Section 11.2, when we refer to the reciprocal basis, that is

$$B^* = \{\vec{e}^{*\alpha_1} \otimes \vec{e}^{*\alpha_2} \otimes \cdots \otimes \vec{e}^{*\alpha_r}\}; \; \forall \alpha_j \in I_n; \; \forall j \in I_r. \tag{11.109}$$

The dot products of each pair of vectors in the basis B^* are defined and notated in an analogous form

$$g^{\alpha_1\beta_1\alpha_2\beta_2\cdots\alpha_r\beta_r}_{\;\circ\;\;\circ\;\;\circ\;\;\circ\;\cdots\;\circ\;\;\circ} = (\vec{e}^{*\alpha_1} \otimes \vec{e}^{*\alpha_2} \otimes \cdots \vec{e}^{*\alpha_r}) \bullet (\vec{e}^{*\beta_1} \otimes \vec{e}^{*\beta_2} \otimes \cdots \vec{e}^{*\beta_r})$$

$$= g^{\alpha_1\beta_1}_{\;\circ\;\;\circ} \cdot g^{\alpha_2\beta_2}_{\;\circ\;\;\circ} \cdots g^{\alpha_r\beta_r}_{\;\circ\;\;\circ}, \tag{11.110}$$

where $g^{\alpha_1\beta_1\alpha_2\beta_2\cdots\alpha_r\beta_r}_{\;\circ\;\;\circ\;\;\circ\;\;\circ\;\cdots\;\circ\;\;\circ}$ are the $(n^r)^2$ elements of a square matrix notated $G^*(r)$, which arise from Formula (11.110), endowing the tensor space with a fundamental metric tensor in contravariant coordinates.

The bilinear form ϕ, defined in (11.104), and notated $\forall T, U \in \left(\overset{r}{\underset{1}{\otimes}} E^n\right)(\mathbb{R})$: $T \bullet U$, is now calculated to work, either with the data tensors in contravariant coordinates (metric tensor in covariant coordinates), or with the data tensors in covariant coordinates (metric tensor in contravariant coordinates):

$$T \bullet U = t^{\alpha_1\alpha_2\cdots\alpha_r}_{\;\circ\;\;\circ\;\cdots\;\circ}\, g^{\;\circ\;\;\circ\;\;\circ\;\;\circ\;\cdots\;\circ\;\;\circ}_{\alpha_1\beta_1\alpha_2\beta_2\cdots\alpha_r\beta_r}\, u^{\beta_1\beta_2\cdots\beta_r}_{\;\circ\;\;\circ\;\cdots\;\circ}$$

$$= t^{\;\circ\;\;\circ\;\cdots\;\circ}_{\alpha_1\alpha_2\cdots\alpha_r}\, g^{\alpha_1\beta_1\alpha_2\beta_2\cdots\alpha_r\beta_r}_{\;\circ\;\;\circ\;\;\circ\;\;\circ\;\cdots\;\circ\;\;\circ}\, u^{\;\circ\;\;\circ\;\cdots\;\circ}_{\beta_1\beta_2\cdots\beta_r}. \tag{11.111}$$

In theory, we can calculate $T \bullet U$ also with other Euclidean contractions for "mixed" data, as indicated in Relation (11.107), though they are certainly infrequent.

In practical examples, which are of major importance in this book, we will use the corresponding fundamental metric tensor matrix expressions:

$$G(r) \equiv [g^{\;\circ\;\;\circ\;\;\circ\;\;\circ\;\cdots\;\circ\;\;\circ}_{\alpha_1\beta_1\alpha_2\beta_2\cdots\alpha_r\beta_r}] = \left(\overset{r}{\underset{1}{\otimes}} G\right)$$

$$= G \otimes G \otimes \cdots \otimes G \text{ (tensor in covariant coordinates)} \tag{11.112}$$

$$G(r) \equiv [g^{\alpha_1\beta_1\alpha_2\beta_2\cdots\alpha_r\beta_r}_{\;\circ\;\;\circ\;\;\circ\;\;\circ\;\cdots\;\circ\;\;\circ}] = \left(\overset{r}{\underset{1}{\otimes}} G^*\right)$$

$$= G^* \otimes G^* \otimes \cdots \otimes G^* \equiv \left(\overset{r}{\underset{1}{\otimes}} G^{-1}\right) \text{ (contravariant).} \tag{11.113}$$

It is also possible, as indicated, that the data tensors be mixed in which case, the fundamental metric tensor will be given in the form of tensor products of the matrices $\{G, G^{-1}, I_n\}$ with the ordering imposed by the pairs of indices $(\alpha_j \beta_j)$ of the fundamental metric tensor. For example, if

$$G(r) \equiv [g^{\alpha_1 \beta_1 \; \circ \; \circ \; \alpha_3 \; \circ \; \circ \; \beta_4 \cdots \; \circ \; \circ}_{\; \; \circ \; \; \circ \; \alpha_2 \beta_2 \; \circ \; \beta_3 \alpha_4 \; \circ \; \cdots \alpha_r \beta_r}]$$

the corresponding matrix would be

$$G(r) = G^{-1} \otimes G \otimes I_n \otimes I_n \otimes \cdots \otimes G \text{ (r matrices).} \qquad (11.114)$$

The dot product $T \bullet U$, is executed in matrix form with the "extended" matrices of the tensors, as in the classic quadratic forms

$$T \bullet U = T^t_{\sigma,1} \bullet G(r) \bullet U_{\sigma,1}. \qquad (11.115)$$

11.11.3 Reciprocal and orthonormal basis

We recommend that the reader study Sections 1.2 and 11.2, before commencing this section.

Next, we study the dot product of two arbitrary basic vectors belonging to two reciprocal basis $\mathcal{B}$ and $\mathcal{B}^*$, respectively.

According to the definitions and the formulation established in (11.108) and (11.110), we have

$$(\vec{e}_{\alpha_1} \otimes \vec{e}_{\alpha_2} \otimes \cdots \vec{e}_{\alpha_r}) \bullet (\vec{e}^{*\beta_1} \otimes \vec{e}^{*\beta_2} \otimes \cdots \vec{e}^{*\beta_r}) = g^{\; \circ \; \beta_1 \; \circ \; \beta_2 \cdots \; \circ \; \beta_r}_{\alpha_1 \; \circ \; \alpha_2 \; \circ \cdots \alpha_r \; \circ}$$

$$= g^{\; \circ \; \beta_1}_{\alpha_1 \; \circ} \cdot g^{\; \circ \; \beta_2}_{\alpha_2 \; \circ} \cdot \cdots \cdot g^{\; \circ \; \beta_r}_{\alpha_r \; \circ}$$

$$= \delta^{\; \circ \; \beta_1}_{\alpha_1 \; \circ} \cdot \delta^{\; \circ \; \beta_2}_{\alpha_2 \; \circ} \cdot \cdots \cdot \delta^{\; \circ \; \beta_r}_{\alpha_r \; \circ} = K,$$

which *shows* the reciprocal character of the bases $\mathcal{B}$ and $\mathcal{B}^*$, because the natural sequences $\{\alpha_1, \alpha_2, \ldots, \alpha_r\}$ and $\{\beta_1, \beta_2, \ldots, \beta_r\}$ satisfy the reciprocity conditions:

$$\text{If } \{\alpha_1, \alpha_2, \ldots, \alpha_r\} = \{\beta_1, \beta_2, \ldots, \beta_r\} \longrightarrow K = 1$$
$$\text{If } \{\alpha_1, \alpha_2, \ldots, \alpha_r\} \neq \{\beta_1, \beta_2, \ldots, \beta_r\} \longrightarrow K = 0. \qquad (11.116)$$

Naturally, one can build the reciprocal basis $(\mathcal{B}_0)^*$, of any other basis $\mathcal{B}_0$ of mixed nature, associated with the tensor space of order r we are dealing with, endowed with such a basis type (which, as has been explained, is not usual).

Finally, we indicate the particular case in which the basis $\{\vec{e}_\alpha\}$ of $E^n(\mathbb{R})$ be *orthonormal*. We will use a *specific notation* for such a situation. The basis of $E^n(\mathbb{R})$ will be notated $\{\vec{h}_\alpha\} \equiv \{\vec{h}_1, \vec{h}_2, \cdots, \vec{h}_n\}$, which automatically implies

$$\vec{h}_\alpha \bullet \vec{h}_\beta = g^{\circ \; \circ}_{\alpha \beta} \equiv \delta^{\circ \; \circ}_{\alpha \beta}, \qquad (11.117)$$

or equivalently, the fundamental metric tensor of $E^n(\mathbb{R})$ is

$$G = I_n. \tag{11.118}$$

Using this notation, the basis associated with the tensor space $\left(\overset{r}{\underset{1}{\otimes}} E^n\right)(\mathbb{R})$, in agreement with Formulas (11.110) and (11.111), satisfies

$$\left(\vec{h}_{\alpha_1} \otimes \vec{h}_{\alpha_2} \otimes \cdots \otimes \vec{h}_{\alpha_r}\right) \bullet \left(\vec{h}_{\beta_1} \otimes \vec{h}_{\beta_2} \otimes \cdots \otimes \vec{h}_{\beta_r}\right) = g_{\overset{\circ}{\alpha_1}\overset{\circ}{\beta_1}} g_{\overset{\circ}{\alpha_2}\overset{\circ}{\beta_2}} \cdots g_{\overset{\circ}{\alpha_r}\overset{\circ}{\beta_r}}$$

$$\equiv \delta_{\overset{\circ}{\alpha_1}\overset{\circ}{\beta_1}} \delta_{\overset{\circ}{\alpha_2}\overset{\circ}{\beta_2}} \cdots \delta_{\overset{\circ}{\alpha_r}\overset{\circ}{\beta_r}} = K,$$

which again satisfies (11.116), revealing that the basis $\mathcal{B}_h = \{\vec{h}_1 \otimes \vec{h}_2 \otimes \cdots \otimes \vec{h}_n\}$ is auto-reciprocal, that is, an orthonormal basis.

The conclusion is that all the associated tensors in this space have *identical* components:

$$t^{\alpha_1 \alpha_2 \cdots \alpha_r}_{\circ \ \circ \ \cdots \ \circ} \equiv t^{\ \circ \ \circ \ \alpha_3 \cdots \alpha_r}_{\alpha_1 \alpha_2 \ \circ \ \cdots \ \circ} = \cdots = t^{\ \circ \ \circ \ \cdots \ \circ}_{\alpha_1 \alpha_2 \cdots \alpha_r},$$

which is a property that extends that of the Euclidean spaces $E^n(\mathbb{R})$, that in basis type $\{\vec{h}_\alpha\}$ their vectors present the same contravariant and covariant coordinates.

The norm of a tensor $T \in \left(\overset{r}{\underset{1}{\otimes}} E^n\right)(\mathbb{R})$ in the basis $\mathcal{B}_h$ will be

$$N(T) = T \bullet T = t^{\alpha_1 \alpha_2 \cdots \alpha_r}_{\circ \ \circ \ \cdots \ \circ} \delta_{\overset{\circ}{\alpha_1}\overset{\circ}{\beta_1}} \delta_{\overset{\circ}{\alpha_2}\overset{\circ}{\beta_2}} \cdots \delta_{\overset{\circ}{\alpha_r}\overset{\circ}{\beta_r}} t^{\beta_1 \beta_2 \cdots \beta_r}_{\circ \ \circ \ \cdots \ \circ}$$

$$= t^{\alpha_1 \alpha_2 \cdots \alpha_r}_{\circ \ \circ \ \cdots \ \circ} \cdot t^{\alpha_1 \alpha_2 \cdots \alpha_r}_{\circ \ \circ \ \cdots \ \circ} = \sum_{\alpha_i=1}^{\alpha_i=n} \left(t^{\alpha_1 \alpha_2 \cdots \alpha_r}_{\circ \ \circ \ \cdots \ \circ}\right)^2 > 0 \text{ if } T \neq \Omega.$$

$$\tag{11.119}$$

Example 11.10 (Metric of Euclidean associated tensors).

1. Show that in a Euclidean space $E^n(\mathbb{R})$ referred to a basis $\{\vec{e}_\alpha\}$ the fundamental metric tensor of which in covariant coordinates is $G = [g_{\overset{\circ}{\alpha}\overset{\circ}{\beta}}]$, the moduli of the associated vectors are identical.

2. Extend this property to the moduli of the associated tensors of $T \in \left(\overset{r}{\underset{1}{\otimes}} E^n\right)(\mathbb{R})$, calculated with the proper metric of the given tensor space, derived from the metric of $E^n(\mathbb{R})$.

Solution:

1. Consider the vector $\vec{V} \in E^n(\mathbb{R})$ referred to a basis $\{\vec{e}_\alpha\}$. The vector components are the column matrices $X \equiv \begin{bmatrix} x^1 \\ x^2 \\ \vdots \\ x^n \end{bmatrix}$ in contravariant coordinates

and $X^* \equiv \begin{bmatrix} x_1 \\ x_2 \\ \vdots \\ x_n \end{bmatrix}$ in covariant coordinates.

We propose

$$N(\vec{V}) \equiv |\vec{V}|^2 \equiv \vec{V} \bullet \vec{V} = X^t G X. \tag{11.120}$$

We know that in matrix form we "lower" the index (it is passed to covariant coordinates) by means of the relation

$$X^* = GX, \tag{11.121}$$

whence, substituting (11.121) into (11.120), the result is

$$N(\vec{V}) \equiv |\vec{V}|^2 \equiv X^t(GX) = X^t X^* = X^t I_n X^*$$
$$= (X^t I_n X^*)^t = (X^*)^t I_n X = (X^*)^t X. \tag{11.122}$$

From (11.121) we obtain $X = G^{-1} X^*$, which when substituted into (11.122), leads to

$$N(\vec{V}) \equiv |\vec{V}|^2 \equiv (X^*)^t G^{-1} X^* = (X^*)^t G^* X^*. \tag{11.123}$$

Relations (11.120), (11.122) and (11.123) show that the moduli of the associated vectors are coincident.

2. The proof for tensors $T \in \left(\overset{r}{\underset{1}{\otimes}} E^n \right) (\mathbb{R})$ belonging to a Euclidean tensor space of order r is done in a analogous form, with the following assumptions. Let $G(r) = \left(\overset{r}{\underset{1}{\otimes}} G \right)$ be the fundamental metric tensor associated with the Euclidean tensor space $\left(\overset{r}{\underset{1}{\otimes}} E^n \right) (\mathbb{R})$, of dimension $\sigma = n^r$. Let $T = [t^{\alpha_1 \alpha_2 \cdots \alpha_n}_{\circ \ \circ \ \cdots \ \circ}]$ be the contravariant components of T and T_σ the extended column-matrix of the T tensor matrix. We know that for obvious reasons, Equation (11.120) is now

$$N(T) = |T|^2 = T \bullet T = T^t_{\sigma,1} G(r) T_{\sigma,1}.$$

Then, (11.121) becomes

$$T^*_{\sigma,1} = G(r) T_{\sigma,1}$$

and (11.122) can be written in reduced form as

$$N(T) = |T|^2 = T^t_{\sigma,1} I_\sigma T^*_{\sigma,1} = (T^*_{\sigma,1})^t I_\sigma T_{\sigma,1}$$

and (11.123) becomes

$$N(T) = |T|^2 = (T^*_{\sigma,1})^t G^*(r) T^*_{\sigma,1},$$

where $G^*(r) = G^{-1}(r)$, which proves the equality of the norm for $T^* = [t_{\alpha_1 \alpha_2 \cdots \alpha_r}^{\circ \ \circ \ \cdots \ \circ}]$.

For the mixed associated tensors of T, of the same species, the process is analogous, we modify only the fundamental metric tensor $G(r)$ associated with the corresponding tensor space. As a model we give the following. If

$$\left(\overset{r}{\underset{1}{\otimes}} E^n \right) (\mathbb{R}) \equiv E^n \otimes E^n_* \otimes E^n_* \otimes E^n \otimes \cdots \otimes E^n_* \otimes E^n,$$

the fundamental metric tensor associated matrix is

$$G(r) = G \otimes G^{-1} \otimes G^{-1} \otimes G \otimes \cdots \otimes G^{-1} \otimes G.$$

□

Example 11.11 (Fundamental metric tensor of $\left(\overset{3}{\underset{1}{\otimes}} E^2 \right) (\mathbb{R})$*).* In a Euclidean space $E^2(\mathbb{R})$ referred to a basis $\{\vec{e}_\alpha\}$, the fundamental metric tensor is given by the matrix $G = \begin{bmatrix} 1 & 1 \\ 1 & 2 \end{bmatrix}$.

1. Calculate the dimension in a basis of the tensor space $\left(\overset{3}{\underset{1}{\otimes}} E^2 \right) (\mathbb{R})$.
2. Obtain the covariant coordinates of the fundamental metric tensor $G(3)$ in the basis of the tensor space.
3. Obtain the contravariant components of the fundamental metric tensor.

Solution:

1. The dimension is given by:

$$\sigma = dim \left(\overset{3}{\underset{1}{\otimes}} E^2 \right) (\mathbb{R}) = n^r = 2^3 = 8.$$

$$B = \{ \vec{e}_\alpha \otimes \vec{e}_\beta \otimes \vec{e}_\gamma \},$$

where $\alpha, \beta, \gamma \in I_2$.

2. The covariant coordinates of the fundamental metric tensor $G(3)$ in the basis of the tensor space are:

$$g^{\circ\circ\circ\circ\circ\circ}_{\alpha\lambda\beta\mu\gamma\nu} = (\vec{e}_\alpha \otimes \vec{e}_\beta \otimes \vec{e}_\gamma) \bullet (\vec{e}_\lambda \otimes \vec{e}_\mu \otimes \vec{e}_\nu) = g^{\circ\circ}_{\alpha\lambda} g^{\circ\circ}_{\beta\mu} g^{\circ\circ}_{\gamma\nu}$$

$$G(3) = [g^{\circ\circ\circ\circ\circ\circ}_{\alpha\lambda\beta\mu\gamma\nu}] = G \otimes G \otimes G = \begin{bmatrix} 1 & 1 \\ 1 & 2 \end{bmatrix} \otimes \begin{bmatrix} 1 & 1 \\ 1 & 2 \end{bmatrix} \otimes \begin{bmatrix} 1 & 1 \\ 1 & 2 \end{bmatrix};$$

$$[g^{\circ\circ\circ\circ\circ\circ}_{\alpha\lambda\beta\mu\gamma\nu}] = G(3) = \left[\begin{array}{cc|cc|cc|cc} 1 & 1 & 1 & 1 & 1 & 1 & 1 & 1 \\ 1 & 2 & 1 & 2 & 1 & 2 & 1 & 2 \\ \hline 1 & 1 & 2 & 2 & 1 & 1 & 2 & 2 \\ 1 & 2 & 2 & 4 & 1 & 2 & 2 & 4 \\ \hline 1 & 1 & 1 & 1 & 2 & 2 & 2 & 2 \\ 1 & 2 & 1 & 2 & 2 & 4 & 2 & 4 \\ \hline 1 & 1 & 2 & 2 & 2 & 2 & 4 & 4 \\ 1 & 2 & 2 & 4 & 2 & 4 & 4 & 8 \end{array} \right].$$

3. The contravariant components of the fundamental metric tensor are:

$$G^{-1} = \begin{bmatrix} 1 & 1 \\ 1 & 2 \end{bmatrix}^{-1} = \begin{bmatrix} 2 & -1 \\ -1 & 1 \end{bmatrix},$$

$$g_{\circ\circ\circ\circ\circ\circ}^{\alpha\lambda\beta\mu\gamma\nu} = \left(\vec{e}^{*\alpha} \otimes \vec{e}^{*\beta} \otimes \vec{e}^{*\gamma}\right) \bullet \left(\vec{e}^{*\lambda} \otimes \vec{e}^{*\mu} \otimes \vec{e}^{*\nu}\right) = g_{\circ\circ}^{\alpha\lambda} g_{\circ\circ}^{\beta\mu} g_{\circ\circ}^{\gamma\nu}$$

$$G^{-1}(3) = G^{-1} \otimes G^{-1} \otimes G^{-1} = \begin{bmatrix} 2 & -1 \\ -1 & 1 \end{bmatrix} \otimes \begin{bmatrix} 2 & -1 \\ -1 & 1 \end{bmatrix} \otimes \begin{bmatrix} 2 & -1 \\ -1 & 1 \end{bmatrix};$$

$$G^*(3) \equiv [g_{\circ\circ\circ\circ\circ\circ}^{\alpha\lambda\beta\mu\gamma\nu}] = G^{-1}(3) = \left[\begin{array}{rr|rr|rr|rr} 8 & -4 & -4 & 2 & -4 & 2 & 2 & -1 \\ -4 & 4 & 2 & -2 & 2 & -2 & -1 & 1 \\ \hline -4 & 2 & 4 & -2 & 2 & -1 & -2 & 1 \\ 2 & -2 & 2 & -2 & -1 & 1 & 1 & -1 \\ \hline -4 & 2 & 2 & -1 & 4 & -2 & -2 & 1 \\ 2 & -2 & -1 & 1 & -2 & 2 & 1 & -1 \\ \hline 2 & -1 & -2 & 1 & -2 & 1 & 2 & -1 \\ -1 & 1 & 1 & -1 & 1 & -1 & -1 & 1 \end{array}\right].$$

□

Example 11.12 (Analysis of a multiform over $\left(\overset{3}{\underset{1}{\otimes}} E^3\right)(Q)$). In a Euclidean space $E^3(Q)$ (Q is the field of the rational numbers), referred to an orthonormalized basis $\{\vec{e}_\alpha\}$ consider the tensor $F : \left(\overset{3}{\underset{1}{\otimes}} E^3\right)(Q) \longrightarrow Q$, which is a multilinear form that associates with each tensor product of three vectors, in $\left(\overset{3}{\underset{1}{\otimes}} E^3\right)(Q)$ a rational scalar of Q.

If the components of $\vec{V}_1, \vec{V}_2, \vec{V}_3$, in the basis $\{\vec{e}_\alpha\}$, are $X = \begin{bmatrix} x^1 \\ x^2 \\ x^3 \end{bmatrix}$, Y, Z the form F is defined as

$$F\left(\vec{V}_1 \otimes \vec{V}_2 \otimes \vec{V}_3\right) = F(X \otimes Y \otimes Z) = x^1 y^1 z^1 + x^2 y^2 z^2 + x^3 y^3 z^3.$$

1. Examine if the given form is or is not a symmetric or alternate multilinear form.
2. We choose as the new basis of $E^3(Q)$ that with vectors

$$\vec{\tilde{e}}_1 = \vec{e}_1 + \vec{e}_2, \quad \vec{\tilde{e}}_2 = \vec{e}_2 + \vec{e}_3 \quad \text{and} \quad \vec{\tilde{e}}_3 = \vec{e}_1 + \vec{e}_3,$$

   which determines the form expression in contravariant coordinates relative to the new basis.
3. Assuming that this form is used to define a *new* "cubic norm" of vectors

$$N(\vec{V}) = F(\vec{V} \otimes \vec{V} \otimes \vec{V}) = f(X, X, X),$$

   are there any vectors $\vec{V} \neq \vec{0}$, of null norm?
   In the affirmative case, give some simple numerical example.

Solution: Any multilinear form has the tensor structure

$$\rho = F(\vec{V_1} \otimes \vec{V_1} \otimes \cdots \otimes \vec{V_r}) = f(X_1 \otimes X_2 \otimes \cdots \otimes \vec{X_r})$$

$$= f_{\alpha_1 \alpha_2 \cdots \alpha_r}^{\circ \circ \cdots \circ} x_{\circ 1}^{\alpha_1 \circ} x_{\circ 2}^{\alpha_2 \circ} \cdots x_{\circ r}^{\alpha_r \circ}; \quad \rho \in Q,$$

that is, a *total* contraction of the tensor

$$F = f_{\alpha_1 \alpha_2 \cdots \alpha_r}^{\circ \circ \cdots \circ} \vec{e}^{*\alpha_1} \otimes \vec{e}^{*\alpha_2} \otimes \cdots \otimes \vec{e}^{*\alpha_r}$$

with the tensor products

$$x_{\circ 1}^{\alpha_1 \circ} x_{\circ 2}^{\alpha_2 \circ} \cdots x_{\circ r}^{\alpha_r \circ} \vec{e}_{\alpha_1} \otimes \vec{e}_{\alpha_2} \otimes \cdots \otimes \vec{e}_{\alpha_r} \in \left(\overset{r}{\underset{1}{\otimes}} E^n \right) (Q).$$

1. Our particular problem presents the form $(r = 3)$ with components

$$\begin{cases} f_{\alpha\beta\gamma}^{\circ\circ\circ} = 0 \text{ if } \alpha \neq \beta; \ \alpha \neq \gamma; \ \beta \neq \gamma \\ f_{111}^{\circ\circ\circ} = f_{222}^{\circ\circ\circ} = f_{333}^{\circ\circ\circ} = 1 \end{cases}$$

and in matrix form it becomes

$$F = [f_{\alpha\beta\gamma}^{\circ\circ\circ}] = \begin{bmatrix} 1 & 0 & 0 \\ 0 & 0 & 0 \\ 0 & 0 & 0 \\ \hline 0 & 0 & 0 \\ 0 & 1 & 0 \\ 0 & 0 & 0 \\ \hline 0 & 0 & 0 \\ 0 & 0 & 0 \\ 0 & 0 & 1 \end{bmatrix},$$

where α is the block row, β the row of each block and γ the column of each block, showing that it is a covariant tensor of order $(r = 3)$, and thus, a multilinear form.

Its *symmetric* and *non-alternate* (non-anti-symmetric) character is also explicit in the matrix representation.

2. The change-of-basis of matrix C is given by

$$\|\vec{e_i}\| = \|\vec{e_\alpha}\| C \rightarrow [\vec{e_1}\vec{e_2}\vec{e_3}] = [\vec{e_1}\vec{e_2}\vec{e_3}] \begin{bmatrix} 1 & 0 & 1 \\ 1 & 1 & 0 \\ 0 & 1 & 1 \end{bmatrix}; |C| = 2 \neq 0;$$

$$C^{-1} = \frac{1}{2} \begin{bmatrix} 1 & 1 & -1 \\ -1 & 1 & 1 \\ 1 & -1 & 1 \end{bmatrix}.$$

We solve this question using two different procedures:

a) First, we find the new covariant coordinates $f^{\circ\circ\circ}_{ijk}$ and then we raise the indices to new contravariant coordinates $f^{ijk}_{\circ\circ\circ}$.

b) First, we find the initial contravariant coordinates, raising the indices to obtain the components $f^{\alpha\beta\gamma}_{\circ\circ\circ}$, and then we perform the change-of-basis to the $f^{ijk}_{\circ\circ\circ}$.

Since the initial frame of reference is orthonormalized, method (b) is more convenient, because the fundamental metric tensor associated with the space $\left(\overset{3}{\underset{1}{\otimes}} E^3_*\right)(Q)$ is $G^*(3) = I_3^{-1} \otimes I_3^{-1} \otimes I_3^{-1} = I_{27}$, so that raising indices is immediate:

$$F_{\sigma,1} = I_{27} \cdot F^*_{\sigma,1},$$

which is equivalent to the identity $f^{\alpha\beta\gamma}_{\circ\circ\circ} = f^{\circ\circ\circ}_{\alpha\beta\gamma}$.

It remains only to change the tensor to the new basis

$$f^{ijk}_{\circ\circ\circ} = f^{\alpha\beta\gamma}_{\circ\circ\circ}\gamma^{i\circ}_{\circ\alpha}\gamma^{j\circ}_{\circ\beta}\gamma^{k\circ}_{\circ\gamma},$$

and in matrix form, for $\sigma = n^r = 3^3 = 27$, the result is

$$\hat{F}_{27,1} = \left(C^{-1} \otimes C^{-1} \otimes \cdots \otimes C^{-1}\right) \bullet F_{27,1}$$

$$= \left(\frac{1}{2}\begin{bmatrix} 1 & 1 & -1 \\ -1 & 1 & 1 \\ 1 & -1 & 1 \end{bmatrix} \otimes \frac{1}{2}\begin{bmatrix} 1 & 1 & -1 \\ -1 & 1 & 1 \\ 1 & -1 & 1 \end{bmatrix} \otimes \frac{1}{2}\begin{bmatrix} 1 & 1 & -1 \\ -1 & 1 & 1 \\ 1 & -1 & 1 \end{bmatrix}\right) \bullet F_{27,1},$$

where

$$F^t_{27} = [1\,0\,0\,0\,0\,0\,0\,0\,0\,|\,0\,0\,0\,0\,1\,0\,0\,0\,0\,|\,0\,0\,0\,0\,0\,0\,0\,0\,1].$$

Once the operations have been performed, we obtain

$$\hat{F}^t_{27,1} = \frac{1}{8}[1\,1\,1\,1\,1\,-3\,1\,-3\,1\,|\,1\,1\,-3\,1\,1\,1\,-3\,1\,1\,|\,1\,-3\,1\,-3\,1\,1\,1\,1\,1],$$

which once condensed, leads to the matrix representation of F by its new contravariant coordinates:

$$[f^{ijk}_{\circ\circ\circ}] = \frac{1}{8}\left[\begin{array}{ccc|ccc|ccc} 1 & 1 & 1 \\ 1 & 1 & -3 \\ 1 & -3 & 1 \\ \hline 1 & 1 & -3 \\ 1 & 1 & 1 \\ -3 & 1 & 1 \\ \hline 1 & -3 & 1 \\ -3 & 1 & 1 \\ 1 & 1 & 1 \end{array}\right].$$

In this situation, the form F operates as $\rho = f^{ijk}_{\circ\circ\circ}x^\circ_i x^\circ_j x^\circ_k$.

3. Since the vector $\vec{V}$ in the initial basis $\{\vec{e}_\alpha\}$ is

$$\vec{V} = [\vec{e}_1 \vec{e}_2 \vec{e}_3] X \equiv [\vec{e}_1 \vec{e}_2 \vec{e}_3] \begin{bmatrix} x^1 \\ x^2 \\ x^3 \end{bmatrix}$$

and assuming that $\forall x^\alpha \in Q$, there must be

$$x^1 = \frac{p^1}{q^1}; \quad x^2 = \frac{p^2}{q^2}; \quad x^3 = \frac{p^3}{q^3}; \quad \text{(arbitrary rational numbers)}$$

and reducing the fractions to common denominator, we obtain $x^1 = \frac{z^1}{q}$; $x^2 = \frac{z^2}{q}$; $x^3 = \frac{z^3}{q}$, and entering these into the form, we must have

$$N(\vec{V}) = f(X, X, X) = (x^1)^3 + (x^2)^3 + (x^3)^3 = \frac{(z^1)^3 + (z^2)^3 + (z^3)^3}{q^3} = 0,$$

which implies

$$(z^1)^3 + (z^2)^3 = (-z^3)^3; \quad \forall z^i \in Z \text{ (relative integers)}.$$

If we let $z^1 = x$; $z^2 = y$; $-z^3 = z$; $x, y, z \in \mathbb{N}$ (natural numbers), we must find some natural solution to the equation:

$$x^3 + y^3 = z^3,$$

which is impossible according to the Fermat theorem (proven by Euler for exponents $n = 3$). There are no vectors of *null* "cubic norm" in Q.

□

Example 11.13 (Tensor metrics: tensor geometries). Consider first a Euclidean space $E_1^2(\mathbb{R})$, referred to a certain basis $\{\vec{e}_\alpha\}$ and the connection fundamental metric tensor of which is given in covariant coordinates by the Gram matrix

$$G_1 \equiv [g_{\alpha\beta}^{\circ\circ}] = \begin{bmatrix} 1 & 2 \\ 2 & 5 \end{bmatrix}$$

and the vectors

$$\vec{a} \ (1, -1) \quad \text{and} \ \vec{b} \ (2, -3), \quad \vec{a}, \vec{b} \in E_1^2(\mathbb{R}).$$

Next, consider another Euclidean space $E_2^3(\mathbb{R})$ referred to a basis $\{\vec{\epsilon}_\gamma\}$, where the Gram matrix associated with its fundamental metric tensor is

$$G_2 \equiv [g_{\gamma\delta}^{\circ\circ}] = \begin{bmatrix} 1 & 1 & -1 \\ 1 & 2 & 0 \\ -1 & 0 & 3 \end{bmatrix},$$

and the vectors:

$$\vec{p} \ (2, -1, 1) \quad \text{and} \ \vec{q} \ (3, -1, 2), \quad \vec{p}, \vec{q} \in E_2^3(\mathbb{R}).$$

1. Examine the proper Euclidean character of $E_1^2(\mathbb{R})$.
2. Obtain the Euclidean modulus of the vectors $\vec{a}$ and $\vec{b}$.
3. Give the dot product $\vec{a} \bullet \vec{b}$ of the two vectors.
4. Calculate the area of the triangle with sides the vectors $\vec{a}$ and $\vec{b}$.
5. Examine the proper Euclidean character of $E_2^3(\mathbb{R})$.
6. Calculate the moduli of the vectors $\vec{p}$ and $\vec{q}$.
7. Calculate the dot product $\vec{p} \bullet \vec{q}$.
8. Calculate the area of the triangle with sides the vectors $\vec{p}$ and $\vec{q}$.
9. Calculate the perimeters of the two mentioned triangles.
10. Obtain the basis $\mathcal{B}_\sigma$ of the tensor space $E_1^2 \otimes E_2^3(\mathbb{R})$.
11. Determine the decomposable tensors $U = \vec{a} \otimes \vec{p}$ and $W = \vec{b} \otimes \vec{q}$, expressing their respective components (column-matrices) in the basis $\mathcal{B}_\sigma$ associated with the tensor space $E_1^2 \otimes E_2^3(\mathbb{R})$.
12. Give the fundamental metric tensor associated with the tensor space $E_1^2 \otimes E_2^3(\mathbb{R})$ and "produced" by the existence of the corresponding tensors G_1 and G_2 of the Euclidean factor spaces. Discuss the Euclidean character of $E_1^2 \otimes E_2^3(\mathbb{R})$.
13. Calculate the moduli of the tensors U and W.
14. Obtain the dot product of the tensors $U, W \in E_1^2 \otimes E_2^3(\mathbb{R})$.
15. Examine the $\cos\theta$ that corresponds to tensors U and W.
16. Obtain the area of the plane triangle of tensor sides U and W, in the tensor space 6-dimensional $E_1^2 \otimes E_2^3(\mathbb{R})$.
17. Obtain the moduli of the tensors:

$$\vec{a} \otimes \vec{a}, \quad \vec{b} \otimes \vec{b}, \quad \vec{a} \otimes \vec{b}, \quad \vec{p} \otimes \vec{p}, \quad \vec{q} \otimes \vec{q}, \quad \vec{p} \otimes \vec{q}.$$

18. Calculate the perimeter of the tensor triangle formed by the tensor vectors U and W.
19. Indicate if it is licit to ask for the dot product $(\vec{a} \otimes \vec{b}) \bullet (\vec{p} \otimes \vec{q})$. In the affirmative case, give such dot product.

Solution:

1. We determine the Gram–Schmidt numbers and the terms of the diagonal matrix:

$$\Gamma_0 = 1; \quad \Gamma_1 = 1; \quad \Gamma_2 = \begin{vmatrix} 1 & 2 \\ 2 & 5 \end{vmatrix} = 1; \quad \hat{g}_{11} = \frac{\Gamma_1}{\Gamma_0} = \frac{1}{1} = 1; \quad \hat{g}_{22} = \frac{\Gamma_2}{\Gamma_1} = \frac{1}{1} = 1,$$

so that there exist canonizing matrices C_1, such that we obtain the Sylvesterian matrix $\hat{G}_1$, by congruence:

$$\exists C_1 | \quad \hat{G}_1 = C_1^t G_1 C_1 = \begin{bmatrix} 1 & 0 \\ 0 & 1 \end{bmatrix}.$$

We have $n = r = \sigma = 2$ (order = rank = Spanish signature), thus G_1 is a positive definite matrix, and thus, $E_1^2(\mathbb{R})$ is a Euclidean space.

2. The Euclidean moduli of the vectors $\vec{a}$ and $\vec{b}$ are:

$$|\vec{a}| = \sqrt{[1\ -1]\begin{bmatrix}1 & 2\\ 2 & 5\end{bmatrix}\begin{bmatrix}1\\ -1\end{bmatrix}} = \sqrt{[-1\ -3]\begin{bmatrix}1\\ -1\end{bmatrix}} = \sqrt{2}$$

$$|\vec{b}| = \sqrt{[2\ -3]\begin{bmatrix}1 & 2\\ 2 & 5\end{bmatrix}\begin{bmatrix}2\\ -3\end{bmatrix}} = \sqrt{[-4\ -11]\begin{bmatrix}2\\ -3\end{bmatrix}} = \sqrt{25} = 5.$$

3. The dot product $\vec{a} \bullet \vec{b}$ is:

$$\vec{a} \bullet \vec{b} = [1-1]\begin{bmatrix}1 & 2\\ 2 & 5\end{bmatrix}\begin{bmatrix}2\\ -3\end{bmatrix} = [-1\ -3]\begin{bmatrix}2\\ -3\end{bmatrix} = 7.$$

4. We extract the covariant coordinates of the vectors $\vec{a}$ and $\vec{b}$ (see special tensors Section 1.4):

$$X^* = G_1 X; \quad X_a^* = \begin{bmatrix}1 & 2\\ 2 & 5\end{bmatrix}\begin{bmatrix}1\\ -1\end{bmatrix} = \begin{bmatrix}-1\\ -3\end{bmatrix}; \quad X_b^* = \begin{bmatrix}1 & 2\\ 2 & 5\end{bmatrix}\begin{bmatrix}2\\ -3\end{bmatrix} = \begin{bmatrix}-4\\ -11\end{bmatrix}.$$

We proceed to create a Cartesian system of three dimensions, adopting an axis $\overline{OZ}$ perpendicular to the OXY (oblique) of our Euclidean space. The new basic vector, over the axis $\overline{OZ}$, will be $\vec{e}_3$ *unitary* ($|\vec{e}_3| = 1$).
In this way, we can calculate the vector product, the modulus of which is $|\vec{a} \wedge \vec{b}| = |\vec{a}||\vec{b}|\sin\theta$, and thus, the sought after area is $S = \frac{1}{2}|\vec{a} \wedge \vec{b}|$:

$$\vec{z}_1 = \vec{a} \wedge \vec{b} = \frac{1}{\sqrt{|G_3|}}\begin{vmatrix}\vec{e}_1 & \vec{e}_2 & \vec{e}_3\\ x_1 & y_1 & z_1\\ x_2 & y_2 & z_2\end{vmatrix} = \frac{1}{\sqrt{\begin{vmatrix}1 & 2 & 0\\ 2 & 5 & 0\\ 0 & 0 & 1\end{vmatrix}}}\begin{vmatrix}\vec{e}_1 & \vec{e}_2 & \vec{e}_3\\ -1 & -3 & 0\\ -4 & -11 & 0\end{vmatrix}$$

$$\vec{z}_1 = -\vec{e}_3; \quad |\vec{z}_1| = |-\vec{e}_3| = 1; \quad S_1 = \frac{1}{2}|\vec{a} \wedge \vec{b}| = \frac{1}{2}|\vec{z}_1| = \frac{1}{2}.$$

5. We proceed in an analogous way to question 1.
   The Gram–Schmidt numbers are

$$\Gamma_0 = 1; \quad \Gamma_1 = 1; \quad \Gamma_2 = \begin{vmatrix}1 & 1\\ 1 & 2\end{vmatrix} = 1; \quad \Gamma_3 = \begin{vmatrix}1 & 1 & -1\\ 1 & 2 & 0\\ -1 & 0 & 3\end{vmatrix} = 1;$$

$$\hat{g}_{11} = \frac{\Gamma_1}{\Gamma_0} = 1; \quad \hat{g}_{22} = \frac{\Gamma_2}{\Gamma_1} = 1; \quad \hat{g}_{33} = \frac{\Gamma_3}{\Gamma_2} = 1$$

$$\exists C_2|\ \hat{G}_2 = C_2^t G_2 C_2 = \begin{bmatrix}1 & 0 & 0\\ 0 & 1 & 0\\ 0 & 0 & 1\end{bmatrix}; \quad n = r = \sigma = 3.$$

G_2 is positive definite and thus, $E_2^3(\mathbb{R})$ is a Euclidean space.

6. The moduli of the vectors $\vec{p}$ and $\vec{q}$ are:

$$|\vec{p}| = \sqrt{[2\ -1\ 1]\begin{bmatrix} 1 & 1 & -1 \\ 1 & 2 & 0 \\ -1 & 0 & 3 \end{bmatrix}\begin{bmatrix} 2 \\ -1 \\ 1 \end{bmatrix}} = \sqrt{[0\ 0\ 1]\begin{bmatrix} 2 \\ -1 \\ 1 \end{bmatrix}} = 1$$

$$|\vec{q}| = \sqrt{[3\ -1\ 2]\begin{bmatrix} 1 & 1 & -1 \\ 1 & 2 & 0 \\ -1 & 0 & 3 \end{bmatrix}\begin{bmatrix} 3 \\ -1 \\ 2 \end{bmatrix}} = \sqrt{[0\ 1\ 3]\begin{bmatrix} 3 \\ -1 \\ 2 \end{bmatrix}} = \sqrt{5}.$$

7. The dot product $\vec{p} \bullet \vec{q}$ is:

$$\vec{p} \bullet \vec{q} = [2\ -1\ 1]\begin{bmatrix} 1 & 1 & -1 \\ 1 & 2 & 0 \\ -1 & 0 & 3 \end{bmatrix}\begin{bmatrix} 3 \\ -1 \\ 2 \end{bmatrix} = [0\ 0\ 1]\begin{bmatrix} 3 \\ -1 \\ 2 \end{bmatrix} = 2.$$

8. We proceed directly, using the vector product $\vec{p} \wedge \vec{q}$, as in question 4. $\vec{p}$ and $\vec{q}$ in covariant coordinates:

$$X_p^* = \begin{bmatrix} 1 & 1 & -1 \\ 1 & 2 & 0 \\ -1 & 0 & 3 \end{bmatrix}\begin{bmatrix} 2 \\ -1 \\ 1 \end{bmatrix} = \begin{bmatrix} 0 \\ 0 \\ 1 \end{bmatrix};$$

$$X_q^* = \begin{bmatrix} 1 & 1 & -1 \\ 1 & 2 & 0 \\ -1 & 0 & 3 \end{bmatrix}\begin{bmatrix} 3 \\ -1 \\ 2 \end{bmatrix} = \begin{bmatrix} 0 \\ 1 \\ 3 \end{bmatrix};$$

$$\vec{z}_2 = \vec{p} \wedge \vec{q} = \frac{1}{\sqrt{|G_2|}}\begin{vmatrix} \vec{e}_1 & \vec{e}_2 & \vec{e}_3 \\ x_1 & y_1 & z_1 \\ x_2 & y_2 & z_2 \end{vmatrix} = \frac{1}{\sqrt{\begin{vmatrix} 1 & 1 & -1 \\ 1 & 2 & 0 \\ -1 & 0 & 3 \end{vmatrix}}}\begin{vmatrix} \vec{e}_1 & \vec{e}_2 & \vec{e}_3 \\ 0 & 0 & 1 \\ 0 & 1 & 3 \end{vmatrix} = -\vec{e}_1;$$

$$|\vec{z}_1| = |-\vec{e}_1| = \sqrt{\vec{e}_1 \bullet \vec{e}_1} = \sqrt{g_{11}} = \sqrt{1} = 1; \quad S_2 = \frac{1}{2}|\vec{p} \wedge \vec{q}| = \frac{1}{2}|\vec{z}_2| = \frac{1}{2}.$$

9. The third side of the triangle of vector sides $\vec{a}, \vec{b}$ is $(\vec{a} - \vec{b})$;

$$X_a - X_b = \begin{bmatrix} 1 \\ -1 \end{bmatrix} - \begin{bmatrix} 2 \\ -3 \end{bmatrix} = \begin{bmatrix} -1 \\ 2 \end{bmatrix};$$

$$|\vec{a} - \vec{b}| = \sqrt{[-1\ \ 2]\begin{bmatrix} 1 & 2 \\ 2 & 5 \end{bmatrix}\begin{bmatrix} -1 \\ 2 \end{bmatrix}} = \sqrt{[3\ 8]\begin{bmatrix} -1 \\ 2 \end{bmatrix}} = \sqrt{13}.$$

P_1(perimeter) $= |\vec{a}| + |\vec{b}| + |\vec{a} - \vec{b}| = \sqrt{2} + 5 + \sqrt{13}$.

The third side of the triangle of vector sides $\vec{p}, \vec{q}$ is $(\vec{p} - \vec{q})$;

$$X_p - X_q = \begin{bmatrix} 2 \\ -1 \\ 1 \end{bmatrix} - \begin{bmatrix} 3 \\ -1 \\ 2 \end{bmatrix} = \begin{bmatrix} -1 \\ 0 \\ -1 \end{bmatrix};$$

$$|\vec{p} - \vec{q}| = \sqrt{[-1 \ 0 \ -1] \begin{bmatrix} 1 & 1 & -1 \\ 1 & 2 & 0 \\ -1 & 0 & 3 \end{bmatrix} \begin{bmatrix} -1 \\ 0 \\ -1 \end{bmatrix}} = \sqrt{2}.$$

P_2 (perimeter)$= |\vec{p}| + |\vec{q}| + |\vec{p} - \vec{q}| = 1 + \sqrt{5} + \sqrt{2}.$

10. $\sigma = $ dimension of $E_1^2 \otimes E_2^3(\mathbb{R}) = r_1 \cdot r_2 = 2 \times 3 = 6.$

Let $\mathcal{B}_\sigma$ be the basis of $E_1^2 \otimes E_2^3(\mathbb{R})$ associated with the respective bases $\{\vec{e}_\alpha\}$ and $\{\vec{\epsilon}_\gamma\}$ of the Euclidean spaces $E_1^2(\mathbb{R})$ and $E_2^3(\mathbb{R})$. We have

$$\mathcal{B}_\sigma \equiv \mathcal{B}_6 = \{\vec{e}_1 \otimes \vec{\epsilon}_1, \vec{e}_1 \otimes \vec{\epsilon}_2, \vec{e}_1 \otimes \vec{\epsilon}_3, \vec{e}_2 \otimes \vec{\epsilon}_1, \vec{e}_2 \otimes \vec{\epsilon}_2, \vec{e}_2 \otimes \vec{\epsilon}_3\}.$$

11. We proceed as follows.

$$[U_{6,1}] = X_a \otimes X_p = \begin{bmatrix} 1 \\ -1 \end{bmatrix} \otimes \begin{bmatrix} 2 \\ -1 \\ 1 \end{bmatrix} \longrightarrow [U_{6,1}]^t = [\,2 \ -1 \ 1 \ -2 \ 1 \ -1\,]$$

$$[W_{6,1}] = X_b \otimes X_q = \begin{bmatrix} 2 \\ -3 \end{bmatrix} \otimes \begin{bmatrix} 3 \\ -1 \\ 2 \end{bmatrix} \longrightarrow [W_{6,1}]^t = [\,6 \ -2 \ 4 \ -9 \ 3 \ -6\,],$$

so that the tensors are

$$U = [\,\vec{e}_1 \otimes \vec{\epsilon}_1 \ \ \vec{e}_1 \otimes \vec{\epsilon}_2 \ \ \vec{e}_1 \otimes \vec{\epsilon}_3 \ \ \vec{e}_2 \otimes \vec{\epsilon}_1 \ \ \vec{e}_2 \otimes \vec{\epsilon}_2 \ \ \vec{e}_2 \otimes \vec{\epsilon}_3\,] \begin{bmatrix} 2 \\ -1 \\ 1 \\ -2 \\ 1 \\ -1 \end{bmatrix}$$

$$W = [\,\vec{e}_1 \otimes \vec{\epsilon}_1 \ \ \vec{e}_1 \otimes \vec{\epsilon}_2 \ \ \vec{e}_1 \otimes \vec{\epsilon}_3 \ \ \vec{e}_2 \otimes \vec{\epsilon}_1 \ \ \vec{e}_2 \otimes \vec{\epsilon}_2 \ \ \vec{e}_2 \otimes \vec{\epsilon}_3\,] \begin{bmatrix} 6 \\ -2 \\ 4 \\ -9 \\ 3 \\ -6 \end{bmatrix}.$$

12. Following the theory, we have

$$G(6) = G_1 \otimes G_2 = \begin{bmatrix} 1 & 2 \\ 2 & 5 \end{bmatrix} \otimes \begin{bmatrix} 1 & 1 & -1 \\ 1 & 2 & 0 \\ -1 & 0 & 3 \end{bmatrix} = \begin{bmatrix} 1 & 1 & -1 & 2 & 2 & -2 \\ 1 & 2 & 0 & 2 & 4 & 0 \\ -1 & 0 & 3 & -2 & 0 & 6 \\ 2 & 2 & -2 & 5 & 5 & -5 \\ 2 & 4 & 0 & 5 & 10 & 0 \\ -2 & 0 & 6 & -5 & 0 & 15 \end{bmatrix}.$$

The Gram–Schmidt numbers of $G(6)$ are

$$\Gamma_0 = 1; \quad \Gamma_1 = 1; \quad \Gamma_2 = \begin{vmatrix} 1 & 1 \\ 1 & 2 \end{vmatrix} = 1; \quad \Gamma_3 = \begin{vmatrix} 1 & 1 & -1 \\ 1 & 2 & 0 \\ -1 & 0 & 3 \end{vmatrix} = 1;$$

$$\Gamma_4 = \begin{vmatrix} 1 & 1 & -1 & 2 \\ 1 & 2 & 0 & 2 \\ -1 & 0 & 3 & -2 \\ 2 & 2 & -2 & 5 \end{vmatrix} = \begin{vmatrix} 1 & 1 & -1 & 2 \\ 0 & 1 & 1 & 0 \\ 0 & 1 & 2 & 0 \\ 0 & 0 & 0 & 1 \end{vmatrix} = 1;$$

$$\Gamma_5 = \begin{vmatrix} 1 & 1 & -1 & 2 & 2 \\ 1 & 2 & 0 & 2 & 4 \\ -1 & 0 & 3 & -2 & 0 \\ 2 & 2 & -2 & 5 & 5 \\ 2 & 4 & 0 & 5 & 10 \end{vmatrix} = \begin{vmatrix} 1 & 1 & -1 & 2 & 2 \\ 0 & 1 & 1 & 0 & 2 \\ 0 & 1 & 2 & 0 & 2 \\ 0 & 0 & 0 & 1 & 1 \\ 0 & 2 & 2 & 0 & 5 \end{vmatrix}$$

$$= \begin{vmatrix} 1 & 1 & 2 \\ 1 & 2 & 2 \\ 2 & 2 & 5 \end{vmatrix} = \begin{vmatrix} 1 & 1 & 2 \\ 0 & 1 & 0 \\ 2 & 2 & 5 \end{vmatrix} = \begin{vmatrix} 1 & 2 \\ 2 & 5 \end{vmatrix} = 1.$$

Finally, using property 2 of Section 1.3.3, we have

$$\Gamma_6 = |G(6)| = |G_1 \otimes G_2| = |G_1|^3 \cdot |G_2|^2 = 1^3 \cdot 1^2 = 1,$$

and thus, all $\hat{g}_{ii} = \frac{\Gamma_i}{\Gamma_{i-1}} = 1$.

There exists a change-of-basis matrix C in the tensor space $E_1^2 \otimes E_2^3(\mathbb{R})$, that can be built by the Gram–Schmidt method, such that

$$\hat{G}(6) = C^t G(6) C \equiv I_6; \quad n = r = \sigma = 6.$$

$G(6)$ is positive definite and thus, $E_1^2 \otimes E_2^3(\mathbb{R})$ connected by the fundamental metric tensor $G(6)$ is a Euclidean tensor space.

13. The moduli of the tensors U and W are:

$$|U| = \sqrt{U_6^t G(6) U_6} = \sqrt{[\,2 \; -1 \; 1 \; -2 \; 1 \; -1\,]\, G(6) \begin{bmatrix} 2 \\ -1 \\ 1 \\ -2 \\ 1 \\ -1 \end{bmatrix}}$$

$$= \sqrt{[\,0 \; 0 \; -1 \; 0 \; 0 \; -3\,] \begin{bmatrix} 2 \\ -1 \\ 1 \\ -2 \\ 1 \\ -1 \end{bmatrix}} = \sqrt{2};$$

we verify that $|U| = |\vec{a}| \cdot |\vec{p}| = \sqrt{2} \cdot 1 = \sqrt{2}$, because $U = \vec{a} \otimes \vec{p}$.

In addition:

$$|W| = \sqrt{W_6^t G(6) W_6} = \sqrt{[\,6 \; -2 \; 4 \; -9 \; 3 \; -6\,]\, G(6) \begin{bmatrix} 6 \\ -2 \\ 4 \\ -9 \\ 3 \\ -6 \end{bmatrix}}$$

$$= \sqrt{\begin{bmatrix} 0 & -4 & -12 & 0 & -11 & -33 \end{bmatrix} G(6) \begin{bmatrix} 6 \\ -2 \\ 4 \\ -9 \\ 3 \\ -6 \end{bmatrix}} = \sqrt{125} = 5\sqrt{5};$$

we verify that $|W| = |\vec{b}| \cdot |\vec{q}| = 5\sqrt{5}$, because $W = \vec{b} \otimes \vec{q}$.

14. The dot product of the tensors U and W is:

$$U \bullet W = U_6^t G(6) W_6$$

$$= \begin{bmatrix} 2 & -1 & 1 & -2 & 1 & -1 \end{bmatrix} \begin{bmatrix} 1 & 1 & -1 & 2 & 2 & -2 \\ 1 & 2 & 0 & 2 & 4 & 0 \\ -1 & 0 & 3 & -2 & 0 & 6 \\ 2 & 2 & -2 & 5 & 5 & -5 \\ 2 & 4 & 0 & 5 & 10 & 0 \\ -2 & 0 & 6 & -5 & 0 & 15 \end{bmatrix} \begin{bmatrix} 6 \\ -2 \\ 4 \\ -9 \\ 3 \\ -6 \end{bmatrix}$$

$$= \begin{bmatrix} 0 & 0 & -1 & 0 & 0 & -3 \end{bmatrix} \begin{bmatrix} 6 \\ -2 \\ 4 \\ -9 \\ 3 \\ -6 \end{bmatrix} = 14.$$

Alternatively, we check that the product of the two real numbers

$$(\vec{a} \bullet \vec{b}) \cdot (\vec{p} \bullet \vec{q}) = 7 \times 2 = 14 \text{ (according to questions 3 and 7).}$$

15. From Questions 13 and 14 we extract the data $|U|, |W|$ and $U \bullet W$, and then

$$\cos \theta = \frac{U \bullet W}{|U| \cdot |W|} = \frac{14}{\sqrt{2} \cdot 5\sqrt{5}} = \frac{14}{5\sqrt{10}};$$

We calculate the cosines:

$$\cos \theta_1 = \frac{\vec{a} \bullet \vec{b}}{|\vec{a}||\vec{b}|} = \frac{7}{\sqrt{2} \cdot 5} \text{ and } \cos \theta_2 = \frac{\vec{p} \bullet \vec{q}}{|\vec{p}||\vec{q}|} = \frac{2}{1 \cdot \sqrt{5}},$$

and then we have

$$\cos \theta_1 \cos \theta_2 = \frac{7}{\sqrt{2} \cdot 5} \cdot \frac{2}{1 \cdot \sqrt{5}} = \frac{14}{5\sqrt{5}} = \cos \theta.$$

16. Since we already have the moduli of the tensors U and W, together with the $\cos \theta$ calculated in the previous question, we calculate the $\sin \theta$:

$$\sin \theta = \sqrt{1 - \cos^2 \theta} = \sqrt{1 - \left(\frac{14}{5\sqrt{10}}\right)^2} = \frac{\sqrt{54}}{5\sqrt{10}} = \frac{3\sqrt{2 \times 3}}{5\sqrt{2 \times 5}} = \frac{3\sqrt{3}}{5\sqrt{5}}$$

and next, the area of the triangle:

$$S = \frac{1}{2}|U||W|\sin\theta = \frac{1}{2}\sqrt{2}\cdot 5\sqrt{5}\cdot\frac{3\sqrt{3}}{5\sqrt{5}} = \frac{3}{2}\sqrt{6}.$$

17. To obtain the moduli of the tensors we proceed as follows.

$$X_{(\vec{a}\otimes\vec{a})} = \begin{bmatrix} 1 \\ -1 \end{bmatrix} \otimes \begin{bmatrix} 1 \\ -1 \end{bmatrix} = \begin{bmatrix} 1 \\ -1 \\ -1 \\ 1 \end{bmatrix};$$

since $\vec{a}\otimes\vec{a} \in \left(\overset{2}{\underset{1}{\otimes}} E_1^2\right)(\mathbb{R})$ we need to know the fundamental metric tensor of the mentioned tensor space in order to calculate the moduli of the tensors then

$$G(4) = G_1 \otimes G_1 = \begin{bmatrix} 1 & 2 \\ 2 & 5 \end{bmatrix} \otimes \begin{bmatrix} 1 & 2 \\ 2 & 5 \end{bmatrix} = \begin{bmatrix} 1 & 2 & 2 & 4 \\ 2 & 5 & 4 & 10 \\ 2 & 4 & 5 & 10 \\ 4 & 10 & 10 & 25 \end{bmatrix};$$

$$|\vec{a}\otimes\vec{a}| = \sqrt{[1\ -1\ -1\ 1]G(4)\begin{bmatrix} 1 \\ -1 \\ -1 \\ 1 \end{bmatrix}} = \sqrt{[1\ 3\ 3\ 9]\begin{bmatrix} 1 \\ -1 \\ -1 \\ 1 \end{bmatrix}} = \sqrt{4} = 2,$$

and we notice that the following holds:

$$|\vec{a}\otimes\vec{a}| = 2 = (\sqrt{2})^2 = |\vec{a}|^2.$$

Similarly, $\vec{b}\otimes\vec{b} \in \left(\overset{2}{\underset{1}{\otimes}} E_1^2\right)(\mathbb{R})$, and then

$$X_{(\vec{b}\otimes\vec{b})} = \begin{bmatrix} 2 \\ -3 \end{bmatrix} \otimes \begin{bmatrix} 2 \\ -3 \end{bmatrix} = \begin{bmatrix} 4 \\ -6 \\ -6 \\ 9 \end{bmatrix};$$

$$|\vec{b}\otimes\vec{b}| = \sqrt{[4\ -6\ -6\ 9]G(4)\begin{bmatrix} 4 \\ -6 \\ -6 \\ 9 \end{bmatrix}}$$

$$= \sqrt{[1\ 6\ 4\ 4\ 4\ 4\ 1\ 2\ 1]\begin{bmatrix} 4 \\ -6 \\ -6 \\ 9 \end{bmatrix}} = \sqrt{625} = 25,$$

and we notice again that the following holds:

$$|\vec{b} \otimes \vec{b}| = 25 = 5^2 = |\vec{b}|^2.$$

Finally, $\vec{a} \otimes \vec{b} \in \left(\overset{2}{\underset{1}{\otimes}} E_1^2 \right) (\mathbb{R})$, whence

$$X_{(\vec{a} \otimes \vec{b})} = \begin{bmatrix} 1 \\ -1 \end{bmatrix} \otimes \begin{bmatrix} 2 \\ -3 \end{bmatrix} = \begin{bmatrix} 2 \\ -3 \\ -2 \\ 3 \end{bmatrix},$$

from which

$$|\vec{a} \otimes \vec{b}| = \sqrt{\begin{bmatrix} 2 & -3 & -2 & 3 \end{bmatrix} \begin{bmatrix} 1 & 2 & 2 & 4 \\ 2 & 5 & 4 & 10 \\ 2 & 4 & 5 & 10 \\ 4 & 10 & 10 & 25 \end{bmatrix} \begin{bmatrix} 2 \\ -3 \\ -2 \\ 3 \end{bmatrix}}$$

$$= \sqrt{\begin{bmatrix} 4 & 11 & 12 & 33 \end{bmatrix} \begin{bmatrix} 2 \\ -3 \\ -2 \\ 3 \end{bmatrix}} = 5\sqrt{2},$$

which confirms again that

$$|\vec{a} \otimes \vec{b}| = 5\sqrt{2} = |\vec{a}| \cdot |\vec{b}|.$$

As a consequence of the obtained results, we can advance the moduli of the rest of the required tensor products, i.e., $|\vec{p} \otimes \vec{p}|, |\vec{q} \otimes \vec{q}|$, and $|\vec{p} \otimes \vec{q}|$, where all the tensors belong to the tensor space $\left(\overset{2}{\underset{1}{\otimes}} E_2^3 \right) (\mathbb{R})$.

We already know that

$$|\vec{p} \otimes \vec{p}| = |\vec{p}|^2 = 1^2 = 1; \quad |\vec{q} \otimes \vec{q}| = |\vec{q}|^2 = (\sqrt{5})^2 = 5; \quad |\vec{p} \otimes \vec{q}| = |\vec{p}||\vec{q}| = \sqrt{5}.$$

18. The third side of the triangle of sides the tensors U and W is the tensor $(U - W)$, because we must not forget that a tensor space is, above all, a *vector* space. From question 11, we get

$$V_6 = [U_6 - W_6] = [U_6] - [W_6] = \begin{bmatrix} 2 \\ -1 \\ 1 \\ -2 \\ 1 \\ -1 \end{bmatrix} - \begin{bmatrix} 6 \\ -2 \\ 4 \\ -9 \\ 3 \\ -6 \end{bmatrix} = \begin{bmatrix} -4 \\ 1 \\ -3 \\ 7 \\ -2 \\ 5 \end{bmatrix};$$

$$|U - W| \equiv |V_6| = \sqrt{V_6^t G(6) V_6}$$

$$= \sqrt{\begin{bmatrix} -4 & 1 & -3 & 7 & -2 & 5 \end{bmatrix} \begin{bmatrix} 1 & 1 & -1 & 2 & 2 & -2 \\ 1 & 2 & 0 & 2 & 4 & 0 \\ -1 & 0 & 3 & -2 & 0 & 6 \\ 2 & 2 & -2 & 5 & 5 & -5 \\ 2 & 4 & 0 & 5 & 10 & 0 \\ -2 & 0 & 6 & -5 & 0 & 15 \end{bmatrix} \begin{bmatrix} -4 \\ 1 \\ -3 \\ 7 \\ -2 \\ 5 \end{bmatrix}}$$

$$= \sqrt{99} = 3\sqrt{11}.$$

Thus, the perimeter of the tensor triangle becomes

$$P = |U| + |W| + |U - W| = \sqrt{2} + 5\sqrt{5} + 3\sqrt{11}.$$

19. The Euclidean dot product of two tensors, *requires* that both belong to *the same* Euclidean space. However:

$$\vec{a} \otimes \vec{b} \in \left(\overset{2}{\underset{1}{\otimes}} E_1^2 \right) (\mathbb{R}) \text{ and } \vec{p} \otimes \vec{q} \in \left(\overset{2}{\underset{1}{\otimes}} E_2^3 \right) (\mathbb{R}),$$

and then, the factor tensors belong to *different* Euclidean spaces, i.e., the question is *not* licit.

Remark 11.2. However, at least in theory, an "exterior connection" *can* be established, *defining* an *exterior* fundamental metric tensor between both linear spaces; this is clearly beyond the boundaries of this book. □

□

Example 11.14 (Common geometric tensors in Euclidean spaces). In the Euclidean ordinary geometric space $E^2(\mathbb{R})$, we consider a convex polygon of m sides $L_i, i \in \{1, 2, 3, \ldots, m\}$, together with the set of $m + 2$ *fixed* points inside it, $\{P, Q, A_i\}$.

With each point A_i we associate a vector $\vec{a}_i$, perpendicular to L_i with direction outside the polygon, and with modulus the length of L_i.

1. a) Calculate the value of the expression

$$\vec{T} = \sum_{i=1}^{i=m} (\vec{p}_i \wedge \vec{a}_i + \vec{a}_i \wedge \vec{q}_i),$$

   where $\vec{p}_i = \vec{PA}_i$; $\vec{q}_i = \vec{QA}_i$; "$\wedge$" is the symbol of the classic cross product.
   b) Idem for the particular case $m = 3$.
2. a) In the Euclidean ordinary geometric space we consider a convex polyhedron of m faces $S_i, i \in \{1, 2, \ldots, m\}$, together with a set of $m + 2$ fixed points interior to it, $\{P, Q, A_i\}$. With each point A_i we associate a vector $\vec{a}_i$, perpendicular to the face S_i, and in the direction outside the polyhedron, and with modulus the value of the area of S_i. Obtain the value of the expression in 1(a).

b) Idem for the particular case $m = 4$.

Solution:

1. a) We assume that each side L_i of the convex polygon is a vector $\vec{L}_i$ directed along all the boundary in the same direction; since the polygon is closed, we have

$$\sum_{i=1}^{i=m} \vec{L}_i = \vec{0}; \quad \sum_{i=1}^{i=m} L_{\circ i}^{\alpha \circ} \vec{e}_\alpha = 0 \rightarrow \sum_{i=1}^{i=m} [\vec{e}_1 \quad \vec{e}_2] \begin{bmatrix} L_{\circ i}^{1\circ} \\ L_{\circ i}^{2\circ} \end{bmatrix} = \vec{0},$$

which implies

$$\begin{cases} \sum_{i=1}^{m} L_{\circ i}^{1\circ} = 0 \\ \sum_{i=1}^{m} L_{\circ i}^{2\circ} = 0 \end{cases} \rightarrow \sum_{i=1}^{m} L_{\circ i}^{\alpha \circ} = 0, \ \alpha \in \{1, 2\}. \tag{11.124}$$

In addition, we have

$$\vec{T} = \sum_{i=1}^{i=m} (\vec{p}_i \wedge \vec{a}_i + \vec{a}_i \wedge \vec{q}_i) = \sum_{i=1}^{i=m} (\vec{p}_i \wedge \vec{a}_i - \vec{q}_i \wedge \vec{a}_i) = \sum_{i=1}^{i=m} (\vec{p}_i - \vec{q}_i) \wedge \vec{a}_i$$

$$= \sum_{i=1}^{i=m} (\vec{PA}_i - \vec{QA}_i) \wedge \vec{a}_i = \sum_{i=1}^{i=m} \vec{PQ} \wedge \vec{a}_i = \vec{PQ} \wedge \sum_{i=1}^{i=m} \vec{a}_i. \tag{11.125}$$

However, $\sum_{i=1}^{i=m} \vec{a}_i = \vec{R}$ is the resultant of the vectors $\vec{a}_i$, and thus $\vec{R}$ is independent of the vectors to be linked to points A_i, or applied to the same point, for example P. In addition, each $\vec{a}_i$ is a vector rotated $90°$ from the corresponding $\vec{L}_i$. Whence, if we denote by $Z = [z_{\circ \alpha}^{i\circ}]$ the rotation matrix, we have

$$\vec{a}_i = a_{\circ i}^{\beta \circ} \vec{e}_\beta = (z_{\circ \alpha}^{\beta \circ} L_{\circ i}^{\alpha \circ}) \vec{e}_\beta. \tag{11.126}$$

Substituting (11.126) into (11.125), we get

$$\vec{T} = \vec{PQ} \wedge \sum_{i=1}^{i=m} (z_{\circ \alpha}^{\beta \circ} L_{\circ i}^{\alpha \circ}) \vec{e}_\beta = \vec{PQ} \wedge z_{\circ \alpha}^{\beta \circ} \left(\sum_{i=1}^{i=m} L_{\circ i}^{\alpha \circ} \right) \vec{e}_\beta,$$

and taking into account (11.124), and replacing this into the last expression we obtain

$$\vec{T} = \vec{PQ} \wedge \vec{0},$$

so that $\vec{T} = \vec{0}$.

b) For the particular case $m = 3$, we also have $\vec{T} = \vec{0}$.

2. a) Since the surface of the polyhedron is closed and convex, the projections of the vectors of the upper and lower faces, over each Cartesian plane XOY, XOZ, YOZ, cancel out, i.e.,

$$\vec{S} = \sum_{i=1}^{i=m} \vec{S}_i = \vec{0};$$

or

$$\sum_{i=1}^{i=m} s_{\circ\ i}^{\alpha\circ} \vec{e}_\alpha = \vec{0},$$

so that in summary, we have

$$\sum_{i=1}^{i=m} s_{\circ\ i}^{\alpha\circ} \vec{e}_\alpha = \vec{0}; \ \alpha \in \{1, 2, 3\}. \tag{11.127}$$

Similarly

$$\vec{T} = \sum_{i=1}^{i=m} (\vec{p}_i \wedge \vec{a}_i + \vec{a}_i \wedge \vec{q}_i) = \vec{PQ} \wedge \sum_{i=1}^{i=m} \vec{a}_i = \vec{PQ} \wedge \sum_{i=1}^{i=m} (s_{\circ\ i}^{\alpha\circ} \vec{e}_\alpha)$$

$$= \vec{PQ} \wedge \left(\sum_{i=1}^{i=m} s_{\circ\ i}^{\alpha\circ} \right) \vec{e}_\alpha,$$

and susbstituting (11.127) into the last expression, we obtain

$$\vec{T} = \vec{PQ} \wedge (0)\vec{e}_\alpha = \vec{PQ} \wedge \vec{0} = \vec{0}.$$

b) Also for $m = 4$, we have $\vec{T} = \vec{0}$.

$\square$

11.12 Exercises

11.1. In a Euclidean space $E^2(\mathbb{R})$ referred to a basis $\{\vec{e}_\alpha\}$ the fundamental metric tensor in covariant coordinates is given by the matrix

$$G = [g_{\alpha\beta}^{\circ\circ}] = \begin{bmatrix} 1 & 1 \\ 1 & 2 \end{bmatrix}.$$

Consider the Euclidean tensor T of matrix representation

$$[t_{\circ\circ\gamma\delta}^{\alpha\beta\circ\circ}] = \begin{bmatrix} 1 & -1 & | & 3 & 1 \\ 1 & 2 & | & 4 & -1 \\ - & - & + & - & - \\ 2 & -5 & | & 1 & 6 \\ 3 & 1 & | & 2 & 2 \end{bmatrix}.$$

1. Obtain the matrix representation T' of the tensor associated with T by its contravariant coordinates.

2. Obtain T'' associated with T by its covariant coordinates.

3. Obtain the tensor T''' associated with T of components $[t^{\circ\circ\gamma\delta}_{\alpha\beta\circ\circ}]$.

4. Obtain the doubly contracted tensor $P = C \begin{pmatrix} \alpha & \beta \\ \gamma & \delta \end{pmatrix} T$.

5. Obtain the tensor $W = C \begin{pmatrix} \gamma & \delta \\ \alpha & \beta \end{pmatrix} T''$.

6. Is $P = W$?

11.2. In the Euclidean vector space $E^3(\mathbb{R})$ referred to a basis $\{\vec{e}_\alpha\}$, the fundamental metric tensor is given by the Gram matrix

$$G = \begin{bmatrix} 2 & 1 & 1 \\ 1 & 2 & 1 \\ 1 & 1 & 2 \end{bmatrix}.$$

Consider the Euclidean tensor of order $r = 3$ represented by

$$[t^{\circ\beta\circ}_{\alpha\circ\gamma}] = \begin{bmatrix} 3 & 0 & 0 \\ 1 & 3 & 0 \\ 0 & 1 & 3 \\ \hline 3 & 1 & 0 \\ 0 & 3 & 1 \\ 0 & 0 & 3 \\ \hline 0 & 1 & 0 \\ 1 & 0 & 0 \\ 0 & 1 & 0 \end{bmatrix}.$$

1. Obtain the representation T' of the totally contravariant associated tensor of T.

2. Obtain the representation T'' of the totally covariant associated tensor of T.

3. We perform in $E^3(\mathbb{R})$ a change-of-basis $\{\vec{e}_\alpha\}$ of matrix

$$C \equiv [c^{\alpha\circ}_{\circ i}] = \begin{bmatrix} 1 & 2 & 3 \\ 2 & 3 & 2 \\ 1 & 2 & 2 \end{bmatrix}.$$

Determine the new matrix representations $\hat{T}, \hat{T}'$ and $\hat{T}''$ of the tensor T and its associated tensor.

4. We first proceed with the matrix $\hat{T}$ of the tensor T in the new basis, and later we determine as associated $\hat{T}'$ and $\hat{T}''$. Verify numerically the results.

11.3. In a Euclidean space $E^3(\mathbb{R})$ the fundamental metric tensor is given by the matrix

$$G = [g_{\alpha\beta}^{\circ\circ}] = \begin{bmatrix} 1 & 0 & 1 \\ 0 & 2 & 0 \\ 1 & 0 & 3 \end{bmatrix}.$$

We consider two Euclidean tensors P and Q of order $r = 3$, with matrix representations

$$[p_{\circ\circ\circ}^{\alpha\beta\gamma}] = \frac{1}{4} \begin{bmatrix} -12 & -24 & -36 \\ -2 & -8 & -14 \\ 4 & 4 & 4 \\ \hline -2 & -8 & -14 \\ 1 & -2 & -5 \\ 2 & 2 & 2 \\ \hline 4 & 4 & 4 \\ 2 & 2 & 2 \\ 0 & 0 & 0 \end{bmatrix} \; ; \; [q_{\alpha\beta\gamma}^{\circ\circ\circ}] = \begin{bmatrix} -8 & -8 & -22 \\ -6 & -6 & -18 \\ -4 & -4 & -14 \\ \hline -6 & -6 & -18 \\ -4 & -4 & -14 \\ -2 & -2 & -10 \\ \hline -4 & -4 & -14 \\ -2 & -2 & -10 \\ 0 & 0 & -6 \end{bmatrix} ,$$

where α is the block row, β is the row of each block, and γ is the column of each block.

1. Are P and Q the same tensor?
2. If the answer is affirmative, give the tensor $Q' = [q_{\alpha\beta\circ}^{\circ\circ\gamma}]$ associated with Q.
3. Execute the Euclidean contraction $V_1 = \mathcal{E}\mathcal{C}(\alpha,\gamma)Q$.
4. Execute the normal contraction $V_2 = \mathcal{C}(\begin{smallmatrix}\alpha\\\gamma\end{smallmatrix})Q'$.
5. Are $\vec{V_1}$ and $\vec{V_2}$ the same vector?

11.4. In a Euclidean space $E^3(\mathbb{R})$ referred to a basis $\{\vec{e}_\alpha\}$ the fundamental metric tensor is given by the Gram matrix

$$G = [g_{\alpha\beta}^{\circ\circ}] = \begin{bmatrix} 3 & 1 & 2 \\ 1 & 1 & 1 \\ 2 & 1 & 2 \end{bmatrix}.$$

Over $E^3(\mathbb{R})$, we consider the tensors S and T of components

$$S = [s_{\circ\delta}^{\gamma\circ}] = \begin{bmatrix} 1 & 0 & -1 \\ 0 & 2 & 0 \\ -1 & 0 & 2 \end{bmatrix}$$

and

$$T = [t^{\circ\,\mu\,\circ}_{\lambda\,\circ\,\nu}] = \begin{bmatrix} 1 & 2 & 0 \\ 0 & -1 & 0 \\ -1 & 1 & 1 \\ \hline 2 & 0 & 5 \\ 3 & 0 & 1 \\ 0 & 0 & 2 \\ \hline -1 & 2 & 0 \\ 2 & 0 & 0 \\ 0 & 0 & 1 \end{bmatrix}.$$

1. Find the covariant coordinates of the tensor $P = S \otimes T$.
2. Representing with $\mathcal{EC}$ the Euclidean contraction of indices of the same species, obtain the contravariant coordinates of the Euclidean tensor $Q = \mathcal{EC}(1,3)T$.
3. a) Starting from the factors, determine the contracted tensor product $Z = \mathcal{EC}(1,4)S \otimes T$.
    b) Starting from the executed product P, determine the contraction $Z = \mathcal{EC}(1,4)P$.
4. Consider in $E^3(\mathbb{R})$ a new basis $\{\vec{e}_i\}$ given by the change matrix:

$$C = \begin{bmatrix} 1 & 2 & 1 \\ 1 & 3 & 1 \\ 1 & 2 & 2 \end{bmatrix}.$$

   a) Obtain the covariant coordinates of the associated tensor of S in the basis $\{\vec{e}_i\}$.
   b) Obtain the contravariant coordinates of the associated tensor of Z in the basis $\{\vec{e}_i\}$.
   c) Obtain the covariant coordinates of the associated tensor of P in the basis $\{\vec{e}_i\}$.

11.5. In the Euclidean space $E^3(\mathbb{R})$ referred to a basis $\{\vec{e}_\alpha\}$ the fundamental metric tensor is given by the matrix

$$G = [g^{\circ\,\circ}_{\alpha\,\beta}] = \begin{bmatrix} 1 & 1 & 1 \\ 1 & 2 & 1 \\ 1 & 1 & 3 \end{bmatrix}.$$

Consider two tensors of order $r = 3$, T and U given by their components:

$$t^{\alpha\,\circ\,\circ}_{\circ\,\beta\,\gamma} = \alpha + \beta + \gamma; \quad u^{\circ\,\circ\,\nu}_{\lambda\,\mu\,\circ} = \lambda + \mu - \nu,$$

and let $P = T \otimes U$.

1. Obtain the contravariant coordinates of $P = \mathcal{C}\begin{pmatrix} \alpha \\ \lambda \end{pmatrix}P$.
2. Obtain the covariant coordinates of $Q = \mathcal{EC}(\alpha,\nu)P$.

3. Execute the dot product $T \bullet U$ directly, determining the matrix of the adequate fundamental metric tensor $G(r)$ that permits the direct calculation, by means of a formula of the type (11.114).

4. Obtain the components of the associated tensor of T, $T' = [t_{\alpha\circ\circ}^{\circ\beta\gamma}]$.

5. Obtain the components of the associated tensor of U, $U' = [u_{\circ\circ\nu}^{\lambda\mu\circ}]$.

6. Obtain the moduli of the tensors T and U.

7. Obtain the moduli of the tensor Q, with the help of the computer.

11.6. In a Euclidean space $E^3(\mathbb{R})$ referred to a basis $\{\vec{e}_\alpha\}$ the fundamental metric tensor is given by the Gram matrix

$$G = [g_{\alpha\beta}^{\circ\circ}] = \begin{bmatrix} 1 & -2 & 0 \\ -2 & 6 & 1 \\ 0 & 1 & 1 \end{bmatrix}.$$

A multilinear form F, applies the tensors $T \in E^3 \otimes E_*^3 \otimes E^3(\mathbb{R})$ in the field $\mathbb{R}$, according to the expression

$$F(T) = 2t_{\circ 2\circ}^{1\circ 1} + 3t_{\circ 1\circ}^{2\circ 2} - 5t_{\circ 1\circ}^{2\circ 3} + 6t_{\circ 2\circ}^{2\circ 1} - 2t_{\circ 2\circ}^{2\circ 3} + 4t_{\circ 1\circ}^{3\circ 1} - t_{\circ 3\circ}^{3\circ 3}.$$

1. Determine the basis $\{\vec{e}^{*\alpha}\}$ reciprocal of the $\{\vec{e}_\alpha\}$, expressed as a linear manifold of the basis $\{\vec{e}_\alpha\}$.

2. Determine the basis $\mathcal{B}^*$ reciprocal of the basis $\mathcal{B}$ of $E^3 \otimes E_*^3 \otimes E^3(\mathbb{R})$ as a function of the vectors in the basis $\mathcal{B}$.

3. Obtain the mnemonic representation of the tensor space to which F belongs.

4. Obtain a matrix representation of the components of the tensor F.

5. Obtain the fundamental metric tensor $G^*(r)$ of the Euclidean tensor space proposed in question 3.

6. Obtain the modulus of F (notated $|F|$).

7. Obtain the fundamental metric tensor $G(r)$ associated with the Euclidean tensor space $E^3 \otimes E_*^3 \otimes E^3(\mathbb{R})$.

8. Obtain the modulus of T (notated $|T|$).

9. Consider the tensor T, with matrix representation

$$[t_{\circ\beta\circ}^{\alpha\circ\gamma}] = \begin{bmatrix} 1 & 1 & 1 \\ 2 & 1 & 2 \\ 1 & 2 & 2 \\ \hline 2 & 2 & 4 \\ 1 & 1 & 1 \\ 2 & 1 & 2 \\ \hline 1 & 0 & 1 \\ 2 & 2 & 1 \\ 1 & 1 & 1 \end{bmatrix},$$

and give the scalar $\rho = F(T)$.

10. Does the quotient $\frac{F(T)}{|F|\cdot|T|}$ suggest something to the reader?

11. We perform in the space $E^3(\mathbb{R})$ a change-of-basis $\{\vec{e_i}\}$ of orthogonal matrix

$$\frac{1}{\sqrt{2}} \begin{bmatrix} 0 & 0 & \sqrt{2} \\ 1 & -1 & 0 \\ 1 & 1 & 0 \end{bmatrix}.$$

Answer all the previous questions, but taking now $\{\vec{e_i}\}$ as the initial basis of $E^3(\mathbb{R})$.

Modular Tensors over $E^n(\mathbb{R})$ Euclidean Spaces

12.1 Introduction

This chapter, after establishing the required mathematical entities, is devoted to defining the orientation tensor, analyzing its Euclidean properties as associated tensors, and its application to Euclidean polar tensors. The new concepts are illustrated with some examples of tensors deduced from the orientation tensor.

12.2 Diverse cases of linear space connections

In Chapter 8, Sections 8.2.1 and 8.2.2 we have established the concept of "oriented linear spaces" and "oriented tensors".

The pseudo-Euclidean spaces $PSE^n(\mathbb{R})$ and the Euclidean spaces $E^n(\mathbb{R})$ are frequently used to build over them tensor spaces $\left(\overset{r}{\underset{1}{\otimes}} PSE^n\right)(\mathbb{R})$ and $\left(\overset{r}{\underset{1}{\otimes}} E^n\right)(\mathbb{R})$, that contain special tensors that are useful in certain applications, as for example the tensor "cross product", "mixed product", "double vector product", etc., extended even in certain cases to dimensions $(n > 3)$ larger than that of the classic geometric space.

What happens if a change-of-basis is performed in the fundamental space $PSE^n(\mathbb{R})$ or $E^n(\mathbb{R})$ of matrix C, such that its relative modulus is $|C| < 0$? Since $|C|$ is negative it can change the orientation of the n-reference frame, which *invalidates* the definition of certain tensors, such as the "cross product".

On the other hand, expressions such as $\sqrt{|G|}$, a factor that accompanies certain pseudo-Euclidean or Euclidean tensors, produce problems when executing changes of basis. If, for example, the fundamental metric tensor of $E^n(\mathbb{R})$ is represented by the Gram matrix $G = [g_{\alpha\beta}^{\circ\circ}]$, it is well known that it changes basis by "congruence", i.e.,

$$\hat{G} = C^t G C,$$

and taking determinants and using the Binet–Cauchy property, we get

$$|\hat{G}| = |C^t G C| = |C^t||G||C| = |C|^2|G|,$$

which leads to

$$\sqrt{|\hat{G}|} = |C|\sqrt{|G|}, \tag{12.1}$$

which is the change-of-basis expression of a modular tensor (modular scalar of weight 1). This leads to a problem if $|C|$ is negative.

Even worse, in the pseudo-Euclidean spaces, it can happen that

$$|\hat{G}| = |C|^2 G \equiv |\text{Sylvesterian}| = \begin{vmatrix} 1 & 0 & 0 & \cdots & 0 & 0 \\ 0 & 1 & 0 & \cdots & 0 & 0 \\ \cdots & \cdots & \cdots & \cdots & \cdots & \cdots \\ 0 & 0 & 0 & \cdots & -1 & 0 \\ 0 & 0 & 0 & \cdots & 0 & -1 \end{vmatrix} = (-1)^{n-\sigma}$$

be negative, where σ is the Spanish signature. Thus, we discover that $|G|$ is negative and so its square root becomes imaginary. This fact has led some authors to define Relation (12.1) in a different form appropriate for the pseudo-Euclidean spaces

$$\sqrt{\text{absolute value of } |\hat{G}|} = |C|\sqrt{\text{absolute value of } |G|},$$

ignoring the tensor or non-tensor nature of the above square root.

Fortunately, in the Euclidean spaces $E^n(\mathbb{R})$ over which we have built the Euclidean tensors of the previous chapter, the fundamental metric tensor matrix, G, is positive definite and so we always have $|G| > 0$. Due to all these considerations, in this chapter we *exclusively* deal with "oriented modular tensors" ($|C| > 0$) over Euclidean spaces ($|G| > 0$).

12.3 Tensor character of $\sqrt{|G|}$

Though in the introduction we have already presented this topic, and we started by saying that it is well known that the fundamental metric tensor changes in a congruent way, etc., it is not a sufficiently rigorous presentation, as is required by its content, so, we propose a more through development.

The fundamental metric tensor satisfies the tensor relations:

$$g_{ij}^{\circ\circ} = g_{\alpha\beta}^{\circ\circ} c_{i\,\circ}^{\circ\alpha} c_{j\,\circ}^{\circ\beta} \quad \text{covariant}$$

$$g_{\circ\circ}^{ij} = g_{\circ\circ}^{\alpha\beta} \gamma_{\circ\alpha}^{i\,\circ} \gamma_{\circ\beta}^{j\,\circ} \quad \text{contravariant,}$$

which expressed to facilitate its treatment in matrix form, become

$$g_{ij}^{\circ\circ} = c_{i\circ}^{\circ\alpha} g_{\alpha\beta}^{\circ\circ} c_{\circ j}^{\beta\circ}$$

$$g_{\circ\circ}^{ij} = \gamma_{\circ\alpha}^{i\circ} g_{\circ\circ}^{\alpha\beta} \gamma_{\beta\circ}^{\circ j}$$

and taking determinants, and considering that

$$[c_{i\circ}^{\circ\alpha}] = C^t; \quad [c_{\circ j}^{\beta\circ}] = C; \quad [\gamma_{\circ\alpha}^{i\circ}] = C^{-1}; \quad [\gamma_{\beta\circ}^{\circ j}] = (C^t)^{-1},$$

we obtain the expressions

$$|g_{ij}^{\circ\circ}| = |C^t||g_{\alpha\beta}^{\circ\circ}||C|$$

$$|g_{\circ\circ}^{ij}| = |C^{-1}||g_{\circ\circ}^{\alpha\beta}||(C^t)^{-1}|$$

and because of the determinant properties, we get

$$|g_{ij}^{\circ\circ}| = |C|^2 |g_{\alpha\beta}^{\circ\circ}| \tag{12.2}$$

$$|g_{\circ\circ}^{ij}| = \frac{1}{|C|^2} |g_{\circ\circ}^{\alpha\beta}|, \tag{12.3}$$

which are a modular tensor of order $r = 0$ and weight $w = 2$, and a modular tensor of order $r = 0$ and weight $w = -2$, respectively.

If we extract the square root, we obtain

$$\sqrt{|g_{ij}^{\circ\circ}|} = |C|\sqrt{|g_{\alpha\beta}^{\circ\circ}|} \tag{12.4}$$

$$\sqrt{|g_{\circ\circ}^{ij}|} = \frac{1}{|C|}\sqrt{|g_{\circ\circ}^{\alpha\beta}|}, \tag{12.5}$$

which are modular tensors of order $r = 0$ and respective weights $w_1 = 1$ and $w_2 = -1$. Formula (12.4) is precisely (12.1), expressed by means of the Gram matrix $|G| \equiv |g_{\alpha\beta}^{\circ\circ}|$. It is convenient to present the relations (12.4) and (12.5) in a reverse form, producing in this way other systems of modular scalars, which will be useful later:

$$\frac{1}{\sqrt{|g_{\circ\circ}^{ij}|}} = |C|\frac{1}{\sqrt{|g_{\circ\circ}^{\alpha\beta}|}} \tag{12.6}$$

$$\frac{1}{\sqrt{|g_{ij}^{\circ\circ}|}} = |C|^{-1}\frac{1}{\sqrt{|g_{\alpha\beta}^{\circ\circ}|}}. \tag{12.7}$$

They are modular tensors of order $r = 0$ and respective weights $w_1 = 1$ and $w_2 = -1$. Formulas (12.2) to (12.7) can also be written in terms of the determinants of the matrices

$$[g_{\alpha\beta}^{\circ\circ}] = G; \quad [g_{\circ\circ}^{\alpha\beta}] = G^{-1}; \quad [g_{ij}^{\circ\circ}] = \hat{G}; \quad [g_{\circ\circ}^{ij}] = \hat{G}^{-1}.$$

12.4 The orientation tensor: Definition

From the fundamental metric tensor G associated with $E^n(\mathbb{R})$ in the basis $\{\vec{e}_\alpha\}$ and the "Levi-Civita ϵ" systems modular tensor in its two simultaneous versions, the contravariant $\epsilon^{\alpha_1\,\alpha_2\,\cdots\,\alpha_n}_{oo\cdotso}$ (known as Levi-Civita tensor density) and the covariant $\epsilon^{oo\cdotso}_{\alpha_1\,\alpha_2\,\cdots\,\alpha_n}$ (known as Levi-Civita tensor capacity), both established and analyzed in Section 8.7, we define the system of scalars called the "orientation tensor" of the following form:

$$\vartheta^{\alpha_1\,\alpha_2\,\cdots\,\alpha_n}_{oo\cdotso} = \frac{1}{\sqrt{|g^{o}_{\alpha\beta}|}}\epsilon^{\alpha_1\,\alpha_2\,\cdots\,\alpha_n}_{oo\cdotso} \tag{12.8}$$

$$\vartheta^{oo\cdotso}_{\alpha_1\,\alpha_2\,\cdots\,\alpha_n} = \sqrt{|g^{o}_{\alpha\beta}|}\,\epsilon^{oo\cdotso}_{\alpha_1\,\alpha_2\,\cdots\,\alpha_n}. \tag{12.9}$$

The complete name of Expression (12.8) is "contravariant orientation tensor of the Euclidean space $E^n(\mathbb{R})$" and that of (12.9) is "covariant orientation tensor of the Euclidean space $E^n(\mathbb{R})$". Evidently, they are *totally anti-symmetric*, because the Levi-Civita tensors are anti-symmetric.

12.5 Tensor character of the orientation tensor

In the title, we have advanced that it is a tensor; however, we need to prove this nature. We start by referring to Formulas (8.71) and (8.72), which express the modular tensor nature of the Levi-Civita tensor density and tensor capacity:

$$\epsilon^{i_1\,i_2\,\cdots\,i_n}_{oo\cdotso} = |C|\epsilon^{\alpha_1\,\alpha_2\,\cdots\,\alpha_n}_{oo\cdotso}\gamma^{i_1\,o}_{o\,\alpha_1}\gamma^{i_2\,o}_{o\,\alpha_2}\cdots\gamma^{i_n\,o}_{o\,\alpha_n} \tag{12.10}$$

$$\epsilon^{oo\cdotso}_{i_1\,i_2\,\cdots\,i_n} = |C|^{-1}\epsilon^{oo\cdotso}_{\alpha_1\,\alpha_2\,\cdots\,\alpha_n}\gamma^{o\,\alpha_1}_{i_1\,o}\gamma^{o\,\alpha_2}_{i_2\,o}\cdots\gamma^{o\,\alpha_n}_{i_n\,o} \tag{12.11}$$

and we also use Formulas (12.4) and (12.7) established in Section 12.3.

We start with the definitions (12.8) and (12.9), which are formulated in the new basis, and we substitute the mentioned formulas, to arrive at the following expressions for the initial basis:

$$\vartheta^{i_1\,i_2\,\cdots\,i_n}_{oo\cdotso} = \frac{1}{\sqrt{|g^{o}_{ij}|}}\epsilon^{i_1\,i_2\,\cdots\,i_n}_{oo\cdotso}$$

$$= \left(|C|^{-1}\frac{1}{\sqrt{|g^{o}_{\alpha\beta}|}}\right)|C|\epsilon^{\alpha_1\,\alpha_2\,\cdots\,\alpha_n}_{oo\cdotso}\gamma^{i_1\,o}_{o\,\alpha_1}\gamma^{i_2\,o}_{o\,\alpha_2}\cdots\gamma^{i_n\,o}_{o\,\alpha_n}$$

$$= \left(\frac{1}{\sqrt{|g^{o}_{\alpha\beta}|}}\epsilon^{\alpha_1\,\alpha_2\,\cdots\,\alpha_n}_{oo\cdotso}\right)\gamma^{i_1\,o}_{o\,\alpha_1}\gamma^{i_2\,o}_{o\,\alpha_2}\cdots\gamma^{i_n\,o}_{o\,\alpha_n}$$

$$= \vartheta^{\alpha_1\,\alpha_2\,\cdots\,\alpha_n}_{oo\cdotso}\gamma^{i_1\,o}_{o\,\alpha_1}\gamma^{i_2\,o}_{o\,\alpha_2}\cdots\gamma^{i_n\,o}_{o\,\alpha_n}.$$

In summary

$$\vartheta^{i_1 i_2 \cdots i_n}_{\quad\;\circ\;\circ\,\cdots\,\circ} = \vartheta^{\alpha_1 \alpha_2 \cdots \alpha_n}_{\quad\;\circ\;\circ\,\cdots\,\circ}\, \gamma^{i_1\;\,\circ}_{\;\circ\,\alpha_1}\gamma^{i_2\;\,\circ}_{\;\circ\,\alpha_2} \cdots \gamma^{i_n\;\,\circ}_{\;\circ\,\alpha_n}, \tag{12.12}$$

which reveals that the orientation tensor $\vartheta^{\alpha_1 \alpha_2 \cdots \alpha_n}_{\quad\;\circ\;\circ\,\cdots\,\circ}$ is a totally contravariant and anti-symmetric *homogeneous* tensor of order $r = n$ (homogeneous is equivalent to saying modular of weight $w = 0$).

In a similar way we prove the tensor nature of $\vartheta^{\;\circ\;\;\circ\,\cdots\,\circ}_{\alpha_1 \alpha_2 \cdots \alpha_n}$:

$$\vartheta^{\;\circ\;\;\circ\,\cdots\,\circ}_{i_1 i_2 \cdots i_n} = \sqrt{|g^{\circ\circ}_{ij}|}\,\epsilon^{\;\circ\;\;\circ\,\cdots\,\circ}_{i_1 i_2 \cdots i_n} = \left(|C|\sqrt{|g^{\circ\circ}_{\alpha\beta}|}\right) |C|^{-1}\epsilon^{\;\circ\;\;\circ\,\cdots\,\circ}_{\alpha_1 \alpha_2 \cdots \alpha_n}\, c^{\;\circ\,\alpha_1}_{i_1\;\,\circ}\, c^{\;\circ\,\alpha_2}_{i_2\;\,\circ} \cdots c^{\;\circ\,\alpha_n}_{i_n\;\,\circ}$$

$$= \left(\sqrt{|g^{\circ\circ}_{\alpha\beta}|}\,\epsilon^{\;\circ\;\;\circ\,\cdots\,\circ}_{\alpha_1 \alpha_2 \cdots \alpha_n}\right) c^{\;\circ\,\alpha_1}_{i_1\;\,\circ}\, c^{\;\circ\,\alpha_2}_{i_2\;\,\circ} \cdots c^{\;\circ\,\alpha_n}_{i_n\;\,\circ}$$

$$= \vartheta^{\;\circ\;\;\circ\,\cdots\,\circ}_{\alpha_1 \alpha_2 \cdots \alpha_n}\, c^{\;\circ\,\alpha_1}_{i_1\;\,\circ}\, c^{\;\circ\,\alpha_2}_{i_2\;\,\circ} \cdots c^{\;\circ\,\alpha_n}_{i_n\;\,\circ}$$

and in summary

$$\vartheta^{\;\circ\;\;\circ\,\cdots\,\circ}_{i_1 i_2 \cdots i_n} = \vartheta^{\;\circ\;\;\circ\,\cdots\,\circ}_{\alpha_1 \alpha_2 \cdots \alpha_n}\, c^{\;\circ\,\alpha_1}_{i_1\;\,\circ}\, c^{\;\circ\,\alpha_2}_{i_2\;\,\circ} \cdots c^{\;\circ\,\alpha_n}_{i_n\;\,\circ}, \tag{12.13}$$

which proves the tensor nature of the orientation tensor $\vartheta^{\;\circ\;\;\circ\,\cdots\,\circ}_{\alpha_1 \alpha_2 \cdots \alpha_n}$, which is a totally covariant and anti-symmetric homogeneous tensor of order $r = n$.

Since the orientation tensor is a totally anti-symmetric homogeneous tensor of order $r = n$, we must not forget that its complete notation is that corresponding to the entities of a linear tensor space. According to the definitions in (12.8) and (12.9) and the declarations established at the end of Section 5, we have:

1. For the contravariant tensor:

$$\Theta = \vartheta^{\alpha_1 \alpha_2 \cdots \alpha_n}_{\quad\;\circ\;\circ\,\cdots\,\circ}\,\vec{e}_{\alpha_1} \otimes \vec{e}_{\alpha_2} \otimes \cdots \otimes \vec{e}_{\alpha_n}$$

$$= \frac{1}{\sqrt{|G|}}\epsilon^{\alpha_1 \alpha_2 \cdots \alpha_n}_{\quad\;\circ\;\circ\,\cdots\,\circ}\,\vec{e}_{\alpha_1} \otimes \vec{e}_{\alpha_2} \otimes \cdots \otimes \vec{e}_{\alpha_n};\; \alpha_i \in I_n;\; \alpha_1 \neq \alpha_2 \neq \cdots \neq \alpha_n.$$

2. For the covariant orientation tensor:

$$\Theta = \vartheta^{\;\circ\;\;\circ\,\cdots\,\circ}_{\alpha_1 \alpha_2 \cdots \alpha_n}\,\vec{e}^{*\alpha_1} \otimes \vec{e}^{*\alpha_2} \otimes \cdots \otimes \vec{e}^{*\alpha_n}$$

$$= \sqrt{|G|}\,\epsilon^{\;\circ\;\;\circ\,\cdots\,\circ}_{\alpha_1 \alpha_2 \cdots \alpha_n}\,\vec{e}^{*\alpha_1} \otimes \vec{e}^{*\alpha_2} \otimes \cdots \otimes \vec{e}^{*\alpha_n};\; \alpha_i \in I_n;\; \alpha_1 \neq \alpha_2 \neq \cdots \neq \alpha_n.$$

12.6 Orientation tensors as associated Euclidean tensors

The ordered development of the determinant $|g^{\circ\circ}_{\alpha\beta}|$ by the elements of its *fixed rows*, with the help of the modular tensor system ϵ, also known as Levi-Civita tensor, can be written as:

$$|g_{\overset{\circ}{\alpha}\overset{\circ}{\beta}}| = g_{1\overset{\circ}{\beta_1}}^{\circ\,\circ} g_{2\overset{\circ}{\beta_2}}^{\circ\,\circ} \cdots g_{n\overset{\circ}{\beta_n}}^{\circ\,\circ} \epsilon^{\beta_1\beta_2\cdots\beta_n}_{\circ\,\circ\,\cdots\,\circ}. \tag{12.14}$$

Executing any permutation in the rows of the previous determinant, is equivalent to multiplying it by $(+1)$, or (-1) depending on the permutation parity; in consequence, Expression (12.14), after permuting its rows, taking into account (8.75) can be written as

$$\epsilon^{\circ\,\circ\,\cdots\,\circ}_{\alpha_1\alpha_2\cdots\alpha_n} |g_{\overset{\circ}{\alpha}\overset{\circ}{\beta}}| = \delta^{1\;2\;\cdots\;n}_{\alpha_1\alpha_2\cdots\alpha_n} g_{1\overset{\circ}{\beta_1}}^{\circ\,\circ} g_{2\overset{\circ}{\beta_2}}^{\circ\,\circ} \cdots g_{n\overset{\circ}{\beta_n}}^{\circ\,\circ} \epsilon^{\beta_1\beta_2\cdots\beta_n}_{\circ\,\circ\,\cdots\,\circ} \tag{12.15}$$

or

$$\epsilon^{\circ\,\circ\,\cdots\,\circ}_{\alpha_1\alpha_2\cdots\alpha_n} |g_{\overset{\circ}{\alpha}\overset{\circ}{\beta}}| = g_{\overset{\circ}{\alpha_1}\overset{\circ}{\beta_1}} g_{\overset{\circ}{\alpha_2}\overset{\circ}{\beta_2}} \cdots g_{\overset{\circ}{\alpha_n}\overset{\circ}{\beta_n}} \epsilon^{\beta_1\beta_2\cdots\beta_n}_{\circ\,\circ\,\cdots\,\circ} \tag{12.16}$$

and since it is

$$|g_{\overset{\circ}{\alpha}\overset{\circ}{\beta}}| = \sqrt{|g_{\overset{\circ}{\alpha}\overset{\circ}{\beta}}|}\sqrt{|g_{\overset{\circ}{\alpha}\overset{\circ}{\beta}}|}, \tag{12.17}$$

we substitute (12.17) into the left-hand side of (12.16), and we pass one factor to the right-hand side, to obtain

$$\epsilon^{\circ\,\circ\,\cdots\,\circ}_{\alpha_1\alpha_2\cdots\alpha_n} \sqrt{|g_{\overset{\circ}{\alpha}\overset{\circ}{\beta}}|} = g_{\overset{\circ}{\alpha_1}\overset{\circ}{\beta_1}} g_{\overset{\circ}{\alpha_2}\overset{\circ}{\beta_2}} \cdots g_{\overset{\circ}{\alpha_n}\overset{\circ}{\beta_n}} \left(\frac{1}{\sqrt{|g_{\overset{\circ}{\alpha}\overset{\circ}{\beta}}|}} \epsilon^{\beta_1\beta_2\cdots\beta_n}_{\circ\,\circ\,\cdots\,\circ} \right). \tag{12.18}$$

If now we substitute into (12.18) Formulas (12.8) and (12.9), which define the orientation tensor, we get

$$\vartheta^{\circ\,\circ\,\cdots\,\circ}_{\alpha_1\alpha_2\cdots\alpha_n} = g_{\overset{\circ}{\alpha_1}\overset{\circ}{\beta_1}} g_{\overset{\circ}{\alpha_2}\overset{\circ}{\beta_2}} \cdots g_{\overset{\circ}{\alpha_n}\overset{\circ}{\beta_n}} \vartheta^{\beta_1\beta_2\cdots\beta_n}_{\circ\,\circ\,\cdots\,\circ}. \tag{12.19}$$

The conclusion is that the contravariant and covariant orientation tensors can be considered as the contravariant and covariant components of a totally anti-symmetric *Euclidean* tensor of order $r = n$, where Formula (12.19) corresponds to the lowering of all contravariant indices, as associated tensors.

So, from this section on, every Euclidean space $E^n(\mathbb{R})$ has *two* important associated tensors to build tensor spaces:

1. the fundamental metric tensor G, and
2. the orientation tensor, ϑ.

These will have important applications in the Euclidean exterior algebras.

12.7 Dual or polar tensors over $E^n(\mathbb{R})$ Euclidean spaces

In Section 8.7.3 we established two types of tensors that are associated with any given anti-symmetric homogeneous tensor, with the names of adjoint or polar tensors of the given tensor. Formulas (8.105) and (8.106) will allow a redefinition of these tensors by simply changing the permutation ϵ tensor in them, to the orientation tensor, which is the one to be used when such tensors are built over Euclidean spaces $E^n(\mathbb{R})$. So, if we have an anti-symmetric tensor $B^{\;\;\circ\;\;\circ\;\cdots\;\circ}_{\beta_1\beta_2\cdots\beta_q}$ of given order q with $(q \leq n)$, the polar tensor of the given tensor is the contravariant tensor

$$D^{\beta_{q+1}\beta_{q+2}\cdots\beta_n}_{\circ\quad\circ\quad\cdots\quad\circ} = \frac{1}{q!}\vartheta^{\beta_1\beta_2\cdots\beta_q\cdots\beta_n}_{\circ\;\circ\;\cdots\;\circ\;\cdots\;\circ} B^{\;\;\circ\;\;\circ\;\cdots\;\circ}_{\beta_1\beta_2\cdots\beta_q}. \qquad (12.20)$$

Similarly, if the given tensor is $A^{\alpha_1\alpha_2\cdots\alpha_p}_{\circ\;\circ\;\cdots\;\circ}$ of order p $(p \leq n)$ and totally anti-symmetric, its polar tensor is

$$D^{\;\;\circ\quad\circ\quad\cdots\quad\circ}_{\alpha_{p+1}\alpha_{p+2}\cdots\alpha_n} = \frac{1}{p!}\vartheta^{\;\;\circ\;\;\circ\;\cdots\;\circ\;\cdots\;\circ}_{\alpha_1\alpha_2\cdots\alpha_p\cdots\alpha_n} A^{\alpha_1\alpha_2\cdots\alpha_p}_{\circ\;\circ\;\cdots\;\circ}. \qquad (12.21)$$

The tensor nature of the polar tensors was analyzed in detail in Section 8.7.3, and nothing need to be added to the proof, with the exception of replacing the orientation tensor ϑ by the permutation tensor ϵ, because the tensor nature of ϑ has been proved in Section 12.5.

In the particular case of a number of indices q in the first case or p in the second, if $p = n-1$, the adjoint or polar tensor has only one index and so, the resulting tensor is a vector. It is customary to give the name "cross product" to the adjoint or polar tensor of the exterior product $\vec{V}_1 \wedge \vec{V}_2 \wedge \cdots \wedge V_{n-1}$.

We initiate the development of the mentioned mode. Since we have assumed that the space is Euclidean, $(E^3(\mathbb{R}))$, and we build the exterior product of two vectors, $\vec{V} \wedge \vec{W} \in \bigwedge_3^{(2)}(\mathbb{R}) \subset \left(\overset{2}{\underset{1}{\otimes}} E_h^3\right)(\mathbb{R})$, we get

$$T = \vec{V} \wedge \vec{W} = (x^\alpha_\circ \vec{e}_\alpha) \wedge (y^\beta_\circ \vec{e}_\beta) = \begin{vmatrix} x^\alpha_\circ & y^\alpha_\circ \\ x^\beta_\circ & y^\beta_\circ \end{vmatrix} \vec{e}_\alpha \wedge \vec{e}_\beta, \quad \alpha, \beta \in I_3; \quad \alpha < \beta,$$

the representation of which as an anti-symmetric tensor is

$$T = t^{\alpha\beta}_{\circ\circ}\vec{e}_\alpha \otimes \vec{e}_\beta = \begin{vmatrix} x^\alpha_\circ & y^\alpha_\circ \\ x^\beta_\circ & y^\beta_\circ \end{vmatrix} (\vec{e}_\alpha \otimes \vec{e}_\beta - \vec{e}_\beta \otimes \vec{e}_\alpha), \quad \alpha, \beta \in I_3; \quad \alpha \neq \beta.$$

The order of this tensor is $p = n-1 = 3-1 = 2$. When extracting the polar tensor, we obtain a covariant tensor, which is called the "cross product" and which unfortunately is also notated $\vec{V} \wedge \vec{W}$ (this causes a confusion between the exterior product of vectors and the cross product).

In this book, the cross product, the development and properties of which correspond to the next chapter, will be notated as $\vec{V} \times \vec{W}$.

Example 12.1 (Tensor product of modular tensors). Let $E^3(\mathbb{R})$ be a Euclidean space over which we consider the corresponding orientation tensor in its contravariant and covariant representations. Show the tensor character of these systems of scalar components, using the tensor product.

Solution:

1. The contravariant orientation tensor is

$$\vartheta^{\alpha\beta\gamma}_{\circ\circ\circ} = \frac{1}{\sqrt{|g^{\circ\circ}_{\alpha\beta}|}}\epsilon^{\alpha\beta\gamma}_{\circ\circ\circ} = \frac{1}{\sqrt{|g^{\circ\circ}_{\alpha\beta}|}} \otimes \epsilon^{\alpha\beta\gamma}_{\circ\circ\circ}. \qquad (12.22)$$

The first factor of (12.22), according to (12.7), is a modular tensor of order $r = 0$ and weight $w = -1$, and the second factor of (12.22), according to (8.71) is a totally anti-symmetric modular tensor of order $r = n = 3$ and weight $w' = 1$. Thus, the tensor product is another modular tensor of order $r = 0 + 3 = 3$ and weight $w = -1 + 1 = 0$, that is, a totally contravariant and anti-symmetric homogeneous $(w = 0)$ tensor of order $r = 3$, which allows its tensor change-of-basis equation to be written as

$$\vartheta^{ijk}_{\circ\circ\circ} = \vartheta^{\alpha\beta\gamma}_{\circ\circ\circ}\gamma^{i\circ}_{\circ\alpha}\gamma^{j\circ}_{\circ\beta}\gamma^{k\circ}_{\circ\gamma}.$$

2. The covariant orientation tensor is

$$\vartheta^{\circ\circ\circ}_{\alpha\beta\gamma} = \sqrt{|g^{\circ\circ}_{\alpha\beta}|}\epsilon^{\circ\circ\circ}_{\alpha\beta\gamma} = \sqrt{|g^{\circ\circ}_{\alpha\beta}|} \otimes \epsilon^{\circ\circ\circ}_{\alpha\beta\gamma}. \qquad (12.23)$$

The first factor of (12.23), according to (12.4), is a modular tensor of order $r = 0$ and weight $w = 1$, and the second factor of (12.23), according to (8.72) is a totally anti-symmetric modular tensor of order $r = n = 3$ and weight $w' = -1$. Thus, the tensor product is another modular tensor of order $r = 0 + 3 = 3$ and weight $w = +1 - 1 = 0$, that is, a totally covariant and anti-symmetric homogeneous $(w = 0)$ tensor of order $r = 3$, which allows its tensor change-of-basis equation to be written as

$$\vartheta^{\circ\circ\circ}_{ijk} = \vartheta^{\circ\circ\circ}_{\alpha\beta\gamma}c^{\circ\alpha}_{i\circ}c^{\circ\beta}_{j\circ}c^{\circ\gamma}_{k\circ}.$$

$\square$

Example 12.2 (Adjoint tensors. Contractions). Over the Euclidean space $E^4(\mathbb{R})$ referred to the basis $\{\vec{e}_\alpha\}$ and for which the interior connection fundamental metric tensor is

$$G = \begin{bmatrix} 3 & 1 & 0 & 0 \\ 1 & 1 & 0 & 0 \\ 0 & 0 & 2 & 1 \\ 0 & 0 & 1 & 1 \end{bmatrix},$$

we build a Euclidean tensor T, of order $r = 2$, the contravariant components of which in the tensor basis associated with the basis $\{\vec{e}_\alpha\}$ are given by the matrix

$$T = [t^{\alpha\beta}_{\circ\circ}] = \begin{bmatrix} 0 & 0 & a & 0 \\ 0 & 0 & 0 & 0 \\ -a & 0 & 0 & 0 \\ 0 & 0 & 0 & 0 \end{bmatrix}; \; a \neq 0.$$

1. Obtain the tensor $S = \text{Adj}\, T$ (adjoint of T).
2. Obtain the totally contracted product *in order* of the first factor indices with those of the second factor of $S \otimes T$.
3. Let $U = T \otimes S - S \otimes T$. Is U an anti-symmetric tensor? If it is, obtain the tensor $L = \text{Adj}\, U$.
4. Obtain the totally contracted product *in order* (of indices 1 with 1 and 2 with 2) of $U \otimes T$.
5. Obtain $\delta^{\alpha\beta\gamma\delta}_{\lambda\mu\nu\rho} t^{\circ\circ}_{\alpha\beta} t^{\circ\circ}_{\gamma\delta}$.

Solution:

$$|G| = \begin{vmatrix} 3 & 1 \\ 1 & 1 \end{vmatrix} \cdot \begin{vmatrix} 2 & 1 \\ 1 & 1 \end{vmatrix} = 2 \times 1 = 2; \quad \sqrt{|g^{\circ\circ}_{\alpha\beta}|} = \sqrt{|G|} = \sqrt{2}.$$

1. We proceed as follows.

$$s^{\circ\circ}_{\gamma\delta} = \frac{1}{2!} \vartheta^{\circ\circ\circ\circ}_{\alpha\beta\gamma\delta} t^{\alpha\beta}_{\circ\circ} = \frac{1}{2!} \sqrt{2} \epsilon^{\circ\circ\circ\circ}_{\alpha\beta\gamma\delta} t^{\alpha\beta}_{\circ\circ}.$$

Since the only non-null components of the tensor $[t^{\alpha\beta}_{\circ\circ}]$ are $t^{13}_{\circ\circ} = -t^{31}_{\circ\circ} = a$, we get

$$s^{\circ\circ}_{\gamma\delta} = \frac{\sqrt{2}}{2} \left(\epsilon^{\circ\circ\circ\circ}_{13\gamma\delta} t^{13}_{\circ\circ} + \epsilon^{\circ\circ\circ\circ}_{31\gamma\delta} t^{31}_{\circ\circ} \right) = \frac{\sqrt{2}}{2} \left[t^{13}_{\circ\circ} - t^{31}_{\circ\circ} \right] \epsilon^{\circ\circ\circ\circ}_{13\gamma\delta}$$

$$= \frac{\sqrt{2}}{2} [a - (-a)] \epsilon^{\circ\circ\circ\circ}_{13\gamma\delta} = \sqrt{2} a \epsilon^{\circ\circ\circ\circ}_{13\gamma\delta};$$

$$s^{\circ\circ}_{24} = \sqrt{2} a \epsilon^{\circ\circ\circ\circ}_{1324} = -\sqrt{2} a \epsilon^{\circ\circ\circ\circ}_{1234} = -\sqrt{2} a$$

$$s^{\circ\circ}_{42} = \sqrt{2} a \epsilon^{\circ\circ\circ\circ}_{1342} = -\sqrt{2} a \epsilon^{\circ\circ\circ\circ}_{1324} = (-\sqrt{2} a)(-1) = \sqrt{2} a,$$

from which we obtain

$$[s^{\circ\circ}_{\gamma\delta}] = \begin{bmatrix} 0 & 0 & 0 & 0 \\ 0 & 0 & 0 & -\sqrt{2}a \\ 0 & 0 & 0 & 0 \\ 0 & \sqrt{2}a & 0 & 0 \end{bmatrix},$$

which gives its unique component $s^{\circ\circ}_{24} = -s^{\circ\circ}_{42} = -\sqrt{2}a$.

2. The totally contracted product is

$$\rho = S \bullet T = s^{\circ\circ}_{\alpha\beta} t^{\alpha\beta}_{\circ\circ} = s^{\circ\circ}_{13} t^{13}_{\circ\circ} + s^{\circ\circ}_{31} t^{31}_{\circ\circ} + s^{\circ\circ}_{24} t^{24}_{\circ\circ} + s^{\circ\circ}_{42} t^{42}_{\circ\circ}$$

$$= 0 \cdot a + 0 \cdot (-a) + (-\sqrt{2}a) \cdot 0 + (\sqrt{2}a) \cdot 0 = 0; \quad S \bullet T = 0,$$

obtaining mutually orthogonal tensors.

3. With the aim of being able to subtract the two tensor product tensors, and later discuss its anti-symmetries, we decide to express the tensor T in covariant coordinates:

$$[t^{\circ\,\circ}_{\alpha\beta}] = [g^{\circ\,\circ}_{\alpha\theta} t^{\theta w}_{\circ\,\circ} g^{\circ\,\circ}_{w\beta}] = \begin{bmatrix} 3 & 1 & 0 & 0 \\ 1 & 1 & 0 & 0 \\ 0 & 0 & 2 & 1 \\ 0 & 0 & 1 & 1 \end{bmatrix} \begin{bmatrix} 0 & 0 & a & 0 \\ 0 & 0 & 0 & 0 \\ -a & 0 & 0 & 0 \\ 0 & 0 & 0 & 0 \end{bmatrix} \begin{bmatrix} 3 & 1 & 0 & 0 \\ 1 & 1 & 0 & 0 \\ 0 & 0 & 2 & 1 \\ 0 & 0 & 1 & 1 \end{bmatrix}$$

$$= \begin{bmatrix} 0 & 0 & 6a & 3a \\ 0 & 0 & 2a & a \\ -6a & -2a & 0 & 0 \\ -3a & -a & 0 & 0 \end{bmatrix}$$

$$U = T \otimes S - S \otimes T; \quad u^{\circ\circ\circ\circ}_{\alpha\beta\gamma\delta} = t^{\circ\circ}_{\alpha\beta} \cdot s^{\circ\circ}_{\gamma\delta} - s^{\circ\circ}_{\alpha\beta} \cdot t^{\circ\circ}_{\gamma\delta}. \tag{12.24}$$

The components $u^{\circ\circ\circ\circ}_{1424}$ and $u^{\circ\circ\circ\circ}_{2324}$, chosen with repeated indices take values

$$u^{\circ\circ\circ\circ}_{1424} = t^{\circ\circ}_{14} \cdot s^{\circ\circ}_{24} - s^{\circ\circ}_{14} \cdot t^{\circ\circ}_{24} = (3a) \cdot (-\sqrt{2}a) - 0 \cdot a = -3\sqrt{2}a^2 \neq 0$$

$$u^{\circ\circ\circ\circ}_{2324} = t^{\circ\circ}_{23} \cdot s^{\circ\circ}_{24} - s^{\circ\circ}_{23} \cdot t^{\circ\circ}_{24} = (2a) \cdot (-\sqrt{2}a) - 0 \cdot a = -2\sqrt{2}a^2 \neq 0$$

and since the tensor U has non-null components with repeated indices, *it is not* an anti-symmetric tensor.

4. Let V be the contraction of the tensor $U \otimes T$, since the only non-null components of T are $t^{13}_{\circ\circ} = -t^{31}_{\circ\circ} = a$, the development of V will be

$$v^{\circ\circ}_{\gamma\delta} = u^{\circ\circ\circ\circ}_{\alpha\beta\gamma\delta} \cdot t^{\alpha\beta}_{\circ\circ} = u^{\circ\circ\circ\circ}_{13\gamma\delta} \cdot t^{13}_{\circ\circ} + u^{\circ\circ\circ\circ}_{31\gamma\delta} \cdot t^{31}_{\circ\circ} = a[u^{\circ\circ\circ\circ}_{13\gamma\delta} - u^{\circ\circ\circ\circ}_{31\gamma\delta}],$$

and substituting (12.24) we have

$$v^{\circ\circ}_{\gamma\delta} = a \left((t^{\circ\circ}_{13} \cdot s^{\circ\circ}_{\gamma\delta} - s^{\circ\circ}_{13} \cdot t^{\circ\circ}_{\gamma\delta}) - (t^{\circ\circ}_{31} \cdot s^{\circ\circ}_{\gamma\delta} - s^{\circ\circ}_{31} \cdot t^{\circ\circ}_{\gamma\delta}) \right)$$

$$= a \left((t^{\circ\circ}_{13} - t^{\circ\circ}_{31}) \cdot s^{\circ\circ}_{\gamma\delta} - (s^{\circ\circ}_{13} - s^{\circ\circ}_{31}) \cdot t^{\circ\circ}_{\gamma\delta} \right)$$

$$= a \left((6a + 6a) \cdot s^{\circ\circ}_{\gamma\delta} - (0 - 0) \cdot t^{\circ\circ}_{\gamma\delta} \right) = 12a^2 \cdot s^{\circ\circ}_{\gamma\delta}.$$

Thus, the matrix representation of the resulting tensor is

$$[v^{\circ\circ}_{\gamma\delta}] = 12a^2 [s^{\circ\circ}_{\gamma\delta}] = \begin{bmatrix} 0 & 0 & 0 & 0 \\ 0 & 0 & 0 & -12\sqrt{2}a^3 \\ 0 & 0 & 0 & 0 \\ 0 & 12\sqrt{2}a^3 & 0 & 0 \end{bmatrix}.$$

5. If we denote by A the expression resulting after the contraction, as $\alpha, \beta, \gamma, \delta$ are dummy indices in A, we can exchange its notation β and α to get

$$A = \delta^{\alpha\beta\gamma\delta}_{\lambda\mu\nu\rho} t^{\circ\circ}_{\alpha\beta} t^{\circ\circ}_{\gamma\delta} \quad \text{and} \quad A' = \delta^{\beta\alpha\gamma\delta}_{\lambda\mu\nu\rho} t^{\circ\circ}_{\beta\alpha} t^{\circ\circ}_{\gamma\delta}.$$

But, on the one hand $\delta^{\beta\alpha\gamma\delta}_{\lambda\mu\nu\rho} = -\delta^{\alpha\beta\gamma\delta}_{\lambda\mu\nu\rho}$ and on the other hand the tensor T in covariant coordinates presents an anti-symmetric matrix, as can be observed in question 3, whence $t^{\circ\circ}_{\beta\alpha} = -t^{\circ\circ}_{\alpha\beta}$, which leads to

$$A' = \delta^{\beta\alpha\gamma\delta}_{\lambda\mu\nu\rho} t^{\circ\circ}_{\beta\alpha} t^{\circ\circ}_{\gamma\delta} = \left(-\delta^{\alpha\beta\gamma\delta}_{\lambda\mu\nu\rho} \right) (-t^{\circ\circ}_{\alpha\beta}) t^{\circ\circ}_{\gamma\delta} = A,$$

and we conclude that when changing $\alpha\beta\gamma\delta$ to $\beta\alpha\gamma\delta$, $\alpha\beta\gamma\delta$ to $\alpha\beta\delta\gamma$, and $\alpha\beta\gamma\delta$ to $\gamma\delta\beta\alpha$, we obtain $A' \to A$, and since $\delta^{\alpha\beta\gamma\delta}_{\lambda\mu\nu\rho}$ requires $\alpha \neq \beta \neq \gamma \neq \delta$ in order not to become null, of the $4! = 24$ possibilities for the indices only $\frac{4!}{2^3} = 3$ permutations remain to be calculated, and then

$$A = 8 \left(\delta^{1234}_{\lambda\mu\nu\rho} t^{\circ\circ}_{12} t^{\circ\circ}_{34} + \delta^{1324}_{\lambda\mu\nu\rho} t^{\circ\circ}_{13} t^{\circ\circ}_{24} + \delta^{1423}_{\lambda\mu\nu\rho} t^{\circ\circ}_{14} t^{\circ\circ}_{23} \right)$$

$$= 8 \left(\delta^{1234}_{\lambda\mu\nu\rho} \cdot 0 \cdot 0 + \delta^{1324}_{\lambda\mu\nu\rho} \cdot 6a \cdot a + \delta^{1423}_{\lambda\mu\nu\rho} \cdot 3a \cdot 2a \right)$$

$$= 8 \left(\delta^{1324}_{\lambda\mu\nu\rho} \cdot 6a^2 - \delta^{1324}_{\lambda\mu\nu\rho} \cdot 6a^2 \right) = 0; \quad \delta^{\alpha\beta\gamma\delta}_{\lambda\mu\nu\rho} t^{\circ\circ}_{\alpha\beta} t^{\circ\circ}_{\gamma\delta} = 0.$$

□

Example 12.3 (Cross products in $E^n(\mathbb{R})$). Consider the Euclidean linear space $E^n(\mathbb{R})$, referred to a basis $\{\vec{e}_\alpha\}$, with respect to which the fundamental metric tensor is given by the matrix $G = [g^{\circ\circ}_{\alpha\beta}]$.

We associate with the linear subspace $S \subset E^n(\mathbb{R})$, generated by the vectors $\beta_0 = \{\vec{x}_1, \vec{x}_2, \ldots, \vec{x}_{n-1}\}$, the vector

$$\vec{w} = \frac{1}{\sqrt{|G|}} \begin{vmatrix} \vec{e}_1 & \vec{e}_2 & \cdots & \vec{e}_n \\ x_{11} & x_{12} & \cdots & x_{1n} \\ \cdots & \cdots & \cdots & \cdots \\ x_{(n-1)1} & x_{(n-1)2} & \cdots & x_{(n-1)n} \end{vmatrix}, \tag{12.25}$$

where $x_{\alpha\beta}$ is the covariant coordinate β of the vector $\vec{x}_\alpha$.

1. Show the invariant or intrinsic character of the expression defining $\vec{w}$.
2. Show that $\vec{w}$ is orthogonal to the linear subspace S, if this is of dimension $(n-1)$. What is $\vec{w}$ when S is not?
3. Setting $|B| \equiv \mathrm{Det}[b^{\circ\circ}_{\alpha\beta}] \equiv \mathrm{Det}[\vec{x}_\alpha \bullet \vec{x}_\beta]$, determine the modulus $|\vec{w}|$ of $\vec{w}$.
4. Analyze the tensor nature of $|\vec{w}|$ and classify it.

Solution:

1. We perform a change-of-basis $||\vec{e}_i|| = ||\vec{e}_\alpha||[c^{\circ\alpha}_{i\circ}]$ in the Euclidean space $E^n(\mathbb{R})$, to examine what happen with the formula of the vector $\vec{w}$.
   Denoting $[c^{\circ\alpha}_{i\circ}] = C$ by the fundamental metric tensor that changes in a congruent form as

$$\hat{G} = C^t G C; \quad |\hat{G}| = |C^t G C| = |G||C|^2 \to \frac{1}{\sqrt{|\hat{G}|}} = |C|^{-1}\frac{1}{\sqrt{|G|}},$$

we proceed to calculate $\vec{w}$ using the available formula, developing the determinant with the help of the Levi-Civita tensor

$$\vec{\hat{w}} = \frac{1}{\sqrt{|\hat{G}|}} \begin{vmatrix} \vec{\hat{e}}_1 & \vec{\hat{e}}_2 & \cdots & \vec{\hat{e}}_n \\ \hat{x}_{11} & \hat{x}_{12} & \cdots & \hat{x}_{1n} \\ \cdots & \cdots & \cdots & \cdots \\ \hat{x}_{(n-1)1} & \hat{x}_{(n-1)2} & \cdots & \hat{x}_{(n-1)n} \end{vmatrix}$$

$$= \left(\frac{1}{\sqrt{|\hat{G}|}}\left(\epsilon^{i_1 i_1 \cdots i_n}_{\circ\,\circ\,\cdots\,\circ}\right)\right)\left(\hat{x}_{1i_1}\hat{x}_{2i_2}\cdots\hat{x}_{(n-1)i_{n-1}}\right)\vec{\hat{e}}_{i_n}$$

$$= \left(|C|^{-1}\frac{1}{\sqrt{|G|}}\right)\left(|C|\epsilon^{\alpha_1\alpha_2\cdots\alpha_n}_{\circ\,\circ\,\cdots\,\circ}\gamma^{i_1\,\circ}_{\circ\,\alpha_1}\gamma^{i_2\,\circ}_{\circ\,\alpha_2}\cdots\gamma^{i_n\,\circ}_{\circ\,\alpha_n}\right)$$

$$\times\left(x_{1\alpha_1}c^{\circ\,\alpha_1}_{i_1\,\circ}\cdot x_{2\alpha_2}c^{\circ\,\alpha_2}_{i_2\,\circ}\cdots x_{(n-1)\alpha_{(n-1)}}c^{\circ\,\alpha_{(n-1)}}_{i_{(n-1)}\,\circ}\right)\left(c^{\circ\,\alpha_n}_{i_n\,\circ}\vec{e}_{\alpha_n}\right)$$

$$= \frac{1}{\sqrt{|G|}}\left(\epsilon^{\alpha_1\alpha_2\cdots\alpha_n}_{\circ\,\circ\,\cdots\,\circ}x_{1\alpha_1}x_{2\alpha_2}\cdots x_{(n-1)\alpha_{(n-1)}}\right)\vec{e}_{\alpha_n}\left(c^{\alpha_1\,\circ}_{\circ\,i_1}\gamma^{i_1\,\circ}_{\circ\,\alpha_1}\right)$$

$$\cdot\left(c^{\alpha_2\,\circ}_{\circ\,i_2}\gamma^{i_2\,\circ}_{\circ\,\alpha_2}\right)\cdots\cdots\left(c^{\alpha_{(n-1)}\,\,\,\circ}_{\circ\,\,\,\,i_{(n-1)}}\gamma^{i_{(n-1)}\,\,\,\circ}_{\circ\,\,\,\,\alpha_{(n-1)}}\right)$$

and since the parentheses represent products of rows by columns of inverse matrices, the products of which belong to the diagonal, they all take value 1, and so

$$\vec{\hat{w}} = \frac{1}{\sqrt{|G|}}\left(\epsilon^{\alpha_1\alpha_2\cdots\alpha_n}_{\circ\,\circ\,\cdots\,\circ}x_{1\alpha_1}x_{2\alpha_2}\cdots x_{(n-1)\alpha(n-1)}\right)\vec{e}_{\alpha_n},$$

which is the development of the determinant

$$\vec{\hat{w}} = \frac{1}{\sqrt{|G|}}\begin{vmatrix} \vec{e}_1 & \vec{e}_1 & \cdots & \vec{e}_1 \\ x_{11} & x_{12} & \cdots & x_{1n} \\ \cdots & \cdots & \cdots & \cdots \\ x_{(n-1)1} & x_{(n-1)2} & \cdots & x_{(n-1)n} \end{vmatrix} = \vec{w}, \text{ i.e., } \vec{\hat{w}} = \vec{w},$$

which shows the intrinsic character of the cross product tensor.

2. We proceed as follows.

$$\vec{w}\bullet\vec{x}_\alpha = \left(\frac{1}{\sqrt{|G|}}\epsilon^{\alpha_1\alpha_2\cdots\alpha_n}_{\circ\,\circ\,\cdots\,\circ}x_{1\alpha_1}x_{2\alpha_2}\cdots x_{(n-1)\alpha_{(n-1)}}\vec{e}_{\alpha_n}\right)$$

$$\bullet(x_{\alpha\gamma}\vec{e}^{*\gamma}) = \frac{1}{\sqrt{|G|}}\epsilon^{\alpha_1\alpha_2\cdots\alpha_n}_{\circ\,\circ\,\cdots\,\circ}x_{1\alpha_1}x_{2\alpha_2}\cdots x_{(n-1)\alpha_{(n-1)}}(x_{\alpha\gamma}\vec{e}_{\alpha_n}\bullet\vec{e}^{*\gamma})$$

$$= \frac{1}{\sqrt{|G|}}\epsilon^{\alpha_1\alpha_2\cdots\alpha_n}_{\circ\,\circ\,\cdots\,\circ}x_{1\alpha_1}x_{2\alpha_2}\cdots x_{(n-1)\alpha_{(n-1)}}(x_{\alpha\gamma}\delta^{\circ\,\gamma}_{\alpha_n\,\circ})$$

$$= \frac{1}{\sqrt{|G|}} \epsilon^{\alpha_1 \alpha_2 \cdots \alpha_n}_{\circ\circ\cdots\circ} x_{1\alpha_1} x_{2\alpha_2} \cdots x_{(n-1)\alpha_{(n-1)}} x_{\alpha\alpha_n}$$

$$= \frac{1}{\sqrt{|G|}} \begin{vmatrix} x_{11} & x_{12} & \cdots & x_{1n} \\ x_{21} & x_{22} & \cdots & x_{2n} \\ \cdots & \cdots & \cdots & \cdots \\ x_{(n-1)1} & x_{(n-1)2} & \cdots & x_{(n-1)n} \\ x_{\alpha 1} & x_{\alpha 2} & \cdots & x_{\alpha n} \end{vmatrix} = 0,$$

because it has two identical rows, and then $\vec{w} \bullet \vec{x}_i = 0$; $\forall \vec{x}_i, 1 \le i \le n-1$. If the vectors $\{\vec{x}_1, \vec{x}_2, \ldots, \vec{x}_{n-1}\}$ are linearly dependent, then $\vec{w} = \vec{0}$ because it is a determinant such that some rows are linear combinations of the rest.

Naturally, we still have $\vec{w} \bullet \vec{x}_i = \vec{0} \bullet \vec{x}_i = 0$.

If the linear subspace S spanned by the vectors in β_0 is of dimension less than $n-1$, then, the vector cross product is the vector $\vec{0}$.

3. Adopting $\{\vec{x}_1, \vec{x}_2, \ldots, \vec{x}_{n-1}, \vec{w}\}$ as a new reference frame, it is known that the mixed product of the basic vectors is

$$\frac{1}{\sqrt{|G'|}} |G'| = \sqrt{|G'|} \tag{12.26}$$

and if we calculate the mixed product, we get

$$\begin{bmatrix} x_1 & x_2 & \cdots & x_{n-1} & \vec{w} \end{bmatrix} = \left(\bigwedge \vec{x}_1 \vec{x}_2 \cdots \vec{x}_{n-1} \bigwedge \right) \bullet \vec{w} = \vec{w} \bullet \vec{w} = |\vec{w}|^2,$$
$$\tag{12.27}$$

and equating both mixed products (12.26) and (12.27), yields

$$|w|^2 = \sqrt{|G'|} = \sqrt{C^t G C} = \sqrt{\begin{vmatrix} x_\alpha \bullet x_\beta & \Omega \\ \Omega & |w|^2 \end{vmatrix}} = \sqrt{|B| \cdot |w|^2} = \sqrt{|B|} \cdot |\vec{w}|$$

and simplifying, we obtain

$$|\vec{w}|^2 = \sqrt{|B|} \cdot |\vec{w}| \to |\vec{w}| = \sqrt{|B|}.$$

4. $|\vec{w}|$ is an intrinsic scalar, that is, a Euclidean tensor of order $r = 0$.

$\square$

Example 12.4 (Tensors obtained from the orientation tensor).

1. Let T be a contravariant and anti-symmetric homogeneous tensor of order $r = n$, built over a $E^n(\mathbb{R})$ the strict component of which, in a basis $\{\vec{e}_\alpha\}$ is $a \in K$.

   Consider a change-of-basis with relative modulus a^{-1}.

   a) Obtain the strict component of T in the new basis.

   b) Obtain the ordinary components of T in both bases.

2. Answer the same questions, assuming T to be covariant.

Solution:

1. a) Let C be the change-of-basis matrix. Then, $|C| = a^{-1}$. Formula (8.74) establishes the expression

$$\delta^{\alpha_1 \alpha_2 \cdots \alpha_n}_{1 \ \ 2 \ \cdots \ n} = \epsilon^{\alpha_1 \alpha_2 \cdots \alpha_n}_{\circ \ \ \circ \ \cdots \ \circ} \quad \text{(Levi-Civita contravariant tensor)}. \quad (12.28)$$

On the other hand, Formula (8.103) establishes a relation between the contravariant anti-symmetric tensors and the generalized Kronecker delta

$$t^{\alpha_1 \alpha_2 \cdots \alpha_n}_{\circ \ \ \circ \ \cdots \ \circ} = \delta^{\alpha_1 \ \alpha_2 \cdots \ \alpha_n}_{(\beta_1 \ \beta_2 \cdots \beta_n)} t^{(\beta_1 \beta_2 \cdots \beta_n)}_{\ \circ \ \ \circ \ \cdots \ \circ}. \quad (12.29)$$

Since the anti-symmetric tensors of order $r = n$ have only one strict component, $(\beta_1\beta_2 \cdots \beta_n) \equiv (12 \cdots n)$, and (12.29) becomes

$$t^{\alpha_1 \alpha_2 \cdots \alpha_n}_{\circ \ \ \circ \ \cdots \ \circ} = \delta^{\alpha_1 \alpha_2 \cdots \alpha_n}_{(1 \ 2 \ \cdots \ n)} t^{(12\cdots n)}_{\ \circ \ \circ \cdots \ \circ},$$

and considering (12.28) and the given data, we get

$$t^{\alpha_1 \alpha_2 \cdots \alpha_n}_{\circ \ \ \circ \ \cdots \ \circ} = \epsilon^{\alpha_1 \alpha_2 \cdots \alpha_n}_{\circ \ \ \circ \ \cdots \ \circ} t^{(12\cdots n)}_{\ \circ \ \circ \cdots \ \circ} = \epsilon^{\alpha_1 \alpha_2 \cdots \alpha_n}_{\circ \ \ \circ \ \cdots \ \circ} \cdot a. \quad (12.30)$$

Similarly, in the new basis we have

$$t^{i_1 i_2 \cdots i_n}_{\circ \ \ \circ \ \cdots \ \circ} = \epsilon^{i_1 i_2 \cdots i_n}_{\circ \ \ \circ \ \cdots \ \circ} \cdot \hat{a},$$

where $\hat{a}$ is the new strict component of the tensor T. Since T is homogeneous, its change-of-basis is

$$t^{i_1 i_2 \cdots i_n}_{\circ \ \ \circ \ \cdots \ \circ} = t^{\alpha_1 \alpha_2 \cdots \alpha_n}_{\circ \ \ \circ \ \cdots \ \circ} \gamma^{i_1 \ \circ}_{\circ \ \alpha_1} \gamma^{i_2 \ \circ}_{\circ \ \alpha_2} \cdots \gamma^{i_n \ \circ}_{\circ \ \alpha_n},$$

and replacing the previous expressions we obtain

$$\left(\hat{a} \cdot \epsilon^{i_1 i_2 \cdots i_n}_{\circ \ \ \circ \ \cdots \ \circ} \right) = \left(a \cdot \epsilon^{\alpha_1 \alpha_2 \cdots \alpha_n}_{\circ \ \ \circ \ \cdots \ \circ} \right) \gamma^{i_1 \ \circ}_{\circ \ \alpha_1} \gamma^{i_2 \ \circ}_{\circ \ \alpha_2} \cdots \gamma^{i_n \ \circ}_{\circ \ \alpha_n}, \quad (12.31)$$

and susbtituting (8.71) which gives the change-of-basis of $\epsilon^{\alpha_1 \alpha_2 \cdots \alpha_n}_{\circ \ \ \circ \ \cdots \ \circ}$, into (12.31), we get

$$\hat{a}|C|\epsilon^{\alpha_1 \alpha_2 \cdots \alpha_n}_{\circ \ \ \circ \ \cdots \ \circ} \gamma^{i_1 \ \circ}_{\circ \ \alpha_1} \gamma^{i_2 \ \circ}_{\circ \ \alpha_2} \cdots \gamma^{i_n \ \circ}_{\circ \ \alpha_n} = a\epsilon^{\alpha_1 \alpha_2 \cdots \alpha_n}_{\circ \ \ \circ \ \cdots \ \circ} \gamma^{i_1 \ \circ}_{\circ \ \alpha_1} \gamma^{i_2 \ \circ}_{\circ \ \alpha_2} \cdots \gamma^{i_n \ \circ}_{\circ \ \alpha_n},$$

and simplifying, we finally obtain

$$\hat{a}|C| = a \rightarrow \hat{a} = |C|^{-1} \cdot a = (a^{-1})^{-1} \cdot a = a^2. \quad (12.32)$$

b) According to (12.30) and (12.32) the ordinary components of T are

$$t^{\alpha_1 \alpha_2 \cdots \alpha_n}_{\circ \ \ \circ \ \cdots \ \circ} = a\epsilon^{\alpha_1 \alpha_2 \cdots \alpha_n}_{\circ \ \ \circ \ \cdots \ \circ}; \quad t^{i_1 i_2 \cdots i_n}_{\circ \ \ \circ \ \cdots \ \circ} = a^2 \epsilon^{i_1 i_2 \cdots i_n}_{\circ \ \ \circ \ \cdots \ \circ}.$$

2. If T is in covariant coordinates we have:

    a) The development is analogous to 1(a), but with the formulas for a covariant tensor.

    The formula corresponding to (12.30) is

$$t^{\circ\ \circ\ \cdots\ \circ}_{\alpha_1\alpha_2\cdots\alpha_n} = a\ \epsilon^{\circ\ \circ\ \cdots\ \circ}_{\alpha_1\alpha_2\cdots\alpha_n}, \tag{12.33}$$

and in the new basis, we will obtain

$$t^{\circ\ \circ\ \cdots\ \circ}_{i_1 i_2\cdots i_n} = \hat{a}\ \epsilon^{\circ\ \circ\ \cdots\ \circ}_{i_1 i_2\cdots i_n}. \tag{12.34}$$

The change-of-basis of the tensor T can be written as

$$t^{\circ\ \circ\ \cdots\ \circ}_{i_1 i_2\cdots i_n} = t^{\circ\ \circ\ \cdots\ \circ}_{\alpha_1\alpha_2\cdots\alpha_n}\, c^{\circ\ \alpha_1}_{i_1\ \circ}\, c^{\circ\ \alpha_2}_{i_2\ \circ}\cdots c^{\circ\ \alpha_n}_{i_n\ \circ},$$

and the Levi-Civita tensor (8.72) gives

$$\epsilon^{\circ\ \circ\ \cdots\ \circ}_{i_1 i_2\cdots i_n} = |C|^{-1}\epsilon^{\circ\ \circ\ \cdots\ \circ}_{\alpha_1\alpha_2\cdots\alpha_n}\, c^{\circ\ \alpha_1}_{i_1\ \circ}\, c^{\circ\ \alpha_2}_{i_2\ \circ}\cdots c^{\circ\ \alpha_n}_{i_n\ \circ},$$

which after replacing terms in (12.33) and simplifying leads to

$$\hat{a}|C|^{-1} = a \rightarrow \hat{a} = |C|a = a^{-1}\cdot a = 1. \tag{12.35}$$

    b) From (12.33) and (12.35) the ordinary components are

$$t^{\circ\ \circ\ \cdots\ \circ}_{\alpha_1\alpha_2\cdots\alpha_n} = a\epsilon^{\circ\ \circ\ \cdots\ \circ}_{\alpha_1\alpha_2\cdots\alpha_n}; \quad t^{\circ\ \circ\ \cdots\ \circ}_{i_1 i_2\cdots i_n} = \epsilon^{\circ\ \circ\ \cdots\ \circ}_{i_1 i_2\cdots i_n}.$$

$\square$

12.8 Exercises

12.1. Consider a Euclidean space $E^4(\mathbb{R})$ referred to a basis $\{\vec{e}_\alpha\}$ and the fundamental metric tensor represented by the Gram matrix

$$G = [g^{\circ\circ}_{\alpha\beta}] = \begin{bmatrix} 1 & 0 & 0 & 1 \\ 0 & 2 & 1 & 0 \\ 0 & 1 & 3 & 2 \\ 1 & 0 & 2 & 5 \end{bmatrix}$$

and four vectors $\vec{V}_1, \vec{V}_2, \vec{V}_3, \vec{V}_4 \in E^4(\mathbb{R})$ represented by the matrix

$$[X_1, X_2, X_3, X_4] = \begin{bmatrix} 1 & 1 & 1 & 4 \\ 2 & 1 & 1 & -2 \\ -1 & 2 & 1 & 1 \\ 0 & 2 & 1 & -1 \end{bmatrix}.$$

1. Determine the exterior vector $V = \vec{V}_1 \wedge \vec{V}_2 \in \bigwedge_4^{(2)}(\mathbb{R})$, giving also the matrix representation of V as an anti-symmetric tensor $\left(V \in (E \otimes E)_h^4(\mathbb{R})\right)$ of second order, in contravariant coordinates.
2. Determine the exterior vector $W = \vec{V}_3 \wedge \vec{V}_4 \in \bigwedge_4^{(2)}(\mathbb{R})$, giving also the matrix representation of W as an anti-symmetric tensor $\left(W \in (E \otimes E)_h^4(\mathbb{R})\right)$ of second order, in contravariant coordinates.
3. Determine the tensor $DV = \text{adj}(V)$, adjoint or polar of the tensor V in matrix representation, by its covariant coordinates and by its contravariant coordinates.
4. Determine the tensor $DW = \text{adj}(W)$, adjoint or polar of the tensor $\vec{W}$ in matrix representation, by its contravariant and its covariant coordinates.
5. Obtain the moduli of V and W.
6. Obtain the dot product of $V \bullet W$ as vectors of its Euclidean tensor space.
7. Obtain the modulus of the tensor DA.
8. Obtain the modulus of the DB.
9. Obtain the dot product $DA \bullet DB$.

12.2. Consider a Euclidean space $E^4(\mathbb{R})$ referred to a basis $\{\vec{e}_\alpha\}$ and fundamental metric tensor of Gram matrix

$$G = [g_{\alpha\beta}^{\circ\,\circ}] = \begin{bmatrix} 2 & 1 & 1 & 1 \\ 1 & 2 & 1 & 1 \\ 1 & 1 & 2 & 1 \\ 1 & 1 & 1 & 2 \end{bmatrix}.$$

Over this space we consider three tensors $T \in \bigwedge_{*4}^{(2)}(\mathbb{R})$ and $\vec{U}, \vec{V} \in E_*^4(\mathbb{R})$; the first one represented by

$$T = \vec{e}^{*1} \wedge \vec{e}^{*2} + 2\vec{e}^{*1} \wedge \vec{e}^{*3} - 2\vec{e}^{*1} \wedge \vec{e}^{*4} + 2\vec{e}^{*2} \wedge \vec{e}^{*3} - 4\vec{e}^{*2} \wedge \vec{e}^{*4} - 4\vec{e}^{*3} \wedge \vec{e}^{*4},$$

the second of unknown components by

$$U = u_1\vec{e}^{*1} + u_2\vec{e}^{*2} + u_3\vec{e}^{*3} + u_4\vec{e}^{*4}.$$

and the third by $\vec{V} = \vec{e}^{*1} - \vec{e}^{*2} + \vec{e}^{*3} - \vec{e}^{*4}$.

We know that the adjoint tensor, $\vec{W}$, of the tensor exterior tensor product $B = T \bigwedge U$, is

$$\vec{W} = \text{adj}(B) = \frac{1}{\sqrt{5}}\left(-6\vec{e}_1 - 4\vec{e}_2 + 7\vec{e}_3 + 5\vec{e}_4\right)$$

and we also know that the dot product $\vec{U} \bullet \vec{V} = -2$.

1. Obtain the matrix representation of T as an anti-symmetric tensor, $T \in \left(\overset{2}{\underset{1}{\otimes}} E\right)_h^4(\mathbb{R})$.

2. Calculate the totally developed literal expression of B.
3. Obtain the components of U.
4. Determine the moduli of the tensors T and U.
5. Determine a vector $\vec{X} \in E_*^4(\mathbb{R})$ such that $T = \vec{V} \wedge \vec{X}$.
6. Verify if the "triple vector product" defined with the data in $E_*^4(\mathbb{R})$ and English notation $\vec{V} \times \vec{X} \times \vec{U}$ is or not $\vec{W}$.
7. Justify the previous result.

12.3. In the Euclidean space $E^2(\mathbb{R})$ referred to the basis $\{\vec{e}_\alpha\}$ and fundamental metric tensor $G = [g^{\circ\circ}_{\lambda\mu}]$, we consider the system of scalars

$$U(\alpha\beta) = \sqrt{g^{\circ\circ}_{\lambda\mu}}\varepsilon(\alpha\beta) \quad \text{and} \quad V(\alpha\beta) = \frac{\varepsilon(\alpha\beta)}{\sqrt{g^{\circ\circ}_{\lambda\mu}}},$$

where $\varepsilon(\alpha\beta)$ is the Levi-Civita ε system of order $n = 2$, associated with $E^2(\mathbb{R})$ ignoring the species.

1. Analyze the tensor nature of the systems $U(\alpha\beta)$ and $V(\alpha\beta)$, for all possible cases, obtaining their (n, r, w).
2. Study the exterior tensor products $U \bigwedge V$ in all licit cases.
3. Obtain the modulus of $U \bigwedge V$.

13

Euclidean Exterior Algebra

13.1 Introduction

In Chapter 9 we have given a detailed exposition of the exterior primal algebra of order r, $\bigwedge_n^{(r)}(K)$, and of the exterior dual algebra $\bigwedge_{n*}^{(r)}(K)$, both built over the vector spaces primal $V^n(K)$ and dual $V_*^n(K)$, giving rise to the concepts of contravariant and covariant coordinates of the respective exterior tensors.

Since in the present chapter there exists only a unique linear space, the Euclidean space $E^n(\mathbb{R})$, which satisfies the identity $E^n(\mathbb{R}) \equiv E_*^n(\mathbb{R})$, we consider $\bigwedge_n^{(r)}(\mathbb{R})$ as a unique linear subspace, i.e., the Euclidean subspace $\left(\overset{r}{\underset{1}{\otimes}} E_h^n \right)(\mathbb{R})$ of the anti-symmetric tensors, referred to a special "exterior" basis that uses only "strict components" for its representation, and in which of special interest is the analysis of the "associated exterior" tensors, that is, the raising or lowering of indices, in its strict components. This will be the central topic of the present chapter, together with some of its practical applications, in the ordinary geometric space (OGS).

We must also mention that this chapter deals with the creation of fundamental connection tensors for exclusive use in exterior algebras.

13.2 Euclidean exterior algebra of order $r = 2$

Consider the vectors $\vec{u}, \vec{v} \in E^n(\mathbb{R})$, expressed in the basis $\{\vec{e}_\alpha\}$ of the Euclidean space

$$\vec{u} = u_\circ^\alpha \vec{e}_\alpha; \quad \vec{v} = v_\circ^\alpha \vec{e}_\beta.$$

Consider the exterior product $\vec{u} \wedge \vec{v}$ as an exterior tensor of order $r = 2$, $\vec{u} \wedge \vec{v} \in \bigwedge_n^{(2)}(\mathbb{R})$. According to Formula (9.4) we have

$$\vec{u} \wedge \vec{v} = (u_\circ^\alpha \vec{e}_\alpha) \wedge (v_\circ^\beta \vec{e}_\beta) = (u_\circ^\alpha v_\circ^\beta) \vec{e}_\alpha \wedge \vec{e}_\beta; \quad \alpha, \beta \in I_n; \quad \alpha \neq \beta, \qquad (13.1)$$

and with the help of (9.5) we also obtain

$$\vec{u} \wedge \vec{v} = \begin{vmatrix} u_\circ^\alpha & v_\circ^\alpha \\ u_\circ^\beta & v_\circ^\beta \end{vmatrix} \vec{e}_\alpha \wedge \vec{e}_\beta; \quad \alpha, \beta \in I_n; \quad \alpha < \beta. \tag{13.2}$$

Let $G \equiv [g_{\alpha\beta}^{\circ\circ}]$ and $G^* = G^{-1} \equiv [g_{\circ\circ}^{\alpha\beta}]$ be the matrices associated with the fundamental metric tensor in covariant and in contravariant coordinates, respectively. According to Formula (11.32) the tensor relations between the primal and dual bases are

$$\vec{e}_\alpha = g_{\theta\alpha}^{\circ\circ} \vec{e}^{*\theta} \quad \text{and also} \quad \vec{e}_\beta = g_{\varphi\beta}^{\circ\circ} \vec{e}^{*\varphi} \tag{13.3}$$

$$\vec{e}^{*\alpha} = g_{\circ\circ}^{\theta\alpha} \vec{e}_\theta \quad \text{and also} \quad \vec{e}^{*\beta} = g_{\circ\circ}^{\theta\beta} \vec{e}_\theta, \tag{13.4}$$

and substituting (13.3) into (13.2), we get

$$\vec{u} \wedge \vec{v} = \begin{vmatrix} u_\circ^\alpha & v_\circ^\alpha \\ u_\circ^\beta & v_\circ^\beta \end{vmatrix} (g_{\theta\alpha}^{\circ\circ} \vec{e}^{*\theta}) \wedge (g_{\varphi\beta}^{\circ\circ} \vec{e}^{*\varphi})$$

$$= \begin{vmatrix} u_\circ^\alpha & v_\circ^\alpha \\ u_\circ^\beta & v_\circ^\beta \end{vmatrix} (g_{\theta\alpha}^{\circ\circ} g_{\varphi\beta}^{\circ\circ}) \vec{e}^{*\theta} \wedge \vec{e}^{*\varphi}; \quad \alpha, \beta, \theta, \varphi \in I_n; \quad \alpha < \beta; \ \theta \neq \varphi.$$

Following the same technique that the one employed in Formula (9.5), in strict components, this is expressed as

$$\vec{u} \wedge \vec{v} = \begin{vmatrix} u_\circ^\alpha & v_\circ^\alpha \\ u_\circ^\beta & v_\circ^\beta \end{vmatrix} \cdot \begin{vmatrix} g_{\theta\alpha}^{\circ\circ} & g_{\theta\beta}^{\circ\circ} \\ g_{\varphi\alpha}^{\circ\circ} & g_{\varphi\beta}^{\circ\circ} \end{vmatrix} \vec{e}^{*\theta} \wedge \vec{e}^{*\varphi}; \quad \alpha, \beta, \theta, \varphi \in I_n; \quad \alpha < \beta; \ \theta < \varphi. \tag{13.5}$$

It must be taken into account that due to the fact that in (13.5) α and β are dummy indices, in reality the product of determinants is a sum of products.

If the data vectors $\vec{u}$ and $\vec{v}$ were to be directly given in covariant coordinates:

$$\vec{u} = u_\theta^\circ \vec{e}^{*\theta}; \quad \vec{v} = v_\varphi^\circ \vec{e}^{*\varphi}.$$

and the exterior product calculated directly, we would have obtained

$$\vec{u} \wedge \vec{v} = \begin{vmatrix} u_\theta^\circ & v_\theta^\circ \\ u_\varphi^\circ & v_\varphi^\circ \end{vmatrix} \vec{e}^{*\theta} \wedge \vec{e}^{*\varphi}; \quad \theta, \varphi \in I_n; \ \theta < \varphi. \tag{13.6}$$

If we denote by $\vec{w}$ the vector $\vec{u} \wedge \vec{v}$, Expression (13.6) allows us to write

$$\vec{w} = \vec{u} \wedge \vec{v} = w_{(\theta\varphi)}^{\circ\circ} \vec{e}^{*\theta} \wedge \vec{e}^{*\varphi} = \begin{vmatrix} u_\theta^\circ & v_\theta^\circ \\ u_\varphi^\circ & v_\varphi^\circ \end{vmatrix} \vec{e}^{*\theta} \wedge \vec{e}^{*\varphi}$$

$$w_{(\overset{\circ}{\theta}\overset{\circ}{\varphi})} = \begin{vmatrix} \overset{\circ}{u}_\theta & \overset{\circ}{v}_\theta \\ \overset{\circ}{u}_\varphi & \overset{\circ}{v}_\varphi \end{vmatrix} ; \ \theta, \varphi \in I_n; \ \theta < \varphi, \tag{13.7}$$

and identifying (13.7) with the strict component of (13.5) we obtain

$$w_{(\overset{\circ}{\theta}\overset{\circ}{\varphi})} = \begin{vmatrix} \overset{\circ}{u}_\theta & \overset{\circ}{v}_\theta \\ \overset{\circ}{u}_\varphi & \overset{\circ}{v}_\varphi \end{vmatrix} = \begin{vmatrix} u_{\overset{\circ}{\alpha}} & v_{\overset{\circ}{\alpha}} \\ u_{\overset{\circ}{\beta}} & v_{\overset{\circ}{\beta}} \end{vmatrix} \cdot \begin{vmatrix} g_{\overset{\circ\circ}{\theta\alpha}} & g_{\overset{\circ\circ}{\theta\beta}} \\ g_{\overset{\circ\circ}{\varphi\alpha}} & g_{\overset{\circ\circ}{\varphi\beta}} \end{vmatrix} ; \ \alpha, \beta, \theta, \varphi \in I_n; \ \alpha < \beta; \ \theta < \varphi.$$
$$\tag{13.8}$$

Notating the strict components of the exterior product as simple exterior tensors, the relation (13.8) can also be notated as

$$w_{(\overset{\circ}{\theta}\overset{\circ}{\varphi})} = w_{\overset{\circ\circ}{}}^{(\alpha\beta)} \begin{vmatrix} g_{\overset{\circ\circ}{\theta\alpha}} & g_{\overset{\circ\circ}{\theta\beta}} \\ g_{\overset{\circ\circ}{\varphi\alpha}} & g_{\overset{\circ\circ}{\varphi\beta}} \end{vmatrix} ; \ \alpha, \beta, \theta, \varphi \in I_n; \ \alpha < \beta; \ \theta < \varphi, \tag{13.9}$$

where (α, β) are dummy indices from which the Einstein convention holds.

Expression (13.9) is the relation between the exterior associated tensors sought after. If an analogous study starting from the vectors $\vec{u} = u_{\overset{\circ}{\alpha}} \vec{e}^{*\alpha}$ and $\vec{v} = v_{\overset{\circ}{\beta}} \vec{e}^{*\beta}$ in covariant coordinates were done, the final result would have led to the following relation, analogous to (13.9):

$$w^{(\overset{\circ}{\theta}\overset{\circ}{\varphi})} = w_{(\alpha\beta)}^{\overset{\circ\circ}{}} \begin{vmatrix} g^{\overset{\theta\alpha}{\circ\circ}} & g^{\overset{\theta\beta}{\circ\circ}} \\ g^{\overset{\varphi\alpha}{\circ\circ}} & g^{\overset{\varphi\beta}{\circ\circ}} \end{vmatrix} ; \ \alpha, \beta, \theta, \varphi \in I_n; \ \alpha < \beta; \ \theta < \varphi. \tag{13.10}$$

From (13.9) and (13.10) we conclude that it is possible to build a "fundamental metric tensor" $G_{\bigwedge_n^{(2)}}$ for the Euclidean exterior space $\bigwedge_n^{(2)}(\mathbb{R})$, the tensors of which have $\binom{n}{2}$ strict components of a total of $\binom{n}{2}^2 = \left(\frac{n!}{2!(n-2)!}\right)^2$ components of the type $\begin{vmatrix} g_{\overset{\circ\circ}{\theta\alpha}} & g_{\overset{\circ\circ}{\theta\beta}} \\ g_{\overset{\circ\circ}{\varphi\alpha}} & g_{\overset{\circ\circ}{\varphi\beta}} \end{vmatrix}$, that together with $G^*_{\bigwedge_n^{(2)}}$, with components of the type $\begin{vmatrix} g^{\overset{\theta\alpha}{\circ\circ}} & g^{\overset{\theta\beta}{\circ\circ}} \\ g^{\overset{\varphi\alpha}{\circ\circ}} & g^{\overset{\varphi\beta}{\circ\circ}} \end{vmatrix}$, permit us to pass directly from "stretched" column vectors built with the tensor strict components of an associated type, to those of another type.

Naturally, all the presented treatment can also be done considering the tensor $T \in \bigwedge_n^{(2)}(\mathbb{R})$ as an arbitrary anti-symmetric tensor, $T \in \left(\overset{2}{\underset{1}{\otimes}} E_h^n\right)(\mathbb{R})$. In fact, let $T \in \bigwedge_n^{(2)}(\mathbb{R})$:

$$T = t_{\overset{\circ\circ}{}}^{(\alpha\beta)} \vec{e}_\alpha \wedge \vec{e}_\beta \text{ with } \alpha, \beta \in I_n \text{ and } \alpha < \beta, \tag{13.11}$$

since each basic vector of $\bigwedge_n^{(2)}(\mathbb{R})$ satisfies

$$\vec{e}_\alpha \wedge \vec{e}_\beta = \vec{e}_\alpha \otimes \vec{e}_\beta - \vec{e}_\beta \otimes \vec{e}_\alpha, \tag{13.12}$$

substituting (13.12) into (13.11), we obtain

$$T = t^{(\alpha\beta)}_{\circ \ \circ}(\vec{e}_\alpha \otimes \vec{e}_\beta - \vec{e}_\beta \otimes \vec{e}_\alpha) \text{ with } \alpha < \beta, \tag{13.13}$$

and since T is a Euclidean anti-symmetric tensor we can decide to lower its indices (α, β), passing (13.13) to the dual basis and using (13.3):

$$
\begin{aligned}
T &= t^{(\alpha\beta)}_{\circ \ \circ}(\vec{e}_\alpha \otimes \vec{e}_\beta - \vec{e}_\beta \otimes \vec{e}_\alpha) \\
&= t^{(\alpha\beta)}_{\circ \ \circ}\left((g^{\circ\circ}_{\theta\alpha}\vec{e}^{*\theta}) \otimes (g^{\circ\circ}_{\varphi\beta}\vec{e}^{*\varphi}) - (g^{\circ\circ}_{\varphi\beta}\vec{e}^{*\varphi}) \otimes (g^{\circ\circ}_{\theta\alpha}\vec{e}^{*\theta})\right) \\
&= t^{(\alpha\beta)}_{\circ \ \circ}(g^{\circ\circ}_{\theta\alpha}g^{\circ\circ}_{\varphi\beta})\left(\vec{e}^{*\theta} \otimes \vec{e}^{*\varphi} - \vec{e}^{*\varphi} \otimes \vec{e}^{*\theta}\right) \\
&= t^{(\alpha\beta)}_{\circ \ \circ}(g^{\circ\circ}_{\theta\alpha}g^{\circ\circ}_{\varphi\beta})\left(\vec{e}^{*\theta} \wedge \vec{e}^{*\varphi}\right); \ \alpha,\beta,\varphi,\theta \in I_n, \ \alpha < \beta; \ \varphi \neq \theta. \tag{13.14}
\end{aligned}
$$

Applying again the same technique as that employed in Formulas (9.5) and (13.5), Expression (13.14) is expressed in strict components, and then

$$T = t^{(\alpha\beta)}_{\circ \ \circ}\begin{vmatrix} g^{\circ\circ}_{\theta\alpha} & g^{\circ\circ}_{\theta\beta} \\ g^{\circ\circ}_{\varphi\alpha} & g^{\circ\circ}_{\varphi\beta} \end{vmatrix}\vec{e}^{*\theta} \wedge \vec{e}^{*\varphi}; \ \alpha,\beta,\theta,\varphi \in I_n; \ \alpha < \beta; \ \theta < \varphi. \tag{13.15}$$

Since T is in the exterior reciprocal basis, its components are the "strict covariant coordinates". So, from (13.15) we conclude that

$$t^{\circ \ \circ}_{(\theta\varphi)} = t^{(\alpha\beta)}_{\circ \ \circ}\begin{vmatrix} g^{\circ\circ}_{\theta\alpha} & g^{\circ\circ}_{\theta\beta} \\ g^{\circ\circ}_{\varphi\alpha} & g^{\circ\circ}_{\varphi\beta} \end{vmatrix}; \ \alpha,\beta,\theta,\varphi \in I_n, \tag{13.16}$$

where $\alpha < \beta$ are dummy indices and $\theta < \varphi$ are free indices.

It is obvious that (13.16) is essentially the same as (13.9).

13.3 Euclidean exterior algebra of order r ($2 < r < n$)

From (9.18), (9.19) and (9.20) we know that the expression of the tensor $\vec{V}_1 \wedge \vec{V}_2 \wedge \cdots \wedge \vec{V}_r \in \bigwedge_n^{(r)}(\mathbb{R})$, decomposable multivector, in strict components is given by

$$
\begin{aligned}
\vec{V}_1 \wedge \vec{V}_2 \wedge \cdots \wedge \vec{V}_r &= \Delta_r \vec{e}_{\beta_1} \wedge \vec{e}_{\beta_2} \wedge \cdots \wedge \vec{e}_{\beta_r} \\
&= |X_{r1} \quad X_{r2} \quad \cdots \quad X_{rr}|\, \vec{e}_{\beta_1} \wedge \vec{e}_{\beta_2} \wedge \cdots \wedge \vec{e}_{\beta_r} \\
&= \begin{vmatrix} x^{\beta_1\circ}_{\circ 1} & x^{\beta_1\circ}_{\circ 2} & \cdots & x^{\beta_1\circ}_{\circ r} \\ x^{\beta_2\circ}_{\circ 1} & x^{\beta_2\circ}_{\circ 2} & \cdots & x^{\beta_2\circ}_{\circ r} \\ \cdots & \cdots & \cdots & \cdots \\ x^{\beta_r\circ}_{\circ 1} & x^{\beta_r\circ}_{\circ 2} & \cdots & x^{\beta_r\circ}_{\circ r} \end{vmatrix}\vec{e}_{\beta_1} \wedge \vec{e}_{\beta_2} \wedge \cdots \wedge \vec{e}_{\beta_r},
\end{aligned}
$$
$$\tag{13.17}$$

where $\beta_j \in I_n$; $\beta_1 < \beta_2 < \cdots < \beta_r$ and the determinants Δ_r are the minors of the matrix $[X_1 \quad X_2 \quad \cdots \quad X_r]_{n,r}$ the columns of which are the data vectors $\vec{V}_1, \vec{V}_2, \ldots, \vec{V}_r$ expressed in the basis $\{\vec{e}_\alpha\}$ of the Euclidean space $E^n(\mathbb{R})$.

Passing (13.17) to the exterior dual basis of the space $\bigwedge_n^{(r)}(\mathbb{R})$, using the relations corresponding to (13.3) and (13.4), i.e.,

$$\vec{e}_{\beta_j} = g_{\overset{\circ}{\theta_i} \overset{\circ}{\beta_j}} \vec{e}^{*\theta_i}; \quad \theta_i, \beta_j \in I_n; \; i \in I_n; \; j \in I_r \tag{13.18}$$

$$\vec{e}^{*\beta_j} = g^{\theta_i \beta_j}_{\circ\ \circ} \vec{e}_{\theta_i}; \quad \theta_i, \beta_j \in I_n; \; i \in I_n; \; j \in I_r, \tag{13.19}$$

we get

$$
\begin{aligned}
\vec{V}_1 \wedge \vec{V}_2 \wedge \cdots \wedge \vec{V}_r &= \Delta_r \vec{e}_{\beta_1} \wedge \vec{e}_{\beta_2} \wedge \cdots \wedge \vec{e}_{\beta_r} \\
&= \Delta_r \left(g_{\overset{\circ}{\theta_1}\overset{\circ}{\beta_1}} \vec{e}^{*\theta_1}\right) \wedge \left(g_{\overset{\circ}{\theta_2}\overset{\circ}{\beta_2}} \vec{e}^{*\theta_2}\right) \wedge \cdots \wedge \left(g_{\overset{\circ}{\theta_r}\overset{\circ}{\beta_r}} \vec{e}^{*\theta_r}\right) \\
&= \Delta_r \left(g_{\overset{\circ}{\theta_1}\overset{\circ}{\beta_1}} g_{\overset{\circ}{\theta_2}\overset{\circ}{\beta_2}} \cdots g_{\overset{\circ}{\theta_r}\overset{\circ}{\beta_r}}\right) \vec{e}^{*\theta_1} \wedge \vec{e}^{*\theta_2} \wedge \cdots \wedge \vec{e}^{*\theta_r},
\end{aligned}\tag{13.20}
$$

where the indices satisfy the conditions $\beta_1 < \beta_2 < \cdots < \beta_r$ and $\theta_1 \neq \theta_2 \neq \cdots \neq \theta_r$.

If we denote by $\{\gamma_1, \gamma_2, \ldots, \gamma_r\}$ the ordered set:

$$\{\theta_1, \theta_2, \ldots \theta_r\} \equiv \{\gamma_1, \gamma_2, \ldots, \gamma_r\}; \quad \gamma_1 < \gamma_2 < \cdots < \gamma_r,$$

the exterior basis that appears in (13.20) can be written as

$$\vec{e}^{*\theta_1} \wedge \vec{e}^{*\theta_2} \wedge \cdots \wedge \vec{e}^{*\theta_r} = \delta^{\theta_1\,\theta_2\,\cdots\,\theta_r}_{(\gamma_1\,\gamma_2\,\cdots\,\gamma_r)} \vec{e}^{*\gamma_1} \wedge \vec{e}^{*\gamma_2} \wedge \cdots \wedge \vec{e}^{*\gamma_r},$$

$$\tag{13.21}$$

where $\theta_1 \neq \theta_2 \neq \cdots \neq \theta_r$; $\gamma_1 < \gamma_2 < \cdots < \gamma_r$, which substituted into (13.20) leads to

$$\vec{V}_1 \wedge \vec{V}_2 \wedge \cdots \wedge \vec{V}_r = \Delta_r \left(\delta^{\theta_1\,\theta_2\,\cdots\,\theta_r}_{(\gamma_1\,\gamma_2\,\cdots\,\gamma_r)} g_{\overset{\circ}{\theta_1}\overset{\circ}{\beta_1}} g_{\overset{\circ}{\theta_2}\overset{\circ}{\beta_2}} \cdots g_{\overset{\circ}{\theta_r}\overset{\circ}{\beta_r}}\right) \vec{e}^{*\gamma_1} \wedge \vec{e}^{*\gamma_2} \wedge \vec{e}^{*\gamma_r},$$

where the term in parentheses that appears in the formula is the development of a minor of the matrix G of the fundamental metric tensor, and Δ_r according to (13.17), is another minor of the data matrix $[X_1 \quad X_2 \quad \cdots \quad X_r]_{n,r}$ that represents to the vectors $\vec{V}_j, 1 \leq j \leq r$. Thus, in summary, we have

$$
\vec{V}_1 \wedge \vec{V}_2 \wedge \cdots \wedge \vec{V}_r =
\begin{vmatrix}
x^{\beta_1 \circ}_{\circ\ 1} & x^{\beta_1 \circ}_{\circ\ 2} & \cdots & x^{\beta_1 \circ}_{\circ\ r} \\
x^{\beta_2 \circ}_{\circ\ 1} & x^{\beta_2 \circ}_{\circ\ 2} & \cdots & x^{\beta_2 \circ}_{\circ\ r} \\
\cdots & \cdots & \cdots & \cdots \\
x^{\beta_r \circ}_{\circ\ 1} & x^{\beta_r \circ}_{\circ\ 2} & \cdots & x^{\beta_r \circ}_{\circ\ r}
\end{vmatrix}
\cdot
\begin{vmatrix}
g_{\overset{\circ}{\gamma_1}\overset{\circ}{\beta_1}} & g_{\overset{\circ}{\gamma_1}\overset{\circ}{\beta_2}} & \cdots & g_{\overset{\circ}{\gamma_1}\overset{\circ}{\beta_r}} \\
g_{\overset{\circ}{\gamma_2}\overset{\circ}{\beta_1}} & g_{\overset{\circ}{\gamma_2}\overset{\circ}{\beta_2}} & \cdots & g_{\overset{\circ}{\gamma_2}\overset{\circ}{\beta_r}} \\
\cdots & \cdots & \cdots & \cdots \\
g_{\overset{\circ}{\gamma_r}\overset{\circ}{\beta_1}} & g_{\overset{\circ}{\gamma_r}\overset{\circ}{\beta_2}} & \cdots & g_{\overset{\circ}{\gamma_r}\overset{\circ}{\beta_r}}
\end{vmatrix}
$$

$$\vec{e}^{*\gamma_1} \wedge \vec{e}^{*\gamma_2} \wedge \ldots \wedge \vec{e}^{*\gamma_r}\gamma_i, \beta_j \in I_n; \; i, j \in I_r;$$

$$\beta_1 < \beta_2 < \cdots < \beta_r; \quad \gamma_1 < \gamma_2 < \cdots < \gamma_r. \tag{13.22}$$

On the other hand, if the data vectors $\vec{V}_j, 1 \leq j \leq r$ are given in covariant coordinates, the data matrix would be $[\, X_1^* \quad X_2^* \quad \cdots \quad X_r^* \,]_{n,r}$, and the minors $\Delta_r^* = |\, X_{r1}^* \quad X_{r2}^* \quad \cdots \quad X_{rr}^* \,|$. Whence the exterior product of the vectors $\vec{V}_j$, directly expressed in the dual basis, would be

$$\vec{V}_1 \wedge \vec{V}_2 \wedge \cdots \wedge \vec{V}_r = \begin{vmatrix} x_{\gamma_1 1}^{\circ\,\circ} & x_{\gamma_1 2}^{\circ\,\circ} & \cdots & x_{\gamma_1 r}^{\circ\,\circ} \\ x_{\gamma_2 1}^{\circ\,\circ} & x_{\gamma_2 2}^{\circ\,\circ} & \cdots & x_{\gamma_2 r}^{\circ\,\circ} \\ \cdots & \cdots & \cdots & \cdots \\ x_{\gamma_r 1}^{\circ\,\circ} & x_{\gamma_r 2}^{\circ\,\circ} & \cdots & x_{\gamma_r r}^{\circ\,\circ} \end{vmatrix} \vec{e}^{*\gamma_1} \wedge \vec{e}^{*\gamma_2} \wedge \cdots \wedge \vec{e}^{*\gamma_r};$$

$$\gamma_1 < \gamma_2 < \cdots < \gamma_r. \quad (13.23)$$

Since Formulas (13.22) and (13.23) represent the same multivector in the same exterior dual basis, its strict components must be identical, which leads to

$$\begin{vmatrix} x_{\gamma_1 1}^{\circ\,\circ} & x_{\gamma_1 2}^{\circ\,\circ} & \cdots & x_{\gamma_1 r}^{\circ\,\circ} \\ x_{\gamma_2 1}^{\circ\,\circ} & x_{\gamma_2 2}^{\circ\,\circ} & \cdots & x_{\gamma_2 r}^{\circ\,\circ} \\ \cdots & \cdots & \cdots & \cdots \\ x_{\gamma_r 1}^{\circ\,\circ} & x_{\gamma_r 2}^{\circ\,\circ} & \cdots & x_{\gamma_r r}^{\circ\,\circ} \end{vmatrix} = \begin{vmatrix} x_{\circ\,1}^{\beta_1\circ} & x_{\circ\,2}^{\beta_1\circ} & \cdots & x_{\circ\,r}^{\beta_1\circ} \\ x_{\circ\,1}^{\beta_2\circ} & x_{\circ\,2}^{\beta_2\circ} & \cdots & x_{\circ\,r}^{\beta_2\circ} \\ \cdots & \cdots & \cdots & \cdots \\ x_{\circ\,1}^{\beta_r\circ} & x_{\circ\,2}^{\beta_r\circ} & \cdots & x_{\circ\,r}^{\beta_r\circ} \end{vmatrix} \cdot \begin{vmatrix} g_{\gamma_1 \beta_1}^{\circ\,\circ} & g_{\gamma_1 \beta_2}^{\circ\,\circ} & \cdots & g_{\gamma_1 \beta_r}^{\circ\,\circ} \\ g_{\gamma_2 \beta_1}^{\circ\,\circ} & g_{\gamma_2 \beta_2}^{\circ\,\circ} & \cdots & g_{\gamma_2 \beta_r}^{\circ\,\circ} \\ \cdots & \cdots & \cdots & \cdots \\ g_{\gamma_r \beta_1}^{\circ\,\circ} & g_{\gamma_r \beta_2}^{\circ\,\circ} & \cdots & g_{\gamma_r \beta_r}^{\circ\,\circ} \end{vmatrix}$$

$$\gamma_i, \beta_j \in I_n; \quad i, j \in I_r; \quad \beta_1 < \beta_2 < \cdots < \beta_r;$$
$$\gamma_1 < \gamma_2 < \cdots < \gamma_r, \quad (13.24)$$

which represents the relation between the strict components in covariant and in contravariant coordinates of the decomposable multivector $\vec{V}_1 \wedge \vec{V}_2 \wedge \cdots \wedge \vec{V}_r$ and are a generalization of Formula (13.8).

Following the same line as that in Section 13.2 and notating the strict components of the exterior product of (13.24) with the general strict notation of exterior tensors, expression (13.24) can also be written as

$$t_{(\gamma_1 \gamma_2 \cdots \gamma_r)}^{\circ\;\circ\;\cdots\;\circ} = t_{\circ\;\circ\;\cdots\;\circ}^{(\beta_1 \beta_2 \cdots \beta_r)} \begin{vmatrix} g_{\gamma_1 \beta_1}^{\circ\,\circ} & g_{\gamma_1 \beta_2}^{\circ\,\circ} & \cdots & g_{\gamma_1 \beta_r}^{\circ\,\circ} \\ g_{\gamma_2 \beta_1}^{\circ\,\circ} & g_{\gamma_2 \beta_2}^{\circ\,\circ} & \cdots & g_{\gamma_2 \beta_r}^{\circ\,\circ} \\ \cdots & \cdots & \cdots & \cdots \\ g_{\gamma_r \beta_1}^{\circ\,\circ} & g_{\gamma_r \beta_2}^{\circ\,\circ} & \cdots & g_{\gamma_r \beta_r}^{\circ\,\circ} \end{vmatrix}, \quad (13.25)$$

where $\beta_1 < \beta_2 < \cdots < \beta_r$ are the dummy indices and $\gamma_1 < \gamma_2 < \cdots < \gamma_r$ are the free indices, which due to the presence of dummy indices is a sum of products.

With Formula (13.25) we end this section. We only mention the reverse formula for raising indices for strict components

$$t_{\circ\;\circ\;\cdots\;\circ}^{(\gamma_1 \gamma_2 \cdots \gamma_r)} = t_{(\beta_1 \beta_2 \cdots \beta_r)}^{\circ\;\circ\;\cdots\;\circ} \begin{vmatrix} g_{\circ\,\circ}^{\gamma_1 \beta_1} & g_{\circ\,\circ}^{\gamma_1 \beta_2} & \cdots & g_{\circ\,\circ}^{\gamma_1 \beta_r} \\ g_{\circ\,\circ}^{\gamma_2 \beta_1} & g_{\circ\,\circ}^{\gamma_2 \beta_2} & \cdots & g_{\circ\,\circ}^{\gamma_2 \beta_r} \\ \cdots & \cdots & \cdots & \cdots \\ g_{\circ\,\circ}^{\gamma_r \beta_1} & g_{\circ\,\circ}^{\gamma_r \beta_2} & \cdots & g_{\circ\,\circ}^{\gamma_r \beta_r} \end{vmatrix}. \quad (13.26)$$

The change-of-basis in $E^n(\mathbb{R})$ is performed exactly the same as in $V^n(\mathbb{R})$, so that there is nothing to add to what has been treated in Chapters 9 and 10 on exterior algebras.

13.4 Euclidean exterior algebra of order r=n

This section refers to the tensors $T \in \bigwedge_n^{(n)}(\mathbb{R})$ of dimension $\binom{n}{n} = 1$, over the Euclidean space $E^n(\mathbb{R})$, where $\bigwedge_n^{(n)}(\mathbb{R})$ is a linear subspace of $\left(\overset{n}{\underset{1}{\otimes}} E_h^n\right)(\mathbb{R}) \subset \left(\overset{n}{\underset{1}{\otimes}} E^n\right)(\mathbb{R})$.

For the exterior tensor in contravariant coordinates, its representation is

$$T = t_{\circ\,\circ\cdots\,\circ}^{(12\cdots n)}\vec{e}_1 \wedge \vec{e}_2 \wedge \cdots \wedge \vec{e}_n; \quad t_{\circ\,\circ\cdots\,\circ}^{(12\cdots n)} \in \mathbb{R} \tag{13.27}$$

and for the exterior tensor in covariant coordinates:

$$T = t_{(12\cdots n)}^{\circ\,\circ\cdots\,\circ}\vec{e}^{*1} \wedge \vec{e}^{*2} \wedge \cdots \wedge \vec{e}^{*n}; \quad t_{(12\cdots n)}^{\circ\,\circ\cdots\,\circ} \in \mathbb{R}. \tag{13.28}$$

As an anti-symmetric tensor $T \in \left(\overset{n}{\underset{1}{\otimes}} E_h^n\right)(\mathbb{R})$, its respective representations (for the non-null components) are

$$T = t_{\circ\,\circ\cdots\,\circ}^{(12\cdots n)}\delta_{(1\;2\;\cdots\;n)}^{\alpha_1\,\alpha_2\cdots\alpha_n}\vec{e}_{\alpha_1} \otimes \vec{e}_{\alpha_2} \otimes \cdots \otimes \vec{e}_{\alpha_n}; \quad \alpha_i \in I_n; \; \alpha_1 \neq \alpha_2 \neq \cdots \neq \alpha_n, \tag{13.29}$$

and for the anti-symmetric covariant $(r = n)$ tensor

$$T = t_{(12\cdots n)}^{\circ\,\circ\cdots\,\circ}\delta_{\alpha_1\alpha_2\cdots\alpha_n}^{(1\;2\;\cdots\;n)}\vec{e}^{*\alpha_1} \otimes \vec{e}^{*\alpha_2} \otimes \cdots \otimes \vec{e}^{*\alpha_n}; \quad \alpha_i \in I_n; \; \alpha_1 \neq \alpha_2 \neq \cdots \neq \alpha_n, \tag{13.30}$$

so that $n!/2$ of them are $t_{(12\cdots n)}^{\circ\,\circ\cdots\,\circ}$ and the other $n!/2$ are $-t_{(12\cdots n)}^{\circ\,\circ\cdots\,\circ}$.

With respect to their relations as associated tensors, it suffices to apply Formula (13.25) with $r = n$ for lowering the indices

$$t_{(12\cdots n)}^{\circ\,\circ\cdots\,\circ} = t_{\circ\,\circ\cdots\,\circ}^{(12\cdots n)}|G| \tag{13.31}$$

and for raising the indices we apply (13.26) with $r = n$ to get

$$t_{\circ\,\circ\cdots\,\circ}^{(12\cdots n)} = t_{(12\cdots n)}^{\circ\,\circ\cdots\,\circ}|G^{-1}|. \tag{13.32}$$

13.5 The orientation tensor in exterior bases

In Chapter 12, Section 12.5 we have established with sufficient detail the "orientation tensor" of an oriented Euclidean space $E^n(\mathbb{R})$, starting from the possibilities offered by modular tensors over Euclidean spaces and we have proved its character as a Euclidean totally anti-symmetric homogeneous tensor of order $r = n$, endowed with its two contravariant and covariant expressions.

As a summary of all that was established in the final part of Section 12.5, we propose a notation for the orientation tensor, valid not only for oriented Euclidean spaces $E^n(\mathbb{R})$, but useful even for pseudo-Euclidean spaces $(PSE^n)(\mathbb{R})$. Let

$$\sqrt{|G_0|} = \sqrt{\text{modulus of } |G|}. \tag{13.33}$$

Its contravariant expression is

$$\Theta = \vartheta^{\alpha_1 \alpha_2 \cdots \alpha_n}_{\circ \; \circ \; \cdots \; \circ} \vec{e}_{\alpha_1} \otimes \vec{e}_{\alpha_2} \otimes \cdots \otimes \vec{e}_{\alpha_n}$$

$$= \frac{1}{+\sqrt{|G_0|}} \epsilon^{\alpha_1 \alpha_2 \cdots \alpha_n}_{\circ \; \circ \; \cdots \; \circ} \vec{e}_{\alpha_1} \otimes \vec{e}_{\alpha_2} \otimes \cdots \otimes \vec{e}_{\alpha_n}$$

$$= \frac{+\sqrt{|G_0|}}{|G_0|} \epsilon^{\alpha_1 \alpha_2 \cdots \alpha_n}_{\circ \; \circ \; \cdots \; \circ} \vec{e}_{\alpha_1} \otimes \vec{e}_{\alpha_2} \otimes \cdots \otimes \vec{e}_{\alpha_n}; \quad \alpha_i \in I_n; \quad \alpha_1 \neq \alpha_2 \neq \cdots \neq \alpha_n$$

and also written with respect to its exterior basis

$$\Theta = \frac{+\sqrt{|G_0|}}{|G_0|} \vec{e}_{\alpha_1} \wedge \vec{e}_{\alpha_2} \wedge \cdots \wedge \vec{e}_{\alpha_n}; \quad \alpha_i \in I_n; \quad \alpha_1 < \alpha_2 < \cdots < \alpha_n. \tag{13.34}$$

Its covariant expression is

$$\Theta = \vartheta^{\circ \; \circ \; \cdots \; \circ}_{\alpha_1 \alpha_2 \cdots \alpha_n} \vec{e}^{*\alpha_1} \otimes \vec{e}^{*\alpha_2} \otimes \cdots \otimes \vec{e}^{*\alpha_n}$$

$$= +\sqrt{|G_0|} \epsilon^{\circ \; \circ \; \cdots \; \circ}_{\alpha_1 \alpha_2 \cdots \alpha_n} \vec{e}^{*\alpha_1} \otimes \vec{e}^{*\alpha_2} \otimes \cdots \otimes \vec{e}^{*\alpha_n}; \quad \alpha_i \in I_n; \quad \alpha_1 \neq \alpha_2 \neq \cdots \neq \alpha_n$$

and also written with respect to its exterior basis

$$\Theta = +\sqrt{|G_0|} \vec{e}^{*\alpha_1} \wedge \vec{e}^{*\alpha_2} \wedge \cdots \wedge \vec{e}^{*\alpha_n}; \quad \alpha_i \in I_n; \quad \alpha_1 < \alpha_2 < \cdots < \alpha_n. \tag{13.35}$$

13.6 Dual or polar tensors in exterior bases

In this section we present exclusively the treatment of polar tensors established over an oriented Euclidean space $E^n(\mathbb{R})$, but the reader, if desired and considering the previous section, can easily extend it to pseudo-Euclidean spaces.

Let $T \in \bigwedge_n^{(r)}(\mathbb{R})$ be a Euclidean tensor of order $r \leq n$ that is completely anti-symmetric, and that is assumed to be given by either of its two representations:

$$t^{(\alpha_1 \alpha_2 \cdots \alpha_r)}_{\circ \; \circ \; \cdots \; \circ} \vec{e}_{\alpha_1} \wedge \vec{e}_{\alpha_2} \wedge \cdots \wedge \vec{e}_{\alpha_r} = t^{\alpha_1 \alpha_2 \cdots \alpha_r}_{\circ \; \circ \; \cdots \; \circ} \vec{e}_{\alpha_1} \otimes \vec{e}_{\alpha_2} \otimes \cdots \otimes \vec{e}_{\alpha_r}$$

or

$$t_{(\alpha_1\alpha_2\cdots\alpha_r)}^{\circ\ \circ\ \cdots\ \circ}\bar{e}^{*\alpha_1}\wedge\bar{e}^{*\alpha_2}\wedge\cdots\wedge\bar{e}^{*\alpha_r}=t_{\alpha_1\alpha_2\cdots\alpha_r}^{\circ\ \circ\ \cdots\ \circ}\bar{e}^{*\alpha_1}\otimes\bar{e}^{*\alpha_2}\otimes\cdots\otimes\bar{e}^{*\alpha_r}.$$

We define the polar tensor of T with the notation A as another Euclidean tensor of order $(n-r)$ obtained as the contracted tensor product of the orientation tensor with T. It is also totally anti-symmetric (if $(n-r)\geq 2$) and of contrary species.

The polar of the contravariant tensor T is

$$a_{\alpha_{r+1}\alpha_{r+2}\cdots\alpha_n}^{\circ\quad\circ\quad\cdots\quad\circ}=\frac{+\sqrt{|G|}}{r!}\epsilon_{\alpha_1\alpha_2\cdots\alpha_r\alpha_{r+1}\alpha_{r+2}\cdots\alpha_n}^{\circ\ \circ\ \cdots\ \circ\quad\circ\qquad\circ\quad\cdots\quad\circ}t_{\circ\ \circ\ \cdots\ \circ}^{\alpha_1\alpha_2\cdots\alpha_r}$$

or in terms of the strict components of T, as $A\in\bigwedge_{n*}^{(n-r)}(\mathbb{R})$:

$$a_{(\alpha_{r+1}\alpha_{r+2}\cdots\alpha_n)}^{\circ\quad\circ\quad\cdots\quad\circ}=\frac{+\sqrt{|G|}}{r!}\epsilon_{\alpha_1\alpha_2\cdots\alpha_r\alpha_{r+1}\alpha_{r+2}\cdots\alpha_n}^{\circ\ \circ\ \cdots\ \circ\quad\circ\qquad\circ\quad\cdots\quad\circ}t_{\circ\ \circ\ \cdots\ \circ}^{(\alpha_1\alpha_2\cdots\alpha_r)}. \qquad (13.36)$$

The polar of the covariant tensor T is

$$a^{\alpha_{r+1}\alpha_{r+2}\cdots\alpha_n}_{\circ\quad\circ\quad\cdots\quad\circ}=\frac{+\sqrt{|G|}}{|G|r!}\epsilon^{\alpha_1\alpha_2\cdots\alpha_r\alpha_{r+1}\alpha_{r+2}\cdots\alpha_n}_{\circ\ \circ\ \cdots\ \circ\quad\circ\qquad\circ\quad\cdots\quad\circ}t^{\circ\ \circ\ \cdots\ \circ}_{\alpha_1\alpha_2\cdots\alpha_r}$$

or in terms of the strict components of T, as $A\in\bigwedge_{n}^{(n-r)}(\mathbb{R})$:

$$a^{(\alpha_{r+1}\alpha_{r+2}\cdots\alpha_n)}_{\circ\quad\circ\quad\cdots\quad\circ}=\frac{+\sqrt{|G|}}{|G|r!}\epsilon^{\alpha_1\alpha_2\cdots\alpha_r\alpha_{r+1}\alpha_{r+2}\cdots\alpha_n}_{\circ\ \circ\ \cdots\ \circ\quad\circ\qquad\circ\quad\cdots\quad\circ}t^{\circ\ \circ\ \cdots\ \circ}_{(\alpha_1\alpha_2\cdots\alpha_r)}. \qquad (13.37)$$

We can observe that the contraction is total. In fact, the order of the tensor A is $(n+r)-(2r)=n-r$, obtained from the order of the tensor product minus the number of indices lost in the contraction.

If $r=n$ the polar tensor is a scalar.

If $r=n-1$ the polar tensor is a vector.

If $r\leq n-2$ the polar tensor is an anti-symmetric tensor.

If we use orthonormalized bases $\{\vec{h}_\alpha\}$ for $E^n(\mathbb{R})$, the factor $+\sqrt{|G|}$, $\frac{\sqrt{|G|}}{|G|}$ takes the value 1 and does not appear in Formulas (13.36) and (13.37), respectively.

It is obvious that the polar tensor can be notated as an anti-symmetric tensor or as an exterior tensor, when $n-r\geq 2$.

If it is a covariant tensor we have

$$A=a_{\alpha_{r+1}\alpha_{r+2}\cdots\alpha_n}^{\circ\quad\circ\quad\cdots\quad\circ}\bar{e}^{*\alpha_{r+1}}\otimes\bar{e}^{*\alpha_{r+2}}\otimes\cdots\otimes\bar{e}^{*\alpha_n}$$

$$=a_{(\alpha_{r+1}\alpha_{r+2}\cdots\alpha_n)}^{\circ\quad\circ\quad\cdots\quad\circ}\bar{e}^{*\alpha_{r+1}}\wedge\bar{e}^{*\alpha_{r+2}}\wedge\cdots\wedge\bar{e}^{*\alpha_n}$$

$$\alpha_i\in I_n;\ \alpha_{r+1}<\alpha_{r+2}<\cdots<\alpha_n$$

and if it is a contravariant tensor:

$$A=a^{\alpha_{r+1}\alpha_{r+2}\cdots\alpha_n}_{\circ\quad\circ\quad\cdots\quad\circ}\vec{e}_{\alpha_{r+1}}\otimes\vec{e}_{\alpha_{r+2}}\otimes\cdots\otimes\vec{e}_{\alpha_n}$$

$$=a^{(\alpha_{r+1}\alpha_{r+2}\cdots\alpha_n)}_{\circ\quad\circ\quad\cdots\quad\circ}\vec{e}_{\alpha_{r+1}}\wedge\vec{e}_{\alpha_{r+2}}\wedge\cdots\wedge\vec{e}_{\alpha_n}$$

$$\alpha_i\in I_n;\ \alpha_{r+1}<\alpha_{r+2}<\cdots<\alpha_n.$$

13.7 The cross product as a polar tensor in generalized Cartesian coordinate frames

Consider the Euclidean space $E^3(\mathbb{R})$, the classic linear space of three dimensions of the ordinary geometric vectors (OGS $\equiv$ ordinary geometric space).

Consider a reference frame of basis $\{\vec{e}_\alpha\}$ and fundamental metric tensor G.

As has been considered in Section 12.7 of the previous chapter, we take two vectors $\vec{V}, \vec{W} \in E^3(\mathbb{R})$ and we build the exterior product $T = \vec{V} \wedge \vec{W}$, where

$$T \in \bigwedge_{3}^{(2)}(\mathbb{R}) \subset \left(\overset{2}{\underset{1}{\otimes}} E^3 \right)_h (\mathbb{R}) \subset \left(\overset{2}{\underset{1}{\otimes}} E^3 \right)(\mathbb{R}).$$

The result obtained in Section 12.7 is

$$T = t^{\alpha\beta}_{\circ\circ} \vec{e}_\alpha \otimes \vec{e}_\beta = \begin{vmatrix} x^\alpha_\circ & y^\alpha_\circ \\ x^\beta_\circ & y^\beta_\circ \end{vmatrix} \vec{e}_\alpha \otimes \vec{e}_\beta, \quad \alpha, \beta \in I_3; \quad \alpha \neq \beta. \tag{13.38}$$

According to Formula (13.35) the corresponding orientation tensor is

$$\Theta = +\sqrt{|G|} \epsilon^{\circ\circ\circ}_{\alpha\beta\gamma} \vec{e}^{*\alpha} \wedge \vec{e}^{*\beta} \wedge \vec{e}^{*\gamma}; \quad \alpha, \beta, \gamma \in I_3; \quad \alpha < \beta < \gamma, \tag{13.39}$$

and, according to (13.36) and (13.38), the cross product tensor will be

$$A = \vec{V} \times \vec{W} = +\sqrt{|G|}\frac{1}{2!}\epsilon^{\circ\circ\circ}_{\alpha\beta\gamma} t^{\alpha\beta}_{\circ\circ} \vec{e}^{*\gamma} = \frac{1}{2}\sqrt{|G|}\epsilon^{\circ\circ\circ}_{\alpha\beta\gamma}\left(x^\alpha_\circ y^\beta_\circ - x^\beta_\circ y^\alpha_\circ \right) \vec{e}^{*\gamma},$$
$$\tag{13.40}$$

and developing the indices $\alpha, \beta, \gamma \in I_3, \alpha \neq \beta \neq \gamma$, we obtain

$$a_1 = \frac{1}{2}\sqrt{|G|}\left[\epsilon^{\circ\circ\circ}_{231}\left(x^2_\circ y^3_\circ - x^3_\circ y^2_\circ \right) + \epsilon^{\circ\circ\circ}_{321}\left(x^3_\circ y^2_\circ - x^2_\circ y^3_\circ \right) \right] = \sqrt{|G|}\begin{vmatrix} x^2_\circ & y^2_\circ \\ x^3_\circ & y^3_\circ \end{vmatrix}$$

$$a_2 = \frac{1}{2}\sqrt{|G|}\left[\epsilon^{\circ\circ\circ}_{132}\left(x^1_\circ y^3_\circ - x^3_\circ y^1_\circ \right) + \epsilon^{\circ\circ\circ}_{312}\left(x^3_\circ y^1_\circ - x^1_\circ y^3_\circ \right) \right] = \sqrt{|G|}\begin{vmatrix} x^3_\circ & y^3_\circ \\ x^1_\circ & y^1_\circ \end{vmatrix}$$

$$a_3 = \frac{1}{2}\sqrt{|G|}\left[\epsilon^{\circ\circ\circ}_{123}\left(x^1_\circ y^2_\circ - x^2_\circ y^1_\circ \right) + \epsilon^{\circ\circ\circ}_{213}\left(x^2_\circ y^1_\circ - x^1_\circ y^2_\circ \right) \right] = \sqrt{|G|}\begin{vmatrix} x^1_\circ & y^1_\circ \\ x^2_\circ & y^2_\circ \end{vmatrix}$$

and then

$$A = \vec{V} \times \vec{W} = a_1\vec{e}^{*1} + a_2\vec{e}^{*2} + a_3\vec{e}^{*3}$$
$$= \sqrt{|G|}\left(\begin{vmatrix} x^2_\circ & y^2_\circ \\ x^3_\circ & y^3_\circ \end{vmatrix} \vec{e}^{*1} - \begin{vmatrix} x^1_\circ & y^1_\circ \\ x^3_\circ & y^3_\circ \end{vmatrix} \vec{e}^{*2} + \begin{vmatrix} x^1_\circ & y^1_\circ \\ x^2_\circ & y^2_\circ \end{vmatrix} \vec{e}^{*3} \right),$$

whish is a tensor in covariant coordinates, or

$$\vec{V} \times \vec{W} = \sqrt{|G|} \begin{vmatrix} \vec{e}^{*1} & x_o^1 & y_o^1 \\ \vec{e}^{*2} & x_o^2 & y_o^2 \\ \vec{e}^{*3} & x_o^3 & y_o^3 \end{vmatrix}. \tag{13.41}$$

With respect to the mixed product, we obtain

$$[\vec{V}\vec{W}\vec{Z}] = \vec{V} \bullet (\vec{W} \times \vec{Z}),$$

and since $\vec{V} = [\vec{e}_1 \vec{e}_2 \vec{e}_3] \begin{bmatrix} x_o^1 \\ x_o^2 \\ x_o^3 \end{bmatrix}$ is in contravariant coordinates, and $\vec{W} \times \vec{Z}$ in

covariant coordinates, the metric tensor for this product is $I_3 = \begin{bmatrix} 1 & 0 & 0 \\ 0 & 1 & 0 \\ 0 & 0 & 1 \end{bmatrix}$,

whence

$$[\vec{V}\vec{W}\vec{Z}] = [x_o^1 x_o^2 x_o^3] I_3 \begin{bmatrix} \begin{vmatrix} y_o^2 & z_o^2 \\ y_o^3 & z_o^3 \end{vmatrix} \sqrt{|G|} \\ \begin{vmatrix} y_o^1 & z_o^1 \\ y_o^3 & z_o^3 \end{vmatrix} \sqrt{|G|} \\ \begin{vmatrix} y_o^1 & z_o^1 \\ y_o^2 & z_o^2 \end{vmatrix} \sqrt{|G|} \end{bmatrix}$$

and operating we obtain

$$[\vec{V}\vec{W}\vec{Z}] = \sqrt{|G|} \begin{vmatrix} x_o^1 & y_o^1 & z_o^1 \\ x_o^2 & y_o^2 & z_o^2 \\ x_o^3 & y_o^3 & z_o^3 \end{vmatrix}, \tag{13.42}$$

which justifies, via a tensor procedure, the formulas in Section 1.4 of "special tensors", with the components of the vectors *in columns*.

13.8 $\sqrt{|G|}$ geometric interpretation in generalized Cartesian coordinate frames

In the OGS referred to an *orthonormal* trihedron, of basis $\{\vec{u}_\alpha\}$, metric tensor $G \equiv I_3$ and $\sqrt{|G|} = 1$, we know, by reasons of metric and analytical geometry, that the volume of the parallelepiped built with the vectors $\vec{V}, \vec{W}, \vec{Z}$ as concurrent edges is

$$V_0 = \text{absolute value of } [\vec{V}\vec{W}\vec{Z}] = \text{absolute value of } \begin{vmatrix} x_o^1 & y_o^1 & z_o^1 \\ x_o^2 & y_o^2 & z_o^2 \\ x_o^3 & y_o^3 & z_o^3 \end{vmatrix}. \quad (13.43)$$

Pre-multiplying (13.43) by itself with the transpose matrix determinant, and taking into account the Binet–Cauchy formula on determinant products, we get

$$V_0^2 = \begin{vmatrix} x_o^1 & x_o^2 & x_o^3 \\ y_o^1 & y_o^2 & y_o^3 \\ z_o^1 & z_o^2 & z_o^3 \end{vmatrix} \cdot \begin{vmatrix} x_o^1 & y_o^1 & z_o^1 \\ x_o^2 & y_o^2 & z_o^2 \\ x_o^3 & y_o^3 & z_o^3 \end{vmatrix} = \begin{vmatrix} \sum_1^3 (x_o^\theta)^2 & \sum_1^3 (x_o^\theta y_o^\theta) & \sum_1^3 (x_o^\theta z_o^\theta) \\ \sum_1^3 (x_o^\theta y_o^\theta) & \sum_1^3 (y_o^\theta)^2 & \sum_1^3 (y_o^\theta z_o^\theta) \\ \sum_1^3 (x_o^\theta z_o^\theta) & \sum_1^3 (y_o^\theta z_o^\theta) & \sum_1^3 (z_o^\theta)^2 \end{vmatrix}$$

$$= \begin{vmatrix} \vec{V} \bullet \vec{V} & \vec{V} \bullet \vec{W} & \vec{V} \bullet \vec{Z} \\ \vec{W} \bullet \vec{V} & \vec{W} \bullet \vec{W} & \vec{W} \bullet \vec{Z} \\ \vec{Z} \bullet \vec{V} & \vec{Z} \bullet \vec{W} & \vec{Z} \bullet \vec{Z} \end{vmatrix}. \quad (13.44)$$

If now we choose a new *arbitrary* basis $\{\vec{e}_\alpha\}$, and we consider the associated fundamental metric tensor

$$G = \begin{bmatrix} \vec{e}_1 \bullet \vec{e}_1 & \vec{e}_1 \bullet \vec{e}_2 & \vec{e}_1 \bullet \vec{e}_3 \\ \vec{e}_2 \bullet \vec{e}_1 & \vec{e}_2 \bullet \vec{e}_2 & \vec{e}_2 \bullet \vec{e}_3 \\ \vec{e}_3 \bullet \vec{e}_1 & \vec{e}_3 \bullet \vec{e}_2 & \vec{e}_3 \bullet \vec{e}_3 \end{bmatrix},$$

we observe that its determinant, after considering Formula (13.44), has an evident geometric interpretation

$$|G| = V_0^2(\vec{e}_1, \vec{e}_2, \vec{e}_3) \quad \text{and} \quad \sqrt{|G|} = V_0(\vec{e}_1, \vec{e}_2, \vec{e}_3), \quad (13.45)$$

i.e., $\sqrt{|G|}$ is precisely the volume of the oblique parallelepiped built with the basic vectors $\{\vec{e}_\alpha\}$ as concurrent edges.

From it, we deduce that Formula (13.42) gives the volume of the parallelepiped built with the (concurrent in O) vectors $\vec{V}, \vec{W}, \vec{Z}$ in an *oblique* arbitrary reference frame.

13.9 Illustrative examples

Example 13.1 (Associated Euclidean exterior tensors). In a Euclidean space referred to a basis $\{\vec{e}_\alpha\}$, the fundamental metric tensor is given by the Gram matrix

$$G = \begin{bmatrix} 1 & -1 & 0 \\ -1 & 2 & 0 \\ 0 & 0 & 1 \end{bmatrix}.$$

Consider a tensor $T \in \left(\overset{2}{\underset{1}{\otimes}} E^3 \right)$ ($\mathbb{R}$) that is given by its cova-contravariant coordinates

$$[t_{\alpha\circ}^{\circ\beta}] = \begin{bmatrix} 1 & 1 & 2 \\ -2 & -1 & 3 \\ -7 & -5 & 0 \end{bmatrix}.$$

1. Obtain its strict covariant coordinates.
2. Obtain its strict contravariant coordinates.
3. Express T as an exterior covariant tensor.
4. Extract the exterior components of T in contravariant coordinates, from the covariant coordinates of the previous question.

Solution: With the aim of maximum information, we use several procedures to solve some questions.

1. *Classic method.* We know, by the theory of Euclidean tensors, that

$$t_{\alpha\beta}^{\circ\circ} = g_{\beta\theta}^{\circ\circ} t_{\alpha\circ}^{\circ\theta} = t_{\alpha\circ}^{\circ\theta} g_{\theta\beta}^{\circ\circ},$$

and in matrix form

$$[t_{\alpha\beta}^{\circ\circ}] = \begin{bmatrix} 1 & 1 & 2 \\ -2 & -1 & 3 \\ -7 & -5 & 0 \end{bmatrix} \begin{bmatrix} 1 & -1 & 0 \\ -1 & 2 & 0 \\ 0 & 0 & 1 \end{bmatrix} = \begin{bmatrix} 0 & 1 & 2 \\ -1 & 0 & 3 \\ -2 & -3 & 0 \end{bmatrix}.$$

Direct method. We first proceed to change notation and notate the dummy and free indices with numbered indices, i.e.,

$$t_{\beta_1\beta_2}^{\circ\;\circ} = g_{\circ\;\beta_1}^{\alpha_1\;\circ} g_{\alpha_2\beta_2}^{\circ\;\circ} t_{\alpha_1\;\circ}^{\circ\;\alpha_2};$$

and in matrix form

$$T_{\sigma,1}' = (I_3 \otimes G) \bullet T_{\sigma,1}: \quad Z_1 = I_3 \otimes G; \quad T_{\sigma,1}' = Z_1 \bullet T_{\sigma,1} \qquad (13.46)$$

$$\sigma = n^r = 3^2 = 9;$$

$$Z_1 = I_3 \otimes G = \begin{bmatrix} 1 & 0 & 0 \\ 0 & 1 & 0 \\ 0 & 0 & 1 \end{bmatrix} \otimes \begin{bmatrix} 1 & -1 & 0 \\ -1 & 2 & 0 \\ 0 & 0 & 1 \end{bmatrix} = \begin{bmatrix} 1 & -1 & 0 & 0 & 0 & 0 & 0 & 0 & 0 \\ -1 & 2 & 0 & 0 & 0 & 0 & 0 & 0 & 0 \\ 0 & 0 & 1 & 0 & 0 & 0 & 0 & 0 & 0 \\ 0 & 0 & 0 & 1 & -1 & 0 & 0 & 0 & 0 \\ 0 & 0 & 0 & -1 & 2 & 0 & 0 & 0 & 0 \\ 0 & 0 & 0 & 0 & 0 & 1 & 0 & 0 & 0 \\ 0 & 0 & 0 & 0 & 0 & 0 & 1 & -1 & 0 \\ 0 & 0 & 0 & 0 & 0 & 0 & -1 & 2 & 0 \\ 0 & 0 & 0 & 0 & 0 & 0 & 0 & 0 & 1 \end{bmatrix}.$$

As $T_{\sigma,1}^t = [\,1\ 1\ 2\ -2\ -1\ 3\ -7\ -5\ 0\,]$, applying (13.46), we get

$$(T_{\sigma,1}')^t = [\,0\ 1\ 2\ -1\ 0\ 3\ -2\ -3\ 0\,],$$

which once condensed leads to

$$t_{\beta_1\beta_2}^{\circ\;\circ} = \begin{bmatrix} 0 & 1 & 2 \\ -1 & 0 & 3 \\ -2 & -3 & 0 \end{bmatrix}.$$

2. *Classic method.* According to the Euclidean tensors theory,

$$t^{\alpha\beta}_{\circ\circ} = g^{\alpha\theta}_{\circ\circ}t^{\circ\beta}_{\theta\circ},$$

and in matrix form

$$[t^{\alpha\beta}_{\circ\circ}] = G^{-1}[t^{\circ\beta}_{\alpha\circ}] = \begin{bmatrix} 2 & 1 & 0 \\ 1 & 1 & 0 \\ 0 & 0 & 1 \end{bmatrix} \begin{bmatrix} 1 & 1 & 2 \\ -2 & -1 & 3 \\ -7 & -5 & 0 \end{bmatrix} = \begin{bmatrix} 0 & 1 & 7 \\ -1 & 0 & 5 \\ -7 & -5 & 0 \end{bmatrix}.$$

Direct method.

$$t^{\beta_1\beta_2}_{\circ\,\circ\,\circ} = g^{\alpha_1\beta_1}_{\circ\,\circ\,\circ}g^{\circ\,\beta_2}_{\alpha_2\,\circ}t^{\circ\,\alpha_2}_{\alpha_1\,\circ}; \quad T''_{\sigma,1} = (G^{-1}\otimes I_3)\bullet T_{\sigma,1}$$

$$Z_2 = G^{-1}\otimes I_3; \quad T''_{\sigma,1} = Z_2\bullet T_{\sigma,1} \qquad (13.47)$$

$$Z_2 = \begin{bmatrix} 2 & 1 & 0 \\ 1 & 1 & 0 \\ 0 & 0 & 1 \end{bmatrix} \otimes \begin{bmatrix} 1 & 0 & 0 \\ 0 & 1 & 0 \\ 0 & 0 & 1 \end{bmatrix} = \begin{bmatrix} 2 & 0 & 0 & 1 & 0 & 0 & 0 & 0 & 0 \\ 0 & 2 & 0 & 0 & 1 & 0 & 0 & 0 & 0 \\ 0 & 0 & 2 & 0 & 0 & 1 & 0 & 0 & 0 \\ 1 & 0 & 0 & 1 & 0 & 0 & 0 & 0 & 0 \\ 0 & 1 & 0 & 0 & 1 & 0 & 0 & 0 & 0 \\ 0 & 0 & 1 & 0 & 0 & 1 & 0 & 0 & 0 \\ 0 & 0 & 0 & 0 & 0 & 0 & 1 & 0 & 0 \\ 0 & 0 & 0 & 0 & 0 & 0 & 0 & 1 & 0 \\ 0 & 0 & 0 & 0 & 0 & 0 & 0 & 0 & 1 \end{bmatrix}.$$

As $T^t_{\sigma,1} = [1\ 1\ 2\ -2\ -1\ 3\ -7\ -5\ 0]$, applying (13.47), the result is

$$\left(T''_{\sigma,1}\right)^t = [0\ 1\ 7\ -1\ 0\ 5\ -7\ -5\ 0],$$

which once condensed leads to

$$[t^{\beta_1\beta_2}_{\circ\,\circ}] = \begin{bmatrix} 0 & 1 & 7 \\ -1 & 0 & 5 \\ -7 & -5 & 0 \end{bmatrix}.$$

3. In the two matrices $[t^{\circ\circ}_{\alpha\beta}]$ and $[t^{\alpha\beta}_{\circ\circ}]$ from the answers to questions 1 and 2, we can notice the anti-symmetric character of tensor T; as such, its strict components define T as an exterior tensor. So, taking the covariant coordinates of the matrix $[t^{\circ\circ}_{\alpha\beta}]$ such that $\alpha < \beta$, that is, the upper triangular matrix of the given matrix, we have

$$T = t^{\circ\circ}_{(12)}\vec{e}^{*1}\wedge\vec{e}^{*2} + t^{\circ\circ}_{(13)}\vec{e}^{*1}\wedge\vec{e}^{*3} + t^{\circ\circ}_{(23)}\vec{e}^{*2}\wedge\vec{e}^{*3}$$

$$= \vec{e}^{*1}\wedge\vec{e}^{*2} + 2\vec{e}^{*1}\wedge\vec{e}^{*3} + 3\vec{e}^{*2}\wedge\vec{e}^{*3} \in \bigwedge\nolimits^{(2)}_{*3}(\mathbb{R}).$$

4. Applying Formula (13.10) to our case ($n = 3$), and remembering the matrix G^{-1}, we have

$$t^{(12)}_{\circ\ \circ} = t^{\circ\ \circ}_{(\alpha\beta)} \begin{vmatrix} g^{1\alpha}_{\circ\circ} & g^{1\beta}_{\circ\circ} \\ g^{2\alpha}_{\circ\circ} & g^{2\beta}_{\circ\circ} \end{vmatrix}$$

$$= t^{\circ\ \circ}_{(12)} \begin{vmatrix} g^{11}_{\circ\circ} & g^{12}_{\circ\circ} \\ g^{21}_{\circ\circ} & g^{22}_{\circ\circ} \end{vmatrix} + t^{\circ\ \circ}_{(13)} \begin{vmatrix} g^{11}_{\circ\circ} & g^{13}_{\circ\circ} \\ g^{21}_{\circ\circ} & g^{23}_{\circ\circ} \end{vmatrix} + t^{\circ\ \circ}_{(23)} \begin{vmatrix} g^{12}_{\circ\circ} & g^{13}_{\circ\circ} \\ g^{22}_{\circ\circ} & g^{23}_{\circ\circ} \end{vmatrix}$$

$$= 1 \cdot \begin{vmatrix} 2 & 1 \\ 1 & 1 \end{vmatrix} + 2 \cdot \begin{vmatrix} 2 & 0 \\ 1 & 0 \end{vmatrix} + 3 \cdot \begin{vmatrix} 1 & 0 \\ 1 & 0 \end{vmatrix} = 1$$

$$t^{(13)}_{\circ\ \circ} = t^{\circ\ \circ}_{(\alpha\beta)} \begin{vmatrix} g^{1\alpha}_{\circ\circ} & g^{1\beta}_{\circ\circ} \\ g^{3\alpha}_{\circ\circ} & g^{3\beta}_{\circ\circ} \end{vmatrix} = 1 \cdot \begin{vmatrix} 2 & 1 \\ 0 & 0 \end{vmatrix} + 2 \cdot \begin{vmatrix} 2 & 0 \\ 0 & 1 \end{vmatrix} + 3 \cdot \begin{vmatrix} 1 & 0 \\ 0 & 1 \end{vmatrix} = 7$$

$$t^{(23)}_{\circ\ \circ} = t^{\circ\ \circ}_{(\alpha\beta)} \begin{vmatrix} g^{2\alpha}_{\circ\circ} & g^{2\beta}_{\circ\circ} \\ g^{3\alpha}_{\circ\circ} & g^{3\beta}_{\circ\circ} \end{vmatrix} = 1 \cdot \begin{vmatrix} 1 & 1 \\ 0 & 0 \end{vmatrix} + 2 \cdot \begin{vmatrix} 1 & 0 \\ 0 & 1 \end{vmatrix} + 3 \cdot \begin{vmatrix} 1 & 0 \\ 0 & 1 \end{vmatrix} = 5,$$

whence

$$T = t^{(\alpha\beta)}_{\circ\ \circ} \vec{e}_\alpha \wedge \vec{e}_\beta = \vec{e}_1 \wedge \vec{e}_2 + 7\vec{e}_1 \wedge \vec{e}_3 + 5\vec{e}_2 \wedge \vec{e}_3 \in \bigwedge\nolimits_3^{(2)}(\mathbb{R}).$$

We observe that the result is the same if we work with the following strict covariant coordinates, given in the solution (point 3):

$$\begin{bmatrix} t^{\circ\ \circ}_{(12)} \\ t^{\circ\ \circ}_{(13)} \\ t^{\circ\ \circ}_{(23)} \end{bmatrix} = \begin{bmatrix} 1 \\ 2 \\ 3 \end{bmatrix}$$

by means of the matrix of an *exterior fundamental metric tensor*:

$$G^{*}_{\bigwedge_3^{(2)}(\mathbb{R})} \equiv G^{-1}_{\bigwedge_3^{(2)}(\mathbb{R})} = \begin{bmatrix} 1 & 0 & 0 \\ 0 & 2 & 1 \\ 0 & 1 & 1 \end{bmatrix} \equiv \begin{bmatrix} 0 & 1 & 0 \\ 0 & 0 & 1 \\ 1 & 0 & 0 \end{bmatrix}^t G^{-1} \begin{bmatrix} 0 & 1 & 0 \\ 0 & 0 & 1 \\ 1 & 0 & 0 \end{bmatrix},$$

where we assume that $\bigwedge_3^{(2)}(\mathbb{R})$ and $\bigwedge_{3*}^{(2)}(\mathbb{R})$ are in dual bases. In fact

$$\begin{bmatrix} 1 & 0 & 0 \\ 0 & 2 & 1 \\ 0 & 1 & 1 \end{bmatrix} \bullet \begin{bmatrix} 1 \\ 2 \\ 3 \end{bmatrix} = \begin{bmatrix} 1 \\ 7 \\ 5 \end{bmatrix},$$

and because it is congruent with G^{-1} it is also positive definite.

$\square$

Example 13.2 (Geometric polar tensors). In the OGS $E^3(\mathbb{R})$, referred to a basis $\{\vec{e}_\alpha\}$, the fundamental metric tensor is represented by the Gram matrix

$$G = [g_{\alpha\beta}^{\circ\circ}] = \begin{bmatrix} 2 & 1 & 0 \\ 1 & 2 & 0 \\ 0 & 0 & 1 \end{bmatrix}.$$

Consider the exterior algebra $\bigwedge_3^{(2)}(\mathbb{R})$ established over that space.

1. Show that for any $T \in \bigwedge_3^{(2)}(\mathbb{R}), \exists \vec{u}, \vec{v} \in E^3(\mathbb{R})$ such that

$$\text{Adjoint of } T = \vec{u} \times \vec{v},$$

   where "$\times$" is the symbol of the ordinary cross product in the OGS.
2. Given the vector $\vec{z} \in E^3(\mathbb{R})$, show that the polar tensor of $(T \wedge \vec{z})$ is the contraction of (Adjoint $T \otimes \vec{z}$).
3. Obtain the polar tensor of the tensor S defined as

$$s_{\circ\circ\circ}^{\mu\nu\gamma} = \delta_{\alpha\beta}^{\mu\nu} r_{\circ\circ\circ}^{\alpha\beta\gamma}; \quad R = T \wedge \vec{z}.$$

4. If the components of T as an anti-symmetric contravariant tensor are

$$[t_{\circ\circ}^{\alpha\beta}] = \begin{bmatrix} 0 & 1 & 2 \\ -1 & 0 & 3 \\ -2 & -3 & 0 \end{bmatrix},$$

   determine the subset of all vectors $\vec{u}, \vec{v} \in E^3(\mathbb{R})$ that satisfy the condition

$$\text{polar of } T = \vec{u} \times \vec{v}.$$

Is it a linear subspace of $E^3(\mathbb{R})$?

Solution:

1. In Example 9.11 of Chapter 9, we established with various details the conditions that an exterior tensor must satisfy to be decomposable, that is, an exterior product of vectors.
   Since $T \in \bigwedge_3^{(2)}(\mathbb{R})$, of order $p = 2$ and dimension $n = 3$, satisfy them, the theorem of point 3 of that example ensures the existence of at least two vectors $\vec{u}, \vec{v} \in E^3(\mathbb{R})$ such that $T = \vec{u} \wedge \vec{v}$. Assuming that these two vectors are known, we write

$$T = t_{\circ\circ}^{(12)} \vec{e}_1 \wedge \vec{e}_2 + t_{\circ\circ}^{(13)} \vec{e}_1 \wedge \vec{e}_3 + t_{\circ\circ}^{(23)} \vec{e}_2 \wedge \vec{e}_3$$

$$= \begin{vmatrix} u_\circ^1 & v_\circ^1 \\ u_\circ^2 & v_\circ^2 \end{vmatrix} \vec{e}_1 \wedge \vec{e}_2 + \begin{vmatrix} u_\circ^1 & v_\circ^1 \\ u_\circ^3 & v_\circ^3 \end{vmatrix} \vec{e}_1 \wedge \vec{e}_3 + \begin{vmatrix} u_\circ^2 & v_\circ^2 \\ u_\circ^3 & v_\circ^3 \end{vmatrix} \vec{e}_2 \wedge \vec{e}_3.$$

Let $A = \text{polar } T$, $|G| = \begin{vmatrix} 2 & 1 & 0 \\ 1 & 2 & 0 \\ 0 & 0 & 1 \end{vmatrix} = 3$ and $+\sqrt{|G|} = \sqrt{3}$;

$$
\begin{aligned}
a^{\circ}_{\gamma} &= \frac{\sqrt{3}}{2!}\,\epsilon^{\circ\circ\circ}_{\alpha\beta\gamma}\,t^{\alpha\beta}_{\circ\circ} = \frac{\sqrt{3}}{2!}\left[\epsilon^{\circ\circ\circ}_{\alpha\beta\gamma}\,t^{(\alpha\beta)}_{\circ\ \circ} + \epsilon^{\circ\circ\circ}_{\beta\alpha\gamma}\left(-t^{(\alpha\beta)}_{\circ\ \circ}\right)\right] \\
&= \frac{\sqrt{3}}{2!}\left[2\epsilon^{\circ\ \circ\ \circ}_{(\alpha\beta)\gamma}\,t^{(\alpha\beta)}_{\circ\ \circ}\right] = \sqrt{3}\,\epsilon^{\circ\ \circ\ \circ}_{(\alpha\beta)\gamma}\,t^{(\alpha\beta)}_{\circ\ \circ};
\end{aligned}
$$

and numerically:

$$
a^{\circ}_{3} = \sqrt{3}\,\epsilon^{\circ\ \circ\ \circ}_{(12)3}\,t^{(12)}_{\circ\ \circ} = \sqrt{3}\begin{vmatrix} u^{1}_{\circ} & v^{1}_{\circ} \\ u^{2}_{\circ} & v^{2}_{\circ} \end{vmatrix}
$$

$$
a^{\circ}_{2} = \sqrt{3}\,\epsilon^{\circ\ \circ\ \circ}_{(13)2}\,t^{(13)}_{\circ\ \circ} = -\sqrt{3}\begin{vmatrix} u^{1}_{\circ} & v^{1}_{\circ} \\ u^{3}_{\circ} & v^{3}_{\circ} \end{vmatrix}
$$

$$
a^{\circ}_{1} = \sqrt{3}\,\epsilon^{\circ\ \circ\ \circ}_{(23)1}\,t^{(23)}_{\circ\ \circ} = \sqrt{3}\begin{vmatrix} u^{2}_{\circ} & v^{2}_{\circ} \\ u^{3}_{\circ} & v^{3}_{\circ} \end{vmatrix},
$$

thus, the tensor is

$$
\begin{aligned}
A &= a^{\circ}_{1}\vec{e}^{*1} + a^{\circ}_{2}\vec{e}^{*2} + a^{\circ}_{3}\vec{e}^{*3} \\
&= \sqrt{3}\left(\begin{vmatrix} u^{2}_{\circ} & v^{2}_{\circ} \\ u^{3}_{\circ} & v^{3}_{\circ} \end{vmatrix}\vec{e}^{*1} - \begin{vmatrix} u^{1}_{\circ} & v^{1}_{\circ} \\ u^{3}_{\circ} & v^{3}_{\circ} \end{vmatrix}\vec{e}^{*2} + \begin{vmatrix} u^{1}_{\circ} & v^{1}_{\circ} \\ u^{2}_{\circ} & v^{2}_{\circ} \end{vmatrix}\vec{e}^{*3}\right).
\end{aligned}
$$

On the other hand, applying Formula (13.41), we have for the cross product of $\vec{u}$ and $\vec{v}$

$$
\vec{u} \times \vec{v} = \sqrt{3}\begin{vmatrix} \vec{e}^{*1} & u^{1}_{\circ} & v^{1}_{\circ} \\ \vec{e}^{*2} & u^{2}_{\circ} & v^{2}_{\circ} \\ \vec{e}^{*3} & u^{3}_{\circ} & v^{3}_{\circ} \end{vmatrix},
$$

which once developed leads to the equality

$$
A = \text{Adjoint } T = \vec{u} \times \vec{v}.
$$

2. According to the statement we have

$$
R = T \wedge \vec{z} = \vec{u} \wedge \vec{v} \wedge \vec{z} = \begin{vmatrix} u^{1}_{\circ} & v^{1}_{\circ} & z^{1}_{\circ} \\ u^{2}_{\circ} & v^{2}_{\circ} & z^{2}_{\circ} \\ u^{3}_{\circ} & v^{3}_{\circ} & z^{3}_{\circ} \end{vmatrix}\vec{e}_{1} \wedge \vec{e}_{2} \wedge \vec{e}_{3} = r^{(123)}_{\circ\circ\circ}\,\vec{e}_{1} \wedge \vec{e}_{2} \wedge \vec{e}_{3}
$$

and we proceed to analyze the polar tensor R:

$$\text{Adjoint } (T \wedge \vec{z}) = \text{Adjoint } R = \frac{\sqrt{|G|}}{3!} \epsilon^{\ \ \ \ }_{\alpha\beta\gamma} r^{\alpha\beta\gamma}_{\ \ \ \circ\circ\circ}$$

$$= \frac{\sqrt{|G|}}{3!} \epsilon^{\ \ \ \circ\circ\circ}_{\alpha\beta\gamma} \left(\epsilon^{\alpha\beta\gamma}_{\circ\circ\circ} \epsilon^{\ \ \ \circ\circ\circ}_{(\alpha\beta\gamma)} r^{(\alpha\beta\gamma)}_{\ \ \ \ \circ\circ\circ} \right)$$

$$= \frac{\sqrt{3}}{3!} \left(3! \epsilon^{\ \ \ \circ\circ\circ}_{(\alpha\beta\gamma)} \right)$$

$$= \sqrt{3} \left(1 \cdot r^{(123)}_{\ \ \ \circ\circ\circ} \right) = \sqrt{3} \begin{vmatrix} u^1_{\circ} & v^1_{\circ} & z^1_{\circ} \\ u^2_{\circ} & v^2_{\circ} & z^2_{\circ} \\ u^3_{\circ} & v^3_{\circ} & z^3_{\circ} \end{vmatrix}. \quad (13.48)$$

In addition we have

$$\text{Contraction (Polar } T \otimes \vec{z}) = \text{Contraction } (A \otimes \vec{z})$$

$$= \text{Contraction } \left[(a^{\circ}_1 \vec{e}^{*1} + a^{\circ}_2 \vec{e}^{*2} + a^{\circ}_3 \vec{e}^{*3}) \otimes (z^1_{\circ} \vec{e}_1 + z^2_{\circ} \vec{e}_2 + z^3_{\circ} \vec{e}_3) \right]$$

$$= [a^{\circ}_1 \ \ a^{\circ}_2 \ \ a^{\circ}_3] \begin{bmatrix} z^1_{\circ} \\ z^2_{\circ} \\ z^3_{\circ} \end{bmatrix} = a^{\circ}_1 z^1_{\circ} + a^{\circ}_2 z^2_{\circ} + a^{\circ}_3 z^3_{\circ}$$

$$= \sqrt{3} \left(\begin{vmatrix} u^2_{\circ} & v^2_{\circ} \\ u^3_{\circ} & v^3_{\circ} \end{vmatrix} z^1_{\circ} - \begin{vmatrix} u^1_{\circ} & v^1_{\circ} \\ u^3_{\circ} & v^3_{\circ} \end{vmatrix} z^2_{\circ} + \begin{vmatrix} u^1_{\circ} & v^1_{\circ} \\ u^2_{\circ} & v^2_{\circ} \end{vmatrix} z^3_{\circ} \right)$$

$$= \sqrt{3} \begin{vmatrix} u^1_{\circ} & v^1_{\circ} & z^1_{\circ} \\ u^2_{\circ} & v^2_{\circ} & z^2_{\circ} \\ u^3_{\circ} & v^3_{\circ} & z^3_{\circ} \end{vmatrix} \quad (13.49)$$

and a comparison of (13.48) and (13.49) leads to

$$\text{Adjoint } (T \wedge \vec{z}) = \text{Contraction (Adjoint } T \otimes \vec{z}).$$

3. The polar tensor of S is

$$\text{Polar } S = \frac{\sqrt{|G|}}{3!} \epsilon^{\ \ \ \circ\circ\circ}_{\mu\nu\gamma} s^{\mu\nu\gamma}_{\ \ \ \circ\circ\circ} = \frac{\sqrt{|G|}}{3!} \epsilon^{\ \ \ \circ\circ\circ}_{\mu\nu\gamma} \left[\epsilon^{\mu\nu\gamma}_{\circ\circ\circ} \epsilon^{\ \ \ \circ\circ\circ}_{(123)} s^{(123)}_{\ \ \ \circ\circ\circ} \right]$$

$$= \frac{\sqrt{|G|}}{3!} \delta^{\mu\nu\gamma}_{\mu\nu\gamma} \epsilon^{\ \ \ \circ\circ\circ}_{(123)} s^{(123)}_{\ \ \ \circ\circ\circ} = \frac{\sqrt{3}}{3!} 3! \cdot 1 \cdot s^{(123)}_{\ \ \ \circ\circ\circ} = \sqrt{3} s^{(123)}_{\ \ \ \circ\circ\circ}$$

$$= \sqrt{3} \delta^{(12)}_{\alpha\ \beta} r^{\alpha\beta3}_{\ \ \ \circ\circ\circ} = \sqrt{3} \left[\delta^{12}_{12} r^{(123)}_{\ \ \ \circ\circ\circ} + \delta^{12}_{21} (-r^{(123)}_{\ \ \ \circ\circ\circ}) \right]$$

$$= \sqrt{3} \cdot 2 \cdot r^{(123)}_{\ \ \ \circ\circ\circ} = 2\sqrt{3} \begin{vmatrix} u^1_{\circ} & v^1_{\circ} & z^1_{\circ} \\ u^2_{\circ} & v^2_{\circ} & z^2_{\circ} \\ u^3_{\circ} & v^3_{\circ} & z^3_{\circ} \end{vmatrix}.$$

4. The given anti-symmetric tensor, expressed as an exterior tensor is

$$T = \vec{e}_1 \wedge \vec{e}_2 + 2\vec{e}_1 \wedge \vec{e}_3 + 3\vec{e}_2 \wedge \vec{e}_3,$$

so that, according to (9.134), the subset of vectors $\vec{u}, \vec{v}$ such that $\vec{u} \times \vec{v} =$ Polar T, is the linear subspace

$$L_1 \equiv t^{(12)}_{\circ\;\circ}x^3 - t^{(13)}_{\circ\;\circ}x^2 + t^{(23)}_{\circ\;\circ}x^1 = 0,$$

which in our case becomes

$$L_1 \equiv 3x^1 - 2x^2 + x^3 = 0,$$

which declares the orthogonality of the vector $\vec{u} \times \vec{v}$ (in covariant coordinates) with the vectors $\vec{x}$ (chosen factors u and v) in contravariant coordinates.

□

Example 13.3 (Adjoint tensors of exterior products). In a Euclidean space $E^3(\mathbb{R})$, referred to a basis $\{\vec{e}_\alpha\}$, the fundamental metric tensor in covariant coordinates is given by the Gram matrix

$$G = \begin{bmatrix} 1 & 1 & -1 \\ 1 & 2 & 0 \\ -1 & 0 & 3 \end{bmatrix}.$$

Consider the vectors $\vec{u}(1,2,3)$ and $\vec{w}(3,2,1)$ defined by its contravariant coordinates.

1. Obtain the strict contravariant and covariant coordinates of the Euclidean tensor $T = \vec{u} \wedge \vec{v}$.
2. Idem of the tensor $A = $ Polar T.
3. Obtain the moduli of the tensors T and A.
4. Is there any relation between the moduli of T and A?

Solution: The matrix of the vectors in contravariant coordinates is $\begin{bmatrix} 1 & 3 \\ 2 & 2 \\ 3 & 1 \end{bmatrix}.$

1. These coordinates are obtained as follows:

$$T = \vec{u} \wedge \vec{v} = \begin{vmatrix} u^\alpha_\circ & v^\alpha_\circ \\ u^\beta_\circ & v^\beta_\circ \end{vmatrix} \vec{e}_\alpha \wedge \vec{e}_\beta$$

$$= \begin{vmatrix} 1 & 3 \\ 2 & 2 \end{vmatrix} \vec{e}_1 \wedge \vec{e}_2 + \begin{vmatrix} 1 & 3 \\ 3 & 1 \end{vmatrix} \vec{e}_1 \wedge \vec{e}_3 + \begin{vmatrix} 2 & 2 \\ 3 & 1 \end{vmatrix} \vec{e}_2 \wedge \vec{e}_3$$

$$= -4\vec{e}_1 \wedge \vec{e}_2 - 8\vec{e}_1 \wedge \vec{e}_3 - 4\vec{e}_2 \wedge \vec{e}_3.$$

and the data in covariant coordinates are

$$[X_1^* X_2^*] = G\,[X_1 X_2] = \begin{bmatrix} 1 & 1 & -1 \\ 1 & 2 & 0 \\ -1 & 0 & 3 \end{bmatrix}\begin{bmatrix} 1 & 3 \\ 2 & 2 \\ 3 & 1 \end{bmatrix} = \begin{bmatrix} 0 & 4 \\ 5 & 7 \\ 8 & 0 \end{bmatrix}$$

$$T = \vec{u} \wedge \vec{v} = \begin{vmatrix} u^{\,\circ}_\alpha & v^{\,\circ}_\alpha \\ u^{\,\circ}_\beta & v^{\,\circ}_\beta \end{vmatrix}\, \vec{e}^{*\alpha} \wedge \vec{e}^{*\beta}$$

$$= \begin{vmatrix} 0 & 4 \\ 5 & 7 \end{vmatrix}\,\vec{e}^{*1} \wedge \vec{e}^{*2} + \begin{vmatrix} 0 & 4 \\ 8 & 0 \end{vmatrix}\,\vec{e}^{*1} \wedge \vec{e}^{*3} + \begin{vmatrix} 5 & 7 \\ 8 & 0 \end{vmatrix}\,\vec{e}^{*2} \wedge \vec{e}^{*3}$$

$$= -20\vec{e}^{*1} \wedge \vec{e}^{*2} - 32\vec{e}^{*1} \wedge \vec{e}^{*3} - 56\vec{e}^{*2} \wedge \vec{e}^{*3},$$

that is,

$$\begin{bmatrix} t^{(12)}_{\circ\,\circ} \\ t^{(13)}_{\circ\,\circ} \\ t^{(23)}_{\circ\,\circ} \end{bmatrix} = \begin{bmatrix} -4 \\ -8 \\ -4 \end{bmatrix};\quad \begin{bmatrix} t^{\,\circ\,\circ}_{(12)} \\ t^{\,\circ\,\circ}_{(13)} \\ t^{\,\circ\,\circ}_{(23)} \end{bmatrix} = \begin{bmatrix} -20 \\ -32 \\ -56 \end{bmatrix}.$$

We can also calculate the covariant coordinates $t^{\,\circ\,\circ}_{(\alpha\beta)}$ from $t^{(\alpha\beta)}_{\circ\,\circ}$, as associated exterior tensor, using (13.16):

$$t^{\,\circ\,\circ}_{(12)} = (-4)\begin{vmatrix} 1 & 1 \\ 1 & 2 \end{vmatrix} + (-8)\begin{vmatrix} 1 & -1 \\ 1 & 0 \end{vmatrix} + (-4)\begin{vmatrix} 1 & -1 \\ 2 & 0 \end{vmatrix} = -20$$

$$t^{\,\circ\,\circ}_{(13)} = (-4)\begin{vmatrix} 1 & 1 \\ -1 & 0 \end{vmatrix} + (-8)\begin{vmatrix} 1 & -1 \\ -1 & 3 \end{vmatrix} + (-4)\begin{vmatrix} 1 & -1 \\ 0 & 3 \end{vmatrix} = -32$$

$$t^{\,\circ\,\circ}_{(23)} = (-4)\begin{vmatrix} 1 & 2 \\ -1 & 0 \end{vmatrix} + (-8)\begin{vmatrix} 1 & 0 \\ -1 & 3 \end{vmatrix} + (-4)\begin{vmatrix} 2 & 0 \\ 0 & 3 \end{vmatrix} = -56,$$

which coincide with the previous results.

2. Because of (13.36), the tensor T, as a contravariant and anti-symmetric tensor of order $r = 2$, has as polar tensor

$$a^{\,\circ}_\gamma = \frac{\sqrt{|G|}}{2!}\epsilon^{\,\circ\,\circ\,\circ}_{\alpha\beta\gamma}t^{\alpha\beta}_{\circ\,\circ} = \frac{\sqrt{|G|}}{2!}\left(\epsilon^{\,\circ\,\circ\,\circ}_{\alpha\beta\gamma} - \epsilon^{\,\circ\,\circ\,\circ}_{\beta\alpha\gamma}\right)t^{(\alpha\beta)}_{\circ\,\circ} = \sqrt{|G|}\epsilon^{\,\circ\,\circ\,\circ}_{(\alpha\beta)\gamma}t^{(\alpha\beta)}_{\circ\,\circ},$$

and then, the covariant coordinates of the tensor A, taking into account that $+\sqrt{|G|} = \sqrt{1} = 1$ and that $\epsilon^{\,\circ\,\circ\,\circ}_{(23)1} = +1;\ \epsilon^{\,\circ\,\circ\,\circ}_{(13)2} = -1;\ \epsilon^{\,\circ\,\circ\,\circ}_{(12)3} = +1,$ are

$$\begin{bmatrix} a^{\,\circ}_1 \\ a^{\,\circ}_2 \\ a^{\,\circ}_3 \end{bmatrix} = \begin{bmatrix} t^{(23)}_{\circ\,\circ} \\ -t^{(13)}_{\circ\,\circ} \\ t^{(12)}_{\circ\,\circ} \end{bmatrix} = \begin{bmatrix} -4 \\ 8 \\ -4 \end{bmatrix}.$$

Similarly, but with Formula (13.37), we obtain

$$a^\gamma_{\,\circ} = \frac{+\sqrt{|G|}}{2!|G|}\epsilon^{\alpha\beta\gamma}_{\circ\,\circ\,\circ}t^{\,\circ\,\circ}_{\alpha\beta} = \frac{\sqrt{|G|}}{2!|G|}\left(\epsilon^{\alpha\beta\gamma}_{\circ\,\circ\,\circ} - \epsilon^{\beta\alpha\gamma}_{\circ\,\circ\,\circ}\right)t^{\,\circ\,\circ}_{(\alpha\beta)} = \frac{+\sqrt{|G|}}{|G|}\epsilon^{(\alpha\beta)\gamma}_{\circ\,\circ\,\circ}t^{\,\circ\,\circ}_{(\alpha\beta)},$$

that is,

$$\begin{bmatrix} a^1_{\circ} \\ a^2_{\circ} \\ a^3_{\circ} \end{bmatrix} = \begin{bmatrix} t^{\circ\,\circ}_{(23)} \\ -t^{\circ\,\circ}_{(13)} \\ t^{\circ\,\circ}_{(12)} \end{bmatrix} = \begin{bmatrix} -56 \\ 32 \\ -20 \end{bmatrix}.$$

We can also calculate the covariant coordinates of the tensor A, a°_{γ}, from the contravariant coordinates $a^{\gamma}_{\circ}$:

$$\begin{bmatrix} a^{\circ}_1 \\ a^{\circ}_2 \\ a^{\circ}_3 \end{bmatrix} = G \begin{bmatrix} a^1_{\circ} \\ a^2_{\circ} \\ a^3_{\circ} \end{bmatrix} = \begin{bmatrix} 1 & 1 & -1 \\ 1 & 2 & 0 \\ -1 & 0 & 3 \end{bmatrix} \begin{bmatrix} -56 \\ 32 \\ -20 \end{bmatrix} = \begin{bmatrix} -4 \\ 8 \\ -4 \end{bmatrix},$$

verifying again the previous results.

3. Notating the tensor T as an anti-symmetric contravariant tensor, its representation is

$$T \in \left(\overset{2}{\underset{1}{\otimes}} E^3 \right) (\mathbb{R}); \quad [t^{\alpha\beta}_{\circ\circ}] = \begin{bmatrix} 0 & -4 & -8 \\ 4 & 0 & -4 \\ 8 & 4 & 0 \end{bmatrix},$$

and as $\sigma = n^r = 3^2 = 9$, "stretched" can be represented as

$$T_{\sigma,1} = [0 \;-4 \;-8 \;4 \;0 \;-4 \;8 \;4 \;0].$$

The connection tensor of $\left(\overset{2}{\underset{1}{\otimes}} \right) (\mathbb{R})$ is

$$G(9) = G \otimes G = \begin{bmatrix} 1 & 1 & -1 \\ 1 & 2 & 0 \\ -1 & 0 & 3 \end{bmatrix} \otimes \begin{bmatrix} 1 & 1 & -1 \\ 1 & 2 & 0 \\ -1 & 0 & 3 \end{bmatrix}$$

$$= \begin{bmatrix} 1 & 1 & -1 & 1 & 1 & -1 & -1 & -1 & 1 \\ 1 & 2 & 0 & 1 & 2 & 0 & -1 & -2 & 0 \\ -1 & 0 & 3 & -1 & 0 & 3 & 1 & 0 & -3 \\ 1 & 1 & -1 & 2 & 2 & -2 & 0 & 0 & 0 \\ 1 & 2 & 0 & 2 & 4 & 0 & 0 & 0 & 0 \\ -1 & 0 & 3 & -2 & 0 & 6 & 0 & 0 & 0 \\ -1 & -1 & 1 & 0 & 0 & 0 & 3 & 3 & -3 \\ -1 & -2 & 0 & 0 & 0 & 0 & 3 & 6 & 0 \\ 1 & 0 & -3 & 0 & 0 & 0 & -3 & 0 & 9 \end{bmatrix},$$

whence

$$|T| = \sqrt{T^t_{\sigma} G(9) T_{\sigma}} = \sqrt{[0 -4 -8 4 0 -4 8 4 0]G(9) \begin{bmatrix} 0 \\ -4 \\ -8 \\ 4 \\ 0 \\ -4 \\ 8 \\ 4 \\ 0 \end{bmatrix}} = \sqrt{1120} = 4\sqrt{70}.$$

In addition $|A| = \sqrt{[a_1 \;\; a_2 \;\; a_3]I_3 \begin{bmatrix} a^1 \\ a^2 \\ a^3 \end{bmatrix}} = \sqrt{[-4 \;\; 8 \;\; -4] \begin{bmatrix} -56 \\ 32 \\ -20 \end{bmatrix}} =$

$\sqrt{224 + 256 + 80} = \sqrt{560} = 4\sqrt{35}.$

4. The relation is $\frac{|T|}{|A|} = \sqrt{2}.$

$\square$

Example 13.4 (Generic expressions of polar tensors).

1. In a Euclidean space $E^2(\mathbb{R})$ referred to the basis $\{\vec{e}_\alpha\}$, obtain in covariant and contravariant coordinates the polar tensor of $P = \vec{a} \wedge \vec{b}$, the exterior product of two vectors $\vec{a}, \vec{b} \in E^2(\mathbb{R})$.

2. In a Euclidean space $E^4(\mathbb{R})$ referred to the basis $\{\vec{e}_\alpha\}$, obtain in covariant and contravariant coordinates the polar tensor of $Q \in \bigwedge_4^{(2)}(\mathbb{R})$, given by its anti-symmetric representation:

$$[q_{\circ\circ}^{\alpha\beta}] = \begin{bmatrix} 0 & a & b & c \\ -a & 0 & d & f \\ -b & -d & 0 & g \\ -c & -f & -g & 0 \end{bmatrix}.$$

Solution:

1. The polar tensor is

$$P = \vec{a} \wedge \vec{b} = \begin{vmatrix} a_1^\circ & b_1^\circ \\ a_2^\circ & b_2^\circ \end{vmatrix} \vec{e}^{*1} \wedge \vec{e}^{*2};$$

and also

$$P = \vec{a} \wedge \vec{b} = \begin{vmatrix} a_\circ^1 & b_\circ^1 \\ a_\circ^2 & b_\circ^2 \end{vmatrix} \vec{e}_1 \wedge \vec{e}_2 \equiv p_{\circ\circ}^{(\alpha\beta)} \vec{e}_\alpha \wedge \vec{e}_\beta.$$

The anti-symmetric tensor, in the contravariant case is

$$P = p_{\circ\circ}^{\alpha\beta} \vec{e}_\alpha \otimes \vec{e}_\beta; \;\; \alpha, \beta \in I_2; \;\; \alpha \neq \beta.$$

The polar tensor A, of P in contravariant coordinates is:

$$a = \frac{+\sqrt{|G|}}{2!} \epsilon_{\alpha\beta}^{\circ\circ} p_{\circ\circ}^{\alpha\beta} = \frac{+\sqrt{|G|}}{2!}(\epsilon_{\alpha\beta}^{\circ\circ} - \epsilon_{\beta\alpha}^{\circ\circ}) p_{\circ\circ}^{(\alpha\beta)}$$

$$= \frac{+\sqrt{|G|}}{2!} \cdot 2\epsilon_{(\alpha\beta)}^{\circ\circ} p_{\circ\circ}^{(\alpha\beta)} = \sqrt{|G|} \epsilon_{(12)}^{\circ\circ} p_{\circ\circ}^{(12)},$$

so that the polar tensor of P is the scalar

$$a = +\sqrt{|G|} \begin{vmatrix} a_\circ^1 & b_\circ^1 \\ a_\circ^2 & b_\circ^2 \end{vmatrix}.$$

Similarly, the polar A of P in covariant coordinates is the scalar

$$a' = \frac{+\sqrt{|G|}}{|G|} \begin{vmatrix} a_1^\circ & b_1^\circ \\ a_2^\circ & b_2^\circ \end{vmatrix}.$$

2. Let B be the polar tensor of Q. From the theory, we obtain

$$b_{\gamma\delta}^{\circ\circ} = \frac{+\sqrt{|G|}}{2!} \epsilon_{\alpha\beta\gamma\delta}^{\circ\circ\circ\circ} q_{\circ\circ}^{\alpha\beta} = \frac{+\sqrt{|G|}}{2!}(\epsilon_{\alpha\beta\gamma\delta}^{\circ\circ\circ\circ} - \epsilon_{\beta\alpha\gamma\delta}^{\circ\circ\circ\circ})q_{\circ\circ}^{\alpha\beta}$$

$$= \frac{+\sqrt{|G|}}{2!} \cdot 2! \epsilon_{(\alpha\beta)\gamma\delta}^{\circ\circ\circ\circ} q_{\circ\circ}^{(\alpha\beta)},$$

and giving values to the indices, and operating only the strict components of tensor B, we get

α	β	γ	δ
1	2	3	4
1	3	2	4
1	4	2	3
2	3	1	4
2	4	1	3
3	4	1	2

$$b_{34}^{\circ\circ} = +\sqrt{|G|}\epsilon_{(12)34}^{\circ\circ\circ\circ} q_{\circ\circ}^{(12)} = +\sqrt{|G|}a; \quad b_{43}^{\circ\circ} = +\sqrt{|G|}(-a)$$

$$b_{24}^{\circ\circ} = +\sqrt{|G|}\epsilon_{(13)24}^{\circ\circ\circ\circ} q_{\circ\circ}^{(13)} = +\sqrt{|G|}(-b); \quad b_{42}^{\circ\circ} = +\sqrt{|G|}b$$

$$b_{23}^{\circ\circ} = +\sqrt{|G|}\epsilon_{(14)23}^{\circ\circ\circ\circ} q_{\circ\circ}^{(14)} = +\sqrt{|G|}c; \quad b_{32}^{\circ\circ} = +\sqrt{|G|}(-c)$$

$$b_{14}^{\circ\circ} = +\sqrt{|G|}\epsilon_{(23)14}^{\circ\circ\circ\circ} q_{\circ\circ}^{(23)} = +\sqrt{|G|}d; \quad b_{41}^{\circ\circ} = +\sqrt{|G|}(-d)$$

$$b_{13}^{\circ\circ} = +\sqrt{|G|}\epsilon_{(24)13}^{\circ\circ\circ\circ} q_{\circ\circ}^{(24)} = +\sqrt{|G|}(-f); \quad b_{31}^{\circ\circ} = +\sqrt{|G|}f$$

$$b_{12}^{\circ\circ} = +\sqrt{|G|}\epsilon_{(34)12}^{\circ\circ\circ\circ} q_{\circ\circ}^{(34)} = +\sqrt{|G|}g; \quad b_{21}^{\circ\circ} = +\sqrt{|G|}(-g).$$

In summary, the tensor B = Polar Q has as covariant matrix representation

$$[b_{\alpha\beta}^{\circ\circ}] = +\sqrt{|G|} \begin{bmatrix} 0 & g & -f & d \\ -g & 0 & c & -b \\ f & -c & 0 & a \\ -d & b & -a & 0 \end{bmatrix}. \tag{13.50}$$

To determine the tensor B in contravariant coordinates $[b_{\circ\circ}^{\alpha\beta}]$, we can consider it as associated with $[b_{\alpha\beta}^{\circ\circ}]$, and raising indices by means of (13.10), we obtain

$$b_{\circ\circ}^{\theta\phi} = b_{\alpha\beta}^{\circ\circ} \begin{vmatrix} g_{\circ\circ}^{\theta\alpha} & g_{\circ\circ}^{\theta\beta} \\ g_{\circ\circ}^{\phi\alpha} & g_{\circ\circ}^{\phi\beta} \end{vmatrix},$$

where the determinants are minors of the matrix G^{-1}. Developing now the first component, we have

$$
b^{12}_{\circ\circ} = b^{\circ\circ}_{12} \begin{vmatrix} g^{11}_{\circ\circ} & g^{12}_{\circ\circ} \\ g^{21}_{\circ\circ} & g^{22}_{\circ\circ} \end{vmatrix} + b^{\circ\circ}_{13} \begin{vmatrix} g^{11}_{\circ\circ} & g^{13}_{\circ\circ} \\ g^{21}_{\circ\circ} & g^{23}_{\circ\circ} \end{vmatrix} + b^{\circ\circ}_{14} \begin{vmatrix} g^{11}_{\circ\circ} & g^{14}_{\circ\circ} \\ g^{21}_{\circ\circ} & g^{24}_{\circ\circ} \end{vmatrix} + b^{\circ\circ}_{23} \begin{vmatrix} g^{12}_{\circ\circ} & g^{13}_{\circ\circ} \\ g^{23}_{\circ\circ} & g^{23}_{\circ\circ} \end{vmatrix}
$$

$$
+ b^{\circ\circ}_{24} \begin{vmatrix} g^{12}_{\circ\circ} & g^{14}_{\circ\circ} \\ g^{22}_{\circ\circ} & g^{24}_{\circ\circ} \end{vmatrix} + b^{\circ\circ}_{34} \begin{vmatrix} g^{13}_{\circ\circ} & g^{14}_{\circ\circ} \\ g^{23}_{\circ\circ} & g^{24}_{\circ\circ} \end{vmatrix}
$$

$$
= + \sqrt{|G|} \left(g \begin{vmatrix} g^{11}_{\circ\circ} & g^{12}_{\circ\circ} \\ g^{21}_{\circ\circ} & g^{22}_{\circ\circ} \end{vmatrix} + (-f) \begin{vmatrix} g^{11}_{\circ\circ} & g^{13}_{\circ\circ} \\ g^{21}_{\circ\circ} & g^{23}_{\circ\circ} \end{vmatrix} + d \begin{vmatrix} g^{11}_{\circ\circ} & g^{14}_{\circ\circ} \\ g^{21}_{\circ\circ} & g^{24}_{\circ\circ} \end{vmatrix} \right.
$$

$$
\left. + c \begin{vmatrix} g^{12}_{\circ\circ} & g^{13}_{\circ\circ} \\ g^{23}_{\circ\circ} & g^{23}_{\circ\circ} \end{vmatrix} + (-b) \begin{vmatrix} g^{12}_{\circ\circ} & g^{14}_{\circ\circ} \\ g^{22}_{\circ\circ} & g^{24}_{\circ\circ} \end{vmatrix} + a \begin{vmatrix} g^{13}_{\circ\circ} & g^{14}_{\circ\circ} \\ g^{23}_{\circ\circ} & g^{24}_{\circ\circ} \end{vmatrix} \right),
$$

and the rest of the components, $b^{13}_{\circ\circ}, b^{14}_{\circ\circ}$, etc. can be developed by the reader, keeping the coefficients in the summands and substituting the minors that correspond to each case.

□

Example 13.5 (Adjoint tensors of order (n-1)). In a Euclidean space $E^4(\mathbb{R})$ referred to a basis $\{\vec{e}_\alpha\}$, of Gram matrix

$$
G \equiv [g^{\circ\circ}_{\alpha\beta}] = \begin{bmatrix} 1 & 1 & 1 & 1 \\ 1 & 2 & 2 & 2 \\ 1 & 2 & 3 & 3 \\ 1 & 2 & 3 & 4 \end{bmatrix},
$$

we consider the exterior product of the vectors $\vec{V}_1, \vec{V}_2, \vec{V}_3$:

$$
T = \vec{V}_1 \wedge \vec{V}_2 \wedge \vec{V}_3,
$$

with contravariant coordinates $\vec{V}_1(0,1,1,1), \vec{V}_2(1,0,1,1)$ and $\vec{V}_3(1,1,0,1)$, and we denote by S the polar tensor of T. With respect to the tensor T, answer the following questions.

1. Since it is a Euclidean totally anti-symmetric contravariant tensor, of third order, obtain its matrix representation in the basis associated with the generic basis $\{\vec{e}_\alpha\}$, as an anti-symmetric tensor.
2. Obtain its covariant coordinates.
3. Obtain the covariant representation of tensor S.
4. Obtain the contravariant representation of tensor S.
5. Generalizing this statement to a Euclidean space $E^n(\mathbb{R})$, of Gram matrix G, where $T = \vec{V}_1 \wedge \vec{V}_2 \wedge \cdots \wedge \vec{V}_{n-1}$, would be the exterior product of $(n-1)$ vectors, and S its polar tensor:

a) Give the expression of the T components as a tensor $T \in \bigwedge_n^{(n-1)}(\mathbb{R})$.

b) Obtain the most simplified version of the components of the polar S, of the tensor T in covariant and contravariant bases.

6. Specify the number of strict components of T and its value in contravariant coordinates.

7. Considering the dot product of the tensor S and a generic vector $\vec{V}_n$, write it if the contravariant coordinates of the vectors are $\vec{V}_i(x_{\circ i}^{1\circ}, x_{\circ i}^{2\circ}, \ldots, x_{\circ i}^{n\circ})$.

8. Idem, but in terms of the covariant coordinates $\vec{V}_i(x_{1i}^{\circ\circ}, x_{2i}^{\circ\circ}, \ldots, x_{ni}^{\circ\circ})$.

Solution:

1. The data matrix is $[\,X_1 \quad X_2 \quad X_3\,] \equiv \begin{bmatrix} 0 & 1 & 1 \\ 1 & 0 & 1 \\ 1 & 1 & 0 \\ 1 & 1 & 1 \end{bmatrix}$, so that applying Formula (13.17), we have

$$T = \vec{V}_1 \wedge \vec{V}_2 \wedge \vec{V}_3 = \begin{vmatrix} 0 & 1 & 1 \\ 1 & 0 & 1 \\ 1 & 1 & 0 \end{vmatrix} \vec{e}_1 \wedge \vec{e}_2 \wedge \vec{e}_3 + \begin{vmatrix} 0 & 1 & 1 \\ 1 & 0 & 1 \\ 1 & 1 & 1 \end{vmatrix} \vec{e}_1 \wedge \vec{e}_2 \wedge \vec{e}_4$$

$$+ \begin{vmatrix} 0 & 1 & 1 \\ 1 & 1 & 0 \\ 1 & 1 & 1 \end{vmatrix} \vec{e}_1 \wedge \vec{e}_3 \wedge \vec{e}_4 + \begin{vmatrix} 1 & 0 & 1 \\ 1 & 1 & 0 \\ 1 & 1 & 1 \end{vmatrix} \vec{e}_2 \wedge \vec{e}_3 \wedge \vec{e}_4$$

$$= 2\vec{e}_1 \wedge \vec{e}_2 \wedge \vec{e}_3 + \vec{e}_1 \wedge \vec{e}_2 \wedge \vec{e}_4 - \vec{e}_1 \wedge \vec{e}_3 \wedge \vec{e}_4 + \vec{e}_2 \wedge \vec{e}_3 \wedge \vec{e}_4.$$

Writing the basic exterior vectors as vectors of the space $\left(\overset{3}{\underset{1}{\otimes}} E_h^3\right)(\mathbb{R})$, we get that T is

$2\,[\vec{e}_1 \otimes \vec{e}_2 \otimes \vec{e}_3 + \vec{e}_2 \otimes \vec{e}_3 \otimes \vec{e}_1 + \vec{e}_3 \otimes \vec{e}_1 \otimes \vec{e}_2 - \vec{e}_2 \otimes \vec{e}_1 \otimes \vec{e}_3 - \vec{e}_3 \otimes \vec{e}_2 \otimes \vec{e}_1 - \vec{e}_1 \otimes \vec{e}_3 \otimes \vec{e}_2]$
$+ [\vec{e}_1 \otimes \vec{e}_2 \otimes \vec{e}_4 + \vec{e}_2 \otimes \vec{e}_4 \otimes \vec{e}_1 + \vec{e}_4 \otimes \vec{e}_1 \otimes \vec{e}_2 - \vec{e}_2 \otimes \vec{e}_1 \otimes \vec{e}_4 - \vec{e}_4 \otimes \vec{e}_2 \otimes \vec{e}_1 - \vec{e}_1 \otimes \vec{e}_4 \otimes \vec{e}_2]$
$- [\vec{e}_1 \otimes \vec{e}_3 \otimes \vec{e}_4 + \vec{e}_3 \otimes \vec{e}_4 \otimes \vec{e}_1 + \vec{e}_4 \otimes \vec{e}_1 \otimes \vec{e}_3 - \vec{e}_3 \otimes \vec{e}_1 \otimes \vec{e}_4 - \vec{e}_4 \otimes \vec{e}_3 \otimes \vec{e}_1 - \vec{e}_1 \otimes \vec{e}_4 \otimes \vec{e}_3]$
$+ [\vec{e}_2 \otimes \vec{e}_3 \otimes \vec{e}_4 + \vec{e}_3 \otimes \vec{e}_4 \otimes \vec{e}_2 + \vec{e}_4 \otimes \vec{e}_2 \otimes \vec{e}_3 - \vec{e}_3 \otimes \vec{e}_2 \otimes \vec{e}_4 - \vec{e}_4 \otimes \vec{e}_3 \otimes \vec{e}_2 - \vec{e}_2 \otimes \vec{e}_4 \otimes \vec{e}_3]$,

and arranging the components of $T = t_{\circ\circ\circ}^{\alpha\beta\gamma}\vec{e}_\alpha \otimes \vec{e}_\beta \otimes \vec{e}_\gamma$ as a column matrix of blocks, with $\alpha = $ row block index, $\beta = $ row of each block, $\gamma = $ column of each block, we have

$$[t_{\circ\circ\circ}^{\alpha\beta\gamma}] = \begin{bmatrix} t_{\circ\circ\circ}^{1\beta\gamma} \\ t_{\circ\circ\circ}^{2\beta\gamma} \\ t_{\circ\circ\circ}^{3\beta\gamma} \\ t_{\circ\circ\circ}^{4\beta\gamma} \end{bmatrix}.$$

The total number of components is $\sigma = n^r = 4^3 = 64$, and the blocks are

$$[t_{\circ\circ\circ}^{1\beta\gamma}] = \begin{bmatrix} 0 & 0 & 0 & 0 \\ 0 & 0 & 2 & 1 \\ 0 & -2 & 0 & -1 \\ 0 & -1 & 1 & 0 \end{bmatrix}$$

$$[t_{\circ\circ\circ}^{2\beta\gamma}] = \begin{bmatrix} 0 & 0 & -2 & -1 \\ 0 & 0 & 0 & 0 \\ 2 & 0 & 0 & 1 \\ 1 & 0 & -1 & 0 \end{bmatrix}$$

$$[t_{\circ\circ\circ}^{3\beta\gamma}] = \begin{bmatrix} 0 & 2 & 0 & 1 \\ -2 & 0 & 0 & -1 \\ 0 & 0 & 0 & 0 \\ -1 & 1 & 0 & 0 \end{bmatrix}$$

$$[t_{\circ\circ\circ}^{4\beta\gamma}] = \begin{bmatrix} 0 & 1 & -1 & 0 \\ -1 & 0 & 1 & 0 \\ 1 & -1 & 0 & 0 \\ 0 & 0 & 0 & 0 \end{bmatrix}.$$

2. The previous contravariant coordinates allow us to know the covariant coordinates, lowering the three indices as a Euclidean tensor associated with an anti-symmetric tensor

$$t_{\beta_1\beta_2\beta_3}^{\circ\,\circ\,\circ} = g_{\alpha_1\beta_1}^{\circ\,\circ}g_{\alpha_2\beta_2}^{\circ\,\circ}g_{\alpha_3\beta_3}^{\circ\,\circ}t_{\circ\,\circ\,\circ}^{\alpha_1\alpha_2\alpha_3},$$

which in matrix form, appears as

$$T_{64,1}(\text{covariant coordinates}) = (G\otimes G\otimes G)\bullet T_{64,1}(\text{contravariant coordinates})$$

followed by the corresponding condensation.
Nevertheless, since in the present chapter we refer to the Euclidean exterior algebra, we execute the lowering of indices by means of Formula (13.25), using only and directly the strict components.
From the previous question the result is:

$$t_{\circ\circ\circ}^{(123)} = 2; \quad t_{\circ\circ\circ}^{(124)} = 1; \quad t_{\circ\circ\circ}^{(134)} = -1; \quad t_{\circ\circ\circ}^{(234)} = 1;$$

whence

$$t_{(123)}^{\circ\,\circ\,\circ} = 2\cdot\begin{bmatrix}1&1&1\\1&2&2\\1&2&3\end{bmatrix} + 1\cdot\begin{bmatrix}1&1&1\\1&2&2\\1&2&3\end{bmatrix} + (-1)\cdot\begin{bmatrix}1&1&1\\1&2&2\\1&3&3\end{bmatrix} + 1\cdot\begin{bmatrix}1&1&1\\2&2&2\\2&3&3\end{bmatrix} = 3$$

$$t_{(124)}^{\circ\,\circ\,\circ} = 2\cdot\begin{bmatrix}1&1&1\\1&2&2\\1&2&3\end{bmatrix} + 1\cdot\begin{bmatrix}1&1&1\\1&2&2\\1&2&4\end{bmatrix} + (-1)\cdot\begin{bmatrix}1&1&1\\1&2&2\\1&3&4\end{bmatrix} + 1\cdot\begin{bmatrix}1&1&1\\2&2&2\\2&3&4\end{bmatrix} = 3$$

$$t_{(134)}^{\circ\,\circ\,\circ} = 2\cdot\begin{bmatrix}1&1&1\\1&2&3\\1&2&3\end{bmatrix} + 1\cdot\begin{bmatrix}1&1&1\\1&2&3\\1&2&4\end{bmatrix} + (-1)\cdot\begin{bmatrix}1&1&1\\1&3&3\\1&3&4\end{bmatrix} + 1\cdot\begin{bmatrix}1&1&1\\2&3&3\\2&3&4\end{bmatrix} = 0$$

$$t_{(234)}^{\circ\,\circ\,\circ} = 2\cdot\begin{bmatrix}1&2&2\\1&2&3\\1&2&3\end{bmatrix} + 1\cdot\begin{bmatrix}1&2&2\\1&2&3\\1&2&4\end{bmatrix} + (-1)\cdot\begin{bmatrix}1&2&2\\1&3&3\\1&3&4\end{bmatrix} + 1\cdot\begin{bmatrix}2&2&2\\2&3&3\\2&3&4\end{bmatrix} = 1,$$

which permits a reconstruction of the anti-symmetric tensor:

$$[t^{\circ\circ\circ}_{\alpha\beta\gamma}] = \begin{bmatrix} t^{\circ\circ\circ}_{1\beta\gamma} \\ t^{\circ\circ\circ}_{2\beta\gamma} \\ t^{\circ\circ\circ}_{3\beta\gamma} \\ t^{\circ\circ\circ}_{4\beta\gamma} \end{bmatrix};$$

the blocks being

$$[t^{\circ\circ\circ}_{1\beta\gamma}] = \begin{bmatrix} 0 & 0 & 0 & 0 \\ 0 & 0 & 3 & 3 \\ 0 & -3 & 0 & 0 \\ 0 & -3 & 0 & 0 \end{bmatrix};$$

$$[t^{\circ\circ\circ}_{2\beta\gamma}] = \begin{bmatrix} 0 & 0 & -3 & -3 \\ 0 & 0 & 0 & 0 \\ 3 & 0 & 0 & 1 \\ 3 & 0 & -1 & 0 \end{bmatrix};$$

$$[t^{\circ\circ\circ}_{3\beta\gamma}] = \begin{bmatrix} 0 & 3 & 0 & 0 \\ -3 & 0 & 0 & -1 \\ 0 & 0 & 0 & 0 \\ 0 & 1 & 0 & 0 \end{bmatrix};$$

$$[t^{\circ\circ\circ}_{4\beta\gamma}] = \begin{bmatrix} 0 & 3 & 0 & 0 \\ -3 & 0 & 1 & 1 \\ 0 & -1 & 0 & 0 \\ 0 & 0 & 0 & 0 \end{bmatrix}.$$

3. Using (13.36), and taking into account that $\sqrt{|G|} = +1$, we have

$$s^{\circ}_{\lambda} = \frac{+1}{3!} \epsilon^{\circ\circ\circ\circ}_{\alpha\beta\gamma\lambda} t^{\alpha\beta\gamma}_{\circ\circ\circ} = \frac{1}{3!} \left(\epsilon^{\alpha\beta\gamma}_{\circ\circ\circ} \epsilon^{\circ\circ\circ}_{\alpha\beta\gamma} \right) \epsilon^{\circ\circ\circ\circ}_{(\alpha\beta\gamma)\lambda} t^{(\alpha\beta\gamma)}_{\circ\circ\circ} = \frac{1}{3!} 3! \epsilon^{\circ\circ\circ\circ}_{(\alpha\beta\gamma)\lambda} t^{(\alpha\beta\gamma)}_{\circ\circ\circ}$$
$$= \epsilon^{\circ\circ\circ\circ}_{(\alpha\beta\gamma)\lambda} t^{(\alpha\beta\gamma)}_{\circ\circ\circ};$$

and giving numerical values to the indices, it is

$$s^{\circ}_{1} = \epsilon^{\circ\circ\circ\circ}_{(234)1} t^{(234)}_{\circ\circ\circ} = (-1) \cdot 1 = -1;$$
$$s^{\circ}_{2} = \epsilon^{\circ\circ\circ\circ}_{(134)2} t^{(134)}_{\circ\circ\circ} = (+1) \cdot (-1) = -1;$$
$$s^{\circ}_{3} = \epsilon^{\circ\circ\circ\circ}_{(124)3} t^{(124)}_{\circ\circ\circ} = (-1) \cdot 1 = -1;$$
$$s^{\circ}_{4} = \epsilon^{\circ\circ\circ\circ}_{(123)4} t^{(123)}_{\circ\circ\circ} = (+1) \cdot 2 = 2;$$

$$S = s_1^{\circ}\vec{e}^{*1} + s_2^{\circ}\vec{e}^{*2} + s_3^{\circ}\vec{e}^{*3} + s_4^{\circ}\vec{e}^{*4} = -\vec{e}^{*1} - \vec{e}^{*2} - \vec{e}^{*3} + 2\vec{e}^{*4}$$

$$= [\,\vec{e}^{*1} \quad \vec{e}^{*2} \quad \vec{e}^{*3} \quad \vec{e}^{*4}\,] \begin{bmatrix} -1 \\ -1 \\ -1 \\ 2 \end{bmatrix}.$$

4. Similarly, using (13.37), and with $\dfrac{+\sqrt{|G|}}{|G|} = 1$, we have

$$s_{\circ}^{\lambda} = \frac{+1}{3!}\epsilon^{\alpha\beta\gamma\delta}{}_{\circ\circ\circ\circ}t^{\circ\circ\circ}_{\alpha\beta\gamma} = \epsilon^{(\alpha\beta\gamma)\delta}{}_{\circ\circ\circ\circ}t^{\circ\circ\circ}_{(\alpha\beta\gamma)};$$

$$s_{\circ}^{1} = \epsilon^{(234)1}{}_{\circ\circ\circ\circ}t^{\circ\circ\circ}_{(234)} = (-1)\cdot 1 = -1;$$

$$s_{\circ}^{2} = \epsilon^{(134)2}{}_{\circ\circ\circ\circ}t^{\circ\circ\circ}_{(134)} = (+1)\cdot 0 = 0;$$

$$s_{\circ}^{3} = \epsilon^{(124)3}{}_{\circ\circ\circ\circ}t^{\circ\circ\circ}_{(124)} = (-1)\cdot 3 = -3;$$

$$s_{\circ}^{4} = \epsilon^{(123)4}{}_{\circ\circ\circ\circ}t^{\circ\circ\circ}_{(123)} = (+1)\cdot 3 = 3;$$

$$S = s_{\circ}^{1}\vec{e}_1 + s_{\circ}^{2}\vec{e}_2 + s_{\circ}^{3}\vec{e}_3 + s_{\circ}^{4}\vec{e}_4 = -\vec{e}_1 - 3\vec{e}_3 + 3\vec{e}_4$$

$$= [\,\vec{e}_1 \quad \vec{e}_2 \quad \vec{e}_3 \quad \vec{e}_4\,] \begin{bmatrix} -1 \\ 0 \\ -3 \\ 3 \end{bmatrix}.$$

5. a) Applying Formula (13.17) for $r = (n-1)$, the result is

$$T = \begin{vmatrix} x^{\alpha_1}{}_{\circ\,1}{}^{\circ} & x^{\alpha_1}{}_{\circ\,2}{}^{\circ} & \cdots & x^{\alpha_1}{}_{\circ\,n-1}{}^{\circ} \\ x^{\alpha_2}{}_{\circ\,1}{}^{\circ} & x^{\alpha_2}{}_{\circ\,2}{}^{\circ} & \cdots & x^{\alpha_2}{}_{\circ\,n-1}{}^{\circ} \\ \cdots & \cdots & \cdots & \cdots \\ x^{\alpha_{n-1}}{}_{\circ\,1}{}^{\circ} & x^{\alpha_{n-1}}{}_{\circ\,2}{}^{\circ} & \cdots & x^{\alpha_{n-1}}{}_{\circ\,n-1}{}^{\circ} \end{vmatrix} \vec{e}_{\alpha_1}\wedge\vec{e}_{\alpha_2}\wedge\cdots\wedge\vec{e}_{\alpha_{n-1}};$$

$$\forall\alpha_i \in I_n;\ \alpha_1 < \alpha_2 < \cdots < \alpha_{n-1},$$

the $\binom{n}{r} = \binom{n}{n-1} = n$ exterior components of which are the strict components, that is, the previous equation can also be written as

$$T = t^{(\alpha_1\alpha_2\cdots\alpha_{n-1})}{}_{\circ\ \circ\ \cdots\ \circ}\vec{e}_{\alpha_1}\wedge\vec{e}_{\alpha_2}\wedge\cdots\wedge\vec{e}_{\alpha_{n-1}};\ \forall\alpha_i \in I_n;\ \alpha_1 < \alpha_2 < \cdots < \alpha_{n-1}.$$

b) Following a process parallel to that in question 3, but generalizing for $r = n-1$, we get

$$S = s_{\lambda}^{\circ}\vec{e}^{*\lambda} = \frac{+\sqrt{|G|}}{(n-1)!}\epsilon^{\circ\ \circ\ \cdots\ \circ}_{\alpha_1\alpha_2\cdots\alpha_{n-1}}t^{\circ\ \circ\ \cdots\ \circ}_{\circ\ \circ\ \cdots\ \circ}{}^{\alpha_1\alpha_2\cdots\alpha_{n-1}}\vec{e}^{*\lambda}$$

$$= \frac{+\sqrt{|G|}}{(n-1)!}(\epsilon^{\alpha_1\alpha_2\cdots\alpha_{n-1}}_{\circ\ \circ\ \cdots\ \circ}\epsilon^{\circ\ \circ\ \cdots\ \circ}_{\alpha_1\alpha_2\cdots\alpha_{n-1}})\epsilon_{(\alpha_1\alpha_2\cdots\alpha_{n-1})\lambda}t^{(\alpha_1\alpha_2\cdots\alpha_{n-1})}_{\circ\ \circ\ \cdots\ \circ}\vec{e}^{*\lambda}$$

$$= \frac{+\sqrt{|G|}}{(n-1)!}(n-1)!\, \epsilon\, {}_{(\alpha_1\alpha_2\cdots\alpha_{n-1})\lambda}^{\quad\circ\ \circ\ \cdots\ \ \circ\ \ \circ}\, t\, {}^{(\alpha_1\alpha_2\cdots\alpha_{n-1})}_{\circ\ \ \circ\ \cdots\ \ \circ}\, \vec{e}^{*\lambda}$$

$$= \sqrt{|G|}(-1)^{n-1}\epsilon\, {}_{\lambda(\alpha_1\alpha_2\cdots\alpha_{n-1})}^{\circ\ \circ\ \ \circ\ \cdots\ \ \circ}\, t\, {}^{(\alpha_1\alpha_2\cdots\alpha_{n-1})}_{\circ\ \ \circ\ \cdots\ \ \circ}\, \vec{e}^{*\lambda},$$

and giving first numerical values to the dummy index λ, followed by the proper values of the α_i (proceeding precisely in this ordered way) we obtain a determinant with the basis $\{\vec{e}^{*\alpha}\}$ of $E^n(\mathbb{R})$ in the first column, and developing by the polar of this column, we obtain

$$S = (-1)^{(n-1)}\sqrt{|G|}\begin{vmatrix} \vec{e}^{*1} & x^{1\circ}_{\circ 1} & x^{1\circ}_{\circ 2} & \cdots & x^{1\ \ \circ}_{\circ(n-1)} \\ \vec{e}^{*2} & x^{2\circ}_{\circ 1} & x^{2\circ}_{\circ 2} & \cdots & x^{2\ \ \circ}_{\circ(n-1)} \\ \cdots & \cdots & \cdots & \cdots & \\ \vec{e}^{*n} & x^{n\circ}_{\circ 1} & x^{n\circ}_{\circ 2} & \cdots & x^{n\ \ \circ}_{\circ(n-1)} \end{vmatrix}, \qquad (13.51)$$

an expression that is considered as the simplest answer to the stated question. By a similar development, we obtain for the tensor S the expression in contravariant coordinates, with data in covariant coordinates:

$$S = (-1)^{(n-1)}\frac{\sqrt{|G|}}{|G|}\begin{vmatrix} \vec{e}_1 & x^{\circ\circ}_{11} & x^{\circ\circ}_{12} & \cdots & x^{\circ\ \ \circ}_{1(n-1)} \\ \vec{e}_2 & x^{\circ\circ}_{21} & x^{\circ\circ}_{22} & \cdots & x^{\circ\ \ \circ}_{2(n-1)} \\ \cdots & \cdots & \cdots & \cdots & \\ \vec{e}_n & x^{\circ\circ}_{n1} & x^{\circ\circ}_{n2} & \cdots & x^{\circ\ \ \circ}_{n(n-1)} \end{vmatrix}. \qquad (13.52)$$

6. We have already mentioned that the number of strict components is $\binom{n}{n-1} = n$. We have also given its value in question 5:

$$t\, {}^{(\alpha_1\alpha_2\cdots\alpha_{n-1})}_{\circ\ \ \circ\ \cdots\ \ \circ} = \begin{vmatrix} x^{\alpha_1\circ}_{\circ 1} & x^{\alpha_1\circ}_{\circ 2} & \cdots & x^{\alpha_1\ \ \circ}_{\circ(n-1)} \\ x^{\alpha_2\circ}_{\circ 1} & x^{\alpha_2\circ}_{\circ 2} & \cdots & x^{\alpha_2\ \ \circ}_{\circ(n-1)} \\ \cdots & \cdots & \cdots & \cdots \\ x^{\alpha(n-1)\circ}_{\circ\quad 1} & x^{\alpha(n-1)\circ}_{\circ\quad 2} & \cdots & x^{\alpha(n-1)\ \ \circ}_{\circ\quad(n-1)} \end{vmatrix};$$

$$\alpha_i \in I_n,\ \alpha_1 < \alpha_2 < \cdots < \alpha_{n-1}. \quad (13.53)$$

7. In this case we have

$$S \bullet \vec{V}_n = (-1)^{n-1}\sqrt{|G|}\begin{vmatrix} \vec{e}^{*1} & x^{1\circ}_{\circ 1} & x^{1\circ}_{\circ 2} & \cdots & x^{1\ \ \circ}_{\circ(n-1)} \\ \vec{e}^{*2} & x^{2\circ}_{\circ 1} & x^{2\circ}_{\circ 2} & \cdots & x^{2\ \ \circ}_{\circ(n-1)} \\ \cdots & \cdots & \cdots & \cdots & \\ \vec{e}^{*n} & x^{n\circ}_{\circ 1} & x^{n\circ}_{\circ 2} & \cdots & x^{n\ \ \circ}_{\circ(n-1)} \end{vmatrix}$$

$$\bullet \left(x^{1\circ}_{\circ n}\vec{e}_1 + x^{2\circ}_{\circ n}\vec{e}_2 + \cdots + x^{n\circ}_{\circ n}\vec{e}_n \right),$$

and taking into account that $\vec{e}^{*\alpha}\bullet \vec{e}_\beta = \delta^\alpha_\beta$, after operating it becomes the determinant:

$$S \bullet \vec{V}_n = (-1)^{n-1}\sqrt{|G|} \begin{vmatrix} x^{1\,o}_{o\,n} & x^{1\,o}_{o\,1} & x^{1\,o}_{o\,2} & \cdots & x^{1\ \ o}_{o\,(n-1)} \\ x^{2\,o}_{o\,n} & x^{2\,o}_{o\,1} & x^{2\,o}_{o\,2} & \cdots & x^{2\ \ o}_{o\,(n-1)} \\ \cdots & \cdots & \cdots & \cdots & \cdots \\ x^{n\,o}_{o\,n} & x^{n\,o}_{o\,1} & x^{n\,o}_{o\,2} & \cdots & x^{n\ \ o}_{o\,(n-1)} \end{vmatrix}.$$

We proceed to pass the first column to the last position by performing the required $(n-1)$ transpositions with each of the remaining columns; after each transposition the determinant changes sign, so that at the end we have

$$S \bullet \vec{V}_n = (-1)^{n-1} \cdot (-1)^{n-1}\sqrt{|G|} \begin{vmatrix} x^{1\,o}_{o\,1} & x^{1\,o}_{o\,2} & \cdots & x^{1\,o}_{o\,n} \\ x^{2\,o}_{o\,1} & x^{2\,o}_{o\,2} & \cdots & x^{2\,o}_{o\,n} \\ \cdots & \cdots & \cdots & \cdots \\ x^{n\,o}_{o\,1} & x^{n\,o}_{o\,2} & \cdots & x^{n\,o}_{o\,n} \end{vmatrix},$$

which leads to

$$S \bullet \vec{V}_n \equiv \left[\vec{V}_1\ \vec{V}_2\ \cdots\ \vec{V}_{n-1}\ \vec{V}_n\right] = \sqrt{|G|} \begin{vmatrix} x^{1\,o}_{o\,1} & x^{1\,o}_{o\,2} & \cdots & x^{1\,o}_{o\,n} \\ x^{2\,o}_{o\,1} & x^{2\,o}_{o\,2} & \cdots & x^{2\,o}_{o\,n} \\ \cdots & \cdots & \cdots & \cdots \\ x^{n\,o}_{o\,1} & x^{n\,o}_{o\,2} & \cdots & x^{n\,o}_{o\,n} \end{vmatrix}, \quad (13.54)$$

where $\left[\vec{V}_1\ \vec{V}_2\ \cdots \vec{V}_{n-1}\ \vec{V}_n\right]$ is the mixed product of those n vectors.

8. It is evident that if we enter in the dot product $S \bullet \vec{V}_n$ with the vector S in contravariant coordinates and the vector $\vec{V}_n$ in covariant coordinates, the final result is

$$S \bullet \vec{V}_n \equiv \left[\vec{V}_1\vec{V}_2\cdots\vec{V}_{n-1}\vec{V}_n\right] = \frac{\sqrt{|G|}}{|G|} \begin{vmatrix} x^{o\,o}_{11} & x^{o\,o}_{12} & \cdots & x^{o\,o}_{1n} \\ x^{o\,o}_{21} & x^{o\,o}_{22} & \cdots & x^{o\,o}_{2n} \\ \cdots & \cdots & \cdots & \cdots \\ x^{o\,o}_{n1} & x^{o\,o}_{n2} & \cdots & x^{o\,o}_{nn} \end{vmatrix}. \quad (13.55)$$

If $n = 3$, in the (OGS) $S \bullet \vec{V}_n$ is the called "mixed product" of three vectors.

□

Example 13.6 (Fundamental exterior tensors). Consider a Euclidean space $E^n(\mathbb{R})$ referred to a basis $\{\vec{e}_\alpha\}$. The fundamental metric tensor is given by the Gram matrix G, in covariant coordinates:

$$G \equiv [g^{o\,o}_{\alpha\beta}].$$

1. Consider the exterior algebra $\bigwedge_n^{(2)}(\mathbb{R})$ of order $r = 2$.

a) Obtain the dimension and the exterior basis of $\bigwedge_n^{(2)}(\mathbb{R})$.

b) Give the generic components of the exterior fundamental metric tensor relative to $\bigwedge_n^{(2)}(\mathbb{R})$.

2. Answer the same questions but for the exterior algebra $\bigwedge_n^{(3)}(\mathbb{R})$.

Solution:

1. For the exterior algebra $\bigwedge_n^{(2)}(\mathbb{R})$.

a) An exterior tensor is given by its strict components with respect to the basis of $\bigwedge_n^{(2)}(\mathbb{R})$:

$$T = t_{\stackrel{(\alpha\beta)}{\circ\;\circ}}\vec{e}_\alpha \wedge \vec{e}_\beta,\ \alpha,\beta \in I_n;\ \alpha < \beta.$$

Thus, the basis of $\bigwedge_n^{(2)}(\mathbb{R})$ is the increasing *ordered* set

$$\{\vec{e}_\alpha \wedge \vec{e}_\beta\}_{\alpha<\beta} \equiv \{\vec{e}_1 \wedge \vec{e}_2, \dots, \vec{e}_{n-1} \wedge \vec{e}_n\}.$$

The dimension σ of $\bigwedge_n^{(2)}(\mathbb{R})$, is the number of vectors in the basis, that is, the number of binary combinations without repetition of the set I_n:

$$\sigma = \binom{n}{2} = \frac{n(n-1)}{2}. \tag{13.56}$$

b) We will use two different notations, to represent the product of the basic vectors of $\bigwedge_n^{(2)}(\mathbb{R})$.

The first notation will carry the expression $g'^{\circ\circ}_{ab}$ to represent the entities of the Gram matrix $G_{\bigwedge_n^{(2)}(\mathbb{R})}$ associated with the space $\bigwedge_n^{(2)}(\mathbb{R})$; a,b,c, etc. represent, with Roman letters, the positions occupied by the basic vectors $\vec{e}_\alpha \wedge \vec{e}_\beta$ in the ordered basis, according to the correspondence indicated in the table:

index a,b	1	2	$\cdots$	$n-1$	n	$n+1$	$\cdots$	$(\sigma-1)$	σ
index α,β	$\vec{e}_1 \wedge \vec{e}_2$	$\vec{e}_1 \wedge \vec{e}_3$	$\cdots$	$\vec{e}_1 \wedge \vec{e}_n$	$\vec{e}_2 \wedge \vec{e}_3$	$\vec{e}_2 \wedge \vec{e}_4$	$\cdots$	$\vec{e}_{n-2} \wedge \vec{e}_n$	$\vec{e}_{n-1} \wedge \vec{e}_n$

On the other hand, the second notation refers to the meaning of the $g'^{\circ\circ}_{ab}$:

$$g'^{\circ\circ}_{ab} = (\vec{e}_\alpha \wedge \vec{e}_\beta)_a \bullet (\vec{e}_\gamma \wedge \vec{e}_\delta)_b, \tag{13.57}$$

where the subindices a and b refer to the position of the Greek letters $\{\alpha,\beta\}, \{\gamma,\delta\}$.

Finally we indicate that the Euclidean spaces have the property indicated in Section 1.3.2, point 5, applicable to four arbitrary vectors of $E^n(\mathbb{R})$, i.e.,

$$(\vec{V}_1 \otimes \vec{V}_3)\bullet(\vec{V}_2\otimes\vec{V}_4) = (\vec{V}_1\bullet\vec{V}_2)\otimes(\vec{V}_3\bullet\vec{V}_4) = (\vec{V}_1\bullet\vec{V}_2)\cdot(\vec{V}_3\bullet\vec{V}_4). \tag{13.58}$$

To clarify the concepts, we give two illustrative models and we apply all the previous properties and notation rules. We remind the reader that $g^{\circ\circ}_{\alpha\beta} \in G$ means $g^{\circ\circ}_{\alpha\beta} = \vec{e}_\alpha \bullet \vec{e}_\beta$, in the metric tensor of $E^n(\mathbb{R})$.

First model. If $n = 5$ and $r = 2$, the scalar $g'^{\circ\circ}_{ab} \equiv g'^{\circ\circ}_{59}$ means the dot product of the fifth basic vector $(\vec{e}_2 \wedge \vec{e}_3)$ and the ninth $(\vec{e}_3 \wedge \vec{e}_5)$, extracted from its table, i.e.,

$$
\begin{aligned}
g'^{\circ\circ}_{59} &= (\vec{e}_2 \wedge \vec{e}_3) \bullet (\vec{e}_3 \wedge \vec{e}_5) \\
&= (\vec{e}_2 \otimes \vec{e}_3 - \vec{e}_3 \otimes \vec{e}_2) \bullet (\vec{e}_3 \otimes \vec{e}_5 - \vec{e}_5 \otimes \vec{e}_3) \\
&= (\vec{e}_2 \otimes \vec{e}_3) \bullet (\vec{e}_3 \otimes \vec{e}_5) - (\vec{e}_2 \otimes \vec{e}_3) \bullet (\vec{e}_5 \otimes \vec{e}_3) \\
&\quad - (\vec{e}_3 \otimes \vec{e}_2) \bullet (\vec{e}_3 \otimes \vec{e}_5) + (\vec{e}_3 \otimes \vec{e}_2) \bullet (\vec{e}_5 \otimes \vec{e}_3) \\
&= (\vec{e}_2 \bullet \vec{e}_3) \otimes (\vec{e}_3 \bullet \vec{e}_5) - (\vec{e}_2 \bullet \vec{e}_5) \otimes (\vec{e}_3 \bullet \vec{e}_3) \\
&\quad - (\vec{e}_3 \bullet \vec{e}_3) \otimes (\vec{e}_2 \bullet \vec{e}_5) + (\vec{e}_3 \bullet \vec{e}_5) \otimes (\vec{e}_2 \bullet \vec{e}_3) \\
&= g^{\circ\circ}_{23} \cdot g^{\circ\circ}_{35} - g^{\circ\circ}_{25} \cdot g^{\circ\circ}_{33} - g^{\circ\circ}_{33} \cdot g^{\circ\circ}_{25} + g^{\circ\circ}_{35} \cdot g^{\circ\circ}_{23} \\
&= 2\left(g^{\circ\circ}_{23}g^{\circ\circ}_{35} - g^{\circ\circ}_{25}g^{\circ\circ}_{33}\right) = 2 \begin{vmatrix} g^{\circ\circ}_{23} & g^{\circ\circ}_{25} \\ g^{\circ\circ}_{33} & g^{\circ\circ}_{35} \end{vmatrix}.
\end{aligned}
\tag{13.59}
$$

Second model. Let $n = 7$ and $r = 2$; the scalar $g'^{\circ\circ}_{ab} \equiv g'^{\circ\circ}_{513}$ means the dot product of the fifth basic vector $(\vec{e}_1 \wedge \vec{e}_6)$ and the thirteenth $(\vec{e}_3 \wedge \vec{e}_5)$, extracted from its table, so that developing as before, we arrive at

$$
g'^{\circ\circ}_{ab} \equiv g'^{\circ\circ}_{513} = (\vec{e}_1 \wedge \vec{e}_6) \bullet (\vec{e}_3 \wedge \vec{e}_5) = 2 \begin{vmatrix} g^{\circ\circ}_{13} & g^{\circ\circ}_{15} \\ g^{\circ\circ}_{63} & g^{\circ\circ}_{65} \end{vmatrix}.
$$

As a practical rule, we observe that the subindices ordered by pairs, first of each factor, second of each factor, etc., are the same as the ordered main diagonal subindices of the final minor; in the models $(2 - 3, 3 - 5)$ and $(1 - 3, 6 - 5)$, respectively.

This result can be generalized for $\bigwedge_n^{(2)}(\mathbb{R})$ by means of the formula

$$
g'^{\circ\circ}_{ab} = (\vec{e}_\alpha \wedge \vec{e}_\beta) \bullet (\vec{e}_\gamma \wedge \vec{e}_\delta)) = 2 \begin{vmatrix} g^{\circ\circ}_{\alpha\gamma} & g^{\circ\circ}_{\alpha\delta} \\ g^{\circ\circ}_{\beta\gamma} & g^{\circ\circ}_{\beta\delta} \end{vmatrix}; \quad \forall g^{\circ\circ}_{\lambda\mu} \in G; \ \alpha < \beta; \ \gamma < \delta,
\tag{13.60}
$$

where the factor 2 is the value of the order $r!$ of the exterior space $\bigwedge_n^{(2)}(\mathbb{R})$.

Formula (13.60) answers the question. The number of elements of $G_{\bigwedge_n^{(2)}(\mathbb{R})} \equiv [g'^{\circ\circ}_{ab}]$ is $\sigma^2 = \left[\binom{n}{2}\right]^2$, but due to the fact that it is a symmetric matrix, we need to calculate only $\binom{\sigma+1}{2} = \frac{\sigma(\sigma+1)}{2}$ terms $g'^{\circ\circ}_{ab}$.

2. For the exterior algebra $\bigwedge_n^{(3)}(\mathbb{R})$.

a) The basis of $\bigwedge_n^{(3)}(\mathbb{R})$ is the increasing ordered set:

$$\{\vec{e}_\alpha \wedge \vec{e}_\beta \wedge \vec{e}_\gamma\}_{\alpha<\beta<\gamma} \equiv \{\vec{e}_1 \wedge \vec{e}_2 \wedge \vec{e}_3, \ldots, \vec{e}_{n-2} \wedge \vec{e}_{n-1} \wedge \vec{e}_n\},$$

of dimension $\sigma = \binom{n}{3}$ basic vectors.

b) Once we have built the corresponding basic table for $\bigwedge_n^{(3)}(\mathbb{R})$, the generic elements of its fundamental metric tensor $G_{\bigwedge_n^{(3)}(\mathbb{R})} = [g'^{\circ\circ}_{ab}]$ are

$$g'^{\circ\circ}_{ab} = (\vec{e}_\alpha \wedge \vec{e}_\beta \wedge \vec{e}_\gamma) \bullet (\vec{e}_\epsilon \wedge \vec{e}_\delta \wedge \vec{e}_\eta) = 3! \begin{vmatrix} g^{\circ\circ}_{\alpha\epsilon} & g^{\circ\circ}_{\alpha\delta} & g^{\circ\circ}_{\alpha\eta} \\ g^{\circ\circ}_{\beta\epsilon} & g^{\circ\circ}_{\beta\delta} & g^{\circ\circ}_{\beta\eta} \\ g^{\circ\circ}_{\gamma\epsilon} & g^{\circ\circ}_{\gamma\delta} & g^{\circ\circ}_{\gamma\eta} \end{vmatrix};$$

$$\alpha < \beta < \gamma; \ \epsilon < \delta < \eta, \qquad (13.61)$$

that has been obtained using the property of Section 1.3.2, point 7:

$$(\vec{V}_1 \otimes \vec{V}_3 \otimes \vec{V}_5) \bullet (\vec{V}_2 \otimes \vec{V}_4 \otimes \vec{V}_6) = (\vec{V}_1 \bullet \vec{V}_2) \cdot (\vec{V}_3 \bullet \vec{V}_4) \cdot (\vec{V}_5 \bullet \vec{V}_6).$$

The number of elements of the matrix $G_{\bigwedge_n^{(3)}(\mathbb{R})}$ is $\sigma^2 \equiv \left[\binom{n}{3}\right]^2$, but we only calculate $\binom{\sigma+1}{2}$ because it is a symmetric matrix.

□

Example 13.7 (Use of exterior metric tensors). Consider a Euclidean space $E^2(\mathbb{R})$ referred to a basis $\{\vec{e}_\alpha\}$. The fundamental metric tensor is given by the Gram matrix G, in covariant coordinates:

$$G \equiv [g^{\circ\circ}_{\alpha\beta}] = \begin{bmatrix} 1 & 2 \\ 2 & 5 \end{bmatrix}.$$

A tensor $T \in \left(\overset{2}{\underset{1}{\otimes}} E^2\right)(\mathbb{R})$ is represented in contra-covariant coordinates by the matrix

$$[t^{\alpha\circ}_{\circ\beta}] = \begin{bmatrix} 2 & 5 \\ -1 & -2 \end{bmatrix}.$$

1. Obtain the strict contravariant coordinates of T.
2. Obtain the strict covariant coordinates of T.

We perform a change-of-basis in $E^2(\mathbb{R})$, of matrix

$$C \equiv [c^{\alpha\circ}_{\circ i}] = \begin{bmatrix} 1 & 3 \\ -1 & -1 \end{bmatrix}.$$

3. Obtain the new strict contravariant coordinates of T.
4. Obtain the new strict covariant coordinates of T.

5. Calculate the modulus of T as exterior contravariant vector in the initial basis.
6. Calculate the modulus of T as exterior covariant vector in the new basis.

Solution:

1. We raise the second index of the Euclidean tensor:

$$t_{\circ\circ}^{\alpha\beta} = t_{\circ\;\theta}^{\alpha\circ}g_{\circ\circ}^{\theta\beta}; \quad G^{-1} = \begin{bmatrix} 5 & -2 \\ -2 & 1 \end{bmatrix};$$

$$[t_{\circ\circ}^{\alpha\beta}] = \begin{bmatrix} 2 & 5 \\ -1 & -2 \end{bmatrix}\begin{bmatrix} 5 & -2 \\ -2 & 1 \end{bmatrix} = \begin{bmatrix} 0 & 1 \\ -1 & 0 \end{bmatrix}.$$

The result is anti-symmetric, with $t_{\circ\;\circ}^{(12)} = 1$, whence $T = t_{\circ\;\circ}^{(\alpha\beta)}\vec{e}_\alpha \wedge \vec{e}_\beta = \vec{e}_1 \wedge \vec{e}_2$.

2. We lower the first index of the tensor

$$t_{\alpha\beta}^{\circ\circ} = g_{\alpha\theta}^{\circ\circ}t_{\circ\beta}^{\theta\circ}; \quad [t_{\alpha\beta}^{\circ\circ}] = \begin{bmatrix} 1 & 2 \\ 2 & 5 \end{bmatrix}\begin{bmatrix} 2 & 5 \\ -1 & -2 \end{bmatrix} = \begin{bmatrix} 0 & 1 \\ -1 & 0 \end{bmatrix},$$

which again shows its anti-symmetry with $t_{(12)}^{\circ\;\circ} = 1$; $T = \vec{e}^{*1} \wedge \vec{e}^{*2}$.

3.

$$C \equiv [c_{\circ i}^{\alpha\circ}] = \begin{bmatrix} 1 & 3 \\ -1 & -1 \end{bmatrix}; \quad C^{-1} \equiv [\gamma_{\circ\alpha}^{i\circ}] = \frac{1}{2}\begin{bmatrix} -1 & -3 \\ 1 & 1 \end{bmatrix}.$$

Applying the formula of the change-of-basis (9.73), we have

$$\hat{t}_{\circ\;\circ}^{(ij)} = t_{\circ\;\circ}^{\alpha\beta)}\begin{vmatrix} \gamma_{\circ\alpha}^{i\circ} & \gamma_{\circ\alpha}^{i\circ} \\ \gamma_{\circ\alpha}^{j\circ} & \gamma_{\circ\alpha}^{j\circ} \end{vmatrix}; \quad \hat{t}_{\circ\;\circ}^{(12)} = t_{\circ\;\circ}^{(12))}\begin{vmatrix} \gamma_{\circ1}^{1\circ} & \gamma_{\circ2}^{1\circ} \\ \gamma_{\circ1}^{2\circ} & \gamma_{\circ2}^{2\circ} \end{vmatrix} = 1 \cdot \begin{vmatrix} -1/2 & -3/2 \\ 1/2 & 1/2 \end{vmatrix} = \frac{1}{2}.$$

4. Applying Formula (9.86), we get

$$\hat{t}_{(12)}^{\circ\;\circ} = t_{(12))}^{\circ\;\circ}|C| = 1 \cdot 2 = 2.$$

5.

$$\begin{aligned}
|T|^2 &= (\vec{e}_1 \wedge \vec{e}_2) \bullet (\vec{e}_1 \wedge \vec{e}_2) = (\vec{e}_1 \otimes \vec{e}_2 - \vec{e}_2 \otimes \vec{e}_1) \bullet (\vec{e}_1 \otimes \vec{e}_2 - \vec{e}_2 \otimes \vec{e}_1) \\
&= (\vec{e}_1 \otimes \vec{e}_2) \bullet (\vec{e}_1 \otimes \vec{e}_2) - (\vec{e}_1 \otimes \vec{e}_2) \bullet (\vec{e}_2 \otimes \vec{e}_1) - (\vec{e}_2 \otimes \vec{e}_1) \bullet (\vec{e}_1 \otimes \vec{e}_2) \\
&\quad + (\vec{e}_2 \otimes \vec{e}_1) \bullet (\vec{e}_2 \otimes \vec{e}_1) \\
&= 2(\vec{e}_1 \bullet \vec{e}_1) \otimes (\vec{e}_2 \bullet \vec{e}_2) - 2(\vec{e}_1 \bullet \vec{e}_2) \otimes (\vec{e}_2 \bullet \vec{e}_1) \\
&= 2 \cdot g_{11}^{\circ\circ}g_{22}^{\circ\circ} - 2 \cdot g_{12}^{\circ\circ}g_{21}^{\circ\circ} = 2|G| = 2 \to |T| = \sqrt{2},
\end{aligned}$$

which reveals that the exterior Gram matrix $G_{\bigwedge_2^{(2)}}$ (for contravaraint vectors) is *the scalar* $2|G|$; so, it is licit to solve it as $|T|^2 = t_{\circ\;\circ}^{(\alpha\beta)}2|G|t_{\circ\;\circ}^{(\alpha\beta)} = 1 \cdot 2(1) \cdot 1 = 2 \to |T| = \sqrt{2}$, that is, applying the general matrix relation $|\vec{V}|^2 = X^t G X$.

6.

$$|T|^2 = t^{\circ\ \ \circ}_{(\alpha\beta)}2|G^{-1}|t^{\circ\ \ \circ}_{(\alpha\beta)} = 1 \cdot 2(1) \cdot 1 = 2; \quad |T| = \sqrt{2}$$

(it is calculated in the initial basis because the modulus is an invariant). We ask the reader to solve this in the new basis, determining the new fundamental metric tensor $\hat{G}^*_{\underset{\wedge_2}{(2)}}$ of $E^2(\mathbb{R})$, using $\hat{t}^{\circ\ \circ}_{(12)}$.

□

Example 13.8 (Change of basis in Euclidean exterior tensors). In a Euclidean space $E^3(\mathbb{R})$ referred to the basis $\{\vec{e}_\alpha\}$, the Gram matrix of the fundamental metric tensor in covariant coordinates is

$$G \equiv [g^{\circ\circ}_{\alpha\beta}] = \begin{bmatrix} 1 & 1 & 0 \\ 1 & 2 & 0 \\ 0 & 0 & 1 \end{bmatrix}.$$

A Euclidean tensor $T \in (E_3 \otimes E^{*3} \otimes E_3)(\mathbb{R})$ is defined by its mixed components referred to the basis $\{\vec{e}_\alpha\}$, according to the block representation matrix

$$[t^{\circ\ \beta\circ}_{\alpha\circ\gamma}] = \begin{bmatrix} 0 & 0 & p \\ 0 & 0 & -p \\ 0 & p & 0 \\ -- & -- & -- \\ 0 & 0 & 2p \\ 0 & 0 & -p \\ -p & 0 & 0 \\ -- & -- & -- \\ -p & -2p & 0 \\ p & p & 0 \\ 0 & 0 & 0 \end{bmatrix},$$

with α the block row, β the row in each block and γ the column in each block.

1. Give the strict contravariant coordinates of T as exterior tensor, if that is possible.
2. Give the strict covariant coordinates of T as exterior tensor, if that is possible.
3. We define a system S of scalar components, by means of the tensor T contracted with the Levi-Civita system, of the following mode, in an arbitrary basis of $E^3(\mathbb{R})$:

$$s(\alpha\beta\gamma) = |G|t^{123}_{\circ\circ\circ}t^{\alpha\beta\gamma}_{\circ\circ\circ}\epsilon^{\phi\theta w}_{\circ\circ\circ}t^{\circ\circ\circ}_{\phi\theta w}.$$

Discuss the tensor nature of S.

4. Assuming that we perform a change-of-basis in $E^3(\mathbb{R})$, of matrix $C = \begin{bmatrix} 1 & 0 & 0 \\ 1 & 2 & 0 \\ 1 & 2 & 3 \end{bmatrix}$, we wish to know the contravariant coordinates of S in the initial and new bases, by means of the simplest possible expressions.

Solution:

1. We lower the index β of the tensor, with the aim of determining its covariant coordinates:

$$t^{\circ\circ\circ}_{\alpha\beta\gamma} = g^{\circ\circ}_{\beta\theta}t^{\circ\theta\circ}_{\alpha\circ\gamma},$$

so that proceeding with the blocks, we get

$$[t^{\circ\circ\circ}_{1\beta\gamma}] = g^{\circ\circ}_{\beta\theta}t^{\circ\theta\circ}_{1\circ\gamma} = \begin{bmatrix} 1 & 1 & 0 \\ 1 & 2 & 0 \\ 0 & 0 & 1 \end{bmatrix} \begin{bmatrix} 0 & 0 & p \\ 0 & 0 & -p \\ 0 & p & 0 \end{bmatrix} = \begin{bmatrix} 0 & 0 & 0 \\ 0 & 0 & -p \\ 0 & p & 0 \end{bmatrix}$$

$$[t^{\circ\circ\circ}_{2\beta\gamma}] = g^{\circ\circ}_{\beta\theta}t^{\circ\theta\circ}_{2\circ\gamma} = \begin{bmatrix} 1 & 1 & 0 \\ 1 & 2 & 0 \\ 0 & 0 & 1 \end{bmatrix} \begin{bmatrix} 0 & 0 & 2p \\ 0 & 0 & -p \\ -p & 0 & 0 \end{bmatrix} = \begin{bmatrix} 0 & 0 & p \\ 0 & 0 & 0 \\ -p & 0 & 0 \end{bmatrix}$$

$$[t^{\circ\circ\circ}_{3\beta\gamma}] = g^{\circ\circ}_{\beta\theta}t^{\circ\theta\circ}_{3\circ\gamma} = \begin{bmatrix} 1 & 1 & 0 \\ 1 & 2 & 0 \\ 0 & 0 & 1 \end{bmatrix} \begin{bmatrix} -p & -2p & 0 \\ p & p & 0 \\ 0 & 0 & 0 \end{bmatrix} = \begin{bmatrix} 0 & -p & 0 \\ p & 0 & 0 \\ 0 & 0 & 0 \end{bmatrix}.$$

In summary:

$$[t^{\circ\circ\circ}_{\alpha\beta\gamma}] = \left[\begin{array}{ccc} 0 & 0 & 0 \\ 0 & 0 & -p \\ 0 & p & 0 \\ \hline 0 & 0 & p \\ 0 & 0 & 0 \\ -p & 0 & 0 \\ \hline 0 & -p & 0 \\ p & 0 & 0 \\ 0 & 0 & 0 \end{array} \right]$$

with $t^{\circ\circ\circ}_{123} = -p$ and $t^{\circ\circ\circ}_{123} = -t^{\circ\circ\circ}_{132} = -t^{\circ\circ\circ}_{213} = t^{\circ\circ\circ}_{231} = t^{\circ\circ\circ}_{312} = -t^{\circ\circ\circ}_{321}$, that is, the tensor is totally anti-symmetric, of strict component $t^{\circ\circ\circ}_{(123)} = -p$.

In addition, $|G| = \begin{vmatrix} 1 & 1 & 0 \\ 1 & 2 & 0 \\ 0 & 0 & 1 \end{vmatrix} = 1$ and $|G^{-1}| = 1$.

Using Formula (13.32) we calculate the strict covariant component

$$t^{(123)}_{\circ\circ\circ} = t^{\circ\circ\circ}_{(123)}|G^{-1}| = (-p)\cdot 1 = -p,$$

and then, in fact, it is possible to express T as a exterior tensor

$$T = (-p)\vec{e}_1 \wedge \vec{e}_2 \wedge \vec{e}_3; \; T \in \bigwedge_3^{(3)}(\mathbb{R}).$$

2. We have already indicated that $t^{\circ\,\circ\,\circ}_{(123)} = -p$, and

$$T = (-p)\bar{e}^{*1} \wedge \bar{e}^{*2} \wedge \bar{e}^{*3}.$$

3. We will establish the tensor relations of the factors of the system S, in the basis $\{\bar{e}_\alpha\}$ and in another basis $\{\bar{e}_i\}$; denoting by C the change-of-basis matrix, we obtain the result. The factor $|G|$ is a determinant of a covariant tensor, because $|G| = |g^{\circ\,\circ}_{\alpha\beta}|$, so that according to point 2 of Example 8.1, we have $|\hat{G}| = |G||C|^2$.

The factor $t^{123}_{\circ\circ\circ} = t^{(123)}_{\circ\circ\circ}$ is the strict component of $T \in \bigwedge^{(3)}_3(\mathbb{R})$; from (9.84) we have

$$\hat{t}^{123}_{\circ\circ\circ} = |C|^{-1}t^{(123)}_{\circ\circ\circ} = |C|^{-1}t^{123}_{\circ\circ\circ}; \quad \hat{t}^{\circ\,\circ\,\circ}_{(123)} = |C|t^{\circ\circ\circ}_{123}.$$

The factor $t^{\alpha\beta\gamma}_{\circ\circ\circ}$ is a homogeneous contravariant tensor of order $r = 3$, and then its tensor relation is

$$\hat{t}^{ijk}_{\circ\circ\circ} = t^{\alpha\beta\gamma}_{\circ\circ\circ}\gamma^{i\,\circ}_{\circ\alpha}\gamma^{j\,\circ}_{\circ\beta}\gamma^{k\,\circ}_{\circ\gamma}.$$

With respect to the system $\epsilon^{\phi\theta w}_{\circ\circ\circ}$ as a modular contravariant tensor we have, according to (8.71):

$$\epsilon^{fzs}_{\circ\circ\circ} = |C|\epsilon^{\lambda\mu\nu}_{\circ\circ\circ}\gamma^{f\,\circ}_{\circ\lambda}\gamma^{z\,\circ}_{\circ\mu}\gamma^{s\,\circ}_{\circ\nu}.$$

Finally, the factor $t^{\circ\circ\circ}_{\phi\theta w}$, as a homogeneous covariant tensor is

$$\hat{t}^{\circ\circ\circ}_{fzs} = t^{\circ\circ\circ}_{\phi\theta w}c^{\circ\phi}_{f\circ}c^{\circ\theta}_{z\circ}c^{\circ w}_{s\circ}.$$

Multiplying all factors in the new basis, we get

$$
\begin{aligned}
s(ijk) &= |\hat{G}|\hat{t}^{123}_{\circ\circ\circ}\hat{t}^{ijk}_{\circ\circ\circ}\epsilon^{fzs}_{\circ\circ\circ}\hat{t}^{\circ\circ\circ}_{fzs}\\
&= |G||C|^2 \cdot |C|^{-1}t^{123}_{\circ\circ\circ} \cdot t^{\alpha\beta\gamma}_{\circ\circ\circ}\gamma^{i\,\circ}_{\circ\alpha}\gamma^{j\,\circ}_{\circ\beta}\gamma^{k\,\circ}_{\circ\gamma} \cdot |C|\\
&\quad \cdot \epsilon^{\lambda\mu\nu}_{\circ\circ\circ}\gamma^{f\,\circ}_{\circ\lambda}\gamma^{z\,\circ}_{\circ\mu}\gamma^{s\,\circ}_{\circ\nu}t^{\circ\circ\circ}_{\phi\theta w}c^{\circ\phi}_{f\circ}c^{\circ\theta}_{z\circ}c^{\circ w}_{s\circ}\\
&= |C|^2|G|t^{123}_{\circ\circ\circ} \cdot t^{\alpha\beta\gamma}_{\circ\circ\circ}\left(\delta^{\phi\circ}_{\circ\lambda}\delta^{\theta\circ}_{\circ\mu}\delta^{w\circ}_{\circ\nu} \cdot \epsilon^{\lambda\mu\nu}_{\circ\circ\circ}\right)t^{\circ\circ\circ}_{\phi\theta w} \cdot \gamma^{i\,\circ}_{\circ\alpha}\gamma^{j\,\circ}_{\circ\beta}\gamma^{k\,\circ}_{\circ\gamma}\\
&= |C|^2|G|t^{123}_{\circ\circ\circ} \cdot t^{\alpha\beta\gamma}_{\circ\circ\circ}\epsilon^{\phi\theta w}_{\circ\circ\circ}t^{\circ\circ\circ}_{\phi\theta w}\gamma^{i\,\circ}_{\circ\alpha}\gamma^{j\,\circ}_{\circ\beta}\gamma^{k\,\circ}_{\circ\gamma},
\end{aligned}
$$

and finally

$$s(ijk) = |C|^2 s(\alpha\beta\gamma)\gamma^{i\,\circ}_{\circ\alpha}\gamma^{j\,\circ}_{\circ\beta}\gamma^{k\,\circ}_{\circ\gamma},$$

which reveals that the system S is a *modular* contravariant tensor of third order, and weight 2, that is,

$$s^{ijk}_{\circ\circ\circ} = |C|^2 s^{\alpha\beta\gamma}_{\circ\circ\circ}\gamma^{i\,\circ}_{\circ\alpha}\gamma^{j\,\circ}_{\circ\beta}\gamma^{k\,\circ}_{\circ\gamma}.$$

4. The contravariant coordinates of S are:
   a) In the initial basis $\{\vec{e}_\alpha\}$:
   From the data, we extract

$$s^{\alpha\beta\gamma}_{\circ\circ\circ} = |G| t^{123\,\alpha\beta\gamma}_{\circ\circ\circ\,\circ\circ\circ} \epsilon^{\phi\theta w}_{\circ\circ\circ} \left(\delta^{(123)}_{\phi\theta w} t^{\circ\circ\circ}_{(123)} \right)$$

$$= 1 \cdot (-p) t^{\alpha\beta\gamma}_{\circ\circ\circ} 3! \cdot (-p) = 3! p^2 t^{(123)}_{\circ\circ\circ} \epsilon^{\alpha\beta\gamma}_{\circ\circ\circ} = -6p^3 \epsilon^{\alpha\beta\gamma}_{\circ\circ\circ}.$$

   b) In the new basis $\{\vec{e}_i\}$ and $|C| = \begin{vmatrix} 1 & 0 & 0 \\ 1 & 2 & 0 \\ 1 & 2 & 3 \end{vmatrix} = 6$:

$$s^{ijk}_{\circ\circ\circ} = |\hat{G}| \hat{t}^{123\,ijk}_{\circ\circ\circ\,\circ\circ\circ} \epsilon^{fzs}_{\circ\circ\circ} \left(\delta^{(123)}_{fzs} \hat{t}^{\circ\circ\circ}_{(123)} \right)$$

$$= |G|\, |C|^2 \cdot |C|^{-1} (-p)\, 3! \left(|C| \cdot (-p) \right) \hat{t}^{ijk}_{\circ\circ\circ}$$

$$= 1 \cdot p^2 \cdot 3! \epsilon^{ijk}_{\circ\circ\circ} \hat{t}^{(123)}_{\circ\circ\circ} = 6\, p^2\, |C|^2 \left(\hat{t}^{(123)}_{\circ\circ\circ} \epsilon^{ijk}_{\circ\circ\circ} \right).$$

Since $|C| = 6$ and $\hat{t}^{(123)}_{\circ\circ\circ} = |C|^{-1} t^{(123)}_{\circ\circ\circ} = \frac{(-p)}{|C|}$, substituting them in the previous expression, we get

$$s^{ijk}_{\circ\circ\circ} = 6\, p^2\, |C|^2 \frac{(-p)}{|C|} \epsilon^{ijk}_{\circ\circ\circ} = -6p^3 |C| \epsilon^{ijk}_{\circ\circ\circ} = -36p^3 \epsilon^{ijk}_{\circ\circ\circ}.$$

We can also answer this question (b) starting from the tensor equation established in the previous question and from the result of 4(a):

$$s^{ijk}_{\circ\circ\circ} = |C|^2 s^{\alpha\beta\gamma}_{\circ\circ\circ} \gamma^{i\circ}_{\circ\alpha} \gamma^{j\circ}_{\circ\beta} \gamma^{k\circ}_{\circ\gamma} = |C|^2 \left(-6p^3 \epsilon^{\alpha\beta\gamma}_{\circ\circ\circ} \right) \gamma^{i\circ}_{\circ\alpha} \gamma^{j\circ}_{\circ\beta} \gamma^{k\circ}_{\circ\gamma}$$

$$= -6p^3 |C| \left(|C| \epsilon^{\alpha\beta\gamma}_{\circ\circ\circ} \gamma^{i\circ}_{\circ\alpha} \gamma^{j\circ}_{\circ\beta} \gamma^{k\circ}_{\circ\gamma} \right) = -6p^3 |C| \epsilon^{ijk}_{\circ\circ\circ} = -36p^3 \epsilon^{ijk}_{\circ\circ\circ}.$$

$\square$

Example 13.9 (Exterior tensor contractions). In the geometric space of vectors $E^3(\mathbb{R})$, with oblique reference frame (basis $\{\vec{e}_\alpha\}$) and fundamental metric tensor

$$G \equiv [g^{\circ\circ}_{\alpha\beta}] = \begin{bmatrix} 1 & 1 & 0 \\ 1 & 5 & 0 \\ 0 & 0 & 1 \end{bmatrix}$$

two Euclidean anti-symmetric tensors T and W, of order $r = n = 3$, are considered.
   The exterior representation of T is

$$T = 2\vec{e}_1 \wedge \vec{e}_2 \wedge \vec{e}_3$$

and its total contraction or inner product with W is $W \bullet T = 72$.

1. Obtain the contravariant exterior expression of W.
2. Obtain the tensor P, the Euclidean contraction of the last two indices of the tensor W, in contravariant coordinates.
3. Obtain in contravariant coordinates the tensor Q, the contracted tensor product of the W by the vector $\vec{V}(1, 1, 1)$ given in covariant coordinates. The contraction is with respect to the third index of W.
4. Obtain the contravariant and covariant coordinates of the polar tensor of Q.
5. We perform a change-of-basis in $E^3(\mathbb{R})$, to the basis $\{\vec{e}_i\}$ with change matrix

$$C = \begin{bmatrix} 1 & 0 & 0 \\ 1 & 2 & 0 \\ 0 & 1 & 2 \end{bmatrix}.$$

Obtain the contravariant and covariant coordinates of the tensors T and W in the new basis.

Solution:

1. Let the strict component of tensor W be $w_{(123)}^{\circ\circ\circ} = h$ and $t_{\circ\circ\circ}^{(123)} = 2$.

$$W \bullet T = w_{\alpha\beta\gamma}^{\circ\circ\circ} t_{\circ\circ\circ}^{\alpha\beta\gamma}$$

$$= w_{123}^{\circ\circ\circ} t_{\circ\circ\circ}^{123} + w_{132}^{\circ\circ\circ} t_{\circ\circ\circ}^{132} + w_{213}^{\circ\circ\circ} t_{\circ\circ\circ}^{213}$$

$$+ w_{231}^{\circ\circ\circ} t_{\circ\circ\circ}^{231} + w_{312}^{\circ\circ\circ} t_{\circ\circ\circ}^{312} + w_{321}^{\circ\circ\circ} t_{\circ\circ\circ}^{321}.$$

Since both tensors are totally anti-symmetric, the sign changes of their components occur simultaneously, so that all summands are positive:

$$W \bullet T = h \cdot 2 + h \cdot 2 + \cdots + h \cdot 2 = 12h; \quad 12h = 72 \rightarrow h = 6; \quad w_{(123)}^{\circ\circ\circ} = 6.$$

According to Formula (13.32), with $|G| = \begin{vmatrix} 1 & 1 & 0 \\ 1 & 5 & 0 \\ 0 & 0 & 1 \end{vmatrix} = 4$, we have

$$w_{\circ\circ\circ}^{(123)} = w_{(123)}^{\circ\circ\circ} |G|^{-1} = \frac{w_{(123)}^{\circ\circ\circ}}{|G|} = \frac{6}{4} = \frac{3}{2}$$

$$W = \frac{3}{2} \vec{e}_1 \wedge \vec{e}_2 \wedge \vec{e}_3.$$

2. To perform contractions we represent the tensors as homogeneous anti-symmetric tensors, which leads to the following Euclidean contraction (of indices of the same species):

$$p_{\circ}^{\alpha} = w_{\circ\circ\circ}^{\alpha\theta\phi} g_{\theta\phi}^{\circ\circ},$$

where

$$[w^{\alpha\beta\gamma}_{\circ\circ\circ}] = \left[\begin{array}{ccc} 0 & 0 & 0 \\ 0 & 0 & 3/2 \\ 0 & -3/2 & 0 \\ \hline 0 & 0 & -3/2 \\ 0 & 0 & 0 \\ 3/2 & 0 & 0 \\ \hline 0 & 3/2 & 0 \\ -3/2 & 0 & 0 \\ 0 & 0 & 0 \end{array}\right] \; ; \; [g^{\circ\circ}_{\beta\gamma}] = \left[\begin{array}{ccc} 1 & 1 & 0 \\ 1 & 5 & 0 \\ 0 & 0 & 1 \end{array}\right].$$

Numerically, we obtain

$$p^1_{\circ} = w^{123}_{\circ\circ\circ}g^{\circ\circ}_{23} + w^{132}_{\circ\circ\circ}g^{\circ\circ}_{32} = 0$$

$$p^2_{\circ} = w^{213}_{\circ\circ\circ}g^{\circ\circ}_{13} + w^{231}_{\circ\circ\circ}g^{\circ\circ}_{31} = 0$$

$$p^3_{\circ} = w^{312}_{\circ\circ\circ}g^{\circ\circ}_{12} + w^{321}_{\circ\circ\circ}g^{\circ\circ}_{21} = 0.$$

$P \equiv \Omega \equiv 0\vec{e}_1 + 0\vec{e}_2 + 0\vec{e}_3$, which is the zero vector of $E^3(\mathbb{R})$.

3.

$$q^{\alpha\beta}_{\circ\circ} = w^{\alpha\beta\theta}_{\circ\circ\circ}v^{\circ}_{\theta};$$

$$[q^{1\beta}_{\circ\circ}] = [w^{1\beta\theta}_{\circ\circ\circ}][v^{\circ}_{\theta}] = \left[\begin{array}{ccc} 0 & 0 & 0 \\ 0 & 0 & 3/2 \\ 0 & -3/2 & 0 \end{array}\right]\left[\begin{array}{c} 1 \\ 1 \\ 1 \end{array}\right] = \left[\begin{array}{c} 0 \\ 3/2 \\ -3/2 \end{array}\right];$$

$$[q^{2\beta}_{\circ\circ}] = [w^{2\beta\theta}_{\circ\circ\circ}][v^{\circ}_{\theta}] = \left[\begin{array}{ccc} 0 & 0 & -3/2 \\ 0 & 0 & 0 \\ 3/2 & 0 & 0 \end{array}\right]\left[\begin{array}{c} 1 \\ 1 \\ 1 \end{array}\right] = \left[\begin{array}{c} -3/2 \\ 0 \\ 3/2 \end{array}\right];$$

$$[q^{3\beta}_{\circ\circ}] = [w^{3\beta\theta}_{\circ\circ\circ}][v^{\circ}_{\theta}] = \left[\begin{array}{ccc} 0 & 3/2 & 0 \\ -3/2 & 0 & 0 \\ 0 & 0 & 0 \end{array}\right]\left[\begin{array}{c} 1 \\ 1 \\ 1 \end{array}\right] = \left[\begin{array}{c} 3/2 \\ -3/2 \\ 0 \end{array}\right].$$

Thus, the results must be stacked in a unique matrix, as *rows*:

$$\left[q^{\alpha\beta}_{\circ\circ}\right] = \left[\begin{array}{ccc} 0 & 3/2 & -3/2 \\ -3/2 & 0 & 3/2 \\ 3/2 & -3/2 & 0 \end{array}\right],$$

and then the result Q is an anti-symmetric tensor of second order ($r = 2$).

4. Let A be the dual or polar of Q:

$$a^{\circ}_{\gamma} = \frac{1}{2!} \theta^{\circ\circ\circ}_{\alpha\beta\gamma} q^{\alpha\beta}_{\circ\circ} = \frac{1}{2!} \sqrt{|G|} \epsilon^{\circ\circ\circ}_{\alpha\beta\gamma} q^{\alpha\beta}_{\circ\circ} = \frac{1}{2} \sqrt{4} \epsilon^{\circ\circ\circ}_{\alpha\beta\gamma} q^{\alpha\beta}_{\circ\circ} = \epsilon^{\circ\circ\circ}_{\alpha\beta\gamma} q^{\alpha\beta}_{\circ\circ};$$

giving numerical values to the indices we get

$$a^{\circ}_1 = \epsilon^{\circ\circ\circ}_{231} q^{23}_{\circ\circ} + \epsilon^{\circ\circ\circ}_{321} q^{32}_{\circ\circ} = 2(1)\left(\frac{3}{2}\right) = 3$$

$$a^{\circ}_2 = \epsilon^{\circ\circ\circ}_{132} q^{13}_{\circ\circ} + \epsilon^{\circ\circ\circ}_{312} q^{31}_{\circ\circ} = 2(-1)\left(-\frac{3}{2}\right) = 3$$

$$a^{\circ}_3 = \epsilon^{\circ\circ\circ}_{123} q^{12}_{\circ\circ} + \epsilon^{\circ\circ\circ}_{213} q^{21}_{\circ\circ} = 2(1)\left(\frac{3}{2}\right) = 3,$$

which shows that A is a vector in covariant coordinates $\vec{A} = 3\vec{e}^{*1} + 3\vec{e}^{*2} + 3\vec{e}^{*3}$, and passing to contravariant coordinates we finally get

$$\begin{bmatrix} a^1 \\ a^2 \\ a^3 \end{bmatrix} = G^{-1} \begin{bmatrix} a_1 \\ a_2 \\ a_3 \end{bmatrix} = \frac{1}{4} \begin{bmatrix} 5 & -1 & 0 \\ -1 & 1 & 0 \\ 0 & 0 & 4 \end{bmatrix} \begin{bmatrix} 3 \\ 3 \\ 3 \end{bmatrix} = \begin{bmatrix} 3 \\ 0 \\ 3 \end{bmatrix}; \quad \vec{A} = 3\vec{e}_1 + 3\vec{e}_3.$$

5. Since T and W are exterior tensors of $\bigwedge^{(3)}_3(\mathbb{R})$, we use Formulas (9.84) and (9.86) to determine its strict components in the new basis.

For T, we have

$$\hat{t}^{(123)}_{\circ\circ\circ} = |C|^{-1} t^{(123)}_{\circ\circ\circ} = \begin{vmatrix} 1 & 0 & 0 \\ 1 & 2 & 0 \\ 0 & 1 & 2 \end{vmatrix}^{-1} \cdot 2 = \frac{2}{4} = \frac{1}{2}$$

$$\hat{t}^{\circ\circ\circ}_{(123)} = |C| t^{\circ\circ\circ}_{(123)} = 4\left(|G| t^{(123)}_{\circ\circ\circ}\right) = 4 \cdot 4 \cdot 2 = 32,$$

and for W, we have

$$\hat{w}^{(123)}_{\circ\circ\circ} = |C|^{-1} w^{(123)}_{\circ\circ\circ} = \frac{1}{4} \times \frac{3}{2} = \frac{3}{8}$$

$$\hat{w}^{\circ\circ\circ}_{(123)} = |C| w^{\circ\circ\circ}_{(123)} = 4 \cdot 6 = 24.$$

□

Example 13.10 (Exterior subspaces). Consider a Euclidean space $E^3(\mathbb{R})$ referred to an orthonormalized basis $\{\vec{u}_\alpha\}$, in which we adopt the following notation criterion, for any pair of generic vectors $\vec{V}, \vec{W} \in E^3(\mathbb{R})$.

In the initial basis $\{\vec{u}_\alpha\}$, we have

$$\vec{V} = ||\vec{u}_\alpha|| X \equiv [\vec{u}_1 \vec{u}_2 \vec{u}_3] \begin{bmatrix} x^1_{\circ} \\ x^2_{\circ} \\ x^3_{\circ} \end{bmatrix};$$

$$\vec{W} = ||\vec{u}_\alpha||Y \equiv [\vec{u}_1 \vec{u}_2 \vec{u}_3] \begin{bmatrix} y_\circ^1 \\ y_\circ^2 \\ y_\circ^3 \end{bmatrix} ;$$

and in another arbitrary basis $\{\vec{e}_i\}$:

$$\vec{V} = ||\vec{e}_i||\hat{X} \equiv [\vec{e}_1 \vec{e}_2 \vec{e}_3] \begin{bmatrix} \hat{x}_\circ^1 \\ \hat{x}_\circ^2 \\ \hat{x}_\circ^3 \end{bmatrix} ;$$

$$\vec{W} = ||\vec{e}_i||\hat{Y} \equiv [\vec{e}_1 \vec{e}_2 \vec{e}_3] \begin{bmatrix} \hat{y}_\circ^1 \\ \hat{y}_\circ^2 \\ \hat{y}_\circ^3 \end{bmatrix} .$$

We perform a change-of-basis in $E^3(\mathbb{R})$, given by the equations

$$\vec{u}_1 = 2\vec{e}_1 - \vec{e}_2 + \vec{e}_3; \quad \vec{u}_2 = \vec{e}_1 + \vec{e}_2 + 3\vec{e}_3; \quad \vec{u}_3 = -\vec{e}_1 + 5\vec{e}_2 - \vec{e}_3,$$

and we know that in the exterior algebra $\bigwedge_3^{(2)}(\mathbb{R})$ associated with $E^3(\mathbb{R})$, there exists a linear space S, of analytical equation

$$S \equiv x_\circ^1 y_\circ^3 - 4x_\circ^2 y_\circ^1 - x_\circ^3 y_\circ^2 + 4x_\circ^1 y_\circ^2 - x_\circ^3 y_\circ^1 + x_\circ^2 y_\circ^3 = 0,$$

that is satisfied by certain multivectors $T = \vec{V} \otimes \vec{W}$ given in contravariant coordinates of $\vec{V}$ and $\vec{W}$.

1. Determine, in contravariant coordinates, the new analytical equation of S that this change produces.
2. Determine the new analytical equation of S in covariant coordinates.
3. Determine the new equation of S in strict contravariant coordinates.
4. Determine the new equation of S in strict covariant coordinates.
5. Determine the modulus of T calculated in the initial and new bases.

Solution: We follow the usual criterion of maximum information, using two different procedures for the solution for questions 1 and 2.

1. (a) *Directly, as anti-symmetric tensors.* Consider the change-of-basis

$$||u^\alpha|| = ||\vec{e}_i||C^{-1}; \quad C^{-1} = \begin{bmatrix} 2 & 1 & -1 \\ -1 & 1 & 5 \\ 1 & 3 & -1 \end{bmatrix}; \quad C = \frac{1}{24} \begin{bmatrix} 16 & 2 & -6 \\ -4 & 1 & 9 \\ 4 & 5 & -3 \end{bmatrix}$$

$$S \equiv x_o^1 y_o^3 - 4x_o^2 y_o^1 - x_o^3 y_o^2 + 4x_o^1 y_o^2 - x_o^3 y_o^1 + x_o^2 y_o^3$$

$$\equiv [x_o^1 x_o^2 x_o^3] \begin{bmatrix} 0 & 4 & 1 \\ -4 & 0 & 1 \\ -1 & -1 & 0 \end{bmatrix} \begin{bmatrix} y_o^1 \\ y_o^2 \\ y_o^3 \end{bmatrix} \equiv X^t A Y,$$

which is a tensor of anti-symmetric matrix A.
Since the equations for the change-of-basis for vectors of $E^3(\mathbb{R})$, are

$$X = C\hat{X}; \quad Y = C\hat{Y},$$

we have

$$S = X^t A Y = (C\hat{X})^t A (C\hat{Y}) = \hat{X}^t (C^t A C)\hat{Y} \equiv \hat{X}^t \hat{A} \hat{Y};$$

and then

$$\hat{A} = C^t A C = \frac{1}{24} \begin{bmatrix} 16 & -4 & 4 \\ 2 & 1 & 5 \\ -6 & 9 & -3 \end{bmatrix} \begin{bmatrix} 0 & 4 & 1 \\ -4 & 0 & 1 \\ -1 & -1 & 0 \end{bmatrix} \frac{1}{24} \begin{bmatrix} 16 & 2 & -6 \\ -4 & 1 & 9 \\ 4 & 5 & -3 \end{bmatrix}$$

$$= \frac{72}{24^2} \begin{bmatrix} 0 & 2 & 6 \\ -2 & 0 & 1 \\ -6 & -1 & 0 \end{bmatrix},$$

whence

$$S = 0; \quad S \equiv \hat{X}^t \hat{A} \hat{Y} = 0 \rightarrow [\hat{x}_o^1 \hat{x}_o^2 \hat{x}_o^3] \begin{bmatrix} 0 & 2 & 6 \\ -2 & 0 & 1 \\ -6 & -1 & 0 \end{bmatrix} \begin{bmatrix} \hat{y}_o^1 \\ \hat{y}_o^2 \\ \hat{y}_o^3 \end{bmatrix} = 0,$$

which operated gives the new analytical expression

$$S \equiv 2\hat{x}_o^1 \hat{y}_o^2 + 6\hat{x}_o^1 \hat{y}_o^3 - 2\hat{x}_o^2 \hat{y}_o^1 + \hat{x}_o^2 \hat{y}_o^3 - 6\hat{x}_o^3 \hat{y}_o^1 - \hat{x}_o^3 \hat{y}_o^2 = 0.$$

(b) *Through the exterior algebra.* Sorting S adequately we get

$$S = 4 \cdot (x_o^1 y_o^2 - x_o^2 y_o^1) + 1 \cdot (x_o^1 y_o^3 - x_o^3 y_o^1) + 1 \cdot (x_o^2 y_o^3 - x_o^3 y_o^2)$$

$$= [411] \begin{bmatrix} x_o^1 y_o^2 - x_o^2 y_o^1 \\ x_o^1 y_o^3 - x_o^3 y_o^1 \\ x_o^2 y_o^3 - x_o^3 y_o^2 \end{bmatrix} = [411] \begin{bmatrix} t_{oo}^{(12)} \\ t_{oo}^{(13)} \\ t_{oo}^{(23)} \end{bmatrix}$$

$$= 4t_{oo}^{(12)} + t_{oo}^{(13)} + t_{oo}^{(23)} = 0, \tag{13.62}$$

which is the exterior equation of S in contravariant coordinates.
From (9.74) we have

$$t^{(\alpha\beta)}_{\circ\ \circ} = t^{(ij)}_{\circ\ \circ} \begin{vmatrix} c^{\circ\alpha}_{i\,\circ} & c^{\circ\beta}_{i\,\circ} \\ c^{\circ\alpha}_{j\,\circ} & c^{\circ\beta}_{j\,\circ} \end{vmatrix},$$

in which the minors belong to C^t, and giving numerical values to the indices, we get

$$t^{(12)}_{\circ\ \circ} = \hat{t}^{(12)}_{\circ\ \circ} \frac{1}{24} \begin{vmatrix} 16 & -4 \\ 2 & 1 \end{vmatrix} + \hat{t}^{(13)}_{\circ\ \circ} \frac{1}{24} \begin{vmatrix} 16 & -4 \\ -6 & 9 \end{vmatrix} + \hat{t}^{(23)}_{\circ\ \circ} \frac{1}{24} \begin{vmatrix} 2 & 1 \\ -6 & 9 \end{vmatrix}$$

$$= \hat{t}^{(12)}_{\circ\ \circ} + 5\hat{t}^{(13)}_{\circ\ \circ} + \hat{t}^{(23)}_{\circ\ \circ}$$

$$t^{(13)}_{\circ\ \circ} = \hat{t}^{(12)}_{\circ\ \circ} \frac{1}{24} \begin{vmatrix} 16 & 4 \\ 2 & 5 \end{vmatrix} + \hat{t}^{(13)}_{\circ\ \circ} \frac{1}{24} \begin{vmatrix} 16 & 4 \\ -6 & -3 \end{vmatrix} + \hat{t}^{(23)}_{\circ\ \circ} \frac{1}{24} \begin{vmatrix} 2 & 5 \\ -6 & -3 \end{vmatrix}$$

$$= 3\hat{t}^{(12)}_{\circ\ \circ} - \hat{t}^{(13)}_{\circ\ \circ} + \hat{t}^{(23)}_{\circ\ \circ} .$$

$$t^{(23)}_{\circ\ \circ} = \hat{t}^{(12)}_{\circ\ \circ} \frac{1}{24} \begin{vmatrix} -4 & 4 \\ 1 & 5 \end{vmatrix} + \hat{t}^{(13)}_{\circ\ \circ} \frac{1}{24} \begin{vmatrix} -4 & 4 \\ 9 & -3 \end{vmatrix} + \hat{t}^{(23)}_{\circ\ \circ} \frac{1}{24} \begin{vmatrix} 1 & 5 \\ 9 & -3 \end{vmatrix}$$

$$= -\hat{t}^{(12)}_{\circ\ \circ} - \hat{t}^{(13)}_{\circ\ \circ} - 2\hat{t}^{(23)}_{\circ\ \circ} .$$

Taking the three relations to (13.62), we obtain

$$S \equiv 4 \cdot \left(\hat{t}^{(12)}_{\circ\ \circ} + 5\hat{t}^{(13)}_{\circ\ \circ} + \hat{t}^{(23)}_{\circ\ \circ} \right) + 1 \cdot \left(3\hat{t}^{(12)}_{\circ\ \circ} - \hat{t}^{(13)}_{\circ\ \circ} + \hat{t}^{(23)}_{\circ\ \circ} \right)$$

$$+ 1 \cdot \left(-\hat{t}^{(12)}_{\circ\ \circ} - \hat{t}^{(13)}_{\circ\ \circ} - 2\hat{t}^{(23)}_{\circ\ \circ} \right) = 6\hat{t}^{(12)}_{\circ\ \circ} + 18\hat{t}^{(13)}_{\circ\ \circ} + 3\hat{t}^{(23)}_{\circ\ \circ},$$

that is,

$$S \equiv 2\hat{t}^{(12)}_{\circ\ \circ} + 6\hat{t}^{(13)}_{\circ\ \circ} + \hat{t}^{(23)}_{\circ\ \circ} = 0, \tag{13.63}$$

that is the equation of S in the new strict contravariant coordinates. Next, we express (13.63) analytically:

$$S \equiv [\,2\ 6\ ] \begin{bmatrix} \hat{t}^{(1\ 2)}_{\circ\ \circ} \\ \hat{t}^{(1\ 3)}_{\circ\ \circ} \\ \hat{t}^{(2\ 3)}_{\circ\ \circ} \end{bmatrix} = [\,2\ 6\ 1\,] \begin{bmatrix} \hat{x}^1_\circ \hat{y}^2_\circ - \hat{x}^2_\circ \hat{y}^1_\circ \\ \hat{x}^1_\circ \hat{y}^3_\circ - \hat{x}^3_\circ \hat{y}^1_\circ \\ \hat{x}^2_\circ \hat{y}^3_\circ - \hat{x}^3_\circ \hat{y}^2_\circ \end{bmatrix};$$

and operating we get

$$S \equiv 2\hat{x}^1_\circ \hat{y}^2_\circ - 2\hat{x}^2_\circ \hat{y}^1_\circ + 6\hat{x}^1_\circ \hat{y}^3_\circ - 6\hat{x}^3_\circ \hat{y}^1_\circ + \hat{x}^2_\circ \hat{y}^3_\circ - \hat{x}^3_\circ \hat{y}^2_\circ = 0,$$

which is the same as the previous result.

2. In the new basis $\{\vec{e}_i\}$ the fundamental metric tensor is

$$\hat{G} = C^t I_3 C = \frac{1}{24} \begin{bmatrix} 16 & -4 & 4 \\ 2 & 1 & 5 \\ -6 & 9 & -3 \end{bmatrix} \frac{1}{24} \begin{bmatrix} 16 & 2 & -6 \\ -4 & 1 & 9 \\ 4 & 5 & -3 \end{bmatrix}$$

$$= \frac{1}{24^2} \begin{bmatrix} 288 & 48 & -144 \\ 48 & 30 & -18 \\ -144 & -18 & 126 \end{bmatrix}$$

$$\hat{G}^{-1} = (C^t C)^{-1} = C^{-1}(C^{-1})^t = \begin{bmatrix} 2 & 1 & -1 \\ -1 & 1 & 5 \\ 1 & 3 & -1 \end{bmatrix} \begin{bmatrix} 2 & -1 & 1 \\ 1 & 1 & 3 \\ -1 & 5 & -1 \end{bmatrix}$$

$$= \begin{bmatrix} 6 & -6 & 6 \\ -6 & 27 & -3 \\ 6 & -3 & 11 \end{bmatrix}.$$

(a) *Directly, as anti-symmetric tensors.* In question 1(a) we have obtained the linear subspace S in the new basis; its analytical expression can be written in matrix form as

$$S = \hat{X}^t \hat{A} \hat{Y} \equiv [\hat{x}_o^1 \hat{x}_o^2 \hat{x}_o^3] \begin{bmatrix} 0 & 2 & 6 \\ -2 & 0 & 1 \\ -6 & -1 & 0 \end{bmatrix} \begin{bmatrix} \hat{y}_o^1 \\ \hat{y}_o^2 \\ \hat{y}_o^3 \end{bmatrix},$$

and remembering the contravariant and covariant relations for vectors of $E^3(\mathbb{R})$, i.e.,

$$\begin{cases} \hat{X}^* = \hat{G}\hat{X} \\ \hat{Y}^* = \hat{G}\hat{Y} \end{cases} \begin{cases} \hat{X} = \hat{G}^{-1}X^* \\ \hat{Y} = \hat{G}^{-1}\hat{Y}^* \end{cases}$$

and that $(\hat{G}^{-1})^t = \hat{G}^{-1}$ because it is a symmetric matrix, we have

$$S = \hat{X}^t \hat{A} \hat{Y} = (\hat{G}^{-1}\hat{X}^*)^t \hat{A}(\hat{G}^{-1}\hat{Y}^*)$$
$$= (\hat{X}^*)^t (\hat{G}^{-1})^t \hat{A}\hat{G}^{-1}\hat{Y}^* = (\hat{X}^*)^t (\hat{G}^{-1} \hat{A}\hat{G}^{-1})\hat{Y}^* \equiv (\hat{X}^*)^t \hat{A}^* \hat{Y}^*,$$

which leads to

$$\hat{A}^* = (\hat{G}^{-1})\hat{A}\hat{G}^{-1} = C^{-1}(C^{-1})^t(C^t AC)C^{-1}(C^{-1})^t = C^{-1}A(C^{-1})^t$$

$$= \begin{bmatrix} 2 & 1 & 1 \\ -1 & 1 & 5 \\ 1 & 3 & -1 \end{bmatrix} \begin{bmatrix} 0 & 4 & 1 \\ -4 & 0 & 1 \\ -1 & -1 & 0 \end{bmatrix} \begin{bmatrix} 2 & -1 & 1 \\ 1 & 1 & 3 \\ -1 & 5 & -1 \end{bmatrix}$$

$$= 3 \begin{bmatrix} 0 & 9 & 7 \\ -9 & 0 & -12 \\ -7 & 12 & 0 \end{bmatrix}.$$

Whence

$$S \equiv (\hat{X}^*)^t \hat{A}^* \hat{Y}^* \equiv [\hat{x}_1^o \hat{x}_2^o \hat{x}_3^o]\, 3 \begin{bmatrix} 0 & 9 & 7 \\ -9 & 0 & -12 \\ -7 & 12 & 0 \end{bmatrix} \begin{bmatrix} \hat{y}_1^o \\ \hat{y}_2^o \\ \hat{y}_3^o \end{bmatrix} = 0,$$

and developing

$$S \equiv 9\hat{x}_1^\circ \hat{y}_2^\circ + 7\hat{x}_1^\circ \hat{y}_3^\circ - 9\hat{x}_2^\circ \hat{y}_1^\circ - 12\hat{x}_2^\circ \hat{y}_3^\circ - 7\hat{x}_3^\circ \hat{y}_1^\circ + 12\hat{x}_3^\circ \hat{y}_2^\circ = 0.$$

If we sort adequately, we obtain the exterior equation of S in new covariant coordinates:

$$S \equiv 9(\hat{x}_1^\circ \hat{y}_2^\circ - \hat{x}_2^\circ \hat{y}_1^\circ) + 7(\hat{x}_1^\circ \hat{y}_3^\circ - \hat{x}_3^\circ \hat{y}_1^\circ) - 12(\hat{x}_2^\circ \hat{y}_3^\circ - \hat{x}_3^\circ \hat{y}_2^\circ)$$

$$= \begin{bmatrix} 9 & 7 & -12 \end{bmatrix} \begin{bmatrix} \hat{x}_1^\circ \hat{y}_2^\circ - \hat{x}_2^\circ \hat{y}_1^\circ \\ \hat{x}_1^\circ \hat{y}_3^\circ - \hat{x}_3^\circ \hat{y}_1^\circ \\ \hat{x}_2^\circ \hat{y}_3^\circ - \hat{x}_3^\circ \hat{y}_2^\circ \end{bmatrix} = \begin{bmatrix} 9 & 7 & -12 \end{bmatrix} \begin{bmatrix} \hat{t}_{(12)}^{\circ \circ} \\ \hat{t}_{(13)}^{\circ \circ} \\ \hat{t}_{(23)}^{\circ \circ} \end{bmatrix};$$

$$S \equiv 9\hat{t}_{(12)}^{\circ \circ} + 7\hat{t}_{(13)}^{\circ \circ} - 12\hat{t}_{(23)}^{\circ \circ} = 0. \tag{13.64}$$

(b) *Through the exterior algebra theory.* We proceed to use its Euclidean character, inside the exterior algebra. Since the relation (13.63) and the matrix $\hat{G}^{-1}$ are known, using Formula (13.26), we get

$$\hat{t}_{\circ \ \circ}^{(i_1 i_2)} = \hat{t}_{(j_1 j_2)}^{\circ \ \circ} \begin{vmatrix} \hat{g}_{\circ \circ}^{i_1 j_1} & \hat{g}_{\circ \circ}^{i_1 j_2} \\ \hat{g}_{\circ \circ}^{i_2 j_1} & \hat{g}_{\circ \circ}^{i_2 j_2} \end{vmatrix},$$

and calculating the components $\hat{t}_{\circ \ \circ}^{(i_1 i_2)}$ we obtain

$$\hat{t}_{\circ \ \circ}^{(12)} = \hat{t}_{(12)}^{\circ \circ} \begin{vmatrix} 6 & -6 \\ -6 & 27 \end{vmatrix} + \hat{t}_{(13)}^{\circ \circ} \begin{vmatrix} 6 & 6 \\ -6 & -3 \end{vmatrix} + \hat{t}_{(23)}^{\circ \circ} \begin{vmatrix} -6 & 6 \\ 27 & -3 \end{vmatrix}$$

$$= 126\hat{t}_{(12)}^{\circ \circ} + 18\hat{t}_{(13)}^{\circ \circ} - 144\hat{t}_{(23)}^{\circ \circ}$$

$$\hat{t}_{\circ \ \circ}^{(13)} = \hat{t}_{(12)}^{\circ \circ} \begin{vmatrix} 6 & -6 \\ 6 & -3 \end{vmatrix} + \hat{t}_{(13)}^{\circ \circ} \begin{vmatrix} 6 & 6 \\ 6 & 11 \end{vmatrix} + \hat{t}_{(23)}^{\circ \circ} \begin{vmatrix} -6 & 6 \\ -3 & 11 \end{vmatrix}$$

$$= 18\hat{t}_{(12)}^{\circ \circ} + 30\hat{t}_{(13)}^{\circ \circ} - 48\hat{t}_{(23)}^{\circ \circ}$$

$$\hat{t}_{\circ \ \circ}^{(23)} = \hat{t}_{(12)}^{\circ \circ} \begin{vmatrix} -6 & 27 \\ 6 & -3 \end{vmatrix} + \hat{t}_{(13)}^{\circ \circ} \begin{vmatrix} -6 & -3 \\ 6 & 11 \end{vmatrix} + \hat{t}_{(23)}^{\circ \circ} \begin{vmatrix} 27 & -3 \\ -3 & 11 \end{vmatrix}$$

$$= -144\hat{t}_{(12)}^{\circ \circ} - 48\hat{t}_{(13)}^{\circ \circ} + 288\hat{t}_{(23)}^{\circ \circ}$$

and substituting these three expressions in (13.63), we have

$$S \equiv 2(126\hat{t}_{(12)}^{\circ \circ} + 18\hat{t}_{(13)}^{\circ \circ} - 144\hat{t}_{(23)}^{\circ \circ}) + 6(18\hat{t}_{(12)}^{\circ \circ} + 30\hat{t}_{(13)}^{\circ \circ} - 48\hat{t}_{(23)}^{\circ \circ})$$

$$+ (-144\hat{t}_{(12)}^{\circ \circ} - 48\hat{t}_{(13)}^{\circ \circ} + 288\hat{t}_{(23)}^{\circ \circ}) = 216\hat{t}_{(12)}^{\circ \circ} + 168\hat{t}_{(13)}^{\circ \circ} - 288\hat{t}_{(23)}^{\circ \circ}$$

and equating to zero and dividing by 24, the result is

$$S \equiv 9\hat{t}_{(12)}^{\circ\ \circ} + 7\hat{t}_{(13)}^{\circ\ \circ} - 12\hat{t}_{(23)}^{\circ\ \circ} = 0,$$

which is again the previous result, Formula (13.64).

3. It has been answered with Formula (13.63).
4. It has been answered with Formula (13.64).
5. In Example 13.6, Formula (13.60) solves the *exterior* fundamental metric tensor associated with $\Lambda_3^{(2)}(\mathbb{R})$. Applying this formula to our problem and with $G = I_3 \equiv \begin{bmatrix} 1 & 0 & 0 \\ 0 & 1 & 0 \\ 0 & 0 & 1 \end{bmatrix}$, for the initial basis we get

$$g'^{\circ\circ}_{12} = (\vec{u}_1 \wedge \vec{u}_2) \bullet (\vec{u}_1 \wedge \vec{u}_3) = 2 \begin{vmatrix} g^{\circ\circ}_{11} & g^{\circ\circ}_{13} \\ g^{\circ\circ}_{21} & g^{\circ\circ}_{23} \end{vmatrix} = 2 \begin{vmatrix} 1 & 0 \\ 0 & 0 \end{vmatrix} = 0$$

$$g'^{\circ\circ}_{11} = (\vec{u}_1 \wedge \vec{u}_2) \bullet (\vec{u}_1 \wedge \vec{u}_2) = 2 \begin{vmatrix} g^{\circ\circ}_{11} & g^{\circ\circ}_{12} \\ g^{\circ\circ}_{21} & g^{\circ\circ}_{22} \end{vmatrix} = 2 \begin{vmatrix} 1 & 0 \\ 0 & 1 \end{vmatrix} = 2$$

$$g'^{\circ\circ}_{22} = (\vec{u}_1 \wedge \vec{u}_3) \bullet (\vec{u}_1 \wedge \vec{u}_3) = 2 \begin{vmatrix} g^{\circ\circ}_{11} & g^{\circ\circ}_{13} \\ g^{\circ\circ}_{31} & g^{\circ\circ}_{33} \end{vmatrix} = 2 \begin{vmatrix} 1 & 0 \\ 0 & 1 \end{vmatrix} = 2$$

and similarly $g'^{\circ\circ}_{13} = g'^{\circ\circ}_{23} = 0$ and $g'^{\circ\circ}_{33} = 0$.

The exterior connection matrix is $G_{\Lambda_3^{(2)}} = \begin{bmatrix} 2 & 0 & 0 \\ 0 & 2 & 0 \\ 0 & 0 & 2 \end{bmatrix}$.

Since $T = t^{(12)}_{\circ\ \circ}\vec{u}_1 \wedge \vec{u}_2 + t^{(13)}_{\circ\ \circ}\vec{u}_1 \wedge \vec{u}_3 + t^{(23)}_{\circ\ \circ}\vec{u}_2 \wedge \vec{u}_3$, we have

$$|T|^2 = [t^{(12)}_{\circ\ \circ} t^{(13)}_{\circ\ \circ} t^{(23)}_{\circ\ \circ}] \begin{bmatrix} 2 & 0 & 0 \\ 0 & 2 & 0 \\ 0 & 0 & 2 \end{bmatrix} \begin{bmatrix} t^{(1)2)}_{\circ\ \circ} \\ t^{(13)}_{\circ\ \circ} \\ t^{(23)}_{\circ\ \circ} \end{bmatrix};$$

$$|T|^2 = 2 \left((t^{(12)}_{\circ\ \circ})^2 + (t^{(13)}_{\circ\ \circ})^2 + (t^{(23)}_{\circ\ \circ})^2 \right).$$

But, T satisfies the equation (13.62) of S, so, one strict component *is not* free. We choose $(t^{(23)}_{\circ\ \circ}) = -(4t^{(12)}_{\circ\ \circ} + t^{(13)}_{\circ\ \circ})$, and substituting it into the expression of $|T|^2$, we get

$$|T| = \sqrt{2}\sqrt{(t^{(12)}_{\circ\ \circ})^2 + (t^{(13)}_{\circ\ \circ})^2 + (4t^{(12)}_{\circ\ \circ} + t^{(13)}_{\circ\ \circ})^2}.$$

In the new basis, we proceed with the minors of $\hat{G} = \frac{1}{24} \begin{bmatrix} 288 & 48 & -144 \\ 48 & 30 & -18 \\ -144 & -18 & 126 \end{bmatrix}$

and with the new strict components of T:

$$T = \hat{t}^{(12)}_{\circ\ \circ}\vec{e}_1 \wedge \vec{e}_2 + \hat{t}^{(13)}_{\circ\ \circ}\vec{e}_1 \wedge \vec{e}_3 + \hat{t}^{(23)}_{\circ\ \circ}\vec{e}_2 \wedge \vec{e}_3$$

of an analogous form but more involved.

□

13.10 Exercises

13.1. Following the theory in Example 13.6 of the present chapter.

1. Determine the matrix of the exterior fundamental metric tensor $G_{\Lambda^{(2)}_2(\mathbb{R})}$ associated with the mentioned algebra, established over a Euclidean space $E^2(\mathbb{R})$ that in the basis $\{\vec{e}_\alpha\}$ presents the Gram matrix

$$G = [g^{\circ\ \circ}_{\alpha\beta}] = \begin{bmatrix} 2 & 1 \\ 1 & 1 \end{bmatrix}.$$

2. Similarly, determine the matrix $G_{\Lambda^{(2)}_3(\mathbb{R})}$ of the algebra established over an $E^3(\mathbb{R})$ that in the basis $\{\vec{e}_\alpha\}$ has as Gram matrix

$$G = [g^{\circ\ \circ}_{\alpha\beta}] = \begin{bmatrix} 3 & 1 & 2 \\ 1 & 1 & 1 \\ 2 & 1 & 2 \end{bmatrix}.$$

3. Idem for $G_{\Lambda^{(2)}_4(\mathbb{R})}$, where

$$G = [g^{\circ\ \circ}_{\alpha\beta}] = \begin{bmatrix} 1 & 0 & 0 & 1 \\ 0 & 2 & 1 & 0 \\ 0 & 1 & 3 & 2 \\ 1 & 0 & 2 & 5 \end{bmatrix}.$$

13.2. In the Euclidean space $E^3(\mathbb{R})$ referred to the basis $\{\vec{e}_\alpha\}$ and the fundamental metric tensor

$$G = [g^{\circ\ \circ}_{\alpha\beta}] = \begin{bmatrix} 2 & 1 & 1 \\ 1 & 2 & 1 \\ 1 & 1 & 2 \end{bmatrix},$$

consider four vectors $\vec{V}_i$, $1 \leq i \leq 4$ by means of the matrix of its contravariant coordinates

$$[X_1 \quad X_2 \quad X_3 \quad X_4] = \begin{bmatrix} 2 & 1 & 5 & 1 \\ -1 & 1 & -2 & -1 \\ 1 & 1 & -2 & -1 \end{bmatrix}.$$

1. Determine the tensor $U = \vec{V}_1 \wedge \vec{V}_2$ by its strict components, as $U \in \Lambda^{(2)}_3(\mathbb{R})$.

2. Determine the tensor $V = \vec{V_3} \wedge \vec{V_4}$ by its strict components, as tensor $V \in \bigwedge_3^{(2)}(\mathbb{R})$.

3. Determine the exterior fundamental metric tensor $G_{\bigwedge_3^{(2)}(\mathbf{R})}$, which connects the mentioned algebra.

4. Give the dot product $U \bullet V$ using the previous metric tensor.

5. Determine the moduli of the tensors U and V.

6. Execute the exterior tensor product $Z = U \wedge \vec{V_4}$.

7. Obtain the tensor $W = $ Polar Z.

8. Obtain the modulus of W.

13.3. Consider a Euclidean space $E^4(\mathbb{R})$ referred to a basis $\{\vec{e}_\alpha\}$, the fundamental metric tensor of which is represented by the matrix

$$G = [g_{\alpha\beta}^{\circ\circ}] = \begin{bmatrix} 1 & 0 & 0 & 1 \\ 0 & 2 & 1 & 0 \\ 0 & 1 & 3 & 2 \\ 1 & 0 & 2 & 5 \end{bmatrix}.$$

Consider two exterior tensors $S, T \in \bigwedge_4^{(2)}(\mathbb{R})$ of strict components $s_{\circ\circ}^{(\alpha\beta)} = \alpha + \beta$ and $t_{\circ\circ}^{(\gamma\delta)} = \delta - \gamma$.

1. Obtain the exterior fundamental metric tensor $G_{\bigwedge_4^{(2)}}$.

2. Obtain the moduli of the tensors S and T.

3. Obtain the dot product $S \bullet T$ of the mentioned tensors.

4. Obtain the tensor $Z = S \wedge T$ exterior tensor product of S and T.

5. Obtain the exterior fundamental metric tensor $\bigwedge_4^{(4)}(\mathbb{R})$.

6. Obtain the modulus of the tensor Z.

7. Obtain the tensor W dual or polar of Z, that is $W = $ Polar Z.

8. Obtain the modulus of W.

9. If we set $\cos\theta = \frac{S \bullet T}{|S| \cdot |T|}$, is it true that $\sin\theta = \frac{|W|}{|S| \cdot |T|}$?

10. We execute in $E^4(\mathbb{R})$ the change-of-basis $\{\vec{e}_i\}$, of associated matrix

$$C \equiv [c_{\circ i}^{\alpha\circ}] = \begin{bmatrix} 1 & 2 & 3 & 1 \\ 1 & 3 & 3 & 2 \\ 2 & 4 & 3 & 3 \\ 1 & 1 & 1 & 1 \end{bmatrix}.$$

Answer again questions 1 to 7, for the basis $\{\vec{e}_i\}$.

Part V

Classic Tensors in Geometry and Mechanics

14

Affine Tensors

14.1 Introduction and Motivation

Although multilinear algebra, tensor theory from the point of view of algebra, has been discussed troughout the previous chapters, we have decided to end this book with a chapter devoted to the most common affine tensors, so that it could act as a bridge to other tensor calculus books or publications.

In those books functional tensors are dealt with, that can be subject to differential and integral processes, and that are built over punctual spaces affine to Euclidean or Hermitian linear spaces, which are their "support" or "reference frame". The position (space coordinates) and the instant (the time) are the usual variables of the mentioned functionals.

From the algebra point of view, nowadays it is possible to have a tensor calculus that is more elaborated, deep and complete that the one its distinguished founders Voigt, Levi-Civita, Ricci, Riemann, Christoffel, Einstein, etc. have given to us, since they unfortunately lacked a sufficiently complete algebraic background, when they started to build such a beautiful mathematical structure.

This chapter aims to show how tensors allow us to solve very interesting practical problems. It is structured in three parts. The first part is devoted to Euclidean tensors in $E^n(\mathbb{R})$, including the projection, the momentum, the rotation and the reflection tensors.

The second part is dedicated to affine geometric tensors or homographies, that include general affinities, homothecies, isometries and products of them.

Finally, the third part is devoted to some important tensors in physics and mechanics, such as the stress and strain tensors, the elastic tensor and the inertial moment tensor.

14.2 Euclidean tensors in $E^n(\mathbb{R})$

Consider an orthonormalized Euclidean space $E^n(\mathbb{R})$ of basis $\{\vec{e}_\alpha\}$. In addition to the basic tensors used in the Euclidean spaces such as the dot product, the cross product, the mixed product of vectors, etc., there exist other tensors that it is convenient to define due to their continuous presence in diverse tensor processes.

14.2.1 Projection tensor

Consider the reference linear space $E^2(\mathbb{R})$, and assume as data one axis (see Figure 14.1), given by its unit vector $\vec{e} = \cos\alpha\vec{e}_1 + \sin\alpha\vec{e}_2$, and a certain vector $\vec{v} = x^1\vec{e}_1 + x^2\vec{e}_2$. We wish to know the vector $\vec{p} = (x^1)'\vec{e}_1 + (x^2)'\vec{e}_2$ orthogonal projection of $\vec{v}$ over the given axis, using tensors. More precisely, we look for a tensor such that its tensor product by a given vector when contracted gives the desired projection.

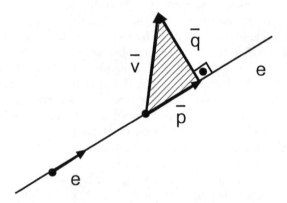

Fig. 14.1. Illustration of the projection tensor.

We denote this executing tensor, which is called a "projector", by $P_{\vec{e}}$:

$$P_{\vec{e}} = \begin{bmatrix} \cos\alpha \\ \sin\alpha \end{bmatrix} [\cos\alpha \quad \sin\alpha] = \begin{bmatrix} \cos^2\alpha & \sin\alpha\cos\alpha \\ \sin\alpha\cos\alpha & \sin^2\alpha \end{bmatrix}.$$

Then, we have

$$\vec{p} = P_{\vec{e}}(\vec{v}); \quad \begin{bmatrix} (x^1)' \\ (x^2)' \end{bmatrix} = \begin{bmatrix} \cos^2\alpha & \sin\alpha\cos\alpha \\ \sin\alpha\cos\alpha & \sin^2\alpha \end{bmatrix} \begin{bmatrix} x^1 \\ x^2 \end{bmatrix}, \qquad (14.1)$$

a matrix product that is none other than the mentioned tensor product contraction.

We define the complementary projector tensor, denoted by $P_{\vec{e}^\perp}$, the tensor that transforms $\vec{v}$ into the vector $\vec{q} = (x^1)''\vec{e}_1 + (x^2)''\vec{e}_2$ such that $\vec{p} + \vec{q} = \vec{v}$.
From the expression $\vec{q} = \vec{v} - \vec{p}$, in matrix form we get

$$\begin{bmatrix} (x^1)'' \\ (x^2)'' \end{bmatrix} = \begin{bmatrix} x^1 \\ x^2 \end{bmatrix} - \begin{bmatrix} \cos^2\alpha & \sin\alpha\cos\alpha \\ \sin\alpha\cos\alpha & \sin^2\alpha \end{bmatrix} \begin{bmatrix} x^1 \\ x^2 \end{bmatrix}$$

$$= \begin{bmatrix} 1-\cos^2\alpha & -\sin\alpha\cos\alpha \\ -\sin\alpha\cos\alpha & 1-\sin^2\alpha \end{bmatrix} \begin{bmatrix} x^1 \\ x^2 \end{bmatrix}$$

$$\vec{q} = P_{\vec{e}^\perp}(\vec{v}); \quad \begin{bmatrix} (x^1)'' \\ (x^2)'' \end{bmatrix} = \begin{bmatrix} \sin^2\alpha & -\sin\alpha\cos\alpha \\ -\sin\alpha\cos\alpha & \cos^2\alpha \end{bmatrix} \begin{bmatrix} x^1 \\ x^2 \end{bmatrix}. \quad (14.2)$$

Evidently, $P_{\vec{e}} + P_{\vec{e}^\perp} = I_2$, as it corresponds to complementary projectors. Similarly, if we deal with the reference linear space $E^3(\mathbb{R})$, we will have the following data and solutions:

$$\vec{e} = \cos\alpha\,\vec{e}_1 + \cos\beta\,\vec{e}_2 + \cos\gamma\,\vec{e}_3; \quad \vec{v} = x^1\vec{e}_1 + x^2\vec{e}_2 + x^3\vec{e}_3;$$

$$\vec{p} = (x^1)'\vec{e}_1 + (x^2)'\vec{e}_2 + (x^3)'\vec{e}_3; \text{ and } \vec{q} = (x^1)''\vec{e}_1 + (x^2)''\vec{e}_2 + (x^3)''\vec{e}_3;$$

$$P_{\vec{e}} = \begin{bmatrix} \cos\alpha \\ \cos\beta \\ \cos\gamma \end{bmatrix} [\cos\alpha \quad \cos\beta \quad \cos\gamma] = \begin{bmatrix} \cos^2\alpha & \cos\alpha\cos\beta & \cos\alpha\cos\gamma \\ \cos\beta\cos\alpha & \cos^2\beta & \cos\beta\cos\gamma \\ \cos\alpha\cos\gamma & \cos\beta\cos\gamma & \cos^2\gamma \end{bmatrix}.$$

Whence, the tensor operates as

$$\vec{p} = P_{\vec{e}}(\vec{v}); \quad \begin{bmatrix} (x^1)' \\ (x^2)' \\ (x^3)' \end{bmatrix} = \begin{bmatrix} \cos^2\alpha & \cos\alpha\cos\beta & \cos\alpha\cos\gamma \\ \cos\alpha\cos\beta & \cos^2\beta & \cos\beta\cos\gamma \\ \cos\alpha\cos\gamma & \cos\beta\cos\alpha & \cos^2\gamma \end{bmatrix} \begin{bmatrix} x^1 \\ x^2 \\ x^3 \end{bmatrix}$$

$$(14.3)$$

and with respect to the *complementary projector*, we have

$$\vec{q} = P_{\vec{e}^\perp}(\vec{v}); \quad \begin{bmatrix} (x^1)'' \\ (x^2)'' \\ (x^3)'' \end{bmatrix} = \begin{bmatrix} \sin^2\alpha & -\cos\alpha\cos\beta & -\cos\alpha\cos\gamma \\ -\cos\alpha\cos\beta & \sin^2\beta & -\cos\beta\cos\gamma \\ -\cos\alpha\cos\gamma & -\cos\beta\cos\gamma & \sin^2\gamma \end{bmatrix} \begin{bmatrix} x^1 \\ x^2 \\ x^3 \end{bmatrix},$$

$$(14.4)$$

where obviously

$$P_{\vec{e}} + P_{\vec{e}^\perp} = I_3. \quad (14.5)$$

A property that is common to projectors is the idempotence: $P_{\vec{e}}^2 = P_{\vec{e}}$.

If we are in the Euclidean space $E^3(\mathbb{R})$, there is also the possibility of *projecting* a vector $\vec{v}$ *over a plane* π that we assume is given in vector form, that is, generated by two unit orthogonal vectors $(\vec{\pi}_1 \bullet \vec{\pi}_2 = 0)$ of π, chosen as the basis of the linear subspace

$$\vec{\pi}_1 = \cos\alpha_1\vec{e}_1 + \cos\beta_1\vec{e}_2 + \cos\gamma_1\vec{e}_3$$
$$\vec{\pi}_2 = \cos\alpha_2\vec{e}_1 + \cos\beta_2\vec{e}_2 + \cos\gamma_2\vec{e}_3.$$

We notate this projection tensor by P_π, and we consider that $P_\pi(\vec{v})$ is the vector sum of the projections $\vec{p}_1$ and $\vec{p}_2$ of $\vec{v}$ over each one of the axes represented by the unit vectors $\vec{\pi}_1$ and $\vec{\pi}_2$. Then, if $P_\pi(\vec{v}) = \vec{p}_\pi$, we have

$$\vec{p}_\pi = P_\pi(\vec{v}) = \vec{p}_1 + \vec{p}_2 = P_{\vec{\pi}_1}(\vec{v}) + P_{\vec{\pi}_2}(\vec{v}); \quad \vec{p}_\pi = (x^1)'_\pi \vec{e}_1 + (x^2)'_\pi \vec{e}_2 + (x^3)'_\pi \vec{e}_3$$

and finally

$$\begin{bmatrix} (x^1)' \\ (x^2)' \\ (x^3)' \end{bmatrix}_\pi = \left(\begin{bmatrix} \cos^2 \alpha_1 & \cos\alpha_1 \cos\beta_1 & \cos\alpha_1 \cos\gamma_1 \\ \cos\alpha_1 \cos\beta_1 & \cos^2 \beta_1 & \cos\beta_1 \cos\gamma_1 \\ \cos\alpha_1 \cos\gamma_1 & \cos\beta_1 \cos\gamma_1 & \cos^2 \gamma_1 \end{bmatrix} \right.$$
$$\left. + \begin{bmatrix} \cos^2 \alpha_2 & \cos\alpha_2 \cos\beta_2 & \cos\alpha_2 \cos\gamma_2 \\ \cos\alpha_2 \cos\beta_2 & \cos^2 \beta_2 & \cos\beta_2 \cos\gamma_2 \\ \cos\alpha_2 \cos\gamma_2 & \cos\beta_2 \cos\gamma_2 & \cos^2 \gamma_2 \end{bmatrix} \right) \begin{bmatrix} x^1 \\ x^2 \\ x^3 \end{bmatrix} \quad (14.6)$$

Some authors call the tensor kP_π with $k \in \mathbb{R}$ and $k \neq 0$, the "displacement tensor".

Example 14.1 (Tensor $S = \sigma P_{\vec{e}}$ in $E^2(\mathbb{R})$). We suspend a weight p at the center of an approximately horizontal cable of length L, which produces, at equilibrium, a vertical displacement f small compared with the length L (see Figure 14.2).

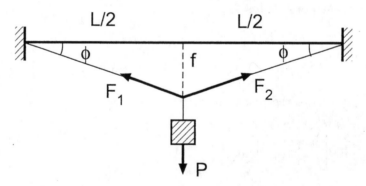

Fig. 14.2. Cable with a suspended weight at its center.

1. Obtain the traction force F produced by the suspension of the weight.
2. We define the cable stress tensor S as the tensor $\sigma P_{\vec{e}}$, where σ is the cable normal stress, A is the cable cross-section, and such that the direction of the e axis coincides with that of the cable in traction. Obtain the stress tensor.
3. Use S to obtain the traction force in a cable section the normal unit vector of which makes an angle θ with e.

Solution:

1. We accept $E^2(\mathbb{R})$ as the vector reference frame for our problem. From the vector force diagram in Figure 14.3, we conclude that

$$\frac{P/2}{F} = \sin\phi$$

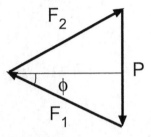

Fig. 14.3. Vector diagram.

and from the geometric characteristic of the suspension, we obtain

$$\frac{f}{L/2} = \sin\phi,$$

thus, we have

$$\frac{P/2}{F} = \frac{f}{L/2} \to F = \frac{PL}{4f}.$$

2. For examining the stresses, we assume that the cable axis coincides with the axis $\overline{OX}$, that is, $\vec{e} = \cos 0°\,\vec{e}_1 + \sin 0°\,\vec{e}_2$ and from Formula (14.1) we have

$$S = \sigma P_{\vec{e}} = \sigma \begin{bmatrix} \cos^2\alpha & \sin\alpha\cos\alpha \\ \sin\alpha\cos\alpha & \sin^2\alpha \end{bmatrix}$$

$$= \frac{F}{A} \begin{bmatrix} \cos^2 0° & \sin 0°\cos 0° \\ \sin 0°\cos 0° & \sin^2 0° \end{bmatrix} = \frac{PL}{4fA} \begin{bmatrix} 1 & 0 \\ 0 & 0 \end{bmatrix}.$$

3. The unit vector $\vec{n}$ normal to the cross section A' to be studied is

$$\vec{n} = \cos\theta\,\vec{e}_1 + \sin\theta\,\vec{e}_2,$$

and then, the stress $\vec{\sigma}'$ at the section A' is $\vec{\sigma}' = S(\vec{n})$:

$$\begin{bmatrix} (\sigma^1)' \\ (\sigma^2)' \end{bmatrix} = \frac{PL}{4fA} \begin{bmatrix} 1 & 0 \\ 0 & 0 \end{bmatrix} \begin{bmatrix} \cos\theta \\ \sin\theta \end{bmatrix} = \frac{PL}{4fA} \begin{bmatrix} \cos\theta \\ 0 \end{bmatrix} \to \vec{\sigma}' = \frac{PL}{4fA}\cos\theta\,\vec{e}_1.$$

□

14.2.2 The momentum tensor

In this section we try to build a tensor, notated $M_{to}(\vec{r}\times)$, such that when transforming $\vec{v}$, it gives as its transform the vector $\vec{r}\times\vec{v}$. Naturally, the reference Euclidean space frame is $E^3(\mathbb{R})$ (orthonormalized).

Since it is a cross product, $\vec{r} = r^1\vec{e}_1+r^2\vec{e}_2+r^3\vec{e}_3$ and $\vec{v} = x^1\vec{e}_1+x^2\vec{e}_2+x^3\vec{e}_3$, the result is

$$\vec{r}\times\vec{v} = \begin{vmatrix} \vec{e}_1 & \vec{e}_2 & \vec{e}_3 \\ r^1 & r^2 & r^3 \\ x^1 & x^2 & x^3 \end{vmatrix} = (r^2x^3 - r^3x^2)\vec{e}_1 + (r^3x^1 - r^1x^3)\vec{e}_2 + (r^1x^2 - r^2x^1)\vec{e}_3.$$

We must have

$$M_{to}(\vec{r}\times)(\vec{v}) \equiv \vec{r}\times\vec{v},$$

whence

$$M_{to}(\vec{r}\times)\begin{bmatrix} x^1 \\ x^2 \\ x^3 \end{bmatrix} \equiv \begin{bmatrix} r^2x^3 - r^3x^2 \\ r^3x^1 - r^1x^3 \\ r^1x^2 - r^2x^1 \end{bmatrix} \rightarrow M_{to}(\vec{r}\times) = \begin{bmatrix} 0 & -r^3 & r^2 \\ r^3 & 0 & -r^1 \\ -r^2 & r^1 & 0 \end{bmatrix}.$$

$$(14.7)$$

In some applications $\vec{r}$ is the position vector of the point A and the vector $\vec{v}$ is linked to it, and in such a case the transformed vector is called a momentum vector (see Figure 14.4) and is notated as

$$\vec{M} = M_{to}(\vec{r}\times)(\vec{v}).$$

$$(14.8)$$

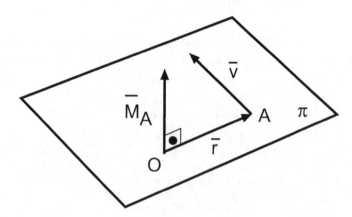

Fig. 14.4. Illustration of the momentum vector.

An interesting property of the momentum vector with respect to the projection tensor is that the successive application of this tensor (established for a unit $\vec{r} = \vec{e}$) is the tensor $(-P_{\vec{e}^\perp})$. In effect, let e be a unit vector, then from (14.7) we have

$$[M_{to}(\vec{e}\times)]^2 = \begin{bmatrix} 0 & -\cos\gamma & \cos\beta \\ \cos\gamma & 0 & -\cos\alpha \\ -\cos\beta & \cos\alpha & 0 \end{bmatrix} \begin{bmatrix} 0 & -\cos\gamma & \cos\beta \\ \cos\gamma & 0 & -\cos\alpha \\ -\cos\beta & \cos\alpha & 0 \end{bmatrix}$$

$$= \begin{bmatrix} \cos^2\alpha - 1 & \cos\alpha\cos\beta & \cos\alpha\cos\gamma \\ \cos\alpha\cos\beta & \cos^2\beta - 1 & \cos\beta\cos\gamma \\ \cos\alpha\cos\gamma & \cos\beta\cos\gamma & \cos^2\gamma - 1 \end{bmatrix}$$

$$= -(I_3 - P_{\vec{e}}) = -P_{\vec{e}^\perp}. \tag{14.9}$$

Finally, we will establish *another* tensor, related to the tensor $M_{to}(\vec{r}\times)$, that is called "*momentum with respect to an axis*", which will be notated $\vec{M}_{(\vec{e})}$.

We define it as the projection vector over an axis $\vec{e}$, of the vector $\vec{M}$.

In this case the data are:

1. A vector $\vec{v}$ and its point A of application.
2. An axis $\vec{e}$.

   Then, we have

$$\vec{M}_{(\vec{e})}(\vec{v}) = P_{\vec{e}}(\vec{M}) = P_{\vec{e}}(M_{to}(\vec{r}\times)(\vec{v}))$$

and if

$$\vec{m} = \vec{M}_{(\vec{e})}(\vec{v}) = m^1\vec{e}_1 + m^2\vec{e}_2 + m^3\vec{e}_3$$

we obtain the matrix expression

$$\begin{bmatrix} m^1 \\ m^2 \\ m^3 \end{bmatrix} = \begin{bmatrix} \cos^2\alpha & \cos\alpha\cos\beta & \cos\alpha\cos\gamma \\ \cos\alpha\cos\beta & \cos^2\beta & \cos\beta\cos\gamma \\ \cos\alpha\cos\gamma & \cos\beta\cos\gamma & \cos^2\gamma \end{bmatrix} \begin{bmatrix} 0 & -r^3 & r^2 \\ r^3 & 0 & -r^1 \\ -r^2 & r^1 & 0 \end{bmatrix} \begin{bmatrix} x^1 \\ x^2 \\ x^3 \end{bmatrix}.$$
$$\tag{14.10}$$

14.2.3 The rotation tensor

We discuss in this section the matrix representation of the rotation tensor, notated $R_{\vec{e}}$, which executes the rotation of a vector $\vec{v}$, with respect to the axis e, an angle θ; the axis is given by a unit vector $\vec{e} = \cos\alpha\vec{e}_1 + \cos\beta\vec{e}_2 + \cos\gamma\vec{e}_3$, which also allows θ to be endowed with a sign, by means of the "right-hand rule", i.e., when taking the axis with the right hand, with the thumb in the direction of the vector $\vec{e}$, the remaining fingers brace the axis in the *positive* (+) sense of the rotation angle θ.

In the proof we will perform a vector process that uses the already established vector tensors (see Figure 14.5).

Consider the vector $R_{\vec{e}}(\vec{v}) \equiv \overline{OB}'$ that results after rotating $\overline{OB} \equiv \vec{v}$ over the axis. We can decompose this vector as the sum of other vectors

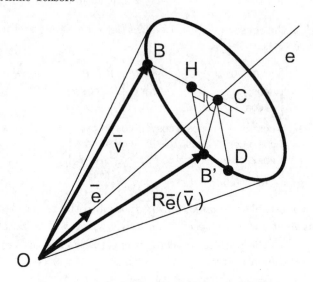

Fig. 14.5. Illustration of the rotation tensor.

$$\overline{OB}' = \overline{OC} + \overline{CH} + \overline{HB}',$$ (14.11)

where:

$\overline{OC}$ is the projection over the axis of the vector $\overline{OB} \equiv \vec{v}$, and we have $\overline{OC} = P_{\vec{e}}(\vec{v})$;

$\overline{CB}$ is the vector complementary projection of $\vec{v}$, that is, $\overline{CB} = P_{\vec{e}^\perp}(\vec{v})$,

$\overline{CH}$ is one part of $\overline{CB}$ with modulus $|\overline{CH}| = |\overline{CB}'| \cos\theta = |\overline{CB}| \cos\theta$, so that

$$\overline{CH} = \overline{CB}\cos\theta = \cos\theta P_{\vec{e}^\perp}(\vec{v});$$

$\overline{CD}$ is a vector orthogonal to the vectors $\vec{e}$ and $\overline{CB}$, which are mutually orthogonal, thus

$$\overline{CD} = \vec{e} \times \overline{OB} = \vec{e} \times \vec{v} = M_{to}(\vec{e}\times)(\vec{v});$$

$\overline{HB}'$ is parallel to $\overline{CD}$ and one part of it, and has modulus

$$|\overline{HB}'| = |\overline{CB}'|\sin\theta = |\overline{CB}|\sin\theta = |\overline{CD}|\sin\theta$$

so that

$$\overline{HB}' = \sin\theta M_{to}(\vec{e}\times)(\vec{v}).$$

As all summands in (14.11) are known, we substitute them to get

$$R_{\vec{e}}(\vec{v}) = P_{\vec{e}}(\vec{v}) + \cos\theta P_{\vec{e}^\perp}(\vec{v}) + \sin\theta M_{to}(\vec{e}\times)(\vec{v}),$$ (14.12)

which can be written, only for tensors, using Formula (14.5):

$$R_{\vec{e}} = P_{\vec{e}} + \cos\theta(I_3 - P_{\vec{e}}) + \sin\theta M_{to}(\vec{e}\times)$$

or

$$R_{\vec{e}} = \cos\theta I_3 + (1 - \cos\theta)P_{\vec{e}} + \sin\theta M_{to}(\vec{e}\times), \qquad (14.13)$$

and in matrix form, using Formulas (14.3) and (14.7), we have

$$R_{\vec{e}} = \cos\theta\begin{bmatrix} 1 & 0 & 0 \\ 0 & 1 & 0 \\ 0 & 0 & 1 \end{bmatrix} + (1 - \cos\theta)\begin{bmatrix} \cos^2\alpha & \cos\alpha\cos\beta & \cos\alpha\cos\gamma \\ \cos\alpha\cos\beta & \cos^2\beta & \cos\beta\cos\gamma \\ \cos\alpha\cos\gamma & \cos\beta\cos\gamma & \cos^2\gamma \end{bmatrix}$$

$$+ \sin\theta\begin{bmatrix} 0 & -\cos\gamma & \cos\beta \\ \cos\gamma & 0 & -\cos\alpha \\ -\cos\beta & \cos\alpha & 0 \end{bmatrix},$$

and remembering that $1 = \cos^2\alpha + \cos^2\beta + \cos^2\gamma$, and operating, we obtain

$$R_{\vec{e}} = \begin{bmatrix} r_{11} & r_{12} & r_{13} \\ r_{21} & r_{22} & r_{23} \\ r_{31} & r_{32} & r_{33} \end{bmatrix}, \qquad (14.14)$$

where

$$r_{11} = \cos^2\alpha + (\cos^2\beta + \cos^2\gamma)\cos\theta$$
$$r_{12} = \cos\alpha\cos\beta(1 - \cos\theta) - \cos\gamma\sin\theta$$
$$r_{13} = \cos\alpha\cos\gamma(1 - \cos\theta) + \cos\beta\sin\theta$$
$$r_{21} = \cos\alpha\cos\beta(1 - \cos\theta) + \cos\gamma\sin\theta$$
$$r_{22} = \cos^2\beta + (\cos^2\alpha + \cos^2\gamma)\cos\theta$$
$$r_{23} = \cos\beta\cos\gamma(1 - \cos\theta) - \cos\alpha\sin\theta$$
$$r_{31} = \cos\alpha\cos\gamma(1 - \cos\theta) - \cos\beta\sin\theta$$
$$r_{32} = \cos\alpha\cos\gamma(1 - \cos\theta) + \cos\beta\sin\theta$$
$$r_{33} = \cos^2\gamma + (\cos^2\alpha + \cos^2\beta)\cos\theta,$$

where $\vec{v} = x^1\vec{e} + x^2\vec{e}_2 + x^3\vec{e}_3$ and $R_{\vec{e}}(\vec{v}) = (x^1)'\vec{e} + (x^2)'\vec{e}_2 + (x^3)'\vec{e}_3$; the tensor $R_{\vec{e}}$ operates in matrix form in the same way as the other tensors, that is,

$$\begin{bmatrix} (x^1)' \\ (x^2)' \\ (x^3)' \end{bmatrix} = R_{\vec{e}}\begin{bmatrix} x^1 \\ x^2 \\ x^3 \end{bmatrix}.$$

Since the rotation is an isometry in $E^3(\mathbb{R})$, it must present an orthogonal matrix, which is to be verified in matrix form using the two tensors $P_{\vec{e}}$ and $M_{to}(\vec{e}\times)$ and their properties:

$$R_{\vec{e}} \bullet R_{\vec{e}}^t = ([\cos\theta I_3 + (1 - \cos\theta)P_{\vec{e}}] + \sin\theta M_{to}(\vec{e}\times))\,([\cos\theta I_3 + (1 - \cos\theta)P_{\vec{e}}]$$
$$- \sin\theta M_{to}(\vec{e}\times))$$
$$= [\cos\theta I_3 + (1 - \cos\theta)P_{\vec{e}}]^2 - \sin^2\theta\,[M_{to}(\vec{e}\times)]^2.$$

Applying Formula (14.9), developing the squares, and taking into account the projector property $P_{\vec{e}}^2 = P_{\vec{e}}$ we get

$$R_{\vec{e}} \bullet R_{\vec{e}}^t = \cos^2\theta I_3 + (1 - 2\cos\theta + \cos^2\theta)P_{\vec{e}}^2 + 2\cos\theta(1-\cos\theta)P_{\vec{e}} - \sin^2\theta[P_{\vec{e}} - I_3]$$
$$= (\sin^2\theta + \cos^2\theta)I_3 + (1 - 2\cos\theta + \cos^2\theta + 2\cos\theta - 2\cos^2\theta - \sin^2\theta)P_{\vec{e}}$$

$$R_{\vec{e}} \bullet R_{\vec{e}}^t = I_3 \rightarrow R_{\vec{e}}^t = R_{\vec{e}}^{-1},$$

which shows that $R_{\vec{e}}$ is a tensor of the associated orthogonal matrix.

14.2.4 The reflection tensor

In the Euclidean space $E^3(\mathbb{R})$, we consider a plane π orthogonal to an axis given by its unit vector $\vec{e}$ (see Figure 14.6).

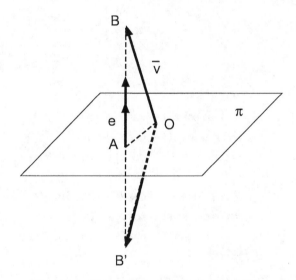

Fig. 14.6. Illustration of the reflection tensor.

The "reflection" tensor, also called the Householder tensor, reflects the image of the vector $\vec{v}$, as if it were a mirror. We denote this tensor by $H_{\vec{e}}$. The vectors $\vec{e}$ and $\vec{v}$ are those already treated in previous sections. Figure 14.6 declares that $\overline{AB} = P_{\vec{e}}(\vec{v})$, and that $\overline{B'B} = 2P_{\vec{e}}(\vec{v})$, and since $\overline{OB}' = H_{\vec{e}}(\vec{v})$, the following vector equation holds:

$$\overline{OB}' + \overline{B'B} = \overline{OB}; \quad \overline{OB}' = \overline{OB} - \overline{B'B}$$

and then

$$H_{\vec{e}}(\vec{v}) = \vec{v} - 2P_{\vec{e}}(\vec{v}) = (I_3 - 2P_{\vec{e}})\vec{v} \qquad (14.15)$$

and taking into account Formula (14.3), the reflection tensor has the matrix representation

$$\vec{v}' = H_{\vec{e}}(\vec{v});$$

$$\begin{bmatrix} (x^1)' \\ (x^2)' \\ (x^3)' \end{bmatrix} = \begin{bmatrix} \sin^2\alpha - \cos^2\alpha & -2\cos\alpha\cos\beta & -2\cos\alpha\cos\gamma \\ -2\cos\alpha\cos\beta & \sin^2\beta - \cos^2\beta & -2\cos\beta\cos\gamma \\ -2\cos\alpha\cos\gamma & -2\cos\beta\cos\gamma & \sin^2\gamma - \cos^2\gamma \end{bmatrix} \begin{bmatrix} x^1 \\ x^2 \\ x^3 \end{bmatrix} \quad (14.16)$$

Tensor $H_{\vec{e}}$ is also an "isometry", that is, of orthogonal matrix, because

$$H_{\vec{e}} \bullet H_{\vec{e}}^t = (I_3 - 2P_{\vec{e}})(I_3 - 2P_{\vec{e}})^t = (I_3 - 2P_{\vec{e}})^2 = I_3 + 4P_{\vec{e}}^2 - 4P_{\vec{e}} = I_3 + 4P_{\vec{e}} - 4P_{\vec{e}} = I_3$$

and so $H_{\vec{e}}^t = H_{\vec{e}}^{-1}$.

Example 14.2 (Properties of tensors $M_{to}(\vec{e}\times), R_{\vec{e}}$ and $H_{\vec{e}}$).

1. Show that the tensor $M_{to}(\vec{e}\times)$ is a contraction of the Levi-Civita tensor permutation and of the unit vector $\vec{e}$.
2. Show that the components of the "axis" of the tensor $M_{to}(\vec{e}\times)$ respond to the expression $\cos\alpha_i = -\frac{1}{2}\epsilon_{ijk}^{\circ\circ\circ} m_{jk}^{\circ\circ}$.
3. Show that $\vec{e}$ is an eigenvector of the rotation tensor $R_{\vec{e}}$. What is its eigenvalue?
4. Given a rotation tensor $R_{\vec{e}}$ by its associated matrix, obtain the value of θ.
5. Show that the determinant $|R_{\vec{e}}| = 1$.
6. From the data tensor $R_{\vec{e}}$, obtain the tensor $M_{to}(\vec{e}\times)$.
7. Show the involutive character of the tensor $H_{\vec{e}}$, that is, when applied twice it returns to the initial state.
8. Show that the determinant $|H_{\vec{e}}| = -1$.

Note: We remind the reader that in orthonormalized bases $\{\vec{e}_\alpha\}$, the tensor indices can appear in contravariant or covariant positions, indistinctly.

Solution:

1. Consider the Levi-Civita permutation tensor of dimension $n = 3$:

$$[\epsilon_{ijk}^{\circ\circ\circ}] = \begin{bmatrix} 0 & 0 & 0 \\ 0 & 0 & 1 \\ 0 & -1 & 0 \\ \hline 0 & 0 & -1 \\ 0 & 0 & 0 \\ 1 & 0 & 0 \\ \hline 0 & 1 & 0 \\ -1 & 0 & 0 \\ 0 & 0 & 0 \end{bmatrix},$$

where i is the block row index, j is the row of each block and k is the column of each block.

Let $\vec{e}$ be the unit vector, $\vec{e} = \cos\alpha\,\vec{e}_1 + \cos\beta\,\vec{e}_2 + \cos\gamma\,\vec{e}_3$ we have

$$[m_{jk}^{\circ\circ}] = [-\epsilon_{\theta jk}^{\circ\circ\circ}x^{\theta}] = [-\epsilon_{1jk}^{\circ\circ\circ}x^1 - \epsilon_{2jk}^{\circ\circ\circ}x^2 - \epsilon_{3jk}^{\circ\circ\circ}x^3]$$

$$= -\begin{bmatrix} 0 & 0 & 0 \\ 0 & 0 & 1 \\ 0 & -1 & 0 \end{bmatrix}\cos\alpha - \begin{bmatrix} 0 & 0 & -1 \\ 0 & 0 & 0 \\ 1 & 0 & 0 \end{bmatrix}\cos\beta - \begin{bmatrix} 0 & 1 & 0 \\ -1 & 0 & 0 \\ 0 & 0 & 0 \end{bmatrix}\cos\gamma$$

$$= \begin{bmatrix} 0 & -\cos\gamma & \cos\beta \\ \cos\gamma & 0 & -\cos\alpha \\ -\cos\beta & \cos\alpha & 0 \end{bmatrix},$$

which is the matrix associated with the tensor $M_{to}(\vec{e}\times)$.

2. We have

$$\cos\alpha_1 = -\frac{1}{2}\left[\epsilon_{123}^{\circ\circ\circ}m_{23}^{\circ\circ} + \epsilon_{132}^{\circ\circ\circ}m_{32}^{\circ\circ}\right] = -\frac{1}{2}\left((-\cos\alpha) + (-1)\cos\alpha\right) = \cos\alpha$$

$$\cos\alpha_2 = -\frac{1}{2}\left[\epsilon_{213}^{\circ\circ\circ}m_{13}^{\circ\circ} + \epsilon_{231}^{\circ\circ\circ}m_{31}^{\circ\circ}\right] = -\frac{1}{2}\left((-1)\cos\beta) + (-\cos\beta)\right) = \cos\beta$$

$$\cos\alpha_3 = -\frac{1}{2}\left[\epsilon_{312}^{\circ\circ\circ}m_{12}^{\circ\circ} + \epsilon_{321}^{\circ\circ\circ}m_{21}^{\circ\circ}\right] = -\frac{1}{2}\left((-\cos\gamma) + (-1)\cos\gamma\right) = \cos\gamma$$

and then
$$\cos\alpha_i\,\vec{e}_i = \cos\alpha\,\vec{e}_1 + \cos\beta\,\vec{e}_2 + \cos\gamma\,\vec{e}_3 = \vec{e}.$$

3. Using Formula (14.13), we have

$$R_{\vec{e}}(\vec{e}) = \cos\theta I_3(\vec{e}) + (1 - \cos\theta)P_{\vec{e}}(\vec{e}) + \sin\theta M_{to}(\vec{e}\times)(\vec{e}),$$

and the projection of the unit vector $\vec{e}$ over $\vec{e}$ is the proper $\vec{e}$, i.e., $P_{\vec{e}}(\vec{e}) = \vec{e}$, and since

$$\begin{bmatrix} 0 & -\cos\gamma & \cos\beta \\ \cos\gamma & 0 & -\cos\alpha \\ -\cos\beta & \cos\alpha & 0 \end{bmatrix}\begin{bmatrix} \cos\alpha \\ \cos\beta \\ \cos\gamma \end{bmatrix} = \begin{bmatrix} -\cos\beta\cos\gamma + \cos\beta\cos\gamma \\ \cos\alpha\cos\gamma - \cos\alpha\cos\gamma \\ -\cos\alpha\cos\beta + \cos\alpha\cos\beta \end{bmatrix} = \begin{bmatrix} 0 \\ 0 \\ 0 \end{bmatrix},$$

we obtain $M_{to}(\vec{e}\times)(\vec{e}) = \vec{0}$, and then

$$R_{\vec{e}}(\vec{e}) = \cos\theta\vec{e} + (1 - \cos\theta)\vec{e} + \vec{0} = \vec{e},$$

which shows it is an eigenvector of eigenvalue $\lambda = 1$.

4. From (14.14) we obtain the trace of the $R_{\vec{e}}$ matrix, which leads to

$$\text{trace}(R_{\vec{e}}) = (\cos^2\alpha + \cos^2\beta + \cos^2\gamma) + 2(\cos^2\alpha + \cos^2\beta + \cos^2\gamma)\cos\theta$$
$$= 1 + 2\cos\theta,$$

from which we get
$$\cos\theta = \frac{\text{trace}(R_{\vec{e}}) - 1}{2}.$$

5. In Section 14.2.3 it was shown that the rotation tensor $R_{\vec{e}}$ has an associated orthogonal matrix, and $|R_{\vec{e}}| = \pm 1$; but choosing the axis $\vec{e} \equiv \vec{e}_3$ (OZ axis), $\cos \alpha = \cos \beta = 0$ and $\cos \gamma = 1$, Formula (14.14) after taking determinants leads to

$$|R_{\vec{e}}| = \begin{vmatrix} 1 & 0 & 0 \\ 0 & \cos\theta & -\sin\theta \\ 0 & \sin\theta & \cos\theta \end{vmatrix} = \cos^2\theta + \sin\theta = +1.$$

6. Transposing the matrix equation (14.13), we have

$$R_{\vec{e}}^t = \cos\theta I_3 + (1 - \cos\theta)P_{\vec{e}} - \sin\theta M_{to}(\vec{e}\times),$$

and subtracting this equation, from (14.13):

$$R_{\vec{e}} - R_{\vec{e}}^t = 2\sin\theta M_{to}(\vec{e}\times) \rightarrow M_{to}(\vec{e}\times) = \frac{1}{2\sin\theta}(R_{\vec{e}} - R_{\vec{e}}^t).$$

7. Because of Formula (14.15), we have

$$H_{\vec{e}}^2 = (I_3 - 2P_{\vec{e}})^2 = I_3^2 + 4P_{\vec{e}}^2 - 4P_{\vec{e}} = I_3 + 4P_{\vec{e}} - 4P_{\vec{e}} = I_3.$$

8. Since $H_{\vec{e}}$ is of orthogonal matrix, $|H_{\vec{e}}| = \pm 1$, and choosing $\vec{e} = \vec{e}_3$, that is, $\cos\alpha = \cos\beta = 0$ and $\cos\gamma = 1$, Formula (14.16) leads to

$$|H_{\vec{e}}| = \begin{vmatrix} 1 & 0 & 0 \\ 0 & 1 & 0 \\ 0 & 0 & -1 \end{vmatrix} = -1.$$

□

Example 14.3 (Reduction of vector systems). In the punctual affine space $E_p^3(\mathbb{R})$ or ordinary geometrical space, we consider the system of vectors $\vec{V}_i$:

$$\vec{V}_1 = 2\vec{e}_1 + 4\vec{e}_2 - 3\vec{e}_3 \text{ lying on the point } A_1(-1, 0, 2)$$
$$\vec{V}_2 = \vec{e}_1 - \vec{e}_2 + 2\vec{e}_3 \text{ lying on the point } A_2(1, 2, 3)$$
$$\vec{V}_3 = 3\vec{e}_1 - 2\vec{e}_2 + \vec{e}_3 \text{ lying on the point } A_3(2, 1, 0)$$
$$\vec{V}_4 = \vec{e}_2 + \vec{e}_3 \text{ lying on the point } A_4(0, 1, 2).$$

1. Obtain the so-called resultant vector $\vec{R} = \sum_{i=1}^{4} \vec{V}_i$ (it is a free vector).
2. Obtain the moment vector of the system with respect to the origin $O(0, 0, 0)$, $\vec{M}_O = \sum_{i=1}^{4} M_{to}(\vec{r}_i \times)(\vec{V}_i)$, where $\vec{r}_i = \overline{OA_i}$ is the position vector of each point A_i, and $\vec{M}_O$ is a free vector, but usually it is assumed to lie on O.
3. Find the moment vector of the system with respect to the point $O'(1, 1, 2)$, $\vec{M}_{O'}$, where now $\vec{r}_i' = \overline{O'A_i}$. The vectors $\vec{M}_O$ and $\vec{M}_{O'}$ must be related.

4. Find a point $E(x, y, z)$ such that $\vec{M}_E = \lambda\vec{R}$, that is, that the moment vector $\vec{M}_E$ be parallel to the resultant. Determine the constant λ.

5. Find the Cartesian equation of the central axis e of the system, the equation of the straight line passing through the point E, with the direction of the resultant vector $\vec{R}$.

6. Show that any point E' on the axis e satisfies the property of the point E: $\vec{M}_{E'} = \vec{M}_E$.

   This common vector $\vec{m} = \vec{M}_E$, is known as the "minimum moment vector" of the system.

7. Show that $\vec{m}$ is the projection of the vector $\vec{M}_O$ over the axis e, i.e., $\vec{m} = P_{\vec{e}}(\vec{M}_O)$.

8. Find the constant ρ called the "rotation radius" of the system:

$$\rho = \frac{|P_{\vec{e}\perp}(\vec{M}_O)|}{|\vec{R}|}.$$

Solution:

1. Summing the four vectors we obtain

$$\vec{R} = 6\vec{e}_1 + 2\vec{e}_2 + \vec{e}_3.$$

2. Using (14.7) we get

$$\begin{bmatrix} m^1 \\ m^2 \\ m^3 \end{bmatrix} = \begin{bmatrix} 0 & -2 & 0 \\ 2 & 0 & 1 \\ 0 & -1 & 0 \end{bmatrix} \begin{bmatrix} 2 \\ 4 \\ -3 \end{bmatrix} + \begin{bmatrix} 0 & -3 & 2 \\ 3 & 0 & -1 \\ -2 & 1 & 0 \end{bmatrix} \begin{bmatrix} 1 \\ -1 \\ 2 \end{bmatrix}$$

$$+ \begin{bmatrix} 0 & 0 & 1 \\ 0 & 0 & -2 \\ -1 & 2 & 0 \end{bmatrix} \begin{bmatrix} 3 \\ -2 \\ 1 \end{bmatrix} + \begin{bmatrix} 0 & -2 & 1 \\ 2 & 0 & 0 \\ -1 & 0 & 0 \end{bmatrix} \begin{bmatrix} 0 \\ 1 \\ 1 \end{bmatrix}$$

$$= \begin{bmatrix} -8 \\ 1 \\ -4 \end{bmatrix} + \begin{bmatrix} 7 \\ 1 \\ -3 \end{bmatrix} + \begin{bmatrix} 1 \\ -2 \\ -7 \end{bmatrix} + \begin{bmatrix} -1 \\ 0 \\ 0 \end{bmatrix} = \begin{bmatrix} -1 \\ 0 \\ -14 \end{bmatrix};$$

$$\vec{M}_O = -\vec{e}_1 - 14\vec{e}_3.$$

3. We know that

$$\vec{r}_i{}' - \vec{r}_i = \overline{O'A_i} - \overline{OA_i} = \overline{O'O} = -\overline{OO'} = -\vec{e}_1 - \vec{e}_2 - 2\vec{e}_3$$

$$\vec{M}_{O'} - \vec{M}_O = \sum_{i=1}^{4} \left[M_{to}(\vec{r}_i{}' \times)(\vec{V}_i) - M_{to}(\vec{r}_i \times)(\vec{V}_i) \right]$$

$$= \sum_{i=1}^{4} M_{to}\left((\vec{r}_i{}' - \vec{r}_i)\times\right)(\vec{V}_i) = \sum_{i=1}^{4} M_{to}(\overline{O'O}\times)(\vec{V}_i)$$

$$= M_{to}(\overline{O'O}\times)\left(\sum_{i=1}^{4} \vec{V}_i\right) = M_{to}(\overline{O'O}\times)(\vec{R});$$

we then have the final formula $\vec{M}_{O'} = \vec{M}_O + M_{to}(\overline{OO'}\times)(\vec{R})$, which in matrix form leads to

$$\begin{bmatrix} -1 \\ 0 \\ -14 \end{bmatrix} + \begin{bmatrix} 0 & 2 & -1 \\ -2 & 0 & 1 \\ 1 & -1 & 0 \end{bmatrix} \begin{bmatrix} 6 \\ 2 \\ 1 \end{bmatrix} = \begin{bmatrix} 2 \\ -11 \\ -10 \end{bmatrix}$$

$$\vec{M}_{O'} = 2\vec{e}_1 - 11\vec{e}_2 - 10\vec{e}_3.$$

4. Applying the last vector formula to the point $E(x, y, z)$, we have

$$\vec{M}_E = \vec{M}_O + M_{to}(\overline{EO}\times)(\vec{R}) = \lambda\vec{R},$$

where

$$\overline{EO} = -\overline{OE} = -x\vec{e}_1 - y\vec{e}_2 - z\vec{e}_3,$$

and in matrix form

$$\begin{bmatrix} m^1 \\ m^2 \\ m^3 \end{bmatrix}_E = \begin{bmatrix} -1 \\ 0 \\ -14 \end{bmatrix} + \begin{bmatrix} 0 & z & -y \\ -z & 0 & x \\ y & -x & 0 \end{bmatrix} \begin{bmatrix} 6 \\ 2 \\ 1 \end{bmatrix} = \lambda \begin{bmatrix} 6 \\ 2 \\ 1 \end{bmatrix},$$

which leads to the system of equations

$$\begin{cases} -y + 2z = 1 + 6\lambda \\ x - 6z = 2\lambda \\ -2x + 6y = 14 + \lambda. \end{cases}$$

Since one of its solutions is

$$E\left(-\frac{22}{41}, \frac{85}{41}, \frac{3}{41}\right); \quad \lambda = -\frac{20}{41},$$

we have

$$\begin{bmatrix} m^1 \\ m^2 \\ m^3 \end{bmatrix}_E = \left(-\frac{20}{41}\right) \begin{bmatrix} 6 \\ 2 \\ 1 \end{bmatrix} = \begin{bmatrix} -\frac{120}{41} \\ -\frac{40}{41} \\ -\frac{20}{41} \end{bmatrix} \rightarrow \vec{M}_E = \frac{1}{41}\left(-120\vec{e}_1 - 40\vec{e}_2 - 20\vec{e}_3\right).$$

5. An equation of the central axis e is

$$e \equiv \frac{x + \frac{22}{41}}{6} = \frac{y - \frac{85}{41}}{2} = \frac{z - \frac{3}{41}}{1} = \mu.$$

6. The parametric equations of this axis give the general expression for an arbitrary point E' on it (for $\mu = 0$, we obtain the point E):

$$E' \begin{cases} x = -\frac{22}{41} + 6\mu \\ y = \frac{85}{41} + 2\mu \\ z = \frac{3}{41} + \mu \end{cases}$$

so that we determine the vector $\vec{M}_{E'}$. According to Formula (14.17), applied to the point E', we get

$$\vec{M}_{E'} = \vec{M}_O + M_{to}(\overline{E'O}\times)(\vec{R});$$

and in matrix form

$$\begin{bmatrix} m^1 \\ m^2 \\ m^3 \end{bmatrix}_{E'} = \begin{bmatrix} -1 \\ 0 \\ -14 \end{bmatrix} + \begin{bmatrix} 0 & \left(\frac{3}{41} + \mu\right) & -\left(\frac{85}{41} + 2\mu\right) \\ -\left(\frac{3}{41} + \mu\right) & 0 & \left(-\frac{22}{41} + 6\mu\right) \\ \left(\frac{85}{41} + 2\mu\right) & -\left(\frac{-22}{41} + 6\mu\right) & 0 \end{bmatrix} \begin{bmatrix} 6 \\ 2 \\ 1 \end{bmatrix}$$

$$= \begin{bmatrix} -1 + \frac{6}{41} + 2\mu - \frac{85}{41} - 2\mu \\ 0 - \frac{18}{41} - 6\mu - \frac{22}{41} + 6\mu \\ -14 + \frac{510}{41} + 12\mu + \frac{44}{41} - 12\mu \end{bmatrix} = \begin{bmatrix} -\frac{120}{41} \\ -\frac{40}{41} \\ -\frac{20}{41} \end{bmatrix};$$

$$\vec{M}_{E'} = \frac{1}{41}(-120\vec{e}_1 - 40\vec{e}_2 - 20\vec{e}_3),$$

that compared with the result $\vec{M}_E$ of question 4, allows us to conclude that effectively it is the vector $\vec{m}$ of the system:

$$\vec{m} = \vec{M}_E = \vec{M}_{E'} = \frac{1}{41}(-120\vec{e}_1 - 40\vec{e}_2 - 20\vec{e}_3).$$

7. The axis unit vector $\vec{e}$ is

$$|\vec{R}| = \sqrt{6^2 + 2^2 + 1^2} = \sqrt{41}; \quad \vec{e} \equiv \frac{6}{\sqrt{41}}\vec{e}_1 + \frac{2}{\sqrt{41}}\vec{e}_2 + \frac{1}{\sqrt{41}}\vec{e}_3.$$

and since the projection tensor $P_{\vec{e}}(\vec{v})$ responds to Formula (14.3), in the present case $P_{\vec{e}}(\vec{M}_O)$ is

$$\begin{bmatrix} m^1 \\ m^2 \\ m^3 \end{bmatrix} = \begin{bmatrix} \frac{36}{41} & \frac{12}{41} & \frac{6}{41} \\ \frac{12}{41} & \frac{4}{41} & \frac{2}{41} \\ \frac{6}{41} & \frac{2}{41} & \frac{1}{41} \end{bmatrix} \begin{bmatrix} -1 \\ 0 \\ -14 \end{bmatrix} = \begin{bmatrix} -\frac{120}{41} \\ -\frac{40}{41} \\ -\frac{20}{41} \end{bmatrix},$$

which implies

$$P_{\vec{e}}(\vec{M}_O) = \vec{m} = \frac{1}{41}(-120\vec{e}_1 - 40\vec{e}_2 - 20\vec{e}_3).$$

8. The complementary projection tensor is given by (14.4), whence $P_{\vec{e}\perp}(\vec{M}_O)$ in matrix form is

$$\begin{bmatrix} \frac{5}{41} & -\frac{12}{41} & -\frac{6}{41} \\ -\frac{12}{41} & \frac{37}{41} & -\frac{2}{41} \\ -\frac{6}{41} & -\frac{2}{41} & \frac{40}{41} \end{bmatrix} \begin{bmatrix} -1 \\ 0 \\ -14 \end{bmatrix} = \frac{1}{41} \begin{bmatrix} 79 \\ 40 \\ -554 \end{bmatrix}$$

and then

$$|P_{\vec{e}\perp}(\vec{M}_O)| = \frac{1}{41}\sqrt{79^2 + 40^2 + 554^2} = \frac{3\sqrt{853 \times 41}}{41}$$

$$\rho = \frac{|P_{\vec{e}\perp}(\vec{M}_O)|}{|\vec{R}|} = \frac{3\sqrt{853 \times 41}}{41\sqrt{41}} = \frac{3\sqrt{853}}{41}.$$

The vectors $\vec{R}$ and $\vec{m}$ applied on any point of the axis e, are equivalent to the data of the vector system, and are its "reduction".   □

14.3 Affine geometric tensors. Homographies

14.3.1 Preamble

The development of the axiomatic properties of the punctual spaces affine to linear spaces corresponds to other books that aim to study analytic geometry, and thus, in the present chapter we assume that the reader is sufficiently familiar with this geometry, to capture with clarity the contents to be discussed. Our punctual real space will be notated $E_p^n(\mathbb{R})$, where E_p refers to a punctual space and n is the dimension of the Euclidean linear space $E^n(\mathbb{R})$ affine to our punctual space.

The punctual space $E_p^n(\mathbb{R})$ is endowed with a reference frame system

$$\{O, \vec{e}_\alpha\} \equiv \{\text{Cartesian axes, of abscissas, of ordinates, of levels, etc.}\}.$$

Such axes are created to "supply support" to the *basic* vectors $\{\vec{e}_\alpha\}$ of the associated Euclidean linear space. While other circumstances are not specified, we assume classical orthonormalized reference systems. Any point $P \in E^n(\mathbb{R})$ has some Cartesian coordinates $X \equiv \begin{bmatrix} x^1 \\ x^2 \\ \vdots \\ x^n \end{bmatrix}$ with a one-to-one correspondence with the position vector

$$\vec{OP} \equiv [\vec{e}_1\ \vec{e}_2\ \ldots\ \vec{e}_n]\, X \equiv x^1\vec{e}_1 + x^2\vec{e}_2 + \cdots + x^n\vec{e}_n \in E^n(\mathbb{R}). \qquad (14.17)$$

Since as in geometric spaces there exist unattainable (at infinity) points P that must be frequently used, a new coordinate is added to the point P, that declares if it is attainable. Thus, any *attainable point* P has two types of coordinates, $P(x^1, x^2, \ldots, x^n)$, the *Cartesian* coordinates and $P(x'^1, x'^2, \ldots, x'^n, t)$, the *homogeneous* coordinates, which are related by the equality

$$\begin{bmatrix} x^1 \\ x^2 \\ \vdots \\ x^n \end{bmatrix} = \begin{bmatrix} x'^1/t \\ x'^2/t \\ \vdots \\ x'^n/t \end{bmatrix}. \qquad (14.18)$$

To transform an attainable point P from Cartesian coordinates to homogeneous coordinates we let $t = 1$. If P is unattainable, it has only homogeneous coordinates, with $t = 0$. To identify an infinite point (a direction), we project such a direction from the origin (see Figure 14.7), by means of a parallel r by

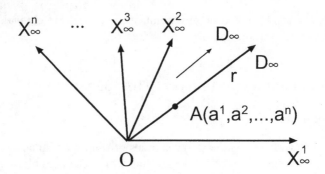

Fig. 14.7. Illustration of how to pass from Cartesian to homogeneous coordinates.

O, and taking an arbitrary attainable point A, on r, we complete its Cartesian coordinates with $t = 0$, in order to get its homogeneous coordinates

$$D_\infty(a^1, a^2, \ldots, a^n, 0).\qquad(14.19)$$

If desired, the point A can be at a unit distance from O, that is, $|\vec{OA}| = 1$. For this, it suffices to divide the coordinates by its modulus, resulting (because the fundamental metric tensor is $G \equiv I_n$) in the expression

$$D_\infty\left(\frac{a^1}{\sqrt{(a^1)^2 + \cdots + (a^n)^2}}, \frac{a^2}{\sqrt{(a^1)^2 + \cdots + (a^n)^2}}, \ldots, \frac{a^n}{\sqrt{(a^1)^2 + \cdots + (a^n)^2}}, 0\right)$$
$$(14.20)$$

which some authors call the *unit vector director* or versor of the given direction D_∞; other authors notate as $D_\infty(\cos\alpha_1, \cos\alpha_2, \ldots, \cos\alpha_n, 0)$ each component of (14.20), by means of the "director cosines" of the unit vector director, $\vec{v}$, because it is obvious that if the mentioned vector is

$$\vec{v} \equiv \vec{OA} \equiv \begin{bmatrix} \vec{e}_1 & \vec{e}_2, & \cdots & \vec{e}_j & \cdots & \vec{e}_n \end{bmatrix} \begin{bmatrix} \dfrac{a^1}{\sqrt{(a^1)^2+(a^2)^2+\cdots+(a^n)^2}} \\ \dfrac{a^2}{\sqrt{(a^1)^2+(a^2)^2+\cdots+(a^n)^2}} \\ \vdots \\ \dfrac{a^n}{\sqrt{(a^1)^2+(a^2)^2+\cdots+(a^n)^2}} \end{bmatrix}\qquad(14.21)$$

then its jth component can be written as

$$\frac{a^j}{\sqrt{(a^1)^2 + (a^2)^2 + \cdots + (a^n)^2}} = \begin{bmatrix} 0 & 0 & \cdots & 1 & \cdots & 0 \end{bmatrix} I_n \begin{bmatrix} \dfrac{a^1}{\sqrt{(a^1)^2+(a^2)^2+\cdots+(a^n)^2}} \\ \dfrac{a^2}{\sqrt{(a^1)^2+(a^2)^2+\cdots+(a^n)^2}} \\ \vdots \\ \dfrac{a^n}{\sqrt{(a^1)^2+(a^2)^2+\cdots+(a^n)^2}} \end{bmatrix}$$
$$= \vec{e}_j \bullet \vec{v} = 1 \times 1 \times \cos\alpha_j = \cos\alpha_j.\qquad(14.22)$$

We have the problematic of the unattainable points as D_∞, because the use of geometric tensors (seldom used in the usual textbooks, which limit their scope to operation over attainable points P) for transforming them is licit and very convenient on some occasions.

14.3.2 Definition and representation

Let $E_p^n(\mathbb{R})$ be the ordinary punctual geometric space, affine to a Euclidean space $E^n(\mathbb{R})$ referred to an orthonormalized basis $\{\vec{e}_\alpha\}$ ($G \equiv I_n$). Assume that a certain linear operator or endomorphism T transforms vectors in $E^n(\mathbb{R})$. An affine mapping $\mathcal{A} : E_p^n(\mathbb{R}) \to E^n(\mathbb{R})$, that associates with each pair of points P and Q of the punctual space $E_p^n(\mathbb{R})$ a vector $\vec{v} \in E^n(\mathbb{R})$, i.e., $\mathcal{A}(P,Q) = \vec{v}$, endows the punctual space $E_p^n(\mathbb{R})$ of an orthonormalized reference frame $\{O; \vec{e}_\alpha\}$, corresponding to the orthonormalized basis of the Euclidean space $E^n(\mathbb{R})$. Moreover, the linear operator T induces in the space $E_p^n(\mathbb{R})$, by means of the affinity $\mathcal{A}$, an "affine punctual transformation" that is denoted by f, and that satisfies the axiomatic properties that follow:

1. $f : E_p^n(\mathbb{R}) \to E_p^n(\mathbb{R})$; f is a linear mapping in $E_p^n(\mathbb{R})$.

$$\forall P, Q \in E_p^n(\mathbb{R}) : f(P) = P'; \ f(Q) = Q'; \ P', Q' \in E_p^n(\mathbb{R}),$$

where *it is not* specified if P' and Q' are or not, attainable.
2. $\mathcal{A}(O, Q') = T[\mathcal{A}(O, Q)] = T[\mathcal{A}(O, P)] + T[\mathcal{A}(P, Q)]$, where P and Q are arbitrary attainable points of $E_p^n(\mathbb{R})$.

If a point $X \in E_p^n(\mathbb{R})$ satisfies the property

$$f(X) = X \tag{14.23}$$

we say that it is a "double" or an "invariant" point.

An affine transformation in $E_p^n(\mathbb{R})$ can have $0, 1, 2, \ldots, n$ double or invariant points, and even a whole affine linear subspace of invariant points, as straight lines, planes, etc.

Since we can notate the affine transformation as a tensor, and we can use it for unattainable points P, we use matrix representations of order $(n + 1)$ and we work in "homogeneous coordinates". The affine geometric tensor transformations are also called *homographies*, and their matrix representation is

$$X' = F \bullet X \to \begin{bmatrix} x'^1 \\ x'^2 \\ \vdots \\ x'^n \\ - \\ t' \end{bmatrix} = \begin{bmatrix} t_{11} & t_{12} & \cdots & t_{1n} & | & a^1 \\ t_{21} & t_{22} & \cdots & t_{2n} & | & a^2 \\ \cdots & \cdots & \cdots & \cdots & | & \cdots \\ t_{n1} & t_{n2} & \cdots & t_{nn} & | & a^n \\ - & - & \cdots & - & + & - \\ b_1 & b_2 & \cdots & b_n & | & b_{n+1} \end{bmatrix} \bullet \begin{bmatrix} x^1 \\ x^2 \\ \vdots \\ x^n \\ - \\ t \end{bmatrix}, \tag{14.24}$$

where F is the affine tensor matrix.

The block of order n of the homography F in (14.24) corresponds to the matrix $T = [t_{ij}]$ of the linear operator T of the Euclidean space $E_n(\mathbb{R})$.

In homogeneous coordinates, the point $O'(a^1, a^2, \ldots, a^n, b_{n+1})$ is the transformed point of the origin O, $(0, 0, \ldots, 0, \ldots, 0, 1)$.

If $b_{n+1} = 0$, the affine point of the origin O is an infinite point O'_∞ of the resulting direction.

If $b_{n+1} \neq 0$, we divide all coordinates by b_{n+1} for the b_{n+1} to become unity, and the point

$$O'\left(\frac{a^1}{b_{n+1}}, \frac{a^2}{b_{n+1}}, \ldots, \frac{a^n}{b_{n+1}}, 1\right)$$

in homogeneous coordinates, is the Cartesian attainable point

$$O'\left(\frac{a^1}{b_{n+1}}, \frac{a^2}{b_{n+1}}, \ldots, \frac{a^n}{b_{n+1}}\right), \tag{14.25}$$

which is the affine transform of the origin O, in this second case.

The homographies of dimension $n = 2$ (in the plane), with an invariant point and a straight line of invariant points, are called *homologies*.

14.3.3 Affinities

If a homography transforms attainable points into attainable points, unattainable points into unattainable points, aligned points into aligned points and in addition maintains the simple ratios of three aligned points A, B, C, then, if

$$(A, B, C) = \frac{\overline{AB}}{\overline{AC}} \Rightarrow (A', B', C') = (A, B, C) \tag{14.26}$$

is called an *affinity* or an affine homographic transformation.

The geometric transformations are classified as "regular" if $|F| \neq 0$ and "singular" or non-regular, if $|F| = 0$.

The matrix representation of affinities is

$$X' = A \bullet X \rightarrow \begin{bmatrix} x'^1 \\ x'^2 \\ \vdots \\ x'^n \\ - \\ t' \end{bmatrix} = \begin{bmatrix} a_{11} & a_{12} & \cdots & a_{1n} & | & \ell^1 \\ a_{21} & a_{22} & \cdots & a_{2n} & | & \ell^2 \\ \cdots & \cdots & \cdots & \cdots & | & \cdots \\ a_{n1} & a_{n2} & \cdots & a_{nn} & | & \ell^n \\ - & - & \cdots & - & + & - \\ 0 & 0 & \cdots & 0 & | & 1 \end{bmatrix} \bullet \begin{bmatrix} x^1 \\ x^2 \\ \vdots \\ x^n \\ - \\ t \end{bmatrix} \tag{14.27}$$

where the transformation matrix is notified by A, and

$$|A| \neq 0, \tag{14.28}$$

i.e., a one-to-one correspondence is required, that is, regularity.

When the block matrix T of A is an orthogonal matrix $(T^t \equiv T^{-1})$, the determinant $|A| = \pm 1$, and the affinity is called *isometry* (direct if $|A| = +1$; and reverse if $|A| = -1$)

The affinities constitute a multiplicative *non-Abelian* group.

Example 14.4 (Affinity). In the ordinary Cartesian system $(O–XY)$ of dimension $n = 2$ consider an affinity such that an invariant point is the point D_∞ in the bisectrix of the second quadrant (affinity direction) and a straight line of invariant points (the affinity axis e) is the straight line $2x - y - 6 = 0$ (Figure 14.8). We also know that in this affinity the straight line r that joins two arbitrary affine points B, B' and D_∞ intersects the axis at the point S such that the simple ratio (S, B', B) is *constant*: $(S, B', B) = k$; $k \in \mathbb{R}$; $k \neq 0$. Such a constant is called the "affinity ratio", which in our case is $k = 3$.

1. Determine the Cartesian equation of the straight line r that joins the points O and D_∞, and obtain the point S, on the axis.
2. Since $O' \in r$ and $(S, O', O) = k$, determine the point O'.
3. Obtain the affinity tensor A.
4. Confirm the above data, obtaining the invariant points of the tensor A resulting from the previous question.
5. Determine the vertices of the triangle $\Delta(O', P', Q')$ affine of the triangle $O(0,0), P(3,0), Q(0,2)$.
6. Confirm that the quotient of the areas σ of the triangles $\Delta(O', P', Q')$ and $\Delta(O, P, Q)$ is the affinity ratio k.
7. Verify that the baricenters G and G' of the triangles above are affine points.

Solution:

1. The point D_∞ has homogeneous coordinates $D_\infty(-1, 1, 0)$. The attainable point O has homogeneous coordinates $O(0, 0, 1)$. Thus, the equation of the straight line r in homogeneous coordinates is

$$r \equiv \frac{X - 0}{-1} = \frac{Y - 0}{1} = \frac{t - 1}{0} \rightarrow Y + X = 0,$$

and in Cartesian coordinates

$$r \equiv y + x = 0.$$

Solving the system

$$\begin{cases} Y + X &= 0 \\ 2X - Y - 6t &= 0 \end{cases}$$

we obtain $S(2, -2, 1)$ in homogeneous coordinates.

2. Since the affinity ratio is $k > 0$, the axis *does not* separate the affine points.

Applying the affinity relation to the O, O' affine points, we have

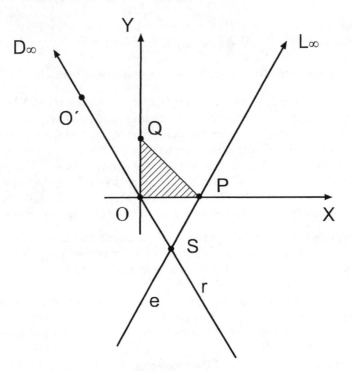

Fig. 14.8. Illustration of Example 14.4.

$$(S, O', O) = \frac{SO'}{SO} \equiv \frac{X_S - X_{O'}}{X_S - X_O} = \frac{Y_S - Y_{O'}}{Y_S - Y_O} = k,$$

Since $S(2, 2, 1)$ and $O(0, 0, 1)$ are in homogeneous coordinates, we get

$$\frac{2 - X_{O'}}{2 - 0} = 3; \quad \frac{(-2) - Y_{O'}}{(-2) - 0} = 3; \rightarrow X_{O'} = -4; \ Y_{O'} = 4,$$

which, in homogeneous coordinates, gives $O'(-4, 4, 1)$.

3. The axis equation in homogeneous coordinates is:

$$e \equiv 2X - Y - 6t = 0,$$

from which we can obtain, for example, the attainable point $P(3, 0, 1)$, and the unattainable point $L_\infty(1, 2, 0)$, etc., because they satisfy its equation. Since D_∞, S, P, L_∞ are invariant points, we know their affine transformed points.

We choose the points O, S, P and their affine transforms, to be substituted into the affine transformation matrix equation (14.27) and get

$$\begin{bmatrix} -4 & 2 & 3 \\ 4 & -2 & 0 \\ 1 & 1 & 1 \end{bmatrix} = A \begin{bmatrix} 0 & 2 & 3 \\ 0 & -2 & 0 \\ 1 & 1 & 1 \end{bmatrix}$$

$$A = \begin{bmatrix} -4 & 2 & 3 \\ 4 & -2 & 0 \\ 1 & 1 & 1 \end{bmatrix} \begin{bmatrix} 0 & 2 & 3 \\ 0 & -2 & 0 \\ 1 & 1 & 1 \end{bmatrix}^{-1} = \frac{1}{3} \begin{bmatrix} 7 & -2 & -12 \\ -4 & 5 & 12 \\ 0 & 0 & 3 \end{bmatrix},$$

which is the matrix of the affinity tensor sought after.
Evidently, $|A| = 1 \neq 0$, which proves the regular character of affinity.
We observe that the matrix A has the format required in Formula (14.27).

4. The straight line e, passing through the points $P(3,0,1)$ and $L_\infty(1,2,0)$ has a homogeneous equation $\frac{x-3}{1} = \frac{y-0}{2} = \frac{t-1}{0} = \lambda$. Thus, the parametric equation of e for its attainable points in homogeneous coordinates is

$$\begin{bmatrix} x \\ y \\ t \end{bmatrix}_e = \begin{bmatrix} 3+\lambda \\ 2\lambda \\ 1 \end{bmatrix},$$

and transforming the points of e, we obtain

$$A \begin{bmatrix} x \\ y \\ t \end{bmatrix}_e = \frac{1}{3} \begin{bmatrix} 7 & -2 & -12 \\ -4 & 5 & 12 \\ 0 & 0 & 3 \end{bmatrix} \begin{bmatrix} 3+\lambda \\ 2\lambda \\ 1 \end{bmatrix}$$

$$= \frac{1}{3} \begin{bmatrix} 9+3\lambda \\ 0+6\lambda \\ 3+0 \end{bmatrix} = \begin{bmatrix} 3+\lambda \\ 2\lambda \\ 1 \end{bmatrix} \equiv \begin{bmatrix} x \\ y \\ t \end{bmatrix}_e,$$

which shows that all points of the axis e are invariant points.
Transforming now $D_\infty(-1,1,0)$, we obtain

$$\frac{1}{3} \begin{bmatrix} 7 & -2 & -12 \\ -4 & 5 & 12 \\ 0 & 0 & 3 \end{bmatrix} \begin{bmatrix} -1 \\ 1 \\ 0 \end{bmatrix} = \frac{1}{3} \begin{bmatrix} -9 \\ 9 \\ 0 \end{bmatrix} = \begin{bmatrix} -3 \\ 3 \\ 0 \end{bmatrix},$$

which in homogeneous coordinates is the unattainable point $D_\infty(-1,1,0)$, an invariant point, as was stipulated in the statement.

5. The transformed point of $O(0,0,1)$ is the point $O'(-4,4,1)$, the transformed point of $P(3,0,1)$ is the point $P'(3,0,1)$ and the transformed point of Q is

$$\begin{bmatrix} x' \\ y' \\ t' \end{bmatrix} = \frac{1}{3} \begin{bmatrix} 7 & -2 & -12 \\ -4 & 5 & 12 \\ 0 & 0 & 3 \end{bmatrix} \begin{bmatrix} 0 \\ 2 \\ 1 \end{bmatrix}$$

$$= \frac{1}{3} \begin{bmatrix} -16 \\ 22 \\ 3 \end{bmatrix} = \begin{bmatrix} -16/3 \\ 22/3 \\ 1 \end{bmatrix} \rightarrow Q'(-16/3, 22/3, 1).$$

6. Since we have

$$\sigma \equiv \text{Area of } \Delta(O, P, Q) = \frac{1}{2} \begin{vmatrix} 0 & 3 & 0 \\ 0 & 0 & 2 \\ 1 & 1 & 1 \end{vmatrix} = \frac{6}{2} = 3.$$

$$\sigma' \equiv \text{Area of } \Delta(O', P', Q') = \frac{1}{2} \begin{vmatrix} -4 & 3 & -16/3 \\ 4 & 0 & 22/3 \\ 1 & 1 & 1 \end{vmatrix}$$

$$= \frac{1}{2}\left(-\frac{64}{3} + 22 + \frac{88}{3} - 12\right) = 9.$$

we get

$$\frac{\sigma'}{\sigma} = \frac{9}{3} = 3 \equiv k.$$

7. The baricenters of both triangles in homogeneous coordinates are

$$X_G = \frac{0+3+0}{3} = 1; \quad Y_G = \frac{0+0+2}{3} = \frac{2}{3} \to G\left(1, \frac{2}{3}, 1\right)$$

$$X_{G'} = \frac{-4+3-16/3}{3} = -\frac{19}{9}; \quad Y_{G'} = \frac{4+0+22/3}{3} = \frac{34}{9} \to G'\left(-\frac{19}{9}, \frac{34}{9}, 1\right)$$

and the affine transform of the baricenter G is

$$\begin{bmatrix} x' \\ y' \\ t' \end{bmatrix} = \frac{1}{3}\begin{bmatrix} 7 & -2 & -12 \\ -4 & 5 & 12 \\ 0 & 0 & 3 \end{bmatrix}\begin{bmatrix} 1 \\ 2/3 \\ 1 \end{bmatrix} = \frac{1}{3}\begin{bmatrix} -4/3 - 5 \\ 10/3 + 8 \\ 3 \end{bmatrix} = \begin{bmatrix} -19/9 \\ 34/9 \\ 1 \end{bmatrix} \equiv G',$$

which shows they are affine points.

□

14.3.4 Homothecies

We give the name homothecy to an affine transformation that has a unique attainable invariant point H (the homothecy center). Any point X and its homothetic one X' are aligned with H, and satisfy the condition

$$(H, X', X) = \frac{\overline{HX'}}{\overline{HX}} = k.$$

The constant k is called the "homothecy ratio"; $k \in \mathbb{R}$, $k \neq 0$.

If $k > 0$ the homothecy is called "direct"; otherwise $(k < 0)$ it is called "inverse".

In the direct homothecies $H \notin \overline{XX'}$ and in the inverse homothecies, $H \in \overline{XX'}$ (segment).

The set of homothecies forms a multiplicative non-Abelian subgroup.

The matrix representation of the homothecy tensor F, in homogeneous coordinates is

$$\begin{bmatrix} x'^1 \\ x'^2 \\ \vdots \\ x'^n \\ - \\ t' \end{bmatrix} = \left[\begin{array}{cccc|c} k & 0 & \cdots & 0 & (1-k)h^1 \\ 0 & k & \cdots & 0 & (1-k)h^2 \\ \cdots & \cdots & \cdots & \cdots & \cdots \\ 0 & 0 & \cdots & k & (1-k)h^n \\ - & - & \cdots & - & - - - - \\ 0 & 0 & \cdots & 0 & 1 \end{array}\right] \bullet \begin{bmatrix} x^1 \\ x^2 \\ \vdots \\ x^n \\ - \\ 1 \end{bmatrix}, \qquad (14.29)$$

where $H(h^1, h^2, \ldots, h^n)$ are the Cartesian coordinates of its homothecy center.

Formula (14.29) shows that the tensor is known as soon as the homothecy center and its ratio, or any other adequate alternative data, are known.

Example 14.5 (Homothecy). In the ordinary Cartesian system $(O-XY)$ of dimension $n = 2$ we consider a homothecy of center $H(1,3)$ and a pair of homothetic points $Q(0,2)$ and $Q'(3,5)$.

1. Obtain the homothecy ratio.
2. Obtain the tensor F of the homothecy.
3. Verify that H is an invariant point.
4. Obtain the homothetic straight line of the straight line $x - y + 6 = 0$.
5. Obtain the homothetic line of the hyperbola $x^2 - y^2 = 4$.

Solution:

1. Applying to our case Formula (14.29), we obtain

$$F = \begin{bmatrix} k & 0 & (1-k)\cdot 1 \\ 0 & k & (1-k)\cdot 3 \\ 0 & 0 & 1 \end{bmatrix},$$

and since the point Q' is the transform of point Q, operating in homogeneous coordinates the points Q and Q', we must have

$$\begin{bmatrix} 3 \\ 5 \\ 1 \end{bmatrix} = \begin{bmatrix} k & 0 & (1-k) \\ 0 & k & (3-3k) \\ 0 & 0 & 1 \end{bmatrix} \bullet \begin{bmatrix} 0 \\ 2 \\ 1 \end{bmatrix} \rightarrow \begin{cases} 3 = 1-k \\ 5 = 2k+3-3k \rightarrow k = -2, \\ 1 = 1 \end{cases}$$

which shows the inverse character of the homothecy.

2. Substituting the value of the homothecy ratio $k = -2$ into the matrix F, we obtain

$$F = \begin{bmatrix} -2 & 0 & 3 \\ 0 & -2 & 9 \\ 0 & 0 & 1 \end{bmatrix},$$

which is the homothecy tensor matrix.

3. We determine the transform of the point $H(1,3,1)$ as

$$\begin{bmatrix} x' \\ y' \\ t' \end{bmatrix} = \begin{bmatrix} -2 & 0 & 3 \\ 0 & -2 & 9 \\ 0 & 0 & 1 \end{bmatrix} \begin{bmatrix} 1 \\ 3 \\ 1 \end{bmatrix} = \begin{bmatrix} 1 \\ 3 \\ 1 \end{bmatrix} \equiv H,$$

which proves its invariant character.

4. From the homothecy equation

$$\begin{bmatrix} x' \\ y' \\ t' \end{bmatrix} = \begin{bmatrix} -2 & 0 & 3 \\ 0 & -2 & 9 \\ 0 & 0 & 1 \end{bmatrix} \begin{bmatrix} x \\ y \\ t \end{bmatrix}$$

we obtain

$$
\begin{bmatrix} x \\ y \\ t \end{bmatrix} = \begin{bmatrix} -2 & 0 & 3 \\ 0 & -2 & 9 \\ 0 & 0 & 1 \end{bmatrix}^{-1} \begin{bmatrix} x' \\ y' \\ t' \end{bmatrix} = \frac{1}{2} \begin{bmatrix} -1 & 0 & 3 \\ 0 & -1 & 9 \\ 0 & 0 & 2 \end{bmatrix} \begin{bmatrix} x' \\ y' \\ t' \end{bmatrix}. \qquad (14.30)
$$

The given straight line in matrix form is $\begin{bmatrix} 1 & -1 & 6 \end{bmatrix} \begin{bmatrix} x \\ y \\ t \end{bmatrix} = 0$ and sub-

stituting (14.30) in it, we obtain

$$
\frac{1}{2} \begin{bmatrix} 1 & -1 & 6 \end{bmatrix} \begin{bmatrix} -1 & 0 & 3 \\ 0 & -1 & 9 \\ 0 & 0 & 2 \end{bmatrix} \begin{bmatrix} x' \\ y' \\ t' \end{bmatrix} = 0 \quad \Leftrightarrow \quad x' - y' - 6t' = 0;
$$

which in Cartesian coordinates is

$$
x - y - 6 = 0.
$$

5. Notating the hyperbola in matrix form and homogeneous coordinates we obtain

$$
\begin{bmatrix} x & y & t \end{bmatrix} \begin{bmatrix} 1 & 0 & 0 \\ 0 & -1 & 0 \\ 0 & 0 & -4 \end{bmatrix} \begin{bmatrix} x \\ y \\ t \end{bmatrix} = 0,
$$

and substituting again (14.30), we get

$$
\frac{1}{4} \begin{bmatrix} x' & y' & t' \end{bmatrix} \begin{bmatrix} -1 & 0 & 3 \\ 0 & -1 & 9 \\ 0 & 0 & 2 \end{bmatrix}^t \begin{bmatrix} 1 & 0 & 0 \\ 0 & -1 & 0 \\ 0 & 0 & -4 \end{bmatrix} \begin{bmatrix} -1 & 0 & 3 \\ 0 & -1 & 9 \\ 0 & 0 & 2 \end{bmatrix} \begin{bmatrix} x' \\ y' \\ t' \end{bmatrix} = 0,
$$

which, once operated, leads to the homothetic hyperbola

$$
\begin{bmatrix} x' & y' & t' \end{bmatrix} \begin{bmatrix} -1 & 0 & -3 \\ 0 & -1 & 9 \\ -3 & 9 & -88 \end{bmatrix} \begin{bmatrix} x' \\ y' \\ t' \end{bmatrix} = 0,
$$

and in Cartesian coordinates

$$
(x)^2 - (y)^2 - 6x + 18y - 88 = 0.
$$

$\square$

14.3.5 Isometries

Although the particular characteristic condition to be satisfied by the "isometries" (the block T_n must be an orthogonal matrix) has already been indicated in Section 14.2.3 in relation to affinities, we devote a brief comment to isometries to explain where that condition comes from.

In the Euclidean space $E^n(\mathbb{R})$ associated with the punctual space $E^n_p(\mathbb{R})$ there exist certain linear operators T that keep invariant the dot product of each pair of vectors and their transforms:

$$\forall \vec{a}, \vec{b} \in E^n(\mathbb{R}) \rightarrow \vec{a} \bullet \vec{b} = T(\vec{a}) \bullet T(\vec{b}). \tag{14.31}$$

In the Euclidean space $E^n(\mathbb{R})$, such operators are called "isometries". Assuming that this Euclidean space is referred to an arbitrary basis $\{\vec{e}_\alpha\}$ and that the connection fundamental metric tensor has an associated matrix G, an isometry T is recognized because its representing matrix T satisfies (14.31) by the following matrix relation:

$$T^t G T = G. \tag{14.32}$$

If the basis $\{\vec{e}_\alpha\}$ of $E^n(\mathbb{R})$ is orthonormalized, then $G \equiv I_n$, and the relation (14.32) can be written as

$$T^t \bullet T = I_n; \text{ or } T^t = T^{-1}, \tag{14.33}$$

proving that the isometry tensor under an orthonormalized basis appears as an orthogonal matrix.

Thus, since in the preamble 14.3.1 we established that our classic reference frames are orthonormalized, the requirement for "any isometry" is that T_n must be an orthogonal matrix.

Next, we deal with the *most outstanding* isometries, first those of dimension $n = 2$ (isometries in a plane) and then, those for $n = 3$ (isometries in a space). There exist other isometries without a proper name, which will not be mentioned here, but we emphasize their existence.

It is important to point out that the isometries constitute a non-Abelian multiplicative group, and we remind the reader about the subdivision formulated in Section 14.2.3, in direct isometries ($|M| = +1$) and inverse isometries ($|M| = -1$).

Translation ($n = 2$)

This isometry is characterized by the fact that the vector difference of two corresponding points is a fixed vector of $E^2(\mathbb{R})$:

If $X' = MX \rightarrow \mathcal{A}(XX') = \vec{t}; \; \vec{t} \in E^2(\mathbb{R})$ (translation vector).

Let $\vec{t} = t^1 \vec{e}_1 + t^2 \vec{e}_2$ be the translation vector, then the matrix representation of this tensor, in homogeneous coordinates, is

$$\begin{bmatrix} x' \\ y' \\ t' \end{bmatrix} = \begin{bmatrix} 1 & 0 & | & t^1 \\ 0 & 1 & | & t^2 \\ - & - & + & - \\ 0 & 0 & | & 1 \end{bmatrix} \begin{bmatrix} x \\ y \\ t \end{bmatrix}, \tag{14.34}$$

which is useful only for translating attainable points.

There exist some straight lines that are invariant. Some authors call them "double straight lines", "invariant straight lines", "guides", etc.

The term "double straight lines" is confusing, because *it is not true* that the points in such straight lines are translation "invariant or double", because this tensor lacks invariant points. Since $|M| = +1$ it is a direct isometry.

Rotation ($n = 2$)

A rotation is a direct isometry in the plane that has a unique invariant point (called its "rotation center" $C(a, b)$), such that if X and X' are two corresponding points ($X' = MX$), the angle $\hat{C}XX' = \theta$ has a constant value, that is called "rotation angle", the sense of which can be "dextrorsum" (the clockwise sense) or "sinextrorsum" (counterclockwise sense).

In this section, instead of proposing the rotation tensor directly, we shall build it before the reader, with the aim of emphasizing the importance of the multiplicative group character of the isometries, and for the reader to be able to do the same in similar circumstances.

We start by assuming that the classic rotation, of center $O(0,0)$ and angle θ, in Cartesian coordinates is known:

$$\begin{bmatrix} x' \\ y' \end{bmatrix} = \begin{bmatrix} \cos\theta & -\sin\theta \\ \sin\theta & \cos\theta \end{bmatrix} \begin{bmatrix} x \\ y \end{bmatrix}. \tag{14.35}$$

Next, we follow the following process:

1. Translation M_1. We translate the center $C(a, b)$ to the origin O.
2. Rotation M_2. We rotate one point an angle θ the sinextrorsum sense, with respect to the actual center using Formula (14.35).
3. Translation M_3. We apply the reverse translation, taking the center C back to its initial position, $C(a, b)$.

Consequently, we have

$$M_1 = \left[\begin{array}{cc|c} 1 & 0 & -a \\ 0 & 1 & -b \\ \hline 0 & 0 & 1 \end{array}\right] ; \quad M_2 = \left[\begin{array}{cc|c} \cos\theta & -\sin\theta & 0 \\ \sin\theta & \cos\theta & 0 \\ \hline 0 & 0 & 1 \end{array}\right] ; \quad M_3 = \left[\begin{array}{cc|c} 1 & 0 & a \\ 0 & 1 & b \\ \hline 0 & 0 & 1 \end{array}\right]$$

and from $M = M_3 \bullet M_2 \bullet M_1$ the result is

$$\begin{bmatrix} x' \\ y' \\ t' \end{bmatrix} = \left[\begin{array}{cc|c} \cos\theta & -\sin\theta & [(1-\cos\theta) \quad \sin\theta]\begin{bmatrix} a \\ b \end{bmatrix} \\ \sin\theta & \cos\theta & [-\sin\theta \quad (1-\cos\theta)]\begin{bmatrix} a \\ b \end{bmatrix} \\ \hline 0 & 0 & 1 \end{array}\right] \begin{bmatrix} x \\ y \\ t \end{bmatrix}, \tag{14.36}$$

which is the matrix representation of the rotation tensor (θ) in sinextrorsum sense.

Its properties include:

- The orthogonal straight line at the center point of the segment $\overline{XX'}$, passes through the rotation center.
- Two corresponding straight lines r and r' form between them the rotation angle θ and are at the same distance from the rotation center.

Central symmetry ($n = 2$)

The central symmetry is a particular case of the homothecy of ratio $k = -1$. Applying Formula (14.29) to our case, we have

$$
\begin{bmatrix} x' \\ y' \\ 1 \end{bmatrix} = \left[\begin{array}{cc|c} -1 & 0 & 2h^1 \\ 0 & -1 & 2h^2 \\ \hline 0 & 0 & 1 \end{array} \right] \begin{bmatrix} x \\ y \\ 1 \end{bmatrix}. \tag{14.37}
$$

The center of this homothecy, the point $H(h^1, h^2)$, that we assume in Cartesian coordinates, is called the "center of symmetry". This symmetry is an isometry because its associated matrix $T \equiv \begin{bmatrix} -1 & 0 \\ 0 & -1 \end{bmatrix}$ is orthogonal.

Since $|M| = +1$, it is a direct isometry.

Some authors classify this isometry as a rotation of $\theta = 180° \equiv \pi$ radians.

Axial symmetry ($n = 2$)

The axial symmetry is characterized by an invariant straight line, which called the "symmetry axis", is the orthogonal straight line at the center of any segment $\overline{XX'}$ joining corresponding points.

Following the method for the rotation listed above, we could obtain the matrix representation of this tensor by the following process: (1) establishment of an axial symmetry of simple axis, (2) general translation and rotation, previous to the symmetry to initially locate the axis in an arbitrary position, executing the symmetry, and (3) reversing the rotation and the translation, in that order. Nevertheless, following the maximum information criterion of this book, we employ another method to establish the tensor.

Assume that the Cartesian equation of the axis is known (see Figure 14.9):

$$
e \equiv Ax + By + C = 0.
$$

If the data were given in the form $e \equiv \frac{x-a}{m} = \frac{y-b}{n}$, we first operate and we write it in the previous form.

We know that the points $M\left(-\frac{C}{A}, 0, 1\right)$ and $N\left(0, -\frac{C}{B}, 1\right)$, which are the intersections of the axis e with the Cartesian axes, are invariant points of the

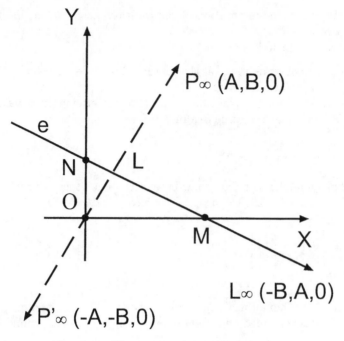

Fig. 14.9. Illustration of the axial symmetry.

axis e with homogeneous equation $AX + BY + Ct = 0$, because they satisfy its equation.

Since the point L_∞ of e is $L_\infty(-B, A, 0)$ because it satisfies its equation, an orthogonal direction will be $P_\infty(A, B, 0)$, because the dot product of both vectors is null in $E^2(\mathbb{R})$:

$$[-B \quad A] \begin{bmatrix} 1 & 0 \\ 0 & 1 \end{bmatrix} \begin{bmatrix} A \\ B \end{bmatrix} = 0,$$

which proves that the two points $P_\infty(A, B, 0)$ and $P'_\infty(-A, -B, 0)$ are symmetric. Entering this information into the fundamental tensor equation ($X' = MX$), we have

$$\begin{bmatrix} -C/A & 0 & -A \\ 0 & -C/B & -B \\ 1 & 1 & 0 \end{bmatrix} = M \begin{bmatrix} -C/A & 0 & A \\ 0 & -C/B & B \\ 1 & 1 & 0 \end{bmatrix},$$

and operating

$$M = \begin{bmatrix} -C/A & 0 & -A \\ 0 & -C/B & -B \\ 1 & 1 & 0 \end{bmatrix} \bullet \begin{bmatrix} -C/A & 0 & A \\ 0 & -C/B & B \\ 1 & 1 & 0 \end{bmatrix}^{-1},$$

which shows that the axial symmetry tensor is

$$
\begin{bmatrix} x' \\ y' \\ t' \end{bmatrix} = \frac{1}{A^2 + B^2} \left[\begin{array}{cc|c} -A^2 + B^2 & -2AB & -2AC \\ -2AB & A^2 - B^2 & -2BC \\ \hline 0 & 0 & A^2 + B^2 \end{array} \right] \begin{bmatrix} x \\ y \\ t \end{bmatrix}. \tag{14.38}
$$

Since the block T of matrix M in (14.38) is

$$
T = \frac{1}{A^2 + B^2} \begin{bmatrix} -A^2 + B^2 & -2AB \\ -2AB & A^2 - B^2 \end{bmatrix};
$$

$$
T \bullet T^t \equiv T^2 = \frac{1}{A^2 + B^2} \begin{bmatrix} (A^2 + B^2)^2 & 0 \\ 0 & (A^2 + B^2)^2 \end{bmatrix} \equiv I_2
$$

it is an orthogonal block and then, M is an isometry.

On the other hand, the determinant $|M| = \frac{-[(A^2 - B^2)^2 + 4A^2 B^2](A^2 + B^2)}{(A^2 + B^2)^3} =$ -1, which proves it is a reverse isometry. This in geometric terms means that we need to move the symmetric figure out of the plane XOY (to the space), to turn it and obtain the prototype figure by superposition. When we reach the mirror symmetry in the space ($n = 3$), we will see that taking the geometric object from the space of dimension $n = 3$ to the space of dimension $n = 4$ to turn it, is not a feasible explanation, so we prefer the tensor properties exposition used up to now, ignoring the imagination of geometry.

Similarity ($n = 2$)

The product of a homothecy by an isometry is called in geometry "similarity".

One of the most common similarities is the homothecy $\times$ rotation, *not* necessarily of *coincident* rotation $C(a, b)$ and homothecy $H(h^1, h^2)$ centers, though this similarity can be reduced to another similarity the centers of which $C(s^1, s^2) \equiv H(s^2, s^2)$ *coincide*, in which case the unique center $C \equiv H \equiv S_0(s^1, s^2)$ is called the "similarity center". We remind the reader that the matrix product carries the *reverse order* to that of its geometric execution.

As the corresponding tensor of a similarity has as associated matrix the product of the matrices associated with the isometry and the homothecy, $S = M \bullet F$, we do not insist further on the matrix representation of similarities. The similarities form a non-Abelian multiplicative group.

The determinant associated with a homothecy in the plane is $|F(n = 2)| > 0$, and then the determinant of S will be

$$|S| > 0 \quad \text{if } |M| = +1 \to \quad \text{(direct similarity)}$$

$$|S| < 0 \quad \text{if } |M| = -1 \to \quad \text{(reverse similarity)}.$$

Example 14.6 (Rotation × Rotation). In the ordinary Cartesian system $(O-XY)$ of dimension $n = 2$ we consider a rotation of center $C_1(a_1, b_1)$ and angle θ_1, followed by a second rotation of center $C(a_2, b_2)$ and angle θ_2.

1. Obtain the resulting isometry tensor.
2. Is this isometry another rotation?
3. Apply this to the numerical cases: $C_1(2, 3)$, $\theta_1 = 60°$ (sinextrorsum) and $C_2(4, 5)$, $\theta_2 = 60°$, (dextrorsum).
4. Idem for the cases $C_1(5, 1)$, $\theta = 45°$ and $C_2(1, 3)$, $\theta_2 = 45°$, both sinextrorsum.

Solution:

1. Using Formula (14.36), we obtain

$$
G_1 = \left[
\begin{array}{cc|c}
\cos\theta_1 & -\sin\theta_1 & a_1(1 - \cos\theta_1) + b_1\sin\theta_1 \\
\sin\theta_1 & \cos\theta_1 & -a_1\sin\theta_1 + b_1(1 - \cos\theta_1) \\
\hline
0 & 0 & 1
\end{array}
\right]
$$

$$
G_2 = \left[
\begin{array}{cc|c}
\cos\theta_2 & -\sin\theta_2 & a_1(1 - \cos\theta_2) + b_1\sin\theta_2 \\
\sin\theta_2 & \cos\theta_2 & -a_1\sin\theta_2 + b_1(1 - \cos\theta_2) \\
\hline
0 & 0 & 1
\end{array}
\right].
$$

The tensor of the indicated isometry $G = G_1 \times G_2$ has associated matrix $G = G_2 \bullet G_1$:

$$
G = \left[
\begin{array}{cc|c}
g_{11} & g_{12} & g_{13} \\
g_{21} & g_{22} & g_{23} \\
\hline
0 & 0 & 1
\end{array}
\right],
\tag{14.39}
$$

where

$$
\begin{aligned}
g_{11} &= \cos(\theta_1 + \theta_2) \\
g_{12} &= -\sin(\theta_1 + \theta_2) \\
g_{13} &= a_1[1 - \cos(\theta_1 + \theta_2)] \\
&\quad + b_1\sin(\theta_1 + \theta_2) + (a_2 - a_1)(1 - \cos\theta_2) + (b_2 - b_1)\sin\theta_2 \\
g_{21} &= \sin(\theta_1 + \theta_2) \\
g_{22} &= \cos(\theta_1 + \theta_2) \\
g_{23} &= -a_1\sin(\theta_1 + \theta_2) + b_1[1 - \cos(\theta_1 + \theta_2)] \\
&\quad - (a_2 - a_1)\sin\theta_2 + (b_2 - b_1)(1 - \cos\theta_2).
\end{aligned}
$$

2. We have the following cases:
   (a) If $\theta_1 + \theta_2 = 0$ or $\theta_1 + \theta_2 = 2\pi$ and $(a_1, b_1) \equiv (a_2, b_2) \rightarrow G \equiv I_3$ (identity).

(b) If $\theta_1 + \theta_2 = 0$ or $\theta_1 + \theta_2 = 2\pi$ and $(a_1, b_1) \neq (a_2, b_2)$ the isometry is a translation, of associated vector

$$\vec{t} = t^1 \vec{e}_1 + t^2 \vec{e}_2 \quad \begin{cases} t^1 = (a_2 - a_1)(1 - \cos\theta_2) + (b_2 - b_1)\sin\theta_2 \\ t^2 = -(a_2 - a_1)\sin\theta_2 + (b_2 - b_1)(1 - \cos\theta_2). \end{cases}$$

(c) If $\theta_1 + \theta_2 = \pi$ it is a central symmetry.

(d) In the remaining cases, it is a rotation of angle $\theta = \theta_1 + \theta_2$.

3. Applying this we get

$$G_1 = \begin{bmatrix} 1/2 & -\sqrt{3}/2 & | & 1 + 3\sqrt{3}/2 \\ \sqrt{3}/2 & 1/2 & | & 3/2 - \sqrt{3} \\ \hline 0 & 0 & | & 1 \end{bmatrix} ; \quad G_2 = \begin{bmatrix} 1/2 & \sqrt{3}/2 & | & 2 - 5\sqrt{3}/2 \\ -\sqrt{3}/2 & 1/2 & | & 5/2 + 2\sqrt{3} \\ \hline 0 & 0 & | & 1 \end{bmatrix} ;$$

$$M = G_2 \bullet G_1 = \begin{bmatrix} 1 & 0 & | & 1 - \sqrt{3} \\ 0 & 1 & | & 1 + \sqrt{3} \\ \hline 0 & 0 & | & 1 \end{bmatrix},$$

which is a *translation* of vector $\vec{t} = (1 - \sqrt{3})\vec{e}_1 + (1 + \sqrt{3})\vec{e}_2$.

4. Applying this we get

$$G_1 = \begin{bmatrix} \sqrt{2}/2 & -\sqrt{2}/2 & | & 5 - 2\sqrt{2} \\ \sqrt{2}/2 & \sqrt{2}/2 & | & 1 - 3\sqrt{2} \\ \hline 0 & 0 & | & 1 \end{bmatrix} ; \quad G_2 = \begin{bmatrix} \sqrt{2}/2 & -\sqrt{2}/2 & | & 1 + \sqrt{2} \\ \sqrt{2}/2 & \sqrt{2}/2 & | & 3 - 2\sqrt{2} \\ \hline 0 & 0 & | & 1 \end{bmatrix} ;$$

$$M = G_2 \bullet G_1 = \begin{bmatrix} 0 & -1 & | & 2 + 3\sqrt{2} \\ 1 & 0 & | & -2 + \sqrt{2} \\ \hline 0 & 0 & | & 1 \end{bmatrix}$$

$$\equiv \begin{bmatrix} \cos 90° & -\sin 90° & | & (2 + \sqrt{2}) \cdot 1 + 2\sqrt{2} \cdot 1 \\ \sin 90° & \cos 90° & | & -(2 + \sqrt{2}) \cdot 1 + 2\sqrt{2} \cdot 1 \\ \hline 0 & 0 & | & 1 \end{bmatrix},$$

which reveals that the isometry M is the rotation

$$M \equiv G, \text{ of center } C(2 + \sqrt{2}, 2\sqrt{2}) \text{ and angle } \theta = 90°.$$

This example shows that the rotations *do not* form a multiplicative group.

$\square$

Example 14.7 (Axial symmetry $\times$ translation). In the ordinary Cartesian plane $(O - XY)$ of dimension $n = 2$, we execute an axial symmetry of axis $e \equiv 2x + y - 2 = 0$ followed by a translation of vector $\vec{t} = 3\vec{e}_1 + 4\vec{e}_2$.

1. Obtain the matrix representation of the resulting isometry.
2. Classify this tensor and indicate if it is one of those discussed in the theory, justifying the answer.
3. Execute this isometry over the triangle that the symmetry axis e determines in the first quadrant.

Solution:

1. Using Formulas (14.38) and (14.34), we obtain

$$M_1 = \frac{1}{2^2 + 1^2} \begin{bmatrix} -2^2 + 1^2 & -2 \cdot 2 \cdot 1 & -2 \cdot 2 \cdot (-2) \\ -2 \cdot 2 \cdot 1 & 2^2 - 1^2 & -2 \cdot 1 \cdot (-2) \\ 0 & 0 & 2^2 + 1 \end{bmatrix} \equiv \frac{1}{5} \begin{bmatrix} -3 & -4 & 8 \\ -4 & 3 & 4 \\ 0 & 0 & 5 \end{bmatrix};$$

$$M_2 = \begin{bmatrix} 1 & 0 & 3 \\ 0 & 1 & 4 \\ 0 & 0 & 1 \end{bmatrix}; \quad M = M_2 \cdot M_1 = \frac{1}{5} \begin{bmatrix} -3 & -4 & 23 \\ -4 & 3 & 24 \\ 0 & 0 & 5 \end{bmatrix}.$$

2. It is an isometry, because its associated matrix has the format of affinities (Formula (14.27) and its block T satisfies the conditions of being an orthogonal matrix:

$$T \cdot T^t = T^2 = \left(\frac{1}{5} \begin{bmatrix} -3 & -4 \\ -4 & 3 \end{bmatrix} \right)^2 = \frac{1}{25} \begin{bmatrix} 25 & 0 \\ 0 & 25 \end{bmatrix} \equiv I_2.$$

The structure of T does not allow us to identify it as a translation or a central symmetry.

In addition, it is not an axial symmetry, because the terms $m_{13} = 23$ and $m_{23} = 24$ do not satisfy the constraints in Formula (14.38).

The final conclusion is that it is one isometry of those mentioned in the introduction to the section 14.3.5, without a proper name.

The present clarification is done with full knowledge of the fact that some authors call it "sliding symmetry", a name that obviously we want to avoid.

As $|M| = -1$, it is a "reverse" isometry.

3. The vertices A, O, B of the mentioned triangle have homogeneous coordinates $A(1, 0, 1)$, $O(0, 0, 1)$, $B(0, 2, 1)$ and their homologous matrices are

$$\frac{1}{5} \begin{bmatrix} -3 & -4 & 8 \\ -4 & 3 & 24 \\ 0 & 0 & 5 \end{bmatrix} \begin{bmatrix} 1 & 0 & 0 \\ 0 & 0 & 2 \\ 1 & 1 & 1 \end{bmatrix} = \frac{1}{5} \begin{bmatrix} 20 & 23 & 15 \\ 20 & 24 & 30 \\ 5 & 5 & 5 \end{bmatrix},$$

which shows that the Cartesian coordinates of the transformed triangle are

$$A'(4, 4), \quad O'(23/5, 24/5), \quad B'(3, 6).$$

$\square$

Example 14.8 (Similarity). In the ordinary Cartesian system $(O-XY)$ of dimension $n = 2$, we consider the following similarity. First, we execute a homothecy F_1 of center $H(-6,4)$ and ratio $k = 1/4$ and then we execute over the homothetic figure a rotation G_2 of center $C(0,0)$ and angle $\theta = 90°$, sinextrorsum.

1. Obtain the matrix representation of the similarity tensor.
2. Classify the mentioned similarity.
3. Obtain the coordinates of the invariant point, called the "similarity center".
4. Determine the points H' and C', homologous forms of the points H and C according to this similarity.
5. Determine the equation of the circumference that passing though the point C' is tangent in C to the straight line HC.
6. Obtain the equation of the circumscribed circumference to the triangle $\Delta(HCH')$; it is obvious that both circumferences pass through C.
7. Show that the other common point of the given circumferences is the similarity center.

Solution:

1.

$$F_1 = \begin{bmatrix} 1/4 & 0 & 3/4 \cdot (-6) \\ 0 & 1/4 & 3/4 \cdot (4) \\ 0 & 0 & 1 \end{bmatrix} = \frac{1}{4} \begin{bmatrix} 1 & 0 & -18 \\ 0 & 1 & 12 \\ 0 & 0 & 4 \end{bmatrix};$$

$$G_2 = \begin{bmatrix} \cos(-90°) & -\sin(-90°) & 0 \\ \sin(-90°) & \cos(-90°) & 0 \\ 0 & 0 & 1 \end{bmatrix} = \begin{bmatrix} 0 & 1 & 0 \\ -1 & 0 & 0 \\ 0 & 0 & 1 \end{bmatrix};$$

$$S = G_2 F_1 = \frac{1}{4} \begin{bmatrix} 0 & 1 & 0 \\ -1 & 0 & 0 \\ 0 & 0 & 1 \end{bmatrix} \begin{bmatrix} 1 & 0 & -18 \\ 0 & 1 & 12 \\ 0 & 0 & 4 \end{bmatrix}; \quad S = \frac{1}{4} \begin{bmatrix} 0 & 1 & 12 \\ -1 & 0 & 18 \\ 0 & 0 & 4 \end{bmatrix},$$

which is the similarity tensor.

2. To classify it we calculate

$$|S| = \frac{4}{4^3} = \frac{1}{16} > 0 \text{ (direct similarity).}$$

3. The double point must satisfy

$$SX = X; \quad (S - I_3)X = \Omega \rightarrow \begin{bmatrix} -1 & -1/4 & 3 \\ -1/4 & -1 & 9/2 \\ 0 & 0 & 0 \end{bmatrix} \begin{bmatrix} x \\ y \\ t \end{bmatrix} = \begin{bmatrix} 0 \\ 0 \\ 0 \end{bmatrix},$$

with solution

$$\frac{x}{\begin{vmatrix} 1/4 & 3 \\ -1 & 9/2 \end{vmatrix}} = \frac{y}{-\begin{vmatrix} -1 & 3 \\ -1/4 & 9/2 \end{vmatrix}} = \frac{t}{\begin{vmatrix} -1 & 1/4 \\ -1/4 & -1 \end{vmatrix}} \rightarrow \frac{x}{66} = \frac{y}{60} = \frac{t}{17},$$

which implies that the similarity center in Cartesian coordinates is $S_0\left(\frac{66}{17}, \frac{60}{17}\right)$.

4. The homologous points are

$$\begin{bmatrix} x' \\ y' \\ t' \end{bmatrix}_{H'} = \frac{1}{4}\begin{bmatrix} 0 & 1 & 12 \\ -1 & 0 & 18 \\ 0 & 0 & 4 \end{bmatrix}\begin{bmatrix} -6 \\ 4 \\ 1 \end{bmatrix} = \begin{bmatrix} 4 \\ 6 \\ 1 \end{bmatrix} \rightarrow H'(4,6)$$

$$\begin{bmatrix} x' \\ y' \\ t' \end{bmatrix}_{C'} = \frac{1}{4}\begin{bmatrix} 0 & 1 & 12 \\ -1 & 0 & 18 \\ 0 & 0 & 4 \end{bmatrix}\begin{bmatrix} 0 \\ 0 \\ 1 \end{bmatrix} = \begin{bmatrix} 3 \\ 9/2 \\ 1 \end{bmatrix} \rightarrow C'(3,9/2).$$

5. The equation of the tangent line at C is $\overline{CH} \equiv \frac{x-0}{-6} = \frac{y-0}{4} : \overline{CH} \equiv$
   $2x + 3y = 0$.
   The equation of the tangent line at C' is $\overline{C'P} \equiv \frac{x-3}{1} = \frac{y-9/2}{\lambda} : \overline{C'P} \equiv$
   $\lambda x - y + (9/2 - 3\lambda) = 0$.
   The equation of the chord $\overline{CC'}$ is $\frac{x-0}{3} = \frac{y-0}{9/2} : \overline{CC'} \equiv 3x - 2y = 0$.
   The set of conics generated by the two tangents and the chord is

$$(2x + 3y)[\lambda x - y + (9/2 - 3\lambda)] + \mu(3x - 2y)^2 = 0,$$

which developed leads to the set

$$(2\lambda+9\mu)x^2+(-3+4\mu)y2+(-2+3\lambda-12\mu)xy+(9-6\lambda)x+(27/2-9\lambda)y = 0. \tag{14.40}$$

Since it is a circumference, it must be

$$\begin{cases} 2\lambda + 9\mu & = -3 + 4\mu \\ -2 + 3\lambda - 12\mu = 0 \end{cases}$$

from which we obtain the parameters

$$\lambda = -2/3; \quad \mu = -1/3,$$

and substituting them into (14.40), we obtain the circumference

$$2x^2 + 2y2 - 6x - 9y = 0.$$

6. The circumscribed circumference to the triangle $\Delta(HCH')$ is

$$\begin{vmatrix} x^2 + y2 & x & y & 1 \\ (-6)^2 + 4^2 & -6 & 4 & 1 \\ 0^2 + 0^2 & 0 & 0 & 1 \\ 4^2 + 6^2 & 4 & 6 & 1 \end{vmatrix} = 0,$$

and operating, the result is

$$x^2 + y2 + 2x - 10y = 0.$$

7. The radical axis associated with the previous circumferences is

$$2(x^2 + y2 + 2x - 10y) - (2x^2 + 2y2 - 6x - 9y) = 0 \rightarrow 10x - 11y = 0,$$

and solving the system of one of them with that of the radical system, we obtain the desired points

$$\begin{cases} x^2 + y2 + 2x - 10y = 0 \\ 10x - 11y = 0 \end{cases} \rightarrow \begin{cases} C(0,0) \\ S_0(66/7, 60/17) \end{cases}$$

which proves the desired result.

□

Translation ($n = 3$)

Nothing needs to be added to what was established for the dimension $n = 2$, with the exception that now the fixed vector $\vec{t} = t^1\vec{e}_1 + t^2\vec{e}_2 + t^3\vec{e}_3$ obviously satisfies $\vec{t} \in E^3(\mathbb{R})$.

In this case, the tensor is represented as

$$X' = MX \rightarrow \begin{bmatrix} x' \\ y' \\ z' \\ t' \end{bmatrix} = \begin{bmatrix} 1 & 0 & 0 & | & t^1 \\ 0 & 1 & 0 & | & t^2 \\ 0 & 0 & 1 & | & t^3 \\ - & - & - & + & - \\ 0 & 0 & 0 & | & 1 \end{bmatrix} \begin{bmatrix} x \\ y \\ z \\ t \end{bmatrix}. \tag{14.41}$$

Central symmetry ($n = 3$)

As has been already indicated in the section dedicated to the central symmetry for ($n = 2$), it is a particular case of homothecy, with $k = -1$, so that applying Formula (14.29), we have

$$\begin{bmatrix} x' \\ y' \\ z' \\ t' \end{bmatrix} = \begin{bmatrix} -1 & 0 & 0 & | & 2h^1 \\ 0 & -1 & 0 & | & 2h^2 \\ 0 & 0 & -1 & | & 2h^3 \\ - & - & - & + & - \\ 0 & 0 & 0 & | & 1 \end{bmatrix} \begin{bmatrix} x \\ y \\ z \\ t \end{bmatrix}. \tag{14.42}$$

The homothecy center is called the "symmetry center", the point $H(h^1, h^2, h^3)$ in Cartesian coordinates, and we have an isometry, because the associated matrix $T \equiv \begin{bmatrix} -1 & 0 & 0 \\ 0 & -1 & 0 \\ 0 & 0 & -1 \end{bmatrix}$ is orthogonal.

Since $|M| = -1$, it is a reverse isometry.

Obviously, H is the center point of all segments $\overline{XX'}$.

Rotation ($n = 3$)

Since in Section 14.2.3 we have done a detailed analysis of the rotation tensor, when we referred to the vector application tensors, in this section devoted to the affine punctual space $E_p^3(\mathbb{R})$, we will use all the available information from there.

Assume that we have the following rotation data:

1. The axis Cartesian equation is

$$\frac{x - a}{\ell} = \frac{y - b}{m} = \frac{z - c}{n},$$

   where $(a, b, c, 1)$ is an attainable point on the axis and (ℓ, m, n, o) is the direction P_∞ of this axis.
2. The rotation angle is θ, the direction of which is given by the "right-hand rule", already established in Section 14.2.3.

The unit vector defining the axis direction is

$$\vec{e} \equiv \frac{\ell}{\sqrt{\ell^2 + m^2 + n^2}}\vec{e}_1 + \frac{m}{\sqrt{\ell^2 + m^2 + n^2}}\vec{e}_2 + \frac{n}{\sqrt{\ell^2 + m^2 + n^2}}\vec{e}_3. \quad (14.43)$$

To execute the rotation G we proceed as follows:

1. We perform a translation M_1 to bring the point (a, b, c) to the Cartesian origin $(0, 0, 0)$ and the whole axis with it.
2. Once the axis passes through O, we perform a rotation $R_{\vec{e}}$ of angle θ with respect to the actual axis, with the help of the corresponding tensor.
3. Once the rotation has been performed, we perform a translation M_2 opposite to the initial one, to reestablish the entire system to its initial position. Naturally, we must have $G = M_2 \cdot R_{\vec{e}} \cdot M_1$.

The indicated matrix M_1 is

$$M_1 = \left[\begin{array}{ccc|c} 1 & 0 & 0 & -a \\ 0 & 1 & 0 & -b \\ 0 & 0 & 1 & -c \\ \hline 0 & 0 & 0 & 1 \end{array}\right];$$

To obtain the matrix associated with tensor $R_{\vec{e}}$ we use Formula (14.14) with the director cosines of our vector $\vec{e}$, written in homogeneous coordinates:

$$R_{\vec{e}} = \frac{1}{\ell^2 + m^2 + n^2}\left[\begin{array}{ccc|c} r_{11} & r_{12} & r_{13} & 0 \\ r_{21} & r_{22} & r_{23} & 0 \\ r_{31} & r_{32} & r_{33} & 0 \\ \hline 0 & 0 & 0 & (\ell^2 + m^2 + n^2) \end{array}\right],$$

where

$$r_{11} = \ell^2 + (m^2 + n^2)\cos\theta$$
$$r_{12} = \ell m(1 - \cos\theta) - n\sin\theta$$
$$r_{13} = \ell n(1 - \cos\theta) + m\sin\theta$$
$$r_{21} = \ell m(1 - \cos\theta) + n\sin\theta$$
$$r_{22} = m^2 + (\ell^2 + n^2)\cos\theta$$
$$r_{23} = mn(1 - \cos\theta) - \ell\sin\theta$$
$$r_{31} = \ell n(1 - \cos\theta) - m\sin\theta$$
$$r_{32} = mn(1 - \cos\theta) + \ell\sin\theta$$
$$r_{33} = n^2 + (\ell^2 + m^2)\cos\theta$$

$$M_2 = \left[\begin{array}{ccc|c} 1 & 0 & 0 & a \\ 0 & 1 & 0 & b \\ 0 & 0 & 1 & c \\ \hline 0 & 0 & 0 & 1 \end{array}\right].$$

The result of $G = M_2 \cdot R_{\vec{e}} \cdot M_1$ is the rotation matrix

$$G = \frac{1}{\ell^2 + m^2 + n^2} \left[\begin{array}{ccc|c} g_{11} & g_{12} & g_{13} & g_{14} \\ g_{21} & g_{22} & g_{23} & g_{24} \\ g_{31} & g_{32} & g_{33} & g_{34} \\ \hline 0 & 0 & 0 & g_{44} \end{array}\right], \tag{14.44}$$

where

$$g_{11} = \ell^2 + (m^2 + n^2)\cos\theta$$
$$g_{12} = \ell m(1 - \cos\theta) - n\sin\theta$$
$$g_{13} = \ell n(1 - \cos\theta) + m\sin\theta$$
$$g_{21} = \ell m(1 - \cos\theta) + n\sin\theta$$
$$g_{22} = m^2 + (\ell^2 + n^2)\cos\theta$$
$$g_{23} = mn(1 - \cos\theta) - \ell\sin\theta$$
$$g_{31} = \ell n(1 - \cos\theta) - m\sin\theta$$
$$g_{32} = mn(1 - \cos\theta) + \ell\sin\theta$$
$$g_{33} = n^2 + (\ell^2 + m^2)\cos\theta$$

$$\left[\begin{array}{c} g_{14} \\ g_{24} \\ g_{34} \\ \hline g_{44} \end{array}\right] = \left[\begin{array}{c} \left[\begin{array}{ccc} s_{11} & s_{12} & s_{13} \\ s_{21} & s_{22} & s_{23} \\ s_{31} & s_{32} & s_{33} \end{array}\right] \left[\begin{array}{c} a \\ b \\ c \end{array}\right] \\ \hline \ell^2 + m^2 + n^2 \end{array}\right], \tag{14.45}$$

where

$$s_{11} = (m^2 + n^2)(1 - \cos\theta)$$
$$s_{12} = -(\ell m(1 - \cos\theta) - n\sin\theta)$$
$$s_{13} = -(\ell n(1 - \cos\theta) + m\sin\theta)$$
$$s_{21} = -(\ell m(1 - \cos\theta) + n\sin\theta)$$
$$s_{22} = (\ell^2 + n^2)(1 - \cos\theta)$$
$$s_{23} = -(mn(1 - \cos\theta) - \ell\sin\theta)$$
$$s_{31} = -(\ell n(1 - \cos\theta) - m\sin\theta)$$
$$s_{32} = -(mn(1 - \cos\theta) + \ell\sin\theta)$$
$$s_{33} = (\ell^2 + m^2)(1 - \cos\theta).$$

The block T of the matrix G is $T_3 = R_{\vec{e}}$, which is an orthogonal matrix, as was proved in the section devoted to the rotation tensor. Thus, G is an isometry.

Since the determinant $|G| = 1$, it is a direct isometry.

Some of its properties are:

- The plane orthogonal to the segment $\overline{XX'}$ passing through its center point contains the rotation axis.
- Two straight lines r and r' that are homologous are at the same distance from the rotation axis.

Remark 14.1. If the rotation axis were given as the intersection of two planes

$$\begin{cases} A_1 x + B_1 y + C_1 z + D_1 = 0 \\ A_2 x + B_2 y + C_2 z + D_2 = 0. \end{cases}$$

and the rotation axis were required in the form

$$\frac{x - a}{\ell} = \frac{y - b}{m} = \frac{z - c}{n},$$

one possible solution for obtaining the rotation matrix G is the formula

$$\frac{x - \frac{1}{\lambda}\begin{vmatrix} (B_1 - C_1) & D_1 \\ (B_2 - C_2) & D_2 \end{vmatrix}}{\begin{vmatrix} B_1 & C_1 \\ B_2 & C_2 \end{vmatrix}} = \frac{y - \frac{1}{\lambda}\begin{vmatrix} (C_1 - A_1) & D_1 \\ (C_2 - A_2) & D_2 \end{vmatrix}}{\begin{vmatrix} C_1 & A_1 \\ C_2 & A_2 \end{vmatrix}} = \frac{z - \frac{1}{\lambda}\begin{vmatrix} (A_1 - B_1) & D_1 \\ (A_2 - B_2) & D_2 \end{vmatrix}}{\begin{vmatrix} A_1 & B_1 \\ A_2 & B_2 \end{vmatrix}}$$

(14.46)

with

$$\lambda = \begin{vmatrix} B_1 & C_1 \\ B_2 & C_2 \end{vmatrix} + \begin{vmatrix} C_1 & A_1 \\ C_2 & A_2 \end{vmatrix} + \begin{vmatrix} A_1 & B_1 \\ A_2 & B_2 \end{vmatrix}.$$

Finally, if the tensor G were the data, the rotation axis can be obtained as the tensor straight line of invariant (double) points, and its angle, as was indicated in Example 14.2 point 4, using the trace of the rotation tensor $R_{\vec{e}}$.   □

Axial symmetry ($n = 3$)

This isometry is a particular case of the rotation, with angle $\theta = 180°$, and then, the only data is the symmetry axis

$$\frac{x - a}{\ell} = \frac{y - b}{m} = \frac{z - c}{n}.$$

The axis unit vector $\vec{e}$ is that in Formula (14.43). We notate this symmetry as $S_{\vec{e}}$. The corresponding tensor matrix is obtained, letting $\theta = 180°$ in Formulas (14.44) and (14.45):

$$S_{\vec{e}} = \frac{1}{\ell^2 + m^2 + n^2} \left[\begin{array}{ccc|c} s_{11} & s_{12} & s_{13} & s_{14} \\ s_{21} & s_{22} & s_{23} & s_{24} \\ s_{31} & s_{32} & s_{33} & s_{34} \\ \hline -n & 0 & 0 & (\ell^2 + m^2 + n^2) \end{array} \right],$$

(14.47)

where

$$s_{11} = \ell^2 - (m^2 + n^2)$$
$$s_{12} = 2\ell m$$
$$s_{13} = 2\ell n$$
$$s_{14} = 2[(m^2 + n^2)a - \ell mb - \ell nc]$$
$$s_{21} = 2\ell m$$
$$s_{22} = m^2 - (\ell^2 + n^2)$$
$$s_{23} = 2mn$$
$$s_{24} = 2[-\ell ma + (\ell^2 + n^2)b - mnc]$$
$$s_{31} = 2\ell n$$
$$s_{32} = 2mn$$
$$s_{33} = n^2 - (\ell^2 + m^2)$$
$$s_{34} = 2[-\ell na - mnb + (\ell^2 + m^2)c].$$

As in the rotation case, it is a direct isometry, where $|S_{\vec{e}}| = 1$. The matrix $S_{\vec{e}}$ is involutive, $S_{\vec{e}}^2 = I_4$.

Other interesting properties are:

- The segments $\overline{XX'}$ of homologous points have an orthogonal line at their center point, which is the symmetry axis.
- Any straight line secant and orthogonal to the axis and any plane orthogonal to the axis are "invariant".

Remark 14.2. The same comments as those for the dimension $n = 2$ and with Formula (14.46) are applicable here.   □

Mirror symmetry ($n = 3$)

This tensor is characterized by having a plane of invariant points, called the symmetry plane. This plane is the plane orthogonal to all segments $\overline{XX'}$, that join homologous points, at its center point. Assume that this symmetry is given by the plane

$$\pi \equiv Ax + By + Cz + D = 0. \tag{14.48}$$

In vector notation, it is the reflection tensor $H\vec{e}$, treated in Section 14.2.4. The plane π is illustrated on Figure 14.6.

Since the data tensor $H_{\vec{e}}$, is a unit vector orthogonal to the plane π, we will use all that has been established in Section 14.2.4, taking a vector $\vec{e}$ from the plane π, that is, the data vector

$$\vec{e} \equiv \frac{A}{\sqrt{A^2 + B^2 + C^2}}\vec{e}_1 + \frac{B}{\sqrt{A^2 + B^2 + C^2}}\vec{e}_2 + \frac{C}{\sqrt{A^2 + B^2 + C^2}}\vec{e}_3, \tag{14.49}$$

which determines $\cos\alpha, \cos\beta$ and $\cos\gamma$. If $\cos\alpha = \frac{A}{\sqrt{A^2+B^2+C^2}}$, then $\sin\alpha = \frac{\sqrt{B^2+C^2}}{\sqrt{A^2+B^2+C^2}}$, with similar conclusions for the remaining director cosines.

Once we have these data, we can enter them in the reflection tensor formula (14.16) to establish the mirror symmetry tensor, notated as S_π, which lacks only the fourth column:

$$S_\pi = \frac{1}{A^2 + B^2 + C^2}\left[\begin{array}{ccc|c} s_{11} & s_{12} & s_{13} & s_{41} \\ s_{21} & s_{22} & s_{23} & s_{42} \\ s_{31} & s_{32} & s_{33} & s_{43} \\ \hline 0 & 0 & 0 & A^2 + B^2 + C^2 \end{array}\right].$$

where

$$s_{11} = -A^2 + B^2 + C^2; \quad s_{12} = s_{21} = -2AB; \quad s_{13} = s_{31} = -2AC$$
$$s_{22} = A^2 - B^2 + C^2; \quad s_{23} = s_{32} = -2BC; \quad s_{33} = A^2 + B^2 - C^2.$$

We know that the points $\left(-\frac{D}{A}, 0, 0, 1\right), \left(0, -\frac{D}{B}, 0, 1\right), \left(0, 0, -\frac{D}{C}, 1\right)$ are the intersections of the plane π with the Cartesian axes, and that they are invariant; thus, transforming them we determine the elements of the fourth column.

From the transformation of the first abscissa, we obtain

$$\frac{(-A^2 + B^2 + C^2)\left(-\frac{D}{A}\right) + 0 + 0 + s_{41}}{A^2 + B^2 + C^2} = -\frac{D}{A} \to s_{41} = -2AD.$$

Similarly, with the other two we obtain

$$s_{42} = -2BD; \quad s_{43} = -2CD,$$

which permits us to complete the mirror symmetry tensor.

The block T determinant is $|T| \equiv |H_{\vec{e}}| = -1$, which requires $|S_\pi| = -1$, and so we have a reverse isometry. In Section 14.2.4 we have proved the involutive character of the reflection tensor $H_{\vec{e}}$, thus, the matrix S_π satisfies $S_\pi^2 = I_4$.

14.3.6 Product of isometries

We give the name "product of isometries" or "decomposable isometries" to those isometries that come from the consecutive executions of several simple isometries, such as those already mentioned.

The corresponding matrix is the product $M = M_k \cdot M_{k-1} \cdots M_2 \cdot M_1$ of the matrices associated with the factor isometries. The resulting isometry can be or not be one of the already studied isometries.

If the determinant $|M| = 1$ the resulting isometry is a direct isometry, otherwise, i.e., $|M| = -1$, we have a reverse isometry.

The set of all isometries M such that $|M| = -1$ are called by some authors "pseudo-isometries", which refers to isometries that are not proper.

We note that the product of an even number of reverse isometries is obviously a direct isometry.

We consider it of interest to mention some concrete decomposable isometries:

- The product of two mirror isometries $S_{\pi_2} \cdot S_{\pi_1}$ is:
  1. A translation, if π_1 and π_2 are parallel.
  2. A rotation if π_1 and π_2 intersect.
- The product of two central symmetries is a translation.
- The product of two axial symmetries $S_{\vec{e}_2} \cdot S_{\vec{e}_1}$, is:
  1. A rotation if the axes $\vec{e}_1$ and $\vec{e}_2$ intersect.
  2. The product of a rotation by a translation if the axes $\vec{e}_1$ and $\vec{e}_2$ cross without intersection.
- The product of three central symmetries, of non-aligned centers, is another central symmetry, of center the fourth vertex of the parallelogram defined by the centers of the given symmetries.
- The product of a rotation by a translation of direction that of the rotation axis, is called helicoidal isometry, though certain authors call it "oblique helicoidal" when the translation direction does not coincide with the rotation axis.

Example 14.9 (Product of transformations). We consider a rotation in the classic geometric space $E_p^3(\mathbb{R})$ of value θ and direct sense (see Figure 14.10). The rotation axis is contained in a plane that passes through the axis $\overline{OZ}$ and makes an angle β with the plane XOZ. That axis passes through the origin O, making an angle α with the axis OZ. Next, we perform a "scaling" (change in the dimensions of the Cartesian axes) of factors k_1, k_2 and k_3, respectively.

1. Obtain the rotation matrix G_1.
2. Obtain the scaling matrix F_2.
3. Obtain the matrix T of the mixed transformation.
4. Give the equation of the transformed sphere of the sphere with equation $x^2 + y^2 + z^2 = 1$, in the following numerical case:

$$\alpha = 45°; \quad \beta = 30°; \quad k_1 = 1; \quad k_2 = 2; \quad k_3 = 3.$$

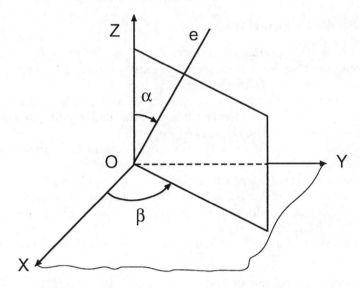

Fig. 14.10. Illustration of a product of transformations.

Solution:

1. The unit vector of the rotation axis is

$$\vec{e} \equiv \sin\alpha\cos\beta\vec{e}_1 + \sin\alpha\sin\beta\vec{e}_2 + \cos\alpha\vec{e}_3.$$

Applying Formulas (14.44) and (14.45), we have the block T_o:

$$T_o = \begin{bmatrix} t_{11} & t_{12} & t_{13} \\ t_{21} & t_{22} & t_{23} \\ t_{31} & t_{32} & t_{33} \end{bmatrix},$$

where

$$t_{11} = \sin^2\alpha\cos^2\beta + (\sin^2\alpha\sin^2\beta + \cos^2\alpha)\cos\theta$$
$$t_{12} = \sin^2\alpha\sin\beta\cos\beta(1 - \cos\theta) - \cos\alpha\sin\theta$$
$$t_{13} = \sin\alpha\cos\alpha\cos\beta(1 - \cos\theta) + \sin\alpha\sin\beta\sin\theta$$
$$t_{21} = \sin^2\alpha\sin\beta\cos\beta(1 - \cos\theta) + \cos\alpha\sin\theta$$
$$t_{22} = \sin^2\alpha\sin^2\beta + (\sin^2\alpha\cos^2\beta + \cos^2\alpha)\cos\theta$$
$$t_{23} = \sin\alpha\cos\alpha\sin\beta(1 - \cos\theta) - \sin\alpha\cos\beta\sin\theta$$
$$t_{31} = \sin\alpha\cos\alpha\cos\beta(1 - \cos\theta) - \sin\alpha\sin\beta\sin\theta$$
$$t_{32} = \sin\alpha\cos\alpha\sin\beta(1 - \cos\theta) + \sin\alpha\cos\beta\sin\theta$$
$$t_{33} = \cos^2\alpha + \sin^2\alpha\cos\theta,$$

which is orthogonal because $T_o \cdot T_o^t = I_3$.

For the rotation

$$G_1 = \begin{bmatrix} & & & | & 0 \\ & T_o & & | & 0 \\ & & & | & 0 \\ -- & -- & -- & + & - \\ 0 & 0 & 0 & | & 1 \end{bmatrix} \quad \text{(in homogeneous coordinates).}$$

2. The scaling is an affine transformation, the matrix of which has the format in Formula (14.27), with the origin $O(0,0,0)$ as unique invariant point:

$$F_2 = \begin{bmatrix} k_1 & 0 & 0 & | & 0 \\ 0 & k_2 & 0 & | & 0 \\ 0 & 0 & k_3 & | & 0 \\ - & - & - & + & - \\ 0 & 0 & 0 & | & 1 \end{bmatrix}.$$

3. $T = F_2 \cdot G_1$.
4. The numerical values are

$$\sin \alpha = \cos \alpha = \frac{\sqrt{2}}{2}; \quad \sin \beta = \frac{1}{2}; \quad \cos \beta = \frac{\sqrt{3}}{2};$$

$$T = \begin{bmatrix} 1 & 0 & 0 & 0 \\ 0 & 2 & 0 & 0 \\ 0 & 0 & 3 & 0 \\ 0 & 0 & 0 & 1 \end{bmatrix} \begin{bmatrix} \frac{3+5\cos\theta}{8} & \frac{\sqrt{3}(1-\cos\theta)-4\sqrt{2}\sin\theta}{8} & \frac{\sqrt{3}(1-\cos\theta)+\sqrt{2}\sin\theta}{4} & 0 \\ \frac{\sqrt{3}(1-\cos\theta)+4\sqrt{2}\sin\theta}{8} & \frac{1+7\cos\theta}{8} & \frac{(1-\cos\theta)-\sqrt{6}\sin\theta}{4} & 0 \\ \frac{\sqrt{3}(1-\cos\theta)-\sqrt{2}\sin\theta}{4} & \frac{(1-\cos\theta)+\sqrt{6}\sin\theta}{4} & \frac{1+\cos\theta}{2} & 0 \\ 0 & 0 & 0 & 1 \end{bmatrix},$$

which leads to

$$T = \begin{bmatrix} \frac{3+5\cos\theta}{8} & \frac{\sqrt{3}(1-\cos\theta)-4\sqrt{2}\sin\theta}{8} & \frac{\sqrt{3}(1-\cos\theta)+\sqrt{2}\sin\theta}{4} & 0 \\ \frac{\sqrt{3}(1-\cos\theta)+4\sqrt{2}\sin\theta}{4} & \frac{1+7\cos\theta}{4} & \frac{(1-\cos\theta)-\sqrt{6}\sin\theta}{2} & 0 \\ \frac{3\sqrt{3}(1-\cos\theta)-3\sqrt{2}\sin\theta}{4} & \frac{3(1-\cos\theta)+3\sqrt{6}\sin\theta}{4} & \frac{3+3\cos\theta}{2} & 0 \\ 0 & 0 & 0 & 1 \end{bmatrix}.$$

However, the application of the operator in this case is simpler if we use the factors.

Since the rotation axis G_1 is one of the sphere diameters, the rotated sphere is the same sphere. Thus, we need to apply only the "scaling" tensor to the sphere:

$$x^2 + y^2 + z^2 - t^2 = 0 \rightarrow [\, x \; y \; z \; t \,] \begin{bmatrix} 1 & 0 & 0 & 0 \\ 0 & 1 & 0 & 0 \\ 0 & 0 & 1 & 0 \\ 0 & 0 & 0 & -1 \end{bmatrix} \begin{bmatrix} x \\ y \\ z \\ t \end{bmatrix} = 0 \text{ (sphere);}$$

$$\begin{bmatrix} x' \\ y' \\ z' \\ t' \end{bmatrix} = F_2 \begin{bmatrix} x \\ y \\ z \\ t \end{bmatrix} \rightarrow \begin{bmatrix} x \\ y \\ z \\ t \end{bmatrix} = F_2^{-1} \begin{bmatrix} x' \\ y' \\ z' \\ t' \end{bmatrix},$$

and substituting this into the sphere equation, we get

$$[\overset{\cdot}{x'} \ \ y' \ \ z' \ \ t'](F_2^{-1})^t \begin{bmatrix} 1 & 0 & 0 & 0 \\ 0 & 1 & 0 & 0 \\ 0 & 0 & 1 & 0 \\ 0 & 0 & 0 & -1 \end{bmatrix} F_2^{-1} \begin{bmatrix} x' \\ y' \\ z' \\ t' \end{bmatrix} = 0,$$

which implies

$$[x' \ \ y' \ \ z' \ \ t'] \begin{bmatrix} 1 & 0 & 0 & 0 \\ 0 & 1/2 & 0 & 0 \\ 0 & 0 & 1/3 & 0 \\ 0 & 0 & 0 & 1 \end{bmatrix} \begin{bmatrix} 1 & 0 & 0 & 0 \\ 0 & 1 & 0 & 0 \\ 0 & 0 & 1 & 0 \\ 0 & 0 & 0 & -1 \end{bmatrix} \begin{bmatrix} 1 & 0 & 0 & 0 \\ 0 & 1/2 & 0 & 0 \\ 0 & 0 & 1/3 & 0 \\ 0 & 0 & 0 & 1 \end{bmatrix} \begin{bmatrix} x' \\ y' \\ z' \\ t' \end{bmatrix}$$

$$[x' \ \ y' \ \ z' \ \ t'] \begin{bmatrix} 1 & 0 & 0 & 0 \\ 0 & 1/4 & 0 & 0 \\ 0 & 0 & 1/9 & 0 \\ 0 & 0 & 0 & -1 \end{bmatrix} \begin{bmatrix} x' \\ y' \\ z' \\ t' \end{bmatrix} = 0 \rightarrow (x')^2 + \frac{1}{4}(y')^2 + \frac{1}{9}(z')^2 = 1,$$

which is an ellipsoid.

□

14.4 Tensors in Physics and Mechanics

Given an arbitrary body in static equilibrium, subject to a set of exterior forces and moments and subject to given supporting conditions, we present the problem of determining the distribution of the interior forces, together with the strains that such stresses produce on this material.

The problem has been faced by substituting the real body by simplified models that allow us to apply tensor methods to obtain results that are close to the real ones. Next, we mention some of the common models.

- *Rigid solid.* A rigid solid is an ideal undeformable solid typical of rational mechanics; by eliminating possible deformations it permits us to calculate the reactions in certain cases (isostatic problems) and not in others (hyperstatic problems).
- *Elastic solid.* This model considers the elastic deformation of the body, which permits us to establish tensor differential equations, frequently in sufficient number for its resolution, but for which integration is the main difficulty.
  It is customary to accept the linearity in the stress–strain relationship (Hooke's law), according to which the object recovers its initial form as soon as the causes of its deformation disappear (linear elasticity).

. In other cases, when the problem nature requires more complicated assumptions, it is necessary to accept that the forces and moments are in-
fluenced by deformation and we enter into the region called "non-linear
elasticity".

- *Plastic solid.* Permanent deformations are accepted.
- *Isotropic solid.* The elastic and plastic properties are the same for all directions.
- *Anisotropic solid.* The elastic and plastic properties depend on the considered direction.

We will consider also the simplifications related to the stress concept.

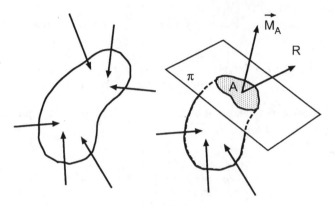

Fig. 14.11. Illustration of the stress concept.

Consider a solid in equilibrium under the external actions and the reactions
acting on its supporting points (see Figure 14.11). Let A be a solid interior
point and π a plane with given orientation passing through A. The solid is
divided into two parts by π. Removing the upper part, but substituting its
effect on point A, in order not to alter its equilibrium, from mechanics we
know (see Example 14.3) that all the actions of the removed upper part on
the point A can be reduced to a single resulting force $\vec{R}$ and a pair of forces,
the moment $\vec{M}_A$. Taking over the plane π and around A a small neighborhood
of area ΔS and assuming the following limits:

$$\lim_{\Delta s \to 0} \frac{\vec{R}}{\Delta s} = \vec{t}; \quad \lim_{\Delta s \to 0} \frac{\vec{M}_A}{\Delta s} = \vec{m} = 0; \quad \text{null pair}$$

in *linear elasticity* the vector $\vec{t}$ (stress vector) is "the stress at the point A
corresponding to the orientation of π" and $\vec{m}$, though it is non-null, is not
considered. However, $\vec{m}$ is considered in *non-linear* elasticity.

If we denote by $\vec{e}$ the unit vector orthogonal to the plane π, the orthogonal
projection of $\vec{t}$ over $\vec{e}$ can be obtained by means of the projection tensor,

notated $\vec{\sigma}$ and called the "normal stress":

$$\vec{\sigma} = P_{\vec{e}}(\vec{t})$$

and the complementary projection vector of the vector $\vec{t}$ is notated $\vec{\tau}$, and is called the "shear stress":

$$\vec{\tau} = P_{\vec{e}^\perp}(\vec{t}) = (I_3 - P_{\vec{e}})(\vec{t}),$$

both at the point A and relative to π.

14.4.1 The stress tensor S

Consider a stress tensor, defined by means of a symmetric matrix

$$S(x,y,z) = \begin{bmatrix} \sigma_x(x,y,z) & \tau_{xy}(x,y,z) & \tau_{xz}(x,y,z) \\ \tau_{xy}(x,y,z) & \sigma_y(x,y,z) & \tau_{yz}(x,y,z) \\ \tau_{xz}(x,y,z) & \tau_{yz}(x,y,z) & \sigma_z(x,y,z) \end{bmatrix},$$

which associates with each point $A(x_0, y_0, z_0)$ of the geometric punctual space $E_p^3(\mathbb{R})$ a *numeric* tensor, then we obtain

$$S = \begin{bmatrix} \sigma_x & \tau_{xy} & \tau_{xz} \\ \tau_{xy} & \sigma_y & \tau_{yz} \\ \tau_{xz} & \tau_{yz} & \sigma_z \end{bmatrix} \quad \text{(stress tensor)}. \tag{14.50}$$

Assuming Cartesian axes with the origin at the point A and parallel to the reference frame, the tensor S assigns to each unit vector $\vec{e} = \cos\alpha\vec{i} + \cos\beta\vec{j} + \cos\gamma\vec{k}$, which gives the orientation of the orthogonal plane π that passes through A, the corresponding stress $\vec{t} = t^1\vec{i} + t^2\vec{j} + t^3\vec{k}$. In summary, the tensor operates as

$$\vec{t} = S(\vec{e}); \quad \begin{bmatrix} t^1 \\ t^2 \\ t^3 \end{bmatrix} = S \begin{bmatrix} \cos\alpha \\ \cos\beta \\ \cos\gamma \end{bmatrix}.$$

For example, the scalars of the first column of the tensor matrix S are none other than the stress components $\vec{t}$ in A, related to the plane YOZ, and $\vec{e}$ is the direction $(1,0,0)$ in this case.

If we calculate the vector $P_{\vec{e}}(\vec{t}) = \vec{\sigma}_e$, we obtain the "normal stress" vector in A to the surface, the modulus of which, according to the projector $P_{\vec{e}}$ is

$$\sigma_e = \vec{e} \cdot \vec{t} = \begin{bmatrix} \cos\alpha & \cos\beta & \cos\gamma \end{bmatrix} \begin{bmatrix} \sigma_x & \tau_{xy} & \tau_{xz} \\ \tau_{xy} & \sigma_y & \tau_{yz} \\ \tau_{xz} & \tau_{yz} & \sigma_z \end{bmatrix} \begin{bmatrix} \cos\alpha \\ \cos\beta \\ \cos\gamma \end{bmatrix},$$

which implies

$$\sigma_e = \sigma_x \cos^2 \alpha + \sigma_y \cos^2 \beta + \sigma_z \cos^2 \gamma + 2\tau_{xy} \cos \alpha \cos \beta$$
$$+ 2\tau_{xz} \cos \alpha \cos \gamma + 2\tau_{yz} \cos \beta \cos \gamma. \tag{14.51}$$

Similarly, if we calculate $(I_3 - P_{(\vec{e})})(\vec{t}) = \vec{\tau}_\pi$, we obtain the "shear stress" vector, the modulus of which can be calculated from the Pythagoras theorem:

$$\tau_\pi = \sqrt{|\vec{t}|^2 - \sigma_e^2}.$$

The "main directions" $\vec{e}_I, \vec{e}_{II}, \vec{e}_{III}$ are those associated with the unit eigenvectors of the matrix S, In such directions, the resulting stress $\vec{t}$ has only a normal stress component $\vec{\sigma}$ and lacks the shear stress component ($\vec{\tau}_\pi = \vec{0}$), that is, $\vec{t} \equiv \vec{\sigma}$. The stresses $\vec{\sigma}_I, \vec{\sigma}_{II}, \vec{\sigma}_{III}$ are called the main stresses.

The characteristic polynomial of the matrix S must satisfy

$$S(\vec{e}) = \sigma \vec{e} \iff \begin{vmatrix} \sigma_x - \sigma & \tau_{xy} & \tau_{xz} \\ \tau_{xy} & \sigma_y - \sigma & \tau_{yz} \\ \tau_{xz} & \tau_{yz} & \sigma_z - \sigma \end{vmatrix} = 0. \tag{14.52}$$

Once the eigenvalues $\sigma_I, \sigma_{II}, \sigma_{III}$ of the symmetric matrix S have been obtained, three mutually orthogonal unit eigenvectors are determined, which supply the orthogonal change-of-basis matrix Q, such that

$$\hat{S} = Q^t S Q \equiv Q^{-1} S Q \equiv \begin{bmatrix} \sigma_I & 0 & 0 \\ 0 & \sigma_{II} & 0 \\ 0 & 0 & \sigma_{III} \end{bmatrix}. \tag{14.53}$$

Some authors resolve the stresses in a certain direction at the point A, starting from the canonized tensor $\hat{S}$ and with the help of interesting geometrical constructions (Mohr diagram), but not from the tensor point of view.

Finally, we comment on other utility of the stress tensor. When varying the directions of the unit vector $\vec{e}$ on the point $\vec{A}$, if we assign it the role of a position vector, the extremity of which is the point $P(x, y, z)$ with respect to the Cartesian system $(A - XYZ)$, it is evident that the set of points covered by its extreme P is the sphere $x^2 + y^2 + z^2 = 1$.

Similarly, we can assign to the extremity P' of the stress vector $\vec{t}$ the coordinates of $P'(X, Y, Z) \equiv P'(kt^1, kt^2, kt^3)$, with the $\vec{t}$ components at a given scale k. Next, we discuss which is the set of points covered by the point P' of $\vec{t}$ corresponding to the extreme points P of $\vec{e}$. Using the tensor relation

$$\begin{bmatrix} X \\ Y \\ Z \end{bmatrix}_{P'} = kS \begin{bmatrix} \cos \alpha \\ \cos \beta \\ \cos \gamma \end{bmatrix}_P$$

we obtain

$$\begin{bmatrix} \cos \alpha \\ \cos \beta \\ \cos \gamma \end{bmatrix} = k^{-1} S^{-1} \begin{bmatrix} X \\ Y \\ Z \end{bmatrix}, \tag{14.54}$$

and since

$$[\cos\alpha \quad \cos\beta \quad \cos\gamma] \begin{bmatrix} \cos\alpha \\ \cos\beta \\ \cos\gamma \end{bmatrix} = 1, \tag{14.55}$$

substituting into (14.54) and (14.55) we get

$$\left(\frac{1}{k}\right)^2 [X \quad Y \quad Z] \left(S^{-1}\right)^t \left(S^{-1}\right) \begin{bmatrix} X \\ Y \\ Z \end{bmatrix} = 1,$$

from which we get the quadratic equation

$$\left(\frac{1}{k}\right)^2 [X \quad Y \quad Z] \left(\begin{bmatrix} \sigma_x & \tau_{xy} & \tau_{xz} \\ \tau_{xy} & \sigma_y & \tau_{yz} \\ \tau_{xz} & \tau_{yz} & \sigma_z \end{bmatrix}^{-1}\right)^2 \begin{bmatrix} X \\ Y \\ Z \end{bmatrix} = 1. \tag{14.56}$$

This quadratic can be classified easily when we subject Formula (14.56) to the change-of-basis in (14.53):

$$\left(\frac{1}{k}\right)^2 [\hat{X} \quad \hat{Y} \quad \hat{Z}] \left(\hat{S}^{-1}\right)^t \left(S^{-1}\right) \begin{bmatrix} \hat{X} \\ \hat{Y} \\ \hat{Z} \end{bmatrix}$$

$$= \left(\frac{1}{k}\right)^2 [\hat{X} \quad \hat{Y} \quad \hat{Z}] \begin{bmatrix} \left(\frac{1}{\sigma_I}\right)^2 & 0 & \\ 0 & \left(\frac{1}{\sigma_{II}}\right)^2 & 0 \\ 0 & 0 & \left(\frac{1}{\sigma_{III}}\right)^2 \end{bmatrix} \begin{bmatrix} \hat{X} \\ \hat{Y} \\ \hat{Z} \end{bmatrix} = 1$$

and operating, we obtain the new Cartesian equation

$$\frac{\hat{X}^2}{(k\sigma_I)^2} + \frac{\hat{Y}^2}{(k\sigma_{II})^2} + \frac{\hat{Z}^2}{(k\sigma_{III})^2} = 1, \tag{14.57}$$

which proves that (14.56) is an ellipsoid, called a "stress ellipsoid", that declares in a visual way the stress state around the point A.

14.4.2 The strain tensor Γ

Assume an orthonormalized reference frame $(O-XYZ)$ in the ordinary punctual space $E_p^3(\mathbb{R})$, and consider a solid body of a continuous elastic material over which several exterior forces and moments act, keeping it in static equilibrium. This originates on its interior some stresses that modify the *relative position* of its particles, a situation known as "elastic deformation". It is convenient to substitute the complex reality of the phenomenon being analyzed by a mathematical model that, close to it, permits us to make acceptable predictions. This section is devoted to the model description.

If a particle initially occupies an arbitrary position $P(x, y, z)$, and after the deformation occupies the position $P'(x', y', z')$, the vector $\overrightarrow{PP'} \equiv \vec{c} \equiv u(x, y, z)\vec{i} + v(x, y, z)\vec{j} + w(x, y, z)\vec{k}$ (see Figure 14.12) is called the "displacement vector" of the given point. In this way we assist in the creation of a *vector field* $\vec{c}$ such that to each concrete point P_0 a displacement vector $\vec{c}_0$ is assigned (which would be numeric if the scalar functions of u, v and w were known).

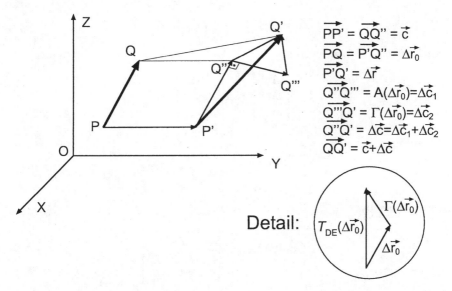

$$\overrightarrow{PP'} = \overrightarrow{QQ''} = \vec{c}$$
$$\overrightarrow{PQ} = \overrightarrow{P'Q''} = \Delta\vec{r_0}$$
$$\overrightarrow{P'Q'} = \Delta\vec{r}$$
$$\overrightarrow{Q''Q'''} = A(\Delta\vec{r_0}) = \Delta\vec{c_1}$$
$$\overrightarrow{Q'''Q'} = \Gamma(\Delta\vec{r_0}) = \Delta\vec{c_2}$$
$$\overrightarrow{Q''Q'} = \Delta\vec{c} = \Delta\vec{c_1} + \Delta\vec{c_2}$$
$$\overrightarrow{QQ'} = \vec{c} + \Delta\vec{c}$$

Detail: $T_{DE}(\Delta\vec{r_0})$ $\Gamma(\Delta\vec{r_0})$ $\Delta\vec{r_0}$

Fig. 14.12. Illustration of the strain tensor.

If the displacement $\overline{QQ'}$ of another point $Q(x + \Delta x, y + \Delta y, z + \Delta z)$, close to P, is known, we can study and compare the changes in the initial distance $|\overline{PQ}|$ of the points P and Q, and the final distance $|\overline{P'Q'}|$ of the new final positions. Similarly, we can analyze the relative position of the vectors $\overline{PQ}$ and $\overline{P'Q'}$, due to the elastic deformation.

The intervening vectors are

$$\Delta\vec{r_0} = \overline{PQ} = \overline{OP} - \overline{OQ} = \Delta x\vec{i} + \Delta y\vec{j} + \Delta z\vec{k}.$$

If we set

$$\Delta r_0 = |\overline{PQ}| = \sqrt{(\Delta x)^2 + (\Delta y)^2 + (\Delta z)^2},$$

where

$$\cos\alpha = \frac{\Delta x}{\Delta r_0}; \quad \cos\beta = \frac{\Delta y}{\Delta r_0}; \quad \cos\gamma = \frac{\Delta z}{\Delta r_0}$$

$$\vec{e} = \cos\alpha\vec{i} + \cos\beta\vec{j} + \cos\gamma\vec{k} \quad \text{(unit vector)},$$

we can assume

$$\Delta\vec{r}_0 = \overline{PQ} = \Delta r_0\vec{e} \tag{14.58}$$

and similarly

$$\Delta\vec{r} = \overline{P'Q'} = \Delta r^1\vec{i} + \Delta r^2\vec{j} + \Delta r^3\vec{k}. \tag{14.59}$$

The tensor that directly transforms the vector $\Delta\vec{r}_0$ into $\Delta\vec{r}$ will be denoted by T_D and called the "strain tensor"

$$\Delta\vec{r} = T_D(\Delta\vec{r}_0) = T_D(\Delta r_0\vec{e}),$$

which in matrix form becomes

$$\begin{bmatrix} \Delta r^1 \\ \Delta r^2 \\ \Delta r^3 \end{bmatrix} = T_D \begin{bmatrix} \Delta x \\ \Delta y \\ \Delta z \end{bmatrix},$$

and also

$$\begin{bmatrix} \Delta r^1 \\ \Delta r^2 \\ \Delta r^3 \end{bmatrix} = \Delta r_0 T_D \begin{bmatrix} \cos\alpha \\ \cos\beta \\ \cos\gamma \end{bmatrix}. \tag{14.60}$$

If we denote by "$\mathcal{J}$" the Jacobian matrix

$$\mathcal{J}\begin{pmatrix} u\ v\ w \\ x\ y\ z \end{pmatrix} \equiv \begin{bmatrix} \frac{\partial u}{\partial x} & \frac{\partial u}{\partial y} & \frac{\partial u}{\partial z} \\ \frac{\partial v}{\partial x} & \frac{\partial v}{\partial y} & \frac{\partial v}{\partial z} \\ \frac{\partial w}{\partial x} & \frac{\partial w}{\partial y} & \frac{\partial w}{\partial z} \end{bmatrix}$$

as in the tensor analysis treatises, the following matrix relation is, as a first approximation, established:

$$T_D = I_3 + \mathcal{J}, \tag{14.61}$$

not considering higher order terms. A tensor field has been created that assigns to each point P_0 a strain tensor T_D, useful for studying the vectors $\Delta\vec{r}_0$ close to P_0.

The conclusions of the mentioned study are separated, in the tensor analysis books, into three steps that are illustrated in Figure 14.12:

1. In the first step the vector $\overline{PQ} \equiv \Delta\vec{r}_0$ undergoes a translation equal to the displacement $\vec{c}$ of the point P, to reach the position $\overline{P'Q''}$. From the figure, we get

$$\overline{PQ} + \overline{QQ'} = \overline{PP'} + \overline{P'Q'}; \quad \Delta\vec{r}_0 + (\vec{c} + \Delta\vec{c}) = \vec{c} + \Delta\vec{r},$$

which is

$$\Delta\vec{r} = \Delta\vec{r}_0 + \Delta\vec{c}. \tag{14.62}$$

2. In the second step we analyze the vector $\Delta \vec{c}$ and we conclude that it is composed of two summand vectors. The first one, the vector $\overline{Q''Q'''}$ represents a rotation of the vector $\Delta \vec{r}_0$ during the deformation. We confirm this by means of a tensor, denoted by $\mathcal{A}$ and called a "rotation tensor", the associated matrix of which is *anti-symmetric*:

$$A = \frac{1}{2}\left(\mathcal{J} - \mathcal{J}^t\right) \equiv \begin{bmatrix} 0 & -w_z & w_y \\ w_z & 0 & -w_x \\ -w_y & w_x & 0 \end{bmatrix}. \qquad (14.63)$$

If $\overline{Q''Q'''} = \Delta \vec{c}_1 = (\Delta c_1)^1 \vec{i} + (\Delta c_1)^2 \vec{j} + (\Delta c_1)^3 \vec{k}$ it operates as $\Delta \vec{c}_1 = A(\Delta \vec{r}_0)$, i.e., in matrix form

$$\begin{bmatrix} (\Delta c_1)^1 \\ (\Delta c_1)^2 \\ (\Delta c_1)^3 \end{bmatrix} = A \begin{bmatrix} \Delta x \\ \Delta y \\ \Delta z \end{bmatrix} = \Delta r_0 A \begin{bmatrix} \cos\alpha \\ \cos\beta \\ \cos\gamma \end{bmatrix}. \qquad (14.64)$$

We insist that this tensor rotates the vector $\Delta \vec{r}_0$, but it *does not* deform it.

3. In the third step we study the second summand of $\Delta \vec{c}$, the vector $\overline{Q'''Q'}$, which represents the true measure variation that $\Delta \vec{r}_0$ undergoes during the process. This summand is determined by the creation of another tensor, denoted by Γ and called a "pure strain tensor". As usual, a new tensor field is created that assigns to each point P_0 its Γ tensor, represented by

$$\Gamma = \frac{1}{2}\left[\mathcal{J} + \mathcal{J}^t\right] \equiv \begin{bmatrix} \epsilon_x & \frac{1}{2}\gamma_{xy} & \frac{1}{2}\gamma_{xz} \\ \frac{1}{2}\gamma_{xy} & \epsilon_y & \frac{1}{2}\gamma_{xz} \\ \frac{1}{2}\gamma_{xz} & \frac{1}{2}\gamma_{yz} & \epsilon_z \end{bmatrix}. \qquad (14.65)$$

Finally, if we set $\overline{Q'''Q'} = \Delta \vec{c}_2 = (\Delta c_2)^1 \vec{i} + (\Delta c_2)^2 \vec{j} + (\Delta c_2)^3 \vec{k}$, the Γ tensor operates as

$$\Delta \vec{c}_2 = \Gamma(\Delta \vec{r}_0); \quad \begin{bmatrix} (\Delta c_1)^1 \\ (\Delta c_1)^2 \\ (\Delta c_1)^3 \end{bmatrix} = \Gamma \begin{bmatrix} \Delta x^1 \\ \Delta x^2 \\ \Delta x^3 \end{bmatrix} = \Delta r_0 \Gamma \begin{bmatrix} \cos\alpha \\ \cos\beta \\ \cos\gamma \end{bmatrix}. \qquad (14.66)$$

As has been indicated, this tensor gives the variation (the increment or decrement) of vector $\Delta \vec{r}_0$, so that, as can be deducted from the detail of Figure 14.12, the vector of *strict deformation* (ignoring rotations) is

$$\Delta \vec{r}_0 + \Gamma(\Delta \vec{r}_0) = (I_3 + \Gamma)(\Delta \vec{r}_0) \equiv \mathcal{T}_{DE}\Delta \vec{r}_0). \qquad (14.67)$$

A new tensor denoted by $\mathcal{T}_{DE}$ and called a "strict deformation tensor" has been created, the associated matrix of which is

$$\mathcal{T}_{DE} = I_3 + \Gamma. \qquad (14.68)$$

The moduli of the vectors $T_{DE}(\Delta\vec{r}_0)$ give the strict measure of the deformation but ignoring rotations and translations.

In summary, we substitute Formulas (14.60), (14.64) and (14.66) into (14.62), which precisely is

$$\Delta\vec{r} = \Delta\vec{r}_0 + (\Delta\vec{c}_1 + \Delta\vec{c}_2),$$

and we obtain the matrix relation

$$\begin{bmatrix} \Delta r^1 \\ \Delta r^2 \\ \Delta r^3 \end{bmatrix} = \Delta r_0 T_D \begin{bmatrix} \cos\alpha \\ \cos\beta \\ \cos\gamma \end{bmatrix} = \Delta r_0 I_3 \begin{bmatrix} \cos\alpha \\ \cos\beta \\ \cos\gamma \end{bmatrix} + \Delta r_0 A \begin{bmatrix} \cos\alpha \\ \cos\beta \\ \cos\gamma \end{bmatrix} + \Delta r_0 \Gamma \begin{bmatrix} \cos\alpha \\ \cos\beta \\ \cos\gamma \end{bmatrix},$$

and from it, the tensor relation

$$T_D = (I_3 + \Gamma) + A = T_{DE} + A, \tag{14.69}$$

a very important expression that relates all tensors that play a role in the total strict elastic deformation.

If we apply only the tensor Γ over the unit vector $\vec{e}$, we obtain the vector of "pure deformation" that corresponds to the mentioned direction. This vector, projected over the direction $\vec{e}$, gives the component "unit enlargement ϵ in the direction of $\vec{e}$", of the pure deformation vector $\Gamma(\vec{e})$:

$$\epsilon = \vec{e} \bullet \Gamma(\vec{e}) = \begin{bmatrix} \cos\alpha & \cos\beta & \cos\gamma \end{bmatrix} \begin{bmatrix} \epsilon_x & \frac{1}{2}\gamma_{xy} & \frac{1}{2}\gamma_{xz} \\ \frac{1}{2}\gamma_{xy} & \epsilon_y & \frac{1}{2}\gamma_{yz} \\ \frac{1}{2}\gamma_{xy} & \frac{1}{2}\gamma_{yz} & \epsilon_z \end{bmatrix} \begin{bmatrix} \cos\alpha \\ \cos\beta \\ \cos\gamma \end{bmatrix}.$$
$$\tag{14.70}$$

The other component of the $\Gamma(\vec{e})$ vector, orthogonal to $\vec{e}$, is called "angular or tangent deformation" and it is represented as $\frac{1}{2}\gamma$. It can be calculated by means of the Pythagoras theorem (in a similar way to that for the stress tensor S in Section 14.4.1), and then, we obtain

$$\frac{1}{2}\gamma = \sqrt{|\Gamma(\vec{e})|^2 - \epsilon^2} \tag{14.71}$$

If we look for the vectors $\vec{\epsilon}$ or $\frac{1}{2}\vec{\gamma}$ we can use the projectors

$$\vec{\epsilon} = P_{\vec{e}}(\Gamma(\vec{e})); \quad \frac{1}{2}\vec{\gamma} = P_{\vec{e}\perp}(\Gamma(\vec{e})). \tag{14.72}$$

We clearly appreciate that, according to the proposal, if the orthonormal basis of $E_p^3(\mathbb{R})$ is $\{\vec{e}_\alpha\}$, the tensor components of Γ can be interpreted as

$$\Gamma_{\alpha\alpha} = \vec{e}_\alpha \bullet \Gamma(\vec{e}_\alpha) \quad \text{and} \quad \Gamma_{\alpha\beta} = \Gamma_{\beta\alpha} = \vec{e}_\alpha \bullet \Gamma(\vec{e}_\beta) = \vec{e}_\beta \bullet \Gamma(\vec{e}_\alpha), \quad \alpha \neq \beta. \tag{14.73}$$

As any symmetric tensor, Γ has real eigenvalues ($\epsilon_I, \epsilon_{II}, \epsilon_{III}$) and the corresponding unit eigenvectors are orthogonal, that represent the directions of maximum or minimum unit deformation (ϵ) and in addition null tangent deformation ($\gamma = 0$). With the given eigenvectors, the orthogonal matrix L is built that canonizes the tensor

$$\hat{\Gamma} = L^t \Gamma L \equiv L^{-1} \Gamma L = \begin{bmatrix} \epsilon_I & 0 & 0 \\ 0 & \epsilon_{II} & 0 \\ 0 & 0 & \epsilon_{III} \end{bmatrix}, \qquad (14.74)$$

all in agreement with the behavior of the stress tensor S.

The directions

$$\vec{e}_a(0, \pm\cos 45°, \pm\cos 45°); \ \vec{e}_b(\pm\cos 45°, 0, \pm\cos 45°); \ \vec{e}_c(\pm\cos 45°, \pm\cos 45°, 0)$$

applied to the tensor $\hat{\Gamma}$ and using the Formulas (14.70) and (14.71), allow us to calculate the maximum tangent deformations ($\frac{1}{2}\gamma_{max}$):

$$\left(\frac{1}{2}\gamma_{max}\right)_a = \sqrt{|\hat{\Gamma}(\vec{e})|^2 - (\hat{\epsilon}_a)^2} = \sqrt{\left(\frac{\epsilon_{II}^2 + \epsilon_{III}^2}{2}\right) - \left(\frac{\epsilon_{II} + \epsilon_{III}}{2}\right)^2}$$

$$= \sqrt{\frac{\epsilon_{II}^2 + \epsilon_{III}^2 - 2\epsilon_{II}\epsilon_{III}}{4}} = \frac{\epsilon_{II} - \epsilon_{III}}{2}. \qquad (14.75)$$

Similarly, for $\vec{e}_b$ and $\vec{e}_c$.

The first invariant of the characteristic polynomial, that is, the trace of Γ, is called the "cubic unit dilatation coefficient" at the point P and is denoted by c_v:

$$c_v = \epsilon_x + \epsilon_y + \epsilon_z = \epsilon_I + \epsilon_{II} + \epsilon_{III}, \qquad (14.76)$$

which represents the unit volume increment of a small volume δV in the neighborhood of P,

$$c_v = \frac{\Delta(\delta V)}{\delta V}, \qquad (14.77)$$

due to elastic deformation.

14.4.3 Tensor relationships between S and Γ. Elastic tensor

In the most general case of anisotropy, Hooke's law can be generalized, assuming that (σ, τ) produce deformations not only in the corresponding directions but in other directions.

In summary, we assume that each of the six stresses $\sigma_x, \sigma_y, \sigma_z, \tau_{xy}, \tau_{xz}, \tau_{yz}$ that constitute the stress tensor S are homogeneous linear functions of the six deformations $\epsilon_x, \epsilon_y, \epsilon_z, \gamma_{xy}, \gamma_{xz}, \gamma_{yz}$ of the tensor Γ. In addition, since the deformation potential function must have coincident cross derivatives, the matrix F must be symmetric, and the generalized Hooke's law, in matrix form becomes

$$
\begin{bmatrix} \sigma_x \\ \sigma_y \\ \sigma_z \\ \tau_{xy} \\ \tau_{xz} \\ \tau_{yz} \end{bmatrix} = \begin{bmatrix} E_{x,x} & E_{x,y} & E_{x,z} & G_{x,xy} & G_{x,xz} & G_{x,yz} \\ E_{x,y} & E_{y,y} & E_{y,z} & G_{y,xy} & G_{y,xz} & G_{y,yz} \\ E_{x,z} & E_{y,z} & E_{z,z} & G_{z,xy} & G_{z,xz} & G_{z,yz} \\ G_{x,xy} & G_{y,xy} & G_{z,xy} & G_{xy,xy} & G_{xy,xz} & G_{xy,yz} \\ G_{x,xz} & G_{y,xz} & G_{z,xz} & G_{xy,xz} & G_{xy,xz} & G_{xy,yz} \\ G_{x,yz} & G_{y,yz} & G_{z,yz} & G_{xy,yz} & G_{xy,yz} & G_{xy,yz} \end{bmatrix} \begin{bmatrix} \epsilon_x \\ \epsilon_y \\ \epsilon_z \\ \gamma_{xy} \\ \gamma_{xz} \\ \gamma_{yz} \end{bmatrix}.
$$

(14.78)

Of the 36 coefficients in this matrix, 21 are not repeated.

The tensor conception of the generalized Hooke's law for an anisotropic elastic solid given in Formula (14.78), appears such that the stress tensor $S = [s_{ij}^{\circ\circ}]$ comes from the contraction of a fourth-order tensor, called the "elastic tensor" $C = [c_{ijk\ell}^{\circ\circ\circ\circ}]$, which represents the elastic properties of the medium, with the deformation tensor $\Gamma = [\gamma_{pq}^{\circ\circ}]$, all of them associated with the Euclidean space $E^3(\mathbb{R})$. The contracted tensor equation is

$$
s_{ij}^{\circ\circ} = c_{ijk\ell}^{\circ\circ\circ\circ}\gamma_{k\ell}^{\circ\circ},
$$

(14.79)

with all indices in covariant position, because they are tensors in orthonormalized bases. The matrix representation of the tensors that appear in (14.79) is

$$
S \equiv [s_{ij}^{\circ\circ}] = \begin{bmatrix} \sigma_x & \tau_{xy} & \tau_{xz} \\ \tau_{xy} & \sigma_y & \tau_{yz} \\ \tau_{xz} & \tau_{yz} & \sigma_z \end{bmatrix}; \Gamma \equiv [\gamma_{k\ell}^{\circ\circ}] = \begin{bmatrix} \epsilon_x & \frac{1}{2}\gamma_{xy} & \frac{1}{2}\gamma_{xz} \\ \frac{1}{2}\gamma_{xy} & \epsilon_y & \frac{1}{2}\gamma_{yz} \\ \frac{1}{2}\gamma_{xz} & \frac{1}{2}\gamma_{yz} & \epsilon_z \end{bmatrix}
$$

$$
C \equiv [c_{ijk\ell}^{\circ\circ\circ\circ}] = \left[\begin{array}{ccc|ccc|ccc} E_{x,x} & G_{x,xy} & G_{x,xz} & G_{x,xy} & G_{xy,xy} & G_{xy,xz} & G_{x,xz} & G_{xy,xz} & G_{xz,xz} \\ G_{x,xy} & E_{x,y} & G_{x,yz} & G_{xy,xy} & G_{y,xy} & G_{xy,yz} & G_{xy,xz} & G_{y,xz} & G_{xz,yz} \\ G_{x,xz} & G_{x,yz} & E_{x,z} & G_{xy,xz} & G_{xy,yz} & G_{z,xy} & G_{xz,xz} & G_{xz,yz} & G_{z,xz} \\ \hline G_{x,xy} & G_{xy,xy} & G_{xy,xz} & E_{x,y} & G_{y,xy} & G_{y,xz} & G_{x,yz} & G_{xy,yz} & G_{xz,yz} \\ G_{xy,xy} & G_{y,xy} & G_{xy,yz} & G_{y,xy} & E_{y,y} & G_{y,yz} & G_{xy,yz} & G_{y,yz} & G_{yz,yz} \\ G_{xy,xz} & G_{xy,yz} & G_{z,xy} & G_{y,xz} & G_{y,yz} & E_{y,z} & G_{xz,yz} & G_{yz,yz} & G_{z,yz} \\ \hline G_{x,xz} & G_{xy,xz} & G_{xz,xz} & G_{x,yz} & G_{xy,yz} & G_{xz,yz} & E_{x,z} & G_{z,xy} & G_{z,xz} \\ G_{xy,xz} & G_{y,xz} & G_{xz,yz} & G_{xy,yz} & G_{y,yz} & G_{yz,yz} & G_{z,xy} & E_{y,z} & G_{z,yz} \\ G_{xz,xz} & G_{xz,yz} & G_{z,xz} & G_{xz,yz} & G_{yz,yz} & G_{z,yz} & G_{z,xz} & G_{z,yz} & E_{z,z} \end{array}\right],
$$

(14.80)

with $1 \leq i,j,k,\ell \leq 3$; where i is the row block index, j is the column block index, k is the row index in each block and ℓ is the column index in each block.

The usual planning of the elastic problem is as follows:

1. The tensor field $S(x,y,z)$ of the stress state is known.
2. We discover the elastic tensor field $C(x,y,z)$, by performing experiments in the laboratory.
3. The tensor field $\Gamma(x,y,z)$ is determined by means of Formula (14.79), which permits us to relate the Jacobian $\mathcal{J}$ partial derivatives.

4. The partial differential equations obtained in the previous step are integrated to discover the displacement tensor field $\vec{c}$.
5. Once $\vec{c}$ is known, the remaining deformation tensors $T_D, \mathcal{A}$ and T_{DE} can be obtained.

In the most common case, that is, the case of *homogeneous isotropic elastic solid*, the following material constants of the material are defined and validated:

1. E, elasticity modulus or Young's modulus.
2. ν, Poisson's ratio.
3. G , transversal elasticity modulus, shear modulus.
4. λ, Lamé's constant.

Some existing relations among them are

$$G = \frac{E}{2(1+\nu)}; \quad \lambda = \frac{\nu E}{(1+\nu)(1-2\nu)}; \quad \nu = \frac{\lambda}{2(\lambda+G)}; \quad E = \frac{(3\lambda+2G)G}{\lambda+G}. \tag{14.81}$$

With these constants we establish the isotropic elastic tensor:

$$C \equiv [c_{ijk\ell}^{\circ\circ\circ\circ}] = \left[\begin{array}{ccc|ccc|ccc} (\lambda+2G) & 0 & 0 & 0 & G & 0 & 0 & 0 & G \\ 0 & \lambda & 0 & G & 0 & 0 & 0 & 0 & 0 \\ 0 & 0 & \lambda & 0 & 0 & 0 & G & 0 & 0 \\ \hline 0 & G & 0 & \lambda & 0 & 0 & 0 & 0 & 0 \\ G & 0 & 0 & 0 & (\lambda+2G) & 0 & 0 & 0 & G \\ 0 & 0 & 0 & 0 & 0 & \lambda & 0 & G & 0 \\ \hline 0 & 0 & G & 0 & 0 & 0 & \lambda & 0 & 0 \\ 0 & 0 & 0 & 0 & 0 & G & 0 & \lambda & 0 \\ G & 0 & 0 & 0 & G & 0 & 0 & 0 & (\lambda+2G) \end{array}\right]. \tag{14.82}$$

Once the contraction in Formula (14.79) is performed, we arrive at the isotropic elastic matrix relation, with the expressions

$$S = 2G\Gamma + \lambda(\text{trace } \Gamma)I_3 \tag{14.83}$$

or

$$\Gamma = \frac{1+\nu}{E}S - \frac{\nu}{E}(\text{trace } S)I_3 \tag{14.84}$$

because

$$\text{trace } S = (3\lambda + 2G)(\text{trace } \Gamma). \tag{14.85}$$

We remind the reader that trace $\Gamma = c_v$ (cubic dilatation coefficient).

Formulas (14.83), (14.84) and (14.85) are equivalent to the Lamé equations:

$$\begin{aligned} \sigma_x &= \lambda c_v + 2G\epsilon_x; \quad \tau_{xy} = G\gamma_{xy} \\ \sigma_y &= \lambda c_v + 2G\epsilon_y; \quad \tau_{xz} = G\gamma_{xz} \\ \sigma_z &= \lambda c_v + 2G\epsilon_z; \quad \tau_{yz} = G\gamma_{yz}. \end{aligned} \tag{14.86}$$

Example 14.10 (Elastic and thermic tensors). Consider a homogeneous, elastic and isotropic solid, of cubic form and edge ℓ, the material of which has a Young's modulus E and a Poisson's ratio ν. Three exterior forces of respective values F_x, F_y and F_z act orthogonally to three contiguous faces of the solid (see Figure 14.13). Due to the elevation of the temperature, thermic stresses are produced inside the solid and are proportional to the thermic gap Δt, where the material proportionality constant is k, where both k and Δt are data.

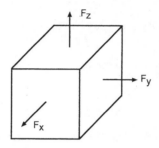

Fig. 14.13. Illustration of the elastic and thermic tensors.

1. Obtain the tensor and matrix representation of the stated problem.
2. Determine the solid unit cubic dilatation coefficient c_v expressing it in terms of the "Bulk modulus", $B = \frac{E}{3(1-2\nu)}$.
3. Assuming that the solid is incompressible, determine its Poisson ratio.
4. Assuming that the solid is liberated from any exterior force, what is the pure deformation Γ tensor to which it is subject due to thermic action?

Solution:

1. Assuming that initially the material is subject to stresses due to the external forces, experimentally we observe that the thermic dilatation relaxes them. This permits us to complete Formula (14.79) by means of the tensor expression
$$s_{ij}^{\circ\circ} = c_{ijk\ell}^{\circ\circ\circ\circ}\gamma_{k\ell}^{\circ\circ} - k_{ij}^{\circ\circ}\Delta t.$$

For an isotropic solid we use the matrix formula (14.83):
$$S = 2G\Gamma + \lambda(\text{trace }\Gamma)I_3 - kI_3\Delta t. \qquad (14.87)$$

2. Developing the diagonal terms, from the matrix equation (14.87) we obtain
$$\sigma_x = 2G\epsilon_x + \lambda(\text{trace }\Gamma) - k\Delta t$$
$$\sigma_y = 2G\epsilon_y + \lambda(\text{trace }\Gamma) - k\Delta t$$
$$\sigma_z = 2G\epsilon_z + \lambda(\text{trace }\Gamma) - k\Delta t,$$

and adding them

$$\sigma_x + \sigma_y + \sigma_z = 2G(\epsilon_x + \epsilon_y + \epsilon_z) + 3\lambda(\text{trace } \Gamma) - 3k\Delta t$$

or

$$\text{trace } S = (2G + 3\lambda)(\text{trace } \Gamma) - 3k\Delta t, \qquad (14.88)$$

which is none other than Formula (14.85) completed with the thermic term.

On the other hand, from Formulas (14.76) and (14.81) we have

$$c_v = \epsilon_x + \epsilon_y + \epsilon_z = \text{trace } \Gamma$$

$$2G + 3\lambda = 2\frac{E}{2(1+\nu)} + 3\frac{\nu E}{(1+\nu)(1-2\nu)} = \frac{(1+\nu)E}{(1+\nu)(1-2\nu)}$$

$$= 3\frac{E}{3(1-2\nu)} = 3B$$

and taking c_v and $(2G + 3\lambda)$ to Formula (14.88) we get

$$\text{trace } S = 3Bc_v - 3k\Delta t \rightarrow c_v = \frac{\text{trace } S}{3B} + \frac{k\Delta t}{B},$$

which in this case is

$$\text{trace } S = \sigma_x + \sigma_y + \sigma_z = \frac{F_x + F_y + F_z}{\ell^2},$$

which finally leads to

$$c_v = \frac{F_x + F_y + F_z}{3\ell^2 B} + \frac{k\Delta t}{B}.$$

3. For it to be incompressible it must be $c_v = 0$, which from Formula (14.76), implies trace $\Gamma = 0$.

Obtaining the trace Γ from Formula (14.87), and taking into account that in the previous question we obtained $(2G + 3\lambda) = \frac{E}{1-2\nu}$, we arrive at

$$\text{trace } \Gamma = \frac{\text{trace } S + 3k\Delta t}{(2G + 3\lambda)} = \frac{1 - 2\nu}{E}(\text{trace } S + 3k\Delta t) = 0,$$

an expression that, if $\nu = \frac{1}{2}$, is null.

4. If there are no exterior forces and moments, the stress tensor is $S = \Omega$, which implies trace $S = 0$. Taking now the first condition to the matrix equation (14.87) and the second to (14.88), we obtain

$$2G\Gamma + \lambda(\text{trace } \Gamma)I_3 - k\Delta t I_3 = \Omega;$$

$$\text{trace } \Gamma = \frac{3k\Delta t}{(2G + 3\lambda)}.$$

In the second question, when we studied the Bulk modulus, we obtained $2G + 3\lambda = \frac{E}{1-2\nu}$, thus, taking the last two expressions to the previous matrix equation, we obtain

$$2G\Gamma + \lambda \left(3k\Delta t \frac{(1 - 2\nu)}{E} \right) I_3 - k\Delta t I_3 = \Omega,$$

and substituting G and λ and operating we get

$$2\frac{E}{2(1+\nu)}\Gamma + \left[\frac{\nu E}{(1+\nu)(1-2\nu)} \left(\frac{3(1-2\nu)}{E} \right) - 1 \right] k\Delta t I_3 = \Omega$$

$$\frac{E}{1+\nu}\Gamma + \left[\frac{3\nu}{1+\nu} - 1 \right] k\Delta t I_3 = \Omega; \quad E\Gamma + (2\nu - 1)k\Delta t I_3 = \Omega,$$

that is,

$$\Gamma = \left(\frac{1 - 2\nu}{E} k\Delta t \right) I_3.$$

□

Example 14.11 (Main stresses). The stress state of a solid at a certain point is

$$\sigma_x = -4; \quad \sigma_y = 2; \quad \sigma_z = 1; \quad \tau_{xy} = 4; \quad \tau_{xz} = \tau_{yz} = 0.$$

1. Obtain the stress tensor S.
2. Obtain the total stress tensor $\vec{t}$ that acts on a plane π the normal to which has the direction of the vector $\vec{r}(1, 2, 3)$ and obtain also the modulus of $\vec{t}$.
3. Determine the "normal component" vector $\vec{\sigma}$ and its modulus σ, corresponding to $\vec{t}$.
4. Determine the "tangent component" vector $\vec{\tau}$ and its modulus τ, acting on the plane π corresponding to $\vec{t}$.
5. Determine the main stresses, vectors $\vec{\sigma}_I, \vec{\sigma}_{II}, \vec{\sigma}_{III}$, and their moduli.
6. Determine the maximum shear stress.
7. Give the orthogonal directions of the planes where the maximum shear stress acts.
8. Give the value of the normal stress corresponding to the maximum shear stress.
9. Obtain the orthogonal direction to a plane such that it has an associated pure shear state ($\sigma = 0$).
10. Obtain the vector $\vec{\tau}$ and its modulus τ in the previous case.

Solution:

1. The stress tensor is

$$S = \begin{bmatrix} -4 & 4 & 0 \\ 4 & 2 & 0 \\ 0 & 0 & 1 \end{bmatrix}.$$

2. The unit vector of direction $\vec{r}$ is

$$\vec{e}_r = \frac{1}{\sqrt{14}}(\vec{i} + 2\vec{j} + 3\vec{k})$$

and then

$$[\vec{t}] = S[\vec{e}_r] = \begin{bmatrix} -4 & 4 & 0 \\ 4 & 2 & 0 \\ 0 & 0 & 1 \end{bmatrix} \frac{1}{\sqrt{14}} \begin{bmatrix} 1 \\ 2 \\ 3 \end{bmatrix} = \frac{1}{\sqrt{14}} \begin{bmatrix} 4 \\ 8 \\ 3 \end{bmatrix} ; \quad \vec{t} = \frac{1}{\sqrt{14}}(4\vec{i} + 8\vec{j} + 3\vec{k})$$

$$|\vec{t}| = \frac{1}{\sqrt{14}}\sqrt{4^2 + 8^2 + 3^2} = \sqrt{\frac{89}{14}}.$$

3. Using Formula (14.3) for the Projector $P_{\vec{e}_r}$, we have

$$P_{\vec{e}_r} = \frac{1}{14} \begin{bmatrix} 1 & 2 & 3 \\ 2 & 4 & 6 \\ 3 & 6 & 9 \end{bmatrix} ;$$

$$P_{\vec{e}_r}(\vec{t}) = \frac{1}{14\sqrt{14}} \begin{bmatrix} 1 & 2 & 3 \\ 2 & 4 & 6 \\ 3 & 6 & 9 \end{bmatrix} \begin{bmatrix} 4 \\ 8 \\ 3 \end{bmatrix} = \frac{29}{14} \left(\frac{1}{\sqrt{14}} \begin{bmatrix} 1 \\ 2 \\ 3 \end{bmatrix} \right)$$

$$\vec{\sigma} = P_{\vec{e}_r}(\vec{t}) = \frac{29}{14}\vec{e}_r = \frac{29}{14\sqrt{14}}(\vec{i} + 2\vec{j} + 3\vec{k}).$$

The previous equality declares that $\sigma = |\vec{\sigma}| = \frac{29}{14}$, though it can also be calculated as a dot product:

$$\sigma = \vec{e}_r \bullet \vec{t} = \frac{1}{\sqrt{14}}[1 \quad 2 \quad 3] \frac{1}{\sqrt{14}} \begin{bmatrix} 4 \\ 8 \\ 3 \end{bmatrix} = \frac{1}{14}(4 + 16 + 9) = \frac{29}{14}.$$

4. Using Formula (14.4) for the "complementary" projector $P_{\vec{e}_r^{\perp}}$, we have

$$P_{\vec{e}_r^{\perp}} = \frac{1}{14} \begin{bmatrix} 13 & -2 & -3 \\ -2 & 10 & -6 \\ -3 & -6 & 5 \end{bmatrix} ;$$

$$P_{\vec{e}_r^{\perp}}(\vec{t}) = \frac{1}{14\sqrt{14}} \begin{bmatrix} 13 & -2 & -3 \\ -2 & 10 & -6 \\ -3 & -6 & 5 \end{bmatrix} \begin{bmatrix} 4 \\ 8 \\ 3 \end{bmatrix} = \frac{9\sqrt{70}}{14\sqrt{14}} \left(\frac{1}{\sqrt{70}} \begin{bmatrix} 3 \\ 6 \\ -5 \end{bmatrix} \right)$$

$$\vec{\tau} = P_{\vec{e}_r^{\perp}}(\vec{t}) = \frac{9\sqrt{5}}{14}\vec{e}_r^{\perp} = \frac{9}{14\sqrt{14}}[3\vec{i} + 6\vec{j} - 5\vec{k}],$$

which declares that $\tau = |\vec{\tau}| = \frac{9\sqrt{5}}{14}$, though it can also be calculated using the Pythagoras theorem:

$$\tau = \sqrt{|t|^2 - \sigma^2} = \sqrt{\frac{89}{14} - \left(\frac{29}{14}\right)^2} = \sqrt{\frac{405}{14^2}} = \frac{9}{14}\sqrt{5}.$$

5. The characteristic polynomial of S is

$$\begin{vmatrix} -4-\sigma & 4 & 0 \\ 4 & 2-\sigma & 0 \\ 0 & 0 & 1-\sigma \end{vmatrix} = 0 \rightarrow (\sigma-1)(\sigma^2 + 2\sigma - 24) = 0,$$

and the corresponding eigenvalues are

$$\sigma_I = 4; \quad \sigma_{II} = 1; \quad \sigma_{III} = -6,$$

which are the moduli of the main stresses; the first two are tensions (positive) and the last one is a compression (negative).
The unit eigenvectors are

$$\text{For } \sigma_I = 4: \quad \begin{bmatrix} -8 & 4 & 0 \\ 4 & -2 & 0 \\ 0 & 0 & -3 \end{bmatrix} \begin{bmatrix} x^1 \\ x^2 \\ x^3 \end{bmatrix} = \begin{bmatrix} 0 \\ 0 \\ 0 \end{bmatrix} \rightarrow \begin{bmatrix} x^1 \\ x^2 \\ x^3 \end{bmatrix} = \begin{bmatrix} 1 \\ 2 \\ 0 \end{bmatrix},$$

which leads to $\vec{a}_I = \frac{1}{\sqrt{5}}(\vec{i} + 2\vec{j})$.

$$\text{For } \sigma_{II} = 1: \quad \begin{bmatrix} -5 & 4 & 0 \\ 4 & 3 & 0 \\ 0 & 0 & 0 \end{bmatrix} \begin{bmatrix} x^1 \\ x^2 \\ x^3 \end{bmatrix} = \begin{bmatrix} 0 \\ 0 \\ 0 \end{bmatrix} \rightarrow \begin{bmatrix} x^1 \\ x^2 \\ x^3 \end{bmatrix} = \begin{bmatrix} 0 \\ 0 \\ 1 \end{bmatrix},$$

which leads to $\vec{a}_{II} = \vec{k}$.

$$\text{For } \sigma_{III} = -6: \quad \begin{bmatrix} 2 & 4 & 0 \\ 4 & 8 & 0 \\ 0 & 0 & 7 \end{bmatrix} \begin{bmatrix} x^1 \\ x^2 \\ x^3 \end{bmatrix} = \begin{bmatrix} 0 \\ 0 \\ 0 \end{bmatrix} \rightarrow \begin{bmatrix} x^1 \\ x^2 \\ x^3 \end{bmatrix} = \begin{bmatrix} 2 \\ -1 \\ 0 \end{bmatrix},$$

which leads to $\vec{a}_{III} = \frac{1}{\sqrt{5}}(2\vec{i} - \vec{j})$.
It can be verified that the change-of-basis associated with the eigenvectors is a direct orthogonal matrix

$$|Q| = \begin{vmatrix} 1/\sqrt{5} & 0 & 2/\sqrt{5} \\ 2/\sqrt{5} & 0 & -1/\sqrt{5} \\ 0 & 1 & 0 \end{vmatrix} = -\frac{1}{5} \begin{vmatrix} 1 & 2 \\ 2 & -1 \end{vmatrix} = 1,$$

that is, the new trihedron is a "direct" trihedron.
The matrix Q produces the "orthogonal canonization" of S as

$$\hat{S} = Q^{-1}SQ \equiv Q^t SQ = \begin{bmatrix} 4 & 0 & 0 \\ 0 & 1 & 0 \\ 0 & 0 & -6 \end{bmatrix}.$$

Finally, the sought after vectors are

$$\vec{\sigma}_I = \frac{4}{\sqrt{5}}(\vec{i} + 2\vec{j}); \quad \vec{\sigma}_{II} = \vec{k}; \quad \vec{\sigma}_{III} = \frac{-6}{\sqrt{5}}(2\vec{i} - \vec{j}).$$

6. According to Formula (14.75) applied to the stress tensor S and directions $\vec{\hat{e}}_a, \vec{\hat{e}}_b, \vec{\hat{e}}_c$, we obtain the following results:

$$(\tau_{max})_a = \frac{\sigma_{II} - \sigma_{III}}{2} = \frac{1 - (-6)}{2} = \frac{7}{2}$$

$$(\tau_{max})_b = \frac{\sigma_{I} - \sigma_{III}}{2} = \frac{4 - (-6)}{2} = 5$$

$$(\tau_{max})_c = \frac{\sigma_{I} - \sigma_{II}}{2} = \frac{4 - 1}{2} = \frac{3}{2}$$

and conclude that the maximum shear stress corresponding to the orthogonal directions $\vec{\hat{e}}_b(\cos 45°, 0, \pm \sin 45°)$ is $\tau_{max} = 5$.

7. The previous orthogonal vectors $(\frac{1}{\sqrt{2}}, 0, \pm\frac{1}{\sqrt{2}})$ are the directions of the maximum shear stress referred to the tensor $\hat{S}$. Since the question refers to the tensor S, we must recover with the change-of-basis Q, the directions to the initial basis $\{\vec{i}, \vec{j}, \vec{k}\}$. In matrix form, the inverse change is

$$X = Q\hat{X};$$

$$X = \begin{bmatrix} 1/\sqrt{5} & 0 & 2/\sqrt{5} \\ 2/\sqrt{5} & 0 & -1/\sqrt{5} \\ 0 & 1 & 0 \end{bmatrix} \begin{bmatrix} 1/\sqrt{2} & 1/\sqrt{2} \\ 0 & 0 \\ 1/\sqrt{2} & -1/\sqrt{2} \end{bmatrix} = \begin{bmatrix} 3/\sqrt{10} & -1/\sqrt{10} \\ 1/\sqrt{10} & 3/\sqrt{10} \\ 0 & 0 \end{bmatrix},$$

so that the orthogonal directions sought after of shear stress $\tau = 5$ are

$$\vec{e}_{b_1} = \frac{1}{\sqrt{10}}(3\vec{i} + \vec{j}) \text{ and } \vec{e}_{b_2} = \frac{1}{\sqrt{10}}(-\vec{i} + 3\vec{j}).$$

The reader can verify, applying the process described in question 4, that in fact a value $\tau = 5$ corresponds to the cited directions.

8. The normal stress corresponding to the maximum shear stresses can be calculated as

$$\sigma_1 = \vec{e}_{b_1} \bullet S(\vec{e}_{b_1}) = \frac{1}{\sqrt{10}}[3 \ \ 1 \ \ 0] \bullet \left(\begin{bmatrix} -4 & 4 & 0 \\ 4 & 2 & 0 \\ 0 & 0 & 1 \end{bmatrix} \frac{1}{\sqrt{10}} \begin{bmatrix} 3 \\ 1 \\ 0 \end{bmatrix} \right)$$

$$= \frac{1}{10}[3 \ \ 1 \ \ 0] \begin{bmatrix} -8 \\ 14 \\ 0 \end{bmatrix} = \frac{1}{10}(-24 + 14 + 0) = \frac{-10}{10} = -1;$$

$$\sigma_2 = \vec{e}_{b_2} \bullet S(\vec{e}_{b_2}) = \frac{1}{\sqrt{10}}[-1 \ \ 3 \ \ 0] \bullet \left(\begin{bmatrix} -4 & 4 & 0 \\ 4 & 2 & 0 \\ 0 & 0 & 1 \end{bmatrix} \frac{1}{\sqrt{10}} \begin{bmatrix} -1 \\ 3 \\ 0 \end{bmatrix} \right)$$

$$= \frac{1}{10}[-1 \ \ 3 \ \ 0] \begin{bmatrix} 16 \\ 2 \\ 0 \end{bmatrix} = \frac{1}{10}(-16 + 6 + 0) = \frac{-10}{10} = -1;$$

thus, we have $\sigma_1 = \sigma_2 = -1$.

The two planes of shear stress $\tau = 5$ undergo normal compressions of value 1.

9. To solve this question we refer it to the reference frame system of tensor $\hat{S}$, and once solved, we return to the initial reference frame, as it was done in question 7.

Let $\vec{e}(\hat{\ell}, \hat{m}, \hat{n})$ be the desired direction; we must have

$$\sigma = \vec{e} \bullet \hat{S}(\vec{e}) = 0; \quad [\hat{\ell} \quad \hat{m} \quad \hat{n}] \begin{bmatrix} 4 & 0 & 0 \\ 0 & 1 & 0 \\ 0 & 0 & -6 \end{bmatrix} \begin{bmatrix} \hat{\ell} \\ \hat{m} \\ \hat{n} \end{bmatrix} = 0; \quad 4\hat{\ell}^2 + \hat{m}^2 - 6\hat{n}^2 = 0.$$

The general solution of this equation, in parametric coordinates, is

$$\begin{cases} \hat{\ell} = \frac{\sqrt{6}}{2}(1 - \lambda^2)\mu \\ \hat{m} = 2\sqrt{6}\lambda\mu \\ \hat{n} = \mu(1 + \lambda^2). \end{cases}$$

We choose $\lambda = \mu = 1$; $(\hat{\ell}, \hat{m}, \hat{n}) = (0, 2\sqrt{6}, 2)$, that is, the vector $\vec{r}(0, \sqrt{6}, 1)$, and we proceed to return it to the initial basis

$$X = Q\hat{X}; \quad \begin{bmatrix} \ell \\ m \\ n \end{bmatrix} = \begin{bmatrix} \frac{1}{\sqrt{5}} & 0 & \frac{2}{\sqrt{5}} \\ \frac{2}{\sqrt{5}} & 0 & \frac{-1}{\sqrt{5}} \\ 0 & 1 & 0 \end{bmatrix} \begin{bmatrix} 0 \\ \sqrt{6} \\ 1 \end{bmatrix} = \frac{1}{\sqrt{5}} \begin{bmatrix} 2 \\ -1 \\ \sqrt{30} \end{bmatrix}.$$

One solution is

$$\vec{r} = 2\vec{i} - \vec{j} + \sqrt{30}\vec{k}; \quad \vec{e}_r = \frac{1}{\sqrt{35}}(2\vec{i} - \vec{j} + \sqrt{30}\vec{k}).$$

10. The vector $\vec{e}_r = \frac{1}{\sqrt{35}}(2\vec{i} - \vec{j} + \sqrt{30}\vec{k})$ has associated stress $\vec{t}_r$:

$$\begin{bmatrix} t_r^1 \\ t_r^2 \\ t_r^3 \end{bmatrix} = \frac{1}{\sqrt{35}} \begin{bmatrix} -4 & 4 & 0 \\ 4 & 2 & 0 \\ 0 & 0 & 1 \end{bmatrix} \begin{bmatrix} 2 \\ -1 \\ \sqrt{30} \end{bmatrix} = \frac{1}{\sqrt{35}} \begin{bmatrix} -12 \\ 6 \\ \sqrt{30} \end{bmatrix},$$

and since $\sigma = 0$, then $\vec{\tau} \equiv \vec{t}_r = \frac{1}{\sqrt{35}}(-12\vec{i} + 6\vec{j} + \sqrt{30}\vec{k})$ and $\tau = |\vec{t}_r| = \frac{\sqrt{12^2 + 6^2 + 30}}{\sqrt{35}} = \sqrt{6}$.

□

Example 14.12 (Mixed stresses). Consider a metal tube for which the scheme is given in Figure 14.14, with radius $R = 30$ cm and width $s = 2$ mm.

Over the ends of the tube two moments are applied:

- A bending moment in the plane YOZ of value $\vec{M}_F = 3\vec{i}$ mT (meter × tonne).
- A torque moment of axis $\overline{OZ}$ and value $\vec{M}_T = 4\vec{k}$ mT.

1. Give the stress tensor S at the point P far from the tube ends.
2. Indicate the points at which the maximum stress occurs.

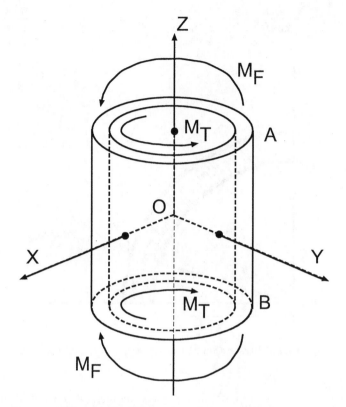

Fig. 14.14. Illustration of the metal tube.

3. Give the value of this maximum stress.
4. Give the unit vector orthogonal to the plane in which this stress occurs.

Solution:

1. We assume as a simplifying assumption that the mean radius of the tube is R, that is, that the width is negligible with respect to the radius. According to this, the tangent stress produced by the torque moment on the plane XOY is

$$\tau = \frac{M_T}{R(2\pi Rs)} = \frac{M_T}{2\pi sR^2},$$

and then the shear stresses on this plane, parallel to the axis $\overline{OX}$ and $\overline{OY}$, as can be seen in Figure 14.15, are

$$\vec{\tau} = \tau_{xz}\vec{i} + \tau_{yz}\vec{j} = -\tau\sin\alpha\vec{i} + \tau\cos\alpha\vec{j} = -\tau\frac{y}{R}\vec{i} + \tau\frac{x}{R}\vec{j},$$

thus, in summary:

$$\tau_{xz} = -\frac{M_T}{2\pi s R^2} \cdot \frac{y}{R} = -\frac{M_T y}{2\pi s R^3}; \quad \tau_{yz} = \frac{M_T}{2\pi s R^2} \cdot \frac{x}{R} = \frac{M_T x}{2\pi s R^3}.$$

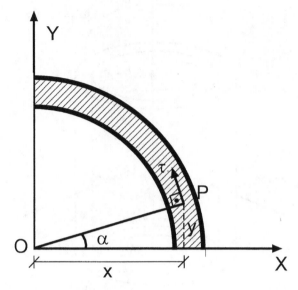

Fig. 14.15. Illustration of the stresses on the metal tube.

On the other hand, due to the bending moment a tension is produced on the plane section XOY, that, according to the theory, is $\sigma_z = \frac{M_F y}{I_x}$. To calculate the inertial moment I_x, of the plane section with respect to the axis $\overline{OX}$ we apply again, as a first approximation, the simplifying assumption of negligible width with respect to the radius. Then, we have

$$I_x = \frac{I_0}{2} = \frac{1}{2}(2\pi R s)R^2 = \pi s R^3,$$

and taking this result to σ_z, we obtain $\sigma_z = \frac{M_F \cdot y}{\pi s R^3}$.

Once these stresses are known, we can determine the stress tensor at the point $P(x, y, z)$ far from the tube ends, at which concentration of stresses takes place due to the proper deformation. The tensor S is

$$S = \begin{bmatrix} \sigma_x & \tau_{xy} & \tau_{xz} \\ \tau_{xy} & \sigma_y & \tau_{yz} \\ \tau_{xz} & \tau_{yz} & \sigma_z \end{bmatrix} = \frac{1}{2\pi s R^3} \begin{bmatrix} 0 & 0 & -M_T y \\ 0 & 0 & M_T x \\ -M_T y & M_T x & 2M_F y \end{bmatrix}.$$

2. By means of the characteristic polynomial, we determine the main stresses and from them we deduce where the maximum stresses occur:

$$\begin{vmatrix} -\sigma & 0 & \tau_{xz} \\ 0 & -\sigma & \tau_{yz} \\ \tau_{xz} & \tau_{yz} & \sigma_z - \sigma \end{vmatrix} = 0; \quad \sigma\left[\sigma^2 - \sigma_z \sigma + (\tau_{xz}^2 + \tau_{yz}^2)\right] = 0.$$

In conclusion, $\sigma[\sigma^2 - \sigma_z\sigma + \tau^2] = 0$ and $\sigma = \frac{\sigma_z}{2} \pm \sqrt{\left(\frac{\sigma_z}{2}\right)^2 + \tau^2}$, and since τ is constant, the maximum value of σ will occur for the $+$ sign and at the tube generatrix AB, for which σ_z is a maximum.

3. The value of σ_z at the generatrix points $\overline{AB}$ $(y = R)$ is

$$(\sigma_z)_{max} \left(\frac{2M_F y}{2\pi s R^3}\right)_{y=R} = \frac{M_F}{\pi s R^2},$$

and the maximum stress becomes

$$\sigma_{max} = \frac{(\sigma_z)_{max}}{2} + \sqrt{\left[\frac{(\sigma_z)_{max}}{2}\right]^2 + \tau^2}$$

$$= \frac{M_F}{2\pi s R^2} + \sqrt{\left(\frac{M_F}{2\pi s R^2}\right)^2 + \left(\frac{M_T}{2\pi s R^2}\right)^2}$$

$$= \frac{1}{2\pi s R^2}\left[M_F + \sqrt{M_F^2 + M_T^2}\right].$$

Numerically, expressing the moments in cm×kg, we get

$$\sigma_{max} = \frac{1}{2\pi \times 0.6 \times 30^2}[3\times10^5 + \sqrt{3^2 + 4^2}\times10^5] = \frac{8 \times 10^4}{339.292} = 235.8 \text{ kg/cm}^2$$

or $\sigma_{max} = 235.8 \times 9.8 \times 10^4 = 2311 \times 10^4$ pascal.

4. The eigenvector corresponding to the eigenvalue σ is

$$\frac{\ell}{\tau_{xz}} = \frac{m}{\tau_{yz}} = \frac{n}{\sigma},$$

and for the σ_{max} is

$$\tau_{xz} = \left(-\tau\frac{y}{R}\right)_{y=R} = -\tau; \quad \tau_{yz} = \left(\tau\frac{x}{R}\right)_{x=0} = 0,$$

so that the eigenvector becomes

$$(\ell, m, n) = (-\tau, 0, \sigma_{max}) = \frac{1}{2\pi s R^2}\left(-M_T, 0, M_F + \sqrt{M_F^2 + M_T^2}\right)$$

$$= \lambda_0(-4, 0, 8).$$

Finally, the unit vector is $\vec{e}_{\sigma_{max}} = \frac{1}{\sqrt{5}}(-\vec{i} + 2\vec{k})$.

□

14.4.4 The inertial moment tensor

Consider the ordinary geometric space referred to the classical orthonormalized system $(O-XYZ)$.

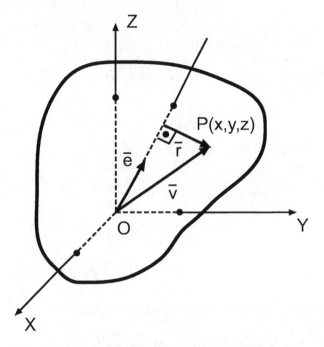

Fig. 14.16. Illustration of the inertial moment tensor.

An homogeneous rigid solid S lies on O, rotates around an axis that passes through O and has instantaneous direction $\vec{e}$.

We consider an elemental mass in the neighborhood of a given point $P(x, y, z)$ of the solid (see Figure 14.16). The sum of all products of the mass times the square of its distance to the axis is called the "inertial moment of the solid with respect to the axis $\vec{e}$" and is denoted by

$$I_e = \int_S r^2 dm. \tag{14.89}$$

If we notate the unit vector $\vec{e}$ as

$$\vec{e} = l\vec{i} + m\vec{j} + n\vec{k}; \quad l^2 + m^2 + n^2 = 1$$

and the position vector of the point P is

$$\vec{v} = \overline{OP} = x\vec{i} + y\vec{j} + z\vec{k},$$

we can calculate the vector $\vec{r}$ using the complementary projector $P_{\vec{e}^\perp}$, Formula (14.20), as

$$\vec{r} = P_{\vec{e}^\perp}(\vec{v}) = \begin{bmatrix} (1 - l^2) & -lm & -ln \\ -lm & (1 - m^2) & -mn \\ -ln & -mn & (1 - n^2) \end{bmatrix} \begin{bmatrix} x \\ y \\ z \end{bmatrix},$$

so that since $P_{\vec{e}^\perp} \equiv P^t_{\vec{e}^\perp}$ we have

$$r^2 = \vec{r} \bullet \vec{r} = [\,x \quad y \quad z\,]\, P^t_{\vec{e}^\perp} \bullet P_{\vec{e}^\perp} \begin{bmatrix} x \\ y \\ z \end{bmatrix} = [\,x \quad y \quad z\,]\, P^2_{\vec{e}^\perp} \begin{bmatrix} x \\ y \\ z \end{bmatrix},$$

and since the projectors are idempotent, we obtain

$$r^2 = [\,x \quad y \quad z\,]\, P_{\vec{e}^\perp} \begin{bmatrix} x \\ y \\ z \end{bmatrix} = [\,x \quad y \quad z\,] \begin{bmatrix} (1-\ell^2) & -\ell m & -\ell n \\ -\ell m & (1-m^2) & -mn \\ -\ell n & -mn & (1-n^2) \end{bmatrix} \begin{bmatrix} x \\ y \\ z \end{bmatrix}$$

or

$$[\,x \quad y \quad z\,] \begin{bmatrix} (m^2+n^2) & -\ell m & -\ell n \\ -\ell m & (\ell^2+n^2) & -mn \\ -\ell n & -mn & (\ell^2+m^2) \end{bmatrix} \begin{bmatrix} x \\ y \\ z \end{bmatrix}.$$

Next, we develop this expression and substitute in Formula (14.89), to obtain

$$\begin{aligned} I_e &= \int_S r^2 dm = \int_S (m^2+n^2)x^2 + (\ell^2+n^2)y2 + (\ell^2+m^2)z^2 \\ &\quad - 2(\ell m xy + \ell n xz + mn yz)dm \\ &= \ell^2 \int_S (y^2+z^2)dm + m^2 \int_S (x^2+z^2)dm + n^2 \int_S (x^2+y^2)dm \\ &\quad - 2\left(\ell m \int_S xydm + \ell n \int_S xzdm + mn \int_S yzdm \right), \end{aligned}$$

where we notate

$$I_x = \int_S (y^2+z^2)dm; \quad I_y = \int_S (x^2+z^2)dm; \quad I_z = \int_S (x^2+y^2)dm;$$

$$P_{xy} = \int_S xydm; \quad P_{xz} = \int_S xzdm; \quad P_{yz} = \int_S yzdm. \tag{14.90}$$

The expressions I_x, I_y, I_z are the inertial moments with respect to the Cartesian axes and because of its essence they are always positive scalars. The expressions P_{xy}, P_{xz}, P_{yz} are called "inertial products" and also "centrifugal moments". If Formulas (14.90) are taken into the expression for I_e, we get

$$\begin{aligned} I_e &= I_x\ell^2 + I_y m^2 + I_z n^2 - 2P_{xy}\ell m - 2P_{xz}\ell n - 2P_{yz}mn \\ &= [\,\ell \quad m \quad n\,] \begin{bmatrix} I_x & -P_{xy} & -P_{xz} \\ -P_{xy} & I_y & -P_{yz} \\ -P_{xz} & -P_{yz} & I_z \end{bmatrix} \begin{bmatrix} \ell \\ m \\ n \end{bmatrix} = \vec{e} \bullet I(\vec{e}). \tag{14.91} \end{aligned}$$

The notation I corresponds to the so-called "inertial moment tensor", and is a symmetric tensor, the matrix representation of which is

$$I = \begin{bmatrix} I_x & -P_{xy} & -P_{xz} \\ -P_{xy} & I_y & -P_{yz} \\ -P_{xz} & -P_{yz} & I_z \end{bmatrix}. \tag{14.92}$$

If the origin O of the Cartesian system coincides with the center of gravity G of the solid S, the inertial moment tensor is denoted by I_G.

If the axis e does not pass through the solid center of gravity, but another parallel axis passes through it, the inertial moment with respect to the passing axis will be denoted by $(I_G)_e$ and its relation to I_e is called the Steiner theorem

$$I_e = m \cdot \Delta^2 + (I_G)_e, \tag{14.93}$$

where m is the solid mass and Δ the distance between the two parallel axes.

If the rotation mentioned at the beginning of this section has angular velocity w around the axis e, the "angular velocity" vector will be

$$\vec{w} = w\vec{e} = wl\vec{i} + wm\vec{j} + wn\vec{k} = w_1\vec{i} + w_2\vec{j} + w_3\vec{k}, \tag{14.94}$$

where w is in radians/sec.

The vector $I(\vec{w})$ is called the "kinetic moment" of the solid S with respect to the axis e and it is denoted by $\vec{h}_e$. If $\vec{h}_e = h_1\vec{i} + h_2\vec{j} + h_3\vec{k}$, we have (remember that $\ell^2 + m^2 + n^2 = 1$)

$$\begin{bmatrix} h_1 \\ h_2 \\ h_3 \end{bmatrix} = I(\vec{w}) = \begin{bmatrix} I_x & -P_{xy} & -P_{xz} \\ -P_{xy} & I_y & -P_{yz} \\ -P_{xz} & -P_{yz} & I_z \end{bmatrix} \begin{bmatrix} w_1 \\ w_2 \\ w_3 \end{bmatrix}$$

$$= wI(\vec{e}) = w \begin{bmatrix} I_x & -P_{xy} & -P_{xz} \\ -P_{xy} & I_y & -P_{yz} \\ -P_{xz} & -P_{yz} & I_z \end{bmatrix} \begin{bmatrix} \ell \\ m \\ n \end{bmatrix}. \tag{14.95}$$

The dot product $\frac{1}{2}\vec{w} \bullet \vec{h}_e$ is denoted by E_c and is called the "kinetic energy" of the solid S due to the rotation $\vec{w}$. We have

$$E_c = \frac{1}{2}\vec{w} \bullet \vec{h}_e = \frac{1}{2}(w\vec{e}) \bullet I(\vec{w}) = \frac{1}{2}(w\vec{e}) \bullet (wI(\vec{e})) = \frac{1}{2}w^2(\vec{e} \bullet I(\vec{e})),$$

and applying Formula (14.91), the result is

$$E_c = \frac{1}{2}I_e w^2. \tag{14.96}$$

As for all the second-order symmetric tensors, the characteristic polynomial of the tensor I presents real eigenvalues, I_1, I_2, I_3, in this case all positive, that are called "main inertial moments".

The unit eigenvectors $\vec{a}_1, \vec{a}_2, \vec{a}_3$, associated with these eigenvalues are the "main directions" and give the maximum and minimum inertial moments, and null inertial products.

With the eigenvector components in columns a orthogonal matrix Q is built, that canonizes the tensor I:

$$\hat{I} = Q^{-1}IQ \equiv Q^t IQ = \begin{bmatrix} I_1 & 0 & 0 \\ 0 & I_2 & 0 \\ 0 & 0 & I_3 \end{bmatrix} ; \quad I_1 > I_2 > I_3 > 0. \tag{14.97}$$

Equation (14.91), which gives the solid inertial moment I_e as a function of the director cosines (ℓ, m, n) of the unit vector $\vec{e}$, can be written as

$$1 = I_x \left(\frac{\ell}{\sqrt{I_e}}\right)^2 + I_y \left(\frac{m}{\sqrt{I_e}}\right)^2 + I_z \left(\frac{n}{\sqrt{I_e}}\right)^2 - 2P_{xy}\left(\frac{\ell}{\sqrt{I_e}}\right)\left(\frac{m}{\sqrt{I_e}}\right)$$
$$-2P_{xz}\left(\frac{\ell}{\sqrt{I_e}}\right)\left(\frac{n}{\sqrt{I_e}}\right) - 2P_{yz}\left(\frac{m}{\sqrt{I_e}}\right)\left(\frac{n}{\sqrt{I_e}}\right),$$

and if we choose a point $A(x, y, z)$ in space, the coordinates of which satisfy the conditions

$$x = \frac{\ell}{\sqrt{I_e}}; \quad y = \frac{m}{\sqrt{I_e}}; \quad z = \frac{n}{\sqrt{I_e}},$$

the end A of the position vector $\overline{OA}$ covers the surface:

$$I_x x^2 + I_y y^2 + I_z z^2 - 2P_{xy}xy - 2P_{xz}xz - 2P_{yz}yz = 1, \tag{14.98}$$

which is known as the "inertial ellipsoid" of the solid S, referred to the axis e.

Example 14.13 (Inertial tensors associated with a solid). Consider an orthogonal parallelepiped, with its edges coinciding at a vertex a, b, c, and assume a material with density $\rho = 1$.

1. Obtain the solid inertial tensor, assuming that its position on the reference frame is $a \in \overline{OX}, b \in \overline{OY}$ and $c \in \overline{OZ}$.
2. Give the inertial tensor I_G, with respect to the system $(G-XYZ)$.
3. Give the inertial moment I_e of the solid with respect to the axis $e \equiv \overline{OG}$, verifying its calculation in the system: (a) $(O-XYZ)$, (b) $(G-XYZ)$.
4. We locate the solid in the reference frame system $(O-XYZ)$ in another position $b \in \overline{OX}; c \in \overline{OY}; a \in \overline{OZ}$. Answer again questions 1, 2 and 3.
5. Are the values of the main inertial moments of the tensors I_{abc}, I_{bca} maintained?

Solution:

1. According to Formulas (14.90), we have:

$$I_x = \int_S (y^2 + z^2)dm = \int \int \int_S (y^2 + z^2)dxdydz$$

$$= \int_0^a dx \int_0^b dy \int_0^c (y^2 + z^2)dz = \frac{abc}{3}(b^2 + c^2)$$

$$I_y = \int_S (x^2 + z^2)dm = \int \int \int_S (x^2 + z^2)dxdydz$$

$$= \int_0^a dx \int_0^b dy \int_0^c (x^2 + z^2)dz = \frac{abc}{3}(a^2 + c^2)$$

$$I_z = \frac{abc}{3}(a^2 + b^2)$$

$$P_{xy} = \int_S (xy)dm = \int \int \int_S (xy)dxdydz$$

$$= \int_0^a xdx \int_0^b xdy \int_0^c dz = \frac{abc}{4}(ab)$$

$$P_{xz} = \frac{abc}{4}(ac)$$

$$P_{yz} = \frac{abc}{4}(bc)$$

and then the inertial tensor, according to Formula (14.92) is

$$I = (abc) \begin{bmatrix} \frac{b^2+c^2}{3} & -\frac{ab}{4} & -\frac{ac}{4} \\ -\frac{ab}{4} & \frac{a^2+c^2}{3} & -\frac{bc}{4} \\ -\frac{ab}{4} & -\frac{bc}{4} & \frac{a^2+b^2}{3} \end{bmatrix}.$$

2. When locating the reference frame system $(G - XYZ)$ with origin at the point $G\left(\frac{a}{2}, \frac{b}{2}, \frac{c}{2}\right)$ and axes parallel to those in $(O - XYZ)$, the limits in the integrals change and since the solid presents three axes of symmetry in G, the inertial products are all null. Summarizing:

$$(I_G)_x = \int_S (y2 + z^2)dm = \int \int \int_S (y2 + z^2)dxdydz$$

$$= \int_{-a/2}^{a/2} dx \int_{-b/2}^{b/2} dy \int_{-c/2}^{c/2} (y2 + z^2)dz = \frac{abc}{12}(b^2 + c^2)$$

$$(I_G)_y = \frac{abc}{12}(a^2 + c^2)$$

$$(I_G)_z = \frac{abc}{12}(a^2 + b^2)$$

and $P_{xy} = P_{xz} = P_{yz} = 0$, which proves that the inertial tensor is

$$I_G = \frac{abc}{12}\begin{bmatrix} b^2 + c^2 & 0 & 0 \\ 0 & a^2 + c^2 & 0 \\ 0 & 0 & a^2 + b^2 \end{bmatrix}.$$

We note that in this case the inertial moments $(I_G)_x, (I_G)_y, (I_G)_z$ are the main inertial moments, and that the main directions are those of the Cartesian axes (for the tensor I_G).

3. (a) The director vector $\overline{OG}$ is $\left(\frac{a}{2}, \frac{b}{2}, \frac{c}{2}\right)$, and so the unit vector will be

$$\vec{e} = \frac{1}{\sqrt{a^2 + b^2 + c^2}}(a\vec{i} + b\vec{j} + c\vec{k}).$$

Using Formula (14.91), we have

$$I_e = (abc)\frac{1}{\sqrt{a^2 + b^2 + c^2}}[a\,b\,c]\begin{bmatrix} \frac{b^2+c^2}{3} & -\frac{ab}{4} & -\frac{ac}{4} \\ -\frac{ab}{4} & \frac{a^2+c^2}{3} & -\frac{bc}{4} \\ -\frac{ab}{4} & -\frac{bc}{4} & \frac{a^2+b^2}{3} \end{bmatrix}\begin{bmatrix} a \\ b \\ c \end{bmatrix}\frac{1}{\sqrt{a^2 + b^2 + c^2}}$$

$$= \frac{abc}{a^2 + b^2 + c^2}$$

$$\left(\frac{b^2 + c^2}{3}a^2 + \frac{a^2 + c^2}{3}b^2 + \frac{a^2 + b^2}{3}c^2 - 2\frac{ab}{4}ab - 2\frac{ac}{4}ac - 2\frac{bc}{4}bc\right);$$

$$I_e = \frac{abc}{a^2 + b^2 + c^2}\left(\frac{2(a^2b^2 + b^2c^2 + a^2c^2)}{3} - \frac{a^2b^2 + b^2c^2 + a^2c^2}{2}\right)$$

$$= \frac{abc}{6}\frac{a^2b^2 + b^2c^2 + a^2c^2}{a^2 + b^2 + c^2}.$$

(b) Using the same formula with the tensor I_G we obtain

$$(I_G)_e = \frac{abc}{12}\frac{1}{\sqrt{a^2 + b^2 + c^2}}[a\,b\,c]\begin{bmatrix} b^2 + c^2 & 0 & 0 \\ 0 & a^2 + c^2 & 0 \\ 0 & 0 & a^2 + b^2 \end{bmatrix}\begin{bmatrix} a \\ b \\ c \end{bmatrix}\frac{1}{\sqrt{a^2 + b^2 + c^2}}$$

$$= \frac{abc}{12}\frac{(b^2 + c^2)a^2 + (a^2 + c^2)b^2 + (a^2 + b^2)c^2}{a^2 + b^2 + c^2} = \frac{abc}{6}\frac{a^2b^2 + b^2c^2 + a^2c^2}{a^2 + b^2 + c^2}.$$

We obtain the same results, as it corresponds to the same axis.

4. (a) Following a parallel process, we obtain

$$I_x = \frac{abc}{3}(c^2 + a^2); \quad I_y = \frac{abc}{3}(b^2 + a^2); \quad I_z = \frac{abc}{3}(b^2 + c^2);$$

$$P_{xy} = \frac{abc}{4}bc; \quad P_{xz} = \frac{abc}{4}ba; \quad P_{yz} = \frac{abc}{4}ca;$$

$$I' = (abc)\begin{bmatrix} \frac{a^2+c^2}{3} & -\frac{bc}{4} & -\frac{ab}{4} \\ -\frac{bc}{4} & \frac{a^2+b^2}{3} & -\frac{ac}{4} \\ -\frac{ab}{4} & -\frac{ac}{4} & \frac{b^2+c^2}{3} \end{bmatrix}.$$

(b)

$$(I_G)_x = \frac{abc}{12}(c^2+a^2); \quad (I_G)_y = \frac{abc}{12}(b^2+a^2); \quad (I_G)_z = \frac{abc}{12}(b^2+c^2);$$

$$I_G' = \frac{abc}{12}\begin{bmatrix} a^2+c^2 & 0 & 0 \\ 0 & a^2+b^2 & 0 \\ 0 & 0 & b^2+c^2 \end{bmatrix}.$$

(c)   i. The director vector $\overline{OG}$ is now (b,c,a), and so the unit vector will be

$$\vec{e}\,' = \frac{1}{\sqrt{a^2+b^2+c^2}}(b\vec{i}+c\vec{j}+a\vec{k})$$

$$I_{e'}' = \frac{abc}{\sqrt{a^2+b^2+c^2}}[b\,c\,a]\begin{bmatrix} \frac{a^2+c^2}{3} & -\frac{bc}{4} & -\frac{ab}{4} \\ -\frac{bc}{4} & \frac{a^2+b^2}{3} & -\frac{ac}{4} \\ -\frac{ab}{4} & -\frac{ac}{4} & \frac{b^2+c^2}{3} \end{bmatrix}\begin{bmatrix} b \\ c \\ a \end{bmatrix}\frac{1}{\sqrt{a^2+b^2+c^2}}$$

$$I_{e'}' = \frac{abc}{6}\frac{a^2b^2+b^2c^2+a^2c^2}{a^2+b^2+c^2} \equiv I_e.$$

ii.

$$(I_G')_{e'} = \frac{abc}{12}\frac{1}{\sqrt{a^2+b^2+c^2}}[b\,c\,a]\begin{bmatrix} a^2+c^2 & 0 & 0 \\ 0 & a^2+b^2 & 0 \\ 0 & 0 & b^2+c^2 \end{bmatrix}\begin{bmatrix} b \\ c \\ a \end{bmatrix}\frac{1}{\sqrt{a^2+b^2+c^2}}$$

$$(I_G')_{e'} = \frac{abc}{6}\frac{a^2b^2+b^2c^2+a^2c^2}{a^2+b^2+c^2} \equiv (I_G)_e.$$

5. We can observe that

$$M^t I M = M^{-1} I M = I',$$

that is, if $M = \begin{bmatrix} 0 & 0 & 1 \\ 1 & 0 & 0 \\ 0 & 1 & 0 \end{bmatrix}$, with $MM^t = I_3$, the result is

$$\begin{bmatrix} 0 & 1 & 0 \\ 0 & 0 & 1 \\ 1 & 0 & 0 \end{bmatrix}(abc)\begin{bmatrix} \frac{b^2+c^2}{3} & -\frac{ab}{4} & -\frac{ac}{4} \\ -\frac{ab}{4} & \frac{a^2+c^2}{3} & -\frac{bc}{4} \\ -\frac{ab}{4} & -\frac{bc}{4} & \frac{a^2+b^2}{3} \end{bmatrix}\begin{bmatrix} 0 & 0 & 1 \\ 1 & 0 & 0 \\ 0 & 1 & 0 \end{bmatrix} = (abc)\begin{bmatrix} \frac{a^2+c^2}{3} & -\frac{bc}{4} & -\frac{ab}{4} \\ -\frac{bc}{4} & \frac{a^2+b^2}{3} & -\frac{ac}{4} \\ -\frac{ab}{4} & -\frac{ac}{4} & \frac{b^2+c^2}{3} \end{bmatrix}.$$

Since the matrices I and I' are orthogonally similar, with passing matrix M, they have the same characteristic polynomial. Thus, the matrices I, I' have the same eigenvalues, and thus, the same "main" inertial moments. In reality, the matrix M is the proper matrix of a rotation around an instantaneous axis that passes through O, as it corresponds to the most general displacement of a solid with a given fixed point. In fact, analyzing the matrix M, we have $|M| = 1$, that is, it is a "direct" rotation.

In Example 14.2, point 2, we obtain the formula giving the value of the rotation angle θ, when its associated matrix is known:

$$\cos\theta = \frac{\text{trace } M - 1}{2} = \frac{0-1}{2} = -\frac{1}{2} \to \theta = 120°; \ \cos\theta = -\frac{1}{2}; \ \sin\theta = \frac{\sqrt{3}}{2}.$$

According to the rotation tensor formula (14.14), identifying the terms m_{11} and m_{12}, we determine the director cosines of the axis e:

$$m_{11} = \cos^2\alpha + (\cos^2\beta + \cos^2\gamma)\cos\theta = \cos^2\alpha + (\cos^2\beta + \cos^2\gamma)(-1/2) = 0$$
$$m_{22} = \cos^2\beta + (\cos^2\alpha + \cos^2\gamma)\cos\theta = \cos^2\beta + (\cos^2\alpha + \cos^2\gamma)(-1/2) = 0.$$

In our case, a simple solution is $(\cos\alpha, \cos\beta, \cos\gamma) = \left(\frac{1}{\sqrt{3}}, \frac{1}{\sqrt{3}}, \frac{1}{\sqrt{3}}\right)$,

which reveals that $M = \begin{bmatrix} 0 & 0 & 1 \\ 1 & 0 & 0 \\ 0 & 1 & 0 \end{bmatrix}$ represents a rotation with respect to the

axis $\frac{x}{1} = \frac{y}{1} = \frac{z}{1}$, the bisectrix of the first quadrant, and angle $\theta = 120°$, an isometry that is equivalent to placing the solid in the position of question 4, with respect to its position in question 1.

□

14.5 Exercises

14.1. Three forces $\vec{F}_1, \vec{F}_2, \vec{F}_3$ with moduli b, $2b$ and $3b$, respectively, act on the vertices of an equilateral triangle, A, B and C (see Figure 14.17):

$$A(-a, 0); \ B\left(0, a\sqrt{3}\right); \ \text{and } C(a, 0),$$

where $\vec{F}_1 \perp \overline{CA}, \vec{F}_2 \perp \overline{AB}$ and $\vec{F}_3 \perp \overline{BC}$, and all moments M_i of each force F_i with respect to the origin O are dextrorsum.

1. Obtain the resultant force $\vec{R}$ of the given system of forces.
2. Obtain the resultant moment of the system with respect to the Cartesian origin O.
3. Obtain the central axis of the given system of forces.

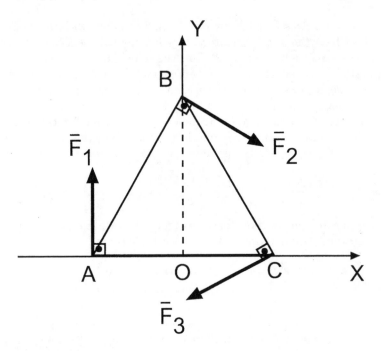

Fig. 14.17. System of forces considered in Exercise 14.1.

4. All the forces are rotated with respect to their application points, in the dextrorsum sense, $90°$. Give the new resultant force $\vec{R}'$.
5. Obtain the new central axis of the given system of forces.
6. Obtain the "center" E of the system.
7. Obtain the minimum moment $\vec{m}_E$ of the system with respect to E.
8. Obtain the mean radius of the minimum moment:

$$\rho = \frac{|\vec{m}_E|}{|\vec{R}|}.$$

14.2. A cube of edge a is referred to a Cartesian system $(O-XYZ)$ the origin of which coincides with the cube center of gravity $(O \equiv G)$ and such that the directions of the axes are parallel to the edges.

We subject the solid to a mixed isometry of the following characteristics:

First, it is subjected to a rotation the axis of which has the direction $\vec{e}\left(\frac{1}{\sqrt{14}}, \frac{2}{\sqrt{14}}, \frac{3}{\sqrt{14}}\right)$, angle $45°$ and sense $(+)$ according to the right-hand rule, and then to a reflection of plane $\pi \equiv \frac{x}{2} - \frac{y}{1} + \frac{z}{3} = a$.

1. Give the coordinates of the cube vertices after the isometry.
2. With the aim of verifying the correct execution of the isometry, determine the lengths of the cube diagonals from the results of the previous question.

3. Assuming that the cube weight is P, we wish to find the required energy to produce the isometry.

14.3. The points of a continuous, isotropic and elastic medium, subject to given exterior forces in equilibrium, undergo a displacement with respect to a reference frame system $(O-XYZ)$ given by the vector

$$\vec{c} = u\vec{i} + v\vec{j} + w\vec{k} = \frac{4}{K}y2\vec{i} + \frac{2}{K}x^2\vec{j} + \frac{1}{K}z^2\vec{k},$$

and the elastic constants of the medium have the following values:

$$K = 10^5 \text{ cms}; \quad E = 2.6 \times 10^5 \text{kg/cm}^2; \quad \nu = 0.3.$$

1. Determine the strain tensor, T_D, at the generic point P.
2. Determine the so-called "rotation" tensor, A, at the point P.
3. Obtain the "strict strain tensor", T_{DE}, at the point P.
4. Obtain the "pure deformation" tensor, Γ, at P.
5. Knowing that the rotation vector $\vec{\theta} = \theta_x\vec{i} + \theta_y\vec{j} + \theta_z\vec{k}$ of the points in a neighborhood of P satisfies the relation (14.7), $M_{to}(\vec{\theta}\times) = A$, obtain the vector $\vec{\theta}$.
6. Obtain the displacement $\vec{c}$, at the point $P_0(1,1,1)$.
7. Obtain the tensors T_D, A, T_{DE}, Γ at the point P_0.
8. Obtain the coordinates of the point P_0', the position of P_0 after the displacement.
9. Obtain the directions of the main strain directions of the tensor Γ at P_0, by means of the corresponding unit vectors.
10. A sphere with center P and elemental radius dr_0 undergoes medium deformation. What is the equation of the transformed surface, with respect to the Cartesian system of the main directions?
11. Obtain the maximum shear strain at P_0.
12. Obtain the strain ε of an element dr_0 with origin at P, that makes identical angles with the three coordinate axes. Give the enlargement at P_0.
13. Using Equations (14.83) give the stress tensor at the point P_0.
14. Obtain the normal and shear stresses in the direction of the enlargement in question 12.

14.4. Consider an orthogonal tetrahedron, homogeneous, with mass m and orthogonal edges a, b, c coincident with the axes $\overline{OX}, \overline{OY}, \overline{OZ}$, respectively, of the ordinary reference Cartesian system.

A system of forces, each one with modulus K acts over each edge and in the directions $\overline{OA}, \overline{OB}, \overline{OC}, \overline{AB}, \overline{BC}, \overline{CA}$.

1. Obtain the resultant moment of the system with respect to the origin O and with respect to the center of gravity G of the tetrahedron.
2. Obtain the torque moment of the system of forces with respect to the axes $\overline{OX}, \overline{OY}$ and $\overline{OZ}$.

3. Obtain the torque moment $M_{\vec{e}}$ of the system with respect to the axis e, represented by the straight line $\overline{OG}$.
4. Obtain the inertial moments I_x, I_y, I_z of the solid with respect to the axes $\overline{OX}, \overline{OY}$ and $\overline{OZ}$.
5. Obtain the inertial products P_{xy}, P_{xz}, P_{yz} of the given tetrahedron.
6. Obtain the inertial tensor of the tetrahedron with respect to the point O.
7. Obtain the inertial moment I_e of the solid with respect to the axis e.
8. If the torque moment $M_{\vec{e}}$ acts during t seconds starting from rest, find the solid angular velocity $|\vec{\theta}_e|$.

References

[1] AKIVIS, M. A., AND GOLDBERG, V. V. *An introduction to linear algebra and tensors.* Dover Publications, New York, 1972.

[2] BAKSALARY, J. K., AND KALA, R. The matrix equation $AXB+CYD = E$. *Linear Algebra Appl. 30* (1980), 141–147.

[3] BALLANTINE, C. S. A note on the matrix equation $H = AP + PA$. *Linear Algebra and Appl. 2* (1969), 37–47.

[4] BARNETT, S. Remarks on solution of $AX + XB = C$. *Electron. Lett. 7* (1971), 385.

[5] BARNETT, S. *Matrices in control theory.* Krieger, Florida, 1984.

[6] BARNETT, S. *Matrices. Methods and Applications.* Oxford University Press, Oxford, 1990.

[7] BISHOP, R. L., AND GOLDBERG, S. I. *Tensor analysis on manifolds.* Dover Publications, New York, 1980.

[8] BORISHENKO, A. I., AND TAPAROV, I. E. *Vector and tensor analysis with applications.* Dover Publications, New York, 1979.

[9] BOURBAKI, N. *Elements de mathematique. Premiere part: Les structures fondamentales de l'analyse. Livre II: Algebre.* Übersetzung aus dem Französischen von D. A. Rajkov. Moskau: Staatsverlag fr physikalisch-mathematische Literatur, 1962.

[10] BOURBAKI, N. *Elements de mathematique. Part 1: Les structures fondamentales de l'analyse. Livre II: Algebre.* Traduit de la ed. francais par G.V. Dorofeev. Moskau: Verlag "Nauka". Hauptredaktion fr physikalisch-mathematische Literatur, 1966.

[11] BOURBAKI, N. *Elements of mathematics. Algebra, Part I:* Chapters 1-3. Hermann, Paris, 1974. Translated from the French.

[12] BOURBAKI, N. *Algebra I. Chapters 1–3.* Elements of Mathematics. Springer-Verlag, Berlin, 1998. Translated from the French, Reprint of the 1989 English translation.

[13] BRILLOUIN, L. *Tensors in mechanics and elasticity.* Academic Press, 1964.

[14] BURBAKI, N. *Algebra: Moduli, koltsa, formy.* Translated from the French by G. V. Dorofeev. Edited by Ju. I. Manin. "Nauka", Moscow, 1966.

[15] BURGESS, H. T. Solution of the matrix equation $X^{-1}AX = N$. *Ann. of Math. (2) 19*, 1 (1917), 30–36.

[16] CHAMBADAL, L., AND OVAERT, J. *Algèbre linèaire et algèbre tensorielle.* Paris: Dunod, 1968.

[17] CHAMBADAL, L., AND OVAERT, J. L. *Cours de mathèmatiques. Algèbre II.* Paris: Gauthier-Villars, 1972.

[18] CULLEN, C. G. *Matrices and linear transformations.* Dover Publications, New York, 1972.

[19] DOBOVIŠEK, M. On minimal solutions of the matrix equation $AX - YB = 0$. *Linear Algebra Appl. 325*, 1-3 (2001), 81–99.

[20] EISELE, J. A., AND MASON, R. M. *Applied matrix and tensor analysis.* John Wiley and Sons, New York, 1970.

[21] FLANDERS, H., AND WIMMER, H. K. On the matrix equations $AX - XB = C$ and $AX - YB = C$. *SIAM J. Appl. Math. 32*, 4 (1977), 707–710.

[22] GOHBERG, I., LANCASTER, P., AND RODMAN, L. *Matrix polynomials.* Academic Press Inc., New York, 1982. Computer Science and Applied Mathematics.

[23] GOLAB, S. *Tensors Calculus.* Elsevier Scientific Publ. Co., Amsterdam-London-New York, 1964.

[24] GOLUB, G. H., AND VAN LOAN, C. F. *Matrix computations,* second ed. Johns Hopkins University Press, Baltimore, 1989.

[25] GOUYON, R. *Calcul tensoriel.* Librairie Vuibert, Paris, 1963.

[26] GREUB, W. *Multilinear algebra,* second ed. Springer-Verlag, New York, 1978. Universitext.

[27] HE, C. N. The general solution of the matrix equation $AXB+CYD = F$. *Acta Sci. Natur. Univ. Norm. Hunan. 19*, 1 (1996), 17–20.

[28] HERMANN, R. *Ricci and Levi-Civita's Tensor Analysis, Paper.* Math. Sci. Press, 1975.

[29] INGRAHAM, M. H., AND TRIMBLE, H. C. On the matric equation $TA = BT + C$. *Amer. J. Math. 63* (1941), 9–28.

[30] JIANG, T., AND WEI, M. On solutions of the matrix equations $X - AXB = C$ and $X - A\overline{X}B = C$. *Linear Algebra Appl. 367* (2003), 225–233.

[31] LANG, S. *Algebra,* third ed. *Graduate Texts in Mathematics, Vol. 211.* Springer-Verlag, New York, 2002.

[32] LI, Y. M. The generalized quaternion matrix equation $AX - DXB = R$. *J. Math. (Wuhan) 22*, 3 (2002), 281–286.

[33] LICHNEROWICZ, A. *Èlèments de calcul tensoriel.* (Collect. Armand Colin, Sect. de Math. 259) Paris: Armand Collin, 1950.

[34] LICHNEROWICZ, A. *Elements of tensor calculus.* Translated by J. W. Leech and D. J. Newman from the 4th revised ed. Methuen's Monographs on Physical Subjects. London: Methuen & Co. Ltd.; New York: John Wiley & Sons Inc., 1962.

[35] LICHNEROWICZ, A. *Einfhrung in die Tensoranalysis.* Mannheim: Bibliographisches Institut, 1966.

[36] LU, C. S. Solution of the matrix equation $AX + XB = C$. *Electron. Lett. 7* (1971), 185–186.

[37] MAC DUFEE, C. C. *The Theory of matrices.* Chelsea Scientific books, New York, 1946.

[38] MARCUS, M. *Finite dimensional multilinear algebra. Part II.* Monographs and Textbooks in Pure and Applied Mathematics. Vol. 23. New York: Marcel Dekker, Inc., 1975.

[39] MARCUS, M. *Introduction to modern algebra,* vol. 47. Monographs and Textbooks in Pure and Applied Mathematics. Marcel Dekker, Inc., New York, Basel, 1978.

[40] MASSEY, W. Cross products of vectors in higher dimensional Euclidean spaces. *Am. Math. Mon. 90* (1983), 697–701.

[41] NIU, H. X. The matrix equation $AX+YA = C$ over an arbitrary division ring. *Qufu Shifan Daxue Xuebao Ziran Kexue Ban 29,* 3 (2003), 39–40.

[42] PARKER, W. V. The matrix equation $AX = XB$. *Duke Math. J. 17* (1950), 43–51.

[43] PENG, Z. Y. The solutions of the matrix equation $AXC + BYD = E$ and their optimal approximation. *Math. Theory Appl. (Changsha) 22,* 2 (2002), 99–103.

[44] PRASOLOV, V. *Problems and theorems in linear algebra.* Translations of Mathematical Monographs. 134. Providence, RI: AMS., 1994.

[45] PRASOLOV, V. *Problems and theorems in linear algebra. (Zadachi i teoremy linejnoj algebry).* Moskva: Nauka. Fizmatlit, 1996.

[46] REISCHER, C., AND SIMOVICI, D. A. On the matrix equation $\mathbf{PAP}^T = \mathbf{B}$ in the two-element Boolean algebra and some applications in graph theory. *An. Şti. Univ. "Al. I. Cuza" Iaşi Secţ. I a Mat. (N.S.) 17* (1971), 263–274.

[47] ROSENBROCK, H. H. *State-space and multivariable theory.* John Wiley & Sons, Inc., New York, 1970.

[48] RUÍZ-TOLOSA, J. R., AND CASTILLO, E. Stretching and condensing tensors. Application to tensor and matrix equations. Tech. Rep. 1, University of Cantabria, Santander, Spain, 2004.

[49] SCHOUTEN, J. A. *Tensor analysis for physicists.* Dover Publications, Inc., New York, 1989.

[50] SHIM, S.-Y., AND CHEN, Y. Least squares solution of matrix equation $AXB^* + CYD^* = E$. *SIAM J. Matrix Anal. Appl. 24,* 3 (2003), 802–808 (electronic).

[51] SOKOLNOKOFF, I. S. *Tensor analysis.* John Wiley and Sons, Inc., New York, 1964.

[52] SYNGE, J. L., AND SCHILD, A. *Tensor calculus.* Dover Publications Inc., New York, 1978. Reprinted from 1949 original.

[53] TIAN, Y. Ranks of solutions of the matrix equation $AXB = C$. *Linear Multilinear Algebra 51,* 2 (2003), 111–125.

[54] WIMMER, H. K. The matrix equation $X - AXB = C$ and an analogue of Roth's theorem. *Linear Algebra Appl.* *109* (1988), 145–147.

[55] YE, Q. K. Solving Lyapunov matrix equation $A^T P + PA = -Q$ by optimization. *Acta Math. Sci. (English Ed.)* *8*, 2 (1988), 145–149.

[56] YOU, X. H., YAN, T., AND MA, S. R. Anti-symmetric solution of the matrix equation $AXB = C$. *J. Nanjing Norm. Univ. Nat. Sci. Ed. 26*, 1 (2003), 6–10.

Index

Addition
  exterior vectors, 324
  of Euclidean tensors, 451
  of modular tensors, 272
  of pseudo-Euclidean tensors, 451
Adjoint tensors, 518
  of exterior products, 547
  of order $(n-1)$, 552
Affine geometric tensors, 597
Affine tensors, 581
Affinities, 600
Affinity, 601
Algebra
  exterior, 315, 369
  exterior dual, over space $V_*^n(K)$, 331
  exterior of order $r = 0$ and $r = 1$, 324
  mixed exterior, 387
  of anti-symmetric tensors, 225
  tensorial anti-symmetric interior, 252
Anti-commutativity
  of the exterior product $\bigwedge$, 331
  of the mixed exterior product $\bigwedge$, 403
Anti-symmetric
  total, 228
Anti-symmetric systems of scalar
  components, 225
Anti-symmetric systems with respect to
  an index subset, 226
Anti-symmetric tensors as functions of
  Kronecker deltas, 301
Anti-symmetry
  in Euclidean tensors, 430
  modular, 280

Applications
  tensor product, 119
  to probability, 24
Associated Euclidean exterior tensors,
  540
Associated Euclidean tensors, 436
Associated properties
  of the orientation tensors, 515
Associated tensors
  with a change-of-basis, 441
Associativity
  of $\otimes$, 38
  of the exterior product, 352
Axial symmetry $(n = 2)$, 609
Axial symmetry $(n = 3)$, 621
Axiomatic properties
  of tensor operations
    in the $\bigwedge_n^{(p,q)}(K)$ algebra, 398
  of tensor operations in exterior
    algebras, 324

Bases
  reciprocal, 335
Bi-stability, by the left and the right,
  414
Bivectors, 360
Building multivectors by contraction,
  376

Capacity
  scalar, 272
Cartesian tensors, 433
Central symmetry $(n = 2)$, 609
Central symmetry $(n = 3)$, 617

Change-of-basis, 6, 116
  in $E^n(\mathbb{R})$, 427
  in a third-order tensor product space,
    65
  in Euclidean exterior tensors, 563
  in exterior algebras, 337, 353
  in heterogeneous tensor spaces, 70
  in homogeneous determinants, 85
  in homogeneous tensor spaces, 72, 73
  in mixed exterior algebras, 404, 407
  in tensor spaces, 65
  of non-tensor nature, 105
  of tensor and non-tensor nature, 103
Classical mechanics, 41
Classification
  of modular tensors, 272
Common geometric tensors in Euclidean
    spaces, 502
Comodular scalars, 342
Comparative tables
  of algebra correspondences, 342
Complements
  on contramodular and comodular
    scalars, 341
Components
  of an anti-symmetric system, 228
  total and strict, 240
Condition
  for $T \in \bigwedge_n^{(2)}(\mathbb{R})$, to be an exterior
    product, 367
Connection
  of spaces primal and dual, 335
  of two linear spaces, 335
  tensor different types, 418
Contracted deltas, 298
Contracted tensor
  products, 276
Contracted tensor product
  of (pseudo-Euclidean) tensors, 455
  of Euclidean tensors, 455
Contraction, 459, 518
  as homomorphism, 55
  multiple, 121
  of ϵ systems, 302
  of exterior r-forms with r-exterior
    vectors, 342
  of generalized Kronecker deltas with
    ϵ systems, 309
  of indices, 92

  of Kronecker deltas of order r, 308
  of tensors, 276
  total symmetric, 217
Contramodular and comodular
    determinants, 363
Contramodular exterior algebras
    $\bigwedge_n^{(n)}(\mathbb{R})$, 361
Contramodular scalars, 341
Contravariant and covariant Euclidean
    tensors direct relation, 488
Coordinates
  contravariant, 5
  covariant, 5
Criteria
  for an exterior tensor $T \in \bigwedge_n^{(n-1)}(K)$
    to be a multivector, 375
  for homogeneous tensor, 73
  for tensor character, 69
    based on quotient law, 122, 277
  for tensor product, 34
Cross product
  as a polar tensor, 538
  in $E^n(\mathbb{R})$, 521
  of modular tensors, 517
  of vectors, 27, 29

Decomposable mixed exterior vectors,
    394
Decomposition
  of an exterior tensor, 373
Delta of order $r < n$ as function of delta
    of order $r = n$, 297
Density
  scalar, 272
Determinants of classic Kronecker
    deltas, 300
Dilatation, 605
Direct n-dimensional tensor endomor-
    phisms, 169
Direct exterior endomorphism, 350
Distributivity
  of $\otimes$, 38
Dot product
  of Euclidean tensors, 483
  of vectors, 27
Dual tensors
  in exterior bases, 536
  over $E^n(\mathbb{R})$ Euclidean spaces, 516

Eigentensors, 159
Einstein
  contraction, 54
  convention, 36
  summation convention, 36
Elastic tensor, 635
Equality
  of Euclidean tensors, 451
  of modular tensors, 272
Equation
  matrix, 22
Equivalent associated tensors, 422
Euclidean
  contraction mnemonic rule, 454
  contraction of indices by the
    Hadamard product, 458
  exterior algebra, 529
    of order $r : (2 < r < n)$, 532
    of order $r = 2$, 529
    of order $r = n$, 535
  homogeneous tensors, 413
  properties
    of the orientation tensors, 515
  tensor algebra, 451
    contraction, 459, 467
    inner product, 474
  tensor character criteria, 427
  tensor contraction, 452
    of order $r = 2$, 457
    of order $r = 3$, 457
    of order $r = 4$, 457
  tensor equality, 451
  tensor metrics, 482
  tensors in $E^n(\mathbb{R})$, 582
Extension to homogeneous tensors, 72
Exterior algebras
  over $V^n(K)$ spaces, 323
Exterior basis
  of a mixed exterior algebra, 397
Exterior homomorphism $H_\otimes$, 347
Exterior homomorphism $H_\wedge$, 347
Exterior homomorphisms, 345
Exterior mappings, 364
Exterior product
  of r linear forms over $V_*^n(K)$, 332
  of r vectors, 319
  of dual exterior vectors, 333
  of exterior vectors, 325
  of mixed exterior vectors, 399

  of mixed tensor, 405
  of r linear forms over $V_*^n(K)$, 332
  of three vectors, 317
  of two vectors, 315
Exterior product of mixed exterior
    vectors, 402
Exterior subspaces, 569
Exterior tensor contractions, 566
Exterior vector mappings, 345
External product
  of Euclidean tensors, 451
  of exterior vector, 325
  of mixed exterior vectors, 399
  of pseudo-Euclidean tensors, 451

Forms
  linear, 51
  quadratic, 51
Formula
  condensation, 15
  extension, 15
Frames
  dual coordinate in affine Euclidean
    spaces, 3
Fundamental exterior tensors, 558
Fundamental metric tensor
  of $\left(\overset{r}{\underset{1}{\otimes}} E^n \right) (\mathbb{R})$, 484, 489
Fundamental tensor G of the connection,
    414

General
  criteria for tensor character, 69
  rules about tensor notation, 48
General matrix representation of
    anti-symmetric tensors example,
    234, 241
General matrix representation of
    symmetric tensors example, 201,
    202
Generalized
  exterior tensor product, 325
  multilinear mappings, 165
Generation
  of anti-symmetric tensors, 232
  of symmetric tensors, 194
Generic expressions
  of polar tensors, 550
Geometric polar tensors, 544

Gram matrix, 4, 414

Homogeneous tensor algebra, 111
Homographies, 597
Homomorphisms
  of metric tensors, 479
  permutation, 133
Homothecy, 604, 605

Inertial tensors associated with a solid,
  651
Inner connection, 483
Inner product, 212
Inner symmetric tensor algebras, 212
Inner tensor product, 120
Intrinsic character
  of symmetry and anti-symmetry, 258
  of tensor anti-symmetry, 236
  of tensor symmetry, 197
Isomers
  of a tensor, 137
  of third-order tensors, 101
Isometries, 606
  decomposable, 623

Kronecker
  delta, 80
  delta generalized, 293
  tensor product, 207

Laws
  cosine law, 27, 28
  quotient law for homogeneous tensors,
    124
  quotient law for modular tensor
    character, 277
Levi-Civita
  tensor systems, 291
Linear combination, 129
Linear space
  oriented, 270

m.t.c. (modular tensor characteristics),
  271
Main modular tensors, 291
Matrix
  algorithms and tensor symmetry, 219
  associated with an operator, 114
  associated with doubly contraction
    homomorphisms, 144

associated with simply contraction
  homomorphisms, 141
equivalent, 20
operation rules for tensor expressions,
  74
permutation, 17
representation
  of a change-of-basis in tensor spaces,
    67
  of permutation homomorphisms,
    127
representation of tensors, 56
Matrix $P_{(i)}$ of third order isomer
  homomorphisms, 129
Metric
  Euclidean tensor, 482
  of Euclidean associated tensors, 487
Metric tensor $G(r)$ for $\left(\overset{r}{\underset{1}{\otimes}} E^n \right) (\mathbb{R})$,
  485
Metric tensor $G_{\bigwedge_n^{(r)}(\mathbb{R})}$ in Euclidean
  exterior algebra, 560
Mirror symmetry $(n = 3)$, 622
Mixed
  anti-symmetric tensor spaces, 387
  exterior algebras, 387
    terminology, 397
  exterior product
    of four vectors, 390
  exterior products, 404
  multivector, 319
  stresses, 644
Mixed exterior vector product, 396
Modular
  quotient law, 281
  tensor character
    of ϵ systems, 292
  tensor nature
    of determinants and its cofactors,
      283
  tensors over $E^n(\mathbb{R})$ Euclidean spaces,
    511
Modular tensor characteristics, 271
Modulus
  relative a change-of-basis, 269
Momentum tensor, 586
Multiform

over $\left(\overset{3}{\underset{1}{\otimes}} E^3 \right) (Q)$, 490

Multilinear mappings, 165
Multiple
  contracted tensor products, 120
  contractions, 121
Multiplication
  by an scalar for exterior vectors, 324
Multivectors, 321

Number of inversion
  of the exterior product, 380
Number of strict components
  of a symmetric system, 192
  of a symmetric system with respect
    to an index set, 191
  of an anti-symmetric system, 229
    with respect to an index subset, 229

Operations
  with modular tensors, 272
Orientation tensor, 514
  in exterior bases, 535
Orthonormal bases in $\left(\overset{r}{\underset{1}{\otimes}} E^n \right) (\mathbb{R})$,
  486

Partial contractions
  of ϵ systems, 303
Permutation, 226
  matrix tensor product, 127
  of indices, 101
Polar tensors
  in exterior bases, 536
  over $E^n(\mathbb{R})$ Euclidean spaces, 516
Polar trihedron, 5
Product
  by a scalar modular tensors, 274
  contracted, 248
  contracted tensor, 113
  direct of matrices, 8
  exterior, 315
  external of matrices, 8
  Hadamard of matrices, 9
  inner, 474
  inner of matrices, 8
  Kronecker of matrices, 8
  Kronecker tensor product, 247
  matrix, 8

mixed exterior, 404
  of axial symmetry by translation, 613
  of exterior vectors, 365
  of rotations, 612
  of symmetric contraction, 215
  of tensors with partial and reverse
    symmetries, contracted, 255
  of transformations, 623
  ordinary of matrices, 8
  scalar, 6
  scalar of matrices, 8
  Schur of matrices, 9
  simply contracted tensor product,
    119
  tensor, 34, 39, 275
  tensorial of matrices, 8
  triple tensor product linear space, 33
Projection tensor, 582
Properties
  axiomatic of modular tensors, 270
  of $H_{\vec{e}}$, 591
  of $M_{to}(\vec{e}\times), R_{\vec{e}}$, 591
  of Euclidean $E^n(\mathbb{R})$ spaces in
    orthonormal bases, 433
  of exterior products of order r, 321
  of matrices, 10
  of matrices with respect to eigenvalues
    and eigenvectors, 12
  square matrices, 11
Pseudo-Euclidean
  tensor algebra, 451
  tensor contraction, 452
Pseudoscalars, 282
Pseudotensors, 269

Quotient space of isomers, 426

Reciprocal bases in $\left(\overset{r}{\underset{1}{\otimes}} E^n \right) (\mathbb{R})$, 486
Rectangular Cartesian tensors, 436
Reduction of vector systems, 593
Reflection tensor, 590
Relation
  between the ϵ systems and the
    alternated multilinear forms, 305
Relative modulus of a change-of-basis,
  269
Rotation
  isomers, 138

tensor, 138, 159, 587
  three dimensions, 618
  Two-dimensions, 608

Scalar invariant, 80
Scalar mappings, 342
Scalar product of four vectors, 28
Scalar tensor product, 120
Scalar triple product, 30
  of vectors, 27
Similarity, 615
  $(n = 2)$, 611
Similitude theorems
  on multilinear mappings, 167, 168
    proof, 172, 176
Solid
  anisotropic, 627
  elastic, 626
  isotropic, 627
  plastic, 627
  rigid, 626
Stress
  main, 640
Strict
  of tensors, 387
Strict components, 199, 238
  of a symmetric mixed tensor, 216
  of a symmetric system, 191
    definition, 191
  of an anti-symmetric mixed tensor,
    256
  of an anti-symmetric system, 228
  of exterior vectors, 318
Strict tensor relationships
  for $\bigwedge_{n*}^{(r)}(\mathbb{R})$ algebras, 339
  for $\bigwedge_{n}^{(r)}(K)$ algebras, 338
Subspaces
  associated with strict components,
    204
Sum
  of exterior vector, 325
  of mixed exterior vectors, 398
Symmetric and anti-symmetric tensors,
  202
Symmetric systems
  of scalar components, 189
  with respect to an index subset, 190
Symmetric tensor algebra, 212
Symmetric tensors

from the tensor algebra perspective,
  206
Symmetric total contractions, 217
Symmetries
  through pure associated tensors, 444
Symmetrized
  tensor associated with a mixed tensor
    extension, 210
  tensor associated with an arbitrary
    pure tensor, 210
  tensors, 211
Symmetry
  axial, 621
  central, 617
  in Euclidean tensors, 430
  mirror, 622
  modular, 280
  total, 190
Systems
  anti-symmetric, 228, 242
  anti-symmetric with respect to an
    index subset, 226
  different classes, 203, 242
  linked, 358
  of determinants with Kronecker
    deltas, 309
  of scalar components, 225
  symmetric, 190

Tensor
  absolute homogeneous, 272
  adjoint or polar, 301
  affine, 581
  analytical representation, 37
  anti-symmetric, 230
    from the tensor algebra perspective,
      246
  anti-symmetries, 257
  anti-symmetrized, 250
    associated with an arbitrary pure
      tensor, 249
  associated in different bases, 448
  associated with a given tensor, 216
  associated with the given tensor, 257
  capacity, 272
  Cartesian, 433
  change-of-basis invariant, 80
  comodular scalar, 272
  concept, 33

contraction, 276
contramodular scalar, 272
density, 272
elastic, 635
first order, 56
fourth order, 78
homogeneous, 47
homogeneous anti-symmetric, 225
homogeneous symmetric, 189
homomorphisms, 111, 113
inertial moment, 647
isomers, 137
isotropic, 80
mappings, 168
modular, 269
modular or pseudotensor, 272
momentum, 586
null, 80
of fifth order, 143
of fourth order, 142
of fourth order mixed, 144
of order ($r = 5$) mixed, 145
of second order, 141
of third order, 141
orientation, 514
projection, 582
proper homogeneous, 272
pseudo escalar, 272
reflection, 590
relative, 269
relative invariant, 272
rotation, 138, 587
rotation tensor and the anti-
    symmetry, 253
scalar capacity, 272
scalar density, 272
second order, 57
second order (matrices), 74
special, 26
strain, 630
stress, 628
third order, 77
transposer, 132
with anti-symmetries, 230
with symmetries, 193
zero order, 80
Tensor and exterior algebras, 369
Tensor character
of polar systems, 301

of subsets of tensor components, 287
of the inner vector's connection in a
    $PSE^n(\mathbb{R})$ space, 416
of the orientation tensor, 514
Tensor criteria
based on contraction, 122
generalization, 69
Tensor geometries, 493
Tensor metrics, 493
Tensor partial Euclidean character
in orthonormal bases, 436
Tensor product, 34, 209
axiomatic properties, 38
Einstein's contraction, 54
of ϵ systems, 305
of Euclidean tensors, 452
of pseudo-Euclidean tensors, 452
of tensors, 50
of vectors in $E^n(\mathbb{R})$, 421
Tensor relationships between S and Γ,
    635
Tensor spaces, 3
anti-symmetric, 243
Tensor total Euclidean character in
    orthonormal bases, 434
Tensors
elastic and thermic, 638
Euclidean homogeneous, 413
in physics and mechanics, 626
obtained from the orientation tensor,
    523
symmetrized, 211
with simultaneous different species,
    302
Terminology, 323, 397
Theorem
Euclidean tensors, 455
first elemental criterion for tensor
    character, 122
general criterion for homogeneous
    tensor character, 122
intrinsic character of anti-symmetry,
    237
intrinsic character of symmetry, 197
of similitude with tensor mappings,
    167
quotient law, 124
second elemental criterion for tensor
    character, 122

tensor contraction, 111, 113
Total and strict components, 200
Total contractions
  of ϵ systems, 304
Translation, 617
  $(n = 2)$, 607
Transposer
  tensor, 132
Transposition, 226
Triple product
  of vectors, 27

Uniqueness
  of associated tensors, 444
Use
  of exterior metric tensors, 561

Vector product of four vectors, 28
Vertical displacement of indices for
    Euclidean tensors, 422
  rule, 424

Weight tensor, 270

Universitext

Aksoy, A.; Khamsi, M. A.: Methods in Fixed Point Theory

Alevras, D.; Padberg M. W.: Linear Optimization and Extensions

Andersson, M.: Topics in Complex Analysis

Aoki, M.: State Space Modeling of Time Series

Audin, M.: Geometry

Aupetit, B.: A Primer on Spectral Theory

Bachem, A.; Kern, W.: Linear Programming Duality

Bachmann, G.; Narici, L.; Beckenstein, E.: Fourier and Wavelet Analysis

Badescu, L.: Algebraic Surfaces

Balakrishnan, R.; Ranganathan, K.: A Textbook of Graph Theory

Balser, W.: Formal Power Series and Linear Systems of Meromorphic Ordinary Differential Equations

Bapat, R.B.: Linear Algebra and Linear Models

Benedetti, R.; Petronio, C.: Lectures on Hyperbolic Geometry

Benth, F. E.: Option Theory with Stochastic Analysis

Berberian, S. K.: Fundamentals of Real Analysis

Berger, M.: Geometry I, and II

Bliedtner, J.; Hansen, W.: Potential Theory

Blowey, J. F.; Coleman, J. P.; Craig, A. W. (Eds.): Theory and Numerics of Differential Equations

Börger, E.; Grädel, E.; Gurevich, Y.: The Classical Decision Problem

Böttcher, A; Silbermann, B.: Introduction to Large Truncated Toeplitz Matrices

Boltyanski, V.; Martini, H.; Soltan, P. S.: Excursions into Combinatorial Geometry

Boltyanskii, V. G.; Efremovich, V. A.: Intuitive Combinatorial Topology

Booss, B.; Bleecker, D. D.: Topology and Analysis

Borkar, V. S.: Probability Theory

Carleson, L.; Gamelin, T. W.: Complex Dynamics

Cecil, T. E.: Lie Sphere Geometry: With Applications of Submanifolds

Chae, S. B.: Lebesgue Integration

Chandrasekharan, K.: Classical Fourier Transform

Charlap, L. S.: Bieberbach Groups and Flat Manifolds

Chern, S.: Complex Manifolds without Potential Theory

Chorin, A. J.; Marsden, J. E.: Mathematical Introduction to Fluid Mechanics

Cohn, H.: A Classical Invitation to Algebraic Numbers and Class Fields

Curtis, M. L.: Abstract Linear Algebra

Curtis, M. L.: Matrix Groups

Cyganowski, S.; Kloeden, P.; Ombach, J.: From Elementary Probability to Stochastic Differential Equations with MAPLE

Dalen, D. van: Logic and Structure

Das, A.: The Special Theory of Relativity: A Mathematical Exposition

Debarre, O.: Higher-Dimensional Algebraic Geometry

Deitmar, A.: A First Course in Harmonic Analysis

Demazure, M.: Bifurcations and Catastrophes

Devlin, K. J.: Fundamentals of Contemporary Set Theory

DiBenedetto, E.: Degenerate Parabolic Equations

Diener, F.; Diener, M.(Eds.): Nonstandard Analysis in Practice

Dimca, A.: Sheaves in Topology

Dimca, A.: Singularities and Topology of Hypersurfaces

DoCarmo, M. P.: Differential Forms and Applications

Duistermaat, J. J.; Kolk, J. A. C.: Lie Groups

Edwards, R. E.: A Formal Background to Higher Mathematics Ia, and Ib

Edwards, R. E.: A Formal Background to Higher Mathematics IIa, and IIb

Emery, M.: Stochastic Calculus in Manifolds

Endler, O.: Valuation Theory

Erez, B.: Galois Modules in Arithmetic

Everest, G.; Ward, T.: Heights of Polynomials and Entropy in Algebraic Dynamics

Farenick, D. R.: Algebras of Linear Transformations

Foulds, L. R.: Graph Theory Applications

Frauenthal, J. C.: Mathematical Modeling in Epidemiology

Friedman, R.: Algebraic Surfaces and Holomorphic Vector Bundles

Fuks, D. B.; Rokhlin, V. A.: Beginner's Course in Topology

Fuhrmann, P. A.: A Polynomial Approach to Linear Algebra

Gallot, S.; Hulin, D.; Lafontaine, J.: Riemannian Geometry

Gardiner, C. F.: A First Course in Group Theory

Gårding, L.; Tambour, T.: Algebra for Computer Science

Godbillon, C.: Dynamical Systems on Surfaces

Godement, R.: Analysis I

Goldblatt, R.: Orthogonality and Spacetime Geometry

Gouvêa, F. Q.: p-Adic Numbers

Gustafson, K. E.; Rao, D. K. M.: Numerical Range. The Field of Values of Linear Operators and Matrices

Gustafson, S. J.; Sigal, I. M.: Mathematical Concepts of Quantum Mechanics

Hahn, A. J.: Quadratic Algebras, Clifford Algebras, and Arithmetic Witt Groups

Hájek, P.; Havránek, T.: Mechanizing Hypothesis Formation

Heinonen, J.: Lectures on Analysis on Metric Spaces

Hlawka, E.; Schoißengeier, J.; Taschner, R.: Geometric and Analytic Number Theory

Holmgren, R. A.: A First Course in Discrete Dynamical Systems

Howe, R., Tan, E. Ch.: Non-Abelian Harmonic Analysis

Howes, N. R.: Modern Analysis and Topology

Hsieh, P.-F.; Sibuya, Y. (Eds.): Basic Theory of Ordinary Differential Equations

Humi, M., Miller, W.: Second Course in Ordinary Differential Equations for Scientists and Engineers

Hurwitz, A.; Kritikos, N.: Lectures on Number Theory

Huybrechts, D.: Complex Geometry: An Introduction

Isaev, A.: Introduction to Mathematical Methods in Bioinformatics

Iversen, B.: Cohomology of Sheaves

Jacod, J.; Protter, P.: Probability Essentials

Jennings, G. A.: Modern Geometry with Applications

Jones, A.; Morris, S. A.; Pearson, K. R.: Abstract Algebra and Famous Inpossibilities

Jost, J.: Compact Riemann Surfaces

Jost, J.: Postmodern Analysis

Jost, J.: Riemannian Geometry and Geometric Analysis

Kac, V.; Cheung, P.: Quantum Calculus

Kannan, R.; Krueger, C. K.: Advanced Analysis on the Real Line

Kelly, P.; Matthews, G.: The Non-Euclidean Hyperbolic Plane

Kempf, G.: Complex Abelian Varieties and Theta Functions

Kitchens, B. P.: Symbolic Dynamics

Kloeden, P.; Ombach, J.; Cyganowski, S.: From Elementary Probability to Stochastic Differential Equations with MAPLE

Kloeden, P. E.; Platen; E.; Schurz, H.: Numerical Solution of SDE Through Computer Experiments

Kostrikin, A. I.: Introduction to Algebra

Krasnoselskii, M. A.; Pokrovskii, A. V.: Systems with Hysteresis

Luecking, D. H., Rubel, L. A.: Complex Analysis. A Functional Analysis Approach

Ma, Zhi-Ming; Roeckner, M.: Introduction to the Theory of (non-symmetric) Dirichlet Forms

Mac Lane, S.; Moerdijk, I.: Sheaves in Geometry and Logic

Marcus, D. A.: Number Fields

Martinez, A.: An Introduction to Semiclassical and Microlocal Analysis

Matoušek, J.: Using the Borsuk-Ulam Theorem

Matsuki, K.: Introduction to the Mori Program

Mc Carthy, P. J.: Introduction to Arithmetical Functions

Meyer, R. M.: Essential Mathematics for Applied Field

Meyer-Nieberg, P.: Banach Lattices

Mikosch, T.: Non-Life Insurance Mathematics

Mines, R.; Richman, F.; Ruitenburg, W.: A Course in Constructive Algebra

Moise, E. E.: Introductory Problem Courses in Analysis and Topology

Montesinos-Amilibia, J. M.: Classical Tessellations and Three Manifolds

Morris, P.: Introduction to Game Theory

Nikulin, V. V.; Shafarevich, I. R.: Geometries and Groups

Oden, J. J.; Reddy, J. N.: Variational Methods in Theoretical Mechanics

Øksendal, B.: Stochastic Differential Equations

Poizat, B.: A Course in Model Theory

Polster, B.: A Geometrical Picture Book

Porter, J. R.; Woods, R. G.: Extensions and Absolutes of Hausdorff Spaces

Radjavi, H.; Rosenthal, P.: Simultaneous Triangularization

Ramsay, A.; Richtmeyer, R. D.: Introduction to Hyperbolic Geometry

Rees, E. G.: Notes on Geometry

Reisel, R. B.: Elementary Theory of Metric Spaces

Rey, W. J. J.: Introduction to Robust and Quasi-Robust Statistical Methods

Ribenboim, P.: Classical Theory of Algebraic Numbers

Rickart, C. E.: Natural Function Algebras

Rotman, J. J.: Galois Theory

Rubel, L. A.: Entire and Meromorphic Functions

Rybakowski, K. P.: The Homotopy Index and Partial Differential Equations

Sagan, H.: Space-Filling Curves

Samelson, H.: Notes on Lie Algebras

Schiff, J. L.: Normal Families

Sengupta, J. K.: Optimal Decisions under Uncertainty

Séroul, R.: Programming for Mathematicians

Seydel, R.: Tools for Computational Finance

Shafarevich, I. R.: Discourses on Algebra

Shapiro, J. H.: Composition Operators and Classical Function Theory

Simonnet, M.: Measures and Probabilities

Smith, K. E.; Kahanpää, L.; Kekäläinen, P.; Traves, W.: An Invitation to Algebraic Geometry

Smith, K. T.: Power Series from a Computational Point of View

Smoryński, C.: Logical Number Theory I. An Introduction

Stichtenoth, H.: Algebraic Function Fields and Codes

Stillwell, J.: Geometry of Surfaces

Stroock, D. W.: An Introduction to the Theory of Large Deviations

Sunder, V. S.: An Invitation to von Neumann Algebras

Tamme, G.: Introduction to Étale Cohomology

Tondeur, P.: Foliations on Riemannian Manifolds

Verhulst, F.: Nonlinear Differential Equations and Dynamical Systems

Wong, M. W.: Weyl Transforms

Xambó-Descamps, S.: Block Error-Correcting Codes

Zaanen, A.C.: Continuity, Integration and Fourier Theory

Zhang, F.: Matrix Theory

Zong, C.: Sphere Packings

Zong, C.: Strange Phenomena in Convex and Discrete Geometry

Zorich, V. A.: Mathematical Analysis I

Zorich, V. A.: Mathematical Analysis II

Druck: Krips bv, Meppel
Verarbeitung: Litges & Dopf, Heppenheim